# EARTH

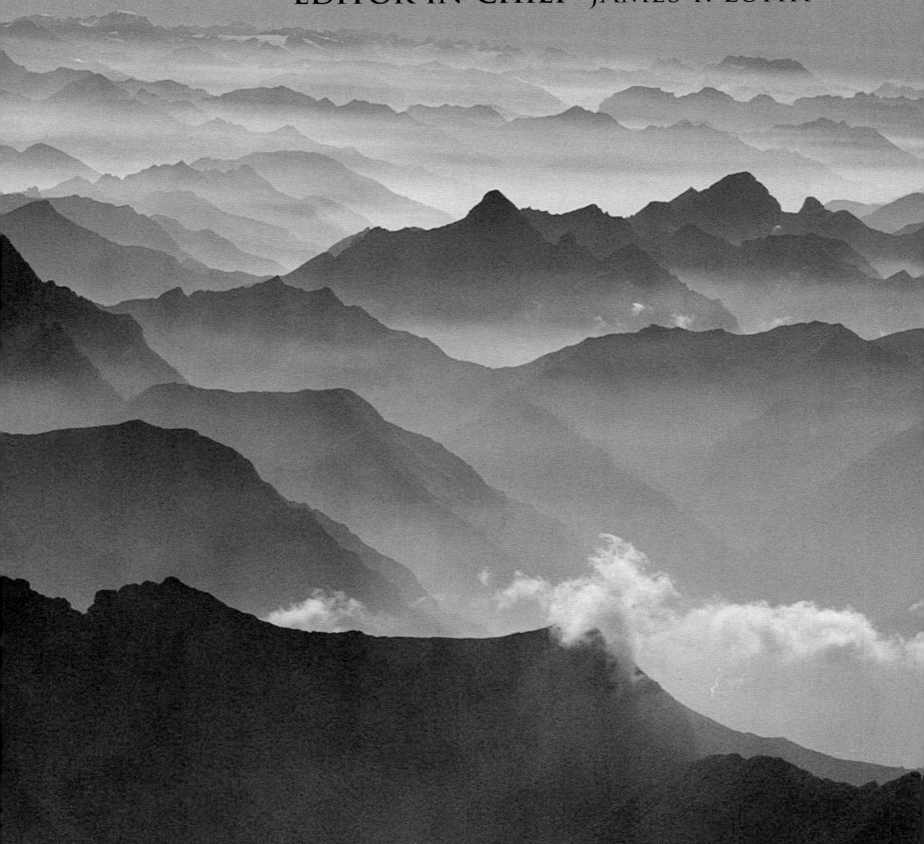

DK
SMITHSONIAN INSTITUTION
EARTH

EDITOR-IN-CHIEF JAMES F. LUHR

LONDON, NEW YORK, MELBOURNE, MUNICH AND DELHI

SENIOR ART EDITOR Caroline Buckingham
ART EDITORS David Ball, Kirsten Cashman, Kenny Grant, Thomas Keenes, Helen Taylor
DESIGNERS Paul Drislane, Lee Riches, Vanessa Thompson
DTP DESIGNER Rajen Shah
PICTURE RESEARCHER Louise Thomas
ILLUSTRATORS Tom Coulson (Encompass Graphics Limited), John Plumer, Planetary Visions Limited
CARTOGRAPHERS David Roberts, Iorwerth Watkins
PRODUCTION CONTROLLERS
Elizabeth Cherry, Sarah Dodd

MANAGING ART EDITOR Philip Ormerod
ART DIRECTOR Bryn Walls

SENIOR EDITOR Peter Frances
PROJECT EDITORS Sophie Bevan, Kim Dennis-Bryan, Sarah Larter, Sean O'Connor, Gill Pitts, David Summers
EDITORS Georgina Garner, Ben Hoare, Giles Sparrow, Nikky Twyman
INDEXER John Noble

MANAGING EDITOR Liz Wheeler
CATEGORY PUBLISHER Jonathan Metcalf

COOLING BROWN
(Rocks, Minerals, Environmental Issues)
ART EDITOR Philip Lord
PROJECT EDITOR Joanna Chisholm
CREATIVE DIRECTOR Arthur Brown
MANAGING EDITOR Amanda Lebentz

SMITHSONIAN PROJECT COORDINATOR
Robyn Bissette

First American Edition, 2003

Published in the United States by
DK Publishing, Inc.
375 Hudson Street
New York, New York 10014

03 04 05 06 07 08 10 9 8 7 6 5 4 3 2 1



Earth / editor-in-chief, James F. Luhr. -- 1st American ed.
p. cm.
Includes index.
ISBN 0-7894-9643-7 (alk. paper)
1.Earth. I. Luhr, James F.

QE501.E27 2003
550 --dc21

2003051573

Color reproduction by Colourscan, Singapore
Printed and bound in Italy by Mondadori

Discover more at
www.dk.com

# CONTENTS

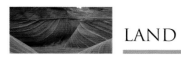

## LAND



## OCEAN



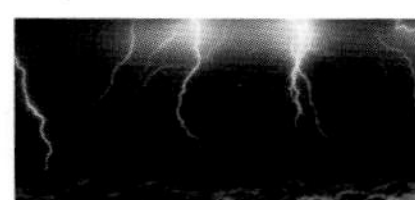

## ATMOSPHERE

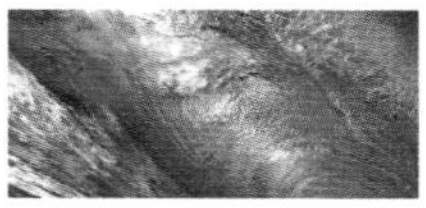

## TECTONIC EARTH

# ABOUT THIS BOOK

*Earth* is divided into five main sections. The first section, PLANET EARTH, is an introduction to the Earth as a whole. The next three sections are about the planet's main environments—the LAND, OCEAN, and ATMOSPHERE. Within these sections, the Earth's features are divided into categories, such as Rivers and Lakes. Introductory pages describe typical features found in each category and explain how they are formed. On the succeeding pages, major features in that category are profiled individually. Thematic panels (see right) appear throughout. The last section, TECTONIC EARTH, is a three-dimensional atlas.

## PLANET EARTH

This overview of our planet is divided into four parts. The History of the Earth is a narrative account of how our planet and the universe as a whole came to be the way they are now. This is followed by The Earth in Space, which sets the Earth in context among the other planets of the Solar System. The Anatomy of the Earth looks at the Earth's structure and materials, while The Changing Earth explains those processes that operate on a global scale.

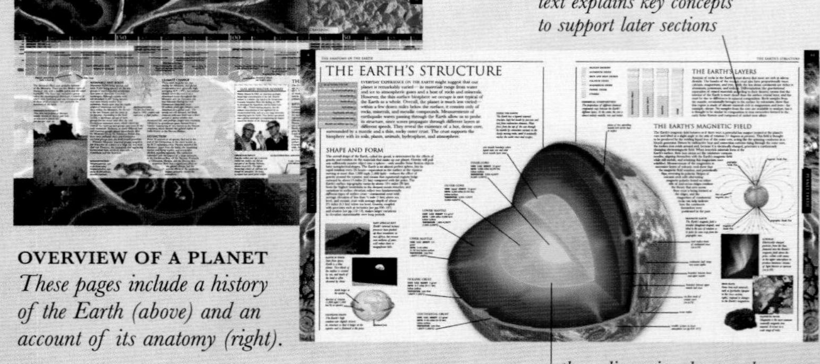


*text explains key concepts to support later sections*

**OVERVIEW OF A PLANET**
*These pages include a history of the Earth (above) and an account of its anatomy (right).*

*three-dimensional artworks reveal structural detail*

## TECTONIC EARTH

The Earth's rocky outer crust is divided into large sections called tectonic plates. It is the movement of these plates, more than any other process, that accounts for the changing shape of the planet's surface. This section contains profiles of the Earth's seven major plates, plus the adjoining minor ones. Supporting maps show human population density and political boundaries.

*locator globe identifies plates being profiled*

*three dimensional model, showing plate boundaries*

*model shows shape of land surface and ocean floor*

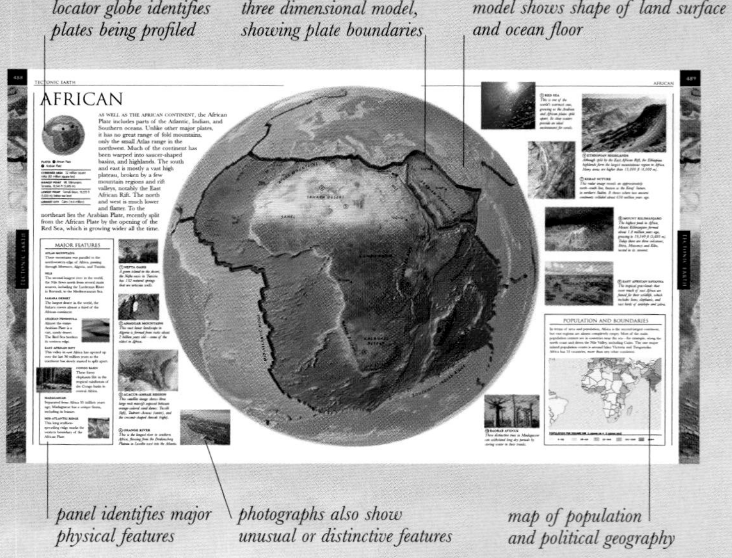


*panel identifies major physical features*

*photographs also show unusual or distinctive features*

*map of population and political geography*

## LAND, OCEAN, AND ATMOSPHERE

These three sections focus on specific features—on the Earth's landmasses, in the five great oceans, and in the layer of gases separating the Earth from space. Within each section, features are grouped by type into smaller sections, and then explained in general and individually profiled. Most profiles describe actual features (such as particular volcanoes), while others characterize general kinds of phenomena (such as types of clouds). Profiles of actual features are presented according to the continent on which they are found, in the following order: the Arctic, North America, Central America, South America, Europe, Africa, north and west Asia, south and east Asia, Australasia, and Antarctica. Within continents, features are arranged approximately in north—south order.

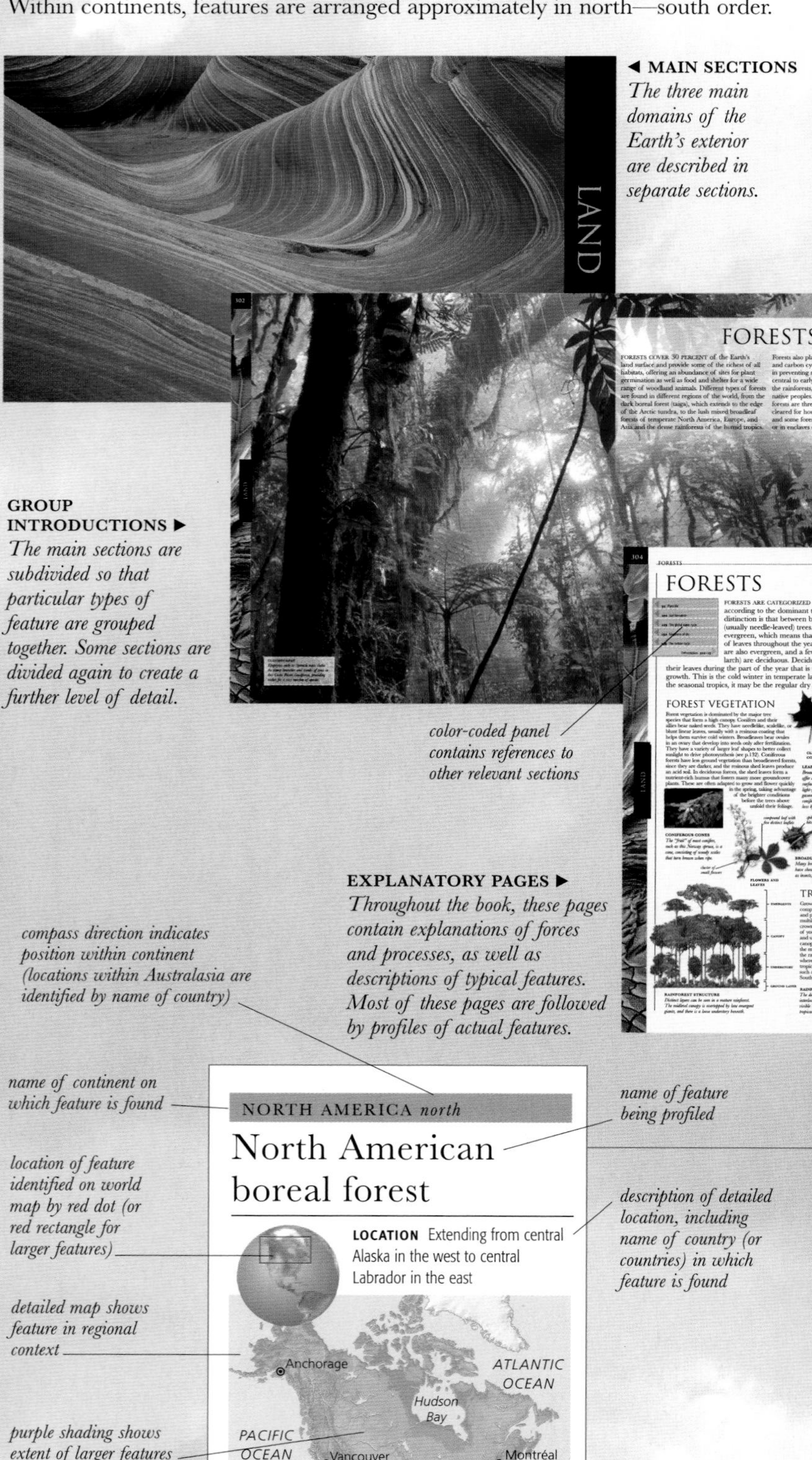


◀ **MAIN SECTIONS**
*The three main domains of the Earth's exterior are described in separate sections.*

**GROUP INTRODUCTIONS** ▶
*The main sections are subdivided so that particular types of feature are grouped together. Some sections are divided again to create a further level of detail.*

*color-coded panel contains references to other relevant sections*

**EXPLANATORY PAGES** ▶
*Throughout the book, these pages contain explanations of forces and processes, as well as descriptions of typical features. Most of these pages are followed by profiles of actual features.*

*compass direction indicates position within continent (locations within Australasia are identified by name of country)*

*name of continent on which feature is found*

*name of feature being profiled*

*location of feature identified on world map by red dot (or red rectangle for larger features)*

*description of detailed location, including name of country (or countries) in which feature is found*

*detailed map shows feature in regional context*

*purple shading shows extent of larger features*

*table of summary information (categories vary between sections)*

*scale bar divided into blocks, each representing 60 miles (100km)*

**LOCATION AND DATA** ▲

# THEMATIC PANELS

**FOREST SIZES**

The technology of satellite imaging has had an enormous impact on ecological monitoring. The comparison of satellite photographs with existing ground surveys allows detailed interpretation and analysis of such images, and enables the accurate mapping of the size of existing forests. The satellite image below, for example, clearly shows the current extent of forest in the Black Hills of South Dakota. It is vital to build up such snapshots of vegetation cover if the effects of factors such as deforestation are to be accurately monitored. Satellite imaging is also used to tr… their path… control m…

**JOSEPH BANKS**

Englishman Joseph Banks (1743–1820) was one of the great botanical explorers and collectors. At age 25, he joined James Cook's expedition to the South Pacific on the *Endeavour*. In 1770, they landed in eastern Australia, where Banks amassed a vast collection of botanical specimens then unknown to Europeans. Appropriately, Captain Cook named their landing place Botany Bay.

▲ **BIOGRAPHY**
*Notable earth scientists, explorers, and others are profiled in this type of feature.*

◀ **SCIENCE**
*These features reveal how scientists have learned about different aspects of the workings of our planet.*

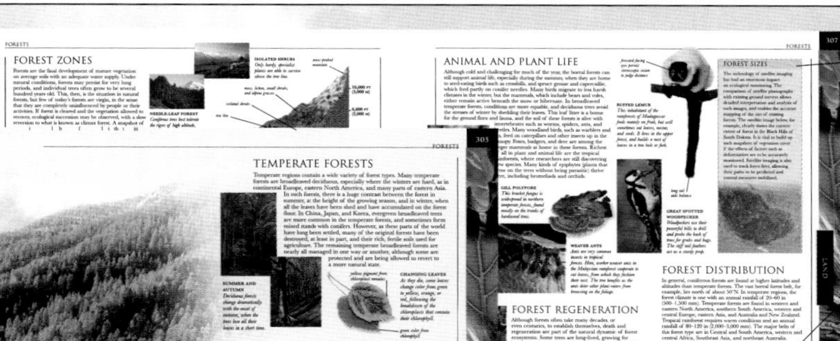

**RAINFOREST FIRES**

Fires are a major threat, even to moist rainforest, and Indonesia has suffered badly, both from natural fires and from those started deliberately as an aid to logging operations or for clearing trees to make way for agricultural land use. Once a forest fire has taken hold, particularly in dry conditions, such as those of the drought in 1997, the proximity of the trees allows it to spread rapidly, destroying not only the trees themselves but virtually all the associated wildlife.

**FIGHTING THE FIRE**
*This villager is attempting to douse a brush fire in Kalimantan, the Indonesian part of Borneo, in 1997.*

◀ ▲ **ENVIRONMENT**
*The ways in which humans are changing the Earth—and how the Earth, in turn, affects humans—are described in double-page features (above) and smaller panels (left) throughout the book. Each double-page article takes a global view of a particular subject.*

*thematic panels (see above) on explanatory pages describe general trends or issues; panels in feature profiles cover localized effects*

*in most sections, a world map shows global distribution of features being described*

*introductory panel defines terms and measurements given in the feature profiles that follow*

▼ **FEATURE PROFILES**
*All profiles contain a text description and a locator or distribution map. Most profiles are also illustrated with a color photograph.*

*profiles are arranged on page in geographical order*

*selected features are described in double-page feature profiles*

## CONTRIBUTORS AND CONSULTANTS

*Earth* is the product of a collaboration between Dorling Kindersley, several contributors (listed below), and staff members of the Smithsonian Institution (listed at the bottom of this panel) working under the direction of James F. Luhr of the National Museum of Natural History.

### CONTRIBUTORS

**Michael Allaby** Atmosphere
**David Burnie** Agricultural Areas, Environment features
**Kim Dennis-Bryan** Soils
**Robert Dinwiddie** The History of the Earth, The Earth in Space, Glaciers, Ocean
**John Farndon** The Changing Earth, Fault Systems, Urban Areas, Industrial Areas, Tectonic Atlas
**Douglas Palmer** The History of the Earth, The Anatomy of the Earth, The Changing Earth, Mountains, Volcanoes, Hot Springs and Geysers
**Clint Twist** Rivers, Lakes, Underground Rivers and Caves, Urban Areas, Industrial Areas, Atmosphere
**Martin Walters** The Changing Earth, Deserts, Forests, Wetlands, Grasslands and Tundra
**Tony Waltham** Igneous Intrusions, Underground Rivers and Caves
**Richard Beatty** Glossary

### CONSULTANTS

**Bruce B. Collette** Ocean
**Douglas H. Erwin** The History of the Earth
**Richard S. Fiske** Igneous Intrusions, Hot Springs and Geysers
**Bevan M. French** Meteorite Impacts
**Andrew K. Johnston** Urban Areas, Industrial Areas
**James F. Luhr** Editor-in-Chief
**Ian G. MacIntyre** Ocean
**Timothy McCoy** The History of the Earth, Rocks, The Earth in Space, Meteorite Impacts
**J. Patrick Megonigal** Soils, Water, Rivers and Lakes, Wetlands, Atmosphere
**William G. Melson** Ocean
**Jeffrey E. Post** Minerals
**Richard Potts** The History of the Earth
**Stanwyn G. Shetler** Life, Deserts, Forests, Grasslands and Tundra, Agricultural Areas
**Lee Siebert** Volcanoes
**Timothy Rose** Underground Rivers and Caves
**Tom Simkin** Tectonic Plates, Plate Boundaries, Volcanoes
**Sorena S. Sorensen** Rocks, Mountains, Glaciers
**Edward P. Vicenzi** Mountains
**Scott L. Wing** The History of the Earth

# DYNAMIC EARTH

With a stable orbit neither too close nor too far from the Sun, the Earth occupies a unique position in the Solar System, with conditions suitable for the origin and evolution of life. Over its 4.56-billion-year history, an unusual combination of events and processes has transformed the Earth into a blue planet. Unlike the other planets, extensive and ever-changing water-filled ocean basins cover most of the Earth's surface and color it blue. Above the ocean surfaces rise continental and island landmasses of varying sizes, the configurations of which also change over geological time.

For billions of years, the oceans have been occupied by abundant life, which has evolved and diversified to occupy the land and air. Nearly all of this life is dependent on a flow of energy from the Sun—and protection from its more damaging radiation by a gaseous atmosphere.

**WOODLAND TO DESERT**
***Heated by the tropical Sun, the Sahara is one of the most inhospitable places on Earth today. However, the Earth's climate is not static: fossils and rock paintings show that over 6,000 years ago, this region's woodlands supported diverse animals and nomadic human hunters.***

A constant dynamic interaction among all of the Earth's processes allows life to flourish in most parts of our planet. From space, the landmasses can be seen to be partly greened by vegetation, but there are also large arid swaths colored yellow-brown by rocks and sands, and others whitened by a permanent covering of ice and snow, where there is little or no plant cover and life barely survives. The solid-and-liquid surface of the Earth is wrapped in the gaseous layer that we know as the atmosphere, with its rapidly fluctuating swirls of water-vapor clouds at lower levels.

The Earth's dynamic behavior is driven by two main sources—one located within the Sun and the other within the Earth. Both are ultimately forms of heat derived from radioactive decay, and both confer enormous benefits on the Earth system and ensure its survival, probably for another 5 billion years. One source, the solar energy striking the Earth's atmosphere, heats the surface, controls weather and hydrologic systems, and drives the processes of erosion and sediment transport. The other heat source, the Earth's core, causes the solid rocks of the mantle (the layer beneath the crust) to churn slowly, much as boiling soup develops rolling convection cells. This allows heat to be more efficiently transferred to the surface, compared with simple conduction through static rock. Mantle convection works with gravity to drive plate-tectonic activity at the Earth's surface.

The Earth's outermost brittle layer of rocks is broken into a mosaic of tectonic plates, and mantle convection is thought to be responsible for the slow but relentless

**THE FORCE WITHIN**

***The Earth's internal heat is most dramatically expressed by erupting volcanoes. Kilauea volcano, on the Pacific island of Hawaii, sits atop a plume of rising heat. This partially melts mantle rocks to form basaltic lava, which erupts through volcanic vents and fissures, and here pours into the sea.***

motions among them. Over geological time, plates can diverge, pouring out huge volumes of lava to create new oceanic crust. Because the Earth does not expand, plate divergence must be balanced by plate convergence, where the margins of cooler and denser oceanic plates descend back into the Earth's interior along deep-sea trenches—the process known as subduction.

Plate movements are responsible for most of our planet's earthquakes and volcanic eruptions, which are concentrated along plate boundaries. Plate-tectonic processes explain the formation, changing size, and distribution of ocean basins, continental landmasses, and islands, as well as most of their prominent features, such as high mountain ranges, ridges, and deep valleys.

Without the tectonic dynamism driven by its inner heat, the Earth's surface would look like that of the Moon: ancient and unchanging except for rare meteorite impacts.

Bodies of rock and ice ranging in size from dust to that of small planets impact upon the Moon, Earth, and other planets in the Solar System. Their craters are readily visible on the Moon, but the Earth's ...

**LAKE NATRON'S FOOD CHAIN**
*The Earth's ecosystems involve complex interactions between the living and nonliving worlds. For example, the water of Tanzania's Lake Natron is highly alkaline because it is fed by springs that rise in areas of volcanic rock. The water supports populations of algae and tiny crustaceans that are eaten by flamingos, which in turn are hunted by birds of prey.*

geological processes tend to obscure and even obliterate them. Fossil and rock records tell us that the continuity of life has been severely affected at specific times in the past by major impacts, including one 65 million years ago that led to the extinction of an estimated 70 percent of all species, including the dinosaurs.

Over time, geological processes—many associated with plate-tectonic activity—have forged materials that have proved to be of great value to human development. These include obvious examples, such as precious metals, ore minerals, and hydrocarbons (oil and natural gas), as well as less obvious materials, such as sand and gravel, which modern societies consume in great volumes. Fresh water is another essential Earth resource that many people have previously taken for granted, but it is now in seriously short supply for a considerable portion of the world's human population.

Most important for life has been the supply of radiant energy from the Sun that bathes the Earth's surface. The Earth's environments range from polar ice caps to hot, dry, tropical deserts, and from dark, pressurized ocean depths to mountaintops with little oxygen but high levels of damaging ultraviolet radiation. From an Earth-bound perspective, these might seem to be extreme environments that are unfavorable to life. However, compared with the Moon and the other planets in the Solar System, which also receive solar energy, the Earth enjoys a unique combination of conditions. These conditions have supported and promoted the evolution of life, which, as far as we can tell, is unique in the Solar System. Some 4 billion years of evolution have produced a wonderful diversity of plants and animals well adapted to the prevailing conditions on Earth, including the colonization of many extreme environments.

The essential difference between the "dead" sterility of the Moon and the vitality of the Earth is the presence of our planet's atmosphere and abundant water. Although the Moon receives much the same amount of energy from the Sun as the Earth does, our planet's carbon-dioxide-rich atmosphere moderates surface temperatures to an average of 59°F (15°C), compared with 34°F (1°C) on the Moon. Consequently, our thick and gas-rich atmosphere allows most of the Earth's water to exist in a liquid state at the surface. In contrast, the Moon's atmosphere is extremely tenuous, and although there is some frozen water among the rock debris at the lunar poles, it is not conducive to the evolution of life.

**ANCIENT FORESTS**
***Trees are some of the largest living things on Earth. These mature stands of sequoia in California are reminiscent of forests of 100–65 million years ago. The trees compete for sunlight, which they combine with carbon dioxide and convert into plant material, while releasing oxygen into the atmosphere and "locking up" carbon.***

Without liquid surface water, the vast majority of Earth's living organisms could not survive.

The atmosphere also protects life from damaging solar radiation, such as ultraviolet light. The composition of the atmosphere has evolved since the Earth's earliest history, when there was no protection from ultraviolet light. At present, the atmosphere is composed largely of nitrogen (78 percent by volume) and oxygen (21 percent), much of which is of biological origin. Importantly, the atmosphere also includes water vapor, which contributes to our weather systems, and a number of other gases, particularly methane and carbon dioxide, which are collectively known as greenhouse gases. By absorbing and re-radiating heat from the surface, the [illegible] of solar energy and raise the temperature of the atmosphere.

The average air temperature at the Earth's surface has been increasing for the past three decades; this phenomenon is called global warming. Over the same interval, human release of carbon dioxide gas through burning of hydrocarbons has also dramatically increased, leading [illegible]

makers to conclude that human activity is a major cause of current global warming. The historical retreat of many mountain glaciers and recent breakups of Antarctic ice shelves are expected to lead to rising sea levels, threatening much of the most fertile and densely populated low-lying coastal land around the world. Dynamic changes in the feedback system among the oceans, atmosphere, and incoming solar radiation have caused significant climate change throughout geological time. Global climates have swung back and forth between glacial icehouse and warm greenhouse states. Despite present human-enhanced global warming, it is debatable whether we are clear of the recent icehouse state of the last 1.8 million years. As we look to the future, one thing is certain: the Earth's system will not be static but will be as dynamic as it has been since the formation of our planet 4.56 billion years ago.

**ATMOSPHERIC ENERGY**
*Vast amounts of energy are transferred in the Earth's atmosphere. A single thunderstorm, for example, releases as much energy as burning 7,000 tons of coal in one hour. This energy transfer is potentially lethal—as when electrical energy builds up and is transferred between storm clouds and the surface as lightning.*

# PLANET EARTH

**ANCIENT LIFE-FORMS**

*Some of the first organisms to inhabit the Earth can still be seen. Over 3 billion years ago, bacteria in shallow seas began to form mounds called stromatolites. Such structures are still forming at Shark Bay, Australia.*

# THE HISTORY OF THE EARTH

OVER THE LAST 200 YEARS, scientists have carefully studied the Earth's rocks and recovered the remains of past animals and plants in order to piece together the history of our planet. The Earth came into existence about 4.56 billion years ago, as a hot, rocky body in our Solar System. Primitive life first appeared in the oceans about 4 billion years ago and has been spreading and diversifying ever since, but its evolutionary path has been far from smooth. The Earth's environments have always been subject to change, as volcanic activity, meteorite impacts, and climate changes have produced often life-threatening and sometimes catastrophic effects. The development of our planet has turned out to be remarkably eventful—and, from the evidence of history, it will continue to be so.

# THE EARTH'S PAST

The Universe 44–45
The Solar System 46–47
Rocks 78–81
Tectonic Plates 106–107
Life 132–35
Tectonic Earth 478–95

GEOLOGISTS USE INFORMATION from many sources to help them reconstruct the Earth's long history and understand how it fits into the even longer evolution of the universe. Valuable information about the early history of the Earth can be gained from extraterrestrial sources—such as meteorites and other bodies in the Solar System. Analysis of rocks, minerals, and fossils found at the surface tells us directly about the crust and provides clues to the nature of deeper layers. This analysis also provides information about climatic and atmospheric changes, geological events (such as the movements of the Earth's crustal plates), and the evolution of life on the planet.

## THE ROCK RECORD

Certain rocks are built up in successive layers (or strata) through the deposition of material by natural processes, with relatively young layers lying on top of older ones. Geologists have matched (or correlated) strata from around the world using distinctive fossils and rock types to produce what is known as a stratigraphical column, stretching over the Earth's entire history. However, the process is complicated by many breaks in the record caused by plate movement and lack of deposition.

**ROCK LAYERS**
*The angular relationship of strata at Siccar Point in Scotland represents a break in the record.*

## GEOLOGICAL TIME

The universe formed about 13–14 billion years ago, and the history of the Earth spans the last 4.56 billion years. To make sense of the immensity of the Earth's history, it has been subdivided by geologists into a hierarchical arrangement. The largest divisions are eons, many hundreds of millions of years long, including the Phanerozoic eon, which extends from 543 million years ago to the present day. Within the Phanerozoic, there are three eras, based on the history of life. Eras are divided into smaller segments called periods, which in turn are divided into epochs. Decisions about where boundaries between the main periods of time should be are ongoing. Evidence from fossils plays an important part in positioning these boundaries. However, international correlation between rock strata can be very difficult because the original sediments may have been deposited in different environments and climatic conditions and therefore contain different fossils.

**AFRICAN SHIELD**
*Diamond-rich sands of the Namibian diamond field cover the Precambrian shield of southern Africa, one of the world's most ancient landmasses.*

**THE DIVISION OF GEOLOGIC TIME**
*Just 250 years ago, the Earth was generally thought to be a few thousand years old. It was not until the 1950s that a near-accurate age of 4.55 billion years was determined. It took another decade to establish that the universe is 13–14 billion years old.*

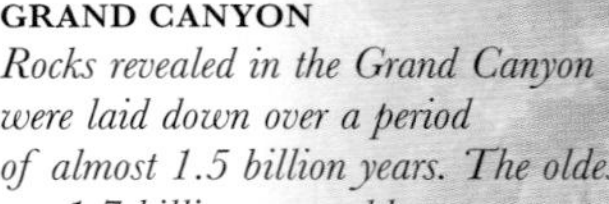

**GRAND CANYON**
*Rocks revealed in the Grand Canyon were laid down over a period of almost 1.5 billion years. The oldest are 1.7 billion years old.*

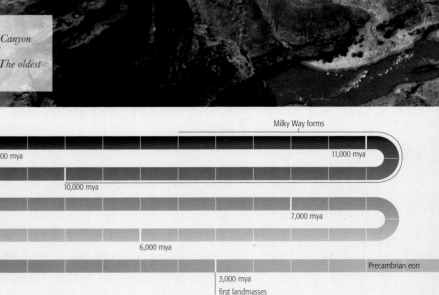

The Big Bang
first galaxies
Milky Way forms
13,000 million years ago (mya)
12,000 mya
11,000 mya
9,000 mya
10,000 mya
8,000 mya
7,000 mya
5,000 mya
6,000 mya
Precambrian eon
4,500 mya
formation of the Earth
4,000 mya
first evidence of life
3,000 mya
first landmasses

# FOSSILS

Fossils are the remains of past life that have been entombed by geological processes in rock strata. They range from pollen grains to the skeletons of giant dinosaurs or whales. Life can be preserved in many ways, and fossils vary from vague traces of past activity (such as burrows or footprints) or body chemistry (biomolecules) to encapsulated bodies (such as insects caught in amber). Most processes of preservation involve considerable loss of information about organisms. Nevertheless, the fossil record shows that life began in the seas approximately 3.8 billion years ago and that from about 543 million years ago, life diversified as it colonized, in succession, land, fresh water, and air. However, the record is highly biased toward marine organisms with hard parts, such as shells.

By studying the processes of burial and fossilization, scientists have been able to search out rarer examples of fossils in which soft parts have been preserved. For example, the beautifully preserved fossils from the Cambrian period found in the Burgess Shale, western Canada (see right), give better insights into the total diversity of past life and its biology.

**PLANT FOSSIL**
*Rosettes of late Carboniferous horsetail (calamite) leaves are preserved here in rock.*

**BURGESS SHALE**
*High in the Canadian Rockies, the World Heritage Site of the Burgess Shale has yielded a remarkable diversity of late-Cambrian marine fossils.*

**PETRIFIED TREE STUMPS**
*Silica replacement of the woody tissues of these Triassic trees from the Chinle formation in Arizona has made them resistant to weathering and erosion.*

**HOW FOSSILS FORM**
*Fossilization is a lengthy and complex process. It requires the rapid covering of an organism's remains by sediment, followed by the deep burial, compression, and chemical changes involved in rock formation. Later uplift and erosion—sometimes combined with excavation—reexpose the fossil at the surface.*

**1. ANIMAL DIES**

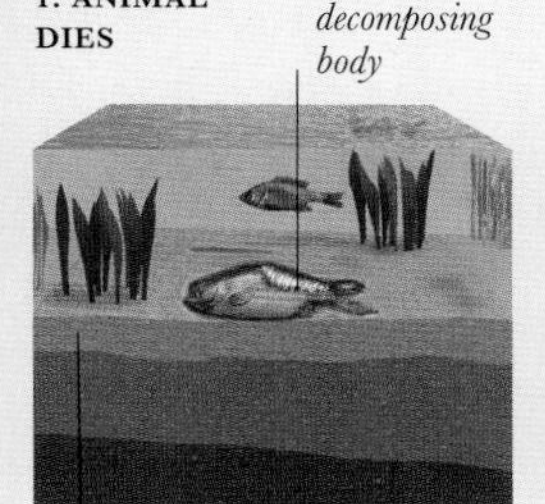


**2. SKELETON BURIED**

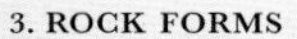

**3. ROCK FORMS**

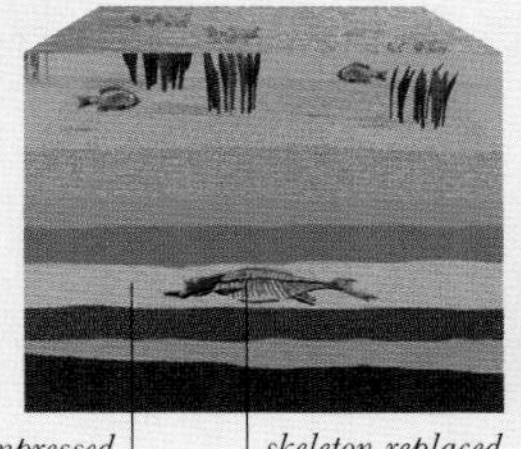


**4. FOSSIL REVEALED**

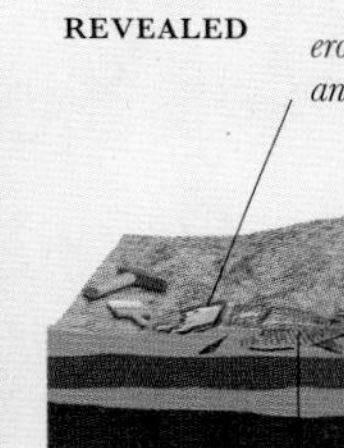


## RADIOMETRIC DATING

Until the advent of radiometric dating at the beginning of the 20th century, there was no reliable method of dating minerals, rocks, and fossils. The discovery of radioisotopes of some elements (such as lead, carbon, and zircon) that occur in minerals, rocks, meteorites, and fossils, together with the measurement of their rates of decay, has made it possible to calculate the age of these materials. However, the limits of radiometric dating depend on the element being used. For example, carbon dating (used for most fossils) cannot provide dates for material that is more than 50,000 years old.

**KEY TO TIMELINE OF EARTH HISTORY**
*The pages that follow contain a visual summary of the history of the Earth, beginning with the formation of the universe. Information is summarized in the following bands:*

EON, ERA, PERIOD, EPOCH — Divisions of geological time. (Decisions on boundaries are ongoing.)

CLIMATE — Average temperatures and composition of the atmosphere. The intensity of the color in this band indicates temperature trends

LIFE — Events and trends in the evolution of flora and fauna

GEOLOGIC EVENTS — Includes maps indicating continental movements

**SYMBOL KEY**
◆ Event occurs at this time
Trend occurs over this period
Event occurs during this period

**MAP KEY**
N. America
Siberia/N. Asia
Baltica/N. Europe
Gondwana
Ice caps

first birds · first primates · early human ancestors · *Homo sapiens*

| EON | Phanerozoic—age of abundant and evident life | | | | | | | | | | | | | | | | | |
|---|---|---|---|---|---|---|---|---|---|---|---|---|---|---|---|---|---|---|
| ERA | Paleozoic—age of ancient life | | | | | | | Mesozoic—age of dominant reptiles | | | | Cenozoic—age of dominant mammals | | | | | | |
| PERIOD | Cambrian | Ordovician | Silurian | Devonian | Carboniferous | | Permian | Triassic | Jurassic | Cretaceous | | Paleogene | | | | | Quaternary | |
| EPOCH | | | | | Mississippian | Pennsylvanian | | | | Lower | Upper | Paleocene | Eocene | Oligocene | Miocene | Pliocene | Pleistocene | Holocene |
| | 543 | 490 | 443 | 418 | 354 | 323 | 290 | 252 | 199.5 | 142 | 99 | 65 | 54.8 | 33.5 | 24 | 5 | 1.8 | 0.01 – 0 |

2,000 mya — first multicelled organisms · 1,000 mya · 543 mya — Phanerozoic eon — Present day · first land plants · first land vertebrates · first dinosaurs

| TIME AFTER BIG BANG | 1 FEMTOSECOND ($10^{-15}$ SECONDS) | 1 PICOSECOND ($10^{-12}$ SECONDS) | 1 NANOSECOND ($10^{-9}$ SECONDS) | 1 MICROSECOND ($10^{-6}$ SECONDS) | 1 MILLISECOND ($10^{-3}$ SECONDS) | 1 SECOND | 1 MINUTE | 1 HOUR | 1 DAY | 1 YEAR | 1 THOUSAND YEARS | 1 MILLION YEARS | 1 BILLION YEARS |
|---|---|---|---|---|---|---|---|---|---|---|---|---|---|
| …PERATURE OF UNIVERSE (K) | | $10^{15}$ $10^{14}$ | $10^{13}$ | $10^{12}$ | $10^{11}$ $10^{10}$ | | $10^{9}$ | $10^{8}$ | $10^{7}$ | $10^{6}$ $10^{5}$ | $10^{4}$ | $10^{3}$ $10^{2}$ 10 | |
| …NSITY OF UNIVERSE ($G/CM^3$) | | $10^{25}$ | $10^{20}$ | $10^{15}$ | $10^{10}$ $10^{5}$ | | $10^{0}$ | | $10^{-5}$ | $10^{-10}$ | $10^{-15}$ $10^{-20}$ | $10^{-25}$ | |
| DIAMETER OF UNIVERSE (M) | | $10^{10}$ | $10^{12}$ | $10^{14}$ | $10^{16}$ | | $10^{18}$ | | $10^{20}$ | $10^{22}$ | | $10^{24}$ | $10^{25}$ |

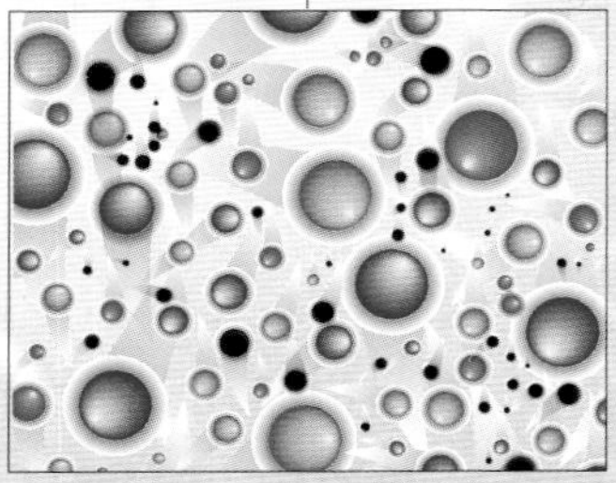

**AFTER THE BIG BANG**
*In its first billionth of a second, the hot, expanding universe was a "soup" of scores of types of particles that varied enormously in their properties.*

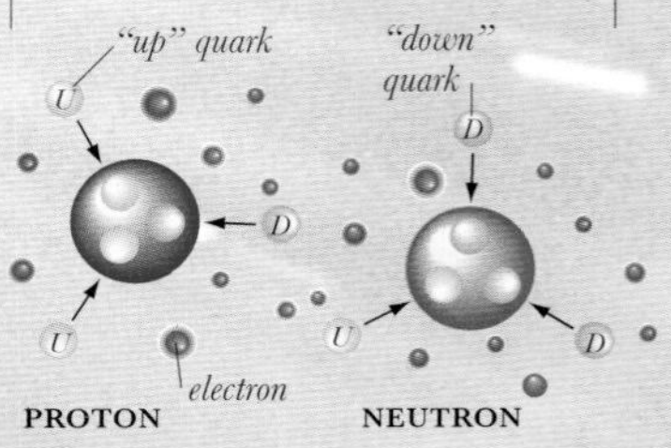

"up" quark
"down" quark
PROTON
electron
NEUTRON

**QUARKS BIND**
*Quarks started combining to form protons and neutrons. One "down" and two "up" quarks made a proton; one "up" and two "down" quarks made a neutron.*

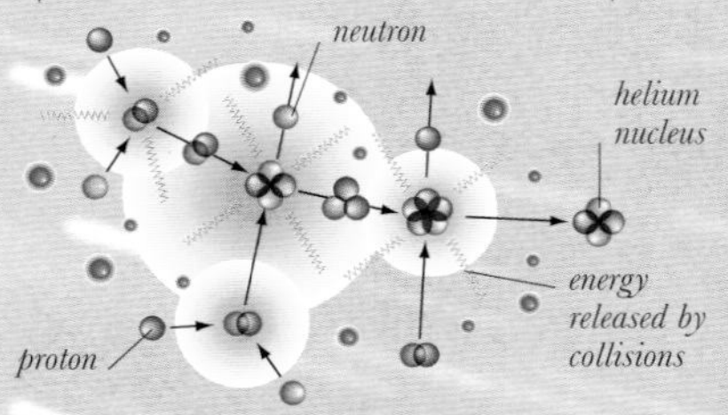

neutron
helium nucleus
energy released by collisions
proton

**BIG BANG NUCLEOSYNTHESIS**
*Between 1 and 180 seconds after the Big Bang, proton and neutron collisions formed the atomic nuclei of some light elements, mainly helium.*

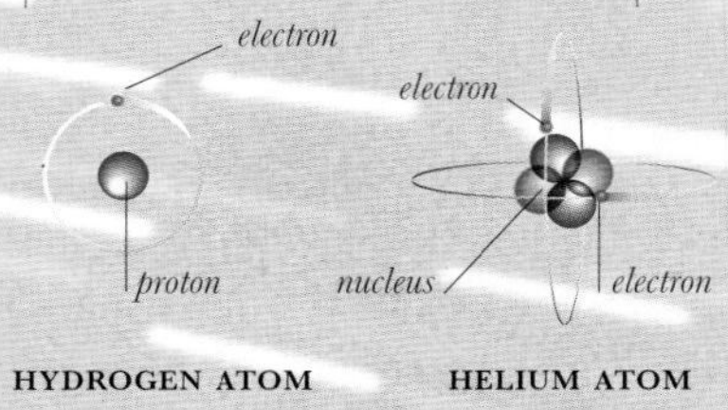

electron
proton
HYDROGEN ATOM
electron
nucleus
electron
HELIUM ATOM

**FIRST ATOMS**
*After about 300,000 years, protons and helium nuclei (consisting of protons and neutrons) began to capture electrons, forming hydrogen and helium atoms.*

# BEFORE THE EARTH

To understand the origins of the Earth, it is necessary to go back to the origin of the universe itself, some 13–14 billion years ago in the event described by the Big Bang theory, when matter, time, energy, and space all came into existence. The early universe was small, hot, and dense. Ever since, it has been expanding and cooling. Within a nanosecond (a billionth of a second), it was hundreds of millions of miles in diameter, with a temperature of tens of trillions of kelvins (K), or degrees above absolute zero (–460°F/–273°C).

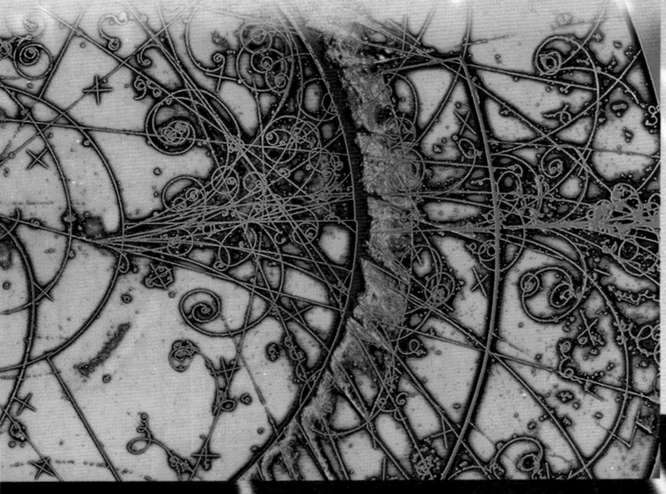

## BUILDING BLOCKS

At this stage, the universe was a seething "soup" of particles created out of energy, together with vast numbers of photons (little packets of radiant energy). Among the most numerous particles were electrons and quarks, but there were also many other kinds that no longer exist today. Although it contained plenty of electrons, the early universe held neither of the other main building blocks of atoms: protons and neutrons. However, after just one microsecond (a millionth of a second), it had cooled sufficiently for vast quantities of protons and neutrons to form as two different types of quarks combined.

Most of the protons were destined to become the nuclei of hydrogen atoms, but about one second after the Big Bang, collisions between protons and neutrons started to form the nuclei of other light elements—helium and tiny amounts of lithium and beryllium. This process, termed Big Bang nucleosynthesis, was completed in three minutes and formed the nuclei of 98 percent of the helium atoms present in the universe today. It also mopped up all the neutrons.

**PARTICLE TRACKS**
*Today, scientists try to simulate what happened in the Big Bang by using particle accelerators to smash subatomic particles together. The resulting tracks can then be studied.*

## THE FIRST ATOMS

For the next few hundred thousand years, the universe continued to expand and cool, but it was still too energetic for atoms to form. If electrons and atomic nuclei came together momentarily, they were quickly split apart by photons, which were themselves trapped in a process of continual collision with the particles. Eventually, after 300,000 years, when the temperature had dropped to about 3,000 K, the protons and other atomic nuclei started capturing electrons permanently, forming the first atoms, which were primarily hydrogen and helium. At the same time, the photons were

### BACKGROUND EVIDENCE

A crucial piece of evidence supporting the Big Bang theory is the existence of Cosmic Microwave Background Radiation (CMBR). This is a faint heat radiation that emanates uniformly from all points in the sky. The only plausible explanation is that this originated in the hot fireball conditions of the early universe, as described by the Big Bang theory.

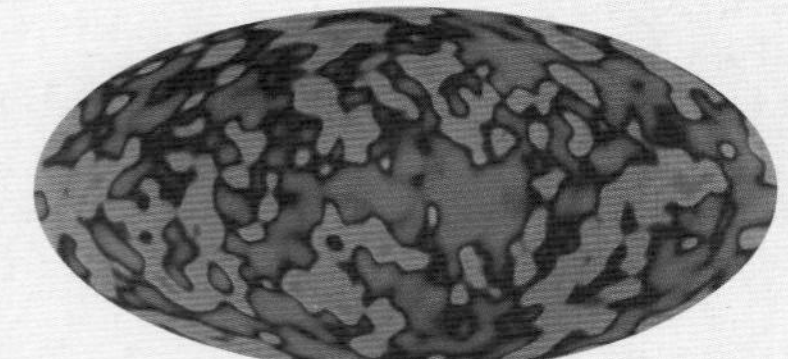

MILLION YEARS AGO: 12,000 | 11,000 | 10,000 | 9,000 | 8,000 | 7,000 | 6,000

TEMPERATURE (K): 9 · 8 · 7 · 6 · 5 · 4

DENSITY (G/CM³): $10^{-27}$ · $10^{-28}$

DIAMETER (M): $10^{26}$

EVENTS: first galaxies formed · formation of the Milky Way · many cycles of star birth and death in the Milky Way · formation of the Solar System (4,560 mya)

released and streamed freely in all directions. At this stage, the universe became transparent, as the earlier "fog" of particles and energy cleared.

## GALAXIES FORM

Gravity now began to cause gas atoms to come together. Over hundreds of millions of years, swirling clouds of hydrogen and helium gas formed and started to extend into long, thin strands. About 13 billion years ago (or 500 million years after the Big Bang), the strands began to clump together to form the first galaxies. Further concentration of matter within galaxies led to the creation of the first stars. The production of energy (including heat and light) in stars began when hydrogen nuclei in their centers started fusing to form helium.

## NEW ELEMENTS

When the first galaxies and stars formed, there were still just four chemical elements in the universe: hydrogen, helium, lithium, and beryllium. The formation of the first stars was highly significant, because it is within these that heavier chemical elements are created from lighter ones, through various processes of fusion. Many of the most common chemical elements on Earth, such as oxygen, silicon, and iron, were made in this way. The very heaviest elements, such as lead, cannot be created in ordinary stars but form only in supernovae—the massive explosions of giant stars in their final death throes. These explosions also distributed new elements throughout galaxies, where they were incorporated into new stars and planets.

**REMNANTS OF A SUPERNOVA**
*Supernovae are huge star explosions that distribute new elements through galaxies. This is the remnant of Cassiopeia A, which exploded about 10,000 years ago.*

## BIRTH OF THE SOLAR SYSTEM

The exact age of our own galaxy, the Milky Way, is not known, but it was probably formed by 10–11 billion years ago. About 4.56 billion years ago, within our galaxy, a clump of gas and dust, known as a nebula, began to condense into what became the Solar System.

Within this nebula, matter coalesced into a denser central region (the proto-Sun) and more diffuse outer regions. Eventually the nebula shrank into a spinning, disklike object, called the protoplanetary disk. Within the disk, dust and ice collided randomly to form ever larger particles. In the center of the disk, as matter was drawn in by gravity, temperatures rose to the point where hydrogen fused to form helium, and a full-fledged star—our Sun—was born.

In other parts of the disk, the predominant matter was solid particles. Closest to the Sun, rocky materials became the main component. In the colder outer regions of the disk, ice particles were more common. As particles throughout the disk became larger, gravity drew them into collisions. This process, known as accretion, caused larger and larger bodies to merge, eventually forming planetesimals. These were the size of boulders, or larger, and composed of rock or rock and ice.

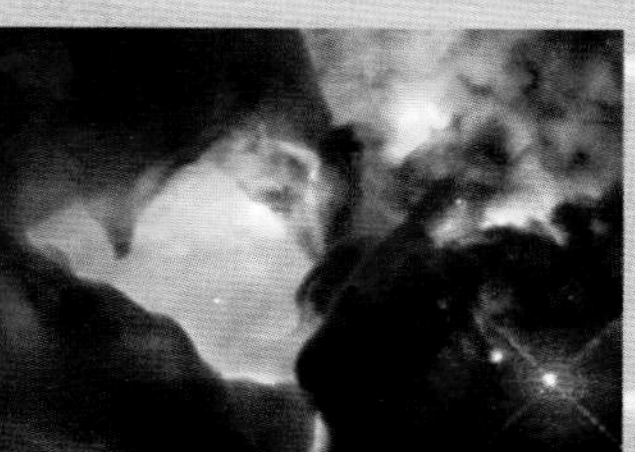

**NEBULA**
*The Solar System condensed out of huge clouds of gas and dust, similar to these of the Lagoon Nebula, a star-forming region in the Milky Way.*

As well as heat and light, the new Sun also emitted a stream of energetic particles known as the solar wind. This "wind" blew volatile gases from the inner areas into the outer areas of the disk. Most of the remaining planetesimals collided to form the four rocky inner planets. They also formed the cores of the outer planets, around which the volatile gases collected.

Not all of the planetesimals merged to form planets. Some "leftovers" remain in the Solar System as two types of bodies: asteroids and comets. Most of the asteroids orbit in a belt between Mars and Jupiter, but some follow Jupiter's orbit, and some others have paths that cross the Earth's orbit. Comets formed from icy planetesimals in the outer edge of the disk.

**FORMATION OF THE SOLAR SYSTEM**
*The Solar System formed in several stages. First a huge nebula (cloud) of gas and dust condensed into a spinning, disklike object. The center of the disk condensed further to form the Sun, while particles in the outer parts of the disk collided and accreted to form the planets.*

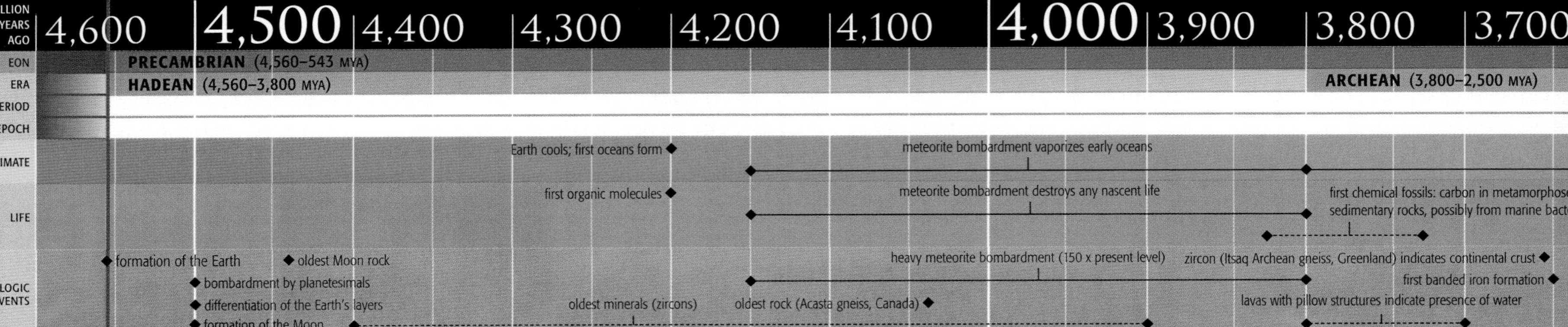


# PRECAMBRIAN EARTH

At one time, the Precambrian eon was seen as a geological *terra incognita*, of unknown age and without fossil remains. Not until 1956, when the age of the Earth was accurately measured, did the immensity of the Precambrian's 4 billion years become glaringly obvious. Since then, fossils have been found, rocks dated, and the formation of the Solar System unraveled. The Earth's early history is now known to be a time of cataclysmic change. Early accretion and differentiation of the core and mantle, along with bombardment from space, culminated in the Moon's formation.

**FORMATION OF THE MOON**
*Around 4.5 billion years ago, a planetary body the size of Mars collided with the Earth, tearing away a large volume of rock. The debris was held by the Earth's gravity, and cooled and coalesced to form the satellite we recognize as the Moon.*

**ZIRCON CRYSTALS**
*Crystals from rocks in Western Australia have been dated at 4.4–3.9 billion years old.*

Continued layering was followed by the formation of a primitive atmosphere, oceans, and possibly life. However, renewed meteorite bombardment, which so scarred the Moon, also devastated the Earth, causing melting of its rocks. Not until about 3.8 billion years ago were an atmosphere and oceans regenerated. The first fossil evidence dates from almost immediately after this time, but these life-forms had to tolerate a lack of oxygen and high ultraviolet radiation because there was no ozone layer to protect them. It took time for the atmosphere and oceans to develop to their present condition. Surface temperatures fell slowly and oxygen levels rose slowly as increasing numbers of photosynthesizing microbes produced oxygen and a protective ozone layer, aided by the emission of water vapor that split to release oxygen and ozone into the upper atmosphere.

**THE EARTH'S LAYERS FORM**
*Early accretion of cosmic material formed a larger, growing body, in which melts formed and migrated. Heavy elements concentrated in the core and lighter ones in the overlying mantle.*

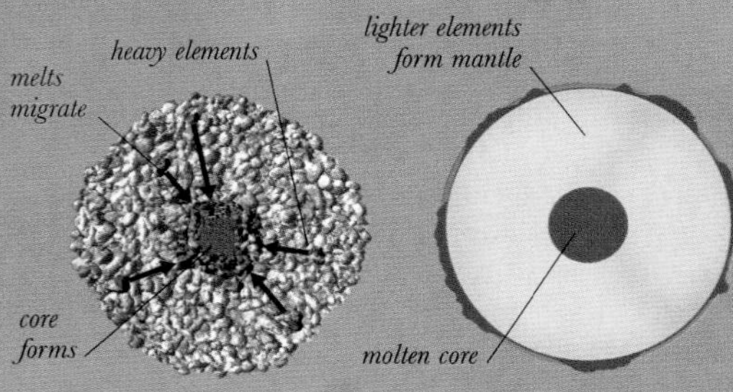


The Precambrian eon is divided into the Hadean, Archean, and Proterozoic eras, during which dynamic processes originating from within the Earth generated new ocean-floor rocks and destroyed them elsewhere, while continents grew and were moved through the processes of plate tectonics (see pp.106–109). Meanwhile, an increasing diversity of marine microorganisms evolved slowly. Major events continued to perturb evolving ecosystems, with large-scale volcanism, impact events, and climate change culminating in runaway glaciations. Although often catastrophic, these upheavals may have stimulated evolution.

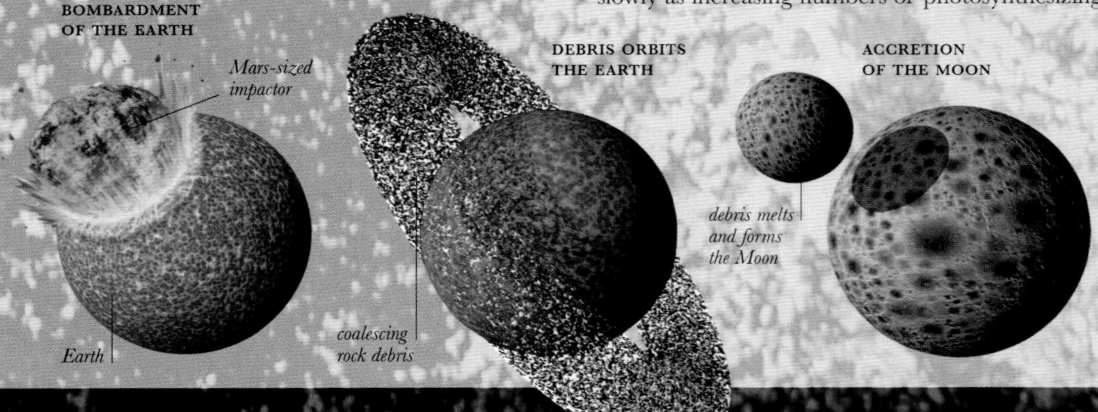


**BOMBARDMENT OF THE EARTH**

**DEBRIS ORBITS THE EARTH**

**ACCRETION OF THE MOON**

500 | 3,500 | 3,400 | 3,300 | 3,200 | 3,100 | 3,000 | 2,900 | 2,800 | 2,700 | 2,60

high surface temperatures falling gradually

temperatures falling rapidly to well below present average

◆ oldest stromatolites (Warrawoona Group, Australia)

◆ microscopic filaments and spheroids

first prokaryotes and eukaryotes (indicated by chemical fossils)

## METEORITE BOMBARDMENT

Dating of the Moon's rocks and the appearance of its heavily cratered surface provide evidence that for some 600 million years after their formation, both the Moon and Earth were subject to intense bombardment from space. The Moon has impact craters, such as the Aitken Basin, up to 1,500 miles (2,500 km) across, and because of its greater size and gravity, the Earth underwent greater bombardment than the Moon. The results were melting of rock and destruction of any nascent life and atmosphere.

**A LIFE-SUSTAINING ATMOSPHERE**

*The Earth's first atmosphere was blasted away by impacts and the solar wind. Volcanism built up a secondary atmosphere by releasing nitrogen, carbon dioxide, and water vapor. The latter was split by ultraviolet light into hydrogen, oxygen, and ozone. The lightest gas, hydrogen, was released into space.*

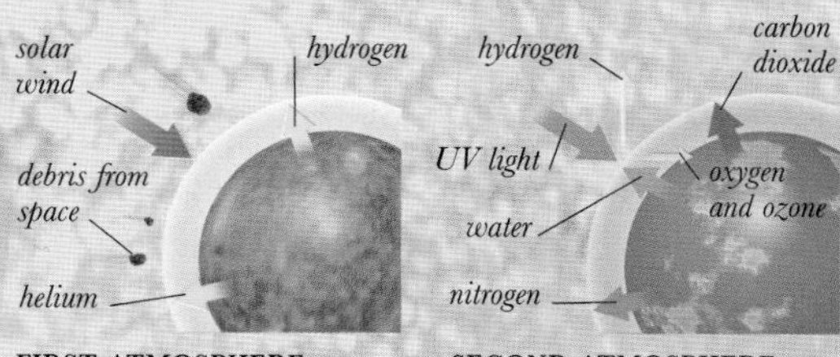


**FIRST ATMOSPHERE** **SECOND ATMOSPHERE**

## OCEANS AND CONTINENTS

The Hadean era, as its name suggests, was hellish: the early Earth's turmoil destroyed any primary atmosphere, oceans, and life. The Archean times that followed saw the water vapor released from volcanoes collecting to form oceans, where salts dissolved to increase salinity gradually. Low-density silicate minerals accumulated to form the Earth's outer crust. But continuing volcanic eruptions resulted in the constant formation of new surface rocks. Since the Earth was no longer expanding through accretion, the cooler and denser rocks, which had been formed previously, sank into the interior to accommodate the new crustal rocks. The Earth's crust was fragmented into a number of plates, with both convergent and divergent margins (see pp.108–109). Low-density rocks accreted into continents and higher-density rocks formed the ocean floor.

## EARLY LIFE

The earliest evidence for life on Earth comes from hydrocarbon residues in metamorphosed sedimentary rocks in Greenland, which date from about 3.8 billion years ago. These residues were derived from living organisms, which were probably aquatic prokaryotes (bacteria) that used light energy from the Sun. Their evolution must have been considerably earlier, perhaps more than 4 billion years ago. It is not known how this early life could have survived the heavy bombardment of the Earth by meteorites that occurred around this time. It is possible that life evolved twice, or was introduced to the Earth by the impacting bodies from space.

By 3.46 billion years ago, microbial photosynthesizers in warm, shallow waters formed mounds known as stromatolites, while other microbes lived off chemicals generated by submarine hot springs. Chemical fossils (see p.28) suggest that the first eukaryotes (organisms with relatively complex cells that contain a nucleus) appeared about 2.7 billion years ago, but the first direct fossil evidence for their existence does not appear until 2.2 billion years ago.

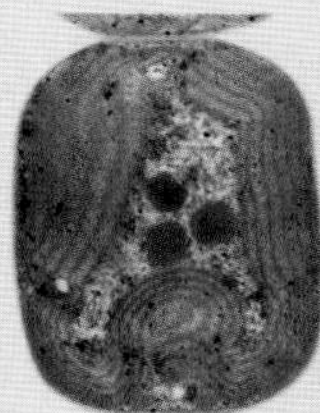

**PROKARYOTES**
*These single-celled organisms have a simple cell that lacks structures (such as a nucleus) found in advanced cells.*

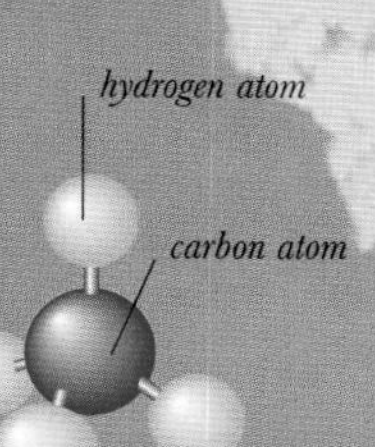


**METHANE ($CH_4$)**
*The simplest organic compound is one carbon atom combined with four hydrogen atoms.*

## CLAIRE PATTERSON

After working on the first atomic bomb and studying radioactivity, US physicist Claire Patterson became interested in calculating the age of the Earth. In 1956, he compared measurements from meteorites and Earth minerals to obtain the age of 4.55 billion years. This was the first accurate estimate.

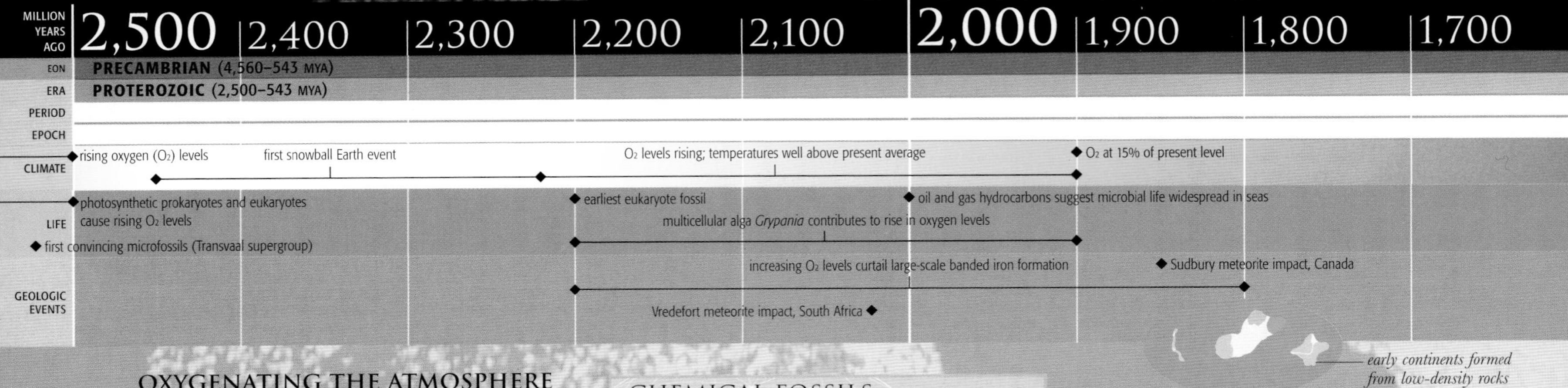


## OXYGENATING THE ATMOSPHERE

About 2,700 million years ago, primitive photosynthesizing microorganisms released increasing volumes of oxygen into the early atmosphere. This oxygen was initially used up oxidizing iron in the oceans, and little entered the atmosphere. By 2.2 billion years ago, oxygen levels were still only about 1 percent of present levels (oxygen comprises 21 percent by volume of today's atmosphere), but by 1,900 million years ago oxygen levels were about 15 percent of present levels. Microorganisms preferring the anoxic (low-oxygen) conditions were forced to adapt by retreating within sediments and below ground.

An important development for evolving life was the formation of the protective ozone layer above the atmosphere to filter out incoming ultraviolet rays that are particularly harmful to DNA. Above a 2-percent oxygen level, an ozone layer will begin to form, so it should have been well established by 1.9 billion years ago. An oxygen-rich atmosphere, protected by an ozone shield, permitted new life-forms to evolve and thrive. Oxygen-tolerant photosynthesizers inherited the Earth.

## MULTICELLULAR LIFE

Coiled filamentous microfossils from Michigan, called *Grypania*, are 2.2 billion years old and provide the first direct fossil evidence for eukaryotes. These organisms may have had multiple eukaryote cells, but the oldest convincing direct fossil evidence for multicells comes in the form of 1.2-billion-year-old fossils of a red alga called *Bangiomorpha* from arctic Canada. The many cells of these microscopic filaments show some specialization, including structures indicating that they reproduced sexually.

By 1,000 million years ago, toward the end of the Proterozoic era and following a prolonged evolutionary gestation, the "big bang" of eukaryotic evolution had started. The evolution of multicellular organisms with sexual reproduction (evidence of fossilized embryos have been found in China dating from about 600 million years ago) paved the way for larger and more diverse organisms. It used to be thought that it was the development of sexual reproduction, with its exchange of genetic material, that was the key to this significant

### CHEMICAL FOSSILS

Even when organisms die and their tissues decay, the organic chemicals (hydrocarbons) of which they are made may survive in sediments, albeit in a degraded form. These are known as chemical fossils. Commonly, these hydrocarbons form mobile oil and gas, but some residues can persist within the rock record. Chemical analysis of these residues can differentiate the organic molecules present, their structural complexity and, broadly, what kind of organisms they are derived from.

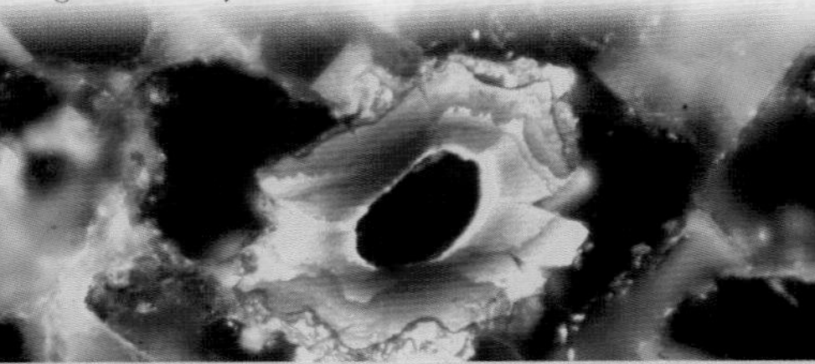

**MODERN STROMATOLITES**
*Known as stromatolites, these mounds of layered sediment with surface films of bacteria are identical to those found in Precambrian sedimentary rocks.*

**BANDED IRON**
*These layered iron deposits were formed by oxidation reactions in the oceans before the formation of the atmosphere.*

1,500 | 1,400 | 1,300 | 1,200 | 1,100 | 1,000 | 900 | 800 | 700

◆ glaciation
◆ glaciation
second snowball Earth event
last snowball Earth event ◆
◆ first multicellular alga fossils (*Bangiomorpha*), showing evidence of sexual reproduction
◆ decline in stromatolites
oldest multicellular (metazoan) body fossils ◆
first skeletons (silica) in unicell algae ◆
◆ major spread of eukaryote population
first Ediacarans ◆
amalgamation of supercontinent of Rodinia, followed by breakup
banded iron formation ◆
banded iron formation ◆
banded iron formation ◆
completion of Rodinian break up ◆
low sea-floor spreading rate ◆

*clustering to form Rodinian supercontinent*

*formation of Gondwanan supercontinent*

diversification. However, it is now known that this innovation alone did not promote eukaryotic diversification of life, as even bacteria exchange genetic material. More important, perhaps, is the ability of multicellular organisms to increase in size beyond the microscopic, with specialization of certain cells for certain tasks within the organism. By 580 million years ago, the earliest large animal fossils (known as the Ediacarans) appear, followed by the first shelled animals, *Cloudina*, in 555-million-year-old marine sediments from Namibia.

**GRYPANIA**
*Coiled ribbons, at least 8 in (20 cm) long, called* Grypania *may be the oldest multicellular eukaryote algae.*

**COMPLEX CELLS EVOLVE**
*Unlike the more primitive prokaryotes, eukaryotes have their nuclear material enclosed in a membrane. Eukaryote cells are usually part of a larger, multicellular organism, while prokaryotes are exclusively single-celled organisms and often incorporate a flagellum for mobility in a watery environment.*

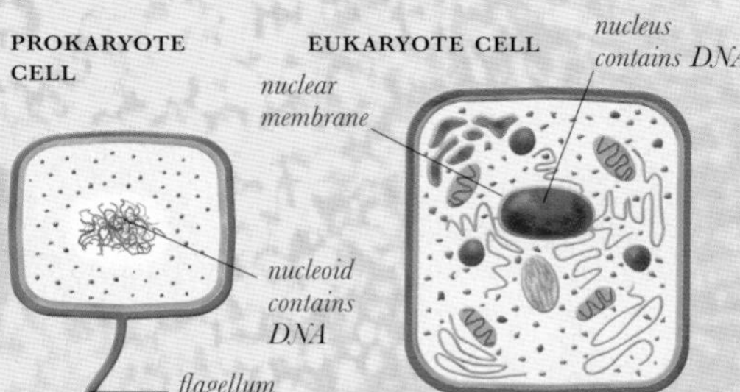


## SNOWBALL EARTH

There is growing evidence for the idea that the Earth suffered runaway glaciation in the Precambrian. The theory claims that ice caps extended from the poles into the tropics, encompassing most of the Earth's surface in what are called snowball events. Proof of glaciation in low latitudes requires evidence that glacial sedimentary rocks were in the tropics at the appropriate time. Measurements of magnetically oriented minerals within Precambrian glacial sediments from all over the globe indicate that 730–580 million years ago there were several snowball events (with glacial retreat in between), and that earlier episodes may have occurred between 2,450 and 2,220 million years ago. Important support for the theory comes from the presence of carbonate sediments, which are typical of low latitudes, immediately above glacially related ones. The appearance of sedimentary iron formations in the oceans suggests that conditions were anoxic, which is consistent with glacial events. The cause of such events is not fully understood but is thought to be related to the clustering of the continents within the tropics, raising the amount of light reflected from the planet and cooling global climates.

**SNOWBALL EARTH**
*Geological evidence shows that low-latitude continents were glaciated; the extent of sea ice cover is less clear.*

## SOFT-BODIED ANOMALIES

The Ediacarans are among the most intriguing Precambrian fossils. A group of sea-dwelling organisms that lived 580–543 million years ago, they are the first known animals. Their bodies were up to 6 ft (2 m) long, and they looked like relatives of jellyfish (see below)—although their relationship to living animal groups is unclear. They had a variety of body shapes, including a flat disk or a frond, while some had attachment disks and curiously quilted or serially divided surfaces. About 100 species are known, all entirely soft-bodied and preserved as sediment molds and infills, suggesting that some lived within the sediment and that their body tissues were tougher than those of jellyfish, which are rarely preserved as fossils. Their first known appearance follows the last snowball event, and their evolution may have been stimulated by its meltdown.

**EDIACARAN FOSSIL**
Spriggina, *an Ediacaran fossil with a head-like structure, is named after Australian geologist Reg Spriggs, who first found the Ediacarans.*

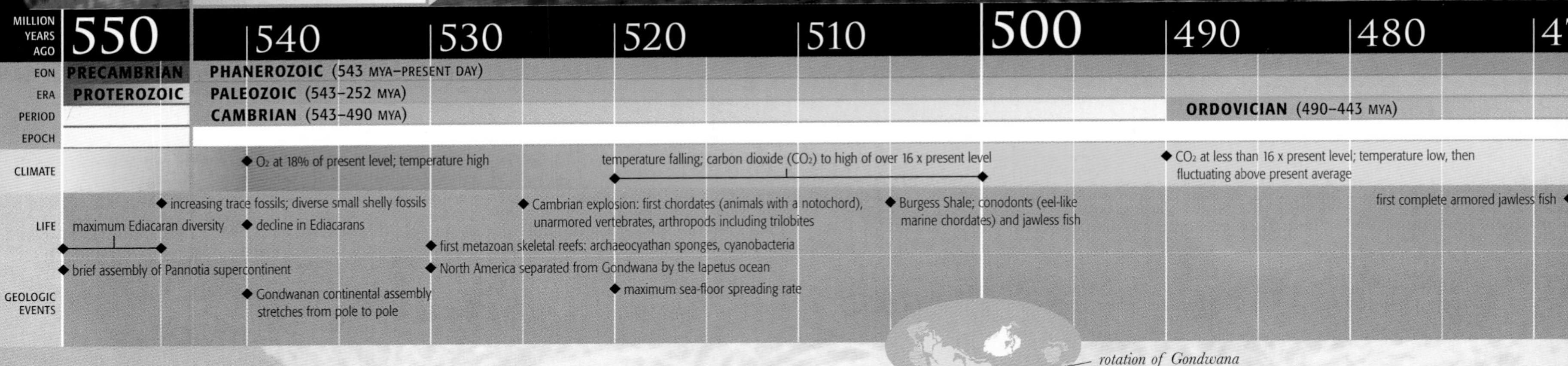


*rotation of Gondwana toward south pole*

# THE PALEOZOIC ERA

Paleozoic means "ancient life", and this era saw the appearance of abundant shelly fossils and the development of land plants. It ended with life's biggest-ever extinction event, at the end of the Permian Period, when 90 percent of all life on the Earth was wiped out.

## PLATE TECTONICS

The Earth's continents are constantly reconfigured by movements of crustal plates, with new oceans opening and the subduction of the old ocean floor (see p.109). During the Paleozoic, most continents were joined to form the supercontinent called Gondwana. Over time, this drifted north from the southern hemisphere to form the even larger supercontinent of Pangaea, which by the end of the Paleozoic stretched from pole to pole.

## MOLECULAR CLOCKS

The molecular clock estimates the age of groups of organisms, without relying on fossils. By measuring genetic similarity between living groups and assuming a constant rate of evolution, the time of their divergence can be calculated. For instance, according to the molecular clock, sharks evolved 528 million years ago, yet fossils first occur in 385-million-year-old strata. The discrepancy may reflect a lack of fossils, or a misunderstanding about rates of evolution.

## THE CAMBRIAN EXPLOSION

In the mid-19th century, fossil evidence seemed to indicate that life began in the earliest Cambrian with the appearance of a variety of sea-dwelling organisms, such as sponges and trilobites, with mineralized shells and skeletons. We now know that life is much more ancient and that shelled animals first appear in late Precambrian rocks. However, it still seems that a large variety of fossil shells appear suddenly at the beginning of the Cambrian. Many of these fossils are very small and represent several different kinds of mollusks and armored arthropods (creatures with paired and jointed limbs)—it is as if there was an explosion in the diversity of life.

Nevertheless, some scientists argue that there was no great explosion of life; rather, it looks that way because a large number of organisms suddenly acquired shells and hard parts, which are much better preserved than soft body parts. Fossil faunas of Cambrian age from China, Canada's Burgess Shale, and northern Greenland show that arthropod diversity was already high, with many different habitats filled by specialized animals. The diversity of these organisms could indicate their presence in the late Precambrian, which would imply that the Cambrian explosion was an artifact of preservation. Nevertheless, within Cambrian times there is fossil evidence for the appearance of most of the major groups of marine invertebrates (those animals that lack a backbone), including a number of groups that subsequently became extinct, such as the trilobites, conodonts, and graptolites. Importantly, from the evolutionary point of view, a recent discovery from China shows that the first fishlike animals with the beginnings of a backbone

**SILURIAN TRILOBITE**
*Marine arthropods called trilobites, with mineralized exoskeletons, evolved in Cambrian times.*

## CHARLES WALCOTT

American paleontologist Charles Walcott (1850–1927) quarried some 70,000 mid-Cambrian fossils from the Burgess Shale in the Canadian Rocky Mountains, which is now a World Heritage Site. The soft-part preservation of its fossils provides insights into marine life of the time. Walcott became director of the US Geological Survey, then secretary of the Smithsonian Institution in Washington, D.C.

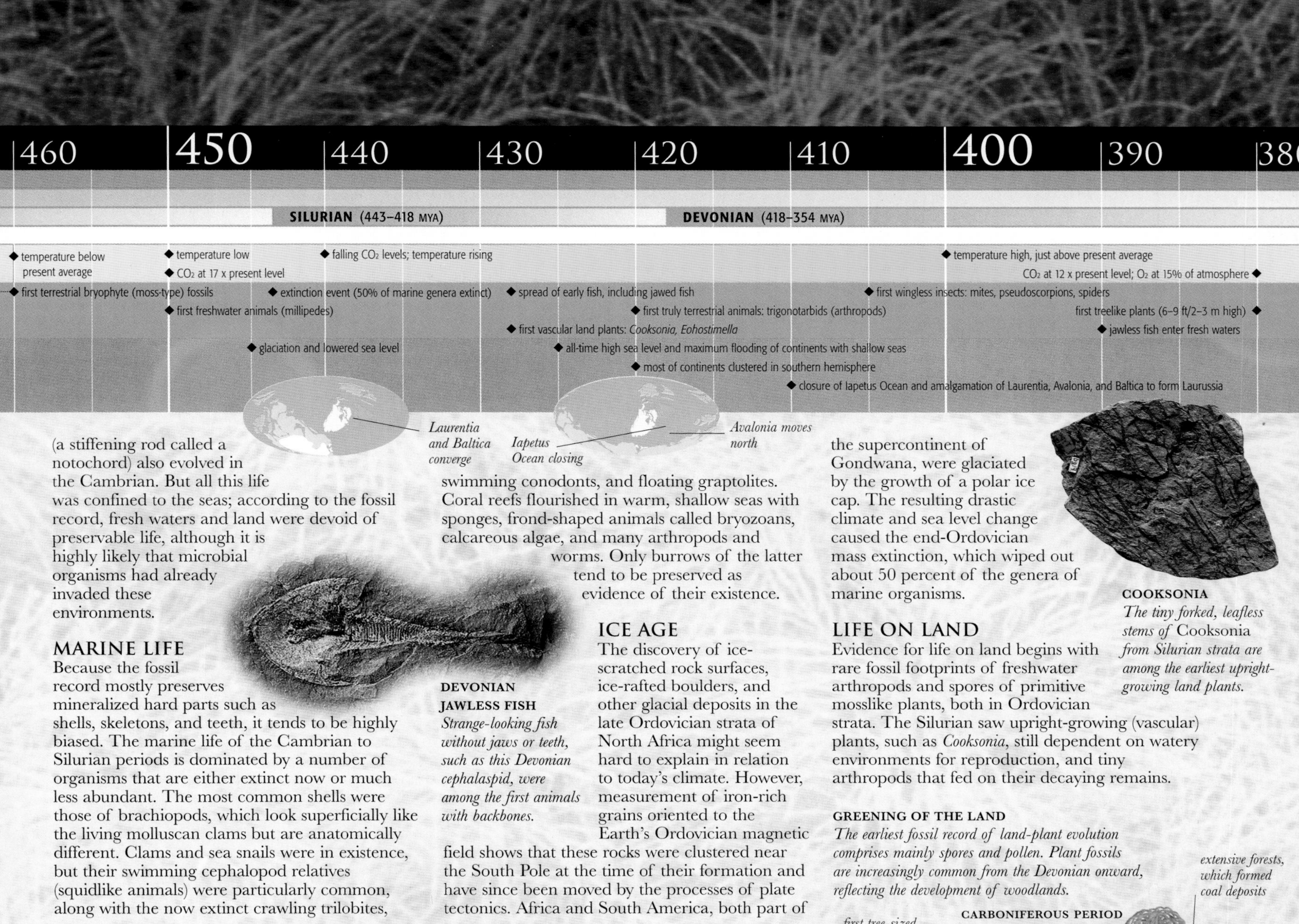


(a stiffening rod called a notochord) also evolved in the Cambrian. But all this life was confined to the seas; according to the fossil record, fresh waters and land were devoid of preservable life, although it is highly likely that microbial organisms had already invaded these environments.

## MARINE LIFE

Because the fossil record mostly preserves mineralized hard parts such as shells, skeletons, and teeth, it tends to be highly biased. The marine life of the Cambrian to Silurian periods is dominated by a number of organisms that are either extinct now or much less abundant. The most common shells were those of brachiopods, which look superficially like the living molluscan clams but are anatomically different. Clams and sea snails were in existence, but their swimming cephalopod relatives (squidlike animals) were particularly common, along with the now extinct crawling trilobites, swimming conodonts, and floating graptolites. Coral reefs flourished in warm, shallow seas with sponges, frond-shaped animals called bryozoans, calcareous algae, and many arthropods and worms. Only burrows of the latter tend to be preserved as evidence of their existence.

**DEVONIAN JAWLESS FISH**
*Strange-looking fish without jaws or teeth, such as this Devonian cephalaspid, were among the first animals with backbones.*

## ICE AGE

The discovery of ice-scratched rock surfaces, ice-rafted boulders, and other glacial deposits in the late Ordovician strata of North Africa might seem hard to explain in relation to today's climate. However, measurement of iron-rich grains oriented to the Earth's Ordovician magnetic field shows that these rocks were clustered near the South Pole at the time of their formation and have since been moved by the processes of plate tectonics. Africa and South America, both part of the supercontinent of Gondwana, were glaciated by the growth of a polar ice cap. The resulting drastic climate and sea level change caused the end-Ordovician mass extinction, which wiped out about 50 percent of the genera of marine organisms.

**COOKSONIA**
*The tiny forked, leafless stems of* Cooksonia *from Silurian strata are among the earliest upright-growing land plants.*

## LIFE ON LAND

Evidence for life on land begins with rare fossil footprints of freshwater arthropods and spores of primitive mosslike plants, both in Ordovician strata. The Silurian saw upright-growing (vascular) plants, such as *Cooksonia*, still dependent on watery environments for reproduction, and tiny arthropods that fed on their decaying remains.

**GREENING OF THE LAND**
*The earliest fossil record of land-plant evolution comprises mainly spores and pollen. Plant fossils are increasingly common from the Devonian onward, reflecting the development of woodlands.*

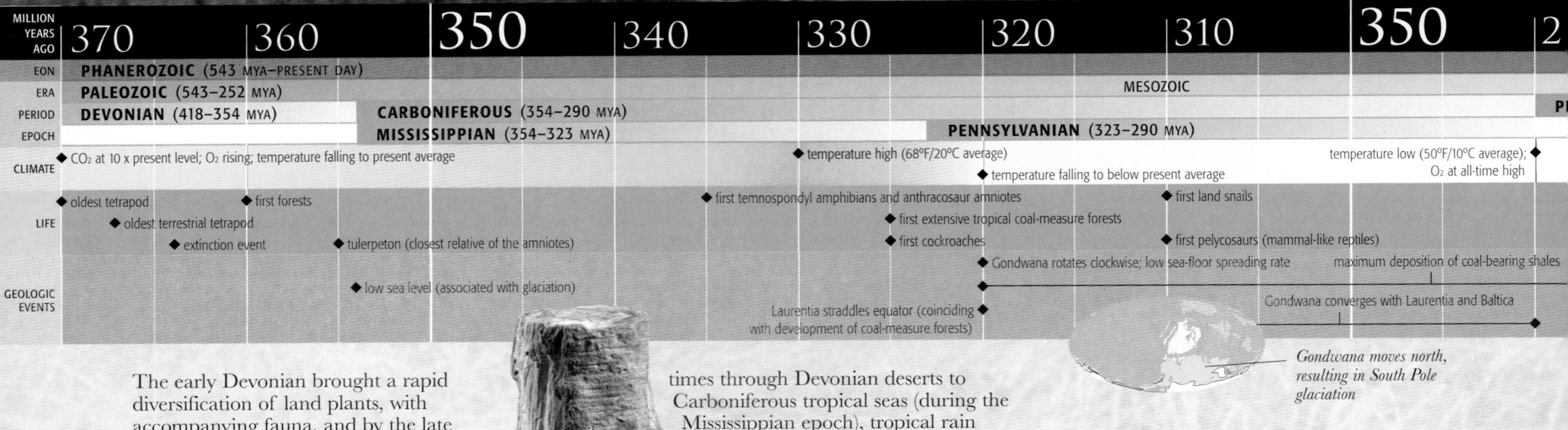


*Gondwana moves north, resulting in South Pole glaciation*

The early Devonian brought a rapid diversification of land plants, with accompanying fauna, and by the late Devonian, the first forests covered low-lying wetlands, which were full of fish. Some of these fish evolved four limbs (making them the earliest tetrapods) and ventured onto land to feed, but they had to return to water to breed.

## CLIMATE CHANGE

Today the sequence of rock strata preserved within individual continents, such as North America, shows a Paleozoic history of drastic climate change, from the cool marine waters of Ordovician times through Devonian deserts to Carboniferous tropical seas (during the Mississippian epoch), tropical rain forests (during the Pennsylvanian epoch), and back to deserts again in the Permian period. However, this sequence of changes is due not, as might be expected, to dramatic changes in global climates, but to the movement of continents through the Earth's climate zones due to plate tectonics.

From Ordovician to Permian times, North America (on the Laurentian Plate) moved from the southern hemisphere across the equator into the northern hemisphere.

**FOSSILIZED TREE STUMP**
*Fossil trees from Antarctica indicate times when the climate of the continent was much warmer than it is today.*

**PERMIAN SANDSTONE**
*Thick piles of sandstone strata were deposited within the vast arid deserts of the Pangaean supercontinent. Structures in the beds show that these were once part of sand dunes.*

## REPTILE RELATIVES

Early land tetrapods were salamander-like animals about 3 ft (1 m) in length. Little is known of their biology except that they had to return to water to breed. By the mid-Carboniferous, both amphibians and more reptilelike groups had evolved. They were 6–10 ft (2–3 m) long and resembled crocodiles and lizards. However, the critical reptilian feature—the ability to lay a membrane-covered (amniote) egg on dry land—is difficult to recognize in the fossils.

From the late Carboniferous into the Permian Period, however, true reptiles were diversifying. The Pangaean supercontinent was occupied by many short-lived reptile groups, some of which were to eventually evolve into the dinosaurs, others into turtles, and yet others into mammals.

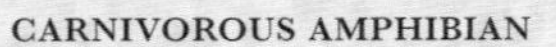

**CARNIVOROUS AMPHIBIAN**
*These early Triassic fossil remains of several meat-eating amphibians were found in South Africa, clustered around those of a plant-eating reptile called* Lystrosaurus.

## EXTINCTION

At the end of the Permian, an estimated 90 percent of all living organisms (60 percent of genera), both marine and land-living, died out in the biggest mass extinction known. Groups that died out include the Paleozoic corals, trilobites, and water scorpions, while the echinoderm crinoids (sea lilies), brachiopods, clams, and sea snails suffered serious losses, as did most life on land.

Geologists have searched for a single cause, especially a major meteorite impact like that which brought about the end-Cretaceous extinction event (see pp.34–35), but without success. The extinction did coincide with a vast outpouring of lavas and volcanic gasses in Siberia (known as the Siberian Traps), which may have altered global climates. Also, there is evidence for widespread anoxia in the oceans, which may have fatally disrupted food chains, causing many groups to die out. The source of this anoxia may have been a sudden release of huge volumes of carbon dioxide gas from frozen masses of methane, known as gas hydrates, which had lain buried as ice in seabed deposits.

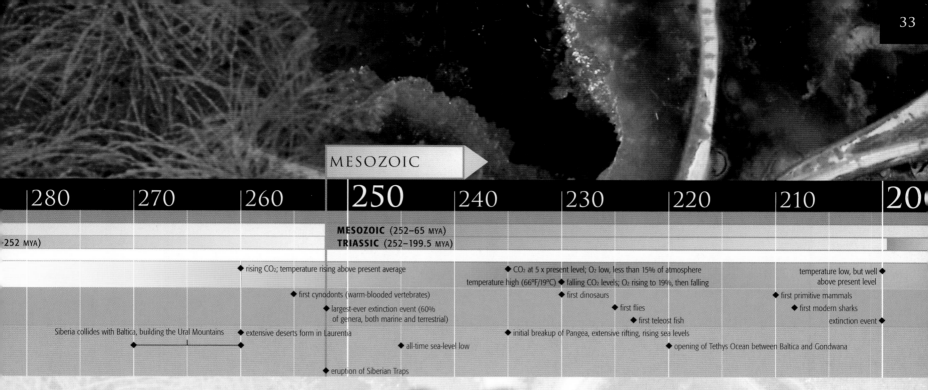

MESOZOIC

280 | 270 | 260 | 250 | 240 | 230 | 220 | 210 | 20

-252 MYA)

**MESOZOIC** (252–65 MYA)

**TRIASSIC** (252–199.5 MYA)

◆ rising $CO_2$; temperature rising above present average

◆ first cynodonts (warm-blooded vertebrates)

◆ largest-ever extinction event (60% of genera, both marine and terrestrial)

Siberia collides with Baltica, building the Ural Mountains

◆ extensive deserts form in Laurentia

◆ all-time sea-level low

◆ eruption of Siberian Traps

◆ $CO_2$ at 5 x present level; $O_2$ low, less than 15% of atmosphere

temperature high (66ºF/19ºC) ◆ falling $CO_2$ levels; $O_2$ rising to 19%, then falling

◆ first dinosaurs

◆ first flies

◆ first teleost fish

◆ initial breakup of Pangea, extensive rifting, rising sea levels

◆ opening of Tethys Ocean between Baltica and Gondwana

temperature low, but well above present level ◆

◆ first primitive mammals

◆ first modern sharks

extinction event ◆

# THE MESOZOIC ERA

Spanning the 187 million years of the Triassic, Jurassic, and Cretaceous periods, the Mesozoic (or "middle life") era lies between two major extinction events, the second one known as the end-Cretaceous extinction. It is often characterized as the Age of the Reptiles, since many major reptile groups were dominant life-forms, including the dinosaurs on land, the ichthyosaurs in the seas, and the pterosaurs in the air.

## MARINE LIFE

The end-Permian extinction and collapse of the Paleozoic reefs caused a major turnover in marine life. Eventually, modern reef corals evolved and marine diversity and food chains were reestablished. The more adaptable molluscan clams and snails gradually took over from brachiopods. Both bony and cartilaginous fish diversified, with some sharks becoming top predators. But the waters were also occupied by new marine reptiles, such as turtles and crocodiles, along with the now-extinct plesiosaurs, dolphinlike ichthyosaurs, and mosasaurs.

**MARINE TURTLES**
*The first turtles with protective shells appeared in the Cretaceous, evolving from the older land-living tortoises of the Triassic.*

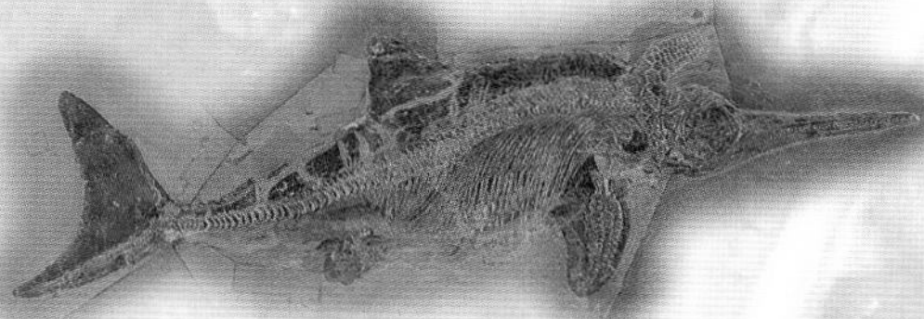

**SEA REPTILES**
*New predators occupied the seas, such as the fast-swimming ichthyosaurs. Shaped like dolphins with toothed, beaklike jaws, these reptiles lived from the Triassic to the Cretaceous.*

## PLANT LIFE

During the late Permian and early Triassic, the dominant Paleozoic plants—such as ferns, club mosses, and horsetails—declined. Newly diversifying groups included conifers, which reproduce by means of seeds borne in cones. They became particularly important in Jurassic and Cretaceous times, along with cycads and cycadlike bennettitaleans. Most of the cycads also had seed-bearing cones and were distributed worldwide, even in polar regions when they were free of ice. Flowering plants (angiosperms) evolved through the Cretaceous. They combined a number of features found in other plants of the Mesozoic, such as flower-type reproductive structures (also found in the bennettitaleans), with the one unique feature that distinguishes flowering plants—unfertilized seeds enclosed in a carpel.

**SURVIVING CYCADS**
*The few living cycads are survivors of a far larger group of seed-bearing plants that were key members of Jurassic and Cretaceous forests.*

## DINOSAURS

For 150 million years, from the late Triassic to the end of the Cretaceous, terrestrial life was dominated by reptiles. One very varied group, known as the dinosaurs, with some 900 genera, diversified from crow-sized bipedal forms to the largest land-living animals ever known—the four-limbed sauropods, at 110 ft (34 m) long.

Dinosaurs were first recognized as a distinct group of reptiles in 1842, their distinguishing characteristic being that they walked with their legs tucked under their bodies, unlike living reptiles, which have splayed legs. Most very large dinosaurs were plant-eaters with long necks to reach the high canopies of Mesozoic trees. However, there were also very large carnivorous predators such as the Jurassic allosaurs, which were up to 40 ft (12 m) long.

**PLACERIAS**
*This late-Triassic 10-ft- (3-m-) long giant from North America was one of last surviving therapsids (mammal-like reptiles).*

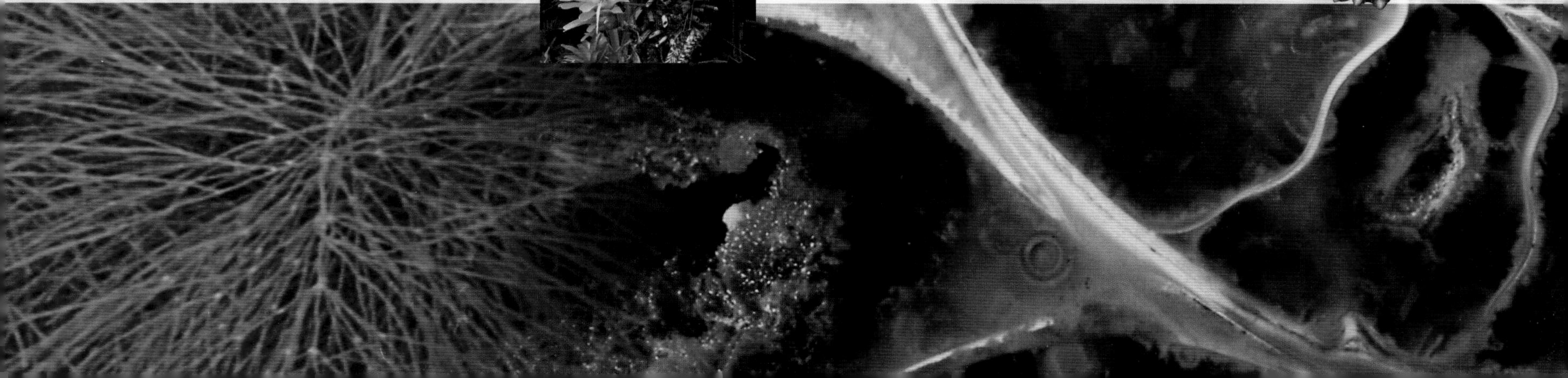

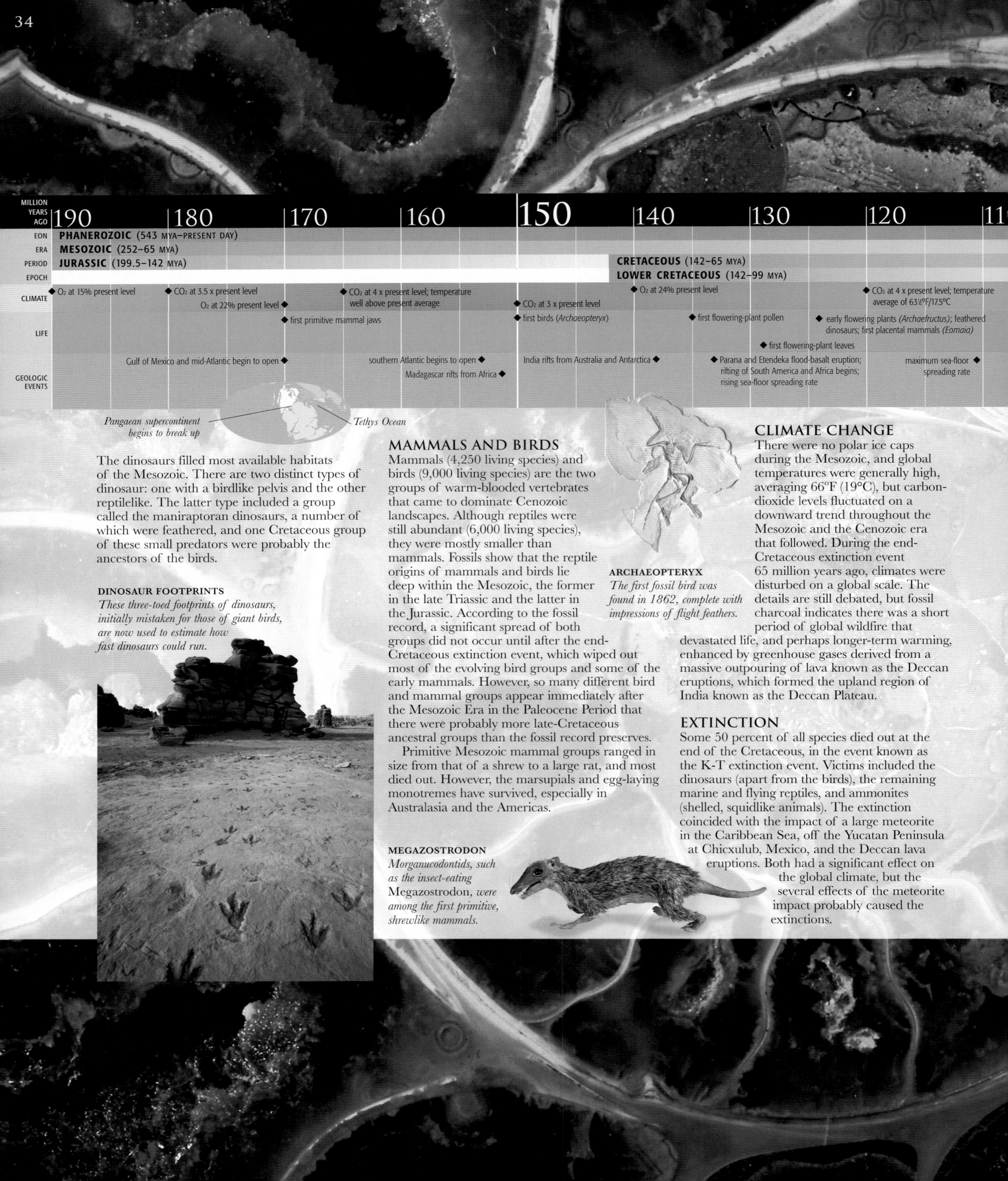


The dinosaurs filled most available habitats of the Mesozoic. There are two distinct types of dinosaur: one with a birdlike pelvis and the other reptilelike. The latter type included a group called the maniraptoran dinosaurs, a number of which were feathered, and one Cretaceous group of these small predators were probably the ancestors of the birds.

**DINOSAUR FOOTPRINTS**
*These three-toed footprints of dinosaurs, initially mistaken for those of giant birds, are now used to estimate how fast dinosaurs could run.*

## MAMMALS AND BIRDS

Mammals (4,250 living species) and birds (9,000 living species) are the two groups of warm-blooded vertebrates that came to dominate Cenozoic landscapes. Although reptiles were still abundant (6,000 living species), they were mostly smaller than mammals. Fossils show that the reptile origins of mammals and birds lie deep within the Mesozoic, the former in the late Triassic and the latter in the Jurassic. According to the fossil record, a significant spread of both groups did not occur until after the end-Cretaceous extinction event, which wiped out most of the evolving bird groups and some of the early mammals. However, so many different bird and mammal groups appear immediately after the Mesozoic Era in the Paleocene Period that there were probably more late-Cretaceous ancestral groups than the fossil record preserves.

Primitive Mesozoic mammal groups ranged in size from that of a shrew to a large rat, and most died out. However, the marsupials and egg-laying monotremes have survived, especially in Australasia and the Americas.

**ARCHAEOPTERYX**
*The first fossil bird was found in 1862, complete with impressions of flight feathers.*

**MEGAZOSTRODON**
*Morganucodontids, such as the insect-eating* Megazostrodon, *were among the first primitive, shrewlike mammals.*

## CLIMATE CHANGE

There were no polar ice caps during the Mesozoic, and global temperatures were generally high, averaging 66°F (19°C), but carbon-dioxide levels fluctuated on a downward trend throughout the Mesozoic and the Cenozoic era that followed. During the end-Cretaceous extinction event 65 million years ago, climates were disturbed on a global scale. The details are still debated, but fossil charcoal indicates there was a short period of global wildfire that devastated life, and perhaps longer-term warming, enhanced by greenhouse gases derived from a massive outpouring of lava known as the Deccan eruptions, which formed the upland region of India known as the Deccan Plateau.

## EXTINCTION

Some 50 percent of all species died out at the end of the Cretaceous, in the event known as the K-T extinction event. Victims included the dinosaurs (apart from the birds), the remaining marine and flying reptiles, and ammonites (shelled, squidlike animals). The extinction coincided with the impact of a large meteorite in the Caribbean Sea, off the Yucatan Peninsula at Chicxulub, Mexico, and the Deccan lava eruptions. Both had a significant effect on the global climate, but the several effects of the meteorite impact probably caused the extinctions.

CENOZOIC

100 | 90 | 80 | 70 | 60 | 50 | 40 | 30 | 20

**CENOZOIC** (65 MYA–PRESENT DAY)

**TERTIARY SUB-ERA: PALEOGENE** (65–23.3 MYA) — **NEOGENE**

**UPPER CRETACEOUS** (99–65 MYA) — **PALEOCENE** (65–54.8) — **EOCENE** (54.8–33.5 MYA) — **OLIGOCENE** (33.5–24) — **MIOCENE** (24–5)

- $CO_2$ at 2 x present level; temperature average of 61°F/16°C
- average temperature of 63°F/17°C; $CO_2$ at 2 x present level; $O_2$ at 27% of present level
- average temperature of 61°F/16°C
- $O_2$ at 23% of present level; temperatures fluctuate around present level
- initial growth of Antarctic ice sheet
- first snakes
- extinction of large marine reptiles
- K-T extinction event
- first primitive whale relative (*Pakicetus*)
- first grasslands and grazing mammals
- first mammalian primate (*Purgatorius*)
- earliest anthropoid (*Eosimias*)
- spread of songbirds
- spread of mammals
- first horses
- molecular divergence of primates from old-world monkeys
- Australia and Antarctica begin to rift; widespread chalk deposition in Tethys Ocean
- building of central Rocky Mountains
- rifting of the North Atlantic continues with intense volcanic activity
- collision of India with Asia begins building of Himalayas
- India begins to move rapidly north
- Chicxulub impact
- Tethys Ocean begins to close
- building of Alps begins
- breakup of Laurasia
- eruption of Deccan Traps, India
- Arabia and Africa rift along Red Sea; Ethiopian flood-basalt eruption
- opening of North Atlantic

*formation of Atlantic and Southern oceans*

*India moves toward Asia*

# THE CENOZOIC ERA

The Cenozoic Era extends from 65 million years ago to the present day (Cenozoic means "recent life"). The era is characterized by a diversification of modern bony fish, flowering plants, pollinating insects, birds, and mammals, including our primate relatives.

## RIFTING AND VOLCANISM

Cenozoic movements of the Earth's plates had a considerable influence on the evolution of life. Mountain building and large-scale uplift down the western flank of the Americas transformed regional climates, while the formation of the Himalayas (see pp.168–69) and the Tibetan Plateau generated the Southeast Asian monsoon (see p.463). Rifting and volcanism opened up the North Atlantic (see pp.402–403) and created the great East African Rift (see pp.148–49), where many of our primate ancestors flourished.

Cenozoic volcanism, associated with rising plumes of heat from within the Earth, led to surface doming and rifting(see p.156). The early Cenozoic saw a region 1,250 miles (2,000 km) wide, from Greenland to Norway, domed by up to 1¼ miles (2 km), leading to rifting and the outpouring of lava, with the Atlantic Ocean flooding north through the rift, a process continuing today in Iceland. Similar doming in Africa led to rifting and the outpouring of the Ethiopian flood basaltic lava (31–28 million years ago) and outpouring of the Columbia River lava in North America (15 million years ago).

**PRESERVED LAVAS AT GIANT'S CAUSEWAY**
*As early Cenozoic lavas erupted in the rift between Norway, Britain, and Greenland, the North Atlantic widened. Evidence of this outpouring can still be seen in Ireland (shown here).*

## LUIS AND WALTER ALVAREZ

Walter Alvarez (b.1940), an American geologist, found high concentrations of the rare element iridium in a marine clay dated to the Cretaceous–Cenozoic boundary. From this finding, in 1980 he developed the hypothesis with his father Luis (1911–88), a Nobel-winning physicist, that the iridium concentration is best explained by an extraterrestrial body, such as a very large meteorite, impacting upon the Earth. They further speculated that a global catastrophe accompanying the impact caused the demise of the dinosaurs.

**CHICXULUB IMPACT**
*Sixty-five million years ago, a meteorite crashed into shallow seas that are now part of the Yucatan Peninsula, Mexico (see p.123). The impact blasted huge volumes of carbonate rock through the atmosphere. On reentry, its radiant heat caused global wildfire.*

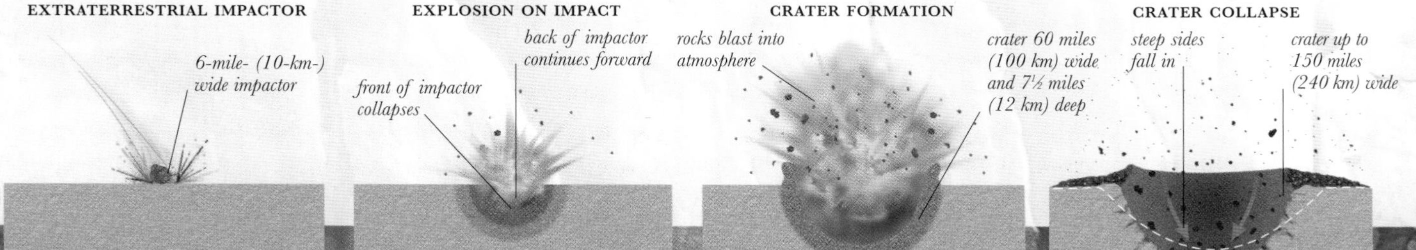

**EXTRATERRESTRIAL IMPACTOR**
*6-mile- (10-km-) wide impactor*

**EXPLOSION ON IMPACT**
*back of impactor continues forward*
*front of impactor collapses*

**CRATER FORMATION**
*rocks blast into atmosphere*
*crater 60 miles (100 km) wide and 7½ miles (12 km) deep*

**CRATER COLLAPSE**
*steep sides fall in*
*crater up to 150 miles (240 km) wide*

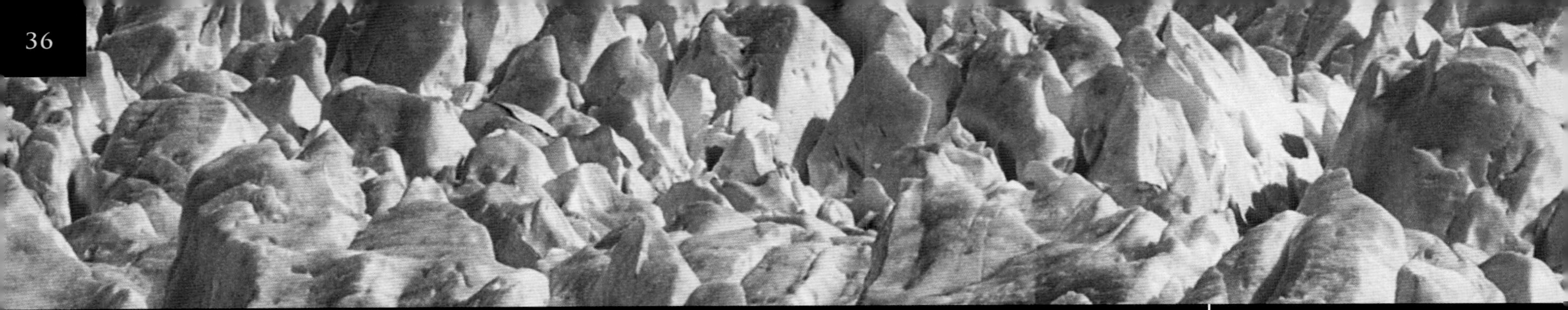

| MILLION YEARS AGO | 19 | 18 | 17 | 16 | 15 | 14 |
|---|---|---|---|---|---|---|

EON **PHANEROZOIC** (543 MYA–PRESENT DAY)

ERA **CENOZOIC** (65 MYA–PRESENT DAY)

PERIOD **TERTIARY SUB-ERA: NEOGENE** (23.3–1.8 MYA)

EPOCH **MIOCENE** (24–5 MYA)

CLIMATE
- temperatures fall ,then fluctuate just above present levels
- temperature high within general ice-house state

LIFE
- spread of primitive apes in Central Africa
- spread of grazing *hipparion* horses in North America and dispersal through Eurasia to Africa
- molecular divergence of orangutan from other great apes
- African forests diminish
- spread of whales and dolphins
- first cattle; spread of snakes, frogs, rats, and mice
- expansion of open grasslands into middle and high latitudes continues throughout Miocene

GEOLOGIC EVENTS
- Australia moves north
- rifting and volcanism in east Africa
- uplift of Tibet
- Columbia River flood-basalt eruption

## MARINE LIFE

Following the end-Cretaceous extinction, life in the seas and oceans gradually took on a modern appearance, from reef-building animals to the top predators. One of the most important developments was the increased diversity of modern bony fish following the extinction of the large predatory marine reptiles. Turtles, a few crocodiles, some iguanas, and snakes remain as marine reptiles, but most are small. The main innovation among vertebrates was the evolution of marine mammals, especially whales and dolphins (cetaceans). Diversifying Eocene mammals evolved a succession of extinct forms, from which modern whales arose 35 million years ago.

## LIFE ON LAND

In early Cenozoic times, mammals took over the habitats of those reptile victims of the end-Cretaceous extinction. However, there were major differences between the emerging mammals, with Australia and South America inheriting the pouched marsupials and egg-laying monotreme mammals, while the rest of the world was soon dominated by placental mammals. The latter give birth to bigger, more advanced young, nurtured for longer within the mother's body via a placenta.

Initially marsupials thrived, evolving both plant-eaters and carnivorous predators, which diversified into forms that were subsequently mimicked by the placental rodents, hippopotamuses, horses, dogs, and big cats. However, as the more advanced placentals evolved and diversified, they tended to displace the marsupials. Formation of a land bridge between North and South America allowed an interchange of placentals and marsupials between the continents. Today the surviving marsupials of the Americas are the opossums, which are still represented by 63 species. But even more marsupials held out for longer on the isolated continent of Australia.

The earliest Cenozoic placental mammals were shrew-like insect-eaters, but soon bigger (sheep-sized) browsing and rooting plant-eaters evolved. By the late Paleocene, there were rhinoceros-sized browsers and some dog-sized carnivores. The number of placental families rose from 21 in the late Cretaceous to 111 in the early Eocene.

The Earth's habitats were soon filled by placental mammals, from otters, seals, and whales in the seas, through the vast range of land mammals—from shrews to giant plant-eaters, such as the extinct 26-ft- (8-m-) long giant rhinoceros *Indricotherium*, and numerous carnivores—to bats in the air. Environmental and climatic changes also affected mammal evolution, especially decreasing forest cover and increasing grasslands, with the coevolution of pollinating insects, songbirds, and grazing mammals. These changes are also thought to have affected the evolution of hominids (higher apes and humans).

**FLOWERING TREES**
*By earliest Cenozoic times, flowering plants had evolved into a diverse range, including the magnolia (above) and laurels that formed the woodlands.*

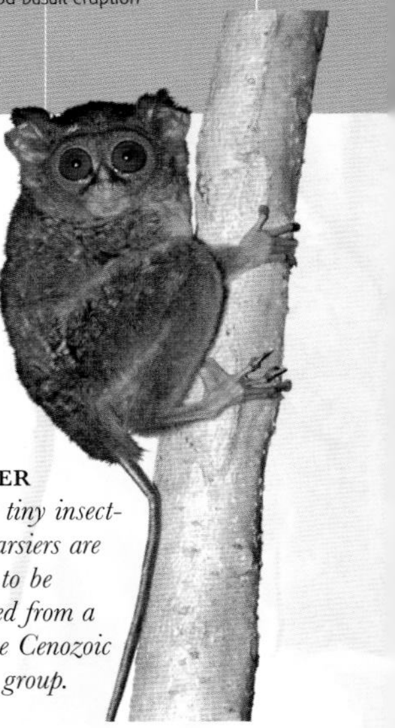

**TARSIER**
*Today's tiny insect-eating tarsiers are thought to be descended from a primitive Cenozoic primate group.*

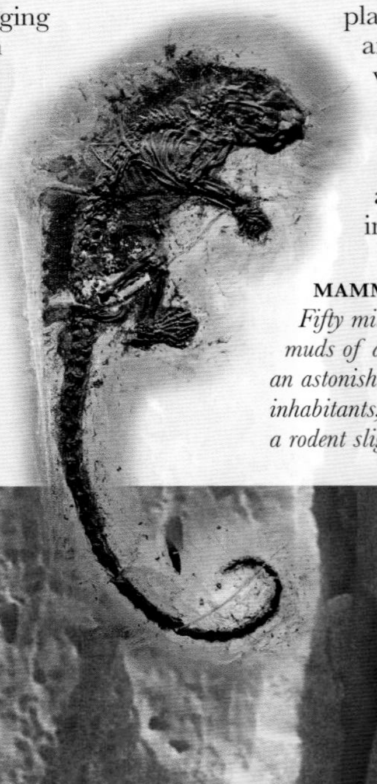

## MOUNTAIN-BUILDING

Cenozoic times saw the formation of four major mountain belts around the world: the Andes (see pp.158–59), the Rocky Mountains (see p.157), the European Alps (see pp.162–63), and the Himalayas (see pp.168–69). The Andes

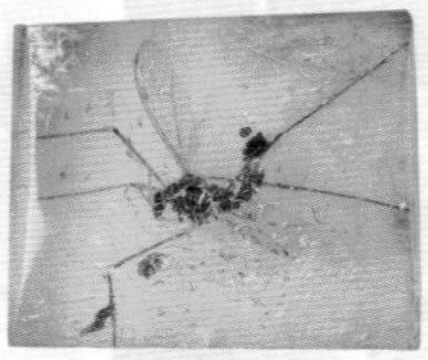

**MAMMALS OF LAKE MESSEL**
*Fifty million years ago, the oil-rich muds of a German lakebed preserved an astonishing diversity of its inhabitants, such as this* Ailuravus, *a rodent slightly larger than a squirrel.*

**AMBER LIFE**
*Since Cretaceous times, amber from some resin-producing trees has trapped and preserved small animals, especially insects.*

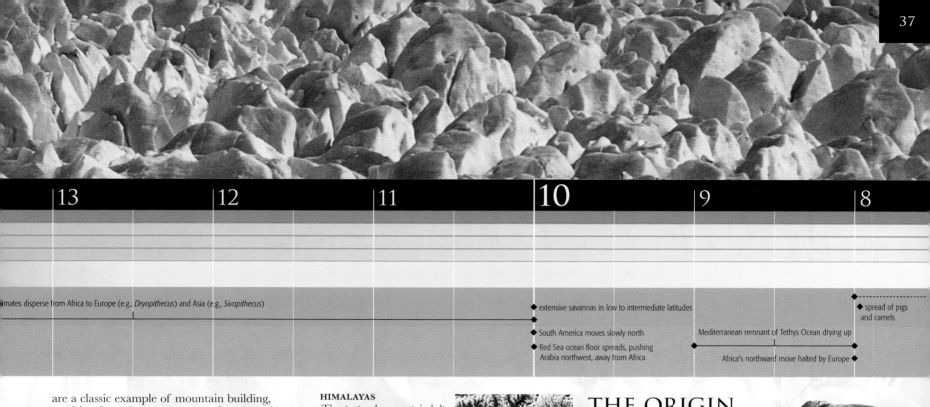

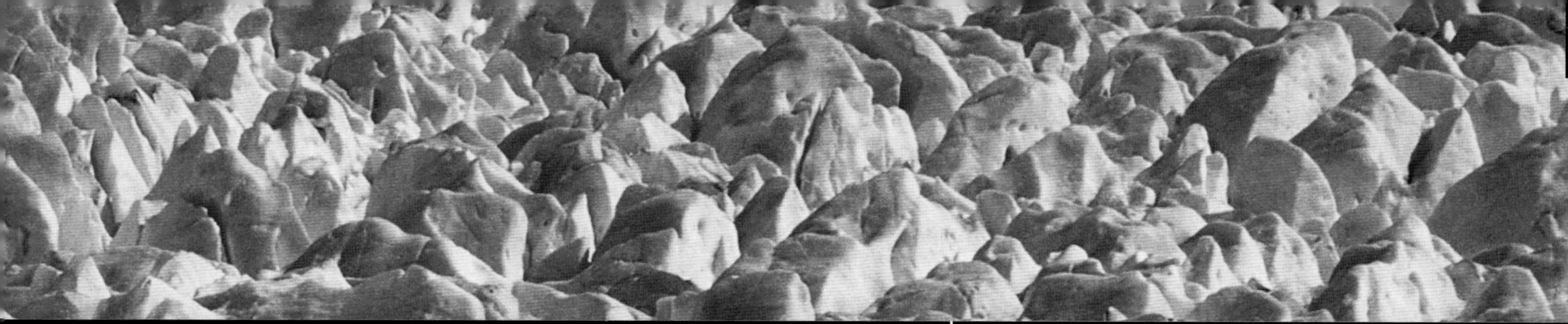

13 12 11 10 9 8

…imates disperse from Africa to Europe (e.g., *Dryopithecus*) and Asia (e.g., *Sivapithecus*)

◆ extensive savannas in low to intermediate latitudes

◆ spread of pigs and camels

◆ South America moves slowly north

Mediterranean remnant of Tethys Ocean drying up

◆ Red Sea ocean floor spreads, pushing Arabia northwest, away from Africa

Africa's northward move halted by Europe ◆

are a classic example of mountain building, resulting from the convergence of an oceanic (Pacific) plate with a continental plate (South America), combined with the subduction of the oceanic plate (see p.109) and frequent earthquakes and volcanoes. By contrast, the Alpine–Himalayan belt resulted from the convergence of the African and Indian plates with Europe and Asia respectively. Africa's northward movement subducted a body of water known as the Tethys Ocean; the intervening rocks were compressed, thickened, and elevated, with intense folding and faulting, leading to the formation of the Alps. The breakaway of India from Africa and Antarctica, which began in the Mesozoic, eventually caused the subduction of the eastern Tethys and its convergence with Asia, beginning 20 million years ago. Again, the intervening rocks were compressed, thickened, and elevated with intense folding and faulting, resulting in the formation of the Himalayas. The continuing northward drive of India is thought to have pushed the deeper part of the Indian plate beneath Tibet, thickening and elevating it without folding. The formation of this high and elongated physical barrier caused significant change in the regional climate, with the development of the Southeast Asian monsoon.

**HIMALAYAS**
*The spectacular mountain belt of the Himalayas, with the high plateau of Tibet to the north and low sediment-filled floodplains to the south, is the result of the convergence of the Indian and Asian plates.*

**SPREADING GRASSLANDS**
*By the Miocene, cooling climates and greater aridity caused forests to break up and grasslands to expand, along with fleet-footed grazing mammals and their predators, such as big cats.*

## THE ORIGIN OF HUMANS

Humans, apes, lemurs, and monkeys are all mammals and were first grouped together as primates in 1758 by Swedish naturalist Carolus Linnaeus (1707–78). The fossil record shows that the earliest primatelike mammals originated as small insect-eating animals, such as the shrewlike *Purgatorius* in the early Paleocene. In the early Cenozoic, a group of tree-climbing, squirrel-like animals (the plesiadapiforms) had evolved in North America and Europe. They were followed by tarsier- and lemurlike primates, which spread into Africa and Asia along with two groups of higher primates, the New World monkeys and the Old World monkeys and apes. It was from the latter group that hominoids evolved in Africa. These included the 18-million-year-old (early Miocene) tailless and monkeylike *Proconsul*, which could climb trees and walk on all fours. By the late Miocene Epoch, the apes diversified and separated into African, European, and Asian branches, some of whose members increased in size.

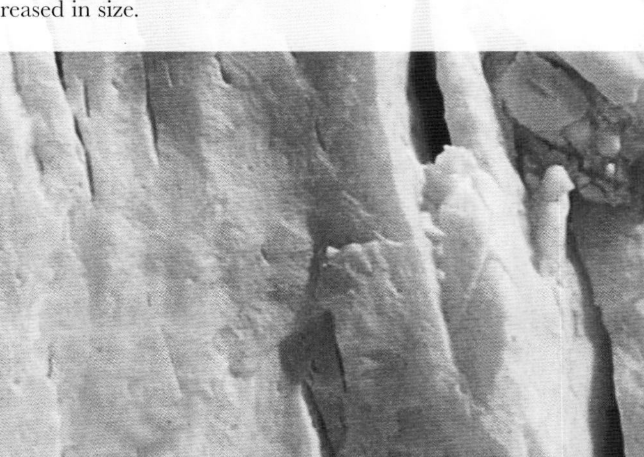

**BIPEDAL BEGINNINGS**
*It seems likely that our ape–human ancestor was a knuckle-walker, like today's apes. However, fossil hominids from 6–7 million years ago appear already to have walked on two feet.*

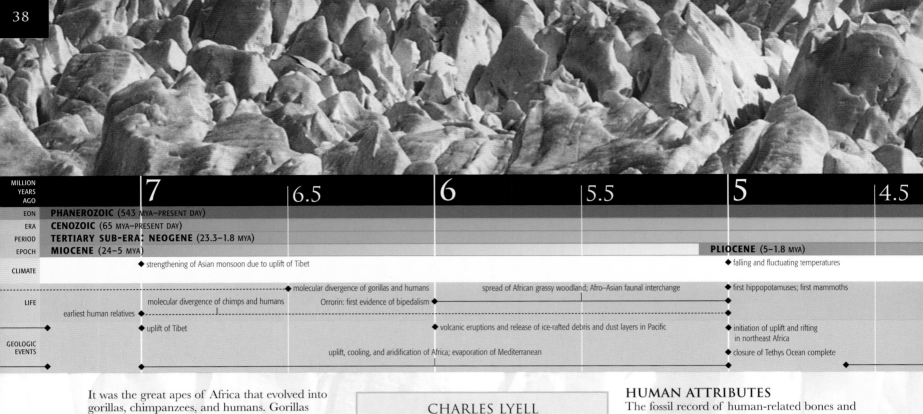

MILLION YEARS AGO: 7 | 6.5 | 6 | 5.5 | 5 | 4.5

EON: **PHANEROZOIC** (543 MYA–PRESENT DAY)

ERA: **CENOZOIC** (65 MYA–PRESENT DAY)

PERIOD: **TERTIARY SUB-ERA: NEOGENE** (23.3–1.8 MYA)

EPOCH: **MIOCENE** (24–5 MYA) | **PLIOCENE** (5–1.8 MYA)

CLIMATE:
- strengthening of Asian monsoon due to uplift of Tibet
- falling and fluctuating temperatures

LIFE:
- molecular divergence of gorillas and humans
- spread of African grassy woodland; Afro–Asian faunal interchange
- first hippopotamuses; first mammoths
- molecular divergence of chimps and humans
- Orrorin: first evidence of bipedalism
- earliest human relatives

GEOLOGIC EVENTS:
- uplift of Tibet
- volcanic eruptions and release of ice-rafted debris and dust layers in Pacific
- initiation of uplift and rifting in northeast Africa
- uplift, cooling, and aridification of Africa; evaporation of Mediterranean
- closure of Tethys Ocean complete

It was the great apes of Africa that evolved into gorillas, chimpanzees, and humans. Gorillas diverged first, about 6–8 million years ago, while the chimpanzees and humans share a common ancestor who, according to the molecular clock (see p.30), lived 5–7 million years ago.

## THE SPREAD OF HOMINIDS

Although the fossil record of our human relatives (the hominids) is very sparse, intensive searching over the last 50 years in southern Africa in general and around the East African Rift Valley has provided some insight into their evolution. A recent discovery of a fossil skull with some human features (belonging to a hominid called *Sahelanthropus*) is dated at 7 million years old. And the 6-million-year-old *Orrorin* leg bones are claimed to indicate upright walking on two legs (bipedalism), although direct evidence does not appear until the Laetoli footprints (see below).

**SKULL RECORD**
*By examining the skulls, it is clear that brain size increases significantly from* Australopithecus *to the large-brained* Homo *species:* H. neanderthalensis *and* H. sapiens.

**AUSTRALOPITHECUS**

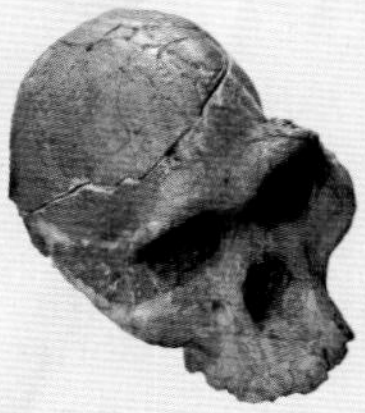

**HOMO NEANDERTHALENSIS**

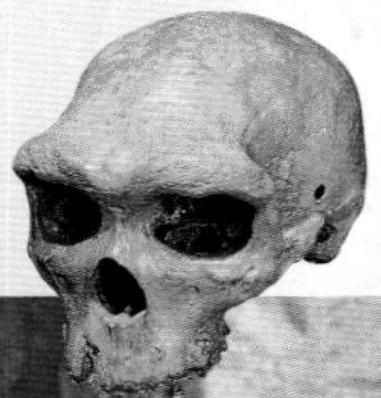

**HOMO SAPIENS**

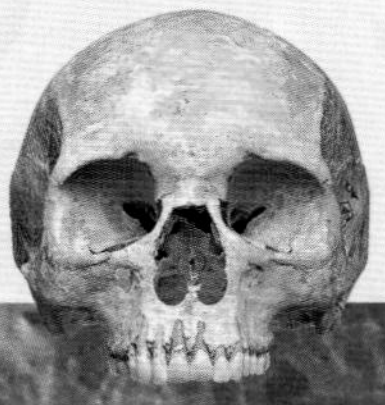

### CHARLES LYELL

A Scottish-born lawyer, Charles Lyell (1797–1875) took up geology and wrote the *Principles of Geology* (1830–33), an important synthesis of geological ideas that greatly influenced the naturalist Charles Darwin. Lyell believed that the interpretation of the past must be based on an understanding of present processes. He also divided the Tertiary into epochs. Lyell initially disagreed with the theory of evolution proposed by Charles Darwin and Alfred Wallace, but finally, if reluctantly, came to accept it.

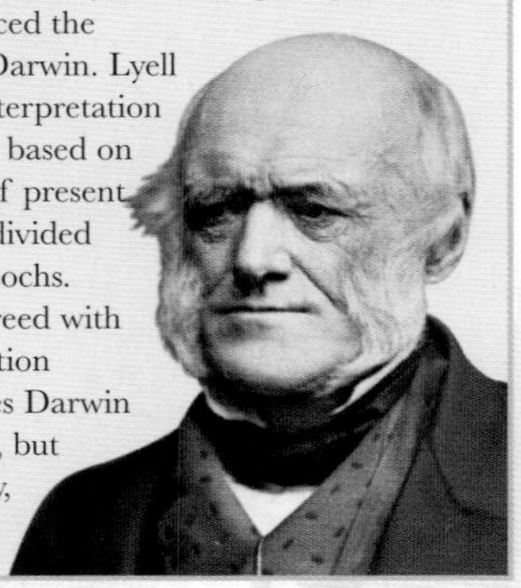

By 3 million years ago, these early human relatives were definitely small (about 3 ft/1 m tall) bipedal apes with small brains (australopithecines). They later split into two branches: large-jawed hominids that fed on plants and small animals, and small-jawed plant-eaters and meat scavengers, which had developing brains and used tools. The latter were the early members of our genus, *Homo*.

**LAETOLI FOOTPRINTS**
*Some 3.6 million years ago in Laetoli, east Africa, two upright-walking human-related adults and a juvenile walked on soft mud, leaving the oldest known footprints.*

## HUMAN ATTRIBUTES

The fossil record of human-related bones and stone tools suggests that human attributes were acquired not all at once but at intervals over at least 4 million years. It used to be thought that upright walking was the fundamental human attribute, but by 4 million years ago, the australopithecines were bipedal. The first primitive stone tools appear about 2.6 million years ago and are thought to be associated with *Homo habilis*. However, it is possible that australopithecines made them, especially since it is now known that chimpanzees use stone tools. The first significant increase in brain size above that of the apes occurs in *Homo habilis*, which had a brain capacity of 40 cubic in (650 cubic cm). By about 1.8–1.9 million years ago, *Homo erectus*, with a larger body, greater mobility, and slightly larger brain size, was the first human relative to migrate from Africa, reaching eastern Asia by about 1.6–1.7 million years ago. The Neanderthals *(Homo neanderthalensis)* and their immediate predecessors were the first human relatives whose brain size reached that of modern humans, at 67–86 cubic in (1,100–1,400 cubic cm). They occupied Europe and Asia from about 230,000 years ago until 30,000 years ago and were successful

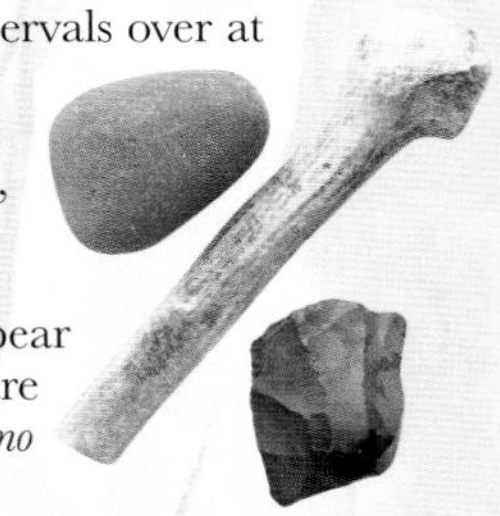

**DEVELOPING SKILLS**
*The fossil record tends to preserve only stone and bone tools, but some of the most primitive tools were probably digging sticks.*

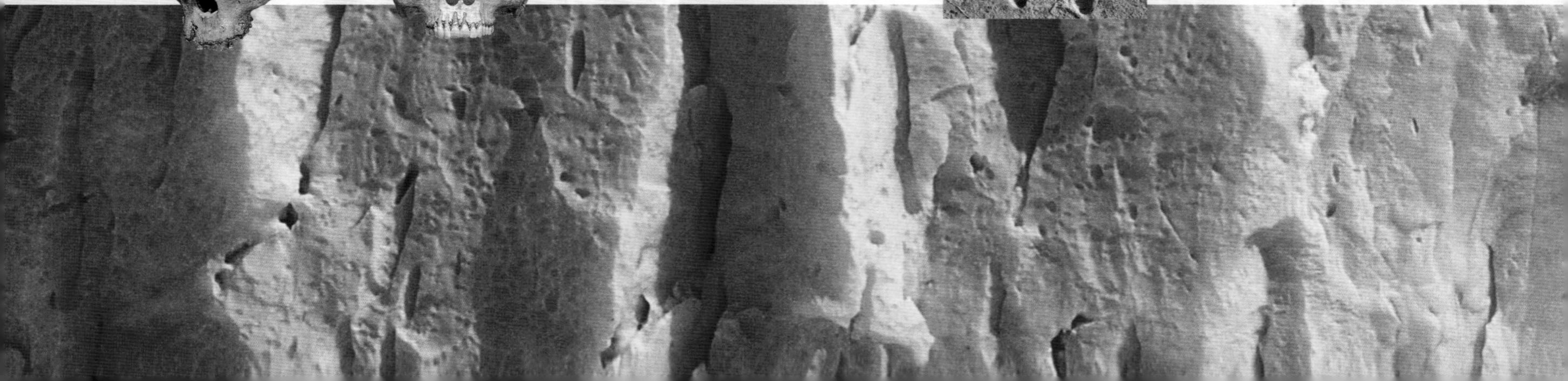

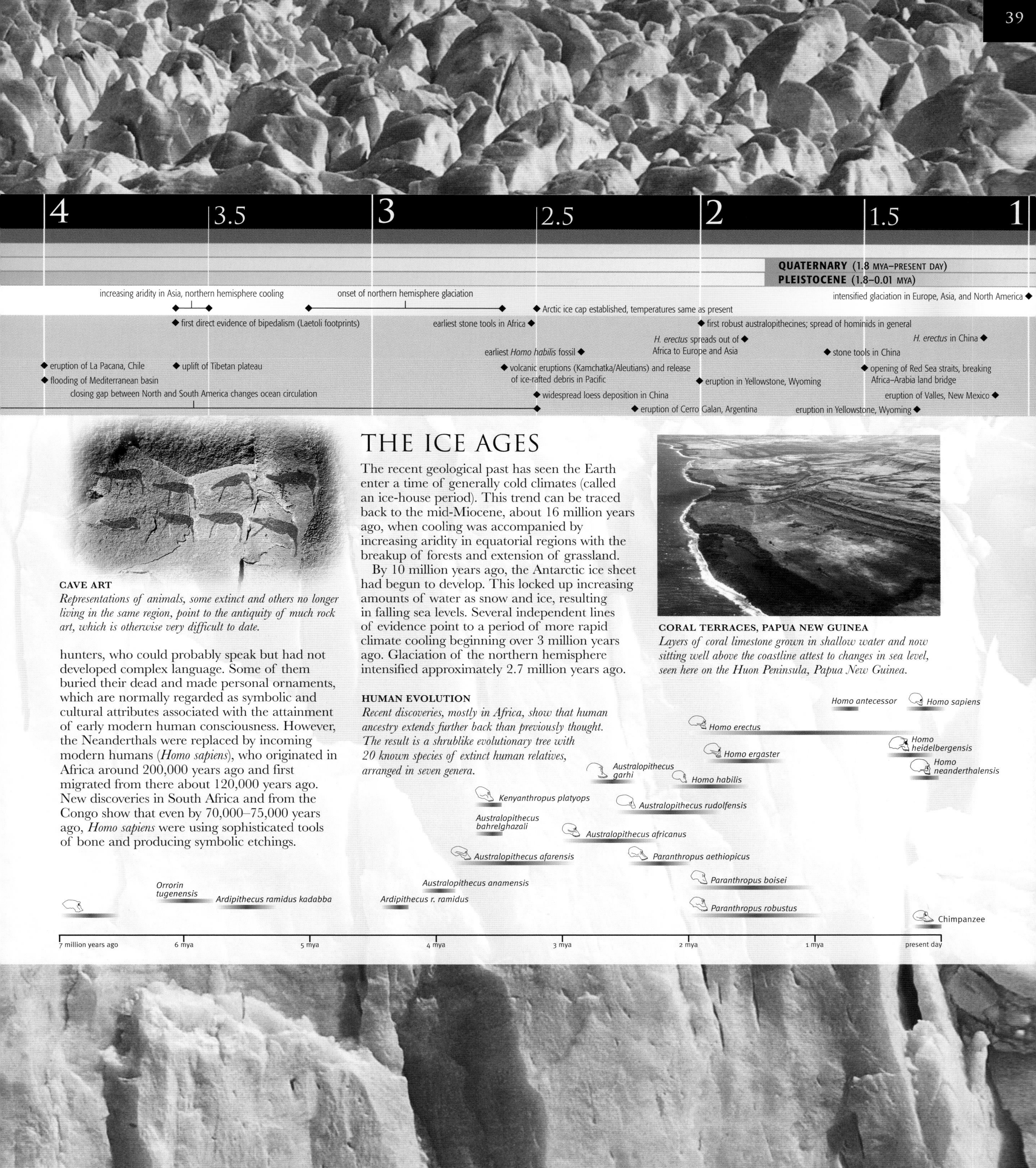



**CAVE ART**
*Representations of animals, some extinct and others no longer living in the same region, point to the antiquity of much rock art, which is otherwise very difficult to date.*

hunters, who could probably speak but had not developed complex language. Some of them buried their dead and made personal ornaments, which are normally regarded as symbolic and cultural attributes associated with the attainment of early modern human consciousness. However, the Neanderthals were replaced by incoming modern humans (*Homo sapiens*), who originated in Africa around 200,000 years ago and first migrated from there about 120,000 years ago. New discoveries in South Africa and from the Congo show that even by 70,000–75,000 years ago, *Homo sapiens* were using sophisticated tools of bone and producing symbolic etchings.

# THE ICE AGES

The recent geological past has seen the Earth enter a time of generally cold climates (called an ice-house period). This trend can be traced back to the mid-Miocene, about 16 million years ago, when cooling was accompanied by increasing aridity in equatorial regions with the breakup of forests and extension of grassland.

By 10 million years ago, the Antarctic ice sheet had begun to develop. This locked up increasing amounts of water as snow and ice, resulting in falling sea levels. Several independent lines of evidence point to a period of more rapid climate cooling beginning over 3 million years ago. Glaciation of the northern hemisphere intensified approximately 2.7 million years ago.

**CORAL TERRACES, PAPUA NEW GUINEA**
*Layers of coral limestone grown in shallow water and now sitting well above the coastline attest to changes in sea level, seen here on the Huon Peninsula, Papua New Guinea.*

**HUMAN EVOLUTION**
*Recent discoveries, mostly in Africa, show that human ancestry extends further back than previously thought. The result is a shrublike evolutionary tree with 20 known species of extinct human relatives, arranged in seven genera.*

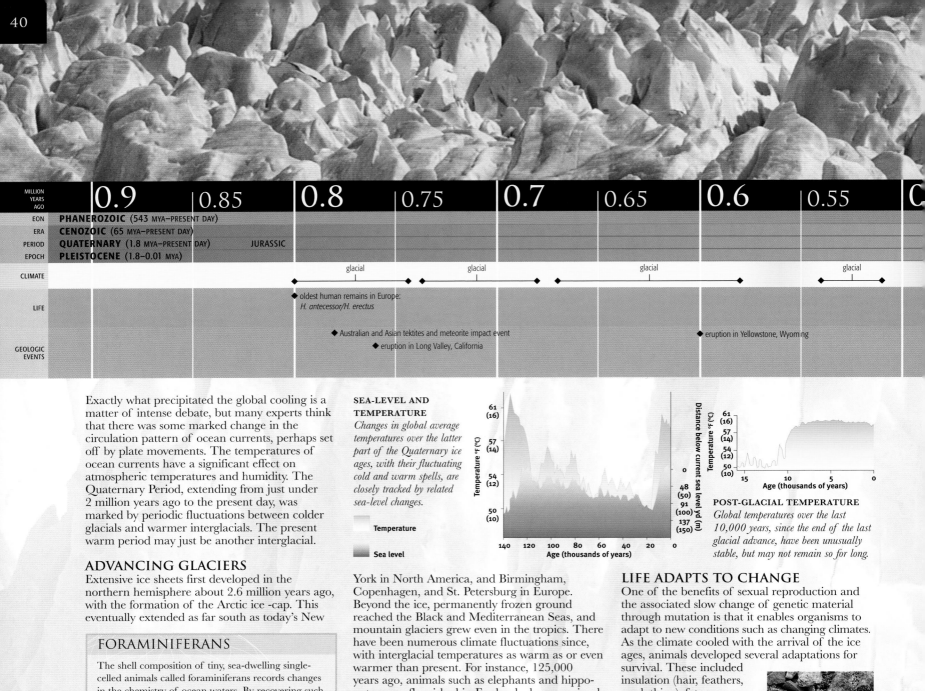

| MILLION YEARS AGO | 0.9 | 0.85 | 0.8 | 0.75 | 0.7 | 0.65 | 0.6 | 0.55 |
|---|---|---|---|---|---|---|---|---|
| EON | PHANEROZOIC (543 MYA–PRESENT DAY) | | | | | | | |
| ERA | CENOZOIC (65 MYA–PRESENT DAY) | | | | | | | |
| PERIOD | QUATERNARY (1.8 MYA–PRESENT DAY) JURASSIC | | | | | | | |
| EPOCH | PLEISTOCENE (1.8–0.01 MYA) | | | | | | | |
| CLIMATE | | | glacial | | glacial | glacial | | glacial |
| LIFE | | | ◆ oldest human remains in Europe: *H. antecessor/H. erectus* | | | | | |
| GEOLOGIC EVENTS | | | ◆ Australian and Asian tektites and meteorite impact event; ◆ eruption in Long Valley, California | | | | ◆ eruption in Yellowstone, Wyoming | |

Exactly what precipitated the global cooling is a matter of intense debate, but many experts think that there was some marked change in the circulation pattern of ocean currents, perhaps set off by plate movements. The temperatures of ocean currents have a significant effect on atmospheric temperatures and humidity. The Quaternary Period, extending from just under 2 million years ago to the present day, was marked by periodic fluctuations between colder glacials and warmer interglacials. The present warm period may just be another interglacial.

## ADVANCING GLACIERS

Extensive ice sheets first developed in the northern hemisphere about 2.6 million years ago, with the formation of the Arctic ice -cap. This eventually extended as far south as today's New York in North America, and Birmingham, Copenhagen, and St. Petersburg in Europe. Beyond the ice, permanently frozen ground reached the Black and Mediterranean Seas, and mountain glaciers grew even in the tropics. There have been numerous climate fluctuations since, with interglacial temperatures as warm as or even warmer than present. For instance, 125,000 years ago, animals such as elephants and hippopotamuses flourished in England where previously and subsequently there were ice sheets. The periodicity of the climate fluctuations was initially about 40,000 years but changed to a 100,000-year cycle about a million years ago due to changes within the global climate cycle.

### FORAMINIFERANS

The shell composition of tiny, sea-dwelling single-celled animals called foraminiferans records changes in the chemistry of ocean waters. By recovering such fossil shells from ocean-floor sediments and analyzing their chemistry over successive generations, the changes in ocean chemistry can be reconstructed. Oxygen-isotope ratios in the calcium carbonate of the shells are measured, and from this information past fluctuations in ocean temperature, ice volume, and hence climate change can be recovered.

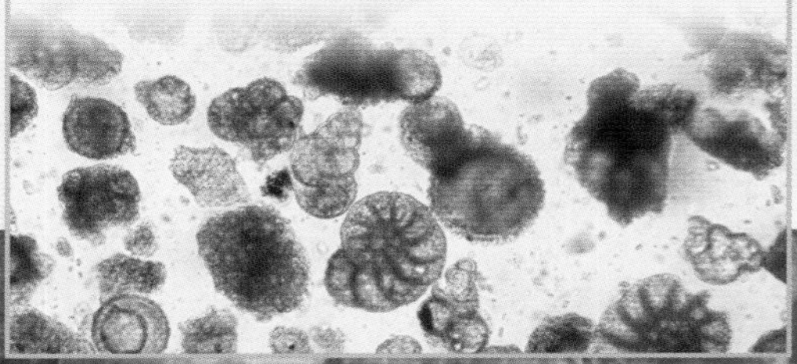

**SEA-LEVEL AND TEMPERATURE**
*Changes in global average temperatures over the latter part of the Quaternary ice ages, with their fluctuating cold and warm spells, are closely tracked by related sea-level changes.*

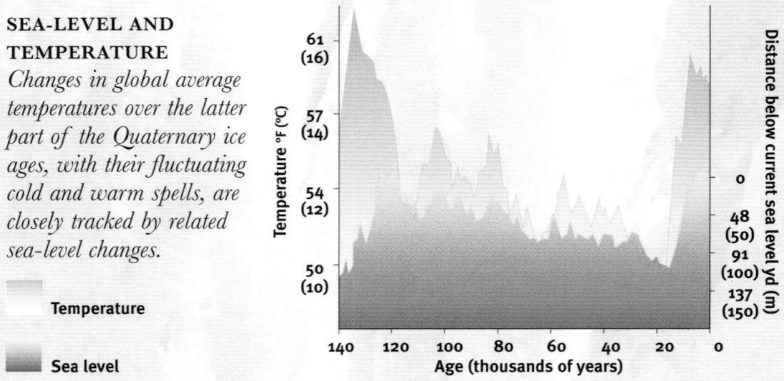

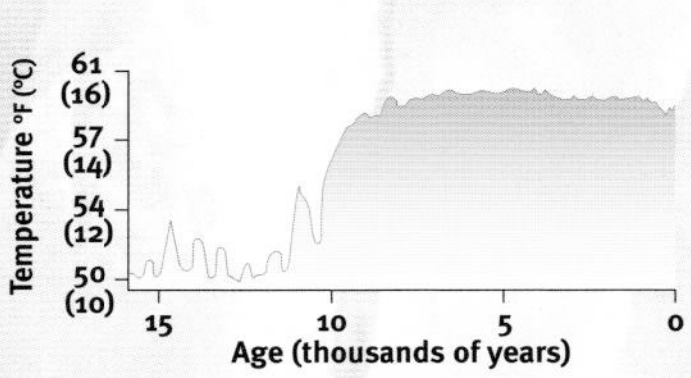

**POST-GLACIAL TEMPERATURE**
*Global temperatures over the last 10,000 years, since the end of the last glacial advance, have been unusually stable, but may not remain so for long.*

**QUATERNARY ICE CAPS**
*Polar ice caps grew and retreated several times during the Quaternary, reaching the maximum extent shown here.*

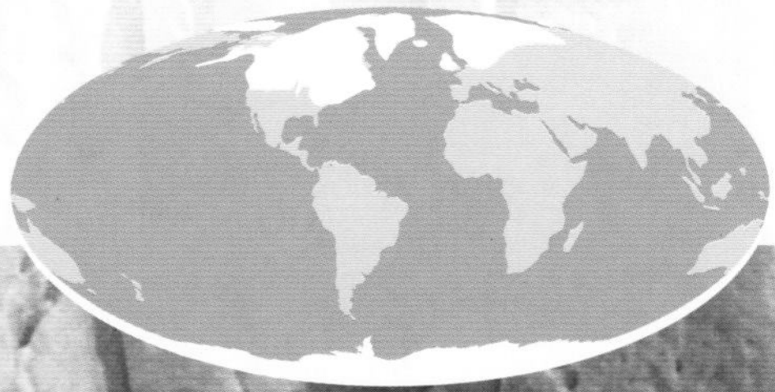

## LIFE ADAPTS TO CHANGE

One of the benefits of sexual reproduction and the associated slow change of genetic material through mutation is that it enables organisms to adapt to new conditions such as changing climates. As the climate cooled with the arrival of the ice ages, animals developed several adaptations for survival. These included insulation (hair, feathers, or clothing), fat reserves for food shortages, and often camouflage for snowy conditions, such as the white coats of Arctic foxes, hares, and polar bears. Typically, cold-adapted

**PURPLE SAXIFRAGE**
*Some flowering plants, such as the Arctic bog cotton and purple saxifrage, are well adapted to tundra conditions.*

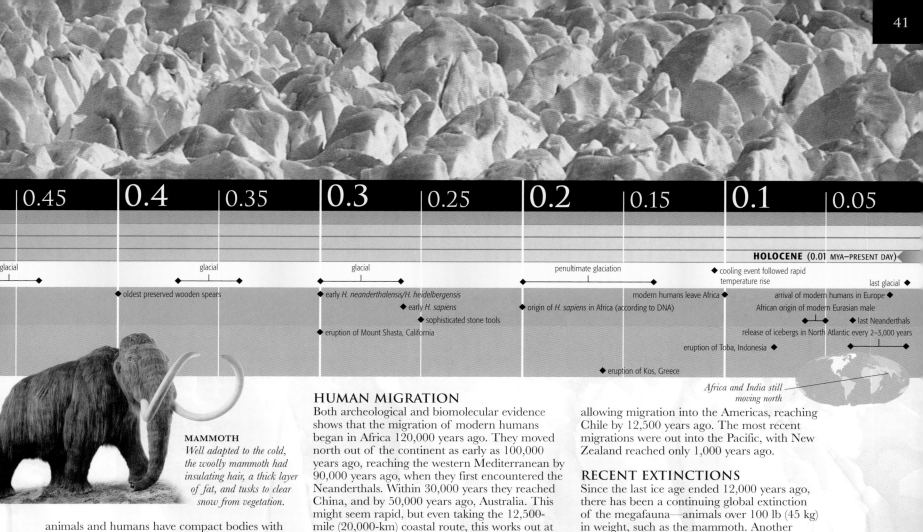

0.45 | 0.4 | 0.35 | 0.3 | 0.25 | 0.2 | 0.15 | 0.1 | 0.05

**HOLOCENE** (0.01 MYA–PRESENT DAY)

glacial

glacial

glacial

penultimate glaciation

cooling event followed rapid temperature rise

last glacial

oldest preserved wooden spears

early *H. neanderthalensis/H. heidelbergensis*

early *H. sapiens*

sophisticated stone tools

eruption of Mount Shasta, California

origin of *H. sapiens* in Africa (according to DNA)

modern humans leave Africa

arrival of modern humans in Europe

African origin of modern Eurasian male

last Neanderthals

release of icebergs in North Atlantic every 2–3,000 years

eruption of Toba, Indonesia

eruption of Kos, Greece

*Africa and India still moving north*

**MAMMOTH**
*Well adapted to the cold, the woolly mammoth had insulating hair, a thick layer of fat, and tusks to clear snow from vegetation.*

animals and humans have compact bodies with small limbs and extremities to decrease their surface area and increase volume for more effective heat conservation. Plants have to withstand not only freezing air temperatures but also very short growing seasons, during which reproduction, seed setting, and dispersal must all take place within a matter of a few months. Frozen soil necessitates shallow roots, so even the woody plants, such as dwarf birch, willow and juniper, tend to be ground-hugging to protect themselves from strong winds.

**CARIBOU MIGRATION**
*Reindeer, also known as caribou, are cold-adapted herd animals that survive in polar regions by migrating over long distances between winter and summer feeding grounds.*

## HUMAN MIGRATION

Both archeological and biomolecular evidence shows that the migration of modern humans began in Africa 120,000 years ago. They moved north out of the continent as early as 100,000 years ago, reaching the western Mediterranean by 90,000 years ago, when they first encountered the Neanderthals. Within 30,000 years they reached China, and by 50,000 years ago, Australia. This might seem rapid, but even taking the 12,500-mile (20,000-km) coastal route, this works out at less than half a mile (1 km) per year.

Ice-age Europe was more of a physical and climatic barrier and was not reached until 40,000 years ago, and the northeast passage through Siberia into Alaska was even more difficult. The Bering land bridge was exposed between 18,000 and 10,200 years ago by lowered sea levels, allowing migration into the Americas, reaching Chile by 12,500 years ago. The most recent migrations were out into the Pacific, with New Zealand reached only 1,000 years ago.

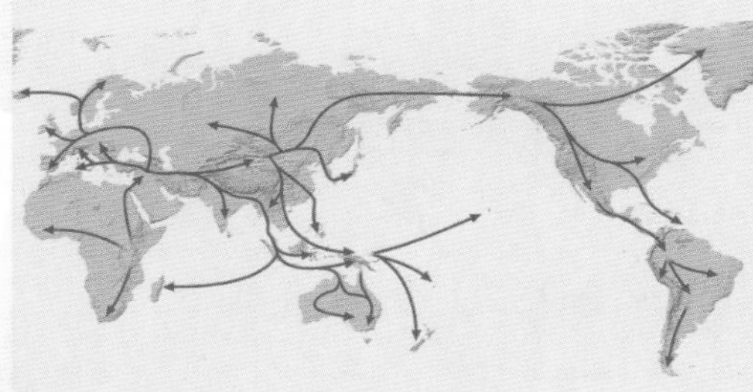

**MIGRATION OF HOMO SAPIENS**
*Through the collection of fossil records, the pathways by which modern humans* (Homo sapiens) *are thought to have achieved global distribution have been reconstructed. The journey is believed to have begun in east Africa about 120,000 years ago.*

## RECENT EXTINCTIONS

Since the last ice age ended 12,000 years ago, there has been a continuing global extinction of the megafauna—animals over 100 lb (45 kg) in weight, such as the mammoth. Another victim was the Neanderthals, whose demise followed the arrival of modern humans about 40,000 years ago. They coexisted for nearly 10,000 years in western Europe, but eventually died out. Analysis of DNA suggests that despite the overlap in time and territory there was no significant interbreeding between the two human species. The exact cause of the megafauna extinction, whether climatic and environmental change, human hunting, or a combination of the two, is still debated. However, recent accurate dating, especially in Australasia, which suffered little climate change, shows that extinctions closely follow the arrival of modern human hunters. The exception is Africa, where modern humans coexisted with large mammals for much longer, but now rising populations are threatening even Africa's big game.

Of course, human influence on the Earth goes far beyond that of the hunter. The encroachment on almost all natural habitats, from the rainforests to the polar wastes, along with the effects of human-enhanced global warming, pose a far greater threat to the planet's flora and fauna, even threatening humankind itself.

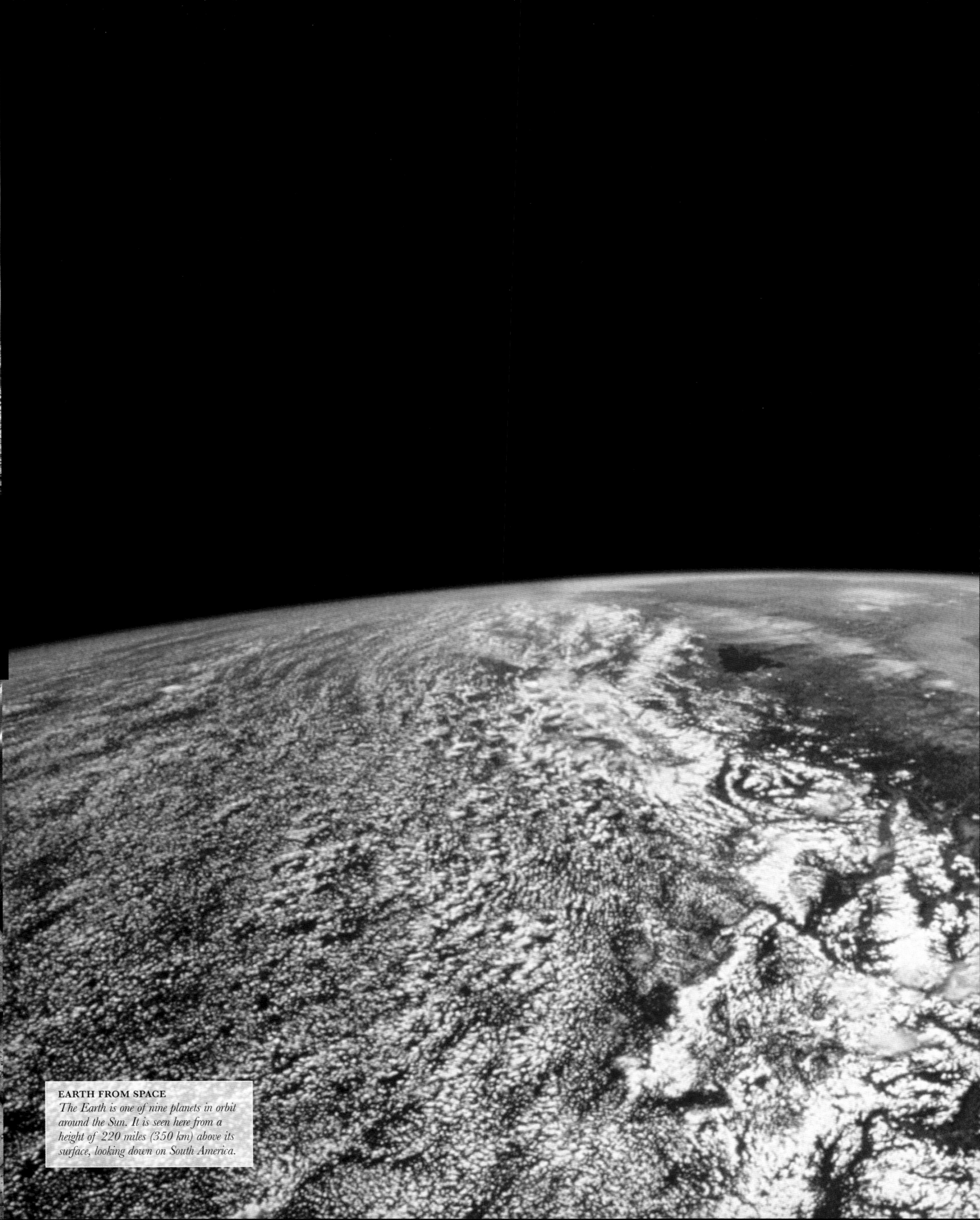

**EARTH FROM SPACE**
*The Earth is one of nine planets in orbit around the Sun. It is seen here from a height of 220 miles (350 km) above its surface, looking down on South America.*

# THE EARTH IN SPACE

SINCE THE LAUNCH OF the first artificial satellites in the 1950s, thousands of views of the Earth have been obtained from space. Although such images can give the initial impression of a self-contained ball of light and color floating in a dark void, they also make us realize that our planet exists within a context. It is just one small world in a single solar system, on the outskirts of a huge galaxy that is just one of billions in the whole universe. The very material that the Earth is made of was originally forged in stars, and the planet's structure and environment have been heavily influenced by its location in space and its interactions with other celestial objects. Today, the most obvious influence comes from the Sun, a vital source of light and energy for our planet, but the Moon also plays a role by creating tides, and the Earth continues to be bombarded by rocks and debris from space. Such factors have altered Earth's history dramatically in the past, and could do so again in the future.

# THE UNIVERSE

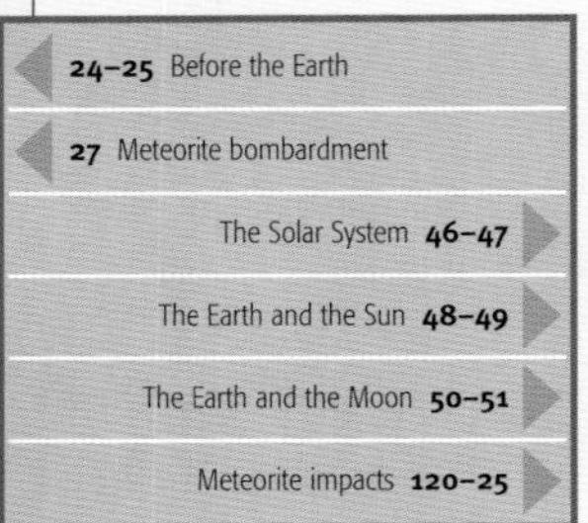
24–25 Before the Earth
27 Meteorite bombardment
The Solar System 46–47
The Earth and the Sun 48–49
The Earth and the Moon 50–51
Meteorite impacts 120–25

THE UNIVERSE IS EVERYTHING that exists, including all matter, energy, and space. Scientists who study the universe as a whole, called cosmologists, are uncertain whether it is finite or infinite, but they have established many other facts. They know that the universe is expanding, that it has no center or edges, and that it came into existence billions of years ago in an enormous explosion called the Big Bang. They also agree that the universe is balanced between two fates: expanding forever or collapsing under its own gravity. The latest research seems to indicate that endless expansion is most likely.

**STAR-FORMING NEBULA**
*This image shows a region of star birth in a nebula (cloud of dust and gas) within the spiral galaxy M33, which is located about 2.7 million light-years from our own galaxy.*

## SIZE AND COMPOSITION

From the Earth, we can see for a maximum of 13–15 billion light-years in any direction (a light-year is the distance light travels in a year). The universe is probably far larger than this, but radiation from objects farther away has not had time to reach us since the Big Bang. Visible matter in the universe is dominated by the simple elements hydrogen and helium, but gravity reveals that normal matter is heavily outweighed by invisible dark matter. The nature of dark matter is unknown, and cosmologists also suspect the existence of dark energy, a mysterious phenomenon that opposes gravity and helps keep the universe expanding.

**SCALE OF THE UNIVERSE**
*The Solar System is just a tiny dot in the Milky Way, itself a speck in the local group of galaxies and nearby universe.*

**HUBBLE DEEP FIELD**
*This Hubble Space Telescope image of a tiny part of the sky reveals hundreds of galaxies.*

**SOLAR SYSTEM**
1.6 light-years

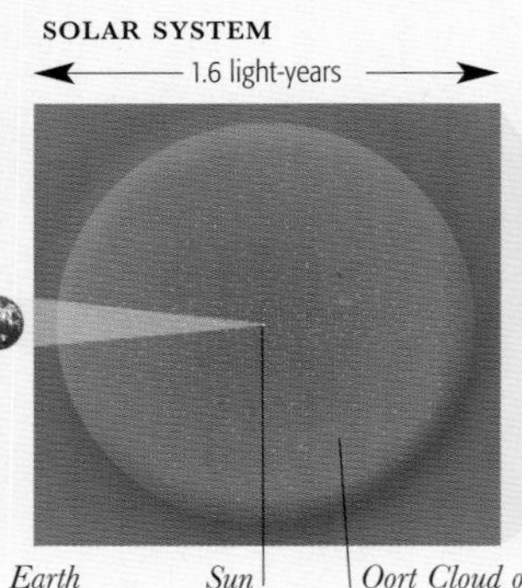

*Earth* *Sun* *Oort Cloud of comets*

**MILKY WAY GALAXY**
100,000 light-years

*Solar System*

**LOCAL GROUP**
5 million light-years

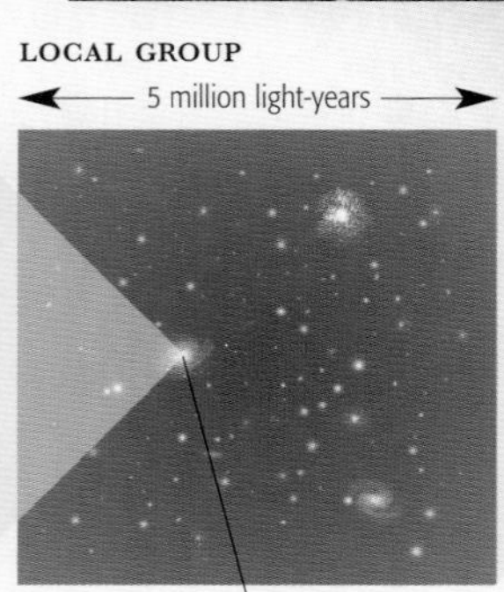

*Milky Way*

**NEARBY UNIVERSE**
500 million light-years

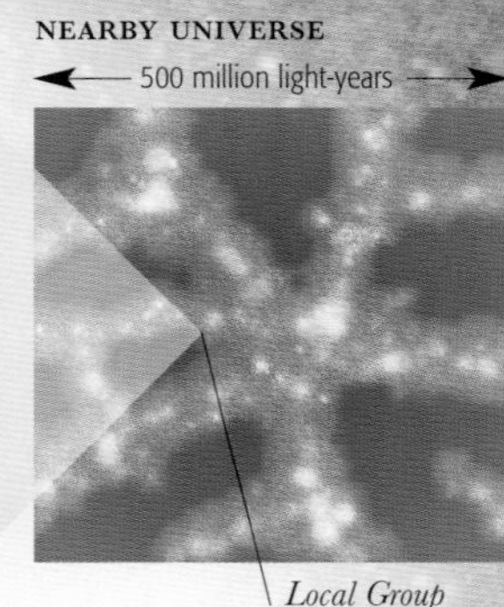

*Local Group*

**BLACK HOLE**
*The bright area in the image to the right is a disk of gas around a black hole at the center of an elliptical galaxy, M87, about 50 million light-years from Earth.*

**BRILLIANT QUASAR**
*About 2 billion light-years away, galaxy 3C 273 (left) has a quasar at its core, formed as a giant black hole sucks in material around it.*

## GALAXIES

Nearly all of the universe's ordinary matter, and some of its dark matter, is gathered into about 100 billion galaxies. Most galaxies contain billions of stars and large clouds of gas and dust. They vary from about 10,000 to 200,000 light-years in diameter, and are usually grouped in clusters of 20 to several thousand. Astronomers identify most galaxies with a combination of letters and numbers. Clusters are grouped in superclusters, and the superclusters form broad bands or filaments, with large voids between them. Since the 1960s, astronomers have puzzled over the identity of remote, highly energetic objects called quasars. Most are now thought to be the result of colossal black holes (regions of extremely compressed matter) forming in the centers of early galaxies. Astronomers now suspect that most large galaxies harbor black holes.

**GALAXY SHAPES**
*Five basic galaxy shapes are recognized: elliptical, lenticular (lens-shaped), spiral, barred spiral (like spiral but with a bar-shaped rather than spherical core), and irregular. Three of these shapes are displayed by the galaxies shown here.*

LENTICULAR GALAXY (NGC 2787)

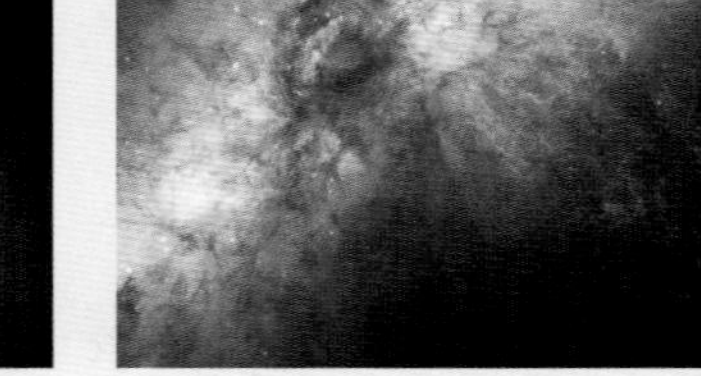

IRREGULAR GALAXY (M82)

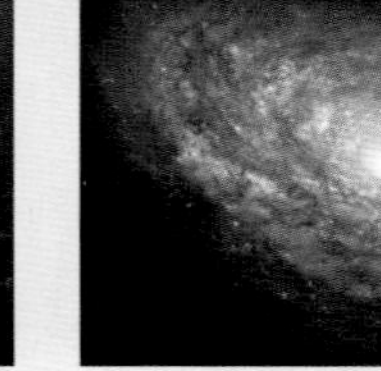

SPIRAL GALAXY (NGC 4414)

# THE MILKY WAY

The Milky Way is our local galaxy, and contains the Sun, the rest of the Solar System, and all the stars visible in the night sky. It is one of a group of about 30 galaxies in our region of the universe, called the Local Group. From careful observation and plotting of the distribution of stars, astronomers believe that the Milky Way is a spiral galaxy. It has a diameter of about 100,000 light-years, contains about 100 billion stars, as well as many large clouds of gas and dust, and is slowly rotating. The Solar System resides in one of the five or six spiral arms of the Milky Way, about two-thirds of the way out from the galaxy's nucleus toward the edge of its disk. Astronomers have established that there are planets orbiting many other stars in the Milky Way. They have also found evidence that there is a huge black hole at the center of the galaxy.

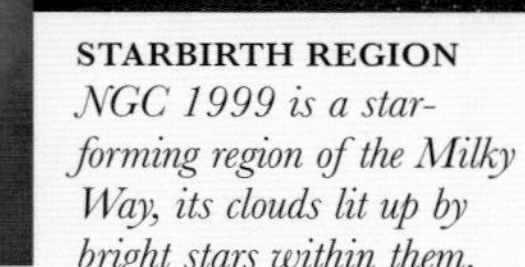

**STARBIRTH REGION** *NGC 1999 is a star-forming region of the Milky Way, its clouds lit up by bright stars within them.*

**MILKY WAY** *A nebulous band of light stretches across the sky as we look across the plane of our galaxy.*

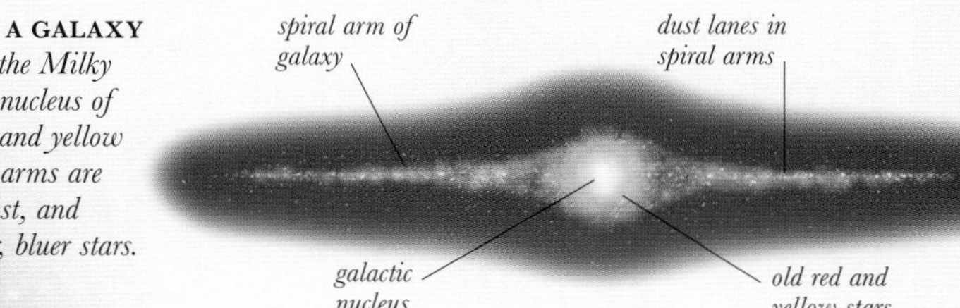


**ANATOMY OF A GALAXY** *At the center of the Milky Way is a dense nucleus of mainly old, red and yellow stars; the spiral arms are richer in gas, dust, and younger, brighter, bluer stars.*

## EDWIN HUBBLE

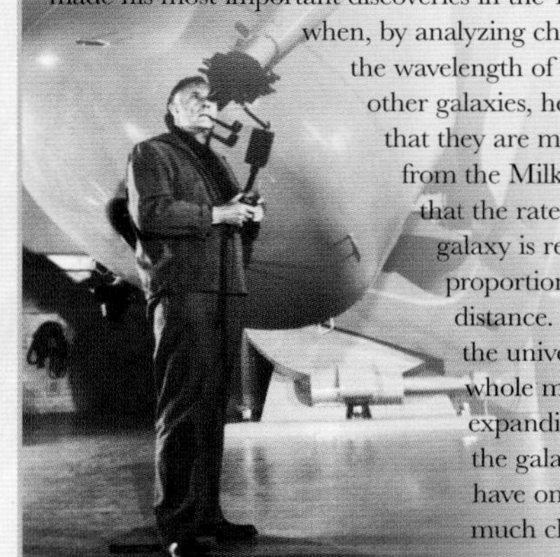

American astronomer Edwin Hubble (1889–1953) is famous for proving that there are galaxies beyond the Milky Way and that the universe is expanding. Hubble made his most important discoveries in the 1920s when, by analyzing changes in the wavelength of light from other galaxies, he showed that they are moving away from the Milky Way, and that the rate at which a galaxy is receding is proportional to its distance. Hence the universe as a whole must be expanding, and all the galaxies must have once been much closer.

**OPEN STAR CLUSTER** *Open star clusters such as the Quintuplet, 25,000 light-years from Earth, are groups of young stars that formed relatively recently.*

**GLOBULAR STAR CLUSTER** *Globular clusters such as Mayall II contain hundreds of thousands of old stars, orbiting in a galaxy's halo or close to its nucleus.*

# STARS

Stars, including the Sun, are hot balls of gas (mostly hydrogen with significant amounts of helium) that generate enormous amounts of energy through nuclear fusion reactions in their cores. The cores of stars are the only places in the universe where elements such as carbon and oxygen can form. Many stars are found in clusters, often surrounded by remnants of the gas clouds from which they collapsed. A star's individual properties depend primarily on its mass, which varies from one star to another. Stars with the highest mass are the hottest and brightest, with surface temperatures of up to 90,000°F (50,000°C), and a blue or blue-white color. These stars quickly use up the hydrogen that fuels their nuclear reactions, so they may shine for no more than a few million years. At the end of their lives, they die in cataclysmic explosions called supernovas. Stars of medium mass, such as the Sun, have surface temperatures of around 7,000–14,000°F (4,000–8,000°C), giving a yellow or orange color, and can shine for billions of years. At the end of their lives, they expand into large, low-density stars called red giants before blowing off most of their matter and finally collapsing to form small white dwarf stars that gradually cool and fade.

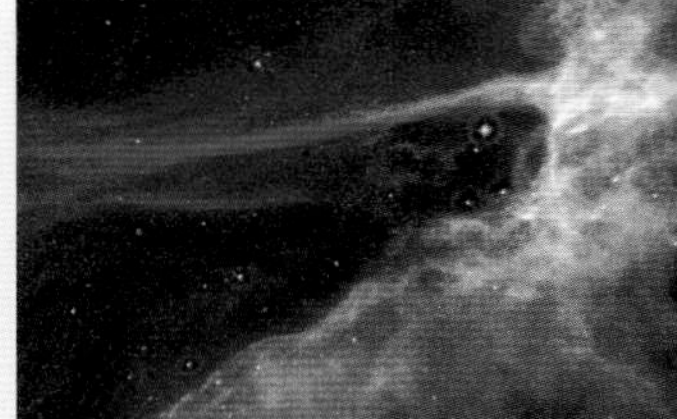

**CYGNUS LOOP** *This blast wave from an ancient supernova heats up the gas clouds in its path and causes them to glow.*

# THE SOLAR SYSTEM

24–25 Before the Earth
44–45 The Universe
The Earth and the Sun 48–49
The Earth and the Moon 50–51
Meteorite impacts 120–25

THE SOLAR SYSTEM CONSISTS OF the Sun and all the objects that orbit it, such as planets, moons, asteroids, and comets. The whole system is thought to have formed 4.56 billion years ago from a spinning cloud of gas and dust. Apart from the Sun, the Solar System can be divided into three main regions. First are the four rocky inner planets (Mercury, Venus, Earth, and Mars). Separated from this region by the asteroid belt are four outer gas giant planets (Jupiter, Saturn, Uranus, and Neptune). Beyond Neptune's orbit is a vast region scattered with small, icy worlds, such as Pluto, and dormant comets. In total, the Solar System is about 15,000 billion km (9,300 billion miles) or 1.6 light years across. The much smaller region containing just the Sun and planets is about 12 billion km (7.5 billion miles) across.

**SUN AND PLANETS**
*The planets are divided into two main groups: the four small inner planets (below) and the four gas giants (right). Pluto (far right) does not fit comfortably into either group.*

MERCURY

VENUS

EARTH

MARS

JUPITER

## THE SUN

The Sun is an enormous ball of hot gas, about 1.4 million km (865,000 miles) across, containing 99 per cent of the Solar System's mass. In its core, nuclear fusion converts hydrogen to helium. This process releases a huge amount of energy that moves outwards by radiation and convection until it reaches the Sun's visible surface, or photosphere, and escapes as radiation. Above the photosphere lie two outer layers – the chromosphere and the corona.

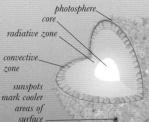

**INSIDE THE SUN**
*The Sun has three inner layers – the core, radiative zone, and convective zone. Light and heat escape at the photosphere.*

*photosphere*
*core*
*radiative zone*
*convective zone*
*sunspots mark cooler areas of surface*
*prominences are loops of gas arching above the photosphere*

## THE INNER PLANETS

The four inner planets have several features in common. They are relatively small, have a structure consisting of a rocky crust, a mantle, and an iron-rich core, and they possess few or no satellites. In other ways they are quite different. Earth is the only planet with an oxygen-rich atmosphere, abundant surface water, and a range of temperatures that ensure most of the water is liquid. These factors have all contributed to the development of life. Mercury has virtually no atmosphere and experiences extremes of temperature. Venus has a thick, carbon-dioxide rich atmosphere that produces extremely high pressures and temperatures at its surface. Mars is cold, with a thin atmosphere and water that exists as ice at its poles and beneath its surface.

### THE INNER PLANETS

Some properties of the inner planets are summarized below. The orbital period (the time a planet takes to orbit the Sun) increases with distance from the Sun. As well as having longer orbits to complete, planets move more slowly when they are further from the Sun.

| Data | Mercury | Venus | Earth | Mars |
|---|---|---|---|---|
| Diameter | 4,875 km (3,029 miles) | 12,104 km (7,521 miles) | 12,756 km (7,928 miles) | 6,780 km (4,213 miles) |
| Average distance from the Sun | 57.9 million km (36 million miles) | 108.2 million km (67.2 million miles) | 149.6 million km (93 million miles) | 227.9 million km (141.6 million miles) |
| Rotation period | 58.6 days | 243 days | 23.93 hours | 24.62 hours |
| Orbital period | 88 days | 224.7 days | 365.26 days | 687 days |
| Surface temperature | -180°C to 430°C (-292°F to 806°F) | 480°C (896°F) | -70°C to 55°C (-94°F to 131°F) | -120°C to 25°C (-184°F to 77°F) |

**INNER PLANET ORBITS**
*All the inner planets orbit the Sun in the same direction and in roughly the same plane. Their orbits are not circular but slightly elliptical (Mercury's orbit is quite markedly elliptical).*

**RELATIVE DISTANCES**
*The distance line below shows the planets' average distances from the Sun roughly to scale.*

Sun Mercury Venus Earth Mars Jupiter Saturn Uranus Neptune Pluto

# THE OUTER PLANETS

The four large outer planets, called the gas giants, are very different from the inner planets, but have many features in common with each other. They all have small, rocky cores and are composed mostly of liquid hydrogen and helium, though Uranus and Neptune also contain large amounts of chemical ices – hydrogen compounds such as water, ammonia, and methane. They have gaseous, often stormy atmospheres, also composed mainly of hydrogen and helium – the atmospheres of Uranus and Neptune contain 2–3 per cent methane as well, which gives them a blue appearance. Each of the gas giants has a ring system made of dust and ice, formed from the fragmented remains of objects that came too close to their powerful gravity. Finally all are orbited by large numbers of moons – several dozen each in the case of Jupiter and Saturn. Beyond the gas giants lies Pluto. The smallest planet, it is an oddity, composed mainly of rock and ice. Its markedly elliptical orbit lies at an angle to the rest of the Solar System, and brings it closer to the Sun than Neptune at times. It also has a huge moon, Charon, and the two objects are often classed as a double planet. Astronomers now think that Pluto is just a large and bright member of the belt of icy objects beyond Neptune's orbit.

**ORBITS OF THE OUTER PLANETS**
*As with the inner planets, the outer planets orbit the Sun in the same direction along elliptical paths. All the orbits are in roughly the same plane, except for Pluto's, which is at an angle to the others.*

### THE OUTER PLANETS

The relationship of increased orbital period with increased distance from the Sun holds with the outer planets as with the inner ones. There is also a drop in surface temperature with increased distance, due to decreasing intensity of solar radiation. All the gas giants except Uranus generate some heat internally.

| Data | Jupiter | Saturn | Uranus | Neptune | Pluto |
|---|---|---|---|---|---|
| Diameter | 142,984 km (88,846 miles) | 120,536 km (74,898 miles) | 51,118 km (31,763 miles) | 49,528 km (30,775 miles) | 2,304 km (1,432 miles) |
| Average distance from the Sun | 778.3 million km (483.6 million miles) | 1,431 million km (889.8 million miles) | 2,877 million km (1,788 million miles) | 4,498 million km (2,795 million miles) | 5,915 million km (3,675 million miles) |
| Rotation period | 9.93 hours | 10.65 hours | 17.24 hours | 16.11 hours | 6.38 days |
| Orbital period | 11.86 years | 29.37 years | 84.1 years | 164.9 years | 248.6 years |
| Temperature | -110°C (-160°F) | -140°C (-220°F) | -200°C (-320°F) | -200°C (-320°F) | -230°C (-380°F) |

SATURN

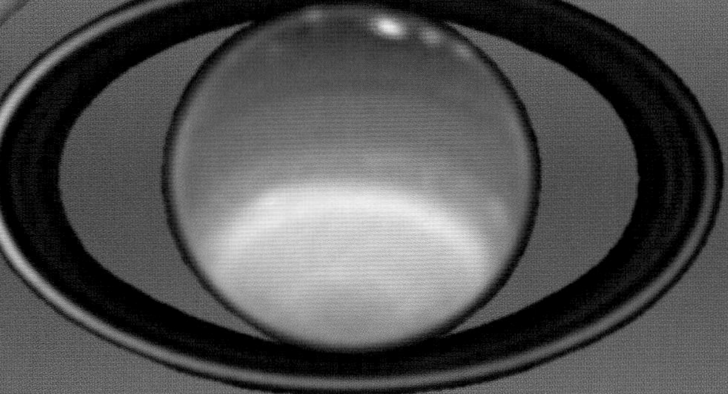

URANUS

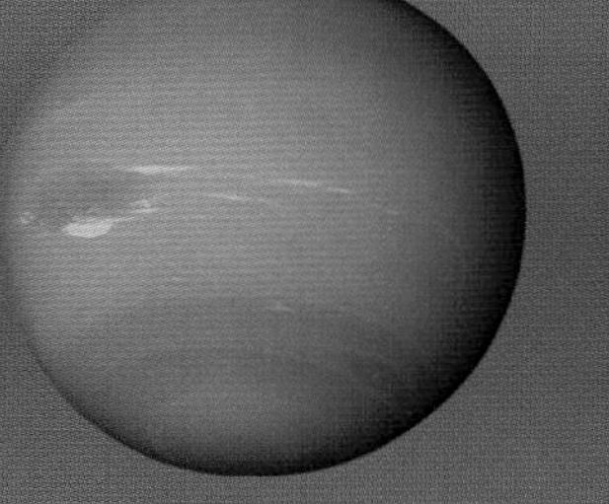

NEPTUNE

PLUTO

## NICOLAUS COPERNICUS

Polish astronomer, cleric, and physician Nicolaus Copernicus (1473–1543) is famous for providing the first detailed and convincing argument that the Earth and planets orbit the Sun and that the Earth spins on its own axis. These ideas opposed the prevailing view of the time, that the Earth was stationary and at the centre of the Universe. Copernicus's heliocentric (Sun-centred) theory of the Universe was published shortly before his death – it took more than a hundred years for his ideas to become widely accepted.

# OTHER SOLAR SYSTEM OBJECTS

In addition to the planets, the Solar System is littered with billions of smaller objects, remnants of its early history. Asteroids are rocky bodies, from a few hundred metres to several hundred kilometres across. Most orbit the Sun in a belt between the inner and outer planets, but others can travel close to Earth. Comets are chunks of ice, frozen gas, and rock particles, usually a few kilometres in diameter. Vast numbers lie dormant in the outer Solar System, mixed with Pluto-like ice dwarfs in a belt just beyond Neptune, or scattered through a more distant spherical shell called the Oort Cloud. When they move close to the Sun and warm up, frozen chemicals vaporize to produce a glowing coma (head) and long tails of dust and ionized gas. Meteoroids are smaller objects, the remains of shattered asteroids and the dust from comets. If they encounter Earth's atmosphere, most burn up as meteors. The few that reach the ground are called meteorites.

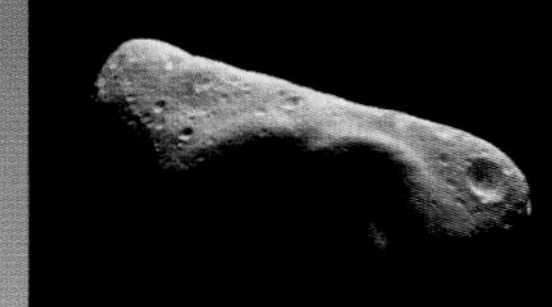

**ASTEROID EROS**
*This 33km- (20-mile-) long object was the first asteroid on which a spacecraft landed.*

**METEOR TRAIL**
*In this photograph, a bright trail can be seen from a meteor, or shooting star, burning up in the atmosphere.*

**COMET HALE-BOPP**
*With its spectacular coma, and tails, this comet was visible for several months in 1996–97. Here it is seen over Mount Whitney, California, USA.*

# THE EARTH AND THE SUN

24–25 Before the Earth
46–47 The Solar System
The Earth and the Moon 50–51
The Structure of the atmosphere 440–41
Energy in the atmosphere 442–43
Climate change 450–51

OF ALL CELESTIAL BODIES, the Sun has the most profound influence on the Earth, affecting our planet in several ways. Most obviously, Earth's movements and orientation in relation to the Sun determine its day–night and seasonal cycles. Most life on Earth ultimately depends on energy derived from sunlight, while solar heating of the atmosphere, oceans, and land helps drive the whole global climatic system. Earth is also affected indirectly by storms, flares, and other intense activity on the solar surface, and the Sun plays a part in the Earth's tides, modifying the effect of the Moon to produce monthly variations in tidal range (see p.428). Finally, variations in Earth's orbit around the Sun are thought to be linked to long-term climate cycles (see p.450).

## THE EARTH'S SEASONS

The Earth orbits the Sun once every 365.25 days, at an average distance of 93 million miles (150 million km). The planet also rotates on its axis in 24 hours, so points on its surface pass from full sunlight to shadow and back, causing day and night. Earth's seasons result from the fact that its spin axis points in a fixed direction throughout its orbit, tilted at 23.5° from vertical. This means that at the summer solstice in June, the northern hemisphere points toward the Sun, and the north polar region is sunlit all day, while the south polar region is in continuous darkness. At the winter solstice in December, the situation is reversed. The Earth's tilt also determines the extent of the tropics. The Sun is overhead at noon on the Tropic of Cancer (23.5°N) in June and on the Tropic of Capricorn (23.5°S) in December.

**MIDNIGHT SUN IN ALASKA**
*During the summer solstice, areas within the Arctic Circle experience a period of continuous daylight.*

**SOLSTICES AND SEASONS**
*The tilt of the Earth's spin axis is responsible for our planet's seasons. At the solstices in June and December, one hemisphere has its longest day, the other its shortest.*

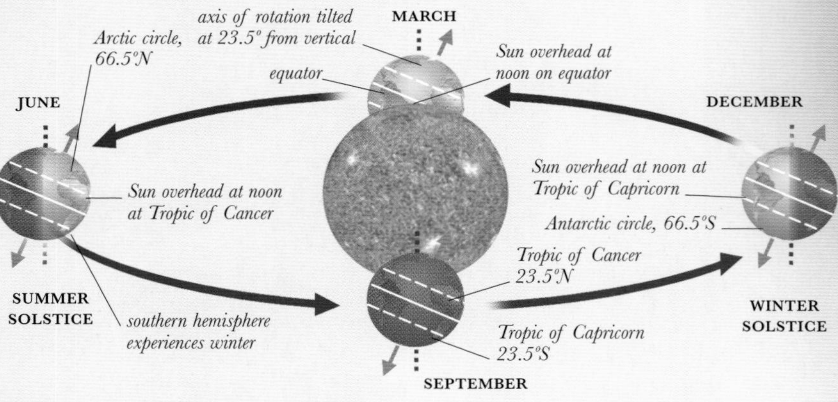


## SOLAR HEATING

The Sun emits various types of radiation, but most is absorbed high in the Earth's atmosphere. Only visible light, some infrared radiation, and a little ultraviolet reach the surface in significant amounts. Light is of huge importance to life on Earth, but infrared is also vital because it heats the atmosphere, oceans, and land. Due to the Earth's curvature, radiant heat falling on tropical regions is more intense than that falling on the poles. High levels of solar radiation in the tropics heat the lower atmosphere there, causing air to rise and move to higher latitudes, where it then cools and sinks. This circulation, taking place in several cells between the equator and poles, combines with the effects of the Earth's rotation to produce global wind patterns (see pp.446–47). Solar radiation also makes the oceans warmer in tropical regions. Warm water is then moved toward the poles by wind-driven currents (see pp.390–91). By transferring heat from the tropics toward the poles, oceanic and atmospheric circulation profoundly influence the Earth's climate.

**SOLAR HEATING**
*Solar heating is greatest in the tropics. Toward the poles, the Sun's rays impinge at an oblique angle and must pass through a greater thickness of atmosphere to reach the ground.*

23.5° angle of tilt
Tropic of Cancer
spin of the Earth
solar radiation
Tropic of Capricorn
axis of rotation
atmosphere

**THE SUN**
*This image, taken by the SOHO spacecraft, shows the seething surface of the Sun in ultraviolet light. A huge prominence of high-temperature plasma (ionized gas) is erupting into the Sun's atmosphere.*

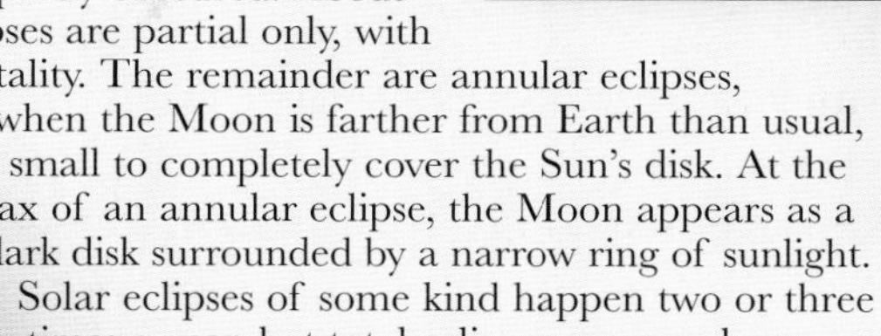

# SOLAR ECLIPSES

An eclipse of the Sun occurs when the Moon partially or totally blocks sunlight from reaching part of the Earth. About 25–30 percent of eclipses are total, with the Moon completely obscuring the Sun for a few moments for viewers located within a limited region called the area of totality. Outside this area is a larger region where viewers see the Sun only partly obscured. About 35 percent of eclipses are partial only, with no area of totality. The remainder are annular eclipses, occurring when the Moon is farther from Earth than usual, and too small to completely cover the Sun's disk. At the climax of an annular eclipse, the Moon appears as a dark disk surrounded by a narrow ring of sunlight. Solar eclipses of some kind happen two or three times a year, but total eclipses occur only once every 18 months on average. During the brief period of totality, the Sun's corona, its tenuous but superheated outer atmosphere, is visible.

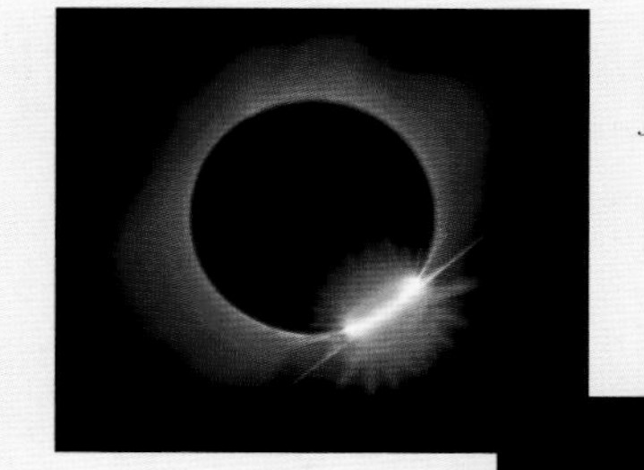

**NEAR-TOTALITY**
*As the Sun emerges from behind the Moon at the end of totality, the beautiful diamond-ring effect may be seen.*

**PARTIAL ECLIPSE**
*During a partial eclipse, the Moon passes across the face of the Sun but does not completely cover it.*

**SOLAR ECLIPSE**
*The shadow cast by the Moon during an eclipse consists of the central umbra (associated with the area of totality) and the penumbra (area of partial eclipse).*

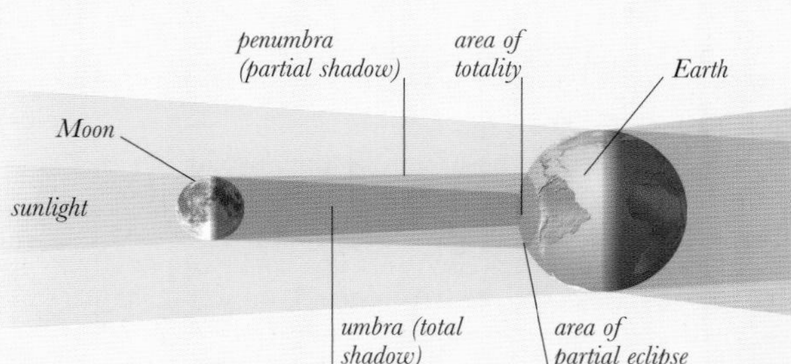


## OBSERVING THE SUN

It is extremely dangerous to look at the Sun through binoculars, a telescope, or smoked glass. The Sun can be viewed safely with the naked eye only during the brief moments of a total eclipse. At other times, it can be observed through filters specifically designed for the purpose (like the ones used by these Mauritanians to view the partial phase of the total eclipse of June 30, 1973) or by projecting the Sun's rays through a homemade pinhole projector onto a screen placed behind the opening.

# SOLAR ACTIVITY

The outer layer of the Sun's atmosphere, the corona, spews out a continuous stream of electrically charged particles that flows out across the Solar System. This solar wind is modified by bursts of activity around the Sun's surface and atmosphere, triggered by intense magnetic activity in these regions. This variable activity includes solar flares and coronal mass ejections (CMEs). When energetic particles from the solar wind and these additional events reach Earth, most are deflected by the planet's magnetic field or magnetosphere (see p.55), but some are channeled into Earth's atmosphere above the magnetic poles. There, the particles interact with gases in the atmosphere to produce beautiful shimmering sheets of light called aurorae, at altitudes of 56–186 miles (90–300 km). When a large burst of particles hits the magnetosphere, the result is a magnetic storm that can interfere with electrical systems on Earth.

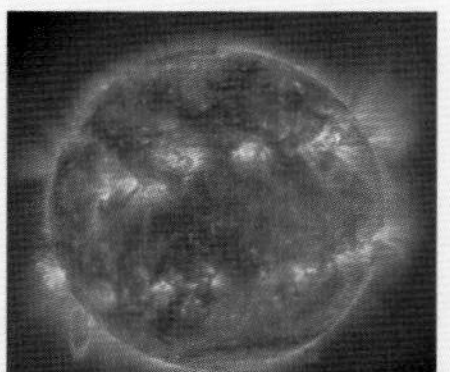

**MAGNETIC FIELDS**
*In this ultraviolet image, wisps of plasma can be seen following the Sun's invisible magnetic field.*

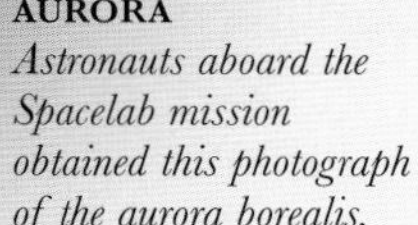

**AURORA**
*Astronauts aboard the Spacelab mission obtained this photograph of the aurora borealis.*

# THE EARTH AND THE MOON

27 Meteorite bombardment
44–45 The universe
46–47 The Solar System
The Earth's structure 54–55
Meteorite impacts 120–25
Tides and waves 428–29

THE EARTH'S MOON IS ONE OF THE LARGEST in the Solar System relative to its planet. This reflects its likely origin in a collision between the infant Earth and a Mars-sized object (see p.26). The Apollo astronauts who visited our satellite found no signs of life there, but analysis of the rocks they brought back has revealed much about the early history of the Earth–Moon system. The Moon's influence on Earth results from its gravity, which is the major cause of tides in Earth's oceans. In the distant past, the Moon orbited much closer to Earth, and its gravitational effects would have been stronger. Indeed, by steadying the Earth's rotation, the Moon may have assisted in the development of a stable climate on Earth, a possible prerequisite for life.

**EARTH AND ITS SATELLITE**
*This image taken by the Galileo spacecraft shows the Earth and the Moon together in space.*

## STRUCTURE OF THE MOON

From measurements made with seismic instruments left on the Moon by the Apollo astronauts, it is known that the Moon has a crust, which is about 37 miles (60 km) thick on the side facing Earth. Beneath the crust is a rocky upper mantle that extends to a depth of about 500 miles (800 km) beneath the surface. Whether or not the Moon has an iron-rich core like Earth's is not known, but if it does, it is probably small. The Moon has no atmosphere, but during the 1990s, the lunar-orbiting Clementine and Lunar Prospector spacecraft discovered that there are probably significant quantities of frozen water, hidden within permanently shadowed craters around its polar regions.

**INSIDE THE MOON**
*The Moon is structurally similar to the rocky planets, but its smaller size means it has cooled faster.*

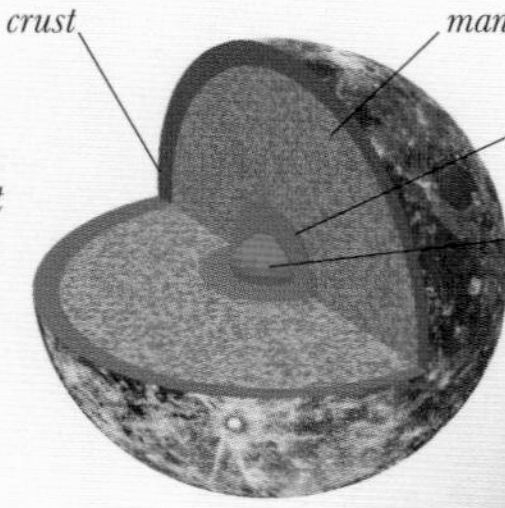

crust
mantle
probable molten outer core
possible small, solid, iron-rich core

**SYNCHRONOUS ROTATION**
*For each orbit of Earth, the Moon spins once on its axis. As a result, it always keeps the same face to Earth.*

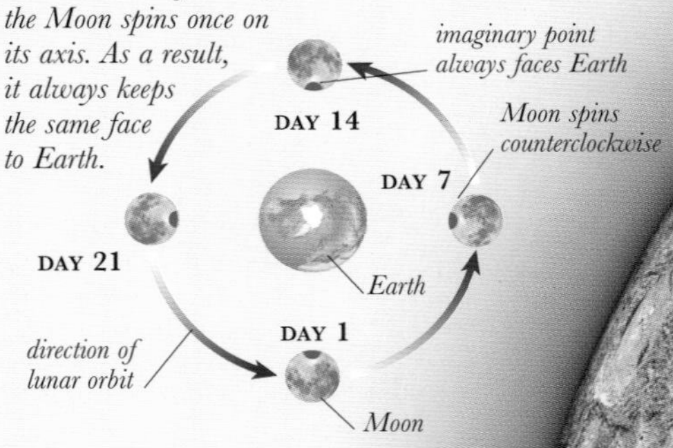

imaginary point always faces Earth
DAY 14
Moon spins counterclockwise
DAY 7
DAY 21
Earth
DAY 1
direction of lunar orbit
Moon

## ORBIT AND PHASES

The Moon orbits Earth at an average distance of 238,900 miles (384,400 km). Both bodies actually orbit their common center of mass, which is located deep within the Earth. With each orbit of 27.3 days, the Moon spins exactly once on its own axis, and as a result it always presents the same face to Earth. This face is called the near side, and the face never seen from Earth is the far side. During each orbit, the angle between the Earth, Moon, and Sun continuously changes. This gives rise to the lunar phases, varying from full moon (when the Earth is located between the Moon and Sun) to new moon (when the Moon lies between the Earth and Sun).

**CHANGING ANGLES**
*During each lunar orbit, the angle between the Earth, Moon, and Sun changes. As a result, the proportion of the Moon's sunlit face seen from Earth also changes.*

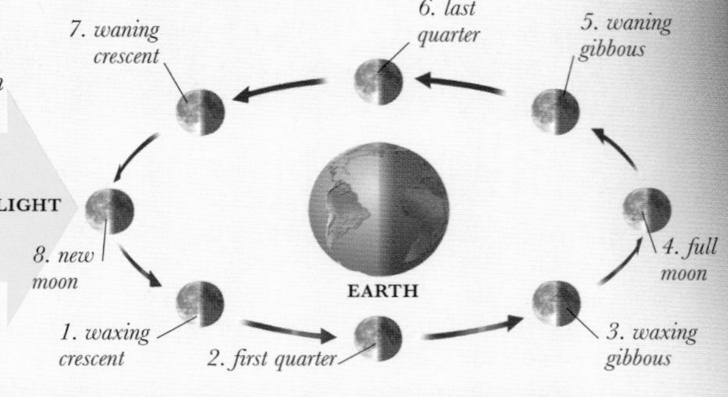



**PHASES OF THE MOON**
*Each lunar month, the appearance of the Moon as seen from Earth passes through the eight phases shown below.*

1. WAXING CRESCENT
2. FIRST QUARTER
3. WAXING GIBBOUS

4. FULL MOON

## APOLLO LANDINGS

Between 1969 and 1972, six pairs of US astronauts landed on the Moon, explored small areas of its surface, left experiments behind, and brought back rock samples. Their work revealed that the Moon is covered by a loose surface layer of rock fragments, called the regolith. The bright lunar highlands proved to be composed mainly of the coarse-grained rock anorthosite, while the maria (or seas) are giant impact craters later filled with basalt from volcanic eruptions.

# LUNAR TIDES

As the Moon and the Earth orbit their common center of mass, two forces are created at the Earth's surface. The first is a pull toward the Moon, which is strongest at points closest to it. This forms a bulge in the oceans toward the Moon. The second force is inertial, resulting from the Earth's orbit around the system's center of mass, and directed away from the Moon to produce a tidal bulge on the opposite side of the Earth's surface. As the Earth spins on its own axis, the bulges sweep over the surface, producing daily fluctuations in sea level called tides. These daily tides are modified by gravitational interactions between the Earth and Sun (see p.428). As a result of frictional drag on Earth's rotation exerted by the tides, the Earth's spin is gradually slowing; at the same time, the Moon is slowly moving away from the Earth.

**TIDAL RANGE**
*Most coastal areas experience one or two tidal cycles a day. The difference between high and low water levels is apparent on this beach.*

**CAUSE OF TIDAL BULGES**
*Gravitational interaction between the Earth and the Moon cause two bulges (exaggerated here) in Earth's oceans. As the Earth spins, the bulges sweep over its surface, though the bulges do not precisely line up with the line between Earth and Moon.*

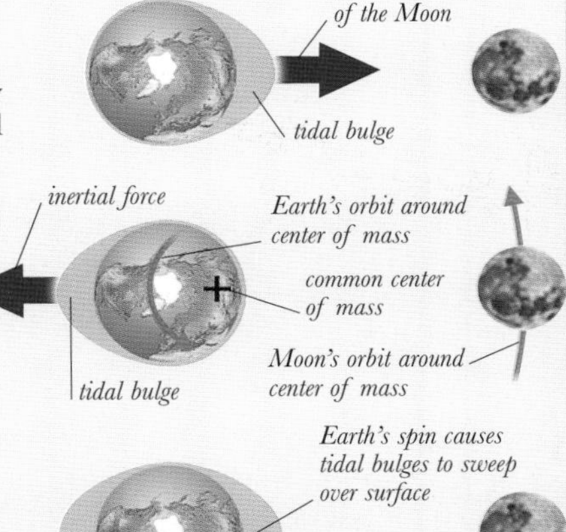


# LUNAR ECLIPSES

A lunar eclipse occurs when the Moon passes through the Earth's shadow as it orbits the planet. This can happen only at a full moon. The Earth's shadow has two components, the umbra (full shadow) and the penumbra (part-shadow). Astronomers recognize three types of lunar eclipse, occurring with roughly equal frequency. In a penumbral eclipse, the Moon passes through the Earth's penumbra, which produces only a slight dimming of the Moon. In a partial eclipse, a portion of the Moon passes through the Earth's umbra, and in a total eclipse the entire Moon passes through the umbra. An eclipse of one kind or another occurs about two to four times each year, and can be seen by anyone on the night side of Earth, even with the naked eye. During a total eclipse, no direct sunlight can reach the Moon, but light rays can be bent toward it by refraction in Earth's atmosphere. Because long-wavelength (red) light is refracted most, the Moon typically takes on a blood-red hue.

**TOTAL ECLIPSE**
*During the total phase of a lunar eclipse, light refracted by Earth's atmosphere still reaches the Moon, turning it red.*

**PARTIAL ECLIPSE**
*In the partial phase of an eclipse, the shape of the Earth is revealed as its shadow creeps across the face of the Moon.*

**ECLIPSE MECHANICS**
*The Earth's shadow consists of the penumbra, where it blocks some of the Sun's rays, and the umbra, where all direct sunlight is blocked. In a total eclipse, the Moon passes through penumbra, umbra, and penumbra again.*

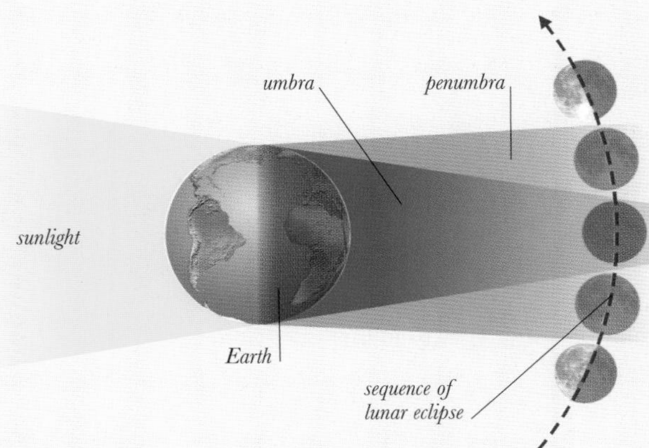


**THE LUNAR SURFACE**
*Notable features of the lunar surface are its bright, heavily cratered highlands, dark maria or seas (huge, ancient impact craters that have filled with lava), and innumerable smaller, younger craters.*

**5. WANING GIBBOUS** **6. LAST QUARTER** **7. WANING CRESCENT** **8. NEW MOON**

**RAW MATERIAL**
*Lava pouring across the island of Hawaii is a vivid demonstration of Earth's hot interior, and of the processes that formed the diverse minerals and rocks at the surface.*

# THE ANATOMY OF THE EARTH

PLANET EARTH IS A COMPLEX structure that is made up of three primary layers: the core, the mantle, and the crust. The density and temperature of the materials that make up the Earth increase approaching the metallic iron core, which has a solid interior and a molten exterior. The brittle outer layer of the planet, however, is made of many different kinds of rock, some up to 4 billion years old. Rocks consist of combinations of minerals, and are formed by various geological processes. These include internal forces, such as the movement of tectonic plates and volcanic eruptions, layering and compression of sediment, and interaction with living organisms and the atmosphere.

# THE EARTH'S STRUCTURE

20–41 The history of the Earth
The core and mantle 56–57
The crust 58–59
Minerals 60–77
Rocks 78–99
Tectonic plates 106–107
Plate boundaries 108–109
Mountains and volcanoes 140–209

EVERYDAY EXPERIENCE ON THE EARTH might suggest that our planet is remarkably varied— its materials range from water and ice to atmospheric gases and a host of rocks and minerals. However, the thin surface biosphere we occupy is not typical of the Earth as a whole. Overall, the planet is much less varied— within a few dozen miles below the surface, it consists only of rocks, minerals, and metallic compounds. Measurements of earthquake waves passing through the Earth allow us to probe its structure, since waves propagate through different layers at different speeds. They reveal the existence of a hot, dense core, surrounded by a mantle and a thin, rocky outer crust. The crust supports the biosphere with its soils, plants, animals, hydrosphere, and atmosphere.

## SHAPE AND FORM

The overall shape of the Earth, called the geoid, is determined by the effects of gravity and rotation on the materials that make up our planet. Gravity will pull any sufficiently massive object into a sphere—only smaller Solar System objects have nonspherical shapes. The Earth is an almost perfect sphere, but its rapid rotation every 24 hours—equivalent to the surface at the equator moving at more than 1,000 mph (1,600 kph)—reduces the effect of gravity around the equator, and means that equatorial regions bulge outward by about 13 miles (21 km) compared with the poles. The Earth's surface topography varies by about 12½ miles (20 km) from the highest mountains to the deepest ocean trenches, and variations in surface elevation reflect two fundamentally different types of surface crust—continental crust with average elevation of less than ⅗ mile (1 km) above sea level, and oceanic crust with average depth of about 2⅔ miles (4.5 km) below sea level. Gravity, coupled with processes such as tectonics (see pp.106–107) and erosion (see pp.112–13), makes larger variations in elevation unsustainable over long periods.

**EAST AFRICAN RIFT**
*Earth's internal tectonic processes have pushed up these mountains in east Africa, but erosion over millions of years will reduce them to insignificant hills.*

**EARTH IN SPACE**
*Seen from space, Earth is a blue planet. Two-thirds of the surface is covered by sea, and much of the land is often obscured by cloud.*

*Earth bulges at the equator*

*direction of rotation (1,000 mph/1,600 km/h) at equator)*

*flattened pole*

**CHANGING SHAPE**
*The Earth's high rotation rate slightly distorts its structure so that it bulges at the equator and is flattened at the poles.*

**INSIDE THE EARTH**
*The Earth has a layered internal structure, kept hot inside by pressure and heat from radioactive elements. Heat flows from the top of the core through the mantle by convection currents in the slowly moving rocks, until it eventually reaches the cooler crust and escapes.*

*core–mantle boundary, where liquid outer core and solid lower mantle layers meet*

INNER CORE
**STATE** Solid **DENSITY** 12 g/cm³
**DEPTH** 3,960 miles (6,370 km) below surface
**TEMPERATURE** 7,200–8,500°F (4,000–4,700°C)

OUTER CORE
**STATE** Liquid **DENSITY** 10 g/cm³
**DEPTH** 3,200 miles (5,150 km) below surface
**TEMPERATURE** 6,300–7,200°F (3,500–4,000°C)

LOWER MANTLE
**STATE** Solid **DENSITY** 5.5 g/cm³
**DEPTH** 1,860 miles (2,990 km) below surface
**TEMPERATURE** 1,800–6,300°F (1,000–3,500°C)

UPPER MANTLE
**STATE** Solid **DENSITY** 3.5 g/cm³
**DEPTH** 3–45 miles (5–70 km) below surface
**TEMPERATURE** Less than 1,800°F (1,000°C)

OCEANIC CRUST
**STATE** Solid **DENSITY** 3 g/cm³
**DEPTH** 0–7 miles (0–11 km) below surface
**TEMPERATURE** Less than 1,800°F (1,000°C)

CONTINENTAL CRUST
**STATE** Solid **DENSITY** 2.7 g/cm³
**DEPTH** 0–45 miles (0–70 km) below surface
**TEMPERATURE** Less than 1,800°F (1,000°C)

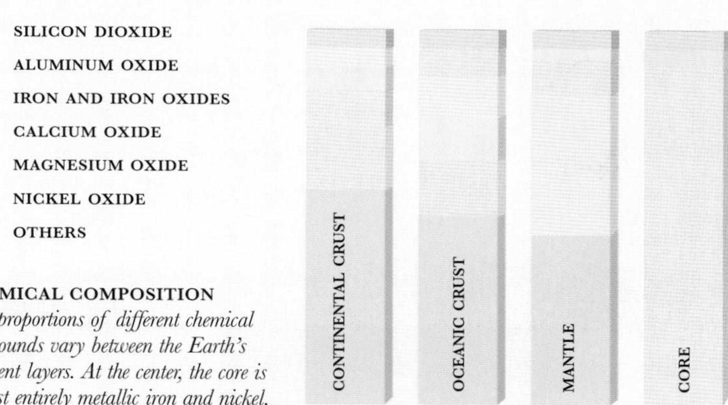


**CHEMICAL COMPOSITION**
*The proportions of different chemical compounds vary between the Earth's different layers. At the center, the core is almost entirely metallic iron and nickel.*

# THE EARTH'S LAYERS

Analysis of rocks in the Earth's crust shows that most are rich in silicon dioxide. The basalts of the oceanic crust also have proportionally more calcium, magnesium, and iron, while the less dense continents are richer in aluminum, potassium, and sodium. Differentiation (the gravitational separation of mixed materials according to their density) means that the interior of the Earth is more dense than the surface crustal layer, and this must be due to differences in mineral composition. Rock samples from the mantle, occasionally brought to the surface by volcanism, show that this region is made of silicate minerals rich in magnesium and iron—for example, olivine. No samples from the core are available to analyze, but it is thought to be similar in composition to iron meteorites formed in the early Solar System and composed of nickel–iron alloys.

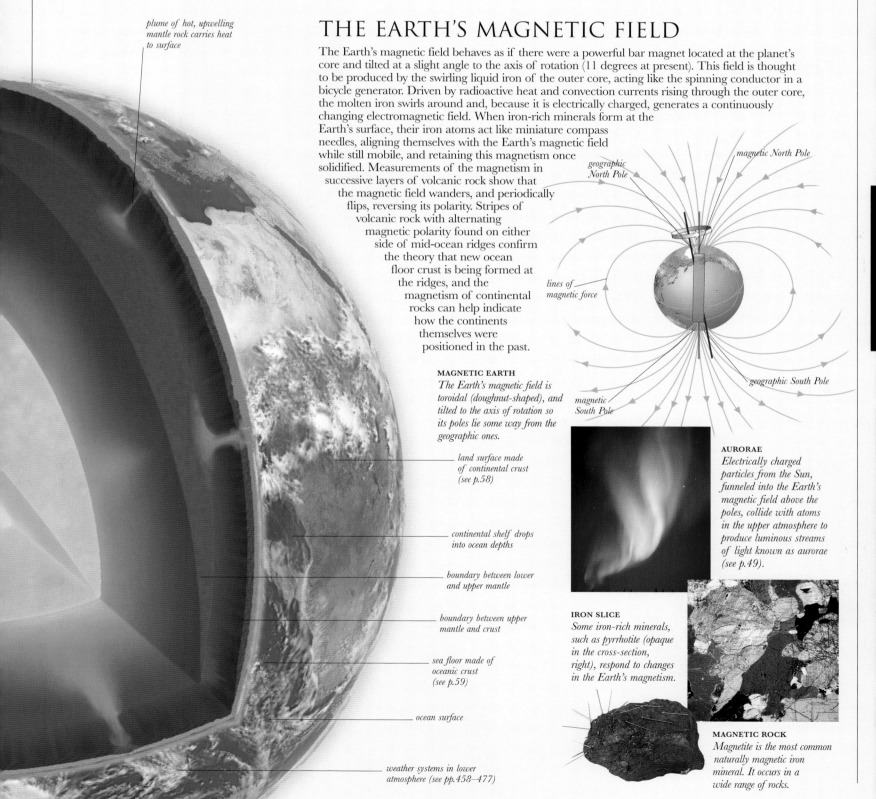

*plume of hot, upwelling mantle rock carries heat to surface*

# THE EARTH'S MAGNETIC FIELD

The Earth's magnetic field behaves as if there were a powerful bar magnet located at the planet's core and tilted at a slight angle to the axis of rotation (11 degrees at present). This field is thought to be produced by the swirling liquid iron of the outer core, acting like the spinning conductor in a bicycle generator. Driven by radioactive heat and convection currents rising through the outer core, the molten iron swirls around and, because it is electrically charged, generates a continuously changing electromagnetic field. When iron-rich minerals form at the Earth's surface, their iron atoms act like miniature compass needles, aligning themselves with the Earth's magnetic field while still mobile, and retaining this magnetism once solidified. Measurements of the magnetism in successive layers of volcanic rock show that the magnetic field wanders, and periodically flips, reversing its polarity. Stripes of volcanic rock with alternating magnetic polarity found on either side of mid-ocean ridges confirm the theory that new ocean floor crust is being formed at the ridges, and the magnetism of continental rocks can help indicate how the continents themselves were positioned in the past.

*geographic North Pole*
*magnetic North Pole*
*lines of magnetic force*
*geographic South Pole*
*magnetic South Pole*

**MAGNETIC EARTH**
*The Earth's magnetic field is toroidal (doughnut-shaped), and tilted to the axis of rotation so its poles lie some way from the geographic ones.*

**AURORAE**
*Electrically charged particles from the Sun, funneled into the Earth's magnetic field above the poles, collide with atoms in the upper atmosphere to produce luminous streams of light known as aurorae (see p.49).*

*land surface made of continental crust (see p.58)*

*continental shelf drops into ocean depths*

*boundary between lower and upper mantle*

*boundary between upper mantle and crust*

*sea floor made of oceanic crust (see p.59)*

*ocean surface*

*weather systems in lower atmosphere (see pp.458–477)*

**IRON SLICE**
*Some iron-rich minerals, such as pyrrhotite (opaque in the cross-section, right), respond to changes in the Earth's magnetism.*

**MAGNETIC ROCK**
*Magnetite is the most common naturally magnetic iron mineral. It occurs in a wide range of rocks.*

# THE CORE AND MANTLE



TOGETHER, THE CORE AND MANTLE make up the vast bulk of the Earth. Except for the uppermost mantle, these regions are far too deep to be studied by direct sampling, so our understanding of them depends on indirect investigations, such as comparisons of rock density, measurement of seismic waves traveling through the Earth, studies of the magnetic field, and laboratory experiments. Together, these data reveal the existence of a metallic core with solid inner and liquid outer layers, separated by a well-defined boundary from the solid rocks of the mantle.

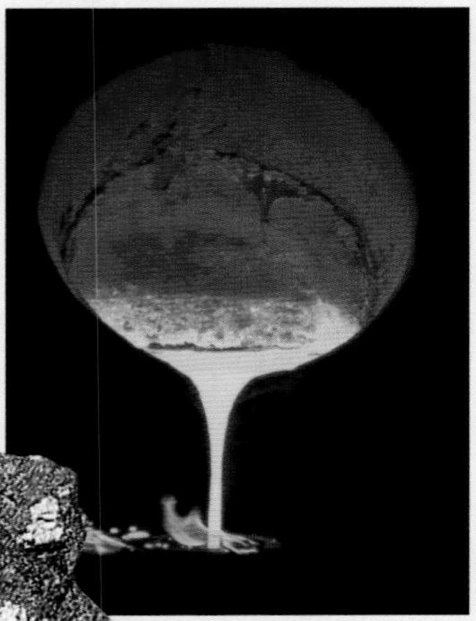

**MOLTEN IRON**
*Metallic iron (seen here being poured in a foundry) has a high density and strongly magnetic character. It probably forms most of the Earth's core, along with a small percentage of nickel and some other lighter element, perhaps sulfur.*

## THE CORE

Comparison of the Earth's overall mass with the density of rocks close to the surface indicates that the core must be significantly denser than the mantle, which is itself denser than the crust. The core is probably dominated by iron and nickel, but about 8–12 percent (by weight) consists of some lighter element, possibly sulfur. The fact that some earthquake waves are not transmitted directly through the core, while others are refracted or bent when they pass through a so-called shadow zone, indicates that there is a solid inner core (only 5 percent by weight of the whole core), surrounded by an outer core of liquid metal. The nature of Earth's magnetic field seems to suggest that it is generated in the swirling liquid of the outer core, while the presence of nickel–iron alloy in the oldest meteorites, formed at the same time and from the same materials as the Earth itself, is more evidence to reinforce this model of the core's composition.

**IRON METEORITE**
*Iron meteorites represent the lowermost mantle and core of large, differentiated asteroids, long since destroyed in violent collisions. The presence of nickel–iron alloys within them is further evidence of Earth's liquid metallic core.*

## THE MANTLE

Mantle material consists mostly of relatively dense silicate minerals, some of which are in high-pressure forms with close-packed atomic structures. Direct evidence for this comes from blocks of upper mantle material called xenoliths, which are occasionally brought to the surface by deep-seated volcanism. These are usually composed of silica-poor peridotite (see p.84), which consists mainly of the mineral olivine (see p.72). The lower mantle is thought to be dominated by denser silicates with the composition of olivine but a more compact structure similar to the denser rare mineral perovskite. The uppermost layers of the mantle are fused to the continental crust, and together crust and upper mantle form the brittle plates of the lithosphere. Just below the lithosphere lies the asthenosphere, where temperatures are almost high enough to melt the rock.

**CORE AND MANTLE**
*During its formation, the Earth differentiated into two main layers—an inner metallic iron core and a thick, solid, and stony silicate mantle, on which a thin outer crust developed. Internal heat from the molten outer core disperses through the solid mantle as rising convection currents.*

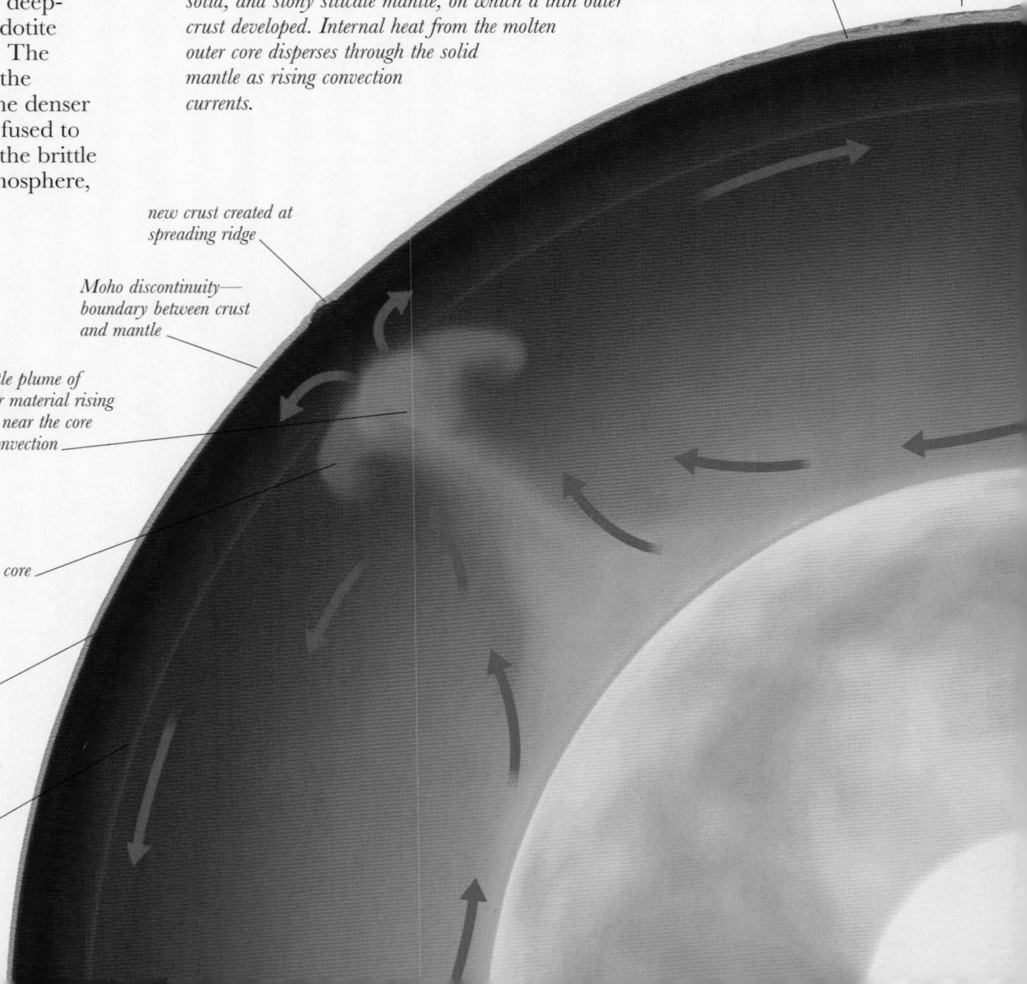

**PERIDOTITE**
*Igneous rocks, such as these in Newfoundland, are the remains of ancient ocean floor. They include coarse-grained, olivine-rich rocks called peridotites, thought to have a composition similar to upper mantle rocks.*

**MANTLE XENOLITHS**
*This xenolith from France is an olivine-rich peridotite caught up in a basaltic lava. It is a rock from the upper mantle that has been carried to the Earth's surface.*

# SEISMIC ACTIVITY

As lithospheric plates move around on the Earth's surface, they produce earthquakes, releasing energy that is transmitted through the Earth as vibrations known as seismic waves. The speed of these waves depends on the properties of the medium they are traveling through—where the rocks are denser, the wave speed increases. Such relationships allow geologists to probe the Earth's deep structure by measuring how long seismic waves take to reach detectors in various locations around the world. At abrupt boundaries, the speed of travel can change dramatically, and seismic waves are refracted just like light waves passing into and out of a fish tank. Such a boundary occurs 1,800 miles (2,900 km) inside the Earth, marking a sharp and substantial increase in density. This is the boundary between the mantle and the core.

## ANDRIJA MOHOROVICIC

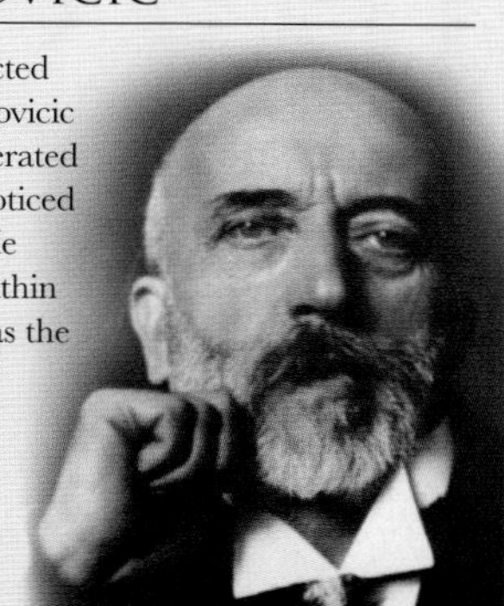

The boundary between crust and mantle was first detected and described by Croatian geophysicist Andrija Mohorovicic (1857–1936). By studying records of seismic waves generated by an earthquake near Zagreb in 1909, Mohorovicic noticed that some waves seemed to arrive earlier than others. He reasoned that they were affected by a density change within the Earth, around 18 miles (30 km) deep. Now known as the Mohorovicic discontinuity, or Moho, its depth changes from 16–56 miles (25–90 km) beneath the continents to 4–6 miles (7–10 km) below the ocean floor. The discontinuity marks the compositional change between the crust and the underlying mantle.

**P- AND S-WAVES**

*Earthquakes produce two types of seismic waves. Compressional or P-waves consist of alternating pulses of stretching and compression through any material. Shear or S-waves shake rocks from side to side and up and down. Because only P-waves can pass through liquids such as molten rock, the two types of waves offer another way of probing the Earth's interior.*

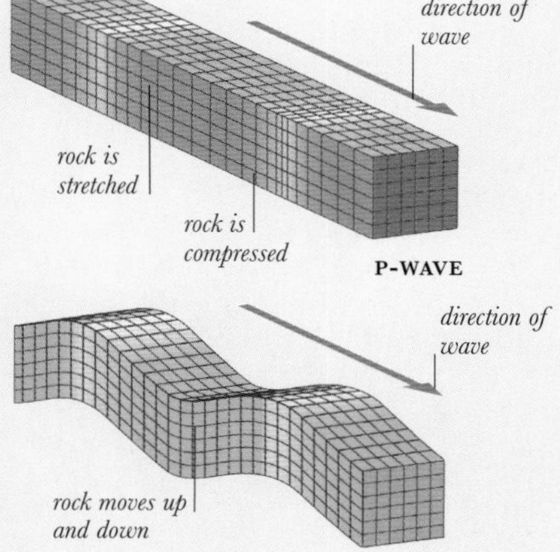


*S-wave travels only through solid rock*

*liquid outer core*

*S-waves are detected up to 105° around the Earth's surface from their origin*

*zone where no S-waves are detected*

*epicenter of earthquake produces seismic waves*

*refraction prevents even P-waves from reaching this shadow zone*

*P-wave travels through solid and molten rock*

*P-waves refracted as they cross between layers*

*P-waves reach areas up to 140° around the Earth's surface from their origin*

**SHOCK WAVES**

*Analysis of the time seismic waves take to be detected at stations around the globe, coupled with the presence of shadow zones where some or all waves cannot penetrate the Earth, supports the idea that the Earth is layered, with a core that is mostly liquid.*

**SEISMIC TOMOGRAPHY**

*Seismic waves alter their speed as they move through rocks under stress. Using a network of detectors, geologists can build up maps of stress patterns in the Earth's crust.*

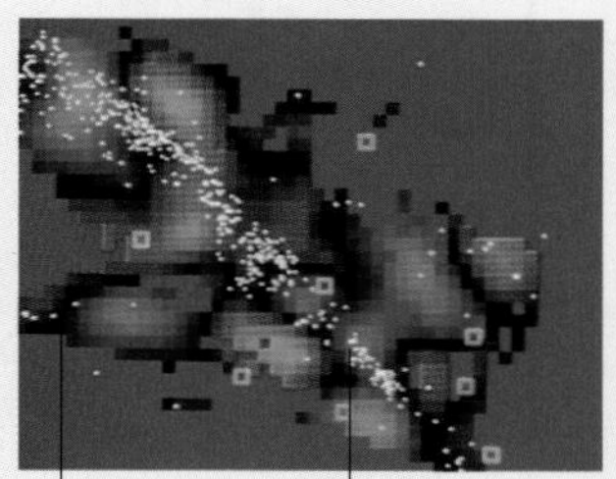

*plot of subsurface rock stresses*

*white dots indicate seismic activity*

# HEAT TRANSFER

The temperature of the Earth's inner core is thought to be around 8,500°F (4,700°C). Heat is convected through the overlying outer core of liquid metal to the base of the mantle, where temperatures are around 6,300°F (3,500°C). At such temperatures, silicate rocks would normally be molten, but the immense pressure deep inside the Earth keeps the rocks of the mantle solid. However, rising heat from the core causes the mantle to slowly circulate, its solid rocks shifting by just a few inches a year, and convecting heat away from the core as they do so. Hot mantle rocks slowly rise toward the surface, cooling and becoming denser until they sink back down to form a convection cell. The top of the cell coincides with the base of the cool and brittle outer lithosphere. This is broken into a series of tectonic plates, the motion of which is thought to be driven in part by the convection below them.

*oceanic crust descending beneath continental crust into mantle (see pp.108–109)*

*crust melts as it descends into mantle*

*upper mantle made primarily from peridotite*

*lower mantle consisting primarily of high-density, magnesium-rich silicate materials*

*molten iron outer core*

*solid iron inner core*

*convection current created by movement of mantle plume*

**FIRES WITHIN**

*Investigation of volcanic eruptions and lava flows, such as this one in Hawaii, helps to improve our understanding of how heat flow is related to igneous activity.*

# THE CRUST

54–55 The Earth's structure
56–57 The core and mantle
Minerals 60–77
Rocks 78–99
Tectonic plates 106–107
Plate boundaries 108–109
Volcanoes 172–93
Ocean tectonics 386–87

THE THINNEST AND OUTERMOST LAYER of the Earth is the crust, with an average thickness of about 18 miles (30 km) below the continents and around 6 miles (10 km) below the oceans. The crust lies on top of the more rigid mantle, and the boundary between the two is marked by the Mohorovicic seismic discontinuity. Overall, the crust is less dense than the mantle, because its rocks are richer in minerals containing relatively light elements such as silicon, aluminum, and calcium. However, there are two distinctly different types of crust—oceanic and continental. Variations in their composition, density, and thickness help to account for differences in their topography, relative age, and history of formation.

## CONTINENTAL CRUST

The continental crust accounts for only about one-third of the Earth's global surface, but it forms all the major landmasses and their fringing shallow seas. Its thickness ranges from 16 to 45 miles (25 to 70 km), with the thickest portions underlying young mountain belts, and it has huge variations in composition—from sedimentary rocks such as sandstone, coal, and limestone (see pp.92–98) through metamorphic rocks such as marble and slate (see pp.87–91) to igneous rocks such as granite and gabbro (see pp.82–86). This variety of rock types has come about mainly because light continental crust is not recycled within the Earth to the same extent as denser oceanic crust. As a result, some continental rocks are up to 4 billion years old, and much of the material now above sea level has been through repeated cycles of erosion, formation into sedimentary rocks, and metamorphosis. Tectonic forces have also subjected the continents to phases of fragmentation, amalgamation, and long-term movement, resulting in the formation of new oceans, mountain-building, widespread volcanism, and occasional coalescence into single landmasses called supercontinents. Throughout all this, new layers of rock have constantly been added to the surfaces and margins of the continents, forming the strata that are the basis for much of our understanding of the geological history of the Earth and, through the fossil record, its life.

**VEGETATION**
*Land vegetation has evolved over more than 450 million years and has played a significant role in altering global climates and forming deposits such as soils and coals (see p.135).*

**FRESH WATER**
*Water precipitated from the atmosphere forms lakes and rivers as it collects and flows back toward the ocean, eroding the landscape through which it flows.*

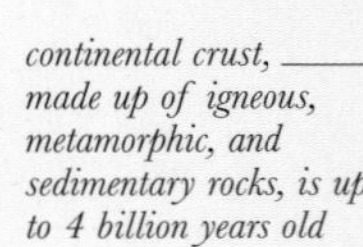

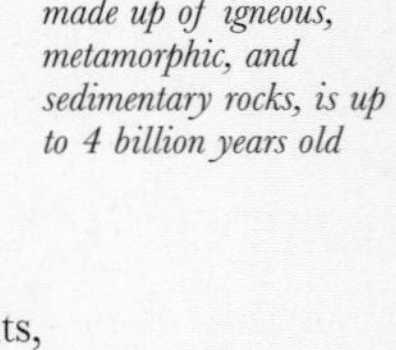

*continental crust, made up of igneous, metamorphic, and sedimentary rocks, is up to 4 billion years old*

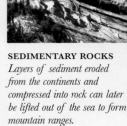

**SEDIMENTARY ROCKS**
*Layers of sediment eroded from the continents and compressed into rock can later be lifted out of the sea to form mountain ranges.*

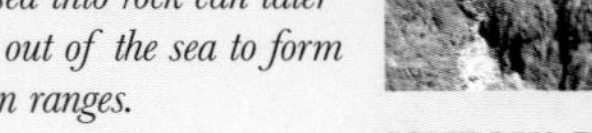

**IGNEOUS ROCKS**
*The other major component of the continental crust, igneous rocks are essentially unchanged since their formation in volcanic eruptions.*

**TIBETAN PLATEAU**
*Regions of high elevation such as this windswept plateau are underlain with continental crust that is far thicker than average.*

**ARAL BASIN**
*Low-lying areas of continental crust are also the thinnest. When surrounded by more elevated regions, they form natural drainage basins.*

**SOIL**
*The addition of organic material and microorganisms to weathered rock debris produces soils, essential to the growth of plants (see p.100).*

*boundary between crust and mantle, known as Mohorovicic discontinuity*

# OCEANIC CRUST

The oceanic crust forms over two-thirds of the Earth's surface, but even the oldest parts of the ocean floor are no more than 200 million years old and covered only by a thin veneer of sediment. Composed of relatively dense basaltic lavas and related rocks, oceanic crust varies from 4 to 7 miles (6 to 11 km) thick. Because of its density, it is less elevated than the lighter continental crustal rocks, with an average depth of about 2 miles (3 km) below sea level. Oceanic crust is continually formed from mantle material within long rifts known as spreading ridges (see p.108). Here, the mantle is rising, heating, and expanding the overlying rocks to form an elevated submarine mountain chain, where oceanic plates on either side of the rift are carried apart as if on opposing conveyor belts. Basaltic lavas erupt from fissures and cones, and cool to form new oceanic floor that is also punctuated by numerous individual volcanoes. These can grow large enough to rise above sea level, and where the crust is moving above a semi-permanent mantle hot spot plume (see p.107), they may form chains of volcanic islands. Elsewhere, in areas called subduction zones, oceanic crust descends into the mantle at the same rate that it is created at the ridges (see p.109).

**ANDESITE CONES, JAVA**
*When the energy released by subduction of oceanic crust partially melts mantle rocks, it can trigger the formation of curved chains of island volcanoes.*

**ICELANDIC RIDGE**
*Because of the density of their basaltic rocks, oceanic crust and mid-ocean ridges do not usually appear above sea level. Iceland is an exception, raised above the waters by exceptional heat flow.*

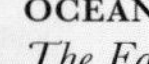

**OCEAN**
*The Earth is the only planet in the Solar System with persistent oceans of liquid water that cycle moisture through the atmosphere to the land and back to the oceans.*

**ATMOSPHERE**
*Above the crust lies the Earth's gaseous atmosphere. Its interaction with oceans and landmasses, and the protection it offers from damaging radiation, are vital for the survival of life.*

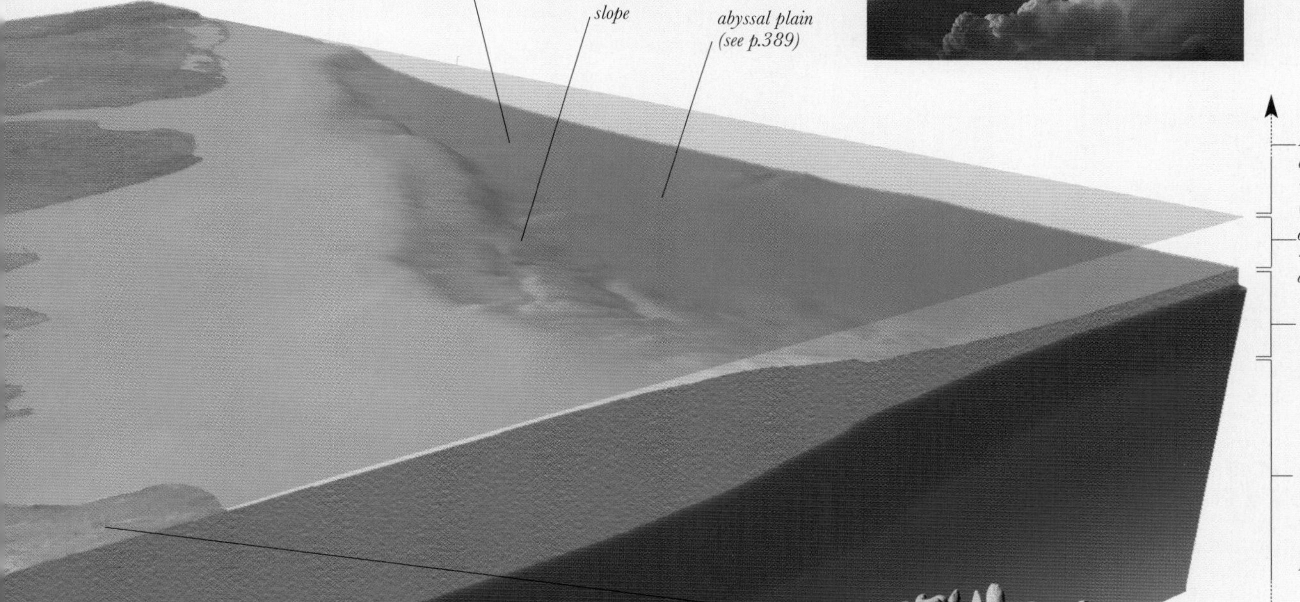

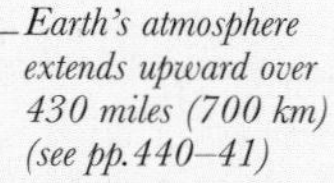

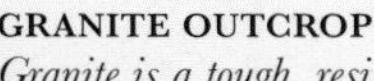

**GRANITE OUTCROP**
*Granite is a tough, resistant igneous rock formed by the slow cooling of magma deep in the continental crust. The granite seen here in the Ahaggar Massif, Algeria, has been exposed by erosion.*

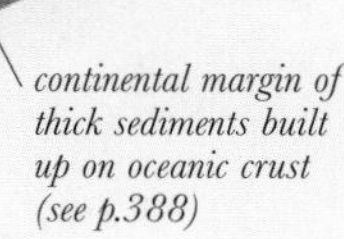

**THE UPPER LAYERS**
*Earth's crust is the uppermost solid layer of a complex outer structure. Fused with the upper mantle, it forms the brittle lithosphere, broken into tectonic plates. These drift on top of the slowly flowing but solid asthenosphere.*

# ISOSTASY

The continental crust is at its thickest beneath young mountain belts, where deep roots of crust stretch down to depths of about 45 miles (70 km)—well into the underlying mantle. Crustal blocks can be thought of as floating on the mantle—a concept known as isostasy. Because the continental crust has a much lower density than the mantle, it is buoyant, in the same way that an iceberg is buoyant in seawater. But the greater the mass of rock above sea level, the more buoyancy is required to support it, and the deeper its roots must go. By comparison, thinner, denser oceanic crust is less buoyant, and so reaches an equilibrium point well below sea level—which is why it is rarely seen as dry land.

**SEA-LEVEL CHANGE**
*Old beaches and cliffs raised above present sea level attest to past changes in the levels of both sea and land.*

# MINERALS

54–55 The Earth's structure
58–59 The crust
Rocks 78–99
Water content and deposits 201
Underground rivers and caves 252–53
Mineral extraction 375–77

MINERALS ARE THE NATURAL building blocks of the rock materials of the Earth and all the solid bodies of the universe. Studying minerals helps us to understand the origin and evolution of the Earth and planets. Most minerals are solid crystalline substances composed of atoms, which are generally arranged in orderly, repeating patterns, giving the mineral its crystalline structure and shape. A few minerals, however, have no such orderly crystalline structure but are amorphous solids similar to glass. Although more than 4,000 minerals have been discovered, only about 30 (known as rock-forming minerals) are common at the Earth's surface.

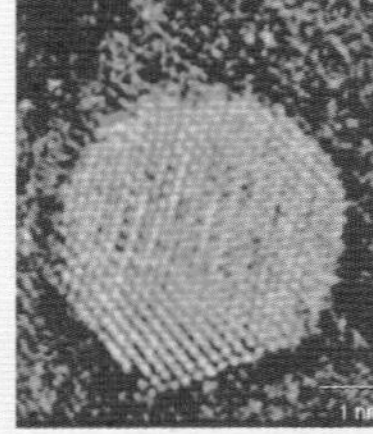

**REGULAR MATRIX**
*Gold atoms grown on a carbon substrate naturally bond and pack together in an orderly cubic structure.*

## MINERAL FORMATION

Individual minerals are distinguished by unique combinations of chemical elements and the arrangement of their atoms. This is determined when a gas or liquid crystallizes into a solid. Generally, the resulting structure is a regular, repeating three-dimensional array, or lattice, of atoms. The lattice grows by the ordered addition of atoms and usually maintains the same shape and composition. Some minerals have well-developed crystal form, but most are aggregates with other textures and habits. Similarly sized mineral grains have a granular texture, while a mineral lacking crystal faces is called massive. Primary minerals are formed at the same time as the host rock; later changes from, for example, erosion or metamorphism, produce secondary minerals.

## CRYSTAL STRUCTURE AND SHAPE

A crystal is a solid such as a mineral with an orderly, repeating atomic structure. With unrestricted growth, it forms a geometric shape with naturally flat planes, called faces, which are often arranged symmetrically. Faces are the external expression of a mineral's internal atomic regularity of structure, and they lie at specific angles one to another, depending on the crystal system to which the mineral belongs. Six main crystal systems—cubic, hexagonal/trigonal, monoclinic, orthorhombic, tetragonal, and triclinic—are defined by their geometric symmetry. Slight misalignment of the structure during growth often produces twinned crystals that are the mirror image of one another. Crystal shape is described using terms such as platy for thin, flat forms and prismatic for columnar crystals.

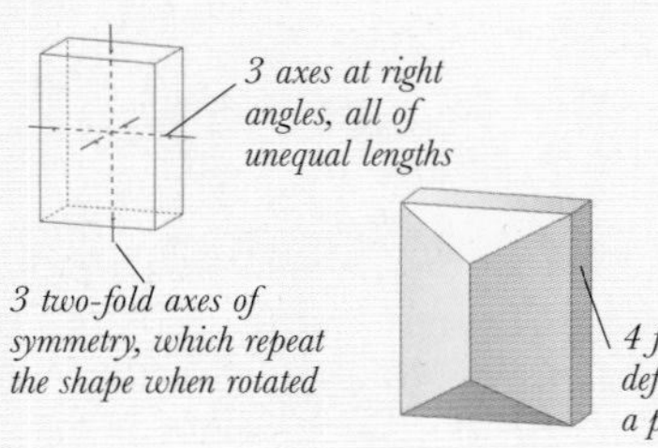

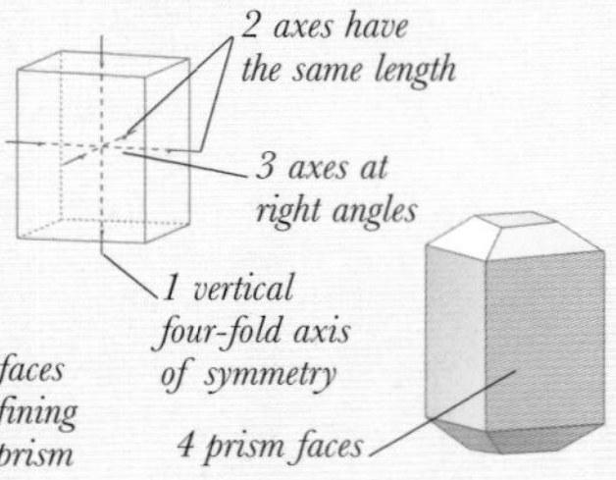

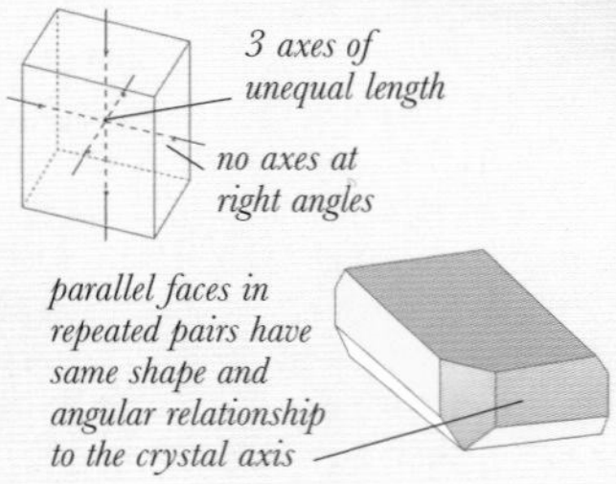

**CRYSTAL SHAPES**
*Crystal systems are defined by the proportions of their axes and the angles between them (see open figures below). Crystals in any one system can assume various shapes (one of which is represented by a solid figure below).*

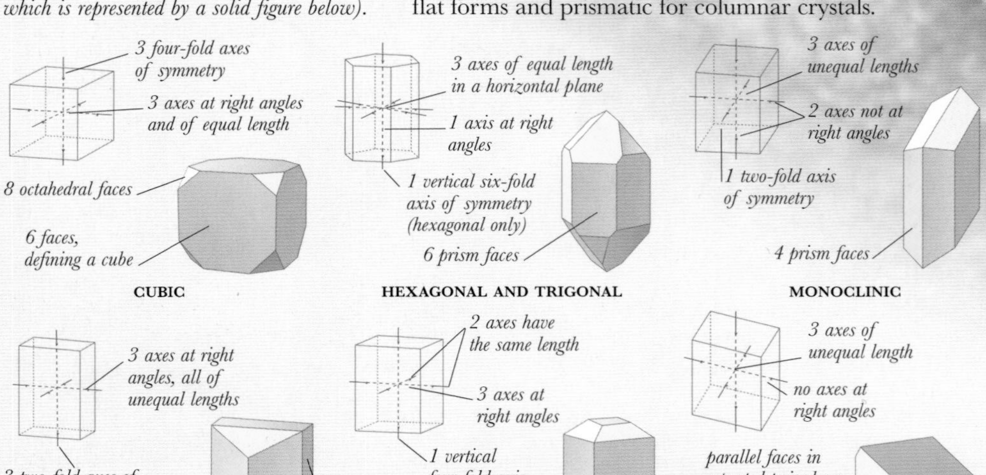


sodium atom
axis of symmetry
chlorine atom

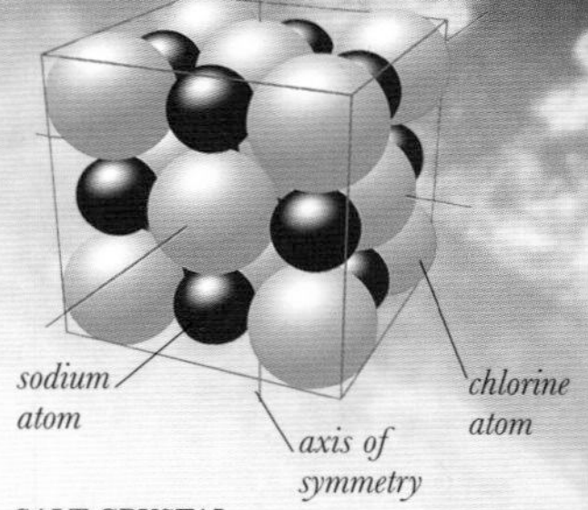

**SALT CRYSTAL**
*Sodium and chlorine atoms pack together to give the mineral halite (salt) a cubic form. Its crystals growing as perfect cubes with six square faces at right angles to one another, as can be seen here.*

SALT CRYSTAL

# MINERAL IDENTIFICATION TESTS

For 2.6 million years, humans and their extinct relatives have exploited particular minerals and rocks to create a variety of objects, from stone tools to swords, plowshares, space shuttles, and exquisite jewelry. To use minerals to their best advantage, it is first necessary to identify their individual characteristics. Over the millennia, many minerals have been found to possess unique combinations of physical properties, such as color, hardness, and specific gravity, which allow them to be picked out even when surrounded by other minerals in a rock. Several of these properties, such as crystal shape, hardness, specific gravity, and cleavage, can be related to actual crystal structures, and are thus predictable. The specific gravity of an open-structured mineral such as graphite, for example, is low, while diamond, with the same composition but a more closely packed structure, has a higher specific gravity. Expert mineralogists can recognize hundreds of different minerals largely on the basis of their appearance and a few tests with a steel point and a hand lens. The identification of other minerals, however, needs optical, chemical, X-ray, or other kinds of analysis.

CALCITE

*cleavage plane*

MUSCOVITE

**CLEAVAGE**
*This describes a plane of weak bonding between atoms in a crystal, along which the crystal may split. Calcite has three cleavage planes, which meet to form rhomboid cleavage fragments, while muscovite mica has well-developed and distinct cleavage in just one plane.*

**FRACTURE**
*Resulting from mineral breakage other than along cleavage planes, fracture frequently shows characteristic forms, such as opal's glasslike, conchoidal (rounded) fracture or kaolinite's irregular one.*

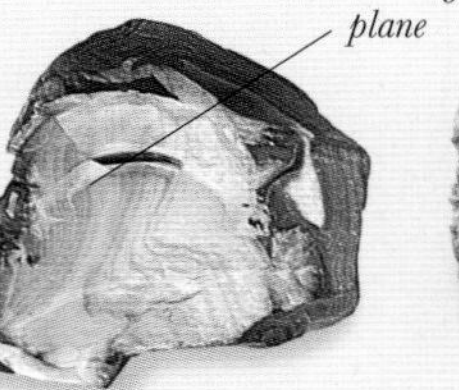


OPAL

KAOLINITE

TALC

DIAMOND

**HARDNESS**
*The strength of a mineral's chemical bonds determines its hardness. This is measured by its resistance to scratching on a 1-to-10 scale, which was invented by Friedrich Mohs in 1822; soft (talc) is 1 and hard (diamond) is 10.*

**SPECIFIC GRAVITY**
*This is determined by the structure and composition of a mineral. The weight of a mineral's atoms relate to its specific gravity. Gold has a higher specific gravity than halite because its atoms are heavier. Specific gravity is measured by comparing the weight of the mineral with an equal volume of water.*

HALITE

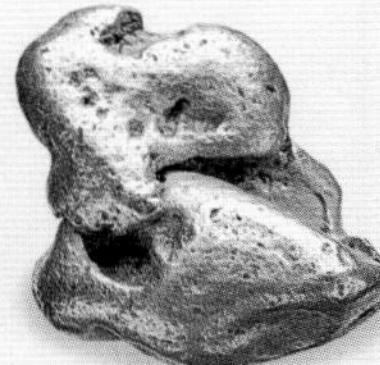

GOLD

AZURITE

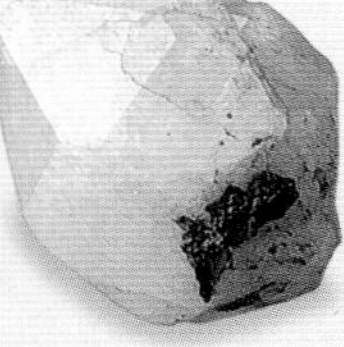

MILKY QUARTZ

**COLOR**
*The composition of a mineral, its atomic structure, often the presence of certain trace elements, and selective light absorption combine to generate the color of a mineral. The element copper in azurite produces its typical blue color. Other minerals are uncolored or white, because no light is absorbed, as in milky quartz.*

**LUSTER**
*The way in which light is absorbed or reflected off the surface of a mineral determines its luster, which is independent of a mineral's color. It may be diagnostic, as in smoky quartz, which shows a vitreous or glassy luster, and galena, which has a metallic one.*

SMOKY QUARTZ

GALENA

QUARTZ

MAGNETITE

**TRANSPARENCY**
*This is the extent to which light can pass through a mineral. It ranges from the water-clear transparency of the rock-crystal form of quartz, to the complete opacity of many metal ore minerals, such as magnetite.*

**STREAK**
*The color of a mineral when crushed to a powder provides its streak, which often differs from the mineral's surface color, especially in certain ore minerals. For example, hematite varies in its color, but its streak is always red.*

HEMATITE

## LIGHT STUDIES

As light passes through many minerals, it is altered by the optical properties of the structure and chemistry of the mineral. By using polarizing microscopes and extremely thin slices of a rock, which transmit light, mineralogists can distinguish the minerals present. Some minerals, however, are opaque and appear black—such as spinel (shown here).

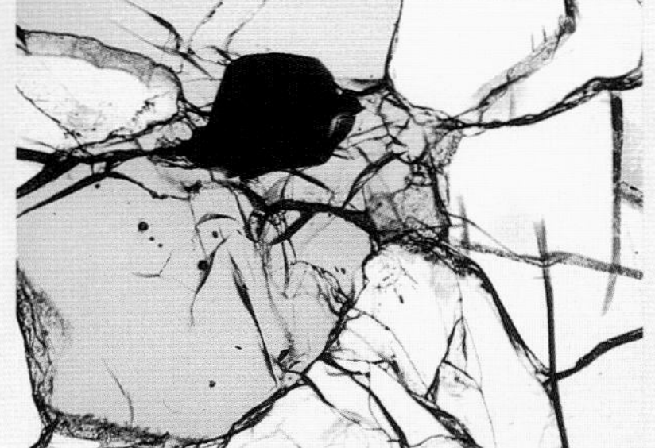

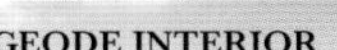

**GEODE INTERIOR**
*Natural rock cavities, or geodes, often become infilled with concentric or radial crystal growths when permeated by mineral-rich solutions. Here, an agate geode has quartz minerals growing in its center.*

# CLASSIFYING MINERALS

Many of the 4,000 or more known natural minerals have similarities in their properties. These have led to their being grouped according to whether they are chemical compounds, such as oxides or silicates, rather than minerals of particular elements—for example, copper or calcium. These groupings are reinforced by their occurrence in similar geological settings. Minerals are chemical compounds with a stability based on an electrical balance between two parts. One positively charged (cationic) part is balanced by a negatively charged (anionic) part, and this is reflected in their chemical formula. The positive part is usually a metal, such as iron (Fe), and is written first. After it comes the negative part, usually a nonmetallic ion, such as oxygen (O) or sulfur (S), or a combination of negatively charged elements, such as sulfate ($SO_4$). Minerals are classified by their anionic groupings, and are commonly divided into the ten groups shown here. Each grouping contains many minerals: for example, the carbonates include over 200 minerals ranging from calcite to azurite.

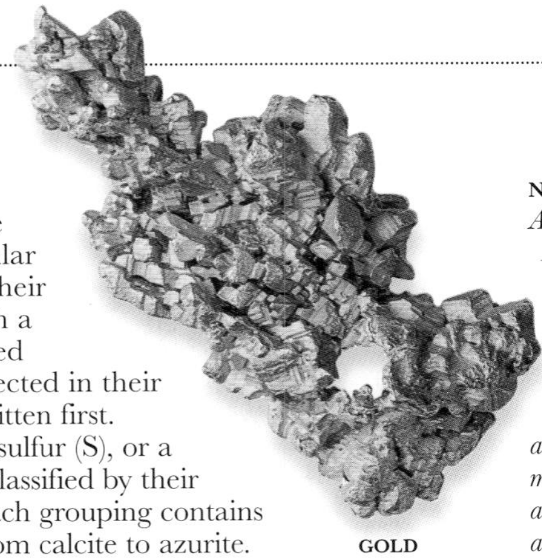

GOLD

**NATIVE ELEMENTS**

*A group of about 20 elements, of which 10 are geologically significant. Native elements include minerals such as gold (Au) and platinum (Pt). In each of these, the single element is free and uncombined. Metallic native elements with relatively low melting-points, such as copper, formed the historical basis for metal technology. The nonmetals sulfur and carbon (especially as diamond) are also economically important.*

STIBNITE

**SULFIDES**

*A group of about 600 minerals, such as stibnite ($Sb_2S_3$) and galena (PbS), in which sulfur (S) combines with metallic and metal-like elements. Most sulfides are opaque, with distinctive colors and characteristic streaks.*

CHROMITE

**OXIDES AND HYDROXIDES**

*Two closely related groups of about 400 minerals, in which metallic elements combine with oxygen ($O_2$) or hydroxyl (OH). Oxides and hydroxides contain some important ore minerals, such as chromite ($FeCr_2O_4$) and goethite (FeO(OH)).*

FLUORITE

**HALIDES AND FLUORIDES**

*Two closely related groups of about 140 minerals, in which fluorine (F), chlorine (Cl), bromine (Br), or iodine (I) combine with elements such as calcium (Ca) or sodium (Na). Fluorite ($CaF_2$) is an example.*

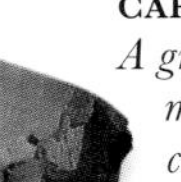

CALCITE

**CARBONATES**

*A group of more than 200 minerals, in which a metal combines with strongly bonded carbonate ($CO_3$) units. Most do not contain water ($H_2O$), but a few do, including azurite. Calcite ($CaCO_3$) is an important carbonate.*

BORAX

**BORATES**

*A group of 125 minerals in which metal elements combine with borate ($BO_3$), as in borax ($Na_2B_4O_5(OH)_4.8H_2O$). Borate units form chains and sheets that are similar to those found in $SiO_4$ groups in silicates.*

ANHYDRITE

**SULFATES**

*A group of about 300 minerals in which metallic elements combine with sulfate ($SO_4$)—for example, anhydrite ($CaSO_4$). Sulfate forms a fundamental structural unit, because sulfur has a strong bond with oxygen.*

WULFENITE

**MOLYBDATES AND TUNGSTATES**

*Two closely related groups of minerals, in which metallic elements combine with molybdate ($MoO_4$) or tungstate ($WO_4$), such as in wulfenite ($PbMoO_4$) and wolframite $(Fe,Mn)WO_4$.*

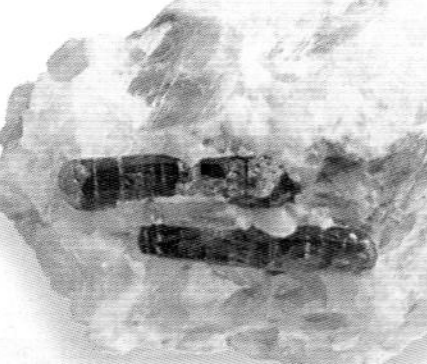

APATITE

**PHOSPHATES AND VANADATES**

*Two closely related groups of more than 400 minerals, in which elements combine with phosphate ($PO_4$) or vanadate ($VO_4$); examples include apatite ($Ca_5(PO_4)_3(F,Cl,OH)$) and carnotite ($K_2(UO_2)2(VO_4)_2.3H_2O$).*

OLIVINE

**SILICATES**

*A group of about 500 minerals in which metallic elements combine with silicon and oxygen ($SiO_4$) to form structural units known as tetrahedra, as in olivine ($(Mg,Fe)_2SiO_4$). These can link singly, in sheets, in chains, or three-dimensionally.*

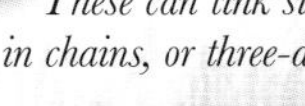

## ABRAHAM GOTTLOB WERNER

Abraham Werner (1749–1817) was a German mineralogist and teacher who attracted students from all over Europe, and even the Americas, to the Freiberg Mining Academy. He produced one of the earliest classification systems for minerals, and was a leading proponent of the Neptunist theory, which claimed that all rocks were laid down sequentially in an ocean created by the Flood described in the Bible. Igneous and metamorphic rocks were deposited, then sedimentary rocks, and finally surface sediments.

# ROCK-FORMING MINERALS

The most abundant elements in the Earth's crust are oxygen and silicon, at 46.6 and 27.7 percent by weight, then aluminum at 8.3 percent and iron at 5.0 percent. Together these elements account for 87.6 percent by weight of the crust. Therefore, silicate minerals, such as olivine, pyroxene, amphibole, feldspar, and quartz, which combine silicon with oxygen, make up more than 90 percent of the rocks of the Earth's surface. They are the main constituents of the most common rocks, especially igneous ones, many metamorphic rocks, and clay-rich sedimentary ones. Nonsilicate minerals contain different groups of chemicals, such as oxygen combined with carbon to form carbonates (as in limestone), which are important sedimentary rocks. Oxygen can also combine with metals to form oxides such as hematite. Most rocks are composed of a few essential minerals, but may also contain small quantities of various other, so-called accessory, minerals.

**RIVER MUD**
*This deposit of river mud (below) is largely made of silicate clay minerals derived from the weathering or alteration and erosion of other silicate minerals such as feldspars.*

**KAOLINITE**
*Many silicate clay minerals, such as kaolinite (above), form minute platy crystals. These produce a single, well-defined cleavage that is parallel to the sheetlike platy surfaces.*

# ORE DEPOSITS

Mineral deposits from which valuable metals can be extracted profitably are known as ores. Market conditions determine whether a naturally occurring concentration of a mineral can be regarded as an ore: the cost of extraction is justified only if it is outweighed by the deposit's market value. Platinum, for example, is so valuable that it need only have a concentration of more than 0.1 parts per million to be exploitable, whereas iron is less valuable and so the ore needs to contain a much higher iron content, perhaps 50 percent, if extracting it from the ground is to be a viable proposition.

**IRON DEPOSITS**
*The sedimentary iron ore deposits of the Hamersley Range in Australia contain about 40 billion tons of ore with a lucrative 55 percent iron concentration. They date from the Precambrian eon.*

**PLACER DEPOSITS**
*Many valuable ore minerals are concentrated in river and shallow-sea sediments, and these can be extracted by panning, washing, and sieving. Here, diamond-rich gravel is being worked in the Moa River, Sierra Leone.*

**HYDROTHERMAL VEINS**
*Many ore deposits are derived from hot, mineral-enriched solutions, which invade the surrounding rocks through any available cracks, cavities, or pores, where they form hydrothermal veins.*

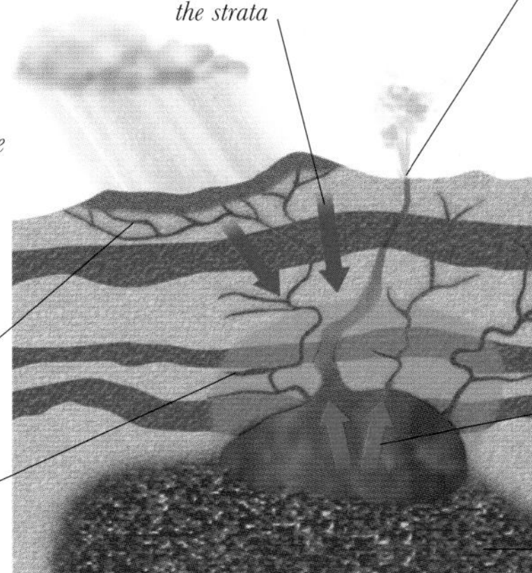


# GEMS

Minerals that have been fashioned for use as personal adornment are known as gems. The most ancient known decorations, some 30,000 years old, were made of bone, shell, and amber, but since stone cutting and polishing technology has developed, most materials selected as gems today are natural, inorganic minerals, commonly crystalline. They are valued as such because they are beautiful, rare, and durable. Out of the 4,000 or more different minerals in the Earth, only 50 or so are typically used as gems. Such is the value of many gems that there has been a very successful technological drive to manufacture synthetic ones, generally using the same chemical compounds as the natural stones themselves.

**OVAL STEP CUT**

**OCTAGONAL STEP CUT**

**BRILLIANT CUT**

**TYPES OF CUT**
*Brilliant cuts are best for delicately colored gems such as sphalerite, to give brightness and "fire," while step cuts show deeply colored stones at their best.*

**CUTTING STONES**
*Gem cutting is designed to enhance a stone's color and beauty. For each gem mineral, selected and angled flat faces (facets) are cut or ground and polished to help reflect light from the stone.*

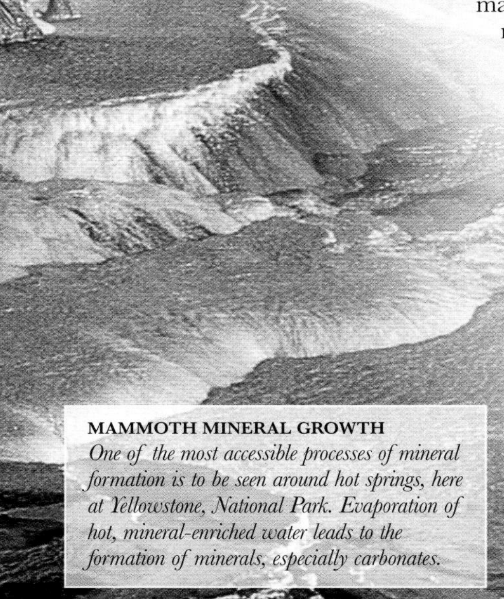

**MAMMOTH MINERAL GROWTH**
*One of the most accessible processes of mineral formation is to be seen around hot springs, here at Yellowstone, National Park. Evaporation of hot, mineral-enriched water leads to the formation of minerals, especially carbonates.*

## MINERAL PROFILES

The pages that follow contain profiles of a selection of minerals. Each profile begins with the following summary information:

**COMPOSITION** Chemical formula

**CRYSTAL SYSTEM** Cubic, tetragonal, orthorhombic, monoclinic, triclinic, or hexagonal/trigonal

**HARDNESS** Resistance to scratching, as measured on Mohs scale, ranging from 1 to 10 and changing in steps of half a unit

**SPECIFIC GRAVITY** The ratio of the mineral's weight to the weight of an equivalent volume of water, expressed to one decimal place

**COLOR** A brief description, including common variations

NATIVE ELEMENTS

## Gold

**COMPOSITION** Au **CRYSTAL SYSTEM** Cubic

**HARDNESS** 2½–3 **SPECIFIC GRAVITY** 19.3

**COLOR** Golden-yellow; golden-yellow streak

The appeal of gold can be traced through several millennia of human history. This precious metal is malleable, soft, and has a relatively low melting point (1,944°F/1,062°C), which means that it can easily be shaped into a variety of forms, from thin sheets to wire; it can be melted and cast, too. Though generally golden, its color depends on impurities: for example, red or pink gold contains traces of copper, and white gold is mixed with silver, platinum, zinc, or nickel, which also makes it harder. Gold is relatively nonreactive, so it retains its luster. Its abundance in the Earth's crust (less than 0.01 parts per million) is low compared with most other metals, and its deposits are scattered around the world. Even low-grade deposits can still be of value because of the high price this mineral commands. It is sometimes found associated with quartz in hydrothermal veins and with copper ores. Weathering and erosion of gold-bearing rocks and the water-borne transport of their debris result in gold and other heavy minerals with high specific gravity being set down as placer deposits. They can be of any geological age, and some are extensive. Most of the world's extracted gold is held as bullion for international financial settlements and increasing amounts as coinage and small bars for investment. Gold is also used to make jewelry and has some more utilitarian applications—it is used to make dental instruments, electronic components, and tarnish-free platings. About 40 percent of world gold production is currently mined by the US, South Africa, and Australia.

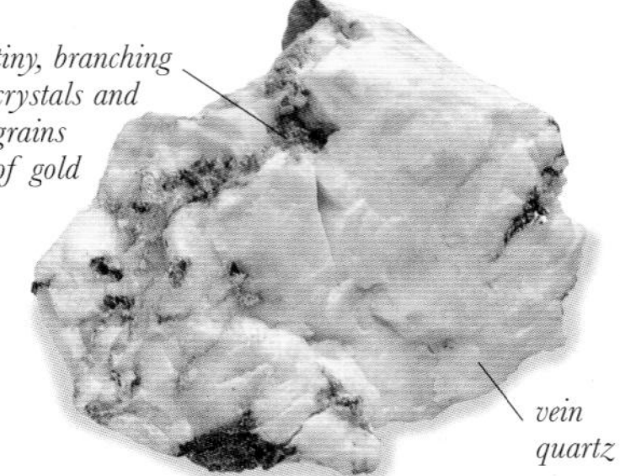


**GOLD IN QUARTZ GROUNDMASS**
*Native gold grains, some as branching, dendritic crystals, are scattered through this vein quartz specimen from a hydrothermal vein.*

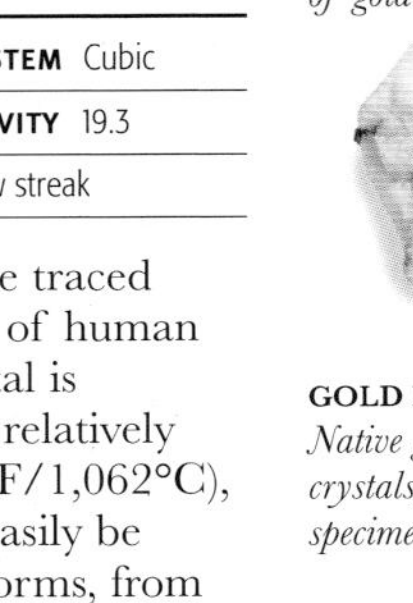


**NATIVE GOLD**
*Isolated nuggets of gold are occasionally found by miners panning placer deposits in streams, but rarely are they as big as this one.*

### THE GOLD RUSH

The pursuit of gold has proved irresistible for many, and this metal has changed lives and redirected history since it was first mined several thousand years ago. During the 19th century, gold rushes followed the discovery of enormous nuggets, such as the Welcome Stranger, which weighed 156 lb (70.7 kg) and was found in Moliagul, Australia, in 1869. Thousands of men abandoned their jobs and families and traveled to gold fields in Australia and South Africa, and in California, British Columbia, the Yukon, and Alaska (as here) in North America. Few succeeded in making their fortunes mining or panning for gold.

NATIVE ELEMENTS

## Silver

**COMPOSITION** Ag **CRYSTAL SYSTEM** Cubic

**HARDNESS** 2½–3 **SPECIFIC GRAVITY** 10.0–11.0

**COLOR** White to gray, tarnishing to black; exhibits grayish-white streak

Native silver, like gold, is a rare, valuable mineral that occurs as nuggets, grains, and wiry branching growths in hydrothermal veins. It is soft and easy to shape; consequently, over the ages, this metal has been used for decorative and utilitarian objects from crowns to cutlery and coins. When polished, it has a bright gray-white metallic color, but it gradually tarnishes black on exposure to the atmosphere. Almost all silver mined today is separated from other minerals, especially large-scale sulfide ore deposits. Mexico and Peru are major producers.

NATIVE ELEMENTS

## Copper

**COMPOSITION** Cu **CRYSTAL SYSTEM** Cubic

**HARDNESS** 2½–3 **SPECIFIC GRAVITY** 8.9

**COLOR** Copper-red, tarnished brown, may stain green; pinkish-red streak

Copper is a relatively abundant metal in the Earth's crust and occurs mostly in sulfide ores such as chalcopyrite. Native copper is found mainly in basaltic lavas, and is formed by the reaction of hydrothermal solutions with iron oxide minerals. It typically occurs as masses or dendrites, and rarely as crystals. Native copper is found as a mineral only in small amounts, while compounds of copper and other elements are more widespread. Historically, copper has been very important in the development of metal technology. Its abundance and ease of working led to wides-pread use from the fifth to third millennia BC, in the Copper Age. Later, copper was combined with tin to make bronze, a tougher and harder alloy; the Bronze Age lasted from 2200 to 800 BC. Copper can easily be drawn into wire and is a very good conductor of heat and electricity, so is widely used in the electrical industry as well as to make pipes, boilers, and roofs (see panel, above).

**COPPER ON GOETHITE BASE**
*Here, dendritic native copper, with its typical branching form, has developed on a goethite mineral surface. The surface of the dendrites has been oxidized to a reddish-brown color.*

### COPPER ROOFS

Native copper is soft and malleable, and can be beaten or rolled into thin but tough and weatherproof sheets. These are readily formed into intricate curves and so are invaluable for covering roofs with complex shapes, such as domes, spires, and cupolas. Although the copper on the roof surface typically weathers to a characteristic green color, the metal will not deteriorate underneath, and the roof will remain in good condition for hundreds of years.

NATIVE ELEMENTS

## Iron

**COMPOSITION** Fe **CRYSTAL SYSTEM** Cubic

**HARDNESS** 4½ **SPECIFIC GRAVITY** 7.3–7.9

**COLOR** Steel-gray to black; gray streak

The Earth's core is predominantly made of metallic iron with some nickel, and it is this iron composition in the core's outer liquid and inner solid parts that gives the planet its magnetic field. Native iron is also found in some meteorites and as a mineral in the Earth's crust, although this is rare, since iron readily combines with oxygen and water to form oxides and other minerals. It is occasionally found in some altered basalts where iron minerals have been reduced to native iron. The large amounts of iron used since the Iron Age in the first millennium BC are derived from smelting of iron minerals such as hematite.

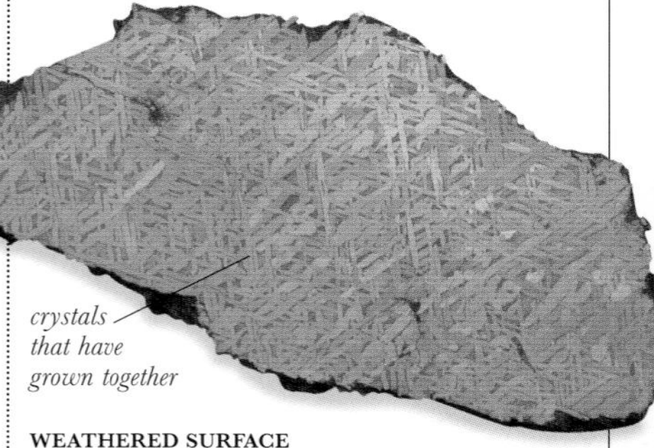


**WEATHERED SURFACE**
*In iron meteorites, iron and nickel combine to form metal alloys with complex crystalline intergrowths. Transit through the atmosphere melts the outer layer to produce a shiny, pitted surface texture.*

NATIVE ELEMENTS

## Sulfur

**COMPOSITION** S **CRYSTAL SYSTEM** Orthorhombic
**HARDNESS** 1½–2½ **SPECIFIC GRAVITY** 2.0–2.1
**COLOR** Bright yellow to brownish-yellow; white streak

Native sulfur typically forms bright yellow crystals, often with prism or pyramid shapes. It is also common as encrusting masses around volcanic vents and fumaroles and in some sedimentary rocks, especially evaporites. The crystals grow from sulfur-rich liquids and gases, such as sulfur dioxide, on the rock surfaces. Sulfur can be distinguished by its color, low melting point (235°F/112.8°C)—it ignites in a candle flame—and relatively low hardness. While native sulfur is unusual as a mineral, the element sulfur is common in many different minerals, especially the sulfide ore minerals and sulfates such as gypsum. The latter occurs in evaporite deposits, especially salt domes around the Gulf of Mexico, dating from the Cenozoic Era. Native sulfur is an extremely important source of the element, which is widely used in the chemical and pharmaceutical industries: for example, in making sulfuric acid, dyes, insecticides, and fertilizers. About 3.5 million tons of native sulfur are obtained globally each year from mining and extracting, especially in the US and Poland, and nearly 50 million tons are recovered each year from other sulfur ores such as pyrite. Sulfur, along with carbon dioxide, chlorine, and fluorine, is released in large quantities in volcanic eruptions—hence its occurrence at or near the craters of volcanoes.

**YELLOW COATING**
*These distinctive sulfur crystals have grown on a rock surface.*

**MINERAL FORMATION**
*These sulfur deposits and other evaporite minerals develop from vents on Anak Krakatoa volcano.*

NATIVE ELEMENTS

## Graphite

**COMPOSITION** C **CRYSTAL SYSTEM** Hexagonal
**HARDNESS** 1–2 **SPECIFIC GRAVITY** 2.1–2.3
**COLOR** Black; gray-black streak

The name graphite is derived from the Greek *grapho* ("to write"), because this soft, opaque, and dark-colored mineral readily marks paper with an erasable gray deposit, a property used to make pencil "lead." Graphite is just one form of carbon found in the natural world—among the others are diamonds, organic-derived charcoal, and coal. Graphite is uncommon, but occurs as scattered grains or flakes in some carbon-rich metamorphic rocks and as veins in pegmatites. Graphite's scaly crystals slide over one another to give a greasy feel, promoting its use as a lubricant in many industries.

NATIVE ELEMENTS

## Diamond

**COMPOSITION** C **CRYSTAL SYSTEM** Cubic
**HARDNESS** 10 **SPECIFIC GRAVITY** 3.5
**COLOR** Mainly colorless, but colored varieties occur; white streak

Perhaps the most famous mineral of all, because of its remarkable hardness and its brilliance when cut as a gemstone, diamond's real value as jewelry has been appreciated only since modern techniques of cutting have fully revealed its "fire" and sparkle. This results from the internal reflection of light after the careful cutting of facets at certain angles to one another. Colored diamonds occur naturally and are highly valued, especially if red, green, or blue, although coloring is usually the result of artificial irradiation. In recent decades, the value of diamond as an abrasive, for cutting metal and other stones, and providing hard-wearing bearings, has outweighed its decorative value. More than 100 million carats (20 tons) are mined each year, mainly in Russia, Australia, South Africa, Botswana, and the Democratic Republic of Congo (see panel, above). Diamonds originate in pipelike intrusions called kimberlites, which have arisen from great depths (90 miles/150 km or more) in the Earth's crust. Within kimberlite pipes, diamonds are formed from carbon at high temperature (5,400°F/3,000°C) and under great pressure. (Since 1955, these conditions have been artificially replicated to make synthetic diamonds.) The diamonds are then moved from their sites of formation by processes of weathering, erosion, and transport. Due to their hardness, they remain largely intact and accumulate as placer deposits within certain sedimentary strata, which form major sources of the mineral. Diamonds can develop after large-scale explosions such as meteorite impacts. Carbon, derived from the terrestrial target rocks, is fused at high pressures and temperatures, resulting in exceedingly small crystals, called microdiamonds, found in deposits that have fallen from the air.

**TRUE GEM**
*This naturally occurring, eight-sided diamond crystal is still embedded in the kimberlite rock in which it was formed by high pressures and temperatures deep within the Earth's interior.*

PEAR-SHAPED

ROUND-SHAPED

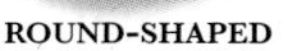

BRILLIANT CUT

**DIFFERENT CUTS**
*As well as occurring in many different colors, diamonds can be cut, or faceted, to enhance their natural "fire."*

### MINING DIAMONDS

The majority of diamonds are now mined on a large industrial scale by just a few major corporations, such as De Beers, using heavy machinery and automated processing. The value of diamonds, however, is so high that there are numerous smaller mining operations around the world, such as this one on the Moa River, Sierra Leone. Often these are in disputed or war-torn territories in less-developed countries, which are too risky for normal mining operations. Here, impoverished miners work in dangerous conditions in the hope of finding diamonds, which they sell cheaply to middlemen or warlords.

**THE OPPENHEIMER DIAMOND**
*This 253.7-carat diamond crystal was found near Kimberley, South Africa, in 1964. It was later given to the Smithsonian Institution.*

SULFIDES

## Pyrite

**COMPOSITION** $FeS_2$ **CRYSTAL SYSTEM** Cubic

**HARDNESS** 6–6½ **SPECIFIC GRAVITY** 4.9–5.2

**COLOR** Pale brass-yellow, darker if tarnished; greenish-black streak

The name pyrite is derived from the Greek *pyros* ("fire"), because this mineral gives off sparks when struck sharply. It is commonly known as "fool's gold," since its golden yellow color can lead to its being mistaken for gold, but it is much harder than gold and not malleable. It usually forms characteristic, well-shaped cubic crystals or twelve-faced crystals called pyritohedrons, with each face having five sides. Pyrite is a relatively common and widespread iron mineral that occurs in a number of different geological settings. In igneous rocks, it is prevalent as an accessory mineral but it may also occur concentrated in a variety of ore deposits. These range from massive, economically important sulfide ore bodies to smaller hydrothermal veins and replacement deposits. Pyrite is also widely distributed in sedimentary rocks, particularly fine-grained mud-rocks deposited in stagnant, low-oxygen conditions. Sulfur-reducing bacteria promote the formation of tiny pyrite crystals within the mud, and these are preserved as the sediments are consolidated into rock. When mud-rock is metamorphosed into slate, the pyrite may grow into larger crystals and sometimes replaces shells or bones of fossils.

**OCTAHEDRAL PYRITE**
*Perfectly formed pyrite crystals are here associated with clusters of small quartz crystals from a hydrothermal vein. The crystals grew from solutions enriched in iron sulfide and silica, respectively.*

SULFIDES

## Sphalerite

**COMPOSITION** ZnS **CRYSTAL SYSTEM** Cubic

**HARDNESS** 3½–4 **SPECIFIC GRAVITY** 3.9–4.1

**COLOR** Yellow–brown to black; variable streak from white to pale yellow to brown

Also known as zincblende, sphalerite is the most common zinc mineral and thus the main source of this metal and rare elements such as cadmium. Its high zinc content (about 67 percent) was, however, discovered only in the 18th century. As a mineral, sphalerite is variable in color and can be difficult to distinguish from minerals such as magnetite. Indeed, the name sphalerite is derived from the Greek *sphaleros* ("treacherous"), because of its lack of identifying features, apart from a resinous luster and well-developed cleavage. Sphalerite frequently occurs with galena and other sulfide minerals in hydrothermal veins, which are mined for their zinc, mainly in China, Australia, and Canada. Sphalerite also occurs in limestone by replacement of the carbonate minerals to form ore bodies with magnetite and pyrrhotite. Transparent brown or green crystals are sometimes found, and these can be faceted as gems but are too soft to be useful jewels.

SULFIDES

## Stibnite

**COMPOSITION** $Sb_2S_3$

**CRYSTAL SYSTEM** Orthorhombic

**HARDNESS** 2 **SPECIFIC GRAVITY** 4.5–4.6

**COLOR** Lead gray; gray streak

This mineral is the major source of the relatively rare metal antimony (only 0.2 parts per million in the Earth's crust), a toxic element used for hardening metal alloys for bearings, lead in batteries, and semiconductors. Stibnite occurs as elongate radiating crystals or massive forms. The latter can be confused with galena, but stibnite's crystal form is distinctive, as is its low melting-point—it liquefies in a match flame. It occurs associated with other sulfides in hydrothermal veins, hot-spring deposits, and within limestone. Most of the world's annual stibnite production comes from China.

long stibnite crystals

quartz and barite matrix

SULFIDES

## Chalcopyrite

**COMPOSITION** $CuFeS_2$ **CRYSTAL SYSTEM** Tetragonal

**HARDNESS** 3½–4 **SPECIFIC GRAVITY** 4.1–4.3

**COLOR** Golden brassy yellow; greenish-black streak

As a copper-iron sulfide, chalcopyrite is the most important copper ore. It occurs as masses and sometimes crystals in many geological settings, such as hydrothermal veins and a number of different metamorphic and igneous rocks. Economically, the most important occurrences of chalcopyrite are in porphyry copper deposits, where veins of sulfide minerals are associated with large igneous intrusions. Pyrite and bornite are commonly present there, too. With its golden color, chalcopyrite is similar to pyrite but is more yellow and has tetragonal, not cubic, crystals.

SULFIDES

## Bornite

**COMPOSITION** $Cu_5FeS_4$ **CRYSTAL SYSTEM** Cubic

**HARDNESS** 3 **SPECIFIC GRAVITY** 5.0–5.1

**COLOR** Brownish bronze, tarnishing purple–blue and black; black streak

A relatively dense and brittle mineral with a characteristic iridescent tarnish, bornite contains about 63 percent copper, 11 percent iron, and 26 percent sulfur. Only occasionally does it form crystals; normally it occurs in masses in pegmatites and other igneous rocks and also in hydrothermal veins with other sulfide minerals such as chalcopyrite. Bornite is relatively common and, because of its high metal content, is an economically important copper mineral. The largest deposits are in Mexico and Butte, Montana. It owes its common name, peacock ore, to its characteristic tarnish of iridescent blue to violet to red.

SULFIDES

## Cinnabar

**COMPOSITION** HgS **CRYSTAL SYSTEM** Trigonal

**HARDNESS** 2–2½ **SPECIFIC GRAVITY** 8.0–8.2

**COLOR** Blood red; deep vermilion-red streak

The very high density of cinnabar distinguishes it from another blood-red mineral, realgar. Cinnabar is the most common and economically important mercury mineral and contains nearly 87 percent mercury. Both massive and crystalline forms occur within fracture zones around active volcanoes and hot springs. Mined since the 7th century BC, cinnabar was a source of black and red pigments and quicksilver (mercury) for mirrors; it is also used in drugs, batteries, electrical instruments, and many chemicals.

SULFIDES

## Realgar

**COMPOSITION** AsS **CRYSTAL SYSTEM** Monoclinic

**HARDNESS** 1½–2 **SPECIFIC GRAVITY** 3.5

**COLOR** Orange-red; orange-red streak

This naturally occurring but rare arsenic sulfide mineral forms grains and well-shaped crystals. The latter have a striking, orange-red color with a resinous luster. Realgar looks similar to cinnabar but is softer and less dense. It typically occurs as a secondary mineral along with yellow orpiment, another arsenic sulfide, in hydrothermal veins and hot-spring deposits. In this context, realgar (70 percent arsenic) is produced by the decomposition of other arsenic minerals such as arsenopyrite. Highly toxic, it was used in medieval medicine and glassmaking, and is now used in fireworks and pesticides.

SULFIDES

## Galena

**COMPOSITION** PbS **CRYSTAL SYSTEM** Cubic

**HARDNESS** 2½ **SPECIFIC GRAVITY** 7.4–7.6

**COLOR** Lead-gray; lead-gray streak

The most common and economically important lead ore (about 87 percent lead), galena forms crystals, masses, and granules. It occurs in a variety of geological settings, from metamorphic rocks to volcanic-related, massive sulfide deposits (including sphalerite and copper ores), and large ore bodies in reef limestones and dolomites associated with hydrothermal fluids. Galena is easily distinguished by its gray color, metallic sheen, high density, and perfect cubic cleavage—it readily breaks into small cubes. Galena was one of the first metal ores to be mined: it was smelted by the Babylonians to make lead vases, and it was much sought-after by the Romans (see panel, above). The word *galena* is Latin for lead ore. Lead derived from galena has been used for printing, in paint, batteries, and in plumbing and electrical wiring, but its toxicity has now greatly reduced its worth, especially for domestic uses. China, Australia, and the US produce half of the world's annual lead output, which comes mostly from galena.

### LEAD AND THE ROMANS

Archaeological finds show that, more than two thousand years ago, the Romans made extensive use of lead for domestic purposes. They discovered that this malleable metal could be cast and beaten into complex shapes, allowing the manufacture of lead vessels and pipes for conducting and storing water around their buildings. Demand for the metal was one reason for the Roman invasion of the British Isles, since there were lead and copper ores in Cornwall. Roman smelting of lead and copper ores between 500 BC and AD 300 caused a small but significant rise in atmospheric pollution, as recorded in Greenland ice cores, before it declined again.

**DUAL PURPOSE**
*Under Emperor Hadrian, the Romans used lead ingots both as a unit of currency and in manufacturing.*

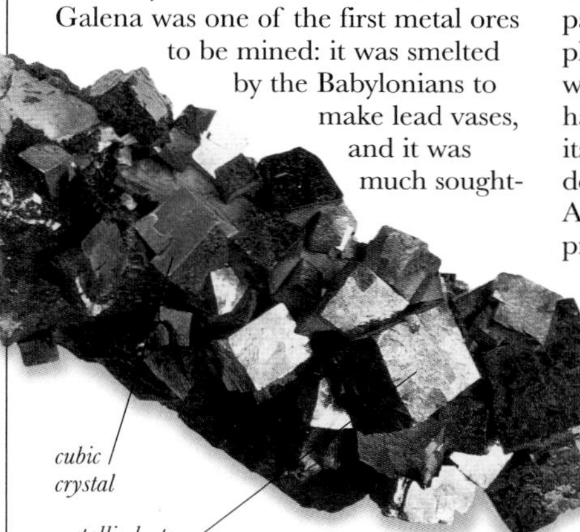


**CRYSTALS OF GALENA**
*Well-formed crystals, such as these cubes of galena, have perfect cubic cleavage and can easily break into even smaller cubes.*

**COMMON LEAD ORE**
*Galena, with its bright luster, is frequently found in massive form (such as here) along with other sulfide minerals in hydrothermal veins.*

OXIDE

## Spinel

**COMPOSITION** $MgAl_2O_4$ **CRYSTAL SYSTEM** Cubic

**HARDNESS** 7½–8 **SPECIFIC GRAVITY** 3.5–4.1

**COLOR** Very variable or colorless; white-brown streak

The composition of this magnesium oxide mineral varies greatly, with iron, chromium, zinc, and manganese atoms replacing some magnesium. Consequently, spinel forms a series of minerals with variable properties, especially color. Some varieties are gem minerals, but large crystals are rare. Red spinel resembles ruby, and can be produced synthetically. Typically, spinel is finely granular, but it also occurs as octahedral crystals in gabbroic igneous rocks and as a result of thermal metamorphism of dolomites. These latter occurrences, along with river sediments derived from them, produce most gem spinel.

OXIDES

## Hematite

**COMPOSITION** $Fe_2O_3$ **CRYSTAL SYSTEM** Trigonal

**HARDNESS** 5–6 **SPECIFIC GRAVITY** 4.9–5.3

**COLOR** Reddish-brown, steel gray to black; red streak

This is probably one of the first minerals to have been exploited by humans. When ground, hematite produces a powder called red ocher, which has been used as a pigment and is found in association with burials dating back some 80,000 years. It is distinguished by its hardness and blood red streak; indeed, its name is derived from the Greek *haima* ("blood"). Widely distributed, hematite is the most important iron ore mineral. It is found in a variety of forms ranging from tabular crystals to bulbous masses with radiating and striated interiors. Hematite occurs in a number of different geological settings, including hydrothermal veins and as a rare component of igneous rocks. It is most common in ancient sedimentary rocks known as banded ironstones (which date from the early Precambrian eon). More than half of the world's annual production of iron ore comes from China, Brazil, and Australia, mostly from hematite.

**MIRROR IMAGE**
*Specular hematite, a crystalline form of hematite, has such a highly reflective crystal face that it was once used for mirrors.*

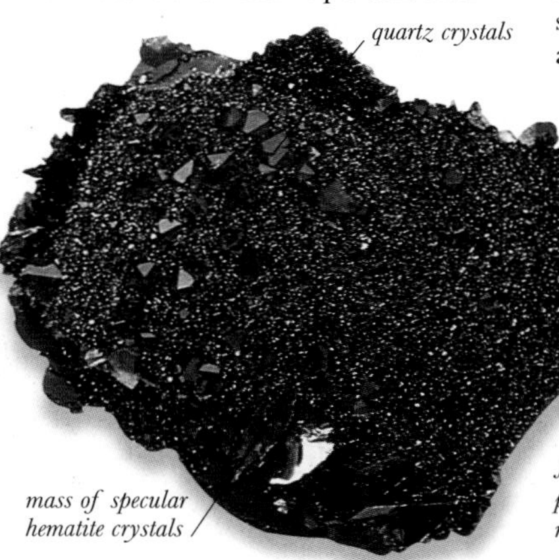


**MIXED CRYSTALS**
*Numerous tiny hematite crystals grow alongside prismatic quartz crystals, as here, on a surface of massive red hematite ore.*

OXIDES

## Magnetite

**COMPOSITION** $Fe_3O_4$ **CRYSTAL SYSTEM** Cubic

**HARDNESS** 5½–6½ **SPECIFIC GRAVITY** 5.2

**COLOR** Black; black streak

Like hematite, magnetite is an important iron ore. It has strong magnetic properties that were recognized by the Chinese in the 11th century BC. Magnetite is also mentioned by the Roman naturalist Pliny the Elder, who suggested that the name was derived from Magnes, a shepherd whose iron-nailed shoes stuck to rocks containing that mineral. Magnetite occurs in massive, granular, and crystalline forms with crystals of eight triangular faces. Widely distributed, it is a common but mostly minor mineral in many igneous rocks, where it sometimes occurs in layers. Magnetite is also often found in metamorphic rocks, some mineral veins, and alluvial placer deposits (see panel, above).

### PANNING STREAMS

Dense and hard minerals such as magnetite resist weathering and erosion. With patience, they can be washed from other placer minerals and sediment in a shallow pan—but this practice is barely viable. Magnetite is shown here being panned in Ivory Coast.

**DIRECTION FINDERS**
*For at least three millennia, humans have used naturally magnetic crystals of magnetite, such as this one, as compasses for finding their way.*

OXIDES

## Ilmenite

**COMPOSITION** $FeTiO_3$ **CRYSTAL SYSTEM** Trigonal

**HARDNESS** 5–6 **SPECIFIC GRAVITY** 4.5–5.0

**COLOR** Black; brownish-red to black streak

Named after the Ilmen Mountains in Russia, ilmenite is an oxide of iron (48 percent) and titanium (52 percent) oxide that occurs in massive form and as tabular crystals. Its appearance is similar to magnetite's, and it is slightly magnetic. It is the world's main source of titanium, occurring as a minor constituent of igneous rocks such as gabbro and diorite, and occasionally in quartz veins and in some metamorphic rocks. Resistant to weathering, ilmenite can accumulate as placer deposits in some river sands, along with magnetite and rutile. Canada, Australia, and South Africa produce some two-thirds of the world's annual output of titanium concentrates, mainly from ilmenite.

OXIDES

## Corundum

**COMPOSITION** $Al_2O_3$ **CRYSTAL SYSTEM** Trigonal

**HARDNESS** 9 **SPECIFIC GRAVITY** 3.9–4.1

**COLOR** Variable, including brown, pink, red (ruby), blue (sapphire), and colorless forms

Several famous and very valuable gem forms, such as ruby (red) and sapphire (blue; see panel below), occur in corundum. When polished, some of these show a six-rayed star of light reflected from tiny, needle-shaped crystals of rutile within the corundum crystal, when viewed at a particular angle. Corundum's natural crystals have rough, barrel, or spindle shapes, and it also occurs with spinel and magnetite as a massive, black, granular form known as emery, which is used as an industrial abrasive. Corundum is found in sodium-rich, granite-related igneous rocks and in their pegmatites, occasionally growing into large crystals. It also develops in a variety of metamorphic rocks, ranging from gneiss and schist to marble. Second in hardness only to diamond, corundum survives erosion to accumulate in alluvial sands and gravels as placer deposits, especially in Sri Lanka and Burma (Myanmar). Synthetic production of gem-quality corundum began in the late 19th century, and by 1902, the Frenchman Auguste Verneuil had produced the first synthetic ruby by fusing aluminum oxide and coloring material with the flame from a blowtorch. Different trace elements vary the color of the synthetic gem: for example, chromium creates ruby red, and iron and titanium makes sapphire blue.

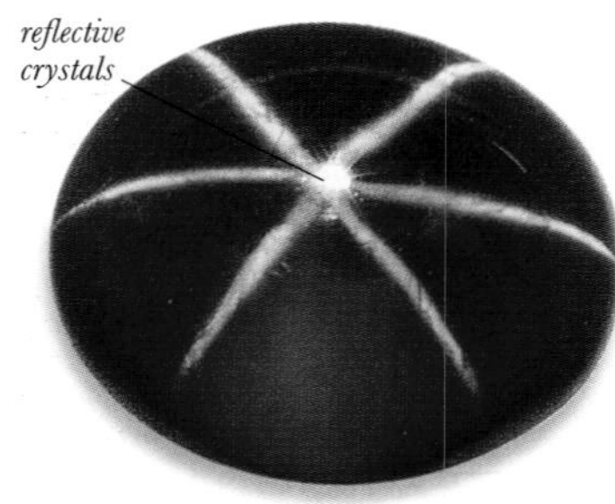


**STAR CABOCHON**
*The rounded shape of a cabochon (a stone cut without facets), such as this ruby, best reveals the star effect from light reflected by tiny rutile crystals.*

### SAPPHIRES

Since the Middle Ages, corundum gems of any color except ruby red have been known as sapphires, but the name is now associated only with blue gems. These can vary from pale to very dark blue to almost black; some gems appear as different shades of blue according to the light. One color description—heavenly blue—has inevitably become associated with tranquility, purity, and peace. Most of the early stones used to come from Sri Lanka and India, where they occurred in pegmatites and placer deposits. They are now found around the world, from the metallic-blue sapphires of Montana, to the dark blue and nearly black stones from Nigeria.

**TREASURED JEWELRY**
*Blue sapphires are much sought-after for their putative spiritual qualities.*

**BRILLIANT CUT SAPPHIRE**

**SAPPHIRE CRYSTAL**
*Blue corundum looks undistinguished as a decorative jewel, but careful cutting reveals the outstanding gem quality of sapphire.*

**CORUNDUM**

OXIDES

## Chromite

**COMPOSITION** $FeCr_2O_4$ **CRYSTAL SYSTEM** Cubic

**HARDNESS** 5½ **SPECIFIC GRAVITY** 4.1–5.1

**COLOR** Black to brownish-black; brown streak

This most important chromium mineral (chromium 73 percent, iron 27 percent) is typically massive or granular, but rare octahedral crystals do occur. Chromite is commonly found as a minor constituent of olivine-rich igneous rocks such as peridotite and serpentinite, but it may be concentrated into layers of sufficient commercial value to exploit. Because chromite is hard, dense, and resistant to erosion and weathering, it accumulates in alluvial sediments as a placer mineral. Its melting point is 3,500°F (1,900°C). When alloyed with nickel, chromium is used for hardening steel; with iron, for making stainless-steel and chromium plating.

OXIDES

## Cassiterite

**COMPOSITION** $SnO_2$ **CRYSTAL SYSTEM** Tetragonal

**HARDNESS** 6–7 **SPECIFIC GRAVITY** 6.8–7.1

**COLOR** Yellow–brown–black–red; white or gray streak

One of the earliest ores to be exploited for metal (it has been mined since 6000 BC), cassiterite can be smelted to produce tin at about 1,000°C (1,800°F). This tin mineral is the world's main source of the metal (78 percent tin), and it occurs in hydrothermal veins and pegmatites associated with granites. Placer deposits of alluvial cassiterite are also found. Cassiterite can be massive or granular or form columnar or pyramidal crystals. China, Indonesia, and Peru are major producers of this tin oxide mineral. Tin is used as a non-toxic, rust-proof coating for steel cans and, alloyed with lead, as a low-melting-point solder for electrical connections.

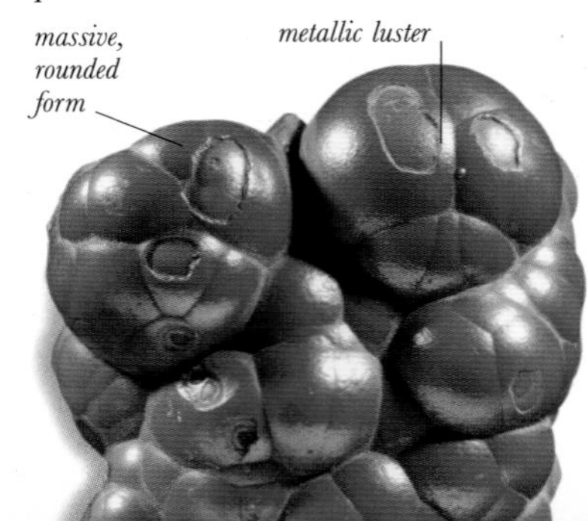


OXIDES

## Rutile

**COMPOSITION** $TiO_2$ **CRYSTAL SYSTEM** Tetragonal

**HARDNESS** 6–6½ **SPECIFIC GRAVITY** 4.2–4.4

**COLOR** Reddish-brown to black; pale brown streak

This mineral was originally mistaken for tourmaline. It was only in 1795 that rutile was found to contain titanium (60 percent), for which it is now the second most important source—ilmenite being the first. Rutile occurs in massive and prismatic crystals, often striated and terminated with pyramidal faces. It grows, too, as fine needles within other minerals such as quartz and corundum. Found in various igneous and metamorphic rocks, rutile can also be produced by the breakdown of minerals such as sphene, and can occur as a placer deposit. Its Latin-derived name (*rutilus*, meaning "golden yellow") was suggested by Abraham Werner (see p.63).

OXIDES

## Uraninite

**COMPOSITION** $UO_2$ **CRYSTAL SYSTEM** Cubic

**HARDNESS** 5–6 **SPECIFIC GRAVITY** 6.5–10.0

**COLOR** Black to dark brown; gray to brownish-black

The radioactive properties of this uranium oxide mineral (also known as pitchblende) were first discovered in 1896 by French physicist Henri Becquerel, and the associated element radium was subsequently identified within it by Polish chemist Marie Curie—both scientists using uraninite specimens from Jáchymov in Bohemia. Uraninite is usually massive and occurs in sulfide-bearing hydrothermal veins. Rare cubic crystals occur in granite pegmatites associated with minerals such as tourmaline and zircon. Sedimentary grains of uraninite are found concentrated in some placer deposits.

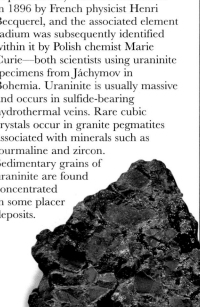

HYDROXIDES

## Goethite

**COMPOSITION** FeO(OH)

**CRYSTAL SYSTEM** Orthorhombic

**HARDNESS** 5–5½ **SPECIFIC GRAVITY** 3.3–4.3

**COLOR** Yellowish to dark brown; yellow-brown streak

First identified as a mineral in its own right in 1806, goethite was named after the German poet J.W. Goethe, who was also a mineral collector. This hydrous iron oxide often occurs as striking globular or stalactitic masses. Internally, it has a radiating, needle-shaped or fibrous structure. Larger crystals are rare but may be bladed or prismatic. Goethite is widespread but of little economic value; it is an oxidation product of iron minerals such as magnetite and pyrite. It is also a hydrothermal mineral and occurs in a form called bog iron ore in some sediments.

HALIDES

## Fluorite

**COMPOSITION** $CaF_2$ **CRYSTAL SYSTEM** Cubic

**HARDNESS** 4 **SPECIFIC GRAVITY** 3.2

**COLOR** Variable, especially blue and green, also colorless; white streak

Beautiful cubic crystals of fluorite in a wide range of colors are sometimes found in hydrothermal mineral veins, either alone or with metal ores, but they are too soft to be worked as gems. With its low melting-point, fluorite is used in smelting iron, and, since ancient Greek times has been used as an ornamental stone. In 1824, German mineralogist Friederich Mohs, who invented the Mohs scale of hardness, discovered the marked fluorescence shown by some fluorite, especially in ultraviolet light.

HALIDES

## Halite

**COMPOSITION** NaCl **CRYSTAL SYSTEM** Cubic

**HARDNESS** 2½ **SPECIFIC GRAVITY** 2.1–2.2

**COLOR** Variable, including pale brown or reddish-brown, and colorless forms; white streak

Of great economic importance as a source of salt, halite (also known as rock salt) is a common mineral in certain sedimentary rocks, where it occurs in massive, granular, or crystal form. Halite is mainly derived from the drying out of seawater in the geological past. The process frequently leaves vast stratified evaporite deposits in which halite is associated with other soluble minerals such as gypsum and anhydrite. Because of halite's low density and ability to recrystallize at low temperatures and pressures, deeply buried halite deposits deform into large dome structures known as diapirs. Although some rock salt is still mined (see panel, above), particularly in the US, Canada, and Germany, most commercial salt is extracted directly from seawater by evaporation.

**CONSISTENT SHAPE**
*Invariably, halite crystallizes as well-formed cubes, with stepped, hopper-shaped faces. It is typically colorless, except when chemical impurities introduce, for example, shades of red.*

### MINING SALT

Since ancient times, salt has been an essential mineral for domestic and industrial purposes. For countries far from the sea, salt mining is a necessity, and many famous salt mines have been excavated over the centuries. Some, such as those in Siberia, have achieved notoriety for the use of slave labor, while others have become famous for the beauty of their cavernous interiors, such as those of Wieliczka, in Poland, and Winsford, England, shown here.

CARBONATES

## Dolomite

**COMPOSITION** $CaMg(CO_3)_2$

**CRYSTAL SYSTEM** Trigonal

**HARDNESS** 3½–4 **SPECIFIC GRAVITY** 2.8–2.9

**COLOR** White, gray to pale brown; white streak

Like its close chemical relative calcite, dolomite is an important rock-forming mineral that has formed in mountain-sized masses. The magnesium in the composition of this calcium and magnesium carbonate generates the small but important differences between dolomite and calcite: for example, dolomite does not dissolve readily in acidic groundwater, and dolomitic carbonate rock does not form solution features such as caves. Most dolomite is a secondary product formed by magnesium-rich solutions interacting with calcium carbonate sediments, especially reef limestones. It can also occur as a primary mineral either alone in magnesium-enriched water or in hydrothermal veins associated with sulfide ores such as sphalerite and galena. Dolomite can be difficult to distinguish from calcite, but it frequently has a characteristic brown color, and develops into curved surfaces made of composite intergrown, or twinned, crystals. It was not recognized as a separate mineral until the end of the 18th century, when Swiss mineralogist H.B. de Saussure named it after the French geologist Déodat de Dolomieu, who had first noticed the distinct properties of dolomite rock. Dolomitic limestone makes good building stone and hardcore. It is also quarried for the manufacture of special cements and refractory linings.

**DOLOMITE MOUNTAINS**
*Massifs such as the Italian Dolomites consist almost entirely of dolomite-rich sedimentary carbonate rocks. These are less prone to chemical weathering than limestone.*

**UNUSUAL FORM**
*Transparent dolomite crystals are rare and, although sometimes attractively colored pink and yellow, are too soft to be used as gemstones.*

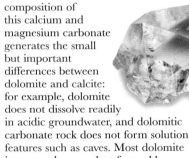

**SADDLE-SHAPED CRYSTALS**
*Dolomite crystals commonly develop with curved surfaces that are composites of overlapping crystals whose orientation changes slightly as they grow.*

CARBONATES

## Rhodochrosite

**COMPOSITION** $MnCO_3$ **CRYSTAL SYSTEM** Trigonal

**HARDNESS** 3½–4½ **SPECIFIC GRAVITY** 3.4–3.7

**COLOR** Rose-red; also brown to gray; often has a brown or black outer crust; white streak

The name of this manganese carbonate mineral, which is often colored a striking rose-pink, is derived from the Greek *rhodokhros* ("rose-colored"). Rhodochrosite generally occurs as crystals or in granular or massive form and sometimes as globular nodules with internal concentric color bands. It is found in hydrothermal veins, with ores of copper, silver, and lead, and in some metamorphic rocks of sedimentary origin. It is a source of manganese, which is used for hardening steel. Argentinian rhodochrosite is made into jewelry.

CARBONATES

## Calcite

**COMPOSITION** $CaCO_3$ **CRYSTAL SYSTEM** Trigonal

**HARDNESS** 3 **SPECIFIC GRAVITY** 2.7

**COLOR** Variable, but generally white or colorless; white streak

One of the most common minerals in the Earth's crust, calcite ranges from massive to granular and crystalline in form. Its crystals grow in many different shapes, from tabular to prismatic, often with well-developed faces. Calcite is found in several types of rock, ranging from sedimentary (limestone) to metamorphic (marble) and igneous (carbonatite lavas). It also occurs in veins, with or without other minerals, and often replaces minerals such as calcium-rich feldspar and pyroxene in igneous rocks. Calcite crystals can grow at the pressures and temperatures found at the Earth's surface, and they readily dissolve in slightly acidic water. This means that calcite is easily leached from rocks and transported in solution to be redeposited elsewhere. At the Earth's surface, this results in the growth of limestone pavements, solution hollows, and cave systems (see pp.252–53). The dissolution of calcite can be reversed: calcium-carbonate-enriched waters deposit calcite in the form of stalactites, stalagmites, and flowstone. Most calcite is associated with sedimentary limestones, especially those of marine origin. Within these rocks, the calcite is deposited as tiny calcium-carbonate crystals and indirectly from the skeletons and shells of sea creatures, especially where reefs are formed. Metamorphism of limestone deep within the Earth's crust produces marble. Apart from its ubiquitous use as a building stone and aggregate, calcite in the form of limestone has been used in great quantities to manufacture lime fertilizer and cement, and in the smelting of metal ores. Clear calcite crystals have also been used in microscope optical systems (see panel, above) due to their refractive properties.

**NAIL-HEAD CALCITE**
*Calcite crystals grow in a variety of shapes, and these have been given common names, such as nail-head calcite, which develops tabular heads on more slender columns.*



**POINTED CRYSTAL**
*High, pyramid-shaped calcite crystals are called dogtooth spar because of their resemblance to teeth.*

### DOUBLE REFRACTION

Light passing through calcite is split into two separate rays because of the atomic structure of this mineral. The same property is shown by all translucent crystals other than those that belong to the cubic system. The rays have different vibration directions and velocities, with the slow ray bent (refracted) from its original path. The measurable difference between the two paths is known as the refractive index and is useful, especially to gemologists, in helping to distinguish between otherwise similar translucent crystals. In calcite, the refractive index is very high (0.172), as can be seen by the apparent doubling of any object viewed through a cleavage fragment where the vibration directions show their maximum divergence.

**CALCITE GROWTH**
*When stained, calcite crystals reveal their growth patterns along with changes in their composition.*

CARBONATES

## Malachite

**COMPOSITION** $Cu_2CO_3(OH)_2$

**CRYSTAL SYSTEM** Monoclinic

**HARDNESS** 3½–4 **SPECIFIC GRAVITY** 3.9–4.0

**COLOR** Pale to dark green; pale green streak

This distinctive green copper mineral is the common weathering product of copper ores. Malachite contains about 57 percent copper and mostly occurs as banded encrustations with bulbous surfaces and an internal, fibrous, radiating habit. It is commonly found in the oxidation zone of other copper mineral deposits, where it is frequently associated with minerals such as goethite and calcite. The ancient Greeks mined malachite to make amulets, which reputedly warded off misfortune. It has since been used as a semi-precious stone.

CARBONATES

## Azurite

**COMPOSITION** $Cu_3(CO_3)_2(OH)_2$

**CRYSTAL SYSTEM** Monoclinic

**HARDNESS** 3½–4 **SPECIFIC GRAVITY** 3.8–3.9

**COLOR** Deep azure blue; pale blue streak

Readily distinguished by its deep blue color, which was widely used as a pigment in early illuminated manuscripts, azurite is compositionally similar to malachite. It develops not only as short, prismatic or tabular crystals but also in massive and earth-like forms. Its occurrence is similar to that of malachite, with which it is often associated as a weathering product of copper ores. Azurite is also mined as a minor copper ore.

BORATES

## Borax

**COMPOSITION** $Na_2B_4O_5(OH)_4.8H_2O$

**CRYSTAL SYSTEM** Monoclinic

**HARDNESS** 2–2½ **SPECIFIC GRAVITY** 1.7

**COLOR** White or colorless; white streak

Borax is one of several relatively rare minerals to contain the element boron. An evaporite mineral, its short prismatic crystals are formed by the evaporation of water in salt lakes, where they then become part of the sediment. Borax can also be found in volcanic exhalations. This massive or crystalline mineral melts easily, is soluble in water, and can crumble to a powder if it dehydrates further. Borax is widely used in the glass, chemical, soap, and food industries.

SULFATES

## Anhydrite

**COMPOSITION** $CaSO_4$

**CRYSTAL SYSTEM** Orthorhombic

**HARDNESS** 3–3½ **SPECIFIC GRAVITY** 2.9–3.0

**COLOR** White or colorless; white streak

Usually in massive form, anhydrite crystals are rare and show three well-developed cleavages at right angles. The cleavage helps to distinguish this evaporite mineral from gypsum, with which it can be confused. Anhydrite's name refers to its composition and is derived from the Greek *anudros* ("without water"). This mineral can be deposited directly from heated seawater (in excess of 108°F/42°C), but is mostly found interbedded with gypsum and halite in evaporite deposits. When ground up, it is used as a soil conditioner and in cement.

SULFATES

## Gypsum

**COMPOSITION** $CaSO_4.2H_2O$

**CRYSTAL SYSTEM** Monoclinic

**HARDNESS** 2 **SPECIFIC GRAVITY** 2.3

**COLOR** White or colorless; white streak

In ancient Greece, gypsum was used as an ornamental stone because it was very easy to carve. The Romans then discovered that heating gypsum to 600°F (300°C) made a plaster that sets hard when mixed with water; such a plaster is still widely used in the building industry. Gypsum ranges from massive to granular (alabaster), fibrous (satin spar), and crystalline (selenite) in form. With low solubility, it is the first mineral to separate from evaporating seawater, followed by anhydrite and other similar minerals. Gypsum also occurs in mineral veins and volcanoes, where sulfuric acid reacts with limestone. In hot deserts, gypsum crystals full of sand grains develop after the evaporation of mineral-enriched waters from flash floods. Such crystals are known as desert roses.

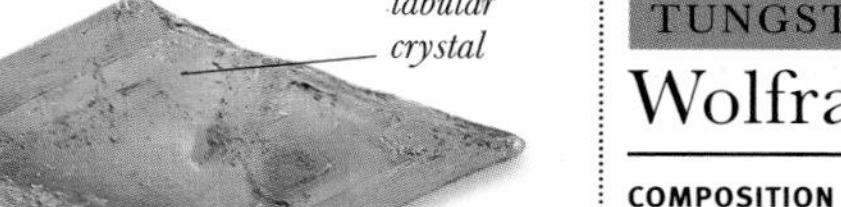

**SELENITE**
*A crystalline form of translucent gypsum, known as selenite, develops into diamond-shaped, tabular crystals in some sedimentary rocks.*

**MINERAL DUNES**
*When inland seas dry up, as here at White Sands, New Mexico, they deposit gypsum as one of several evaporate minerals. Gypsum grains are then picked up by the wind and blown into dune formations.*

TUNGSTATES

## Wolframite

**COMPOSITION** $(Fe,Mn)WO_4$

**CRYSTAL SYSTEM** Monoclinic

**HARDNESS** 5–5½ **SPECIFIC GRAVITY** 7.0–7.5

**COLOR** Gray to brownish-black; brownish-black streak

This mineral is the most economically important tungsten ore. Originally known as wolfram, tungsten is symbolized as an element by the letter W, for wolframite. Tungsten typically forms tabular or prismatic crystals, yet can be massive or granular, too. It is mainly found in granite-related pegmatites and quartz veins but also occurs in some hydrothermal veins associated with other ore minerals such as sphalerite, cassiterite, and galena. Wolframite can accumulate in placer deposits as well. It used to be considered a waste material because it did not melt—the element tungsten was not recognized until 1781. Because of its exceptionally high melting point (6,000°F/3,387°C), tungsten is used today for electric filaments and steels for high-speed tools.

MOLYBDATES

## Wulfenite

**COMPOSITION** $PbMoO_4$

**CRYSTAL SYSTEM** Tetragonal

**HARDNESS** 3 **SPECIFIC GRAVITY** 6.5–7.0

**COLOR** Shades of orange-yellow to red or brown; white streak

A striking lead-and-molybdate mineral, wulfenite can be found in massive and crystalline forms. Its crystals develop as beautiful square tablets of orange-yellow to reddish-brown. This secondary mineral forms in the oxidized zone of ore deposits of lead and molybdenum, and may be associated with other lead minerals, such as anglesite and cerussite. When present in sufficient quantities, it is mined as a lead ore, since its composition is around 61 percent lead oxide. Wulfenite was named after F. X. Wülfen, the Austrian mineralogist who first described it in 1785.

PHOSPHATES

## Apatite

**COMPOSITION** $CA_5(PO_4)_3(F,Cl,OH)$

**CRYSTAL SYSTEM** Hexagonal

**HARDNESS** 5 **SPECIFIC GRAVITY** 3.1–3.2

**COLOR** Variable, including green, brown, and white; white streak

Difficult to distinguish from minerals such as tourmaline, apatite's name is Greek for "deceit," because of its misleading appearance, especially in crystalline form. It is also found in massive and granular states. Apatite occurs in hydrothermal veins, pegmatites, and metamorphosed limestone. In sedimentary rocks, it is formed from organic materials. It is used as a fertilizer and in the chemical industry.

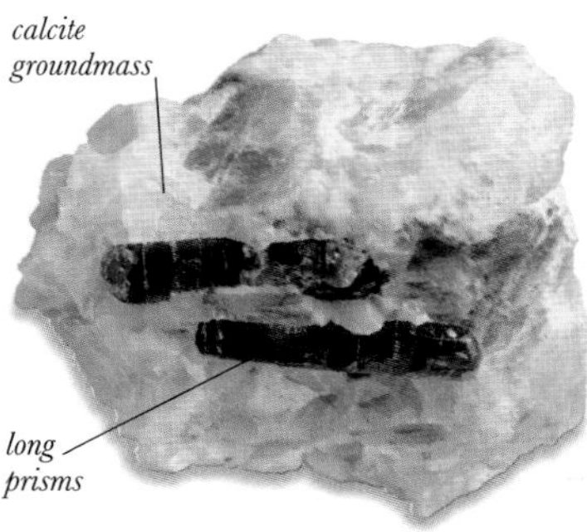


PHOSPHATES

## Turquoise

**COMPOSITION** $CuAl_6(PO_4)_4(OH)_8.5H_2O$

**CRYSTAL SYSTEM** Triclinic

**HARDNESS** 5–6 **SPECIFIC GRAVITY** 2.6–2.8

**COLOR** Sky blue to bluish-green to greenish-gray; greenish-white streak

A phosphate mineral, often with a distinctive, pale blue-green color, turquoise was one of the first gemstones to be mined. Normally massive or granular, it is rarely crystalline. Turquoise is secondary in origin, and occurs in highly altered veins within igneous rocks and metamorphosed sedimentary rocks rich in aluminosilicates. Sometimes it is highly porous and full of cracks. Its name is derived from Turkey, where during the Ottoman Empire it was traded as a precious stone, especially with Persia. Today, localities in Mexico, the US, China, and Iran produce turquoise. Imitation turquoise was first created in France in 1972 using the Gilson process, but it is not a true synthetic, since its properties differ from the natural forms. Nevertheless, much turquoise sold today is synthetic.

SYNTHETIC TURQUOISE

NATURAL TURQUOISE

**NATURAL AND SYNTHETIC**
*Turquoise is much in demand for jewelry, but it can be extremely difficult to tell the difference between synthetic (above) and natural (right) stones.*

**BLUE-STREAKED ROCK**
*Natural turquoise often occurs in thin veins and is highly porous and full of cracks, so it is unusual to obtain gem-quality pieces of large size.*

VANADATES

## Carnotite

**COMPOSITION** $K_2(UO_2)_2(VO_4)_2.3H_2O$

**CRYSTAL SYSTEM** Monoclinic

**HARDNESS** 2 **SPECIFIC GRAVITY** 4.0–5.0

**COLOR** Lemon to greenish-yellow; yellow streak

Containing about 63 percent uranium oxide, carnotite is highly radioactive and is mined for its uranium content. It is deposited in sedimentary rocks from vanadium- and uranium-enriched waters. Rarely crystalline, carnotite typically has a powdery or earthy texture. It is hard to distinguish from other uranium minerals except by using X-ray diffraction.

SILICATES

## Olivine group

**COMPOSITION** $(Mg,Fe)_2SiO_4$

**CRYSTAL SYSTEM** Orthorhombic

**HARDNESS** 6½–7 **SPECIFIC GRAVITY** 3.2–4.4

**COLOR** Clear yellowish-green, also yellow-brown–black; white streak

Commonly found in igneous rocks, olivine is the most abundant of all silicate minerals. It actually comprises a group of minerals whose composition ranges from magnesium-rich forsterite to iron-rich fayalite. Perfect crystals are rare, but there is a green gem form (see right). This was first brought to Europe by returning Crusaders from the Red Sea island of Zebirget, where it had been mined for 3,500 years. Olivine's primary origin is in silica-poor intrusive igneous rocks (such as gabbro and peridotite) and extrusive rocks (such as basalt). Dunite is a rare intrusive rock mainly made of olivine, and forsterite crystals up to 5½ in (14 cm) long have been found in dunite pegmatites in the Urals, Russia. The metamorphism of magnesium-rich dolomitic limestone produces green forsterite marbles. Olivine was first recognized in 1772 in a meteorite. It is readily altered by hydrothermal solutions and prolonged weathering to create serpentinite minerals. It is used to make heat-resistant glass.

**SANDS OF TIME**
*Weathering and erosion of Hawaii's basalt lavas have produced these dark green sands, which consist mostly of olivine.*

**GEM OLIVINE**
*Perfect crystals of green, transparent olivine, known as peridot, are rare and so especially prized as gems.*

**OLIVINE IN BASALT**
*Through a microscope, a transparent section of basalt rock, as here, reveals fissured yellow olivine crystals as a main constituent mineral.*

**UNEXPECTED PROPERTIES**
*Although a fairly hard mineral, olivine is susceptible to change and so is usually full of cracks.*

SILICATES

## Sillimanite

**COMPOSITION** $Al_2SiO_5$

**CRYSTAL SYSTEM** Orthorhombic

**HARDNESS** 6½–7½ **SPECIFIC GRAVITY** 3.2–3.3

**COLOR** White-gray or colorless; white streak

A hard aluminosilicate mineral, sillimanite is similar in structure and mode of formation to kyanite and andalusite. Typically found in regional metamorphic rocks, sillimanite forms long, fibrous, prismatic crystals with slightly variable hardness, and in some pegmatites it can be of gem quality. It is second in density to kyanite, and is named after the late-18th-century American chemist Benjamin Silliman.

SILICATES

## Garnet group

**COMPOSITION** $X_3Y_2Si_3O_4$ where X = Ca, Mg, Fe, or Mn; Y = Al ,Fe, Ti, or Cr

**CRYSTAL SYSTEM** Cubic

**HARDNESS** 6–7½ **SPECIFIC GRAVITY** 3.5–4.3

**COLOR** Variable, including red-brown, yellow-green, and black; white streak

Within this group of silicate minerals, some elements can replace others in the atomic structure, resulting in great variation in color. Crystals are common and may grow to a few inches at relatively high temperatures. The kind of garnet found within various types of rock is linked to the chemistry of the host rock, especially in metamorphic and some igneous rocks: for example, magnesium-rich reddish pyrope is common in peridotites and associated serpentinites, while the rare green uvarovite garnet is found in some chrome-bearing serpentinites. Garnet minerals are particularly common in metamorphic rocks such as schist, gneiss, and eclogite. They also occur in placer deposits, because of their durability. Garnet has been used as a gemstone since Greek and Roman times but was especially popular in the 19th century, when Bohemian garnets—particularly pyropes—were traded extensively throughout Europe and North America. Since the 1960s, synthetic garnet has been produced mainly for industrial use—as abrasives and as bearings in fine instruments—and as gemstones.

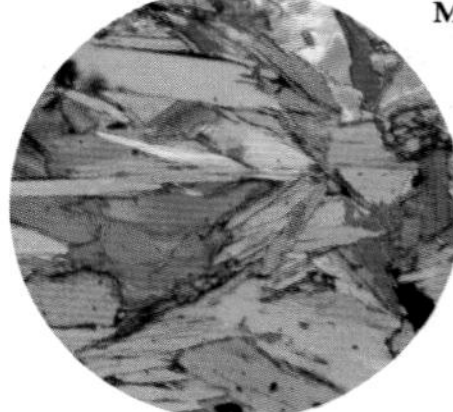

**METAMORPHIC SCHIST**
*This thin section of garnet reveals typical spherical garnet crystals, despite rock compression.*

**GROSSULAR CRYSTALS**
*Intergrown garnet crystals with well-developed, four-sided rhombic faces, and colored orange by calcium, here form the variety known as grossular garnet.*

**DIFFERENT ELEMENTS**
*The variety of environments in which garnet grows affects its composition and leads to a considerable range of gem coloring.*

SILICATES

## Zircon

**COMPOSITION** $ZrSiO_4$

**CRYSTAL SYSTEM** Tetragonal

**HARDNESS** 7½ **SPECIFIC GRAVITY** 4.6–4.7

**COLOR** Variable, including brown, green, blue, and black; white streak

One of the oldest minerals on the Earth, zircon's hardness has allowed it to survive in many types of rocks (see panel, below). In particular, it occurs as an accessory mineral in granites and alkali-rich intrusive igneous rocks such as syenite and certain pegmatites. Because of its resistance to weathering and erosion, zircon accumulates in alluvial placer deposits. It can also tolerate considerable metamorphism and is widely found in regional metamorphic terrains. For a long time zircon was confused with diamond, especially in gems from Sri Lanka. It is also the main source of the element zirconium, which is used in nuclear reactors and as an abrasive. South Africa, the US, and Australia are major producers.

**RESILIENT CRYSTALS**
*Zircon's well-formed crystals, which are formed in igneous rocks, survive considerable metamorphism without deformation of their shape.*

**STEP-CUT ZIRCON GEM**

### ZIRCON DATING

The resistance of zircon crystals to chemical and physical change makes them ideal for use in radiometrically dating very old rocks. Some radioactive elements, such as uranium, decompose over time, forming what are known as daughter isotopes. The rate of this decay has been established. This means that, by detecting single zircon crystals (or even zones within crystals) in old rocks and measuring the concentration of daughter isotopes in them against their original content, it is possible to calculate how much decay has taken place and thereby estimate the age of the rock.

SILICATES

## Andalusite

**COMPOSITION** $Al_2SiO_5$

**CRYSTAL SYSTEM** Orthorhombic

**HARDNESS** 6½–7½ **SPECIFIC GRAVITY** 3.1–3.2

**COLOR** Pale pink–gray–brown–green; white streak

Sometimes found in massive form, andalusite generally forms prismatic crystals with a square cross-section. This aluminosilicate mineral usually occurs in thermally metamorphosed rocks formed under low pressure, such as hornfel. It is also found in some pegmatites, where gem-quality crystals may develop, and occasionally in placer deposits. It is used in the production of porcelain and other heat-resistant materials.

**CHIASTOLITE**
*This variety of andalusite contains carbon inclusions in a dark cross-shape.*

SILICATES

## Kyanite

**COMPOSITION** $Al_2SiO_5$ **CRYSTAL SYSTEM** Triclinic

**HARDNESS** 5½–7 **SPECIFIC GRAVITY** 3.5–3.7

**COLOR** Blue, white, or gray; white streak

Flat, blade-shaped crystals are typical of kyanite. Like its polymorphs andalusite and sillimanite, kyanite is found in regionally metamorphosed rocks, but it has a denser structure. Consequently, it is found in schist and gneiss formed at high pressures and temperatures. Kyanite also occurs in pegmatites, sometimes as deep blue gem-quality stones—its name deriving from the Greek *kuanos* ("dark blue").

**STEP CUT**

**FLAWED BEAUTY**
*Despite their closely packed structure, kyanite gems commonly have pressure cracks.*

**ROCK MATRIX WITH CRYSTALS**

SILICATES

## Topaz

**COMPOSITION** $Al_2SiO_4(OH,F)_2$

**CRYSTAL SYSTEM** Orthorhombic

**HARDNESS** 8 **SPECIFIC GRAVITY** 3.5–3.6

**COLOR** Variable yellow–brown–blue or colorless; white streak

Characteristically, topaz forms prismatic crystals of variable color with vertically lined faces, but it can also occur in massive or granular forms. It is mostly found in pegmatites associated with granite, along with tourmaline, beryl, and apatite. Topaz also occurs in placer deposits and is well known as a gemstone. The discovery of large deposits in Brazil has provided spectacularly large crystals, including one weighing 596 lb (271 kg), and another that is the biggest cut blue topaz (21,005 carats). Another beautiful topaz, found in 1740, was set in the Portuguese crown in the mistaken belief that it was a diamond (the so-called Braganza "diamond" of 1,640 carats). The first-century Roman naturalist Pliny the Elder thought that the name topaz was derived by the Romans from the Red Sea island Topazos. The mineral found there was long mistaken for topaz and is now known to be olivine.

**COLOR RANGE**
*Commonly colored yellow to brown, topaz is occasionally found in other colors, such as blue.*

**GIANT TOPAZ**
*This 22,892-carat topaz gemstone was cut from a crystal that was found in Brazil. It is one of the world's largest gemstones.*

**PEAR-SHAPED TOPAZ**
*Tiny internal flaws demonstrate that this topaz gem has been cut from a natural crystal.*

SILICATES

## Staurolite

**COMPOSITION** $(Fe,Mg,Zn)_2Al_9(Si,Al)_4O_{22}(OH)_2$

**CRYSTAL SYSTEM** Monoclinic

**HARDNESS** 7–7½ **SPECIFIC GRAVITY** 3.7–3.8

**COLOR** Reddish-brown to brown-black; white streak

This hard aluminosilicate mineral is associated with kyanite and mica in regionally metamorphosed phyllite, schist, and gneiss. The composition of staurolite is complex, with considerable substitution to give a number of varieties, such as Zambian lusakite, which has been mined as a blue pigment. Staurolite is often found as well-formed prismatic crystals, which are generally opaque and frequently twinned in the form of a distinctive cross (see panel, right)—the name staurolite being Greek for "cross stone." Crystals of staurolite are frequently larger than those that surround them, in which case they are referred to as porphyroblasts. Staurolite is also found in placer deposits.

**DOMINANT ROLE**
*In places, staurolite is such a common mineral that its crystals comprise a significant proportion of the rock in which it is embedded.*

### RELIGIOUS SYMBOLS

The frequent occurrence of staurolite as twinned cruciform crystals has led to their use as amulets for Christians over the centuries. Some of the best specimens were found in the mica schists of the Swiss Alps and were sold as *Baseler taufstein* ("Basel baptismal stone"). Because they were so readily available, staurolite amulets were not expensive and became widely worn.

SILICATES

## Epidote group

**COMPOSITION** $X_2Y_3Si_3O_{12}OH$ where X = Ca; Y = Al, Fe, or Mg

**CRYSTAL SYSTEM** Orthorhombic and monoclinic

**HARDNESS** 5–7 **SPECIFIC GRAVITY** 3.2–4.5

**COLOR** Variable, including green-black; gray streak

In this group of complex silicates, magnesium and iron (as in piedmontite) can be replaced by aluminum and iron (as in epidote), depending on the chemical environment within which the mineral is developing. Epidote is mineralogically close to clinozoisite, and both minerals typically form columnar, prismatic, and needle-shaped crystals, but they may also be massive. While the natural forms are commonly black to green in color, gem varieties may be yellow, dark green, or brown. This mineral is rarely cut as a gem, however, since there is a well-developed cleavage along which the crystals can easily break. The name epidote is Greek for "an addition," because this mineral was long mistaken for tourmaline until the French mineralogist René Haüy distinguished epidote as a separate mineral in 1801. Epidote is a relatively common calcium aluminosilicate in low- and medium-grade metamorphic rocks, especially those derived from igneous rocks such as basalt (amphibolite) and from limestone (marble). It also occurs in veins in igneous rocks. A new valuable blue variety of the epidote group is tanzanite, found in Arusha, Tanzania, in 1967 by an Indian tailor, Manuel d'Souza.

**FASHIONING EPIDOTE**
*Cut epidote gems such as this one are unusual, because their crystals are difficult to shape due to their perfect cleavage.*

**PROBLEMS OF IDENTITY**
*The elongate and lined prisms of epidote crystals look extremely similar to those of tourmaline, for which they can at times be mistaken.*

SILICATES

## Beryl

**COMPOSITION** $Be_3Al_2Si_6O_{18}$

**CRYSTAL SYSTEM** Hexagonal

**HARDNESS** 7½–8 **SPECIFIC GRAVITY** 2.6–2.8

**COLOR** Variable, including blue (aquamarine), green (emerald), yellow (heliodor), white, and colorless; white streak

Beryl is best known for its gem varieties, especially green emerald and pale blue-green aquamarine, both of which are highly valued. As well as occurring as a crystal, this hard beryllium mineral is also found in a massive form, which may be mistaken for quartz. Beryl crystals are usually marked by characteristic longitudinal lines and commonly develop in pegmatitic veins, especially in granite. The mineral also occurs in metamorphic schist and gneiss, and sedimentary placer deposits. Beryl crystals can be very large, with specimens up to 20 ft (6 m) long and weighing 1.5 tons being found in Maine; such crystals can be mistaken for apatite, but beryl is harder. The largest aquamarine found was a hexagonal prism discovered in Brazil in 1910; it weighed 243 lb (110.5 kg) and measured 19 x 16 in (48 x 40 cm). By comparison, emerald crystals rarely grow more than a few inches long and are commonly flawed with cracks and tiny crystals of other minerals, but this has not detracted from their value. Emeralds have in the past been particularly associated with Ottoman, Persian, and Mogul rulers, although the stones actually originated in Colombia, South America, and were sold by indigenous peoples and Spanish colonialists (see panel, above). Beryl is the main source of beryllium, which is one of the lightest metals known and an important ingredient of special metal alloys. More than 80 percent of the world's annual beryl production comes from the US.

**LARGE CRYSTALS**
*Beryl crystals, such as this fine gem-quality prismatic one grown from a metamorphic rock groundmass, can be extremely long and heavy.*

### BERYL MINING

Emeralds were mined by the ancient Egyptians as long ago as 1300 BC, and for a thousand years the mines of Sikait and Zabara, in Egypt, were Europe's principal source of these gems. When the Spanish invaded the Americas, however, they discovered that Colombian emeralds were widely traded by native peoples from Mexico to Chile. In 1573, the Spanish captured the Muzo beryl mine, in Colombia, which then supplanted the Egyptian source for Europe. Nevertheless, the value of beryl gems is so great that they are still mined wherever they occur, regardless of the quantity.

AQUAMARINE

EMERALD

**POPULAR CUT STONES**
*Aquamarine and emerald are among the best-known of several attractive and valuable gem varieties of the mineral beryl.*

SILICATES

## Tourmaline group

**COMPOSITION** $(Ca,Na,K)(Al,Fe,Li,Mg,Mn)_3(Al,Cr,Fe,V)_6Al_6(BO_3)_3Si_6O_{18}(OH,F)_4$

**CRYSTAL SYSTEM** Trigonal

**HARDNESS** 7 **SPECIFIC GRAVITY** 3.0–3.2

**COLOR** Variable, including yellow, green, blue, pink, brown, and black; white streak

Color variation may occur along the length of a single tourmaline crystal. Tourmaline crystals exhibit a piezo-electric effect, in which an electric current applied to one end of a crystal has its charge reversed at the other end (for example, from positive to negative)—a property that has been used in electronics. Tourmaline is a hard, silicate gem mineral with long, lined, prismatic crystals. These have a triangular cross-section with curved faces. Crystals commonly occur in radiating or parallel clusters in granite pegmatites and metamorphic schist and gneiss, especially those that have been altered by boron bearing hydrothermal solutions.

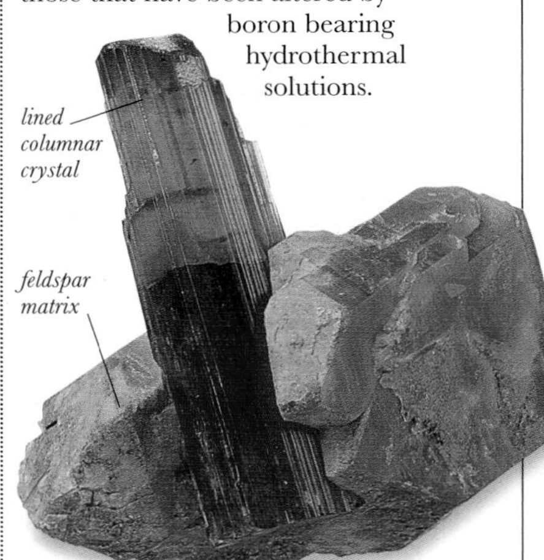


SILICATES

## Pyroxene group

**COMPOSITION** $X_2Si_2O_6$ where X = Mg, Fe, Mn, Li, Ti, Al, Ca, or Na

**CRYSTAL SYSTEM** Orthorhombic and monoclinic

**HARDNESS** 5–6½ **SPECIFIC GRAVITY** 3.2–4.0

**COLOR** Pale to dark brownish-green or bronze; greenish-white streak

Pyroxenes are some of the most widespread of all rock-forming silicate minerals. They can be broadly divided into clinopyroxenes, which have monoclinic crystals, and orthopyroxenes, which have orthorhombic crystals. Well-formed crystals are typically short, dark-colored prisms, some of which occur as gems: for example, green enstatite, reddish-brown hypersthene, and greenish-diopside. Clinopyroxenes (such as aegirine and gabbro) contain combinations of calcium, sodium, aluminum, iron, or lithium, while orthopyroxenes (such as hypersthene and enstatite) have very little calcium but may contain magnesium and iron. Pyroxene minerals are commonly found as masses, grains, or better-defined crystals within quartz-poor igneous rocks such as basalt, gabbro, and pyroxenite. Calcium-rich clinopyroxene may also occur in some metamorphosed limestone, while orthopyroxene may be found in stony meteorites. Pyroxene can easily be confused with amphibole. Pyroxene crystals, however, have two sets of cleavages at 90° to one another, while amphibole has two sets at about 120°. The green-colored pyroxene jadeite is usually massive with a very compact structure. These features give it great strength, which is why it has been used to make objects such as ax blades and carved ornaments.

**AUGITE PHENOCRYST**
*Crystals that grow larger than their surrounding ones are called phenocrysts. Here, a well-formed, short, stubby prismatic phenocryst of augite has developed within a finer-grained rock matrix.*

**DIABASE**
*The pyroxenes are the brown grains in this thin section of diabase igneous rock.*

**GABBRO**
*The pale and golden-yellow grains in this thin section of gabbro are clinopyroxene; the black and white grains are feldspar.*

**AEGIRINE CRYSTAL**
*Named after Aegir, the Scandinavian sea god, this very dark green clinopyroxene mineral was first discovered and described in Norway in 1835.*

DIOPSIDE

ENSTATITE

**DIFFERENT GEM TYPES**
*Olive green diopside is a gem-quality clinopyroxene, while emerald green enstatite is an orthopyroxene.*

SILICATES

## Amphibole group

**COMPOSITION** $X_2Y_5(OH)_2(Si,Al)_8O_{22}$ where X = Ca or Na; Y = Mg, ,Fe, or Al

**CRYSTAL SYSTEM** Monoclinic

**HARDNESS** 5–6 **SPECIFIC GRAVITY** 3.0–3.6

**COLOR** Variable, including white–green–black; pale green streak

Amphiboles are a group of rock-forming silicate minerals that have a complex atomic structure in which the elements magnesium, iron, calcium, aluminium, and sodium replace one another. Amphibole resembles pyroxene except for its 120° cleavages and its hydroxyl (–OH) groups, which form an integral part of its composition.

**HORNBLENDE CRYSTALS**
*Long, prismatic crystals and green-black coloring are characteristic of hornblende crystal growth.*

**CRYSTALLINE RIEBECKITE**
*The crystalline form, shown here, of this amphibole mineral is not a danger to health, unlike riebeckite in its fibrous form—asbestos.*

One of the most important amphibole minerals is hornblende, which contains calcium, sodium, magnesium, and iron. It is widespread in both igneous rocks and metamorphic rocks, such as amphibolite, while other amphiboles, such as tremolite, actinolite, and nephrite (jade), are found mainly in metamorphic rocks. Another amphibole, called riebeckite (also known as crocidolite or blue asbestos), is now notorious for its long-term toxicity and link with lung diseases such as asbestosis and mesothelioma cancer.

**ANDESITE MAGNIFIED**
*Amphibole crystals appear brown in this thin section of the volcanic lava andesite.*

SILICATES

## Talc

**COMPOSITION** $Mg_3Si_4O_{10}(OH)_2$

**CRYSTAL SYSTEM** Monoclinic

**HARDNESS** 1 **SPECIFIC GRAVITY** 2.6–2.8

**COLOR** White, gray–green; white streak

This soft silicate mineral (a Mohs scale hardness value of one, the lowest possible) has a soapy or greasy feel. Talc is formed by the metamorphism of magnesium-silicate minerals such as pyroxene, amphibole, or olivine. Talc usually occurs as foliated or granular masses and only rarely as crystals. Massive talc is known as steatite or soapstone and can easily be carved for decorative purposes. The heat-resistant properties of talc are extensively exploited for the manufacture of fire-resistant materials, especially ceramics. China and the US produce more than a third of the world's annual output of talc.

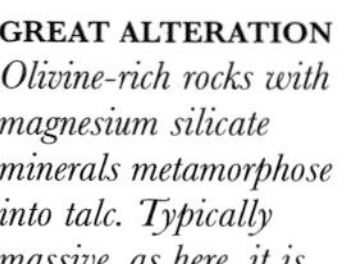


**GREAT ALTERATION**
*Olivine-rich rocks with magnesium silicate minerals metamorphose into talc. Typically massive, as here, it is cut through by lines of fracture and stress.*

**TALC QUARRIES**
*Uplift and erosion have brought talc rocks to the Earth's surface, where they can be extracted by open-cast quarrying, as here in the Shetland Isles, UK.*



SILICATES

## Mica group

**COMPOSITION** $XY_{2-3}(OH)_2(Si,Al)_{4-5}O_{10}$ where X = K, Na, or Ca; Y = Al, Mg, Fe, or Li

**CRYSTAL SYSTEM** Monoclinic

**HARDNESS** 2–4 **SPECIFIC GRAVITY** 2.7–3.3

**COLOR** Variable, including white, pink, green, and brown–black; white streak

The micas are an important group of complex silicate minerals that are characterized by their distinctive atomic structure, which consists of silicon and oxygen tetrahedra linked together to form sheets with relatively weak bonds between the sheets. Cleavage within crystals is thus well developed between successive sheets. Crystals are tabular in shape and often hexagonal in outline. They occur as scattered flakes or booklike aggregates of flakes. Mica is widespread throughout many igneous, metamorphic, and sedimentary rocks. Composition varies from white micas rich in potassium and aluminum (such as muscovite) to darker micas rich in magnesium and iron (such as biotite) and rarer varieties such as pink-colored, lithium-rich micas (such as lepidolite). Muscovite is characteristic of quartz-rich igneous rocks such as granite and granitic pegmatites, as well as many metamorphic rocks, especially phyllite and schist. It also survives weathering and erosion well and is an important component of sedimentary rocks, from sandstone through siltstone to shale. Biotite is more typical of a wider range of igneous rocks, from granite through to diorite along with their pegmatites, as well as a great variety of metamorphic rocks. Dark mica is less commonly found in sediments, because it is more readily altered and weathered into residual minerals. Some micas are extensively used in industry, because they are good thermal and electrical insulators. The US and China produce more than half of the annual global output of mica.

**BIOTITE**
*This dark brown-colored but slightly translucent mica mineral has a perfect cleavage between flakes.*

vitreous luster

**TYPICAL MUSCOVITE**
*White-colored muscovite crystals here exhibit their characteristic tabular form, with each crystal having a single, well-developed cleavage parallel to the largest (tabular) face.*

**CRYSTALS OF LEPIDOLITE**
*Found in granite and pegmatites (as here), this mica mineral contains lithium and is usually distinguished by its lilac to pink coloring.*

### MICA WINDOWS

In some pegmatites, mica crystals grow to great size—one found in the Urals, Russia, was 54 square feet (5 square meters) in area and 1.6 ft (0.5 m) thick. Like slate rocks, micas can be split along the well-developed cleavage plane into very thin sheets, which are transparent and can readily be cut. They have been widely used as substitutes for glass in industrial furnaces and domestic stoves, because mica has a very high melting-point and does not easily break. Wherever available and cheap, mica sheets have also been used as window panes in domestic buildings, especially in rural areas where glass was difficult to obtain or was too expensive.

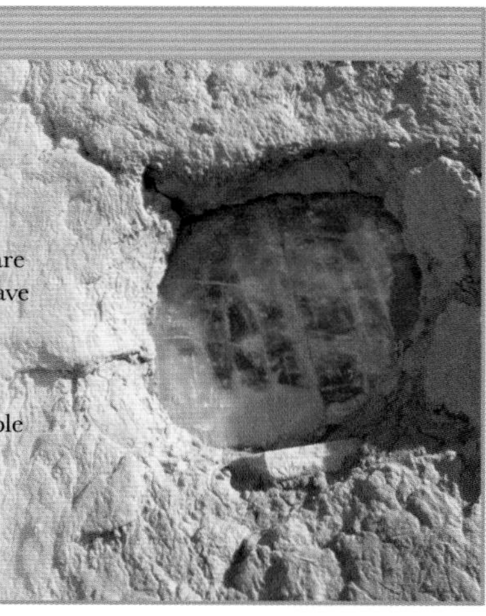

SILICATES

## Chlorite group

**COMPOSITION** $(Mg,Fe,Mn,Al)_{4-6}(Si,Al)_4O_{10}(OH,O)_2$

**CRYSTAL SYSTEM** Monoclinic

**HARDNESS** 2–3 **SPECIFIC GRAVITY** 2.6–3.3

**COLOR** Green, also yellow-brown; white streak

This widespread group of minerals includes chamosite and clinochlore. They are similar to mica in composition, structure, and cleavage, except that their cleavage flakes are brittle, not elastic. Chlorite is commonly massive or fine-grained; crystals do occur, often in scaly aggregates. It is found in igneous rocks as well as low-grade metamorphic rocks and sediments. The name chlorite is derived from the Greek *khloros* ("green"). Green chamosite develops in certain iron-rich sediments and has been mined as an iron ore.

SILICATES

## Zeolite group

**COMPOSITION** Complex and variable aluminosilicates

**CRYSTAL SYSTEM** Cubic, orthorhombic, and monoclinic

**HARDNESS** 3½–5½ **SPECIFIC GRAVITY** 2.0–2.5

**COLOR** White, yellow, pink, brown, or colorless; white streak

There is considerable substitution of sodium and calcium, and occasionally barium and potassium, in the chemical composition of this complex group of aluminosilicates, which includes scolecite, analcime, and natrolite. Zeolite contains water that can be expelled on heating and then regained on exposure to moist air, without destroying the structure. Typically found in basalt and occasionally sedimentary rocks, it occurs as fibers, needles, prisms, or complex cubes. It has been used for water softening and extensively as industrial catalysts.

SILICATES

## Kaolinite group

**COMPOSITION** $Al_2Si_2O_5(OH)_4$

**CRYSTAL SYSTEM** Triclinic and monoclinic

**HARDNESS** 2–2½ **SPECIFIC GRAVITY** 2.6–2.7

**COLOR** White or yellow; white streak

In detail, kaolinite is made up of microscopic, flaky-layered crystals with perfect basal cleavages, like mica, even though this clay mineral mostly looks like an earthy mass. This cleavage confers important properties on kaolinite and other related clay minerals such as illite and dickite: they are pliable and tend to act as lubricants. Kaolinite is formed through intense weathering or hydrothermal action on aluminosilicate minerals such as feldspar. It has been used in the manufacture of porcelain since the 6th century in China, from where the name derives. It is also used as a filler in paper and paint. More than half of the annual output of kaolinite is produced by the US and Uzbekistan.

SILICATES

## Vermiculite

**COMPOSITION** $(Mg,Ca)_{0.7}(Mg,Fe,Al)_6(Al,Si)_8O_{22}(OH)_4.8H_2O$

**CRYSTAL SYSTEM** Monoclinic

**HARDNESS** 1½ **SPECIFIC GRAVITY** 2.3

**COLOR** Yellow–brown; white streak

The unusual properties of vermiculite have led to its widespread use in the manufacture of thermal and sound-insulation materials, especially since it is not carcinogenic like asbestos. This chemically complex aluminosilicate mineral is similar in structure and appearance to mica, forming flaky-layered crystals with perfect basal cleavage. When heated, however, it expands greatly along the cleavage, to form worm-like masses—hence its name, which is derived from the Latin *vermiculus* ("little worm"). South Africa produces about half the annual vermiculite output.

SILICATES

## Feldspar group

**COMPOSITION** $X(Al,Si)_4)_8$ where X = Na, K, Ca, or Ba

**CRYSTAL SYSTEM** Monoclinic and triclinic

**HARDNESS** 6–6½ **SPECIFIC GRAVITY** 2.5–2.8

**COLOR** Variable, including white, pink, green, blue, brown, and colorless; white streak

One of the most important groups of rock-forming aluminosilicate minerals, feldspars are widely distributed throughout igneous, metamorphic, and sedimentary rocks. They are found in one of two types: potassic feldspars (such as orthoclase and microcline); and plagioclase feldspars (such as albite and labradorite), in which calcium and sodium are substituted for one another. The formation of feldspar minerals in igneous rocks depends on the chemical composition and temperature of the original melt (magma). Low-temperature potassic feldspars develop in silica-rich magmas that result in granites and rhyolite lavas; while plagioclase feldspars grow in more silica-poor, higher-temperature magmas that result in gabbros and basaltic lavas. Typically, feldspar forms tabular or prismatic crystals, many of which are twinned in highly distinctive patterns.

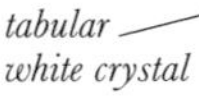


**ALBITE CRYSTALS**
*These white albite crystals exhibit the typical tabular form of many feldspar minerals, although a close look reveals repeated twinning.*

MICROCLINE

MOONSTONE ORTHOCLASE

POLISHED LABRADORITE

YELLOW ORTHOCLASE

**DIFFERENT FELDSPARS**
*Varying enormously in appearance, feldspar may occur as opaque microcline crystals, gem-quality orthoclase, or reflective and decorative labradorite.*

SILICATES

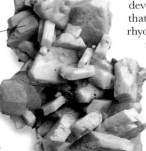

## Serpentine group

**COMPOSITION** $(Mg,Fe,Ni)_3Si_2O_5(OH)_4$

**CRYSTAL SYSTEM** Monoclinic

**HARDNESS** 2½–4 **SPECIFIC GRAVITY** 2.5–2.6

**COLOR** Pale to dark green, also brown, yellow, or white; white streak

One of the serpentine minerals—the fibrous chrysotile asbestos—has been identified as carcinogenic (see panel, right), but the common, nonfibrous forms of this widespread group of magnesium silicates are deemed safe. Such nonfibrous forms are either massive or scaly. Massive serpentine is used as an ornamental stone, and a nickel-bearing variety (garnierite) is mined for nickel in New Caledonia. Serpentine is essentially a secondary mineral, formed from magnesium-rich orthopyroxenes or olivine, and may be abundant enough to form masses called serpentinite bodies.

**CHRYSOTILE ASBESTOS**
*This fibrous form of the serpentine group is a health hazard, because its numerous fibers can readily be separated from the solid mineral.*

### ASBESTOS

Until recently, asbestos was regarded as a very useful fire- and chemical-resistant material. Mixed with cement, it was molded into pipes and sheets, and added to paints, in the home and in numerous industries. Asbestos is made from a fibrous serpentinite (chrysotile) or a fibrous amphibole (crocidolite), but both these minerals are carcinogenic. Inhalation or contact can cause long-term damage that may even result in death many years later.

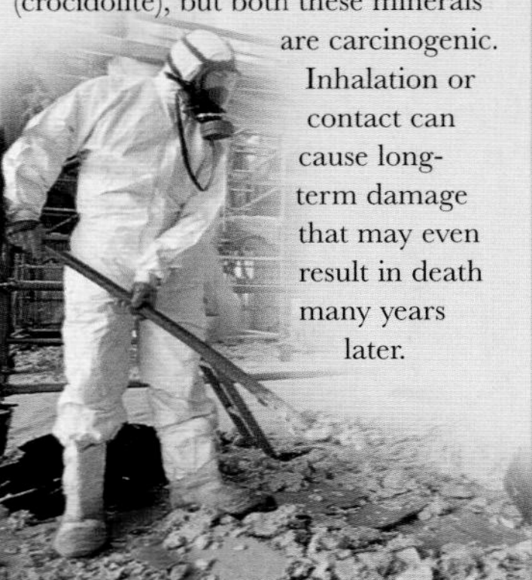

SILICATES

# Quartz

**COMPOSITION** $SiO_2$ **CRYSTAL SYSTEM** Trigonal

**HARDNESS** 7 **SPECIFIC GRAVITY** 2.65

**COLOR** Very variable, from white to black, or colorless; white streak

Due to its resistance to weathering and erosion, this common, rock-forming, silica mineral accounts for most of the grains in sediments found in deserts, river deposits, and coastal areas. When compressed or cemented, such sediment forms quartz sandstones, which are an important component of sedimentary rock. Quartz minerals are common in most sedimentary, metamorphic, and igneous rocks, except for those igneous rocks derived from silica-poor magmas (such as gabbro) and their metamorphosed equivalents (such as pyroxenite). They develop mostly in the continental crust, and range from well-developed crystals to surface encrustations and zoned cavity linings (such as in chalcedony and agate). Crystals are usually six-sided prisms, whose faces are often covered with horizontal lines and are topped with pyramidal faces. Giant crystals weighing several tons are known. In igneous rocks, quartz is very common in granite and rhyolite, and in pegmatites or hydrothermal veins related to them. Widespread throughout schist and gneiss, it may form pure quartz veins and be an important constituent of intrusive mineral veins.

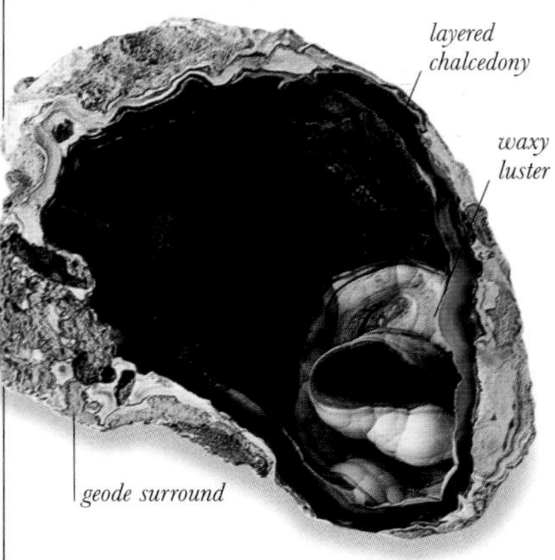


**CRYSTALLINE CAVITY**
*Some rocks contain a cavity, or geode, which here is layered with a microcrystalline form of quartz known as chalcedony.*

**CITRINE CRYSTAL**
*This hexagonal prism with its pyramidal faces is the characteristic form of quartz, and is seen here in a yellow variety known as citrine.*

## QUARTZ ARTIFACTS

Differences in quartz-related minerals can be very great and have resulted in the formation of a large number of colored varieties. Because they are relatively common, are hard, and can be cut and polished, a number have been adopted as precious stones by different cultures over the millennia. These include rock crystal, which is pure, colorless, and transparent quartz; purple (amethyst) and yellow (citrine) varieties; multi-colored agates; and fine-grained or massive forms (such as blood red jasper and black-and-white-striped onyx). Reflective bands of tiger's-eye or agate with internal mineral growths can also be found.

**EGYPTIAN CHEST ORNAMENT**
*The quartz gems shown here are typical of those used in South America and Egypt for decoration.*

**ZONED CRYSTALS**
*The color zones in this amethyst formed as the crystals grew, with traces of iron coloring the purple sectors.*

quartz

amethyst

**COLORED BANDS**
*The layered structure of this sardonyx is emphasized by its coloration. Sardonyx forms in lava cavities and is a variety of chalcedony.*

parallel bands

**QUARTZ OUTCROP**
*These quartz-bearing rocks at Shining Rock Wilderness, North Carolina, stand out against the rest of the landscape; such rocks are very resistant to surface stresses, such as weathering and erosion.*

## PETRIFIED FORESTS

Woody plant tissues are frequently fossilized by quartz and related silica minerals such as opal. Silica-enriched groundwater fills the porous tissue with mineral particles, which preserve the detailed cell structure of the original plant (such as these trees at Blue Mesa, Arizona). Such cell information useful to paleobotanists when identifying a plant and the conditions under which it grew.

# ROCKS

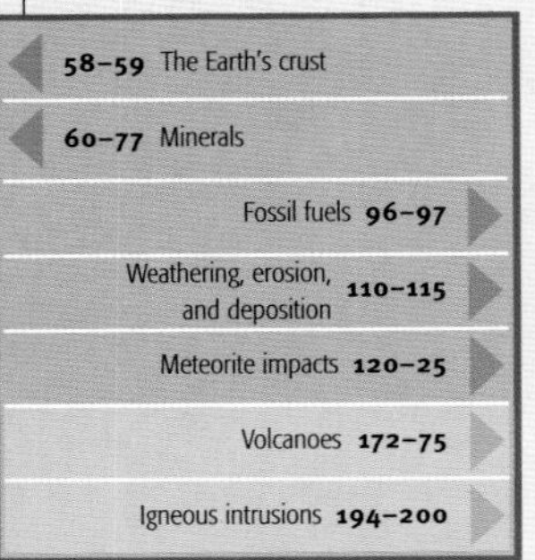
58–59 The Earth's crust
60–77 Minerals
Fossil fuels 96–97
Weathering, erosion, and deposition 110–115
Meteorite impacts 120–25
Volcanoes 172–75
Igneous intrusions 194–200

THE EARTH'S CRUST IS made of solid materials known as rocks, which are naturally occurring assemblages of minerals. Most of the Earth's rocks are concealed beneath soil and vegetation, but in some places they are exposed at the surface, where they may form features such as volcanoes and granite mountains. The huge diversity of the Earth's rocks has developed over thousands of millions of years through the geological processes of igneous activity, changes in form known as metamorphism, and the formation of sediments and sedimentary rocks (including organic materials such as coal). Although most of the Earth's rocks are formed within the planet, meteorites originating in outer space are also found on its surface.

## MINERALS AND ROCKS

Rocks are made of minerals, and although they vary enormously across the Earth, rocks all form by three basic processes. Igneous rocks solidify from a hot, molten state. Some cool slowly underground and produce the relatively large crystals of plutonic rocks, such as granite. Others erupt at the surface to form volcanic rocks such as glass, the rapidly quenched product of the molten liquid. Metamorphic rocks are transformed from older rocks of any type by heat and pressure within the Earth. Processes that take place at the Earth's surface, such as erosion and transport, lead to the formation of sedimentary rocks.

**IGNEOUS ROCK**
*Formed by the slow crystallization of magma deep underground, granite consists of interlocking crystals of feldspar, quartz, and mica.*

**PLUTONIC IGNEOUS ROCK FORMATION**
*In a plutonic igneous rock (see p.79), the first minerals to grow form large crystals. Crystals that form later are smaller and less distinct.*

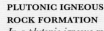

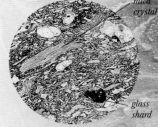

**MAGNIFIED IGNIMBRITE**
*This rock formed as magma cooled rapidly after a volcanic eruption. It consists of crystals, rock fragments, and glass shards welded together.*

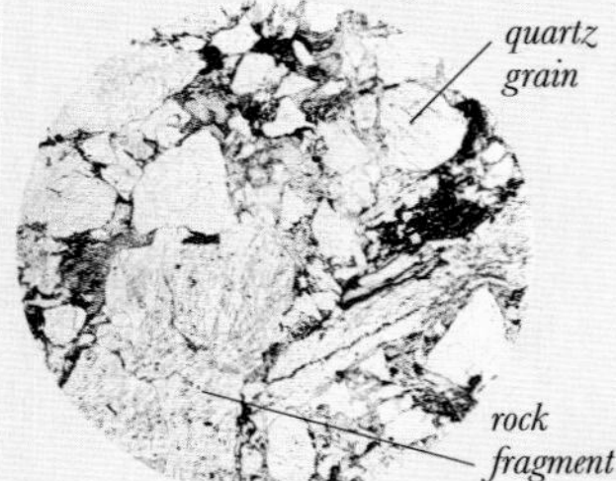


**MAGNIFIED SANDSTONE**
*Sedimentary sand grains of different shapes, sizes, and minerals are held together by a fine-grained matrix, or mineral groundmass, in this sandstone.*

**SEDIMENTARY ROCK AGGREGATIONS**
*When mineral grains are deposited by wind or water, they form loose sediments. Over time, these are compacted to form solid rock.*

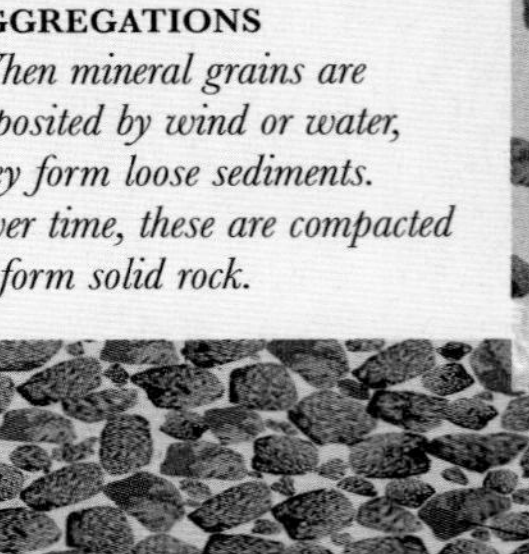

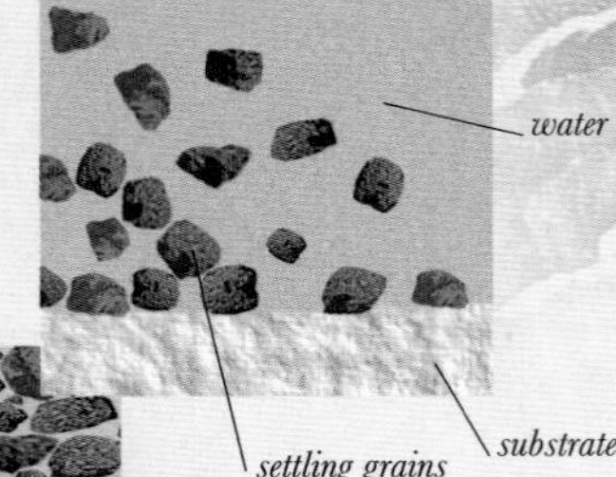


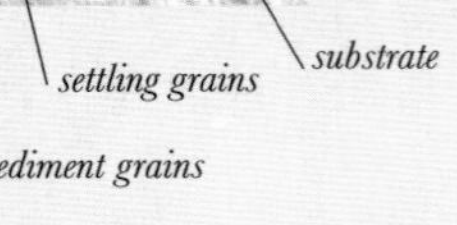

# TYPES OF ROCKS

Earth's numerous rock types can be grouped, according to the way they formed, into igneous, metamorphic, and sedimentary rocks, plus a small number of meteorites. Igneous rocks, which were once molten, are divided into extrusive (or volcanic) rocks, such as lavas, formed on the Earth's surface, and intrusive (or plutonic) rocks, such as granite, formed beneath the surface. Metamorphic rocks form from any preexisting rocks transformed by heat and pressure. They vary from low-grade metamorphic rocks such as slate (derived from shale) to high-grade schist and gneiss (from shale and sandstone). Metamorphic rocks often have distinctive planar textures called schistosity, in which platy minerals are aligned in parallel. Sedimentary rocks are formed from sediment through weathering, erosion, dissolution, transport of particles, biochemical precipitation, and deposition. They can broadly be divided into those transported by water, wind, or gravity. Land-based (terrestrial) deposits include river sediments, lake deposits, organic deposits such as coal, residual deposits from intense weathering, and volcanic sediments. Marine deposits include those formed grain by grain, as well as biochemical deposits such as carbonates.

**VOLCANIC DEBRIS**
*Igneous rock material erupted from the Novarupta volcano in Alaska is here being transported and redeposited to form river sediments.*

**STRATIFIED SEDIMENTARY ROCK**
*Layers of chalk sediment are transformed into limestone through burial, compaction, and recrystallization. In Normandy, France (right), folding and uplift have exposed the rocks to erosion, depositing new sediment on the seabed.*

**HOW ROCKS ARE MADE**
*Igneous, metamorphic, and sedimentary rocks form in a self-perpetuating cycle (below). Volcanic activity creates rocks at the Earth's surface. Erosion of all surface rocks produces sediments, which burial transforms into sedimentary rocks. Metamorphism induces further changes, and uplift and erosion expose them all.*

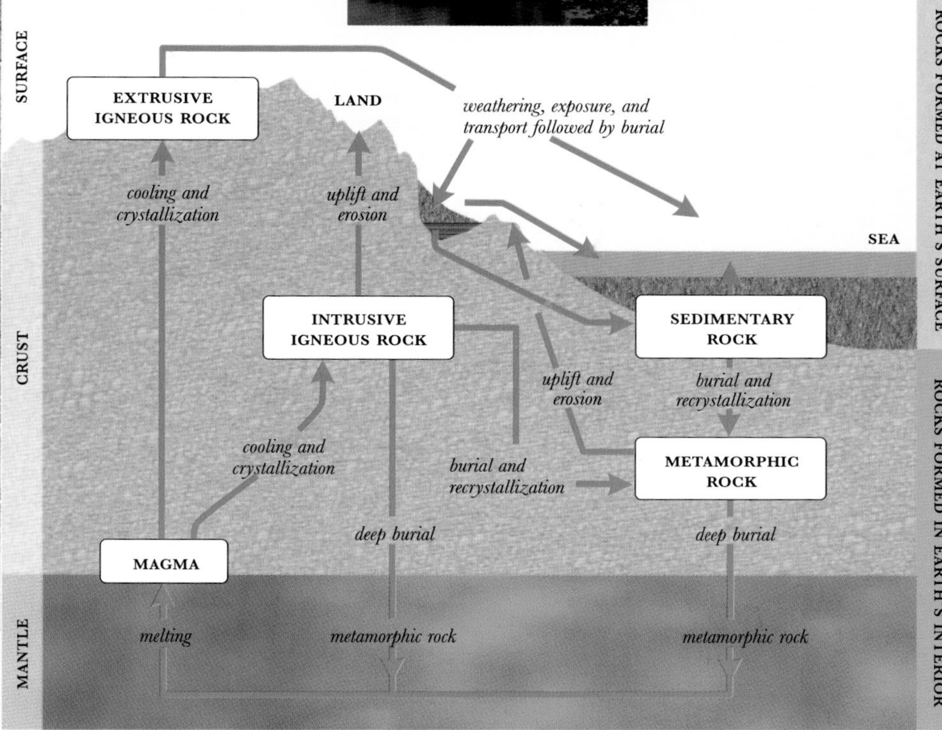


**METAMORPHIC ROCK**
*Bedding planes separating layers of ancient seabed sediments in Pembrokeshire, Wales, have been tilted and flattened by tectonic forces, with cleavage forming parallel to the layers.*

**IGNEOUS ROCK**
*As molten rock, such as this Hawaiian lava, hardens quickly, it forms a glassy crust, because rapid cooling leaves no time for crystals to grow.*

**BANDED SANDSTONE**
*The original sedimentary layering (bedding) of sandstone, such as here at Petra, Jordan, can be obscured by later weathering and chemical staining to produce a false bedding.*

# IGNEOUS ROCKS

**IGNEOUS INTRUSION**
*This vertical intrusive diabase dike in the Cuillin Hills, UK, acted as a feeder for the extrusive basaltic lavas that form the cliffs at top left.*

Rocks that solidify from a hot, molten state are known as igneous and are broadly divided into extrusive rocks, such as those produced by volcanoes, and intrusive bodies, such as plutons and dikes. They vary in composition from basalt to granite and in texture from rapidly cooled glasses and tiny crystals to slowly cooled coarse grains. Igneous rocks include the basalts (volcanic) and underlying gabbros (plutonic) that create all of the oceanic crust, basaltic island volcanoes (such as those of Iceland), and many hazardous subduction-zone volcanoes (such as Fuji, see p.190), as well as expansive plutons of granite (such as the North American Sierra Nevada) and gabbro (such as the Skaergaard, see p.196). Plutonic rocks also form smaller bodies, which are classified (in decreasing size) as plutons, stocks, sills (roughly horizontal), and dikes (roughly vertical). The term pegmatite refers to silica-rich (granitic) plutonic bodies with giant crystals, which are sources of both gems and industrial minerals rich in rare elements.

**IGNEOUS ROCK FORMATION**
*Molten rock (magma) collects in underground chambers within the Earth's crust. It may solidify in place to form an intrusion, or it may erupt at the surface to form an extrusive igneous rock.*

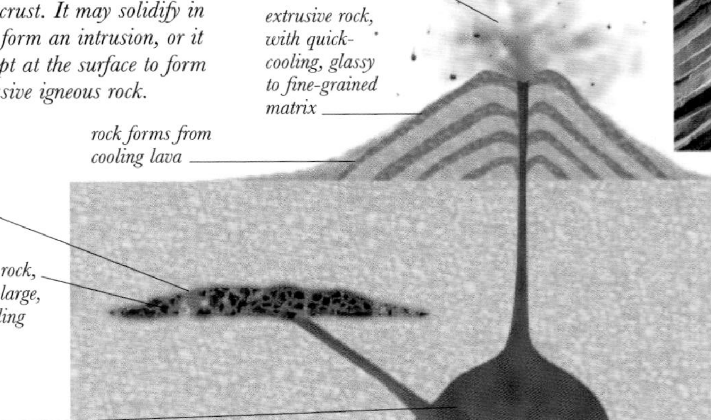


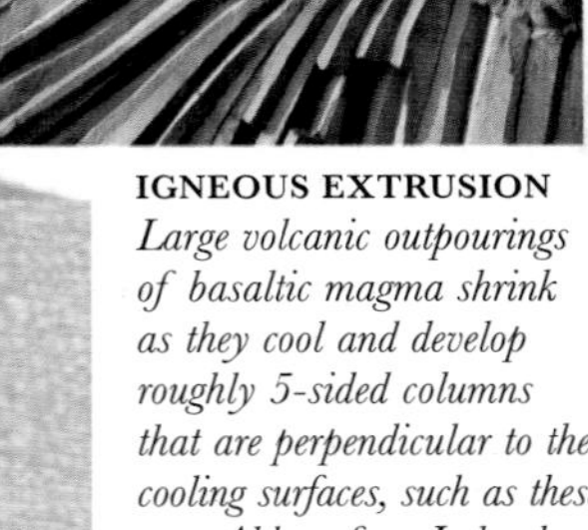

**IGNEOUS EXTRUSION**
*Large volcanic outpourings of basaltic magma shrink as they cool and develop roughly 5-sided columns that are perpendicular to the cooling surfaces, such as these near Aldeyarfoss, Iceland.*

# METAMORPHIC ROCKS

All rocks can be changed by heat and pressure. Such changes are called metamorphism. The most significant transformations occur deep within the Earth, but local changes can be produced nearer the surface. Strong pressure creates dynamic metamorphism, typified by the compression of mud-rocks into slates. On a smaller scale, such pressure may be localized—for example along fault planes, where rocks are distorted by the Earth's movements. Intense heat produces thermal metamorphism, which ranges from localized heating of rocks around a small igneous body to more widespread heating that produces mile-wide zones of recrystallization, called thermal aureoles, around major, deep-seated plutonic intrusions. Together, heat and pressure on a large scale create regional metamorphism, which is graded from low to high, with increasing pressure and temperature triggering the formation of assemblages of new minerals, and so converting the original rock texture into a metamorphic texture, often dominated by planar foliation.

**SPECTACULAR FOLDING**
*Intense folding of sedimentary rocks, seen here in the Ugab valley, Namibia, also initiates dynamic metamorphism through compression and recrystallization.*

**THERMAL METAMORPHISM**
*Where large bodies of igneous rock such as granite are formed, they radiate sufficient heat into the surrounding rocks to alter their mineralogy, a process known as thermal metamorphism.*

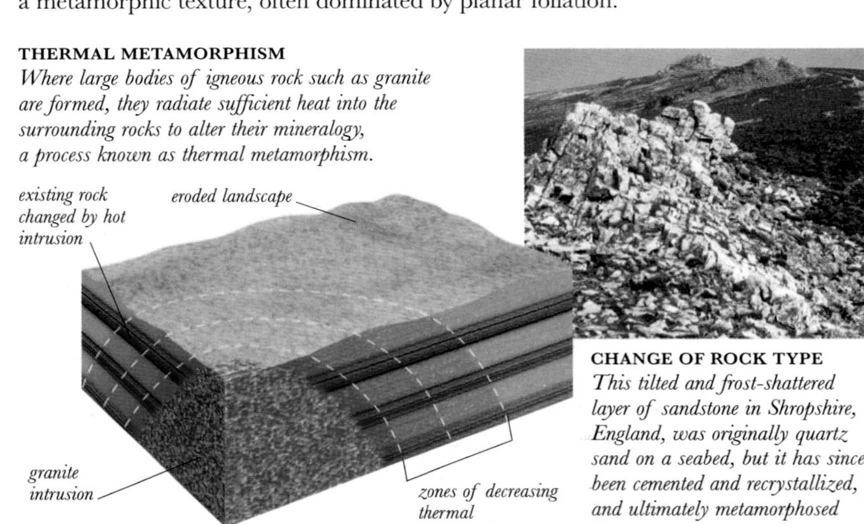


**CHANGE OF ROCK TYPE**
*This tilted and frost-shattered layer of sandstone in Shropshire, England, was originally quartz sand on a seabed, but it has since been cemented and recrystallized, and ultimately metamorphosed into quartzite.*

**DYNAMIC METAMORPHISM**
*The Earth's movements can apply pressure to sedimentary rocks and transform them through dynamic metamorphism by orientating mica minerals to form slaty cleavage.*

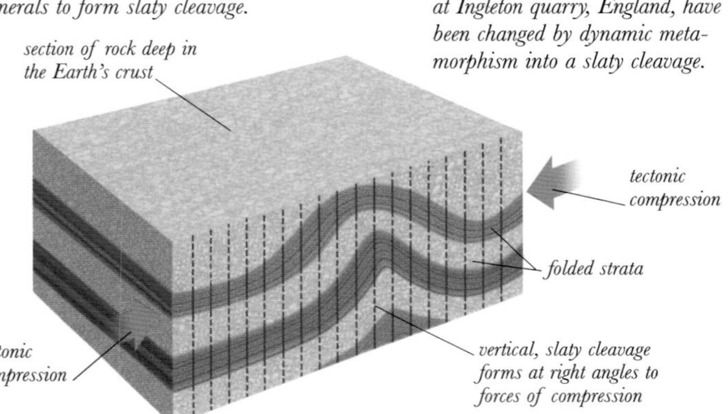

**CREATION OF SLATE**
*Clay minerals of seabed muds at Ingleton quarry, England, have been changed by dynamic metamorphism into a slaty cleavage.*

# SEDIMENTARY ROCKS

For at least 3.8 billion years of the Earth's history, sediments have accumulated on the surface of the planet. Laid down layer upon layer and separated by bedding planes, they accumulate in sequences, with the oldest layer at the bottom. With deep burial, they are compacted (lithified) into sedimentary rock, occasionally disturbed by folding or faulting, and returned to the surface again. These processes were first demonstrated by Nicolaus Steno (see panel, below). The Earth's depositional environments vary enormously, producing many kinds of sediments and sedimentary rocks. The sands and muds from ancient rivers, deserts, and seas, along with the deposits associated with glaciers and volcanoes, are all preserved in the Earth's stratigraphic record. The original sediment is mainly mineral and rock debris from weathering, erosion, and transport by water (for example, sandstone), wind (loess), glacial and mass movement (tillite), biochemical activity (limestone), or evaporation (salt deposits). Once laid down, sediments may be altered by a variety of processes. Water is driven off, mainly due to compaction, and further changes over time transform the soft sediment into hard and brittle layered rocks such as sandstone, limestone, and shale.

**QUARTZ SANDSTONE**
*At a magnification of 25x, the individual quartz grains of this sandstone can be seen to be held together by tiny but well-formed crystals of a quartz cement—the small, angular, surface protrusions.*

**WATER-LAID SEDIMENTARY ROCKS**
*The alternation of limestone and shale deposits at Lyme Regis, southern England, show that conditions of sedimentation kept changing from carbonate to that of fine-grained mud deposition.*

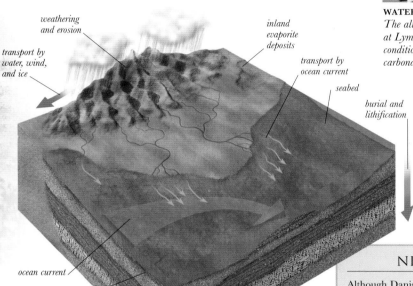

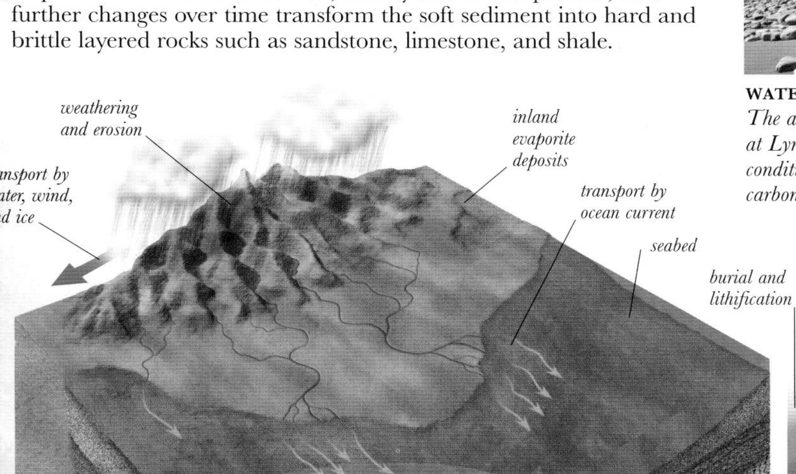


**SEDIMENTARY ROCK FORMATION**
*The creation of sedimentary rocks occurs in overlapping stages. Rocks are weathered and eroded on land to create sediments that ultimately are transported to the sea. There, further transport may occur before deposition, burial, and lithification, the process that turns loose sediment into hard sedimentary rock.*

**ANCIENT FRESHWATER DEPOSITS**
*Many old sedimentary rocks, like these in Bryce Canyon, Utah, are derived from sediments originally deposited in rivers and lakes.*

## NICOLAUS STENO

Although Danish, Nicolaus Steno (1638–86) studied in Italy and became physician to the Grand Duke of Tuscany. While there, he demonstrated the principle of superposition and showed that sedimentary rock strata were originally laid horizontally on top of one another by the deposition of particles from a fluid. Consequently, in any series of strata, younger layers lie on older ones. He also ascertained that tilted and deformed strata result from displacement by the Earth's movements after deposition, and proved that fossils are the remains of once-living organisms.

# METEORITES

Natural objects from space made of iron and stone crash through the Earth's atmosphere all the time. They range from dust-sized particles to massive lumps weighing many tons. Such meteorites have fallen throughout the history of the Earth, and they are important because they record the birth of the Solar System 4.56 billion years ago and its subsequent history. Records of meteorite falls first appear in cuneiform writings, in about 1900 BC, while fossilized meteorites recently found in Sweden fell to the Earth more than 450 million years ago. Meteorites are made of the same elements as the Earth and Sun. These elements combine to form more than 300 different minerals, including many not found on the Earth. The abundance and types of minerals found in the meteorites are used to classify them into stones (both chondrites and achondrites), stony irons (such as pallasites), and irons.

*alloy of nickel and iron metal*

**METEORITE SECTION**
*Cut in half, this spindle-shaped meteorite (above) with its black surface shows a gray metallic interior made of intergrown crystals of nickel–iron alloy typical of iron meteorites.*

**ANTARCTIC FIND**
*In recent decades, a large number of extremely well-preserved meteorites (right) have been collected from the surface of the Antarctic ice sheet, where they have accumulated over hundreds of years.*

## ROCK PROFILES

The pages that follow contain profiles of a selection of rock types. Each profile begins with the following summary information:

**ORIGIN** (igneous) Intrusive or extrusive; (metamorphic) dynamic, thermal, or regional; (sedimentary) marine, freshwater, windblown, glacial, organic, terrestrial, residual, or volcanic; (meteorites) extraterrestrial

**COLOR** A brief description, including common variations

**GRAIN SIZE** Fine grains are less than 0.004 in (0.1 mm) in diameter; medium, 0.004–0.08 in (0.1–2 mm); coarse, more than 0.08 in (2 mm)

# IGNEOUS AND METAMORPHIC ROCKS

Apart from the core, the vast bulk of the Earth is made up of igneous rocks and metamorphic rocks formed by the cooling and crystallization of molten silicate rocks and their transformation in the solid state to new textures and assemblages of minerals. Igneous rocks can be classified according to their composition (especially the relative proportions of silica and iron-magnesium minerals). They can also be grouped according to grain size (from fine to coarse, reflecting how quickly they cooled) and color (silica-rich rocks, dominated by quartz and feldspar, tend to be pale-colored, while silica-poor rocks tend to be colored darker by iron-magnesium minerals, such as olivine, pyroxenes, and amphiboles). Also invaluable is texture (how the crystal grains relate to one another). Other characteristics, such as the mode of formation and presence of certain minerals or assemblages of minerals (for example, chlorite, kyanite, and garnet), are particularly useful in categorizing metamorphic rocks, because they reflect the temperature and pressure conditions under which they were formed.

## BUILDING WITH GRANITE

The mineral composition and structure of granite give it good load-bearing properties, so that it can be used for a variety of building purposes, such as road-building (shown here, in Egypt). Granite can be a massive rock without any planes of weakness, but cracks, called joints, may form as the rock cools and pressure is released. These joints can be exploited in quarrying; otherwise, granite has to be drilled and blasted to extract blocks. The surface of rough-hewn granite can be susceptible to chemical weathering, but when polished, granite is resistant to such weathering. The relative coarseness of the grain means that it is not suitable for fine carving.

IGNEOUS

## Granite

**ORIGIN** Intrusive **GRAIN SIZE** Mostly coarse

**COLOR** White–red, pale green-blue, gray–black

Typically, granite is a tough rock that forms pronounced topographical features and is resistant to erosion. It forms by the slow cooling of magma deep in the Earth's crust, and is composed of silica-rich minerals such as quartz, feldspar, and smaller amounts of mica and amphibole (generally hornblende). Granite's color varies from white to red, pale green or blue, even black, depending on the colors of the minerals and the texture. Most crystals are a fraction of an inch in size and interlock with one another, but some minerals such as feldspar grow bigger, up to several inches. Granite is composed of about 70 percent silica ($SiO_2$) and more than 10 percent quartz, and forms large crystals. Overall, granite does not vary greatly in its mineral composition, except near the margins of granite intrusions, where blocks called xenoliths, which are derived from the surrounding rocks, can be found in various stages of assimilation. Granite forms large intrusive masses called batholiths, with volumes of tens to thousands of cubic miles, yet it can also occur as a lens, vein, or dike measured in yards. Because granite forms between one and several miles beneath the Earth's surface, it is exposed only after long periods of erosion or tectonic movements of the Earth's crust. Feldspars in granite break down to form kaolin-rich clays on the Earth's surface.

**MAGNIFIED SECTION**
*Brown biotite crystals set in a matrix of quartz and feldspar (both clear-colored) are typical of plutonic igneous rock granite.*

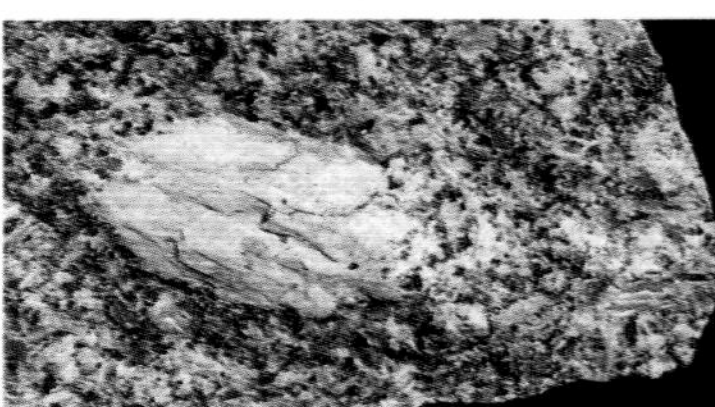

**GRANITE PORPHYRY**
*A large feldspar crystal set in smaller crystals gives this granite from southwest England a texture known as porphyritic.*

**TEXTURES AND COLORS**
*Different mineral compositions cause the pink and red coloring in the two coarse-grained granites shown here, while a medium-grained granophyre has more amphiboles and dark micas than usual.*

*ferro-magnesian minerals create dark color*

**GRANOPHYRE**

*biotite mica*

**PINK GRANITE**

*orthoclase feldspar crystals*

**RED GRANITE**

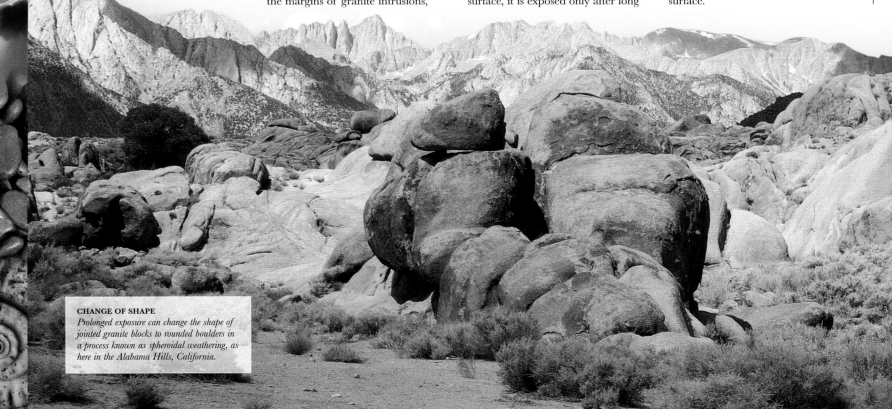

**CHANGE OF SHAPE**
*Prolonged exposure can change the shape of jointed granite blocks to rounded boulders in a process known as spheroidal weathering, as here in the Alabama Hills, California.*

IGNEOUS

## Pegmatite

**ORIGIN** Intrusive **GRAIN SIZE** Coarse

**COLOR** White, pale pink, and pale gray

After most of a granitic intrusion has cooled and crystallized, the residual liquid part of the magma solidifies into extremely coarse-grained rocks called pegmatites. Pegmatites typically occur as sheet- or lens-shaped bodies at the margins of granitic plutons. Like granites, most pegmatites are composed of quartz, feldspars, and micas. Pegmatites are an important source of industrial minerals and rare elements such as boron, beryllium, and lithium. They also form gem minerals such as tourmaline and beryl.

IGNEOUS

## Granodiorite

**ORIGIN** Intrusive **GRAIN SIZE** Coarse

**COLOR** White, gray, and black

Granodiorite is the most voluminous of all intrusive igneous rocks in the continental crust, and its composition is thought to approximate the average composition of all rocks in the continental crust. With the addition of iron and magnesium, granite grades into granodiorite and then diorite. Like granite, granodiorite forms large intrusions within the continental crust, but it is slightly darker in color as a result of its composition: it has more plagioclase (over 65 percent) than potassic feldspar and slightly more (up to 40 percent) dark minerals such as biotite mica. A tough rock, granodiorite is widely quarried, cut, and crushed for use as facings for buildings, curbstones, and road surfaces.

IGNEOUS

## Diabase

**ORIGIN** Intrusive **GRAIN SIZE** Medium

**COLOR** Dark green, gray, and black

Broadly similar in composition to gabbro, diabase typically comprises pyroxene, plagioclase feldspar (generally labradorite), and iron-titanium oxides such as ilmenite. Varieties of diabase range from those with some quartz to those with large amounts of olivine or hornblende. Diabase usually occurs as sheetlike or circular, plug-shaped intrusions related to a volcano. In some cases, the rocks surrounding such intrusions have been thermally metamorphosed. This suggests that many diabase intrusions have acted as pathways for magma feeding extrusive lavas or further intrusive and extensive dikes and sills. Diabase intrusions occasionally include exotic blocks (called xenoliths) of coarse-grained igneous or metamorphic rocks that have been raised from great depths. These are of particular interest because they provide samples of rocks from the base of the continental crust and possibly the upper mantle, which are not normally available. The mineral grains of diabase interlock, to produce a very compact rock with high crushing strength. It is therefore often used as aggregate in load-bearing layers of roadbeds.

**MINERAL CONTENT**
*The characteristic medium-grained texture and dark color of diabase rocks are derived from their abundant amphibole and mica minerals.*

**DIABASE DIKES**
*This roadcut in Jordan reveals a network of branching and rejoining dark diabase dikes in a surrounding pink host rock.*



IGNEOUS

## Gabbro

**ORIGIN** Intrusive **GRAIN SIZE** Coarse

**COLOR** Gray–black

Named after a town in Tuscany, Italy, by Christian Leopold von Buch (see panel, right), gabbro is a coarse-grained, plutonic rock with a high percentage (20–65 percent) of dark minerals, especially pyroxene and olivine. As these dark minerals increase at the expense of feldspar, gabbro grades into peridotite. The light-colored minerals are mostly plagioclase feldspar (35–80 percent). Gabbro is found in continental and oceanic crusts, but it is a more important constituent of the latter, where it forms as part of what is known as the ophiolite sequence. Below ocean-floor sediments, there are pillow lavas, sheeted dikes, and then gabbro. This has developed from the slow cooling of basaltic magma at depth. Gabbro can form plutons, or sheetlike bodies. Gravity segregation of the crystallizing minerals has produced some layered gabbros such as the Bushveld Complex of South Africa, which contains economically important minerals such as chromite. Gabbro is a tough, compact rock and is used as an ornamental facing stone, especially when it contains an iridescent variety of labradorite feldspar.

### CHRISTIAN LEOPOLD VON BUCH

One of the most eminent geologists of all time, Christian Leopold von Buch (1774–1852) studied mineralogy with Abraham Gottlob Werner (see p.63) at Freiberg, where one of his fellow students was the famous geologist and geographer Alexander von Humboldt (see p.159). Von Buch traveled widely and published on many geologically related topics. His early studies focused on volcanism, but his later interests were in fossils and stratigraphy. He effectively defined the Jurassic system of strata in his 1839 book *Über den Jur in Deutschland.*

**GABBRO IN ANTARCTICA**
*Layered gabbro intrusions in Antarctica are here cut through by dikes, which are compositionally similar but finer-grained.*

**TRANSFORMATION OF MINERALS**
*White plagioclase feldspar and dark minerals such as pyroxene (shown here) typically crystallize to form gabbro from slow-cooling, silica-poor magmas.*

IGNEOUS

## Diorite

**ORIGIN** Intrusive **GRAIN SIZE** Coarse

**COLOR** Black and white

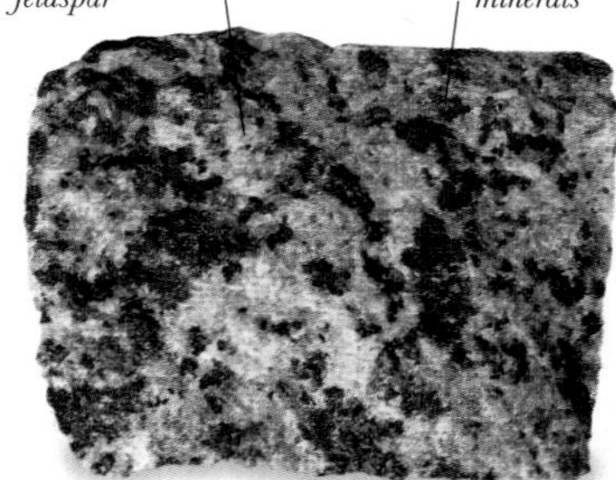


Less silica-rich than granodiorite, diorite is the intrusive equivalent of andesitic lava and has little or no quartz (less than 20 percent). It is composed essentially of plagioclase feldspar and has more dark (mafic) minerals, such as amphibole and pyroxene (up to 50 percent), than granodiorite, of which it is a compositional variant. As a result, diorite is darker in color and can be difficult to distinguish from gabbro, which it also grades into. Diorite often contains fragments (xenoliths) of other rocks, especially gabbro. This tough rock is used for ornamental purposes and as road metal.

IGNEOUS

## Lamprophyre

**ORIGIN** Intrusive and extrusive

**GRAIN SIZE** Fine–coarse **COLOR** Dark gray

The name lamprophyre is derived from Greek and means "glistening porphyry," referring to its texture. The rapidly cooled matrix is typically very fine-grained, but it is studded with larger phenocrysts of mica or amphibole (or both). These two water-bearing minerals with strong cleavages give lamprophyre its glistening porphyritic texture. The restriction of feldspars to the matrix and the presence of phenocrystic mica and amphibole attest to the exceptionally high water content of lamprophyric magmas. Lamprophyre typically forms late-stage dikes and sills cutting across granitic plutons. Volcanic types are also known. (These include the scoria shown below—this is a type of volcanic glass from a late-Pleistocene cinder cone in Mexico.)

IGNEOUS

## Lamproite

**ORIGIN** Intrusive and extrusive **GRAIN SIZE** Coarse

**COLOR** Dark gray

Lamproites are volcanic igneous rocks that contain diamonds. They are rich in potassium but silica-poor, containing 45–55 percent by weight of silicon dioxide. Lamproite generally overlaps in composition with kimberlite and lamprophyre, and may be subdivided into olivine and leucite lamproite. Major minerals include olivine, leucite, and phlogopite mica, as well as accessory minerals such as calcite, apatite, serpentine, spinel, and diamond—but only olivine lamproite contains diamonds. Lamproite typically occurs as intrusive dikes or extrusive flows, but may also form in diatremes or pipes on the margins of ancient shields or cratons. A lamproite pipe in the Kimberley region of Australia produced 40 percent by volume of the world's diamonds at its peak.

IGNEOUS

## Kimberlite

**ORIGIN** Intrusive **GRAIN SIZE** Medium–coarse

**COLOR** Bluish-gray to greenish-gray and black

The world's primary source of diamonds is kimberlite pipes. These steep-sided, deep volcanic conduits often occur in clusters, which merge at depth. Kimberlite is a dark-colored, originally coarse-grained, ultramafic rock with variable composition. It is often fragmented during the process of intrusion. The predominant minerals are olivine, mica, pyrope garnet, orthopyroxene, and a variety of accessory minerals, including diamond. The matrix is generally altered to serpentine, chlorite, and carbonate minerals. Kimberlite pipes also often include exotic rocks, known as xenoliths, which are derived from the Earth's crust and upper mantle.

dark matrix

xenolith

### MINING KIMBERLITE

Diamonds are rare but exceedingly valuable minerals within kimberlite, so it is worth processing many tons of rock to extract a few diamonds. Because of the near-vertical walls of kimberlite pipes, the process of mining them for their diamonds, as here in Kimberley, South Africa, has resulted in some of the biggest and deepest artificial holes in the world.

**A SOURCE OF DIAMONDS**
*Diamond-bearing kimberlite forms volcanic pipes, less than a half-mile wide, that extend to great depths.*

IGNEOUS/METAMORPHIC

## Peridotite

**ORIGIN** Intrusive

**GRAIN SIZE** Coarse **COLOR** Dark green to black

Peridotite is the main rock of the Earth's upper mantle. Most peridotite is a dark-colored, dense rock that slowly cooled at depth, and it is often found layered with gabbro. It is mainly composed of olivine and pyroxene, along with garnet or spinel, and more rarely amphibole and phlogopite. Peridotite occurs as sheetlike rocks associated with other mafic to ultramafic rocks typical in oceanic crust sequences, or in large massifs within mountain belts. It is also found as xenoliths brought up by alkali-rich basaltic magmas.

**CANADIAN COASTLINE**
*These layered ultramafic rocks at Lobster Cove, Canada, include peridotite.*

IGNEOUS

## Basalt

**ORIGIN** Intrusive and extrusive **GRAIN SIZE** Fine

**COLOR** Grayish-black to black when fresh

The most common igneous rock on the Earth's surface, basalt forms the rock floor of most of the oceans. It also occurs in continental settings as extensive plateau basalts, such as in the Columbia River region of the northwest US (see p.176). In both settings, basalt is found as intrusive and extrusive bodies of rock. This fine-grained, dark-colored equivalent of gabbro is dominated by pyroxene, olivine, plagioclase, and accessory iron-titanium oxide minerals such as magnetite and ilmenite. Basalt may include some large crystals, typically olivine or plagioclase. It also commonly contains former gas bubbles, known as vesicles, which may become filled with zeolite, carbonate, or agate by a secondary process. Basalt weathers to pale green, brown, or gray or, where oxidized, to red. It occurs mostly as extrusive lavas, but also forms intrusions such as dikes and sills. On cooling, it may form distinctive, jointed columns. The occurrence of intrusions at sites such as the basaltic lavas of the Giant's Causeway in Northern Ireland and the nearby Fingal's Cave on the Scottish island of Staffa (see left) has fascinated scientists and artists alike for hundreds of years. When fresh, basalt makes a high-grade aggregate, and it has been used as an ornamental stone since ancient Egyptian times.

fine-grained matrix

small crystals

**FINE-GRAINED TEXTURE**
*The crystalline texture of basalt is so fine-grained that individual crystals are not visible to the unaided eye, but they give basalt a tough, compact structure.*

**FINGAL'S CAVE**
*The dramatic scenery surrounding Fingal's Cave in Scotland inspired the classical composer Felix Mendelssohn to write his now-famous Overture.*

**AMYGDALOIDAL BASALT**
*Amygdaloidal means almond-shaped and refers to the gas bubbles commonly found in basaltic lavas, which are often infilled with secondary minerals.*

IGNEOUS

## Dacite

**ORIGIN** Mostly extrusive **GRAIN SIZE** Fine–coarse

**COLOR** Pale to medium shades of gray and pink, through to brown

Named after a Roman province in Romania, dacite is a light- to medium-colored volcanic rock with about 65 percent by weight of silica. It occurs as extrusive lavas, small intrusions such as domes and plugs, or pumice and ash, having been formed by the rapid cooling of viscous magma extruded at 1,500– 1,700°F (800–900°C). In 1980, Mount St. Helens, Washington, erupted dacite pumice and ash, as did Mount Pinatubo, in the Philippines, in 1991. Typically porphyritic with phenocrysts of plagioclase, quartz, hornblende, or biotite mica, colored varieties have been used historically as ornamental stones, and currently as aggregates.

IGNEOUS

## Obsidian

**ORIGIN** Mostly extrusive **GRAIN SIZE** Fine

**COLOR** Greenish-black to black when fresh

Rapid cooling of highly viscous, hot, rhyolitic lava before individual minerals have had time to form crystals gives obsidian its overall glassy texture. Typically, this dense, opaque to transparent rock is dark-colored, due to the presence of very fine-grained iron-titanium oxide minerals such as magnetite. Small crystals of feldspar or quartz may form, and the rock sometimes shows flow banding with alternating glassy and devitrified layers. Obsidian occurs in a number of different volcanic settings, ranging from glassy surfaces to voluminous lavas (such as the 1,300-year-old Big Obsidian Flow in Newberry, Oregon), small intrusions, and the outer layers of lava domes. Obsidian has been used in jewelry since ancient times: for example, the black eyes of the young King Tutankhamen's gold funerary mask were made from polished obsidian.

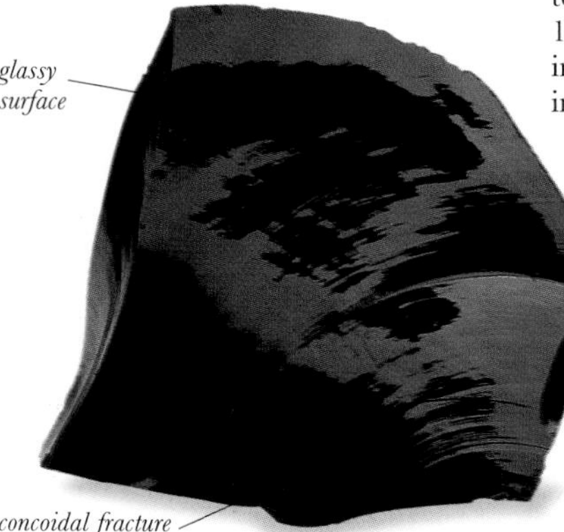


**REFLECTIVE IMAGE**
*The smooth, rounded, cracked surface of obsidian is typically dark-colored and slightly translucent, with a glassy texture similar to that of manufactured glass.*

### OBSIDIAN TOOLS

For prehistoric peoples, obsidian was ideal for making very sharp blades and points because of its glassy texture, conchoidal fracture, and considerable hardness. Broken shards of obsidian were extremely sharp—though fragile—but if damaged they could be reworked to give new, equally sharp edges. Obsidian, for example, was often used in the technically sophisticated Clovis-point weapons adopted by hunters in the Americas from the tenth millennium BC.

IGNEOUS

## Andesite

**ORIGIN** Mostly extrusive **GRAIN SIZE** Fine–coarse

**COLOR** Variable, including brown-purple to gray

Named after the Andes Mountains of South America, andesite is a fine-grained to porphyritic volcanic rock common in many of the world's subduction-related volcanic arcs. Andesite is typically about 60 percent silica by weight and erupts at temperatures of 1,750–1,800°F (950–1,000°C). Plagioclase feldspar commonly forms prominent phenocrysts accompanied by pyroxenes, iron-titanium oxides, and, in some andesites, the hydrous minerals amphibole or biotite (or both). Andesitic magma can erupt explosively, but typically the gas propellant bleeds away from the magma prior to eruption as sluggish and viscous block-lava flows.

IGNEOUS

## Rhyolite

**ORIGIN** Mostly extrusive **GRAIN SIZE** Fine

**COLOR** White to pale gray, green, pink, and brown

This light-colored, often banded rock is rich in sodium, potassium, and silica (70–78 percent by weight). It is primarily found in continental settings. Rhyolite contains a fine-grained matrix, which may be glassy and sometimes includes visible phenocrysts (larger crystals) of feldspar, quartz, or mica. Color banding and parallel phenocrysts within the rhyolite indicate the direction of the magma flow in the conduit through which the lava was released. Silica-rich rhyolite magma is relatively cool, emerging at temperatures as low as 1,300–1,500°F (700–800°C). Due to its high silica content, rhyolitic magma is extremely viscous, which causes it to flow slowly and to retain its gases, because gas bubbles cannot effectively rise through the viscous liquid. If rhyolite has a low gas content as it nears the surface, it erupts to form thick, sluggish lava flows that tend to pile up around the vent, forming domelike constructions. More commonly, rhyolite is rich in gas as it nears the surface and erupts explosively. The once-dissolved volatiles, mainly water and carbon dioxide, expand rapidly into the atmosphere. Expansion froths the magma and rips it apart to form pumice and ash, and propels the fragmented magma out of the vent at high speed. The hot eruption column entrains and heats surrounding air, which can cause the plume to rise buoyantly into the stratosphere. In other cases, the margins of the plume collapse downward to form destructive pyroclastic flows. Prehistoric rhyolitic eruptions, such as those that have occurred at Yellowstone in the past 2.1 million years, have ejected up to several thousand cubic miles of magma.

**TWO CRYSTAL SIZES**
*Although characteristically fine-grained and pale-colored, rhyolite may contain a variety of slightly larger crystals within its matrix.*

**THICK LAVA FLOWS**
*Silica-rich rhyolite lavas are viscous and so do not easily flow any great distance after expulsion. Here, in the Rhyolite Hills, Iceland, they have therefore formed very thick flows.*

IGNEOUS

## Volcanic deposits

**ORIGIN** Extrusive **GRAIN SIZE** Fine–coarse

**COLOR** Very variable, including white, gray–black

**VOLCANIC DEBRIS**
*Fine ash through to pebble-sized lapilli are among debris erupted from volcanoes that falls back to Earth.*

The material erupted by volcanoes varies according to the magma from which it was derived. As recognized by James Hutton (see panel, right), these deposits range from effusive lavas to a variety of other forms. These other forms are collectively known as volcanic deposits, and they include scoria, pumice, ash-fall, pyroclastic-flow, and debris-flow deposits. The nature of each volcanic deposit depends on the chemical composition, gas content, and temperature of its magma during eruption. Silica-poor magmas such as basalt have low viscosities and high temperatures.

COARSE ASH

LAPILLI (PUMICE)

FINE ASH

Consequently, basaltic magmas are very fluid and, on eruption, develop a wide range of fluidal shapes. Clots of basalt thrown through the air, called bombs, can develop aerodynamic shapes (see below). Fast-moving basaltic lava flows can form blocks with jagged, spiny surfaces known by the Hawaiian name a'a, while slower-moving basalts have smooth and ropy textures known by the Hawaiian name pahoehoe. When basalt erupts below water, it forms stacked, bulbous masses called pillow basalts. Magmas richer in silica (andesite, dacite, and rhyolite) are more viscous, which inhibits their ability to degas. Nonetheless, many silica-rich magmas do lose their gases during ascent, and erupt to form sluggish and thick block-lava flows. Other silica-rich magmas, however, reach the surface with high contents of water, carbon dioxide, and other gases. These explosively fragment the magma into pumice and ash, sending eruptive clouds high into the atmosphere, where they are blown downwind and begin to fall back to Earth. The largest, densest particles fall closest to the vent, and finer and less dense particles drop at progressively greater distances, forming thinner deposits. The margins of the eruptive column can collapse to form pyroclastic flows (see panel, above). Such flows can travel rapidly downhill, coming to rest as accumulations of pumice, ash, and rock fragments called pyroclastic-flow deposits. When stacked, the pressure in the hot interior bonds the deposit into welded tuff.

gas-bubble vesicle

**PUMICE**
*This gas-frothed laval rock is filled with numerous cavities, or vesicles, and is so light that it can float.*

**PYROCLASTIC LAYERS**
*These layers of ash and pebble-sized rock and pumice pieces have piled up around the volcano on the Italian island of Lipari.*

### PYROCLASTIC FLOWS

Our knowledge of volcanic phenomena such as pyroclastic flows has generally come from observations made during or following important recent eruptions. These dramatic and deadly phenomena were only dimly perceived prior to the eruptions of the Caribbean volcanoes Mount Pelée and Soufrière St. Vincent in 1902, when pyroclastic flows (also known by the French term *nuées ardentes*, or "glowing clouds") were observed and photographed as they descended canyons on the flanks of the volcanoes. Volcanologists came to realize that pyroclastic flows are hot mixtures of gas, pumice, ash, and old rock fragments that move downhill as turbulent, gravity-driven flows. Insight from these eruptions led to the realization that many ancient deposits of pumice and ash (such as those from the eruption of Vesuvius in AD 79, see p.186), which were formerly thought to have fallen through the air, were actually pyroclastic-flow deposits.

**PYROCLASTIC-FLOW DEPOSIT**
*Layers of ash with different-sized particles, such as these from around Herculaneum, Italy, are now interpreted as deposits of pyroclastic flows.*

### JAMES HUTTON

One of the most important Earth scientists of the 18th century, James Hutton (1726–79) was born in Scotland and educated in chemistry and medicine in Edinburgh, Paris, and Leiden. He studied volcanic rocks in particular and showed that both intrusive and extrusive rocks are of igneous origin. His *Theory of the Earth*, published in 1785–88, was particularly influential.

**VOLCANIC BOMB**
*Molten, low-viscosity basaltic magma blasted through the air can assume aerodynamic forms, such as shown by this streamlined volcanic bomb.*

*spindle-shaped*

*piled droplets*

**LAVA STALAGMITE**
*When fluid, molten basaltic lava drips into a lava tube, it can cool drop by drop. As the drops land on the ground, they accumulate to form a stalagmite-like structure.*

**PAHOEHOE LAVA FLOW**
*Hot, fluid basalt forms a plastic skin, which can stretch and fold as the lava moves within it. The skin gradually hardens to a ropy-textured, glassy surface, as seen here at Kilauea volcano, Hawaii.*

METAMORPHIC

## Skarn

**ORIGIN** Thermal **GRAIN SIZE** Fine–coarse

**COLOR** Most commonly green and red, but also gray, brown, black, and white

Formed from high-temperature metamorphism and alteration of carbonate rock, skarn contains a variety of minerals rich in calcium, magnesium, and iron (such as garnet and pyroxene), as well as sulfide minerals of copper, lead, iron, zinc, and tungsten in commercially valuable quantities. Skarn deposits are found adjacent to porphyry deposits within the thermal contact zone of large igneous intrusions (such as granites) that have invaded carbonate rocks. Mineral-enriched solutions react with the surrounding rocks to form new silicate minerals and ores.

METAMORPHIC

## Serpentinite

**ORIGIN** Dynamic **GRAIN SIZE** Fine–coarse

**COLOR** Mainly green with black streaks

This generally green rock is made of the mineral serpentine and is commonly banded, blotched, or streaked with colors ranging from black to bright green and occasionally red. Serpentinite is also typically transected by veins of chrysotile asbestos. It is mainly composed of a silica-poor, hydrated magnesium-silicate, which is in turn the alteration product of ferro-magnesian minerals, especially olivine—serpentinite being derived from the igneous rock peridotite, which is rich in olivine. Other minerals present include pyroxene and iron oxides. Serpentinite is a relatively dense but soft metamorphic rock, and it is often fibrous in texture and readily deformed. The presence of serpentinite within mountain belts provides evidence of major tectonic events in the geological past, which brought slices of ocean-floor rocks or even the mantle itself into contact with the continental crust. Most serpentinite is found in convergence zones between tectonic plates, such as within the Alpine–Himalayan mountain belt. These are often the only remaining traces of the ocean basin that once existed between two continental masses that are now firmly joined together (or sutured). Substantial exposures of serpentinite include those found at the Bay of Islands, in Newfoundland, Canada, and those in Oman, on the Arabian Peninsula. For example, the parent rocks of the Oman serpentinites were originally formed on the Tethys Ocean floor in mid-Cretaceous times. The closure of the Tethys Ocean resulted in part of the sea-floor rock being thrust by faults onto the Arabian continent.

**DEEP-SEA CHANGES**
*These serpentinite rocks at the Lizard Peninsula, in far southwestern England, were metamorphosed 400 million years ago.*

**ORNAMENTAL STONE**
*The attractive coloration of serpentinite rock such as this, with its greens and blacks, has led to its widespread use as an ornamental stone.*

METAMORPHIC

## Marble

**ORIGIN** Thermal **GRAIN SIZE** Fine–coarse

**COLOR** Mainly white, pink, green, brown, and black

Over millennia, marble has been valued for its smooth texture, color, and ease of working for both sculpture and building purposes. It is the product of thermal or regional metamorphism of limestone. Heat and pressure cause the carbonate minerals to recrystallize into compact and often homogenous marble of various colors. Superficially, it may be confused with quartzite (another metamorphic rock), but marble can be scratched by a steel point, and it also reacts to weak hydrochloric or acetic acid. Recrystallization tends to remove any cracks or cleavage, but traces of folds—often highly contorted—may be preserved, indicating that the rock flowed during metamorphism. Also, marble that is the product of low-grade metamorphism may preserve traces of fossils and sedimentary rocks. The close intergrowth of the carbonate minerals gives some marble a sugar-like texture that is suitable for carving (see panel, below). It is composed mainly of calcite but may contain other minerals rich in magnesium, calcium, and iron, such as serpentine, dolomite, phlogopite mica, and the amphibole tremolite, which can produce distinctive colors and textures. Decorative marble is often particularly valued for specific color and textural variations as a result of such mineral "impurities" in the matrix. As proportions of these accessory minerals increase, marble grades into other calc-silicate metamorphic rocks such as skarn. Marble is widespread through regional metamorphic terrains wherever there were original limestones. It is therefore often interlayered or found close to phyllite, quartzite, and schist. In addition, it may be occur in the thermal metamorphic zones around large igneous intrusions.

### MARBLE SCULPTURE

Because marble can be carved readily, it has been used since preclassical times for sculpture and in buildings such as the Parthenon, in Greece, and the Taj Mahal, in India. The pure white marbles of Greece and Italy have been reshaped by the greatest sculptors, from Praxiteles to Michelangelo, to produce world-renowned statues. Unfortunately, marble is prone to attack by acid rain, which, since the Industrial Revolution, has seriously damaged marble stonework on many buildings worldwide.

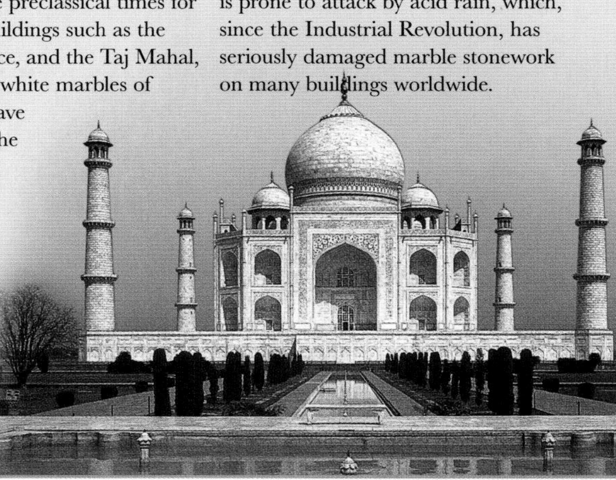

**RENOWNED**
*The Taj Mahal, in India, was built of white marble and semi-precious stones in 1632–54.*

**MARBLE TEXTURE**
*The crystalline calcium carbonate in marble gives it a compact texture that is suitable for intricate carving.*

**FLAWLESS ROCK**
*Some of the best-quality marble, such as that found in Carrara, Italy, has a pure white color and can be cut into almost any form.*

METAMORPHIC

## Hornfels

**ORIGIN** Thermal **GRAIN SIZE** Fine–medium

**COLOR** Variable, including white and shades of gray, green, blue, and black

The heat given off by large igneous intrusions can metamorphose surrounding sedimentary rocks into a new rock type called hornfels. This hard, compact stone has a fine matrix of interlocking mineral grains. Some varieties, such as pyroxene hornfels, however, are medium grained and completely recrystallized. The color of hornfels depends on the composition of the original rock, but it generally looks spotted, because of the random growth of the metamorphic minerals.

**GARNET HORNFELS**
*Rounded garnet crystals have here grown larger than the other minerals of the matrix and so have given this hornfels a distinctive spotted appearance.*

The new mineral growths (the spots) on, for example, garnet hornfels are technically known as porphyroblasts, and the texture they produce is termed porphyroblastic. Their size and shape depend on the extent of metamorphism and the minerals growing within them. Generally, the rocks are too fine-grained to allow individual minerals to be distinguished in a hand specimen. Some porphyroblasts have diagnostic shapes: for example, chiastolite grows in long prisms with a distinct cross shape of inclusions within each grain. These inclusions are seen in the direction perpendicular to the prism's long axis. Heating from an intrusion to temperatures within the range 600–1,800°F (300–1,000°C) causes solid-state changes to occur, with the progressive development of metamorphic minerals depending on the temperature. A succession of hornfels is therefore found, from low-temperature, spotted slate with porphyroblasts, which cannot be identified by eye in a hand specimen, through cordierite hornfels, which may be in direct contact with the intrusion or pass into a higher-grade pyroxene hornfels, which is in direct contact. The thermal contact zone around even very large granite intrusions may be surprisingly narrow (3 ft/1 m or so thick), showing that the heat from the intrusion dissipated rapidly away from the contact. The name hornfels is an old German miners' term meaning "horn rock."

**CHIASTOLITE HORNFELS**
*Long, square-sectioned needles of chiastolite here grow randomly through a fine-grained matrix.*

**PYROXENE HORNFELS**
*The pyroxene crystals in this hornfels specimen can be distinguished only with the aid of a microscope.*

METAMORPHIC

## Migmatite

**ORIGIN** Regional **GRAIN SIZE** Medium–coarse

**COLOR** Dark gray, white, and pink

Migmatite is formed at great depth in the continental crust and consists of light-colored (leucocratic) and dark-colored layers. Unlike gneiss, which it superficially resembles, migmatite shows evidence of partial melting either by the *in situ* production of granite from a preexisting rock or by the fine-scale injection of granite veins. The latter appears as small clusters of crystals, bigger lenses, or veins, which may be layered or highly folded. Folded structures in migmatites indicate that deformation is common in partially molten bodies of rock. In 1907, Finnish geologist Jakob Johannes Sederholm gave migmatite its name, which is derived from the Greek *migma* ("mixture").

METAMORPHIC

## Slate

**ORIGIN** Regional **GRAIN SIZE** Fine

**COLOR** Generally gray, also tinged green or purple

This compact rock is characterized by the way in which it may be split into thin sheets along parallel planes of cleavage. This property—along with slate's waterproof quality—have led to its widespread use as a building material, especially as roofing slate (see panel, below). The formation of slates occurs during low-grade metamorphism of mud-rocks (fine-grained, clay-rich sediments such as mudstones, shales, siltstones, and fine-grained volcanic deposits such as layered tuffs). When mud-rock is subjected to compression, it is reduced in volume as water is driven out and the minute clay mineral grains are all reoriented at right angles to the pressure direction. This can happen irrespective of any original bedding planes within the sediment or any shape they are folded into. Cleavage formation can destroy any original sedimentary structures or fossils entombed within the rock. When the cleavage is closely parallel to the original sedimentary bedding, however, fossils may be preserved, even if considerably distorted. Since the cleavage is generally a product of the same forces that produce folding, there is an important geometrical relationship between the cleavage and fold geometry. Like minerals of the mica group, the clay minerals have a tabular (or platy) shape and a well-developed plane of cleavage. When they are orientated parallel to one another, the whole rock acquires the same parallel slaty cleavage, as it is known. Mica can be a constituent of slate; both primary and synmetamorphic mica grains define some slaty cleavage. Slate may vary enormously in color but typically is gray, purplish, or greenish-gray. Quarrymen have used the word slate in a much wider sense to describe any fissile rock that can be used as roofing material: for example, Stonesfield slate from Oxfordshire, England, is an unmetamorphosed platy Jurassic limestone.

**BLACK SLATE**
*Two successive deformations have produced this slate—the second, weaker one having realigned some of the minerals diagonally.*

*poorly developed cleavage*

*brachiopod shell*

**FOSSIL-IMPRINTED SLATE**
*Occasionally, fossils like this Paleozoic shell are preserved within slates. They can be difficult to identify if they are highly flattened and distorted.*

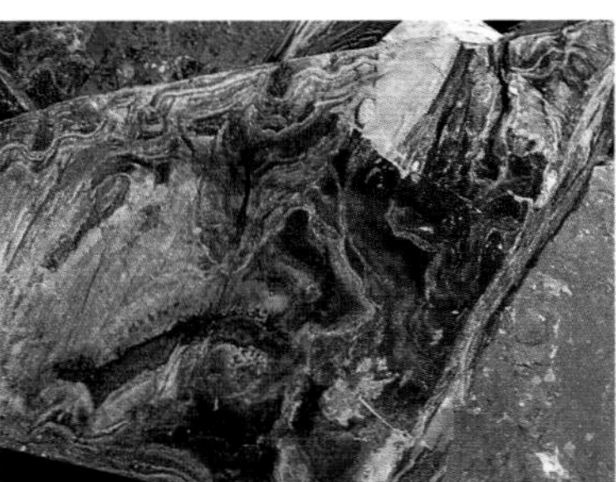

**COLOR BANDING**
*Cleavage surfaces of slates are often coated with mineral deposits such as the metals iron and copper, as here in Pembrokeshire, UK.*

### SLATE ROOFS

When highly compacted and cleaved into slates, mud-rocks make ideal waterproof roofing materials. They can be split into thin sheets, even as fine as 3/16 in (5 mm), which are not too heavy and can be trimmed to shape. Good-quality slate shingles can last for hundreds of years. Impurities such pyrite (iron sulfide), which are common and can be hard to detect in the slate, gradually erode and can make the slate porous. The craft of hand- splitting and trimming slates is arduous, labor-intensive, and unhealthy (because the dust is harmful), and is no longer practiced to the extent it once was— most commercial "slates" are now preformed and made of reconstituted mineral material.

**WEATHER RESISTANCE**
*Well-fitting slates protect this mountain house in Piedmont, Italy, against rain and snow.*

METAMORPHIC

## Phyllite

**ORIGIN** Dynamic, low-grade regional

**GRAIN SIZE** Fine–medium

**COLOR** Greenish-gray to plain gray

With progressive metamorphism, slate grades into cleaved phyllite with a slight increase in grain size, and then into schist. Phyllite's cleavage planes have a characteristic sheen resulting from the reflection of light from the individual grains, which are barely detectable by the unaided eye. A single cleavage is generally dominant but is sometimes corrugated or folded by subsequent deformation. Phyllite therefore does not split as easily or uniformly as slate and cannot be used in the same way as a building material. As in slate, the cleavage is produced by the reorientation of clay mineral grains in response to confining pressure. Phyllite represents a greater amount of recrystallization of original clay minerals than slate to form minerals such chlorite and mica (especially muscovite). The name phyllite is derived from the Greek for "leaf," referring to the parallel cleavage.

METAMORPHIC

## Quartzite

**ORIGIN** Thermal and dynamic

**GRAIN SIZE** Fine–medium

**COLOR** Mostly white to gray

Superficially, quartzite may be confused with marble, although it is considerably harder than both the latter and most other metamorphic rocks. Consequently, quartzite often forms resistant features such as ridges in a landscape. Produced by the metamorphism of sandstone, it is generally pale-colored or white but may be in various shades of brown or gray, depending on its minor mineral constituents, especially magnetite and pyrite. Quartzite is predominantly made of quartz grains that have recrystallized. With high-grade metamorphism, however, individual grains within sandstones may be lengthened by the combination of pressure, heat, and stress, which may flatten or shear the rock. This is best seen in conglomerates, where the grains are very large. When sandstone grains recrystallize, they overgrow their original grain boundaries and closely intermesh to form compact and brittle quartzite. Original sedimentary structures and fossils may be destroyed in the process, though with low levels of metamorphism, traces of some sedimentary structures may persist, as well as outlines of some fossils. The main variation in quartzite is due to original compositional variations such as the presence of feldspar, mica, or pyrite. Because it is brittle, quartzite is often fractured by a series of joints that may act as conduits for mineral-rich fluids. Due to their relative hardness and chemical stability, eroded quartzite fragments (clasts) persist longer in sediments than most other rocks.

**HARD-WEARING**
*Quartzite is a hard, tough, and compact rock because it is almost entirely composed of quartz grains that are interlocked.*

**SHATTERED QUARTZITE**
*These broken, angular blocks of quartzite in Ramon, Israel, show how brittle this rock is, yet it is very resistant to weathering and erosion.*

METAMORPHIC

## Schist

**ORIGIN** Regional **GRAIN SIZE** Medium–coarse

**COLOR** Variable, including white and shades of gray, green, blue, brown, and black

The essential feature of schist is its parallel planes of similarly oriented minerals within the rock. This originates in the same way as cleavage, with minerals growing with a preferred orientation in response to a powerful pressure and deforming stress on the rock. Minerals that are tabular (such as mica or chlorite) or blade- or needle-shaped (such as kyanite or sillimanite) can all contribute to the formation of schistosity. Clay minerals are absent, having been recrystallized into mica and chlorite. In general, schists are all produced by relatively high pressures and temperatures typical of large-scale regional metamorphism. The result is that schist typically has medium- to coarse-grained texture, but may frequently contain larger grains of particular minerals called porphyroblasts. Some minerals tend to grow within specific pressure and temperature ranges, and consequently can be used as indicators of the conditions under which their host schist originally developed. The schist rock name is then modified by the indicator mineral—for example, biotite or garnet schist (see below). The dominant minerals for most common schists include quartz and mica plus various other metamorphic minerals such as garnet, kyanite, and staurolite. The name schist is an old German miners' term and is derived from the Greek *skhizein* (" split" or "divide"). In some mountainous regions where slates are not available, schist is used as roofing material, since it can be finely split.

**FOLDED SCHIST**
*The parallel planes in this specimen have been compressed into corrugations by later movements of the Earth.*

*shiny surface*

*crumpled cleavage*

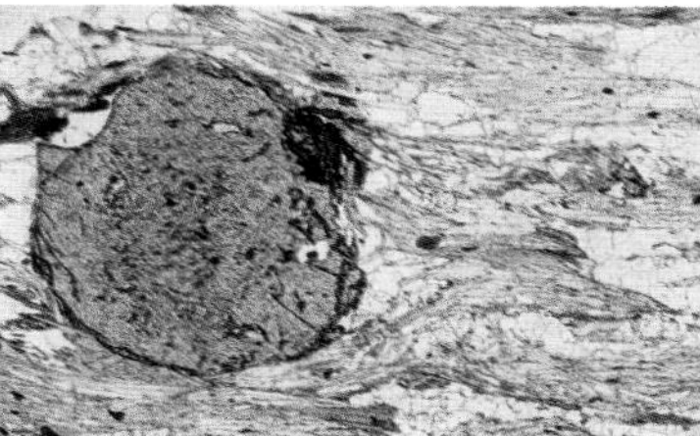

**MICROSCOPIC VIEW**
*As they grow, rounded garnet crystals form islands around which mica grains wrap, as seen in this highly magnified thin section.*

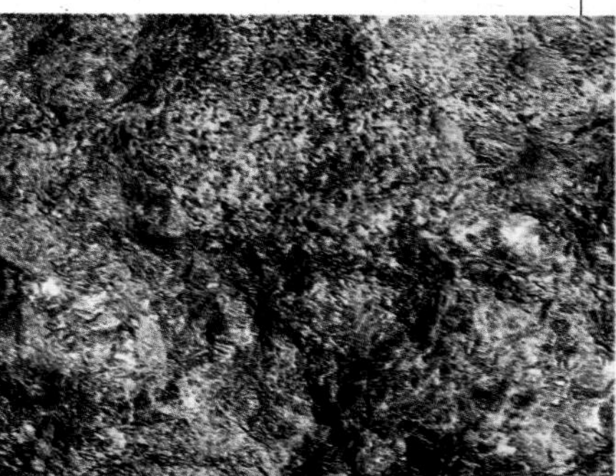

**GARNET SCHIST**
*Large, rounded garnets indicate that this schist developed at relatively high temperatures and pressures deep within the continental crust.*

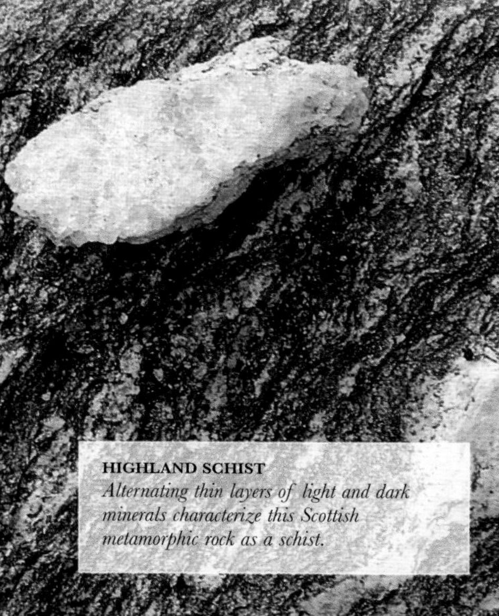

**HIGHLAND SCHIST**
*Alternating thin layers of light and dark minerals characterize this Scottish metamorphic rock as a schist.*

METAMORPHIC

## Gneiss

**ORIGIN** Regional **GRAIN SIZE** Medium–coarse
**COLOR** Pink–gray

With its characteristic alternation of thin light and dark layers, this silica-rich metamorphic rock has a feldspar content of more than 20 percent. It is the product of high-grade metamorphism at temperatures approaching the melting points of the constituent minerals, thus allowing some of them to recrystallize and produce the porphyroblastic textures seen in gneiss. Although gneiss is typically an overall pink or gray, its coloring may vary within the rock, depending on its composition. It is composed of layers, often in different colors and sometimes with a wavy, planar pattern of orientated biotite grains. Pale layers are dominated by white quartz and plagioclase feldspar, although pink feldspar may also be present. Darker bands include biotite and hornblende and tend to have parallel orientations. Some minerals, such as garnet or clustered associations of feldspar and quartz, form eye-shaped lumps as gneiss is strongly deformed (see panel, right), much as peanuts in hot taffy keep their shape as the confection is stretched. Gneiss may be derived from sedimentary or igneous rocks. The more compact varieties of gneiss are used as building stone, especially those with spectacular folds and color bands and other metamorphic textures when cut and polished. Because of their high crushing strength, these compact varieties are also used as aggregates. The name gneiss probably comes from a Slavonic word for "nest"; it was in use in medieval times by miners in Central Europe. The word was adopted—along with schist—by German mineralogist and teacher Abraham Gottlob Werner (see p.63) and his followers in the 18th century.

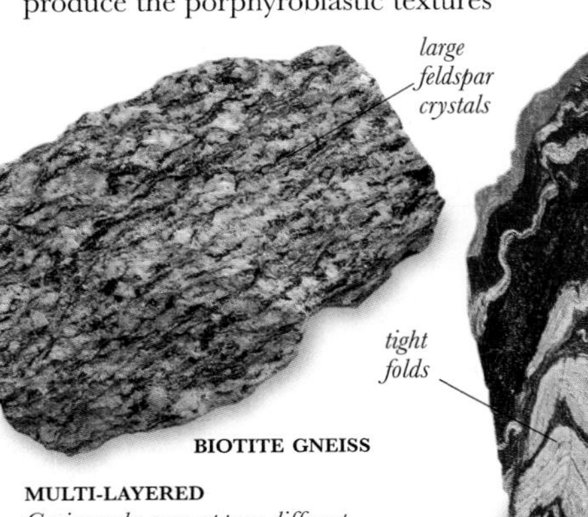


**MULTI-LAYERED**
*Gneiss rocks may appear different depending on the rocks they originated from, yet they share characteristic alternating light and dark layers.*

tight folds

**PRECAMBRIAN GNEISS**

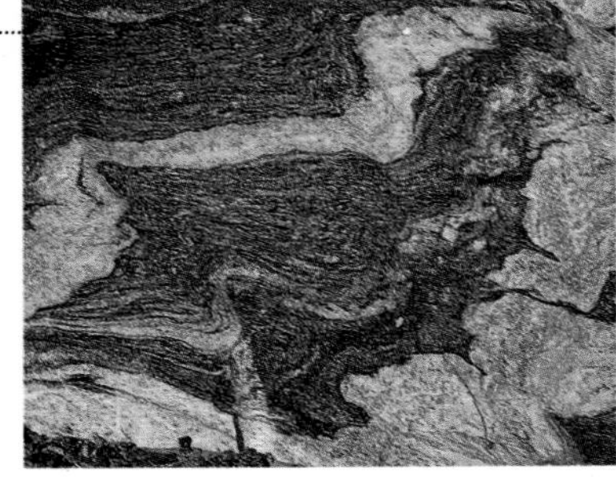

**FOLDED GNEISS**
*Alternating pale- and dark-colored layers are intensely metamorphosed and folded in this ancient gneiss from Namibia, in southwest Africa.*

### GARNET AUGEN GNEISS

Garnet is a common mineral in rocks, such as schist and gneiss, found in areas of regional metamorphism. Generally, the garnet is the iron-rich, pink variety almandine. It tends to grow as distinctive porphyroblastic "augen" (eye-shaped grains). Sometimes the garnet crystals rotate during growth to produce what are known as snowball garnets, which have internal spiral trails of small inclusions of other minerals that are visible with a microscope. Garnet minerals are useful to geologists because their occurrence within regionally metamorphosed rocks indicates that the rocks formed within certain pressure and temperature ranges. At pressures and temperatures equivalent to depths of 60 miles (100 km) beneath the Earth's surface, redistribution of aluminum among the minerals of upper-mantle peridotites leads to a progressive conversion of spinel and pyroxene into garnet.

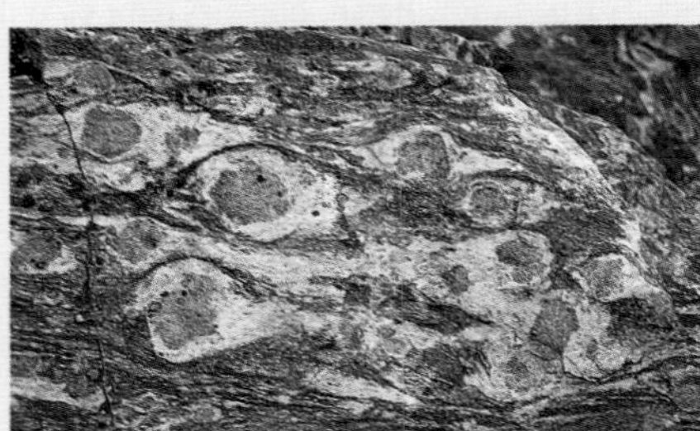

METAMORPHIC

## Eclogite

**ORIGIN** Thermal and dynamic
**GRAIN SIZE** Medium–coarse
**COLOR** A mixture of green and red

One of the densest silicate rocks known, eclogite is typically massive and formed by very high pressures and temperatures in excess of 900°F (500°C) associated with regional metamorphism at considerable depth in the Earth's crust. Although rare at the surface, it occurs as exotic blocks within serpentinite or kimberlite, as well as in mountain belts that preserve metamorphic rocks that formed in ancient subduction zones. Eclogite is formed when slabs of dense ocean-floor rocks such as basalt subduct beneath the edge of a continent or another oceanic slab. On reaching depths of 50–56 miles (80–90 km), pressure and temperature conditions cause the silica-poor, dark silicates such as amphibole to lose their water and so transform into the denser minerals of eclogite. Eclogite provides key information about the process of subduction. This rock can be striking when cut and polished, showing bright green and red from its main constituent minerals—green pyroxene (omphacite) and red garnet, and other high-pressure minerals such as kyanite.

red garnet
green pyroxene

**TYPICAL COLORATION**
*The characteristic and attractive eclogite mixture of red garnet and green omphacite pyroxene is clearly visible in this hand specimen.*

**DEEP SAMPLE**
*Dark red garnet crystals have grown as large porphyroblasts within the green pyroxene matrix of this eclogite hand specimen.*

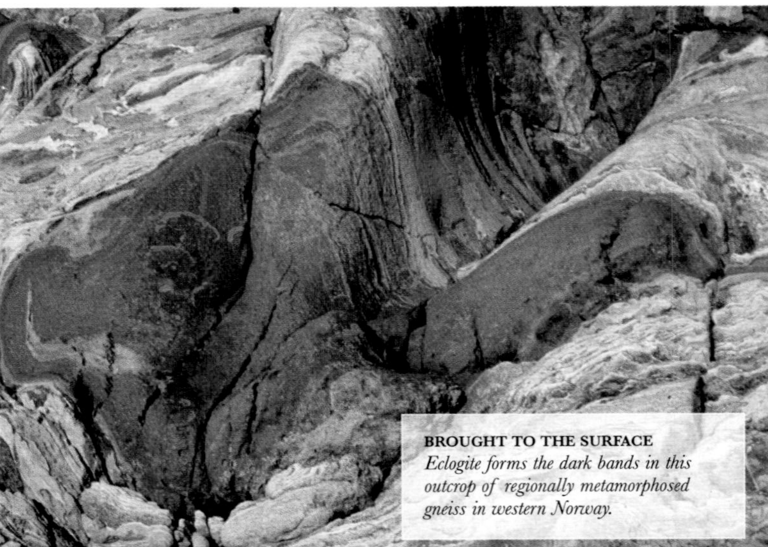

**BROUGHT TO THE SURFACE**
*Eclogite forms the dark bands in this outcrop of regionally metamorphosed gneiss in western Norway.*

METAMORPHIC

## Granulite

**ORIGIN** Regional **GRAIN SIZE** Medium–coarse
**COLOR** Brownish-gray to dark gray

This tough, massive rock is formed at the base of the continental crust under high pressure and temperature. Typically, granulite may be banded and display what is known as a granoblastic texture, in which most of the minerals form a mosaic of roughly rounded, interlocking crystals. They have variable mineralogical compositions including feldspar, quartz, and anhydrous minerals (meaning those without water) rich in iron and magnesium, such as pyroxene and garnet. Granulite forms at temperatures above 1,300°F (700°C) and over a wide range of pressures equivalent to crustal depths of 4–28 miles (7–45 km) below sea level. It is thought that much of the lower part of the continental crust is composed of granulite, which surfaces only in deeply eroded mountain belts.

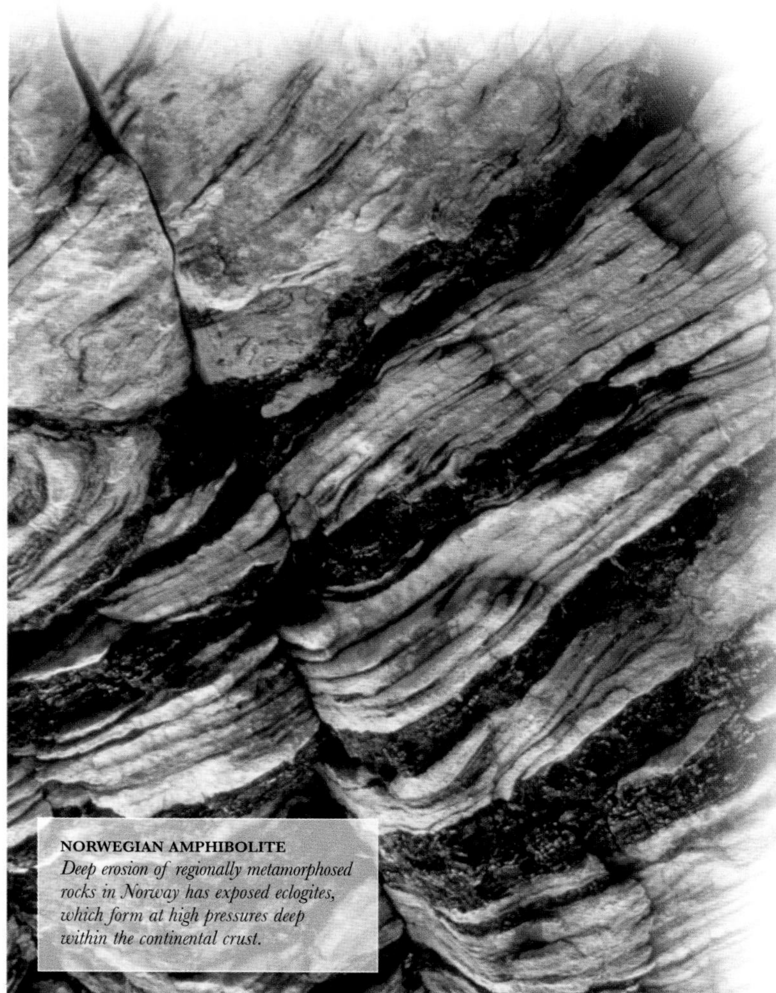
**NORWEGIAN AMPHIBOLITE**
*Deep erosion of regionally metamorphosed rocks in Norway has exposed eclogites, which form at high pressures deep within the continental crust.*

METAMORPHIC

## Amphibolite

**ORIGIN** Regional **GRAIN SIZE** Coarse
**COLOR** Green–black

Amphibolite is one of the few metamorphic rock types that are named after a single mineral (amphibole). This mineral is extremely important in metamorphic rocks that are derived from extrusive rocks (lavas), such as basalt or andesite, and intrusive rocks, such as gabbro and diorite. This is because the mineral structure of amphibole can incorporate a large variety of elements—indeed, some geologists have described amphibole as a "garbage can" or "sponge" because it may contain so many elements. Despite changes in its composition, amphibole remains a stable mineral during the metamorphism of mafic to intermediate igneous rocks. This property also explains why amphibolite is found in the higher-temperature parts of both thermal and regional metamorphic terrains worldwide, provided the original rocks were of basaltic to andesitic composition. Amphibolite is formed at moderate temperatures (900–1,300°F/500–700°C) and across a wide range of pressure conditions, which typically occur 6–19 miles (10–30 km) below sea level. The major mineral components of amphibolite are 50 percent or more hornblende amphibole (and sometimes other amphiboles, such as actinolite or tremolite) and plagioclase feldspar, as well as a variety of other minerals such as biotite and garnet. Some of these, especially garnet, may grow as porphyroblasts—relatively large crystals surrounded by smaller ones. The presence of large numbers of amphibole crystals aligned in parallel planes may give the rock a distinctive, cleavagelike pattern known as foliation. Sometimes amphibolite is divided into alternating layers of dark amphibole and light feldspar. Amphibolite's strength and compactness mean that it can also be used as a high-grade aggregate with good load-bearing properties for road construction.

**MINERAL COMPOSITION**
*Garnet crystals set in a matrix of amphibole and plagioclase crystals help identification of this metamorphic rock as an amphibolite.*

METAMORPHIC

## Mylonite

**ORIGIN** Dynamic **GRAIN SIZE** Fine
**COLOR** Variable, including gray, brown, and black

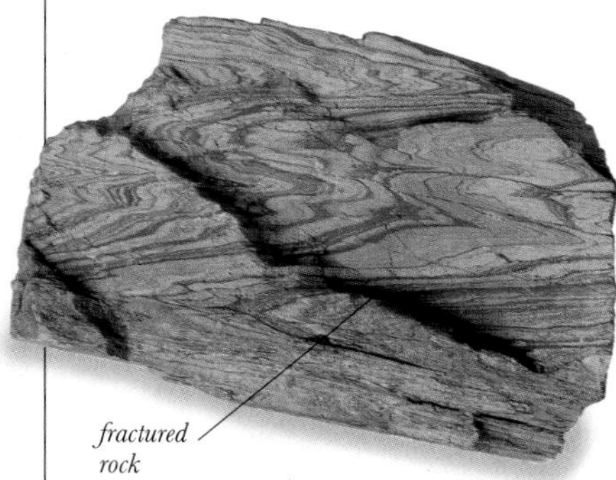


Mylonite is a banded or laminated metamorphic rock in which the original texture or fabric of the parent rock is destroyed by intense shearing deep within a fault zone. The parent rock can be of almost any kind, and its fabric is replaced by a kind of roughly developed foliation—a deformed streaky pattern known as cataclastic texture. The grains are reduced in size from those of the parent rock, and may recrystallize into a hard, compact, and often splintery rock with orientated crystal growth and occasionally eye-shaped augen porphyroblasts.

METAMORPHIC

## Jadeitite

**ORIGIN** Dynamic **GRAIN SIZE** Fine
**COLOR** Shades of green, gray, white, pink, and mauve

Jadeitite is a very hard and tough metamorphic rock formed largely from the pyroxene-group mineral jadeite, which is a sodium alumino-silicate mineral. It is usually massive and colored in various shades of green from white to almost black, and including a rare mauve-colored variety, as well as the very rare emerald-green Imperial, Kingfisher, or quetzal jade. Impurity-free jadeitite is white. The finest-quality jadeitite is translucent. Jadeitite's toughness is due to a strong network of closely interlocking prismatic crystals that formed at high pressure in crustal depths of 9–18 miles (15–30 km)—that is, in the lower part of the continental crust—but at relatively low temperatures (about 750°F/400°C). These conditions, which can be attained only in subduction zones, transform basalt into a blue-colored schist that contains the blue-colored amphibole glaucophane, plus lawsonite or epidote. Jadeite was first distinguished as a mineral in 1863, when French chemist Augustine Damour analyzed an object from the New World, which he thought was the same mineral as Chinese jade. In fact, it transpired that Chinese jade was not the sodium alumino-silicate mineral jadeite. Instead, it is a massive form of the calcium-magnesium-iron silicate mineral amphibole, sometimes referred to as nephrite actinolite.

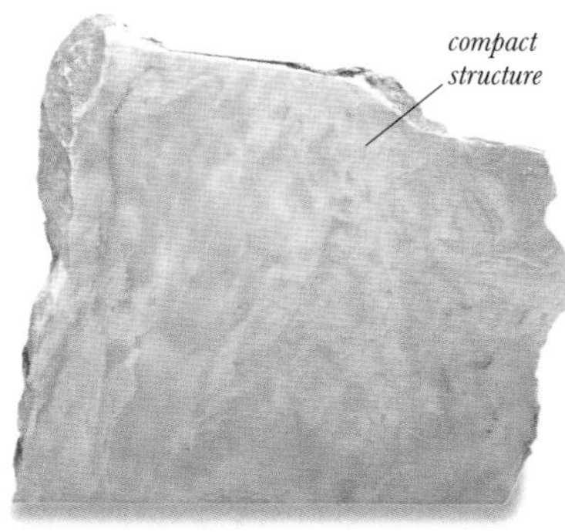

**COLOR VARIATIONS**
*Although commonly green, the color of jadeitite is extremely variable. The lavender coloring is here caused by manganese and iron impurities.*

### JADE ARTIFACTS

Because of its great toughness and beauty, the rock jadeitite, which is made of the mineral jadeite, has been valued for thousands of years, especially in China and Mexico. Wood found with jadeite artifacts in Mexico, for example, has been carbon-dated to 1500 BC. Initially, jadeitite was probably used for utilitarian purposes; it was certainly made into high-quality axes. Its rarity, combined with its durability, attractive colors, and capacity to be carved into intricate, three-dimensional forms without breaking, have since resulted in jadeitite being adopted more extensively, for ornamental and ceremonial purposes. Jadeitite is mined commercially in Burma (Myanmar), Russia, and Guatemala. The name jadeite is a derivation of *le jade*, which is a French corruption of the Spanish *pietra de ijada* ("stone of the loin or kidney").

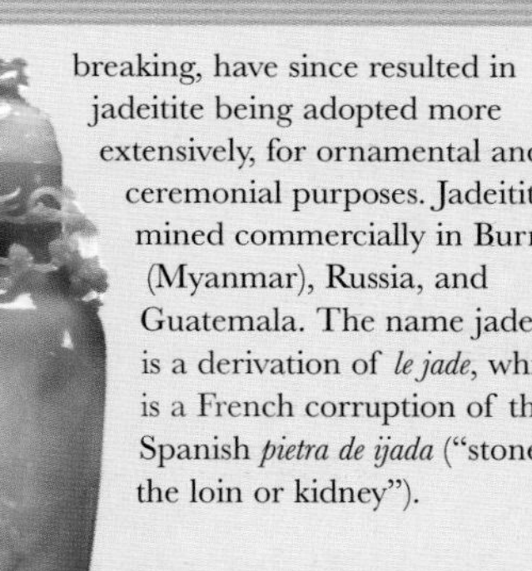
**ENHANCING THE COLORS**
*Expert carvers of jade combine natural color variations in their designs, such as the pale green and lavender seen here.*

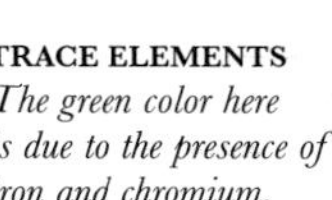

**TRACE ELEMENTS**
*The green color here is due to the presence of iron and chromium.*

# SEDIMENTARY ROCKS

These rocks are formed from deposits that accumulate under the influence of gravity, layer by layer, on the Earth's surface, and are laid down by water, wind, and ice. Those deposited in seas and oceans can be grouped as marine deposits, while those formed on land are terrestrial. Their categorization has arisen from the practicalities of geological mapping and economic use as well as scientific understanding. A broad division is made into: clastic (or fragmental) rocks; and chemical and biochemical deposits. Clastic rocks are subdivided by grain size from coarsest (conglomerates) to finest (mudstone), with further identification by composition. Sandstones, for example, may be separated into micaceous sandstone or arkose. Sedimentary rocks of chemical and biochemical origin are categorized according to their chemistry. Most are marine and biochemical, especially limestones (predominantly calcium carbonate) along with siliceous cherts, while those of organic carbon composition include peats and coals. Purely chemical deposits range from evaporites (such as gypsum), through iron deposits to phosphorites and dolomites, which are formed from limestones after deposition.

SEDIMENTARY

## Breccia

**ORIGIN** Glacial and terrestrial

**COLOR** Variable, depending on mineralogy

**GRAIN SIZE** Coarse

Fragments of angular rock distinguish breccia from round-grained conglomerate. In breccia, fossils are rare, and layering is not generally evident. Although this rock has a variable composition, it typically comprises immature sediment that has not been transported far. Consequently, it tends to form from sediments derived from terrestrial erosion or mass movement such as high-angle scree slopes or cliff bases (known as talus). Some frost-shattered or glacial moraine debris is made up of breccia, but the former is not normally consolidated as a rock, and the latter tends to show signs—such as ice scratches—of its glacial origin. The term breccia is also used to describe rock fragments produced by fault movement, and these are often cemented with minerals derived from hydrothermal fluids and may be of some economic value.

SEDIMENTARY

## Conglomerate

**ORIGIN** Marine, freshwater, glacial, and terrestrial

**COLOR** Variable, depending on mineralogy

**GRAIN SIZE** Coarse

Typically containing pebble-sized grains but also including boulders more than 3 ft (1 m) across, conglomerates are the coarsest-grained sedimentary rocks. They are formed through high-energy processes of sedimentation such as glaciation, gravitational movement, and rapid water flow in environments such as mountain streams and wave-battered beaches. The grains are rounded by friction and generally set in a finer-grained matrix. Composition varies enormously and depends on the parent rocks and the amount of weathering and erosion—immature conglomerates can include some soft rocks, such as mudstones or limestone. Fossils are rarely present.

SEDIMENTARY

## Arkose

**ORIGIN** Freshwater and terrestrial

**COLOR** Variable, including white, yellow, and brown

**GRAIN SIZE** Medium–coarse

This particular kind of sandstone contains a mixture of both quartz and feldspar grains (more than 25 percent of the rock), sometimes with a clay matrix and a calcite cement. Arkose generally forms where feldspar grains are particularly abundant—in deposits derived from the weathering and erosion of metamorphic and igneous rocks. Compared with quartz, the survival of feldspar grains in sedimentary environments is relatively limited, since the feldspar is chemically weathered to form clay minerals, especially in hot and humid climates. Arkose therefore forms around high-latitude mountain belts, where uplift and high rates of erosion expose feldspar-rich rocks such as granite. It is also found in rivers or alluvial plains at the base of mountains.

SEDIMENTARY

## Graywacke

**ORIGIN** Marine **COLOR** Mostly gray

**GRAIN SIZE** Fine–coarse

Analysis of graywacke's sedimentary structures reveal much about evolving marine basins of sedimentation. Graywacke is a mixture of quartz, feldspar, other minerals, and rock fragments, set in a clay matrix—the clay being mostly secondary in origin, derived from the alteration of mineral grains and rock fragments. Grains within graywacke vary in size and shape, and may be poorly arranged, with a rough grading from coarse at the bottom to fine at the top. Graywacke deposits are often well-defined layers, many square miles in area and up to 3 ft (1 m) or so thick. Interlayered with shale, they have sharp bases passing up through finer, rippled layers into shale above. Graywacke deposits are laid down by turbulent, sediment-laden currents in deep marine waters. Such submarine flows can erode the sea bed, leaving elongate hollows called flute marks. Thick piles of graywackes can survive subduction of ocean-floor rocks and be thrust onto continental margins.

**MIXED COMPOSITION**
*A typical graywacke like this one consists of mineral grains and fragments held together by clay particles.*

**PEMBROKESHIRE COASTLINE**
*These Paleozoic graywacke deposits in Abereiddy Bay, Wales, were originally laid down on the sea bed some 440 million years ago.*

SEDIMENTARY

## Siltstone

**ORIGIN** Marine, freshwater, and glacial

**COLOR** Variable, including gray, brown, reddish-brown

**GRAIN SIZE** Fine–medium

This clastic sedimentary rock typically has fine internal layering and sometimes small ripples. Its particles are smaller than fine sand but larger than clay grains. Most silt is quartz, but clay is often present. As a sediment, siltstone tends to behave more as sand than clay, and is produced by prolonged weathering, erosion, and transport of sediment in both marine and terrestrial environments, including glacial ones. Being relatively fine-grained, siltstone often contains fossils.

SEDIMENTARY

## Sandstone

**ORIGIN** Marine, freshwater, windblown, and glacial

**COLOR** Variable, including white, yellow, brown to red-black

**GRAIN SIZE** Medium

Accounting for 10–15 percent of all sedimentary rock in the Earth's crust, sandstone typically occurs as stratified layers of sand-sized particles, held together by various mineral cements. Typically porous, sandstone forms economically important reservoirs for water and hydrocarbons. Sandy sediment accumulates particle by particle, through the actions of wind and water, on land and at sea. It is deposited in different environments, from high mountain rivers and lakes to the bottom of the deep ocean. The environment in which a sandstone deposit formed is reflected in many of its characteristics, from its large-scale geometry to smaller structures and its composition. For example, river-lain sandstones are dozens to hundreds of miles long, up to several miles wide, and up to hundreds of yards thick. Locally, they appear as meandering channels, with curved lower boundaries and flat upper surfaces.

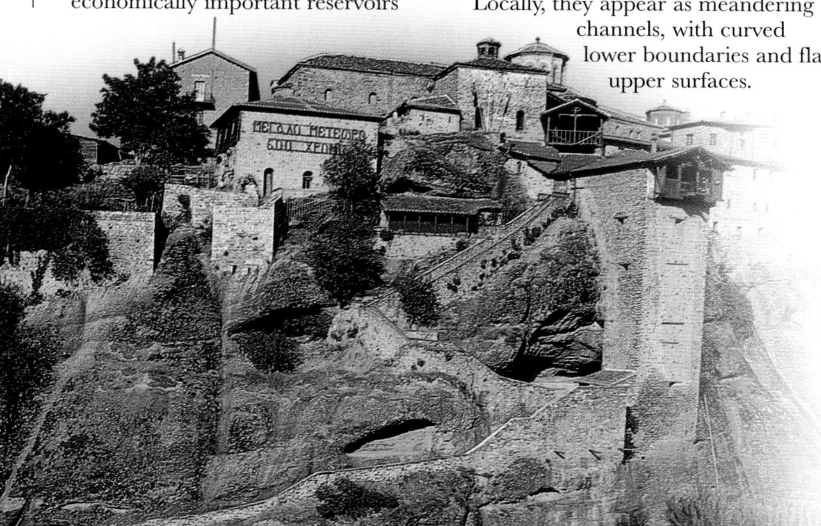

### DINOSAUR FOOTPRINTS

The fossil record of many organisms ranging from worms to dinosaurs is preserved by their trace fossils, such as burrows or footprints, left in soft, sandy or muddy sediment, as here in Namibia. Where the traces are quickly infilled by younger sediment, they can be protected from destruction, and on burial and lithification may be preserved. Subsequent uplift, erosion, and weathering of ancient sedimentary strata may then reveal the fossil traces once again. Analysis of these can tell scientists a great deal about the behavior of the long-vanished animals.

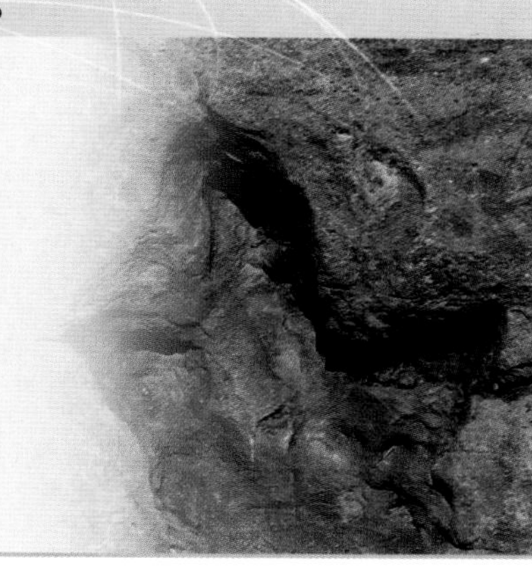

Sequences of rippled layers with well-rounded, similar-sized grains reflect the original river's current strength, varying from coarse, high-energy basal layers passing up into finer, low-energy tops and then silts and muds. Technically, the term sandstone refers to the grain size of the particles, which may be composed of any rock fragment or mineral combination. In practice, most sandstones are rich in quartz, because quartz is the most common residual mineral in sedimentary environments. Sandstone can be classified according to its grain size (fine, medium, and coarse) and composition: for example, arkose (or arkosic sandstone) is feldspar-rich. The size and roundness of the grains reflect how far they have traveled, and their composition may indicate the geology of their source rocks.

**CLIFF-TOP RETREAT**
*These Greek monasteries at Metéora perch on top of a cliff made of sandstone, which has resisted erosion over several millennia.*

**FOSSILIFEROUS**
*An ancient, sand-laden river carrying plant debris, which has been buried, will later turn into sandstone of the kind shown here.*

SEDIMENTARY

## Claystone

**ORIGIN** Marine, freshwater, and glacial

**COLOR** White through to pale shades of gray, yellow, brown, and red

**GRAIN SIZE** Fine

This common sedimentary rock is typically composed of particles less than 0.0002 in (0.004 mm) across—mostly derived from the weathering of feldspar and other minerals. The particles in claystone are predominantly silicate clay minerals, such as kaolinite, organic material, or minute calcium carbonate crystals—hence the name of the rock. Claystone is similar in grain size and composition to shale but lacks its layered structure and ability to be split. Fine-grained clay particles can be transported great distances by wind and water.

SEDIMENTARY

## Mudstone and shale

**ORIGIN** Marine, freshwater, and glacial

**COLOR** White, also pale shades of gray, yellow, brown, and red

**GRAIN SIZE** Fine

The most abundant of all sedimentary rocks, mudstone and shale are variable in composition but commonly include silt and clay minerals, organic material (of economic value in oil shale), iron oxides, and minute crystals of minerals such as pyrite and gypsum. When the grains occur randomly within the sediment, the resulting rock is a blocky mudstone with little or no layering. However, where the grains have been oriented by the steady flow of a current or by compression to produce layering along which the rock splits, the resulting rock is called a shale. Mudstone and shale generally contain well-preserved fossils.

SEDIMENTARY

## Loess

**ORIGIN** Windblown and glacial

**COLOR** Pale gray to brown

**GRAIN SIZE** Fine

Lifted by the wind from dry land surfaces, loess is a fine-grained silt and clay-sized material that is carried as airborne dust, frequently over very great distances, before being redeposited and eventually compacted to form sedimentary rock. Dry, windy landscapes lacking in vegetation, such as arid and semiarid continental interiors, are typical source regions for loess, which is finer-grained than sand. There are huge loess deposits in eastern Europe and much of Asia, all of which are derived from areas that were on the edges of ice sheets during the last ice age. Fluctuations in loess sediment can help provide a detailed record of climate change.

SEDIMENTARY

## Dolomite

**ORIGIN** Marine

**COLOR** Variable, but mostly white or pale shades of yellow, gray, or brown

**GRAIN SIZE** Fine–coarse

To distinguish it from the mineral of the same name, dolomite is also known as dolostone. This carbonate rock is mostly formed by post-depositional chemical replacement of original limestone. Calcium ions are exchanged for magnesium ions by percolating magnesium-enriched water. The original carbonate mineral is converted by this process of dolomitization into the magnesium carbonate mineral dolomite. Typically, dolomite forms massive rocks in which the original bedding and other sedimentary structures are destroyed by the process of dolomitization.

SEDIMENTARY

## Limestone

**ORIGIN** Marine and freshwater

**COLOR** Variable, but mostly white or pale shades of yellow, gray, or brown

**GRAIN SIZE** Fine–coarse

After mudstone and sandstone, limestone is the next most abundant sedimentary rock. It forms extensive, thick, multiple layers within the continental crust, and contains carbonate minerals, especially calcite (50 percent or more), which recrystallize during lithification into a hard and brittle rock. Carbonate sedimentation often involves animals or plants, and especially microorganisms. These change the local chemical environment, transforming tiny carbonate crystals (generally the mineral aragonite) into carbonate muds on shallow sea beds. A wide range of organisms—from the microscopic, single-celled foraminifera to reef-builders (such as corals) and gigantic vertebrates (for example, dinosaurs and whales)—remove calcium from their surroundings to build their shells and bones (the latter are made of calcium phosphate). Following death, carbonate remains may accumulate in sufficient quantities to form sediments such as ocean-bed muds and reef limestones. In shallow tropical seas with high evaporation rates, carbonate may also be deposited directly on the sea bed. There, the currents roll the carbonate-accreting grains around like snowballs (up to a fraction of an inch in diameter), forming carbonate sands called oolites, which are later lithified to form rocks. Shallow tropical sea carbonates build up the most important limestone strata. In the past, such seas have flooded extensive parts of the Earth's continents, building huge reefs and associated limestone sediments: for example, during Lower Carboniferous (Mississippian) times, they covered much of the ancient continent of Laurentia (now roughly North America) as well as Europe. Limestone rocks can dissolve to form karst and cave features and then, on evaporation, be redeposited to form travertine deposits around hot springs and geysers (see p.201), and cave carbonate deposits such as flowstones and stalagmites (see p.253).

oolitic grains

**OOLITIC LIMESTONE**
*Rounded grains (oolites) of calcium carbonate rolled by seabed currents have here been cemented by carbonate mud to form rock.*

fossil shell

**FOSSILIFEROUS LIMESTONE**
*A deposit of shells that has accumulated in sufficient quantity may turn into limestone rock. This one has a matrix of carbonate minerals.*

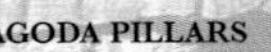

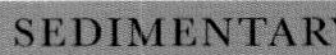

**PAGODA PILLARS**
*This limestone in Lunan, southern China, which dates from the Permian Period, has been largely dissolved away by chemical weathering.*

SEDIMENTARY

## Evaporites

**ORIGIN** Marine and freshwater

**COLOR** Generally white or pale shades of yellow and gray, through to red

**GRAIN SIZE** Fine–coarse

Minerals such as halite (rock salt) and gypsum form extensive sedimentary deposits beside shallow seas or, in arid areas, within saltwater lakes and seas. Evaporation triggers the growth of salt crystals from brines derived from the weathering of local rocks. Thus, ancient evaporite deposits can be good indicators of past climates, geography, and seawater composition. Seawater evaporation produces a sequence of evaporite minerals regulated by their solubility, from calcite through gypsum to halite and potassium and magnesium salts such as sylvite (bitter salt) and carnallite. Some evaporite deposits are economically important for the production of salt, gypsum, and potash, which is an important constituent of fertilizer. Because of their solubility, evaporite deposits may be dissolved and redeposited in other layers.

crystalline texture

**ROCK SALT**
*Although crystalline halite formed by evaporation of seawater can look like marble or quartzite, it differs in being soluble and much softer.*

**DEVIL'S GOLF COURSE**
*In Death Valley, California, salt evaporites have formed pressure-rimmed pans as growing crystals push against one another while they develop.*

SEDIMENTARY

## Iron formations

**ORIGIN** Marine, freshwater, and terrestrial

**COLOR** Reddish-brown to red

**GRAIN SIZE** Fine

Sedimentary iron formations are of great economic importance, with their iron minerals deposited in a variety of environments (see panel, right). The most common sediment-related iron minerals include oxides (such as hematite, magnetite, and goethite), carbonates (such as siderite), and sulfides (such as pyrite and marcasite). Iron-rich strata may show sedimentary features such as rounded grains and current ripples. Fossil remains, especially plants, are common in Phanerozoic iron deposits. Most iron deposits are chemical in origin and most important are the banded iron formations of the Precambrian Eon, which form the bulk of the world's iron reserves, with iron contents in excess of 50 percent by weight. Typically hundreds of yards thick, they first occur around 3.8 billion years ago, and, apart from those associated with the late Precambrian ice ages, are rare in the last 1.8 billion years of the geological record. The most extensive have thinly bedded layers of chert alternating with layers of magnetite and hematite, and are products of chemical sedimentation processes. Others are associated with volcanism.

specular hematite

chert

**BANDED IRONSTONE**
*Such alternating layers of red chert and black hematite, which is rich in iron, are typical of marine deposits dating from the Precambrian.*

### ORE DEPOSITS

Iron is the most important metal in the modern world, and is mostly sourced from sedimentary ores. Precambrian banded-iron deposits in the Lake Superior region have produced most of the iron ore mined in North America over the last 120 years. Iron's availability was an important factor in the rapid industrialization of that continent at the end of the 19th century. Australia is now a major producer of iron ore, from similar deposits, as here in the Hamersley Range.

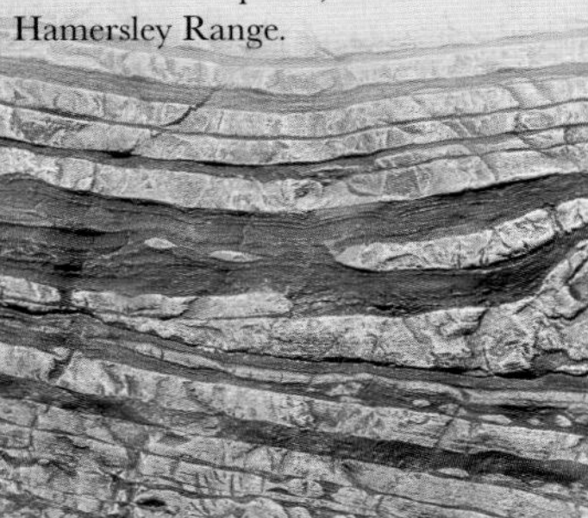

SEDIMENTARY

# Organic deposits

**ORIGIN** Marine, freshwater, and terrestrial

**COLOR** Generally dark brown to black

**GRAIN SIZE** Fine

This group of sedimentary deposits includes the economically important hydrocarbons (such as peat, coal, oil, and asphalt). In addition, it includes other related organic products that are preserved in rocks, such as plant-derived amber resin. Essentially, organic deposits are made of carbon-related materials derived from the post-mortem breakdown of organic tissues—both plant and animal. All life consists of complex organic compounds, mainly long chains of carbon atoms (aliphatic molecules) such as carbohydrates, proteins, and lipids. In the process of photosynthesis, plants on land and algae in the sea convert carbon dioxide into organic tissue using light energy from the Sun. In recent decades, however, an independent process of chemosynthesis, provided by hydrothermal fluids, has been found to be widespread within the deep ocean.

Following the death of an organism, its tissues decay and are broken down, often forming rings of carbon atoms (aromatic molecules). The carbon molecules may be dissolved, forming humic or fulvic acids, or be stored in sediments. The latter may develop into petroleum and natural gas (methane) in marine sediments, depending on a succession of geological events. On land, organic matter is stored in soils and aquatic deposits such as peat, coal, and gas (methane), depending on geological constraints of burial pressure and temperature, as realized by William Smith (see panel, above). Having low densities, petroleum and gas often migrate from their source rocks and may return to the surface and be lost, or they may be stored underground in porous reservoir rocks such as sandstone. Peat and coal, however, are made of accumulated plant materials in layers up to several yards thick, interbedded with fossil soil layers and other rocks such as sandstone, shale, and limestone, depending on the original depositional site. Since the Carboniferous Period, tropical forests and swamps have been extensive enough to form coal deposits, the grade of which depends on the burial depth of the plant-bearing layer. With increasing pressure and temperature, peat-type deposits are progressively transformed into lignite, bituminous coal, and anthracite, which have progressively higher carbon contents and calorific values and lower moisture contents and volatility.

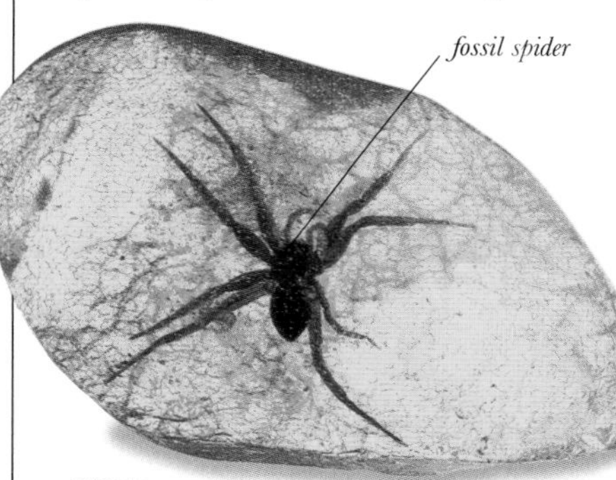


**AMBER**
*Resin from some kinds of ancient pines trapped invertebrates and, on hardening, became part of the sedimentary rock record known as amber.*

**TAR PITS**
*Natural seepages of oil occasionally appear at the surface, where their volatile compounds are lost, leaving heavy tars, as here in Trinidad.*

## WILLIAM SMITH

The son of a rural English blacksmith, William Smith (1769–1839) earned his living as a civil engineer and land surveyor, building canals and sinking shafts for coal. From this practical experience of rock sequences, Strata Smith, as he was nicknamed, independently developed techniques for mapping and correlating strata. Single-handedly, he produced the first detailed geological map of England, Wales, and parts of Scotland in 1815, and provided cross-sections of the correct vertical succession of strata, which he identified from the color, texture, and composition of the rocks. In doing so, Smith laid down the principles of geological mapping.

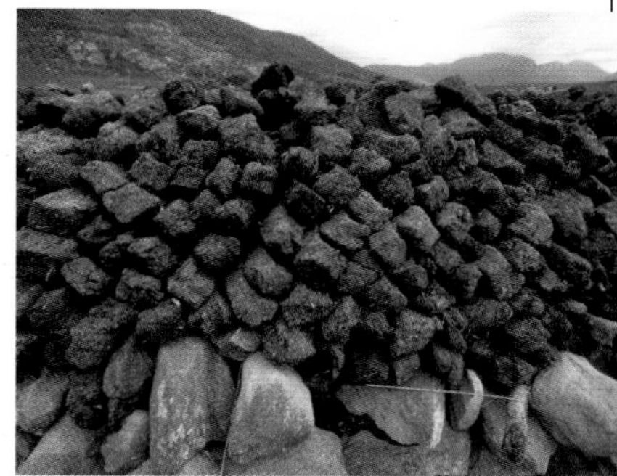

**PEAT FUEL**
*Organic plant deposits of bogs and swamps have over the centuries been extensively cut and dried, as here near Loch Cloon, in County Kerry, Ireland. The blocks are then used for domestic fuel and garden soil mix—and even in power stations.*

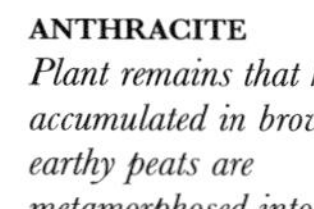
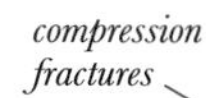

**ANTHRACITE**
*Plant remains that have accumulated in brown, earthy peats are metamorphosed into anthracite by deep burial and elevated temperature.*

**COASTAL COALS**
*In Portugal, erosion and open-pit mining have exposed these thin, dark coal seams with paler soil layers and other mixed sediments.*

**FLOODED FIELD**
*Drenched by crude oil, two men struggle to cap a Kuwaiti oil well that was set on fire during the Gulf War of 1991. The Middle East contains over half the world's oil reserves.*

# FOSSIL FUELS

Despite the growing use of renewable resources, fossil fuels still supply more than four-fifths of the world's energy needs. Oil and gas often occur together, and are extracted using sophisticated drilling technology that maximizes shrinking reserves. The third fossil fuel—coal—is much more abundant than oil and gas, although it is harder to extract, and also more polluting when used. The aftereffects of coal mining dominate the landscape in some parts of the world.

## OIL AND GAS EXTRACTION

When the world's first oil well opened in Pennsylvania in 1859, it struck oil at a depth of just 69 ft (21 m). A century and a half later, such easily accessible deposits are extremely rare. In the relentless quest for more oil, today's oil wells can be more than 16,500 ft (5,000 m) deep, and instead of drilling directly downward, drill bits can be steered underground at almost any angle. Called directional drilling, this enables a single oil rig to collect oil from an area of more than 40 square miles (100 square km).

Compared to coal mining, oil extraction creates very little surface spoil. One of oil's few waste products is natural gas, which has now become an important fuel in its own right. But with great distances often separating production and consumption, the overall impact of oil and gas extraction is high. Oil installations can pollute groundwater on land, and numerous accidents by oil-carrying tankers underline the pollution hazard at sea. Oil and gas extraction can also intrude into environmentally sensitive areas.

## COAL MINING

Coal was the fuel that powered the Industrial Age. In energy terms, its place has since been taken by oil, yet it remains an essential resource, particularly in the developing world. Of all the fossil fuels, coal is the simplest to use, but because it is a sedimentary rock, it is also the hardest and most dangerous to extract.

The earliest coal mines were shallow pits, excavated where coal seams were near the surface. When these became depleted, miners were forced to dig downward to find seams, which they could follow underground. By the early 1600s, some mines in Europe were already more than ³⁄₅ mile (1 km) long, and they generated waste piles that have become a hallmark of mining country worldwide. With underground mining, subsidence and flooding are constant risks, and the older a mine is, the greater these problems become.

Much of today's coal is produced by an alternative method: surface or open-pit mining. In this, the excavation is typically carried out by power shovels or self-propelled augers, which remove the coal by digging or boring through it. Compared to underground mines, surface mines are cheaper and safer, but they alter the landscape even more. In many mines, coal makes up less than 10 percent of the extracted rock. The rest is waste, which is returned to the ground. Because of its lack of soil and high mineral content, such waste is a difficult habitat for plants. Fortunately, in recent years, researchers have bred plant strains that can tolerate this human-made environment. With these, and careful landscaping, waste piles can be stabilized and their visual impact much reduced.

**SURFACE MINING**
*The Arkwright coal mine, near Derby, is one of England's largest open-pit mines. Coal from this region powered the world's earliest locomotives.*

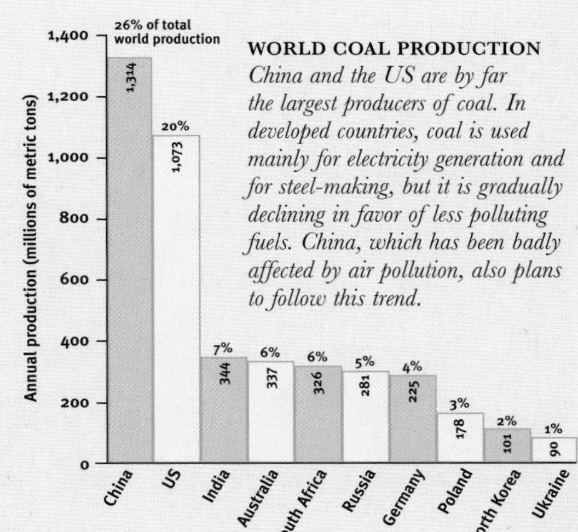


**WORLD COAL PRODUCTION**
*China and the US are by far the largest producers of coal. In developed countries, coal is used mainly for electricity generation and for steel-making, but it is gradually declining in favor of less polluting fuels. China, which has been badly affected by air pollution, also plans to follow this trend.*

## FUTURE SUPPLIES

The Earth's accessible coal reserves amount to at least 1 trillion tons, which is enough to last several centuries if consumption rates remain unchanged. However, coal is a highly polluting fuel, so environmental considerations are likely to limit its use. There is enough natural gas to last at least another century, but for oil the picture is less clear. Oil exploration constantly finds new reserves, while improving technology increases the efficiency of extraction. But even so, some experts believe that commercially viable oil reserves may be exhausted within the next 50 years.

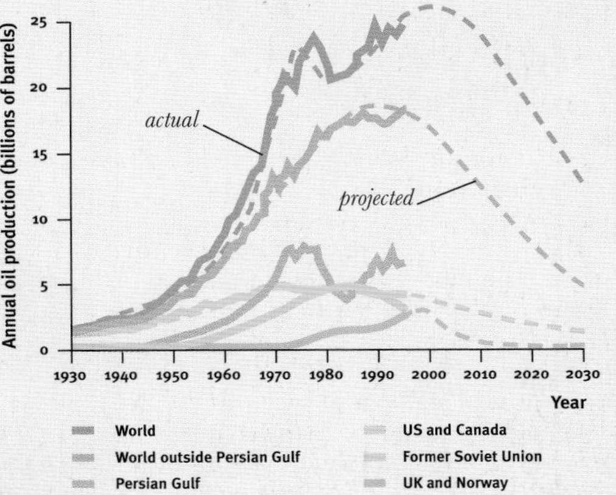


**WORLD OIL PRODUCTION**
*The world's oil producers appear to be reaching the top of a bell-shaped curve, which means that oil supplies will soon begin to decline. For economic reasons, a switch to other fuels will be needed long before oil finally runs out.*

SEDIMENTARY

## Bauxite

**ORIGIN** Terrestrial

**COLOR** Variable, including white, gray, yellow, and red

**GRAIN SIZE** Fine–medium

When exposed at the surface, many minerals are not stable in normal processes of weathering, especially in humid tropical regions. Decaying vegetation releases organic acids, which percolate down to the bedrock, causing chemical changes known as leaching. This removes soluble constituents of minerals, leaving a residue to form new minerals. The most immobile residual elements are silicon, aluminum, and iron, usually in the form of oxide or hydroxide minerals in the rock bauxite. Bauxite is an ore containing over 50 percent aluminum oxide and is found in tropical limestone areas, where all the limestone has been dissolved away. Small mineral residues from the original limestone are left behind, concentrated, and form the insoluble bauxite minerals by additional chemical leaching. Bauxite also occurs as a residual deposit through the leaching of alumina and silica-rich minerals from igneous, metamorphic, and sedimentary rocks. Bauxite is a rock composed primarily of aluminum hydroxide minerals (for example, gibbsite) along with clay minerals, silica, and iron hydroxide minerals (for example, limonite). Layers of bauxite may be dozens of yards thick and take millions of years to develop. It is the principal ore of aluminum, with Australia producing nearly a third of the global ore production. Bauxite is a useful indicator of past climates, because it records periods of tropical climate in the geological past.

**VARYING LEVELS**
*Hydrated aluminum oxide minerals make up the large grains and the intervening cement in this bauxite ore hand specimen.*

**SURFACE EXTRACTION**
*Bauxite deposits in Jamaica are formed by the intense tropical weathering of surface rocks, and can therefore conveniently be quarried.*

SEDIMENTARY

## Phosphate deposits

**ORIGIN** Marine and freshwater

**COLOR** Greenish-gray to brownish-gray

**GRAIN SIZE** Fine–medium

Animal bones and the feces of carnivorous animals are the source of most phosphate deposits. Such deposits (known as phosphorites) are typically layered accumulations (with more than 20 percent of phosphate minerals, especially apatite). Present-day phosphate deposits formed when animal remains accumulated relatively quickly in marine sediments and were then altered by chemical reactions. Phosphate deposits are commonly associated with limestones, and some are of economic value.

**GUANO DEPOSITS**
*Guano deposited by fish-eating birds on islands such as this one in the Galápagos has in the past proved to be an economical source of phosphate.*

SEDIMENTARY

## Septarian concretion

**ORIGIN** Marine, freshwater, and terrestrial

**COLOR** Grayish-yellow to brown

**GRAIN SIZE** Fine

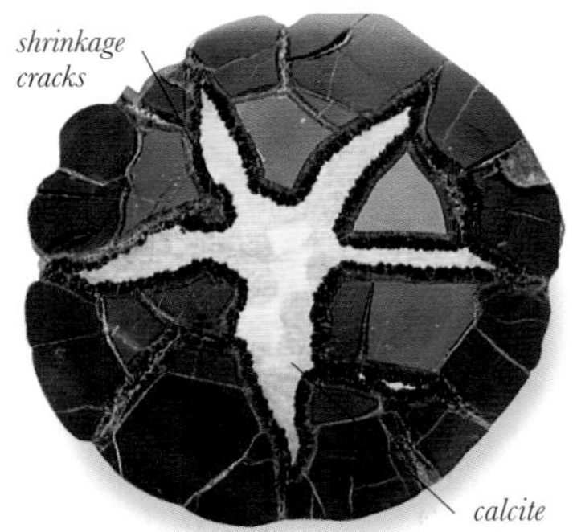


Concretions are typically formed within sediments. They vary in their composition, shape (round or disk-shaped to irregular), and size (inches to yards), but are essentially the same material as the host sediment, which has been cemented (concreted) by other minerals, commonly carbonate, silica, or iron oxide. A septarian concretion has internal cracks filled with crystals such as calcite. The infilling mineral also hardens or cements the concretion, often before compression and flattening, and it is therefore more resistant to weathering and erosion than the surrounding sediment.

SEDIMENTARY

## Chert

**ORIGIN** Marine

**COLOR** Variable, but mostly gray to black

**GRAIN SIZE** Extremely fine

Chert is an extremely fine-grained (cryptocrystalline) form of silica. Also known as flint, it develops in a variety of sedimentary environments both as isolated nodules and in more continuous layers and massive beds, especially in limestone, banded ironstone, and above ocean-floor basalts. Most chert is formed by the accumulation of silica in the sediment by the decay of silica-rich organisms (such as glass sponges and radiolarians), hydrothermal activity on the ocean floor, and—during the Precambrian eon—due to changes in ocean chemistry, which created sedimentary iron formations. In some cases, further chemical changes within the sediment form silica gels, some of which harden into chert, often around fossil remains and before compaction of the wet sediment. Due to its hardness and resistance to erosion, chert may accumulate to form significant deposits such as beach conglomerates. It has also provided an invaluable source of weapons and tools for humans (see panel, above).

**FLINT CORE**
*Flint nodules can form in sedimentary rocks, especially limestone, and may sometimes include organic fossils not normally preserved.*

### FLINT TOOLS

The glassy texture of chert, along with its hardness, toughness, and brittleness, have proved to be suitable qualities for stone tools such as axes, blades, and points, and so this rock was an important resource in prehistoric times. First crafted by early peoples around 2.6 million years ago, chert has continued to be used even in more recent times, as part of the firing mechanism of flintlock muskets.

SEDIMENTARY

## Nodules

**ORIGIN** Marine, freshwater, and terrestrial

**COLOR** Variable, depending on mineralogy

**GRAIN SIZE** Fine–coarse

Minerals such as calcite, quartz, pyrite, and apatite may crystallize in spherical aggregates, to develop nodular growths in many sedimentary environments. Most nodules form within sediment, while others grow near or at the sediment–water interface. Abundant ferro-manganese nodules containing valuable metals such as titanium, chromium, cobalt, copper, and zinc form on the deep sea floor, but so far are too expensive to recover. Other nodules may preserve rare fossils or soft tissues.

# METEORITES

Meteorites are rocks found on the Earth's surface that have an extra-terrestrial origin. They continuously fall to the Earth from space, with a small number derived from the Moon and Mars. They can be divided into three broad groups: stony meteorites, iron meteorites, and a mixed stony-iron group. By far the most common are stony meteorites (subdivided into chondrites and achondrites), which are made of pyroxene, olivine, and plagioclase silicate minerals with some nickel–iron metal. By contrast, the irons are mainly nickel–iron alloys. Stony irons such as pallasites are rare; they contain mixtures of silicates and nickel–iron alloy. Dating from the early stages of the formation of the Solar System, meteorites provide scientists with invaluable evidence about the minerals that were originally aggregated in the Earth.

METEORITES

## Pallasites

**ORIGIN** Extraterrestrial **GRAIN SIZE** Coarse

**COLOR** Yellow to green and metallic gray

When sliced open, pallasites reveal a mixture of rounded, yellowish-green olivine crystals in a shiny gray groundmass. This kind of stony-iron meteorite, which accounts for about 1 percent of meteorite falls, was named after German naturalist Peter Pallas who, in 1776, provided the first detailed description of a meteorite that he had found in Siberia. Pallasites typically comprise one part magnesium-rich olivine crystals to two parts metal. Where the metal is even more abundant, it shows a pattern called Widmanstätten structure. Pallasites are thought to originate at the core–mantle boundary of long-destroyed asteroids, where the thick, olivine-rich silicate mantle material mixes with the underlying metallic iron core. Similarities in metal composition (among elements such as gallium, germanium, and nickel) suggest there is a link between some pallasites and some iron meteorites. However, compositional differences between pallasites indicate that they are derived from more than one asteroid.

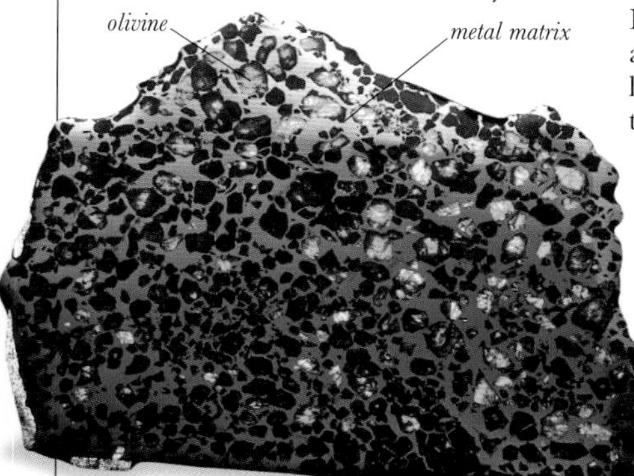


METEORITES

## Achondrites

**ORIGIN** Extraterrestrial **GRAIN SIZE** Coarse

**COLOR** Gray

Achondrites are rare, constituting only about 3 percent of all known meteorites. Despite their scarcity, they are made of the types of igneous rocks most familiar to humans. Achondrites formed in the outer layer (or crust) of an asteroid, which is similar to the crust of the Earth. During the asteroid's early history, the Earth was molten. When it was in this state, dense metal sank to the center to form a core, and lighter silicate melts rose to the surface, erupted in lava flows and fountainlike volcanoes, and eventually solidified and cooled.

**ROUGH SURFACE**
*The apparently chaotic mixture of medium and coarse grains on the surface of this achondrite in fact conceals its crystalline igneous texture.*

Many asteroids were also molten during the Solar System's early history and underwent similar differentiation. The igneous origin of achondrites was first recognized in 1825 by the German Gustav Rose, who was a professor of mineralogy at Berlin University. With Alexander von Humboldt (see p.159), Rose traveled widely in Asiatic Russia and Siberia, collecting minerals and rocks including meteorites. Composed of a variety of silicate minerals, including olivine, pyroxene, and plagioclase, achondrites typically contain far less metal than the more common chondrites. Achondrites are divided into several groups. The eucrites—the most prevalent group—are basalts. Composed essentially of pyroxene and plagioclase, they form first when chondrites melt. They resemble basalts found in abundance on the Earth, particularly on ocean floors. The related diogenites formed when the basaltic melt stagnated in large magma chambers and tiny pyroxene crystals grew and sank through to melt to form a layer of pyroxene. Howardites—rocks composed of eucrites and diogenites mixed together by impact—testify to a long history of bombardment in space. A variety of other meteorites are also classified as achondrites, including angrites, aubrites, ureilites, and acapulcoites. While these groups are considerably smaller, they suggest that melting was a common process that modified many asteroids during the early history of the Solar System.



METEORITES

## Chondrites

**ORIGIN** Extraterrestrial

**GRAIN SIZE** Medium–coarse

**COLOR** Brownish-gray with black crust

First analyzed by French chemist Antoine Lavoisier in 1772, chondrites are by far the most common type of meteorite, forming about 86 percent of meteorites that have fallen to the Earth. Whether this reflects their true abundance in space is less certain. Chondrites have provided the oldest and most precise dates (of 4.56 billion years) for rocks of the Solar System. Since they contain the same ratio of elements as the Sun, it is likely that they formed within a few million years of the birth of the Solar System. Chondrite meteorites are essentially cosmic sediments. They take their name from their major constituent, chondrules. These are rounded spheres formed when molten silicates cooled rapidly in the early solar nebula. Chondrites comprise olivine, pyroxene, and plagioclase from chondrules, along with metal and sulfide. A long history is recorded in chondrite meteorites after their initial formation. This history includes more than 60 million years of heating within asteroids, subsequent impacts that heated parts of asteroids, fragmentation that sent chondrites on their journey toward the Earth, and their eventual arrival on our planet.

**CHANGES DURING TRANSIT**
*As chondrites move through the atmosphere, their exteriors melt to form black, glassy crusts. The insides remain a mass of silicate mineral grains.*

### MARTIAN METEORITES

The possibility of life on Mars cannot be discounted, even though recent claims of microscopic organisms and hydrocarbon residues being preserved in a Martian meteorite known as ALH 84001 (shown here) have now been questioned. There is no doubt, however, that extraterrestrial hydrocarbons are preserved in some chondrites. Magnetite grains and a microbelike tubule were found within carbonate minerals from ALH 84001, and the tubule was thought to be a minute, rod-shaped, bacterial fossil. Nevertheless, some critics believe that its structure resulted from mineral formation at high temperature and was therefore unlikely to have occurred within bacterial cells. Also, the complex structured hydrocarbons had Earth-derived, radiocarbon isotopes and so may be contaminants rather than from the planet Mars.

METEORITES

## Irons

**ORIGIN** Extraterrestrial **GRAIN SIZE** Coarse

**COLOR** Metallic gray with brown-black crust

About 5 percent of meteorites are irons, which are composed of iron–nickel metal alloy (which is 5–20 percent nickel by weight). When the cores of ancient asteroids cooled, the metal separated into two minerals—kamacite and taenite—whose intergrowth forms this meteorite's Widmanstätten structure (see Pallasite, above). Because Earth's metallic core cannot be reached, iron meteorites provide the only material evidence for its probable composition.

# SOILS

Weathering 110–11
Deposition 114–15
Mass movement 116–17
Nutrient cycles 135
The world's agricultural areas 344–53

A THIN LAYER OF SOIL covers most of the Earth's land surfaces. This complex mixture of unconsolidated, weathered rock and organic material is essential to all terrestrial life. Plants need soil in order to grow, and to supply them with nutrients and water. Millions of microorganisms, which are responsible for the recycling of organic waste, live in soil and contribute to its fertility. Soil is also an important filter for much of the water that enters rivers and lakes. Different types of soils support different ecosystems. Knowledge of these soil types allows people to cultivate land for crops and to plan construction of buildings and transportation networks.

## SOIL TYPES

Soils develop in different places, from different rocks and under different climates, and so they have varying properties. Texture is particularly important—soils can be coarse-, medium-, or fine-grained, depending on the relative amounts of sand, clay, and silt they contain. Fine-grained clay and silt soils have small air spaces and retain water, so in regions with heavy rainfall, they tend to become waterlogged. In contrast, coarse-grained, sandy soils have large air spaces and are well drained, but can be infertile, as nutrients are easily washed away. Soil acidity, or pH, is measured on a scale from 0 (most acidic) to 14 (most alkaline). Acidity determines nutrient availability and influences the activity of soil organisms. Soils are classified in various ways, but a common system is the United States Department of Agriculture (USDA) classification, in which soils are classified according to the type of climate and vegetation under which they formed.

### PROTECTING SOIL

When land is cultivated, the natural vegetation is removed, leaving the soils exposed to erosion by both wind and water. In order to minimize the effects of erosion, farmers may adopt various agricultural practices, including contour plowing. By following the lay of the land, wind resistance is reduced, and the soil's natural drainage is better conserved.

## SOIL FORMATION

All soils consist of a mixture of inorganic rock fragments and decomposing organic material, or humus. These materials are arranged in a series of layers, or horizons, that together form a soil profile. As a soil matures and deepens over time, progressively more layers (the O, A, B, and C horizons) develop. These are the combined result of processes involving climate, vegetation, soil organisms, and topography.

**DEVELOPMENT OF A WOODLAND SOIL**
*The complex, layered structure of a woodland soil takes time to develop. Initially, the parent rock is broken down by weathering, forming a regolith. Plants then establish, and organic material starts to build up on the surface. As the mineral and organic content mix, the different layers form, with humus-rich layers at the surface and mineral-rich layers at the base.*

**BREAKDOWN OF ORGANIC MATTER**
*The fungi on this fallen birch not only absorb nutrients from the tree, but also aid in its breakdown into humus, which can then be incorporated into the soil.*

**SOIL DEVELOPMENT**
*This fine-grained clay soil is developing from an underlying mudstone, which is being broken down by, among other things, plant roots forcing the rock apart.*

### SOIL TYPE PROFILES

The pages that follow contain profiles of the 12 main soil types in the USDA system. Each profile begins with the following summary information:

**AREA** Area of the world covered by the particular soil type

**DEPTH** Shallow (0–3 ft/0–1 m), deep (3–30 ft/1–10 m), or variable

**PH** Slightly, moderately, or strongly acidic or alkaline; neutral; variable

TROPICAL AND MID-LATITUDES

## Undeveloped soils

**DISTRIBUTION** Widely distributed, except in northern tundra regions. Extensive in the Sahara Desert, north Africa, and throughout the Middle East

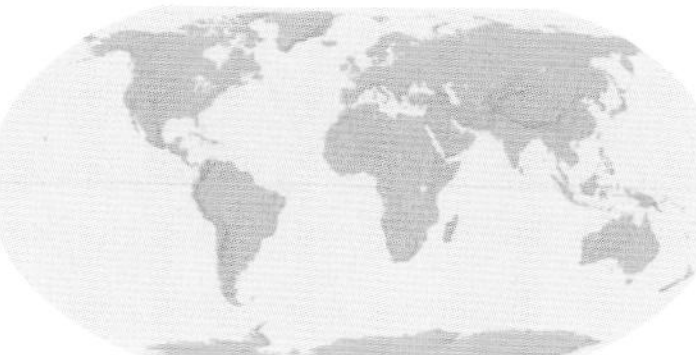

**AREA** 8.2 million square miles (21.1 million square km)

**DEPTH** Shallow **PH** Variable

Very young, shallow soils without any distinct layers are classified as entisols. They lack a vertical profile because there has been insufficient time for one to develop. Some form a thin covering on sloping ground where the rate of soil formation is equivalent to that of erosion; others are deeper, fine-grained deposits of loess (see p.93) or alluvium. Given sufficient time, entisols will mature and develop into another type of soil.

ENTISOL ON SLOPE

TROPICAL AND MID-LATITUDES

## Young soils

**DISTRIBUTION** Widely distributed, except in polar regions. Most extensive in Europe and the floodplains of large rivers, such as the Ganges

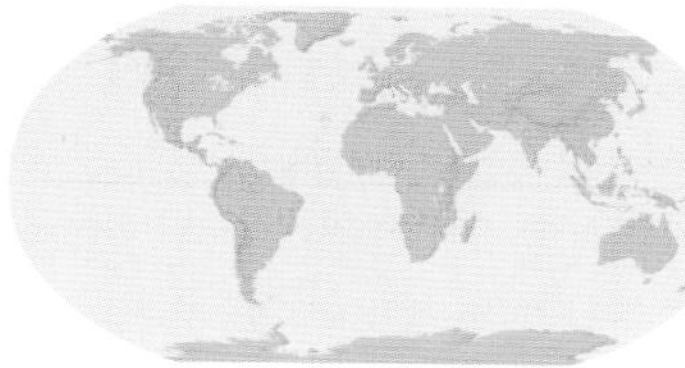

**AREA** 4.95 million square miles (12.8 million square km)

**DEPTH** Variable **PH** Variable

A young soil with a poorly developed vertical profile is called an inceptisol. Soils of this type are found in a range of environments and vary widely in fertility. Horizon development is limited by their youth and in some cases also by climate, but as small amounts of organic material collect in the uppermost layer, it darkens and becomes distinct from the layer below. Inceptisols that form in alluvial deposits are important for growing rice.

INCEPTISOL PROFILE

WORLDWIDE

## Volcanic soils

**DISTRIBUTION** Located in areas of volcanic activity, such as the Pacific's Ring of Fire, and oceanic islands such as the Galápagos, Iceland, and the Canary Islands

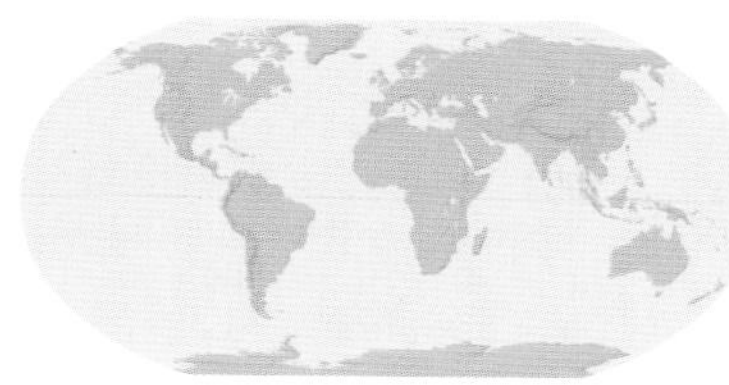

**AREA** 0.35 million square miles (0.9 million square km)

**DEPTH** Often deep **PH** Neutral

Volcanic soils, referred to as andisols, are usually only a few thousand years old. Since they develop from volcanic ash and other volcanic material, they have many unique qualities, including retention of phosphorus in a form that makes it unavailable to plants. Despite this, andisols are usually very fertile and large areas are extensively cultivated (see p.184). Andisols have remarkable water-holding abilities, especially if they contain large amounts of volcanic ash. However, their lack of bulk when dry makes them susceptible to erosion. Where a series of volcanic eruptions has occurred over time, ancient soil profiles, called paleosols, may be preserved beneath the newer ones.

**LAYERED SOIL**
*The alternating upper layers in this soil are caused by repeated burial of the existing profile by volcanic ash.*

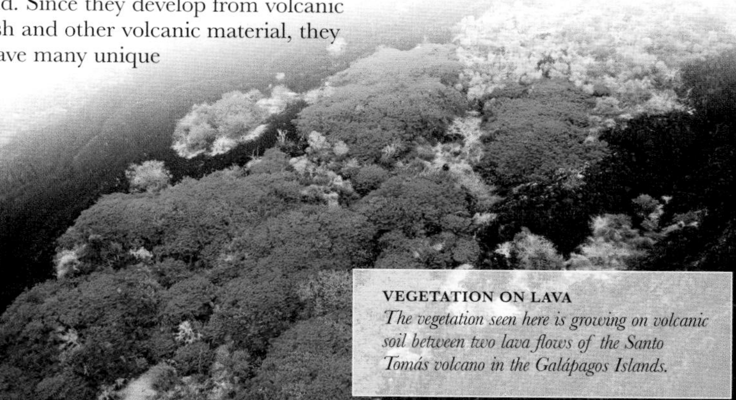

**VEGETATION ON LAVA**
*The vegetation seen here is growing on volcanic soil between two lava flows of the Santo Tomás volcano in the Galápagos Islands.*

POLAR REGIONS

## Frozen soils

**DISTRIBUTION** Restricted to high latitudes in the northern hemisphere, with isolated pockets in some mountainous areas

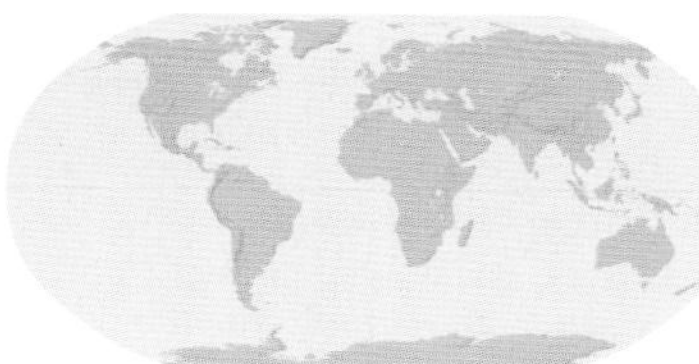

**AREA** 4.35 million square miles (11.3 million square km)

**DEPTH** Deep **PH** Moderately to slightly acidic

Frozen soils, also known as gelisols, are limited to extremely cold polar regions and high mountains, where permafrost is present within 3 ft (1 m) of the surface. The major factor controlling formation of gelisols is climate. Low temperatures not only maintain the permafrost, (see p.342) but also impede decomposition of organic material and downward movement of water. This restricts the soil's development, resulting in a buildup of organic material at the surface. During the warmer summer months, the sun is sufficiently strong to melt the ice held in the surface layers of the soil, but at night it freezes again. This cyclical freeze and thaw of the groundwater is called cryoturbation. It causes the soil to heave and churn, resulting in considerable structural modification, including distortion of soil horizons and the formation of characteristic stone polygons on the soil surface in some areas. Gelisols, due to their location in high latitudes, are not agriculturally important.

**FRUTICOSE LICHEN**
*This is one of the few plants that can survive the freezing conditions of the Arctic tundra. It is a staple food of the caribou.*

**LOW TEMPERATURES**
*In this Alaskan gelisol, microbial activity is restricted by low temperatures, causing organic matter to build up at the surface.*

**CRYOTURBATION**
*The pale horizon in this gelisol has been disrupted by episodes of freezing and thawing of the soil water, creating an uneven upper boundary.*

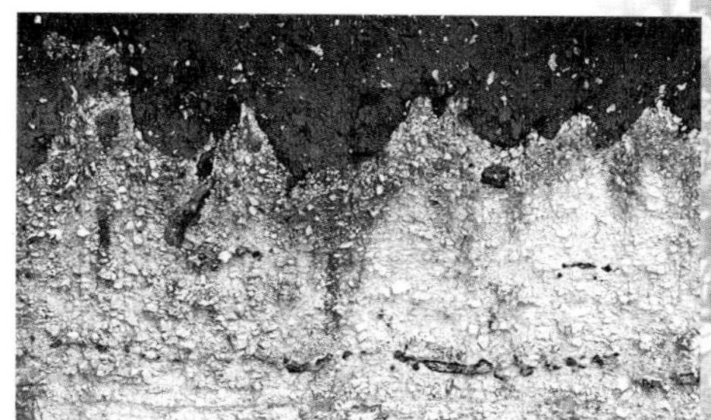

**STONE POLYGONS IN ICELAND**
*These stones have been sorted from finer material by frost heave caused by freezing and thawing of the soil above the permafrost.*

MID-LATITUDES

## Cool temperate, acidic soils

**DISTRIBUTION** Concentrated mainly in the northern hemisphere, covering large areas of eastern Canada and Scandinavia

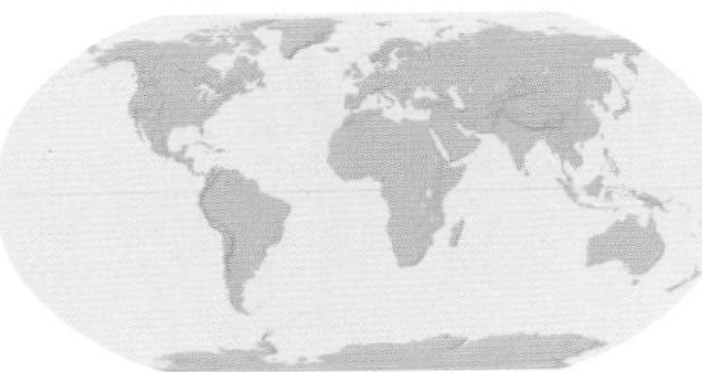

**AREA** 1.3 million square miles (3.4 million square km)

**DEPTH** Deep **PH** Very strongly acidic

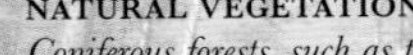

Coniferous forests in cool, wet regions, such as Scandinavia, usually grow on soils known as acidic spodosols. Also called podzols, these soils develop from sandy parent materials, and so are coarse-textured, allowing water to pass rapidly downward. This movement of water removes humus and water-soluble bases (alkaline chemical compounds) from the surface layers by a process known as leaching. As a result, the upper layers of the soil become increasingly acidic and deficient in minerals. Spodosols have well-developed vertical profiles, with striking horizons of brown, white, red, and orange. The characteristic light-colored upper horizon, created by leaching,

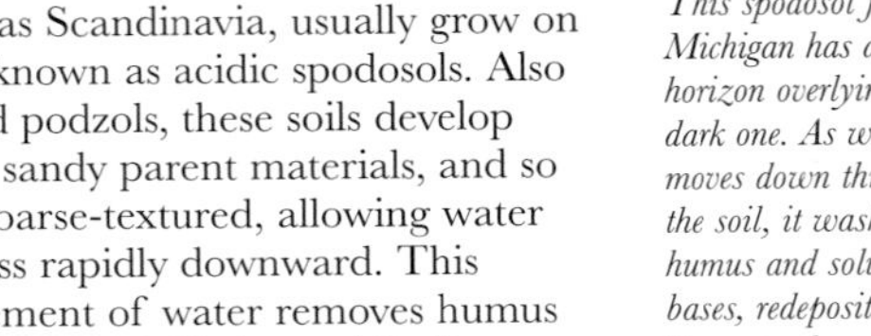

**VERTICAL PROFILE**
*This spodosol from Michigan has a light horizon overlying a dark one. As water moves down through the soil, it washes out humus and soluble bases, redepositing them at a lower level.*

is referred to as an eluviated A horizon (also called an E horizon). This overlies a darker layer, which may be reddish in color due to the accumulation of humus and oxides of aluminum and iron washed in from the horizon above. The oxides are carried downward by acidic organic compounds that leach from the slowly decomposing foliage of conifers such as pines, which are commonly associated with these soils. The iron and humus carried by the water precipitate into a solid form some distance below the horizon, creating a brightly colored layer called a spodic horizon. Although not important for agriculture, spodosols can be made productive by the addition of lime (to reduce the acidity level) and fertilizers.

**NATURAL VEGETATION**
*Coniferous forests, such as this one between Boda and Orrefors in Sweden, are the natural vegetation cover that develops on spodosols.*

**SOIL NEMATODES**
*Millions of nematodes live in the spaces between soil particles.*

TROPICS AND MID-LATITUDES

## Moderately weathered soils

**DISTRIBUTION** Widely distributed, but especially prevalent in temperate areas with humid climates, such as southern Europe

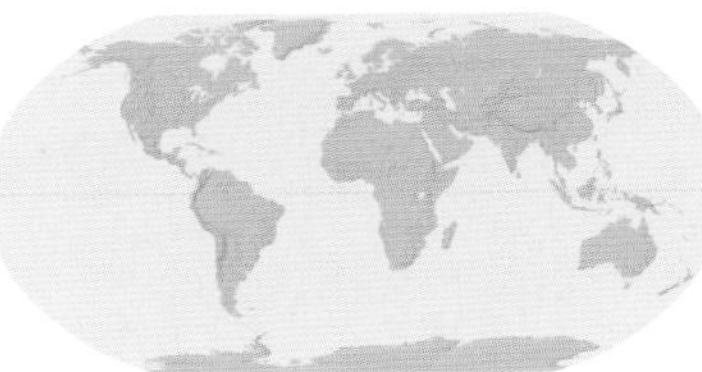

**AREA** 4.9 million square miles (12.6 million square km)

**DEPTH** Deep **PH** Moderately to slightly acidic

Well-developed forest soils in which clays have accumulated at lower levels are called alfisols. The clay horizon is a sign of advanced age compared to most other soils, and is overlain with a deep organic layer. Alfisols are usually associated with deciduous woodlands that generate plentiful organic matter in the form of shed leaves. Alfisols are fertile and easy to farm, so they are important for agriculture.

**ALFISOL PROFILE**

TROPICS

## Weathered tropical and subtropical soils

**DISTRIBUTION** Mainly confined to tropical regions, except for southeast US (excluding Florida) and parts of Korea and Japan.

**AREA** 4.2 million square miles (11 million square km)

**DEPTH** Deep **PH** Strongly to moderately acidic

Formed over a long period, these soils, termed ultisols, are strongly acidic due to intense leaching and chemical weathering. They have well-developed profiles with a yellow or red A horizon overlying a darker B horizon where clays and iron oxides accumulate. Although not naturally very fertile due to leaching, these soils can be productive if well-managed.

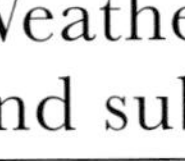

**ULTISOL PROFILE**
*Ultisols, such as this one from North Carolina, have a shallow A horizon overlying a darker, deep B horizon.*

TROPICS

## Highly weathered tropical soils

**DISTRIBUTION** Mainly found in stable, humid environments of tropical regions in Africa and South America

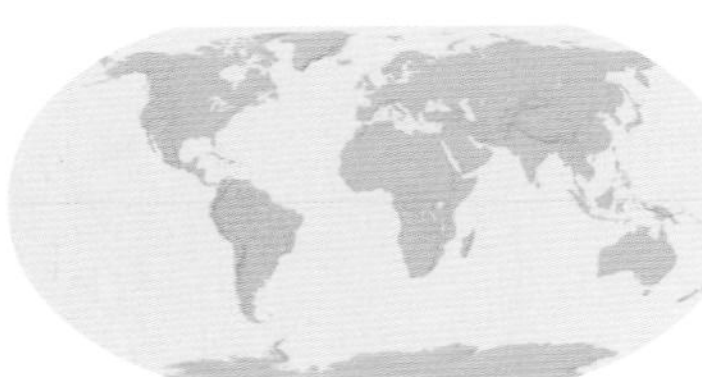

**AREA** 3.8 million square miles ( 9.8 million square km)

**DEPTH** Very deep **PH** Very strongly acidic

Tropical, acidic soils are called oxisols. They are deep, lack horizons, and are colored bright red due to extensive weathering that leaves behind iron and aluminum oxides. Subsistence farming rapidly exhausts oxisols. Once deserted, they are vulnerable to severe erosion.

**OXISOL EXPOSED BY A ROAD CUTTING**

MID-LATITUDES

## Organic soils

**DISTRIBUTION** Mostly in the northern hemisphere in countries with cool, wet climates. Also in parts of the Everglades, Florida, and coastal mangroves of Asia

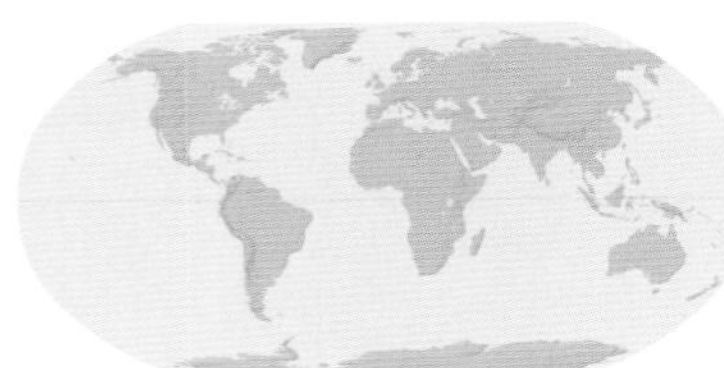

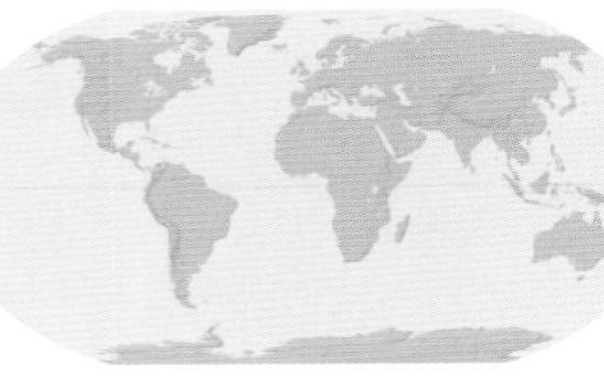

**AREA** 0.6 million square miles (1.5 million square km)

**DEPTH** Deep **PH** Moderately acidic

Organic soils rich in carbon are also known as peat or muck, and technically as histosols. They generally occur on flat land with poor natural drainage, and in places where continuous waterlogging slows the rate of organic deposition. Low temperatures slow the process still further so that organic material accumulates faster than it is broken down. Histosols have surface horizons that are more than 16 in (40 cm) thick, and nearly a third of their mass is of organic origin. Many are acidic with few nutrients available to plants. Drainage of histosols leads to shrinkage and wind erosion.

**HISTOSOL PROFILE**
*Organic matter in this soil from Michigan decomposes when the water table is low.*

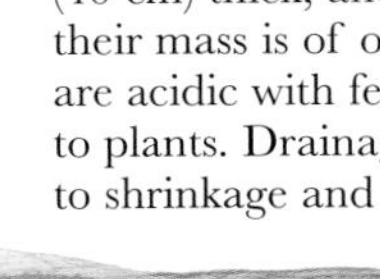

**ORGANIC SOIL**
*Waterlogging, caused by impeded drainage, is typical of peat soils, such as this one in County Mayo, Ireland.*

MID-LATITUDES

## Grassland soils

**DISTRIBUTION** Found around the world under what was once natural grassland. Absent from Africa except in the extreme north

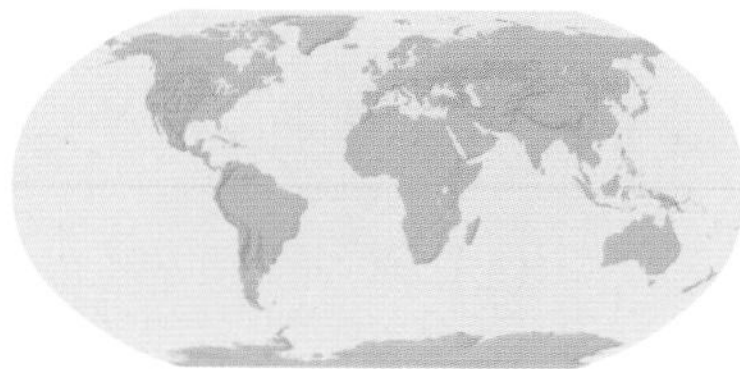

**AREA** 3.48 million square miles (9 million square km)

**DEPTH** Deep **PH** Slightly acidic

The soils of temperate grasslands, such as the steppes of Russia, the pampas of South America, and the prairies of North America, are known as mollisols. These are mature, well-drained soils with well-developed profiles characterized by a deep, dark A horizon about $\frac{3}{4}$–3 ft (0.25–1 m) deep. The dark color reflects a high organic content, mainly derived from plant roots that have been broken down and mixed with the mineral particles by the many millions of organisms living in the soil. These include bacteria, fungi, earthworms, ants, and termites, as well as larger animals such as moles, gophers, and prairie dogs. These soils also contain a high content of nutrients such as magnesium, calcium, potassium, and sodium, allowing plants to obtain these important substances very easily. πIn more arid areas, calcification may occur; this process results in a concentration of calcium carbonate in the B horizon. Mollisols are the most productive of all soils and have long been used by humans for growing crops. Today, they are cultivated to grow many of the world's cereals, especially wheat. However, fine-grained mollisols that develop from windblown loess deposits (see p.93) can be subject to severe erosion when natural vegetation is removed (see panel, below).

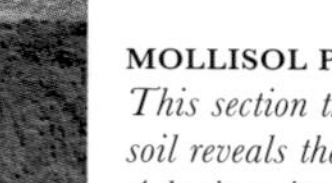

**MOLLISOL PROFILE**
*This section through a grassland soil reveals the humus-rich, dark A horizon interwoven with grass roots.*

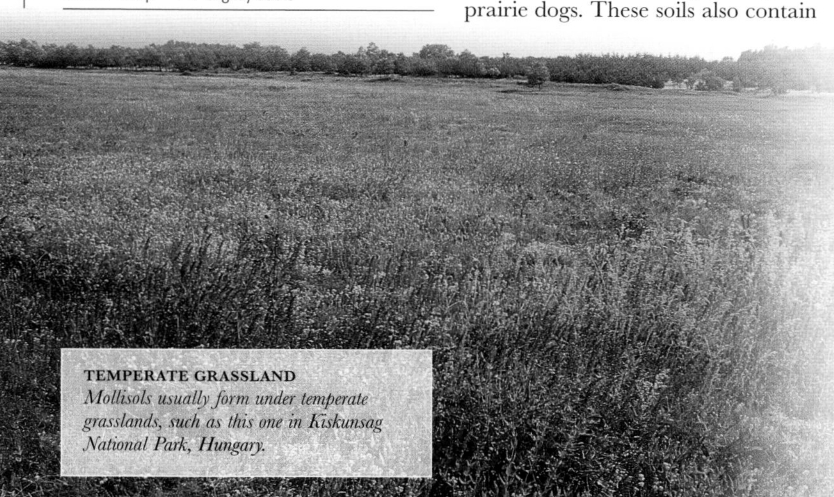

**TEMPERATE GRASSLAND**
*Mollisols usually form under temperate grasslands, such as this one in Kiskunsag National Park, Hungary.*

### FROM GRASSLAND TO DUST BOWL

The fine-grained soil of the American Great Plains is stabilized by hardy grasses that protect it from heavy rain and droughts. In the early 1900s, the area now known as the Dust Bowl was settled by people with little knowledge of farming fragile soils. They plowed millions of hectares of Colorado, New Mexico, Texas, and Oklahoma to grow cotton and wheat, depleting the soil of moisture and nutrients. In the 1930s, drought prevented their crops from growing, and much of the topsoil was blown away in dust storms. Due to the Great Depression, cash was in short supply, and thousands of people were forced out of their homes. Since 1935, the federal government has taken steps to prevent further erosion, such as planting windbreaks, but the soils remain vulnerable during droughts.

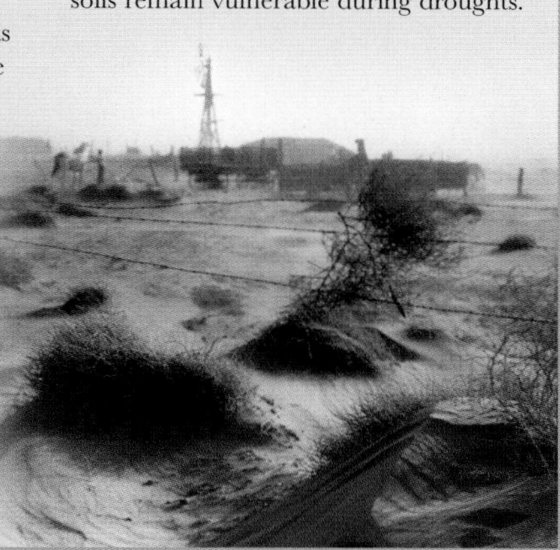

TROPICS AND MID-LATITUDES

## Desert soils

**DISTRIBUTION** In arid regions such as the Great Basin area of the western US, the Middle East, southwest South America, and much of Australia

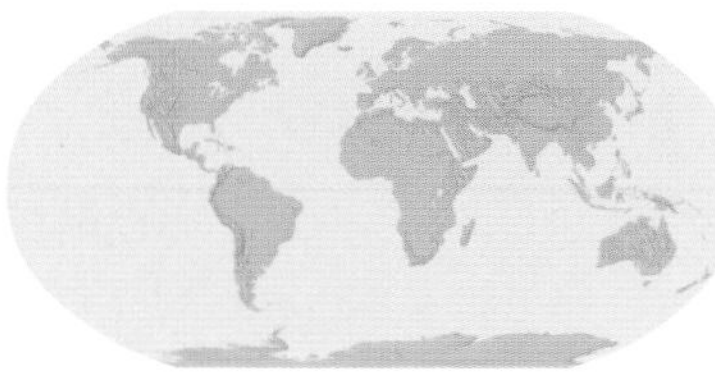

**AREA** 6.1 million square miles (15.7 million square km)

**DEPTH** Variable **PH** Neutral to slightly alkaline

The soils characteristic of desert areas with hot, dry climates are known as aridisols. They are alkaline mineral soils with poorly developed vertical profiles, although a salt layer may accumulate at or near the surface. These salts are dissolved in water that is drawn upward through the soil by capillary action and then redeposited as the water evaporates. Because there is little rainfall in these areas, vegetation is sparse, and there is little organic material in the surface horizon. Scarcity of vegetation also means that there are few plant roots to anchor the soil particles, so many aridisols are subject to wind erosion. Where erosion has taken place, new soils start to form, and so aridisols are often associated with undeveloped soils (see p.101). Although aridisols are dry for at least six months of the year, they never develop the deep vertical cracks typical of swelling soils (see right). Under irrigation, these soils can be productive and may be used to cultivate crops such as wheat and potatoes.

**ARIDISOL PROFILE**
*Upward movement of water may transport salt to the surface, as seen here in Nevada.*

**DRY CLAY PAN**
*During droughts, virtually all of the surface water evaporates, leaving dry clay pans, such as this one in the Namib Desert, surrounded by sand.*

TROPICS AND MID-LATITUDES

## Swelling soils

**DISTRIBUTION** On all continents, in areas with marked wet and dry seasons, including the Deccan Plateau of India, eastern Australia, Sudan, and Ethiopia

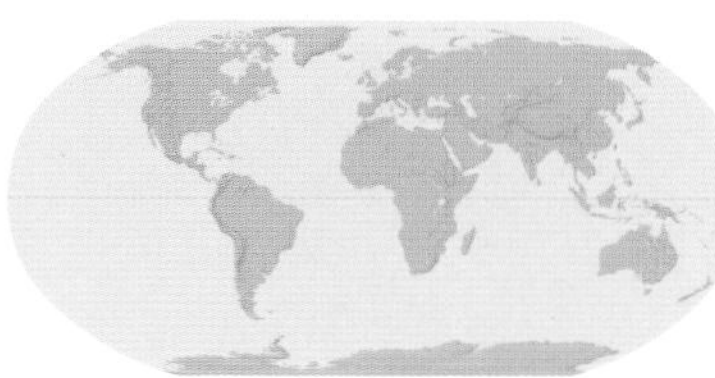

**AREA** 1.2 million square miles (3.2 million square km)

**DEPTH** Deep **PH** Neutral to slightly acidic

Swelling soils, or vertisols, are found mainly in semiarid tropical regions that experience marked seasonal rainfall. They are formed from parent rock rich in clay minerals (see p.63), and the high clay content causes vertisols to shrink or swell as the moisture level falls or rises. In dry periods, vertical cracks up to 1$\frac{1}{2}$ ft (0.5 m) deep may develop. The soil movement not only impedes development of horizons in the profile but can also be a major obstacle to engineering works. The natural vegetation cover for these soils is grass or woodland.

**VERTISOL PROFILE**
*This vertisol from southern Idaho shows a dry, crumbly profile with deep cracks. After rain, its character will alter dramatically.*

**SURFACE CRACKS**
*This vertisol in Utah shows the deep cracks that develop during dry periods. After heavy rain, the soil swells and the cracks disappear.*

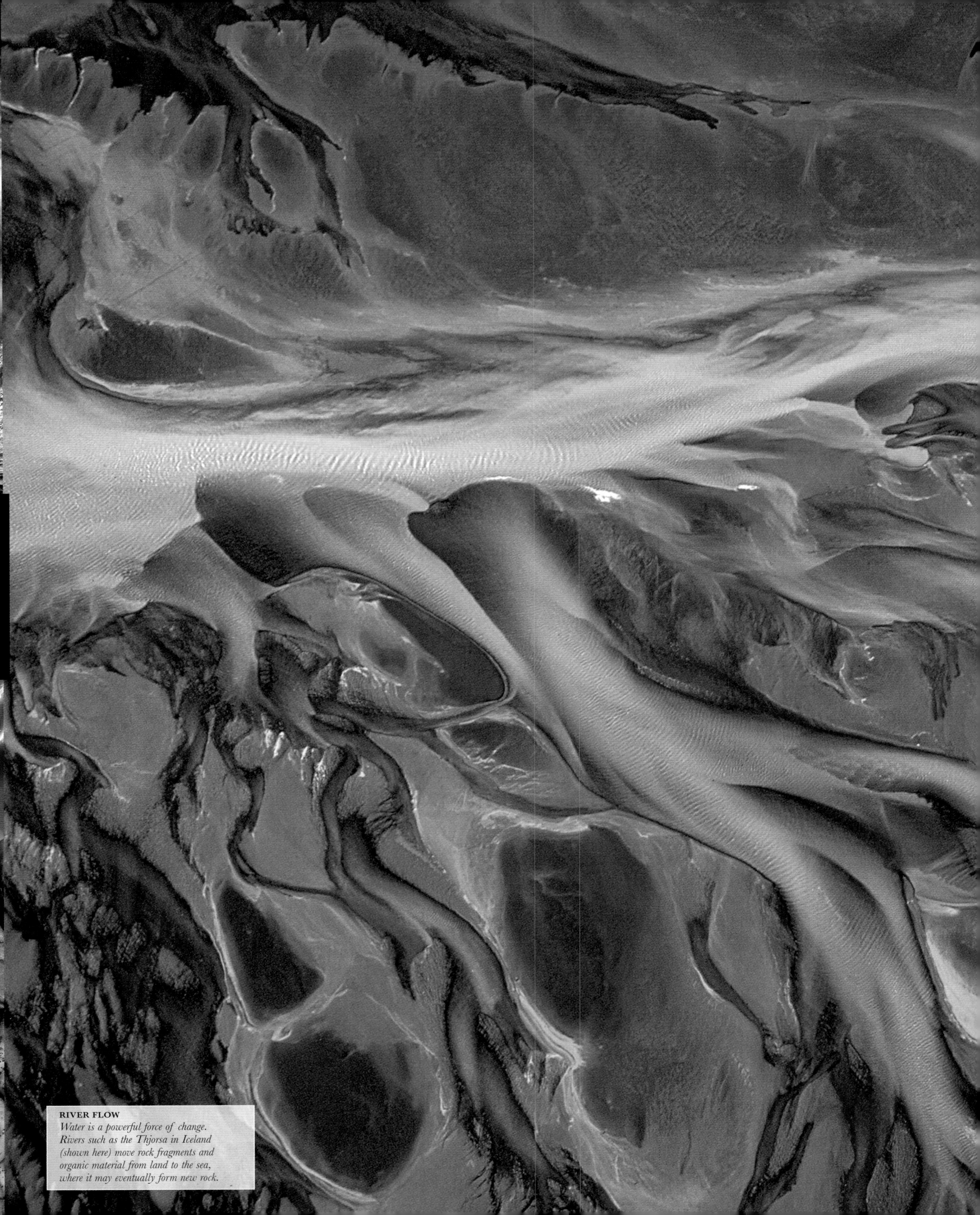

**RIVER FLOW**
*Water is a powerful force of change. Rivers such as the Thjorsa in Iceland (shown here) move rock fragments and organic material from land to the sea, where it may eventually form new rock.*

# THE CHANGING EARTH

THE SURFACE OF THE EARTH is constantly changing under the influence of forces within the planet, surface processes, and even the impact of objects from outer space. The effects of such change are often discernible only after many thousands of years, although some changes such as landslides and meteorite impacts have an instant effect. The relentless movement of the tectonic plates, which form the Earth's surface, can open new oceans, create mountains, and change continents. Wherever the Earth's materials are exposed at the surface, they are subjected to the processes of weathering, erosion, and transport by wind, water, and ice. These processes can sculpt rock and create sediment, which is deposited on land and beneath the sea. Water covers most of the Earth's surface and is a significant agent of change. The Earth is also the only planet known to support life. Living things both adapt to the conditions on the surface and shape the environment around them.

# TECTONIC PLATES

28–35 The Earth's past
58 The crust
Plate boundaries 108–109
Fault systems 142–53
Mountains 154–71
Volcanoes 172–73
Ocean tectonics 386–87
Tectonic Earth 478–95

THE COOL, BRITTLE outer rock layer of the Earth is called the lithosphere, and it is broken into seven large, continent-sized tectonic plates and about a dozen much smaller plates. These irregularly shaped segments fit together like a jigsaw puzzle and cover the entire surface of the Earth. Tectonic plates have an outer layer of crustal rocks that is inseparable from an underlying layer of upper-mantle rocks. Throughout geological time, the plates have probably moved across the surface of the Earth, and so continents have been shuffled around, oceans have opened and closed, and mountains and volcanoes have formed.

### ALFRED WEGENER

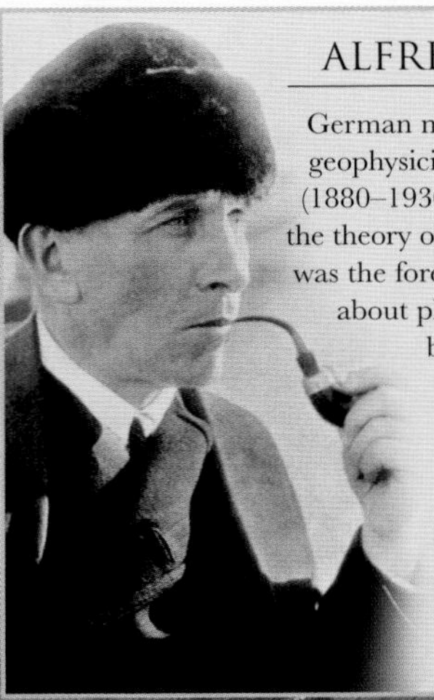

German meteorologist and geophysicist Alfred Wegener (1880–1930) was responsible for the theory of continental drift, which was the forerunner of current ideas about plate tectonics. Impressed by the complementary shapes of the coastlines of Africa and South America, he believed that all the continents once joined together in a supercontinent called Pangaea, which split in the Mesozoic Era.

## THE EARTH'S PLATES

Seven major plates, which are up to 60 miles (100 km) thick, cover 94 percent of the Earth. They carry the major continents, most of which were once part of the ancient landmass of Pangaea (see p.30). The Pacific Plate is by far the largest in area at 42 million square miles (108 million square kilometers). It is followed in size by the African Plate; the Eurasian Plate, which includes Europe and Asia; the Australian Plate; the North American Plate; Antarctica; and the South American Plate. The remaining surface area is accounted for by about a dozen much smaller plates, such as the Juan de Fuca Plate, which is located off the coast of the northwest US. The Pacific Plate is constructed entirely of oceanic crust and mantle, while, at present, the rest of the major plates contain both continental and oceanic material. Plate size is constantly changing as some expand and others get smaller. These processes occur at plate boundaries (see pp.108–109). A balance between the two processes is maintained, since the size of the Earth is constant. Most plate boundaries occur within the oceans or adjacent to continents, but some are marked on land by mountain belts, such as the Himalayas, which is where the Eurasian and Indian plates meet, or volcanic island arcs, such as the Aleutians in Alaska, which mark the subduction of the Pacific Plate beneath the North American Plate.

**TECTONIC PLATES TODAY**
*At present the Earth's surface is broken into seven major tectonic plates and about a dozen minor ones. The boundaries of the plates are marked by features such as mountains, ocean ridges, and deep-sea trenches.*

**WESTERN AUSTRALIA**
*The Precambrian rocks of Western Australia (right) are more than 3.4 billion years old. They form a stable continental mass that is part of the Australian Plate.*

**CASCADE RANGE**
*When two plates converge, mountains rise. Subduction of the Juan de Fuca Plate beneath the North American Plate has generated the Cascade Range in the northwestern US, including Mount Adams on the horizon.*

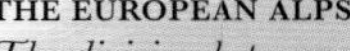

**THE EUROPEAN ALPS**
*The division between the African and Eurasian plates is marked by the high peaks of the European Alps.*

# HOW PLATES MOVE

The Earth's tectonic plates are known to be in constant motion, shifting by up to 6 in (15 cm) each year. However, the mechanism that causes them to move has been hotly debated. Some scientists have argued that organized convection cells in the mantle move the tectonic plates. The relatively cool lithospheric plates are underlain by mantle rock that is hot enough to flow, and this rock is thought to churn in three-dimensional cells, similar to the much more rapid convective cells in a pan of boiling soup. At their upper margins, mantle convection cells are thought to act like conveyer belts, dragging along the overlying plates. Other scientists discount the idea of organized three-dimensional convection cells in the mantle, and instead argue that gravity pushes actively growing plates down and away from high, thermally buoyant spreading ridges. Furthermore, once an oceanic plate segment enters a subduction zone, gravity acts to pull the cool and relatively dense plate downward into the hot mantle.

**PLATE MOTION**
*Convection currents within the Earth's mantle are one possible explanation for the movement of the Earth's lithospheric plates. Hot mantle is thought to rise, cool, and then sink again in a circular motion, moving the plates along as it does so.*

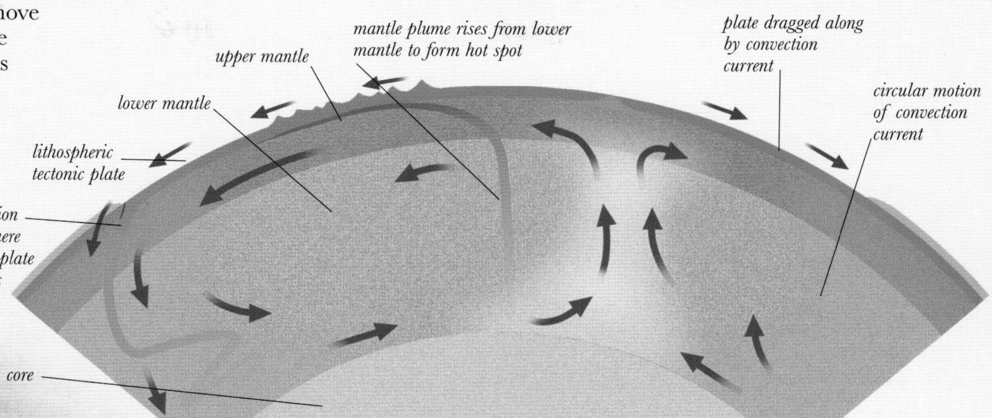


# PAST PLATE MOVEMENT

Several scientific breakthroughs have led to the reconstruction of past movements of the Earth's tectonic plates. Mapping of plate boundaries, the discovery of sea-floor spreading (see p.108), and measurement of the direction and rate of movement of the plates has enabled scientists to recreate their history. There is also much geological evidence to give clues to the positions of the plates in the past. For instance, the distribution of ancient mountains, such as the Caledonian belt—which today has remnants on either side of the Atlantic (see pp.158 and 160), and unusual deposits and fossils, have helped to pin down past plate positions. Fossils of the marine reptile *Mesosaurus* have been found in both Africa and South America, suggesting that the two continents were once joined. Mapping and drilling of the ocean floor beginning in the 1950s and 1960s provided further evidence of plate movement, as it was discovered that the oldest known sea-floor rocks were just 180 million years old. Thus the ocean floor is relatively young, and any older rocks have disappeared, swallowed back within the Earth. Plate movements from the present back to the Permian Period show that over the course of about 200 million years, today's continents evolved from a continental cluster called Pangaea, which stretched from pole to pole (see p.30). Late Precambrian models have also been reconstructed, and include a group of continents called Rodinia. Little information exists for earlier tectonic distributions.

**FOSSIL FOUNDATIONS**
*Fossils are a key to past plate reconstructions. Identical fossils have been found on either side of the Atlantic. Other species, such as this* Gasosaurus, *are isolated within stable continents.*

**MADAGASCAN PRIMATE**
*The lemurs of Madagascar, including this ring-tailed lemur, were isolated as a species when the island split from Africa in the late Cretaceous Period.*

## MEASURING PLATE MOVEMENT

Radio telescope arrays are helping to provide data about the movement of tectonic plates that is accurate to a fraction of an inch. The radio telescopes of the Very Long Baseline Array, including this one at Mauna Kea Observatory in Hawaii, provide accurate measurements of their location by measuring radio signals from distant quasars. By comparing data from other radio telescopes on other plates around the world, the relative plate motions can be determined.

# PLATE BOUNDARIES



THERE ARE THREE MAIN TYPES of boundaries between the Earth's tectonic plates. At two of these types of boundaries, plates change in size. Plate growth occurs at divergent boundaries along sea-floor spreading ridges, while plate consumption occurs at convergent boundaries, which are marked by deep-ocean trenches and subduction zones. Divergent and convergent zones form 80 percent of plate boundaries; the remaining 20 percent are transform boundaries, where two plates slide past one another with no significant change in the size of either plate.

**VOLCANIC CHAIN**
*Iceland is split by the divergent plate boundary between the North American and Eurasian plates. Volcanoes have formed parallel to the boundary.*

## DIVERGENT BOUNDARIES

Where plates move apart, they form divergent (or constructive) boundaries. An example of a divergent boundary is the Mid-Atlantic Ridge, which stretches beneath the ocean between the North and South Poles, although it rises above sea level in Iceland. Divergent boundaries are marked by slowly spreading rift valleys (see p.156). As two plates move away from each other, the mantle rises and partially melts to form basaltic magma. When the molten basalt erupts on the sea floor, it solidifies as it comes into contact with ocean water, forming oceanic crust, a process that is repeated as the plates continue to diverge. New oceanic crust is created at divergent boundaries at a rate of an inch or two per year, in a process called sea-floor spreading. Over many millions of years, this widening process can lead to the opening of new oceans. Because the Earth remains the same size, lithosphere is consumed by subduction (see opposite page) at the same rate that it is created. The straight lines of divergent boundaries are disrupted by the Earth's curvature, and transform boundaries (where two plates slide past each other) form. These are neither constructive nor destructive boundaries.

*direction of plate movement*
*spreading ridge*
*mid-ocean rift, where molten basalt reaches the sea floor*
*divergent plate boundary*
*plates move past each other at transform boundaries like trains on opposite tracks*
*lithosphere*
*asthenosphere*
*rising magma from mantle*
*diverging ocean plate*
*oceanic crust formed when magma cools and solidifies*

**DIVERGENT AND TRANSFORM BOUNDARIES**
*Divergent plate boundaries are offset by transform boundaries. Plates slide past one another along transform boundaries, and, since the lithosphere is neither destroyed nor created, they are also known as conservative plate boundaries.*

### MAGNETIC REVERSALS

The rate at which new crust is created along a spreading ridge can be measured by studying the magnetic properties of the rocks on either side. As upwelling magma cools, iron-rich minerals within it, orientated by the Earth's magnetic field, are set in position. These reflect the direction and inclination of the magnetic poles at the time the rocks formed. Their magnetism can be mapped by magnetometers, and these reveal symmetrical patterns on either side of a ridge, which show the periodic reversals of the Earth's magnetic field. In this image, bands of color identify rocks of normal and reversed magnetization.

**VOLCANIC ACTIVITY**
*At Geysir in Iceland (left), groundwater, superheated by hot rocks below the Mid-Atlantic Ridge, regularly blasts out at the surface in the form of a steam geyser.*

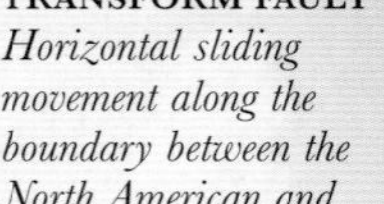

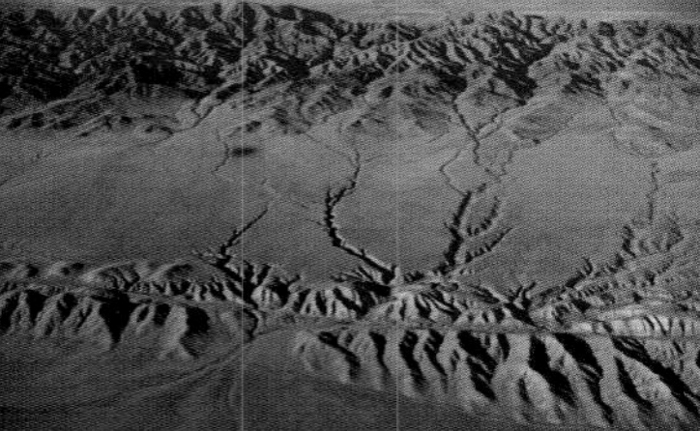

**TRANSFORM FAULT**
*Horizontal sliding movement along the boundary between the North American and Pacific plates occurs on either side of the San Andreas fault.*

## CONVERGENT BOUNDARIES

Convergent boundaries (also known as destructive boundaries) occur where plates move toward one another. Continental crust is thicker and less dense than oceanic crust. If continental and oceanic plate boundaries converge—for example, as is happening off the coast of South America—a subduction zone forms, at which the denser oceanic crust descends beneath the continental margin. This leads to uplift of the continental plate, and the formation of volcanic mountains, such as the Andes. Where two oceanic plates converge (see p.155), the older of the two will be subducted beneath the younger, since it is cooler and denser. At these boundaries, volcanic island arcs form (see p.387). Convergent boundaries involving oceanic plates are marked by deep-sea trenches, where the ocean floor reaches depths of up to 7 miles (11 km). At subduction zones, earthquakes occur deep within the Earth as the tectonic slab descends into the mantle. The plate also begins to bleed fluids, which trigger partial melting in the mantle. The resulting magma can ascend to the surface in a volcanic eruption. When continental plates converge, no subduction occurs, since the plates are of similar density. Instead, the pressure leads to the formation of mountain ranges such as the Himalayas (see pp.154–55).

**VALLE DE LA LUNA, ANDES**
*The Andes have been formed by crustal compression as a result of the convergence of the Nazca and South American plates.*

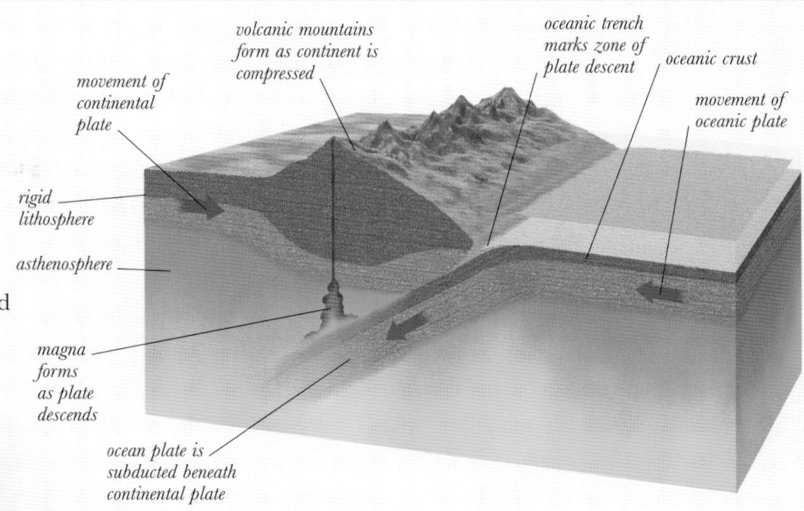


**SUBDUCTION ZONE**
*When an oceanic plate collides with a continental plate, it sinks into the mantle and eventually melts. The collision leads to the formation of an ocean trench and a chain of volcanic mountains on land.*

## WORLD BOUNDARIES

Six of the seven largest tectonic plates are predominantly continental, but the largest of all, the Pacific Plate, is oceanic. Asia and North America, in particular, have grown by the addition of numerous smaller plates, and are known as accretionary terranes. These are geologically welded to the main plates. Well-defined zones of seismic and volcanic activity are distributed along the current plate boundaries, as shown here.

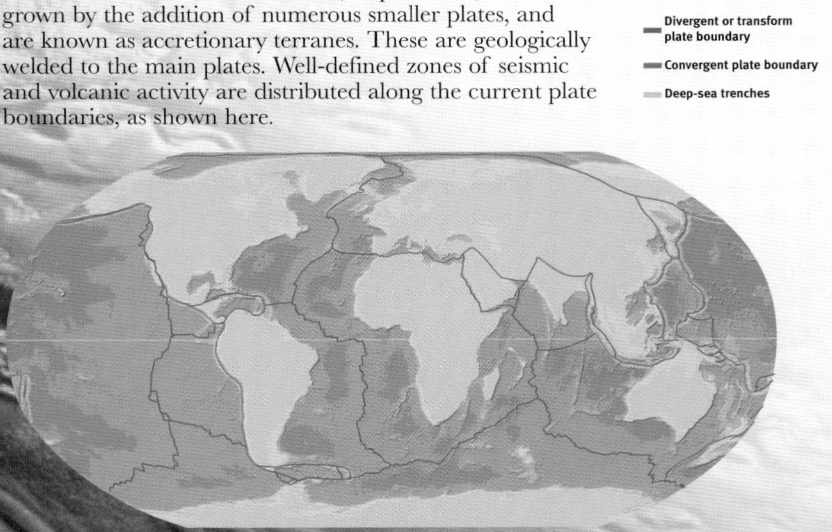

Divergent or transform plate boundary
Convergent plate boundary
Deep-sea trenches

### JOHN TUZO WILSON

In 1965, Canadian geophysicist John Tuzo Wilson (1908–93) demonstrated that sea-floor spreading ridges are segmented and offset by what he called transform-fault boundaries. Wilson also put forward the theory of hot spots (see p.387) and their part in the formation of volcanic island chains such as Hawaii. His discoveries were a key contribution to the acceptance of the theory of plate tectonics.

# WEATHERING

78–103 Rocks
102–103 Soils
Erosion 112–13
Mass movement 116–17
Karst landscapes 253
Deserts 284–85
Precipitation and clouds 466–69

WHEN ROCKS ARE EXPOSED at the Earth's surface they are subjected to a group of physical, chemical, and biological processes collectively known as weathering. Over long periods—sometimes millions of years—weathering can lead to the disintegration of rocks, making them susceptible to later transport and erosion. Many rocks are formed in conditions that are very different from those on the Earth's surface. When rocks are exposed, they react in various ways with the atmosphere, water, and living organisms, depending on their structure and composition. The speed at which weathering occurs also depends on climatic conditions—rocks in warm, humid environments tend to weather more quickly than those in arid or cold conditions.

**SHATTERED ROCK**
*Exactly how desert boulders shatter, as this one in Arizona has, is debated. Cycles of heat expansion and contraction and frost shattering may play a role in this type of weathering.*

## PHYSICAL WEATHERING

The process whereby rocks break apart without substantial change to their chemical structure is known as physical or mechanical weathering, and is largely due to changes in pressure or temperature. All rocks have natural areas of weakness, along which cracks can appear. During the processes of uplift (see p.79) and erosion (see pp.112–13), which expose rocks at the surface, much of the heat and pressure that surrounded them is removed, so they cool and shrink, causing them to fracture. Water invades the resulting cracks, and when it freezes, it expands by almost one-tenth of its volume. This exerts immense pressure on the surrounding rock and enlarges the fracture. If a cycle of freezing and thawing is established, the rock eventually breaks up. Called frost shattering or heaving, such freeze–thaw processes are particularly visible on exposed mountainsides. Similar processes of expansion and contraction may also occur in deserts, where the fluctuation between daytime rock surface temperatures, which can reach up to 140°F (60°C), and nighttime temperatures, which can be freezing, may result in rock fracture.

**FROST-SHATTERED ROCK**
*The fracture of rocks by freeze–thaw action in exposed mountain areas, such as this one in Wales, is a physical weathering process.*

# CHEMICAL WEATHERING

Physical weathering exposes rock surfaces to other processes, including chemical weathering. This type of weathering can lead to the breakdown of some minerals within the rock, and the formation of new ones. In warm, wet climates, chemical weathering occurs far more rapidly than in arid conditions. The process is enhanced by rainfall, which combines with carbon dioxide in the atmosphere to form an acidic solution. As this oozes through soils, it takes some minerals, particularly calcium carbonate, into solution in a chemical reaction that is reversible. For example, limestone readily dissolves in acidic water, which can hollow out underground passages in the rock. Inside these passages, the limestone may be redeposited from solution to create rock formations such as stalactites or flowstone (see pp.252–53). The chemical weathering of limestone produces a distinctive type of landscape called karst. Groundwater and seawater can also lead to reactions with other minerals. For instance, many rock-forming silicate minerals, such as olivine, combine with water. At or just below the Earth's surface, micas, feldspars, and some amphiboles may also combine with water to form clay minerals, such as kaolinite (china clay). These are a principal component of the Earth's soil and sediment, so chemical weathering is a hugely important geological process. Many rocks also contain iron-rich minerals, which react in the atmosphere to form iron oxide residues that are characterized by their rust color.

**SALT WEATHERING**
*Seawater has lashed against this rock, penetrated it, and then evaporated to leave behind salt crystals, which have expanded to produce holes in the surface.*

**KARST LANDSCAPE**
*Extensive weathering of limestone has led to the creation of spectacular scenery in Guilin, China. The hills and towers are all that remain of the original rock.*

## ACID RAIN

Chemical weathering by acid rain, which is so effective in forming cave systems, also attacks limestone buildings and sculptures (see below). Since the Industrial Revolution, the process has accelerated, particularly in areas of atmospheric pollution. Nitrogen oxides and sulfur dioxide are released as a result of burning fossil fuels and vehicle emissions, and compounds react in the atmosphere to form acid rain. Damage is also caused by acid-rain runoff into groundwater, rivers, and lakes.

**LICHEN**
*In the soil-free Antarctic landscape, primitive lichens obtain nutrients from rock surfaces. Acids within the lichens dissolve the outer parts of the rock.*

# BIOLOGICAL WEATHERING

Living things, especially microorganisms, play an important role in the weathering of rocks, reacting with them to cause biochemical decay. Lichens tolerate extreme environmental conditions, and were probably the first organisms to colonize the land. Their acids dissolved the minerals within rocks, breaking them down to form soil. This addition of organic material to inorganic rock debris was an important stage in the formation of materials that could support the growth of the first land plants. In recent years it has been discovered that microbial life exists far below the ground, living on the surface of mineral grains. This evidence suggests that biological weathering may play a far more important role in the alteration of rock-forming minerals than has been recognized previously. Macroscopic plants, such as trees, also play a significant role in the formation of soils as they release organic acids that attack rocks. They also contribute to physical weathering as their roots and seeds penetrate and grow within the cracks in rocks, forcing them to open further. It could also be argued that humans have been one of the most influential organisms to have brought about weathering of landscapes, by clearing land for agriculture and construction.

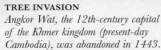

**PLANT COLONIZATION**
*Even recent lava flows, such as this example in Hawaii, can be colonized by plants. These succulents derive nutrition from their harsh environment.*

**TREE INVASION**
*Angkor Wat, the 12th-century capital of the Khmer kingdom (present-day Cambodia), was abandoned in 1443. Since then it has been overtaken by the growth of the encroaching jungle.*

# EROSION

78–103 Rocks
110–11 Weathering
Deposition 114–15
Erosion and transport 213
Underground erosion 252
Glacial erosion 265
Wind and dunes 285
Erosional and depositional coastlines 432

THE PROCESSES BY WHICH soil and rock material are loosened, dissolved, and removed from any part of the Earth's surface are known collectively as erosion (from the Latin *erodere*, which means "to gnaw"). They include corrasion, which is the mechanical erosion of rocks by materials carried within water, wind, and ice, and the movement of rock material as sediment or in solution. The processes of erosion combine to lower the Earth's surface. Without significant erosion, the rock debris that results from weathering (see pp.110–11) would accumulate where it formed. The key element of erosion is transport, which carries debris away. There are two primary ways in which material is transported: by the movement of air, water, or ice, and by the forces of gravity.

### W. M. DAVIS

US physical geographer William Morris Davis (1850–1934) was a pioneering student of landscape erosion and erosive cycles. He envisioned a cycle in which juvenile landscapes evolved to maturity and old age, with characteristic landforms for each stage. Davis's model saw rivers cutting steep V-shaped valleys, which matured into broader valleys with meandering rivers and floodplains, which then became more prolonged "old-age" lowland landscapes close to sea level. Over the last few decades, the reality has been shown to be more complex.

## WATER EROSION

Flowing water is continually reshaping the landscape, and the effects of its erosive power are evident all around us. Water is capable of lifting and transporting loose rock fragments (sediment), which then act abrasively on rock, wearing it away. It can also wash away soil and dissolve minerals, carrying them downstream before depositing them elsewhere (see pp.114–15). When rainwater falls on slopes, it runs downhill, cutting small channels called rills, which can deepen with further rainfall to form gullies. These channels can eventually join streams and rivers. Rivers are important erosive agents, carving their way through the Earth's surface to create valleys and canyons, and carrying sediment toward the sea. The volume of sediment and the distance it is transported depend upon the size of the particles a river carries and the energy of the flow. A major flood has huge erosive potential. Seawater is also a powerful erosive force, as waves constantly bombard coastlines, reshaping them and creating dramatic rock formations, such as caves and stacks (see p.432).

**AUSABLE RIVER**
*A powerful, fast-flowing river, such as this one in New York State, transports abrasive particles, and can quickly cut a steep gorge through rock. In other settings, rivers may erode the rock more slowly.*

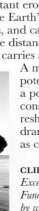

**CLIFF EROSION**
*Exceptionally high tidal ranges in the Bay of Fundy, Canada, and sediment-laden attack by waves, have cut through the cliff face to produce these free-standing rocks.*

### ACCELERATED EROSION

High winds, the shifting course of the Yellow River (see p.233), and overcultivation of land in northwest China's arid interior have led to the erosion of the vast layer of fertile loess (silt) that blankets the landscape. The loess was deposited by winds from the arid Arctic regions of northern Asia during the last ice age. Today, lack of vegetation to bind the surface allows strong winds to lift these light particles into the air, where they are carried in suspension as dust storms over large distances. Millions of tons of China's loess are stripped away by wind and water erosion every year.

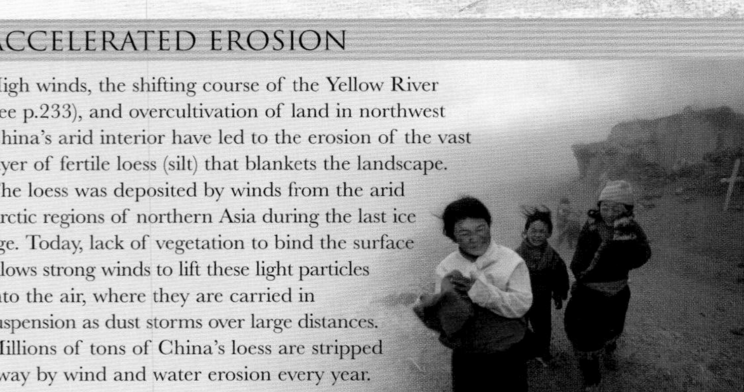

## WIND EROSION

Moving air has the power to carry huge volumes of sediment. Wherever there is a lack of vegetation or moisture to hold surface particles together, loose grains can be picked up and transported by wind. Material that is carried by wind is in itself an effective agent of erosion, sculpting rocks into a variety of shapes. These include ventifacts, which are rocks that have been blasted by particles carried within the air flow, and yardangs, which are wind-shaped desert ridges (see p.284). By removing loose material, the wind exposes rock to further weathering and erosion, with the result that land surfaces are actively transformed by these two processes. Until the Earth's landscapes were covered by vegetation in the Permian Period, wind was probably as significant an agent of erosion as water. Today, wind erosion occurs primarily in arid or desert regions. Only the tiniest grains, about the size of a particle of silt or clay, are lifted high into the wind flow. The transport of larger grains takes place in a process called saltation, in which grains are lifted up to 3 ft (1 m) into the air and carried along briefly before landing back on the ground. As each particle bounces back to the surface, it collides with other fragments that are too large to be lifted, and shunts them along in a creeping motion. This results in the buildup of ripples and dunes.

**VENTIFACT**
*Grains carried by wind have sandblasted this Antarctic erratic boulder. Known as ventifacts, such structures are normally associated with hot deserts.*

**ERODED ROCK**
*Wind erosion has stripped away the softer strata of this sedimentary rock in the Sahara Desert, emphasizing its constituent layers.*

## GLACIAL EROSION

Glaciers form in areas where ice and snow remain year-round and where, over the course of a year, more accumulates in winter than melts in summer. At present, 10 percent of the Earth's land surface, about 5.75 million square miles (14.9 million square km), is covered by glacial ice that is actively eroding the landscape. However, 18,000 years ago, during the last ice age, this proportion was closer to 30 percent. Rock-laden ice is one of the most effective abrasive forces in the natural world. The ice removes loose rock debris and incorporates it into its flow, which then carves, scours, and scrapes the surfaces it creeps over, dramatically modifying landscapes. These effects become visible only when the ice melts. The classic land-forms of glacial erosion include pyramid-shaped mountain peaks with steep rock walls called glacial horns—one of the best examples of which is Switzerland's Matterhorn (see p.163); scooped-out cirques separated by sharp rock ridges called arêtes (see p.265); and broad U-shaped valleys (see p.265). Many rock surfaces are covered with telltale traces of glacial erosion in the form of scratches and grooves.

**DELICATE ARCH**
*The flow of seawater millions of years ago, ice wedging, and collapse along joints contributed to the formation of this remarkable stone arch in Utah.*

**GLACIAL GROOVE**
*Smooth, rounded rock outcrops covered in parallel scratches and grooves, such as these in northern Norway, are typical of erosion by rocks embedded in glacial ice.*

**GLACIAL VALLEY**
*This broad valley in northern England's Lake District, surrounded by steep, bare rock walls that contain a flat valley floor without a major river, is a typical result of glacial erosion.*

# DEPOSITION

92–99 Sedimentary rocks
112–13 Erosion
110–11 Weathering
Mass movement 116–17
Mountains and volcanoes 154–93
Rivers 210–35
Glaciers 262–69
Depositional coasts 433

THE PRODUCT OF THE PROCESSES of weathering and erosion is sediment, which is moved across the Earth's surface by the fluid motion of water, wind, ice, or mass movement (see pp.116–17). As this movement slows down—for example, if a river reaches flatter ground—particles carried within the flow are deposited, with denser fragments falling out first. Sediment is deposited in layers in natural "traps," such as river deltas, or basins on land or in the sea. This accumulation of layers of sediment represents the initial stage of sedimentary rock formation (see p.81). Deposition is also an important process in the formation of fertile land for cultivation.

## LAND RECLAMATION

Where land is at a premium for human habitation, such as around Hong Kong's harbor, materials are removed from other locations and deposited to form new land. In Hong Kong, material has been removed from mountain tops and deposited in the sea, supplemented by rock that has been transported from elsewhere in the world. Land reclamation has increased the space for building in Hong Kong, but has also narrowed one of the world's finest natural harbors.

## WATER-BORNE DEPOSITION

Water plays a major role in the transportation and deposition of sediment. The particles carried by water can range from tiny grains of clay to boulders, depending on the speed and depth of the flow. Water is also capable of dissolving certain minerals and rocks, especially limestones, and transporting the mineral constituents in solution. Some minerals may subsequently come out of solution to form new minerals. Rivers carry sediment from the steeper parts of their courses, depositing it within lakes and on floodplains or in deltas. Floodplains are important sites of deposition: for instance, the Mississippi floodplain in the US (see p.217) is 9 miles (15 km) wide, and is characterized by fertile deposits of sand, silt, and mud. Much river-borne material is also carried toward the sea and deposited in deltas (see p.215). The deltas of large rivers, such as the Nile (see p.229) and the Ganges (see p.234), are sites of significant deposition. As it enters the sea, river-borne sediment is redistributed, contributing to the formation of continental shelves. Sediment can also be transported into the ocean by a high-energy flow called a turbidity current (see p.388).

**RIVER TRANSPORT**
*Pebbles and boulders of various sizes are transported downstream by high-energy waters that also smooth their shape.*

**ANCIENT SANDFLATS**
*Washed by waves and currents, layers of sand were deposited on an ancient beach. The layers were buried and compacted to form this sandstone in Utah.*

## SEDIMENT SAMPLING

By taking samples of the layers of sediment at a deposition site, scientists can reconstruct the conditions that prevailed in the past. From these samples, it is possible to determine the structure of the deposits. This is done by measuring the thickness of each layer of sediment, identifying fossils, measuring the size and shape of grains, and analyzing mineral composition, all of which can be used to track the site's geological history.

**WHITEHAVEN BEACH, AUSTRALIA**
*These white sands are constantly shifting in the shallow tropical sea, but some will eventually be deposited on continental shelves.*

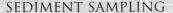

# WIND-BORNE DEPOSITION

Air cannot transport large particles of sediment. However, atmospheric wind and storm systems can be so energetic that they carry enormous volumes of dust, silt, and fine sand. As wind subsides, it deposits the material it is carrying. A strong wind can carry sediment over vast distances—particles from the Sahara Desert in Africa have been found deposited on the opposite side of the Atlantic, in Barbados. Most wind-formed landscapes are situated within the world's hot, arid equatorial deserts, although there are significant mid-latitude deserts, such as the Gobi in Asia (see p.298). The most striking desert landforms caused by wind-borne deposition are dunes, which are usually between 15 and 100 ft (5 and 30 m) high. Some dunes are concentrated in huge sand seas called ergs, which have an area of more than 12,000 square miles (32,000 square km). Dunes vary in shape, and include ripples and ridges, which are transverse and parallel to wind direction; crescent-shaped barchans; domes; and star-shaped formations (see p.285). Narrow zones of sand dunes also form along coastlines, as beach sands are blown inland and deposited. One of the Earth's most significant wind deposits is a thick layer of fine, fertile silt called loess. It covers large areas of the Earth's surface, particularly in China, and was deposited during the last ice age (see p.112).

**LAYERED SANDSTONE**
*Distinctive layers of large-scale dune deposits, dating back to the Jurassic Period, are evident in these sandstones in Utah.*

**AFRICAN SANDSTORM**
*Wind-driven sandstorms, such as this massive system over Chad in central Africa, can transport and deposit desert particles over many thousands of miles.*

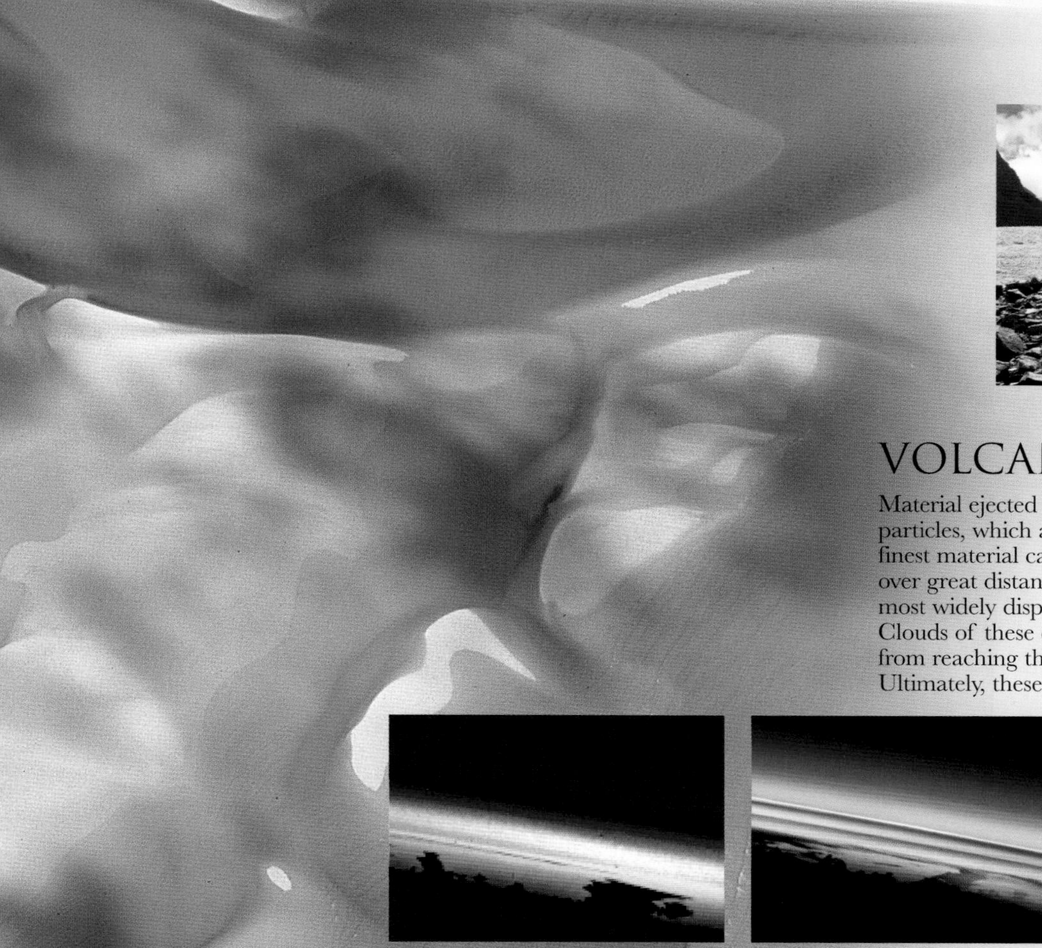

# GLACIAL DEPOSITION

Rock material eroded by mountain glaciers and ice sheets is called till, and consists of particles that range from tiny fragments to boulders. This detritus is carried downhill by the glacier, and then deposited over surrounding landscapes. The features produced by till deposits are moraines, which range from linear trains of debris to wide sheets of rock that can be molded into many different shapes (see p.265). Till that is deposited at the sides of glaciers is known as a lateral moraine, and creates ridges parallel to ice flow. If two glaciers merge, a medial moraine is formed from the till, while ridges formed at right angles to the front edge of a glacier are called terminal moraines. These deposits remain as distinctive landforms when glaciers recede. Meltwaters reshape till and generate sediments called fluvioglacial deposits that consist primarily of sand and gravel. Rivers that flow beneath glaciers channel these deposits into sinuous ridges called eskers.

**MEDIAL MORAINE**
*A medial moraine, such as this one in Switzerland, is formed when two glaciers merge. Their lateral moraines combine to form a central band of till deposits.*

# VOLCANIC DEPOSITION

Material ejected by volcanoes into the atmosphere includes ash and aerosol particles, which are deposited in layers over surrounding landscapes. The finest material can be spewed high into the atmosphere and distributed over great distances before it is finally deposited. The longest-lasting and most widely dispersed volcanic particles are droplets of sulfuric acid. Clouds of these droplets absorb incoming solar radiation, preventing it from reaching the Earth's surface, which then affects climate (see p.451). Ultimately, these droplets fall out of the atmosphere at the poles, where they are buried by snow. These deposits have been detected within cores drilled from polar ice sheets.

**DEPOSITS IN THE ATMOSPHERE**
*The left-hand image, taken from the Space Shuttle in 1984, shows a clear atmosphere. The right-hand image was taken less than two months after the eruption of Mount Pinatubo in 1991, which released nearly 20 million tons of aerosol particles into the atmosphere. The two distinct dark layers are aerosol particles, some of which eventually fell to Earth.*

# MASS MOVEMENT

112–13 Erosion
Landslides 118–19
Fault systems 142–43
Earthquakes 150–51
Volcanoes 172–93
Glaciers 260–81
Tsunamis 428

ROCK MATERIALS MAY APPEAR TO BE STRONG and capable of maintaining mountains and cliffs high above their surroundings. However, all rocks have finite strengths, and the pull of gravity, combined with the processes of weathering (see pp.110–11) and erosion (see pp.112–13), alters the shape of the Earth's surface. A steep cliff may survive for millennia, but can quickly collapse if it is undercut by a river. A hillside may seem to have stable, shallow slopes, but will fail suddenly and collapse, sometimes with disastrous results, if the forces that bind it are altered by rain, vegetation clearance, or a geological event such as an earthquake (see pp.118–19). This is called mass movement, and the speed of the motion depends on the angle of the slope and the material involved.

## SLOPE STABILITY

The slope of a pile of material has a natural angle of repose that is maintained by friction between the individual particles. This is true of both slopes that are consolidated (in which the particles are joined) and those that are unconsolidated (in which they are loose). If the angle is steepened or disturbed, the pile will collapse until it reaches a more stable position. Masses of material will also start to move, under the influence of gravity, if the cohesive strength is altered by the presence of water, additional weight, or sudden movement such as an earthquake. Any of these factors can drastically alter the equilibrium of the mass and cause it to collapse. Vast amounts of unstable unconsolidated and consolidated material are found all over the Earth's surface both on land and beneath the sea. Much of this is in slopes that are close to their maximum angle of repose and, if triggered, will fail. The inhabitants of high mountains and coasts with steep cliffs are all too familiar with the processes of mass movement and the potential instability of slopes.

**SAND DUNES**
*If disturbed, sand grains will roll down the slope of a sand dune until they reach their natural angle of repose.*

**ANGLES OF STABILITY**
*When loose materials are poured, they form natural piles. Large particles form slopes with higher angles of repose than those made of fine particles.*

## ROCK MASS MOVEMENT

Several factors and processes contribute to the downward movement of rock material under the force of gravity. Rock that is weakened by weathering, particularly the process of frost shattering (see p.110), can break away from a slope and tumble down it. Weathering can decrease the strength of a rock mass until it too becomes unstable and moves down a slope. A rock slide occurs when a block breaks off and moves along a bedding plane or a joint. The most devastating rock mass movement is a rock avalanche, which can be triggered by an earthquake or torrential rain. This type of mass movement is violent, sudden, and rapid, moving at speeds of up to 200 mph (320 km/h), and contains many pieces of rock. The largest rock avalanches are the result of the collapse of volcanoes, such as the one that occurred during the eruption of Mount St. Helens in 1980 (see p.177). Evidence of rock mass movement can be seen in talus or scree slopes, which accumulate at the bottom of cliffs or mountains. Rocks can also form part of unconsolidated mass movement (see opposite page).

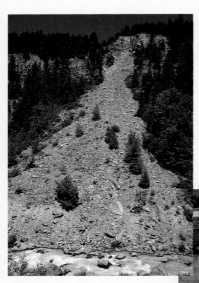

**ROCK FALL**
*Avalanches of rock debris, such as this one above the Hood River in Oregon, are common in steep, hilly areas prone to heavy rainfall.*

**CLIFF SLIDE**
*Cliffs that are weakened by the bombardment of waves are prone to rock falls, especially in areas where movement along active faults can trigger them, such as the coast of California.*

# UNCONSOLIDATED MASS MOVEMENT

Soil and sediment consisting of uncemented grains can move downhill in various ways. The addition of water further affects this movement. Unconsolidated mass movements include creep, earthflows, mudflows, and debris avalanches. Soil creep and solifluction (see below) are slow movements (fractions of an inch per year) that usually occur over gentle slopes. These processes are common in permanently frozen ground (permafrost), particularly during the seasonal thaws, when saturated debris moves downhill over surfaces lubricated by water, fine sediment, or a combination of the two. Debris flows and mudflows containing soil, rocks, and water can occur suddenly and move very rapidly. Lahars occur on the sides of volcanoes, and can be triggered by an eruption or rainfall that remobilizes ash and volcanic debris to create a fast-moving and often catastrophic flow. Such rapid movements can pose substantial threat to life, since they have the potential to envelop everything in their path. Mass movements of unconsolidated material can include simple, unconfined hillside flows, those that are confined by valleys and eventually spread out onto open ground, and highly destructive large-scale flows with enough energy to override topography.

**MUDFLOW**
*Heavy rainfall on this cliff on the Isle of Wight, UK, has saturated and liquified the clay within it, causing it to collapse and resulting in a mudflow that has spread across this beach.*

**SOLIFLUCTION**
*The slow flow of water-saturated soil from higher to lower ground is known as solifluction. The tilt of these old building piles in Svalbard, in the Arctic, is evidence that solifluction has occurred.*

## SUBMARINE MASS MOVEMENT

The large blocks of rock detected on the submarine slopes of oceanic volcanoes (such as those of the Hawaiian Islands in the Pacific and the Canary Islands in the Atlantic) suggest that these unstable edifices have repeatedly collapsed and swept down their submarine slopes. Imaging of the ocean floor surrounding the Canary Islands has revealed blocks of rock ⅔ mile (1 km) wide on the underwater slopes, extending 50 miles (80 km) out from the islands. Geologists have suggested that the Cumbre Vieja volcano on La Palma in the Canary Islands (see right), which is the most active volcano in the group, could collapse and cause a huge submarine mass movement, triggering a tsunami with the potential to devastate the eastern seaboard of the US.

# TYPES OF SLOPES

The development, form, and decline of slopes is a fundamental feature of landscape evolution and depends on a number of elements. These include the relative hardness of the materials forming the slope, and the elements that impact upon it, such as wind and rain. For example, resistant rocks such as granite can form perpendicular cliffs, while softer rock such as shale forms slopes that are more gradual. Free-standing cliffs may form as a result of undercutting by waves or stream erosion, glacial excavation, differential weathering, or collapse along vertical joints. Then, sloping pediments of debris build up and protect the lower parts of the cliffs from further weathering or erosion. In arid and semiarid environments, prominent rock cliffs such as mesas (see p.285) are common, but in areas where the climate is more varied and there is more rainfall, fragmented debris may dominate the slope—unless another geological process, such as glaciation, removes it.

**BUTTES**
*Vertical-faced mesas or smaller buttes, such as these in Monument Valley, Utah, rising above gently sloping pediments, are typical of slope development under arid conditions.*

**SCREE**
*Rock debris that has been loosened by freeze–thaw action (see p.110) tumbles down a hillside, piling up to form scree slopes on the sides of valleys like this one in England.*

**AVALANCHES**
*An avalanche is a downward movement of snow, ice, or debris in a mountainous area, such as Alaska (seen here). It can be triggered by sudden noises or thawing.*

# LANDSLIDES

On January 13, 2001, 100,000 tons of loose volcanic rock slid down a hillside and smashed through the suburb of Santa Tecla, San Salvador. This was one of many landslides produced when a major earthquake hit El Salvador, leading to extensive damage and fatalities. But landslides are not triggered solely by natural causes. Human activity can also set them off, when tree-felling or construction work inadvertently undermines or destabilizes steeply sloping ground.

## SLOPES THAT FAIL

Landslides are mass movements (see p.116) that occur when gravity overcomes friction and pulls soil or rock downhill. During a landslide, part of a slope shears away along a basal failure plane, and initially moves as a coherent block. In the Santa Tecla landslide, the moving material soon broke up, but in some landslides it stays more or less intact as it shifts. An example of this occurred on the coast of Dorset, England, in 1839. Coastal cliffs gave way, carrying clifftop farmland with them, but the fields remained usable, and crops were grown on them for several years.

One of the most common causes of landslides is heavy rainfall, which saturates the ground. Earthquakes and volcanic eruptions are much more violent triggers. Human activities can increase both the risk of landslides and the danger they present.

In mountainous regions, forest clearance is a frequent cause of landslides, because it increases water runoff and accelerates erosion. Road construction is another hazard, because it cuts into hillsides, destabilizing rock. Urbanization also causes problems. In Malaysia, a landslide demolished a 12-story apartment block in 1994, while in Hong Kong, landslides have killed nearly 500 people in the last 50 years. Most of these fatalities have been due to poor construction techniques on steep slopes, aggravated by the region's heavy summer rainfall.

**VAIONT DAM DISASTER**
*Following a landslide, water from the Vaiont reservoir destroyed the town of Longarone in 1963. A second overspill—also caused by a landslide—occurred in 1966.*

Although some landslides strike without any warning, in many cases ground movements show that an area is at risk. During the early 1960s, several small landslips occurred during the construction of the Vaiont Dam in the Italian Alps. During the fall of 1963, rising water levels saturated a clay layer, triggering a massive landslide that plunged into the reservoir below. The dam held, but displaced water surged over its crest, devastating a town in its path. All together, more than 2,000 people lost their lives.

The most catastrophic landslides in recorded history took place during a major earthquake in the Gansu region of China in 1920, which triggered collapses of steep cliffs of fine-grained loess soils. Ten large cities and many villages were buried, and an estimated 200,000 people lost their lives during a single day.

### LARGEST LANDSLIDES

Most of the largest landslides and debris flows in the last 100 years (listed below) have occurred in tectonically active zones, which are subject to earthquakes and eruptions.

| Location | Date | Cause | Fatalities |
|---|---|---|---|
| Gansu, China | 1920 | Earthquake | 200,000+ |
| Vaiont, Italy | 1963 | Dam construction | 2,600 |
| Mount Huascarán, Peru | 1970 | Earthquake | 18,000 |
| Mount St. Helens, Washington | 1980 | Volcanic eruption | 60 |
| Nevado del Ruiz, Colombia | 1985 | Volcanic eruption | 23,000 |
| Huaraz, Peru | 1991 | Lake drainage | 5,000 |
| Honduras and Nicaragua | 1998 | Hurricane | 2,000+ |
| Santa Tecla, San Salvador | 2001 | Earthquake | 500 |

## MUD ON THE MOVE

Landslides move at speeds ranging from a slow, barely perceptible creep to more than 60 mph (100 km/h) on steep slopes. Both debris avalanches and water-saturated mass movements (mudflows) can travel many dozens of miles, causing widespread devastation and smothering everything in their paths. Mudflows are common in the western US during El Niño years (see p.449), and they were responsible for many of the fatalities caused by Hurricane Mitch when it stalled over Central America in 1998, shedding considerable quantities of rain.

Mudflows in volcanic terrain, called lahars, can occur in association with eruptions, or later, as rains remobilize loose ash deposits. Fine ash is particularly dangerous, because it easily forms a fluid mass. When a lahar comes to a halt, the transported mud or ash sets like cement, making escape almost impossible. In 1985, a lahar poured down the flanks of Nevado del Ruiz, a volcano in Colombia (see p.180) and struck the town of Armero, 30 miles (50 km) to the east. Most of its 29,000 inhabitants were asleep, and fewer than a third survived.

Mudflows on this scale are impossible to prevent, but in the future, improved communications may reduce the cost in human lives. A satellite-based system, now under trial in the Caribbean, will gather rainfall data and flash warnings to home computers or even cellular phones.

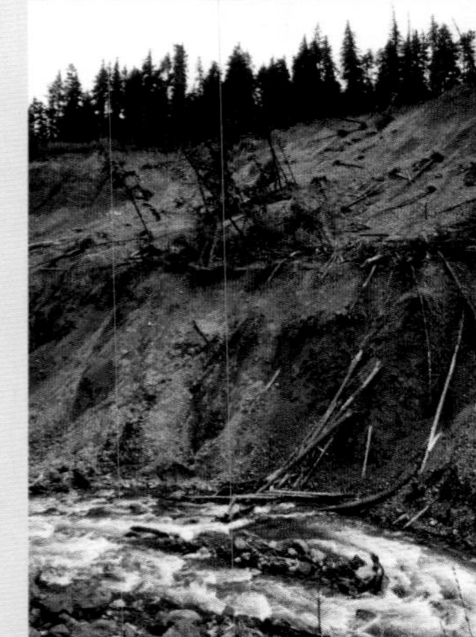

**EXPOSED TO THE ELEMENTS**
*Clearing a forested area aggravates natural erosion, leading to frequent landslides, as has happened here in Cabin Creek in the Cascade Range of western North America.*

**FATEFUL FALL**
*The landslide that hit Santa Tecla, El Salvador, flattened all the buildings its path, while others just a short distance away escaped harm.*

# METEORITE IMPACTS

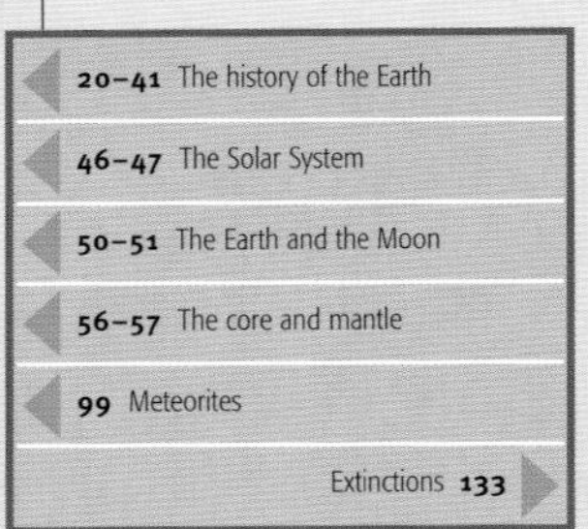

20–41 The history of the Earth
46–47 The Solar System
50–51 The Earth and the Moon
56–57 The core and mantle
99 Meteorites
Extinctions 133


THE EARTH IS regularly struck by debris from space. In the last billion years, it has been hit by about 130,000 meteorites large enough to produce a crater at least ⅔ mile (1 km) wide. Although this kind of impact site is visible on the Moon, the surface of the Earth is so geologically active that evidence of many impacts has disappeared. Until recently, it was believed that the Earth was hit by large meteorites only very early in its history. However, scientists now realize that the Earth is being struck continually, and that a number of giant craters have been preserved. So far, more than 160 impact sites have been identified worldwide, with new craters discovered every year.

**WOLFE CREEK CRATER, AUSTRALIA**
*This Australian impact structure shows features typical of meteorite craters: the near-circular depression is surrounded by a well-preserved blanket of debris.*

## CAUSES OF IMPACTS

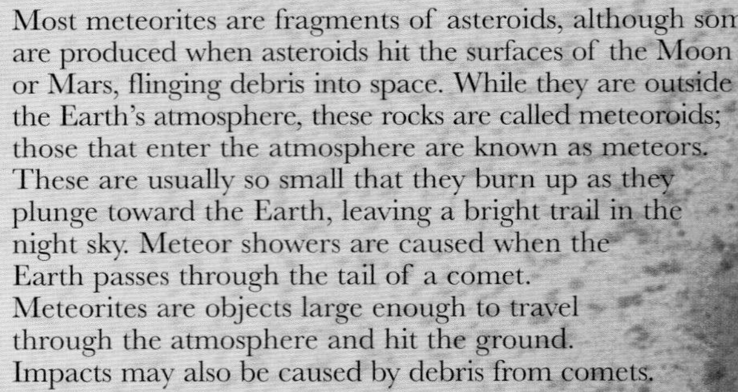

Most meteorites are fragments of asteroids, although some are produced when asteroids hit the surfaces of the Moon or Mars, flinging debris into space. While they are outside the Earth's atmosphere, these rocks are called meteoroids; those that enter the atmosphere are known as meteors. These are usually so small that they burn up as they plunge toward the Earth, leaving a bright trail in the night sky. Meteor showers are caused when the Earth passes through the tail of a comet. Meteorites are objects large enough to travel through the atmosphere and hit the ground. Impacts may also be caused by debris from comets. If a meteorite is the size of a house, or even larger, the force of the impact blasts out a crater.

**METEOR TRAIL**
*Meteors, which are also called shooting stars, are visible in the night sky when they burn up as they travel through the Earth's atmosphere.*

**NAMIBIAN METEORITE**
*This 60-ton rock discovered at Hoba West, Namibia, is the largest meteorite to have been found on Earth.*

### EUGENE SHOEMAKER

US scientist Eugene Shoemaker (1928–97) was an expert on meteorites. His work confirmed the first identification of an impact crater on the Earth: Meteor Crater in Arizona (see p.123). In the process, he and Edward Chao discovered coesite, a silica mineral produced under the extreme pressure and temperature of an impact. The presence of coesite is now taken as a key indicator of an impact. With his wife Carolyn and friend David Levy, Shoemaker discovered Comet Shoemaker-Levy 9, which hit Jupiter in 1994.

## IDENTIFYING IMPACT SITES

The most obvious indication of a meteorite impact is a large, circular crater, and geologists may begin their search for new sites by scouring satellite images of the Earth's surface for such features. Although craters erode over time, scars called astroblemes ("star wounds") remain. Once spotted, each site is investigated for evidence to confirm impact. Usually, a meteorite vaporizes on impact, but shattered or melted remnants may still exist. These fragments can resemble ordinary rock, but if rich in iridium, osmium, or platinum they probably originate from a meteorite. Geologists also look for damage to rocks around the impact site. These effects are called shock metamorphism, and include shatter cones, tektites, diaplectic glass (a natural glass formed during a meteorite impact), and high-pressure silica minerals such as stishovite and coesite.

**TEKTITE**
*Tektites are beads of glass that are formed when droplets of silica-rich molten material are expelled from a crater during a meteorite impact.*

**SHOCKED QUARTZ**
*This image of a quartz grain reveals a key sign of a meteorite impact—shock lamellae. These are deformed layers within the quartz, indicated by bright colors.*

**METEOR CRATER, ARIZONA**
*With its distinctive simple bowl shape, Meteor Crater is the result of the most recent significant meteorite impact on Earth, about 50,000 years ago. It was the first major impact structure to be identified* *(see p.123)*.

# IMPACT CRATERS

When a meteorite strikes the Earth, the impact sends shock waves through the ground, squeezing the surrounding rock to two or three times its usual density. The compressed rock then springs back and shatters into fragments, hurling chunks upward and outward, along with pieces of the meteorite that have not vaporized. The result is a bowl-shaped crater that is much larger than the meteorite. Rock fragments called ejecta are blasted far beyond the crater. Other fragments fall back into the hollow to form rocks called breccia. Small craters form in just a few seconds, and even giant ones form in a few minutes. Meteorite craters vary widely in appearance from small, cup-shaped bowls to giant depressions filled with multiple ridges and mounds. They are divided into two groups: simple and complex (see below).

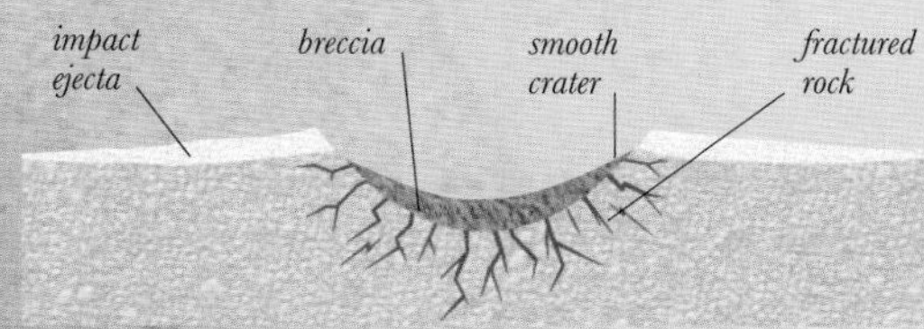


**SIMPLE CRATER**
*Typically no more than 2 1/2 miles (4 km) in diameter, simple craters are smooth bowls that are wider than they are deep. They often have a steep, well-defined rim, and there may be a layer of breccia at the bottom of the crater that is thicker at the center.*

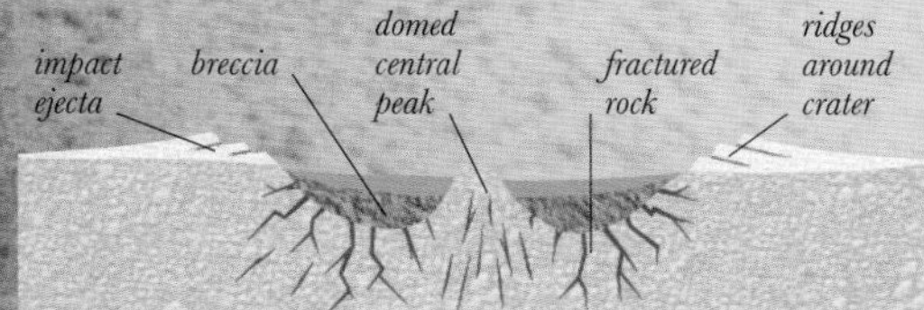


**COMPLEX CRATER**
*Complex craters are larger than simple craters, and they are usually shallower. This type of crater often has rings of ridges. Many also have a domed central peak, which was raised as the compressed rock sprang back after impact.*

# AFTER IMPACT

The potential energy of a significant meteorite impact on the Earth is over 100 million megatons—more than the world's entire nuclear arsenal. Studying nuclear explosions has given scientists some idea of what might happen after the impact of a large meteorite about 6 miles (10 km) in diameter. At the moment of collision, an immense wave of heat and pressure roars outward, flattening and incinerating everything over a vast area. Massive amounts of debris are blasted high into the atmosphere and blown around the world by strong winds. Hot ash rains down, starting forest fires, while dust clouds linger in the atmosphere for months, blocking out the Sun and turning the Earth into a dark and frozen planet. When the sky finally clears, the carbon dioxide that has flooded the atmosphere creates a greenhouse effect, warming the global climate by an average of 27°F (15°C).

**DINOSAUR EXTINCTION**
*The global effects of a large meteorite may be so devastating that scientists suspect they are the cause of many of the mass extinctions that have occurred in the Earth's history, including the death of the dinosaurs 65 million years ago.*

## SHATTER CONES

Among the most persuasive signs of a meteorite impact are shatter cones. These are rock structures that have distinctive fractures resembling horsetails. The fractures converge in a cone shape, which can range in size from 1 in (2.5 cm) to over 6 ft (2 m) in length. Fractures such as these are only caused by sudden, intense pressure on existing rock. In impact craters, most shatter cones point upward, which indicates that the impact on the rock came from above.

## METEORITE IMPACT PROFILES

The pages that follow contain profiles of the most significant impact craters in the world. Each profile begins with the following summary information:

**DATE OF IMPACT** Approximate date on which the meteorite hit the Earth

**TYPE** Simple or complex

**DIAMETER** Width across the crater

ARCTIC

## Haughton

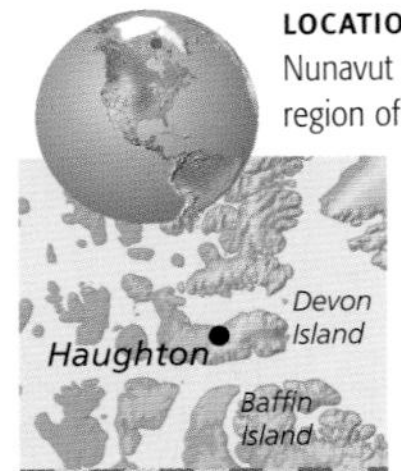


**LOCATION** On Devon Island, Nunavut Territory, in the Arctic region of northwest Canada

**DATE OF IMPACT** 23 million years ago

**TYPE** Complex

**DIAMETER** 15 miles (24 km)

Most craters that are as old as the Haughton impact site have undergone significant erosion. However, because it lies within the Arctic Circle, where much of the water is frozen solid, Haughton has escaped the worst effects of weathering. The lack of vegetation covering the site makes it easy for geologists to study the crater both on the ground and from space using satellites. Studies have shown that the giant meteorite that struck here penetrated 1 mile (1.7 km) into the ground—as far as the buried ancient rocks of the continental crust. As it crashed into the Earth, it threw into the air a huge shower of shattered fragments of this ancient crust, which can now be found on the surface.

**ARCTIC IMPACT**
*The Haughton meteorite struck within the hilly landscape of Devon Island in Arctic Canada.*

### MARS TEST

Rocky, arid, and icy cold, Haughton Crater resembles conditions found on the surface of Mars. Scientists have been studying the crater to see what it can tell them about the geology of the Red Planet and also how astronauts could survive on Mars.

NORTH AMERICA *northeast*

## Manicouagan

**LOCATION** Within Réservoir Manicouagan, north of the Laurentian Mountains, Quebec

**DATE OF IMPACT** 212 million years ago

**TYPE** Complex

**DIAMETER** 60 miles (100 km)

The Manicouagan Crater is one of the largest and most intact impact structures on the Earth's surface. Its width is exceeded by only four other craters in the world and matches that of the Popigai Crater in Russia (see p.125). It also one of the Earth's oldest craters, caused by the impact of a huge meteorite in the Triassic Period. The site is notable for its circular central plateau, which is surrounded by the ring-shaped Lake Manicouagan. The crater has been the subject of a great deal of speculation about whether the impact caused a wave of global extinction. Paleontologists identify the boundary of the Triassic and Jurassic periods as one of the times in the Earth's history when vast numbers of species disappeared in a short span of time, an incident known as the T-J event. One theory to explain this is that the Manicouagan meteorite impact threw so much rock and dust into the atmosphere that its effect on the planet's climate destroyed many living organisms. However, some scientists believe that the impact happened too early to be responsible for this, and that other cataclysmic events caused the mass extinctions.

**RIVERS AND LAKE**
*Manicouagan Crater has become a focus for numerous surrounding rivers and streams, which flow into its lake.*

NORTH AMERICA *northeast*

## Sudbury

**LOCATION** North of Lake Huron and the town of Sudbury, Ontario

**DATE OF IMPACT** 1,840 million years ago

**TYPE** Complex

**DIAMETER** 125 miles (200 km)

The Sudbury Crater is such an unusual shape that it took geologists a long time to identify it as a meteorite crater. The area is one of the Earth's richest deposits of nickel-copper sulfide minerals, and was initially thought to be of volcanic origin. Now, however, scientists are certain that it is an impact site. The object that smashed into the Earth here produced a crater about 12 miles (20 km) deep. What makes the crater so unusual is the fact that it is elliptical and is nearly twice as long as it is wide. The crater is now believed originally to have been circular—its elliptical shape was produced by post-impact deformation of the Canadian Shield, the stable continental rocks in which the impact took place.

**NICKEL MINING**
*The highly productive mines that lie within the Sudbury Crater yield over 1,000 tons of nickel and copper ores every day.*

NORTH AMERICA *central*

## Manson

**LOCATION** Around the town of Manson, northwest of Des Moines, Iowa

**DATE OF IMPACT** 74 million years ago

**TYPE** Complex

**DIAMETER** 23 miles (37 km)

Completely buried up to 295 ft (90 m) beneath glacial deposits, the Manson impact site was thought to be a volcanic structure, but studies in the 1960s revealed its true origin. Initially, the crater was dated at about 66 million years old, so scientists began to wonder if it might be linked to the extinction of the dinosaurs, which occurred at a similar time. In the 1990s, scientists drilled about a dozen research cores to find out more about Manson. They found that the structure consists of a ring-shaped moat around a central peak. They also revised the date of impact to 74 million years ago, well before the demise of the dinosaurs. The effects of the impact must been enormously destructive, killing most wildlife within 600 miles (1,000 km).

NORTH AMERICA *east*

## Chesapeake Bay

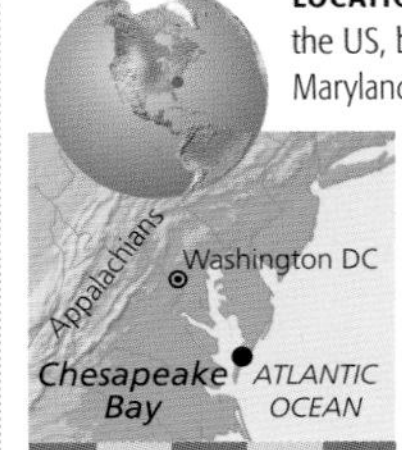


**LOCATION** On the Atlantic coast of the US, beneath Chesapeake Bay, Maryland, and Virginia

**DATE OF IMPACT** 35 million years ago

**TYPE** Complex

**DIAMETER** 53 miles (85 km)

One of the largest impact craters on Earth, Chesapeake Bay was identified by chance in the 1980s when teams drilling the sea bed for water sources found samples of jumbled rocks, and an ocean survey ship found shattered quartz and microtektites. Subsequent seismic and gravity surveys then revealed the huge extent of the crater.

**SATELLITE IMAGE OF CHESAPEAKE BAY**

NORTH AMERICA *west*

## Meteor Crater

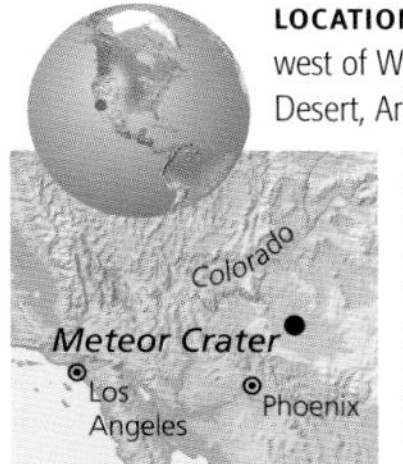


**LOCATION** East of Flagstaff and west of Winslow, in the Painted Desert, Arizona

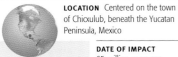

**DATE OF IMPACT** 50,000 years ago

**TYPE** Simple

**DIAMETER** 3/4 mile (1.2 km)

A vast dish carved into the Arizona desert, Meteor Crater is perhaps the Earth's most famous meteorite crater. Its shape has been remarkably well preserved, and it is comparatively young for an impact site. Scientists estimate that the meteorite that crashed here was about 150 ft (50 m) across and weighed about 300,000 tons. It probably hurtled into the ground at a speed of 40,000 mph (65,000 km/h), creating an immense force as it hit the surface—equivalent to about 20 million tons of TNT exploding. Fragments of the nickel-iron meteorite that impacted are scattered all around the crater, as are tiny metallic balls, which were created as the meteorite vaporized on impact. It was the presence of nickel-iron that led Daniel Moreau Barringer (see panel, right) to conclude that the crater was the result of a meteorite impact.

Further crucial evidence was discovered in 1960 by Eugene Shoemaker, Edward Chao, and Daniel Milton, who found coesite and stishovite at the site. These minerals form only under enormous pressure and at high temperatures, and are key signs of an extraterrestrial impact. Shoemaker's demonstration three years later of the similarities between Meteor Crater and craters made by nuclear tests in Nevada clinched the argument.

**CRATER AND CANYON**
*The Colorado Plateau is scarred by two dramatic geological structures: Meteor Crater (left) and the Grand Canyon (right).*

### DANIEL MOREAU BARRINGER

Meteor Crater is also called Barringer Crater after the mining engineer who was determined to prove its impact origin, Daniel Moreau Barringer (1860–1929). He found it hard to convince geologists of his theory, which he put forward in 1902, and lost a fortune searching for the meteorite he was convinced was buried there. It was later realized that the meteorite had vaporized on impact.

**ARIZONA'S GIANT IMPACT SITE**
*Meteor Crater is a dramatic, cup-shaped bowl in the Arizona desert. It was the first meteorite impact site on the Earth's surface to be positively identified.*

NORTH AMERICA *southeast*

## Chicxulub

**LOCATION** Centered on the town of Chicxulub, beneath the Yucatan Peninsula, Mexico

**DATE OF IMPACT** 65 million years ago

**TYPE** Complex

**DIAMETER** 100–150 miles (160–240 km)

Buried deep beneath the limestone of Mexico's Yucatan Peninsula, Chicxulub is one of the largest meteorite craters yet found on the Earth. Some estimates put its width at over 185 miles (300 km). At the end of the Cretaceous Period, 65 million years ago, a comet or an asteroid that may have been more than 6 miles (10 km) wide crashed into the Earth at this site. The impact was cataclysmic: fires raged over the surface, giant tsunamis radiated across the oceans, and the planet was rocked by massive earthquakes. Many scientists believe that the devastating global effects of the Chicxulub impact caused the extinction of the dinosaurs (see p.34) and two-thirds of other animal species. Both the size of the catastrophe and its timing are right, and the evidence is very convincing. The Chicxulub Crater is a complex crater with several ring-shaped ridges and a central mound like that of the Manicouagan Crater (see opposite page). To explain its shape, scientists have suggested that the impact deformed the solid rock of the Yucatan Peninsula rapidly, almost like a fluid.

Rings rippled out from the point of impact, and at the center, a towering mound, at least twice the height of Mount Everest, piled up and then subsided gradually. Because it is buried so deep beneath the surface, Chicxulub was discovered almost by accident when an oil company was prospecting in the area in the 1980s.

**CATASTROPHIC CHICXULUB**
*Magnetic and gravity surveys were used to create this vertically exaggerated computer image of the crater hidden about 2/3 mile (1 km) beneath the Yucatan Peninsula. The top of the image represents north.*

### K-T BOUNDARY

About 65 million years ago, two-thirds of the Earth's animal species died out suddenly. Scientists link this event to a thin layer of clay that dates from the same time, which is sandwiched between Cretaceous (K) and Tertiary (T) rocks. Called the K-T boundary, this clay contains high concentrations of iridium, an element rare on the Earth but abundant in meteorites. It is likely that this layer represents debris from the Chicxulub impact that was suspended in the atmosphere. This would have blocked out the Sun's rays, turning the world ice-cold and spelling doom for the dinosaurs.

TOWN BUILT WITHIN A CRATER
*Within the Ries Crater lies the medieval walled town of Nördlingen, which was built on the dried sediments of the lake that once filled the impact site.*

EUROPE *northwest*

## Ries

**LOCATION** In southern Bavaria, northwest of Munich, western Germany

**DATE OF IMPACT** 15 million years ago

**TYPE** Complex

**DIAMETER** 15 miles (24 km)

The Ries Crater is perhaps the most studied impact structure in Europe. It is a flat, circular basin with a narrow rim around the edge. The Ries meteorite is thought to have been about ⅔ mile (1 km) across. Like some other impact sites, the Ries Crater is not alone in the landscape. Just 22 miles (36 km) to the southwest is another, less perfectly shaped basin called the Steinheim Crater, which measures only 2.4 miles (3.8 km) across. This was created by a much smaller meteorite that was barely 300 ft (100 m) long. Radioactive dating suggests that both these meteorites struck at the same time. It is possible that they were the result of a single meteorite that disintegrated as it came through the atmosphere, but the craters are so far apart that this seems unlikely. Some scientists think it more probable that two meteorites swept into the atmosphere together. When the larger Ries meteorite struck, it penetrated the limestone and marl sediments at the surface and plunged right through to the crystalline crustal basement over 2,000 ft (600 m) below the surface of the Earth. The shock melted surface rocks, which were ejected from the site, cooling into spheres of transparent green glass (tektites) that were scattered across the landscape of Bohemia and Moravia in the Czech Republic. Clear examples of this green glass, called moldavite, are today cut to make semi-precious stones. Soon after the impact, both the Ries and the Steinheim craters filled with water to become lakes, but over the next 2 million years they became choked with sediment and eventually formed dry land.

AFRICA *west*

## Bosumtwi

**LOCATION** Beneath Lake Bosumtwi, northwest of Accra and south of Kumasi, Ghana, west Africa

**DATE OF IMPACT** One million years ago

**TYPE** Complex

**DIAMETER** 6 miles (10 km)

Hidden deep within rainforest, and drowned beneath the waters of the lake that fills its crater, Bosumtwi has been more awkward to study than other impact sites. Moreover, the heat and humidity of the region have resulted in rapid erosion of many of its features. However, scientists have managed to confirm that the crater is complex, with a raised central dome buried beneath the lake floor. Tektites of a similar age found along the Ivory Coast are now thought to have been flung out by the Bosumtwi impact.

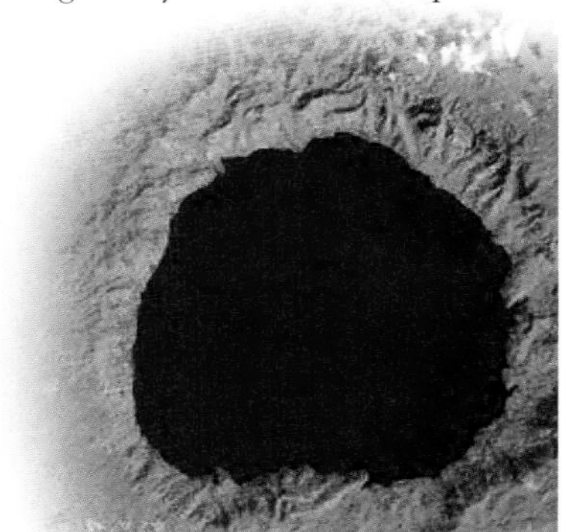

LAKE BOSUMTWI

AFRICA *south*

## Vredefort

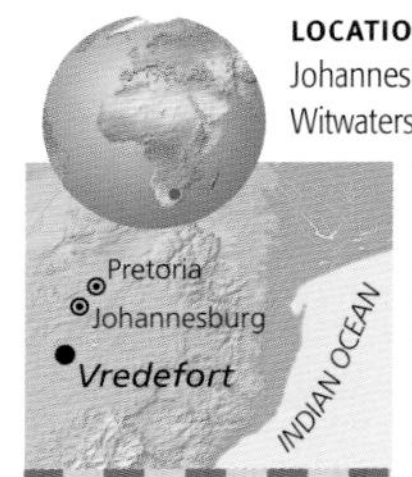


**LOCATION** Southwest of Johannesburg, east of the Vaal River, Witwatersrand Basin, South Africa

**DATE OF IMPACT** 2,020 million years ago

**TYPE** Complex

**DIAMETER** 180 miles (300 km)

The Vredefort impact site is marked by the oldest and largest known crater on the Earth's surface It is also one of the few multi-ringed craters in the world. Such structures are common on the Moon but are rare on the Earth, since most would have formed in the early days of the planet's history and have been destroyed by geological processes. It was initially thought that the dome in the center of the Vredefort Crater was the result of a volcanic explosion. However, since the mid-1990s, mounting mineralogical evidence, accompanied by the discovery of dramatic distortion of rocks within the dome, has convinced scientists that it is the site of a meteorite impact. Shatter cones are found frequently in the bed of the nearby Vaal River, while rocks that lie flat outside the crater have been forced into almost vertical positions within the basin. The extraterrestrial body that hit Vredefort was one of the biggest meteorites ever to have hit the Earth, measuring about 6 miles (10 km) across.

### MINING CRATERS

Many impact sites are also the locations of valuable mineral ores, which have subsequently been exploited by mining companies. At Sudbury Crater in Canada (see p.122), the rich nickel and copper ores were formed by the heat of the meteorite's impact. The older sediments beneath Vredefort are rich in gold, which is preserved beneath the crater. Meteorite craters also make very good traps for oil, as it works its way toward the surface through cracks formed by an impact.

GOLD MINING IN VREDEFORT
*These mines yield over $7 billion worth of gold annually, and contain the largest concentration of the precious metal in the world.*

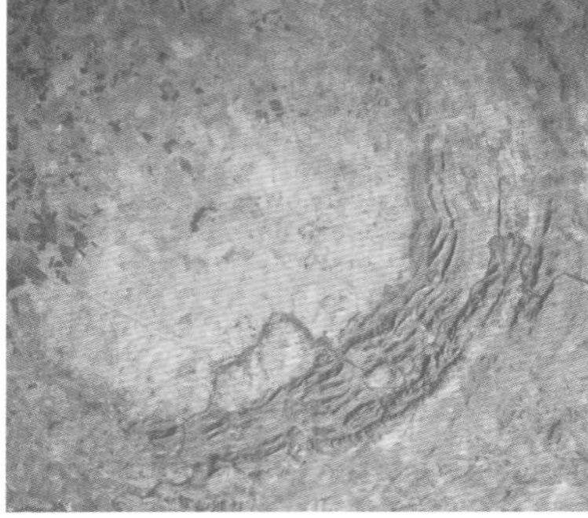

EARTH'S OLDEST CRATER
*This satellite view of the Witwatersrand Basin in South Africa clearly reveals the circular wrinkles at the edge of the giant Vredefort structure.*

ASIA *west*

## Kara and Ust Kara

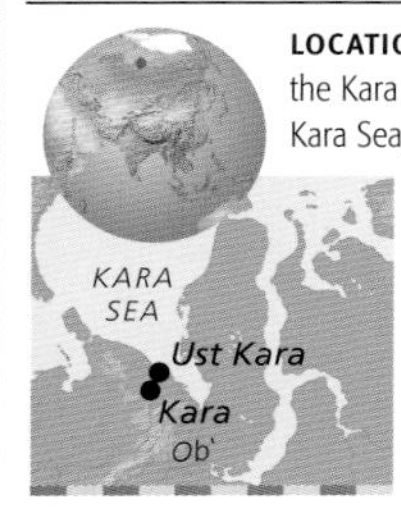


**LOCATION** Close to the mouth of the Kara River, near the coast of the Kara Sea, northwest Siberia, Russia

**DATE OF IMPACT** 70 million years ago

**TYPE** Complex

**DIAMETER** 40–75 miles (65–120 km)

Discovered in the 1970s, the Kara Crater is situated in an area of bleak, barely inhabited tundra. It was initially believed that there was just one crater here, about 40 miles (65 km) wide. Then more impact outcrops were spotted to the west, by the coast of the Kara Sea (see p.400). Some geologists suggested that this was another crater, named Ust Kara, that formed at the same time. However, the results of scientific expeditions in 2001 concluded that the two sites probably form one vast crater. Its age once led scientists to suggest the crater as an alternative candidate to Chicxulub (see p.123) for the K-T event that might have wiped out the dinosaurs, but it is too old.

ASIA *west*

## Tunguska

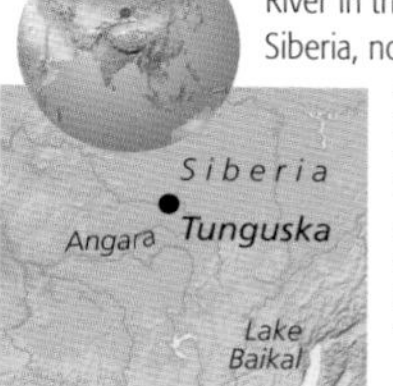


**LOCATION** Near the Tunguska River in the forests of eastern Siberia, north of Lake Baikal, Russia

**DATE OF IMPACT** 1908

**TYPE** No crater has been discovered

**DIAMETER** Unknown

Tunguska is one of the Earth's great mysteries. On June 30, 1908, local people witnessed a fireball streaking through the sky before vanishing below the horizon. A huge explosion followed, felling trees over an area of 850 square miles (2,200 square km). There have been several expeditions to the Siberian site to try to discover the cause of the explosion; one theory is that it was due to a comet vaporizing before it hit the surface.

DEVASTATION AT TUNGUSKA

ASIA *northeast*

## Popigai

**LOCATION** Near the town of Popigai, east of the Popigai River, northern Siberia, Russia

**DATE OF IMPACT** 35 million years ago

**TYPE** Complex

**DIAMETER** 62 miles (100 km)

The vast Popigai Crater is one of the Earth's largest impact sites, and was probably caused by a 3-mile- (5-km-) wide meteorite. The force of the impact was sufficient to turn graphite into microscopic diamonds, and ejecta discovered as far away as Massignano, Italy, may have originated there. The timing of the Popigai collision coincides closely with the impact at Chesapeake Bay (see p.122). This has led to the conjecture that at the time of the impact, the Earth may have been subjected to a comet shower.

POPIGAI IMPACT SITE SEEN FROM SPACE

ASIA *southwest*

## Lonar

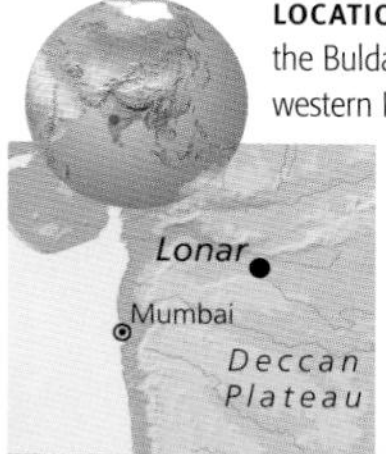


**LOCATION** Northeast of Mumbai, in the Buldana district of Maharashtra, western India

**DATE OF IMPACT** 50,000 years ago

**TYPE** Simple

**DIAMETER** 1 1/10 mile (1.8 km)

Filled with a lake and surrounded by temples and dense woodland, Lonar is undoubtedly one of the world's most spectacular meteorite impact sites. According to Hindu legend, the crater was created by Lord Vishnu when he destroyed the demon Lavanasur. The site was first recognized as a meteorite impact crater by the American geologist G.K. Gilbert in 1896. The crater is about 560 ft (170 m) deep, with a rim that is 65 ft (20 m) high. Estimates suggest the meteorite that struck here was about 200 ft (60 m) long. The Lonar Crater is remarkable in that the impact occurred in hard basalt, and therefore the shape of the crater is exceptionally well preserved. The salty, alkaline chemistry of the lake means that it is inhabited by unique species of flora and fauna.

**SALT LAKE**
*Surrounded by a rim thrown up around the crater by the meteorite's impact, Lonar's lake is one of the most saline on Earth, and is an almost perfect circle.*

AUSTRALASIA *Australia*

## Woodleigh

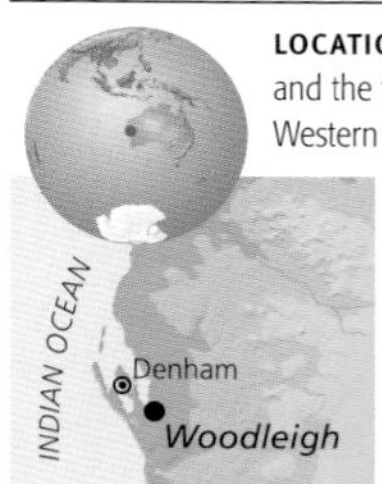


**LOCATION** South of Shark Bay and the town of Denham, Western Australia

**DATE OF IMPACT** 250–364 million years ago

**TYPE** Complex

**DIAMETER** 37 miles (60 km)

Woodleigh was discovered by mining prospectors in the 1970s, but it wasn't until a 1997 survey revealed a telltale dome that a meteorite impact was suspected. Shocked quartz, which was drilled from the dome in 1999, confirmed the site's origins. An impact this large, from a meteorite about 2 miles (3 km) wide, probably caused major devastation to the planet's living organisms.

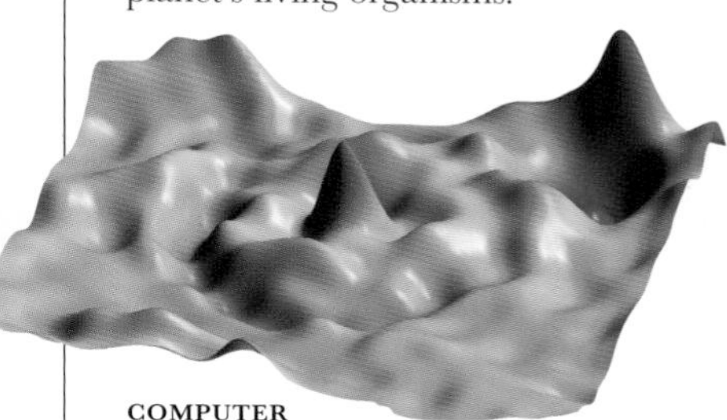

COMPUTER IMAGE OF WOODLEIGH CRATER

AUSTRALASIA *Australia*

## Gosses Bluff

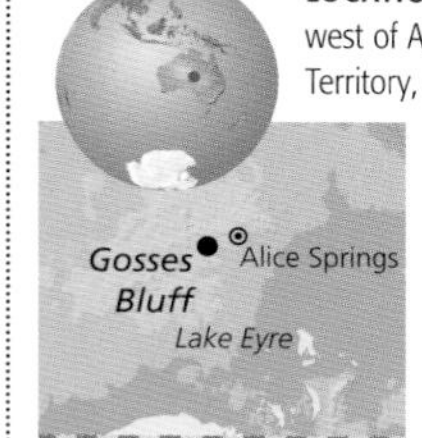


**LOCATION** Near Hermannsburg, west of Alice Springs, Northern Territory, Australia

**DATE OF IMPACT** 142 million years ago

**TYPE** Complex

**DIAMETER** 14 miles (22 km)

One of the best-preserved meteorite craters in the world, Gosses Bluff is an imposing sandstone ring in the northern Australian desert. The site is very important to the Western Arrerent Aboriginal people, featuring in some of their Dreamtime stories. The crater formed when an asteroid 2/3 mile (1 km) wide smashed into the surface of the Earth. The meteorite is thought to have penetrated 3 miles (5 km) into the ground and vaporized. The ground then rebounded, forming the crater's central dome. Over millions of years, some of the crater's original features have eroded and weathered, and the entire land surface of the region is about 1 1/4 miles (2 km) lower than it was at the time of impact. However, in addition to the striking outer ring, there remains a 2 3/4-mile- (4.5-km-) wide inner ring, which is overlain with breccias and the remnants of the central dome.

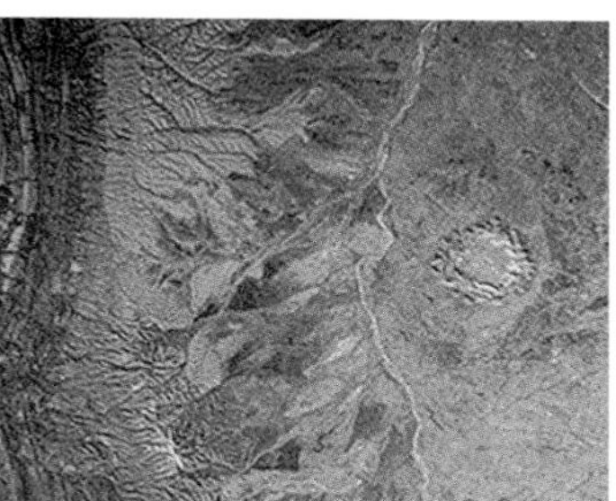

**CENTRAL UPLIFT**
*The true shape of the Gosses Bluff Crater, with its ring of sandstone hills and its slightly raised central area, becomes clearly visible in this photograph from space.*

**SANDSTONE CIRCLE**
*Rearing out of the desert like a castellated wall, the outer ring of the Gosses Bluff Crater is a striking reminder of the massive asteroid impact millions of years ago.*

# WATER

Water resources 130–31
Rivers 212–13
Lakes 236
Freshwater quality 242–43
Glaciers 262–65
Ocean water 384–85
Polar oceans 394–95
Precipitation and clouds 466–73

THE EARTH'S OUTER LAYERS are dominated by water. More than two-thirds of its surface is covered with liquid water; if frozen water, in the form of ice, is also included, this proportion rises to more than four-fifths. Water is essential to life because it is an excellent solvent and it can move or flow easily. Living organisms require not merely the presence of water, but a continuing supply for the maintenance of life. Humans are typical of most animals, being 62 percent water; soft-bodied aquatic creatures such as jellyfish are more than 98 percent water.

**SURFACE TENSION**
*The film that forms on water is strong enough to support the weight of insects such as the common back-swimmer. This predatory bug can hang down from the surface of ponds.*

## PROPERTIES OF WATER

Pure water has no color, taste, or smell. Its freezing point, at which it turns to solid ice, is 32°F (0°C), and its boiling point, where it changes into water vapor, is 212°F (100°C). Water's density, or mass per volume, is exactly 1 kg per liter, or about 1 lb per pint. This is high for a liquid—that is, water is relatively heavy—and the figure is used as a standard for comparing densities of other substances. Each molecule of water is made of two hydrogen atoms and one oxygen atom (written $H_20$). A pinhead-sized drop of water contains about one billion billion molecules. These attract each other powerfully, especially at the surface of the water, where their mutual attraction forms a strong "skin," known as surface tension. Since water is difficult to break by tension, it can "creep" into small holes and along narrow cracks. Its ability to change state easily, and to spread or flow even through rock, means that water is always on the move through global and local cycles (see p.129).

## SALT WATER

Salt water is found in seas and oceans, coastal lagoons, and inland lakes with few or no outlets, such as the Dead Sea (see p.251). In the latter, minerals are washed into drainage basins, where they accumulate at high concentrations because pure water evaporates relatively quickly. The main dissolved minerals in seawater are sodium (10.8 parts per thousand) and chloride (19.35 ppt), with smaller amounts of other salts (see p.384). The freezing point of salt water is 28.6°F (–1.9°C) at 35 ppt. Overall salinity (or salt concentration) is highest in the tropics (above 35 ppt), where evaporation occurs fastest, and lowest near the poles (below 30 ppt), where there is slower evaporation and melting ice. Seawater is slightly less saline than the body fluids of most marine invertebrates, so water tends to enter their bodies by osmosis; to counteract the inflow, they have body structures such as gills and nephridia (tiny tubes) that are able to pump water out. Marine vertebrates, including most fish, face the opposite problem. Seawater is two to three times more concentrated than their body fluids, so they lose water through their gills and other surfaces. To counteract this, they drink seawater and rapidly filter out the salts, mainly through the kidneys and other specialized organs, to produce small volumes of highly concentrated urine.

**SOCKEYE SALMON**
*Salmon hatch in fresh water, migrate to the sea to mature, then return to their home river to breed. Their physiology changes to enable them to cope with varying salinity.*

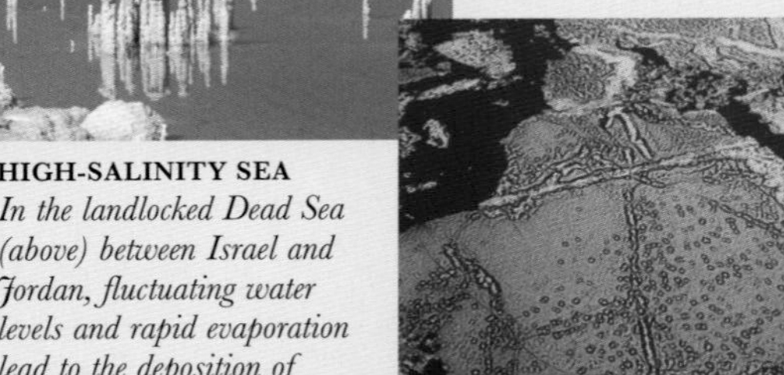

**HIGH-SALINITY SEA**
*In the landlocked Dead Sea (above) between Israel and Jordan, fluctuating water levels and rapid evaporation lead to the deposition of minerals in pillarlike shapes.*

**LOW-SALINITY PACK ICE**
*Pack ice is a typical feature of low-salinity seas, such as the Beaufort Sea off northern Canada (left). In winter, the seawater freezes easily to form a solid sheet, which breaks up in summer.*

**RJUKANDI WATERFALL**
*A mass of water flows over this Icelandic waterfall. As water moves in this way, it releases a large amount of erosive energy.*

# FRESH WATER

Of all the water on Earth, about 3 percent is fresh—that is, it contains far lower concentrations of dissolved minerals than salt water. Almost four-fifths of this fresh water is locked up in polar regions as glacier ice; the other fifth lies under the surface, as groundwater in rocks. Only 0.3 percent of fresh water is in liquid form at the surface, in rivers, lakes, and wetlands, yet this is the form most useful for land animals and many aquatic creatures and plants. Although it may look pure, natural fresh water usually holds measurable amounts of at least 25 dissolved minerals, such as silica, calcium, potassium, magnesium, and iron. The relative terms "hard" and "soft" are used to describe water with higher or lower levels of these minerals, respectively. Rainwater is naturally slightly acidic. It absorbs carbon dioxide gas from the air and from the soil, and becomes a weak form of carbonic acid. This is almost insignificant compared to acid rain formed by atmospheric pollution (see p.111), but it is enough to dissolve chemically rocks such as limestones, where cracks gradually widen to form potholes and caverns (see p.252). The temperature of fresh water varies hugely. Far underground, it may be warmed by hot rocks until it is superheated. It may then burst forth as a geyser.

**ACIDIC WATER**
*Bogs are isolated from alkaline groundwater and depend entirely on rainwater, which is relatively mineral-poor. As a result, they are naturally acidic, a condition that favors the growth of sphagnum moss, but few other plant species.*

**MINERAL TERRACE**
*At Pamukkale in southwest Turkey, hard water reaches the surface as hot springs. Minerals in the water precipitate out and become solid, forming terraces (below).*

**HOT WATER**
*The temperature of hot springs (above) may exceed boiling point. Yet life survives here, especially in the form of thermophilic, or heat-loving, microorganisms such as blue-green algae.*

# ICE

At any one time, almost four-fifths of all fresh water is frozen solid as ice. Ice exists in numerous forms, including mountain glaciers, polar ice sheets, icebergs, periglacial landforms (see p.342), and mountaintop coverings. The mass of floating ice in the Arctic Ocean is, on average, 16–23 ft (5–7 m) thick. The ice covering the landmass of Antarctica has accumulated over thousands of years to become more than 2.8 miles (4.5 km) thick in places. Ice is slightly less dense than liquid water, so icebergs and ice cubes float (although only one-eighth to one-tenth of their bulk is exposed above the surface). As water cools toward its freezing point, it becomes most dense at 39°F (4°C). Below this temperature, it starts to expand again. This property is helpful to aquatic life. In a pond or lake, the densest water at 39°F (4°C) sinks to the bottom, which enables animals to survive beneath the ice that forms at the colder surface.

**ICEBERG**
*As an iceberg warms, chunks break off and it floats at different levels in the water, leaving a stacked series of grooves eroded by the waves.*

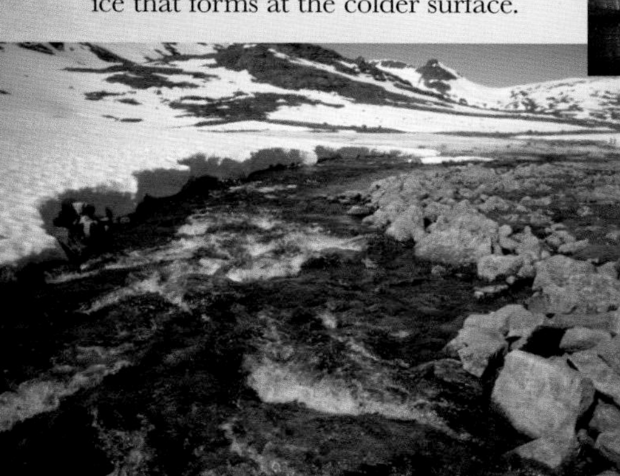

**MELTWATER**
*Rising temperatures in spring melt ice and snow on higher ground, producing a surge of water along rivers. The tips of glaciers also release meltwater.*

# WATER IN THE ATMOSPHERE

The proportion of water vapor in the atmosphere is known as absolute humidity. It varies with air temperature and pressure, from almost zero to four percent by volume. Often a more useful measure is relative humidity. This compares the actual amount of water vapor in a given volume of air to the maximum amount which that air could hold. When relative humidity is 100 percent, the air is completely saturated and cannot hold any more moisture. At this point, no further evaporation can take place. Water also exists in the atmosphere in both liquid and solid states. When water vapor condenses around particles of dust, it forms either tiny droplets or ice crystals, depending on the air temperature. Large masses of droplets or crystals become visible in the form of clouds (see pp.466–73). Another feature of water droplets and ice crystals is that they float easily in even the weakest air currents, and so are continually colliding and merging. If they grow sufficiently large, they fall as either rain or snow. Sometimes, powerful winds create updrafts that blow raindrops upward into supercooled clouds, where they rapidly freeze, only to fall again as pellets of frozen rain (or hailstones).

**SNOWSTORM**
*Windblown snow collects on the sides of the Grand Canyon, Arizona, during a storm. In low atmospheric temperatures—for example, in winter in temperate regions or at high altitudes—clouds may be made of ice crystals, and precipitation takes the form of snow.*

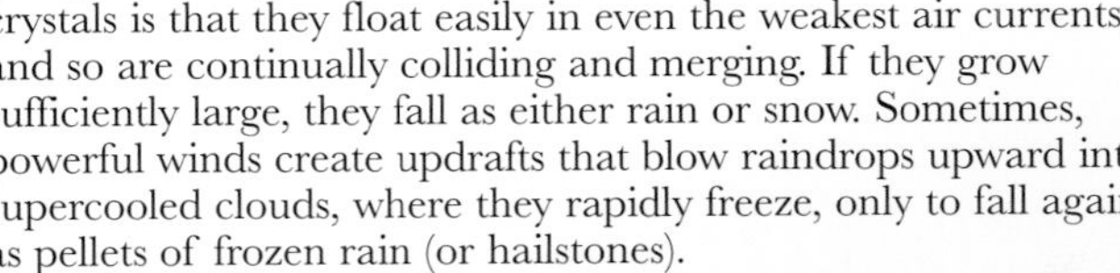

**ALTOCUMULUS**
*These mid-altitude clouds occur when a large air mass is forced to rise, leading to cooling and condensation over a wide area. They can be gray or white or both.*

**CIRRUS**
*The wispy strands of cirrus clouds are made of ice crystals blown into shape by the wind. They form when an air mass cools at high altitude and becomes saturated, forming ice instead of water droplets.*

PLANET EARTH

**RAINSTORM OVER BOHOL SEA**
*Banks of moisture-laden clouds cause a squall off the southern Philippines. Clouds contain the same mixture of water vapor and droplets as steam.*

# THE GLOBAL WATER CYCLE

Water represents only 0.2 percent of Earth's weight, but the global water cycle, or hydrological cycle, is one of the world's most important systems. It is responsible for the continuous circulation of water around the planet. The Sun provides a massive input of energy to Earth in the form of solar radiation, especially infrared radiation or heat (see p.442). This warms liquid water, which evaporates into invisible water vapor and then dissipates into the atmosphere. About 22 percent of the solar radiation that reaches Earth heats oceans, seas, lakes, and rivers, changing liquid water into water vapor. Only pure water evaporates; any dissolved minerals and other substances in it are left behind, mainly as salts in the sea. This process is, in effect, solar-powered distillation (purification) of water. Water vapor is also given off by plants during transpiration, and by animals in exhaled air and perspiration. As water vapor rises into higher, cooler regions of the atmosphere, it condenses to form clouds. Eventually, this returns to the land as precipitation. Here, it is warmed by the Sun—and so the global water cycle continues.

**PRECIPITATION**
*Rain, hail, sleet, snow, frost, and dew form a vital part of the global water cycle. They return water in the atmosphere to the Earth's surface, and in the process release the latent energy stored in water vapor.*

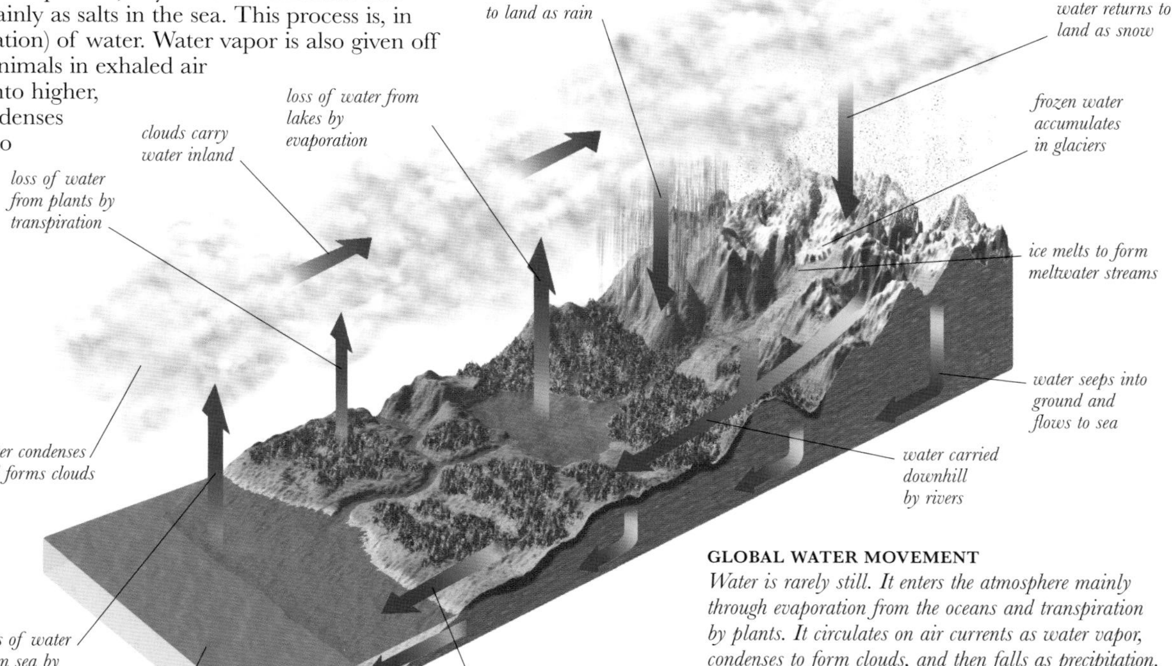


**GLOBAL WATER MOVEMENT**
*Water is rarely still. It enters the atmosphere mainly through evaporation from the oceans and transpiration by plants. It circulates on air currents as water vapor, condenses to form clouds, and then falls as precipitation. Back on the land, water moves as rivers and glaciers, and it soaks into the soil and rocks; it is also stored in lakes, oceans, and the tissues of living things.*

**BLUE-AND-WHITE PLANET**
*From space, the Earth appears mainly blue and white. These areas are the visible forms of water, in seas and oceans, and in clouds. Water is the chief weather-determining component of the atmosphere.*

**FRESHWATER SPRING**
*The Ozark Plateau has many springs, including Greer Springs, Missouri (above), which releases an average flow of 264 million gallons (a billion liters) of water each day.*

# THE LOCAL WATER CYCLE

Many interconnected local water cycles together make up the global water cycle. On a local scale, water movement is determined by factors such as the amount and type of precipitation, the contours of the landscape, and the geology (the composition and layering of soil and rock). Water can percolate through porous or permeable rocks such as chalks, but not through impermeable types such as clay. So regions with porous soils and rocks tend to have less surface drainage, in the form of streams and rivers, because rain and other precipitation soaks straight down. Percolating water accumulates above poorly permeable layers, saturating rocks and creating groundwater deposits. A rock formation that conducts groundwater is known as an aquifer, which people can draw upon for water wells. The upper level of groundwater saturation is called the water table, and usually rises and falls with varying precipitation through the seasons. Where the surface of the land intersects an aquifer, water may escape from it as a spring. Alternatively, the water can percolate unseen below the surface, through to rivers, lakes, and wetlands.

## PAVED LANDSCAPE

Today, vast areas of land are covered with waterproof layers, such as asphalt and concrete, that greatly alter local water cycles. Water can no longer percolate easily into porous soils and rocks to replenish groundwater supplies. Instead, it flows into drains and channels and rushes away into larger watercourses. This can lower the water table, cause springs to dry up, and deplete local groundwater supplies needed for irrigation and human use. The sudden rush of surface water after a storm may also cause local flooding.

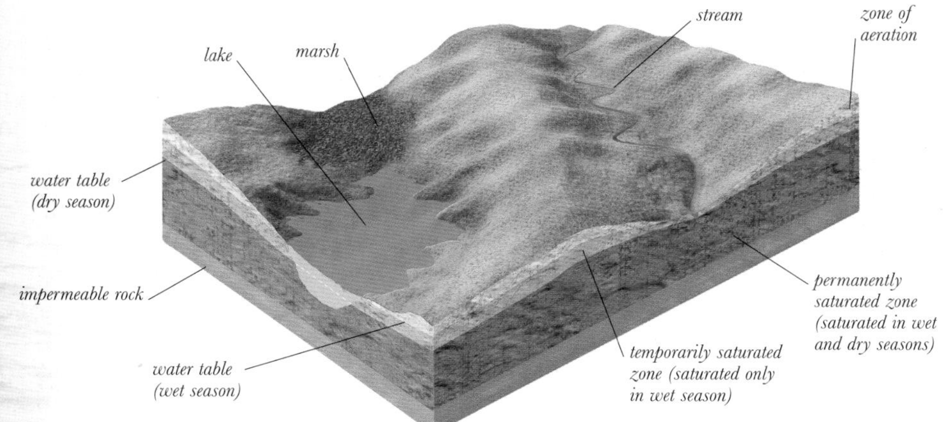


**LOCAL WATER MOVEMENT**
*In a given area, water may exist at the surface, such as in a stream, marsh, or lake. Other water is underground. In the zone of aeration, air fills the spaces between particles of soil and rock. Beneath this lie zones of temporary and permanent water saturation, which vary according to the rainfall. The water table is the level below which the ground is saturated.*

# WATER RESOURCES

Every year, about 9,500 cubic miles (40,000 cubic km) of water evaporates from the oceans and falls on land. This yearly flow is about ten times that of the Amazon River, and—in theory—it is enough to support at least five times the current number of people on Earth. Most of the world's fresh water, however, is difficult to reach, and the remainder is unevenly spread. As demand for water grows, the world's resources are coming under increasing strain.

## SURFACE WATER

Water supplies come from two major sources. Some is removed from rivers and lakes (surface water), and some from aquifers—natural reservoirs of water in porous rock (groundwater). A small amount is also produced by the process of desalinization, but this is an expensive procedure that relatively few countries can afford. Surface water has the advantage of being accessible, and, because it is quickly replenished, it can be treated as a renewable resource. Unfortunately, it is also easily polluted—a growing problem in regions with high populations and poor hygiene.

Globally, domestic use accounts for less than a tenth of total water consumption, and industry uses about 25 percent. The biggest consumer by far is agriculture, which takes about 60 percent, much of it from rivers. Most of this water is for irrigation, and once it is has been used, it often evaporates instead of draining back into natural waterways. In arid parts of the world, from the American West to China, irrigation has—at times—slowed major rivers to a trickle.

Cutting household consumption can help to conserve water, but reducing use by agriculture can have an even greater impact. By drip-feeding plants with moisture, microirrigation is able to reduce water consumption by more than 250 percent. But this is financially feasible only for high-value crops. For large-scale cultivation—such as cereal farming—growing drought-resistant varieties can yield much greater savings.

**HUMAN-MADE OASIS**
*A golf course contrasts with desert scenery near Las Vegas, Nevada. The fairways are irrigated from the Colorado River—an example of recreational water use.*

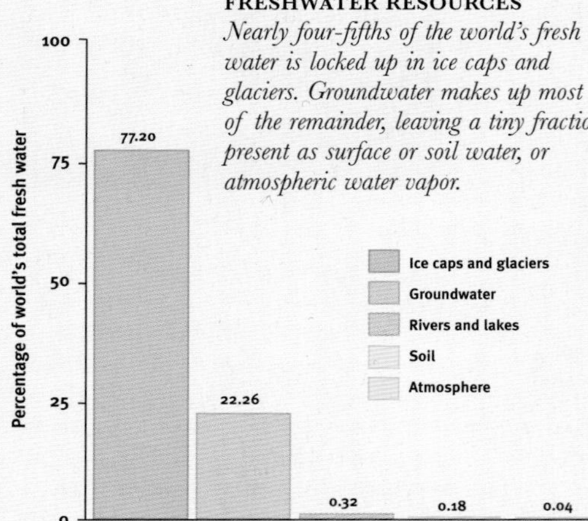


**FRESHWATER RESOURCES**
*Nearly four-fifths of the world's fresh water is locked up in ice caps and glaciers. Groundwater makes up most of the remainder, leaving a tiny fraction present as surface or soil water, or atmospheric water vapor.*

## GROUNDWATER

Some of the world's earliest civilizations depended on groundwater for their survival, and aquifers now underpin life in many dry parts of the world. Groundwater is harder to reach than surface water but is less prone to pollution, and it is often present in places where surface water is scarce. The drawback is that it replenishes itself very slowly. If too much water is used, the water table sinks and wells run dry. The drop in the water table can be precipitous: in southern India, falls of more than 82 ft (25 m) have been recorded in a single year.

In the American Great Plains, declining water tables threaten some of the most productive farmland in the world. Here, farming relies on the Ogallala Aquifer, an immense reservoir of fossil meltwater dating back to the last ice age. Large-scale use of this water began in the 1970s, and without conservation, it may be depleted within the next 50 years. Australian farmers could face similar problems in the Great Artesian Basin—a huge aquifer where boreholes currently tap about 343 million gallons (1.3 billion liters) a day.

**TESTING TIME**
*In Pakistan, a technician and a farmer check water from a newly sunk artesian well. In such wells, pressurized groundwater flows up to the surface.*

## WATER WARS

When water is in short supply, the potential for conflict grows. Within national borders, this can lead to protracted legal disputes, but when it develops between countries, it may even cause wars. More than 250 of the world's largest river basins straddle more than one country, and many experts believe that the threat of "water wars" is likely to increase as the human population grows. According to international law, water cannot be owned outright, but at present there is little to prevent upstream nations from exploiting supplies in ways that harm their neighbors. Likely flash-points for water conflict include the Middle East, southern Africa, and Central Asia—all regions where low rainfall brings about water stress.

In 2003—the International Year of Fresh Water—a UN report predicted that the average supply of water per person would fall by over a third in 20 years, and that 7 billion people could face a shortage of water by 2050 unless urgent action is taken now. With static water resources, conservation will become increasingly important, but an even bigger challenge will be to make sure that existing supplies are shared fairly.

**NATURAL CYCLE**
*In many parts of India, the monsoon climate means that most of the year's rain falls in a four-month period, after which reservoirs dry out all too quickly.*

# LIFE

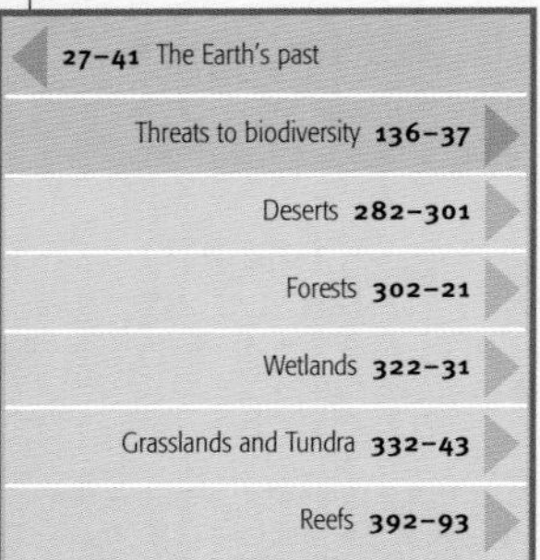
27–41 The Earth's past
Threats to biodiversity 136–37
Deserts 282–301
Forests 302–21
Wetlands 322–31
Grasslands and Tundra 332–43
Reefs 392–93

LIVING ORGANISMS HAVE been present on Earth for about 3.8 billion years—or some two-thirds of the Earth's history. Over the course of about 2 billion years, microorganisms drastically changed the composition of gases in the atmosphere, in particular by adding oxygen. Since then, organisms have spread through the seas and over most of the land, adapting to changing conditions. Their very presence has also affected the nature of the planet so that the Earth today is the result of myriad intimate interactions between the nonliving and living.

## WHAT IS LIFE?

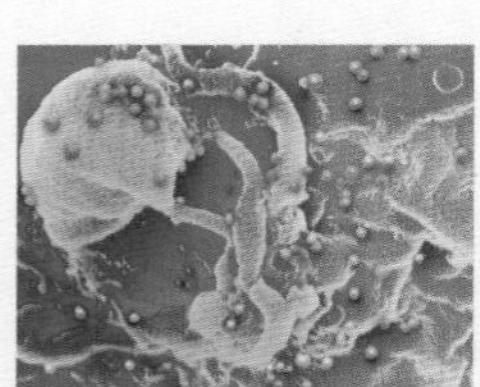

**AIDS VIRUS**
*Like all viruses, HIV (seen here in red) cannot function in the absence of other organisms. It must take over living cells in order to replicate itself.*

Organisms as diverse as zebras and grasses are easy to identify as living. But this task is more difficult with tiny, hard-cased seeds that can lie dormant for centuries, or with many microscopic organisms. Several key features help to define life. At some stage in its life cycle, an organism has an active internal chemistry, in which substances dissolved in water are mobilized and change through chemical reactions. To power these chemical processes, the organism takes in and expends energy. It also takes in nutrients and raw materials so that it can grow. Organisms also reproduce—they make more of their kind. Certain nonliving things, such as mineral crystals, show some of these features, including growth, but not all. Viruses lie at the boundary between living and nonliving. These microorganisms can be subjected to freezing, boiling, and crystallizing and remain totally inactive, but when conditions are favorable, they burst into life. Virus particles enter the cells of other organisms and take over the cellular machinery to make copies of themselves, which are released when the cell bursts open. However, viruses cannot replicate on their own. They are packets of chemicals that simply copy themselves, without truly being alive.

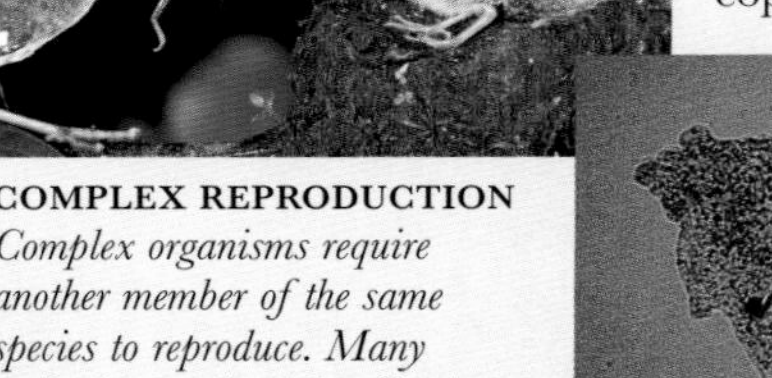

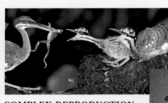

**COMPLEX REPRODUCTION**
*Complex organisms require another member of the same species to reproduce. Many animals care for their offspring.*

**SIMPLE REPRODUCTION**
*Simple organisms and some plants can produce offspring identical to themselves. The single-celled amoeba shown here can divide into two cells, both of which can feed and grow back to full size, and split again, all in a few hours.*

### KINGDOMS OF LIFE

Of the different systems scientists use to classify living things into groups, the most widely used is that of five kingdoms. This groups living things on the basis of the nature of their cells or how they obtain energy and nutrients. However, this system is continually under debate, and some recent classifications propose around 30 kingdoms encompassed within 3 superkingdoms.

**Animals**

Animals are complex, multicellular organisms that gain energy by ingesting other organisms, either in part or whole, as food. Many animals can move from place to place, and they sense and react to their surroundings quickly. All animals can move at least some part of their body. Animals are either vertebrates (with a backbone) or invertebrates (without a backbone).

VERTEBRATE

INVERTEBRATE

**Plants**

Plants are multicellular organisms that gain energy from sunlight. Special pigments absorb light energy and use it to join simple low-energy chemicals to make more complex high-energy ones, especially sugars, by the process of photosynthesis.

FLOWERING PLANT

NONFLOWERING PLANT

**Fungi**

Fungi gain energy and nutrients from other living or recently dead organisms by breaking down their tissues into simple chemicals outside the fungal body and then absorbing them. Some fungi, such as yeasts, are microscopic; others, such as mushrooms and toadstools, form large fruiting bodies.

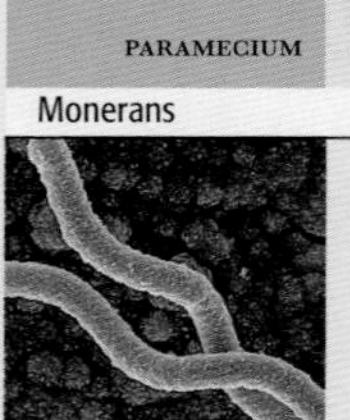

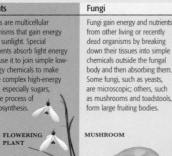

MUSHROOM

**Protists**

Protists are single-celled organisms with an ordered internal structure. Some protists ingest food like animals; others gain energy from sunlight, like plants.

PARAMECIUM

**Monerans**

The simplest living organisms are single cells with little internal structure. The two main groups of monerans are bacteria and cyanobacteria (blue-green algae).

BACTERIUM

## DIVERSITY

The bewildering number and variety of living things is organized using a system of classification that begins with kingdoms (see panels, below). Each kingdom is divided into major groups called phyla, which are split in turn into classes, orders, families, genera, and species. The species as the basic unit is usually defined not in terms of appearance, but by reproduction. An animal species includes all organisms that can breed with each other to produce viable offspring. The insects, class Insecta, are thought to include more species than all other animal groups and all other kingdoms added together—probably about 8 million. To date, just over one million of those insect species have been discovered and described.

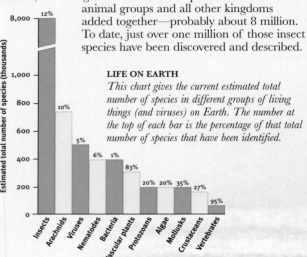

**BROWN KIWI**
*The number of species fluctuates with scientific knowledge. Kiwis were once regarded as two species, brown and spotted, but recent studies suggest that there are actually four.*

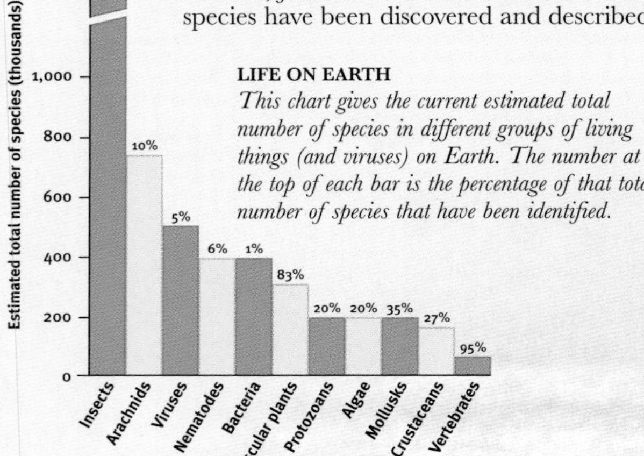


**LIFE ON EARTH**
*This chart gives the current estimated total number of species in different groups of living things (and viruses) on Earth. The number at the top of each bar is the percentage of that total number of species that have been identified.*

# EVOLUTION

Organisms evolve by becoming better adapted to their environment. The environment includes physical conditions and other organisms (which interact by eating or being eaten by one another, as well as in other ways). However, the environment does not stay constant. Climates fluctuate greatly through time, and other living things are themselves evolving. Within a species, individuals vary slightly in their features. Those with features that better adapt them to the current environment are more likely to survive and leave offspring. If the features are genetically determined—inherited by their offspring—then over generations the useful features spread, and the species changes, or evolves. If a species is found in two different habitats, each subgroup becomes adapted to its own habitat, and one species eventually becomes two, in an evolutionary process known as speciation.

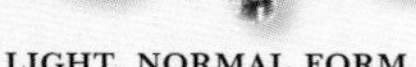

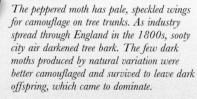

**LIGHT, NORMAL FORM**

**DARK, OR CITY, FORM**

**PEPPERED MOTH**

*The peppered moth has pale, speckled wings for camouflage on tree trunks. As industry spread through England in the 1800s, sooty city air darkened tree bark. The few dark moths produced by natural variation were better camouflaged and survived to leave dark offspring, which came to dominate.*

**EXTINCT IN THE WILD**

*The Hawaiian koki'o, a species of tree unique to the island of Molokai, has been extinct in the wild since 1918. Ten grafted plants remain in cultivation, but the species of Hawaiian honeycreeper that fed on the plant's nectar is now extinct.*

# EXTINCTIONS

The vast majority of species that ever lived on Earth are now extinct—they have died out completely. The fossil record shows that creatures such as ammonites, trilobites, and dinosaurs once thrived in huge numbers but no longer exist. Extinction is a natural part of evolution and the turnover of species. In the face of changing climates, food sources, and competitors, some species adapt and thrive, while others perish. Through the Earth's history, there have been rapid major changes resulting in five mass extinctions. The most severe was the "Great Dying" at the end of the Permian Period, about 252 million years ago, when over 75 percent of land species and over 90 percent of marine species died out. The last mass extinction was 65 million years ago. Between mass extinctions, the natural background rate of extinction among organisms is estimated at about one species per century. However, that rate has rocketed since humans spread over the Earth and altered it for their own needs, and a sixth mass extinction is currently underway. Today, it is estimated that one species per day, perhaps even one species every 20 minutes, is being lost.

**COEXISTING SPECIES**

*All life relies on water. Here, springbok, gemsbok, Burchell's zebra, and ostrich gather at a drying water hole in the Kalahari Desert, southwest Africa.*

# BIOMES AND ECOSYSTEMS

Ecologists recognize several levels of organization among living things. A biome is a broad category, taking in similar assemblages of plants and animals, across whole regions and continents. For example, the coniferous forest biome, in which the trees have needlelike leaves that can withstand long, cold winters, stretches over vast areas of the northern continents. A biome can be divided into several smaller-scale, more specific communities of plants and animals living in a certain area (known as a habitat). A community could be as large as a huge lake, or as small as a rotting tree stump (a microhabitat). So the temperate forest biome includes communities such as woods of oak, beech, birch, maple, and many other deciduous trees. Biomes are typically named after their dominant plants – for example, the grassland biomes include communities based on elephant grass, pampas grasses, cotton-grasses, and others. Biomes are determined mainly by environmental factors such as temperature, rainfall, and wind or water currents. Also important are the topography and soil type. Each biome has not only its characteristic plants, but also typical species of animals. An ecosystem is a functional rather than descriptive term. It refers to the ways living organisms interact with each other and with the physical and chemical factors of their surroundings. An ecosystem is never totally self-contained – a foraging animal can move from a forest to nearby grassland, for example.

**SEA OTTER COMMUNITY**
*Sea otters live in the forests of kelp that lie off the coasts of California and Alaska. The sea otter's diet includes red and purple sea urchins, whose main food is the kelp. If the sea otter population declines, then the sea urchin population grows unchecked, which prevents the kelp fronds from developing.*

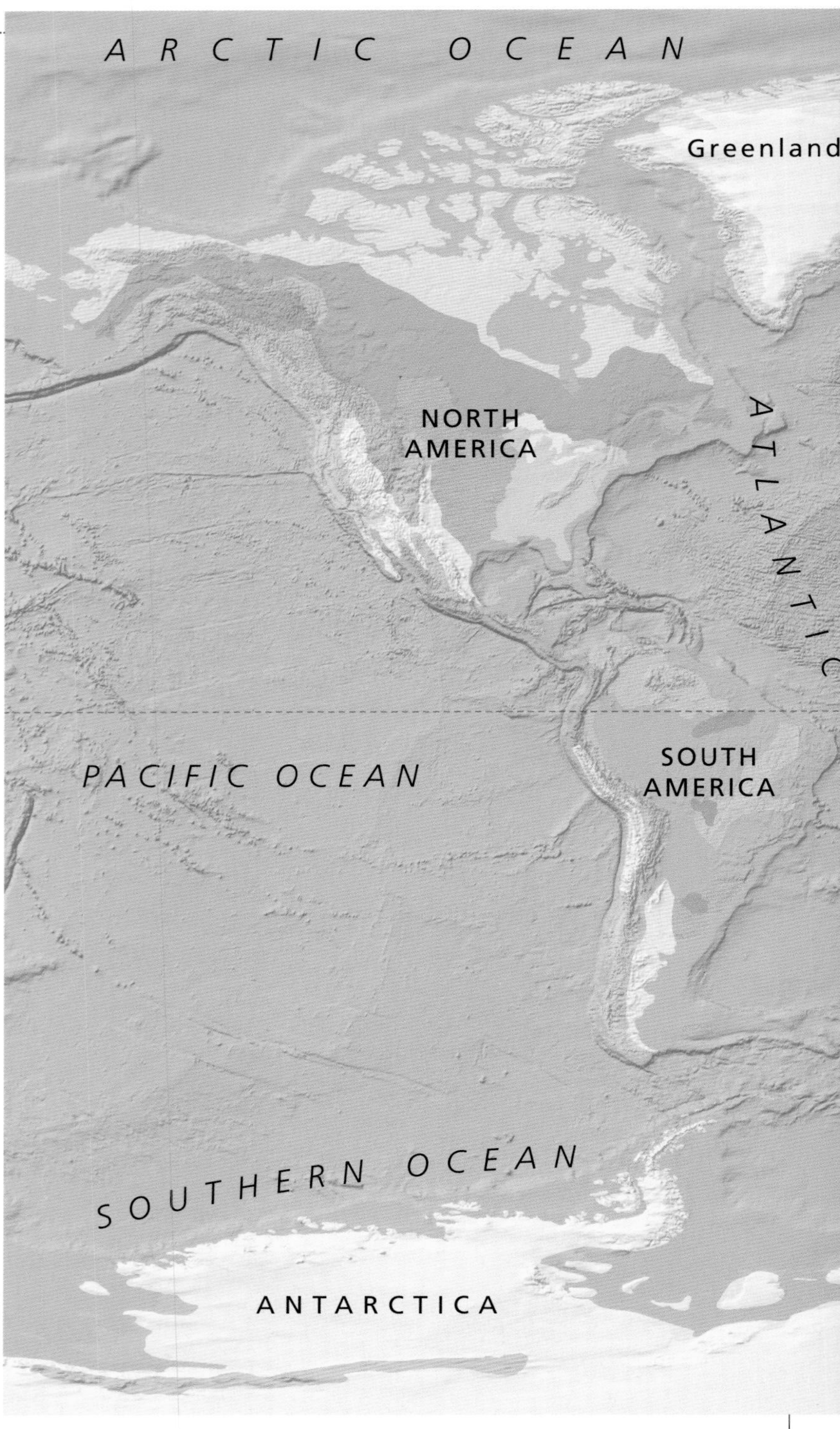


# BIOGEOGRAPHY

The study of how living things are distributed around the world, why they live there, and especially how they got there, is known as biogeography. Each species is adapted to its own habitat, which may be continuous, like a vast tract of grassland, or discontinuous, like a group of islands in the ocean. Species can spread through a continuous habitat relatively easily. They have more difficulty overcoming geographic or habitat barriers—as when a lowland species tries to cross a mountain range, although this may not be a problem to animals that can fly. The geographical history of the Earth, with its changing pattern of bridges and barriers caused by plate movements and volcanic activity, has greatly affected present-day species distribution. In some places, what is now the ocean bed was exposed long ago by lower sea levels. It was colonized by terrestrial plants and animals, and used as a land bridge for species to spread from their original area to another region with a similar habitat. Other types of geographical barriers encourage new species to evolve. This occurs, for example, when members of a terrestrial species arrive at an oceanic island on a floating mass of matted vegetation. The island has its own specialized habitat, to which some members of the species adapt. They gradually evolve into a new species, distinct from the founder species. Continued island-hopping may lead to a group of related but distinct species that are adapted to different habitats.

GROUND FINCH

CACTUS FINCH

**DARWIN'S FINCHES**
*The Galápagos Islands were colonized a few million years ago by finches blown from South America. They evolved into several species that exploit different food sources.*

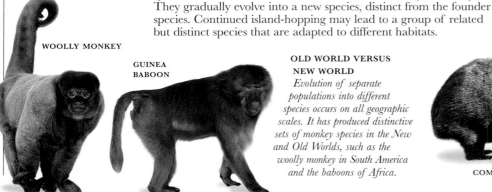

WOOLLY MONKEY

GUINEA BABOON

**OLD WORLD VERSUS NEW WORLD**
*Evolution of separate populations into different species occurs on all geographic scales. It has produced distinctive sets of monkey species in the New and Old Worlds, such as the woolly monkey in South America and the baboons of Africa.*

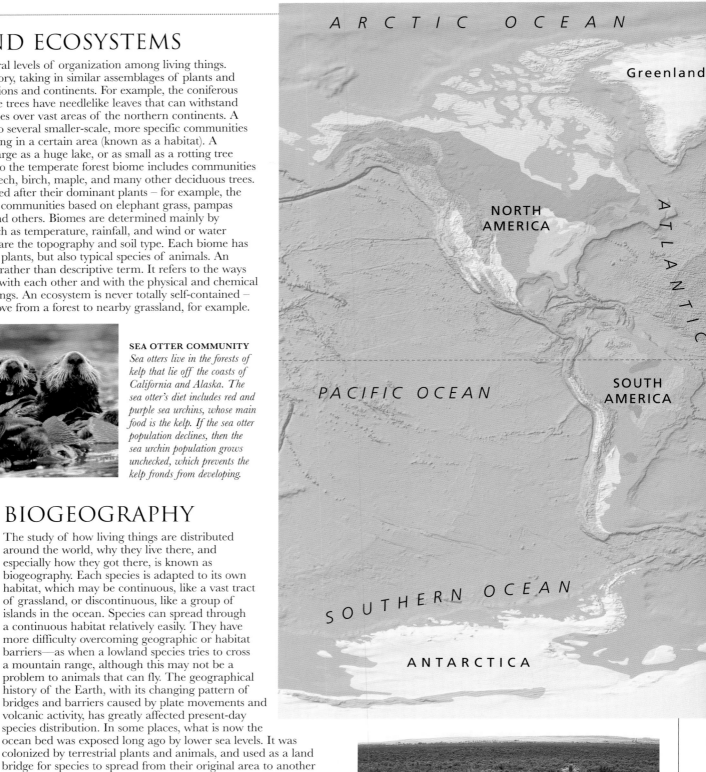

ACORN BANKSIA

COMMON WOMBAT

**ENDEMIC SPECIES**
*Australia has been an island for over 40 million years. Its wildlife has evolved in unique directions with thousands of endemic species (those that occur naturally nowhere else). These include more than 170 species of marsupials, such as the common wombat, and 75 species of the plant genus* Banksia.

ARCTIC OCEAN
ASIA
EUROPE
AFRICA
PACIFIC OCEAN
INDIAN OCEAN
Australia
OCEAN
SOUTHERN OCEAN
ANTARCTICA

**BIOMES OF THE WORLD**
*This map shows the distribution of the Earth's major biomes. The biome distribution shown here is the pattern that would exist if human-made changes to the planet, resulting from urbanization, industrialization, the spread of agriculture, and logging, had not occurred. Only the major wetland areas are shown here, since the rest are too small to show up on a map of this scale, and the size of many of them varies with the seasons. Similarly, only the largest lakes and inland seas have been shown, and no rivers have been mapped.*

TROPICAL FOREST
TEMPERATE FOREST
CONIFEROUS FOREST
TROPICAL GRASSLAND
TEMPERATE GRASSLAND
TUNDRA
WETLAND
DESERT
MOUNTAIN
POLAR

# NUTRIENT CYCLES

Certain chemicals are vital to living things, being common constituents of the complex organic molecules found in living cells. One of the most important is the element carbon, on which all life is based. Other important chemicals include nitrogen and phosphorus for plant growth, and the metals magnesium (which is part of the pigment chlorophyll that captures sunlight energy in plants) and iron (found in the blood of many animals). But these chemicals are finite and must be reused. So the Earth can be regarded as a giant self-contained ecosystem where the same nutrients are recycled, going around on an endless variety of pathways through countless organisms. In the carbon cycle, carbon dioxide from the air is incorporated into carbon compounds in the living tissues of plants and phytoplankton by photosynthesis. The carbon is subsequently returned to the atmosphere as a by-product of respiration by the plants and phytoplankton, and the animals that ate them. The carbon cycle also includes important inorganic components and processes (see below).

**NITROGEN FIXATION**
*Some species of plants, such as the acacias shown here, have nitrogen-fixing bacteria living in their roots. The bacteria take nitrogen gas from the air and reduce it to nitrate, a nutrient for the plant.*

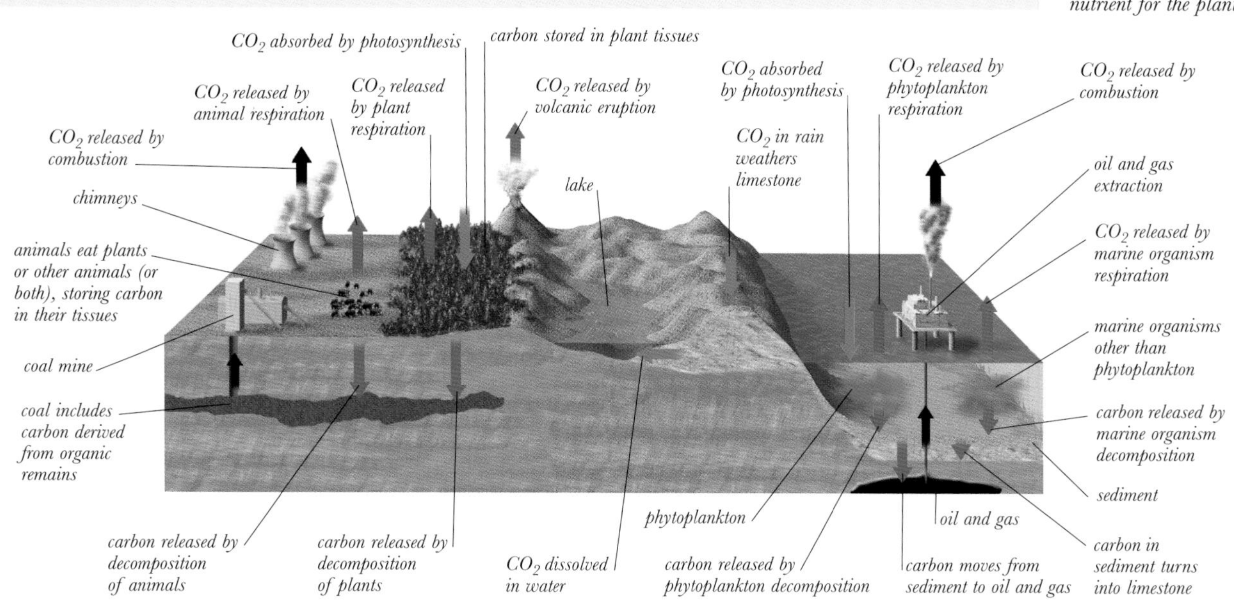


**THE CARBON CYCLE**
*The pathways by which carbon is cycled through ecosystems are complex. As organisms die and their cells rot, some of the carbon in them enters the soil on land or the sediments on the sea floor. Here it can be used by microorganisms, or taken in as mineral substances such as carbonates through roots, for new plant growth. In some cases carbon is transformed into fossil fuels such as coal or oil. In addition to the natural routes, which include volcanic eruptions, humans add carbon dioxide ($CO_2$) to the atmosphere, mainly through the burning of fossil fuels.*

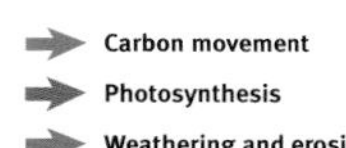

# THREATS TO BIODIVERSITY

Although no one knows how many species of living things exist on the Earth, one fact is certain: the Earth's biodiversity—or biological richness—is currently undergoing a steep decline. According to some estimates, more than 5 percent of the planet's species are disappearing each decade, which is the highest rate for 65 million years. Mass extinctions happened long ago on several occasions, triggered by natural events. In today's case, humans are largely to blame.

**LIFE IN ISOLATION**
*Giant tortoises evolved on remote islands, such as the Galápagos and Seychelles. Hunting, and competition from introduced animals, have made many island races extinct.*

## A DOWNWARD TREND

The current decline in biodiversity affects life on all fronts. According to the International Union for the Conservation of Nature and Natural Resources (IUCN), nearly one in four mammal species is under threat, while the figure is one in eight for birds. Plants fare less badly, but even so, more than 7,000 species are listed as endangered. However, these figures represent only the tip of the iceberg, because the data is far from complete. Every one of the world's 4,763 mammal species has been assessed, for example, yet less than 0.1 percent of invertebrate animals, which number more than 1 million species, have been evaluated in the same detail.

The forces threatening biodiversity are much easier to establish. Habitat destruction is by far the most important cause, because it sweeps away not only individual species but also the ecosystems on which they depend. Habitats can also be disrupted by introduced species, by pollution, and by the direct exploitation of living things—particularly through hunting and collecting. Pollution can have localized effects, but it can also cause habitat disruption on a worldwide scale.

INTRODUCED SPECIES

UNKNOWN

HABITAT DISRUPTION

HUNTING

**CAUSES OF RECENT EXTINCTION**
*Three main factors are behind the more than 400 cases of animal extinction, dating back to the year 1600. In the last 50 years, habitat destruction has become the most important cause of biodiversity decline.*

## BIODIVERSITY HOT SPOTS

For climatic and geographic reasons, biodiversity is unevenly spread. Some large regions, such as the Arctic, contain relatively few species, while some much smaller regions boast considerable biological riches. These areas, known as biodiversity hot spots, include ecologically isolated parts of continents, as well as islands such as the Galápagos. Together, two dozen hot spots—making up less than a fiftieth of the Earth's land surface—are home to over a third of the world's vertebrate animals and flowering plants.

Hot spots are concentrated in the tropics, where habitat destruction can do disproportionate damage. Tropical rainforests and coral reefs are particularly vulnerable, and are home to many declining species. But the fall in biodiversity is not just a tropical problem: after two centuries of economic growth, many of the world's industrialized countries have pushed their natural inhabitants to the very edge of survival.

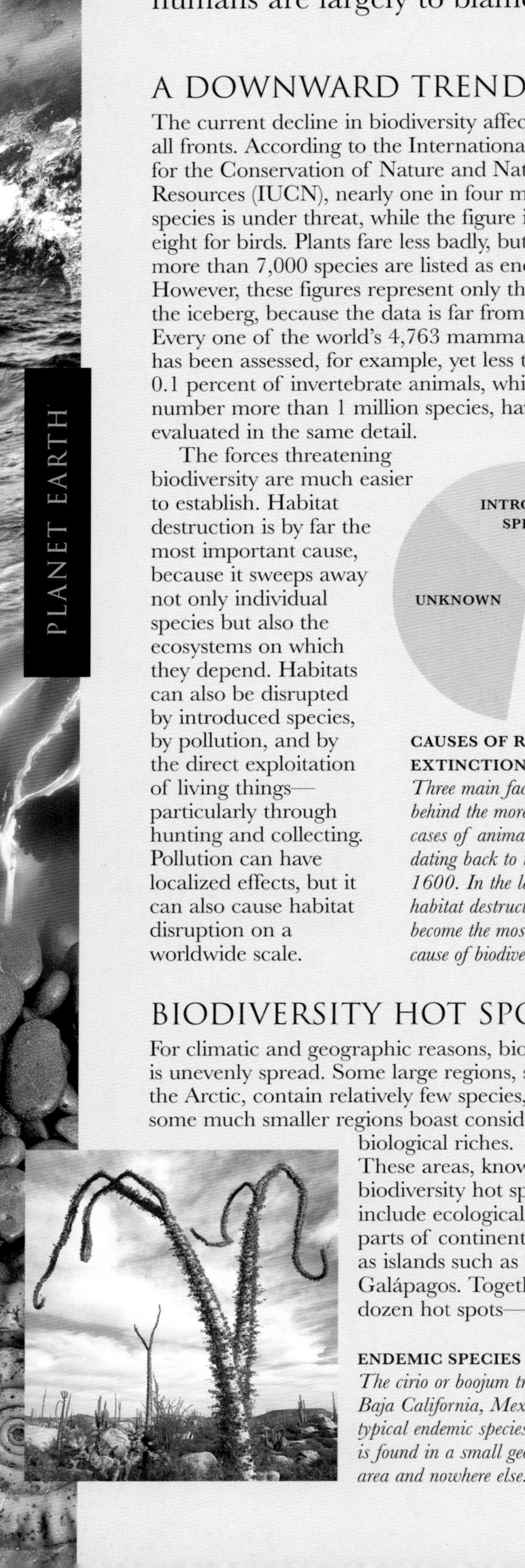

**ENDEMIC SPECIES**
*The cirio or boojum tree, from Baja California, Mexico, is a typical endemic species—one that is found in a small geographical area and nowhere else.*

## PRESERVING BIODIVERSITY

Biodiversity matters because it is a reservoir of resources on which we all depend. Today's crop plants and farm animals originally came from the wild, and their wild relatives still contain a huge reserve of potentially useful genes. In the past, plants have supplied us with hundreds of useful drugs, and it is likely that they will offer many more in the future. A high level of biodiversity is generally believed to stabilize ecosystems and to provide vital environmental services, such as water purification and soil formation. Without these, human survival may be put at risk.

Where species are facing the immediate risk of extinction, emergency methods can be used to help them survive. These include breeding species in captivity so they can be released at a later date. However, this form of protection is extremely expensive and raises difficult questions about which species should be selected. It is easier, for example, to raise funds for saving tigers than vultures, even though vultures play an equally important part in natural food chains. Most ecologists believe that the key to stabilizing biodiversity lies in preserving fully functioning habitats. Many also advocate that hot spots should have priority when resources are limited.

**WILDLIFE CRISIS**
*In the last 30 years, the number of black rhinoceros has fallen by 96 percent—a decline that is entirely due to unauthorized hunting.*

**THEMES AND VARIATIONS**
*This collection of crickets and katydids is from one of the world's richest habitats—the tropical forests of Costa Rica. Despite decades of study, new species are discovered every year.*

# LAND

**MOUNT ST. HELENS, CASCADE RANGE**
*In May 1980, a violent eruption tore apart the snow-capped peak of Mount St. Helens in Washington State. Such events are a reminder of the hugely powerful, and often devastating, internal workings of the Earth.*

# MOUNTAINS AND VOLCANOES

THE EARTH'S ROCKY ELEVATIONS, mountains and volcanoes, have an enduring fascination. Mountain belts form imposing barriers and are liable to violent extremes of weather. They are also associated with catastrophic events such as landslides and avalanches. Until recently, some people saw mountains and volcanoes as the homes of wrathful gods who vent their anger without warning, shaking the ground, and spewing fire, rocks, and ash into the air. But mountains and volcanoes are just the most visible sign of the tectonic forces shaping the Earth—elsewhere, fault systems and earthquake zones follow the outlines of the tectonic plates themselves, and igneous intrusions mark areas where molten material has penetrated existing rock. Since scientific investigation of these phenomena began in the 18th century, they have lost many of their superstitious associations, but our fascination with them continues, and they remain impressive reminders of the often spectacular power of Earth's continuing evolution.

# FAULT SYSTEMS



THE RELENTLESS MOVEMENT of the Earth's tectonic plates puts the brittle rock of the crust under so much strain that it can buckle or even fracture altogether. When rock fractures, it can result in huge blocks slipping past one another. These cracks in the Earth's surface are called faults, and often extend through the brittle upper crust for many miles. Many faults are the result of past movement and are inactive. However, there are active fault systems all around the Earth, and massive amounts of energy can be unleashed in a few seconds as rocks lurch past each other in an earthquake.

## FAULTS AND JOINTS

Faults are cracks in rocks across which there has been displacement. They range from tiny fissures to sets of interconnecting faults that are many miles long. Some active faults move in a slow, continuous manner called creep. More often, they move in a series of spasmodic jumps, which can release energy in the form of an earthquake (see opposite page). Over millions of years, the cumulative effect of countless small movements along the fault can displace rocks across a fault a substantial distance up, down, or horizontally. Not all rock fractures are faults: joints are cracks along which there is no evidence of movement. These are usually caused by tectonic activity, but may also be the result of erosion (see pp.112–13), which strips away the overlying layer of rock, releasing pressure and allowing the surface of rocks to expand and crack. They can also be formed as igneous rock cools. Joints are exploited by the processes of weathering (see pp.110–11), which may exaggerate the shape of the original fracture. They usually occur close to the surface.

**LIMESTONE JOINTS**
*Sedimentary rocks, such as this limestone in England, have particularly marked joints. When the limestone is exposed at the surface, its strong pattern of crisscross joints is further etched by acidic rainfall. Such striking formations are called limestone pavements.*

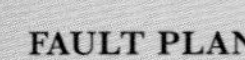

**FAULT PLANE**
*The huge uplift of rock on one side of a fault over a period of thousands of years has created this dramatic precipice in Greece. Scratch marks called slickensides can be seen on the face of the cliff.*

## MEASURING EARTHQUAKES

Earthquakes are recorded on a seismograph, which charts the magnitude of each vibration. They were traditionally measured on the Richter scale. Each step in the scale, which begins at 0 and has no upper limit, represents a 30-fold increase in the energy released by an earthquake. The greatest magnitude recorded, 9.5, was from the 1960 Chilean earthquake. Many scientists now use the moment-magnitude scale, which combines seismograph readings with measurements of rock movement.

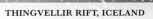

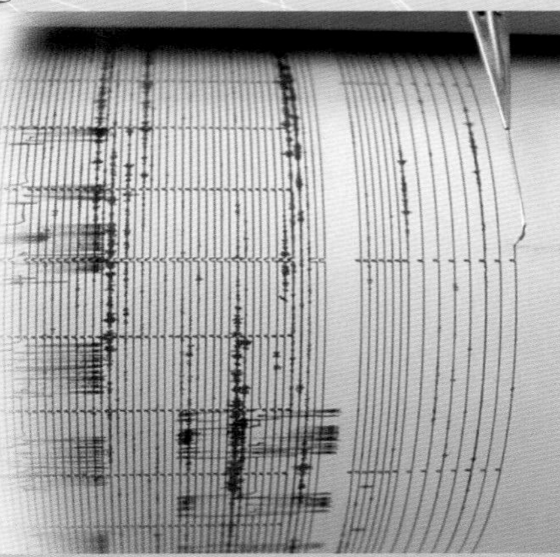

## CHARLES RICHTER

In 1935, American seismologist Charles Richter (1900–85) devised the scale for measuring earthquakes that bears his name. Before this, seismologists had been able to compare earthquakes only by using the Mercalli scale, a rule-of-thumb guide to the severity of their effects. The Richter Scale, devised with Beno Gutenberg (1880–1960), measured the magnitude of each earthquake precisely, from the variations in movement it produced on a seismograph (see below). Richter also mapped out the zones in North America that were vulnerable to earthquakes, but he was scornful about the idea of predicting them.

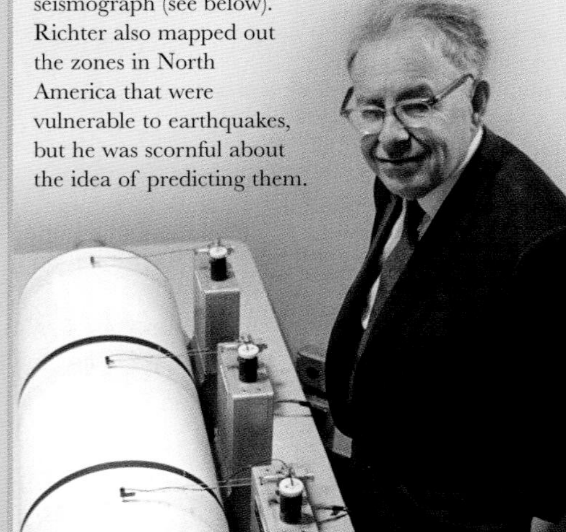

**THINGVELLIR RIFT, ICELAND**
*This exposure of fault lines parallel to the Thingvellir Rift provides dramatic evidence of tectonic activity. The rift lies on the boundary separating the American and Eurasian plates.*

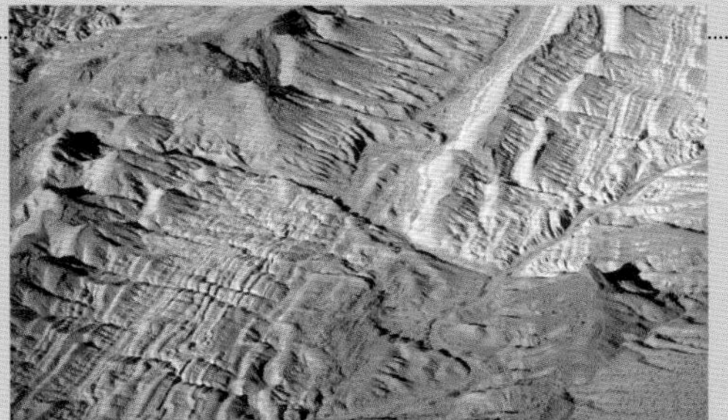

**STRIKE-SLIP FAULT, NEVADA**
*The displacement of the two sides of this right-lateral strike-slip fault can be seen in the white areas on either side of the crack, which were once continuous vertical layers.*

# TYPES OF FAULTS

Faults are created by the compression or extension of the Earth's crust as its tectonic plates move (see pp.108–09). The slope or angle of a fault to the horizontal is called the dip. Faults that show vertical movement are called dip-slip faults, and may be either normal or reverse. Normal faults occur where tension in the Earth's crust pulls the rock apart, and allows a block to slip down the fault plane, which is the surface along which the crack occurs. Parallel sets of normal faults form rift valleys (see p.156), and are associated with divergent plate boundaries. Reverse faults, which are known as thrust faults if they are particularly shallow, occur where compression in the crust pushes one block of rock up over another. If two blocks slide past each other horizontally, the fracture is classified as a strike-slip fault, an example of which is the San Andreas Fault (see pp.144–45). If the opposite block in a strike-slip fault moves left, it is left-lateral; if it moves right, it is right-lateral. If a strike-slip fault is combined with compression or tension, the blocks can slide diagonally, creating an oblique-slip fault.

**FAULT LINE, UZBEKISTAN**
*The dramatic effects of fault displacement on the landscape are visible in this escarpment on the outskirts of Tashkent, Uzbekistan, which is the result of ancient earthquakes.*

**FAULT MOVEMENT**
*Faults are classified by their direction of movement, which reflects surrounding tectonic stresses, and whether they shorten or lengthen the original surfaces.*

*movement along fault plane*
*fault plane*
*surface lengthened by faulting*
*down-faulted block*

**NORMAL DIP-SLIP FAULT**

*upthrust, overhanging block*
*surface shortened by faulting*

**REVERSE THRUST DIP-SLIP FAULT**

*opposite block (viewed from either side of the fault) moves left*
*extension of crust*

**STRIKE-SLIP (LEFT-LATERAL)**

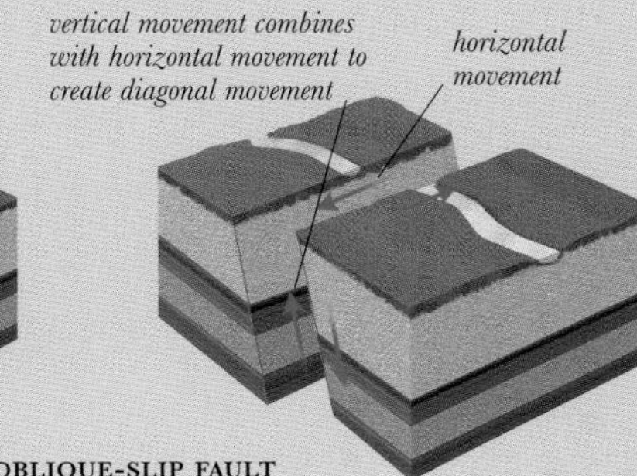


**OBLIQUE-SLIP FAULT**

# EARTHQUAKES

A sudden slip on a fault releases energy in the form of an earthquake, which radiates outward as seismic waves (see p.57). The Earth's major earthquake zones coincide with faults between tectonic plates. The immense forces generated as these plates grind together trigger most large earthquakes. This happens as the rocks on either side of a fault lock, and stress builds up. The fault eventually ruptures, sending shock waves shuddering through the planet. The Pacific Ocean subduction zones are the source of 80 percent of the world's earthquakes. Almost every day, there are several hundred minor earthquakes around the world. Disturbances such as the massive 1964 Alaskan earthquake, which measured 9.2 on the Richter Scale, and triggered a devastating tsunami, are rare. However, there are about 20 earthquakes the size of the 1999 event in Izmit (see p.147) every year.

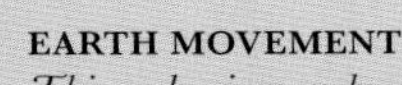

**EARTH MOVEMENT**
*This radar image shows the ground displacement from an earthquake. The closely packed contours near the black line of the fault indicate the greatest movement.*

## FAULT SYSTEMS PROFILES

The pages that follow contain profiles of a selection of the world's most significant fault systems. Each profile begins with the following summary information:

**TYPE** Normal dip-slip, reverse dip-slip, reverse thrust dip-slip, strike-slip (left-lateral), strike-slip (right-lateral), or oblique slip

**LENGTH/AREA** Extent of fault line or complex

**ACTIVITY** Active or inactive

NORTH AMERICA *west*

# San Andreas Fault

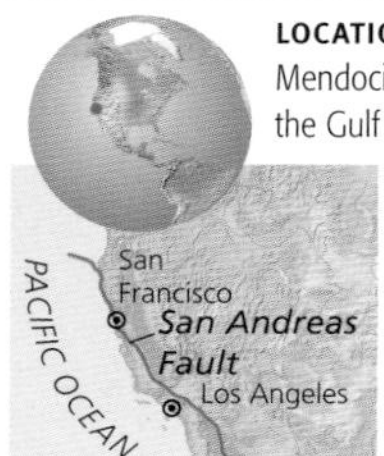


**LOCATION** Extending from Cape Mendocino, northern California, to the Gulf of California

**TYPE** Strike–slip (right-lateral)

**LENGTH** 800 miles (1,200 km)

**ACTIVITY** Active

**SLIPPING SIDEWAYS**
*Images from space reveal the gash across the land created by the San Andreas fault as the Pacific and North American plates move in opposite directions.*

Slicing across California's coastal region, the San Andreas Fault is one of the Earth's most famous faults. It is a strike-slip fault in which the rocks move sideways in opposite directions, although, because it forms a plate boundary, the San Andreas is also a transform fault (see p.108). To the west of the fault is the Pacific Plate, which extends from the edge of California almost as far as the eastern edge of Asia. To the east is the North American Plate, which forms the bulk of the continent. As the Pacific Plate rotates, coastal California is sliding slowly northwest past the rest of North America. Over the past 20 million years the Pacific Plate has moved about 350 miles (560 km) relative to North America, averaging about ½ in (1 cm) a year. Plate movement seems to be accelerating, and over the past century the fault has shifted almost 2 in (5 cm) annually. The San Andreas Fault is not one long crack in the Earth's crust, but consists of four major segments and several minor sections. Along some segments of the fault, the strain of the plate movement is released as minor tremors. In major sections, the plates

## BAILEY WILLIS

One of the leading figures in American geology in the early 20th century, Bailey Willis (1857–1949) was a pioneer in the study of California's San Andreas Fault. Willis studied how San Francisco's Golden Gate Bridge might be affected by earthquakes and led the battle for more stringent safety standards for the city's structures.

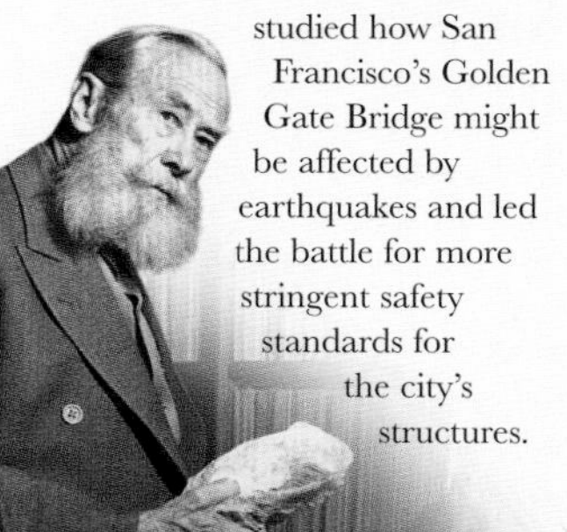

## THE LAST "BIG ONE"

At 5:12 a.m. on April 18, 1906, the Pacific Plate lurched approximately 20 ft (6 m) northward along a 267-mile (430-km) stretch of the northern San Andreas Fault. In just a few seconds, centuries of pent-up energy were released as a massive earthquake that devastated San Francisco. Although the earthquake destroyed many buildings, the worst damage resulted from the fires that raged through the city in its aftermath.

**VISIBLE FAULT LINE**
*The San Andreas Fault scars the Californian landscape. In addition to the very obvious fault line itself, it is often possible to see hills, valleys, and streams truncated and offset on opposite sides of the fault.*

lock together for many years, allowing the buildup of stresses that are eventually unleashed as massive earthquakes. These disturbances typically occur in one section of the fault at a time. In 1857, a sudden movement along the segment of the fault in the Transverse Ranges, which separate Central and Southern California, resulted in a severe earthquake that opened a crack 218 miles (350 km) long. In 1906, movement in the northern section of the fault caused an earthquake, estimated to have measured 8.3 on the Richter scale, that devastated San Francisco (see panel, left). An earthquake with its epicenter at Loma Prieta rocked San Francisco again in 1989. It is feared that pressure is building up in the southern section of the fault, and when it finally unlocks, it will unleash an earthquake nicknamed "the Big One." This earthquake is expected to happen by 2032. Most of California's population lives close to the San Andreas Fault, and some communities are built right over it.

**FLOODED FAULT**
*Water exploits the cracks in the landscape created by faults. Rivers flow along them and lakes form in them. In the case of Tomales Bay, the sea has flooded an inlet created by the San Andreas Fault.*

## EARTHQUAKE WATCH

In 1985, the US Geological Survey predicted that an earthquake would strike the small Californian community of Parkfield in 1993. The village was fitted with equipment to monitor the potential disturbance, although it did not happen. However, Parkfield remains the most closely observed earthquake zone in the world, as scientists probe the ground to measure factors such as strain in the rocks, heat flow, and geomagnetism.

**DEATH VALLEY, CALIFORNIA**
*The lowest point in the Basin and Range—and the US—is Death Valley, at 282 ft (86 m) below sea level. Surrounded by sharp mountain peaks, the basin contains several extremely active faults.*

NORTH AMERICA *west*

# Basin and Range

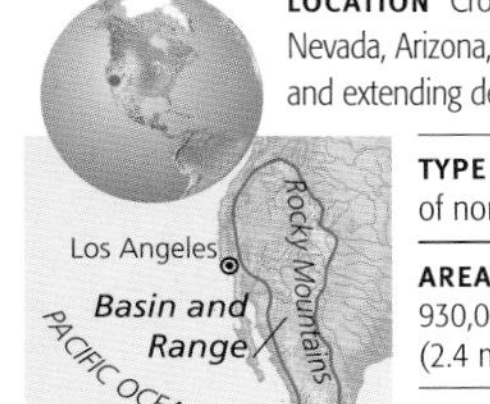

**LOCATION** Crossing California, Utah, Nevada, Arizona, Oregon, and Texas, and extending deep into Mexico

**TYPE** Complex pattern of normal dip-slip faults

**AREA** 930,000 square miles (2.4 million square km)

**ACTIVITY** Active

The Basin and Range country of the southwestern US and northern Mexico has one of the most distinctive landscapes in the world. More than 100 long, roughly parallel mountain ranges run from north to south through the region, each separated from the next by a wide, flat, dry desert basin. On each side, the mountain ranges are bounded by faults that slope into the valleys. This extraordinary landscape is the result of extensive stretching of the North American Plate, which began about 20 million years ago. The crust was pulled, cracked, and thinned, and it is now almost twice its original width. As this happened, the crust fractured, and fault after fault opened up. Basins were created as the land dropped down between parallel faults. As the basin blocks slid down, the mountains were pushed up, reaching heights of 11,500 ft (3,500 m) in places. Although there are a variety of fault types in the region, the majority of the faults are normal dip-slip faults, created as the Earth's crust was wrenched apart. The faults are all steep, typically descending into the crust at an angle of 60° but flattening at depth. As the mountain ranges rose, they were immediately affected by weathering and erosion (see pp.110–13). Water, wind, and ice have worn down the peaks and deposited debris in the intervening valleys. Sometimes, so much debris has collected in the basins that the solid bedrock is buried thousands of yards below the surface.

**SNOW-CAPPED SIERRA NEVADA**
*The uplifted blocks forming California's Sierra Nevada are so high that some of the peaks are permanently snow-capped.*

**LINEAR VALLEYS**
*In the Great Basin of southern Nevada, the pattern of alternating parallel mountain ridges and deep basins is clearly visible.*

EUROPE *north*

## Midland Valley

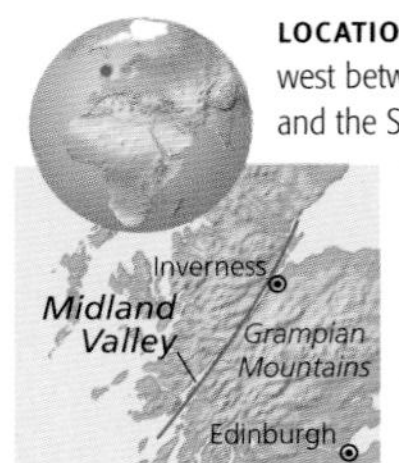


**LOCATION** Running from east to west between the Southern Uplands and the Scottish Highlands, Scotland

**TYPE** Strike-slip (right-lateral)

**LENGTH** 56 miles (90 km)

**ACTIVITY** Inactive

The Midland Valley is bounded on either side by two very ancient strike-slip faults that divide Scotland in two. On the southern side of the valley lies the Southern Upland Fault, while to the north is the Highland Boundary Fault. The origin of these faults dates back more than 540 million years to a time when northern Scotland was part of the ancient continent of Laurentia and was separated from the rest of the British Isles, which were situated on the ancient continent of Avalonia. During the course of the next 200 million years, Laurentia moved gradually toward Avalonia, carrying northern Scotland toward the rest of Britain. As the continents converged, the enormous pressure formed the Scottish mountains in an event called the Caledonian Orogeny (see p.156). About 400 million years ago, Laurentia began to swing away to the southeast, ripping Scotland apart again, and creating not only the two great faults on either side of Midland Valley, but also another fault north of the Highlands, Glen Mor (also known as Great Glen), which forms the long basin filled by Loch Ness.

**LOCH LOMOND**
*Seen from Duncryne Hill on its southern shore, Loch Lomond is the largest of Scotland's lochs, and is crossed by the Highland Boundary Fault.*

**HIGHLAND BOUNDARY**
*The Scottish Highlands rise along the northwestern margin of the Midland Valley, where the Highland Boundary Fault creates a dramatic shift in the region's geology.*

### COAL MINING

About 320 million years ago, Scotland sat on the equator and had a tropical climate. Swampy forests flourished on the ancient land where the Midland Valley now lies. Over hundreds of millions of years, the remains of the plants from these forests have compressed to form Scotland's richest coal deposits, which have been exploited for over 900 years.

EUROPE *north*

## Moine Thrust

**LOCATION** Extending northwest from the south of the Isle of Skye to the Moine Peninsula, Scotland

**TYPE** Reverse thrust dip-slip

**LENGTH** 112 miles (180 km)

**ACTIVITY** Inactive

**NORTHWEST HIGHLAND MOUNTAIN**
*At 3,484 ft (1,062 m) high, An Teallach is one of Scotland's many spectacular mountains. It is the site of intense geological study, because layers of the Moine Thrust belt are clearly exposed here.*

Thrust faults are shallow reverse faults that are created as Earth's upper crust is squeezed, forcing one block up to overhang another. The Moine Thrust is not a single thrust fault, but a formation called a thrust belt. These are created when plate movement forces layers of crust up and over each other, and then pulls them back, a motion that happens again and again. The discovery of the Moine Thrust in 1907 was a milestone in the history of geology—it was the first thrust belt to be identified. It formed between 410 and 430 million years ago, as Scotland was compressed by tectonic movement. Thrust belts are now recognized as features at the edges of mountain ranges all around the world.

EUROPE *north*

## North Sea Basin

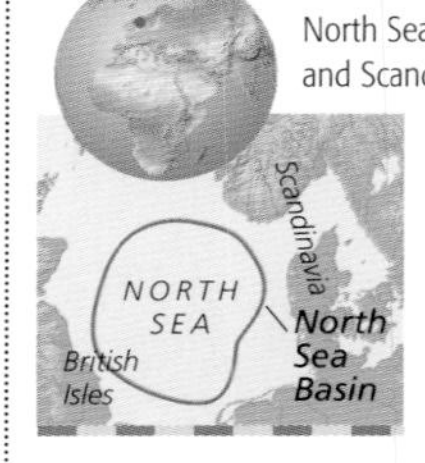


**LOCATION** Lying beneath the North Sea between the British Isles and Scandinavia

**TYPE** Normal dip-slip

**LENGTH** 155 miles (250 km)

**ACTIVITY** Inactive

As geologists prospected the rocks beneath the North Sea for oil in the 1960s, they discovered not only a rift valley but a whole series of faults. Rift valleys form when two normal faults pull in opposite directions, so that as the Earth's crust stretches apart, a block drops down between them (see p.156). The rift valley under the North Sea dates from the Jurassic Period, about 200 million years ago, when the supercontinent of Pangea began to break up, opening up the Atlantic Ocean and the North Sea. It overlies a complex series of much older faults dating from about 50 million years earlier. By studying these structures using seismic surveys that probe into the solid bedrock, geologists are discovering more about how continents diverge, and also the way that rich oil deposits accumulate in down-thrown fault features such as those in the North Sea Basin.

**DRILLING FOR OIL**
*The rich oil fields beneath the North Sea are the largest natural oil and gas deposits in Europe.*

LAND

EUROPE *eastern*

## North Anatolian Fault

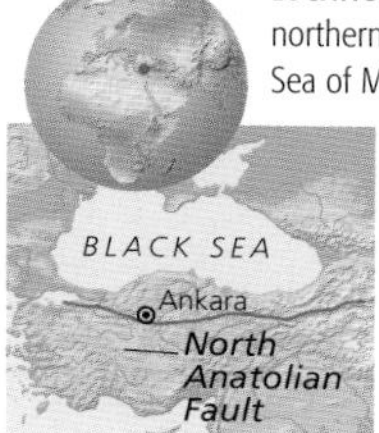


**LOCATION** Extending across northern Turkey and beneath the Sea of Marmara

**TYPE** Strike-slip (right-lateral)

**LENGTH** 600 miles (1,000 km)

**ACTIVITY** Active

The North Anatolian Fault is one of the most energetic earthquake zones in the world. Turkey is set on a minor tectonic plate that is being squeezed westward as the Arabian and Eurasian plates move together. The Anatolian Plate is grinding past the Eurasian Plate at ½ in (1 cm) to 8 in (20 cm) each year, setting off earthquakes as it moves. Since the terrible Erzincan earthquake of 1939, there have been seven earthquakes along this fault measuring more than 7.0 on the Richter scale, each of which has happened at a point progressively farther west. Seismologists studying this pattern believe that earthquakes occur in "storms" over a number of decades, and that one earthquake triggers the next. By analyzing the stresses caused along the fault by each earthquake, they were able to forecast a disturbance that struck the town of Izmit with such devastating effect in August 1999. It is thought that the chain is not complete, and that an earthquake will strike even farther west along the fault—perhaps in the heavily populated city of Istanbul.

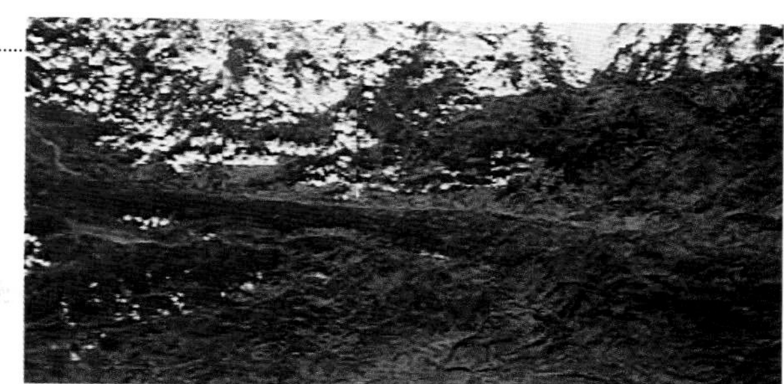

**SEEN FROM SPACE** *The great scar across the Turkish landscape made by the North Anatolian Fault is clearly visible when viewed from space.*

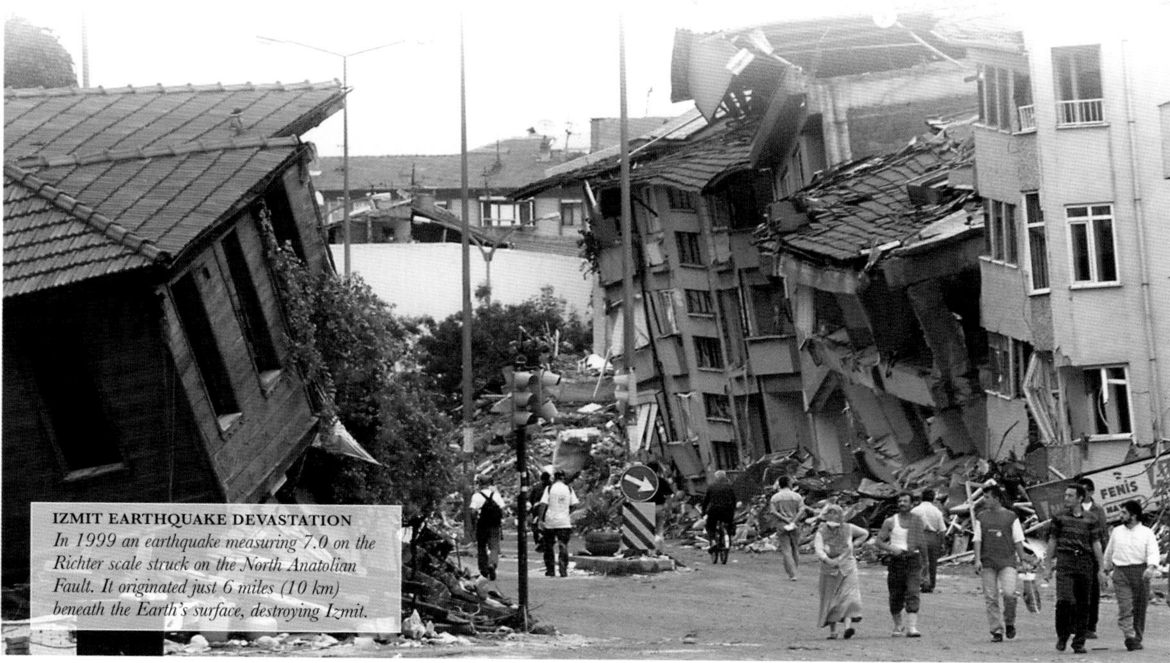


**IZMIT EARTHQUAKE DEVASTATION** *In 1999 an earthquake measuring 7.0 on the Richter scale struck on the North Anatolian Fault. It originated just 6 miles (10 km) beneath the Earth's surface, destroying Izmit.*

EUROPE *central*

## Rhine Rift

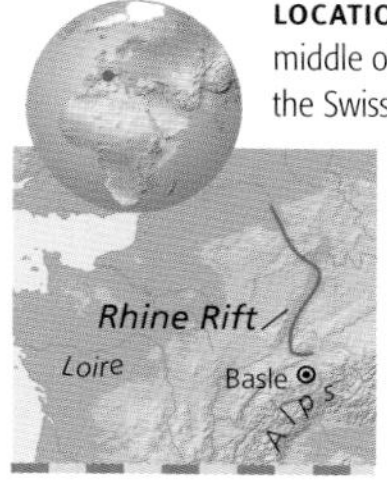


**LOCATION** Extending through the middle of northwest Europe, from the Swiss Alps to the North Sea

**TYPE** Reverse thrust dip-slip

**LENGTH** 820 miles (1,320 km)

**ACTIVITY** Active

The Rhine Rift is a striking reminder of the tectonic forces that formed the European continent. Rift valleys are usually created when continents diverge, causing land to drop down within the gap as the crust is pulled apart. However, the Rhine Rift is thought to have formed as the two halves of what is now northwestern Europe knitted together between 40 and 50 million years ago. As this happened, a long strip of rock called a graben (from the German word for "ditch") dropped between the two continental massifs: the Vosges in France and the German Black Forest. It was initially assumed that the Rhine Rift was inactive. However, scientists have recently detected signs of movement beneath the surface. Faults within stable continental interiors rarely rupture, but if they do, the subsequent earthquake can be sudden and extremely violent. Northwest Europe's worst earthquake occurred in 1356 in the Upper Rhine Valley, destroying the Swiss city of Basle, and knocking down buildings as far as 125 miles (200 km) away. Some geologists believe the cause of the earthquake was not the Rhine Rift, but a fault in the Alps. However, Basle may yet fall victim to an earthquake unleashed by the Rhine Rift—an active fault has been found southwest of the city. In 1992 an earthquake along the Peel Fault, which is farther down the Rhine, rocked the town of Roermond in the Netherlands.

**FAULTED FOREST** *Germany's Black Forest lies on a horst block, which is a section of rock that has been left upstanding after the land on either side of it faulted and dropped downward.*

**KAISERSTUHL, GERMANY** *To the north of Basle is Kaiserstuhl, which is an ancient complex of volcanoes. These were created along the plate margin that also caused the Rhine Rift to form.*

AFRICA *east*

# East African Rift

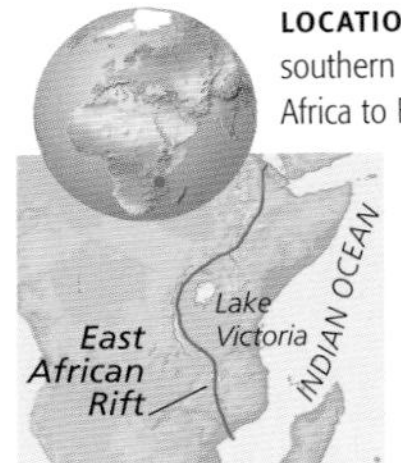


**LOCATION** Extending from the southern Red Sea through East Africa to Beira in Mozambique

**TYPE** Normal dip-slip

**LENGTH** 4,000 miles (6,400 km)

**ACTIVITY** Active

One of the longest fault systems in the world, the East African Rift is part of a huge set of fissures in the Earth's crust called the East African Rift System, which threatens to split Africa in two. The East African Rift is the primary branch of the system, which extends from Jordan in the north, through the Dead Sea and the Red Sea (see p.414), and along the length of East Africa to the mouth of the Zambezi River (see p.231). The average width of the valley is 30 miles (50 km), although at its widest point in the Danakil Desert it is nearly 300 miles (480 km) across. Its steep walls rise 2,955 ft (900 m) above the valley floor, but in some places, such as Kenya's Mau Escarpment, the cliffs soar to 8,860 ft (2,700 m). The valley splits into two branches in Kenya and Tanzania. The western arm extends from the north shore of Lake Nyasa (see p.247) along the western borders of Tanzania, forming an arc that is dominated by Lake Tanganyika (see p.248). The eastern branch includes lakes Manyara (see p.247) and Natron. Most of the lakes in the East African Rift are extremely deep, with Tanganyika descending to 4,800 ft (1,471 m). The East African Rift System lies along the boundaries of three tectonic plates, the Arabian, the African–Nubian, and the African–Somalian. It started to form when the plates began to diverge about 100 million years ago. As the Earth's crust stretched, volcanoes erupted on the surface; they dot the length of the rift today, and include Erte Ale in Ethiopia and Ol Doinyo Lengai in Tanzania (see p.188). About 20 million years ago a series of huge faults opened, and the land dropped between the three plates to form the East African Rift. The plates meet at a triple junction under the Afar Triangle in Ethiopia, where the Red Sea merges with the Gulf of Aden. This may be the beginning of the world's next major ocean, as hot mantle material doming under the junction splits the plates apart. It is thought that the three plates will eventually separate, and that the Horn of Africa will break off from the rest of the continent to form an island.

**RIFT VALLEY WILDLIFE**
*The East African Rift is home to a huge array of endemic species of African wildlife, among which are leopards.*

**VOLCANIC LANDSCAPE**
*Where the East African Rift meets the Red Sea, at the Ardoukoba Volcanic Field in Djibouti, the terrain is dominated by volcanoes. A fault-bounded rift extends seaward in the center of the picture.*

**RED SEA CORAL REEFS**
*The Red Sea was formed by the divergence of the African and Arabian plates. Stunning coral reefs have formed in the sea's warm shallows.*

**STEEP VALLEY CLIFFS**
*At Losiolo Maralal, Kenya, the valley floor has dropped away to leave sheer escarpments above the Suguta Valley, one of the hottest places on the Earth.*

## EARLY HUMANS

In 1974, American paleontologist Don Johanson found some of the most important hominid remains ever discovered, an almost complete skeleton of a 3-million-year-old *Australopithecus afarensis*. The skeleton, nicknamed Lucy, is that of a drowned woman. It was preserved in volcanic mud, where minerals from the surrounding liquid slowly replaced the calcium in the bones. The land turned to desert, and the ground was then washed away by rain to reveal the skeleton millions of years later.

## THE LEAKEY FAMILY

No name is more closely linked to the wealth of early hominid discoveries in the East African Rift than that of Leakey. Louis Leakey (1903–72) and his wife Mary (1913–96) discovered the first skull of an *Australopithecus* in Olduvai Gorge, Tanzania, in 1959. They also discovered that the remains of the earliest humans, *Homo habilis* and *Homo erectus*. Later, Mary Leakey found 3.5-million-year-old footprints in Tanzania, proving that these human ancestors walked upright. The couple's son, Richard, has also made significant discoveries in the same field.

LAND

**EAST AFRICAN RIFT AT THE RED SEA**
*This image shows the northern edge of the East African Rift and its extension between the African and Arabian plates. The resulting rifts have been filled by the Gulf of Suez (left) and the Gulf of Aqaba (right).*

**MOVING EARTH**
*Japan's 1995 Kobe earthquake had a magnitude of 6.9—strong enough to bend train tracks, topple bridges, and make elevated sections of the city's freeways collapse.*

# LIVING WITH EARTHQUAKES

Earthquakes are the most dangerous of all natural hazards. Unlike storms and volcanic eruptions, they strike within seconds, giving no opportunity for escape. Quite apart from the toll in lives, they cause damage that can take years to repair. Earthquakes cannot be prevented, but current research aims to make life safer in earthquake zones. The chief goal—which still remains elusive—is a reliable way of forecasting exactly when and where future earthquakes will strike.

## PREDICTING EARTHQUAKES

Forecasting earthquakes is notoriously difficult because they are so complex. Hundreds of variables are involved, and no two fault systems are exactly the same. However, two kinds of data are particularly revealing: the past record of earthquakes in a given area, and the present buildup of stress within a fault system's rocks.

The past record gives average frequencies for earthquakes, which can be used as a guide to future earthquakes. For example, if there has previously been a major earthquake every 100 years, there is a 10 percent probability of one in any given decade, if earthquakes are randomly spaced. But recent research into fault-system stress suggests that earthquakes are not random. Far from relieving stress, one earthquake can actually increase the likelihood of another one elsewhere. In 1999, this discovery was successfully used to predict a major aftershock in Turkey, three months after the devastating earthquake at Izmit, which claimed 25,000 lives (see p.147).

### LARGEST EARTHQUAKES

The ten most powerful earthquakes of the last 100 years are here ranked by moment magnitude–a measurement that reflects seismic energy more accurately than the Richter scale.

| Location | Date | Magnitude |
|---|---|---|
| Chile | May 22, 1960 | 9.5 |
| Prince William Sound, Alaska | March 28, 1964 | 9.2 |
| Andreanof Islands, Aleutian Islands | March 9, 1957 | 9.1 |
| Kamchatka | November 14, 1952 | 9.0 |
| Coastal Ecuador | January 31, 1906 | 8.8 |
| Rat Islands, Aleutian Islands | February 4, 1965 | 8.7 |
| India–China border | August 15, 1950 | 8.6 |
| Kurile Islands | October 13, 1963 | 8.5 |
| Banda Sea, Indonesia | February 1, 1938 | 8.5 |
| Kamchatka | February 3, 1923 | 8.5 |

## EARTHQUAKE-PROOFING

The Izmit earthquake vividly illustrated a well-established fact: it is not earthquakes that kill, but buildings. Earthquakes have a relatively minor impact in the open—many people even fail to notice them—but when they shake rigid structures, the results can be catastrophic. In Izmit (see right), where the initial shake had a magnitude of 7.4, many new apartment buildings collapsed completely, while older ones nearby were either unaffected or toppled over while remaining intact. Subsequent investigations showed that many of the failures were due to inadequate design or poor building materials.

**SELECTIVE DESTRUCTION**
*The Izmit 1999 earthquake in Turkey destroyed many new apartments, and reinforced the need for better-designed buildings in earthquake-prone areas.*

**BUILT TO LAST**
*Because of its earthquake-proof design, the Transamerica Building successfully weathered the 1989 earthquake in San Francisco.*

In California and Japan—two of the world's most populous earthquake zones—strict building codes have saved many lives. Here, high-rise buildings are designed to be slightly elastic, so that the energy from a shock is dissipated harmlessly, instead of fracturing supporting columns and floors. Foundations are equipped with mechanical dampers—structures that function in much the same way as shock absorbers in cars. Using this kind of technology, skyscrapers such as San Francisco's Transamerica Building (see above) have emerged from major shocks unscathed.

Almost as important is the prevention of fire. After San Francisco's 1906 earthquake, fire swept through its central business district (see p.144); similarly, in 1923, thousands of houses were burned in Tokyo, causing almost as much loss of life as the earthquake itself. Although the use of nonflammable building materials greatly reduces this danger, risks still remain. In 1995, more than 5,000 people died when an earthquake hit Kobe (see p.152)—a city widely believed to be one of the most earthquake-proof in the world.

## TSUNAMIS

When earthquakes occur beneath the sea, the sudden slippage of the sea bed can produce tsunamis, or tidal waves. These radiate out from the earthquake site, traveling across oceans at speeds of up to 500 mph (800 km/h). In the open ocean, tsunamis are rarely more than 3 ft (1 m) high, but as they approach coasts, their amplitude increases, creating a wall of water that can tower 65 ft (20 m) above the normal high-tide level.

Like earthquakes, tsunamis cannot be prevented, but monitoring systems can warn when a tsunami is on its way. In the Pacific Basin—the world's most tsunami-prone region—an ocean-wide warning system was established in 1948. Based in Hawaii, the system receives information about earthquakes around the Pacific Rim and issues alerts when tsunamis are detected. With satellite technology, several hours' warning can often be given, allowing coastal areas to be evacuated in time.

ASIA *east*

## Nojima Fault

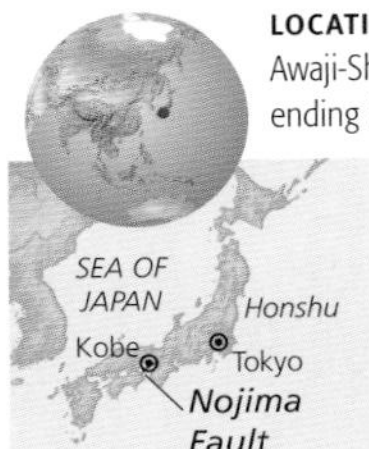


**LOCATION** Extending the length of Awaji-Shima island, Osaka Bay, ending beneath Kobe, Japan

**TYPE** Strike-slip (right-lateral)

**LENGTH** 37 miles (60 km)

**ACTIVITY** Active

Early on the morning of January 17, 1995, Kobe, Japan's sixth-largest city, was hit by the worst earthquake to strike the country since the 1923 Tokyo disaster. The earthquake lasted for barely 20 seconds, but it was so powerful—reaching 6.9 on the Richter scale—and so close to the city that its effects were devastating. About 6,300 people were killed, and 300,000 people were left homeless. Scientists began an urgent search for the origins of the earthquake, since the disaster had occurred in a region of the country that was not prone to serious seismic disturbances. The major fault in the region is the Median Tectonic Line (MTL), which extends beneath the sea off the eastern coast of Japan. The Nojima Fault, which is part of the Arima Takatsuki Line, which in turn is a branch of the MTL, was found to be the cause of the Kobe disturbance. The earthquake began just 9–12 miles (15–20 km) beneath the surface of the Earth, and within 12 miles (20 km) of the city itself, which is why its effects were so catastrophic. During the earthquake, the ground on the south side of the Nojima Fault moved 5 ft (1.5 m) to the right and slipped 4 ft (1.2 m) downward.

### KOBE EARTHQUAKE

The Kobe earthquake was traced to the Nojima Fault on Awaji-Shima island, which is one of a series of faults. Southern Japan sits on the eastern edge of the Eurasian Plate, where the Philippine Plate is being subducted along the Nankai Trough. However, the Eurasian Plate is moving in a different direction. The plate boundary is marked by strike-slip faults that trigger seismic disturbances.

**CRACKED EARTH**
*The movement along the Nojima Fault during the Kobe earthquake was marked at the surface by a visible rupture and cracks that scarred the surrounding landscape.*

### REBUILDING

The Kobe earthquake destroyed 100,000 buildings and damaged an equal number. The repair cost for the infrastructure alone was estimated at $150 billion. Yet rebuilding work began immediately, and within a year all railroads and almost all roads had opened again. The people who suffered the most were the poor, who could not afford to rebuild their wooden houses, which were destroyed by the earthquake.

**KOREAN FAULT ZONE**
*The active fault zone that has recently been discovered in South Korea begins in the heavily populated southeast corner of the peninsula and extends northwest from its edge.*

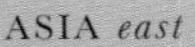

ASIA *east*

## Southeast Korea

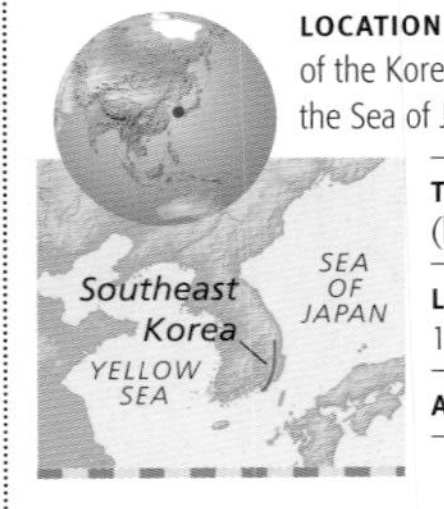


**LOCATION** On the southeast edge of the Korean Peninsula, between the Sea of Japan and the Yellow Sea

**TYPE** Strike-slip (left-lateral)

**LENGTH** 124 miles (200 km)

**ACTIVITY** Active

Until recently, Korea was not believed to be a major earthquake zone. However, over 30 active faults have now been mapped in the southeast of the peninsula and beneath the surface of the surrounding ocean. Unlike neighboring Japan, Korea is not located next to any plate boundaries, situated instead within the relatively stable interior of the Eurasian Plate. Nevertheless, there has been at least one minor earthquake here every year for the past 2,000 years. The possibility that Korea may unexpectedly be hit by a major earthquake like the one that devastated Kobe (see Nojima Fault, left) has focused attention on the peninsula's fault network—particularly since many nuclear and industrial facilities are situated in the region. The major fault is the Yangsan, which runs 125 miles (200 km) inland in a northeasterly direction from the peninsula's tip. The Yangsan Fault is a strike-slip fault that is moving by up to $1/25$ in (1 mm) every year. Most of the Korean faults were originally opened up by the way the Pacific Plate interacts with the Eurasian Plate and the Okhotsk Plate. The resulting Korean faults were all originally right-lateral strike-slip faults. But the way the Pacific Plate is now being subducted under the Okhotsk and Eurasian plates has reversed the movement of the faults, and they are now left-lateral faults.

AUSTRALASIA *New Zealand*

## Great Alpine Fault

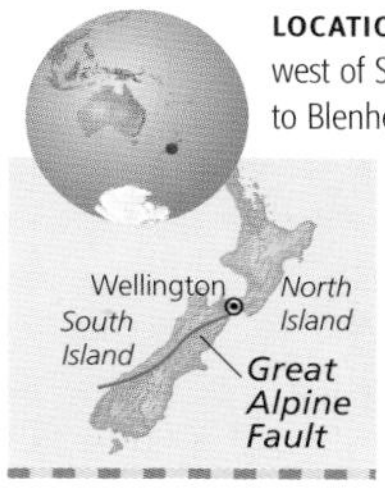

**LOCATION** Extending along the west of South Island from Fiordland to Blenheim, New Zealand

**TYPE** Strike-slip (right-lateral)

**LENGTH** 310 miles (500 km)

**ACTIVITY** Active



The Great Alpine Fault of New Zealand is one of the Earth's most remarkable faults. At its northern and southern limits it is easily visible both from space and from the Earth's surface. New Zealand is located on the boundary between the Australian and Pacific plates. In the ocean to the east of New Zealand's North Island, the Pacific Plate is being subducted under the Australian Plate, and magma that is produced by the process forms North Island's active volcanic zone. The Great Alpine Fault lies on South Island, marking the line where the two plates meet, and cutting right through the middle of New Zealand's towering Southern Alps (see p.170). Only the northwest of South Island sits on the Australian Plate. The rest of the island lies to the east of the fault, on the Pacific Plate.

**FRACTURED EARTH**
*The activity of the Great Alpine Fault is clearly revealed in this field on South Island, where a large crack has opened across the Earth's surface.*

Along this section of the plate boundary, the Australian Plate is being subducted, bending at an angle of about 40°. As a result of this plate convergence, the Southern Alps are being thrust over the Australian Plate, lifting them up at about 1/4 in (7 mm) a year. As the Australian Plate descends beneath the Pacific Plate, it generates earthquakes deep under the southeast part of South Island. The Pacific Plate is also rotating counterclockwise. This creates horizontal movement along the Great Alpine Fault, and, because the Pacific Plate is moving south, it is a right-lateral fault. Sideways movement is so extensive that the Great Alpine Fault is classified as a transform boundary (see p.108). The origins of the fault, and thus the Southern Alps, date back about 26 million years to a time geologists call the Kaikoura Orogeny. At this time, New Zealand was barely above sea level and there were no mountains, but as the Pacific and Australian plates twisted and thrust together, the Southern Alps were forced up and the Great Alpine Fault opened.

### HAROLD WELLMAN

Harold Wellman (1908–99) was an English geologist who discovered the Great Alpine Fault. During World War II, the New Zealand Geological Survey sent him to South Island to look for the mineral mica. While there, he discovered a line across the ground where rocks changed from granite to schist. Wellman realized that the rocks had been displaced over 310 miles (500 km) sideways along this line. He had found one of the Earth's major faults.

**MOUNTAIN FAULT**
*The straight line of the Great Alpine Fault strikes across the landscape of New Zealand's South Island. The line of the fault becomes visible as it emerges in the Southern Alps.*

# MOUNTAINS

106–09 Tectonic plates
142–53 Fault systems
172–93 Volcanoes
Glaciers 262–81
Forests 302–321
Ocean tectonics 386–87
Mountain climates 457
Precipitation and clouds 466–73

MOUNTAINS ARE ROCK MASSES that have been elevated high above their surroundings, mostly by the Earth's plate-tectonic processes. The origin of a mountain belt can generally be traced to one of the Earth's plate boundaries, past or present. Most mountains can be classified by the type of plate boundary involved in their formation (see pp.108–09). A broad classification can be based on whether the plate boundary is divergent or convergent. The plates involved may be both continental, both oceanic, or of differing types. Once they have been uplifted by tectonic forces, mountains are sculpted into peaks, valleys, and other features by the processes of weathering and erosion.

**ANTARCTIC FOLD**
*Intense folding and metamorphism have transformed the white layers seen in this picture of Antarctica's Royal Society Range from limestone into marble, and deformed them into an overturned anticline.*

## FOLDING

Rocks near the Earth's surface are usually hard and brittle, and when subject to pressure they may break, as happens during an earthquake. However, when rocks are deeply buried, heated, and compressed over a long period, they become ductile and can deform, leading to folding, and making them susceptible to the geological processes associated with mountain building. Horizontal sedimentary rock strata buckle and dislocate into a variety of curved shapes. Upfolds, known as anticlines, are separated by downfolds known as synclines. These processes can produce folds from a fraction of an inch to many miles long. The hills and valleys flanking mountain belts are typically symmetrical folds with vertical planes of symmetry. By contrast, deep within mountain belts, where temperatures are greater, folds may become overturned to such an extent that the fold becomes almost flat (recumbent), and the tip may become detached and slide or be forced farther along a fault plane.

**VERTICAL STRATA, NAMIBIA**
*The effect of differential stress (unequal pressure in different directions) on rock strata can be powerful enough to rotate them from a horizontal to a vertical position.*

**SYMMETRICAL FOLD**
*limb* 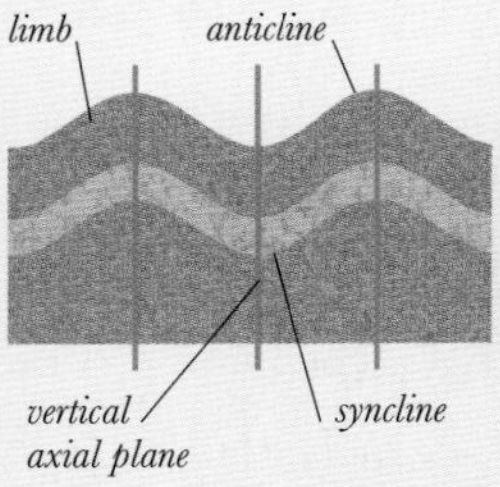
*anticline*
*vertical axial plane*
*syncline*

**ASYMMETRICAL FOLD**
*shallow limb*
*steep limb*
*steeply inclined axial plane*

**OVERTURNED FOLD**
*very gentle limb* 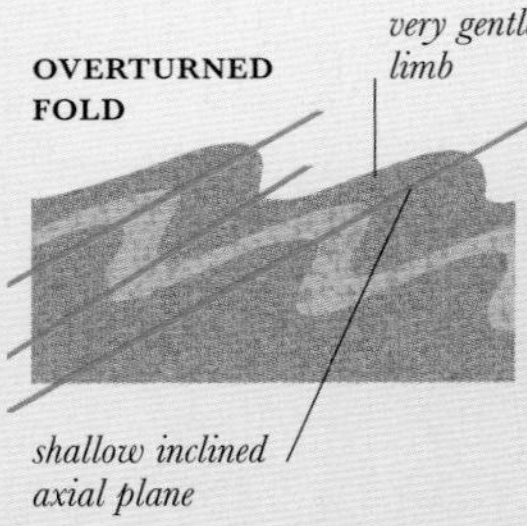
*shallow inclined axial plane*

**RECUMBENT FOLD** 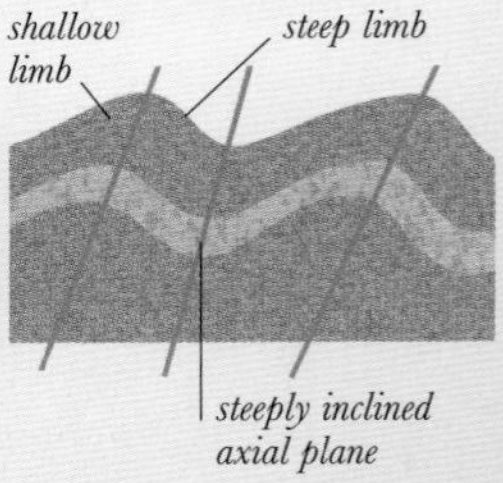
*overturned fold limb*
*thrust fault*
*very shallow axial plane*

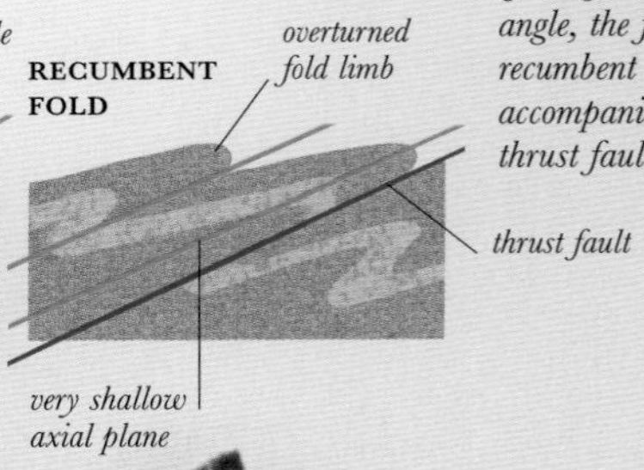

**FOLD SHAPES**
*Folds with limbs (sides) whose axial planes are inclined at the same angle are called symmetrical folds. If the axis of a fold is tilted to one side, and one limb has deformed so that it is steeper than the other, the fold is asymmetrical. When the fold axis is inclined at 45° or less to the horizontal, the fold is overturned. If the fold axis lies at a shallow angle, the fold is known as recumbent and is often accompanied by low-angle thrust faulting.*

**ANDEAN PEAKS**
*Although they began forming millions of years ago, the Andes, seen here in Patagonia, are among the Earth's youngest mountains.*

# CONTINENTS IN COLLISION

Some of the Earth's most impressive mountain ranges, such as the Himalayas (see pp.168–69), have been formed by continents colliding directly with each other. Most tectonic plates are a combination of upper mantle rocks with dense oceanic crust and lighter but thicker continental crust. When an ocean basin is consumed by subduction, two continental masses move toward each other and meet, forming a convergence zone. Their similar density means neither will readily subduct into the mantle. Subduction cannot continue, even though the plates carrying the continents continue to converge. Instead, the continental margins, with their thick layers of sediments washed off the continents, and then the continents themselves collide to form vast folded and faulted mountain belts in which crustal rocks are thickened, deformed, and metamorphosed. Within such a belt, slivers of oceanic crust and even upper mantle rocks may be preserved, testifying to the earlier subduction.

*mountain range formed by continental collision*

*plateau behind mountain range*

*overlying plate*

*major fault*

*subducted oceanic crust*

*suture zone*

*plate attached to subducted sea floor*

**MOUNT EVEREST**
*Mount Everest's weight is supported by a root of thickened continental crust below it that floats on the mantle. This balancing act is called isostasy.*

**PLATE MOVEMENT**
*As continents come together, sea floor from one plate is subducted beneath the other, creating volcanic mountains. As the continental crust blocks finally collide, a second range is created by compression and uplift, with the previous mountains forming a plateau.*

# OCEAN–CONTINENT COLLISIONS

Where the oceanic part of a plate converges with the edge of a continent, a subduction zone is formed, as the denser oceanic slab of crust and mantle is driven beneath the less dense continental crust. This type of subduction forms a continental volcanic arc, such as the Cascade Range of western North America. Sediments scraped from the ocean floor by the subduction process are faulted onto the continental margin, which is further compressed and thickened by folding. Such subduction is also responsible for the Ring of Fire around the Pacific Ocean, and contributes to the building of the young mountain belt extending from the South American Andes (see pp.158–59) through the Western Cordillera of North America and the mountains of the Aleutian Islands to Japan and beyond.

*continental plate*

*mountain range*

*subducting oceanic plate*

*deep ocean trench formed*

**ANDEAN-STYLE MOUNTAINS**
*An oceanic plate subducts below the overriding continental plate. The continent's edge thickens, buckles, and is built up further by volcanism.*

**ANDES AND PACIFIC OCEAN**
*This image, taken from space, shows the Chilean Andes, with the Pacific Ocean to the right of the range.*

# OCEAN–OCEAN COLLISIONS

When converging plates are both capped by oceanic crust of similar density, a volcanic island arc (see p.387) can be produced. Typically, this consists of mountainous islands with a parallel trench running offshore along one side of the chain. The trench marks where the plate that is older, cooler, and therefore slightly denser is subducting beneath the other one. Hot molten rock (magma) rises through the overriding plate to erupt at the surface, building up gradually into a chain or arc of mountainous volcanic islands such as the Marianas, Vanuatu, or Solomon Islands in the Pacific and the Antilles and Scotia arc in the Atlantic. Island arcs and their associated sediments can sometimes become joined or accreted to continental margins, forming so-called exotic terranes. Such terranes are thought to have contributed to new mountains on the Pacific margin of North America's Western Cordillera.

*converging oceanic plate*

*arc of mountainous volcanic islands*

*subduction zone*

*subduction of older and cooler oceanic plate*

**VOLCANIC ISLAND ARC**
*Subduction of the older plate triggers volcanic activity, creating an island arc and forming new continental crust.*

**JAPANESE ISLANDS**
*Plate convergence below Japan generates frequent and often damaging earthquakes, fault movements, and occasional volcanic eruptions.*

# RIFT MOUNTAINS

Where continental crust is pulled apart at divergent boundaries, the brittle surface rocks do not stretch very far before fracturing along parallel fault lines. Between the faults, the central block subsides, forming a rift valley flanked by escarpments that can reach mountainous dimensions, as in parts of the East African Rift (see pp.148–49). If rifting continues, plates may split in two, with a new ocean between them. Where oceanic plates diverge, they generate sea-floor spreading (see pp.386–87), leading to the formation of submarine mountain ranges quite different in structure from those on land. As the plates are pulled apart, molten rock rises from the upper mantle to erupt along the axis of the sea-floor spreading ridge. These ridges form the Earth's longest mountain chain.

**RIFT VALLEY**
*The East African Rift is flanked by mountains, including volcanoes, that rise over 6,600 ft (2,000 m) above the valley floor.*

steep fault scarps
parallel faults
down-faulted central block
magma rises through fractures in crust
young deposits preserved in rift valley
plates pull apart
gentle back slopes

**RIFTING**
*A rift valley, flanked by parallel faults, is formed where rocks are pulled apart on a regional rather than a continental scale.*

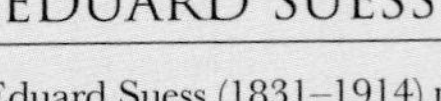

## EDUARD SUESS

Austrian geologist Eduard Suess (1831–1914) pioneered studies of the folds, faults, and geological structure of the European Alps. Suess's acclaimed four-volume book *The Face of the Earth* (*Das Antlitz der Erde,* 1885–1909) was an encyclopedic overview of the Earth's mountains, their structure and formation, and a history of the oceans. He was also the geologist who first put forward the idea—later disproved—that there could have been land bridges that joined the continents of South America, Africa, India, and Australia. Suess named this once-vast continent Gondwanaland; that name is still used for an ancient supercontinent.

# EPEIROGENIC MOVEMENTS

The Earth's crust can be subjected to large-scale movement without extensive deformation and folding. This movement can be either downward, which results in the creation of large sedimentary basins, or upward to form elevated domes or plateaus. Both trends are referred to as epeirogenic movements. These events tend to be gradual, and usually affect large regions, such as the Colorado Plateau in the US. An area of about 15,500 square miles (40,000 square km), the plateau is cut off from the Great Plains by the Rocky Mountains, and has been stable for at least 500 million years. Epeirogenic uplift in the last 65 million years has led to increased erosion by the Colorado River, excavating deep canyons in the plateau.

**UTAH GOOSENECKS**
*At Goosenecks State Park in Utah, epeirogenic uplift has led to the formation of a deep canyon by the San Juan River.*

# OROGENIC BELTS

Over geological time, plate movements have produced innumerable episodes of mountain-building around the world, a process known as orogenesis. The resulting linear arrays of mountains, shown below, are called orogenic belts, with the events that form them known as orogenies. They are usually identified by a geographical name, such as the Caledonian orogeny, which was first identified in Scotland (known as Caledonia to the Romans). Due to later plate movements, most ancient orogenic belts are no longer located where they formed; for instance, the Caledonian belt was originally an extension of the North American Appalachians.

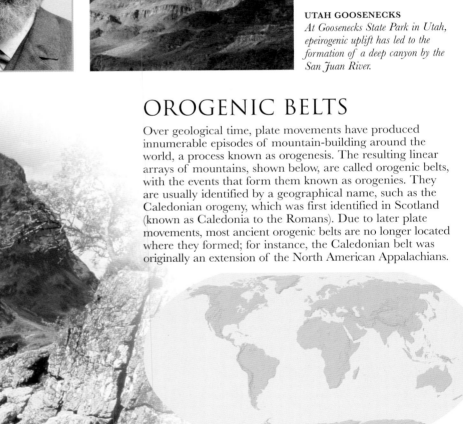

**ANCIENT MOUNTAIN ROOTS**
*The Scottish hills now barely qualify as mountains, but they are the eroded remnants of once higher mountains.*

## MOUNTAIN PROFILES

The pages that follow contain profiles of some of the world's mountain belts. Each profile includes the following summary information.

**HIGHEST PEAK** Name and height above sea level

**LENGTH** Distance along the belt's longest axis

**AGE** Geological era within which the mountain belt formed

**FORMATION TYPE** Continent–continent, ocean–continent, ocean–ocean, or rift

NORTH AMERICA *west*

# Rocky Mountains

**LOCATION** Stretching along the western side of North America from Alaska to New Mexico

**HIGHEST PEAK** Mount McKinley, Alaska 20,321 ft (6,194 m)

**LENGTH** 3,700 miles (6,000 km)

**AGE** Mesozoic–Cenozoic

**FORMATION TYPE** Ocean–continent

The Rocky Mountains are part of one of the largest mountain belts on Earth, the great Western Cordillera of North America, which extends from above the Arctic Circle to the Sierra Madre Occidental in Mexico. Strictly speaking, the Rocky Mountains are the high peaks of the cordillera's eastern ranges, which also include the Canadian Rockies, and the Northern, Central, and Southern Rockies in the US. Most of the mountains peak above 3,000 ft (1,000 m), with more than 100 summits above 10,000 ft (3,000 m). Within the belt is North America's continental divide, to the west of which rivers flow into the Pacific, while in the east they flow into the Arctic and Atlantic oceans. Geologically, the history of the cordillera is extremely complex, with large amounts of uplift taking place in the last 15 million years. The range includes intensely deformed zones with extensive folds and faults, such as the Canadian Rockies; belts of volcanic activity and lava plateaus, which are evident in Yellowstone, Wyoming; and highly metamorphosed areas with granite intrusions, including Grand Teton. Features such as Wyoming's Great Divide Basin are also part of the chain. The Alaskan ranges are snow-covered and heavily glaciated. An abundance of wildlife inhabits the cordillera, including mountain goats and sheep, bears, wolves, cougars, and moose.

**AMERICA'S ICON** *The bald eagle is the only species of eagle endemic to North America.*

## FEATURES OF THE WESTERN CORDILLERA

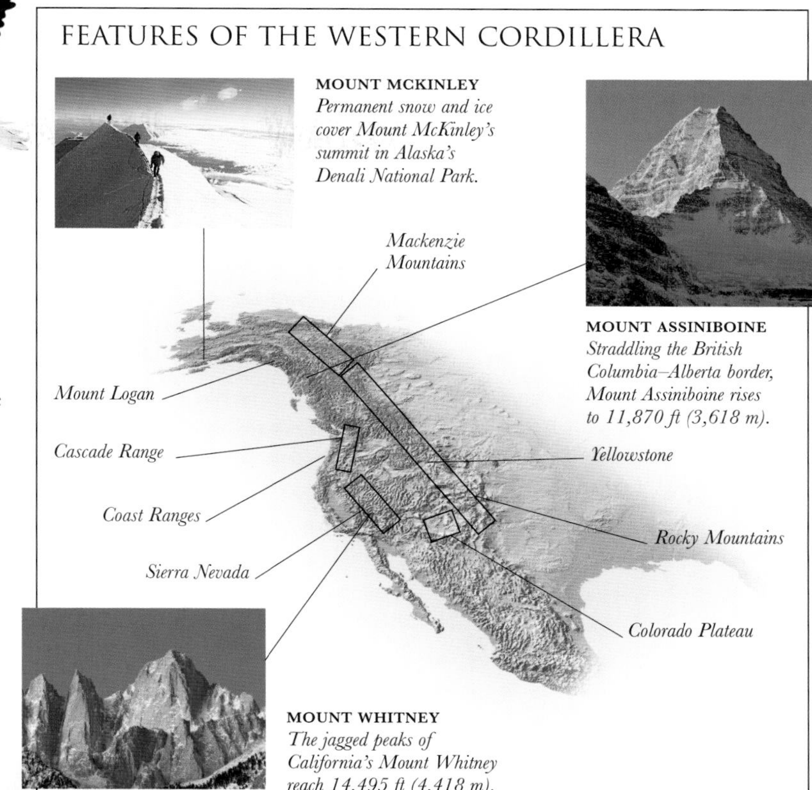


**MOUNT MCKINLEY** *Permanent snow and ice cover Mount McKinley's summit in Alaska's Denali National Park.*

**MOUNT ASSINIBOINE** *Straddling the British Columbia–Alberta border, Mount Assiniboine rises to 11,870 ft (3,618 m).*

**MOUNT WHITNEY** *The jagged peaks of California's Mount Whitney reach 14,495 ft (4,418 m).*

**ROCKY MOUNTAIN LANDSCAPE** *The headwaters of the Snake River, with its flat alluvial plain and terraces, lie in the Teton Range in northwest Wyoming. They are overlooked by Grand Teton peak, which reaches a height of 13,770 ft (4,197 m).*

## ALASKAN OIL PIPELINES

The discovery of oil and gas in Alaska has led to the construction of huge pipelines that overcome the barrier presented by the high mountains of Alaska. However, the pipelines cross environmentally sensitive wilderness areas, creating a physical obstacle for migrating animals such as caribou. Although burying pipelines might appear to solve such difficulties, it can actually cause further problems, since they may fracture. This is also a region of permafrost, which is unstable. In places, pipelines are elevated to allow wildlife to pass safely beneath.

NORTH AMERICA *east*

## Appalachians

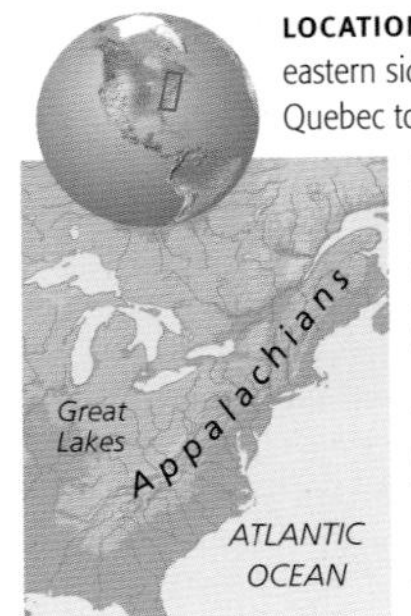


**LOCATION** Running along the eastern side of North America, from Quebec to Alabama

**HIGHEST PEAK** Mount Mitchell, North Carolina 6,684 ft (2,037 m)

**LENGTH** 1,500 miles (2,400 km)

**AGE** Paleozoic–Mesozoic

**FORMATION TYPE** Continent–continent

A much-eroded, ancient, and complex mountain belt, the Appalachians were formed during a series of plate separations and collisions. Today, the chain consists of a series of ranges that increase in geological complexity from west to east. The slightly folded and uplifted plateaus in the west give way to parallel ridges and valleys, followed by the highly deformed Blue Ridge. The belt becomes less elevated closer to the Atlantic coast, where metamorphosed sediments and volcanic rocks are overlain by the younger accumulations of the coastal plain. The Appalachians are the North American continuation of Europe's Caledonian range (see p.160).

**APPALACHIAN AMPHIBIAN**
*There are 34 native varieties of salamanders resident in the Appalachian forests, including this marbled salamander.*

**TENNESSEE MOUNTAINS**
*The forest-clad slopes of much of the Appalachian range disguise a geologically complex area of ancient metamorphosed rocks.*

NORTH AMERICA *south*

## Sierra Madre Oriental

**LOCATION** In eastern Mexico – extending from north to south, and bending west at the northern end

**HIGHEST PEAK** Cerro Peña Nevada, Mexico 11,955ft (3,644m)

**LENGTH** 700 miles (1,100km)

**AGE** Mesozoic–Cenozoic

**FORMATION TYPE** Ocean–continent

Much of Mexico is elevated above 3,000ft (1,000m), with the three Sierra Madre ranges and the Mexican Volcanic Belt forming the country's mountain backbone. A continuation of the Western Cordillera, the Sierra Madre Oriental was formed by compressional tectonic forces and folding about 50–70 million years ago, which produced a long, dramatic series of alternating anticlines and synclines in limestone.

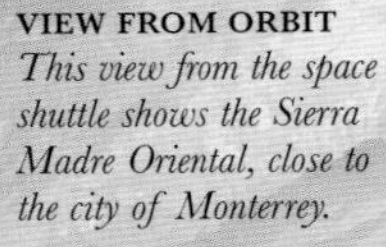

**VIEW FROM ORBIT**
*This view from the space shuttle shows the Sierra Madre Oriental, close to the city of Monterrey.*

**LIMESTONE CRESTS**
*The sharp peaks of the Sierra Madre Oriental are formed of limestone, which was elevated by tectonic forces.*

SOUTH AMERICA *west*

## Andes

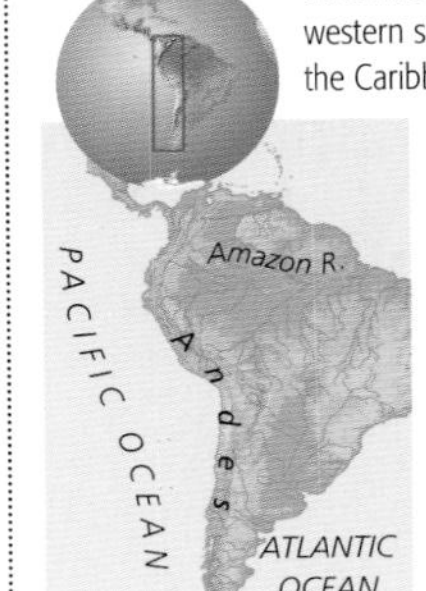

**LOCATION** Running down the western side of South America from the Caribbean Sea to Cape Horn

**HIGHEST PEAK** Aconcagua, Argentina 22,834 ft (6,959 m)

**LENGTH** 4,500 miles (7,200 km)

**AGE** Mesozoic–Cenozoic

**FORMATION TYPE** Ocean–continent

The longest mountain chain in the world (apart from the Mid-Atlantic Ridge beneath the Atlantic Ocean), the Andes is also one of the Earth's most spectacular and active mountain belts. Within it lies one of the most arid places on Earth, Chile's Atacama Desert (see p.290); the world's highest navigable body of water, Lake Titicaca (see p.241); the source of the world's greatest river, the Amazon (see pp.222–23); and the highest city in the world, La Paz, in Bolivia, which is situated at an altitude of 11,910 ft (3,630 m). The whole range is frequently affected by earthquakes and eruptions from 183 active

### FEATURES OF THE ANDES

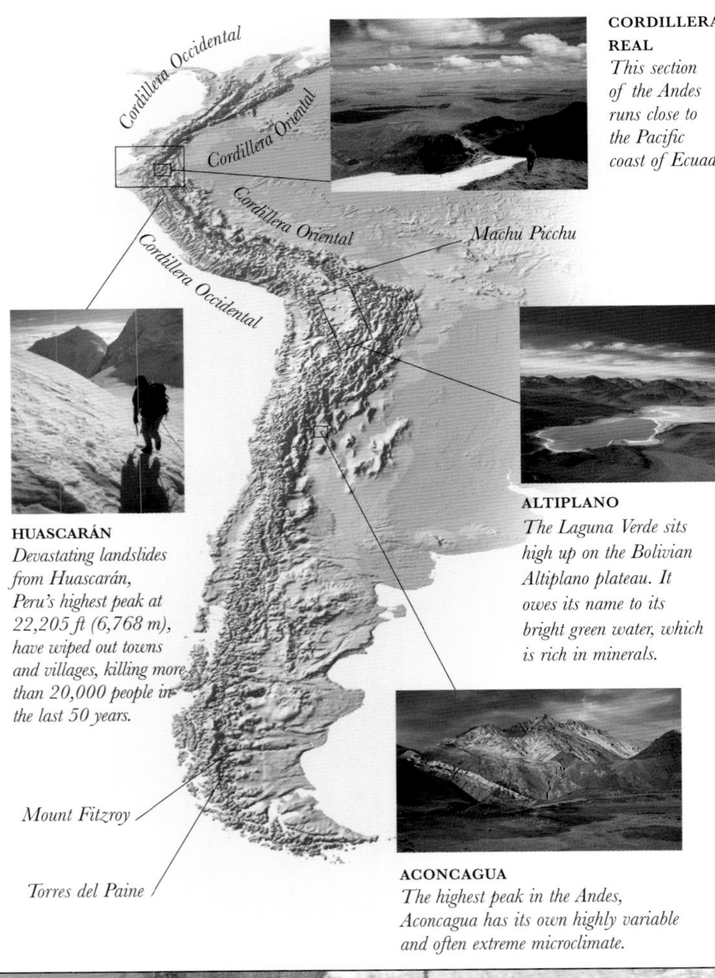


**CORDILLERA REAL**
*This section of the Andes runs close to the Pacific coast of Ecuador.*

**HUASCARÁN**
*Devastating landslides from Huascarán, Peru's highest peak at 22,205 ft (6,768 m), have wiped out towns and villages, killing more than 20,000 people in the last 50 years.*

**ALTIPLANO**
*The Laguna Verde sits high up on the Bolivian Altiplano plateau. It owes its name to its bright green water, which is rich in minerals.*

**ACONCAGUA**
*The highest peak in the Andes, Aconcagua has its own highly variable and often extreme microclimate.*

**JUTTING PEAKS**
*The tooth-shaped peaks of the Fitzroy Range in Patagonia are the stumps of much higher mountains that have been steadily eroded by glaciers.*

LAND

**HIGH FLIER**
*With the largest wing area of any bird, the Andean condor can fly at altitudes of 18,000 ft (5,500 m).*

volcanoes, including the highest on Earth. The mountains ascend abruptly from sea level on the western, Pacific coast of South America to altitudes of over 21,300 ft (6,500 m). The belt forms a formidable physical, meteorological, and biological barrier, ranging from 60 miles (100 km) wide at the southern tip of the continent to 435 miles (700 km) wide in the central region. In Peru, Bolivia, and Ecuador, between the Cordillera Occidental range in the west and the Cordillera Oriental in the east, lies the Altiplano (also known as the Puña), which is the world's most densely populated high plateau. The farmers who inhabit the Altiplano exist in one of the most hostile environments on Earth for human life: the high altitude means that the air is thin, and the weather is extreme. The southern sector of the range, which runs through Patagonia and Tierra del Fuego, has been glaciated as far as sea level and is characterized by numerous fjords. Geologically, the mountains are the result of the continuing eastward subduction of the Nazca Plate beneath the South American Plate. The subduction zone is marked by the Peru–Chile Trench off South America's Pacific coast, which plunges to more than 26,250 ft (8,000 m) below sea level. The present topography of the Andes is largely the result of deformation that has taken place during the past 50 million years, with significant elevation of the mountains occurring over the past 10 million years. As such, the belt is young in geological terms, and mountain-building continues in the present. Minerals such as gold, silver, platinum, and copper are all mined in the Andes.

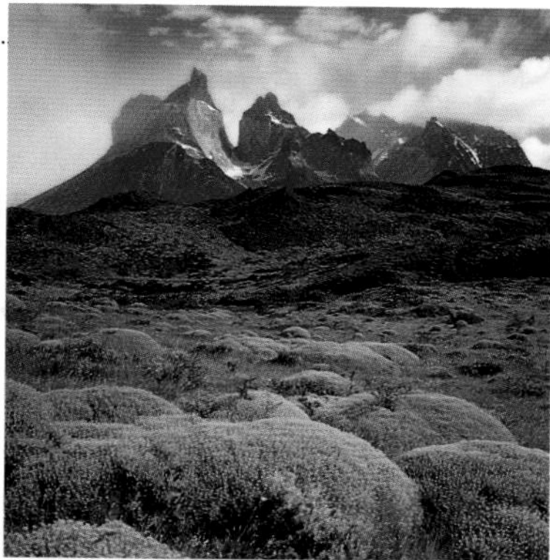

**GRANITE HORNS**
*The granite peaks of the Torres del Paine soar to 10,006 ft (3,050 m) above the glacial lakes of Patagonia in southern Chile.*

## ALEXANDER VON HUMBOLDT

German aristocrat Alexander von Humboldt (1769–1859) began to explore South America in 1799 with the French botanist Aimé Bonplan (1773–1858). The pair spent five years studying the continent's geology and botany. Humboldt was the first to climb Chimborazo, Ecuador's highest peak, and describe its environment.

## CITY OF THE INCAS

Situated high in the Peruvian Andes, Machu Picchu is a remarkably preserved Inca city. The exact function of the settlement is unknown, but it may have been a royal estate built by the Inca emperor Pachacuti in the 15th century. Located on a prominent ridge, partly surrounded by steep cliffs, the site includes houses, religious architecture, tombs, and agricultural terraces. When the Spanish conquistadors arrived in South America in the 16th century, Machu Picchu was abandoned. The city remained hidden for centuries, until 1911, when it was found by US archaeologist Hiram Bingham.

EUROPE *east*

## Urals

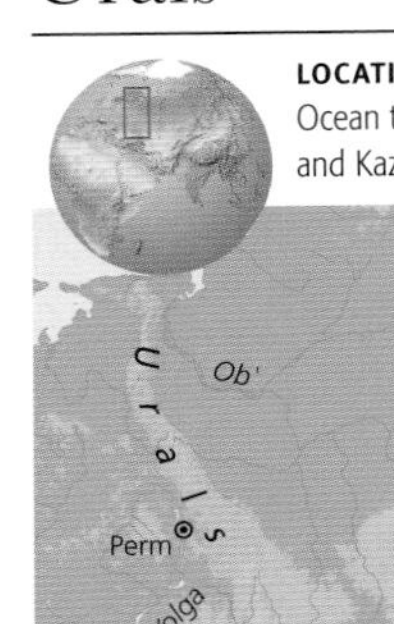


**LOCATION** Running from the Arctic Ocean to the border between Russia and Kazakhstan

**HIGHEST PEAK** Narodnaya, Russia 6,214 ft (1,894 m)

**LENGTH** 1,500 miles (2,400 km)

**AGE** Paleozoic

**FORMATION TYPE** Continent–continent

The ancient, extensively eroded Urals form a wall-like, linear mountain chain traditionally regarded as the boundary between Europe and Asia. Only 125 miles (200 km) wide at its broadest point, it is a narrow range. The Urals represent the seam along which the Baltic and Russian continents collided about 300 million years ago, after the vast Uralian Ocean had been completely subducted under the Russian continent. The movement that formed the mountains was part of the assembly of the supercontinent Pangea, which once covered two-fifths of the world's surface. Following subduction of the ocean crust, the compressive forces of convergence crumpled sediments on the leading edges of the two colliding continents, thickening the crust and elevating the mountains. In 1841 British geologist Roderick Murchison mapped the stratification of deposits in the western Urals. In doing so he defined the Permian Period (named after the Russian region of Perm), which extended from about 290 million to 252 million years ago and saw the extinction of much of the marine life of the period. Sulfide mineral deposits, including copper, iron, and zinc, were formed by hydrothermal vents on the ancient ocean floor. These are now preserved in the southern Urals and form an ore belt that is 1,200 miles (2,000 km) long. The area has been one of the most important industrial regions of modern times. The central and southern Urals are heavily forested, while the northern mountains feature alpine meadows and tundra.

**THICKLY FORESTED MOUNTAINS**
*Much of the Urals are covered in taiga, the Russian name for the coniferous forest that is indigenous to Eurasia and stretches from the Baltic Sea to the Pacific Ocean (see p.316).*

### PLATINUM MINING

Pure platinum is an extremely rare and valuable natural mineral. It was once confused with silver, but it actually has a much higher melting point of 3,222°F (1,772°C). With a specific gravity of 21.5, platinum is the densest naturally occurring mineral. Nuggets of platinum have been found in placer deposits (see p.63), especially at Nizhniy Tagil in the Urals, where one of the largest accumulations of the mineral in the world is found.

EUROPE *northwest*

## Caledonian Mountains

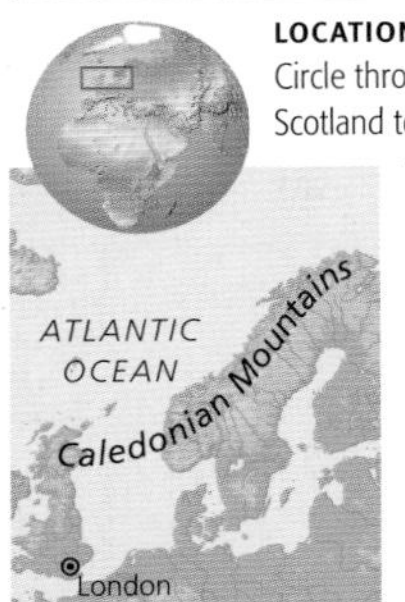


**LOCATION** Running from the Arctic Circle through Scandinavia and Scotland to northwest Ireland

**HIGHEST PEAK** Galdhøpiggen, Norway 8,100 ft (2,469 m)

**LENGTH** 1,625 miles (2,600 km)

**AGE** Paleozoic

**FORMATION TYPE** Continent–continent

The Caledonian Mountains are the European section of a much longer mountain chain that continues as the Appalachians (see p.158) on the opposite side of the Atlantic Ocean. This ancient mountain belt is one of the most intensely studied in the world. This is especially true of its British ranges, studies of which generated many formative ideas about mountain-building in the late 19th century. However, none of the pioneers of geology could have guessed at the extraordinary history of the Caledonian Mountains. Five hundred million years ago, North America, Scotland, and northwest Ireland were part of the ancient continent of Laurentia, and were separated from Avalonia (England, Wales, and southwest Ireland) by the Iapetus Ocean, all of which were located in the southern hemisphere. About 490 million years ago, Avalonia began to move north, subducting the Iapetus Ocean in the process. Eventually, Avalonia impacted upon Laurentia, a convergence that resulted in a period of mountain-building. Much later, as the North Atlantic Ocean began to open 80 million years ago, North America split from northwest Ireland and Scotland, which converged with the rest of what is now the British Isles. The Caledonian Mountains are continuing to rise by a tiny amount each year. The range includes the North West Highlands and Grampian mountains of Scotland (which are the most elevated peaks in the British Isles) and the Jotunheimen of Norway. The rugged landscape of the range today features craggy, snow-capped peaks, glens, lochs, and fjords.

**BUACHAILLE ETIVE MOR**
*This striking triangular peak descends to Glen Coe, in the Scottish Grampian Mountains.*

**NORWEGIAN FJORDLAND**
*With their precipitous mountainsides, the deep fjords of Norway's Atlantic coast are evidence of the incredible erosive power of glaciers.*

EUROPE *west*

# Pyrenees

**LOCATION** Between France and Spain, extending from the Atlantic Ocean to the Mediterranean Sea

**HIGHEST PEAK** Aneto, Spain 11,168 ft (3,404 m)

**LENGTH** 270 miles (435 km)

**AGE** Late Paleozoic–Cenozoic

**FORMATION TYPE** Continent–continent

The Pyrenees form a natural barrier between France and the Iberian Peninsula (Spain and Portugal). Two-thirds of the belt is situated in Spain, and there are relatively few natural passes between the mountains. Nevertheless, human predecessors (*Homo antecessor*) managed to cross the mountains into northern Spain approximately 800,000 years ago. The limestone caves of the Pyrenees are particularly famous for preserving some of the most recent Neanderthal remains in Europe, which are approximately 29,000 years old. The walls of the caves also contain paintings by the earliest modern humans, dating back about 15,000 years. The geological history of the Pyrenees is long, complex, and intimately associated with that of the Iberian Peninsula. During the Cretaceous Period, approximately 142 million years ago, Iberia became a microcontinent, separating from the major landmasses during the breakup of the supercontinent of Pangea. As the Atlantic Ocean opened, the Iberian plate was wedged between Europe and North Africa, leading to the initial development of the Pyrenees. Then, about 60 million years ago, rotation of the African and Eurasian plates meant that the Iberian Plate again converged with southern France. The resulting thickening and folding of the crust further uplifted the Pyrenees. The peninsula's motion ended about 10 million years ago. During the last ice age, the Pyrenees experienced extensive glaciation. The mountains are characterized by jagged peaks that are steepest on the French side of the range. In the west, the slopes are heavily wooded, but the granite peaks of the east are much starker with less vegetation. Dramatic, semicircular glaciated hollows known as cirques (see p.262) are a prominent feature of the Pyrenees, and the mountains also contain some of Europe's most spectacular waterfalls, the highest of which, at the Cirque de Gavarnie, plunges 1,385 ft (422 m). The wildlife of the Pyrenees includes abundant bird species, among them vultures (including the rare lammergeier), eagles, capercaillie, and ptarmigan. Mammals include chamois, wild boar, lynx, and a tiny population of brown bears.

## FEATURES OF THE PYRENEES

**CIRQUE DE GAVARNIE** *This sheer rock wall is the site of the highest waterfall in Europe.*

**HAUTES-PYRÉNÉES** *Straddling the French–Spanish border, the Hautes-Pyrénées include the highest peaks in the belt.*

**PYRENEAN IBEX** *Once common throughout Europe, wild goats were depicted by early humans. The Pyrenean ibex is a surviving species, though it is declining dramatically.*

**FRINGED PINKS** *Thousands of varieties of wild flower grow in the Pyrenees, including these delicate fringed pinks, which are found in the Spanish mountains.*

## WATCHING SPACE

The Earth's increasingly polluted atmosphere mars our view of the universe. Building astronomical observatories as high as possible, and away from heavily populated areas, minimizes the effects of atmospheric pollution and light interference. The Pic du Midi Observatory is situated above the clouds in the Pyrenees.

**ORDESA NATIONAL PARK, SPAIN** *Rising high above the tree line, horizontal rock strata form precipitous cliffs above valleys that were once filled with glacial ice. Only the high peaks stood out above the glaciers at the height of the last ice age.*

EUROPE *south*

# Alps

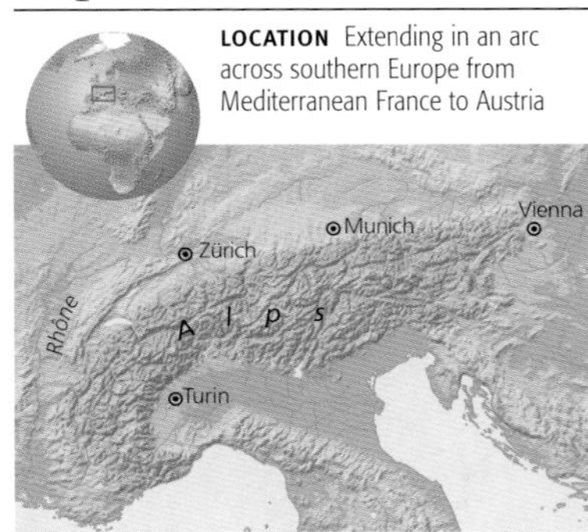


**LOCATION** Extending in an arc across southern Europe from Mediterranean France to Austria

**HIGHEST PEAK** Mont Blanc, France 15,771 ft (4,807 m)

**LENGTH** 650 miles (1,050 km)

**AGE** Mesozoic–Cenozoic

**FORMATION TYPE** Continent–continent

**EDELWEISS**
*This perennial alpine flower has tufted leaves and flowers protected by thick, woolly bracts. These protect the edelweiss from the freezing temperatures that are common close to the snow line where it grows.*

**ALPINE MEADOW**
*Retreating glaciers allow specialized alpine plants, especially grasses and saxifrages, to colonize newly exposed rocky soils within a few years.*

The largest mountain chain in Europe, the Alps contain some of the most intensely studied mountains in the world. Studies of the Alps in the 18th and 19th centuries led to early theories about mountain formation and the nature of large-scale folding and thrusting. Between 60 and 120 miles (100–200 km) wide, the Alps form a curved fold belt, with several peaks rising to over 13,000 ft (4,000 m). The west of the chain includes the Maritime and Pennine Alps; the central ranges include the Bernese (or Central) Alps; and the eastern ranges include the Dolomites and the Austrian Alps. In the past, the mountains have formed a physical obstacle to human migrations, although roads were first built through the belt's natural passes in Roman times. The Alps were generated by the convergence of the African and Eurasian plates. This process began about 90 million years ago, when the African plate began to move north, subducting the ancient Tethys Ocean and uplifting the sediment on the sea floor to form faulted recumbent fold structures called nappes. Folding and thrusting continued, with a major period of mountain building taking place 33–15 million years ago. The current landscape of the Alps is the result of glaciation during the last 2 million years. Primary features include U-shaped and hanging valleys (see p.265), and lakes such as Como (see p.245) and Garda, which have been naturally dammed by glacial moraines. The longest remaining glacier in the Alps is the Aletsch (see p.275), which extends for 25 km (16 miles).

## ALBRECHT PENCK

German geomorphologist Albrecht Penck (1859–1945) was director of the Berlin Institute of Oceanography and Geography. Penck studied the development of landforms and surface deposits in the Alps and the boulder clay of the North German Plain to determine the extent to which glaciation had affected them. This led him to propose that four main episodes of glaciation had taken place in the area. He also argued that the elevation of mountains is limited by the balance between uplift and erosion, a theory that is well illustrated by the Alps. The mountains continue to be uplifted by approximately 1/50 in (0.5 mm) every year, a rise that is compensated for by similar levels of erosion. Penck's most famous work is *The Morphology of the Earth's Surface* (1894).

## ICE MUMMY

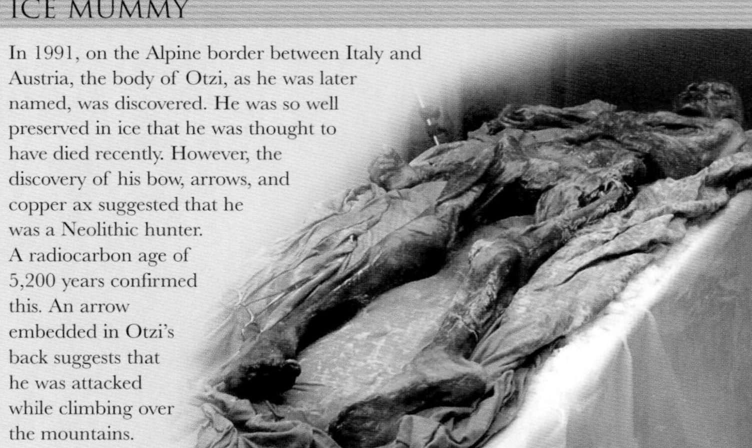

In 1991, on the Alpine border between Italy and Austria, the body of Otzi, as he was later named, was discovered. He was so well preserved in ice that he was thought to have died recently. However, the discovery of his bow, arrows, and copper ax suggested that he was a Neolithic hunter. A radiocarbon age of 5,200 years confirmed this. An arrow embedded in Otzi's back suggests that he was attacked while climbing over the mountains.

## FEATURES OF THE ALPS

**MONT BLANC**
*The highest mountain in the Alps, Mont Blanc was first climbed in 1786.*

**NORTH FACE OF THE EIGER**
*At 13,025 ft (3,970 m) high, with exposed, rocky, clifflike walls, the Eiger was not climbed until 1858, and its north face defied all attempts until 1938.*

**GROSSGLOCKNER**
*A classic glacial horn in the Hohe Tauern range, Grossglockner is the highest mountain in the Austrian Alps, at 12,461 ft (3,798 m).*

**FORMIDABLE MATTERHORN**
*With its pyramid-shaped peak and precipitous rock walls sculpted by glacial ice, the Matterhorn, 14,692 ft (4,478 m) high, is seen as the epitome of the alpine mountain form.*

EUROPE *southwest*

## Sierra Nevada

**LOCATION** In southern Spain, running west–east between Granada and the Mediterranean Costa del Sol

**HIGHEST PEAK** Mulhacén, Spain 11,421 ft (3,481 m)

**LENGTH** 63 miles (100 km)

**AGE** Mesozoic–Cenozoic

**FORMATION TYPE** Continent–continent

Southern Spain's snow-capped Sierra Nevada mountain range forms a dramatic backdrop to the Mediterranean beaches of the Costa del Sol. The range is the most southerly part of the more extensive Betic Cordillera, which runs southwest to northeast from Cadíz to Alicante, and continues offshore to the Balearic Islands of Majorca, Minorca, and Ibiza. The formation of the Sierra Nevada was contemporary with that of the Atlas Mountains (see opposite) and the European Alps (see pp.162–63). It resulted from the northward movement of the African Plate during the Cenozoic Era. The Sierra Nevada continues to rise by about 1/8 in (3 mm) every 100 years. The southern folds of the range, known as Las Alpujarras, are deep valleys that provide fertile farming land.

**ARID BADLANDS**
*Deforestation and overgrazing of fragile, semiarid lands have resulted in erosion and the formation of semidesert badlands in the Sierra Nevada.*

EUROPE *central*

## Carpathians

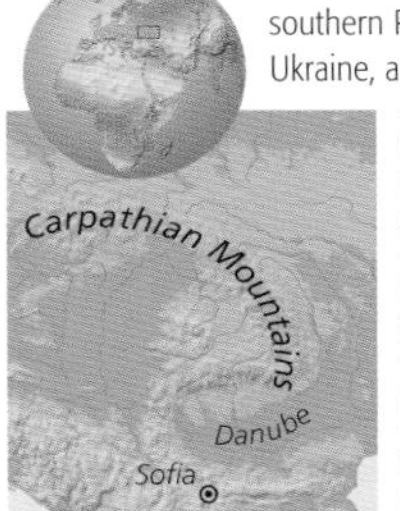


**LOCATION** Forming an arc from southern Poland through Slovakia, Ukraine, and into central Romania

**HIGHEST PEAK** Gerlachovka, Slovakia 8,711 ft (2,655 m)

**LENGTH** 900 miles (1,450 km)

**AGE** Mesozoic–Cenozoic

**FORMATION TYPE** Continent–continent

**APOLLO BUTTERFLY**
*A huge conservation program was undertaken to preserve this species from extinction in the Pieniny range of the Polish Carpathians.*

The Carpathians are an extension of the European Alps, and they contain a complex mixture of ancient rocks covered by younger sediments that are often highly folded and faulted. Many of the rocks that form the range were once deposits in the Mesozoic Tethys Ocean. The highest Carpathian peaks are the granitic Tatra Mountains in Slovakia, where steep, rocky walls rise above U-shaped valleys. In Romania, where more than half of the mountains are situated, the chain breaks up into several ranges, the loftiest being the Transylvanian Alps. Broad, fertile valleys, lakes, and limestone plateaus are all features of the range.

**STEEP-SIDED VALLEY**
*Pasture land is grazed by sheep at the base of sharp, towering Carpathian peaks in Slovakia.*

EUROPE *south*

## Dinaric Alps

**LOCATION** Running along the coast of the Adriatic Sea from Slovenia into northern Albania

**HIGHEST PEAK** Durmitor, Yugoslavia 8,274 ft (2,522 m)

**LENGTH** 470 miles (750 km)

**AGE** Mesozoic–Cenozoic

**FORMATION TYPE** Continent–continent

The northwest–southeast trend of the Dinaric Alps parallels Italy's Apennines, from which they are separated by the Adriatic Sea. Deep gorges have been cut into the mountains by rivers such as the Neretva. The soluble nature of the limestone that forms the mountains has led to the development of the region's famous karst landscape (see p.253). Below ground, extensive cave systems have developed, and in places cavern collapse has formed large hollows called poljes.

**DURMITOR, YUGOSLAVIA**

EUROPE *east*

## Caucasus Mountains

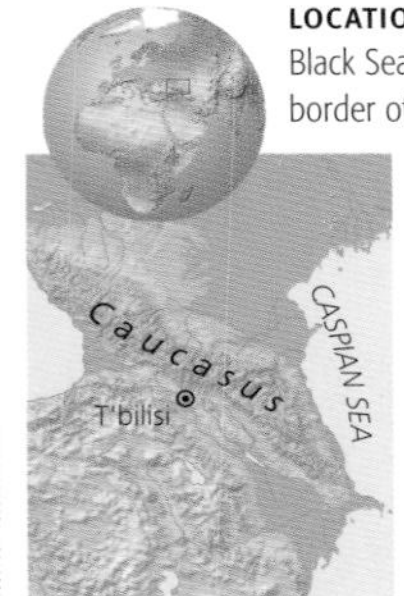


**LOCATION** Running between the Black Sea and Caspian Sea along the border of Russia and Georgia

**HIGHEST PEAK** El'brus, Russia 18,510 ft (5,642 m)

**LENGTH** 750 miles (1,200 km)

**AGE** Late Paleozoic–Cenozoic

**FORMATION TYPE** Continent–continent

One of Europe's natural boundaries in the east, the Caucasus Mountains consist of two parallel ranges separated by a central trough. The southeastern peaks are mainly limestone, which has formed karst landscapes. A higher range, formed of igneous rocks, dominates in the northwest. The mountains contain one of the world's deepest caves, Snezhnaya in Georgia, which is 4,823 ft (1,470 m) deep.

**PONTIC MOUNTAINS, TURKEY**
*On the southern shore of the Black Sea, the Lesser Caucasus fold structure continues westward as the Pontic Mountains.*

**BELOLAKAYA**
*This peak attracts climbers, hikers, and tourists. It is one of several high mountains in the western Caucasus, the tallest of which is Dombay-Ul'gen, at 13,262 ft (4,042 m).*

AFRICA *northwest*

# Atlas Mountains

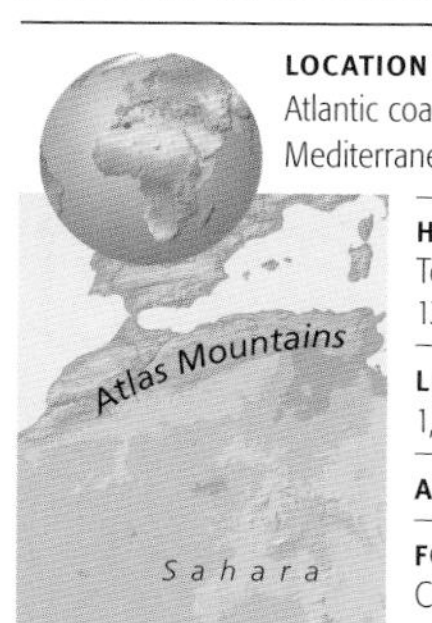


**LOCATION** Extending from the Atlantic coast of Morocco to the Mediterranean east coast of Tunisia

**HIGHEST PEAK** Toubkal, Morocco 13,665 ft (4,165 m)

**LENGTH** 1,500 miles (2,400 km)

**AGE** Mesozoic–Cenozoic

**FORMATION TYPE** Continent–continent

The Atlas Mountains form a high barrier between the Mediterranean Sea and the Sahara. Many of the tallest peaks preserve evidence of glaciation during the last ice age, with cirques (see p.262) and periglacial features (see p.342) evident. The mountains do not form a continuous chain, but are a series of ranges that are interspersed with plateaus, basins, and gorges. The most prominent range is the High Atlas, which runs inland from Agadir on Morocco's west coast for 400 miles (650 km) and contains several snow-capped peaks, including the highest, Toubkal. The eastern ranges of the Atlas form less elevated plateaus. In the north, the mountain slopes receive ample rainfall and are forested with cedar, pine, and oak. The more southerly ranges are more arid, with salt flats occurring in those areas of the Atlas close to the Sahara. A powerful earthquake in 1960 destroyed Agadir and confirmed the presence of major active faults separating the Atlas from the North Saharan Basin. Like the European Alps (see pp.162–63) and Spain's Sierra Nevada (see opposite), the Atlas Mountains were built by plate convergence between Africa and Europe, beginning about 60 million years ago. However, as is also true for Central Europe, there are folds and faults in the Moroccan ranges that were created during an era known as the Variscan orogeny, a period of mountain-building that occurred about 300 million years ago. Younger sedimentary rocks, including many limestones that were originally deposited on the floor of the Tethys Ocean, eventually covered these folds. The northernmost ranges were formed by folding that began 60 million years ago. Further movement of the African Plate, approximately 20 million years ago, deformed rocks in central Tunisia and Algeria, and elevated the central plateaus. The mountains are rich in minerals, including coal, iron and gold.

**ATLAS MONKEY**
*The Barbary ape is actually a monkey that has become adapted to living in the rocky cliffs and high-altitude oak and cedar forests of the Atlas range.*

## WATER SOURCES

In the arid regions of the Atlas Mountains, water is a scarce commodity, and many communities depend on meltwater streams formed by the winter snows. In places, the water is collected in deep cisterns cut into the rock centuries ago. With the advent of global warming, there is a risk that diminishing snowfalls will result in water shortages for the inhabitants.

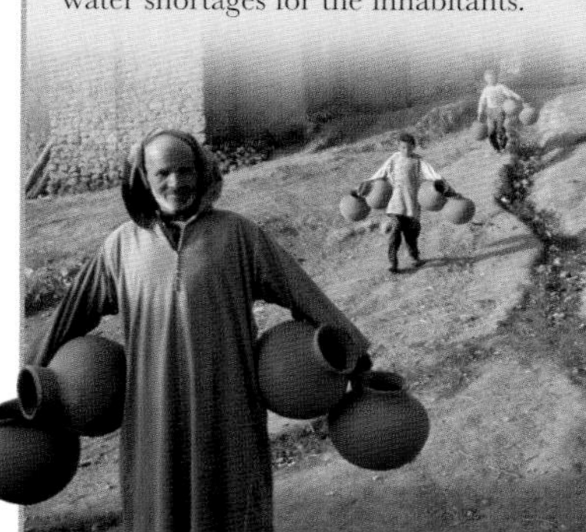

## FEATURES OF THE ATLAS MOUNTAINS

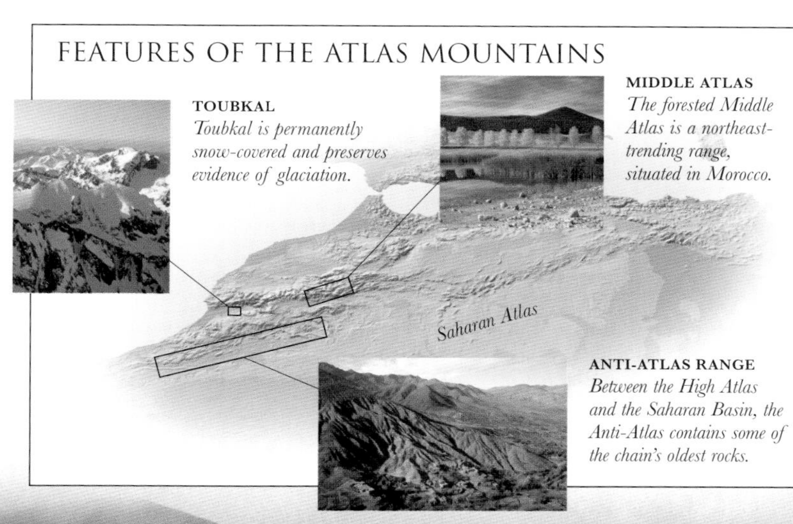


**TOUBKAL**
*Toubkal is permanently snow-covered and preserves evidence of glaciation.*

**MIDDLE ATLAS**
*The forested Middle Atlas is a northeast-trending range, situated in Morocco.*

**ANTI-ATLAS RANGE**
*Between the High Atlas and the Saharan Basin, the Anti-Atlas contains some of the chain's oldest rocks.*

**BARREN SLOPES**
*The inhabitants of ancient settlements continue a precarious existence in the arid parts of the Atlas. Survival depends on local microclimates and the availability of water for plant growth and domestic uses.*

AFRICA *south*

## Drakensberg Plateau

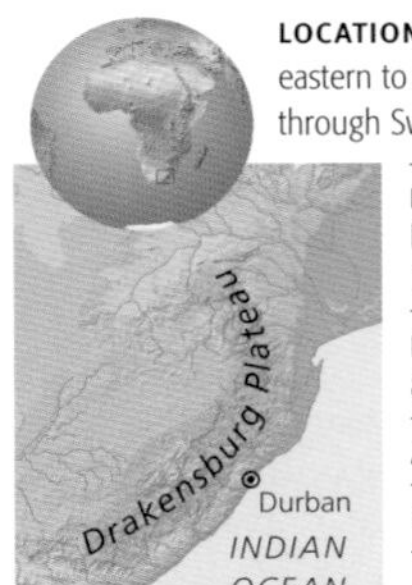


**LOCATION** Extending from north-eastern to southern South Africa, through Swaziland and Lesotho

**HIGHEST PEAK** Ntlenyana, Lesotho 11,424 ft (3,482 m)

**LENGTH** 800 miles (1,290 km)

**AGE** Mesozoic–Cenozoic

**FORMATION TYPE** Rift

A vast, westward-dipping escarpment, the Drakensberg range is one of the world's largest elevated plateaus. It separates a narrow coastal plain from southern Africa's vast interior of savanna and desert. The east of the plateau is dominated by the Great Escarpment, which soars to over 10,000 ft (3,300 m), high enough for frost action to fracture the rocks. In places, the sheer scarp face rises to over 7,000 ft (2,100 m) from its base to the crest. The Drakensberg Plateau consists of sedimentary rocks capped with volcanic deposits. The base of the plateau was built up by the non-marine Karoo sediments laid down about 250–300 million years ago. They begin with glacial deposits (the Dwyka tillite) and pass up into terrestrial coals. The summit of the plateau contains remnants of the ancient continent called Gondwanaland, which were originally close to sea level before being elevated prior to the breakup of the landmass 200 million years ago. A rising hot spot domed the region, stretching and rifting the crust as Africa and South America separated. Subsequently, basalts erupted through volcanic fissures in the underlying sediment, a process that continued for about 50 million years. These layers of basalt originally covered an area of approximately 800,000 square miles (2 million square km), a region that includes much of southern Africa and parts of what is now Antarctica. The basaltic deposits were originally 4,900 ft (1,500 m) thick, but erosion has heavily dissected them. The resulting landscape of the Drakensberg Plateau is incredibly varied, including steep sandstone cliffs, free-standing pinnacles, torrential waterfalls, and extensive caves. Major peaks include Mont-aux-Sources (which is the origin of three rivers, including the Tugela) and Giant's Castle, so named because its steep, rocky cliffs resemble battlements. Rivers rising in the mountains, such as the Tugela and Orange, have carved impressive gorges through the rock. The diverse and inaccessible landscape, which also includes extensive high grasslands, provides habitats for an astonishing array of flora and fauna, with about 300 species of indigenous plant life. Drakensberg's name is taken from the Afrikaans for "dragon's mountain," but it is also known by its Zulu name of Ukhahlamba.

**ABOVE THE CLOUDS** *The Great Escarpment of the Drakensberg rises high above the low clouds of the coastal plains. The clouds bring moisture to the lower slopes, while the top of the plateau is much drier.*

**ROCKY MAMMAL** *Related to early hoofed mammals, the rock hyrax inhabits rocky slopes in colonies of up to 40 for protection against predators.*

### FEATURES OF THE DRAKENSBERG PLATEAU

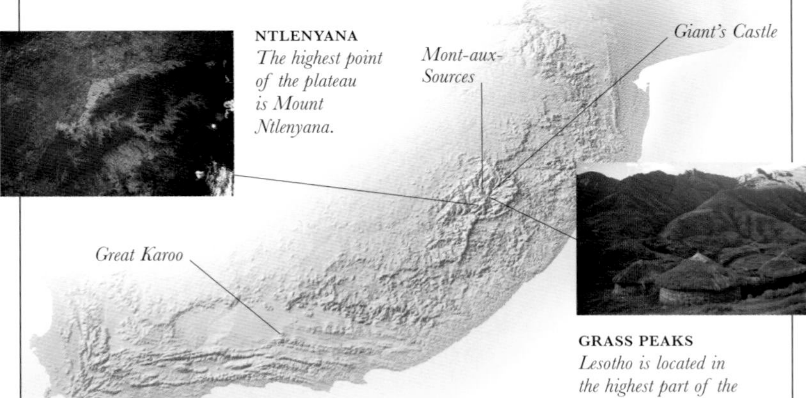


**NTLENYANA** *The highest point of the plateau is Mount Ntlenyana.*

**GRASS PEAKS** *Lesotho is located in the highest part of the Drakensberg Plateau.*

**BLYDE RIVER CANYON** *The Three Rondavels (also known as the Three Sisters) form part of the Blyde River Canyon in the Transvaal, South Africa. The cliffs of the canyon rise to 2,600 ft (800 m).*

ASIA *west*

## Zagros Mountains

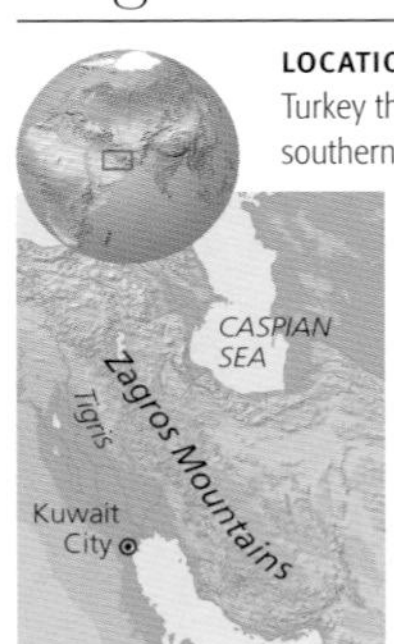


**LOCATION** Running from eastern Turkey through northern Iraq to southern Iran and the Persian Gulf

**HIGHEST PEAK** Damavand, Iran 18,603 ft (5,671 m)

**LENGTH** 950 miles (1,530 km)

**AGE** Mesozoic–Cenozoic

**FORMATION TYPE** Continent–continent

The Zagros Mountains run parallel to the valley of the Euphrates River and the Persian Gulf. Part of an unstable fold belt, the range is vulnerable to earthquakes—it includes a complex zone of thrust faulting and volcanism. In the center of the region lies an area of salt domes up to 5,000 ft (1,500 m) high. Many of the Zagros Mountains are perennially snow-capped.

### OIL-RICH HILLS

The sediments of the western foothills of the Zagros Mountains are a substantial source of hydrocarbons. Folding has produced anticlines, which trap petroleum in reservoirs beneath impermeable rock. However, some oil seeps to the surface, where volatile gases burn off.

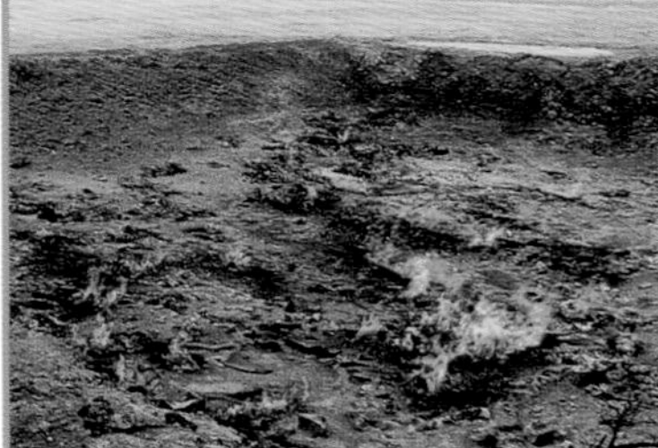

**PARS SNOW FIELDS** *Although on the same latitude as southern Turkey in the Mediterranean, the Pars region of the Zagros Mountains rises to over 16,400 ft (5,000 m), and is covered with snow in winter.*

ASIA *central*

## Altai Mountains

**LOCATION** Running from eastern Kazakhstan through Russia and Mongolia to western China

**HIGHEST PEAK** Belukha, Russia 14,783 ft (4,506 m)

**LENGTH** 1,250 miles (2,000 km)

**AGE** Late Paleozoic

**FORMATION TYPE** Continent–continent

The Altai mountain system forms the northern rim of the Dzungarian Basin depression in western China. Much of the range exceeds 10,000 ft (3,000 m), and there are numerous glaciated peaks over 13,000 ft (4,000 m) in the Mongolian sections. The highest peak, Mount Belukha, lies in the north at the headwaters of the Ob′ (see p.233) and Irtysh rivers in Russian Siberia.

**SNOW LEOPARD** *This endangered species of big cat is ideally suited to the cold mountains of the Altai, where it hunts small mammals and birds.*

**MONGOLIAN HIGHLANDS** *The Altai Mountains rise above the cold steppe grasslands of Mongolia, home to nomads and their herds of horses, camels, goats, and sheep.*

LAND

## ANCIENT INHABITANTS

Southern Africa's rock art was first described in the mid-18th century by European travelers. Engravings and paintings of hunting scenes, containing beautiful representations of animals and human hunters, are widespread throughout the continent. It is very difficult to date the rock art, but it is now evident it was produced by modern humans who inhabited the African continent over 40,000 years ago.

ASIA *central*

# Tien Shan Mountains

**LOCATION** To the north of the Tarim Basin between western China, Kyrgyzstan, and Kazakhstan

**HIGHEST PEAK** Pik Pobedy, Kazakhstan 24,407 ft (7,439 m)

**LENGTH** 1,400 miles (2,250 km)

**AGE** Late Paleozoic–Cenozoic

**FORMATION TYPE** Continent–continent

With more than 30 peaks of 20,000 ft (6,000 m) or more in height, the Tien Shan mountain belt is an imposing physical barrier. Not until 1856 did a Russian explorer, Peter Semonyov, become the first European to survey the Tien Shan, whose Chinese name translates as "celestial mountains." The mountain belt consists of a series of parallel, east–west trending ranges around Ozero Issyk-Kul', which is one of the largest mountain lakes in the world. The western Alai range includes numerous lakes, and the surrounding peaks average about 15,000 ft (4,600 m) in height. In the east, the mountains of the Tien Shan become lower and narrower, but then rise again to over 16,400 ft (5,000 m) in the Bogda Shan range, with Bogda Feng in China ascending to 17,864 ft (5,445 m). The east–west trending folds of the range expose lower Paleozoic rocks that have been folded, faulted, and uplifted in a series of tectonic events that included early subduction and subsequent continent–continent collision. These disturbances continue in the present, with massive earthquakes associated with major faults in the region. The Tien Shan range separates two major topographical depressions, the southerly Turpan Depression and the northern Dzungarian Basin. The floor of Turpan Pendi is 505 ft (154 m) below sea level. When the English explorer Charles Howard-Bury visited the area in June 1913, he marveled at the diversity and profusion of wild flowers, commenting that he had never seen "such a luxuriant flora" as he discovered in the alpine meadows of the Tien Shan. The mountains are the natural habitat of many domesticated ornamental and edible plants, including fruit trees, roses, tulips, and onions. The Tien Shan's animal life includes the native white-clawed bear.

**KHAN TENGRI**
*Inylchek, the largest glacier in the Tien Shan, flows 39 miles (62 km) down the western slopes of the Khan Tengri massif.*

**GLACIER COVER**
*Peaks that rise above 13,000 ft (4,000 m) in the Tien Shan are capped by glaciers, covering an area of about 3,670 square miles (9,500 square km).*

**MOUNTAIN FARMS**
*Fertile pastures flank the Tien Shan Mountains of Kyrgyzstan and Kazakhstan. Herds of goats, sheep, and horses are farmed in the region.*

ASIA *central*

## Himalayas

**LOCATION** Running southeast from northern Pakistan and India across Nepal to Bhutan

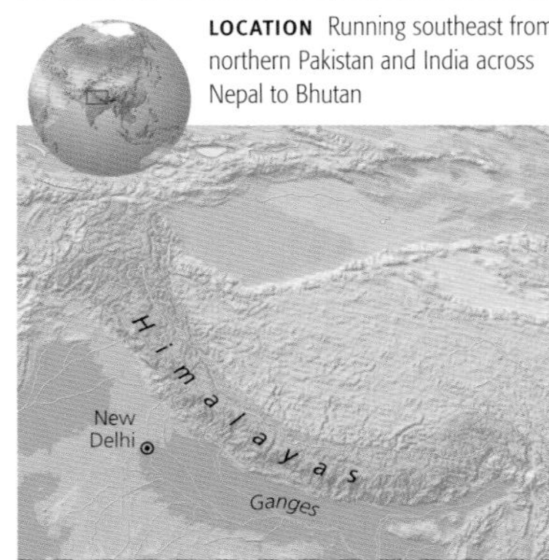


**HIGHEST PEAK** Mount Everest, Nepal 29,035 ft (8,850 m)

**LENGTH** 2,400 miles (3,800 km)

**AGE** Cenozoic

**FORMATION TYPE** Continent–continent

### FEATURES OF THE HIMALAYAS

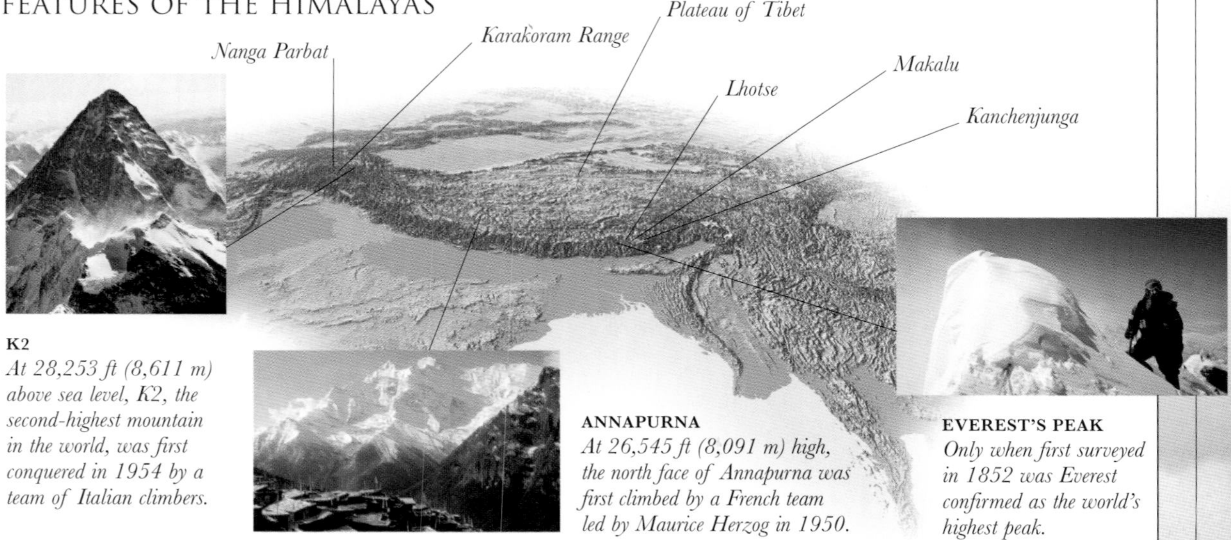

Nanga Parbat
Karakoram Range
Plateau of Tibet
Lhotse
Makalu
Kanchenjunga

**K2**
*At 28,253 ft (8,611 m) above sea level, K2, the second-highest mountain in the world, was first conquered in 1954 by a team of Italian climbers.*

**ANNAPURNA**
*At 26,545 ft (8,091 m) high, the north face of Annapurna was first climbed by a French team led by Maurice Herzog in 1950.*

**EVEREST'S PEAK**
*Only when first surveyed in 1852 was Everest confirmed as the world's highest peak.*

In addition to being the highest mountains on Earth, the Himalayas are also one of the youngest mountain belts. Formed within the last 50 million years, the belt is largely composed of deformed and metamorphosed rocks from the Indian crustal plate. Mountain building began with the collision of the Indian continental plate and Southeast Asia. This convergence caused India's northern edge to shorten by about 1,200 miles (2,000 km), thickening the Indian crust to 34 miles (55 km) beneath the Himalayas and 43 miles (70 km) below Tibet. The elevation of the Himalayas transformed the region's climate, establishing the annual Southeast Asian monsoon, with its regular torrential rains. Between 155 and 218 miles (250 and 350 km) wide, the mountains form a virtually impassable barrier, although one species, the bar-headed goose, regularly flies across the peaks from Central Asia to the Indian subcontinent. A series of ranges run parallel along the length of the Himalayan chain. In the south, the mountains rise from the flat, fertile Ganges Plain to the low, gently folded Siwalik Range. This range contains numerous fossils of early land-living mammals, including the first apes, which have been found to be 5–25 million years old. A major

### NORGAY AND HILLARY

New Zealander Edmund Hillary (born 1919) and Nepali Tenzing Norgay (1914–86) were the first people to reach the summit of Everest, on May 29, 1953. Hillary went on to lead the first overland crossing of Antarctica, and Tenzing to train guides at the Mountaineering Institute in Darjeeling, India.

thrust fault separates the Siwaliks from the central Himalayas, which rise up to 16,400 ft (5,000 m) high. This geologically complex range includes slices of Precambrian and Paleozoic rocks covered with younger sediments. Brought to the surface by great thrust faults, these rocks have been deeply eroded, and frequent landslides still occur in the region. To the north are the Great Himalayas, which soar to more than 23,000 ft (7,000 m) above sea level. These mountains are formed of Upper Paleozoic rocks, slices of the Tethys Ocean floor, and granite intrusions thrust over early Cenozoic sediments. Many peaks in this range reach over 26,000 ft (8,000 m) and are covered with permanent snow and glaciers. Two major rivers, the Indus (see p.235) and the Brahmaputra, have their sources in these mountains.

**HIMALAYAN POPPIES**
*Many varieties of poppies collected in the Himalayas in the 19th century are today familiar garden plants all over the world.*

Beyond and to the north of the Great Himalayas lies the vast Plateau of Tibet. The Himalayas are still rising, at rates of up to 1/16 in (4 mm) a year, growth that is offset by the processes of weathering and erosion. Because the belt is so vast, plant species are varied, with tea cultivated on the lower slopes and oak, rhododendrons, and pine forests growing on the higher slopes. The highest peaks are inhospitable, with little or no vegetation.

**PLATEAU OF TIBET**
*With an average altitude of 16,400 ft (5,000 m), the Tibetan Plateau, seen in this picture taken from the space shuttle, is the highest and largest plateau in the world.*

## MOUNTAIN OF GARBAGE

Since the summit of Mount Everest was first reached in 1953, several hundred people have ascended the mountain, with more than 100 perishing in the attempt. For many, climbing Everest is now merely another extreme sport, but the business provides a valuable income for the local population. However, the disposal of waste by climbing parties is now a major concern. Debris, including empty gas bottles, does not rot away at high altitude and is difficult to bury.

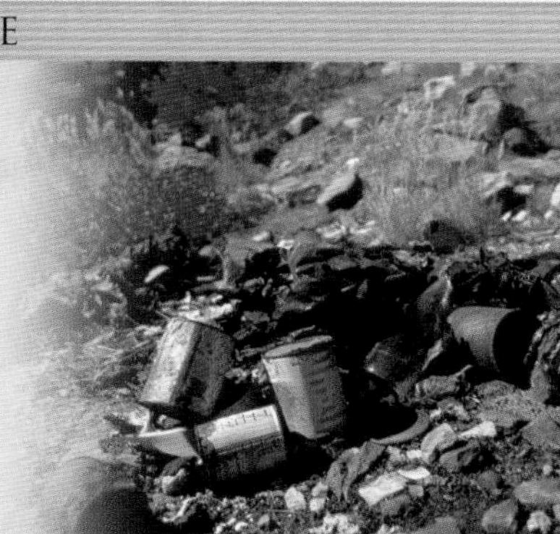

**FROM OCEAN TO SKY**
*Many Himalayan rocks were originally sediments on the Tethys Ocean floor. Now they lie within the highest mountains in the world, many of which rise over 26,000 ft (8,000 m) above sea level.*

AUSTRALASIA *Australia*

## Great Dividing Range

**LOCATION** Running from the Cape York Peninsula, Queensland, along Australia's eastern coast to Tasmania

**HIGHEST PEAK** Mount Kosciuszko, Australia 7,310 ft (2,228 m)

**LENGTH** 2,250 miles (3,600 km)

**AGE** Paleozoic

**FORMATION TYPE** Ocean–continent

To the first Europeans who arrived there, the elevated eastern margin of the Australian continent, now known as the Great Dividing Range, seemed to be a single mountain chain rising abruptly from the narrow coastal plain. However, approached from Cairns in the north, the chain forms an uplifted plateau with a great scarp face set back inland. The Great Dividing Range actually consists of a complex of tablelands and lowlands, rising to higher, alpinelike peaks in the south. The landscape ranges from green hills covered with lush tropical rainforest in Queensland, to snow-covered mountains in New South Wales. The mountains are the source of several major rivers, including the Murray (see p.235) and the Darling. The wildlife inhabiting the range is as diverse as the landscape, and includes kangaroos and platypuses. The highest peak, Mount Kosciuszko, was first noted by Polish-born explorer Paul Strzelecki in 1829, and is named after a Polish national hero.

**BLUE MOUNTAINS**
*Trees cling to the rugged slopes of the Three Sisters, an outcrop in the Blue Mountains, New South Wales.*

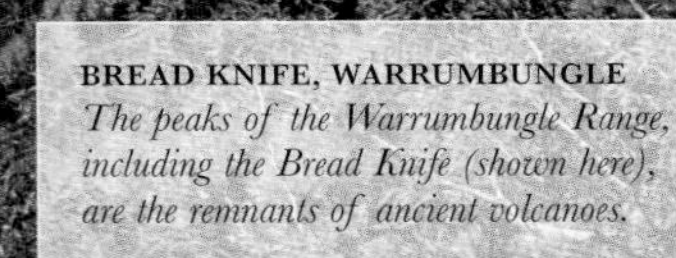

**BREAD KNIFE, WARRUMBUNGLE**
*The peaks of the Warrumbungle Range, including the Bread Knife (shown here), are the remnants of ancient volcanoes.*

AUSTRALASIA *New Zealand*

## Southern Alps

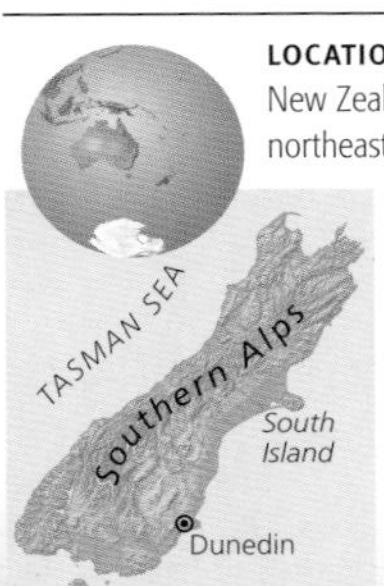


**LOCATION** Running the length of New Zealand's South Island, from northeast to southwest

**HIGHEST PEAK** Mount Cook, New Zealand 12,284 ft (3,744 m)

**LENGTH** 310 miles (500 km)

**AGE** Cenozoic

**FORMATION TYPE** Ocean–continent

A spectacular range of glaciated mountains, valleys, and fjords, the Southern Alps dominate New Zealand's South Island. The range is at its highest near the center, where the snow fields around Mount Cook feed glaciers such as the Tasman Glacier (see p.280), which flows for 17 miles (27 km). The Southern Alps were formed by the convergence of the Australian and Pacific plates. The relatively young range—it was formed in the last 60 million years—is still tectonically active and continues to experience uplift. Year-round rain from westerly winds is among the heaviest in the world, with up to 26 ft (8 m) falling in a year. Thus the western slopes of the range are clad in forest, while in the more arid rain shadow to the east, tussocky grasses grow, providing pasture for summer grazing by highland sheep.

**SOUTHERN PEAK**
*Mount Cook is just one of many peaks above 10,000 ft (3,000 m) in the Southern Alps.*

**ALPINE PARROT**
*The kea, a large, opportunistic parrot, has prospered better than its relative, the endangered and flightless kakapo.*

**MITRE PEAK, FIORDLAND**
*South Island's southwest coast is deeply indented with glacial fjords and valleys flanked by steep, triangular peaks.*

ANTARCTICA

# Transantarctic Mountains

**LOCATION** Extending across Antarctica from Oates Land to the Antarctic Peninsula

**HIGHEST PEAK** Vinson Massif, Antarctica 16,066 ft (4,897 m)

**LENGTH** 2,200 miles (3,500 km)

**AGE** Late Paleozoic–Cenozoic

**FORMATION TYPE** Continent–continent

The curved belt of the Transantarctic Mountains separates the highly elevated, ancient region of Greater Antarctica in the east from the younger and lower Lesser Antarctica in the west. More than three-fourths of the continent is covered by a deep ice sheet (see pp.278–79) but several mountain peaks over 13,000 ft (4,000 m) high emerge from it. The belt extends northwest through the continent and includes many volcanoes, some still active, including Deception Island. It is thought to mark a region along the western edge of Greater Antarctica that has been subjected to repeated compression and extension events over very long periods. The mountains continue offshore to form the South Sandwich volcanic island arc and the southwestern end of the Pacific Ring of Fire.

## FEATURES OF THE TRANSANTARCTIC MOUNTAINS

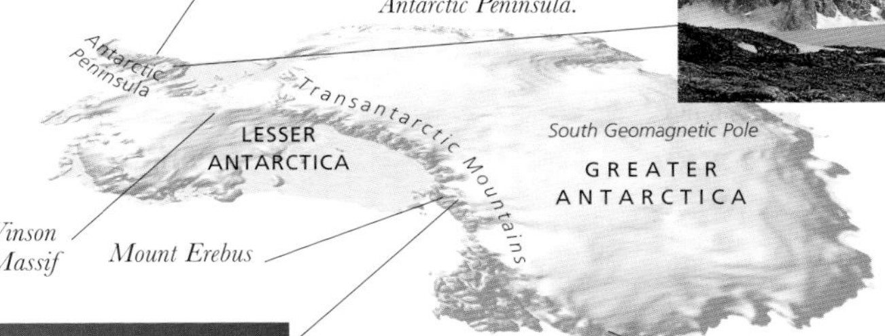

**MT. WILLIAM RANGE**
*The Mount William Range is situated on the Antarctic Peninsula.*

**ROYAL SOCIETY RANGE**
*The Royal Society Range rises from the Scott Coast, in Victoria Land.*

**DRY VALLEY**
*With temperatures permanently below freezing and lying in a rain shadow, some parts of Antarctica's Dry Valley have seen no rainfall for more than 2 million years.*

## PAST CLIMATES

Antarctica's Alexander Island has yielded a fossilized conifer tree that was alive in the mid-Cretaceous Period, approximately 100 million years ago. The tree is surrounded by fossils of its foliage and those of liverworts and ferns. It was entombed as it grew by sediments from floods. At the time, Antarctica was in a polar position, and the presence of such coniferous woodlands is convincing evidence of a much warmer climate during the period.

**DEEP IN SNOW**
*Forming the northeast coast of the Ross Ice Shelf, the mountains of Queen Maud Land hold back the East Antarctic Ice Sheet, except where glaciers spill through.*

# VOLCANOES

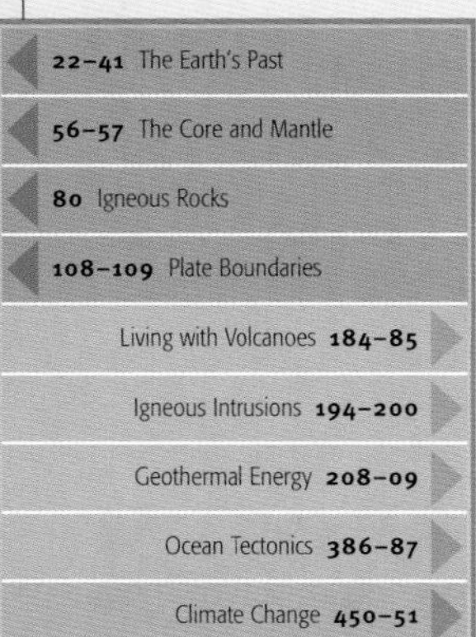
22–41 The Earth's Past
56–57 The Core and Mantle
80 Igneous Rocks
108–109 Plate Boundaries
Living with Volcanoes 184–85
Igneous Intrusions 194–200
Geothermal Energy 208–09
Ocean Tectonics 386–87
Climate Change 450–51

VOLCANIC ERUPTIONS OF MOLTEN ROCK from within the Earth are awesome demonstrations of the pent-up heat energy stored within our planet. The world is dotted with volcanoes and their rock products, both ancient and modern—reminders of the Earth's long history of volcanism. The variety of materials volcanoes produce, their styles of eruption, and the edifices they create all reflect the geological environments within which they develop. Active volcanoes are found on land and, above all, in the deep sea, and their distribution—which is far from random—reflects the continuing dynamism of the planet.

## ERUPTING MAGMA

Volcanism is the process by which magma—molten rock from inside the Earth—rises through the Earth's crust onto the surface. Magma chambers within the crust are created by localized melting and upward migration of partially molten crust and mantle rock. This melting requires special conditions, such as the ascent through convection of solid mantle rock beneath oceanic spreading ridges, the addition of seawater at subduction zones, or the presence of rising mantle plumes beneath hot-spot volcanoes. Magma travels from the chamber up to the Earth's surface either through fractures in the crust or by literally melting a path through the surrounding rock. A volcano is both the vent at the Earth's surface through which volcanic materials erupt and the edifice created by their accumulation around the vent. As well as magma that erupts as lava, volcanic products include gas, ash, and fragmented rocks (see p.175). These can pour gently out onto the Earth's surface or be blasted through the atmosphere, to be spread globally by high-level winds (see p.174).

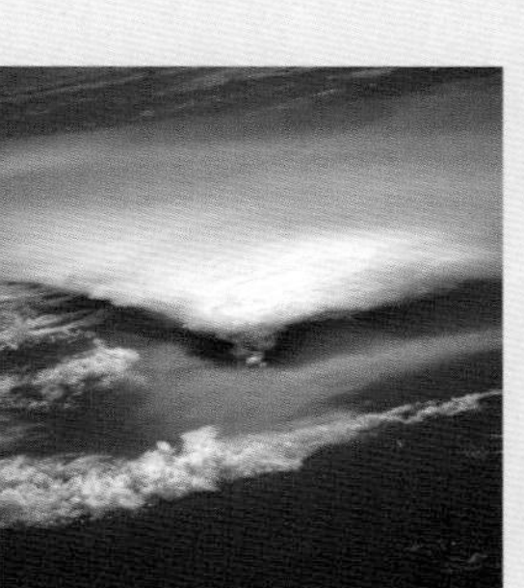

**VISIBLE FROM SPACE**
*The scale and power of volcanic eruptions, such as this one at Rabaul, blast gas and debris high into the stratosphere. They can be seen even from space.*

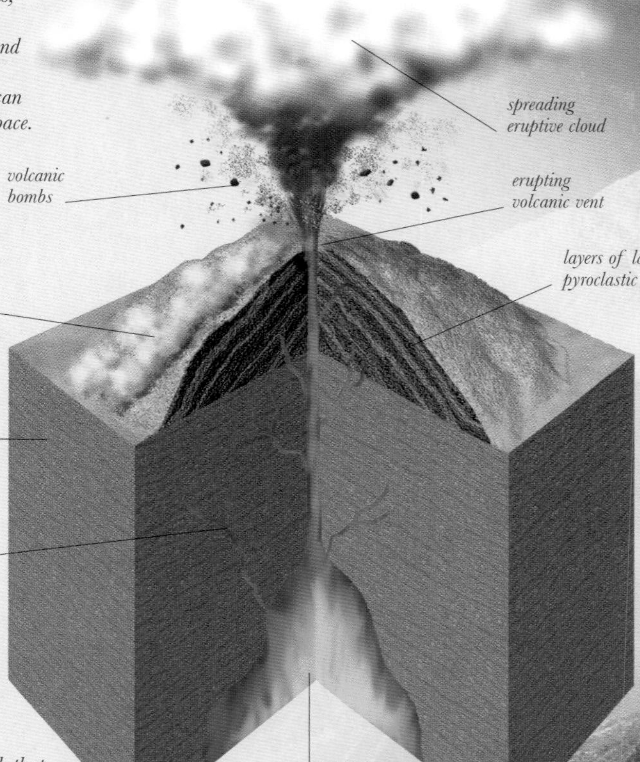

spreading eruptive cloud
volcanic bombs
erupting volcanic vent
layers of lava and pyroclastic materials
pyroclastic flow
surrounding country rock
magma-filled cracks and fissures
magma chamber

**VOLCANISM**
*A typical volcanic system is based around a central "plumbing" network that transfers magma from its crustal chamber to the surface.*

**VOLCANIC COMPLEX**
*The Tengger volcanic complex in Java (foreground) consists of five overlapping volcanoes. In such cases, the vents can have intricate underground connections.*

# LAVAS AND VOLCANOES

Volcanic material erupted at the Earth's surface varies according to the properties of the magma from which it is derived. Differences in chemical composition, gas content, and temperature all affect the way in which magmas erupt, and therefore the types of landforms and volcanoes they produce. Magmas that have high silica content and low temperature are very viscous and slow-moving. If they reach the surface with high gas content, their eruptions are highly explosive. If they lose their gas during ascent, they instead erupt as lava. Silica-rich lavas are thick, slow-moving block lavas that do not travel far from their source. They are associated with subduction at destructive plate margins and the assimilation of rock materials from continental crust. Basaltic lavas, including pahoehoe and 'a'a, have low silica content and high temperatures. They are extremely fluid and can travel at high speeds—up to 60 mph (100 km/h)—over great distances. They are associated with hot-spot activity, such as that under Hawaii, as well as ocean-floor spreading, where they produce pillow lavas (see pp.306–07).

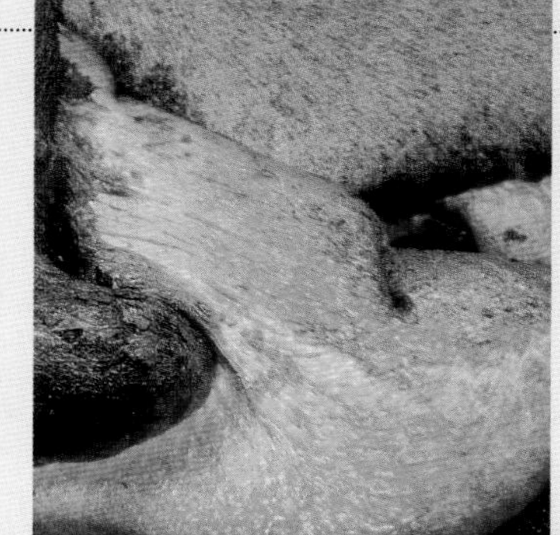

**PAHOEHOE (KILAUEA)**
*Known by its Hawaiian name, this fluid basaltic lava cools to form a glassy skin. It can have a flat, hummocky, or ropy surface produced by further flow after the skin has formed.*

**A'A (KILAUEA)**
*Pronounced "ah-ah," this fast-flowing basaltic lava has a rough texture, formed as the surface cools and fractures.*

**BLOCK LAVA (NEVADO DE COLIMA)**
*Silica-rich, viscous lavas, such as andesites and rhyolites, typically flow as piles of large, fractured blocks.*

**SHIELD VOLCANO (MAUNA KEA)**
*Fluid basaltic lavas that can flow easily across the ground tend to build massive volcanoes with low, angled slopes and a shield-shaped profile.*

**STRATOVOLCANO (MOUNT FUJI)**
*The flanks of these large, often steep-sloped volcanoes are built up of alternating layers of more viscous lava and pyroclastic deposits (see p.175).*

**DOME VOLCANO (MOUNT ST. HELENS)**
*Lava domes are rounded, steep-sided mounds of silica-rich lava that is too viscous to flow very far from the vent before it cools and crystallizes.*

**CINDER CONE (CERRO NEGRO)**
*Steep, straight-sided cinder cones are made of loose, angular, volcanic rock fragments that fall from the eruption cloud and accumulate above a vent.*

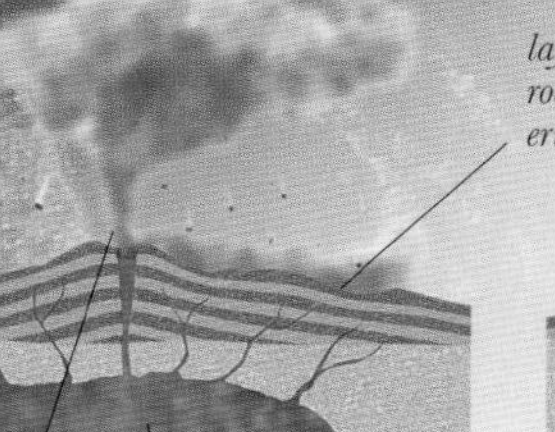

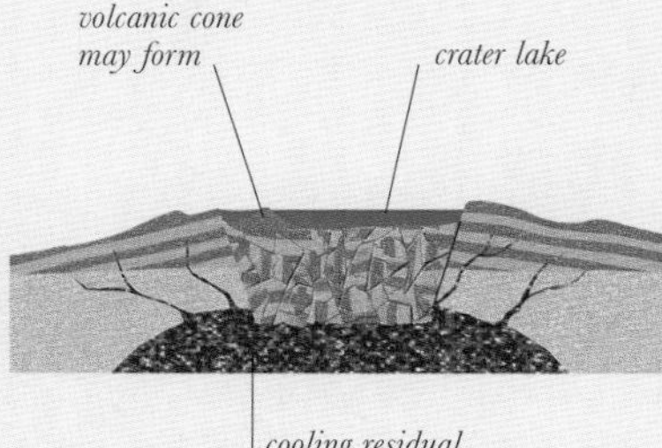


**CALDERA FORMATION**
*A caldera is a broadly circular depression formed when a volcano's magma chamber is emptied As the chamber becomes depleted, and support of the overlying rock is removed, the volcano's summit collapses, forming a caldera.*

# ERUPTION STYLES

Volcanologists traditionally have recognized different styles of volcanic eruption, although these styles can often mark different phases in a single event. Strombolian eruptions are spasmodic and discrete explosive bursts, a few seconds long, which eject pyroclastic materials (see opposite page) a relatively short distance into the air. High gas pressures fragment magma within the vent, but with frequent release, no sustained eruption cloud develops. Vulcanian eruptions are more violent, producing highly fragmented ash and rock. Eruptive columns can rise 6–12 miles (10–20 km), with ash spread over huge areas. Plinian eruptions expel large columns of pyroclasts and gas at high velocities (hundreds of yards per second) up to 30 miles (45 km), well into the stratosphere. Collapse of such columns can produce deadly pyroclastic flows, containing gas and pyroclasts of all sizes, traveling at high speeds and covering large distances. In contrast, Hawaiian eruptions produce copious low-viscosity lavas, associated with shield volcanoes. Where similar eruptions occur in shallow water, they are known as Surtseyan. These eruptions can be highly explosive as a result of water invading the vent, expanding as it turns to steam, and fragmenting the magma to blast out glassy fragments.

**PLINIAN ERUPTION (MAYON)**
*These eruptions produce massive clouds of expanding gas that carry volcanic materials high into the atmosphere. They are named after Pliny the Younger, who described the eruption of Vesuvius in AD 79.*

**PHREATIC ERUPTION (SURTSEY)**
*Phreatic eruptions are steam-driven explosions produced when water, either underground or on the surface, is heated by magma or hot rocks and expands rapidly.*

**FISSURE ERUPTION (KILAUEA)**
*In 1971, lava erupted from this fissure, which extends southwest from Halemaumau crater at Kilauea volcano, Hawaii, and covered the surrounding landscape in basalt.*

**LAHAR (MT. PINATUBO)**
*The saturation of volcanic ash by water, from heavy rain, glacial meltwater, or drainage of crater lakes, can generate catastrophic mud or debris flows, called lahars, that devastate surrounding landscapes.*

**STROMBOLIAN ERUPTIONS**
*Intermittent, small-scale explosive eruptions like this one in Hawaii are produced as gas bubbles through a magma column in a vent.*

## MONITORING VOLCANOES

Many millions of people live near dangerously active volcanoes, often because their agriculture relies on rich volcanic soils (see pp.184–85). Advance notice of when eruptions might occur is highly desirable, but it also has to be consistently accurate for local populations to heed the warnings. Unlike major earthquakes, which can happen with little or no warning, large-scale volcanic eruptions are often preceded by warning signs. Volcanologists have developed a number of techniques for monitoring the evolution of magmas beneath volcanoes and anticipating when volcanic activity is likely to develop into a potentially dangerous phase. These include measuring and interpreting the seismic activity that accompanies eruptions, analysis of changes in gases expelled from volcanic vents, and remote sensing by satellite to measure any changes in the elevation or shape of the volcanic edifice.

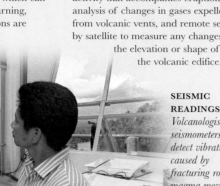

**SEISMIC READINGS**
*Volcanologists use seismometers to detect vibrations caused by fracturing rock, magma movement, or bursting gas bubbles prior to eruptions.*

# VOLCANIC DEPOSITS

Many different types of volcanic material can be preserved in the Earth's rock record, and their analysis reveals a great deal about the nature of past volcanism. Although lavas are the best-known volcanic products and can vary considerably in composition, texture, and volume (see p.173), fragmented volcanic rock materials, or pyroclasts, are in many ways the most interesting and have the greatest impact on life. They are produced by explosive eruptions of parent magmas that evolve into gaseous columns containing fragmented rock, minerals, and glassy shards. Deposits are classified by particle size. The smallest, usually called ash, is fine enough to stay aloft for days or weeks and can be distributed globally by high-altitude winds. Fragments can also combine with hot gases to produce glowing pyroclastic flows rolling across landscapes at up to 125 mph (200 km/h). At the other end of the scale, some volcanoes have been known to eject rock fragments bigger than a house up to several miles. Tephra is a collective term for fragments of volcanic rock and lava of all sizes that are blasted into the air.

**LAYERED DEPOSITS**
*Pyroclastic eruptive material is deposited in a way similar to sediments and accumulates as stratified layers.*

**ASH BURIAL**
*Volcanic deposits of all sizes can smother and bury landscapes and buildings, as here on Montserrat, and totally extinguish life.*

**LAVA BOMBS**
*Large fragments of ejected molten rock take on aerodynamic shapes as they travel through the air.*

# VOLCANO DISTRIBUTION

The global distribution of volcanoes clearly relates to the Earth's tectonic plates. About 80 percent of volcanoes above sea level occur at convergent plate boundaries, 5 percent occur at divergent boundaries, and mantle hot spots account for the rest, found away from plate boundaries. The volcanic chains around the Pacific (called the Ring of Fire) include volcanic island arcs, formed by ocean–ocean plate convergence, and Andean-style continental arcs, produced by ocean–continent convergence. Continental plate divergence generates rift volcanoes, such as those of eastern Africa, while oceanic plate divergence forms ridges, such as that of the mid-Atlantic, and isolated volcanic islands.

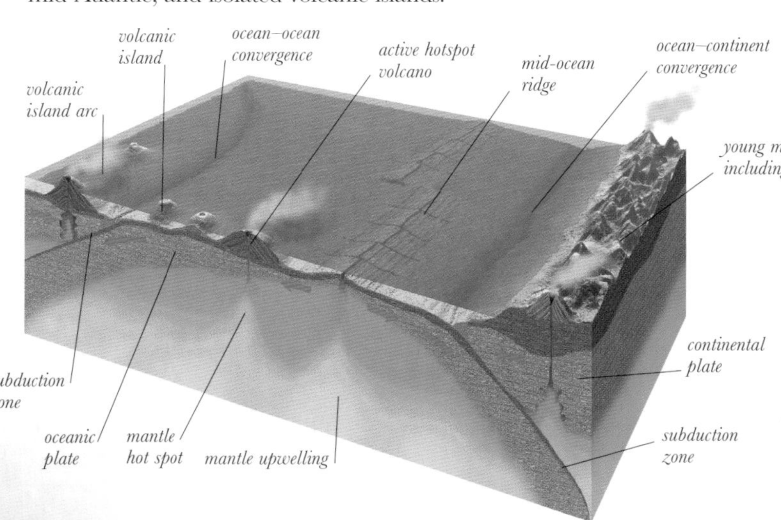


**VOLCANOES AND PLATES**
*Volcanic activity is closely associated with the movement of tectonic plates. Most active volcanism is found at undersea mid-ocean ridges where crust is being pulled apart. It is also produced by convergent plate boundaries, where subduction occurs, and rising mantle hot spots.*

## VOLCANO PROFILES

The pages that follow contain profiles of many of the the world's most significant and best-known volcanoes. Each profile begins with the following summary information:

**HEIGHT** Elevation above sea level of each volcano's highest point. Due to volcanic activity, this figure is likely to change over time

**TYPE** Stratovolcano, shield volcano, cinder cone, caldera, fissure vent, maar (vent formed by steam explosions), tuff cone, lava dome, flood basalt, complex

**LAST ERUPTION** Date of the last confirmed eruption, or, in the case of ancient volcanoes, an estimate based on rock records

NORTH AMERICA *northwest*

## Novarupta

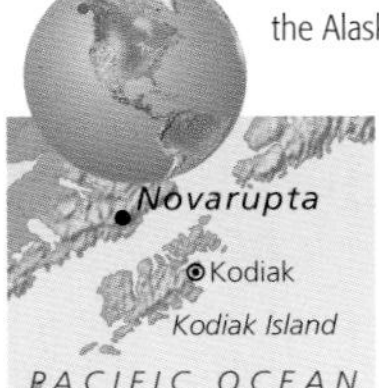


**LOCATION** In the Katmai region of the Alaskan Peninsula

**HEIGHT** 2,760 ft (841 m)

**TYPE** Caldera, lava dome

**LAST ERUPTION** 1912

Although Novarupta is the lowest volcano in the Katmai area of Alaska, it was formed, in 1912, by the biggest volcanic eruption of the 20th century. Within 60 hours, some 7 cubic miles (30 cubic km) of silica-rich rhyolitic magma blasted out of Novarupta. Most of it fell back to Earth as ash, while the remainder formed a pyroclastic flow—the renowned Valley of Ten Thousand Smokes (VTTS) ash flow (see p.204). At the end of the eruption, two calderas formed. An indistinct shallow depression about 1.2 miles (2 km) wide formed at Novarupta and was followed by the extrusion of a lava dome, 215 ft (65 m) high and 1,250 ft (380 m) wide, at the vent of the VTTS ash flow. A much larger collapse caldera, about 3 miles (5 km) wide, was formed at Katmai volcano 10 miles (16 km) to the east, as its underground magma reservoir drained toward Novarupta. Traces of ash that settled after the 1912 eruption have been found in Greenland ice cores.

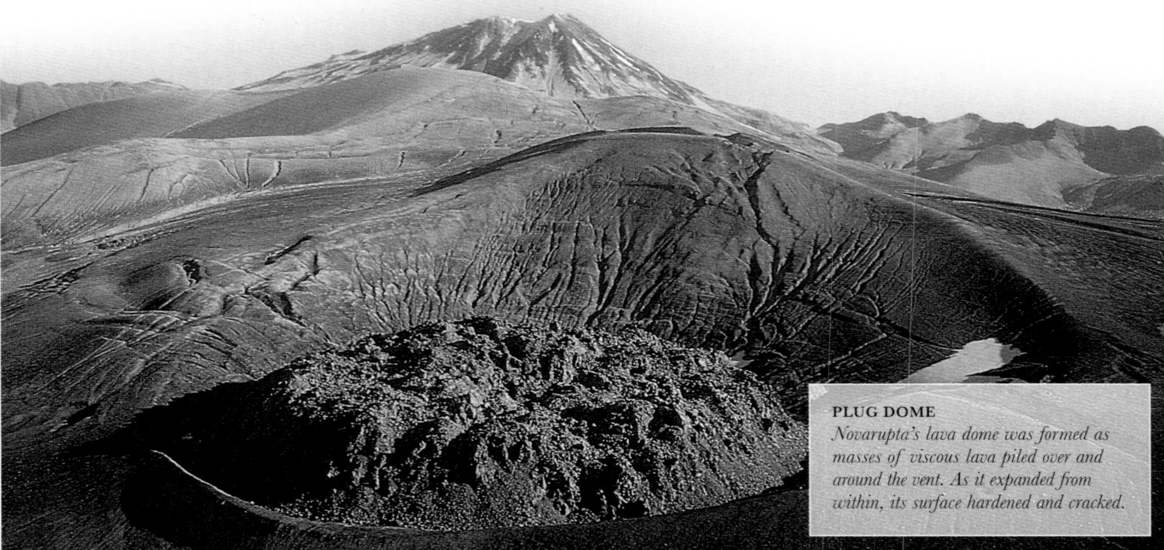

**PLUG DOME**
*Novarupta's lava dome was formed as masses of viscous lava piled over and around the vent. As it expanded from within, its surface hardened and cracked.*

NORTH AMERICA *west*

## Mount Rainier

**LOCATION** In the northern Cascade Range, Washington State

**HEIGHT** 14,410 ft (4,392 m)

**TYPE** Stratovolcano

**LAST ERUPTION** About 1825

Heavily glaciated, with an ice-filled summit crater complex, Mount Rainier is the highest peak in the Cascade range. The collapse of a much larger structure about 5,700 years ago produced great lahars that flowed more than 60 miles (100 km) to Puget Sound. The present-day cone, capped by overlapping craters, formed during a major explosive eruption about 2,200 years ago.

**MOUNT RAINIER CAPPED WITH SNOW**

NORTH AMERICA *west*

## Columbia River Plateau

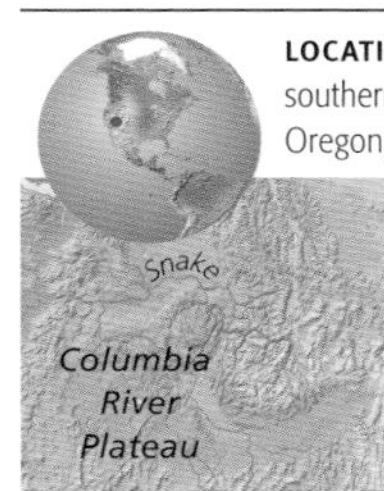


**LOCATION** Ranging across southern Washington, northern Oregon, and Idaho

**HEIGHT** Up to 6,600 ft (2,000 m)

**TYPE** Flood basalt

**LAST ERUPTION** About 15 million years ago

Between 17 and 15 million years ago, about 40,000 cubic miles (170,000 cubic km) of basaltic lava poured out of deep crustal fissures to create the Columbia River Plateau. Many of the flows are over 160 miles (100 km) long and up to 100 ft (30 m) thick, and some cooled to form distinctive shrinkage columns. More than 95 percent of the basalts may have been erupted in less than 2 million years. The origin of the eruption is thought to have been a rising mantle plume or hot spot beneath the continental crust, which produced partial melting in the upper mantle. The increased heat flow caused the crustal rocks above to dome and stretched, creating deep cracks that tapped the magma below and allowed it to flow out at the surface.

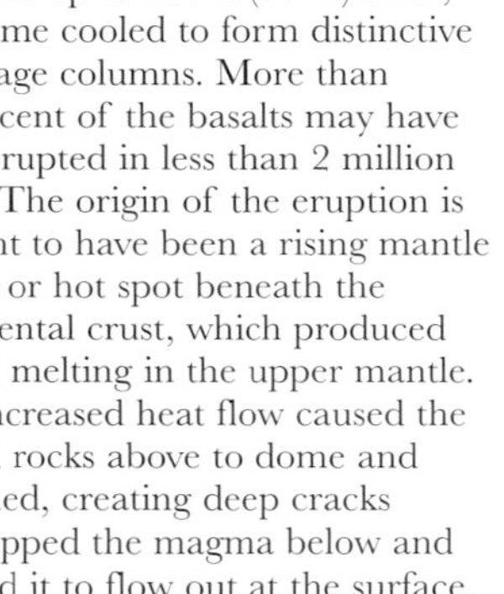

**TWIN SISTERS**
*These basalt pillars were isolated about 15,000 years ago when catastrophic floods carved great channels in the Plateau.*

NORTH AMERICA *west*

## Crater Lake

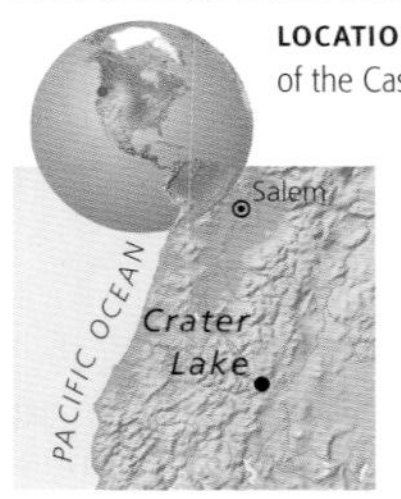


**LOCATION** In the southern part of the Cascade Range, Oregon

**HEIGHT** 8,159 ft (2,487 m)

**TYPE** Caldera

**LAST ERUPTION** About 2290 BC

Crater Lake, the second-deepest lake in North America at 1,934 ft (589 m) deep, fills a caldera 9 miles (14 km) wide and is surrounded by a spectacular rock wall. The caldera marks the site of a complex of overlapping volcanoes, known as Mount Mazama. This complex was built up 420,000–40,000 years ago and erupted 6,850 years ago in one of the biggest eruptions of the Holocene (the last 10,000 years). A huge Plinian blast showered ash as far away as Alberta, Canada, and pyroclastic flows traveled up to 40 miles (65 km) from the volcano. A series of now lake-covered lava domes formed on the caldera floor a few hundred years after its formation.

**VOLCANIC DEPOSITS**
*Crater Lake's "Pinnacles" are pyroclastic flow deposits that have eroded into dramatic spires.*

**FLOODED CALDERA**
*Crater Lake fills an almost perfectly circular caldera. The lake's surface is broken by Wizard Island, a cinder cone that erupted out of the caldera floor.*

NORTH AMERICA *west*

# Mount St. Helens

**LOCATION** In the northern region of the Cascade Range, Washington State

**HEIGHT** 8,363 ft (2,549 m)

**TYPE** Stratovolcano

**LAST ERUPTION** 1991

The strikingly symmetrical cone of Mount St. Helens had long been admired as North America's Mount Fuji. However, on May 18, 1980, in one of the world's best-documented eruptions, the top 4,600 ft (1,400 m) of the volcano was destroyed, leaving a horseshoe-shaped caldera. Magma intrusion caused the north flank to bulge outward and then collapse in a debris avalanche, uncorking an explosive eruption. The avalanche traveled about 16 miles (25 km) at speeds of up to 250 ft (75 m) per second, filling a valley with debris up to 640 ft (195 m) deep. An explosive blast of hot gases and magma traveled ahead of the debris avalanche, flattening more than 10 million trees over about 230 square miles (600 square km). Later, collapse at the base of the eruptive column produced numerous pyroclastic flows, at temperatures of about 1,300°F (700°C). Formed in nine eruptive phases beginning 50,000–40,000 years ago, Mount St. Helens has been the most active volcano in the Cascade Range during the Holocene Epoch (the last 10,000 years). The modern volcano has been constructed within the last 2,500 years as basaltic, andesitic, and dacitic magma erupted from both summit and flank vents.

**BEFORE ERUPTION**
*The volcano's conical peak was flanked by evergreen forests and grassy meadows.*

**AFTER ERUPTION**
*Seared by a lethal combination of red-hot gases, rocks, and ash, the landscape to the north of the volcano resembled a desolate moonscape.*

## RATES OF RENEWAL

Catastrophic volcanic eruptions can destroy the plant and animal life of a region. Mount St. Helens provided an invaluable opportunity to monitor the regeneration of wildlife after an eruption. Studies revealed that recolonization occurs at a much faster rate than scientists had estimated. Mats of river algae (see below) and sprouting shrubs were harbingers of regrowth, followed by birds, small mammals, and, eventually, elk.

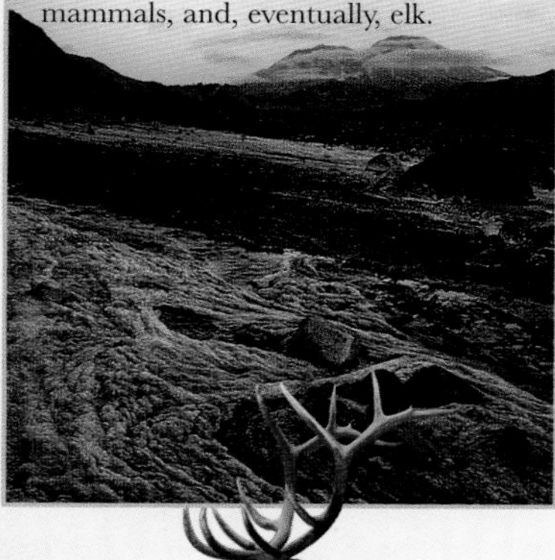

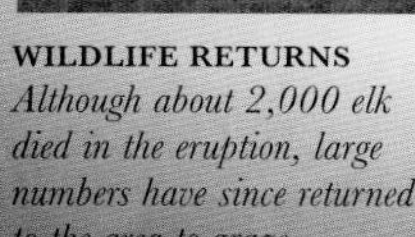

**WILDLIFE RETURNS**
*Although about 2,000 elk died in the eruption, large numbers have since returned to the area to graze.*

**PLINIAN ERUPTION**
*In 1980, Mount St. Helens produced a Plinian-style eruption, blasting an eruptive column of gas, ash, and pumice fragments 15 miles (24 km) into the atmosphere.*

**THE MOUNTAIN EXPLODES**
*As the north side of the volcano collapsed, a rock avalanche was set in motion (1). Pressure inside the volcano was released, and clouds of ash blasted out (2–3). These eventually caught up with and overtook the avalanche (4).*

PACIFIC OCEAN *central*

## Kilauea

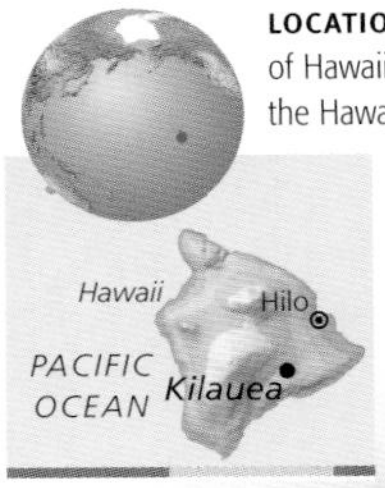

**LOCATION** On the southeast coast of Hawaii, the most southeasterly of the Hawaiian Islands

**HEIGHT** 4,009 ft (1,222 m)

**TYPE** Shield volcano

**LAST ERUPTION** 2003

Kilauea is the most active of the overlapping volcanoes that have built up the island of Hawaii over the past million years. With its massive low-angle slope profile and wide caldera, it is a typical shield volcano built up by frequent summit and flank basaltic lava flows interspersed with lava-lake activity within the crater. About 90 percent of Kilauea's surface is formed by lava that is less than 1,100 years old. Since 1983, during a period of constant eruption, flows have covered over 40 square miles (100 square km) and destroyed up to 200 houses. Most of Kilauea is hidden beneath the sea. Hawaii is part of a chain of volcanic islands built up over the past 75 million years as the Pacific Plate passed over a mantle hot spot at a rate of about 3½ in (9 cm) per year.

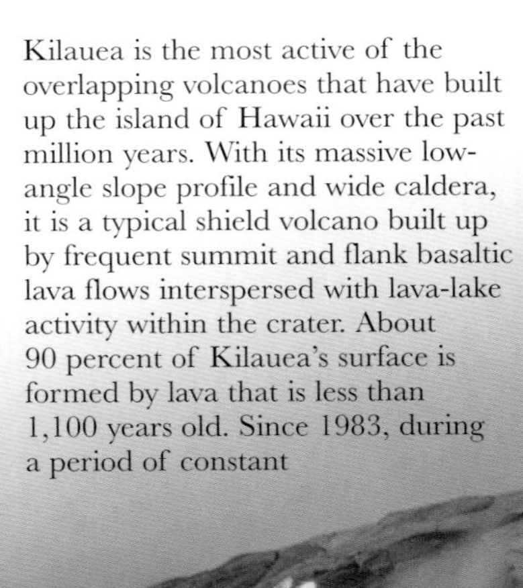

**LAVA FIELD**
*Low-viscosity, basaltic lava streams down the flanks of Kilauea. Lava flows can eventually reach the coast, where they cascade into the sea.*

### SAMPLING LAVA

Sampling a volcano's lava is one way to monitor changes in its behavior. Volcanologists studying Kilauea throw hammer-headed cables into lava tubes or, as shown here, take hand samples directly from lava flows or vent spatter. It is important to take hot lava samples, to obtain rapidly quenched glass for analysis.

PACIFIC OCEAN *central*

## Mauna Loa

**LOCATION** In the center of Hawaii, the most southeasterly of the Hawaiian Islands

**HEIGHT** 13,681 ft (4,170 m)

**TYPE** Shield volcano

**LAST ERUPTION** 1984

**SUMMIT CRATERS**

Rising nearly 5½ miles (9 km) from the ocean floor, and covering about half the island of Hawaii, Mauna Loa is the largest active volcano in the world. Almost 90 percent of its surface is basaltic lava less than 4,000 years old, erupted from flank fissures and overflowed from a summit lava lake. There have also been large-scale submarine avalanches, one of which traveled nearly 60 miles (100 km).

NORTH AMERICA *south*

## Paricutín

**LOCATION** In the Mexican Volcanic Belt, of southwestern Mexico, inland from the Pacific coast

**HEIGHT** 9,216 ft (2,809 m)

**TYPE** Cinder cone

**LAST ERUPTION** 1952

**VOLCANO ABLAZE**

Paricutín is the best-known volcano of about 1,000 volcanic centers in the Michoacán–Guanajuato volcanic field. It is a cinder cone that started to erupt from a cornfield in 1943 and grew to 1,102 ft (336 m) within a year. The eruption continued until 1952, adding an extra 288 ft (88 m) to its height. Paricutín provided volcanologists with the rare chance to witness and document the birth, growth, and death of a volcano.

NORTH AMERICA *south*

## Colima

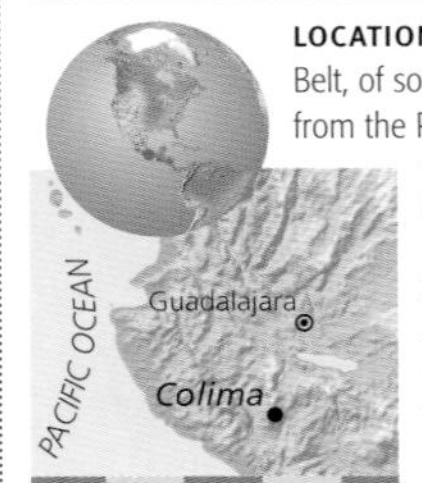


**LOCATION** In the Mexican Volcanic Belt, of southwestern Mexico, inland from the Pacific coast

**HEIGHT** 12,664 ft (3,860 m)

**TYPE** Stratovolcano

**LAST ERUPTION** 2003

Colima is the most prominent complex in the western Mexican Volcanic Belt. It is made up of the 14,173-ft (4,320-m) Nevado de Colima and the younger 12,664-ft (3,860-m) Volcán de Colima, an active stratovolcano that grew on the south flank of Nevado in a southward-moving pattern typical of the belt. Volcán de Colima is Mexico's most active volcano—its active cone lies within a 3-mile- (5-km-) wide caldera formed by the collapse of an ancestral cone.

Major slope failures at both Nevado and Volcán de Colima have formed a thick apron of debris-avalanche deposits. One of the biggest known occurred about 18,000 years ago, traveling 70 miles (120 km) to the Pacific coast. The volcanic complex began to grow in late Pleistocene times, and recorded eruptions date back to the 16th century, with occasional major explosive eruptions. The most recent, in 1913, destroyed the summit and left a steep-walled crater that was refilled and then overtopped by the growth of a lava dome that has fed five block lava flows since 1961.

**SHEER FLANKS**
*The steep profile of the Volcán de Colima has been shaped by viscous block-lava flows.*

**GAS SAMPLING**
*Volcanologists collect samples from a fumarole field at the rim of Colima's crater. Monitoring composition and temperature of the gases can help scientists understand and predict the volcano's behavior.*

NORTH AMERICA *south*

## Popocatépetl

**LOCATION** In the Mexican Volcanic Belt, within the Puebla region of south-central Mexico

**HEIGHT** 17,887 ft (5,452 m)

**TYPE** Stratovolcano

**LAST ERUPTION** 2003

Popocatépetl is North America's second-highest volcano, towering over Mexico city to its northwest. With its symmetrical shape, summit glaciers, and frequent eruptions during recorded history, it has an imposing presence. From the Pleistocene Epoch onward, the volcano has grown to 17½ miles (28 km) across. The summit crater is 2,200 ft (670 m) in diameter and up to 1,480 ft (450 m) deep. Since Popocatépetl's formation, at least three previous cones have been destroyed by gravitational collapse. These have produced massive debris-avalanche deposits to the south. Three major Plinian eruptions have occurred over the last few thousand years, most recently in AD 800, producing pyroclastic flows and large-volume lahars. The Aztecs recorded frequent historical eruptions before the arrival of Western explorers in the 15th century, and gave the volcano its name, meaning "smoking mountain."

**STEEP-WALLED CRATER**
*Popocatépetl's deep, oval-shaped crater has near-vertical walls. On the crater floor, a succession of sporadically erupting lava domes has evolved.*

### FERTILE SOILS

Volcanic products are rich in minerals and glass that can produce very fertile and well-drained soils. In highly populated areas, such as here in the shadow of Popocatépetl, the fertile soil is of great benefit to local farmers, despite the potential risks.

NORTH AMERICA *south*

## El Chichón

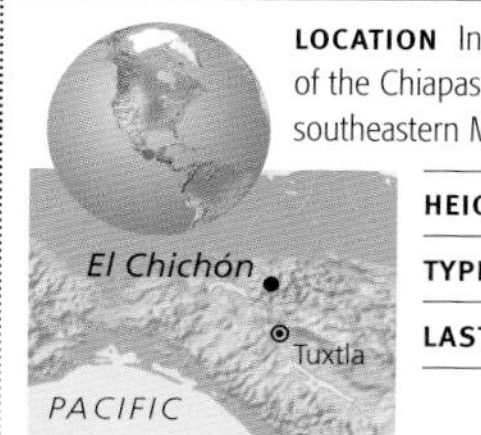


**LOCATION** In an isolated part of the Chiapas region of southeastern Mexico

**HEIGHT** 3,478 ft (1,060 m)

**TYPE** Lava dome

**LAST ERUPTION** 1982

This lava-dome and tuff-cone complex was relatively unknown until 1982, when it produced a series of highly explosive eruptions. Pyroclastic flows swept outward for more than 5 miles (8 km), destroying 9 villages and killing more than 2,000 people. The magma was extremely sulfur-rich, and sulfuric acid droplets forming in the stratosphere produced brilliant sunsets around the globe.

**ACIDIC POST-ERUPTION CRATER LAKE**

CENTRAL AMERICA

## Masaya

**LOCATION** Between the west coast of Lake Nicaragua and the Pacific coast, southwestern Nicaragua

**HEIGHT** 2,083 ft (635 m)

**TYPE** Caldera

**LAST ERUPTION** 2001

**ESCAPING GASES**

One of Nicaragua's most active volcanoes, Masaya is an 7-mile- (11-km-) wide caldera, with more than a dozen vents, surrounded by walls 985 ft (300 m) high. The twin cones of Masaya and Nindiri have been the source of frequent recorded eruptions. A major Plinian eruption of basaltic tephra occurred about 6,500 years ago. Since then, frequent lava flows have covered much of the caldera floor, and in AD 1670 one overtopped the northern rim.

CENTRAL AMERICA

## Arenal

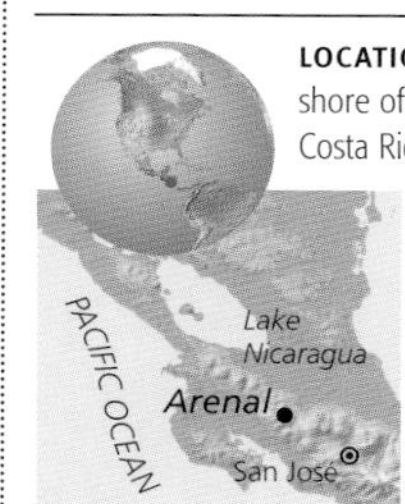


**LOCATION** On the southeastern shore of Lake Arenal in central Costa Rica

**HEIGHT** 5,436 ft (1,657 m)

**TYPE** Stratovolcano

**LAST ERUPTION** 2003

Arenal is the youngest and one of the most active volcanoes in Costa Rica. The 1968 Vulcanian eruption signaled the start of the current long-term eruptive period. It threw large bombs up to 3 miles (5 km) from the crater and generated pyroclastic flows (one of which killed 70 people) before releasing viscous lava flows from vents at the summit and the upper part of its western slope.

**INCANDESCENT ROCK AVALANCHES**

CENTRAL AMERICA

## Soufrière Hills

**LOCATION** Occupies the southern half of the island of Montserrat in the eastern Caribbean Sea

**HEIGHT** 3,002 ft (915 m)

**TYPE** Stratovolcano

**LAST ERUPTION** 2003

Soufrière Hills is a complex volcano, with a series of lava domes forming its summit area. Although swarms of seismic shocks had been felt at 30-year intervals throughout the 20th century, the volcano was thought to be inactive. Then, in an eruption beginning in 1995, pyroclastic flows and mudflows overran and destroyed Plymouth, Montserrat's capital, burying it under several yards of pyroclastic deposits. The southern end of the island was evacuated and, inevitably, the population suffered major disruption. As the activity has continued, Soufrière Hills has become one of the world's most closely monitored volcanoes.

**ERUPTION CLOUD**
*Soufrière Hills produces Plinian eruptions, sending huge clouds of ash high into the air and smothering the surrounding area.*

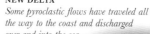

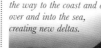

**NEW DELTA**
*Some pyroclastic flows have traveled all the way to the coast and discharged over and into the sea, creating new deltas.*

SOUTH AMERICA *northwest*

## Nevado del Ruiz

**LOCATION** In the northern region of the Cordillera Central range, central Colombia

**HEIGHT** 17,457 ft (5,321 m)

**TYPE** Stratovolcano

**LAST ERUPTION** 1991

The modern volcano of Nevado del Ruiz sits within the caldera of the older Ruiz volcano. Its cone consists of a series of lava domes, with the summit occupied by the Arenas crater, 3,300 ft (1 km) wide and 790 ft (240 m) deep. The volcano's flanks are shaped by large landslides. Although Nevado del Ruiz is only about 300 miles (500 km) north of the equator, its position high in the Andes means that it is capped throughout the year by large volumes of snow and ice. Because of this, heat released by eruptions has, during recorded history, caused summit glaciers to melt, creating devastating lahars. In 1985, Nevado del Ruiz was the scene of South America's deadliest volcanic eruption and one of the worst volcanic disasters of modern times (see panel, right). A relatively small volume of hot ejecta, spewed across the snow- and ice-covered summit, proved to be a lethal combination.

**BROAD MASS**
*Nevado del Ruiz is a wide, sprawling volcano that covers over 77 square miles (200 square km).*

### LAHAR DISASTER

On November 13, 1985, pyroclastic flows, released by a small eruption, melted the summit ice cap, triggering a series of devastating lahars. Channeled down narrow valleys for up to 60 miles (100 km), they wiped out the town of Armero and killed more than 23,500 people. When the flows stopped, they set like concrete, offering victims little chance of escape.

SOUTH AMERICA *northwest*

## Galeras

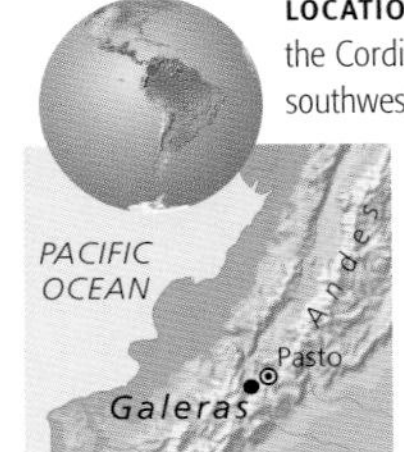


**LOCATION** At the southern end of the Cordillera Central range, southwest Colombia

**HEIGHT** 14,029 ft (4,276 m)

**TYPE** Complex volcano

**LAST ERUPTION** 2000

Now a modern cone positioned within an older caldera, Galeras has been active for more than a million years. Major eruptions and weakening of the edifice on many occasions have led to debris avalanches, pyroclastic flows, and widespread air-fall deposits. Situated next to the city of Pasto, the volcano is closely monitored.

**PASTO'S VOLCANIC BACKDROP**

SOUTH AMERICA *northwest*

## Cotopaxi

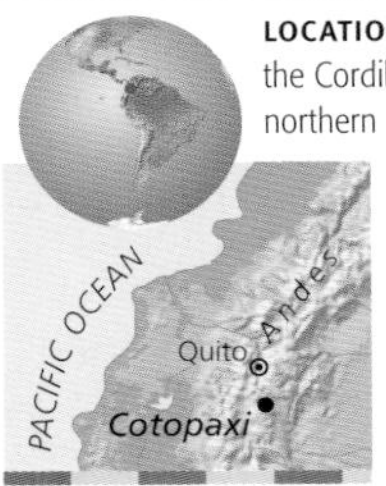


**LOCATION** At the southern end of the Cordillera Central range, northern Ecuador

**HEIGHT** 19,393 ft (5,911 m)

**TYPE** Stratovolcano

**LAST ERUPTION** 1940

The glacier-clad Cotopaxi is the highest active volcano on Earth. One of Ecuador's most active volcanoes, it has a steep-sided, almost perfectly symmetrical cone capped with deep craters nested within the summit.

**VOLCANIC DEBRIS**
*A huge block of volcanic rock lies on a flank of Cotopaxi, where it was probably deposited by a large debris avalanche or powerful lahar.*

The present cone, scarred by lava flows and lahars, has been constructed over the last 5,000 years. Cotopaxi has a long history of explosive eruptions, including that of 1534, which put an end to a battle between the Incas and the Spanish. In one of its most violent recorded eruptions, in 1877, an eruptive column collapsed, generating pyroclastic flows and lahars that devastated nearby valleys and flowed over 60 miles (100 km) into the Pacific Ocean.

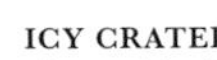

**ICY CRATER**
*Situated in the northern Andes, Cotopaxi is the highest active volcano in the world. Its peak is clad with glaciers and a constant snow covering.*

SOUTH AMERICA *southwest*

## Cerro Azul–Quizapu

**LOCATION** In the southern Andes, central Chile, close to the border with Argentina

**HEIGHT** 12,428 ft (3,788 m)

**TYPE** Stratovolcano

**LAST ERUPTION** 1967

The steep-sided Cerro Azul volcano is located at the southern end of a volcano chain called the Descabezado Grande–Cerro Azul eruptive system. It has a 1,640-ft- (500-m-) wide summit crater and several flank cinder cones, including the three La Resoloma craters to the west and Los Hornitos to the southwest. Quizapu is a vent on the northern flank that formed in 1846 with the emission of a large volume of lava. In 1932, Quizapu was the site of one of the largest explosive eruptions of the 20th century, ejecting a large volume of tephra and creating a crater up to 2,300 ft (700 m) wide and 490 ft (150 m) deep.

**QUIZAPU CRATER**

SOUTH AMERICA *east*

## Paraná Plateau

**LOCATION** Across southern Brazil, eastern Paraguay, northern Argentina, and northwest Uruguay

**HEIGHT** Up to 3,300 ft (1,000 m)

**TYPE** Fissure vent

**LAST ERUPTION** About 120 million years ago

The eruption about 120 million years ago of more than 290,000 square miles (750,000 square km) of basaltic lava in the Paraná region of South America coincided with the splitting of that continent from Africa and the opening of the south Atlantic Ocean. Heat from within the mantle formed the Walvis hot spot under the Paraná region, in eastern South America, and Namibia in southwest Africa, which at the time were connected. Doming of the crust over a 625-mile- (1,000-km-) wide zone developed deep and extensive fractures, through which lava poured onto the Earth's surface. Continuing volcanic activity led to further rifting between the continents and the formation of a new ocean basin. The hot spot is today marked by the volcanic island of Tristan da Cunha, which last erupted in 1961.

ATLANTIC OCEAN *north*

## Grímsvötn

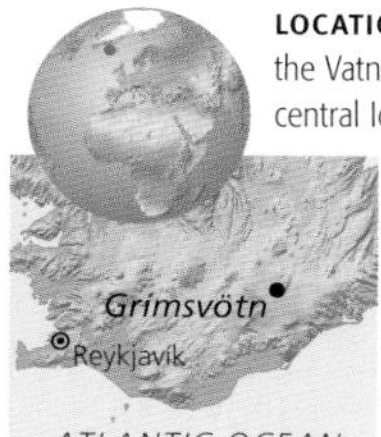


**LOCATION** Lying largely beneath the Vatnajökull ice cap of central Iceland

**HEIGHT** 5,659 ft (1,725 m)

**TYPE** Caldera

**LAST ERUPTION** 1998

Iceland's most active volcano in recorded history, Grímsvötn is largely covered by the Vatnajökull ice cap. Only the southern rim of its 5-mile- (8-km-) wide caldera is exposed above the ice, and heat from the volcano frequently causes floods of melted glacier ice (jökulhlaups). Long fissures extending from the volcano include the well-known Laki fissure, which produced the largest lava flow in recorded history. In 1783 about 3½ cubic miles (15 cubic km) of lava was erupted over a period of seven months. More than 9,000 people died from starvation and disease in what became Iceland's worst natural disaster.

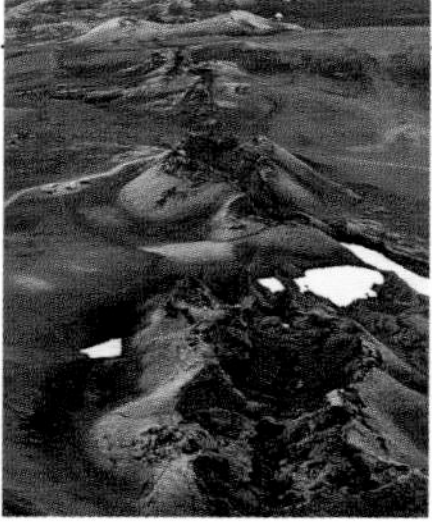

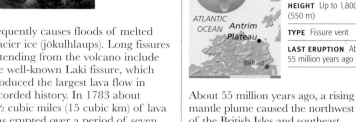

**LAKI FISSURE**
*The Laki fissure system is 17 miles (27 km) long and contains over 140 craters.*

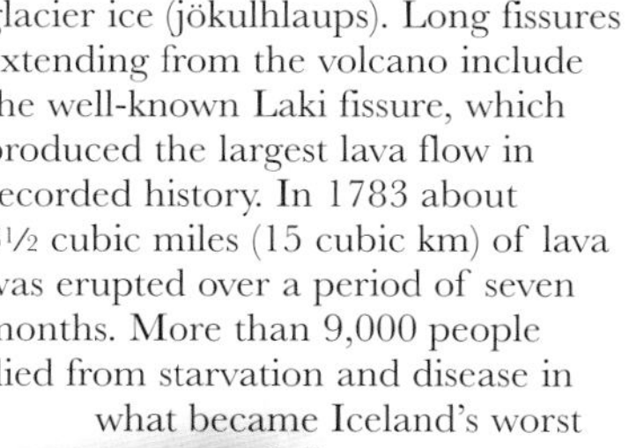

**ERUPTING DOME**
*During the last eruption, in 1998, a plume 6 miles (10 km) high erupted from the caldera.*

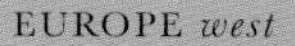

EUROPE *west*

## Antrim Plateau

**LOCATION** On the coast of County Antrim, in northeastern Northern Ireland

**HEIGHT** Up to 1,800 ft (550 m)

**TYPE** Fissure vent

**LAST ERUPTION** About 55 million years ago

About 55 million years ago, a rising mantle plume caused the northwest of the British Isles and southeast Greenland to dome upward. Rifting and sea-floor spreading erupted a pile of basaltic lavas up to 1¼ miles (2 km) thick, with flows up to 26 ft (8 m) thick. The lava flows are interbedded with pyroclastic material, and plant fossils in the beds show a humid temperate climate. The Giant's Causeway, a group of thousands of mostly hexagonal basalt columns, is world-famous for its spectacular appearance.

**THE GIANT'S CAUSEWAY**

EUROPE *central*

## West Eifel Field

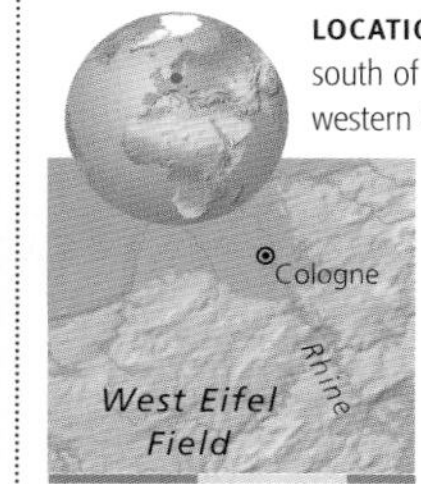


**LOCATION** In the Rhine Valley, south of the city of Cologne, western Germany

**HEIGHT** 1,968 ft (600 m)

**TYPE** Maars, cinder cones

**LAST ERUPTION** 7,000–11,000 years ago

The West Eifel Field is associated with the Rhine Rift (see p.147). Over the last 730,000 years, it has seen alternating episodes of Strombolian and phreatic activity, producing some 240 cinder cones, maars, and small stratovolcanoes. They cover an area of about 230 square miles (600 square km), with the last eruption occurring 7,000–11,000 years ago. More than 70 maars, up to 3,300 ft (1 km) in diameter and 650 ft (200 m) deep, were formed between 12,500 and 10,000 years ago, when the end of the last ice age produced abundant meltwater.

**MAAR MINING**

EUROPE *west*

## Chaîne des Puys

**LOCATION** In the Auvergne region of central France, in the northern Massif Central

**HEIGHT** 4,803 ft (1,464 m)

**TYPE** Cinder cones, lava domes, maars

**LAST ERUPTION** About 4040 BC

The Chaîne des Puys was extensively studied by 19th-century scientists and played an important role in the history of geology. It is a series of cinder cones, lava domes, and maars aligned north to south. The maars are craters formed by explosive eruptions, rimmed by ejected debris, and often filled by lakes. Their formation began about 70,000 years ago, with pyroclastic flows and long lava flows, and finally ended about 6,000 years ago. The most recent volcanic activity occurred near Besse-en-Chandesse. It has been radiocarbon-dated at around 4040 BC and included powerful explosions that created Lac Pavin Maar.

**VOLCANO CHAIN**
*A series of about 90 cinder cones, maars, and lava domes extends across nearly 35 miles (50 km) of France's northern Massif Central.*

EUROPE *south*

## Stromboli

**LOCATION** One of the volcanic Aeolian Islands off the north coast of Sicily in the Mediterranean

**HEIGHT** 3,038 ft (926 m)

**TYPE** Stratovolcano

**LAST ERUPTION** 2003

The small island of Stromboli is an active volcano rising about 9,850 ft (3,000 m) from the ocean floor. It is one of the few almost continuously erupting volcanoes on Earth, having been active since records began more than 2,500 years ago. Stromboli's active summit vents are located in the Sciara del Fuoco, a curved scarp that was formed by a slope failure about 5,000 years ago. The volcano gave its name to one of the main types of volcanic eruption (see p.174). Most of its activity is moderate in scale, consisting of brief, but explosive, ejections of glowing lava fragments to heights seldom more than 490 ft (150 m) and, less frequently, lava flows. Periods of more intense activity can produce prolonged eruptions from fountains and ejection of large bombs.

**INCANDESCENT DISPLAY**
*This long-exposure photograph shows a typical small-scale explosive eruption ejecting glowing rock particles into the air.*

EUROPE *south*

## Mount Etna

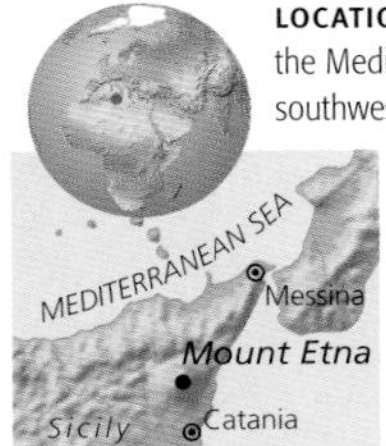


**LOCATION** On the eastern side of the Mediterranean island of Sicily, southwest of the Italian mainland

**HEIGHT** 10,991 ft (3,350 m)

**TYPE** Stratovolcano

**LAST ERUPTION** 2003

Etna is Europe's highest volcano, and has been almost continuously active since observations were first recorded, about 2,500 years ago. About 2.5 million years old, Etna probably originates from a mantle hot spot. The volcano has a tall cone with a base circumference of about 95 miles (150 km) and a surface area of about 620 square miles (1,600 square km). Etna's most prominent feature is the Valle del Bove, a 6-mile- (10-km-) wide horseshoe-shaped caldera open to the southeast. The summit is a complex structure of old and new craters, and currently there are four active craters. The sides of the volcano are scored by deep fissures that radiate from the summit. Up to 200 small ash cones are present on its flanks, with three larger cones near the summit. Etna produces basaltic lava that has low viscosity, and lava flows extend to the foot of the volcano on all sides, reaching the sea over a broad area on the southeast flank. The summit craters are in a permanent state of low-level activity, with lava fountains erupting and phreatic explosions (see pp.174–75). Although Etna has not been considered an especially explosive volcano, at times it can be spectacular. Etna's largest and most famous eruption, in 1669, lasted 122 days. Earthquakes opened an 8-mile- (12-km-) long fissure and outpourings of basaltic lavas rapidly overwhelmed nearby villages (see panel, right). The 1986 summit eruption produced lava fountains that jetted up to 1 mile (1.6 km) above the crater, and a column of ash and gas that rose 6 miles (10 km) into the atmosphere. More recently, eruptions in both 2001 and 2002 again brought Etna to the world's attention.

### DIVERTING LAVA FLOWS

Efforts to divert destructive lava flows on Etna date as far back as 1669, when lava overwhelmed a number of villages and threatened the town of Catania. A party was sent to try to breach a levee and allow the lava to spread in a different direction, away from the town. However, this action met with opposition from neighboring villagers, and the lava flow was eventually left to enter the town. In recent years, methods have involved the use of industrial equipment to build earth dams and concrete barriers, and even the use of explosives. However, most lava-diversion attempts meet with only temporary success.

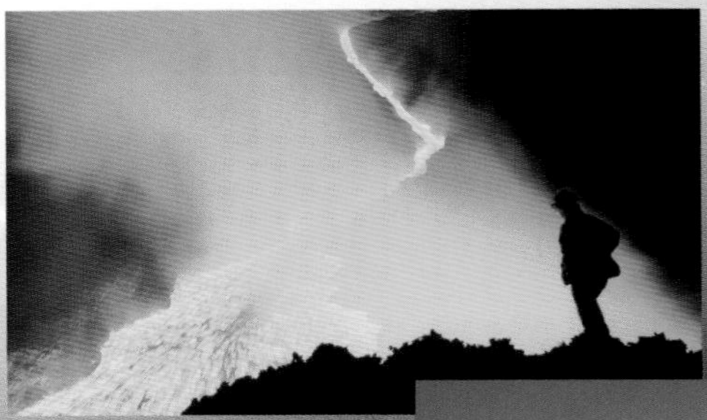

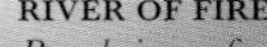

**RIVER OF FIRE**
*Broad rivers of alkali basaltic lava flow down the lava field of the Valle del Bove on Etna's southeast flank in January 1992.*

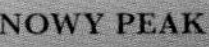

**SNOWY PEAK**
*Close to Etna's summit, the desolate wasteland of solid lava and ash is covered with snow for much of the year. Its fertile lower slopes are widely cultivated.*

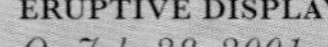

1 2 3 4

**ERUPTIVE DISPLAY**
*On July 28, 2001, explosions from Etna's newly formed Lago cone hurled clouds of ash high into the air (1 and 2). Later, molten materials were discharged from two separate vents simultaneously (3 and 4).*

**STROMBOLIAN ACTIVITY ON ETNA**
*Fountains of fluid lava and violent ejection of partially molten volcanic bombs, seen in this long-exposure photograph, are typical of Strombolian eruptions (see p.174).*

**OUT OF THE ASHES**

*Planted in pits designed to trap water, vines flourish in arid but fertile volcanic soil on Lanzarote, the easternmost of the Canary Islands.*

# LIVING WITH VOLCANOES

Volcanoes and the materials that they produce are essential for life on Earth. Without them, surface water would not have formed, and living things could not have evolved. Volcanoes also raise minerals and melts from deep in the Earth's crust, helping to create fertile soil. But this beneficial role comes at a price, because eruptions are among the most deadly hazards in the world. That danger cannot be removed, but with modern technology, the risk it carries can be reduced.

## VOLCANOES AS NEIGHBORS

Although volcanoes have always been feared, people recognized their beneficial effects long before anything was known about the chemistry of soils. In Sicily, for example, the lower slopes of Mount Etna have been farmed for several thousand years, and were an important source of agricultural produce in classical times. On the Indonesian island of Java, weathered volcanic soil currently supports one of the densest populations on the Earth, with an average of nearly 2,100 people per square mile (800 per square kilometer). In contrast, the Amazon Basin, which is situated at almost exactly the same latitude as Java and enjoys a similar climate, is nonvolcanic, and suffers from extremely poor soils, which are difficult to cultivate successfully.

Mount Etna has erupted more than 200 times in recorded history, but despite this turbulent past, it has inflicted relatively little loss of human life. Etna's immense size means that most farms are distant from the summit, and the mountain's lava streams often follow predictable paths that people can avoid. Many volcanoes in populated regions, however, are prone to violently explosive eruptions (see p.175). Java's volcanoes fall into this category, as do those of Mount Pinatubo, in the Philippines. When the latter erupted in 1991, the death toll was high, and thousands of farms were buried by ash.

**STANDING SENTINEL**
*A volcano rises above rice fields in Kagawa Prefecture, Shikoku. More than 60 volcanoes are recorded as active in the Japanese archipelago.*

## WARNING SIGNS

Modern volcanology began in the early 1900s, when an observatory was built on Kilauea—the world's largest active volcano—in Hawaii. Before this, people relied on a mixture of observation and superstition to judge when a volcano might erupt. Seismic tremors provided some clues, as did the sudden escape of gas, but just as much attention was paid to unusual behavior in wild animals. Unfortunately, these portents often turned out to be unreliable—a fact vividly demonstrated in AD 79, when Mount Vesuvius in Italy erupted, burying Pompeii and its inhabitants under several yards of ash (see p.186).

Volcanologists now gather various kinds of data to predict volcanic activity. One of the most useful is the physical deformation that occurs when magma rises up through a volcano's interior. Sudden expansion—such as that witnessed an Mount St. Helens in 1980—is one of the surest signs that an eruption is imminent. Until recently, deformation measurements had to be taken manually, using instruments fixed to a crater's rim, but measurements can now be done remotely by satellite-based radar without any human presence at the crater.

**CLOSE ESCAPE**
*These houses in Shimabara, Japan, were damaged after Mount Unzen erupted in 1993—about 200 years after the last devastation.*

## MONITORING VOLCANOES

Seismic activity is the most widely monitored parameter at active volcanoes—a warning sign is an increase or, just as ominously, a sudden stop in movement. Such information is gathered by automated seismic sensors, and is then transmitted to research stations for analysis. Seismic sensing provided advance warning of the Mount Pinatubo eruption, although, because of its immense scale, not everyone was able to escape.

Moving magma causes changes in a volcano's magnetic field, and vent-gas analysis provides a further line of evidence, because the concentration of gases changes as hot magma approaches the surface. Originally developed in Japan, vent-gas analysis can signal increased volcanic activity months before a volcano explodes, and after an eruption, it also provides confirmation that volcanic activity is in decline. As yet, a fail-safe method of predicting eruptions does not exist, but with these monitoring techniques, that goal may not be far off.

**GAS ANALYSIS**
*A volcanologist collects gas samples on the island of Vulcano, off the north coast of Sicily. The last major eruption here was in 1890.*

Annual emission rate (metric tons per day)
2,500
2,000
1,500
1,000
500
0
1980 1981 1982 1983 1984 1985 1986 1987 1988
Year

**GAS EMISSIONS**
*The toxic acidic gas sulfur dioxide is emitted by active volcanoes. At Mount St. Helens, sulfur dioxide emissions reached peak levels during the 1980 eruption, and then gradually subsided. The measurements were made by a sampling airplane.*

EUROPE *south*

## Vesuvius

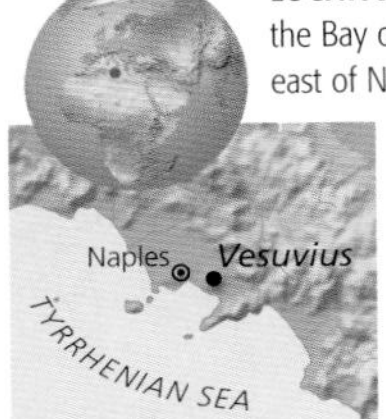


**LOCATION** On the eastern coast of the Bay of Naples, 7 miles (12 km) east of Naples in southwest Italy

**HEIGHT** 4,203 ft (1,281 m)

**TYPE** Complex volcano

**LAST ERUPTION** 1944

Vesuvius is a frequently active volcano overlooking the Bay of Naples. Its cone, with a base circumference of about 45 miles (70 km), sits within the caldera of an older volcano, Monte Somma, that formed about 17,000 years ago. Vesuvius is best known for its Plinian eruptions, such as the one in AD 79 that destroyed Pompeii and

Herculaneum (see panel, below). Since the modern cone formed, there have been eight major eruptions, accompanied by large pyroclastic flows and surges, and high eruptive columns capable of carrying pumice, ash, and bombs dozens of miles into the air. In 1631, during the largest eruption since AD 79, pyroclastic flows reached as far as the sea. More recently, Vesuvius erupted in 1906 and 1944, causing fatalities on both occasions. Today, about 3 million people live within the eruptive range of Vesuvius, and evacuation plans are in place.

**GIANT PROFILE**
*The city of Naples sits in the shadow of Vesuvius, whose fertile slopes are dotted with villages and vineyards.*

**LAST MAJOR ERUPTION**
*The 1944 eruption of Vesuvius claimed the lives of 27 people. The average repose time between eruptions is 50 years, so the next one is overdue.*

### DESTRUCTION OF POMPEII

Most of the victims of the eruption that destroyed Pompeii in AD 79 died very suddenly, smothered by fast-moving, high-temperature pyroclastic surges. A town of 20,000 people, Pompeii was buried under 10 ft (3 m) of materials that effectively "fossilized" the inhabitants and building remains.



EUROPE *southeast*

## Santorini

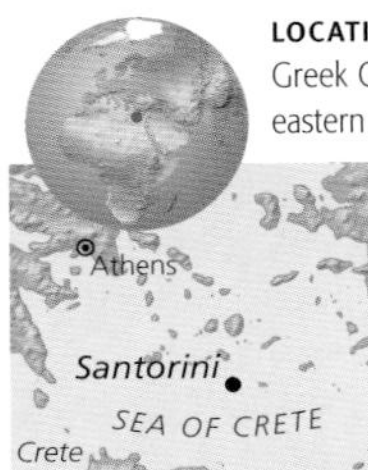


**LOCATION** One of the volcanic Greek Cyclades Islands in the eastern Mediterranean Sea

**HEIGHT** 1,204 ft (367 m)

**TYPE** Caldera

**LAST ERUPTION** 1950

**SANTORINI'S FLOODED CALDERA**

Santorini is a complex of overlapping volcanoes and calderas forming a circular group of islands with steep, inward-facing walls. The most recent caldera-forming eruption occurred in about 1640 BC, when about 14–16 cubic miles (58–68 cubic km) of material was ejected, and the current sea-filled caldera was formed. It is thought that this eruption might have caused the collapse of the Minoan civilization. New islands in the center of the caldera were formed by a series of eruptions beginning in 197 BC.

EUROPE *south*

## Vulcano

**LOCATION** One of the volcanic Aeolian Islands off the north coast of Sicily in the Mediterranean Sea

**HEIGHT** 1,640 ft (500 m)

**TYPE** Stratovolcano

**LAST ERUPTION** 1890

The island of Vulcano is the source of the word "volcano"—according to Roman legend, Vulcano was home to Vulcan, the god of fire, and his forges. It also gives its name to the Vulcanian style of eruption, which is characterized by moderate-to-violent eruptive columns dominated by viscous and solid ejecta. Although it is only a small island (with an area of 8.5 square miles/22 square km), Vulcano is constructed of four volcanic complexes that have developed over the last 120,000 years. The Piano caldera, in the southern half of the island, is overlapped to the north by the Fossa cone, a modern volcano that last erupted in 1888–90. Vulcanello, on the northern tip of the island, is a partly submerged volcano with a lava platform, which now forms a low peninsula. The Vulcano complex also includes a combination of hot springs, mud pots, and active fumaroles.

**NEARBY VILLAGE**
*A busy seaside town on the neighboring island of Lipari sits in the shadow of the summit of the active Vulcano.*

**FUMAROLE GASES**
*Vulcano has become a popular site with volcanologists, who monitor the gases emitted by the many steaming fumaroles on the volcano's flanks.*

ATLANTIC OCEAN *east*

## Pico de Teide

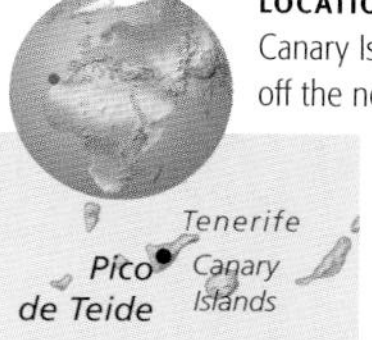


**LOCATION** On Tenerife, one of the Canary Islands in the Atlantic Ocean, off the northwest coast of Africa

**HEIGHT** 12,188 ft (3,715 m)

**TYPE** Stratovolcano

**LAST ERUPTION** 1909

The island of Tenerife was formed by a complex of overlapping volcanoes that remain active today. Pico de Teide lies within the 10-mile- (17-km-) wide Las Cañadas caldera. An eruption in 1492 was witnessed by Christopher Columbus on his way to the New World.

FLANK CONE

AFRICA *east*

## Erta Ale

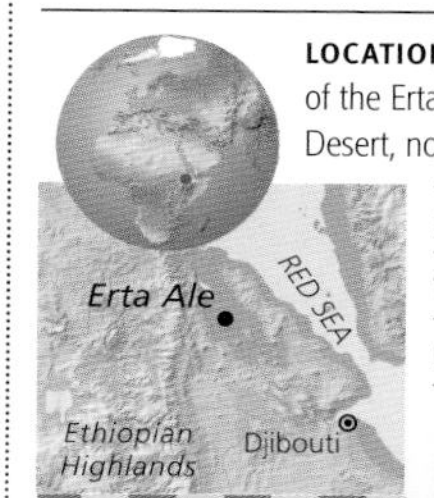


**LOCATION** The prominent feature of the Erta Ale range in the Danakil Desert, northeastern Ethiopia

**HEIGHT** 2,011 ft (613 m)

**TYPE** Shield volcano

**LAST ERUPTION** 2003

The remote Erta Ale is a shield volcano 30 miles (50 km) wide that rises to over 2,000 ft (600 m) from below sea level. Its 1-mile- (1.6-km-) wide elliptical summit crater contains circular, steep-sided pit craters. Another, larger, elongated depression parallel to the Erta Ale range is located to the southeast of the summit and is bounded by curvilinear fault ridges. Recent basaltic lava flows have erupted from these fault fissures into the caldera and have overflowed its rim. Erta Ale is Ethiopia's most active volcano, and its summit pit craters are renowned for their perpetually churning lava lakes. These have been active since at least 1967, but possibly since 1906, which would make it one of the longest known eruptions in recorded history. Recently, fissure eruptions have occurred on the volcano's northern flank.

PIT CRATER
*The smaller of the two summit craters, Erta Ale's active south pit crater is about 500 ft (150 m) across its almost perfectly proportioned diameter.*

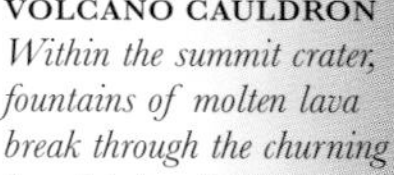

VOLCANO CAULDRON
*Within the summit crater, fountains of molten lava break through the churning lava lake's solidified, black surface crust.*

LAND

AFRICA *east*

## Mount Kilimanjaro

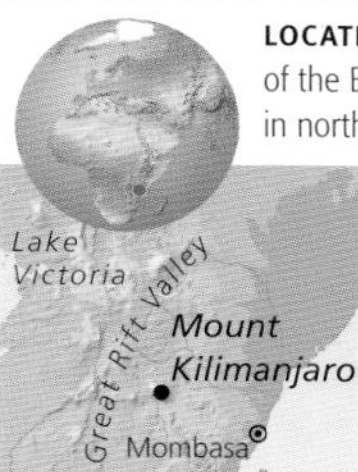


**LOCATION** At the southern end of the East African Rift's eastern fork, in northeastern Tanzania

**HEIGHT** 19,340 ft (5,895 m)

**TYPE** Stratovolcano

**LAST ERUPTION** Unknown

Africa's highest mountain, Kilimanjaro is one of several volcanoes whose formation is associated with the eastern part of the East African Rift (see pp.148–49). An elongated complex cone, more than 30 miles (50 km) long, it consists of three volcanoes. The mountain was mainly constructed during the Pleistocene Epoch (1.8 million–10,000 years ago) but includes a group of nested summit craters that are apparently younger. Kibo is the highest and central cone and erupted between the other two—Shira and Mawenzi. It is topped by a glacier-covered subsidence caldera with an inner crater called the Ash Pit. The 2¼-mile- (3.5-km-) wide caldera gives the summit its broad, elongated profile. Most of the 250 flanking lava and cinder cones are less than 330 ft (100 m) high. Recent volcanic activity has been mostly confined to fumaroles around Kibo's crater. Lying just 200 miles (320 km) north of the equator and yet high enough to be snow-capped, Kilimanjaro supports a unique ecosystem adapted to a wide range in temperature.

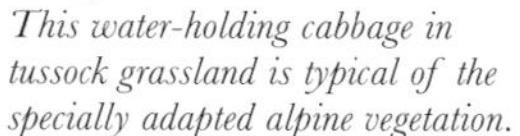

UNIQUE VEGETATION
*This water-holding cabbage in tussock grassland is typical of the specially adapted alpine vegetation.*

### GLOBAL WARMING

Global warming is transforming one of Africa's most iconic landmarks. Recent studies have suggested that since 1912, more than 80 percent of the ice and snow at its summit has melted, and that by 2020 what remains may have vanished. As well as reshaping a dramatic skyline, the consequences for communities that rely on the meltwater could be huge.

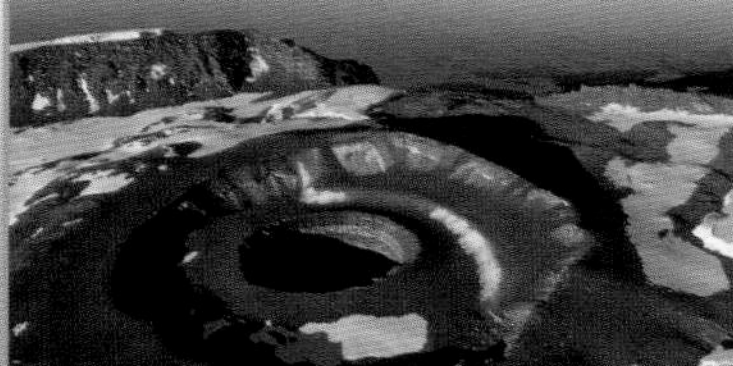

MASSIVE MOUNTAIN
*Rising from the savanna plains of eastern Africa, the massive Kilimanjaro is one of the most famous and most imposing volcanoes in the world.*

AFRICA *central*

## Lake Nyos

**LOCATION** One of the craters on the Oku Volcanic Field of northwestern Cameroon

**HEIGHT** 9,878 ft (3,011 m)

**TYPE** Maar

**LAST ERUPTION** Unknown

Lake Nyos, which lends its name to this volcano, lies within the crater of one of 29 maars in what is known as the Oku Volcanic Field. This, in turn, is part of a zone of crustal weakness, called the Cameroon Volcanic Line, which encompasses the Cameroon Mountain stratovolcano and extends for about 1,000 miles (1,600 km), with half its length submerged in the Atlantic Ocean. The Lake Nyos crater might have been created by an explosive phreatic eruption about 500 years ago. Its rim, which is up to 1 mile (1.6 km) wide, is made up of fragments of basaltic ejecta containing large shattered blocks of granite. Lake Nyos was brought to the world's attention by a catastrophe in 1986 that killed about 1,800 people (see panel, above). This release of a great cloud of noxious gas was caused by the lake's water being saturated with carbon dioxide seeping from underground springs. It is thought that the gas could have been released from the water at the bottom of the lake by a landslide or earthquake. Since 1990, a team of French scientists has been working to degas the lake, and a series of pipes has been installed to try to prevent future buildups of carbon dioxide.

### DEADLY GASES

In 1986 the release of a suffocating carbon-dioxide cloud from Lake Nyos killed around 1,800 people, along with over 6,000 cattle. Up to ¼ cubic mile (1 cubic km) of the gas, traveling at nearly 30 mph (50 km/h), was channeled down surrounding valleys for about 14 miles (23 km). As it moved, the heavy gas hugged the ground, displacing the air and asphyxiating all humans and animals in its path.

**CRATER LAKE**
*The crater lake is more than 650 ft (200 m) deep, and during the rainy season, excess water often floods over the rim and down the nearby valleys.*

AFRICA *central*

## Nyiragongo

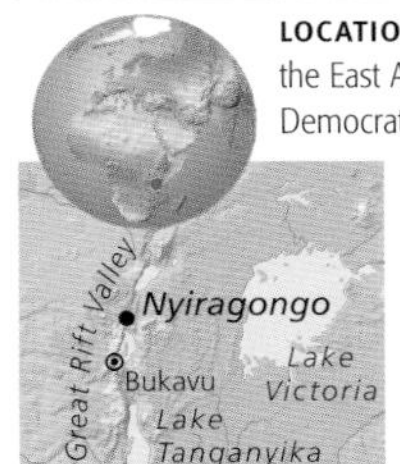


**LOCATION** At the southern end of the East African Rift's western fork, Democratic Republic of Congo

**HEIGHT** 11,384 ft (3,470 m)

**TYPE** Stratovolcano

**LAST ERUPTION** 2003

One of the most active volcanoes in the world, Nyiragongo is famous for the lava lake in its summit crater, first discovered by German explorer Adolf von Gotzen in 1894. On January 10, 1977, lava broke through the crater walls and traveled at up to 60 mph (100 km/h) toward the city of Goma, killing 50 to 100 people. The lake started to build up again in 1982, and Goma was partially overrun by lava flows in January 2002.

**CHURNING LAVA**

AFRICA *east*

## Ol Doinyo Lengai

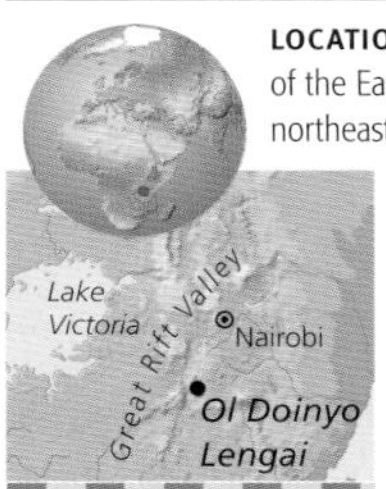


**LOCATION** At the southern end of the East African Rift's eastern fork, northeastern Tanzania

**HEIGHT** 9,482 ft (2,890 m)

**TYPE** Stratovolcano

**LAST ERUPTION** 2003

The cone-building stage of this 370,000-year-old symmetrical volcano ended about 15,000 years ago. Its summit consists of two craters. The inactive older southern crater is covered with vegetation and volcanic ash. Activity in the northern crater is centered around many hornitos (stacks pushed up from an underlying lava flow) and small cones. Ol Doinyo Lengai is most renowned for being the only volcano known to have erupted compositionally unique carbonatite lavas and tephra in recorded history. With a viscosity close to that of water (due to low silica content), its lava is the most fluid in the world, and also the coolest, with temperatures up to only 1,100°F (590°C). This lava flows black during the day and glows a deep red at night, and when it comes in contact with water, a chemical reaction turns it white. Long-term lava effusion has been punctuated with some ash eruptions, and strong explosive eruptions were recorded in 1917, 1940, 1960, and 1966.

**MOUNTAIN OF GOD**
*Ol Doinyo Lengai is known to the Masai as the "Mountain of God." Local villages rely on tourists who visit the volcano.*

**ACTIVE CRATER**
*The volcano's crater has gradually filled up with ash and rock. Its volume has been estimated at 14 cubic miles (60 cubic km).*

**STEEP PROFILE**
*The imposing, steep-sided form of Ol Doinyo Lengai rises from the flat, wide plain, south of Tanzania's Lake Natron.*

AFRICA *east*

## Piton de la Fournaise

**LOCATION** On the French island of Réunion in the Indian Ocean, off the east coast of Madagascar

**HEIGHT** 8,632 ft (2,631 m)

**TYPE** Shield volcano

**LAST ERUPTION** 2002

One of the biggest and most active volcanoes in the world, Piton de la Fournaise originated from a mantle hot spot under the Indian Ocean. Much of its 530,000-year eruption record has overlapped with that of the nearby Piton des Neiges shield volcano. Eastward slumping of the volcano produced three successive calderas, formed 250,000, 65,000, and less than 5,000 years ago. A 1,300-ft- (400-m-) high cone lies in the center of the youngest of the three, Caldera de l'Enclos Fouque. The summit of the cone comprises two craters—Bory and the larger Dolomieu, from which most of the recent eruptions have originated. More than 150 eruptions have been recorded since the 17th century, producing fluid basaltic lava flows. The volcano's activity is closely monitored by an observatory on its slopes.

### VOLCANO TOURISM

Every year, over 300,000 tourists are drawn to the isolated island of Réunion especially to visit Piton de la Fournaise. As one of the world's most active volcanoes, it offers the chance to witness at first hand the awe-inspiring scene of an eruption. Hiking paths wind through the dramatic landscape, and lookout points offer safe viewing platforms.

**FURNACE PEAK**
*An active cone in the summit crater of the volcano belches fumes as molten lava wells up inside. In English, Piton de la Fournaise translates as "Furnace Peak."*

ASIA *north*

## Siberian Traps

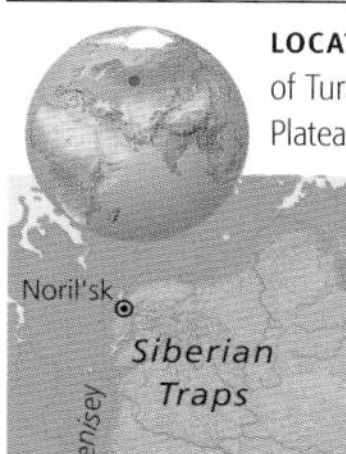


**LOCATION** Centered on the town of Tura on the central Siberian Plateau, northern Russia

**HEIGHT** Up to 1,650 ft (500 m)

**TYPE** Fissure vent

**LAST ERUPTION** About 250 million years ago

One of the largest and most enigmatic of flood basalt effusions, the Siberian Traps cover more than 115,000 square miles (300,000 square km) of Arctic wilderness. About 360,000 cubic miles (1.5 million cubic km) of lava erupted through the Earth's crust within about a million years. This outpouring does not appear to be connected to rifting, although it may have been the result of a mantle plume. The timing of the formation of the Siberian Traps coincides with the Permo-Triassic extinction event—the largest extinction in the Earth's history (see pp.32–33). It has been argued that gases associated with such a massive eruption could have had enough effect on global climates to disrupt the food chain severely by damaging plant growth on a global scale.

ASIA *northeast*

## Kliuchevskoi

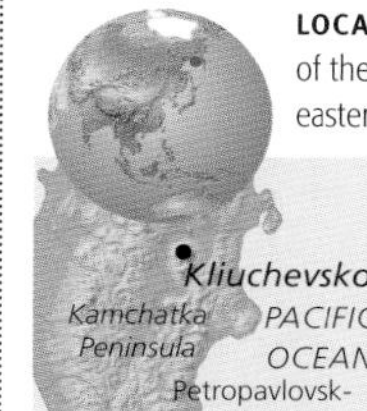


**LOCATION** Near the eastern coast of the Kamchatka Peninsula, eastern Siberia, Russia

**HEIGHT** 15,863 ft (4,835 m)

**TYPE** Stratovolcano

**LAST ERUPTION** 2003

Symmetrical, snow-capped Kliuchevskoi is the highest and most active of a chain of volcanoes along the eastern side of the Kamchatka Peninsula. Since its formation about 6,000 years ago, it has frequently produced explosive and effusive eruptions, without any extended periods of inactivity. Numerous cones and craters have been formed during the past 3,000 years by more than 100 flank eruptions. However, most recorded eruptions have originated from the 2,300-ft- (700-m-) wide summit crater, which is frequently reshaped. Many of Kliuchevskoi's eruptions have been viewed from space (see far right).

**SYMMETRICAL PEAK**
*Kliuchevskoi is a perfectly symmetrical stratovolcano formed by layers of pyroclastic materials and lava flows.*

**ERUPTION FROM SPACE**
*Photographed by astronauts on the space shuttle* Endeavour, *the eruption on September 30, 1994 sent a massive ash plume to a height of 11 miles (18 km).*

LAND

ASIA *east*

## Mount Fuji

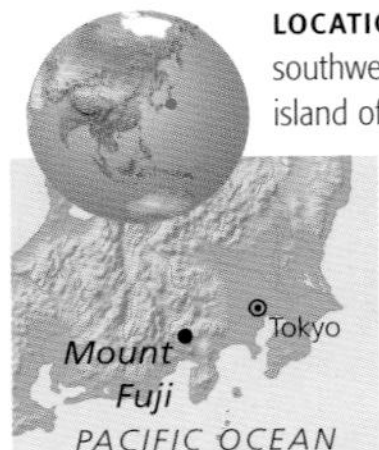

**LOCATION** About 55 miles (90 km) southwest of Tokyo, on the Japanese island of Honshu

**HEIGHT** 12,388 ft (3,776 m)

**TYPE** Stratovolcano

**LAST ERUPTION** 1708

Japan's highest mountain, Mount Fuji is a symbol of its homeland, and one of a chain of volcanoes along the western margin of the Pacific Ocean. It is a typical stratovolcano, with a steep-sided, symmetrical cone built up of layers of lava flows alternating with ash and other debris. Growth of the present volcano began 11,000 years ago on top of older volcanic remnants. Within 3,000 years, about 80 percent of the current bulk of the volcano had built up in outpourings of basaltic lava. During the long eruptive history that followed, lava flows and violent eruptions were emitted from the 2,300-ft- (700-m-) wide summit crater and numerous flank vents, building up more than 100 cones. Some of these lava flows blocked river drainage to the north, forming the Fuji Five Lakes area, which has become a popular tourist area. The last eruption in 1707 was the largest in historical times. It formed a large new crater on the east flank and deposited ash on Tokyo, over 55 miles (90 km) to the northeast.

**ICONIC BEAUTY**
*The picturesque symmetrical cone of Mount Fuji has been celebrated by hundreds of artists. On a clear day, the mountain can be seen from over 90 miles (150 km) away.*

### ANNUAL FESTIVAL

The Fujiyoshida Fire Festival, held at the end of August each year, is the highlight of Mount Fuji's tourist season. Shrines are dedicated to the goddess of Mount Fuji to give thanks for a safe climbing season. The mountain is the sacred epicenter of the country, and climbing it is treated by many as a religious experience.

ASIA *east*

## Baitoushan

**LOCATION** Straddling the mid-section of the border between China and North Korea

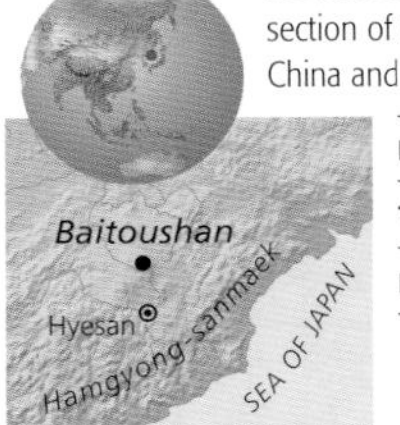


**HEIGHT** 9,003 ft (2,744 m)

**TYPE** Stratovolcano

**LAST ERUPTION** 1702

The Baitoushan (or Changbaishan) stratovolcano, 38 miles (60 km) wide, was built up over an older shield volcano. In around AD 1000, one of the largest eruptions in recorded history formed its summit caldera, 3 miles (5 km) wide and 2,790 ft (850 m) deep. This is now filled by Lake Tianchi, or Sky Lake. Tephra were deposited more than 1,000 miles (1,600 km) away in northern Japan. Only four eruptions have been recorded since the 15th century.

SKY LAKE IN THE BAITOUSHAN CALDERA

ASIA *east*

## Mount Unzen

**LOCATION** On the Shimabara Peninsula, to the east of Nagasaki on the Japanese Island of Kyushu

**HEIGHT** 4,921 ft (1,500 m)

**TYPE** Complex volcano

**LAST ERUPTION** 1996

The huge Unzen volcanic complex includes three large stratovolcanoes within a 25-mile- (40-km-) long rift valley. The Mayu-yama lava dome complex formed about 4,000 years ago and was the source of a massive debris avalanche on May 21, 1792. The avalanche generated a tsunami up to 180 ft (55 m) high that devastated the coastline and killed more than 14,500 people. Subsequent eruptive activity has been restricted to the summit and flanks of the Fugen-dake volcano. The most recent period of sustained volcanic activity, during 1990–95, was centered around the growth of a new summit lava dome. It has been estimated that during this time, more than 10,000 pyroclastic flows were generated by the dome collapsing. Many of these flows swept up to 3 miles (5 km) down the local river valleys, destroying hundreds of homes and causing many fatalities.

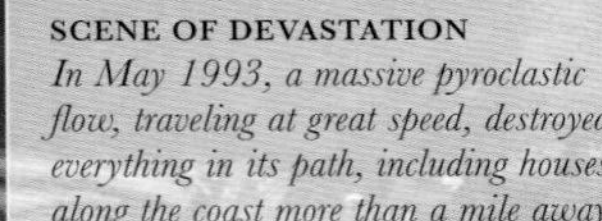

**SCENE OF DEVASTATION**
*In May 1993, a massive pyroclastic flow, traveling at great speed, destroyed everything in its path, including houses along the coast more than a mile away.*

### MAURICE AND KATIA KRAFFT

French volcanologists Maurice and Katia Krafft became famous for filming the hazardous nature of pyroclastic flows. Tragically, in 1991, while filming on Mount Unzen, they were killed by an incandescent ash flow that swept suddenly down the mountain.

ASIA *southeast*

## Mount Pinatubo

**LOCATION** On the west coast of the island of Luzon in the Philippines

**HEIGHT** 4,875 ft (1,486 m)

**TYPE** Stratovolcano

**LAST ERUPTION** 1993

In June 1991, Mount Pinatubo, a relatively unknown complex of lava domes, erupted in the second most violent volcanic explosion of the 20th century. A series of Plinian eruptions sent about 2 cubic miles (10 cubic km) of rock and ash 25 miles (40 km) into the atmosphere. Huge pyroclastic flows incinerated land up to 11 miles (17 km) from the volcano, and lahars destroyed villages up to 38 miles (60 km) away. The top 820 ft (250 m) of the mountain collapsed, leaving a summit caldera. Although monitoring had given enough warning for a timely evacuation of the local population, 300 people were killed directly by the eruption, partly due to the simultaneous arrival of Typhoon Yuna. As well as causing widespread disruption (see panel, right) the ash darkened the sky for a number of days. A month after the eruption, fine aerosol particles had circled the Earth, and had even reduced global temperatures. Lahars redistributing the deposits of the eruption continue to pose a threat to local populations.

### ESCAPING THE ASH

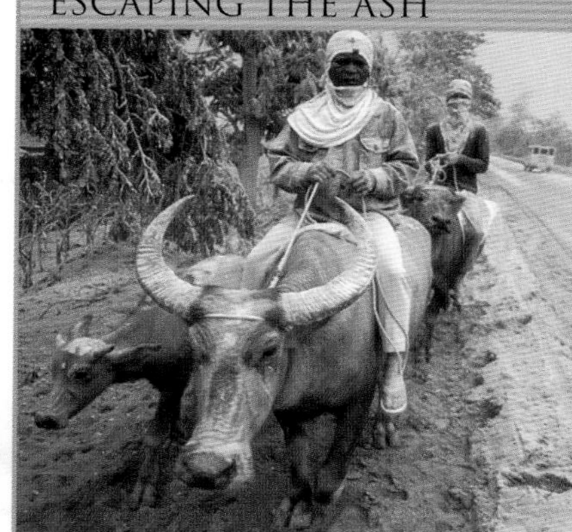

Volcanic ash causes severe disruption both on the ground and in the air. At least 14 airliners flew into emissions from the 1991 eruption and several had to make emergency landings with badly damaged engines. Fine ash can also cause health problems for local populations. The eruption filled the air and covered the ground with huge amounts of ash. A major increase in respiratory illnesses such as bronchitis was reported at the time, and some believe that ash contamination could still be killing people today.

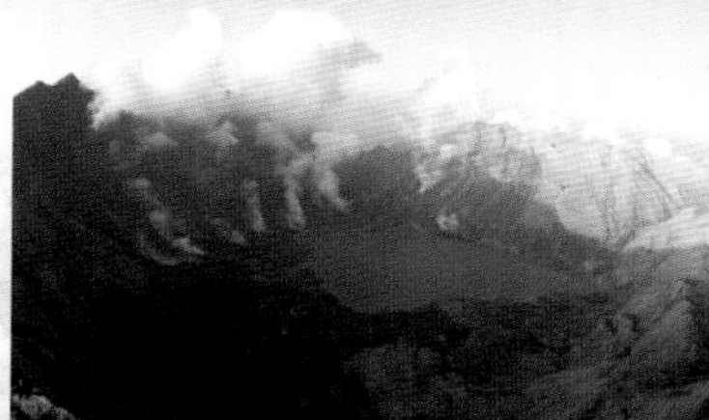

**CALDERA LAKE**
*The 1½-mile- (2.5-km-) wide summit caldera created by the 1991 eruption is partially filled by a lake.*

**RACING THE FLOW**
*A lucky individual makes a dramatic escape from a pyroclastic flow, traveling at speeds of more than 50 mph (80 km/h), released by the 1991 eruption.*

ASIA *southeast*

## Tambora

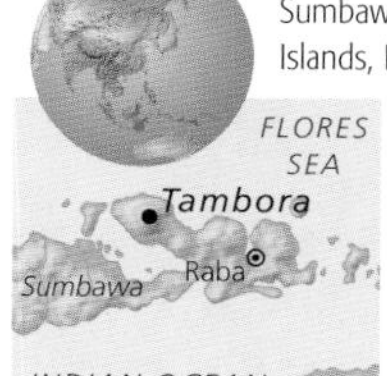


**LOCATION** On the island of Sumbawa, one of the Lesser Sunda Islands, Indonesia

**HEIGHT** 9,350 ft (2,850 m)

**TYPE** Stratovolcano

**LAST ERUPTION** About 1967

The 1815 eruption of Tambora was the largest in recorded history. It ejected more than 30 cubic miles (150 cubic km) of volcanic materials, producing a caldera 4 miles (6 km) wide and 4,100 ft (1,250 m) deep.

**MASSIVE CALDERA**

More than 10,000 people were killed by pyroclastic flows, and many thousands more were victims of subsequent famine and disease. Worldwide circulation of the dust and gas produced global climate change and a series of highly colored sunsets famously painted by J.M.W. Turner. The year 1816 was to be known as the "year without a summer."

ASIA *southeast*

## Taal

**LOCATION** South of Pinatubo on the west coast of the island of Luzon in the Philippines

**HEIGHT** 1,312 ft (400 m)

**TYPE** Caldera

**LAST ERUPTION** 1977

One of the most active and powerful volcanoes in the Philippines, Taal has erupted more than 30 times since records began in 1572. Its caldera, 12½ miles (20 km) wide, is filled by Lake Taal. The 3-mile- (5-km-) wide Volcano Island in the north of the lake has been the focus of all of the volcano's recorded eruptions. This complex of stratovolcanoes, tuff rings, and cones has produced massive pyroclastic flows and surges. A 1911 eruption killed 1,300 people, and 200 died in 1965. Since 1991, Taal has showed signs of increased activity.

**LAKE TAAL, THE FLOODED CALDERA**

ASIA *southeast*

## Merapi

**LOCATION** On the Indonesian Island of Java, just north of the city of Yogyakarta

**HEIGHT** 9,737 ft (2,968 m)

**TYPE** Stratovolcano

**LAST ERUPTION** 2003

Lying in one of the world's most densely populated areas, Merapi is one of Indonesia's most active volcanoes. The upper part of the volcano is unvegetated due to frequent eruptions of pyroclastic flows and lahars, which have accompanied the growth and collapse of the steep-sided summit lava dome. The dome occupies an unsupported position, at the western edge of the summit, which leaves it prone to collapse when it swells. The collapses release pyroclastic flows that have devastated cultivated lands and caused many fatalities throughout recorded history. Eruptions of Merapi frequently produce hot, pyroclastic flows, formed by gravitational collapse from lava flows and domes, and sometimes known as Merapi-type pyroclastic flows. Merapi was one of a dozen "decade" volcanoes chosen in 1991 for scientific monitoring as part of the United Nations International Decade for Natural Disaster Reduction.

**STEEP SLOPES**
*Merapi's upper flanks are partly bare of vegetation, due to frequent activity.*

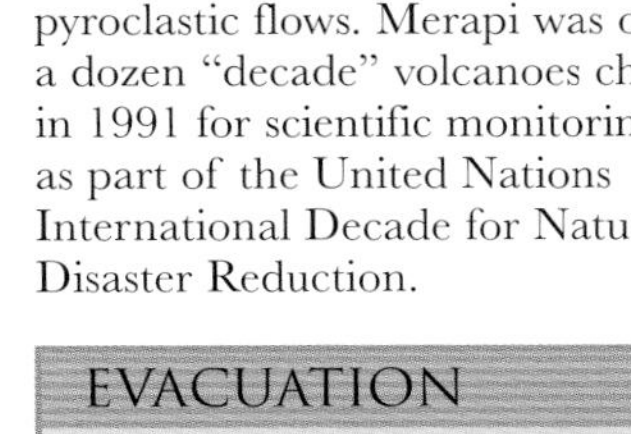

**LAVA DOME**
*A volcanologist examines an unstable lava dome as it emits steam and gas from a fumarole vent.*

### EVACUATION

Merapi's proximity to one of the world's most densely populated areas poses a constant threat. In such situations, advance warning of eruptions and effective crisis management are important.
In February 2001, volcanologists monitoring the volcano issued a maximum alert, and thousands of villagers, already affected by ash falls, had time to flee to safety.

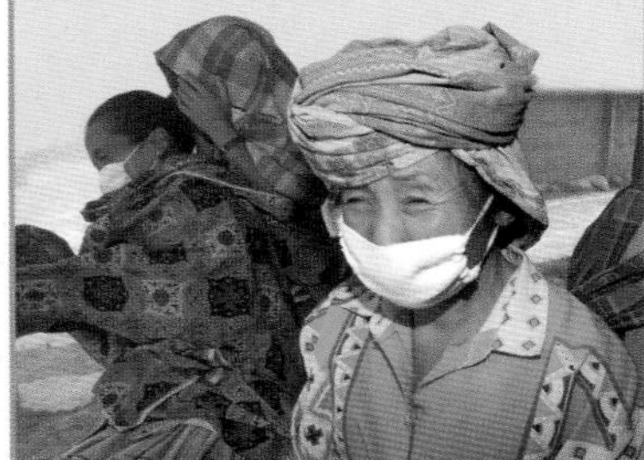

ASIA *southeast*

## Krakatau

**LOCATION** A volcanic island in the Sunda Strait between Sumatra and Java, Indonesia

**HEIGHT** 2,667 ft (813 m)

**TYPE** Caldera

**LAST ERUPTION** 2001

An eruption in around AD 416 caused the ancestral Krakatau volcano to collapse, and formed a 4-mile- (7-km-) wide caldera. The regrown volcano was dramatically transformed in 1883 by one of the most notorious volcanic eruptions in recorded history. Earthquakes during the late 1870s preceded a series of eruptions, culminating in a massive explosion on August 27, 1883 that was heard on Rodrigues Island, 2,890 miles (4,653 km) away in the Indian Ocean. An eruptive column rose 16 miles (25 km), showering ash and pumice over the region and blotting out sunlight for two days. Pyroclastic flows traveled up to 25 miles (40 km) across the sea and destroyed coastal communities in Sumatra. Situated in one of the busiest shipping lanes in the world, Krakatau's eruption was seen and logged by several vessels. The caldera collapse displaced enormous volumes of seawater and generated a series of tidal waves, or tsunamis, that devastated coastal settlements and killed more than 36,000 people. Two-thirds of the island was consumed in the collapse. Four decades later, a post-collapse cone, known as Anak Krakatau, appeared within the caldera.

**EARLY COLONIZER**
*As Krakatau's vegetation has been repeatedly eliminated, species like this morning glory provide excellent models of tropical vegetation succession.*

**CHILD OF KRAKATAU**
*Anak Krakatau ("Child of Krakatau") has frequently erupted since 1928. Since its formation, it has been rising out of the sea and continues to grow with increased activity.*

**STROMBOLIAN DISPLAY**
*Krakatau typically exhibits fairly mild Strombolian or Vulcanian eruptions. These produce basaltic lava flows and fountains or ash and lava bombs.*

LAND

ASIA *southeast*

## Rabaul

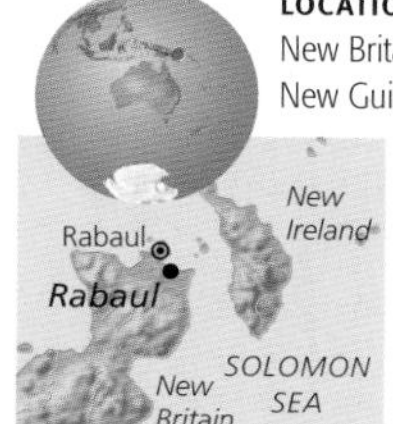


**LOCATION** On the east end of New Britain Island, Papua New Guinea

**HEIGHT** 2,257 ft (688 m)

**TYPE** Caldera

**LAST ERUPTION** 2003

Two major caldera-forming eruptions, about 7,100 and 1,400 years ago, formed the 9-mile- (14-km-) wide Rabaul caldera and Blanche Bay, a broad, sheltered harbor for the city of Rabaul. The outer flanks of the volcano are made up of pyroclastic-flow deposits. Three small volcanoes lie outside the caldera to the north and east. Several pyroclastic cones on the caldera floor have explosively erupted in recorded history, including the Vulcan cone, formed during a large eruption in 1878. The most recent period of unrest began in 1971 with frequent seismic activity. During 1984, up to several hundred tremors a day were recorded, and part of the harbor was elevated by over 3 ft (1 m). In 1994, the powerful and simultaneous eruption of the Vulcan and Tavurvur cones forced the temporary abandonment of Rabaul, the largest city in New Britain. Between 1995 and 2003, the volcano remained active with intermittent ash emissions.

**BILLOWING ASH CLOUD**
*During activity in 2000, Rabaul's Tavurvur cone erupted a cloud of ash and debris that traveled more than 1 mile (1.5 km) into the atmosphere.*

AUSTRALASIA *New Zealand*

## Ruapehu

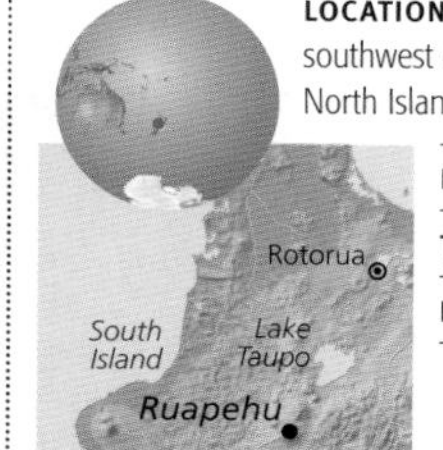


**LOCATION** About 25 miles (40 km) southwest of Lake Taupo on the North Island of New Zealand

**HEIGHT** 9,176 ft (2,797 m)

**TYPE** Stratovolcano

**LAST ERUPTION** 1997

Ruapehu is the tallest mountain on New Zealand's North Island, and one of its most active volcanoes. A 24-cubic-mile (100-cubic-km) volcanic complex, it is surrounded by a ring of volcanic debris produced by a series of eruptions between about 22,600 and 10,000 years ago. The active vent near its summit contains a crater lake that formed about 3,000 years ago. There have been about 50 eruptions since the mid-19th century. Explosive phreatic eruptions from this crater lake have frequently produced lahars, one of which destroyed a railroad bridge and train in 1953, killing 151 people.

**EXPLOSIVE OUTPUT**
*During an eruption in 1996, the snow-covered Ruapehu released huge clouds of ash that covered surrounding ski fields and disrupted air traffic.*

ANTARCTICA *central*

## Mount Erebus

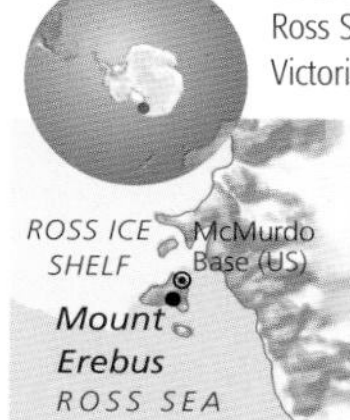


**LOCATION** On Ross Island in the Ross Sea, just off the Scott Coast, Victoria Land

**HEIGHT** 12,447 ft (3,794 m)

**TYPE** Stratovolcano

**LAST ERUPTION** 2003

Mount Erebus is the most southerly volcano on Earth to have been active in recorded history. Overlooking McMurdo Sound, it is one of three major volcanoes on Antarctica's Ross Island. Its summit has been modified by successive generations of caldera formation. A summit plateau at about 2,000 ft (3,200 m) marks the rim of the youngest caldera, within which the modern cone was constructed. The elliptical crater at the summit, which is 2,000 ft (600 m) wide and 360 ft (110 m) deep, is unusual in having a small but persistent lava lake on its floor. The lava is also of unusual alkaline composition. High-temperature anomalies associated with the lava lake can even be detected from space by satellites. The glacier-clad volcano was erupting when first sighted by Captain James Ross in 1841 (see panel, right), and Strombolian activity was also seen in December 1912, when the volcano was visited by members of Scott's Antarctic expedition. More recently, fluxes in the emission of sulfur dioxide from the volcano have been measured as part of a global program to try to correlate such gas emissions with the eruptive process. Continuous lava-lake activity with minor explosions, punctuated by occasional larger Strombolian explosions that eject bombs onto the crater rim, have been documented since 1972.

**INSIDE EREBUS ICE**
*A rappeler descends into an ice cave near the crater of Mount Erebus. These caves were formed by volcanic heat rising to the surface from below.*

### JAMES CLARK ROSS

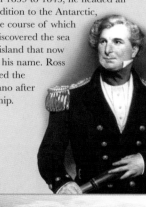

In 1831 the British naval explorer James Clark Ross (1800–62) discovered the magnetic north pole while on an expedition in the Arctic. From 1839 to 1843, he headed an expedition to the Antarctic, in the course of which he discovered the sea and island that now bear his name. Ross named the volcano after his ship.

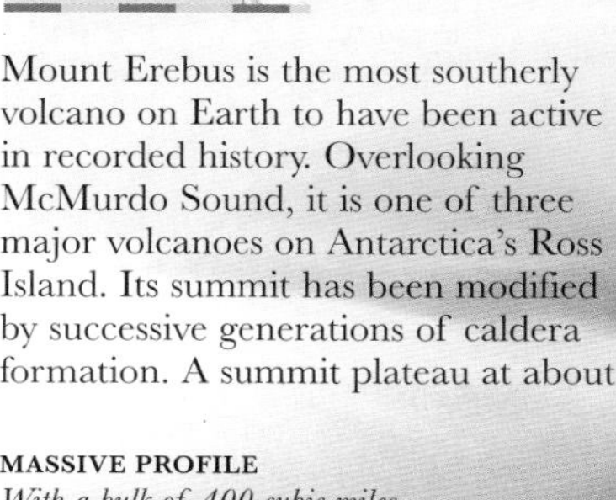

**MASSIVE PROFILE**
*With a bulk of 400 cubic miles (1,700 cubic km), Mount Erebus is one of the world's most impressive volcanoes.*

# IGNEOUS INTRUSIONS

56–57 The core and mantle
58–59 The crust
63 Ores
78–81 Rocks
82–91 Igneous and metamorphic rocks
172–75 Volcanoes
Deserts 284–87

AN IGNEOUS INTRUSION FORMS when magma—molten or partially molten rock—invades cracks in existing rock or stagnates within the Earth's crust and solidifies to form new rock. Magma is generated in the crust and upper mantle, and because it is hot and contains water and gases, it migrates upward. A small proportion erupts from volcanoes at the surface (see p.173). However, most magma cools and crystallizes beneath the surface to create igneous intrusions. Because they are formed underground, these rock formations, which include batholiths, dikes, and sills, are seen as features of the landscape only after erosion has stripped away the rocks that cover them.

## INTRUDING MAGMA

Igneous intrusions form when cooling, crystallization, and loss of gases cause a once fluid and buoyant magma to solidify. Most rocks that form intrusions originate as either granitic or basaltic magma. Granitic magmas have a low density, and small volumes (a few miles across) rise slowly through the Earth's upper crust, pushing existing rock (termed country rock) aside. However, very large granite masses, which are perhaps 1,000 times larger, are formed when magma digests and incorporates heated country rock. This surrounding rock begins to melt, and elements of it are assimilated into the magma, which then sets. Such processes generally take place deep within the Earth's crust, and granite intrusions are typically found in areas of active mountain building. Basaltic magmas are less dense and more fluid. They inject into cracks and zones of weakness in country rock, and they are most commonly seen as dikes or sills (see opposite page). These intrusions have the same composition as extruded basaltic lava, but contain larger crystals. The crystals within basaltic magma can settle at different rates to form distinctive layers of rock that have different abundances of minerals.

**INTRUSIVE CONTACT**
*Here, dark, sedimentary country rock has been assimilated into an intruding magma, which has crystallized to form this pink granite.*

**PEGMATITE**
*The very large crystals in this pegmatite (see p.83) in Utah were formed by an intruding granite magma that was rich in water and gases.*

**LAYERED INTRUSION**
*These igneous rocks with differing compositions were created in layers as dark, heavy crystals settled and cooled in still-molten magma.*

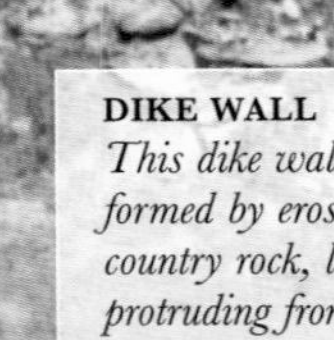

**DIKE WALL**
*This dike wall in Colorado has been formed by erosion of the surrounding country rock, leaving strong igneous rock protruding from the ground.*

## CRYSTAL STRUCTURE

All igneous rocks are formed by cooling of magma, so their primary characteristic is a mosaic of interlocking crystals. The tight jigsaw-puzzle pattern of crystal boundaries gives igneous rocks their high strength. Crystal size depends largely on the rate of cooling. Intrusive magmas lose heat slowly into the surrounding rocks, so they have larger crystals than extrusive rocks, which cool rapidly in air or water. The minerals of igneous rocks are mostly silicates (see pp.72–77); in this microscopic view of granite, the feldspars and quartz are gray, while the biotite appears as several different strong colors.

# INTRUSIVE ROCK BODIES

Different types of igneous intrusions can be named according to their shape, which is related to the type of rock that forms them. Granitic magmas are usually viscous, so they tend to form huge batholiths that have an area of at least 40 square miles (100 square kilometers). Basaltic magmas are more fluid, so they flow into thin gaps and cool vertically as dikes, which run through the rock strata, or they set horizontally as sills along bedding planes (the boundaries between layers of sedimentary rock). Because rock cools more slowly underground than it does on the surface, the rocks that form igneous intrusions have larger crystals than those formed when lava solidifies at the surface. Those that form dikes are medium-grained, and are usually diabase (see p.83), while those that form larger intrusive bodies, such as batholiths and stocks, are coarse-grained gabbro or granite (see pp.82–83). Ring dikes, which are shaped like vertical tubes, are formed as magma fills cracks created by its upward pressure. Cylindrical plugs are formed from magma that has cooled inside volcanic vents. Laccoliths are large, lens-shaped intrusions that are usually formed of gabbro. Lopoliths are lenses and sheets that are also primarily gabbro, and are found in basins that have sagged under the weight of the intrusion.

**TYPES OF IGNEOUS INTRUSIONS**

*Most igneous intrusions are classified by their size and shape. The intrusions shown here range from relatively small horizontal sills and vertical dikes, to conical ring dikes, volcanic plugs, lens-shaped laccoliths, and huge batholiths, which can be more than 60 miles (100 km) in length.*

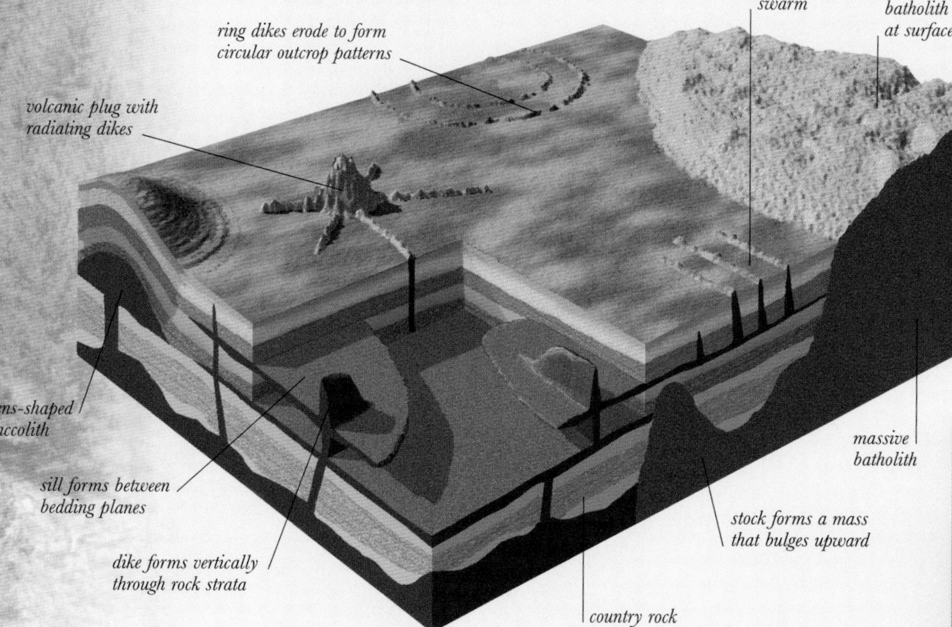

**SIERRA NEVADA, NORTH AMERICA**

*Almost the entire Sierra Nevada range is a massive batholith, which was formed by multiple phases of granitic magma intrusion. This cooled beneath the surface of the Earth before being exposed by subsequent erosion.*

**SILL AND DIKE**

*The pale, banded, metamorphosed sedimentary rocks in this hillside exposure in Oregon are cut by two small intrusions of darker diabase. One is a dike that has filled a tension fissure cutting across the beds (bottom); the other is a sill that has injected along a single bedding plane (top).*

**EXPOSED DIKE**

*The typically hard rock of an igneous intrusion forms the core of this rocky crag in Colorado. Debris eroded from the margins forms the more gentle slopes, which contrast with the sharp profile of the intrusion.*

## ECONOMIC IMPORTANCE

Intrusive igneous rocks are so strong that they are widely quarried to make hard rock aggregate, which is required in huge quantities for use in the manufacture of concrete and to build roads. Some types of granite are also highly valued as large blocks of cut rock, known as dimension stone. Strong, unfractured, and light in color, granite is sawn and polished to make thin slabs of glossy stone for covering building exteriors. The granite in this Sardinian quarry is cut out by powerful water jets and traditional wire saws, without using explosives.

## IGNEOUS INTRUSION PROFILES

The pages that follow contain profiles of some of the world's igneous intrusions. Each profile begins with the following summary information:

**AGE** When the igneous intrusion was formed

**TYPE** Batholith, dike, dike swarm, ring dike, sill, plug, stock, layered intrusion

**ROCK** The main rock type that makes up the intrusion

ARCTIC

## Skaergaard

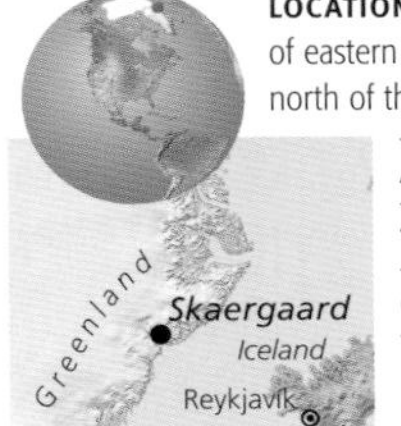


**LOCATION** In the coastal mountains of eastern Greenland, just to the north of the Arctic Circle

**AGE** 60 million years

**TYPE** Layered intrusion

**COMPOSITION** Gabbro

Dissected by deep, glaciated valleys and fjords, and exposed in the soil-free mountains of Greenland, the Skaergaard intrusion is the world's most intensively studied large igneous body. Where it is exposed at the surface, the intrusion is about 6 miles (10 km) across. It is notable for its vast sequence of layered rocks, which were formed by differentiation of a mass of basaltic magma. Crystals of dense, silicate minerals sank through the liquid remnants of the original magma, but were interrupted by convection currents as they did so. When this happened, the magma composition changed, with bands of rock forming from different minerals as the layers accumulated and cooled progressively. The dominant gabbro layers near the base of the intrusion contrast with the small layer of granophyre (see p.82), which was the last rock to crystallize at the end of the batholith's molten life.

**CLEAN ROCK**
*The Arctic conditions lead to frost shattering (see p.110), which leaves bare outcrops of igneous rock. This makes Skaergaard the perfect place to study magmas and rocks.*

**COLOR-CODED**
*Color changes identify the mineral layers within the Skaergaard gabbro, and lumps of coarse, pegmatitic material are remnants of olivine-rich layers that fragmented before they solidified.*

NORTH AMERICA *northwest*

## Mackenzie Dikes

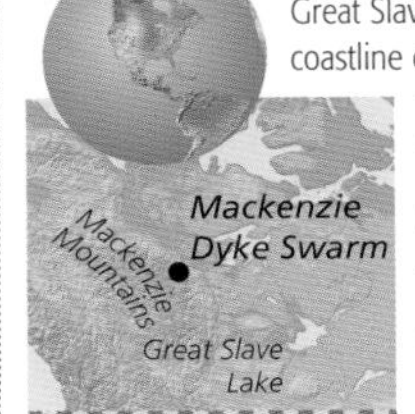


**LOCATION** Stretching between Great Slave Lake and the Arctic coastline of northern Canada

**AGE** 1,260 million years

**TYPE** Dike swarm

**COMPOSITION** Diabase

Hundreds of parallel intrusions form the largest dike swarm in the world, which is exposed in the ice-scoured lowlands of the Canadian Shield. The dikes formed when magma intruded into fissures that opened along a zone of tension above an active mantle plume. Flood basalts formed where huge volumes of magma spilled onto the surface, and layered-gabbro intrusions have been detected at depth. The result is a trinity of forms that record the formation of new crust in the Earth's distant past.

**MACKENZIE DIKE SWARM**

**HUDSON RIVER VALLEY**
*At the top of the sill is a colonnade of diabase columns. Trees grow on the weathered fragments of these structures, which have fallen to the base of the cliff.*

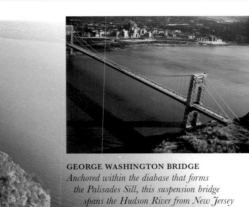

**GEORGE WASHINGTON BRIDGE**
*Anchored within the diabase that forms the Palisades Sill, this suspension bridge spans the Hudson River from New Jersey to Manhattan in New York City.*

NORTH AMERICA *northeast*

## Palisades Sill

**LOCATION** Formed along the New Jersey side of the Hudson River, northward from New York City

**AGE** 190 million years

**TYPE** Sill

**COMPOSITION** Diabase

For 50 miles (80 km), the Hudson River flows beneath the dark wall of the Palisades Sill. Nearly horizontal, this dramatic outcrop is a sheet of intrusive diabase. The sill is about 1,100 ft (330 m) thick, and intruded into now-eroded Triassic sandstones at shallow depth. Where the base of the sill is exposed, there is about 50 ft (15 m) of fine-grained diabase that chilled rapidly against the contact rock. Above this is a 18-ft- (5-m-) thick layer that is predominantly olivine (see p.72). This was the first mineral to crystallize, and its dense crystals sank through magma that was still 95 percent liquid. The Palisades Sill is a clear example of the differentiation of magma, with rocks of varying compositions created from the same source.

LAND

NORTH AMERICA *west*

# Devil's Tower

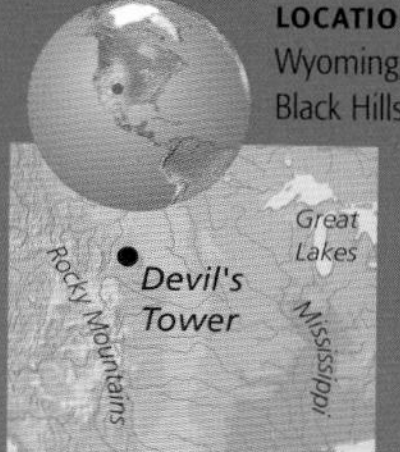


**LOCATION** In the Great Plains of Wyoming, just northwest of the Black Hills of Dakota

**AGE** 40 million years

**TYPE** Plug

**COMPOSITION** Phonolite

The Devil's Tower originated as a volcanic vent, but it owes its spectacular appearance to erosion, which has removed the surrounding sedimentary rock to expose igneous rocks that cooled and solidified underground. The original volcano lost its gas pressure by exploding to the surface through a cover of sediment, but much of the magma remained in an underground plug that was nearly 1,000 ft (300 m) in diameter. As it cooled, the magma contracted and fractured, forming polygonal joints. When the cooling fronts moved inward, these fractures evolved into remarkably uniform columns. (A similar phenomenon occurs when mud dries to form polygonal cracks.) The tower reaches a height of 867 ft (264 m). Horizontal columns also intruded from the edge of the plug, but never extended far. Most of this fringe of horizontal columns has been eroded, though where it is not masked by debris, some can be seen in the flared base of the tower. The phonolite that forms the Devil's Tower is derived from continental crust; it contains less silica than the other rhyolitic magmas, and is distinguished by its small crystals of dark green aegirine, a pyroxene mineral that is rich in sodium.

**SUMMIT OF THE TOWER**
*The domed peak of the intrusion exposes the tops of the columns that form the mass of the Devil's Tower.*

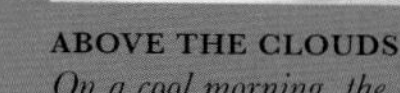

**ABOVE THE CLOUDS**
*On a cool morning, the Devil's Tower rises through a layer of cloud to mimic the situation about 15 million years ago, when much less of the intrusion was projecting above the surrounding land.*

## ROCK CLIMBING

The splendid columnar joints of the Devil's Tower intrusion provide perfect routes for the serious climber who enjoys long crack lines in very strong rock. However, natural cooling is never perfect, and some columns do merge or split, so climbers have to choose which joints they pursue with care.

**WYOMING'S WONDER**
*The striking structure of the Devil's Tower has distinguished it as a sacred place for some American Indians. It appeared in the movie* Close Encounters of the Third Kind.

**THE WINGED ROCK**
*Though the Navajo name suggests that Ship Rock flew to its present site, the geological wings are the remarkable dikes that now stand high as walls across the desert of New Mexico.*

NORTH AMERICA *southwest*

## Ship Rock

**LOCATION** In the desert of New Mexico, west of the San Juan Mountains

**AGE** 30 million years

**TYPE** Plug

**COMPOSITION** Lamprophyre

Though it now stands 1,700 ft (500 m) above the surrounding plains, Ship Rock formed underground when magma cooled within the feeder pipes of a volcano. The lavas and pyroclastic deposits of the volcano have since disappeared, and subsequent erosion of the underlying soft shales has steadily lowered the surface of the surrounding plains. The resistant rock of the plug is far less worn, and stands high above the desert. The upper part of Ship Rock is a lava breccia that was created when rock was shattered by explosive eruptions within the volcano's vent. When this happened, the rocks that constitute today's summit were probably less than 3,300 ft (1,000 m) below the Earth's surface. The lamprophyre that forms the intrusion is similar to diabase, except that it contains more mica and less feldspar. It also contains small masses (xenoliths) of peridotite and eclogite. This suggests that the magma had a source deep within the upper mantle. In addition to feeding the main conduit, magma also ascended into radiating fissures, where it cooled to form the dikes that distinguish Ship Rock. These were exposed as the adjacent surface eroded, resulting in spectacular dike walls. The rock ribs are about 10 ft (3 m) thick and stand 65 ft (20 m) high, and the longest dike stretches for nearly 2 miles (3 km).

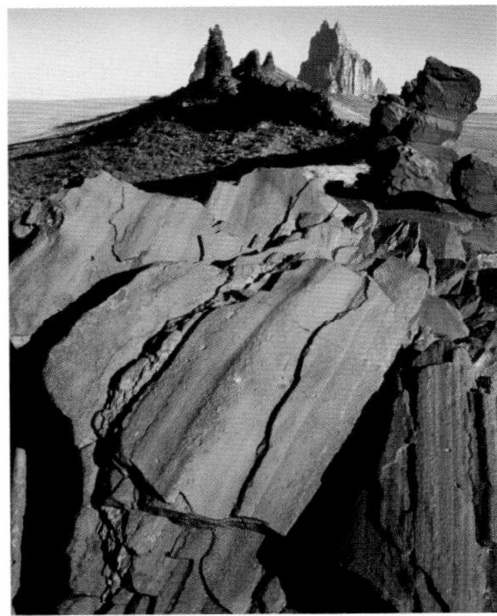

**VOLCANIC REMNANTS**
*The rock exposures in the desert around Ship Rock are the volcanic remnants of intrusions that were once buried deep beneath the surface of the Earth.*

NORTH AMERICA *west*

## El Capitan and Half Dome

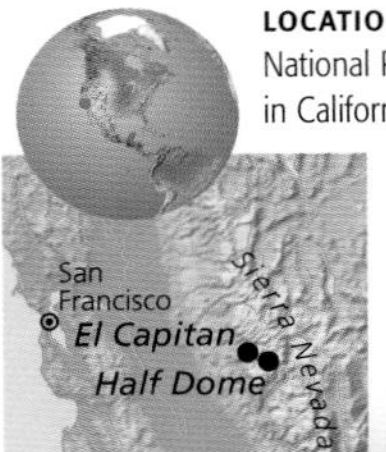


**LOCATION** Within Yosemite National Park, in the Sierra Nevada in California

**AGE** 82 million years

**TYPE** Batholith

**COMPOSITION** Granite and granodiorite

Known as the Incomparable Valley, Yosemite cuts deep into the largest and strongest of the Sierra Nevada granitic intrusions, and is bounded by the huge, steep vertical cliffs that make it so famous. Like all batholiths, the Sierra Nevada is not formed from a single type of granite. It consists of a series of intrusions that were emplaced over 50 million years during the Cretaceous Period, when the region sat above an active convergent plate boundary. Two of the batholith's most distinctive outcrops are El Capitan and Half Dome. The composition of the intrusions varies, since different magmas invaded different areas of country rock. El Capitan—Yosemite's highest rock wall—is formed of a true granite, but Half Dome—the area's steepest wall—is formed of granodiorite, which contains less alkali feldspar. Both are massive rocks with few weaknesses, as is typical of the largest batholiths.

**EL CAPITAN**
*Towering above Lower Yosemite Valley, this rock wall is a result of erosion, which has shaped this part of the huge Sierra Nevada batholith.*

**HALF DOME**
*Yosemite's most famous landmark is the rounded end of a granite ridge between two deep valleys. One side of Half Dome was broken off at a major joint during glacial sculpting of the region.*

SOUTH AMERICA *south*

## Chuquicamata

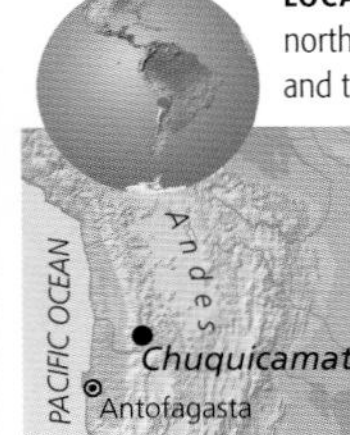


**LOCATION** In the Atacama Desert of northern Chile, between the Andes and the Pacific Ocean

**AGE** 33 million years

**TYPE** Stock

**COMPOSITION** Granodiorite

**DESERT COPPER MINE**

The porphyritic granodiorite intrusion at Chuquicamata is associated with the largest of several porphyry copper ore bodies that are found throughout the western mountains of North and South America. Metal sulfides are dispersed throughout the stock, and half a million tons of copper are mined every year from over 4 billion tons of ore. Mining has created an open pit that is over 2½ miles (4 km) long and 2,650 ft (810 m) deep.

SOUTH AMERICA *east*

## Sugar Loaf

**LOCATION** Overlooking the entrance to Guanabara Bay, Rio de Janeiro, Brazil

**AGE** 800 million years

**TYPE** Batholith

**COMPOSITION** Granite

The Sugar Loaf is solid granite, and is actually a tiny part of a huge batholith that stretches along the Brazilian coast on either side of Guanabara Bay. It was formed by partial melting within a belt of metamorphic gneisses (see p.90) at the center of a plate convergence zone that was active about 800 million years ago in late Precambrian times. Since then, erosion has left stumps of granite that form the magnificent peaks enclosing Rio de Janeiro. Sugar Loaf is not the highest promontory in the area (it rises to only 1,295 ft/395 m), but its rounded profile makes it distinctive. The rounded outline is typical of exposed masses of strong, homogenous, joint-free granite that were once at the core of a massive batholith. When intrusive granite crystallizes, it is confined by horizontal and vertical stress, due to the weight of the rock above it. Once exposed, these stresses are lost, and the rock relaxes, opening curved joints as it does so. Erosion strips away these shells of rock, each time exposing a smoothly curved surface.

### TOURISM

Rock that is almost white, bold curves, and towering cliffs are the characteristic features of mountains cut into the granites that form large batholiths. The resulting landscapes are often tourist attractions. Few are finer than Rio de Janeiro's two dramatic granite peaks, the Sugar Loaf and Corcovado, which can be ascended by cable car or tram.

**BRAZIL'S GRANITE ICON**
*The smooth, rounded structures of the Sugar Loaf and the peninsula that extends to it are both formed from the huge mass of granite at the core of a batholith.*



EUROPE *north*

## Ardnamurchan

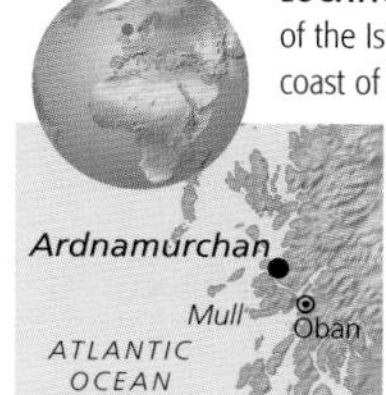


**LOCATION** A peninsula to the north of the Isle of Mull on the west coast of Scotland

**AGE** 60 million years

**TYPE** Ring dike complex

**COMPOSITION** Gabbro

A complex of gabbroic ring dikes forms the Point of Ardnamurchan, which is about 5 miles (8 km) across and is the most westerly point of the British mainland. A decline in pressure over a large magma chamber caused conical fractures to form. Magma intruded into these fissures to form dikes. Subsequent cone-shaped sheet intrusions then formed when magma pressure increased. Above these deep intrusions, huge volcanic craters erupted lavas and pyroclastic deposits.

**POINT OF ARDNAMURCHAN**

EUROPE *north*

## Castle Rock

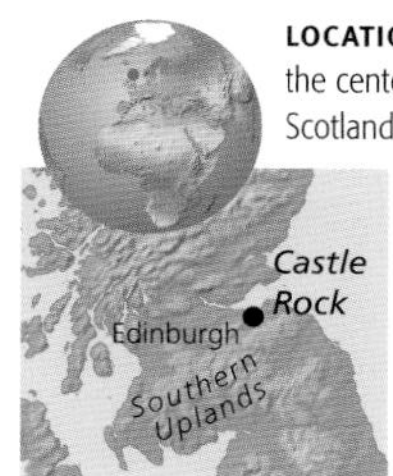


**LOCATION** Beneath the castle in the center of the city of Edinburgh, Scotland

**AGE** 325 million years

**TYPE** Plug

**COMPOSITION** Diabase

Edinburgh's great crag of black diabase, which is now crowned by a castle and walled courtyard, is the exposed feeder pipe of an ancient volcano. The entire volcanic edifice has long been lost to erosion, and its only remnant is the basaltic magma that solidified deep beneath the eruption crater. The magma cooled more slowly than the extruded, air-cooled basaltic lavas and so formed large crystals, which are typical of diabase. About one million years ago, during the last ice age, ice sheets swept over Castle Rock; however, the tough diabase plug remained in place. Ice flowed around the crag and deposited till debris (see p.115) in its unscoured lee. This resulted in the formation of the long eastern rampart that now supports the High Street of Edinburgh's Old Town. The modern landform is a classic example of a feature known as a crag-and-tail.

**EDINBURGH'S ROYAL ROCK**
*The hard black diabase of Castle Rock's intrusive plug forms a crag far steeper than the ancient volcano that once existed here.*

EUROPE *north*

## Fair Head Sill

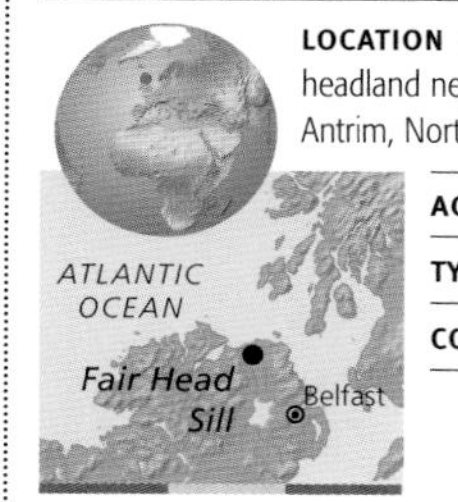


**LOCATION** Forming a coastal headland near Ballycastle, County Antrim, Northern Ireland

**AGE** 58 million years

**TYPE** Sill

**COMPOSITION** Diabase

**FAIR HEAD SILL**

The great prow of Fair Head is the exposed edge of a 260-ft- (80-m-) thick diabase sill, which intruded into Carboniferous sandstones as lavas poured across the rest of the region. The sill is scored by columnar cooling fractures, and is notable for its huge sandstone xenoliths.

EUROPE *north*

## Whin Sill

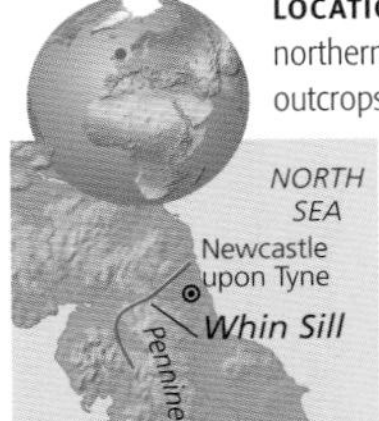


**LOCATION** Underlying England's northern Pennine hills, with outcrops at the edges

**AGE** 295 million years

**TYPE** Sill

**COMPOSITION** Diabase

Whin is an Old English quarryman's word for hard black stone. This aptly describes the diabase that forms Whin Sill. In the late Carboniferous Period, basaltic magma rose from beneath the Earth's crust and spread to intrude along shallow bedding planes in sedimentary rocks. The intrusion is complex; it cuts across the bedding to form sheets at multiple levels and appears to have been fed by more than one magma chamber. Whin Sill is therefore not a single sill, but actually a collection of cross-cutting sills and flat, lens-shaped intrusions. Nevertheless, at many outcrops it is a uniform and unbroken sheet of diabase that is approximately 100 ft (30 m) thick.

**COLONNADED ESCARPMENT**
*Rough columnar jointing in the diabase adds texture to the long escarpment cliffs wherever Whin Sill emerges above the surface.*

### HADRIAN'S WALL

An escarpment that extends across the Pennines marks the northern edge of Whin Sill. It was used as a natural line of defense in AD 122, when the Roman emperor Hadrian built a wall to keep the Scots out of the Roman province of Britannia.

AFRICA *south*

## Great Dike

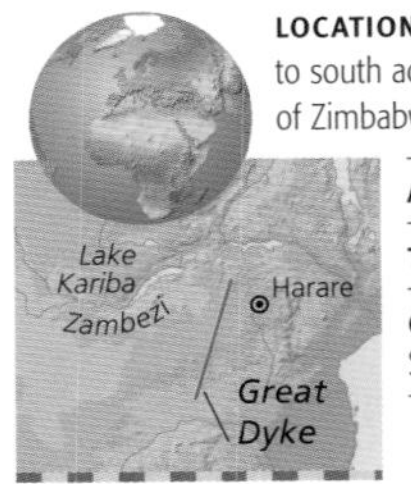


**LOCATION** Stretching from north to south across almost the whole of Zimbabwe

**AGE** 2,600 million years

**TYPE** Layered intrusion

**COMPOSITION** Serpentinite

Called a dike because it is long and thin at the surface, most of this intrusion is about 3 miles (5 km) wide, and so is too large to be a true dike. It consists mainly of peridotite (see p.84) and dunite, which have altered to become serpentinite (see p.87), though its upper layers consist of gabbro. Within the intrusion, crystal settling has created chromite bands that are the world's largest source of chromium ore. The mineral bands show a synclinal structure (see p.154), suggesting that the igneous body is actually a lopolith. However, its base is hidden, and its margins are both major faults, so Great Dike may be best classified as a layered intrusion.

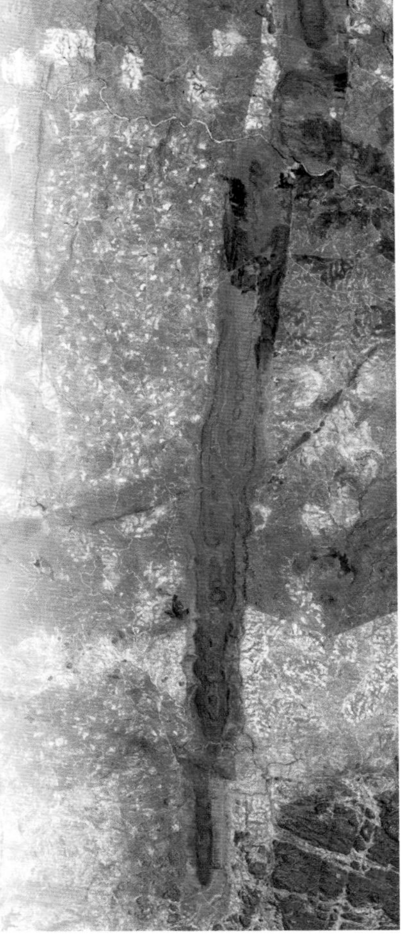

**LINEAR INTRUSION**
*This satellite image shows the long, thin range of wooded hills that marks the line of the Great Dike across the plains of Zimbabwe.*



AFRICA *west*

## Aïr Mountains

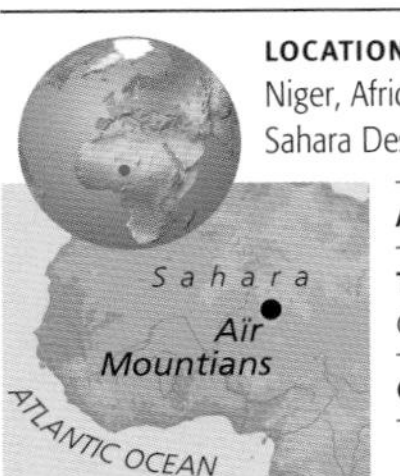


**LOCATION** In the northern part of Niger, Africa, within the southern Sahara Desert

**AGE** 410 million years

**TYPE** Ring dike complexes

**COMPOSITION** Granite

The major ring dike complexes of the Aïr Mountains were formed where three continental plates collided. Each complex is about 37 miles (60 km) in diameter and contains one or more ring dikes that are 65–650 ft (20–200 m) thick. Before the plates welded together to form the huge basement block of northern Africa, their boundaries were subduction zones (see p.108–09), melting rock and generating magma. Around the junction, granitic magmas reached the surface, creating huge explosive volcanoes, whose lavas and pyroclastic deposits have since eroded. Further eruptive phases caused declines in the underground magma pressure, resulting in large caldera collapses. Beneath the surface, gigantic blocks of rock subsided inside huge conical ring fractures. Then more magma rose through these new fractures. Some of the new magma fed younger volcanoes, but the magma that solidified within the fissures created ring dikes. Erosion has brought the modern surface down to a level that cuts through the ancient intrusive rings.

**DESERT DIKES, AÏR MASSIF**
*Barren rock crags within a sea of sand dunes are the eroded stumps of old dikes and plugs in the Sahara's Aïr Mountains.*

### MINING URANIUM

Granitic magmas contain traces of uranium. When erosion exposed the Aïr dikes, mountain streams carried the uranium as soluble oxides into sedimentary basins. There, low-oxygen conditions in nearly stagnant lakes caused precipitation of the uranium. Today it is mined from the exposed basins in Niger's desert, but its original source was the Aïr intrusions.

ANTARCTICA *north*

## Salvesen Mountains

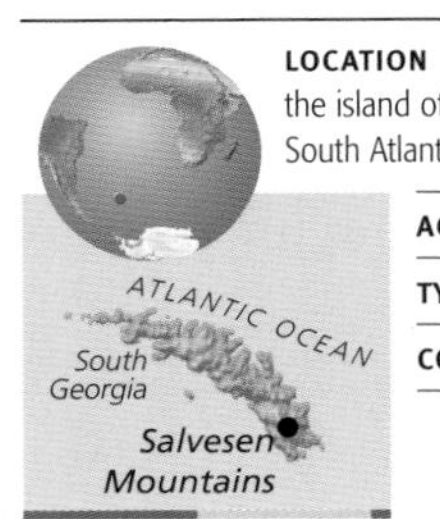


**LOCATION** At the southern tip of the island of South Georgia, in the South Atlantic

**AGE** 127 million years

**TYPE** Stock

**COMPOSITION** Granite

The roots of the island of South Georgia are intrusive igneous rocks, though they form only a small part of the modern outcrops. Stocks of white Cretaceous granite are intruded into Jurassic basaltic lavas and diabase dikes, both of which are black, so they create striking color contrasts in exposures across the Salvesen Mountains, the most jagged of the island's icebound mountain ranges. Both the granite and the basalt are remnants of rising magma formed on a divergent plate boundary where the southern Atlantic opened. The main ranges of South Georgia, famously crossed by Ernest Shackleton in 1916, are less rugged and precipitous than the Salvesen Mountains, since they are made of folded sandstones. These were formed from sand deposition, the sediment for which was derived from erosion of the igneous rocks and rifting continental blocks.

# HOT SPRINGS AND GEYSERS

52 Heat transfer
63 Sulfur
128–29 Water
170–91 Volcanoes
Geothermal energy 208–209
Ocean tectonics 386–87

HOT SPRINGS, SPOUTING GEYSERS, fumaroles, boiling mud pots, and many mineral deposits are all evidence of the Earth's internal heat. They all depend upon the underground heating of water by hot rocks. Once hot water and steam have been generated, they are forced to the surface, often charged with dissolved minerals and gas. Hot springs produce warm water steadily, while geysers release intermittent bursts, often with copious amounts of steam. Fumaroles are vents where only steam and other gases escape.

## GEOTHERMAL SYSTEMS

The flow of internally generated heat up to the Earth's surface is almost constant over the globe, but in some places the heat flow is much higher and produces hot springs and associated geothermal phenomena. These zones of high heat-flow tend to occur at the margins of the Earth's tectonic plates and are normally associated with volcanism. Many active volcanoes are underlain by reservoirs of partially molten rock, centers of heat at abnormally shallow depths. Any groundwater coming into contact with these hot rocks is heated, but because it is under pressure at depth, the boiling point is raised above normal and the water becomes superheated. Heated water tends to rise toward the surface, losing pressure as it does so. Flashing over into steam may then propel the water upward with such force that it gushes forth above ground as a geyser.

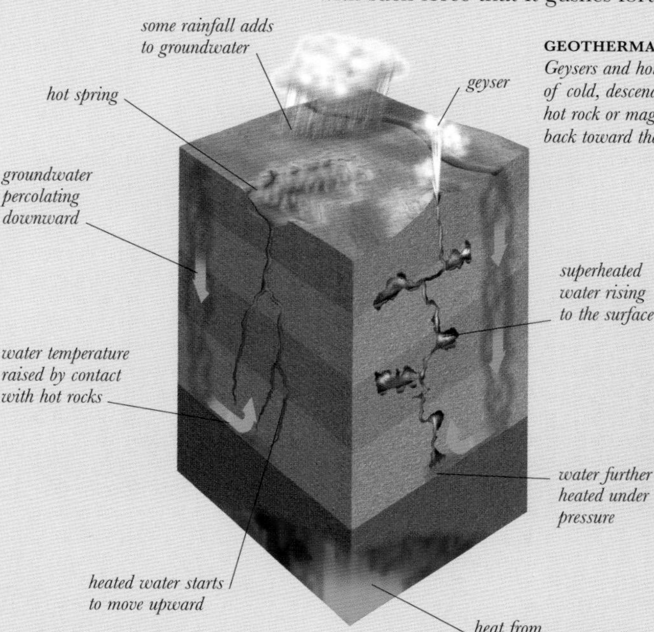


**GEOTHERMAL SYSTEMS**
*Geysers and hot springs are generated by the cycle of cold, descending surface water being heated by hot rock or magma, expanding, and being driven back toward the surface.*

**USING GEOTHERMAL ENERGY**
*Many hot springs, such as these in Kusatsu, Japan, are harnessed for heat, power, and domestic uses.*

## WATER DEPOSITS

Much of the water that feeds hot springs and geysers originates as rainwater, infiltrating below ground through joints or cracks to the local water table. Where groundwater is heated by a geothermal source, it rises, drawing in additional cold groundwater toward the heat source as it does so. While underground, the heated water dissolves soluble minerals from the rocks, but, on reaching the surface, evaporation and cooling cause some of these dissolved minerals to be re-precipitated. The resulting deposits are dominated by silica, calcium, sodium, and potassium carbonates and chlorides, but include other minerals and metals, such as copper and lead, depending upon the local rocks.

**WHITE DOME GEYSER**
*Hot underground rock turns groundwater to steam and boiling water, which spurts out of this geyser in Yellowstone National Park.*

**BUBBLING CAULDRON**
*The constantly boiling waters of Punchbowl Spring, Yellowstone, are contained within a raised rim of carbonate deposits (travertine), about 2½ ft (75 cm) high, which is shaped like a punch bowl.*

**SULPHUR CRYSTALS**
*Hot gases expelled from volcanic fumaroles often leave deposits, such as these sulphur crystals on Vulcano, in the Aeolian Islands.*

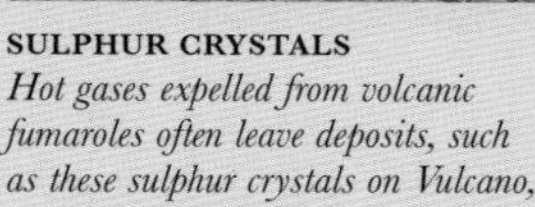

### HOT SPRINGS AND GEYSERS PROFILES

The pages that follow contain profiles of a selection of the world's hot springs and geysers. Each profile begins with the following summary information:

**TYPE** Hot spring, mud pool, geyser, or fumarole

**FREQUENCY** Interval between periods of activity

NORTH AMERICA *northwest*

# Yellowstone

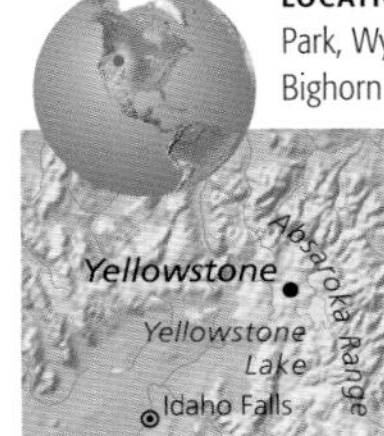


**LOCATION** In Yellowstone National Park, Wyoming, between the Bighorn Basin and Snake River Plain

**TYPE** Geysers, hot springs

**FREQUENCY** **Old Faithful** 45–110 min, **Steamboat Geyser** irregular

Yellowstone displays some of the most spectacular hydrothermal activity in the world, especially from its dramatic geysers (about 200 in number), of which Old Faithful is the most famous. There are also thousands of fumaroles, hot springs, and boiling mud pools. This thermal activity is a reminder that Yellowstone is the site of one of the largest explosive volcanic eruptions in recent history, which led to the collapse of an ancient volcano, about 600,000 years ago, to form a giant caldera. The magma chamber that fed the original volcano is still present, about

**SURVIVING THE WINTER**
*For many generations, American bison have trekked to Yellowstone's hot springs and pools in winter, to benefit from the warmer temperatures and to feed on the plants that grow there.*

3¾ miles (6 km) below Yellowstone, and is the heat source for the present geothermal activity. Plentiful rainfall keeps its hydrothermal system well charged with water. The geysers form where the underground circulation of hot water is restricted. Under pressure at depth, water is superheated without boiling. As it rises, pressure decreases, allowing steam to form, which further accelerates upward movement. Water

**TERRACED WATERFALLS AND POOLS**
*Each day, evaporation deposits two tons of carbonate mineral at Mammoth Hot Springs, which enlarges the ever-changing Minerva Terrace.*

and steam then gush out at the surface as geysers. Gushing ceases when the hot-water reservoir is exhausted. The period between gushes depends on the time it takes for the underground system to recharge. Famously, Old Faithful produces columns of steam and water 100–180 ft (30–55 m) high every 67 minutes on average. At Yellowstone, the water temperature at 660 ft (200 m) depth is 392°F (200°C). This hot underground water supplies not only geysers but also springs. The rising hot water dissolves soluble minerals from the surrounding rocks, but, on surfacing, evaporation and cooling reverse the process, and mineral deposits form, such as those at Minerva Terrace. At Mammoth Hot Springs, terraces of limy flowstone (see p.253) contain beautiful, variably colored pools at different temperatures, their colors produced by species of temperature-sensitive algae. Boiling mud pools are normally gray, but in Yellowstone's Fountain Paint Pot these too are multicolored, and up to 66 ft (20 m) in diameter.

**ERUPTING GEYSER**
*Castle Geyser is one of Yellowstone's 200 or so active geysers, which are a constant reminder that the region is still geologically active.*

**THE LODGEPOLE PINE**
*The dominant tree of Yellowstone is the lodgepole pine, so called because its straight trunk is ideal for use as tepee poles. Its needle-shaped leaves point downward so the snow slips off them in winter.*

## LIFE AT HIGH TEMPERATURES

Prismatic Pool hot spring (shown below) is host to aquatic algae and other microorganisms that can survive high temperatures. Investigation of their tolerance of extreme conditions is providing insights into how life may have originated in the extremely harsh environments of the early Earth, and whether life can exist in the even more difficult conditions found on other planets. Thermophiles (heat-loving organisms) are among the most primitive forms of life. Over 4 billion years ago, the Earth's surface was hot, and the atmosphere was charged with carbon dioxide, nitrogen, and methane. Similar conditions are found in hot springs today, making them natural laboratories for studying life's origins.

**OLD FAITHFUL**
*After the famous geyser jets a spout of hot water around 165 ft (50 m) high into the air every 45–110 minutes, water and steam droplets hang in the air for about four minutes.*

NORTH AMERICA *northwest*

## Valley of Ten Thousand Smokes

**LOCATION** Near Katmai, in the Aleutian Range at the northeast end of the Alaskan peninsula

**TYPE** Fumaroles

**FREQUENCY** Extinct

In 1912, about 13 miles (22 km) of the Ukak valley was filled by the largest pyroclastic flow of modern times as it poured from vents near Mount Katmai (see p.176). The area is called the Valley of Ten Thousand Smokes because, for 15 years afterward, vaporized water from the stream and wet sediments continued to escape the surface through small holes and cracks.

THE SMOKING VALLEY

NORTH AMERICA *west*

## Fly Geyser

**LOCATION** On the flat, high plains adjacent to the Black Rock Desert, northwest Nevada

**TYPE** Geyser

**FREQUENCY** Continuous

This geyser has three outlets and is surrounded by hot-water pools contained by terraces of travertine (calcium carbonate). Over the years, the geyser has built up three rocky pinnacles of mineral deposits that continuously spout water. Some of the pools have been enlarged, allowing them to cool enough for bathing. Lying at an altitude of 4,270 ft (1,300 m), the Fly Geyser region receives an annual rainfall of less than 11¾ in (300 mm), but this, combined with runoff from the surrounding mountains, is sufficient to feed the local aquifer. The adjacent Black Rock Desert is the dry bed of a lake that covered about 8,660 square miles (22,420 square km) during the last ice age. Its sediments are floored by Pleistocene basaltic lavas, and the presence of the geyser shows that there must still be an associated magma chamber or intrusion at shallow depth, acting as a heat source.

**THREE-VENTED GEYSER**
*Travertine terraces surround Fly Geyser's continuously spouting trio of cones, which jet from the ancient dry lake floor of the Black Rock Desert.*

LAND

NORTH AMERICA *west*

## Soda Springs

**LOCATION** At the head of the Bear River valley, southeastern Idaho

**TYPE** Geyser

**FREQUENCY** Hourly

It is claimed that Soda Springs is the only geyser in the world to be made by human activity. On November 30, 1937, a well was drilled for hot water to supply a swimming pool. At about 330 ft (100 m) down, the drill hit a reservoir containing a pressurized mixture of carbon dioxide and hot water. The drill-hole provided an escape route, sending a geyser 165 ft (50 m) into the air. The borehole was capped and fitted with a valve, which restricts activity to once an hour. This part of the Bear River valley contains numerous other hot springs, such as Steamboat Springs, so called because of the steampipe-like noise that the hot water makes as it gushes forth. In the 19th century, migrants traveling the Oregon Trail stopped here to bathe and to wash their clothes.

**MANMADE GEYSER**
*Idaho's Soda Springs geyser is perhaps the only human-made geyser in the world, capped in such a way that it erupts every hour, on the hour.*

SOUTH AMERICA *west*

## El Tatio geysers

**LOCATION** Northeast of San Pedro de Atacama in the Chilean Andes, South America

**TYPE** Geysers, hot springs, and fumaroles

**FREQUENCY** Variable

At an altitude of 14,176 ft (4,321 m), the El Tatio geysers, hot springs, and fumaroles are the highest in the world. The area of thermal activity covers 4 square miles (10 square km) of a rift-valley floor. Water surfaces at temperatures of about 187°F (86°C)—the boiling point for water at this altitude. The water reservoir is within volcanic rocks capped by impermeable layers; faults conduct the hot water to the surface. The exact heat source is unknown, but is likely to be magma or an igneous intrusion beneath the valley.

FUMAROLES AT EL TATIO

ATLANTIC OCEAN *north*

## Geysir

**LOCATION** In the Haukadular rift valley, 50 miles (80 km) east of Reykjavik, southwest Iceland

**TYPE** Geyser

**FREQUENCY** 8–10 hours

The term "geyser" is derived from this feature, first described in 1294. Although there has been no volcanic activity here for about 10,000 years, temperatures below ground may reach 464°F (240°C). In 1915, Geysir became dormant, but earthquake activity in 1935 reactivated it for a time. After earthquakes in June 2000, Geysir currently erupts every 8–10 hours.

WATER BUBBLING IN GREAT GEYSIR

ATLANTIC *north*

## Strokkur Geyser

**LOCATION** In the Haukadalur rift valley, 50 miles (80 km) east of Reykjavik, southwest Iceland

**TYPE** Geyser

**FREQUENCY** 8 min

Strokkur, whose name means "churn" in Icelandic, is one of the most famous geysers in Iceland. It became active following an earthquake in 1789, and erupted regularly until 1896, when another earthquake blocked its flow. In 1963 the conduit was cleaned out by local people, and it has been active ever since. The eruption of the geysers in this area is driven by rising hot and pressurized water flashing into steam. Water at a depth of 75 ft (23 m) is at a temperature of about 248°F (120°C), but the weight of the water in the conduit above prevents it from boiling. At 52 ft (16 m), the temperature of the water may rise above boiling point and raise the water in the conduit slightly, setting off a chain reaction. The pressure decrease allows more water to boil and flash into steam, which in turn drives the water upward with increasing velocity to erupt at the surface. When the water in the upper part of the conduit has erupted, an action lasting just a few minutes, the steam phase is exhausted and the cycle begins again. With nearby Geysir (see opposite), Strokkur is one of Iceland's leading tourist attractions. There are a number of hot springs in the vicinity, as well as steam vents and colorful algal deposits.

**REGULAR CYCLE**
*Strokkur's eight-minute repeat cycle owes its regularity to human intervention. In 1963, the conduit was cleaned to ensure regular eruptions.*

**WELLING UP**
*The interval between Strokkur's eruptions depends on the time its reservoir channels take to refill and raise their temperature back to the critical point.*



EUROPE *west*

## Chaudes-Aigues

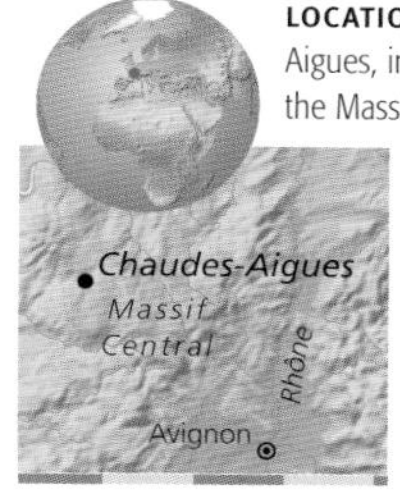


**LOCATION** In the town of Chaudes-Aigues, in the Auvergne region of the Massif Central, France

**TYPE** Hot springs

**FREQUENCY** Continuous

The name Chaudes-Aigues literally means "hot waters." The inhabitants of this small medieval town still enjoy central heating developed from the world's first geothermal system, established in the 14th century. First used by the Romans in the time of Nero, over 30 hot springs provide the hottest spring waters in Europe, at 180°F (82°C). Today Chaudes-Aigues is a spa town, famed for the curative powers of its hot, mineral-enriched spring waters. The hot springs are probably related to volcanic heat below the Chaîne des Puys (see p.181) and the famous spring waters of Volvic, even though Chaudes-Aigues lies 60 miles (100 km) to the south. The Chaîne des Puys includes about 90 cinder cones and maars (explosion craters formed without basaltic lava flows), many aligned along northeast–southwest fissures. The last volcanic activity occurred about 6,000 years ago.

EUROPE *south*

## Lardarello

**LOCATION** In the central region of the Colline Metallifere (Metalliferous Hills), Tuscany, central Italy

**TYPE** Hot springs

**FREQUENCY** Continuous

One of the world's largest geothermal fields and one of the oldest centers of geothermal power production, Lardarello has been generating electrical power since 1904. The heat source is located some distance below the surface in metamorphic rocks. Above, permeable limestones covered by impermeable shales and clays are folded into a dome, forming a trap for hot water and steam. Faults through the arch of the dome channel water to the surface, where they form hot springs.

**USING STEAM TO PRODUCE ELECTRICITY**

EUROPE *south*

## Solfatara

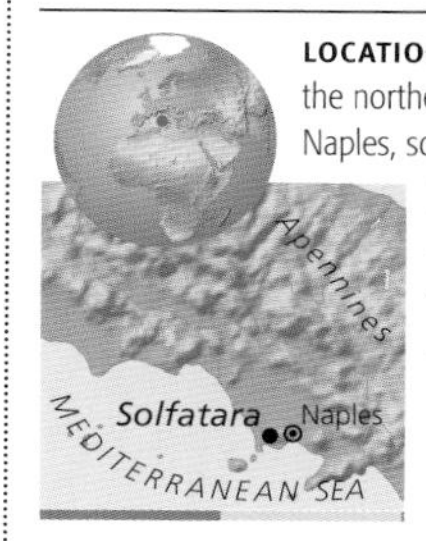


**LOCATION** West of Vesuvius, on the northern shore of the Bay of Naples, southern Italy

**TYPE** Hot springs, fumaroles

**FREQUENCY** Continuous

Solfatara is a crater lying within an ancient volcanic caldera called the Phlegrean Fields. About 40 hot springs and numerous fumaroles venting volcanic gases litter the caldera. Those of Solfatara's crater were believed by the Romans to cure a host of medical conditions, but are now the site of intense scientific research into the frequency and intensity of earthquakes, and changes in gas composition. The Bay of Naples, including the port of Pozzuoli, has risen and subsided by several yards over the last few millennia, accompanied by countless earth tremors (15,000 were detected between 1982 and 1984 alone). These movements affect the aquifer below Solfatara and water levels in the various wells of the region. The present Solfatara well was built in the early 19th century to extract alum (potassium aluminum sulfate) from the spring water. Deep magma chambers heat the overlying hydrothermal system, whose temperature varies from 390° to 480°F (200° to 250°C) if there is a rapid inflow of seasonal rain.

**STEAMING FUMAROLES**
*Bocca Grande ("Large Mouth"), shown here, is the largest fumarole at Solfatara. It was described by Virgil as the entrance to Hell.*

EUROPE *south*

## Lipari

**LOCATION** One of the Aeolian Islands, in the Mediterranean Sea, 22 miles (35 km) off northern Sicily

**TYPE** Fumaroles

**FREQUENCY** Continuous

Lipari is one of the Aeolian Islands, an active volcanic arc to the north of Sicily. The volcanoes are the result of subduction of the African plate beneath the southern edge of the Eurasian plate. This stimulates partial melting that feeds the volcanoes at the surface. In the past, volcanic activity on Lipari has produced highly viscous lavas, often with a glassy crust. Today, deep pockets of magma at depth produce vigorous fumarolic activity at the surface, and monitoring shows a wide range of temperatures over time, from about 392°F (200°C) to as much as 1,202°F (650°C). This is associated with increasing gas flow from depth and thermal expansion, which opens new conduit cracks. There is concern that volcanic activity is moving north from the nearby island of Vulcano and could cause an eruption near Lipari.

AFRICA *east*

## Bogoria

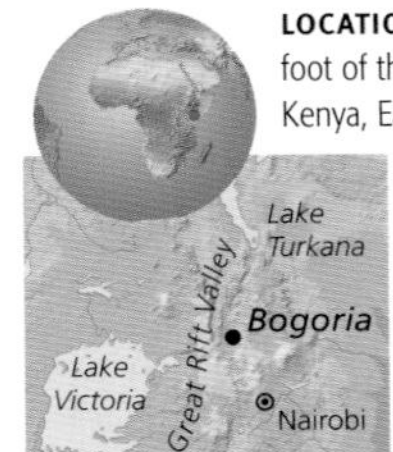


**LOCATION** On Lake Bogoria, at the foot of the Laikipia Escarpment, Kenya, East Africa

**TYPE** Hot springs and geysers

**FREQUENCY** Continuous

Hydrothermal features are rare in Africa but are present on the western and southern sides of Lake Bogoria. Here, groundwater is heated at depth by igneous intrusions buried below the valley's sediments and volcanic infill. As it rises, the hot water dissolves soluble salts (especially sodium carbonate) from the highly alkaline volcanic rocks. The hot springs feed soda lakes, which are home to alkali-tolerant algae and brine shrimp, upon which flamingos feed.

LAKE BOGORIA GEYSER AND HOT SPRINGS

ASIA *northeast*

## Nagano

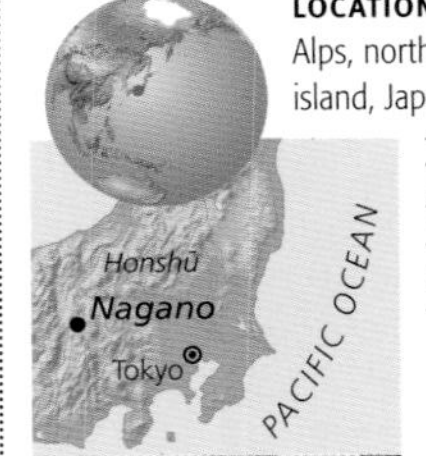


**LOCATION** In the central Japan Alps, northwest of Tokyo, Honshu island, Japan

**TYPE** Hot springs and fumaroles

**FREQUENCY** Continuous

The hot springs and fumaroles of the resort of Yamanouchi near Nagano are famous for their unusual inhabitants—Japanese macaque monkeys. The fumaroles expel sulfur-dioxide-enriched steam, and the hot springs, with their mineral-enriched waters, feed a series of pools. In the 1960s, the macaques discovered the benefits of immersing themselves in the hot water, and soon it became part of their behavioral culture, passed on from generation to generation. Today, they bathe in two specially excavated pools; people bathe elsewhere.

**JAPANESE MACAQUES AT NAGANO**
*About 250 Japanese macaques use the hot springs at Nagano, where snow may persist for four months of the year.*

ASIA *northeast*

## Beppu

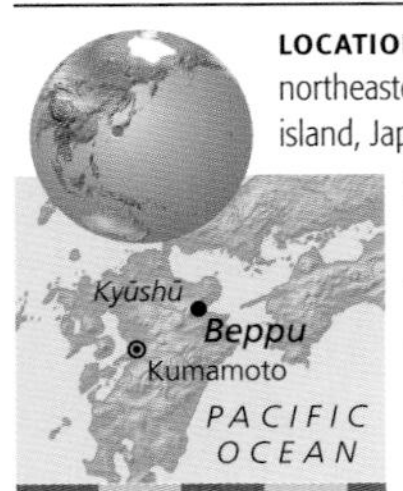


**LOCATION** In Beppu city, on the northeastern shore of Kyushu island, Japan

**TYPE** Geyser, hot springs, mud pools, and fumaroles

**FREQUENCY** 20–25 min

The active volcanism of the Japanese islands means that the country is also rich in hot springs, which are now used for a variety of domestic, social, and industrial purposes. Beppu city is Japan's largest spa resort, with many fumaroles and mudpools, 2,849 springs, and the Tatsumaki Jigoku geyser. The volume of hot water that gushes from this hydrothermal system exceeds 36 million gallons (136,000 kiloliters) each day, giving it the second-highest output in the world. Beppu has a number of hot springs, or jigokus (a Buddhist word meaning "burning hell"), including Blood Pond Jigoku, where red mud emerges with the boiling water; Ocean Jigoku, with its blue water; and Tatsumaki Jigoku geyser, which spouts water 66 ft (20 m) every 25 minutes. The springs are situated around two faults, Kannawa to the north and Asamigawa to the south. Their heat source is associated with 100,000-year-old buried lava domes that previously generated volcanism in the area. The abundant hot water, which surfaces at temperatures of 104–212°F (40–100°C), has been harnessed for domestic and industrial use, processing food, breeding fish, and producing geothermal power.

**BEPPU'S STEAM POWER**
*These chimneys are releasing not pollutants but excess steam from the geothermal energy plant, which powers local industry and heats homes.*

### COMMUNITY BATHING

The curative and restorative properties of Japan's hot springs have been appreciated for hundreds of years, and bathing in the hot springs is something of a national recreation. Beppu's hot baths range from domestic tubs in private houses, to free municipal facilities for poorer citizens, to luxurious private baths where the rich can relax, meet friends, and gossip.

**HOT SPRING AT BEPPU**
*Blood Pond Jigoku is so called because the water is charged with red mud. The red color is caused by iron oxide.*

AUSTRALASIA *New Zealand*

## Waimangu

**LOCATION** In Waimangu Valley, near Rotorua, on the North Island of New Zealand

**TYPE** Fumaroles

**FREQUENCY** Continuous

The volcanic belt of North Island, New Zealand, was the site of several very large explosive eruptions in the past 10,000 years. Its volcanoes, near the southwest end of the Pacific's Ring of Fire (see p.175), are associated with subduction of the Pacific plate. Waimangu Valley is a bleak area of raw volcanic craters, boiling pools, and occasional geysers. It lies 6 miles (10 km) southwest of Lake Tarawera and close to Lake Rotomahana. One geyser, which appeared in 1902 and lasted until 1905, spouted water 1,475 ft (450 m) into the air at intervals of between 5 and 30 hours and was the highest geyser plume ever recorded. Above the Waimangu Valley, there used to be scattered geysers and a series of tiered pink and white terraces, known as the White Terraces, which contained water-filled silica deposits. Maoris had traditionally used the site for bathing and cooking, and Waimangu was later regarded as the Yellowstone of its day, but early on the morning of June 10, 1886, all of these features were obliterated by a violent eruption of nearby Mount Tarawera. Within a couple of hours, a massive volume of basaltic scoria was blasted from a fissure 11 miles (17 km) long. Fragments of the scoria fell over a huge area, some landing 18 miles (30 km) away. The eruption also buried three Maori villages, killing 155 people. One of these villages has now been re-excavated, in a way similar to Pompeii (see p.186), while the cliffs above Lake Rotomahana still steam perpetually. The cratered fissure is still visible as a chasm on Mount Tarawera.

**INFERNO CRATER**
*One of the attractions of Waimangu Reserve is this water-filled crater, which is a reminder of the 1886 eruption of Mount Tarawera.*

**WARBRICK TERRACE**
*The mineral deposits and hot-water pools of Warbrick Terrace are a reminder of the world-famous White Terrace, destroyed in 1886.*

AUSTRALASIA *New Zealand*

## Rotorua

**LOCATION** In the town of Rotorua, near Lake Rotorua, on the North Island of New Zealand

**TYPE** Geysers, hot springs, and fumaroles

**FREQUENCY** Virtually continuous

Rotorua is a modern lakeshore resort that lies in an extremely fertile region of volcanic hills and lakes around Lake Rotorua. Settled since the 14th century by the Arawa group of Maori tribes, the area lies in the north of the Taupo Volcanic Zone, which encompasses active geysers, fumaroles, hot springs, and boiling mud pools. This hydrothermal activity is all that remains of one of the most violent volcanic eruptions of the last 2,000 years. Around AD 180, extensive pyroclastic flows from the Taupo Volcanic Zone covered the area, devastating all life. The event passed without human record as the Maoris had not yet colonized the island. Today, Geyser Flat has seven active geysers; the most spectacular is Prince of Wales Feathers, a triple geyser that spouts to a height of 40 ft (12 m) and is always followed by an eruption of the Pohutu geyser, which rises to 100 ft (30 m).

**INDICATOR GEYSER**
*The Prince of Wales Feathers geyser resembles an emblem on the prince's coat of arms. It was named following a royal visit in the early 1900s.*

### GEOTHERMAL POWER

Wairakei, near Rotorua, was the second geothermal field to be developed after Lardarello (see p.205). Water heated to boiling point by an igneous intrusion rises through a permeable volcanic ash aquifer. Wairakei is liquid-dominated and benefits from high pressure. Since the geothermal aquifer has been punctured by boreholes, the rising water crosses its flash-point and turns to steam on its way to the surface.

The regular sequential timing of the geysers suggests that they are linked. They also have associated boiling mud pools, formed by steam leaking from underground geothermal reservoirs. Lady Knox geyser lies 15 miles (24 km) to the southeast. It is artificially encouraged to spout to a considerable height by a daily dose of soap suds.

**WAI-U-TAPU MUD POOLS**
*A mixture of black sulfide with white silica and kaolin clay gives these mud pools their distinctive color, while rising gas generates the bubbles.*

**CHAMPAGNE POOL**
*The hot waters of Champagne Pool are bordered with carbonate deposits and give off clouds of sulfurous gas, which fill the air with a rotten-egg smell.*

**WINTER WARMTH**
*Bathers enjoy a refreshing dip at the Svartsengi geothermal power plant in Iceland, which is one of the world's leading producers of such energy.*

# GEOTHERMAL ENERGY

The Earth's internal heat represents a vast and practically inexhaustible source of energy, yet such geothermal energy is exploited only to a limited extent. This may change if economically viable methods of harnessing it are found. Geothermal energy is mainly produced by natural radioactive decay of uranium, thorium, and potassium. Unlike energy from fossil fuels, it also generates little carbon dioxide, so it does not contribute significantly to global warming (see pp.452–53).

## RESERVOIRS IN ROCKS

The Earth's core is at a temperature of about 7,200°F (4,000°C). Heat constantly flows through the mantle and crust (see p.55) and escapes through the atmosphere to cold outer space. Over most of the planet's surface, such heat flow cannot be felt, let alone put to work. However, in areas where the heat energy of the Earth has been concentrated, such as active volcanic zones, geothermal energy can be exploited. Here, intense heat can boil groundwater that is close to the surface, creating underground reservoirs of superheated water or steam. If the water escapes upward, it can emerge in hot springs, geysers, or fumaroles.

Hot springs were used to heat houses in Roman times, about two millennia ago, and the practice still continues in Iceland and Japan. This simple method of tapping geothermal energy does have some disadvantages. The water from hot springs often gives off sulfurous gases, and it can only be used locally, because its temperature soon falls if it is piped far from its source.

A much more flexible way of exploiting geothermal energy is to harness it for the generation of electricity. The world's first such geothermal power plant opened at Larderello, Italy, in 1904, and installations now operate in more than 20 countries worldwide. Most of these power plants tap into existing reservoirs of steam or water, which drive the turbines. The largest plant—at the Geysers in northern California—generates more than 1,700 megawatts (MW) of electrical power, which is sufficient to supply about half a million homes.

**NORTHERN HEAT**
*Although it is not far from the Arctic Circle, tomato plants can be grown in this geothermally heated greenhouse in Hveragerdhi, Iceland.*

### LEADING GEOTHERMAL PRODUCERS

Geothermal energy currently provides for less than 0.02 percent of the world's annual energy needs. With the exception of Iceland, most producers are located in the volcanically active Pacific Rim.

| Country | Production (MW) | Global proportion (%) |
|---|---|---|
| US | 2,228 | 27.9 |
| Philippines | 1,909 | 23.9 |
| Italy | 785 | 9.8 |
| Mexico | 755 | 9.5 |
| Indonesia | 589 | 7.4 |
| Japan | 547 | 6.9 |
| New Zealand | 437 | 5.5 |
| Iceland | 170 | 2.1 |
| El Salvador | 161 | 2.0 |
| Costa Rica | 142 | 1.8 |

## LOW-LEVEL HEAT

Even in regions in which the Earth's heat flow is not concentrated, geothermal heat can be used as a way of saving energy. Through a closed system of underground pipes, a heat pump can transfer geothermal heat between a house and the ground beneath it, but it does not create the heat. During the winter, the ground temperature is often higher than the air in unheated rooms. As a result, water in the pipes collects heat from the ground and releases it indoors. In the summer, the situation is reversed. The pump acts as an air-conditioner, because the water absorbs heat as it flows indoors, and releases it back into the ground.

This way of exploiting geothermal energy works best in areas with continental climates, in which there is a large difference between winter and summer temperatures. In such conditions, a geothermal heat pump can cut electricity or fuel use by up to 50 percent—saving money as well as reducing carbon dioxide emissions.

## UNTAPPED RESOURCES

At present, geothermal sources meet only a tiny fraction of the world's total energy requirements, and most of this energy comes from regions in which there is geothermally heated water underground. However, much more widespread than these hydrothermal reservoirs are hot, dry rocks. If ways can be found to extract their heat, geothermal energy production could soar.

Several major engineering problems have to be overcome before this goal is realized. One is that zones of hot, dry rock are typically 1¼ miles (2 km) or more underground. Once a borehole has been drilled to this depth, the heat energy has to be collected and brought to the surface. In experimental wells, water has been used to fracture the rock, creating fissures that fill with steam. The steam then travels to the surface, where its energy can be utilized—or that, at least, is the theory. However, despite its promise, energy production from hot, dry rock has yet to become a commercial reality.

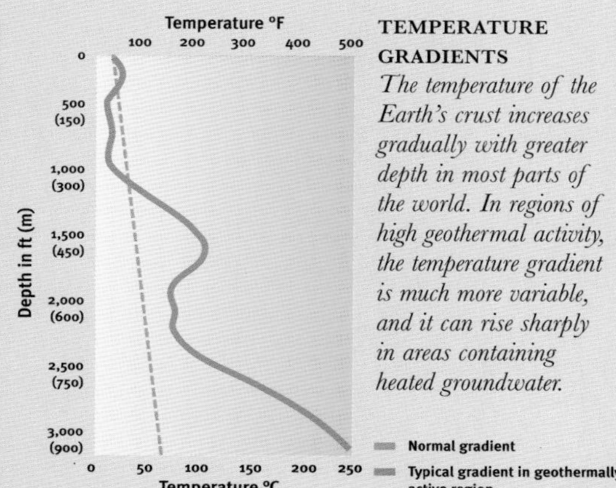


**TEMPERATURE GRADIENTS**
*The temperature of the Earth's crust increases gradually with greater depth in most parts of the world. In regions of high geothermal activity, the temperature gradient is much more variable, and it can rise sharply in areas containing heated groundwater.*

**SHAPING THE SURFACE**
*Meandering across the Kenyan countryside, the present channel of the River Mara bypasses land scarred by old meanders and oxbow lakes.*

# RIVERS AND LAKES

RIVERS FLOW AND LAKES FORM in all but the world's hottest, coldest, and driest places. They hold a small, but vital, proportion of the Earth's surface fresh water: lakes hold about one-half of one percent, and river channels contain much less—about one-fortieth of one percent. This tiny fraction is, however, disproportionately significant, because rivers shape the land. Rivers are the most powerful erosive force on Earth—across the continents, the dominant landscape is one of hill slopes and river valleys. Given time, rivers can wear down mountains and carry them to the sea. Powered by the force of gravity, the world's rivers deliver about 20 billion tons of land surface to the oceans every year. Many rivers are old in geological terms, and have increased in size with age, forming systems that drain continents. Some have hollowed out vast caves and run largely underground. By comparison, most lakes are very young, temporary accumulations of water that will inevitably shrink and disappear.

# RIVERS

112 Water erosion
114 Waterborne deposition
126–29 Water
Dams 226–27
Lakes 236–51
Freshwater quality 242–43
Underground rivers and caves 252–59
Wetlands 322–31

RAIN OR MELTED SNOW that does not evaporate flows downhill across the land surface. This surface flow is channeled by small irregularities into tiny rivulets that merge and flow down gullies into streams. These accumulate water and increase in size along their courses, creating valleys and becoming rivers that receive the inflow of smaller rivers, known as tributaries. The nature of a particular river at any point on its course depends on its rate of descent, the amount of water it contains, and the underlying geology of the land. These factors dictate the rates of erosion and deposition and, consequently, the degree to which the river modifies the landscape.



## DRAINAGE BASINS

A river and its tributary streams form a system that collects all the runoff (surface flow) within its drainage basin. The mountains, or other high ground, that separate the drainage basins of different river systems are called the watershed divide. A notable example is the Continental Divide along the Rocky Mountains in North America, which separates eastward- and westward-flowing rivers. River systems form different patterns of drainage on the landscape, depending on the rock type, slope of the land, and movements of the Earth's crust. The most common pattern is dendritic, where tributary streams join a main river like the branches of a tree. The points where rivers meet are called confluences. In addition to the tributaries, groundwater seeping into the river channel may account for up to 30 percent of a river's total volume. Most of the water carried in a river eventually flows into the sea.

*glaciers are source of some rivers*

*rain-fed tributaries flow into river*

*watershed divide between drainage basins*

*lakes provide water storage within drainage basin*

*forests reduce amount of runoff*

**CATCHMENT AREAS**
*Within a drainage basin, the total amount of water collected by a river system is influenced by the constancy of its sources, the type (or absence) of vegetation cover, and how much water is diverted for use by humans.*

**HEADWATERS TUMBLE**
*High in the mountains of Jotunheimen National Park in Norway, meltwater flows over bare rock on its way to becoming a river. Headwaters such as these need not necessarily flow through permanent channels.*

**YELLOWSTONE FALLS**
*The upper courses of many rivers—such as the Yellowstone River in Wyoming (shown here)—include dramatic waterfalls, rapids, and gorges.*

# EROSION AND TRANSPORT

The size and amount of material carried by rivers varies enormously. Even the slowest, smoothest flow of water will wash away small particles of rock. The turbulent flow of a river, with churning currents and sudden changes in velocity, has much greater power. The flow of water erodes the banks and bed by friction, suction, and solution. This flow also carries with it large quantities of solid material—the products of weathering or the river's own processes. It is this material, known as a river's sediment load, that usually accounts for most of its erosive force, through the abrasion of sand, gravel, pebbles, and boulders rolling and bouncing along the riverbed. Most solid material is, however, transported in suspension as fine particles of clay and silt.

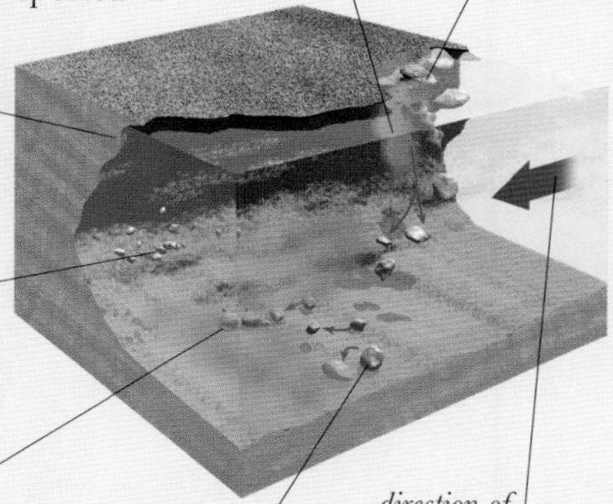

sediment cloud
undercut bank collapses
bank undercut by suspended sediment
clay and silt particles in suspension
colliding pebbles may shatter
larger particles roll or bounce along riverbed
direction of water flow

**RIVER EROSION**
*Small particles are washed away, suspended in the water flow. Larger particles become concentrated on the riverbed where the current is strongest, and multiple impacts gradually reduce their size.*

**POTHOLES**
*Potholes, such as these in South Africa, form if hard boulders are "trapped" in depressions of softer rock and then swirled around by turbulent currents.*

**EROSION BY UNDERCUTTING**
*This section of canyon wall in Zion National Park, Utah, has been undercut by the river—not by the gentle flow pictured, but by winter floodwaters crashing through the gorge.*

# DEPOSITION

When the sediment-carrying capacity of a river decreases, deposition occurs. This can happen for a number of reasons—for example, because the current lessens where the river bends, widens, or flattens. Erosion and deposition are closely interlinked processes and may take place simultaneously on opposite banks of the same river bend. The most prominent features produced by deposition are point bars along riverbanks, and riffles, sandbars, and islands in midstream. The floodplains and deltas seen in the lower courses of rivers are depositional features on a much grander scale.

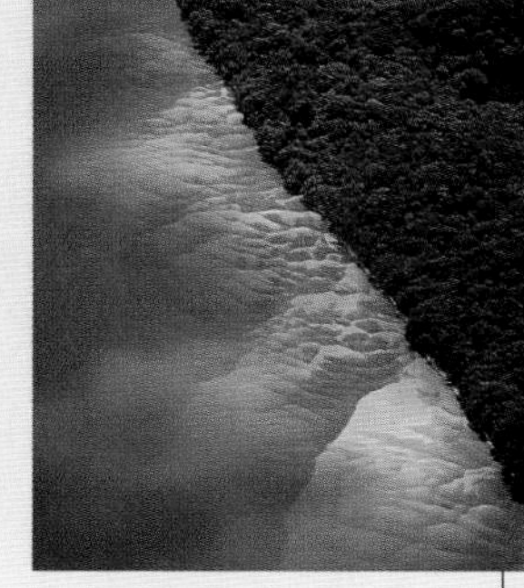

**SANDBANKS**
*Point bars are deposited along a river's banks below the surface, where the water flow is slowest.*

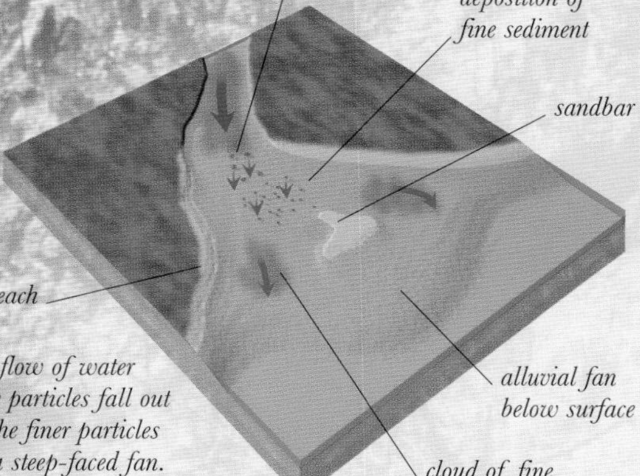

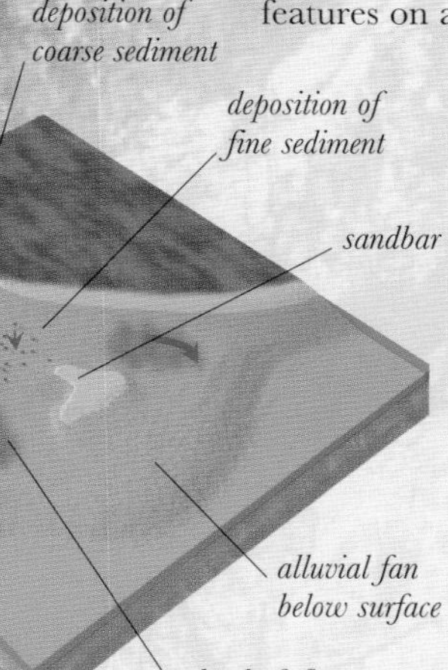


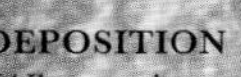

**DEPOSITION**
*When a river widens, the flow of water spreads and slows. Coarse particles fall out of suspension first, with the finer particles deposited downstream in a steep-faced fan. As the flow spreads out, a sandbar is deposited in the middle where the current is weakest.*

**DRIED-UP RIVER**
*In summer, when water levels fall and evaporation increases, rivers such as the Yellow River shrink, revealing beds of dried mud.*

**EROSION AND DEPOSITION**
*Where the current is strongest, a river's bank is undercut; where the water moves more slowly on the other side, deposition occurs.*

LAND

# UPPER COURSES

In most rivers, the upper courses are characterized by steep descent and the inflow of many small tributaries. The river is usually narrow and fast-flowing, and it erodes a distinctive V-shaped valley that zigzags between interlocking spurs. The rate of flow along the upper course may be highly variable, depending on seasonal factors such as snowmelt or heavy rainfall, and most mountain rivers are subject to regular flooding. Most of the valley-cutting takes place when the river is in flood stage. Flooding also maintains small floodplains where the river crosses flatter ground. When crossing a high, sloping plateau, a river may cut a deep gorge and become incised, particularly if the plateau is being uplifted.

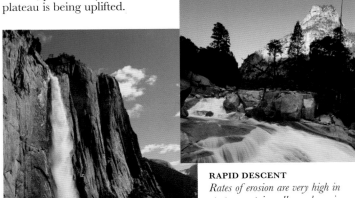

**RAPID DESCENT**
*Rates of erosion are very high in steep mountain valleys where rivers race downhill.*

**WATERFALL**
*At the base of most waterfalls, such as this one in California, the waters erode a plunge pool, which is enlarged by the scouring action of rock fragments from the cliff face.*

## PREDICTING FLOODS

In Great Britain, the Environmental Protection Agency uses radar sensors carried by aircraft to determine precisely the height of land along the margins of rivers that are prone to flooding. Using this information, scientists run computer simulations that predict the effects on the land of a river rising to a particular height. Areas that are more vulnerable to flooding can be identified and the flood defenses planned accordingly.

**FLOOD RISK**
*This computer-generated image shows the effects of a 5 ft- (1.5 m-) rise in the level of a river in northern England.*

predicted flood area

river channel

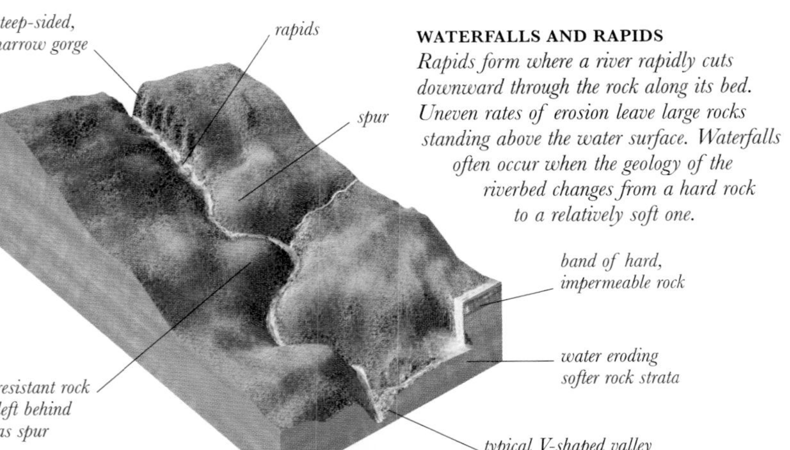


**WATERFALLS AND RAPIDS**
*Rapids form where a river rapidly cuts downward through the rock along its bed. Uneven rates of erosion leave large rocks standing above the water surface. Waterfalls often occur when the geology of the riverbed changes from a hard rock to a relatively soft one.*

**FERTILE RIVER VALLEY**
*This combination of cone-shaped hills and a flat valley floor is typical of rivers flowing through limestone valleys in the tropics and subtropics.*

## LOWER COURSES

Rivers generally flow down gentler gradients in their lower courses compared to their upper reaches. Few such rivers follow one straight channel: some braid into numerous channels separated by temporary islands; others meander in great looping curves. A meandering river may cut rapidly downward, forming an incised meander, or it may wander from side to side, creating a wide, flat valley with steep sides. Meanders that are cut off as the river shifts its course form distinctive oxbow lakes. Deposition rates are high along lower courses and are often linked to seasonal flooding. Sediments deposited by floodwaters may build up into deep beds. Flooding also produces natural levees (raised banks) along stretches of riverbank.

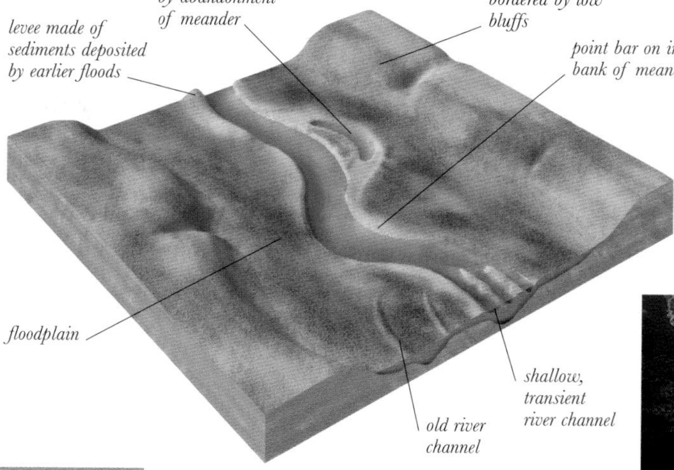

*levee made of sediments deposited by earlier floods*
*oxbow lake, formed by abandonment of meander*
*wide river valley bordered by low bluffs*
*point bar on inside bank of meander*
*floodplain*
*old river channel*
*shallow, transient river channel*

**FLOODPLAINS**
*A floodplain landscape (left) is subject to constant change as the river shifts its meandering course, leaving behind levees, oxbow lakes, and abandoned channels. Because floodplains are flat and regularly covered with rich alluvial deposits, they are attractive both for agriculture and for human settlement. Floodplains are some of the most densely populated areas in the world.*

**BRAIDING ON LOWER COURSE**
*This pattern of braided channels on a river in Denali National Park, Alaska, will remain stable until the flow of the river changes.*

### CLEARING WATERWAYS

Rivers that are used for navigation often need to be dredged to clear them of sediment and vegetation. Some large river highways have to be dredged constantly in order to keep important navigation channels open. Here, a dredger is clearing a deepwater channel in the bed of the Mississippi—a river particularly prone to such problems, as it carries half a billion tons of sediment to the sea each year.

## DELTAS

Many rivers flow into the sea through a delta—a wide, flat region where the river branches into numerous distributary channels, and where new land is being created as the river deposits any material it is still carrying in suspension. The smallest particles may be carried a considerable distance out to sea, as fresh water slowly mixes with seawater. A typical delta takes the form of a broad fan that slopes gently seaward. Its shape depends on the balance between the river's flow, the tide, currents, and wave action. In some cases, the river's flow or coastal currents may be too strong for a delta to form. Some rivers form deltas when they enter lakes, but these lacustrine deltas, as they are known, have a steeper slope because water mixing takes place more rapidly.

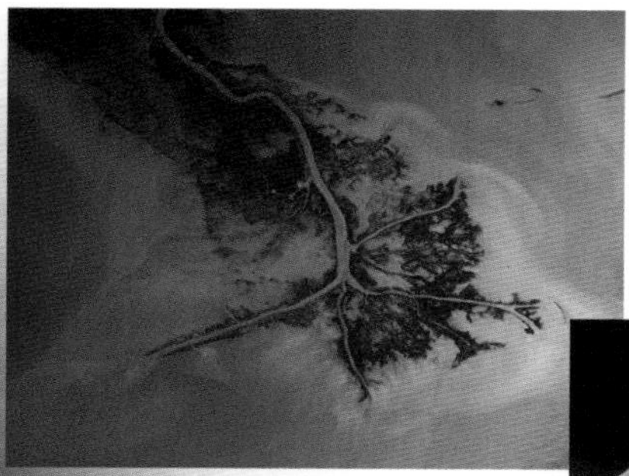

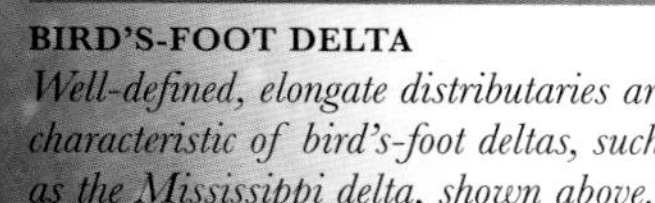

**BIRD'S-FOOT DELTA**
*Well-defined, elongate distributaries are characteristic of bird's-foot deltas, such as the Mississippi delta, shown above.*

**FAN DELTA**
*In this satellite image, the fertile Nile delta contrasts with the surrounding desert sands. Constant wave action keeps the delta's coastline smooth and distinct.*

### RIVER PROFILES

The pages that follow contain profiles of the world's main rivers. Each profile begins with the following summary information:

**LENGTH** Distance from source to mouth

**VERTICAL FALL** Height descended by river from source to mouth

**BASIN AREA** Area drained by the main river and all of its tributaries

**MAIN TRIBUTARIES** Subsidiary rivers that flow into the main river

NORTH AMERICA *northwest*

## Mackenzie

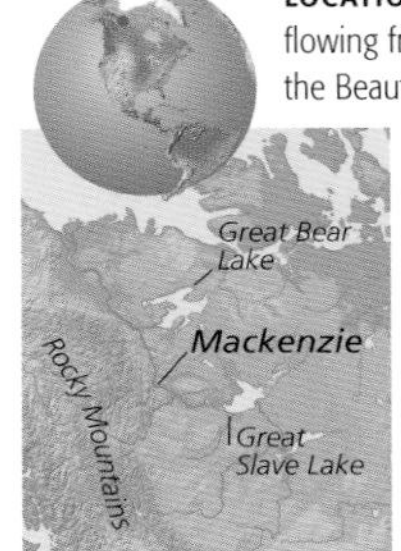


**LOCATION** In northeast Canada, flowing from Great Slave Lake into the Beaufort Sea in the Arctic Ocean

**LENGTH**
1,060 miles (1,705 km)

**VERTICAL FALL**
512 ft (156 m)

**BASIN AREA**
700,000 square miles (1.8 million square km)

**MAIN TRIBUTARIES**
Liard, Great Bear, Peel

The basin of the Mackenzie River is the largest in Canada and the second-largest in North America. From the western end of Great Slave Lake, the river flows northwest, mainly over ground subject to permafrost, and into the sea through a seasonally marshy delta. In winter the river freezes along its entire length. For the most part it is a broad river, 1–4 miles (1.5–6.5 km) wide, which has steep gravel banks and frequently braids into channels. Mackenzie waters leaving Great Slave Lake are clear, any sediment having been deposited in the lake. The waters of the main tributary, the Liard, are very muddy, and the two flows do not completely mix for some 200 miles (320 km) downstream of their confluence. Midway along its course, the Mackenzie meets Great Bear River, which drains Great Bear Lake.

**CARIBOU**
*The Mackenzie lowlands, which are covered with a mosaic of forests, swamps, shrubland, and grassy tundra, are home to herds of caribou.*

**PINGO ON MACKENZIE RIVER DELTA**
*Located in the most isolated corner of Canada's Northwest Territories, the delta is a constantly changing wilderness blanketed by snow and ice for seven months of the year.*

**FROZEN RIVER IN WINTER**
*Ice usually forms in mid-October, and parts of the river remain frozen until early June. Tributaries often thaw before the main river, causing flooding.*

NORTH AMERICA *northeast*

## Hudson

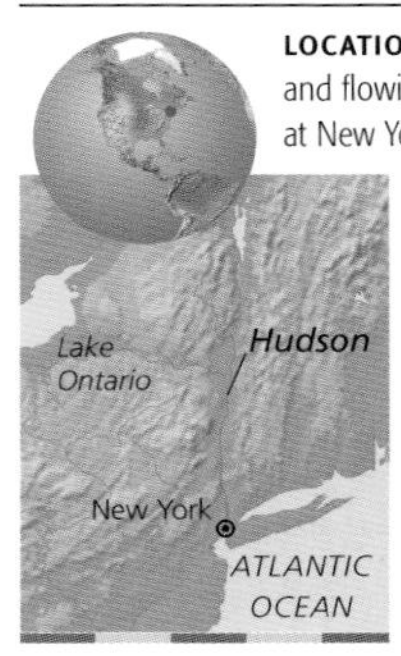


**LOCATION** In New York State, and flowing into the Atlantic Ocean at New York City

**LENGTH**
315 miles (507 km)

**VERTICAL FALL**
4,323 ft (1,371 m)

**BASIN AREA**
13,370 square miles (34,628 square km)

**MAIN TRIBUTARIES**
Mohawk

From its source in Lake Tear of the Clouds, a postglacial lake in the Adirondack Mountains, the Hudson flows southeast to its confluence with the Mohawk near Albany, then turns southward for the last 150 miles (240 km) of its course. The lower Hudson Valley (and that whole section of eastern coastline) was drowned by the sea at the end of the last ice age, leaving a submarine canyon about 110 miles (180 km) long. Tidal influence extends as far as Albany, which has a 4½-ft (1.4-m) high tide. A prominent feature of the lower valley is the Palisades (see p.196), a series of steep cliffs where the river cuts into hard rock. Since its exploration by Henry Hudson in 1609, the river has been commercially important. Nineteenth-century improvements, notably the Erie Canal, connected the Hudson with the Great Lakes, and the rapidly extending western "frontier." It remains a major river highway.

**HUDSON PASSING MANHATTAN ISLAND**
*The shipping piers seen here along Manhattan's Lower West Side underline the Hudson's importance as a transportation route.*

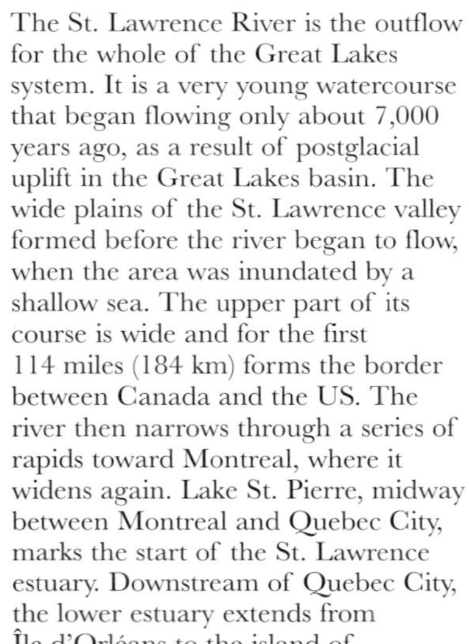

NORTH AMERICA *northeast*

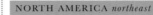

## St. Lawrence

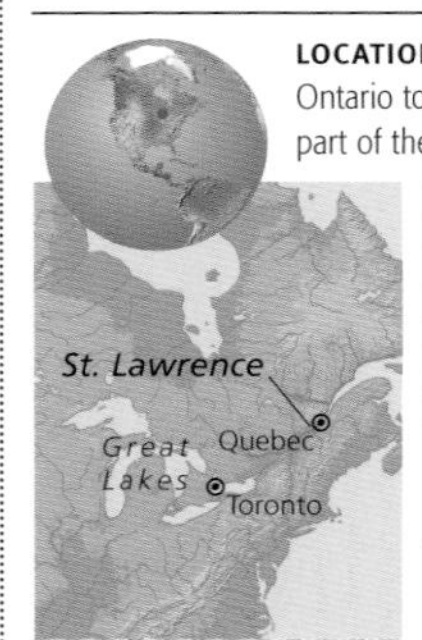


**LOCATION** Flows from Lake Ontario to the Atlantic Ocean; forms part of the Canada–US border

**LENGTH**
760 miles (1,244 km)

**VERTICAL FALL**
246 ft (75 m)

**BASIN AREA**
500,000 square miles (1.3 million square km)

**MAIN TRIBUTARIES**
Ottawa, St. Maurice

The St. Lawrence River is the outflow for the whole of the Great Lakes system. It is a very young watercourse that began flowing only about 7,000 years ago, as a result of postglacial uplift in the Great Lakes basin. The wide plains of the St. Lawrence valley formed before the river began to flow, when the area was inundated by a shallow sea. The upper part of its course is wide and for the first 114 miles (184 km) forms the border between Canada and the US. The river then narrows through a series of rapids toward Montreal, where it widens again. Lake St. Pierre, midway between Montreal and Quebec City, marks the start of the St. Lawrence estuary. Downstream of Quebec City, the lower estuary extends from Île d'Orléans to the island of Anticosti, where the river enters the Atlantic at the Gulf of St. Lawrence. Extensive engineering during the 20th century has transformed the river into the St. Lawrence Seaway.

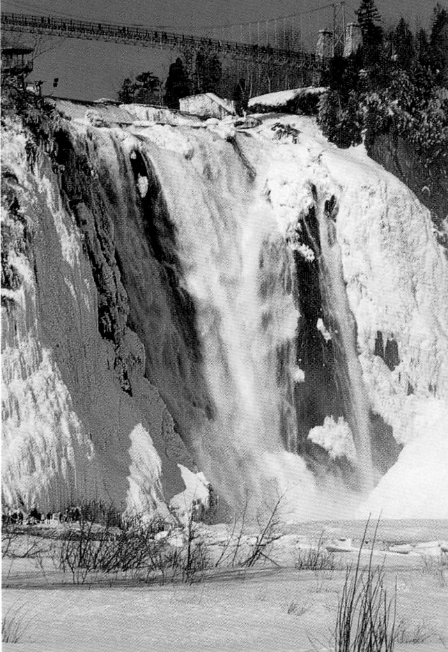

**FROZEN FALLS IN WINTER**
*Water plunging over the Montmorency Falls on the Montmorency River in Quebec freezes as it falls, creating curtains of frozen spray.*

NORTH AMERICA *central*

# Mississippi

**LOCATION** In the US, flowing from near the Canadian border in Minnesota to the Gulf of Mexico

**LENGTH** 2,350 miles (3,780 km)

**VERTICAL FALL** 1,475 ft (450 m)

**BASIN AREA** 1.2 million square miles (3.2 million square km)

**MAIN TRIBUTARIES** Missouri, Ohio, Arkansas, Tennessee

Together with its many tributaries, the Mississippi River forms a vast and ancient river system. Its drainage basin extends over almost the whole of the United States from the Rocky Mountains to the Appalachians. The western tributaries drain the Great Plains, while the eastern ones drain the Appalachian Plateau. The source of the Mississippi is Lake Itasca in northern Minnesota, which the river leaves as a clear 10-ft- (3-m-) wide stream that winds eastward through extensive marshes, fed by other lakes and streams. It then turns southward, skirting the higher ground around the western end of Lake Superior. Downstream of Minneapolis, the Mississippi begins to collect the upper-course tributaries that give rise to its name, which means "Father of Waters" in the native Algonkian language. The influx of the Missouri River near St. Louis brings huge quantities of sediment that change the character of the river. Downstream, at Cairo, Illinois, the inflow of the Ohio River marks the original beginning of the Mississippi delta, which has been inching southward for the last 100 million years. Flowing through a gap between the Cumberland and Ozark plateaus, the Mississippi follows a broad valley cut through the alluvial deposits by meltwater at the end of the last ice age. Although the Missouri ("Big Muddy"), 2,560 miles (4,120 km) long, is considered the Mississippi's main tributary, the Ohio—length 975 miles (1,570 km)—contributes the greater volume of water. Only along its lower course, south of Cairo, does the Mississippi achieve the full majesty of "Ol' Man River," more than 1 mile (1,600 m) wide and carrying about half a billion tons of sediment to the sea each year. The delta has shifted over an area of about 200 square miles (320 square km) during the last 6,000 years, and forms a branching extension into the Gulf of Mexico, southeast of New Orleans. It is currently shrinking, due to a combination of natural and human-induced factors.

**MILK RIVER MEETING THE MISSOURI**
*The Milk River gathers some of its distinctive pale sediment as it loops through southern Canada. It meets the Missouri in Montana.*

**MINNESOTA MARSHES**
*The low-lying and fast-eroding shoreline of Lake Minnetonka reveals the delicate balance between land and water in the wetlands of Minnesota.*

## FLOOD CONTROL

Levees for protection against floods were first built on the Mississippi in the early 18th century. Long stretches are now ramparted in this way. The variable waters of the Ohio cause most floods, but in 1993, heavy rains over the northern Great Plains caused devastating floods upstream from Cairo. With little warning, East Dubuque and other towns on the upper Mississippi and lower Missouri were inundated by a so-called "100-year" event.

**SLOUGHS ON MISSISSIPPI RIVER**
*This upstream view of the river in northern Iowa shows narrow, dead-end sloughs (backwaters) between a network of sandbars that have been stabilized by vegetation.*

NORTH AMERICA *west*

# Colorado

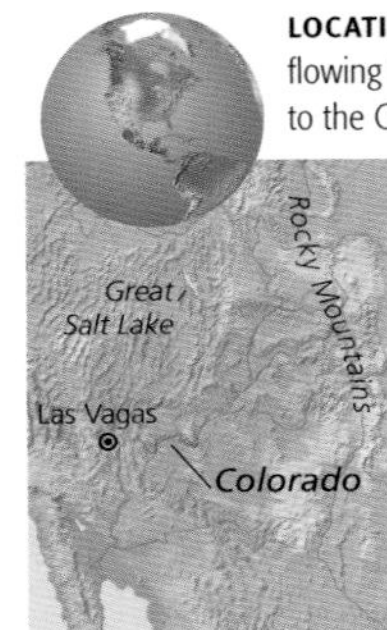


**LOCATION** Mainly within the US, flowing from the Rocky Mountains to the Gulf of California

**LENGTH**
1,450 miles (2,333 km)

**VERTICAL FALL**
14,200 ft (4,320 m)

**BASIN AREA**
244,000 square miles (632,000 square km)

**MAIN TRIBUTARIES**
Green, Little Colorado, Gila

The Colorado, which drains the arid southwestern quarter of the US, is the world's most formidable canyon-cutter. In response to the tilting uplift of the Colorado Plateau, the river has cut more than 1,000 miles (1,600 km) of deep canyons, the most spectacular stretch of which is the 220-mile-(350-km-) long Grand Canyon in northern Arizona. The Colorado River rises from snowmelt high in the Rocky Mountains in northern Colorado, on the westward slopes of the Continental Divide. It flows southwest across the Colorado Plateau into Utah, where the confluence with the Green River in the Canyonlands region brings waters from the northernmost reaches of its drainage basin in Wyoming. Downstream in northern Arizona, the main trunk of the Colorado's branching canyon system—the Grand Canyon—reaches 18 miles (29 km) in width and cuts down through layers of sedimentary rock that record 2 billion years of geological history. Below the canyon, after exiting the plateau, the course turns southward, forming the California–Arizona state line. Along its lower course the Colorado is a slow, meandering river, laden with fine silt and subject to flooding. The last 80 miles (128 km) run through Mexico to a shrinking delta. The Colorado River is the most managed river in the world. Large-scale engineering began in the northern part of the drainage basin in the 1920s, with great tunnels cut beneath the Continental Divide to divert some of the river's water eastward to the Great Plains.

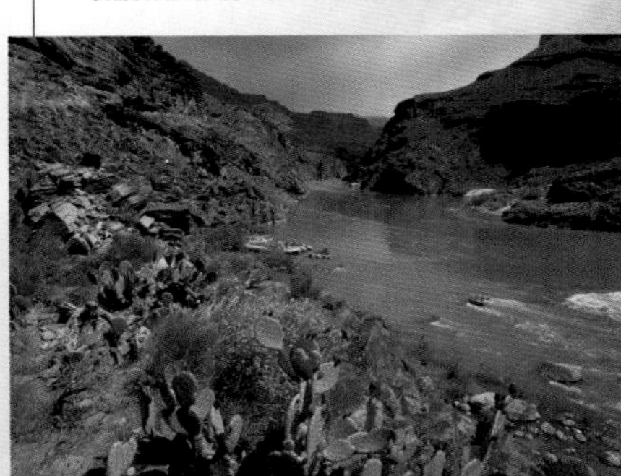

**CACTI BY THE COLORADO**
*The climate of the Colorado Plateau is hot and arid. Only cacti and low desert shrubs can grow—there are no trees, even at the river's edge.*

**THE GRAND CANYON**
*Created by the river, the magnificent rock cliffs of the Grand Canyon are preserved by the aridity of the desert climate. At the base of the canyon, the river has cut down to 1.7 billion-year-old rocks.*

Subsequent attention has focused on the lower course. The 726-ft- (221-m-) high Hoover Dam created Lake Mead in 1935, and provides both electricity and irrigation water. Downstream, considerable amounts of water are diverted to the cities of Las Vegas, Phoenix, Tucson, San Diego, and Los Angeles. The Imperial Dam diverts most of the water that remains along the All-American Canal to California's Imperial Valley. Although the flow at the mouth of the Colorado is now much reduced, it has in the past effected great changes upon the coastline. The Gulf of California once extended much farther north, until the Colorado changed its course in prehistoric times. Sediment deposited on the river's banks dammed the Gulf, creating the current shoreline and the great Salton Sink, a stretch of former seabed that extends inland some 300 miles (480 km). In 1905 a breached irrigation channel allowed the Colorado's waters to flood into the sink, creating the 70-mile- (112-km-) long Salton Sea.

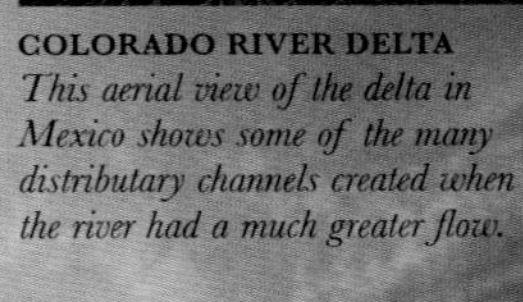

**COLORADO RIVER DELTA**
*This aerial view of the delta in Mexico shows some of the many distributary channels created when the river had a much greater flow.*

## REGULATING THE COLORADO

The 710-ft- (216-m-) high Glen Canyon Dam is located in Arizona about 110 miles (175 km) upstream of the Grand Canyon. It was constructed in the 1960s to store water from winter rains to generate electricity later in the year, when the river's natural flow diminishes. By restricting the flow of the Colorado, the dam has had an alarming effect on the environment along riverbanks in the Grand Canyon. Instead of being scoured clean by regular winter floods, parts of the canyon are now starting to fill with sediment. In the 1990s, attempts were made to redress the balance by releasing artificial floods from the dam.

**CUTTING POWER OF THE COLORADO**
*The erosive power of the Colorado River is clearly evident in this picture, which shows a deep channel cut through the colorful sedimentary rocks of the Grand Canyon.*

NORTH AMERICA *southwest*

## Rio Grande

**LOCATION** Flows from the Rocky Mountains to the Gulf of Mexico; is part of the US–Mexico border

**LENGTH** 1,885 miles (3,034 km)

**VERTICAL FALL** 12,000 ft (3,650 m)

**BASIN AREA** 172,000 square miles (445,000 square km)

**MAIN TRIBUTARIES** Chama, Conchos, Pecos, Salado

Rising high in the San Juan Mountains of southern Colorado, the fifth-longest river in North America flows through New Mexico into Texas. From El Paso on the high plains to Brownsville on the delta, the Rio Grande defines the border—1,250 miles (2,000 km) long—between Texas and Mexico. Fed by springs below the San Juan snowfield, its upper course flows eastward along pine-forested canyons, then turns south into New Mexico, where the river has cut the Rio Grande gorge and White Rock Canyon. The middle course flows through basin-and-range terrain, skirting the San Andres Mountains, before descending through three canyons, over 1,500 ft (450 m) deep, onto the Gulf Coastal Plain. In the US, these canyons form Big Bend National Park. With much of its water being diverted for irrigation, the Rio Grande winds sluggishly across the coastal plain to the Gulf of Mexico. So much of the river's water is diverted that during long dry periods, surface flow at the delta disappears.

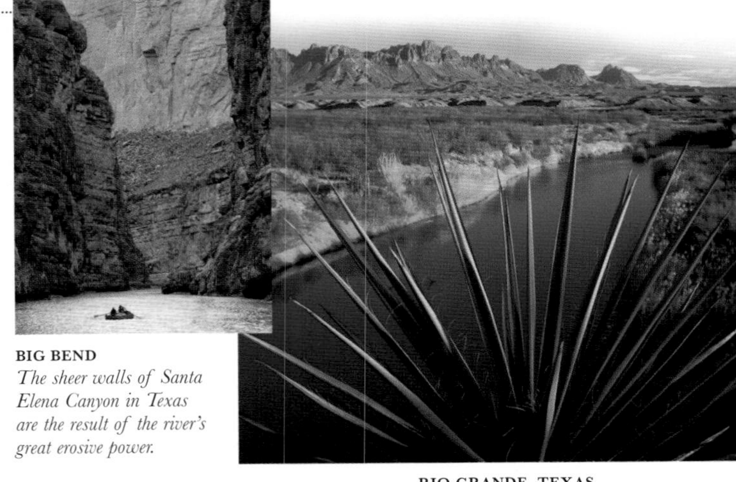

**BIG BEND**
*The sheer walls of Santa Elena Canyon in Texas are the result of the river's great erosive power.*

**RIO GRANDE, TEXAS**
*Upstream of the Big Bend canyons, with the Chisos Mountains rising in the distance, the Rio Grande provides water to the arid landscape of the high plains.*

SOUTH AMERICA *southeast*

## Plate River

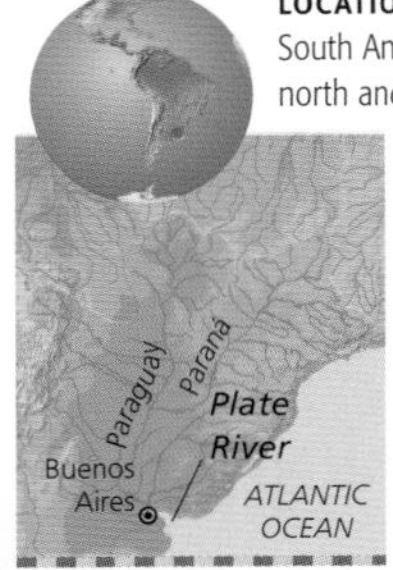


**LOCATION** On the eastern coast of South America, with Uruguay to the north and Argentina to the south

**LENGTH** 180 miles (290 km)

**VERTICAL FALL** Nil

**BASIN AREA** 1.6 million square miles (4.2 million square km)

**MAIN TRIBUTARIES** Paraná, Uruguay

Not strictly a river, nor a marine gulf, the Plate River is a huge, funnel-shaped estuary formed by the mouth of the Uruguay River coinciding with the delta of the Paraná River. Extending over 13,500 square miles (35,000 square km), and 136 miles (219 km) wide where it meets the Atlantic, the Plate receives the sediment load of three great rivers—the Uruguay, the Paraná, and the Paraguay (the major tributary of the Paraná)—that have a combined length of nearly 5,000 miles (8,000 km). As a result, the Plate is choked by sediment that accumulates into great shoals, where the water depth can be reduced to as little as 5 ft (1.5 m). Constant dredging is necessary to maintain deep-water channels to the ports of Montevideo and Buenos Aires. The catchment area of the Plate extends over parts of Argentina, Bolivia, Brazil, Paraguay, and Uruguay. The Paraná River flows some 2,485 miles (4,000 km) from the highlands of central Brazil to the Plate. Along its upper course, the most notable feature is the Guaira Falls, which carries several times the volume of Niagara Falls. Downstream, the Iguaçu River brings water still churning from a drop of 270 ft (82 m) over the spectacular Iguaçu Falls. The Paraguay River runs 1,585 miles (2,550 km) from its source in the Mato Grosso region of Brazil, through the Pantanal wetlands, to its confluence with the Paraná. The river has a shallow gradient—most of its catchment area lies below 650 ft (200 m)—and it is subject to seasonal flooding. The Uruguay River rises on the western slopes of southern Brazil's coastal range and loops inland along a course of about 1,600 miles (1,000 km) to the Plate estuary.

**PARANÁ RIVER DELTA**
*This false-color satellite image shows an area of marshland to the north of Buenos Aires, Argentina, where the Paraná and Uruguay rivers meet.*

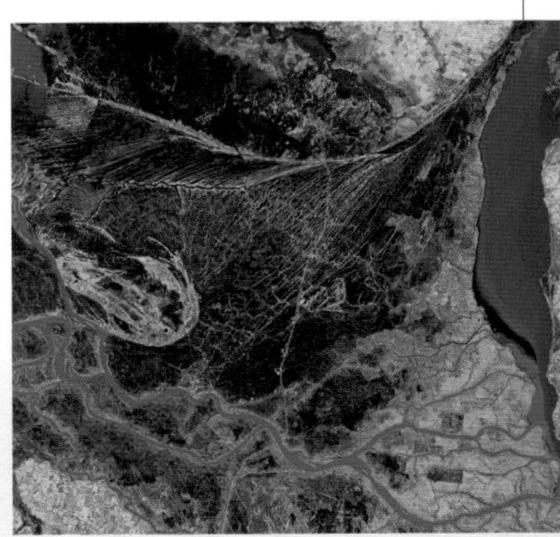

**DEVIL'S THROAT AT IGUAÇU FALLS**
*Upstream from its confluence with the Paraná, the Iguaçu River plunges over a great horseshoe-shaped waterfall into a steep-sided canyon called Devil's Gorge.*

SOUTH AMERICA *north*

# Orinoco

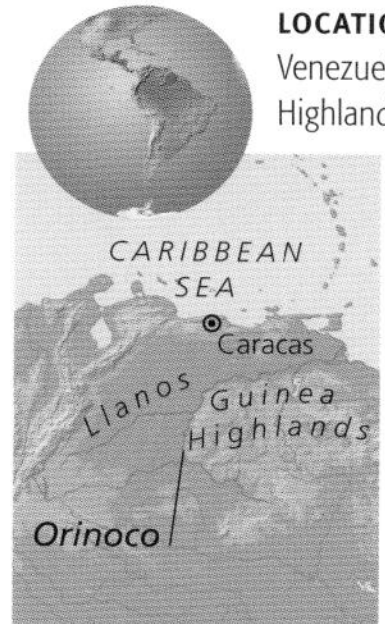


**LOCATION** Flows through Venezuela from the Guiana Highlands to the Atlantic Ocean

**LENGTH**
1,337 miles (2,151 km)

**VERTICAL FALL**
3,523 ft (1,074 m)

**BASIN AREA**
366,000 square miles (948,000 square km)

**MAIN TRIBUTARIES**
Guaviare, Meta, Arauca, Apure, Caroni

The second-longest river in South America, the Orinoco flows in a great arc through Venezuela. Its drainage basin covers about 80 percent of that country and about 25 percent of Colombia. The upper reaches contain two remarkable features: Angel Falls, the world's highest uninterrupted waterfall, cascades down Auyan Tebui (Devil's Mountain); and Casiquiare Channel, 221 miles (355 km) long, diverts some of the Orinoco's waters southward into the Rio Negro, a tributary of the Amazon. On the east edge of the Guiana Highlands, the Orinoco flows through a series of boulder-strewn rapids (Region de los Raudales) before turning northeastward. The lower course is marked by numerous swamps (notably the Llanos wetlands, see p.328) before branching into a delta some 270 miles (436 km) from the Atlantic coast. The main channel discharges at Boca Grande ("Big Mouth"). The rainfall that feeds the Orinoco is very seasonal, and flooding along the lower course is frequent and extensive. At the end of the dry season in April, the river at Ciudad Bolívar may be 100 ft (30 m) lower than at the height of the rains in July.

**ANGEL FALLS, VENEZUELA**
*Before joining the Orinoco (via the Caroni), the Churun River plunges 3,212 ft (979 m) over Angel Falls at Canaima, Venezuela.*

**WATTLED JAÇANA**
*Also known as the lily-trotter, this bird has large, spidery feet that work like snowshoes, spreading its weight and enabling it to walk on floating vegetation.*

*elongated toe*

**ORINOCO DELTA**
*The delta covers about 14,000 square miles (36,500 square km) and is home to the now rare Orinoco crocodile. Many of the islands are used for cattle ranching and growing cacao.*

## JIMMIE ANGEL

Angel Falls is named after American aviator Jimmie Angel (1899–1956), who first sighted the falls from the air in 1933. Four years later, he landed on top of Devil's Mountain, but damaged his plane in the process and had to trek back through rainforest to the nearest settlement. Angel, who achieved near-legendary status in Venezuela, died in Panama from injuries received in a flying accident.

SOUTH AMERICA *north*

# Amazon

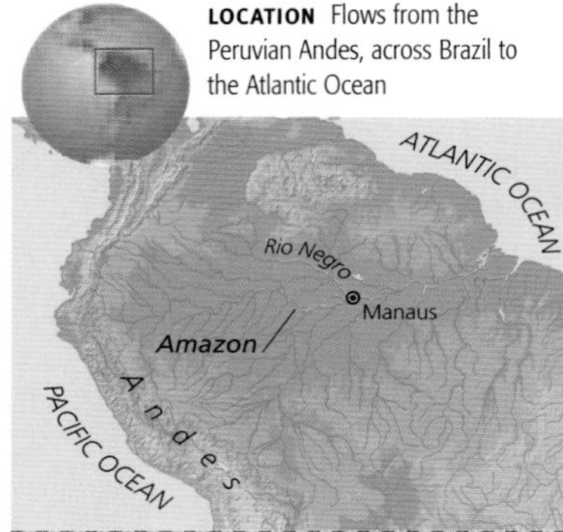


**LOCATION** Flows from the Peruvian Andes, across Brazil to the Atlantic Ocean

**LENGTH** 3,990 miles (6,430 km)

**VERTICAL FALL** 18,000 ft (5,500 m)

**BASIN AREA** 2.7 million square miles (7.1 million square km)

**MAIN TRIBUTARIES** Juruá, Madeira, Negro

The Amazon is the world's greatest river, measured both by the size of its basin and by the volume of water it discharges. On a world scale, the Amazon delivers about 20 percent of all the water reaching the sea from rivers. Its nearest rivals, the Congo and Yellow rivers, each deliver only about 4 percent of the total. Some 10 miles (16 km) wide at Manaus, while still 1,000 miles (1,600 km) from the sea, it is truly a giant river. The Amazon flows from mountainsides and glacier-fed lakes high in the Andes of southern Peru, within about 100 miles (160 km) of the Pacific Ocean. The upper course, known as the Marañón River, flows north, and then turns east, descending into the Amazon Basin—a vast depression sinking under the weight of material eroded from surrounding highlands. Here, it meanders for more than 2,000 miles (3,200 km) from Iquitos to Belém. Marshy areas along the banks are liable to seasonal floods, but most of the basin is covered by dry-footed equatorial rainforest that experiences daily convectional thunderstorms. Along its middle and lower course, the Amazon receives many tributaries from the Guiana Highlands to the north, and the Brazilian Central Plateau in the south. These sources—Andean glaciers, daily rains, and the numerous tributaries—contribute to the 203 billion gallons (770 billion liters) of water that the Amazon pours into the Atlantic every hour. The flow of the river is so strong that there is no delta; instead, it discharges through a wide, mangrove-fringed estuary that straddles the equator. Despite its size, the Amazon delivers less sediment to the ocean than the Mississippi, which has only one-tenth of the Amazon's flow.

**AMAZON HEADWATERS**
*Many of the Amazon's Andean tributaries freeze in winter, as seen here in the Bolivian Cordillera Real.*

**BRAIDED CHANNELS NEAR MANAUS**
*Along its middle course, the river frequently braids between strings of low islands that constantly shift as the river channel alters following seasonal rains.*

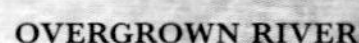

**OVERGROWN RIVER**
*The river's placid backwaters are covered in mats of vegetation, such as these giant water lilies, which provide a valuable habitat for many animals.*

**GIANT RIVER OTTER**
*This highly endangered mammal lives in small family groups along quiet stretches of the river throughout the Amazon Basin.*

## DEFORESTATION

The Amazon Basin cradles the greatest expanse of rainforest on Earth—an irreplaceable resource that is being destroyed at an alarming rate. The Amazonian forests (see p.311) face a number of threats—logging, large-scale agriculture, and the slash-and-burn subsistence farming of internal migrants drawn to the Amazonian frontier. Stripped of cover, and deprived of the long-term release of the nutrients locked within the trees, Amazonian soils deteriorate rapidly into barren wastes, and add to the river's sediment load. Despite widespread concern and publicity, the deforestation continues, and some experts predict a catastrophic collapse of the entire forest ecosystem within the next 30–50 years.

**AMAZON TRIBUTARY**
*This view of the Cuieiras River (a tributary of the Rio Negro) is typical of the Amazon Basin—wide, unruffled rivers branching through a seemingly endless lush rainforest.*

EUROPE *west*

## Thames

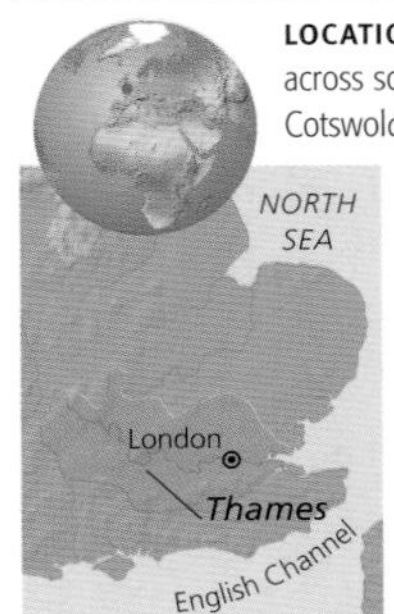


**LOCATION** In Great Britain, flowing across southern England from the Cotswold Hills to the North Sea

**LENGTH** 210 miles (336 km)

**VERTICAL FALL** 356 ft (108 m)

**BASIN AREA** 5,200 square miles (13,600 square km)

**MAIN TRIBUTARIES** Colne, Kennet, Wey

The longest river in Great Britain, the Thames arises from springs in the Cotswolds, a ridge of limestone hills in south-central England. The upper course winds eastward along the northern edge of the Chiltern Hills until it breaks through at the Goring gap near Oxford. Here, where it is known locally as the Isis, the river is about 100 ft (30 m) wide. The lower course flows southeast along the wide, shallow Thames valley, which is characterized by gravel terraces and clays deposited by successive pulses of postglacial meltwater. The river is tidal for the lower 90 miles (145 km) of its length, and the tidal range at London, the lowest bridging point, is about 23 ft (7 m). Although London's inhabitants have a comfortable relationship with the Thames, the river in fact poses a considerable threat. Heavy rains combined with exceptionally high tides have flooded the city on many occasions. Below London, the Thames widens into an estuary that is about 5 miles (8 km) wide, and then discharges into the North Sea.

**THE THAMES BARRIER, LONDON**
*The Thames Barrier is a series of huge steel gates that can be raised to protect the city against flooding at high tide.*

EUROPE *southwest*

## Tagus

**LOCATION** Flowing across the Iberian Peninsula from eastern Spain to the Atlantic coast of Portugal

**LENGTH** 626 miles (1,007 km)

**VERTICAL FALL** 5,216 ft (1,590 m)

**BASIN AREA** 31,505 square miles (81,600 square km)

**MAIN TRIBUTARIES** Gallo, Zezere

Rising in the Sierra de Albarracin about 90 miles (140 km) east of Madrid, the Tagus is one of several rivers that drain the Meseta Plateau. Initially fast-flowing, through limestone gorges, the river slows and widens as it flows westward, past the city of Toledo, through increasingly arid landscapes. For a short distance it forms part of the Spain–Portugal border, and the final 170 miles (275 km) of its course flow through Portugal. The estuary at Lisbon is one of the world's finest natural harbors. The Tagus has been widely harnessed for both power and irrigation by both countries. Upstream of the border section, a dam at Alcántara in Spain has created an artificial lake, large by European standards, that extends back nearly 100 miles (160 km) along the river's middle course.

**TAGUS RIVER PASSING TOLEDO, SPAIN**
*The oldest parts of this city, founded before the time of the Romans, are perched high on a rocky outcrop that overlooks a bend in the river's gorge.*

EUROPE *west*

## Severn

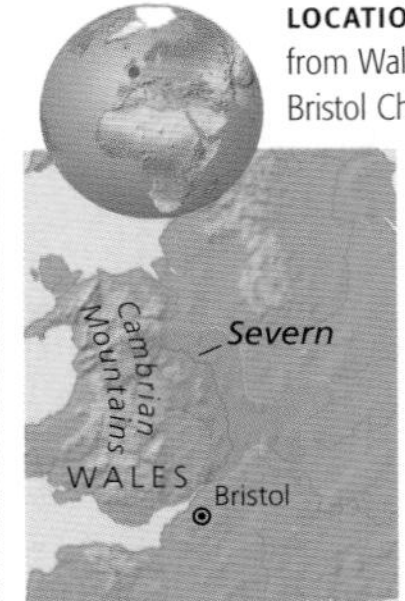


**LOCATION** In Great Britain, flowing from Wales through England to the Bristol Channel and the Irish Sea

**LENGTH** 180 miles (290 km)

**VERTICAL FALL** 2,000 ft (600 m)

**BASIN AREA** 4,350 square miles (11,266 square km)

**MAIN TRIBUTARIES** Avon, Stour, Teme, Usk

Rising on Plynlimon mountain in western Wales, the upper course of the Severn descends 1,500 ft (455 m) in its first 15 miles (24 km) into the Vale of Powys. After cutting the Ironbridge Gorge through hills on the English–Welsh border, the river follows a semicircular course through central England. One of its middle-course tributaries, the Warwickshire Avon, flows through William Shakespeare's hometown of Stratford-upon-Avon. South of Gloucester the river widens and meanders to an estuary that discharges into the Bristol Channel below Newport. Locally, the lower course of the Severn is noted for its tidal bore, a wave that moves at about 15 mph (25 km/h), reaching a height of nearly 10 ft (3 m).

EUROPE *west*

## Seine

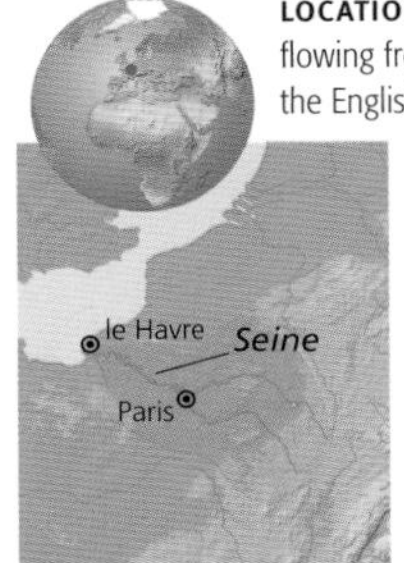


**LOCATION** In northern France, flowing from the Burgundy region to the English Channel at Le Havre

**LENGTH** 480 miles (780 km)

**VERTICAL FALL** 1,545 ft (471 m)

**BASIN AREA** (32,000 square miles (83,000 square km)

**MAIN TRIBUTARIES** Aube, Marne, Oise

The Seine defines what is often called the Paris Basin. Upstream of the city, the river and its tributaries have cut down through limestone to create a series of fertile plateaus that form the region known as the Île de France. Downstream of Paris, the Seine flows through chalk hills, and below Rouen it follows a series of wide meanders before widening into its estuary.

**ALEXANDRE III BRIDGE, PARIS**

EUROPE *west*

## Loire

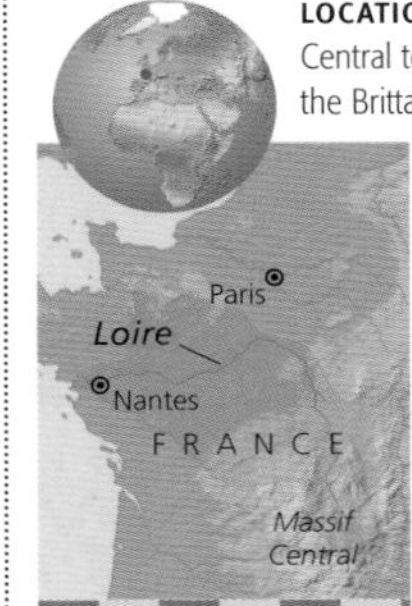


**LOCATION** Flows from the Massif Central to the Atlantic coast south of the Brittany peninsula in France

**LENGTH** 634 miles (1,020 km)

**VERTICAL FALL** 4,500 ft (14,850 m)

**BASIN AREA** 45,000 square miles (117,000 square km)

**MAIN TRIBUTARIES** Allier, Maine, Vienne

The longest river in France, the Loire rises among volcanic peaks in the Cevennes mountains at the south-east edge of the Massif Central. Its source is only about 100 miles (160 km) from the Mediterranean Sea and the mouth of the Rhône. Initially flowing north along a series of faulted depressions in the massif, the Loire then meanders through central France along a shallow, steep-sided valley. The river pivots west near the city of Orléans, passing Tours and Angers. Below Nantes, the Loire widens to an estuary that discharges into the Bay of Biscay at St. Nazaire. The river's flow is very irregular, and parts of the middle course, which is studded by strings of sandy islands, run almost dry in summer. In autumn and spring, heavy rains can cause flooding, especially in the lower reaches.

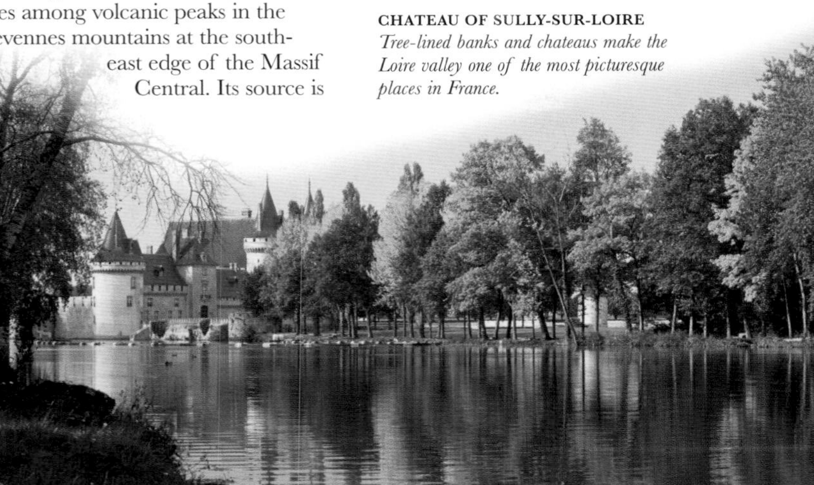

**CHATEAU OF SULLY-SUR-LOIRE**
*Tree-lined banks and chateaus make the Loire valley one of the most picturesque places in France.*

LAND

EUROPE *northwest*

## Rhine

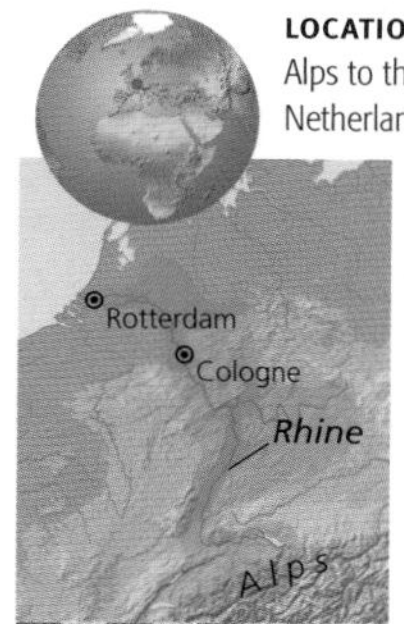


**LOCATION** Flowing from the Swiss Alps to the North Sea coast of the Netherlands

**LENGTH** 820 miles (1,320 km)

**VERTICAL FALL** 7,690 ft (2,339 m)

**BASIN AREA** 85,000 square miles (220,000 square km)

**MAIN TRIBUTARIES** Main, Moselle, Neckar

The most commercially important river of western Europe, the Rhine rises high in central Switzerland. Its headwaters, the Vorderrhein and the Hinterrhein, are fed by snowmelt that reaches a maximum in June and July. The upper Rhine flows north into Lake Constance (see p.245) through a small delta. From the western end of the lake, the river continues westward, plunging over the 100-ft- (30-m-) high Rhine Falls in northern Switzerland, and along the southern edge of the Black Forest mountains, before swinging north near Basel into the two-stage Rhine Valley. The upper part of the valley is a broad and ancient rift valley (see p.147) about 20 miles (32 km) wide—with the Vosges mountains to the west and the Black Forest to the east—the sides of which contain the river's floodplain. Just below Basel, for about 100 miles (160 km), the Rhine forms part of the border between Germany and France. Downstream, the German section of river has been straightened during the last two centuries by cutting through its many meanders. After passing Mainz, the Rhine enters the lower valley, a narrow, twisting gorge, 90 miles (145 km) long, cut between the Hunsrück and Eifel hills on the western side, and the Taunus and Westerwald on the eastern side. South of Bonn, the valley widens onto a plain, and the river's lower course passes the cities of Cologne, Düsseldorf, and Duisburg. The Rhine estuary begins below Emmerlich near the German–Dutch border, and the main stream enters the North Sea at Rotterdam. The River Meuse follows a parallel course to the sea and adds its waters to their combined estuary.

**THE RHINE NEAR KAUB**
*The snaking course of the narrow lower valley is overlooked by hundreds of castles, some of which date back many centuries.*

### INDUSTRIAL RIVERS

The Rhine is heavily used by commercial vessels, such as these ships passing the huge Bayer chemical plant in Leverkusen, Germany. Most of the commercial traffic is concentrated in the heavily industrialized section between Cologne and the sea. Despite strict controls over both shipping and disposal of factory waste, the water is heavily polluted.

EUROPE *west*

## Rhône

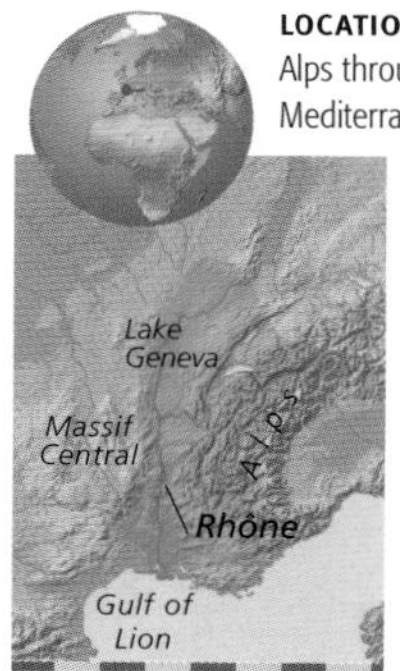


**LOCATION** Flowing from the Swiss Alps through France to the Mediterranean Sea

**LENGTH** 505 miles (813 km)

**VERTICAL FALL** 6,000 ft (1,800 m)

**BASIN AREA** 37,000 square miles (96,000 square km)

**MAIN TRIBUTARIES** Ardèche, Arve, Isère, Saône

The Rhône is the only major European river that flows into the Mediterranean Sea. Descending from the Rhône glacier in southern Switzerland, its upper course flows westward across the Valais region, through gorges that follow structural faults and along glaciated valleys. The river then turns north across the Alps and forms a delta as it enters Lake Geneva (see p.245). The Rhône exits the southwestern end of the lake and follows a zigzag westward course through the Jura mountains of eastern France. Near Lyons it receives the waters of its main tributary, the Saône, and turns southward. Its lower course follows a trough between the Alps and the Massif Central. Receiving tributaries on both banks, the Rhône proceeds through a series of gorges cut through rocky outcrops, separated by sediment-filled basins with terraced sides that are often inundated by seasonal flooding. As it approaches the sea, the river braids over an increasingly marshy landscape. The Rhône delta starts near Arles, where the river diverts into two main channels that flow across the Camargue wetland (see p.329) and into the sea west of Marseilles.

**HEADWATERS OF THE RHÔNE**
*The ice-cold waters of the river cross the Vallée de la Clarée near Briançon in the French Alps.*

**CHAMOIS**
*This agile antelope grazes along the upper reaches of the Rhône in the French and Swiss Alps. In winter it descends to lower levels to find food.*

**RHÔNE AT AVIGNON**
*Along its lower reaches, the Rhône can present a placid, tranquil image that is quite at odds with its wild and racing Alpine origins.*

EUROPE *central*

## Oder

**LOCATION** Flowing from the Czech Republic to the Baltic Sea in Poland

**LENGTH** 550 miles (886 km)

**VERTICAL FALL** 2,080 ft (633 m)

**BASIN AREA** 46,000 square miles (119,000 square km)

**MAIN TRIBUTARIES** Neisse, Warta

The Oder rises in the mountainous northeast of the Czech Republic, but nine-tenths of its drainage basin lies within Poland, and a stretch of its lower course forms part of the Polish–German border. The river exits the mountains through a depression known as the Moravian Gate. It is a very wide river, up to 6 miles (9.5 km) across in places. It is also very shallow, with an average depth of about 3 ft (90 cm). The river freezes for about one month in winter. As it approaches the sea, the main channel braids into several parallel streams. Below the city of Szczecin, it enters a lagoon and then flows into the Bay of Pomerania.

# DAMS

Dams are some of the largest engineered structures on Earth. They supply almost a fifth of the world's electricity, and they significantly reduce the risk of floods and droughts. However, in the industrialized world, the construction of very large dams is almost at a standstill, largely because suitable sites have already been exploited. In developing countries, such as India and China, it continues apace.

## DAMS AND DRAINAGE

Earth-filled dams date back to the time of ancient Egypt, but concrete ones are a much more recent innovation. Dam-building became a matter of national prestige in the 20th century. The US, for example, built a series of massive dams in the Colorado River basin (see p.218), and the former Soviet Union dammed its major rivers and supervised the building of the Aswan High Dam, a rock-filled barrage straddling the Nile River in Egypt (see p.229). These huge projects were followed by others in Africa and South America. One of them—the Itaipú Dam, on the border between Brazil and Paraguay—currently houses the largest hydroelectric plant in the world.

Electricity production is the chief reason for building mega-dams, but they also smooth out seasonal variations in water flow, preventing potentially dangerous floods. In China, this is one of the reasons for building the Three Gorges project, which will regulate the Yangtze River (see p.232). When complete, it will stop disastrous floods, which over the centuries have claimed many lives.

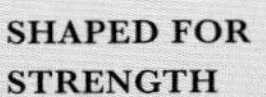

**SHAPED FOR STRENGTH**
*The Hoover Dam spans the lower reaches of the Colorado River. At 726 ft (221 m), it is the tallest dam in the US.*

### LARGEST HYDROELECTRIC DAMS

The world's hydroelectric dams have a combined capacity of about 700,000 megawatts (MW), but they exploit only about a quarter of this energy, owing to practical and environmental considerations.

| Name | Location | Max power (MW) |
|---|---|---|
| Itaipú | Brazil–Paraguay | 14,000 |
| Guri | Venezuela | 10,000 |
| Tucuruí | Brazil | 8,370 |
| Grand Coulee | US | 6,494 |
| Sayano-Shushensk | Russia | 6,400 |
| Krasnoyarsk | Russia | 6,000 |
| Churchill Falls | Canada | 5,428 |
| La Grande-2 | Canada | 5,328 |
| Bratsk | Russia | 4,500 |
| Moxoto | Brazil | 4,328 |

## ECOLOGICAL EFFECTS

The Three Gorges project epitomizes the environmental conundrum that dams can pose. The power they generate is renewable and theoretically pollution-free. These are major advantages for a country such as China, which still relies heavily on coal. But with extremely large dams, these benefits are offset by some serious drawbacks. By blocking sediment flow,

**DIVIDED BY DAMS**
*The erection of dams shown on the map above has limited the movements of the rare Indus River dolphin. The dolphins are now confined to isolated pockets between dams, in which they interbreed and so jeopardize their survival.*

dams deprive floodplains of fertile silt, which wetland plants need to become established on exposed soils. Instead, the silt accumulates in reservoirs, which reduces their working life.

For freshwater wildlife, large dams have mixed effects. Reservoirs create new habitats for animals, but dams form barriers that can be difficult to pass. This is a problem for migratory fish, and also for other species that normally stay on the move. Riverside plants decline when sediment levels drop, although reservoirs often make ideal homes for floating water plants. Unfortunately, these floating plants include invasive species such as water hyacinth, a problematic weed in warm parts of the world.

## GAINS AND LOSSES

Dams make existing water supplies easier to exploit. In the American southwest, many farms depend on water from the Colorado Basin reservoirs, as do rapidly growing cities such as Phoenix and Las Vegas. In this region, dams attract plenty of environmental criticism, but current lifestyles there would be unsustainable without the water that dams provide.

In other parts of the world, small-scale dams are vital for rural life. In southern India, for example, thousands of hand-built reservoirs—which are known locally as tanks—capture rainwater from the annual monsoon. Without them, agriculture would be impossible during the dry season. But dams can also displace people and threaten their livelihoods. When the Three Gorges project is finally completed, an estimated 1.3 million people will lose their homes, and their farmland will disappear under a lake 373 miles (600 km) long.

**FLOODED FOREST**
*Rotting vegetation in newly flooded reservoirs releases methane—a gas that adds to the greenhouse effect.*

**LIQUID RESERVES**
*The Glen Canyon Dam on the Colorado River generates hydroelectric power. The water it traps is also used for drinking and irrigation.*

EUROPE *central*

## Danube

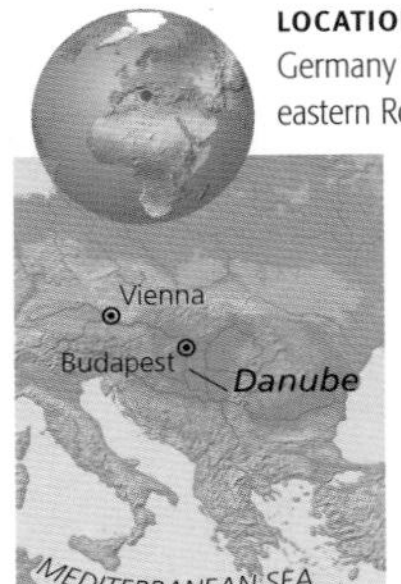


**LOCATION** Flowing from southern Germany to the Black Sea coast in eastern Romania

**LENGTH** 1,770 miles (2,860 km)

**VERTICAL FALL** 2,174 ft (662 m)

**BASIN AREA** 315,000 square miles (816,000 square km)

**MAIN TRIBUTARIES** Drava, Sava, Tisza

Europe's second-longest river, the Danube, rises in Germany from the confluence of the Brege and Brigach rivers. It flows eastward into Austria, descending through the Hungarian Gate to the Little Alföld plain. After passing Budapest, the river flows southward across the Great Hungarian Plain and between the Carpathian and Balkan mountains. Here the river has cut the deepest gorge in Europe, the Iron Gate, with sheer walls 2,600 ft (800 m) high. The Danube's lower course passes across the Wallachian Plain. As it nears the Black Sea, the river is deflected northward by the ancient and eroded Dobruja Hills, before turning east once more. It discharges into the sea through three main distributaries, in a delta that has an area of about 1,900 square miles (5,000 square km).

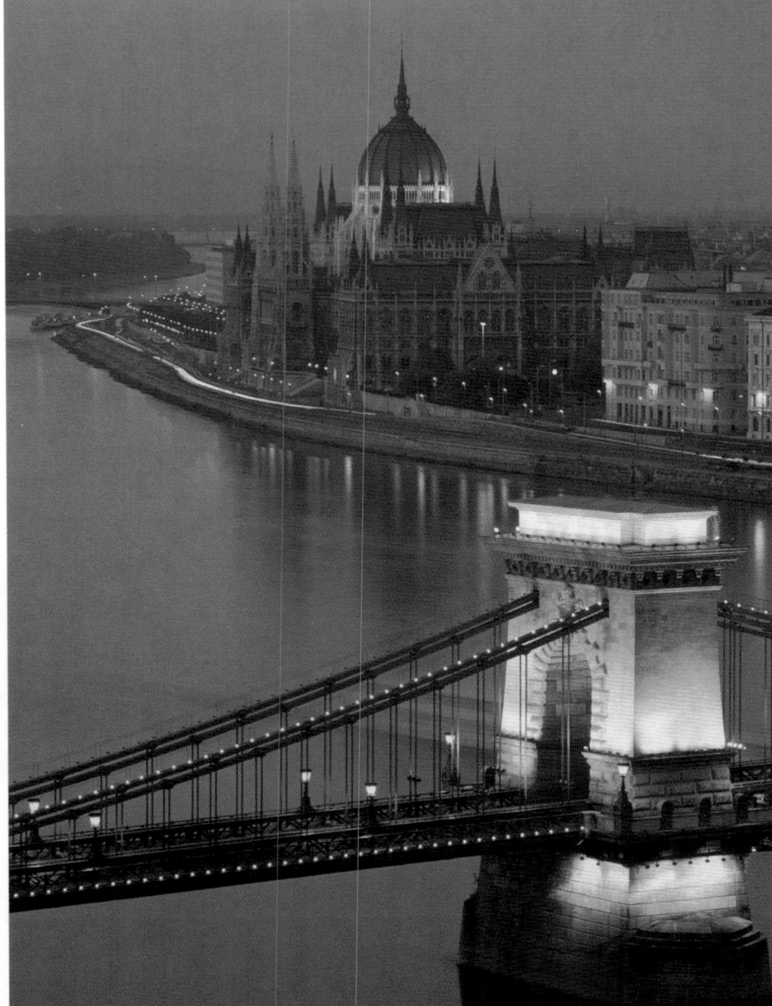

**THE DANUBE PASSING BUDAPEST**
*The Chain Bridge, seen here in front of the Hungarian Parliament Building, joins what were once two separate cities—Buda and Pest.*

### FLOODING

During the summer of 2002, unusually heavy summer rains across central Europe caused the Danube to overtop its banks in many places. Millions of people living along its course were affected, and thousands of homes were inundated, as here in the Slovak capital, Bratislava. Downstream, the Hungarian capital, Budapest, was also affected. Upstream of Bratislava, in Austria, the Danube reached an all-time high as the region experienced its heaviest rainfall ever recorded. The town of Linz and part of Vienna were flooded, despite an excellent system of flood prevention.

LAND

EUROPE *southeast*

## Volga

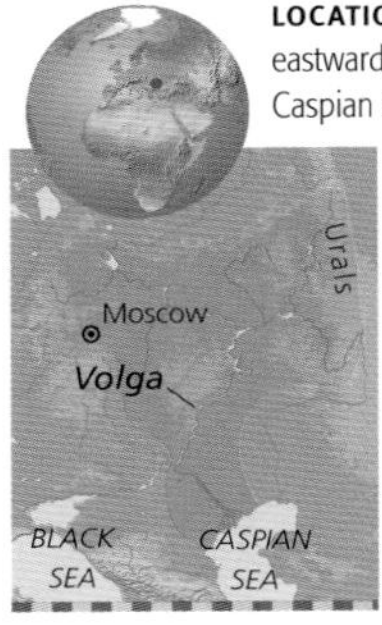


**LOCATION** In Russia, flowing eastward then southward to the Caspian Sea

**LENGTH** 2,293 miles (3,690 km)

**VERTICAL FALL** 848 ft (258 m)

**BASIN AREA** 525,000 square miles (1.4 million square km)

**MAIN TRIBUTARIES** Oka, Kama

The Volga is Europe's longest river and one of Russia's most important transportation routes. The river rises 748 ft (228 m) above sea level in the Valdai Hills about 250 miles (400 km) northwest of Moscow, and its drainage basin extends eastward to the Urals. Its upper course runs northeast through swamp forest, lakes, and reservoirs, then turns southeast along a narrow valley between the Danilov and Uglich highlands. Near the city of Nizhniy Novgorod, the Volga receives the waters of the Oka and increases considerably in size. The river continues its southeast course through damp forest until it passes Kazan, where it turns southward and flows through a progressively more arid landscape. Downstream of Kazan the river, swollen by the inflow of the Kama, flows between the Volga Hills and the Turgay Plateau. The most notable feature of its middle course is the Samara Bend—a remarkable loop 100 miles (160 km) long, around a piece of high ground. Near Volgagrad the river splits in two. The Volga and its main distributary, the Akhtuba, follow a parallel course into the Caspian Depression where, downstream of Astrakhan, they discharge through a wide delta into the Caspian Sea (see p.250) about 100 ft (30 m) below sea level. The Volga freezes for about 100 days each year. Its importance for transport has been enhanced by the construction of canals, notably the Volga–Baltic Waterway completed during the 1960s. Numerous hydroelectric dams have also been built across its course.

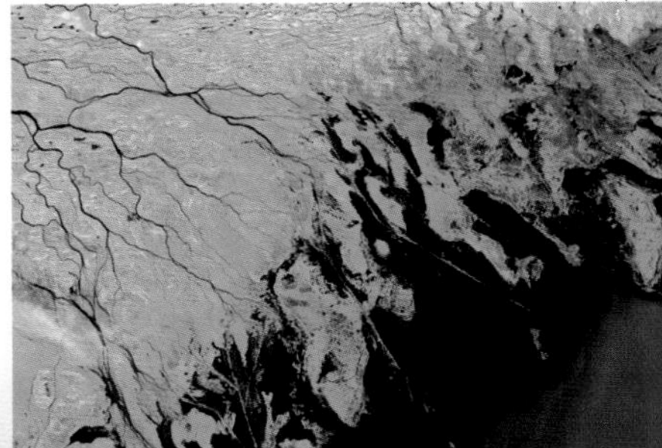

**VOLGA DELTA**
*The delta (shown here in a satellite image) is lush and well-watered compared with the barren desert landscape that surrounds it.*

**VOLGA PASSING NIZHNIY NOVGOROD**
*Ice-free for about 200 days each year, the Volga has for centuries formed a river highway through the heart of European Russia.*

AFRICA *northeast*

# Nile

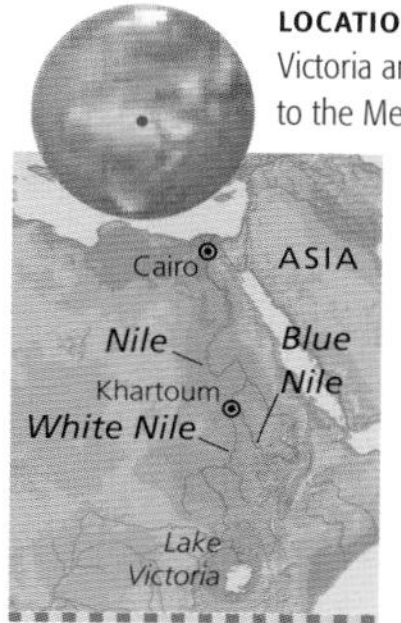


**LOCATION** Flows north from Lake Victoria and the Ethiopian highlands to the Mediterranean coast in Egypt

**LENGTH** 4,132 miles (6,648 km)

**VERTICAL FALL** 3,725 ft (1,135 m)

**BASIN AREA** 1.3 million square miles (3.4 million square km)

**MAIN TRIBUTARIES** White Nile, Blue Nile

The longest river in the world, the Nile is one of the few that are recognized as having more than one main source. The longest stream of the Nile—the White Nile—flows from the northern end of Lake Victoria (see p.247) as the Victoria Nile and over the Murchison Falls into the northern part of the Great Rift Valley (see pp.148–49). The river crosses the northern end of Lake Albert (exiting as the Albert Nile) and then descends through a series of gorges to the flat, marshy plain of the Sudd (see p.331). At Khartoum, the White Nile is joined by the waters of the Blue Nile, with its source at Lake Tana in the Ethiopian Highlands. From Khartoum to southern Egypt, the Nile passes through a series of six great cataracts and then enters a long, narrow valley between desert escarpments before it widens to a broad triangular delta north of Cairo. This delta region, and a long, narrow green ribbon along the Nile, make up the 3 percent of Egypt's land area that supports cultivation. Thanks to its two sources, the Nile provides both a constant source of water and an annual source of fertile silt. The White Nile, flowing from equatorial rain-fed lakes, ensures the Nile's constancy, but contributes only 16 percent of its volume downstream of Khartoum. The Blue Nile contributes 84 percent of the lower river's volume, and accounts for all of its variability. The regular annual flooding of the Nile, which delivers fine, black, fertile silt across the floodplain, is caused by winter rains and snowmelt in the Ethiopian Highlands. The flood season begins in April at Khartoum and gets progressively later in the river's more northerly reaches—the high water of the floods at Cairo occurs in October. The completion of the Aswan High Dam in 1970 created a huge reservoir, Lake Nasser, which extends across the Egyptian–Sudanese border. The dam has significantly moderated the downstream flow of the Nile.

nostrils on top of snout

eyes on top of head

muscular, vertically flattened tail

**NILE CROCODILE**
*This dangerous, powerful reptile is found along the entire length of the Nile and White Nile, and also on rivers in central and western Africa.*

**THE UPPER NILE VALLEY**
*As it flows through verdant forest-lined rapids in Uganda, the Victoria Nile is a foaming, turbulent, and fast-flowing river fed by heavy year-round rainfall across Lake Victoria's drainage area.*

**FERTILE NILE VALLEY**
*The banks of the Nile near Luxor are lined with fields. The farmed area extends only as far as the escarpment in the background.*

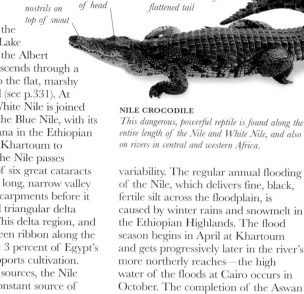

## ABU SIMBEL

The damming of the Nile to create Lake Nasser in the 1960s threatened to submerge several important relics of Egypt's ancient civilization, including the temple of Pharaoh Ramses II at Abu Simbel, built in the 13th century BC. Engineers and archaeologists cooperated to carefully dismantle the temple and reassemble it at a new site some 200 ft (60 m) above the waters of the lake. A smaller temple dedicated to Nefertiti, Ramses' favorite wife, was moved to the same site.

**THE NILE DELTA**
*This satellite image shows the Nile's enormous influence on Egypt. Most of the population lives in the delta region, or scattered along the valley.*

AFRICA *west*

# Congo

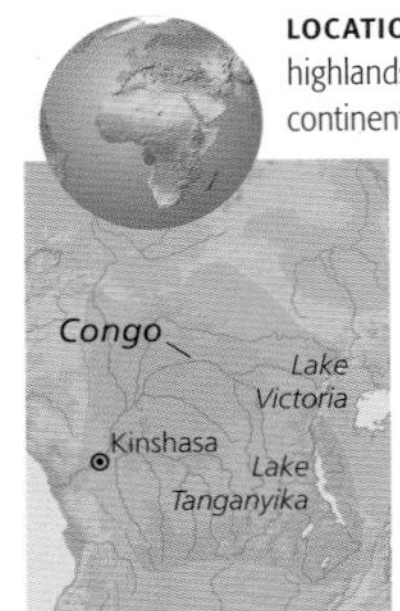


**LOCATION** Flows from the highlands of east Africa across the continent to the Atlantic Ocean

**LENGTH**
2,900 miles (4,670 km)

**VERTICAL FALL**
5,780 ft (1,760 m)

**BASIN AREA**
1.4 million square miles (3.5 million square km)

**MAIN TRIBUTARIES** Kwa, Lualaba, Sangha, Ubangi

The Congo is second only to the Amazon in the volume of water it carries to the oceans. It is the second-longest river in Africa, and sweeps in a great counterclockwise loop that drains most of central Africa. The Congo's headwaters are fed by rains on the western slopes of the East African Plateau near the southern end of Lake Tanganyika (see p.248) in Zambia. The upper course collects several smaller rivers and flows through rapids and small lakes before descending over the seven-stage Boyoma Falls upstream of Kisangani and entering the Congo Basin—the vast, near-circular depression that extends over most of the river's catchment area. The river widens considerably over the lowlands, reaching a width of 8 miles (13 km) in places, with strings of islands dividing the flow into several channels. Most of the basin is covered by tropical rainforest, and there are extensive marshes along the river margins and between the many tributary streams. This rain marsh, as it is known, is subject to flooding at all times of the year, especially in the western half of the basin. When tributaries that usually reach their highest level at different times happen to reach peak flow simultaneously, flooding can be extensive. The Congo leaves its basin flowing southwest through a narrow gorge, known as the Chenal or Corridor, that it has cut through the Bateke Plateau. It then widens and flows into the Malebo Pool, a lake 17 miles (27 km) wide. The city of Brazzaville is located on the northern shore, while Kinshasa stands on the southern shore. Upstream of the pool, the Congo is navigable for more than 1,000 miles (1,600 km) of its length. Downstream of Malebo, rapids and waterfalls, which long hindered European exploration of the river, block access to the sea. Over about 200 miles (320 km), the river descends nearly 1,000 ft (300 m) and then continues along a gorge and through mangrove forests to the Atlantic. The flow of the Congo is so strong and constant that no delta has formed; instead, the flow continues along a submarine canyon, depositing its sediment in a fan on the ocean floor. Many of the features on the Congo (including the Malebo Pool, Boyoma Falls, and the town of Kisangani) were formerly named after British explorer Henry Stanley, who, together with his French rival Pierre Brazza, was among the first Europeans to explore this region of Africa.

**BOYOMA FALLS**
*Fish are a great natural bounty for the people of the Congo Basin. In swirling rapids like this one upstream of Kisangani, fish are harvested in conical traps hung from sturdy wooden scaffolding.*

**LIVING BETWEEN RIVER AND RAINFOREST**
*The Congo's banks are lined with villages like this. The villagers benefit from plentiful food and good transportation, but they live with the risk of flooding.*

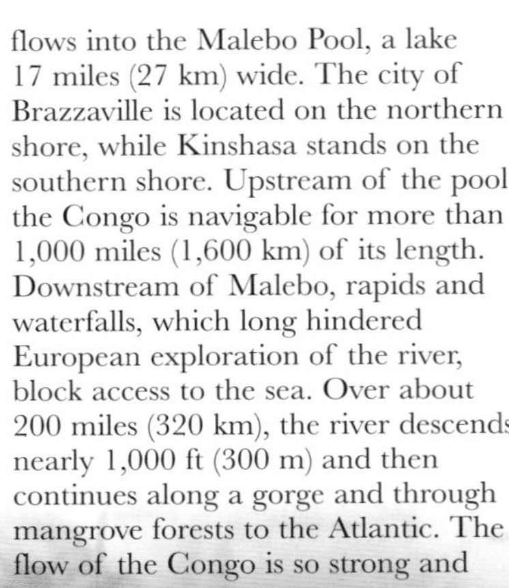

**UPPER CONGO**
*The upper reaches of the Congo are a hot, wet, sparsely populated and little-explored region of the African continent.*

LAND

AFRICA *west*

## Niger

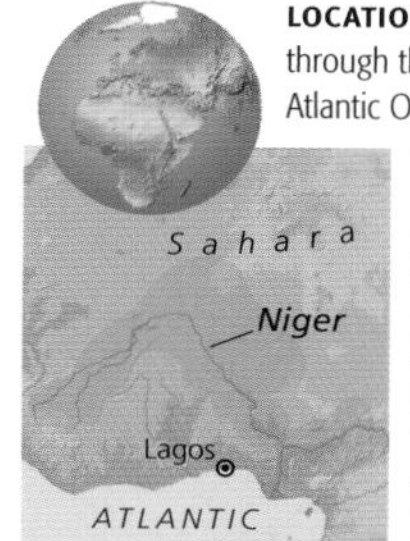


**LOCATION** Flows from Guinea through the southern Sahara to the Atlantic Ocean in Nigeria

**LENGTH** 2,600 miles (4,180 km)

**VERTICAL FALL** 2,800 ft (850 m)

**BASIN AREA** 730,000 square miles (1.9 million square km)

**MAIN TRIBUTARIES** Bani, Benue, Kaduna

The major river of West Africa, the Niger rises in the Fonta Djallon highlands of northern Guinea. It flows northeast over the lowlands of the Sahel, then curves through the Sahara Desert before turning southeast to pass through Niger and Nigeria. The river flows through dense rainforest before reaching the sea, forming a delta in the Gulf of Guinea.

**NIGER DELTA**
*The delta where the Niger meets the sea is dwarfed by its inland delta, but is still the largest in Africa and is mainly covered in dense mangrove forest.*

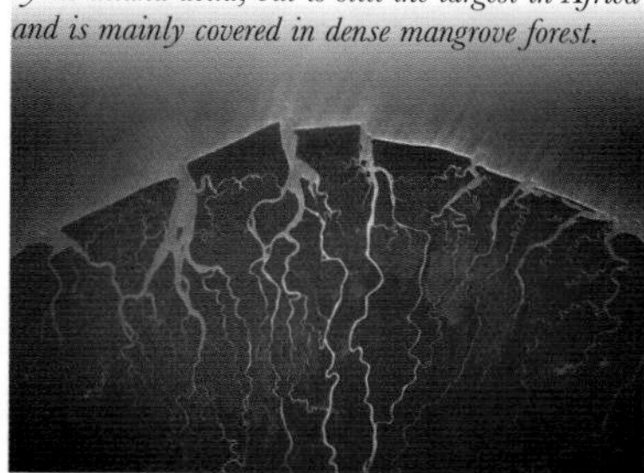

AFRICA *southeast*

## Zambezi

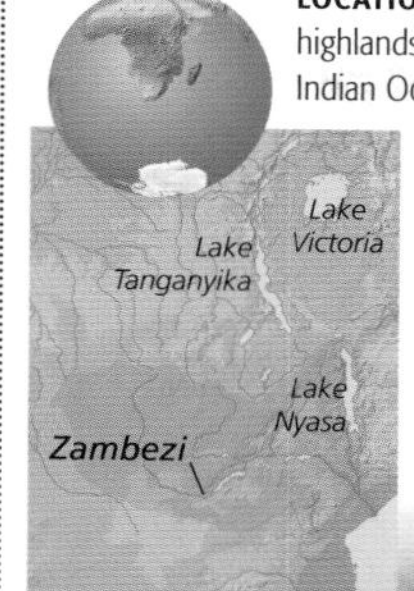


**LOCATION** Flows from the highlands of central Africa to the Indian Ocean in Mozambique

**LENGTH** 2,200 miles (3,540 km)

**VERTICAL FALL** 4,800 ft (1,460 m)

**BASIN AREA** 500,000 square miles (1.3 million square km)

**MAIN TRIBUTARIES** Kafue, Shire

Africa's fourth-largest river rises as a small spring in the Kalene Hills of southwestern Zambia, and then flows eastward, gaining water from many small tributaries. The Zambezi attains a width of about 5,500 ft (1,700 m) before plunging over the spectacular Victoria Falls, 355 ft (108 m) high. Downstream of the falls, where the river forms the border between Zambia and Zimbabwe, its deep, wide valley has been flooded by a lake 175 miles (280 km) long, which formed following the construction of the Kariba Dam in 1959 to generate electricity for the region.

**HIPPOPOTAMUS**
*Plant-eating hippopotamuses live in the shallows along the Zambezi's edge and can easily capsize canoes and other small craft.*

**VICTORIA FALLS IN ZAMBIA**
*More than double the height and width of Niagara Falls (see p.238), the local name for this magnificent waterfall, Mosi-oa-Tuya, translates as "the smoke that thunders."*

ASIA *west*

## Euphrates

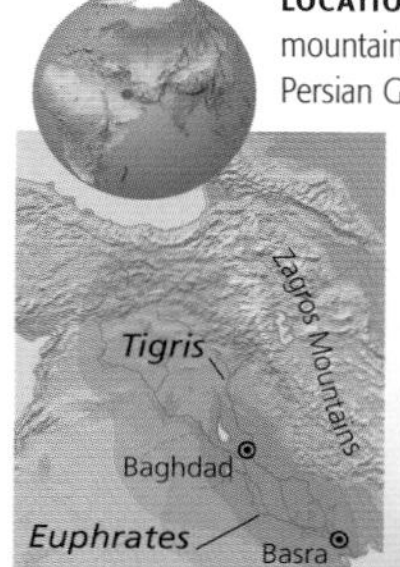


**LOCATION** Flows from the mountains of eastern Turkey to the Persian Gulf

**LENGTH** 1,740 miles (2,800 km)

**VERTICAL FALL** 8,800 ft (2,680 m)

**BASIN AREA** 430,000 square miles (1.1 million square km)

**MAIN TRIBUTARIES** Tigris

Together with its main tributary and near twin, the Tigris (which is 1,180 miles/1,890 km long), the Euphrates forms the largest river system in western Asia. Its headwaters originate in the highlands of eastern Turkey. The Tigris rises about 50 miles (80 km) to the east, and collects much of its downstream waters from the Great and Little Zab rivers, which flow from the Zagros Mountains in Iran. For most of their length, the rivers follow nearly parallel courses to the sea, giving rise to the old name for Iraq: Mesopotamia ("land between rivers"). As they descend from the highlands, the two rivers follow deep, converging valleys that define a triangle of desert known as Al Jazirah ("the Island"). Near Baghdad they come within 50 miles (80 km) of each other, only to diverge again. In their lower courses, the rivers' channels are branching and highly unstable. Their combined floodplain is dissected by numerous abandoned channels and irrigation canals. The two rivers meet at Basra and then follow a single course, known as the Shatt al 'Arab, through extensive reed-filled marshes to their outflow at the head of the Persian Gulf.

### WARS OVER WATER

In the rain-starved Middle East, water is the most precious of all natural resources. The Ataturk Dam across the Euphrates is one of several that have been constructed in Turkey. These have greatly restricted the downstream flow of the river, and the governments of both Syria and Iraq have protested strongly about Turkey's "theft" of the water.

**THE EUPHRATES VALLEY**
*Upstream of its confluence with the Tigris, the Euphrates flows along a high-sided valley, the floor of which has been leveled by successive floods and course changes.*

ASIA *east*

# Yangtze

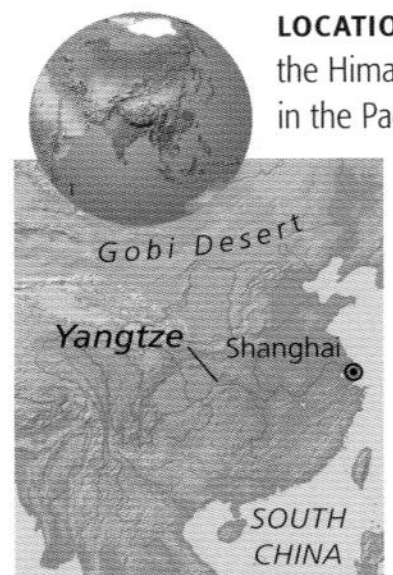


**LOCATION** In China, flowing from the Himalayas to the East China Sea in the Pacific Ocean

**LENGTH**
3,900 miles (6,300 km)

**VERTICAL FALL**
18,000 ft (5,480 m)

**BASIN AREA**
760,000 square miles (2 million square km)

**MAIN TRIBUTARIES**
Han Shui, Ya-Lung Chiang

The Yangtze is the longest river in Asia and the third-longest in the world. It rises in the Kunlun Mountains, south of the Takla Makan desert, and flows southeastward across the Tibetan Plateau along narrow, steep valleys. Over the first 1,000 miles (1,600 km), the Yangtze descends at an average rate of about 10 ft per mile (2 m per km). The river then turns eastward across Sichuan Province before flowing through the famous Yangtze Gorges. The three gorges have a combined length of about 56 miles (90 km) and are no more than 600 ft (180 m) wide, with rock walls towering 1,200–2000 ft (350–600 m) above the river. At this point, the Yangtze reaches depths of more than 500 ft (150 m), making it the deepest river in the world, and it is here that the world's largest dam project is under construction. The river exits the gorges onto a lake-strewn plain and loops south along a wide valley, exiting onto the southern edge of the North China Plain. Along parts of the lower course, the normal river level can be several yards (meters) above that of the plain, kept safely in check by natural and artificial levees. The waters of the Yangtze are essential to this densely populated rice-growing area, where the summer drought can last for eight weeks. Conversely, when the rains finally arrive, the Yangtze frequently overtops its banks, flooding huge areas. After meandering northeastward, the river divides into several distributaries and forms a delta. Shanghai is located between the outflow of the main southern distributaries. The Yangtze has always been China's most commercially important river, and is linked by the Grand Canal, 1,000 miles (1,600 km) long, to the Yellow River and Beijing to the north.

**YANGTZE HEADWATERS**
*In the Tibet Autonomous Region of China, the upper course of the Yangtze River flows through icy mountain scenery that is almost barren of vegetation.*

## FLOOD LEVEES

Constant maintenance is essential to keep the levees along the Yangtze in good condition and safely above the level of the river. Records reveal that there have been catastrophic floods at least 100 times in the last 2,000 years. The floods of 1932 and 1954, though not as high as some in the river's history, inundated the cities of Nanjing and Wuhan, killed about 300,000 people, and made 40 million homeless.

**YANGTZE RIVER DOLPHIN**
*This endangered mammal is found in the lower Yangtze. Overhunting and pollution have greatly reduced its numbers.*

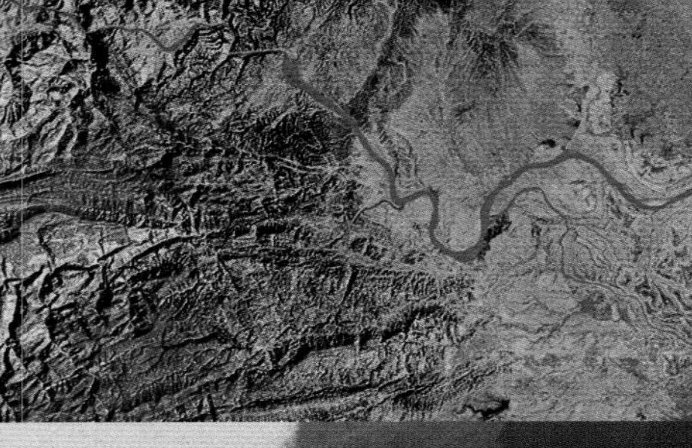

**THE YANGTZE GORGES**
*At the top left of this image, the river can be seen snaking through the Yangtze Gorges before widening and looping across the downstream plain.*

**YANGTZE FISHERMEN**
*The crew of a traditional fishing boat casts nets on the Yangtze as the river flows along one of the many valleys it has cut through the undulating hills of Sichuan Province.*

ASIA *east*

## Yellow River

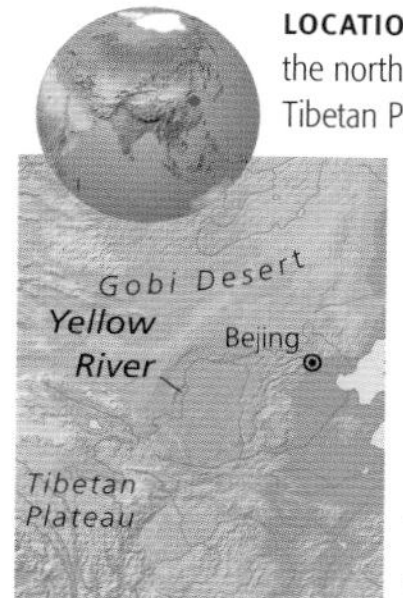


**LOCATION** In China, flowing across the north of the country from the Tibetan Plateau to the Yellow Sea

**LENGTH**
3,390 miles (5,460 km)

**VERTICAL FALL**
15,000 ft (4,500 m)

**BASIN AREA**
740,000 square miles (1.9 million square km)

**MAIN TRIBUTARIES**
Wei He, Fen He

Also known as the Huang-He or Huang-Ho, the Yellow River is China's second-longest river and the muddiest river in the world. It carries 30 times more silt per cubic yard (meter) of water than the Nile, and it delivers almost as much volume of sediment to the sea as the Ganges but transported in a much smaller overall flow. Rising on the Tibetan Plateau, the Yellow River follows an erratic zigzag course across northern China. Initially fast-flowing over barren upland, the river descends onto a high plateau covered by loess deposits (see p.93) that date from the last ice age and are several hundred yards (meters) thick in places. It is here that the river collects its heavy load of sediment. The river continues southward onto the North China Plain, then eastward to its current delta, north of the Shandong peninsula. Since the Yellow River is subject to flooding, most of the lower reaches are ramparted by levees. Heavy flooding can lead to changes in the river's course, and its local nickname is "Ungovernable." Before 1854, the river entered the Yellow Sea south of the Shandong peninsula. When the levees fell into disrepair in the 1850s, floods caused the Yellow River to shift its course some 200 miles (320 km) farther north.

**JAPANESE CRANE**
*This endangered water bird is famous for its elegant and intricate courtship dance, and is considered to be a symbol of happiness and good luck.*

**YELLOW RIVER AT LAJIA**
*In Qinhai Province, western China, the Yellow River flows through narrow valleys and across small floodplains with fields of rapeseed and barley.*

**HOKOU FALLS**
*The yellow color of the river, caused by suspended particles of loess, shows as it falls between Shanxi and Shaanxi provinces in central China.*

ASIA *northwest*

## Ob′

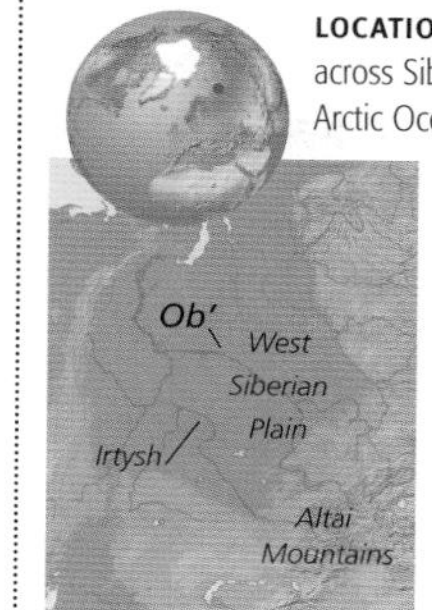


**LOCATION** In Russia, flowing across Siberia to the Kara Sea in the Arctic Ocean

**LENGTH**
2,300 miles (3,700 km)

**VERTICAL FALL**
Exact source unknown

**BASIN AREA**
1.2 million square miles (3 million square km)

**MAIN TRIBUTARIES**
Chulym, Irtysh, Tom

The Ob′ drains the sixth-largest catchment area in the world, and some authorities combine it with the Irtysh to form a single river 3,360 miles (5,400 km) long. The Ob′ rises in the foothills of the Altai Mountains and flows northward through progressively colder and drier forest landscapes, and then across open tundra, where it separates into two channels, the Greater Ob′ and Lesser Ob′. The streams recombine before flowing into an elongated delta that has been partly submerged beneath the sea in the Gulf of Ob′. The river is frozen for up to 200 days per year.

ASIA *south*

## Irrawaddy

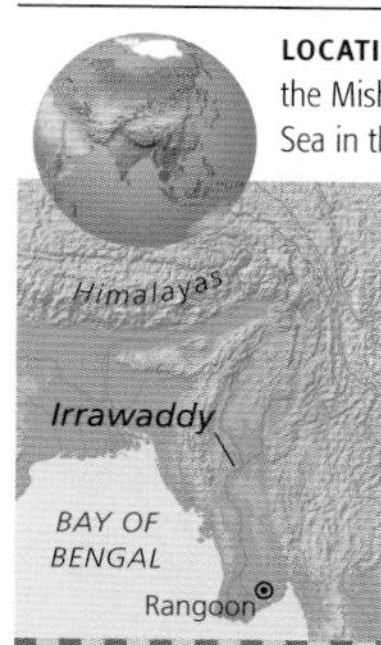


**LOCATION** In Burma, flows from the Mishmi Hills to the Andaman Sea in the Indian Ocean

**LENGTH**
1,300 miles (2,100 km)

**VERTICAL FALL**
Exact source unknown

**BASIN AREA**
158,000 square miles (411,000 square km)

**MAIN TRIBUTARIES**
Chindwin

Rising from the meltwaters of east Himalayan glaciers, the Irrawaddy flows southward to Mandalay and its confluence with the Chindwin. It then meanders through the dry central region of Burma, where it provides much-needed irrigation water. It follows a narrow valley east of the Arakan Yoma highlands, forming a triangular delta west of the Gulf of Thailand. Rangoon is located on one of the numerous distributaries.

**PANNING FOR GOLD AT PAGAN, BURMA**
*In northern Burma, the Irrawaddy and many of its tributaries contain placer deposits, carried down from the mountains when the rivers are in flood.*

ASIA *south*

## Mekong

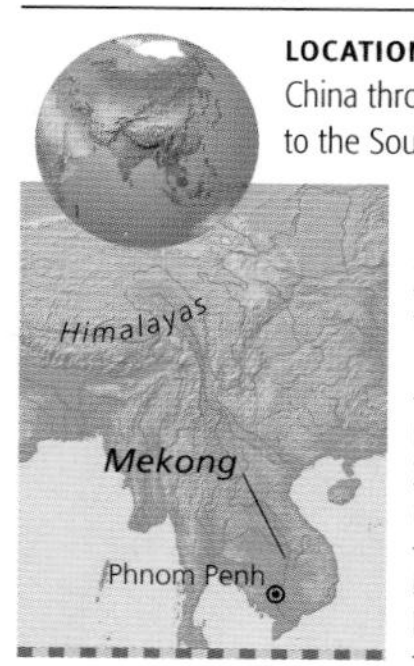


**LOCATION** Flows from western China through Laos and Cambodia to the South China Sea

**LENGTH**
2,500 miles (4,000 km)

**VERTICAL FALL**
16,250 ft (4,950 m)

**BASIN AREA**
305,000 square miles (795,000 square km)

**MAIN TRIBUTARIES**
Kang, Mun, Srepok

The longest river in Southeast Asia, the Mekong rises on the Tibetan Plateau, where it flows along deep valleys and is part of the Burma–Laos border. Downstream it forms part of the Thai–Laos border, flowing through Cambodia before forming a delta near Hô Chi Minh city in Vietnam.

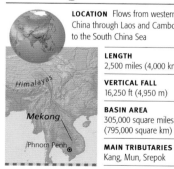

**RAPIDS ON THE MEKONG RIVER**
*The border of Thailand and Laos sees the Mekong pass through spectacular rapids. Local people fish in swirling pools by the riverside.*

ASIA *northwest*

## Ganges

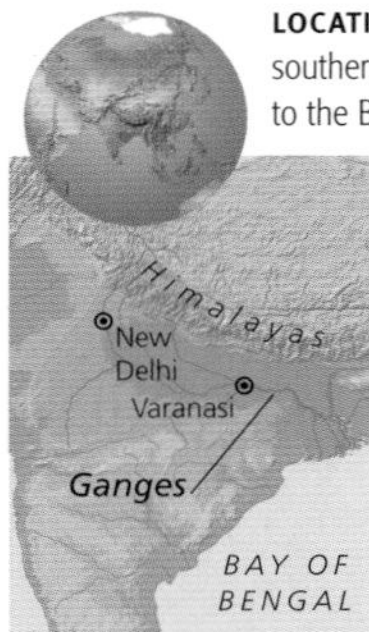


**LOCATION** Flows along the southern edge of the Himalayas to the Bay of Bengal in India

**LENGTH**
1,557 miles (2,506 km)

**VERTICAL FALL**
10,000 ft (3,030 m)

**BASIN AREA**
625,000 square miles (1.6 million square km)

**MAIN TRIBUTARIES**
Brahmaputra, Ghaghar, Yamuna

The Ganges River, sacred to the Hindu religion, drains about a quarter of India, and delivers more sand and silt to the sea than any other river in the world. Boosted at the head of its delta by the inflow of its largest tributary, the Brahmaputra, the Ganges discharges about 2 billion tons of sediment each year. Rising in the southern Himalayas, on the Indian side of the border with China, the Ganges is formed by the confluence of the Alaknanda and Bhagirathi rivers. The upper course flows westward, then southward, and descends onto the north Indian plain about 70 miles (110 km) east of Delhi. The river's middle course meanders sluggishly eastward along a wide floodplain covered with sandy alluvial deposits. Over a distance of more than 1,000 miles (1,600 km), the river falls only about 600 ft (180 m). After passing Varanasi and Patna, the Ganges skirts the northern slopes of the Rajmahal Hills and turns southeast onto the Bengal lowlands. Here, the river divides into many distributaries, creating a network of interlacing channels that stretches 250 miles (400 km) across the northern coast of the Bay of Bengal. The main western distributary is the Hooghly, which flows through Calcutta. Most of the Ganges delta is in Bangladesh, and Dhaka is situated at the northeastern corner. In total, the Ganges delta covers an area of about 22,000 square miles (57,000 square km). The seaward fringes, known as the Sundarbans, are densely forested in places, and are one of the last refuges of the Bengal tiger.

**UPPER COURSE OF THE GANGES**
*In the mountainous north of India, every small area of riverside lowland, overlooked by Himalayan foothills, is intensely cultivated.*

**HEADWATERS OF GANGES**
*The Bhagirathi Valley in Uttar Pradesh, India, is overlooked by the mountain glaciers that are the source of the Ganges.*

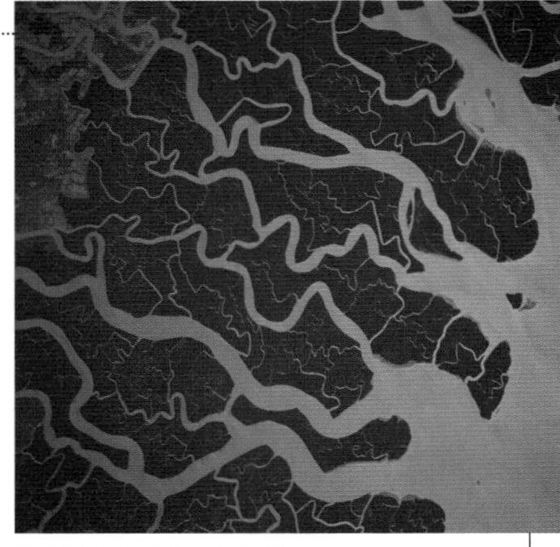

**MOUTHS OF THE GANGES**
*This satellite view shows some of the river's thousands of silt-laden distributaries winding between forested areas of consolidated ground.*

### FLOODING IN BANGLADESH

The Bengal Basin, within which the Ganges delta lies, covers 97,000 square miles (252,000 square km) and is prone to seasonal flooding caused by monsoon rains. However, an even greater flood threat is posed by the tropical cyclones that sometimes sweep up the Bay of Bengal from the Indian Ocean. Storm surges inundate much greater areas than ordinary riverine floods and can extend over 75 percent of Bangladesh's land area, most of which is less than 33 ft (10 m) above sea level. Despite the danger of flooding, the area has a population of about 90 million, and the rich alluvial soils are intensively farmed. The floods of October 1998 affected over a third of the population. Homes were lost, crops and cattle were destroyed, and roads and bridges were washed away.

**GANGES AT VARANASI**
*The Ganges River is considered sacred by Hindus, and millions of pilgrims visit the city of Varanasi each year to immerse themselves in the holy waters.*

ASIA *northwest*

## Indus

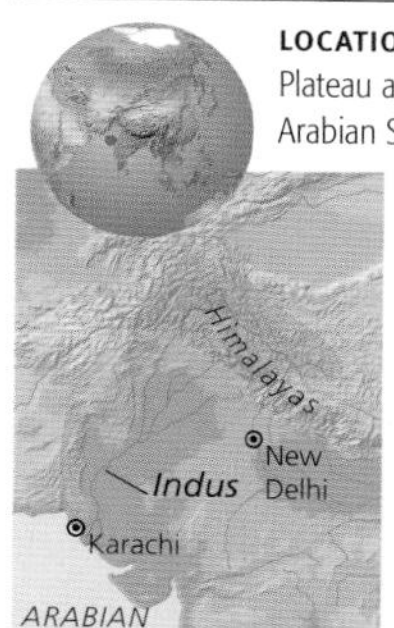


**LOCATION** Flows from the Tibetan Plateau across the Himalayas to the Arabian Sea

**LENGTH** 1,800 miles (2,900 km)

**VERTICAL FALL** 16,000 ft (4,848 m)

**BASIN AREA** 450,000 square miles (1.2 million square km)

**MAIN TRIBUTARIES** Chenab, Kabul, Jhelum, Sutlej

The greatest of the trans-Himalayan rivers, the Indus flows from headwaters fed by snowmelt high in southwest China. The upper course flows northwest through the disputed territory of Jammu and Kashmir, along a wide, faulted valley, between the foothills of the Karakoram and Zaskar mountains. The Indus then turns southwest into Pakistan, where it crosses the Himalayas south of the Hindu Kush, flowing through gorges, the floors of which are more than 12,000 ft (3,650 m) below the surrounding peaks. After its confluence with the Kabul River between Peshawar and Islamabad, the Indus crosses the Potwar Plateau and descends onto the Punjab Plain, where it receives many tributaries flowing from the southern foothills of the Himalayas. Skirting the northern fringes of the Thar Desert, the last 300 miles (480 km) of the river's course meander across a wide floodplain, and downstream of Hyderabad it splits into several distributaries, forming a delta known as the "Mouths of the Indus." The city of Karachi is located at the northern end of the delta.

**INDUS RIVER IN KASHMIR**
*The river flows sedately past sandy terraces that correspond to higher levels in the past, when the river was swollen by snowmelt, and more turbulent.*

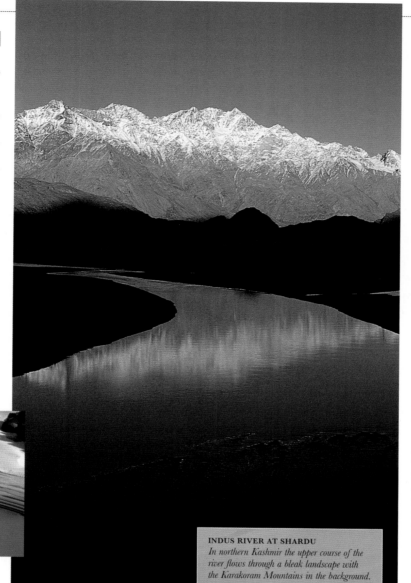

**INDUS RIVER AT SHARDU**
*In northern Kashmir the upper course of the river flows through a bleak landscape with the Karakoram Mountains in the background.*

### EARLY CIVILIZATION

The Indus—like the Nile, Euphrates, and Yellow rivers—gave rise to the first civilizations. Over 4,000 years ago, people of this region lived in cities of baked-brick houses, paved streets, and sewers. Some of them used animal-decorated seals like this one, written in an as yet undeciphered script.

AUSTRALASIA *southeast*

## Murray

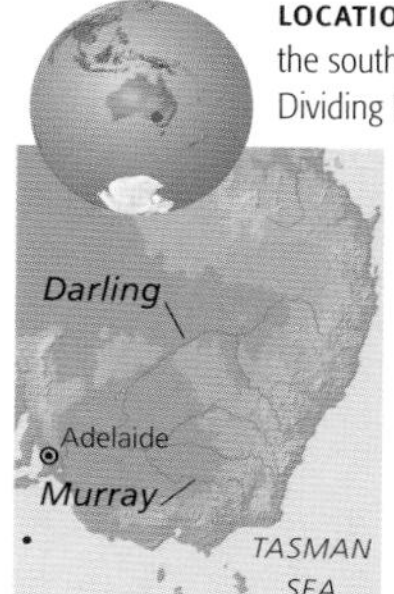


**LOCATION** In Australia, flows from the southern end of the Great Dividing Range to the Indian Ocean

**LENGTH** 1,610 miles (2,590 km)

**VERTICAL FALL** 2,050 m (6,750 ft)

**BASIN AREA** 412,000 square miles (1.1 million square km)

**MAIN TRIBUTARIES** Darling, Murrumbidgee

The Murray, with its major tributary, the Darling, forms Australia's only major river system—one that drains the temperate southeastern corner of an otherwise mainly arid continent. Their combined catchment area contains more than 80 percent of the country's irrigated farmland. From a source high in the Australian Alps, about midway between Canberra and Melbourne, the Murray flows northeast and then turns westward, descending into the Murray Valley near the town of Wagga Wagga. The valley is flanked by the Riverina Plains of New South Wales to the north and the plains of northern Victoria to the south. Along its middle course the river receives its main tributaries, which bring waters from the northern part of the catchment area. For the last 400 miles (650 km) of its course, the Murray flows through South Australia along a series of deep gorges separated by shallow lakes. The river exits into the ocean through the Coorong Lagoon (see p.331), into Encounter Bay south of Adelaide. The level of the Murray is erratic, especially during droughts, and the river dried up completely three times between 1800 and 1950. Recently, its flow has been increased by diverting water from the south-flowing Snowy River into its headwaters, as part of the Snowy Mountains hydroelectric program.

**GALAH**
*The galah is the most widespread cockatoo in Australia and is common around the Murray River.*

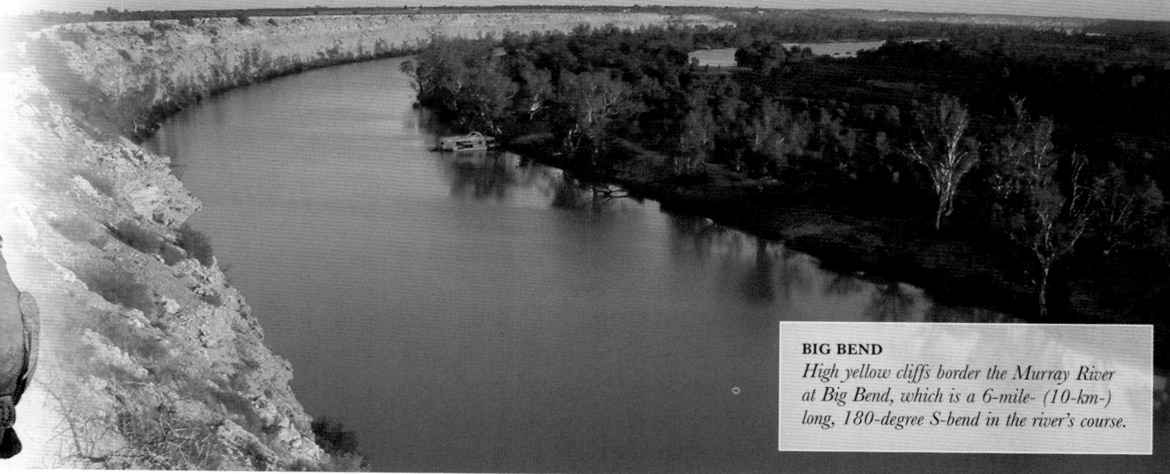

**BIG BEND**
*High yellow cliffs border the Murray River at Big Bend, which is a 6-mile- (10-km-) long, 180-degree S-bend in the river's course.*

LAND

# LAKES

110–11 Erosion by water
112–13 Deposition by water
126 The water cycle
212–35 Rivers
Water quality 242–43
Underground rivers 252–59
Wetlands 322–31

A LAKE FORMS when surface water collects in a depression with no immediate outlet. The water level rises until an outlet is formed, by which time a substantial amount of water may have accumulated. Lakes vary greatly in size and depth: some are no deeper than the average pond, while others are more than two-thirds of a mile (1 km) deep, and some have beds below sea level. Some lakes are millions of years old, but most are much younger, and date back to the end of the last ice age. Not all lakes consist of fresh water. Some, mainly in arid regions, are saline, with levels of salinity varying from slight to six times that of seawater.

## LAKE FORMATION

Some lakes, such as Baikal and Tanganyika (see p.248), formed in depressions along fault lines or in areas of rifting (see p.156). Others, such as the huge Caspian Sea (see p.250), were once part of an ocean, but have become cut off from the larger body of seawater. A few lakes have been formed by volcanic action, when lava dams a river or a caldera is formed. Most, however, are the result of glaciation, and this is reflected in their distribution. Most North American lakes, for example, are located in areas that 25,000 years ago were under a thick ice sheet. Glaciers create lakes in two main ways: valley floors are hollowed and deepened by glacial erosion and later filled in by water; and the ridges of moraines deposited by glaciers (see p.265) form dams across valleys that trap water behind them.

**FAULT LAKE**
*Loch Ness in Scotland occupies a long depression in an ancient structural fault line, running east–west, that cuts the country in two.*

**KETTLE LAKES**
*Kettle lakes, seen here in central Chile, form when large blocks of melting ice become surrounded by glacial debris. Such lakes have no outlet.*

**CALDERA LAKE**
*A near-circular caldera lake on the Kamchatka Peninsula in eastern Siberia is enclosed by the rim of an extinct volcano. The rounded outline is characteristic of this type of lake.*

## INFLOW AND OUTFLOW

Lakes receive much of their inflow from surface water that may carry a considerable amount of sediment. Most lakes have a single outflow into a river system—those without a defined outflow are described as unchanneled. Some lakes in arid regions have no outflow at all—the inflow is balanced by evaporation. Geologically speaking, most lakes are short-lived, due to infilling by sediment and lowering of the water level through the down-cutting action of the outflow river.

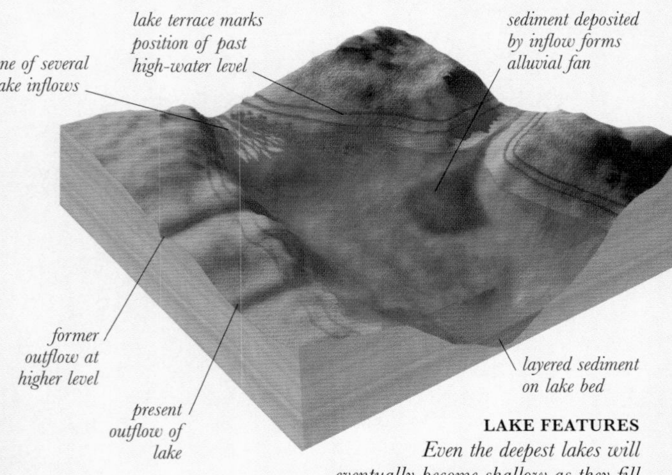


**LAKE FEATURES**
*Even the deepest lakes will eventually become shallow as they fill with sediment. Deposition is greatest where rivers flow into the lake. Infilling may sometimes give rise to wetlands.*

**MORAINE LAKE**
*Peyto Lake, in Alberta, Canada, formed in a glaciated valley when meltwater became trapped by moraines deposited across the valley during successive stages of glaciation.*

### LAKE PROFILES

The pages that follow contain profiles of the world's main lakes. Each profile begins with the following summary information:

**AREA** Area covered by the lake
**MAXIMUM DEPTH** Deepest part of the lake
**ELEVATION** Height of the lake above sea level
**OUTLET** Main river flowing out of the lake

NORTH AMERICA *northwest*

## Great Bear Lake

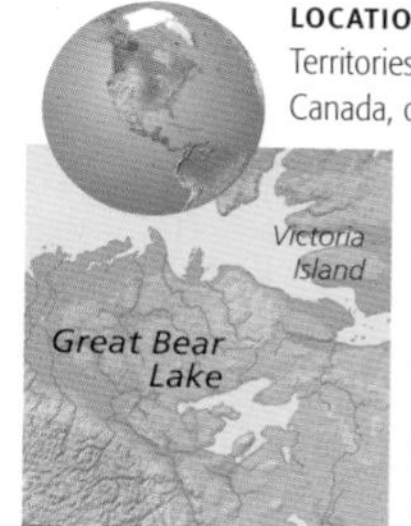


**LOCATION** In the Northwest Territories, in far northwestern Canada, on the Arctic Circle

**AREA** 12,024 square miles (31,150 square km)

**MAXIMUM DEPTH** 1,464 ft (446 m)

**ELEVATION** 610 ft (186 m)

**OUTLET** Great Bear River

Great Bear Lake is the largest lake entirely within Canadian territory and the fourth-largest in North America. Its northern shores lie within the Arctic Circle, and it is the largest inland body of water to exist at such a high latitude. The lake is irregular in shape and branches into three arms at the western end. The water is deepest near the eastern shoreline. The lake was formed by the scouring action of glaciers as they flowed from the North American ice sheet down river valleys. Lakes formed in this way are often called piedmont lakes. When the glaciers melted at the end of the last ice age, their waters became trapped behind moraines, forming a huge lake that has since partially drained, leaving as remnants Great Bear Lake, Great Slave Lake (see below), and the numerous small lakes that lie between them. The lake is situated in gently rolling wilderness, with evergreen forest along its southern shores and tundra extending to the north. Its waters are often cited as an example of a pristine lake environment, although it has far fewer fish species than lakes of comparable size in more temperate regions because its waters are much colder. The only settlement on its shores is the former mining center of Port Radium, named after the radioactive element, which was shipped from the mines here during World War II to make the first atomic bombs. Great Bear Lake receives inflow from the Whitefish River on its northwestern arm, and from other smaller rivers. The Great Bear River exits from the western end of the lake, flows westward, and delivers its waters into the Mackenzie River about 300 miles (480 km) upstream from the delta. Great Bear Lake and the Great Bear River both freeze over for about seven months each year.

**BROWN BEAR**
*The grizzly or brown bear is a top predator of forests around the lake's southern shores.*

**ECHO BAY ON THE NORTH SHORE**
*In summer, even the remotest northern bays are free of ice. In winter, the frozen lake and the shore are covered in snow.*

NORTH AMERICA *northwest*

## Great Slave Lake

**LOCATION** In the Northwest Territories of Canada, east of the Mackenzie Mountains

**AREA** 11,027 square miles (28,568 square km)

**MAXIMUM DEPTH** 2,016 ft (614 m)

**ELEVATION** 512 ft (156 m)

**OUTLET** Mackenzie River

Great Slave Lake is the second-largest lake in Canada and the main reservoir for the Mackenzie River system. The lake is irregular in shape and has a rocky, mainly forested shore with wide bays and numerous islands. It is located along the boundary between the ancient rocks of the Canadian Shield and the newer strata underlying the Great Plains. At Great Slave Lake this division is marked by a large inlet that extends northward from the northern shore. The lake's waters support a small commercial fishery. Other developments around its shores include the gold-mining town of Yellowknife, on the northern arm of the lake, and several small tourist resorts. The main inflow is the Slave River, which enters the southern shore bringing water from Lake Athabasca and the Peace River. The Mackenzie River exits from the southwestern end of the lake and is the sole outflow. The lake freezes in winter and is ice-free for only four months of the year.

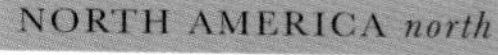

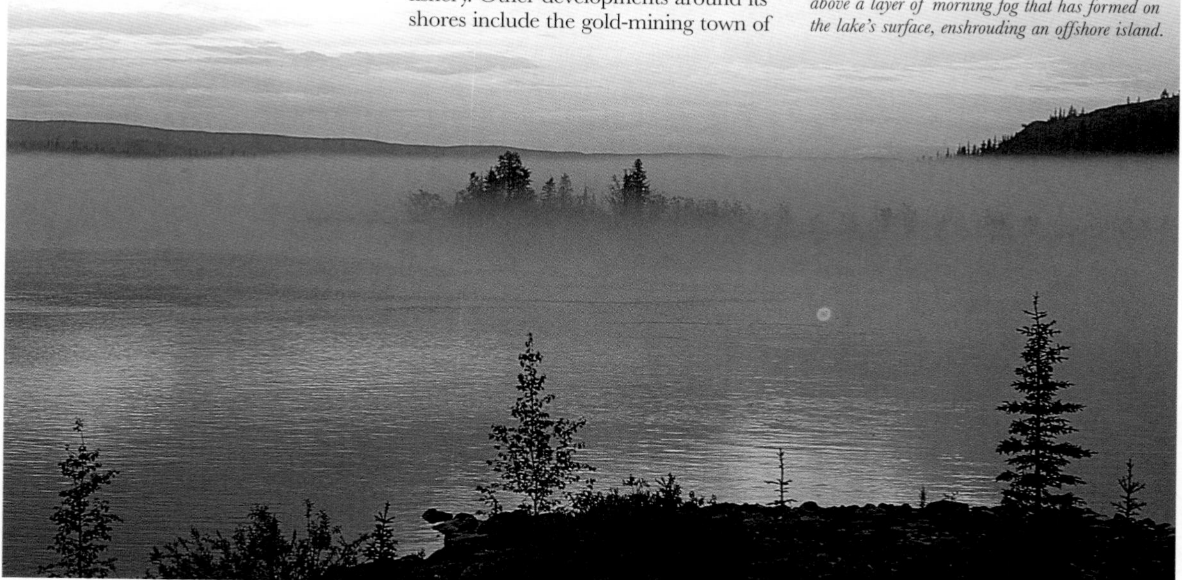

**LAKE COASTLINE**
*The shores of Great Slave Lake are covered with dense coniferous forest, the tops of which stand out above a layer of morning fog that has formed on the lake's surface, enshrouding an offshore island.*

NORTH AMERICA *north*

## Lake Winnipeg

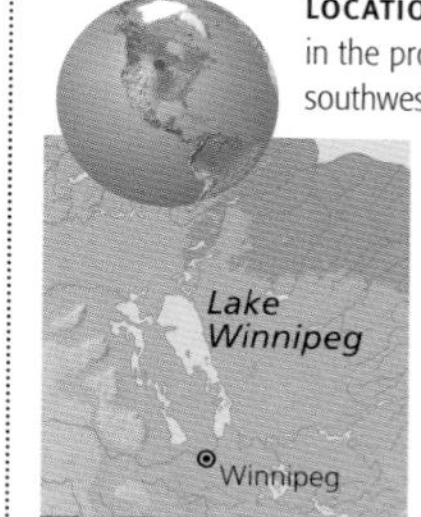


**LOCATION** In south-central Canada, in the province of Manitoba, southwest of Hudson Bay

**AREA** 9,167 square miles (23,750 square km)

**MAXIMUM DEPTH** 118 ft (36 m)

**ELEVATION** 712 ft (217 m)

**OUTLET** Nelson River

Lake Winnipeg is a shallow lake at the eastern edge of the Great Plains. It is the remnant of a much larger glacial lake that also included lakes Manitoba and Winnipegosis. Much of its water comes from the Saskatchewan River, which flows from the Canadian Rockies to the west. Another inflow, the Red River, drains the high plains. The Nelson River flows from the northeast end of the lake into Hudson Bay.

**ROCKY SHORELINE**

NORTH AMERICA *northeast*

# Great Lakes

**LOCATION** On the border of the US and Canada; Lake Michigan is entirely within the US

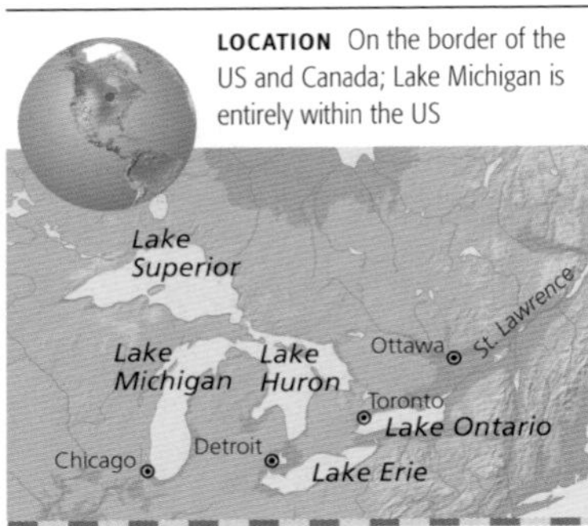


**AREA Superior** 51,159 square miles (82,367 square km), **Michigan** 36,024 square miles (58,000 square km), **Huron** 37,000 square miles (59,570 square km), **Erie** 16,037 square miles (25,820 square km), **Ontario** 11,807 square miles (19,010 square km)

**MAXIMUM DEPTH Superior** 1,333 ft (406 m), **Michigan** 923 ft (281 m), **Huron** 749 ft (228 m), **Erie** 210 ft (64 m), **Ontario** 735 ft (224 m)

**ELEVATION Superior** 601 ft (183 m), **Michigan** 580 ft (177 m), **Huron** 578 ft (176 m), **Erie** 571 ft (174 m), **Ontario** 246 ft (75 m)

**OUTLET** St. Lawrence River

The Great Lakes form the largest system of freshwater lakes in the world, with a combined surface area of 95,500 square miles (244,000 square km). The system drains from west to east, discharging into the sea from Lake Ontario through the St. Lawrence River (see p.216). The gradient between the lakes is generally very shallow. Superior drains into Michigan through the St. Mary's River. Michigan has almost the same level as Huron, and the two are connected by the Straits of Mackinac, 4 miles (6 km) wide. Huron drains through the St. Clair River, Lake St. Clair, and the Detroit River into Erie, which is drained in turn by the Niagara River into Lake Ontario. The descent between Erie and Ontario is greater than between any of the others, and includes the spectacular Niagara Falls, 167 ft (51 m) high. The Great Lakes were formed during the last ice age, when successive episodes of glaciation eroded a landscape of mountains and river valleys. When the ice melted, Superior, Michigan and Huron were all part of one huge lake (known as Lake Algonquin), which drained into the Ottawa River. Post-glacial uplift altered the drainage patterns, and the lakes began to take on their present shape about 10,000 years ago. The present shorelines became stable about 3,000 years ago, but the western end of Lake Erie is still sinking at about $\frac{1}{10}$ in (2 mm) per year. Only the western and northern shores of Superior are rocky; the other shores are covered by thick glacial deposits, with concentric ridges of terminal moraine at the southern ends of Michigan and Huron. Numerous cities, including Buffalo, Chicago, Cleveland, Detroit, Milwaukee, and Toronto, are situated around the lakes.

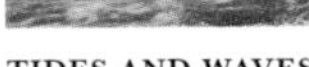

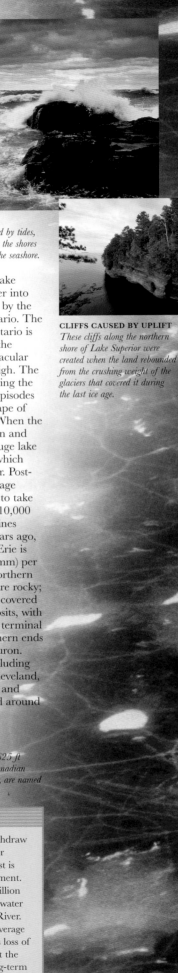

**TIDES AND WAVES**
*Although the Great Lakes are unaffected by tides, the wind can produce waves that buffet the shores of Lake Superior like breakers against the seashore.*

**CLIFFS CAUSED BY UPLIFT**
*These cliffs along the northern shore of Lake Superior were created when the land rebounded from the crushing weight of the glaciers that covered it during the last ice age.*

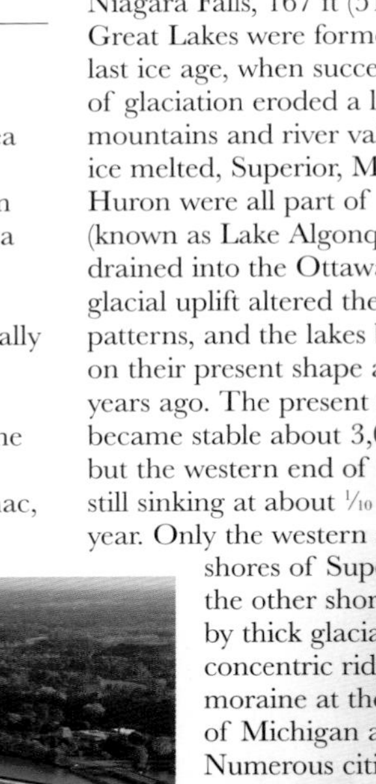

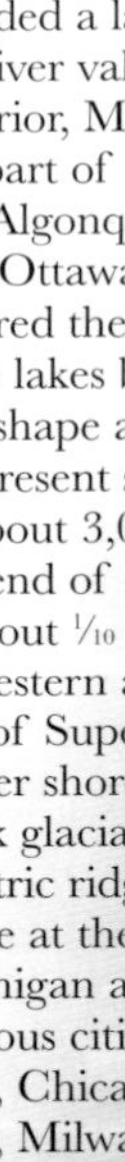

**NIAGARA FALLS**
*The Horseshoe Falls, 2,625 ft (800 m) wide, on the Canadian side of the Niagara River, are named for their distinctive shape.*

## URBAN SHORELINE

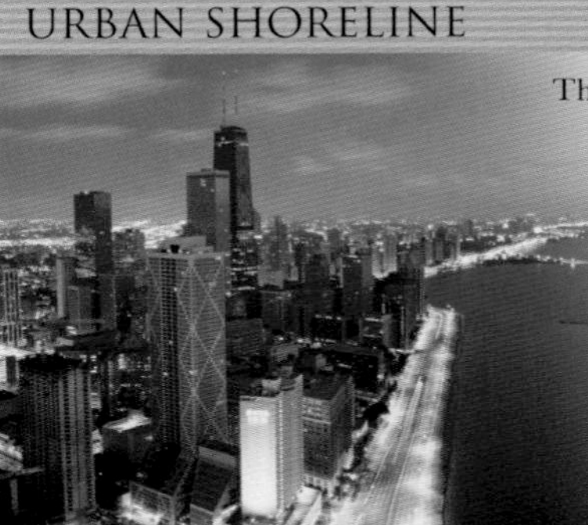

The cities around the Great Lakes withdraw huge amounts of water each day for municipal and industrial use. Most is returned to the lakes after treatment. Chicago, however, sends 2.9 billion gallons (10.9 billion liters) of water every day to the Mississippi River. Combined with lower-than-average rainfall in recent years, this loss of water is beginning to affect the level of the lakes. The long-term effects of this fall remain unclear.

**CLEARING A PATH**
*During the winter months, the Coast Guard uses powerful icebreakers to keep the Great Lakes shipping routes open—in this case, Thunder Bay on Lake Superior.*

NORTH AMERICA *east*

## Lake Seneca

**LOCATION** In west-central New York state south of Lake Ontario

**AREA** 67 square miles (174 square km)

**MAXIMUM DEPTH** 650 ft (198 m)

**ELEVATION** 443 ft (135 m)

**OUTLET** Seneca River

Lake Seneca is one of the famed Finger Lakes, a series of long, narrow lakes that lie roughly parallel with each other on the undulating uplands to the south of Lake Ontario. The lakes were formed more than a million years ago, during the last stages of the most recent ice age, when a series of shallow glaciated valleys running on a north–south axis became blocked at both ends by glacial moraines. Seneca is the second-largest of the Finger Lakes and is by far the deepest, with parts of its bed lying 180 ft (54.5 m) below sea level. The lake receives water from streams draining the surrounding hills, including Catherine Creek and the Keuka Lake outlet. Its outflow, the Seneca River, exits at the northeastern corner and, having received the waters of Lake Cayuga, flows into the Oswego River, which eventually discharges into Lake Ontario.

NORTH AMERICA *west*

## Lake Tahoe

**LOCATION** In the Sierra Nevada mountains between Nevada and California

**AREA** 193 square miles (499 square km)

**MAXIMUM DEPTH** 1,640 ft (500 m)

**ELEVATION** 6,200 ft (1,900 m)

**OUTLET** Truckee River

The fifth-deepest lake in the world, Lake Tahoe originated in a deep depression between two faults that subsequently became dammed by lava flow. Although fed by small streams all around its shores, the main inflow is at the southern end, through the Upper Truckee Marshes. The Truckee River flows out of the northwestern end of the lake into Nevada. Lake Tahoe is a popular tourist destination.

DEEP BLUE WATERS OF LAKE TAHOE

NORTH AMERICA *southeast*

## Lake Okeechobee

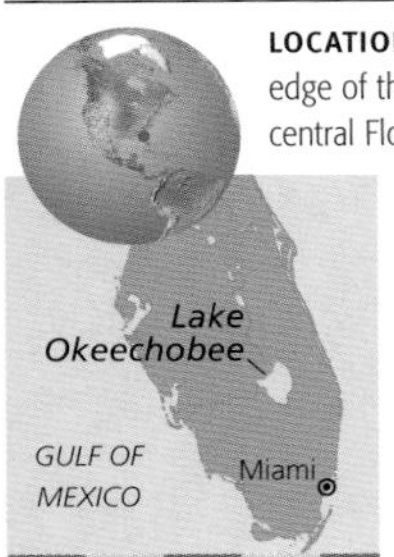


**LOCATION** At the northern edge of the Everglades in central Florida

**AREA** 732 square miles (1,895 square km)

**MAXIMUM DEPTH** 15 ft (4.5 m)

**ELEVATION** 13 ft (4 m)

**OUTLET** Unchanneled

Lake Okeechobee is the main reservoir of fresh water feeding the Everglades' "river of grass"—although it no longer flows with enough force to keep the tide at bay in southern Florida. The lake lies in a shallow, sloping depression that was once part of the sea bed, and is dammed along its southern shore by an accumulation of peat. The lake receives water from the Kissimmee River and from Lake Istokpoga. There is no channeled outflow—the lake's waters flow south through the Everglades on a broad front. Since the 1950s, numerous canals have been constructed to divert the lake's water for agriculture, and to meet the increasing demand of resort towns along Florida's eastern coastline.

**OKEECHOBEE PLANT LIFE**
*The marshy areas around the lake are colonized by a variety of aquatic plants, such as these waterlilies. Some are native; others are introduced.*

NORTH AMERICA *west*

## Great Salt Lake

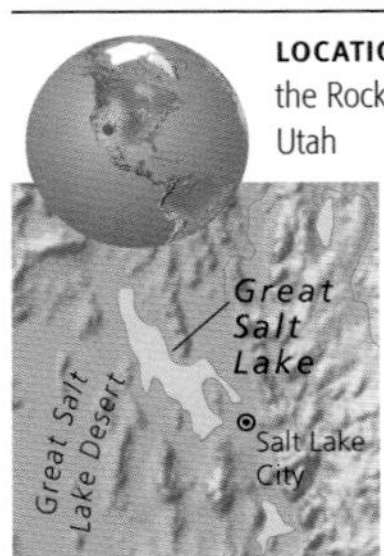


**LOCATION** To the west of the Rocky Mountains, in northern Utah

**AREA** 1,930 square miles (5,000 square km)

**MAXIMUM DEPTH** 39 ft (12 m)

**ELEVATION** 4,200 ft (1,280 m)

**OUTLET** None

The Great Salt Lake is the largest inland body of salt water in the western hemisphere, and at 25 percent salinity is about five times saltier than the sea. It is a remnant of the prehistoric Lake Bonneville, which covered most of the floor of the Great Basin (see p.288) during the last ice age. The present lake is surrounded by great tracts of sand, salt flats, and salt marsh and receives most of its inflow from the Bear Creek and Weber rivers, which bring waters from the Wasatch Range along the eastern shore. The area of the lake is extremely variable—between 1964 and 1987, it more than doubled in size, reaching a maximum area of nearly 2,300 square miles (6,000 square km). A causeway constructed in the 1950s divides the lake into two parts. The smaller northern section is both shallower and saltier than the rest of the lake.

**SALT LAKE SHALLOWS**
*Because the lake lies in a shallow basin, small changes in its level have large effects on its surface area, exposing or submerging large areas of its bed.*

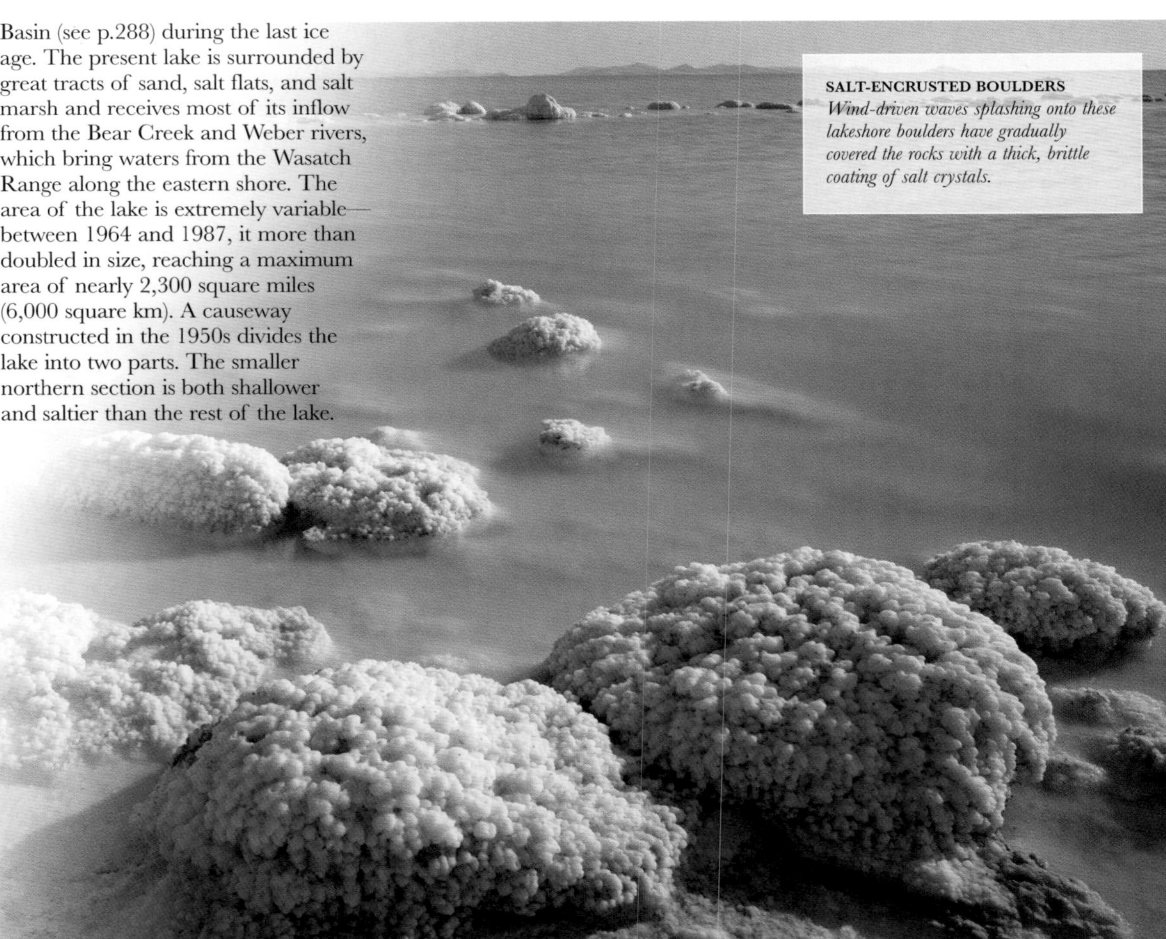

**SALT-ENCRUSTED BOULDERS**
*Wind-driven waves splashing onto these lakeshore boulders have gradually covered the rocks with a thick, brittle coating of salt crystals.*

SOUTH AMERICA *northwest*

## Lake Titicaca

**LOCATION** In the central Andes, on the Altiplano plateau, bordered by Peru and Bolivia

**AREA** 3,386 square miles (8,772 square km)

**MAXIMUM DEPTH** 923 ft (281 m)

**ELEVATION** 12,516 ft (3,812 m)

**OUTLET** Desaguadero River

Lake Titicaca is the largest freshwater lake in South America, and the highest lake of any significant size in the world. The lake sits in a structural depression at the northern end of the Altiplano plateau that separates the Eastern and Western cordilleras of the Andes. Roughly oval in shape, the lake is 120 miles (195 km) long and about 50 miles (80 km) wide. It is divided into two unequal parts by a narrow strait known as the Tiquina. The lake bottom slopes down from west to east. The lake is widest at the northwest, and the deepest parts are near the northern end, where the shoreline is defined by the Bolivian Cordillera Real mountains. The southeastern part of the lake, below the Tiquina, is much shallower, with an average depth of about 30 ft (9 m). Titicaca receives inflow from across the northern part of the Altiplano and is fed by more than 20 rivers, the largest of which is the Ramis. Each year the level of the lake fluctuates by up to 13 ft (4 m), with the highest levels occurring after the summer rains between December and March. Evaporation in the arid mountain climate removes about 95 percent of the lake's inflow. The sole outflow, the Desaguadero, exits from the shallow southern end of the lake and flows 185 miles (300 km) south to Lake Poopó, which has no outlet to the sea. Evidence of terraces made by former shorelines shows that Titicaca was much larger during the period immediately after the last ice age. Stretches of the present shoreline are covered by marshes and beds of totora reeds. These reeds are still used by some of the people living around the lakeshore, to make boats, houses, and floating islands in the traditional manner. Three fish species have evolved in the waters of the lake, which because of the high altitude contain about one-third less oxygen than fresh water at sea level. These endemic species supported a small local fishery, but their population was devastated by the introduction of trout for commercial fisheries during the 1930s. The shipping route across Lake Titicaca is a vital link in the railroad network that connects land-locked Bolivia with the sea.

**STRAITS OF TIQUINA**
*This view looking eastward across the straits into the deeper section of the lake shows the bleak, rocky landscape that makes up much of its northern shoreline.*

**VICUÑA**
*Adapted to high-altitude living, the vicuña grazes on the grasslands of the Altiplano. Its domesticated cousin, the alpaca, is raised for the fibers of its coat.*

### INCA SETTLEMENTS

The Inca, who flourished until about 500 years ago, considered Lake Titicaca to be the sacred and ancestral home of their agriculture and their civilization. According to legend, their sun god, Manco Capac, sprang from the waters of the lake and founded the Inca dynasty at Cusco. The ruins of Inca buildings are still to be found around the lakeshores and on some of the islands, providing ample evidence of great skill at building with large blocks of unmortared stone.

**REED BOATS ON UROS ISLAND**
*Boats made from slim bundles of totora reeds are moored to an island made of larger bundles of reeds, forming a thick floating platform on which houses are built.*

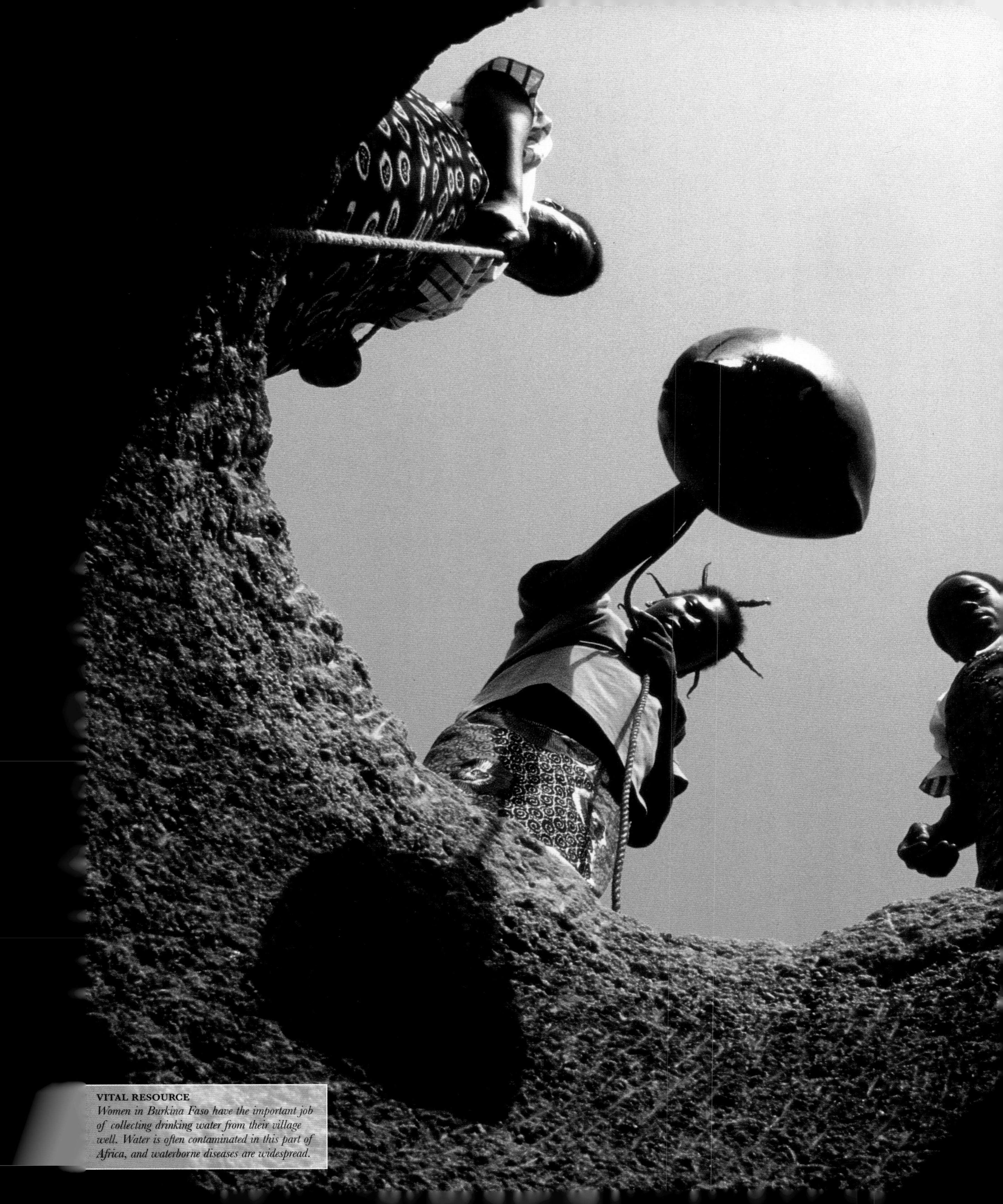

**VITAL RESOURCE**
*Women in Burkina Faso have the important job of collecting drinking water from their village well. Water is often contaminated in this part of Africa, and waterborne diseases are widespread.*

# FRESHWATER QUALITY

Throughout human history, water has played a dual role: as a life-giving liquid and as a resource for waste disposal. Without careful management, these two uses can conflict, potentially dangerously. In the developed world, water pollution reached crisis proportions around the mid-twentieth century, but is now increasingly controlled. In developing countries, poor water quality remains a major health threat.

## NATURAL PURIFICATION

In nature, pure fresh water is rare. Groundwater is normally purer than surface water, because it is filtered by porous rock. It contains dissolved minerals, but very little organic matter. Surface water also contains minerals, but even when it looks clean, it often teems with living things and their dead remains. The vast majority of its living inhabitants are microorganisms, such as bacteria. Most of the bacteria need oxygen to break down and purify organic matter in the water, and this process is most effective when the water contains high levels of dissolved air. Natural purification is also helped by water movement, because this stirs up sediment and dissolves more air. However, if water becomes overloaded with organic matter—for example, from human sources—the normal waste-disposal system may break down.

**WARNING SIGN**
*In this river, industrial pollution is very evident. Other kinds of water pollution can be much harder to detect.*

## EUTROPHICATION

In fresh water, two mineral nutrients—nitrogen and phosphorus—are essential for the growth of microorganisms. These nutrients are often scarce, and the growth of microorganisms stops once they have been used up. When untreated sewage and agricultural runoff enter the water, nitrogen and phosphorus levels can suddenly climb, and the microorganism population booms. This creates more living matter, which uses up oxygen. As oxygen levels plunge, fish and other animals die, and the waterway fills with organic sludge. This process, which is known as eutrophication, is prevented by treating sewage before it is discharged, and by controlling fertilizer use.

**ALGAL BLOOM**
*Lush growths of algae are a sign of eutrophication, which causes oxygen levels in water to fall.*

## POLLUTION

Untreated waste is also hazardous because it may contain disease-carrying microorganisms. Normally, clean and waste water are kept apart during water treatment and waste management, but if this separation is breached—for example, through lack of sanitation—waterborne diseases can spread. To prevent this, water is periodically tested for indicator organisms, such as the bacterium *Escherichia coli*, a common resident of the human intestine. If large numbers of *E. coli* are detected in drinking water, immediate action is needed to trace their source.

Water pollution is caused not only by microorganisms but also by industrial waste, which affects water quality in different ways. Synthetic organic chemicals can enter freshwater food chains, and from there they can reach drinking water supplies. These chemicals include pesticides, herbicides, and pharmaceutical drugs, as well as substances that are utilized in manufacturing. Industrial waste can also contain inorganic pollutants, such as acids, zinc, and lead, many of which are known to be toxic. Safety limits for synthetic organics, which are relatively recent "inventions," are much more difficult to gauge, especially since thousands of new organics are synthesized every year.

Heat is another form of pollution that affects many urban rivers. One of the chief sources is industrial effluent, particularly water that is used to cool power stations. Heat affects the growth of plants and animals. It also reduces the amount of oxygen that water can carry—something that affects much aquatic life.

**WASTE DISPOSAL**
*Water offers a cheap and easy way for industry to dispose of waste.*

## FUTURE TRENDS

The increasing global population and the spread of industrialization mean that water pollution will remain an acute problem in many parts of the world. According to UN figures released in 2003, about 6 million tons of waste are dumped each day into rivers, lakes, and streams—a figure that is likely to increase. In developed countries, economic growth has helped to bring about improvements in water quality, but many developing countries are finding it difficult to follow this trend.

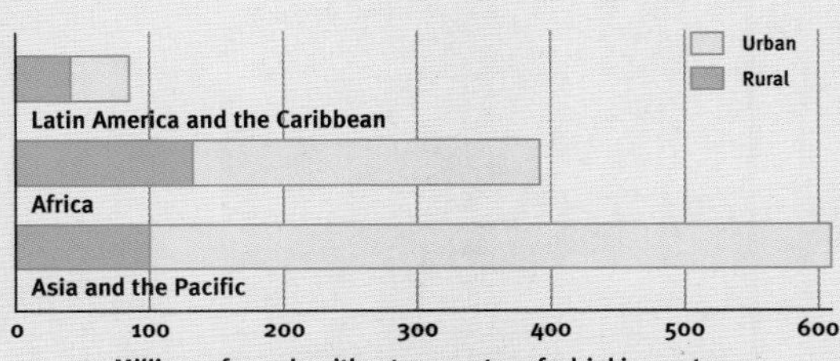


**ACCESS TO CLEAN WATER**
*Within the developing world, the number of people who are without access to clean drinking water varies. In population terms, Africa fares worst, with poor water quality affecting at least 50 percent of the population.*

CENTRAL AMERICA

## Lake Nicaragua

**LOCATION** In southwestern Nicaragua, only 9 miles (15 km) from the Pacific coastline

**AREA** 3,146 square miles (8,150 square km)

**MAXIMUM DEPTH** 230 ft (70 m)

**ELEVATION** 105 ft (32 m)

**OUTLET** San Juan River

The largest body of fresh water in Central America, Lake Nicaragua is oval in shape and was once a wide bay on the ocean coastline. The present lake was formed when volcanic activity on the Pacific Rim (see p.175) produced a narrow ridge that extended from the Cordillera de Guanacaste mountain range northward, cutting off the bay from the sea. This 12-mile- (19-km-) wide ridge now forms the lake's western shoreline. The bay's salt water has been gradually freshened by rain-fed streams flowing down the surrounding hills, combined with a steady outflow to the sea. This process has produced a unique combination of animal life, and Lake Nicaragua is the only lake in the world to have freshwater species of marine animals such as sharks and swordfish. The lake is dotted with hundreds of islands, some of which contain active volcanoes. Among the many rivers flowing into the lake is the Tipitat River, which drains Lake Managua, located to the northwest. Lake Managua was formed and freshened by the same processes that created Lake Nicaragua. The San Juan River flows from the southeastern corner of the lake through dense rainforest to the Caribbean Sea. Locally, the lake is famed for its apparent tides, which are in fact standing waves that oscillate across the lake surface—caused by winds blowing up the San Juan valley.

**TOUCAN**
*The chestnut-mandibled toucan nests high in the forest canopy on hillsides around the lake.*

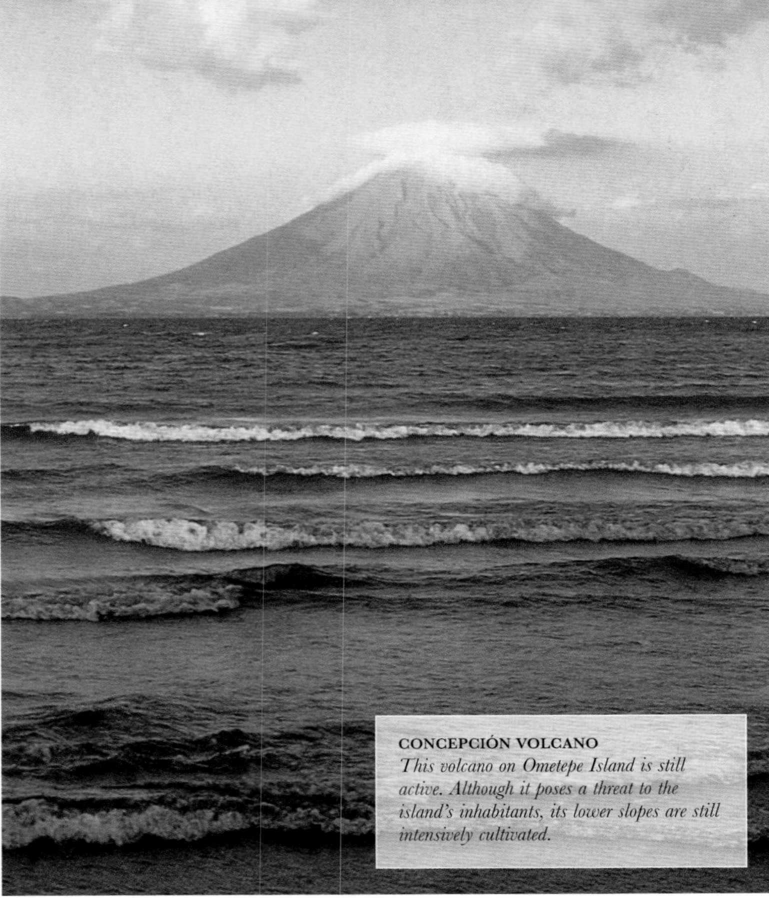

**CONCEPCIÓN VOLCANO**
*This volcano on Ometepe Island is still active. Although it poses a threat to the island's inhabitants, its lower slopes are still intensively cultivated.*



EUROPE *north*

## Lake Ladoga

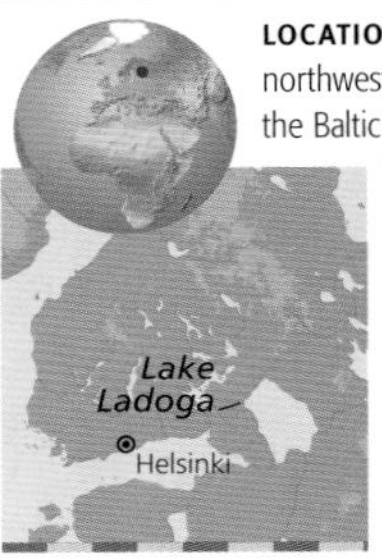


**LOCATION** In the Karelia region of northwestern Russia, to the east of the Baltic Sea

**AREA** 6,823 square miles (17,675 square km)

**MAXIMUM DEPTH** 755 ft (230 m)

**ELEVATION** 16 ft (5 m)

**OUTFLOW** Neva River

The largest lake in Europe, Ladoga lies in a glacially modified depression. The water is deepest near the high, rocky cliffs of the northern shores. In the south, the lake is much shallower, with a low shoreline. Ice starts to form in December, and the entire lake freezes between January and May. Ladoga is fed by several rivers and two smaller lakes nearby. The Neva River flows from the southwestern corner of the lake and then through St. Petersburg about 25 miles (40 km) to the west.

**BATTERED SHORELINE**
*The bleak, rocky shores of Lake Ladoga are sometimes pounded by wind-driven waves up to 15 ft (4.5 m) high.*

EUROPE *northwest*

## Loch Ness

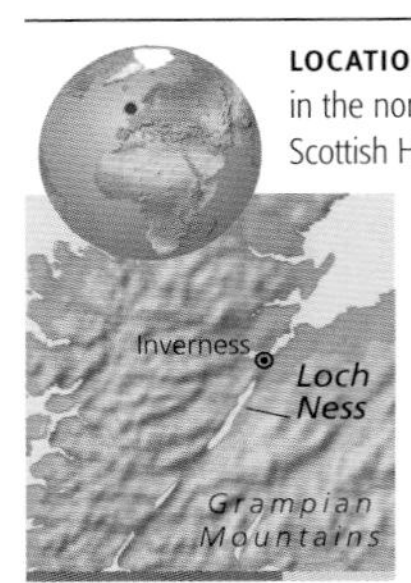

**LOCATION** In Inverness, Scotland, in the northwestern part of the Scottish Highlands

**AREA** 22 square miles (56 square km)

**MAXIMUM DEPTH** 755 ft (230 m)

**ELEVATION** 52 ft (16 m)

**OUTFLOW** Ness River

Famous for "Nessie," the monster that is said to inhabit its depths, Loch Ness is a long, narrow lake at the northern end of the Great Glen, near the town of Inverness. The glen is an ancient, faulted depression that bisects Scotland diagonally between the Grampian Mountains and the North West Highlands. The fault accounts both for the lake's shape and for its depth. Loch Ness is 22 miles (39 km) long, but only about 1 mile (1.5 km) wide. Its average depth is more than 330 ft (100 m) below the level of the sea, which is 6 miles (10 km) away. Despite many investigations, no conclusive evidence of a monster has ever been found.

**URQUHART CASTLE ON LOCH NESS**

EUROPE *northwest*

## Lake Windermere

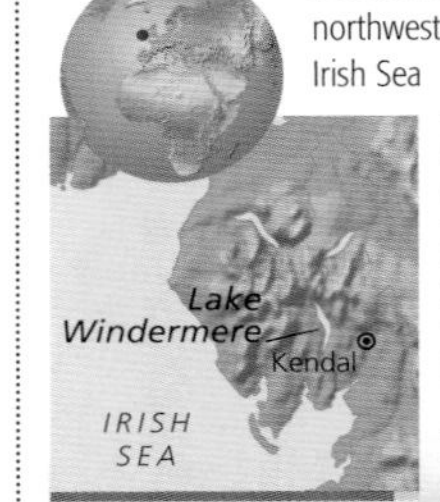


**LOCATION** In the Lake District of northwestern England, east of the Irish Sea

**AREA** 6 square miles (15 square km)

**MAXIMUM DEPTH** 210 ft (64 m)

**ELEVATION** 128 ft (39 m)

**OUTFLOW** Leven River

Windermere, the largest lake in England, formed at the end of the last ice age in a deep, glaciated valley that runs north–south through the Cumbrian Mountains. It is a long, narrow lake, deepest at the northern end. Fed by rain and snow from the surrounding highland, Windermere also receives waters from other nearby lakes, including Grasmere. Water flows out through the Leven River.

**LAKE DISTRICT**
*Windermere is the largest of 17 major lakes that comprise England's Lake District.*

EUROPE *northwest*

## Lough Derg

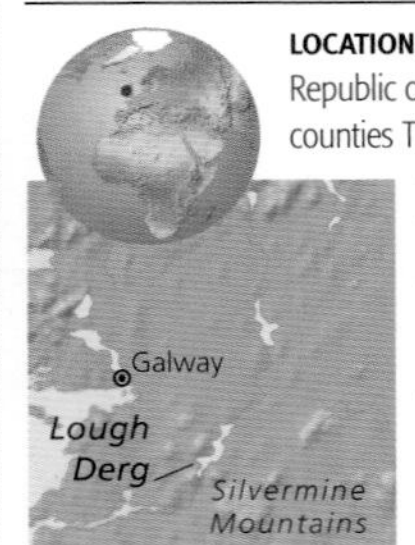


**LOCATION** In the southwestern Republic of Ireland at the boundary of counties Tipperary, Galway, and Clare

**AREA** 46 square miles (118 square km)

**MAXIMUM DEPTH** 118 ft (36 m)

**ELEVATION** 112 ft (34 m)

**OUTFLOW** Shannon River

Lough Derg is located at the southern extremity of Ireland's central plain, about 18 miles (30 km) upriver from the mouth of the Shannon River. The irregularly shaped lake lies in a glaciated section of the Shannon valley. The narrow southern arm, the deepest part of the lake, was formed during the last ice age, when ice was forced into the neck of the gorge through which the river now exits the lake. An island on Lough Derg—called Saint Patrick's Purgatory because St. Patrick is reputed to have stayed there in AD 445—is of religious significance.

EUROPE *central*

## Lake Constance

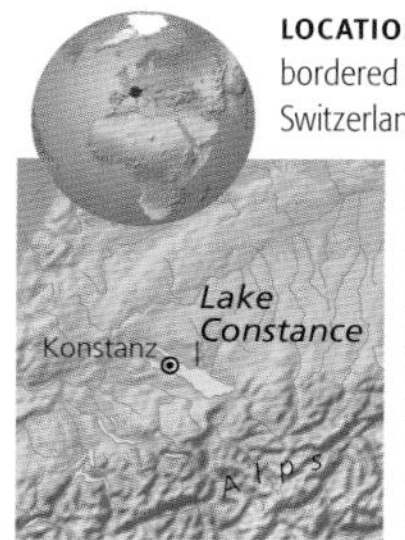


**LOCATION** In the northern Alps, bordered by Austria, Germany, and Switzerland

**AREA** 209 square miles (541 square km)

**MAXIMUM DEPTH** 827 ft (252 m)

**ELEVATION** 1,295 ft (395 m)

**OUTFLOW** Rhine River

Lake Constance occupies the lower part of the valley created by the Rhine Glacier during the last ice age. Measuring 40 miles (64 km) long and 8½ miles (13.5 km) wide, it is central Europe's second-largest freshwater lake, and supplies drinking water to about 4.5 million people. The marshes along the lake margins support large numbers of waterfowl and shorebirds. The Rhine River enters the lake at the southeastern end, on the Swiss–Austrian border. At the western end of the lake, where the glacier spread out over low foothills, the lake splits into two arms. The Rhine exits through the southern arm at the German–Swiss border.

BATHHOUSE ON LAKE CONSTANCE

EUROPE *central*

## Lake Geneva

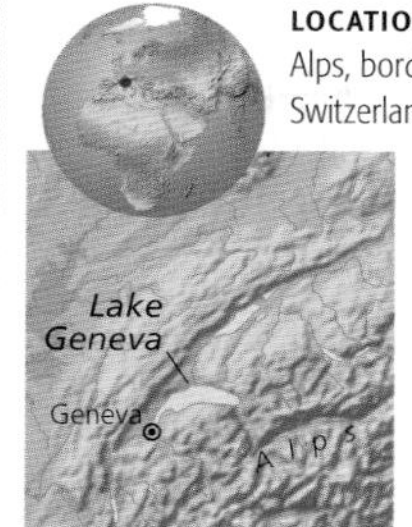


**LOCATION** In the northwestern Alps, bordered by France and Switzerland

**AREA** 224 square miles (581 square km)

**MAXIMUM DEPTH** 1,018 ft (310 m)

**ELEVATION** 1,221 ft (372 m)

**OUTFLOW** Rhône River

Lake Geneva is the largest of the Alpine lakes, forming a broad crescent, with the Rhône River entering from the southeastern tip and exiting from the southwestern tip. Although the lake is only 8½ miles (14 km) across at its widest point, strong winds can cause violent storms that are more characteristic of an inland sea. The northern shoreline is densely populated, the main cities being Geneva and Lausanne. The south-facing slopes along the lakeshore near Lausanne form the largest wine-producing area in Switzerland.

VITICULTURE ON THE NORTHERN SHORE

EUROPE *central*

## Lake Como

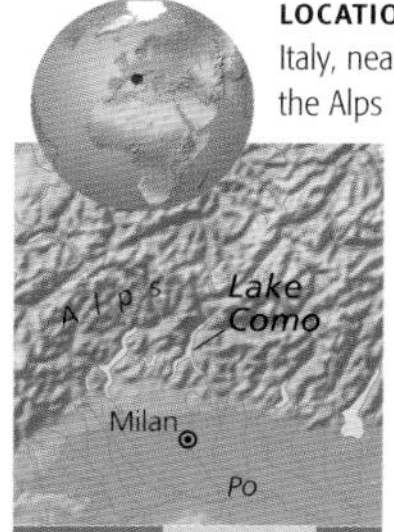


**LOCATION** In Lombardy, northern Italy, near the southern edge of the Alps

**AREA** 56 square miles (146 square km)

**MAXIMUM DEPTH** 1,346 ft (410 m)

**ELEVATION** 650 ft (198 m)

**OUTFLOW** Adda River

Lake Como is the deepest of the Alpine lakes. It has three branches, each about 15 miles (25 km) long and about 2½ miles (4 km) wide. Its inverted Y-shape is defined by deep, faulted valleys that were produced during the formation of the Alps. The valley floors have been depressed and eroded by subsequent glaciations, which accounts for the lake's great depth; although it is an Alpine lake, parts of its bed are more than 660 ft (200 m) below the level of the Mediterranean Sea. Despite being surrounded by mountains—up to 8,000 ft (2,400 m) high at the the northern end, descending to about 2,000 ft (600 m) in the south—the lake enjoys a mild Mediterranean climate and is a popular tourist destination. Lake Como receives most of its water from the snow-fed Adda River, which flows from the eastern Alps and enters the lake's northern arm. The lake also receives the inflow of numerous other rivers and streams. The Adda is the lake's sole outflow—it exits from the southeastern arm and is a tributary of the Po River.

LAKE COMO AND BELLAGIO
*This view up the northern arm of the lake shows a high wedge of rock known as Bellagio (right center). During the last ice age, this outcrop caused the lake-forming glacier to split into two arms.*

ANCIENT RESORT
*Como's spectacular mountain setting and mild climate have made it a popular resort for at least 2,000 years. Some towns around the lake date back to the Roman period.*

AFRICA *north*

## Lake Chad

**LOCATION** On the southern fringes of the Sahara Desert, bordered by Cameroon, Chad, Niger, and Nigeria



**AREA** Highly variable

**MAXIMUM DEPTH** 33 ft (10 m)

**ELEVATION** 920 ft (280 m)

**OUTFLOW** Bahr el Ghazal

Lake Chad is a large, shallow, oval lake, with a highly variable surface area. The level of the lake fluctuates annually, according to patterns of rainfall in the western Sahel. Generally it covers between 3,800 and 10,000 square miles (10,000 and 25,000 square km). Changes in level have a major impact because the lake is shallow, averaging only about 13 ft (4 m). In 1970, when Lake Chad was at its maximum, it was the fourth-largest lake in Africa, and since then it has shrunk dramatically. This is due to short-term drought and increased use of the lake's waters for irrigation, but there is also a long-term downward trend. Lake Chad is all that remains of a much larger lake, which 10,000 years ago was about 165 ft (50 m) deeper. The present lake is divided in two by a sandy ridge known as the Great Barrier that runs east–west and is breached by a single channel. The northern part of the lake is the deepest, and has a shoreline of inlets, islets, and peninsulas, formed by dunes up to 100 ft (30 m) high. The southern part is shallower, and around the inflow of the Chari River the shoreline supports marshes. The Chari flows from the Ubangi Plateau to the south and provides 95 percent of the lake's inflow. The rain-fed water of the Chari also accounts for the lake's continuing freshness in a desert climate. If the level falls, it is the deeper, northern portion that is most likely to dry out completely, because the Great Barrier cuts it off from the Chari's inflow. When levels are at their highest, excess water flows westward along the Bahr el Ghazal.

**DRY LAKE BASIN**
*The Chad Basin has been drying out for centuries. During the drought of the 1970s and 1980s, the northern part of the lake almost dried out completely.*

**SHRINKING LAKE**
*The shallow southern part of the lake is becoming colonized with beds of papyrus reeds that allow windblown sand to accumulate more quickly.*

AFRICA *east*

## Lake Turkana

**LOCATION** In the Great Rift Valley, bordered by Ethiopia and Kenya



**AREA** 2,606 square miles (6,750 square km)

**MAXIMUM DEPTH** 360 ft (110 m)

**ELEVATION** 1,180 ft (360 m)

**OUTFLOW** None

Also known as Lake Rudolf, Turkana is a narrow lake, 165 miles (265 km) long, running through the middle of the eastern arm of Africa's Great Rift Valley. It is the remnant of a larger lake that 5,000 years ago was three times as long, about 260 ft (80 m) deeper, and drained northeastward into the White Nile. The present lake is deepest in the center and at the southern end, where it is dammed by a lava flow. The lake has three islands, all of which are extinct volcanoes. The southern shores, once part of the lakebed, consist of mud flats and dunes. The Omo River, flowing in at the northern end, brings water from the southern Ethiopian Highlands, and accounts for almost all the lake's inflow. With no outflow, Lake Turkana loses water mainly through evaporation in the hot, arid climate. The rate of evaporation is increased by strong winds, which often blow across the lake. Lake Turkana is subject to seasonal and periodic changes in level, but the overall trend is downward. Since 1900, the level of the lake has dropped by about 50 ft (15 m).

**VOLCANIC PEAK**
*The steep slopes of South Island, a large volcano (visible in the distance in this picture), descend to the deepest part of Lake Turkana.*

### FISHING AT TURKANA

The waters of Lake Turkana are rich in fish species, and fishing the lake has been the mainstay of some lakeside populations for thousands of years. The lake continues to produce about 1,000 tons of fish each year, but stocks of the main food fishes—Nile perch, tilapia, and tiger fish—are dwindling. To maintain the size of the catch, an increasing number of species are being harvested for food. Sports fishermen also use the lake.

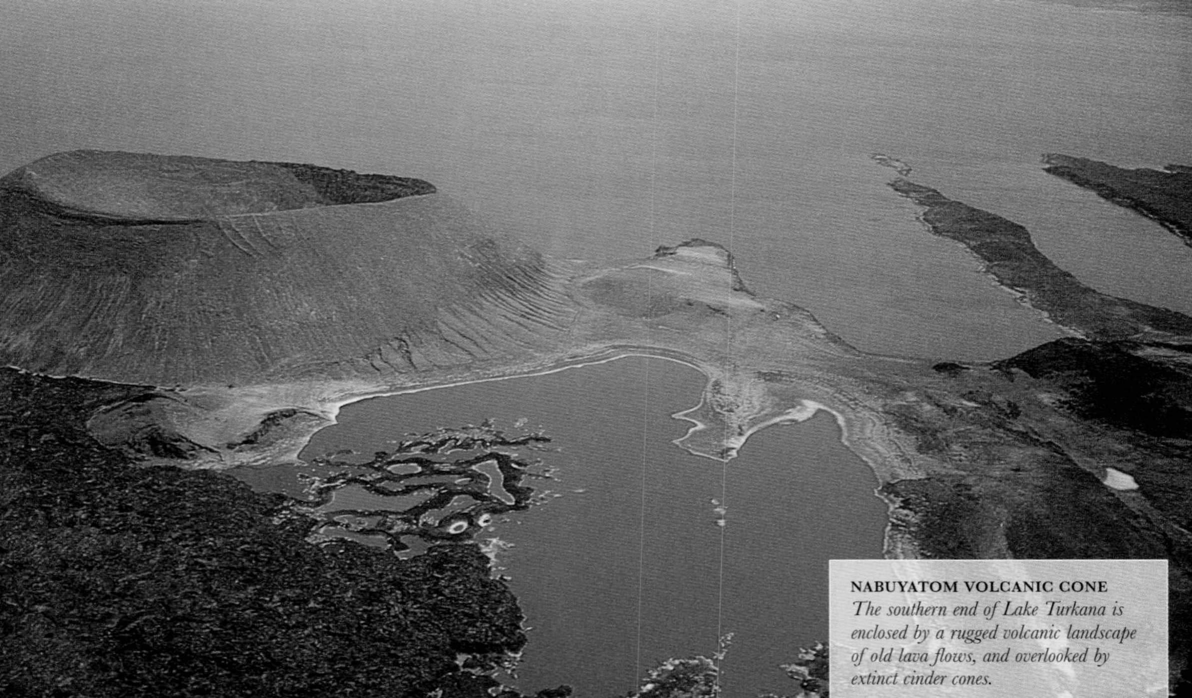

**NABUYATOM VOLCANIC CONE**
*The southern end of Lake Turkana is enclosed by a rugged volcanic landscape of old lava flows, and overlooked by extinct cinder cones.*

LAND

AFRICA *east*

# Lake Victoria

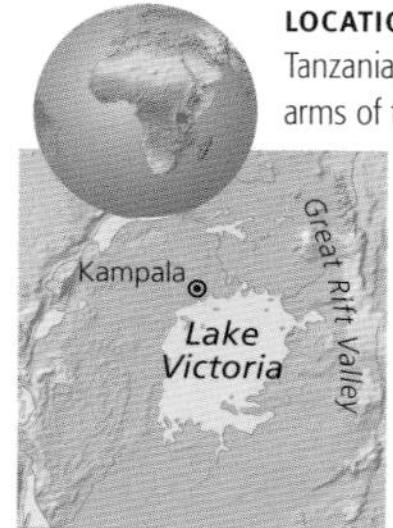


**LOCATION** Bordered by Kenya, Tanzania, and Uganda between the arms of the Great Rift Valley

**AREA** 26,250 square miles (68,000 square km)

**MAXIMUM DEPTH** 270 ft (84 m)

**ELEVATION** 3,725 ft (1,134 m)

**OUTFLOW** Nile River

Lake Victoria is the largest lake in Africa, the second-largest freshwater lake in the world, and the main reservoir on the Nile River. The lake is roughly square in shape and is situated in the middle of the plateau that separates the western and eastern arms of the Great Rift Valley. It formed about 2 million years ago, when tilting of the crust between the arms diverted surface water into a shallow depression on the plateau's surface. Lake Victoria has no major inflows—its level is maintained by year-round heavy rainfall across the lake's surface and runoff from the surrounding upland. Its shores are one of the most densely populated parts of Africa, and the cities of Kampala, Kisumu, and Mwanza are situated around the lake. Many of the lake's numerous islands are also inhabited. The northern shore has numerous long, thin peninsulas; between them are areas of swamp that extend far inland and provide a refuge for the rare sitatunga antelope. Lake Victoria is about 2 million years old—which is ancient by lake standards—and several hundred endemic species, mainly cichlid fishes, have evolved. The introduction of the Nile perch (a food fish endemic to the Nile and Lake Turkana) into the northern part of lake has proved an ecological disaster for local species. The perch is a voracious predator that has now spread to all parts of the lake, and the native fish population has been catastrophically reduced.

**TRANSPORTATION ROUTES**
*Steamships crisscross the lake between cities, but life around the lakeshore mainly relies on even older forms of agriculture and transportation.*

**COLLECTING FRESH WATER**
*Lake Victoria's extensive shoreline provides ready access to water for millions of people.*

AFRICA *east*

# Lake Nakuru

**LOCATION** In western Kenya, in the eastern arm of the Great Rift Valley

**AREA** 15 square miles (40 square km)

**MAXIMUM DEPTH** 10 ft (3 m)

**ELEVATION** 5,780 ft (1,760 m)

**OUTFLOW** None

Lake Nakuru is a small, shallow, salty and alkaline lake—a remnant of a much larger lake that has shrunk to almost nothing over the last 10,000 years. At least once during the 20th century, Lake Nakuru has dried out completely. It is popular with tourists because of the hundreds of bird species that congregate there—particularly the spectacular flamingos.

**FLAMINGOS FLOCKING ON THE LAKE**

AFRICA *east*

# Lake Manyara

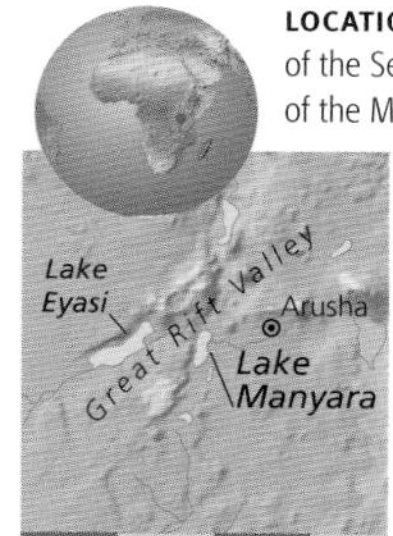


**LOCATION** In Tanzania, southeast of the Serengeti Plain and northwest of the Masai Steppe

**AREA** 85 square miles (230 square km)

**MAXIMUM DEPTH** Not known

**ELEVATION** 3,100 ft (945 m)

**OUTFLOW** None

Lake Manyara lies at the southern end of the eastern arm of the Great Rift Valley, beneath the sheer rock walls of the valley's western escarpment. The lake has no perennial surface water sources, just a few occasional streams. Most of its inflow comes from beneath its shallow bed, having percolated down from the volcano-strewn Ngorongoro Highlands. This groundwater is saturated with minerals derived from volcanic rock. Despite there being extensive salt and soda deposits along areas of the lakeshore, its water is tolerated by catfish and bream, and by hippopotamuses. The Lake Manyara National Park, on the western shore, was established in the 1960s specifically for the protection of elephants, but is more famous for its tree-climbing lions. Visitors can also see more than 300 bird species.

AFRICA *east*

# Lake Nyasa

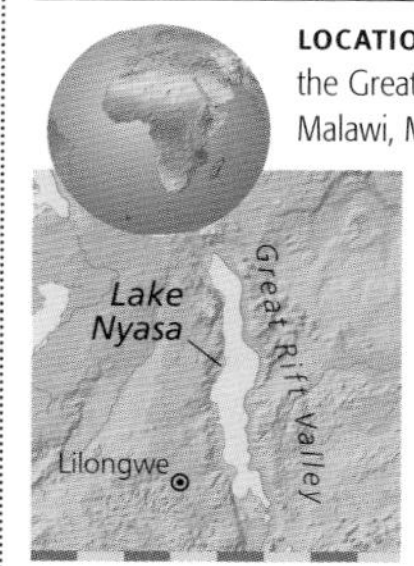


**LOCATION** At the southern end of the Great Rift Valley, bordered by Malawi, Mozambique, and Tanzania

**AREA** 2,470 square miles (6,400 square km)

**MAXIMUM DEPTH** 2,318 ft (706 m)

**ELEVATION** 1,640 ft (500 m)

**OUTFLOW** Shire River

Lake Nyasa is the most southerly of the Great Rift Valley lakes and was formed by the same rifting process that produced Lake Tanganyika (see p.248), which lies about 220 miles (350 km) to the northwest. The long, narrow lake basin is delimited mainly by faults. Lake Nyasa is deepest in the north, where the bottom drops away sharply from the western shore. It becomes increasingly shallow toward the southern end, where its only outlet, the Shire River, flows sluggishly out of Monkey Bay to join the Zambezi River. In contrast, it receives water from no fewer than 14 rivers. Lake Nyasa has more fish species than any other lake in world, with perhaps as many as 1,000 different cichlid fishes living in its surface waters. However, as with other deep tropical lakes, there is no vertical circulation of water. The deep waters lack oxygen and are saturated with noxious gases such as hydrogen sulfide, methane, and carbon dioxide, produced by decomposition of vegetation that has sunk to the bottom.

**STILL WATERS**
*A placid, sand-fringed lagoon on the northern shore gives no clue that Lake Nyasa is the second-deepest lake in Africa.*

AFRICA *east*

## Lake Tanganyika

**LOCATION** At the southern end of the Great Rift Valley, between Burundi, Dem. Rep. of Congo, and Tanzania

**AREA** 12,350 square miles (32,000 square km)

**MAXIMUM DEPTH** 4,800 ft (1,471 m)

**ELEVATION** 2,538 ft (773 m)

**OUTFLOW** Lukuga River

The second-deepest lake in the world, Tanganyika is a long, narrow lake extending 410 miles (660 km) along a block-fault depression (graben) formed by the rifting process in East Africa about 15 million years ago (see pp.148–49). It has steep-sided, forested shorelines, with the Mitumba Mountains along the western shore towering about 6,600 ft (2,000 m) above the lake. The eastern shore is dominated by the highlands of the East African Plateau. There is lakeside lowland only where rivers enter or leave the lake. A subsurface ridge extends across the middle of the lake, dividing it into two extremely deep basins. Year-round heavy rain washes in large quantities of sediment that have accumulated in layers up to 3 miles (5 km) deep. The main inflows are the Ruzizi River, which enters at the north end, descending 2,000 ft (600 m) from Lake Kivu; the Malagarasi River on the eastern shore; and the Kalambo River, which enters on the southern tip, its waters having plunged over a 705-ft- (215-m-) high waterfall before reaching the lake. The single outflow, the Lukuga, exits midway along the western shore and adds its flow to the headwaters of the Congo River. Sometimes the outflow becomes blocked by floating vegetation, causing local flooding.

**LEMON CICHLID**
*This species of cichlid, found only in the southern part of the lake, is one of dozens of species of cichlids that are unique to Lake Tanganyika.*

### HENRY MORTON STANLEY

The British-born American journalist and explorer Henry Morton Stanley (1841–1904) famously "found" the "missing" British missionary-explorer David Livingstone in 1871 at Ujiji on the eastern shore of Lake Tanganyika. They then explored the northern end of the lake, looking for the source of the Nile. In 1874 Stanley organized and led a further expedition that crossed Africa coast to coast, from west to east.

**EASTERN SHORE**
*The densely forested slopes of the Mohale Mountains in Tanzania descend smoothly to the water's edge—there are no beaches or lakeshore lowlands.*

ASIA *northeast*

## Lake Baikal

**LOCATION** In Russia, south of the Central Siberian Plateau, near Mongolia

**AREA** 12,160 square miles (31,500 square km)

**MAXIMUM DEPTH** 5,716 ft (1,741 m)

**ELEVATION** 1,497 ft (456 m)

**OUTFLOW** Angara River

Baikal is the deepest lake in the world, containing about 20 percent of the world's surface fresh water. The lake is 395 miles (636 km) long and about 30 miles (50 km) wide. It formed in a deep and ancient block fault some 25 million years ago, which also makes it the oldest lake in the world. The fault appears to be still active, as the lake is widening by about 1 in (2.5 cm) per year. On its western side, Baikal is lined by mountains; the eastern shoreline has gentler slopes. The deepest parts are in the center and near the southern end. Layers of sediment on the lake bed are up to 4 miles (7 km) thick in places. The continual descent of cold water keeps the bottom levels well oxygenated, and animal life is found at all depths. The lake receives inflow from many rivers, mainly along its eastern shore, and these include the Barguzin and the Selenga, which forms a delta as it enters the lake. Baikal also receives water below its surface from hot freshwater vents in the lake bed. Irkutsk is situated near the southern end of the lake on its single outflow, the Angara River. The lake is ice-free for about six months each year.

**BAIKAL SEAL**
*The world's only freshwater seal is just one of the 1,500 or so plant and animal species that are unique to the lake.*

**STORM CLOUDS OVER LAKE**
*During the winter months, the surrounding mountains channel fierce storms along Lake Baikal.*

### FISHING

Lake Baikal provides an important source of food to the people living on its shores, despite remaining frozen much of the time. Local fishermen employ traditional methods to fish through a hole cut in the winter ice. The major food fish is the Omul, but 56 species of fish occupy the lake, and more than half of them are only found here.

**ARAL SEA SHORELINE**
*The volume of water entering the Aral Sea from the Syr Darya River has shrunk considerably because of the demand for water for crop irrigation.*

ASIA *west*

# Aral Sea

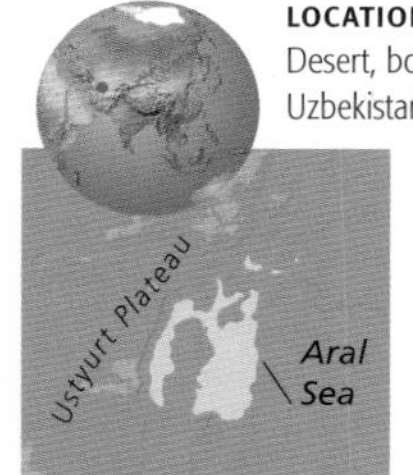


**LOCATION** North of the Kara Kum Desert, bordered by Kazakhstan and Uzbekistan

**AREA** 10,000 square miles (26,000 square km)

**MAXIMUM DEPTH** 190 ft (58 m)

**ELEVATION** 121 ft (40 m)

**OUTFLOW** None

Once the world's fourth-largest inland body of water, the Aral Sea is now drying up rapidly, making it one of the world's most conspicuous examples of man-made desertification. Fifty years ago it was a wide, shallow, saline sea—healthy and full of life. In the mid-20th century, its fisheries accounted for 3 percent of the total catch of the former Soviet Union. Now, its area is greatly reduced, and there are no fish because water salinity has tripled. The Aral Sea was cut off from the oceans during the last ice age but maintained its size in the arid Central Asian climate because of the inflow of two mighty rivers—the Syr Darya, flowing down from the Tien Shan Mountains into the northern end, and the Amu Darya, bringing water from the Pamirs and entering along the southern shore, where it formed a large delta. In the 1930s, increasingly large amounts of water were diverted from these rivers to provide irrigation water for vast fields of cotton. While cotton production soared, the lake began to shrink, slowly at first and then alarmingly quickly. During the last 35 years, the level of the Aral Sea has dropped by more than 66 ft (20 m) and less than 40 percent of its surface area remains. The western shores are defined by the high Ustyurt Plateau, and the drop in level is most apparent along the northern and eastern shores, where great expanses of former lake bed are now exposed. The shrinking of the Aral Sea has reduced the humidity of the local climate, so that summers are now hotter and winters are colder. Millions of tons of dust and mineral salts have been blown away by the wind, adversely affecting the health of 5 million people who have to breathe the dust-laden air.

## WATER PIPELINES

Toxic dust blowing from the desert created by the evaporation of the Aral Sea has contaminated water supplies over a wide area. Muynoq is located near the Amu Darya delta, once the region's greatest source of fresh water. It was a thriving port, but is now 35 miles (56 km) from the coast of the Aral Sea. Today, its inhabitants must rely on water brought from distant reservoirs along pipelines such as the one shown below.

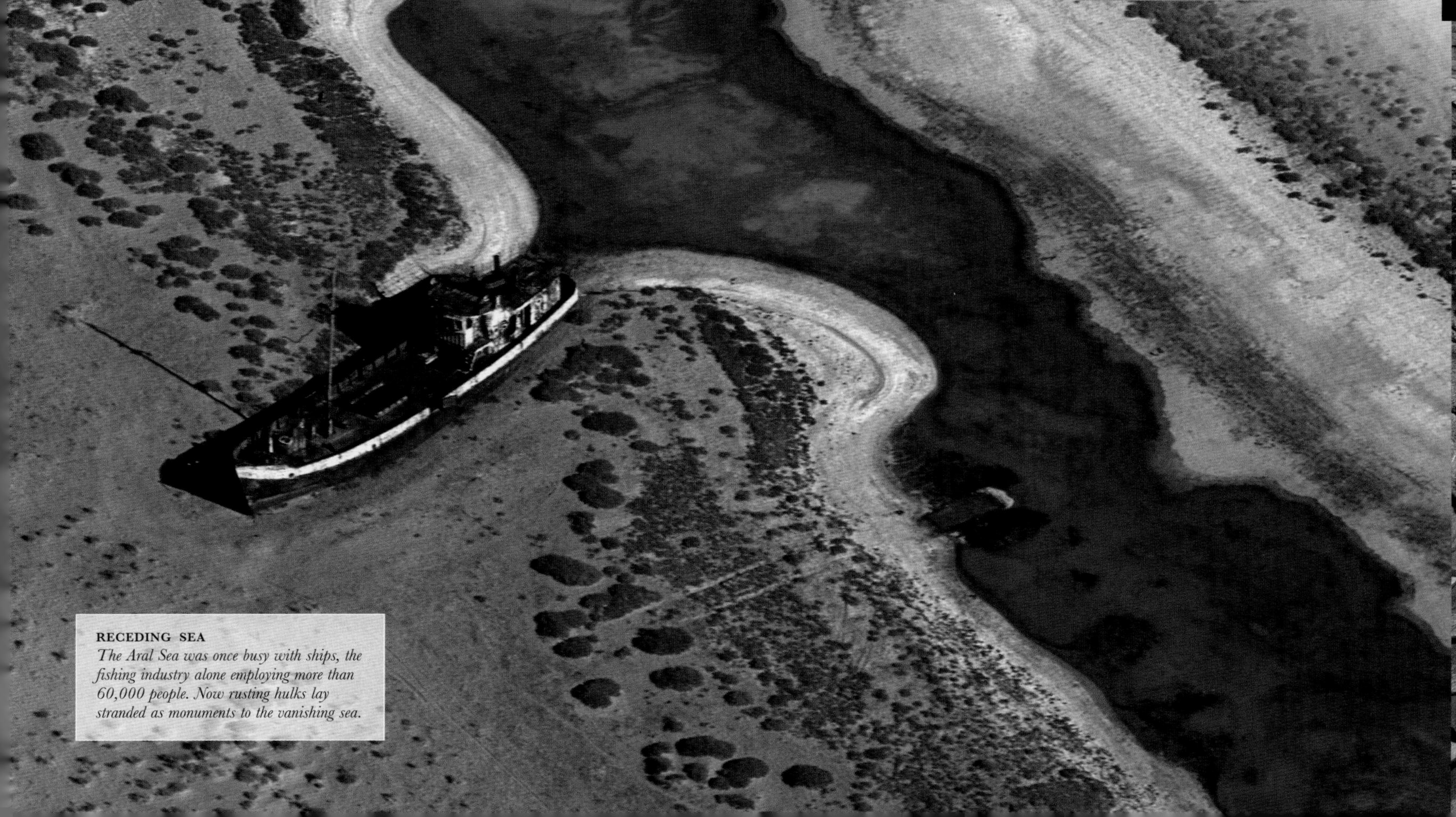

**RECEDING SEA**
*The Aral Sea was once busy with ships, the fishing industry alone employing more than 60,000 people. Now rusting hulks lay stranded as monuments to the vanishing sea.*

ASIA *west*

## Caspian Sea

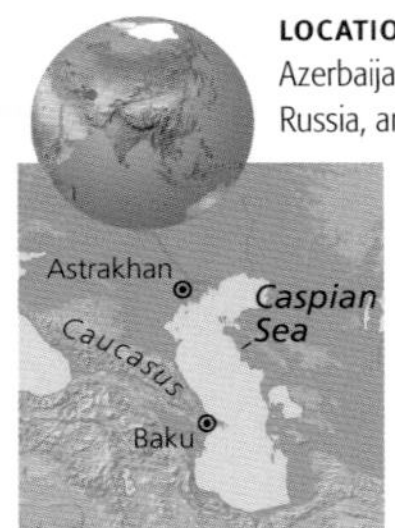


**LOCATION** On the borders of Azerbaijan, Iran, Kazakhstan, Russia, and Turkmenistan

**AREA** 143,000 square miles (371,000 square km)

**MAXIMUM DEPTH** 3,120 ft (950 m)

**ELEVATION** 100 ft (30 m) below sea level

**OUTFLOW** None

The world's largest inland body of water, the saline Caspian Sea occupies an ancient, tilted depression in the Earth's surface. The water is deepest at the southern end, where the Elburz Mountains line its shores. Cut off from the Mediterranean Sea by falling water levels during the last ice age, the Caspian has no outflow, and owes its continuing existence to the Ural and Volga rivers, which supply much of its inflow. The eastern shore is defined by the cliffs of the arid Ustyurt Plateau, and the Kara Kum Desert lies to the southeast. The water's salinity varies from 1 percent in the north, near the Volga inflow, to about 20 percent in the shallow Kara Bogay Bay on the eastern shore, which is almost totally cut off from mixing with the rest of the Caspian's waters. (Ocean salinity averages about 3.5 percent.) For thousands of years, the water level of the Caspian Sea has experienced dramatic and unpredictable fluctuations. For example, the water rose 8 ft (2.5 m) between 1978 and 1995.

### OIL EXTRACTION

The Caspian Sea has proven oil reserves that equal those of the US, and the region is rapidly being transformed by the oil industry. Oil extraction was formerly centered on Baku in Azerbaijan, on the Caspian's southwestern shoreline. It has now expanded to include newly developed oil and gas fields in Kazakhstan at the northern end of the Caspian.

**STERLET**
*This smaller relative of the caviar-producing sturgeon is one of the Caspian's most valuable food fish.*

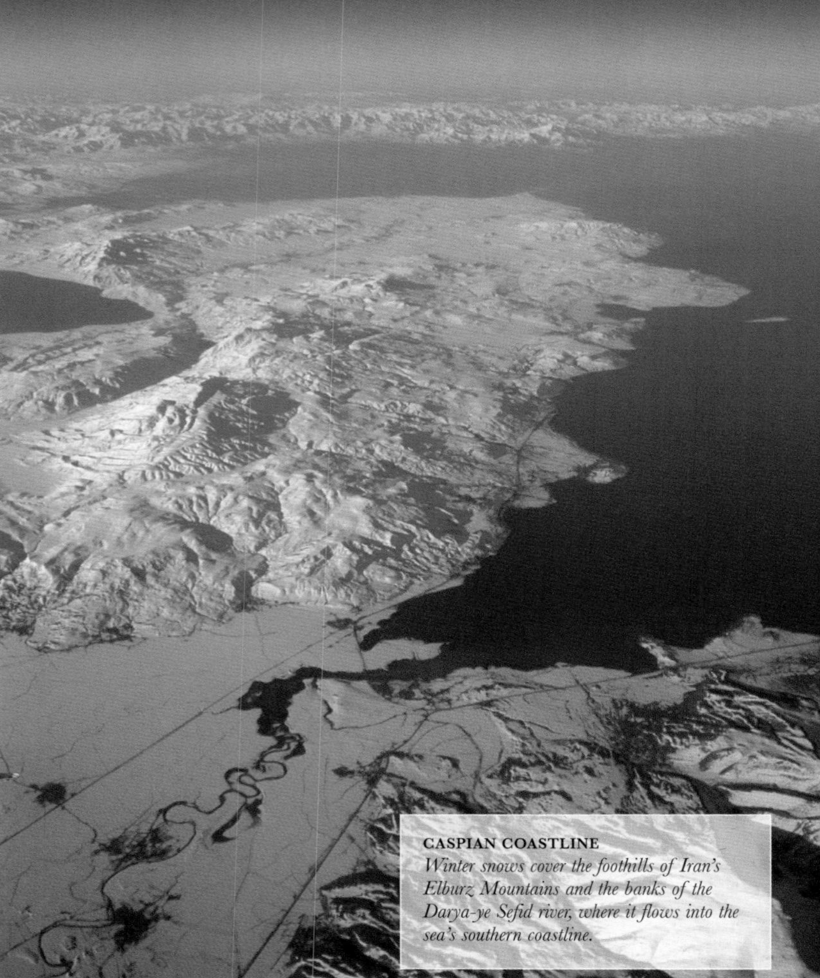

**CASPIAN COASTLINE**
*Winter snows cover the foothills of Iran's Elburz Mountains and the banks of the Darya-ye Sefid river, where it flows into the sea's southern coastline.*



ASIA *central*

## Lake Balkhash

**LOCATION** In southeastern Kazakhstan, south of the Central Kazakh Uplands

**AREA** 7,025 square miles (18,200 square km)

**MAXIMUM DEPTH** 82 ft (25 m)

**ELEVATION** 1,115 ft (340 m)

**OUTFLOW** None

Balkhash is a long, narrow, curved lake, situated in an east–west depression. The lake is divided into two parts by the Sarymsck Peninsula, which juts out from the southern shore. The western part of the lake is wide and shallow, and the water freshness is maintained by the inflow of the Ili River near the lake's southern end. To the east of the peninsula, the lake is narrower and deeper. There is little mixing between the two parts of the lake, and the water in the eastern part has much higher levels of dissolved minerals because it receives only the inflow of minor rivers. The level of the lake fluctuates seasonally by about 10 ft (3 m), and during the last 100 years Balkhash's area has varied by up to 50 percent—from 6,000 square miles (15,600 square km) to 9,000 square miles (23,400 square km). There are valuable copper deposits on the northern shore that are processed in the nearby town of Balkhash.

**DIVIDED LAKE**
*The Sarymsck Peninsula, across the middle of Lake Balkash, is visible near the center of this space shuttle image.*

ASIA *west*

## Sea of Galilee

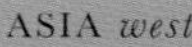

**LOCATION** In northern Israel, close to the borders with Syria and Lebanon

**AREA** 64 square miles (166 square km)

**MAXIMUM DEPTH** 157 ft (48 m)

**ELEVATION** 696 ft (212 m) below sea level

**OUTFLOW** Jordan River

Known also as Lake Tiberias, and renowned from the Bible, the Sea of Galilee is a small, shallow, pear-shaped lake situated in a deep, rifted depression that is an extension of Africa's Great Rift Valley. The Jordan River flows southward along the rift and through the Sea of Galilee. North of the lake, around the inflow of the Jordan, is the narrow, fertile Gennesaret Plain, with the Golan Heights to the northeast in Syria. The western shores of the lake lie beneath the sheer wall of the rift, and the southern shores are confined by a volcanic ridge. The flow of the Jordan into the lake has diminished recently because water is being diverted for irrigation purposes upstream of the lake. As a result, the lake's waters are slowly becoming saline.

**SOUTHEASTERN SHORELINE**
*The Sea of Galilee is seen here from the southeastern shore. The wall of the rift is visible across the water to the west.*

ASIA *west*

## Dead Sea

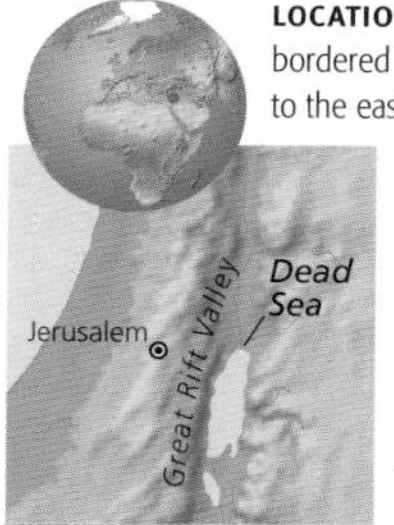


**LOCATION** North of the Red Sea, bordered to the west by Israel and to the east by Jordan

**AREA** 313 square miles (810 square km)

**MAXIMUM DEPTH** 1,084 ft (330 m)

**ELEVATION** 1,329 ft (405 m) below sea level

**OUTFLOW** None

The Dead Sea lies in a depressed piece of continental crust (graben) that separates two tectonic plates, the Sinai Subplate and the Arabian Plate. To the north and south the basin is bordered by igneous intrusions called sills. Fed by the Jordan River, and having no outflow, the Dead Sea is the lowest lake in the world—its surface is the lowest point on the Earth's land surface, at 1,320 ft (400 m) below sea level. It is also one of the saltiest lakes, with an average salinity of about 32 percent. The Dead Sea achieved its greatest extent about 10,000 years ago, and it has been shrinking slowly ever since. It is currently about 50 miles (80 km) long and up to 11 miles (18 km) wide. Inflow from the Jordan and the occasional rain shower are insufficient to balance the very high rates of evaporation. Recently the rate of shrinkage has increased because of the diminished flow of the Jordan. During the last quarter of the 20th century, the area of the lake declined by about 20 percent, and the level of the lake is currently dropping at a rate of about 39 in (1 m) per year. The Lisan Peninsula, which once divided the lake into northern and southern parts, now forms the southern shoreline. What was the southern part of the lake now consists only of evaporation pans for salt and mineral production. In the deepest part of the lake, where the bottom waters are very dense and saline, there is little vertical mixing. The deepest water in the Dead Sea is many thousands of years older than the surface waters.

**DEAD SEA COASTLINE**
*The paler sediments seen here on the barren shoreline of the Dead Sea indicate that the water level was once much higher.*

**LAKESIDE SALT DEPOSITS**
*As the waters of the Dead Sea evaporate, they leave behind an unearthly landscape dotted with mounds of salt crystals and pools of extremely saline liquid.*

ASIA *southeast*

## Qinghai Hu

**LOCATION** In the Nan Mountains of north-central China, east of the Qaidam Pendi basin

**AREA** 1,772 square miles (4,460 square km)

**MAXIMUM DEPTH** 125 ft (38 m)

**ELEVATION** 10,500 ft (3,200 m)

**OUTFLOW** None

Qinghai Hu is a shallow, slightly saline lake situated in a broad depression on a high plateau and ringed by mountains. It is the largest lake in China and is also known as Koko Nor, and alternatively as the Blue Lake because of its intense color. The area of the lake varies seasonally, and approaches 2,300 square miles (6,000 square km) when water levels are at their highest. Terraces around the shores indicate much higher levels during the last ice age, when Qinghai Hu probably drained westward into the Tarim Basin.

**QINGHAI HU SHORELINE**

ASIA *southeast*

## Lake Toba

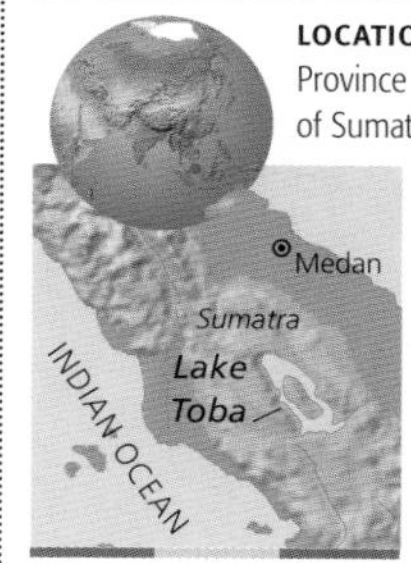


**LOCATION** In North Sumatra Province on the Indonesian island of Sumatra

**AREA** 425 square miles (1,100 square km)

**MAXIMUM DEPTH** 1,737 ft (529 m)

**ELEVATION** 2,970 ft (905 m)

**OUTFLOW** Asahan River

Lake Toba is the largest caldera lake in the world and is situated high in the Barisan Mountains of northern Sumatra. The lake is oval in shape—56 miles (90 km) long by 19 miles (30 km) wide and the cliffs of the enclosing caldera rim rise up to 3,940 ft (1,200 m) above the surface. It is over 1,518 ft (460 m) deep. Lake Toba is situated on an active fault line that runs the length of Sumatra, and was formed when prolonged lava flow emptied and collapsed a subterranean chamber, producing a caldera that soon filled with water. A subsequent volcanic eruption between 30,000 and 75,000 years ago produced the large island of Samosir, with an area of 250 square miles (640 square km), in the middle of the lake. The shape of Samosir has been modified by further volcanic activity. At the northern end of the lake, water enters via the Sipiso-Piso waterfall. The Asahan River exits from the southern end of the lake and flows into the Indian Ocean.

**SAMOSIR ISLAND**
*Samosir, which has two extinct volcanoes, occupies about one-third of the lake. The rich volcanic soils on the island are intensively cultivated.*

AUSTRALASIA *Australia*

## Lake Eyre

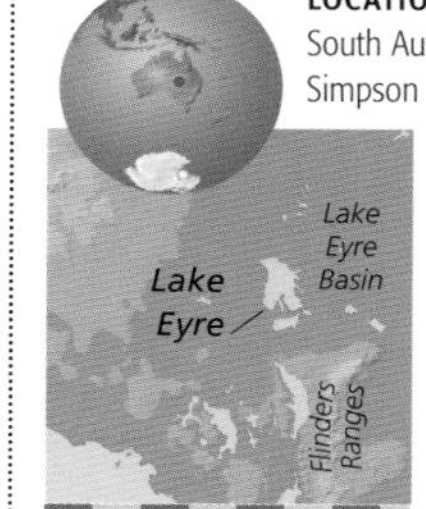


**LOCATION** In the state of South Australia, south of the Simpson Desert

**AREA** 3,745 square miles ( 9,700 square km)

**MAXIMUM DEPTH** 18 ft (5.5 m)

**ELEVATION** 53 ft (16 m) below sea level

**OUTFLOW** None

A salt lake of variable size, Lake Eyre is the lowest point on the Australian continent. It receives water from ephemeral streams, fed by erratic rainfall over much of central Australia. Thought to have once dried out completely, the lake reached its present size following exceptionally heavy rains at the end of the 20th century.

**DRY LAKEBED**
*This dried-out area of lakebed is covered by thick deposits of gypsum overlain by a thin crust of salt.*

# UNDERGROUND RIVERS AND CAVES

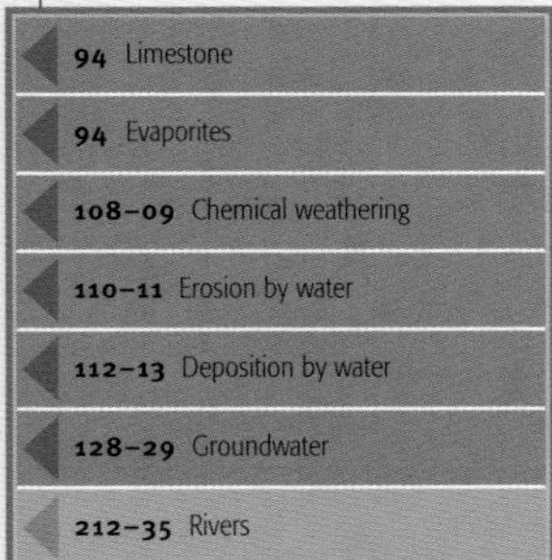
94 Limestone
94 Evaporites
108–09 Chemical weathering
110–11 Erosion by water
112–13 Deposition by water
128–29 Groundwater
212–35 Rivers

BENEATH THE SURFACE in many of the Earth's limestone regions are large passages and caves. These are formed as limestone dissolves in slightly acidic water. As groundwater seeps downward, tiny fractures gradually enlarge into open passages. When streams and rivers are swallowed by caves, erosion continues more rapidly. Underground rivers scour out huge waterfall shafts and dissolve great tunnels. Subsequently, chambers are modified by rock-fall from unsupported ceilings, and seepage water creates deposits of sparkling calcite. While young, active caves have noisy streams and cascades; old, abandoned caves are silent wonderlands decorated with stalagmites and stalactites.

## LIMESTONE IN SOLUTION

Limestone is rock mostly composed of calcium carbonate (see p.94), which dissolves in naturally acidic water. Rainwater absorbs carbon dioxide from the air and from microscopic soil organisms to become weak carbonic acid. As it seeps downward, it can dissolve large amounts of limestone. If that water then drips out into a cave, it loses carbon dioxide into the air and deposits excess calcite (calcium carbonate) as stalactites and stalagmites (see opposite) in order to maintain its chemical equilibrium.

**STRAW STALACTITE**
*Tiny crystals of calcite can be seen here growing inside a water drop at the end of a cave straw (miniature stalactite) just 5mm (1/5in) in diameter.*

## TRACING STREAMS

Underground streams can be tracked through large caves, but not where they are lost in narrow fissures. Then their courses are traced using fluorescent dyes, such as fluorescein. Even when diluted by many converging cave streams, this dye can be detected in minute quantities at a spring to prove the underground flow path from a distant sinkhole.

## UNDERGROUND EROSION

When streams pass underground, two main types of cave passage are formed—canyons and tubes. Underground streams and waterfalls above the water table cut deep, twisting canyon passages, by removal of limestone in solution and sediment abrasion of their floors. Below the water table, caves are full of slowly moving water, which dissolves limestone walls, floors, and ceilings equally, creating tubular tunnels. Many caves descend steeply to the water table and then continue sideways, as flooded passages, until they intersect the surface, where the emerging water forms a spring (also known as a resurgence).

**LIMESTONE LANDSCAPE**
*Sinking streams, bare rock pavements, and scattered sinkholes (or dolines) form the visible half of a limestone landscape. The unseen half is the network of cave passages, some with rivers, and others that are now dry and contain calcite deposits.*

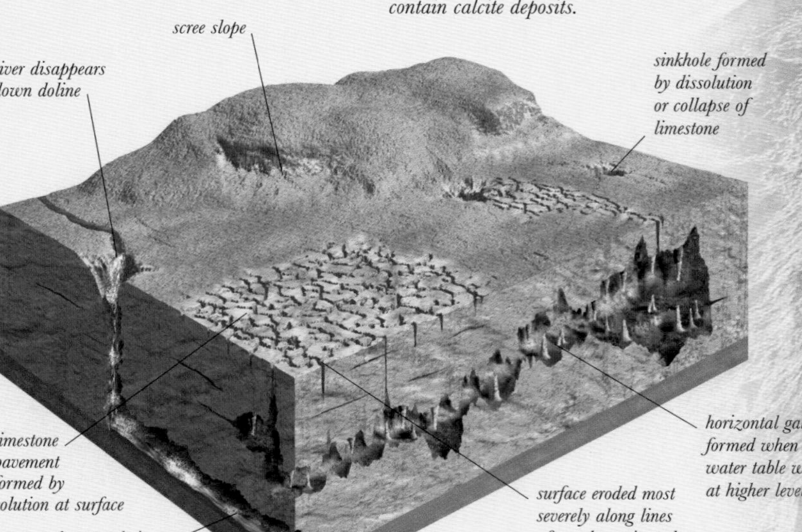


LAND

## CAVE DEPOSITS

Sand, mud, calcite, gypsum, and guano are all common cave deposits, but the most widespread and beautiful are those made of calcite. Groundwater saturated with calcium carbonate deposits calcite in the form of stalactites where water drips from a cave roof. Stalagmites form where the water hits the cave floor. Banks of solid calcite on the walls and floors of caves are called flowstones. A rimstone pool forms on the surface of a flowstone deposit where water collects behind a wall of calcite. More calcite may be deposited as water flows over and down the wall. Most cave calcite contains traces of uranium; the radioactive decay of its atoms can be used to determine the age of deposition, and thus to date the cave.

**ARAGONITE CRYSTALS**
*In Australia's Jenolan Caves, tiny crystals of aragonite (a form of calcium carbonate) grow under thin films of water derived from condensation or seepage.*

**CAVE PEARLS**
*These calcite spheres in Golconda Cavern, England, are cave pearls, which grew by rolling around in a pool disturbed by dripping water.*

## KARST LANDSCAPES

Karst is defined as a limestone landscape with underground drainage. It is distinguished by extensive patches of bare rock, because the weathering of limestone produces little or no soil. Rock outcrops exposed to rainfall are fretted by grooves called karren, formed as weakly acidic rainwater runs across them. They start as narrow fissures on joints or meandering grooves, and evolve into deep gullies between pinnacles of remnant limestone. Sinkholes or dolines formed by collapses in the surface can merge to form large sunken regions, equivalent to valleys in other landscapes. As the sinkholes expand, the residual hills between them become cones, or can be undercut to form towers. The weathering of karst landforms relies on abundant rainfall and lush vegetation (to provide carbon dioxide through its decay), so the most mature karst and the largest caves form in wet, tropical areas.

**LIMESTONE PAVEMENT**
*This pavement in England's Pennines was scraped clean by glaciers during the last ice age and then etched with deep fissures by postglacial rainwater.*

**TOWER KARST SCENERY**
*Isolated limestone towers with vertical cliffs overlook the wide plain around Yangshuo, in Guangxi, southern China. They represent the ultimate mature karst landscape; tower karst is also known by its Chinese name, fenglin.*

**PINNACLE KARST**
*With razor edges honed by tropical rainfall, spectacular pinnacles and blades of limestone rise clear of the rainforest canopy on the karst hills of Gunung Mulu National Park, Sarawak, Malaysia.*

**LEVIATHAN CAVE**
*This cave in Nevada was formed by a river that has now disappeared. The impressive arch shape is due to later collapse of the roof.*

### UNDERGROUND RIVERS AND CAVES PROFILES

The pages that follow contain profiles of some major underground rivers and caves. Each profile begins with the following summary information:

| | |
|---|---|
| **LENGTH** | Total length of all the passages within the cave system |
| **DEPTH** | Vertical distance from the cave entrance to its deepest known point |

NORTH AMERICA *west*

## Carlsbad Cavern

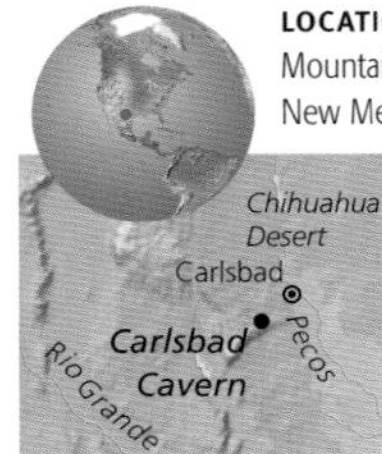


**LOCATION** In the Guadalupe Mountains of southeastern New Mexico

**LENGTH** 31 miles (50 km)

**DEPTH** 1,037 ft (316 m)

Left high and dry in the mountains of the Chihuahua Desert, Carlsbad Cavern has no underground river and almost no dripping water, but is a gigantic relic of long-finished karst solutional processes. It was formed in Permian limestones that were invaded by fluids rising from the hydrocarbon reservoirs in the Delaware Basin, just to the east. These fluids were rich in corrosive sulfuric acid that dissolved the limestone, removing it in solution on an unusually large scale. The Big Room, which was formed about 4 million years ago, displays evidence of this very unusual origin. Although rarely noticed by visitors, there are thick banks of chalky white gypsum left by the chemical reaction between the sulfuric acid and the limestone. Only later was Carlsbad Cavern invaded by percolating rainwater, which both continued the excavation by solution, and also deposited the massive calcite stalactites and stalagmites. Just inside its gaping entrance, the spacious Bat Cave once held thick deposits of bat guano, most of which was mined out in the 1800s to use as fertilizer. It was the miners who first explored the extensive lower passages, which contain many classic cave features. The descending Main Corridor is floored by huge limestone blocks, called breakdown, that have fallen from the faulted roof, totally changing its shape from the original solutional cave. Lower down in Carlsbad Cavern, the Boneyard is a complex of small, unmodified chambers and tunnels that show how the limestone was dissolved in random shapes when the cave was full of water. Beyond this, the famous Big Room is not so much a great chamber as a giant, twisting passage that is over 400 ft (120 m) wide in places. Most of Carlsbad Cavern is decorated with a profusion of calcite deposits, and the Big Room contains some very large stalagmites within its high, open space. The Big Room is the end of the cave section open to the public, but large tunnels descend beneath it, ending at the Lake of the Clouds, named after the huge bulbous stalactites that hang over a deep pool where the cave meets the present-day water table.

**BATS AT THE CAVE ENTRANCE**
*Every evening, hundreds of thousands of Mexican free-tailed bats leave their daytime roost in Bat Cave to feed in the surrounding countryside.*

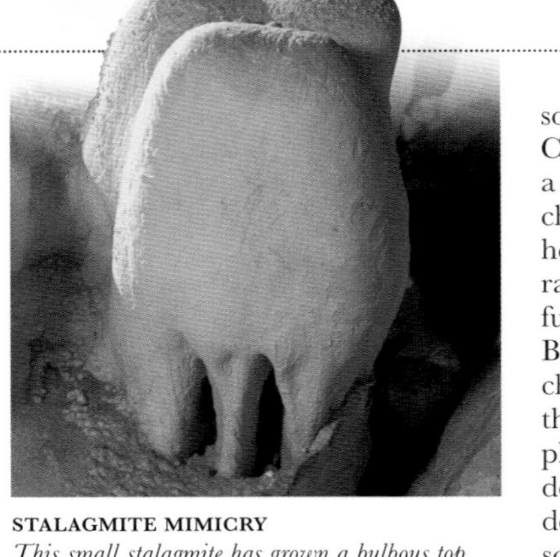

**STALAGMITE MIMICRY**
*This small stalagmite has grown a bulbous top with stalactite ribs hanging below, so that it is now always known as the "bashful elephant."*

**DRIPSTONE DECORATIONS**
*Minor irregularities in the cave wall cause drips of water to deposit calcite, creating little flowstone terraces. Water dripping off the terraces deposits more calcite as curtains and fringes of stalactites.*

**PAINTED GROTTO**
*In this corner of Carlsbad Cavern's Big Room, calcite is colored by natural pigments of iron oxides and hydroxides. The Big Room has an area of 357,480 square feet (33,210 square meters).*

**PAPOOSE ROOM**
*This chamber gets its name from an American Indian word for a small child. The cavity was originally created by sulfuric acid; the stalactites were later formed by carbonic acid.*

NORTH AMERICA *east*

## Mammoth Cave

**LOCATION** In southwest Kentucky, beneath the Chester Upland

**LENGTH** 352 miles (563 km)

**DEPTH** 377 ft (115 m)

By far the longest cave system known in the world, Mammoth Cave has a huge network of passages on many levels. The nearly horizontal Carboniferous limestone in which it occurs carries water from sinkholes on the Pennyroyal Plateau to springs along the Green River. There are many cave passages, each with a long, meandering canyon that passes into an elliptical tube beneath the water table. Downcutting of the Green River has caused the water table to fall in successive stages, leaving many large, dry, abandoned tubes. The highest level of caves contains sediments that are 2.3–3.5 million years old; these survive beneath the protective sandstone ridges of the Chester Upland. Most of the youngest caves drain through the ridges, where they connect with some of the older caves, while a few pass beneath the intervening Houchins and Doyel valleys. Except where shafts drop down through the limestone beds, most caves follow the same bedding planes within the rock sequence. Consequently, there are numerous chance passage intersections, which have linked the caves into the one extraordinarily long system.

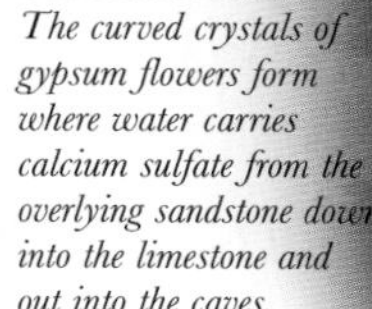

**GYPSUM FLOWER**
*The curved crystals of gypsum flowers form where water carries calcium sulfate from the overlying sandstone down into the limestone and out into the caves.*

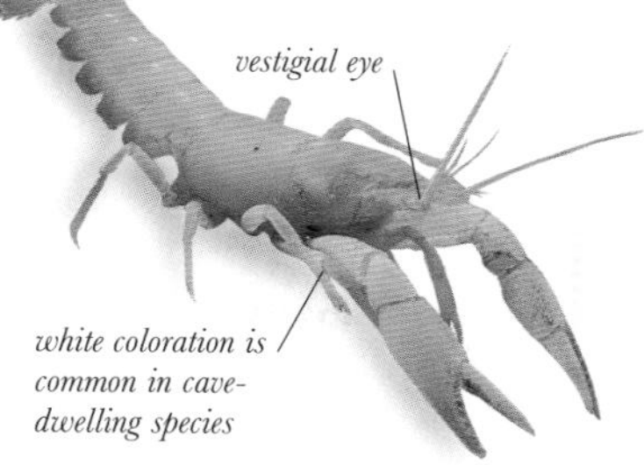


**EYELESS CAVE-DWELLER**
*Due to the lack of light, this Indiana eyeless crayfish has no need for eyes or color pigment, but instead has extended sensory feelers.*

**ABANDONED TUBULAR CAVE**
*The many long, tubular tunnels of Mammoth Cave were formed below the water table before it fell to a lower level.*

CENTRAL AMERICA

## Windsor Caves

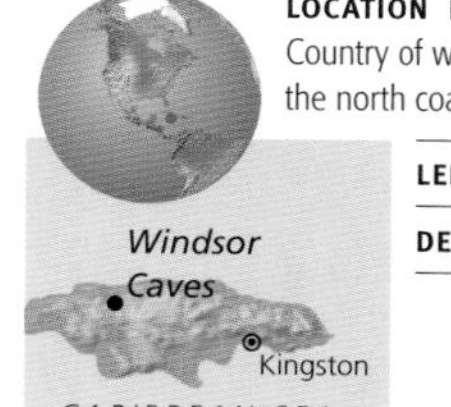


**LOCATION** Beneath the Cockpit Country of western Jamaica, close to the north coast

**LENGTH** 2 miles (3 km)

**DEPTH** 65 ft (20 m)

The forest-covered hills in Jamaica's Cockpit Country are often likened to an upturned egg carton. They constitute a famous example of cone karst and are drained entirely underground. Windsor Great Cave is one of the few caves known in the karst; it is a high-level remnant of an ancient drainage system, now largely choked with calcite deposits. There is a small cave stream, but the main drainage lies below the water table at greater depths within the limestone.

**WINDSOR GREAT CAVE**

NORTH AMERICA *south*

## Sistema Huautla

**LOCATION** In the Sierra Mazateca, part of the Sierra Madre del Sur, in Oaxaca, Mexico

**LENGTH** 39 miles (56 km)

**DEPTH** 4,840 ft (1,475 m)

The limestone highlands of the Sierra Mazateca contain one of the world's finest collections of caves. Nita Nanta is the highest entrance to one of a dozen caves that converge in Sistema Huautla; each is a narrow stream canyon broken by a staircase of waterfall shafts. To the south, the giant dolines of San Agustin drain into shafts that feed the underground river. The combined waters then drop down into a largely unexplored flooded zone. The main passage loops far below the water table before emerging from underwater passages in Santo Domingo Canyon. As the canyon has cut deeper, it has triggered changes throughout the cave system. Fresh stream passages have been formed under now-abandoned tunnels. The oldest passages include chambers, modified by massive roof collapse, near the foot of the San Agustin shafts.

**EXPLORING THE HUAUTLA CAVES**
*Cascading streamways and lashing waterfalls make exploration in Huautla both challenging and exciting for cavers.*

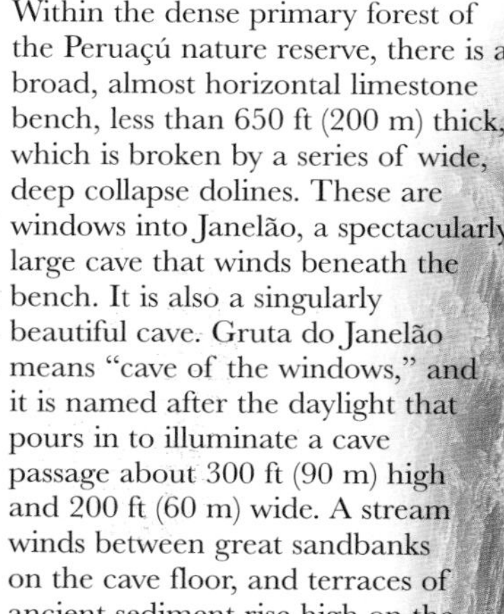

SOUTH AMERICA *east*

## Gruta do Janelão

**LOCATION** In the middle São Francisco valley of Minas Gerais, eastern Brazil

**LENGTH** 3 miles (5 km)

**DEPTH** 500 ft (150 m)

Within the dense primary forest of the Peruaçú nature reserve, there is a broad, almost horizontal limestone bench, less than 650 ft (200 m) thick, which is broken by a series of wide, deep collapse dolines. These are windows into Janelão, a spectacularly large cave that winds beneath the bench. It is also a singularly beautiful cave. Gruta do Janelão means "cave of the windows," and it is named after the daylight that pours in to illuminate a cave passage about 300 ft (90 m) high and 200 ft (60 m) wide. A stream winds between great sandbanks on the cave floor, and terraces of ancient sediment rise high on the walls, which suggest that a far larger river flowed through the cave in the distant past. Giant stalactites hang from the cave roof—one has been measured as 92 ft (28 m) long and is perhaps the longest stalactite in the world. There are also massive banks of calcite flowstone on the walls and floors of the cave. The whole site is the product of karst processes that have been active for a very long time, but the exact age of the cave is not yet known.

**CAVE OF THE WINDOWS**
*Daylight from holes in the roof (and the presence of a person sitting on a sandbank at the lower right in this photograph) make it possible to appreciate the sheer size of the cave.*

EUROPE *northwest*

## Ease Gill Caverns

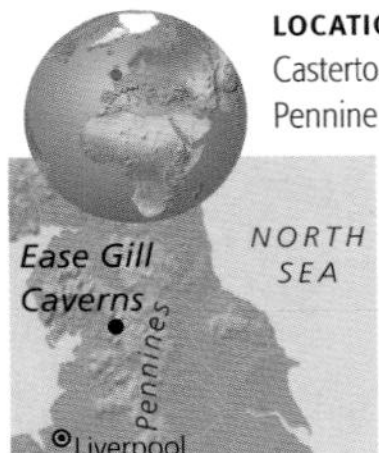


**LOCATION** Beneath Leck and Casterton Falls in the northern Pennine karst, England

**LENGTH** 44 miles (70.5 km)

**DEPTH** 692 ft (211 m)

Dozens of streams drain through this rambling series of caves, before resurfacing at Leck Beck Head. The present-day inlet passages are long, clean-washed stream canyons, interrupted by vertical waterfall shafts; some intersect with old, abandoned tunnels. The large, high-level tunnels, which date from the Pleistocene Epoch, have calcite decorations and are partly choked with in-washed sediments.

**CAVE SHAFT**
*This deep, vertical shaft with its waterfall is created where a cave stream drops from one level to another. It forms a staircaselike descent through the limestone beneath Leck Fell in Yorkshire.*

EUROPE *west*

## Lascaux

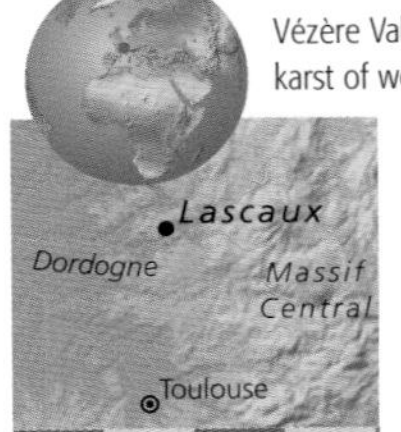


**LOCATION** In the side of the Vézère Valley within the Dordogne karst of western France

**LENGTH** 500 ft (150 m)

**DEPTH** 30 ft (10 m)

This tiny remnant of an ancient cave passage was long ago left high and dry in a limestone hillside. The passage was later truncated as erosion wore away the hillside, leaving an open entrance in the side of the Vézère Valley. Lascaux was found by people of the Solutrean and Magdalenian cultures in the Old Stone Age (or late Paleolithic), about 17,000 years ago. They walked into the open cave and into its dark zone, where they painted on the upper walls and arched ceilings of the first five small chambers, using natural pigments of ocher, iron oxide, and charcoal. They produced beautiful figures of horses, aurochs, bulls, and stags, and even created narrative strips of hunting scenes. The cave was rediscovered in 1940, when local boys pursued their dog down a small hole in the rocks and soil that masked the entrance. Its wealth of Paleolithic art is regarded as the finest in the world.

### PRESERVATION

When the Lascaux cave was opened to visitors, the ancient paintings were damaged by new calcite deposition induced by environmental changes and were discolored by algae from spores carried in through the enlarged entrance. Preservation and visitor access could not coexist. Visitors to Lascaux now see an exact replica of the cave and its paintings.

**PALEOLITHIC CAVE ART**
*The perfect proportions of the animals within the ceiling frieze in Lascaux's Great Hall of the Bulls reflect the superb skills of bygone cave artists.*

EUROPE *west*

## Vercors

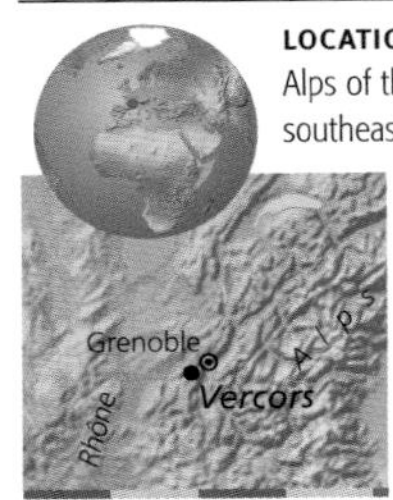


**LOCATION** In the limestone lower Alps of the Provence Alpes region, southeast France

**LENGTH** Numerous long caves

**DEPTH** 4,170 ft (1,271 m)

With an area of about 400 square miles (1,000 square km), Vercors is the largest karst area in Europe. Formed in Cretaceous limestone more than 1,300 ft (400 m) thick, it contains many long, deep, and spectacular caves. Their passages are a mix of large, old, dry tunnels with calcite deposits, and narrow streamways with many fine waterfalls. The most famous cave is the very deep Gouffre Berger. Narrow shafts descend to the Grand Galerie, which is decorated with large stalagmites. The tunnel follows the slope of the limestone beds for 2 miles (3 km) to a flooded zone behind the resurgence in the Cuves de Sassenage.

**GOURNIER CAVE LAKE, VERCORS**
*Just inside the resurgence entrance to the Gournier Cave, a placid lake lies beneath a profusion of calcite stalactites and flowstone cascades.*

EUROPE *west*

## Grotte Casteret

**LOCATION** In the crest of the Spanish Pyrenees, within the Ordesa National Park, Spain

**LENGTH** 1,600 ft (500 m)

**DEPTH** 115 ft (35 m)

Grotte Casteret is a limestone cave in which water seeps in as a liquid and then freezes on entering the cave. Such ice caves, as they are known, are rare because they rely on delicately balanced microclimates. Though high in the Pyrenees, the ground is not frozen, so water can seep into the limestone. However, in winter cold air becomes trapped in the cave and remains there for most of the summer. Water freezes almost as soon as it drips into the cave, creating huge icicles and columns of glass-clear ice. The floor of the main chamber is a sheet of ice, which overflows in a frozen cascade to another ice floor in a lower chamber.

**CURTAINS OF ICE**
*Glass-clear icicles drape the walls of Grotte Casteret wherever seepage water emerges from the limestone fissures.*

**ICE CAVE**
*This grand ice column in Casteret's main chamber is a rare and transient feature. If the cave warms, it will melt; if it cools, groundwater will be unable to enter and replenish the ice.*

EUROPE *south*

## Skocjanske Jame

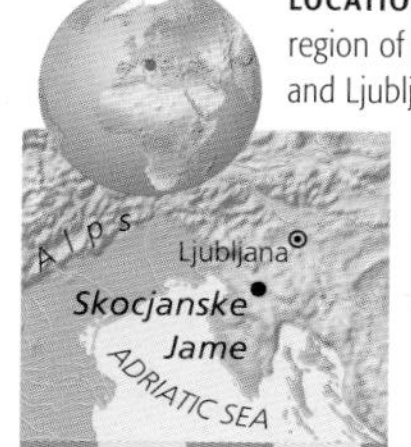


**LOCATION** In the heart of the Kras region of Slovenia, between Trieste and Ljubljana

**LENGTH** 3⅗ miles (5.8 km)

**DEPTH** 800 ft (250 m)

With a tourist pathway built high on the walls of its huge river passage, Skocjanske Jame is one of the most spectacular show caves in the world. Beside the village of Matuvan, the Reka River sinks through two giant dolines and runs beneath cliffs over 330 ft (100 m) high into the yawning cave entrance. There the underground river flows in a deep canyon cut down into a wider ancient passage, which now survives as high-level ledges adorned with calcite deposits. The abandoned gallery, which was the river's outlet before the canyon developed, is now blocked by massive calcite deposits and debris from a gigantic collapsed doline. The cave is the largest river sink in the Kras—the limestone region that gives its name to karst landscapes around the world.

**UNDERGROUND RIVER** *The first part of Skocjanske Jame's river passage receives light filtering in from the cave's giant entrance.*

**SINKHOLE ENTRANCE** *The deep entrance dolines or sinks of this cave system are crossed by the Reka River before it finally sinks underground.*

EUROPE *southeast*

## Optimisticheskaya

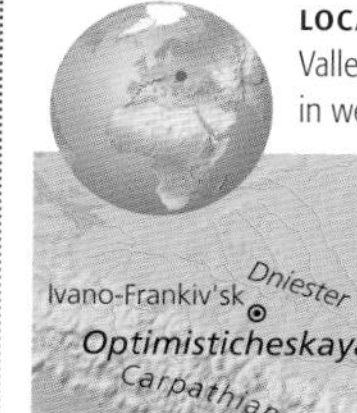

**LOCATION** Beneath the Dniester Valley, northwest of the Black Sea, in western Ukraine

**LENGTH** 132 miles (212 km)

**DEPTH** 50 ft (15 m)

Limestone is not the only rock that can be dissolved to form caves. For more than a million years, groundwater has drained through a thin bed of gypsum lying beneath a limestone and clay cover in the area of the Ukraine called the Podil's'kyy flatlands. Water, flowing along joints and faults in the gypsum below the water table, dissolved the walls to make each into a long, narrow fissure. These intersect with each other to create a seemingly endless underground maze, which has been left dry by the falling water table. The Optimisticheskaya network is accessible only through a single entrance, dug through the floor of a doline in 1966 by local cavers. Removal of gypsum in solution continues today, creating another generation of caves to the south of Optimisticheskaya, where the gypsum still lies below the water table.



AFRICA *south*

## Cango Caves

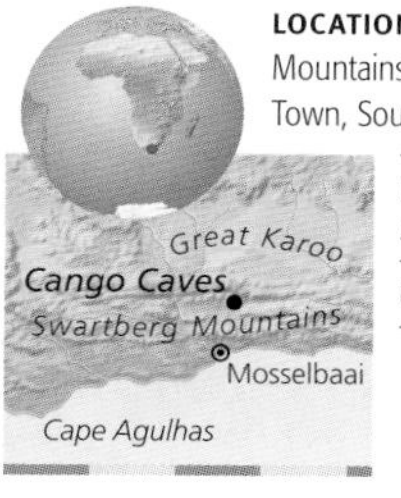


**LOCATION** In the Swartberg Mountains to the east of Cape Town, South Africa

**LENGTH** 3⅓ miles (5.3 km)

**DEPTH** 200 ft (60 m)

Isolated in a narrow outcrop of Precambrian limestones in the highlands near Oudtshoorn, the Cango Caves are remarkable for their wealth of calcite deposits. They are a remnant of a major channel below the water table that was drained when nearby surface valleys were cut to lower levels. Subsequently calcite deposition has occurred on a massive scale, leaving a spectacular array of flowstone and large stalagmites. Bushmen left paintings on the walls of the entrance chamber thousands of years ago, but the entrance was lost, and rediscovered by local farmers only in 1780.

**LARGE DRIPSTONE DEPOSIT**

**BOTHA'S HALL**
*The chambers of Cango Caves are sections of a larger main passage, separated from neighboring sections by walls and columns of calcite.*

ASIA *west*

## Krubera

**LOCATION** High on the Arabika Massif of the Western Caucasus Mountains in Georgia

**LENGTH** 3 miles (5 km)

**DEPTH** 5,610 ft (1,710 m)

Krubera was established as the world's deepest known cave following exploration in January 2001. Thick Mesozoic limestones form spectacular karst across the summits of the Western Caucasus, and explorations in the 1980s found a number of deep cave systems, including Krubera, draining the glaciated Ortobalangan Valley. Krubera is a steeply descending cave system with a single active streamway of narrow, meandering canyons, broken by nearly 40 major shafts and numerous smaller cascades. The bottom chamber is blocked by boulders that prevent access into the main river passage, which must exist to drain all the caves of the western Arabika. The proven resurgence of the cave water is at springs on the shore of the Black Sea, 1,740 ft (530 m) below the level of the bottom of the known cave. In early 2003, the depth record passed to Gouffre Mirolda in France, now known to be 5,685 ft (1,733 m) deep.

ASIA *southeast*

## Nanxu Arch

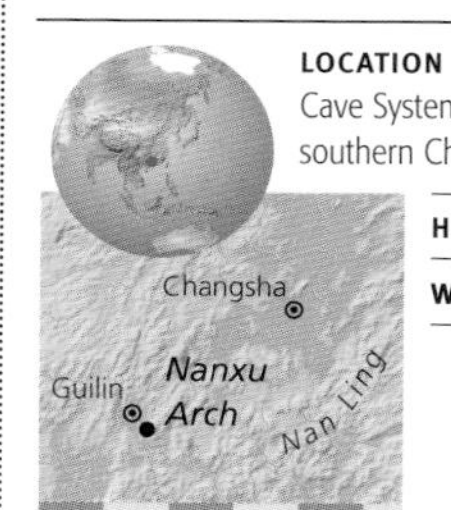


**LOCATION** Beside the Guanyuan Cave System, on the Li River in southern China

**HEIGHT** 200 ft (60 m)

**WIDTH** 130 ft (40 m)

The beautiful rock arch standing above Nanxu is the last surviving remnant of a large, old cave passage that once passed through the towering limestone hills in the karst of southeastern China. The arch's smooth profile is a result of surface weathering after the cave passage was exposed. The small river that formed the old cave now sinks just below the arch, into the younger Guanyuan Cave System.

**LIMESTONE ARCH**

ASIA *southeast*

# Gunung Mulu

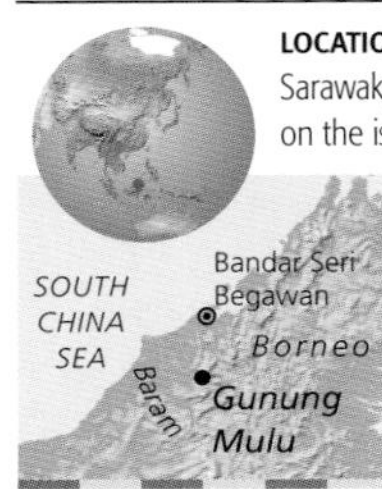


**LOCATION** In the rainforests of Sarawak, near the border with Brunei on the island of Borneo, Malaysia

**LENGTH** Over 200 miles (320 km) in total

**DEPTH** 1,542 ft (470 m)

In 1978, British cavers discovered a series of gigantic caves with some of the largest passages in the world, all beneath the rainforests of the Gunung Mulu National Park. Deer Cave (Gua Payau) has a single passage over 330 ft (100 m) high and wide for its entire 3,500-ft- (1-km-) long route through a limestone ridge. Good Luck Cave (Lubang Nasib Bagus) has a long stream passage that leads into Sarawak Chamber—the world's largest underground space, 2,300 ft (700 m) long and over 1,000 ft (300 m) wide beneath an arched limestone roof over 330 ft (100 m) high. Clearwater Cave (Gua Air Jernih) has over 69 miles (110 km) of passages that have already been mapped. The lower levels have a major river draining through massive canyon passages, while the upper levels are huge, abandoned tubes decorated with calcite deposits. All these caves have been formed by the copious flows of rainwater draining from the sandstone mountain of Gunung Mulu (Mount Mulu) into the steep limestone ridges of Api and Benarat, and out onto the shale lowlands of the Melinau River. Wherever a river met the limestone, a cave formed and then matured into a large tunnel. As the ground surface lowered, the rivers created successive newer caves at lower levels and abandoned the older passages above. Dating of sediments in Clearwater Cave shows that the caves have evolved over a period of 2 million years. Throughout this period, surface erosion has progressively lowered the sandstone hills and shale lowlands. In contrast, erosion of the limestone has occurred largely underground, and minimal surface lowering has left the cavernous hills as ever-higher features of the changing landscape.

## NESTING BIRDS

Thousands of swiftlet nests, made of moss and saliva, are glued to the walls of the Mulu caves, and swiftlets constantly fly through the darkness, navigating by echolocation with distinctive clicks. Local people use swiftlet nests to make birds'-nest soup, but due to conservation measures in the cave, the birds here are safe from all predators except voracious racer snakes, which are excellent climbers.

**SKYLIGHT**
*Daylight filtering through the forest canopy beams down a hole in the roof of a gallery in the tourist section of Clearwater Cave.*

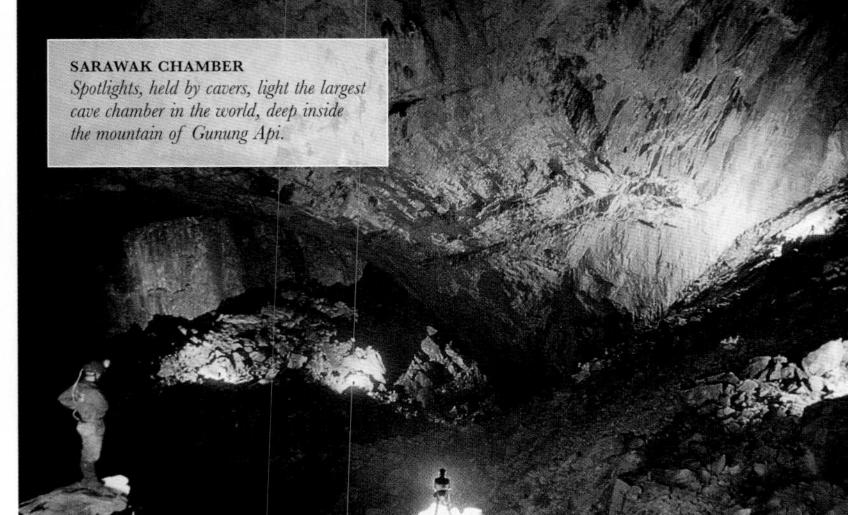

**SARAWAK CHAMBER**
*Spotlights, held by cavers, light the largest cave chamber in the world, deep inside the mountain of Gunung Api.*

**BATS LEAVING DEER CAVE**
*Every evening a seemingly endless procession of bats trails out of the cave entrance en route to feed on insects living in the nearby rainforest.*

**DEER CAVE, SARAWAK**
*The world's largest passage is lit by daylight from its southern entrance. The cave floor is covered by stream sediments and guano from the 2 million bats roosting overhead.*

ASIA *east*

## Akiyoshi-do

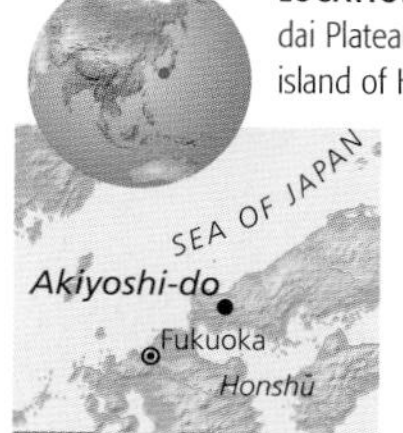


**LOCATION** Beneath the Akiyoshi-dai Plateau, at the western tip of the island of Honshu, Japan

**LENGTH** 5½ miles (8.7 km)

**DEPTH** 450 ft (137 m)

The rolling limestone plateau of Akiyoshi-dai contains a host of caves, the longest and most important being Akiyoshi-do. A stream enters the cave via two sinkholes and then passes through a short flooded section, before entering a passage with tall chambers that were created by successive collapses of the cave roof. The main passage is part of the lowest cave level, and the wide stream canyon has been carved into the floor of an original tubular cave that appears to be at least 100,000 years old. A layer of volcanic ash within the cave sediments is dated to about 34,000 years ago. The famous Hyakumai pools lie in the large stream passage that leads to the tall resurgence entrance in the foot of a cliff.

**THE CASCADE OF HYAKUMAI-ZAI**
*Each terraced rimstone pool is retained by a calcite barrier that is deposited by the overflow of calcium-carbonate-saturated water.*

AUSTRALASIA *Australia*

## Koonalda Caves

**LOCATION** Beneath the Nullarbor Plain, close to the coast of South Australia

**LENGTH** 4,000 ft (1,200 m)

**DEPTH** 260 ft (80 m)

The flat, barren landscape of the Nullarbor Plain is broken by more than 130 large dolines, formed by surface collapse. Koonalda is one of these, where a very large cave passage is accessible below the debris. The Nullarbor Plain is made of Miocene limestone that has been drained by underground passages since it was uplifted about 14 million years ago. The caves formed as widely spaced tunnels, filled with slowly moving water at or below the water table. During later phases of uplift, the water table fell, leaving some of these passages dry. Further collapse of limestone ceilings has left high chambers, some of which have broken out at the surface, creating the collapse dolines that are the caves' only entrances. Koonalda has four chambers in its main passage, besides the one that forms its entrance. Between the collapse features, the stable sections of cave roof arch over long lakes of clear, brackish water. At the end of the third lake, the roof dips, and the cave continues as a submerged tunnel.

**DOLINE ENTRANCE**
*The 130-ft- (40-m-) wide entrance of Koonalda Caves provided easy access for Aboriginal flint miners more than 20,000 years ago.*

ASIA *southeast*

## Ha Long Bay

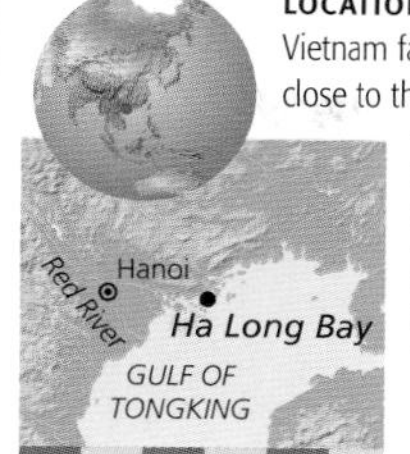


**LOCATION** On the coast of Vietnam facing the Gulf of Tongking, close to the Chinese border

**LENGTH** Numerous short caves

**DEPTH** All less than 230 ft (70 m)

With its 1,969 limestone islands, Ha Long Bay provides one of the world's most distinctive and spectacular landscapes. Most of the islands are fringed by vertical cliffs, and many are just individual towers taller than they are wide. Since invasion by the sea, marine erosion has undercut the cliffs, so that the island profiles are even steeper than those of the inland towers well-known across the border in southern China. There are hundreds of caves within the islands of Ha Long Bay, but none is long or deep because they are limited by the sizes of the islands. Some caves are remnants of ancient systems, now decorated with calcite deposits. Other younger caves have natural tunnels at sea level, which lead through the limestone hills into drowned dolines—hidden inland lakes that have no other access to the outside world.

**HANG BO NAU**
*A vista of limestone hills and islands is framed by the mouth of the Bo Nau Cave, the remnant of an ancient passage on Island 268 in Ha Long Bay.*

ASIA *southeast*

## Tham Hinboun

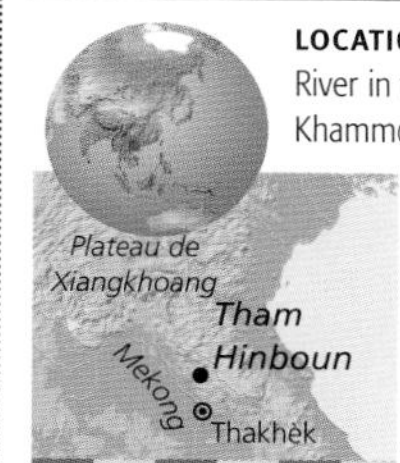


**LOCATION** East of the Mekong River in the limestone mountains of Khammouan, Laos

**LENGTH** 7¾ miles (12.4 km)

**DEPTH** 130 ft (40 m)

The Hinboun River flows through this cave for a distance of over 4.5 miles (7 km), and is the world's longest navigable underground river. The cave originated with a switchback route, following fractures and bedding in the limestone. Subsequently, the high passages were eroded by retreating waterfalls, while the lower ones were removed by solution of their underwater ceilings. Today the cave is an almost horizontal tunnel passing between two open karst basins.

AUSTRALASIA *Australia*

## Jenolan Caves

**LOCATION** In the western flank of the Blue Mountains, in New South Wales, Australia

**LENGTH** 6⅔ miles (10.6 km)

**DEPTH** 280 ft (85 m)

Australia's best-known tourist caves were discovered around 1838 by escaped convict and bushranger James McKeown. They lie in a narrow outcrop of Silurian limestone that is crossed by the Jenolan River. Grand Arch, at just 400 ft (120 m) long, is all that remains of an ancient cave; but to either side the limestone contains a host of smaller cave passages with streams still flowing through the lower levels. Some Jenolan caves were formed in Permo-Carboniferous times before being completely filled with Permian carbonates and gravels. Later streams have re-excavated a few of the old caves, or cut new ones through the limestone.

**JUBILEE CAVE**
*Delicate calcite crystals that grow in pools of calcium-carbonate-saturated water are a feature of this richly decorated part of the Jenolan Caves.*

**GRAND COLUMN**
*This large stalagmite in River Cave has developed an apron of stalactites and curtains below its overgrown top.*

LAND

**WALL OF ICE**
*As the Hubbard Glacier meets the sea in Disenchantment Bay, ice detaches from its 300-ft- (90-m-) high terminal wall. These chunks of ice form icebergs, which then float away into the Gulf of Alaska.*

# GLACIERS

ABOUT ONE-TENTH OF the Earth's land surface is covered in the ice masses known as glaciers, and they hold over 75 percent of the world's fresh water. Glaciers are among the most beautiful of natural phenomena, but they also carve out landscapes and transport vast amounts of rock as they advance and retreat. Glaciers come in many different forms and sizes. For example, some glaciers carry pinnacles of ice, streams of water, even lakes on their surfaces. Others have volcanoes underneath them, colonies of worms living inside them, or long floodwater channels running through them. In recent years, glaciers have become important geological features to study. If all glaciers were to melt, sea levels would rise significantly. Therefore the behavior of glaciers in response to global warming is being closely monitored.

# GLACIERS

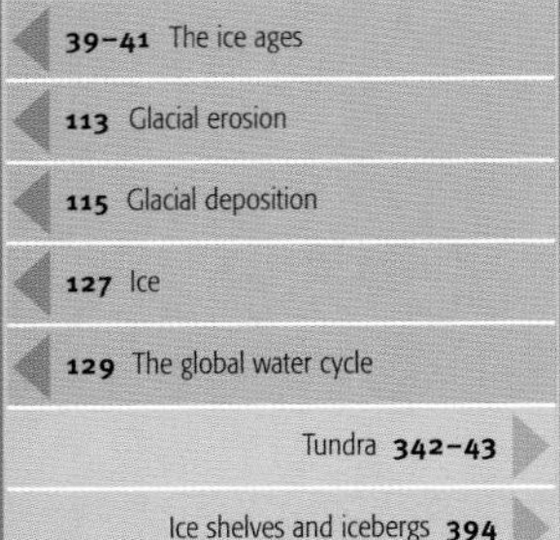
39–41 The ice ages
113 Glacial erosion
115 Glacial deposition
127 Ice
129 The global water cycle
Tundra 342–43
Ice shelves and icebergs 394

A GLACIER IS AN ACCUMULATION of many layers of snow, transformed over time and by their own weight into a mass of ice that persists from year to year. This ice mass slowly deforms and begins to move, most of it traveling downhill. A small glacier may be no bigger than a football field, but the largest, called ice sheets, can be continent-sized. Glaciers are an important reservoir of fresh water, and can act like time capsules, holding information—in the form of trapped air, chemicals, and dust—that tells us much about changing climatic conditions over hundreds of thousands of years.

## GLACIER STRUCTURE

Most glaciers except for the very largest descend from a mountain area and flow down a valley. The top of the glacier usually lies in a bowl-shaped area called a cirque. The ice below the cirque is often broken up by deep cracks called crevasses, formed in response to stress as the ice grinds over uneven bedrock. Over time, rocks fall onto the glacier or become attached to the moving ice and are plucked off the bedrock. They are deposited at the melting terminus (snout) of the glacier, leaving a debris pile called a terminal moraine, or at the sides (forming lateral moraines). Where glaciers join, rocks carried on their edges merge to form a surface stripe called a medial moraine. Broadly similar processes occur in larger glaciers, although these are less constrained by topography (see p.264).

**UPSTREAM FEATURES**
*The Bagley Ice Field in Alaska, 5 miles (8 km) wide, is typical of the upper areas of a glacier, which are composed of clean ice fractured by crevasses.*

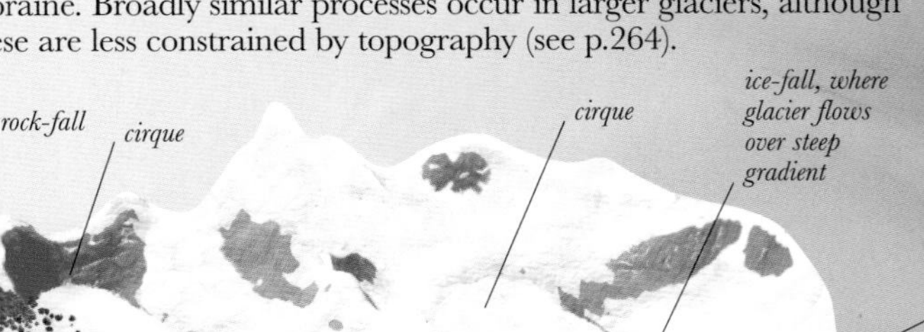


**MERGING GLACIERS**
*Two glaciers, originating from separate cirques, merge and continue down a valley. A wide medial moraine marks where they have joined. At the terminus, a lake of meltwater forms behind the terminal moraine.*

**MELTWATER STREAM**
*A glacier's surface may be broken up by streams of meltwater and studded with boulders.*

**TERMINUS**
*Glaciers can be hundreds (or even thousands) of yards deep. This is most apparent at a glacier's terminus, where an enormous cliff or mound of ice can be seen, as here in Harriman Fiord, Alaska.*

**MELTWATER CAVES**
*Ice caves, such as this one in Alaska's Muir Glacier, are formed by meltwater streams eroding vertical sinkholes, tunnels, and finally large chambers within the ice.*

# GLACIER ICE

Newly fallen snow contains masses of air between individual snowflakes. As more settles onto a previous fall, the flakes beneath are compressed and deformed, by partial melting and refreezing, into smaller, rounder ice particles. These adhere to each other, initially forming small ice granules and then a material consisting of larger granules, known as firn. As more snow falls, further compression occurs, causing the firn to weld into solid, dense glacial ice.

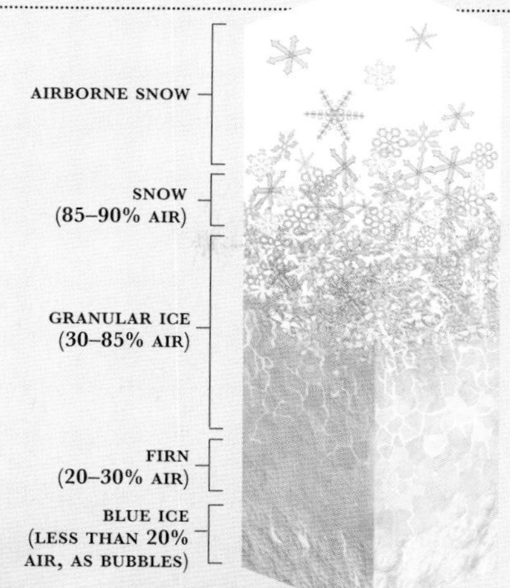

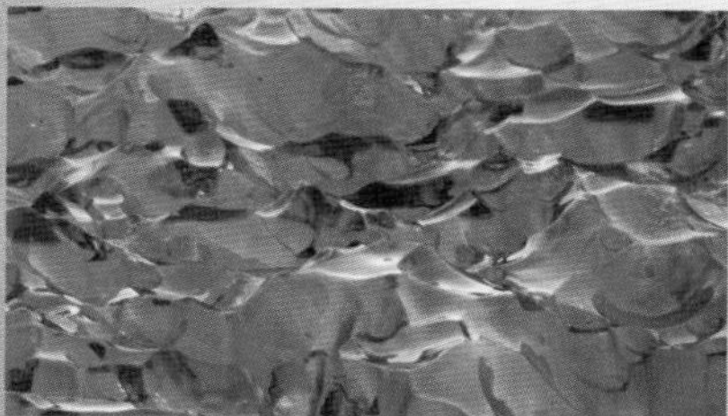

**BLUE ICE**
*Dense glacial ice owes its intense color to its crystalline structure, which absorbs all except the shortest (bluest) wavelengths of visible light. Blue ice is stronger and contains less air than white ice.*

AIRBORNE SNOW
SNOW (85–90% AIR)
GRANULAR ICE (30–85% AIR)
FIRN (20–30% AIR)
BLUE ICE (LESS THAN 20% AIR, AS BUBBLES)

**STAGES OF FORMATION**
*As glacial ice is formed, there is a gradual decrease in the proportion of air to snow or ice. Depending on the location of the glacier, the process can take anywhere from 5 to 3,000 years.*

# MOVEMENT

In the upper parts of a glacier, known as the accumulation area, there is a net gain of ice each year, as some of each winter's snowfall remains until the next winter. From here, ice flows downhill due to gravity. The mechanism of ice movement varies. In some glaciers (called warm-based), a thin layer of water forms under the glacier, allowing the ice to slide over the bedrock. In others (cold-based), the base of the glacier is frozen to the bedrock, so ice can move only by slow deformation. Eventually the ice reaches the lower parts of the glacier, where it melts and evaporates or, in some glaciers, breaks off as icebergs. This lower zone of the glacier, where there is a net loss of ice over the year, is called the ablation area. The section between the accumulation and ablation areas, where ice gains and losses are in balance, is called the equilibrium zone. The difference between the glacier's overall ice gains and losses is called its mass balance.

**CALVING ICE**
*Some glaciers end in lakes or the sea, where huge chunks of ice calve (break off) and float away as icebergs.*

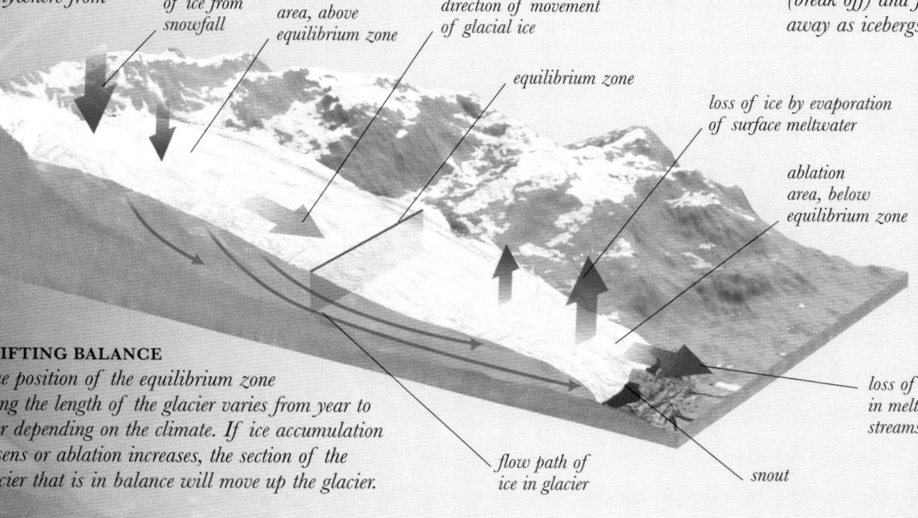


**SHIFTING BALANCE**
*The position of the equilibrium zone along the length of the glacier varies from year to year depending on the climate. If ice accumulation lessens or ablation increases, the section of the glacier that is in balance will move up the glacier.*

# ADVANCE AND RETREAT

Over a period of years, if more ice forms on a glacier than melts, the glacier must grow. It may widen, thicken, or advance (or any combination of the three). Conversely, if more ice is lost than forms, the glacier shrinks. The ice in its lower parts may thin and stagnate, or the snout may retreat, or both. Since 1900, most glaciers around the world have retreated, although within this time there have also been periods of advance, both by groups of glaciers and by individual glaciers (especially by surge-type glaciers with cyclical patterns of rapid advance and slow retreat). The retreat is probably linked to global warming, but scientists are divided over whether human activity is the main cause.

**RETREATING GLACIER**
*The Blomstrandenbreen Glacier in Svalbard, Norway, has retreated dramatically in the past 80 years, as the photo shown above, taken in 1922, and the photo on the left, taken in 2002, demonstrate.*

**ADVANCING GLACIER**
*When a glacier advances, it simply brushes aside objects such as trees. Here, the Taku Glacier, which advanced more than 4½ miles (7 km) between 1899 and 1989, pushes through a hundred-year-old forest near Juneau, Alaska.*

# TYPES OF GLACIERS

Glaciers vary enormously in their size and form. The largest, called continental ice sheets, and somewhat smaller glaciers called ice caps, almost totally submerge the landscapes on which they lie. These glaciers are slightly dome-shaped. Driven by gravity, the ice in them flows outward from their centers toward their edges, where channels of ice flow, called ice streams and outlet glaciers, are visible. Outlet glaciers are bounded by ice-free ground, while ice streams are surrounded by slower-moving ice. The other main types of glaciers are found in areas where the ice flow is constrained by mountains. An ice field is a huge expanse of ice at high altitude, partly hemmed in by mountains, that gives rise to numerous outlet glaciers. A cirque glacier is a small glacier mainly confined to a depression on the flank of a mountain. Where ice flows out of a cirque and extends far down a valley, the result is a valley glacier. This can be dozens or even hundreds of miles long. When a valley or outlet glacier flows into a flat, lowland area, it may spread out into a wide expanse of ice called a piedmont lobe.

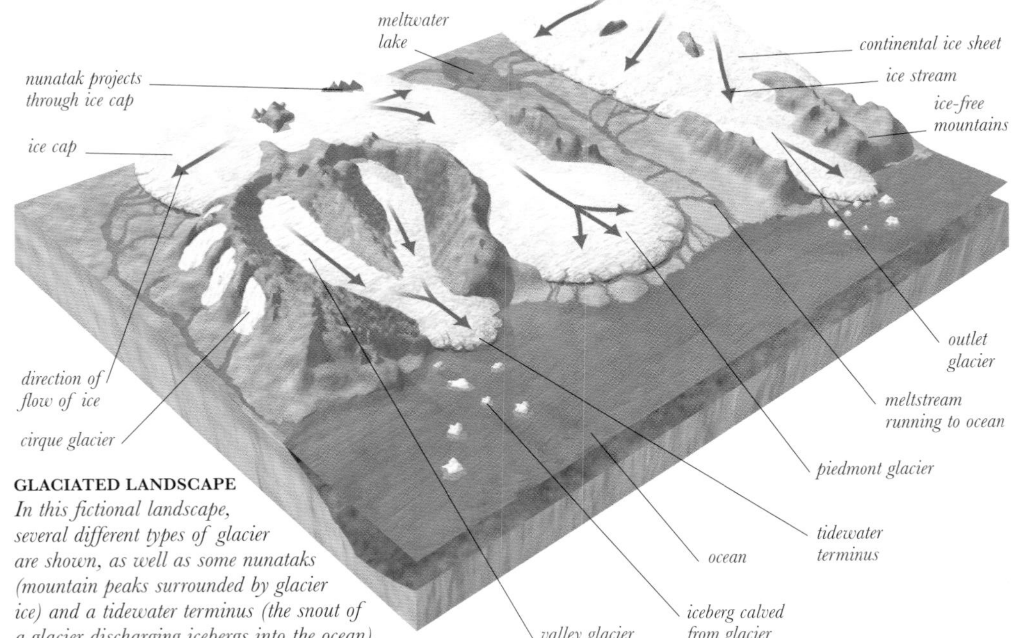


**GLACIATED LANDSCAPE**
*In this fictional landscape, several different types of glacier are shown, as well as some nunataks (mountain peaks surrounded by glacier ice) and a tidewater terminus (the snout of a glacier discharging icebergs into the ocean).*

**ICE-SHEET MARGIN**
*Thin meltwater streams can be seen here on the surface of the Greenland Ice Sheet at one of the points where it meets the sea.*

**OUTLET GLACIERS**
*Several outlet glaciers from the Spitsbergen Ice Cap on Svalbard, Norway, flow through lines of nunataks toward the sea.*

**PIEDMONT GLACIER**
*This satellite image shows Alaska's Malaspina Glacier, which consists almost entirely of an enormous piedmont lobe, the world's largest.*

**VALLEY GLACIER**
*The Aletsch, a 16-mile- (25-km-) long valley glacier, flows through the Swiss Alps. The upper part of the glacier is shown here.*

**CALIFORNIAN GLACIATED LANDSCAPE**
*McGee Creek in the Sierra Nevada is a U-shaped valley sculpted by a glacier during the last ice age. An array of cirques, horns, and arêtes can be seen among the mountains in the background. In the foreground are distinct lateral moraines—ridges formed from rock debris deposited at the side of the glacier.*

## GLACIAL EROSION

Glaciers erode mountainous regions into characteristic shapes, some of which only become apparent when the glaciers eventually melt and disappear (as at McGee Creek, opposite below). Erosion occurs partly by the plucking of rocks from valleys and mountainsides as they freeze to glacial ice, and partly by the moving mass of ice and its rock load grinding or abrading the bedrock. At the head of a glacier, the ice typically carves out a cirque. Where two cirques back up against each other, ridges called arêtes are formed; three or more cirques can combine to produce a pyramidal peak called a horn. Downstream, a glacier moving through a valley will usually deepen and broaden the valley from a V-shape into a U-shaped profile. Another common feature is the hanging valley. This is seen where glacial erosion has occurred faster in the main valley than in a smaller tributary valley. When the glaciers melt, the tributary is left suspended high above the more deeply carved main valley. Landscapes like the one shown in the illustration on the right can be seen today in many mountainous regions of Europe and North America, such as parts of Norway, northern Great Britain, and the Rocky Mountains in Canada and the US.

**CIRQUE AND ARETE**
*This deeply eroded cirque, with a long ridge, or arête, at its crest on the far side, sits above Gryllefjord in Norway.*

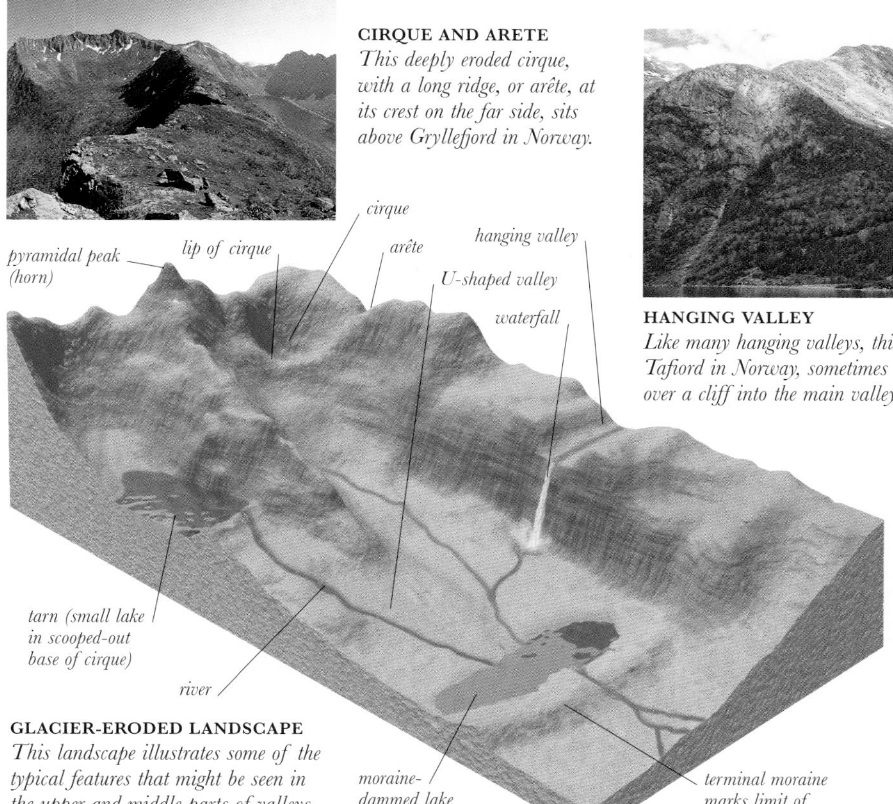


**GLACIER-ERODED LANDSCAPE**
*This landscape illustrates some of the typical features that might be seen in the upper and middle parts of valleys that were once filled with glaciers.*

**HANGING VALLEY**
*Like many hanging valleys, this one in Muldalen, above Tafjord in Norway, sometimes discharges a waterfall over a cliff into the main valley below.*

**ERODED VALLEY**
*The floor of this valley has been deeply eroded by past glaciation.*



## DEPOSITION

When glaciers melt, they deposit material on the landscape. Much of this matter is a mixture of particles, ranging in size from sand grains to boulders, called till. The deposits come in many forms. For example, a terminal moraine is a broad ridge of till marking the farthest point to which a glacier advanced. An esker is a twisting ridge of sand and gravel that snakes its way across the landscape. It is formed from material deposited by a stream that once flowed under the glacier. A drumlin is a long, cigar-shaped mound of till that was smoothed in the direction of the glacier's flow. An erratic is a large, rounded boulder that a glacier carried far from its source in the mountains and eventually deposited at its terminus. When an ice block detaches from a melting glacier and is partially buried by deposits from meltwater streams, it melts and creates a depression in the layer of till called a kettle hole. If the depression fills with water, it becomes a kettle lake. When a meltwater channel becomes choked with coarse material, it causes the river to divide into a series of diverging and converging segments, known as braided streams.

**KETTLE LAKE**
*This kettle lake lies in Dolma La Pass, one of the routes to Mount Kailas, the sacred mountain of Tibet, which is known locally as Ghang Rimpoche, or "jewel of snow."*

**ERRATIC**
*This erratic boulder in Yorkshire, England, was transported by a glacier and then deposited on younger rock.*

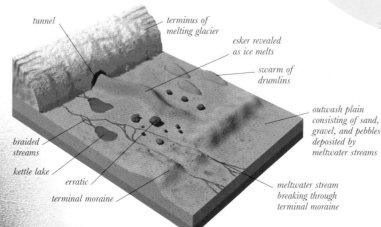

tunnel
terminus of melting glacier
esker revealed as ice melts
swarm of drumlins
outwash plain consisting of sand, gravel, and pebbles deposited by meltwater streams
braided streams
kettle lake
erratic
terminal moraine
meltwater stream breaking through terminal moraine

**DEPOSITIONAL FEATURES**
*A melting glacier typically leaves several of the features seen here. Over large parts of lowland North America and Europe, deposits such as these can still be discerned in the landscape, providing clues to the region's glacial history.*

## WORLD DISTRIBUTION

Glaciers can form only on areas of land where there is some snowfall and very low temperatures persist for much of the year. The only places that satisfy both these conditions are close to the poles or in high mountain areas. The world's largest glaciers are massive ice sheets that cover Antarctica, while in high northern latitudes, huge areas of glaciation exist over Greenland, parts of Iceland, Norway, and Alaska. Closer to the equator, glaciers occur only at high altitudes, such as in the Alps and Himalayas. In the tropics, some areas of glaciation are found in the northern Andes and on the highest mountains in Africa, Mexico, and Indonesia.

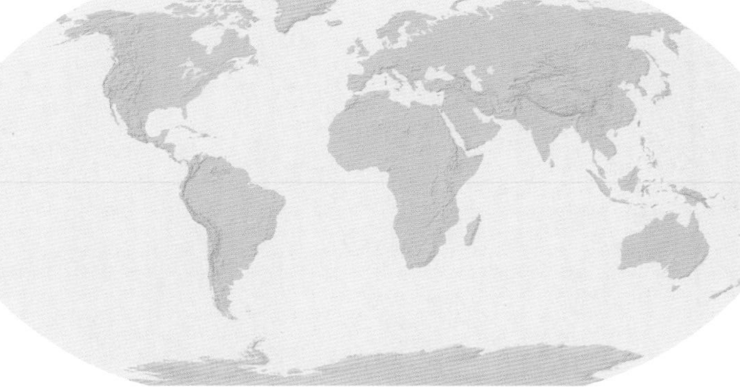

### GLACIER PROFILES

The pages that follow contain profiles of a selection of the world's glaciers. Each profile begins with the following summary information:

**TYPE** Cirque glacier, valley glacier, piedmont glacier, outlet glacier, ice cap, ice field, or ice sheet

**TERMINUS** Terrestrial snout, terrestrial lobe, tidewater (sea-calving) front, or lake-calving front (cirque, valley, piedmont, and outlet glaciers only)

**AREA** Surface area (all glacier types)

**LENGTH** Distance from source to terminus (all glacier types except ice caps and ice sheets)

**STATUS** Advancing, retreating, static, or uncertain (since 1990)

NORTH AMERICA *northwest*

## Trapridge Glacier

**LOCATION** In the St. Elias Mountains, 53 miles (85 km) west of Kluane Lake, Yukon, Canada

**TYPE** Cirque glacier

**TERMINUS** Terrestrial snout

**AREA** 2 square miles (5 square km)

**LENGTH** 2 miles (3 km)

**STATUS** Advancing

Trapridge is a surge-type glacier, undergoing periodic rapid transfers of ice from the upper to the lower parts, which increase in thickness and then surge forward. This is followed by a quiet phase, as ice builds up again in the upper glacier and thins or retreats lower down. Trapridge, which last surged in the 1940s, is unusual because its higher parts rest on a bed of till—the glacier's movement can be partly attributed to deformation of the soft till beneath it. Surges may happen when water drainage beneath the upper parts of a glacier is blocked by increasing pressure. This leads to an extension of the water layer under the ice, allowing the ice to slip rapidly over the bedrock.

**READY TO SURGE?**
*The bulge that is developing at its terminus may be a sign that the Trapridge Glacier will surge soon.*

NORTH AMERICA *northwest*

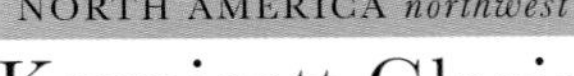

## Kennicott Glacier

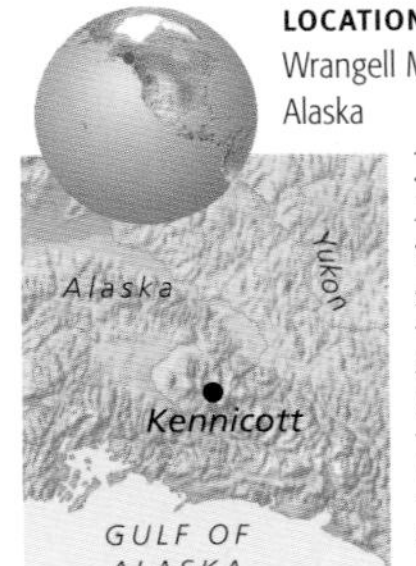


**LOCATION** In the southern Wrangell Mountains, south-central Alaska

**TYPE** Valley glacier

**TERMINUS** Terrestrial snout

**AREA** 97 square miles (250 square km)

**LENGTH** 29 miles (47 km)

**STATUS** Retreating

This glacier pours down from Alaska's Wrangell Mountains, passing the old mining town of Kennicott on the way. Early last century, the glacier was so large that it loomed above the town. Since then the glacier's terminus has retreated, leaving mounds of silt, while its lower parts have become covered in a thick layer of rock debris. Several side valleys branch off from the Kennicott Valley, and in spring many of them fill with lakes as the glacier blocks meltwater streams from the valleys. As temperatures rise, channels beneath the glacier grow, connecting with the lakes and releasing the water.

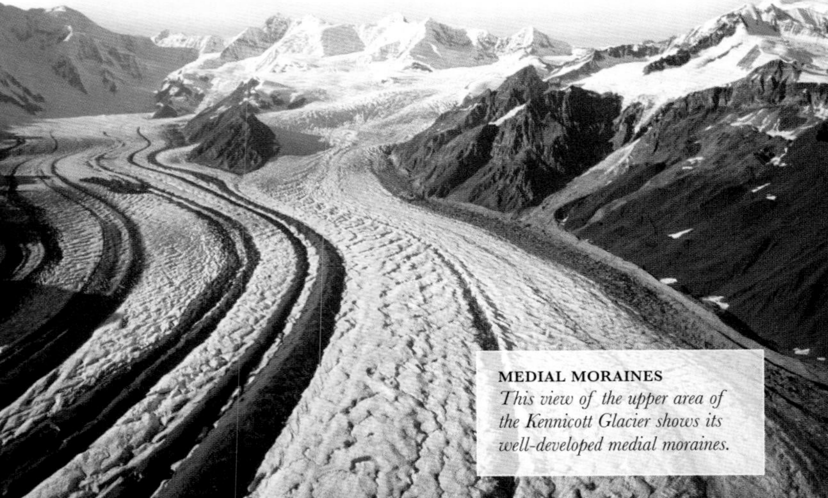

**MEDIAL MORAINES**
*This view of the upper area of the Kennicott Glacier shows its well-developed medial moraines.*

NORTH AMERICA *northwest*

## Black Rapids Glacier

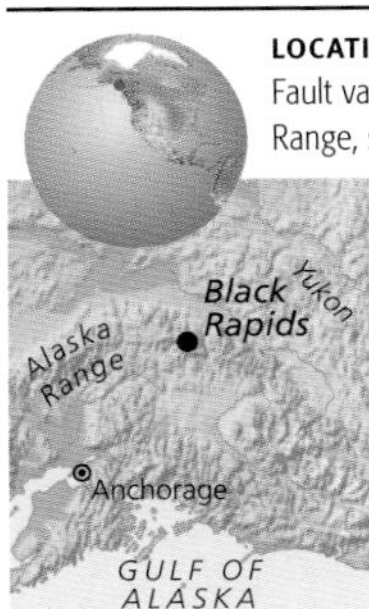


**LOCATION** Occupies the Denali Fault valley in the east-central Alaska Range, south-central Alaska

**TYPE** Valley glacier

**TERMINUS** Terrestrial snout

**AREA** 58 square miles (150 square km)

**LENGTH** 29 miles (47 km)

**STATUS** Retreating

Sometimes called the Galloping Glacier, the Black Rapids is a surge-type glacier (see Trapridge, above). In 1936–37, the terminus advanced 4 miles (6.5 km) in three months. The photograph below reveals a typical feature of surge-type glaciers: dark, looped medial moraines on the surface.

**LOOPED (OR S-SHAPED) MORAINES**

NORTH AMERICA *northwest*

## Hubbard Glacier

**LOCATION** In Canada's St. Elias Mountains, descending through southeastern Alaska

**TYPE** Valley glacier

**TERMINUS** Tidewater (sea-calving) front

**AREA** 1,350 square miles (3,500 square km)

**LENGTH** 76 miles (122 km)

**STATUS** Advancing

The Hubbard is the largest tidewater glacier in North America, discharging huge amounts of ice into Alaska's Disenchantment Bay. Unusually, it has been advancing in fits and starts for more than a century, and it now threatens to block off an arm of the bay called Russell Fiord, turning it into a lake. The glacier has twice dammed the fjord already, in 1986 and 2002, causing a dramatic rise in its water level, before the waters eventually broke through the ice dam in a torrential flood. There are fears that if the fjord becomes permanently blocked, it will overflow, inundating land traditionally used by native American peoples.

**THE THREAT TO RUSSELL FIORD**
*This aerial view shows, at bottom right, the narrow mouth of the Russell Fiord, which flows into Disenchantment Bay, on the left.*

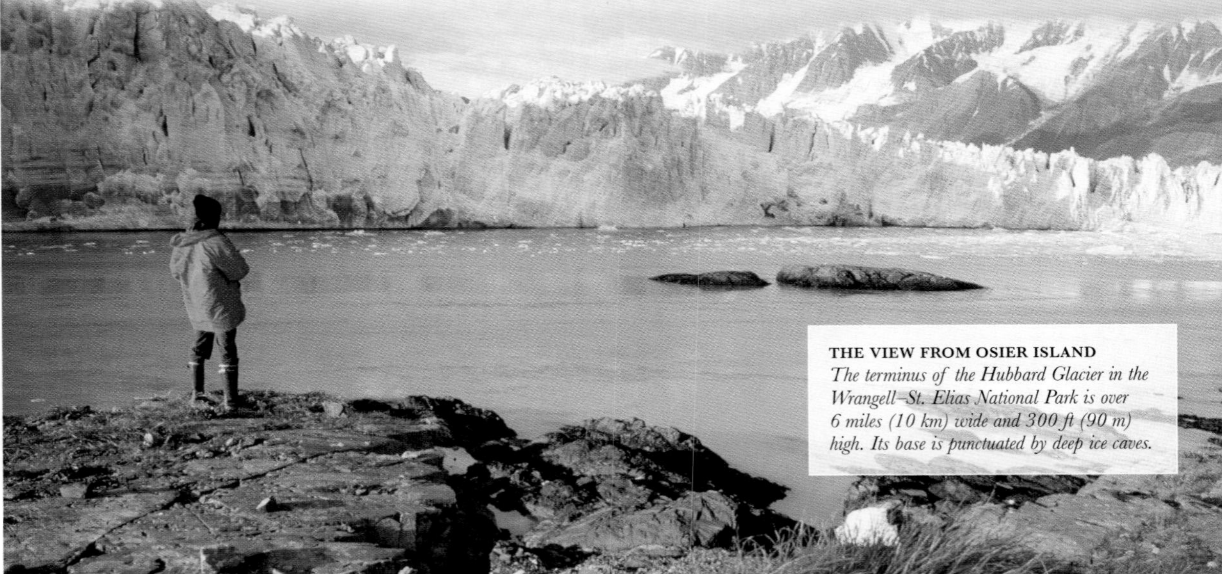

**THE VIEW FROM OSIER ISLAND**
*The terminus of the Hubbard Glacier in the Wrangell–St. Elias National Park is over 6 miles (10 km) wide and 300 ft (90 m) high. Its base is punctuated by deep ice caves.*

NORTH AMERICA *northwest*

## Bering Glacier

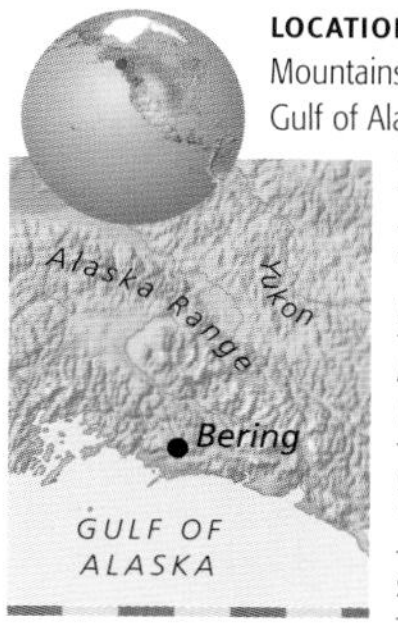

**LOCATION** In the Chugach Mountains, descending toward the Gulf of Alaska, Alaska

**TYPE** Piedmont glacier

**TERMINUS** Lake-calving piedmont lobe

**AREA** 2,000 square miles (5,200 square km)

**LENGTH** 112 miles (180 km)

**STATUS** Retreating

The Bering Glacier is the next-largest glacier in the world after the Antarctic and Greenland ice sheets. It can be thought of as three different glaciers combined into one. The upper region, called the Bagley Ice Field, is an ice basin about 56 miles (90 km) long, almost completely filling several interconnecting valleys in the eastern Chugach Mountains. This is where most of the Bering Glacier's ice forms. The ice field gently slopes down toward the west from an altitude of about 6,900 ft (2,100 m). From its western end, a wide trunk of ice (which is an outlet glacier) turns southwest, feeding the lower area of the Bering, which consists of a piedmont lobe 29 miles (47 km) wide. This lobe is fringed by a lake, Vitus Lake, into which the glacier calves icebergs. From the melting edge of the glacier and the lake, streams flow a short distance into the Gulf of Alaska. The Bering Glacier occasionally makes a rapid forward advance—in 1995 it surged about ½ mile (750 m) in only two weeks. But over the long term, the glacier is in retreat; during the last century, it is estimated to have lost about 3 percent of its surface area. Parts of the lower glacier are covered in rock debris. This is the result of melting, which causes rock particles carried by the glacier to become concentrated at its surface. In some areas, the combination of a fine layer of rock debris with melting ice underneath has even allowed some vegetation to grow on top of the glacier. The retreat of the glacier has revealed the remains of an ancient spruce forest that it overran about 1,500 years ago, as well as shells of marine organisms, indicating that part of the land that the glacier currently sits on was once under the ocean.

**ICE COLLAPSE**
*This view of the stagnating lower part of the Bering Glacier shows the vegetation growing on its surface and a craterlike area of ice collapse.*

**SURFACE TEXTURE**
*Parts of the lower and middle regions of the glacier have a highly fractured, patterned surface.*

### AERIAL MAPPING

In recent years, researchers in Alaska have started using aircraft-borne laser devices to create precise contour maps of glacier surfaces such as the Bering Glacier's. As the aircraft flies over the ice, a laser beam is used to measure the distance from aircraft to glacier. Simultaneously, the aircraft's position is determined using GPS (the global positioning system). When the heights measured were compared with those given on old US Geological Survey charts, it was revealed that most of Alaska's glaciers have thinned in the past 50 years.

**SNAKING TO THE SEA**
*Just a small part of the Bering Glacier's extensive terminus is seen here, with Vitus Lake in the foreground. The sinuous medial moraines are strikingly evident.*

NORTH AMERICA *northwest*

## Worthington Glacier

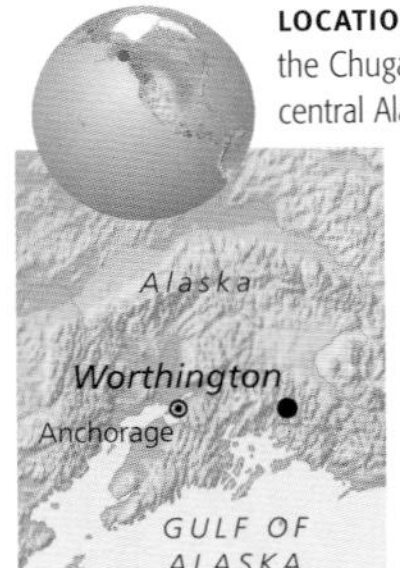


**LOCATION** In the central part of the Chugach Mountains, south-central Alaska

**TYPE** Valley glacier

**TERMINUS** Terrestrial snout

**AREA** 4 square miles (10 square km)

**LENGTH** 4 miles (7 km)

**STATUS** Retreating

A small valley glacier, the Worthington descends from 6,500 ft (2,000 m) and terminates close to a major highway. The changing shape of boreholes drilled through the glacier to the bedrock has revealed that at least 20 percent of the glacier's movement is due to internal deformation—the rest is due to slippage over the bedrock.

DOUBLE-ARMED TERMINUS

NORTH AMERICA *northwest*

## Columbia Glacier

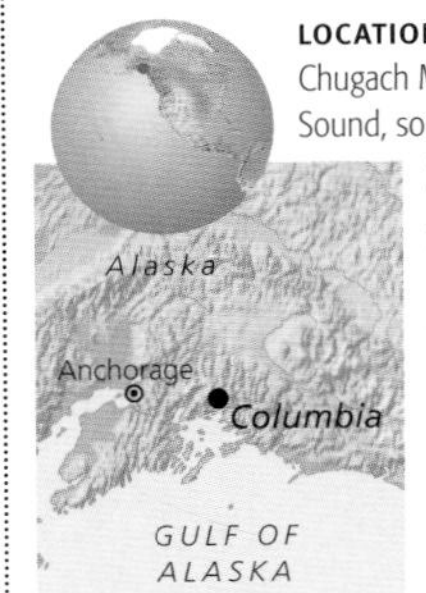


**LOCATION** Descending from the Chugach Mountains to Prince William Sound, south-central Alaska

**TYPE** Valley glacier

**TERMINUS** Tidewater (sea-calving) front

**AREA** 400 square miles (1,050 square km)

**LENGTH** 32 miles (52 km)

**STATUS** Retreating

One of the fastest-moving glaciers in North America, the Columbia is also one of the most rapidly retreating. Since the 1970s, the velocity of the ice near its terminus has quadrupled to an astonishing 80 ft (24 m) or so per day. However, this fast forward movement is more than canceled out by the rapid loss of ice at the terminus. As a result, since around 1982, the terminus has retreated about 9 miles (14 km) and has thinned considerably. Scientists are not convinced that the glacier's fast retreat can be attributed solely to global warming, since other nearby glaciers are not retreating as rapidly. Instead, they point to special factors related to the bedrock channel beneath the front of the glacier. This is very deep and slants backward, and as a result, the front of the glacier is almost afloat on meltwater, explaining why it moves forward so quickly. The high ice velocity in turn increases the rate at which icebergs are calved, and causes ice to be lost faster than it can be replenished by the upper parts of the glacier.

CONCEALED TERMINUS

*The bulk of this glacier's 3-mile- (5-km-) wide terminus is hidden beneath the waters of Prince William Sound. Calved icebergs are a recognized hazard to shipping in the area.*

SURFACE DEBRIS

*Parts of the lower glacier are covered in a thin layer of rock debris that has concentrated at the surface as the ice has melted.*

NORTH AMERICA *northwest*

## Margerie Glacier

**LOCATION** In the St. Elias Mountains, descending to Glacier Bay, southeastern Alaska

**TYPE** Valley glacier

**TERMINUS** Tidewater (sea-calving) front

**AREA** 14 square miles (35 square km)

**LENGTH** 14 miles (22 km)

**STATUS** Advancing

The Margerie Glacier is one of several tidewater glaciers that originate from the eastern end of the St. Elias Mountains. They then descend to discharge ice into a large inlet of the Gulf of Alaska called Glacier Bay, which became a national park in 1980. The terminus of the glacier is in Tarr Inlet at the northern end of the Bay. At one time, about 100 years ago, the Margerie was a mere tributary of a much larger glacier, the Grand Pacific Glacier, which filled the whole of Tarr Inlet. Earlier still, when the English naval officer Captain George Vancouver discovered Glacier Bay in 1794, the entire bay was filled with ice. Since then the Grand Pacific has shrunk and Margerie has become a calving glacier in its own right. Today, it is one of the most spectacular glaciers in the Bay, towering about 265 ft (80 m) above the sea surface, with another 400 ft (120 m) or so underwater. Unlike most other Alaskan glaciers, it has advanced slightly in recent years.

WALL OF BLUE ICE

*The Margerie Glacier regularly discharges large chunks of blue ice from its 1-mile- (1.5-km-) wide terminus into Glacier Bay, with thunderous cracking noises.*

NORTH AMERICA *northwest*

## Mendenhall Glacier

**LOCATION** Descending from the Juneau Ice Field, Coast Mountains, southeastern Alaska

**TYPE** Outlet glacier

**TERMINUS** Lake-calving front

**AREA** 40 square miles (100 square km)

**LENGTH** 15 miles (27 km)

**STATUS** Retreating

The Mendenhall is an outlet glacier from the Juneau Ice Field, a vast area of ice that straddles the border between Alaska and British Columbia. The glacier terminates in a lake 2 miles (3 km) long, just a few miles north of the city of Juneau. Recently, the terminus has been retreating at a rate of about 130 ft (40 m) a year, while at the same time the lake has greatly increased in size. It is anticipated that the glacier will finally recede out of the lake in about 2015. Once the glacier can no longer calve icebergs, its rate of retreat should slow.

SURFACE MELTWATER

*Streams, such as the one shown here, can develop where surface ice melts faster than it can be absorbed by the glacier.*

NORTH AMERICA *northwest*

# Malaspina Glacier

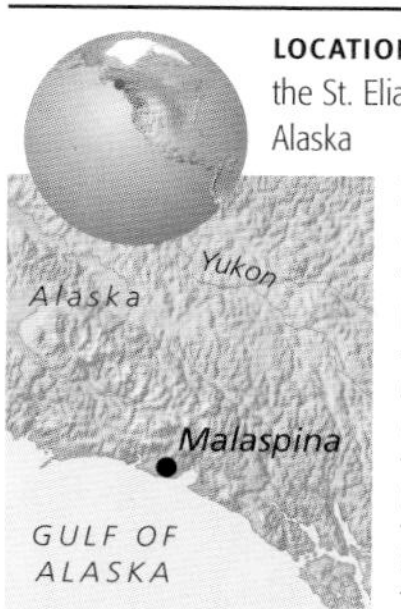


**LOCATION** On the coastal side of the St. Elias Mountains, southeastern Alaska

**TYPE** Piedmont glacier

**TERMINUS** Terrestrial lobes

**AREA** 1,500 square miles (3,900 square km)

**LENGTH** 34 miles (55 km)

**STATUS** Retreating

Named after an Italian, Alessandro Malaspina, who explored this part of Alaska in 1791, the Malaspina Glacier is famous for containing the world's largest piedmont lobe. It is also the second-largest glacier in North America. It consists of three lobes, each fed by different outlet glaciers that flow out of a large ice field lying high up in the St. Elias Mountains behind the Malaspina. The largest of these feeder glaciers is the Seward Glacier (seen in the image below as the blue ribbon of ice entering from the top). This feeder is itself 2½ miles (4 km) wide at its lower end—it gives rise to the largest, central lobe of the Malaspina. To the west, the Agassiz Glacier (the blue stream at the top left of the image) feeds a smaller western lobe. To the east, the Hayden and Marvine glaciers give rise to a small eastern lobe. Seismic surveys have shown that the Malaspina is up to 2,000 ft (600 m) thick, and that, roughly in the middle of the glacier, ice at its base is eroding the bedrock at a depth of about 1,000 ft (300 m) below sea level. From there, the ice is actually forced upward toward the lobe-shaped terminus of the glacier, where it melts and deposits the rock debris as till. All around the 60-mile- (100-km-) long terminus of the glacier is a region of stagnating ice, thickly overlain with rock debris. Surrounding this is a wide area of till studded with meltwater lakes and kettle holes. At its southern end, the glacier almost reaches sea level at Sitkagi Bluffs on the Gulf of Alaska. There, a series of terminal moraines protects it from contact with the open ocean. If the sea level were to rise sufficiently to connect the glacier to the ocean, the glacier would start calving icebergs and would probably retreat significantly. A remarkable feature of the Malaspina, and some other coastal glaciers in Alaska, is the occurrence of ice worms. These small segmented worms are found in colonies of millions of individuals near the ice surface during the day; at night they burrow deep into the ice. The worms are thought to survive by feeding on green algae that grow near the surface of the glacier.

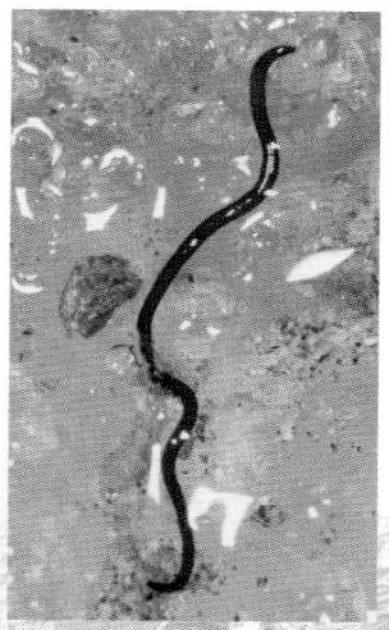

**ICE WORM**
*The tiny Ice Worm is one of very few species of animal known to be able to survive in glacier ice. This one is about ½ in (12 mm) long.*

**ALTERNATING CURRENTS**
*Much of the Malaspina Glacier's surface is covered in a complex pattern of contorted medial moraines. These are an indication of surge behavior—that is, patterns of alternating fast and slow ice flow.*

**TERMINAL LOBE**
*In this false-color satellite image of the Malaspina Glacier, taken by NASA's Landsat 7 in 2000, the ice and snow show as blue and white, the red areas are rock, and the brown patches are bands of vegetation.*

**CAPE YORK, NORTH GREENLAND**
*In this sunset view of the Greenland coast, the ice sheet is visible in the central background, with an outlet glacier descending to the frozen sea. Calved icebergs are locked in the sea ice.*

ARCTIC

## Greenland Ice Sheet



**LOCATION** Covering 85 percent of the island of Greenland in the Arctic Circle

**TYPE** Ice sheet

**AREA** 668,000 square miles (1.73 million square km)

**STATUS** Retreating

Despite its name, Greenland is mostly covered in a thick layer of ice. A single glacier, the Greenland Ice Sheet, makes up most of this ice, although smaller ice caps occur at the edges. The ice sheet is the northern hemisphere's largest glacier, with an average thickness of 5,900 ft (1,790 m), and a volume of 0.62 million cubic miles (2.6 million cubic km). Its surface is slightly domed, reaching about 10,800 ft (3,290 m) above sea level, slightly to the southeast of the island's center. A second, lower dome lies in the southeast corner. From these domes, ice moves slowly toward the edges of the ice sheet. Around most of its periphery, the ice is constrained by coastal mountains, so there are few places where it meets the sea along a broad front. As a result, Greenland has no ice shelves, but in many places, large outlet glaciers flow through valleys between the mountains and discharge enormous quantities of icebergs into the sea. One of the main outlets, Jakobshavn Glacier in west Greenland, is the fastest-flowing glacier in the world. At its terminus, the ice flows at speeds of about 3 ft (1 m) per hour, producing over 20 million tons of icebergs per day. An important factor operating on the ice sheet is albedo—the tendency of light-colored surfaces to reflect radiation (see p.443). The ice sheet has a high albedo that keeps its summer temperature far lower than it would be otherwise. There are concerns, however, that the ice sheet is melting due to global warming. It is so large that it comprises almost 10 percent of the world's fresh water reserves, and if it melted completely, the oceans would rise by 20–23 ft (6–7 m). ICEsat, a NASA satellite launched in December 2002, is monitoring both the Greenland and Antarctic ice sheets (see pp.278–79), using lasers that measure surface height changes as small as ⅖ in (1 cm) per year. ICEsat should be able to confirm whether the ice sheets are growing or shrinking, and whether they are causing sea-level change.

### WEATHER STATION

As part of the Greenland Climate Network project, 18 automatic weather stations have been established around the ice sheet. Instruments collect data on air temperature, humidity, atmospheric pressure, solar radiation, wind speed and direction, and heat flows in the snow cover overlying the ice sheet and at the snow/air interface. Data is transmitted hourly via satellite links to research centers in the US. The collected data may establish whether the climate around the ice sheet is getting warmer, and the extent to which this is linked to ice thinning.

**BEACHED ICEBERGS**
*Most North Atlantic icebergs originate from the Greenland Ice Sheet. The icebergs enrich the sea around Greenland with fresh water and oxygen, attracting an abundance of fish.*

**ICEBERG MAKER**
*Egi Glacier is a typical iceberg-calving glacier in Greenland. Its terminus rises 330 ft (100 m) above the sea and extends 3,000 ft (900 m) below water level. When the icebergs fall into the sea, they can cause tidal waves 30 ft (10 m) high.*

NORTH AMERICA *northwest*

## Athabasca Glacier

**LOCATION** Extending from the Columbia Ice Field, in the Rocky Mountains of Alberta

**TYPE** Outlet glacier

**TERMINUS** Terrestrial snout

**AREA** 2.5 square miles (6 square km)

**LENGTH** 4 miles (6 km)

**STATUS** Retreating

The Athabasca is one of more than 20 outlet glaciers from the Columbia Ice Field, the largest body of ice in the Rocky Mountains, with a surface area of 135 square miles (350 square km). Remarkably, the water runoff from this ice field flows into three different oceans: the Pacific, Atlantic, and Arctic. The top part of the Athabasca Glacier, which descends from the ice field at an altitude of about 8,860 ft (2,700 m), consists of three huge ice-falls like a series of giant steps. Farther down, the glacier descends gently and spreads out to a broad snout. The depth of the ice over the whole glacier varies from about 300 to 1,000 ft (90–300 m). The Athabasca has been steadily shrinking since about 1850 and is estimated to have lost 30 percent of its mass since 1900. Currently, its rate of retreat is about 50 ft (15 m) a year. Where the glacier has paused in its retreat, distinct ridges of large terminal moraines are clearly visible.

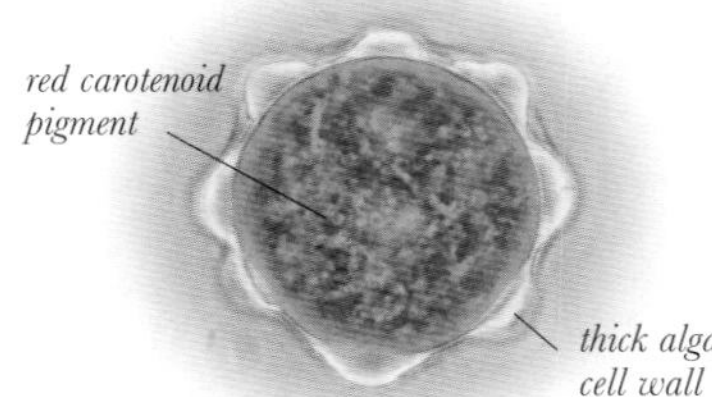


**WATERMELON SNOW**
*The patches of "pink snow" found on some glaciers, including the Athabasca, are actually colonies of bright red algae.*

**EVIDENCE OF RETREAT**
*Piles of moraine surround the snout and sides of this glacier. High up on the horizon is the Columbia Ice Field.*

NORTH AMERICA *northwest*

## Kaskawulsh Glacier

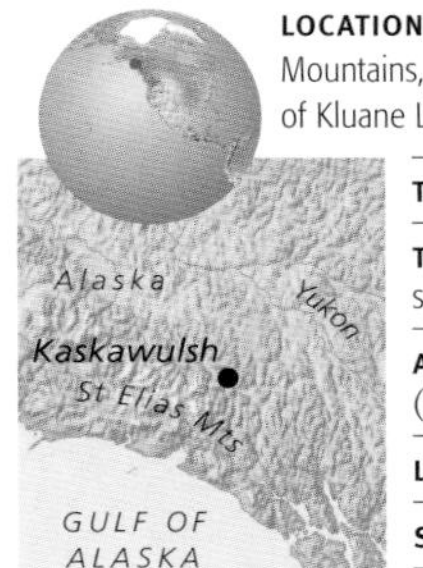


**LOCATION** In the St. Elias Mountains, 33 miles (53 km) south of Kluane Lake, Yukon

**TYPE** Valley glacier

**TERMINUS** Terrestrial snout

**AREA** 58 square miles (150 square km)

**LENGTH** 31 miles (50 km)

**STATUS** Retreating

Arguably the most spectacular glacier in Canada, the Kaskawulsh is notable for its sheer size and its impressive medial moraines (see below). These formed as a result of the merging of tributary glaciers into the main glacier trunk. The snout gives rise to the Slims River, which maintains the level of the Yukon's largest lake, Kluane Lake.

**AERIAL VIEW OF KASKAWULSH GLACIER**

NORTH AMERICA *northwest*

## Palisade Glacier

**LOCATION** In the southern part of the Sierra Nevada range, central California

**TYPE** Cirque glacier

**TERMINUS** Terrestrial snout

**AREA** 1 square mile (3 square km)

**LENGTH** 1¼ miles (2 km)

**STATUS** Retreating

The Palisade is one of about 100 small glaciers in the Sierra Nevada that have survived since the last ice age. At its top is a horizontal crack, called a bergschrund, which formed when the glacier fractured and its main mass slipped down, leaving a wall of ice frozen to the rock.

**THE HEAD OF THE PALISADE GLACIER**

SOUTH AMERICA *west*

## Pastoruri Glacier

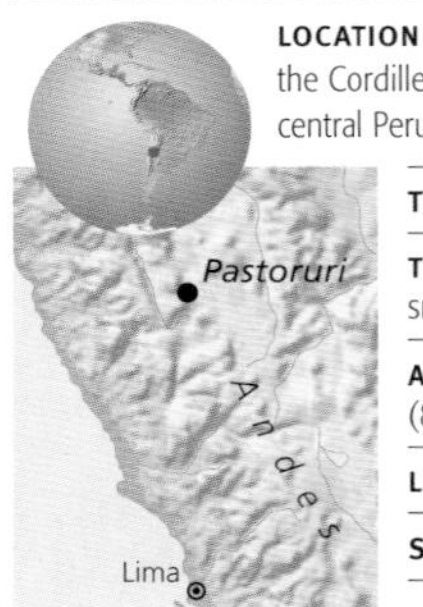


**LOCATION** In the southern part of the Cordillera Blanca, Andes, north-central Peru

**TYPE** Cirque glacier

**TERMINUS** Terrestrial snout

**AREA** 3 square miles (8 square km)

**LENGTH** 2½ miles (4 km)

**STATUS** Retreating

The tropical regions of South America still harbor a few areas of glaciation, such as the Pastoruri Glacier in Peru. Occupying the flank of a 17,200-ft- (5,240-m-) high Andean peak, the glacier has steep, clifflike edges and is usually covered in soft snow, with a few heavily crevassed areas. It is much used by ice-climbers and snowboarders.

**PASTORURI GLACIER, LOOKING DOWNHILL**

SOUTH AMERICA *southwest*

## Southern Patagonian Ice-field

**LOCATION** In the Patagonian Andes, along the Chile–Argentina border, southwest of Patagonia

**TYPE** Ice-field

**AREA** 5,020 square miles (13,000 square km)

**LENGTH** 225 miles (360 km)

**STATUS** Retreating

The Southern Patagonian Ice Field is the largest glaciated region in the southern hemisphere outside of Antarctica. It is the main remnant of a much larger area of glaciation (about 40 times bigger in area) that extended over much of the southern Andes about 10,000 years ago. Like all ice fields, it is a single, extensive, mostly level-surfaced mass of ice that covers most of a mountain region, apart from the highest peaks and ridges. Monte Fitzroy at 11,073 ft (3,375 m) and Cerro Pirámide at 11,089 ft (3,380 m) are two of the main peaks that the ice field surrounds. At its widest point, the ice field is about 56 miles (90 km) across, and its average width is some 25 miles (40 km). On its western side, nearly 50 significant outlet glaciers flow out of it, reaching sea level in the rugged fjords on the coast of Chile. Many of these fjords are choked with ice. On the eastern side, which is mainly in Argentina, large outlet glaciers calve icebergs into extensive lakes at about 650–900 ft (200–270 m) above sea level. Three of these lakes (Lago Martin, Lago Viedma, and Lago Argentino) can be seen in the image below. The lakes occupy valleys that were filled by large glaciers during the last ice age. Comparison of various aerial and satellite photographs of the ice field taken since 1945 has shown that, with a few exceptions, most of its outlet glaciers have retreated over the past 50 years.

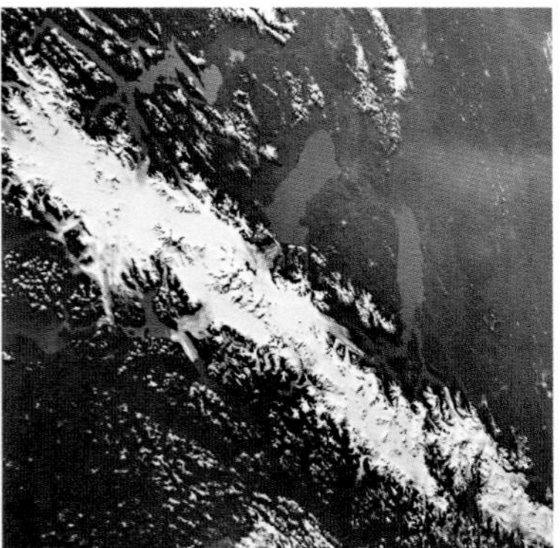

**ICE CAP REMNANT**
*This satellite photograph showing the vast extent of the Southern Patagonian Ice Field was taken looking toward the northeast.*

**LAKE-CALVING FRONT**
*Ice-block falls from this 200-ft (60-m) wall can produce waves big enough to drown people on the nearby shore.*

SOUTH AMERICA *southwest*

# Perito Moreno Glacier

**LOCATION** Extending from the Southern Patagonian Ice Field, southern Argentina

**TYPE** Outlet glacier

**TERMINUS** Lake-calving front

**AREA** 80 square miles (200 square km)

**LENGTH** 12 miles (20 km)

**STATUS** Static

**GLACIER TERMINUS**
*When the Perito Moreno Glacier advances, it forms a dam between Lago Argentino (on the right in this photograph) and Brazo Rico (on the left).*

Perito Moreno is an impressive outlet glacier from the Southern Patagonian Ice Field (see opposite), descending to the east of the ice field, where it calves icebergs into the largest lake in Argentina, Lago Argentino. Unlike most outlet glaciers in the region, its terminus has scarcely changed its position in the last 90 years. During that period, the glacier has made repeated temporary advances, each time blocking off an arm of the lake, called Brazo Rico. Whenever this happens, a separate lake has been created, with a rising water level that floods a large area of grassland. After six to 12 months, but always in summer, the lake bursts through the ice dam, discharging about a billion tons of water in 24 hours and flooding the shores of Lago Argentino.

EUROPE *northwest*

# Vatnajökull Ice Cap

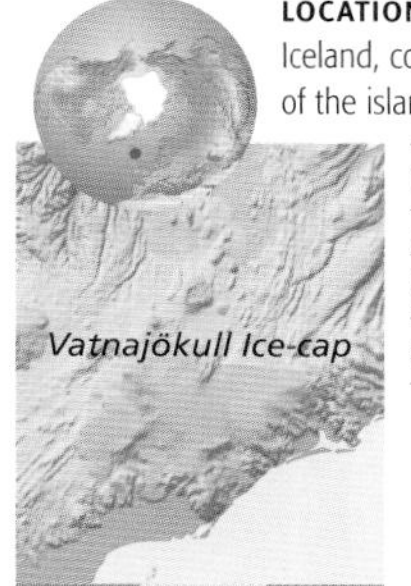


**LOCATION** In southeastern Iceland, covering about 8 percent of the island

**TYPE** Ice cap

**AREA** 3,100 square miles (8,100 square km)

**STATUS** Retreating

Vatnajökull, a classic ice cap, is the biggest glacier in Europe and one of several ice caps on Iceland. Roughly ellipse-shaped, the glacier completely covers the mountainous terrain it sits on and is slightly domed. Its average thickness is about 1,300 ft (400 m). Around its edges are numerous large outlet glaciers that drain the main body of ice toward the sea. Iceland lies in a region of high tectonic activity, and there are several active volcanoes and volcanic fissures under the western parts of Vatnajökull. In a few areas, heat from these melts the base of the ice-cap, creating subglacial lakes. In October 1996, an eruption from one of the volcanic fissures sent a huge plume of ash and steam some 3,000 m (10,000 ft) up into the air. Water from the melting ice cap soon filled the caldera of Grímsvötn, a subglacial volcano (see p.181), and then overflowed as a massive outburst flood under an outlet glacier (see panel below).

**BREIDAMERKURJÖKULL**
*This is the largest outlet glacier of the Vatnajökull Ice Cap, seen here calving ice into a small lake.*

## FLOODING HAZARD

Glaciers can cause catastrophic floods, and because the problem is so common in Iceland, glaciologists have adopted an Icelandic word, jökulhlaup ("glacier flood"), to refer to the phenomenon. Icelandic jökulhlaups result from subglacial volcanic activity, but in other parts of the world, a more common cause is the bursting of lakes (which have either overfilled or were previously ice-dammed), as glaciers melt and disintegrate. The Vatnajökull flood of 1996 (see above) destroyed many roads and bridges below the glacier.

**SUBGLACIAL ERUPTION**
*In December 1998, an eruption at the edge of Grímsvötn sent a plume of ash and steam 33,000 ft (10,000 m) up into the air above the ice cap.*

EUROPE *north*

## Kongsvegen Glacier

**LOCATION** In western Spitsbergen, Svalbard, 775 miles (1,250 km) north of the Norwegian mainland

**TYPE** Outlet glacier

**TERMINUS** Tidewater (sea-calving) front

**AREA** 40 square miles (105 square km)

**LENGTH** 12 miles (20 km)

**STATUS** Advancing slightly

Svalbard is an archipelago under Norwegian jurisdiction in the Arctic Ocean. Its largest island, Spitsbergen, is about 80 percent covered in glaciers, of which one of the best studied is the Kongsvegen. This is classified as a polythermal glacier, which means that there are large temperature variations within it—a cold surface layer of ice, 160–520 ft (50–160 m) thick, overlies a much warmer layer of ice at its base. Ice flow in polythermal glaciers tends to increase during the summer over the lower part of the glacier. This is thought to be caused by a buildup of meltwater beneath the glacier, lubricating its flow. In recent years, the terminus of Kongsvegen has been retreating. However, because it is a surge-type glacier (see p.263), this does not necessarily mean that the amount of ice in the glacier is falling. Indeed, detailed mass-balance studies conducted since 1987 have indicated that the glacier has actually slightly gained in ice mass overall.

**DIRTY ICE CLIFF**
*This partially grounded cliff of ice, visible at one side of the glacier's terminus, contains medial moraines and englacial rock debris.*

**MERGED GLACIER TERMINUS**
*At its terminus, the Kongsvegen merges with another glacier, the Kronebreen. The merged glaciers calve icebergs into Kongsfjorden along a front 2½ miles (4 km) wide.*



EUROPE *north*

## Jostedalsbreen Ice Field

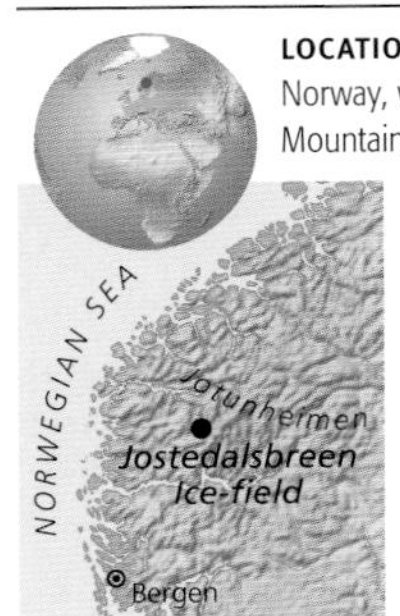


**LOCATION** In southwestern Norway, west of the Jotunheimen Mountains and north of Sognafjord

**TYPE** Ice field

**AREA** 88 square miles (487 square km)

**LENGTH** 60 miles (97 km)

**STATUS** Advancing

Jostedalsbreen is the largest glacier on the European mainland, a remnant of a vast ice sheet that covered the whole of Norway until about 10,000 years ago. It survives in southwest Norway primarily because of high regional snowfall rather than particularly cold temperatures, and this is reflected by a high rate of melting at the snouts of its 50-some outlet glaciers.

EUROPE *north*

## West Svartisen Ice Cap

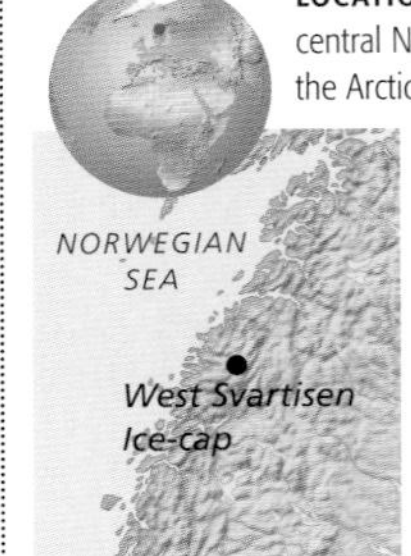


**LOCATION** In the coastal area of central Norway, at the latitude of the Arctic Circle

**TYPE** Ice cap

**AREA** 85 square miles (201 square km)

**STATUS** Uncertain

The West Svartisen Ice Cap occupies a high plateau near the Norwegian coast, with most of its surface lying above 3,600 ft (1,100 m). Like Jostedalsbreen, it owes its existence more to high snowfall than to low temperature. The ice cap's surface is featureless and slightly domed, but at its edges are a number of outlet glaciers. One of these, Engabreen, is notable on several counts. First, it is the lowest-lying glacial arm on the European mainland. Second, its mass balance has been studied since 1970 and shows a considerable positive balance (gain of ice); this has been accompanied by an advance of its terminus, as well as an increase in width and thickness of the glacier. The third is the laboratory beneath its surface (see panel, right).

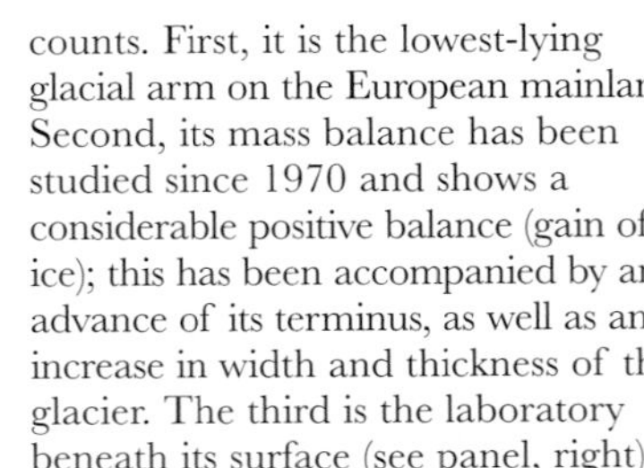

**ENGABREEN GLACIER**
*This arm of the West Svartisen Ice Cap descends to a lake that is just a few yards above sea level.*

### GLACIER LABORATORY

Tunnels have been drilled in the bedrock under the Engabreen Glacier as part of a hydroelectric power project. At the side of these tunnels, a laboratory has been excavated under the glacier, where the ice is about 660 ft (200 m) thick. This has allowed researchers to install instruments that continuously record the stress on the bedrock and the speed, temperature, and pressures within the ice as it slips across the rock. The results are helping scientists develop mathematical models for predicting glacier movement.

EUROPE *west*

## Glacier d'Argentière

**LOCATION** In the Savoie Alps, 8 miles (15 km) northeast of Mont Blanc, France

**TYPE** Valley glacier

**TERMINUS** Terrestrial snout

**AREA** 4 square miles (10 square km)

**LENGTH** 6 miles (10 km)

**STATUS** Static

The Glacier d'Argentière is one of numerous glaciers that descend from the flanks of the Mont Blanc Massif, a spectacular mountain chain that lies on the borders of France, Italy, and Switzerland. Surrounded by some of the highest mountains in the Alps, the glacier descends from around 13,100 ft (4,000 m) to 4,750 ft (1,450 m), via a series of steep ice-falls in its lower parts. The glacier terminus, which is about 1½ miles (2 km) above the resort of Argentière, has stayed more or less in the same position since 1970, although it shows a general pattern of retreat going back to the 1850s, at the end of the Little Ice Age (see p.451). Attempts have been made to tap into the subglacial drainage of meltwater to provide hydroelectric power. To do this, tunnels were excavated under the glacier in the 1960s and a water intake put in place. The project was not entirely successful, due to changes in the course of the meltwater drainage, and the tunnels have now been closed. However, the excavations did enable some scientific studies to be made via further tunnels bored into the ice using hot-water hoses. One study measured the rate at which ice movement was wearing down the marble bedrock at the glacier's base. The rate was 1½ in (3.6 cm) per year, although this is not thought to be representative of abrasion rates on glacial beds in general.

### ALPINE PURSUITS

The glaciers around the Mont Blanc Massif are heavily used by alpine sports enthusiasts. Cross-country skiing is possible on the upper parts of some glaciers—including the Glacier d'Argentière, which is so smooth that light aircraft can land on it—although care is needed to avoid the crevasses. Other popular pursuits include snow-boarding, ski-mountaineering, and ice-climbing. To facilitate this, nighttime refuges have been set up in numerous, otherwise highly inhospitable, locations around the region.

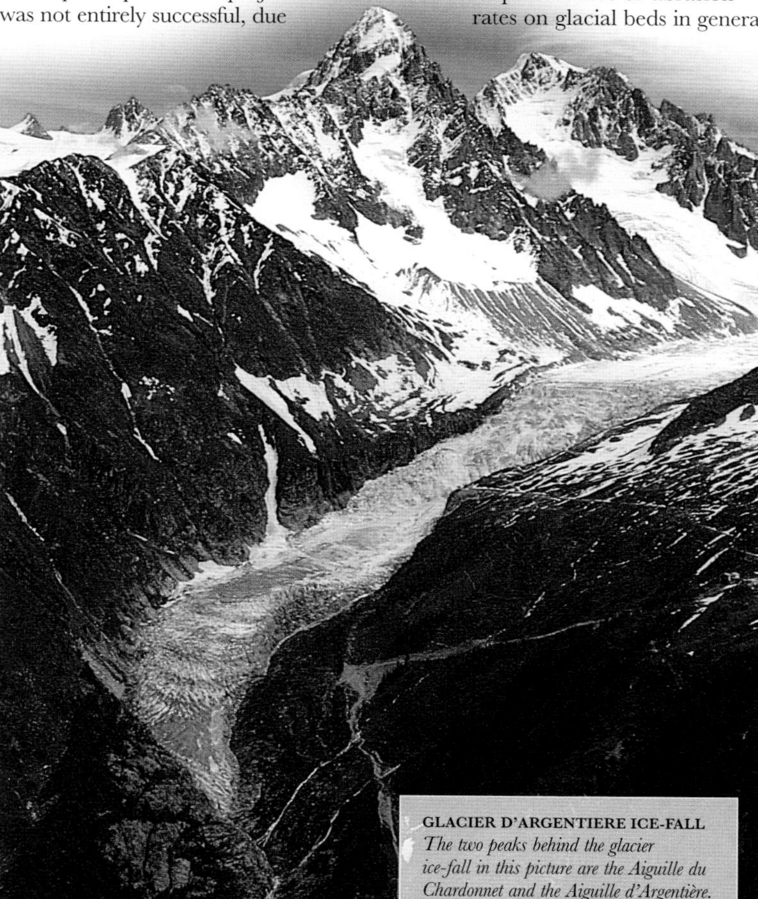

**GLACIER D'ARGENTIERE ICE-FALL**
*The two peaks behind the glacier ice-fall in this picture are the Aiguille du Chardonnet and the Aiguille d'Argentière.*

EUROPE *west*

## Mer de Glace

**LOCATION** In the Savoie Alps, 6 miles (10 km) northeast of Mont Blanc, France

**TYPE** Valley glacier

**TERMINUS** Terrestrial snout

**AREA** 15 square miles (40 square km)

**LENGTH** 7½ miles (12 km)

**STATUS** Retreating

The Mer de Glace, so called because it resembles a sea of frozen waves, has distinct alternating bands of light and dark ice, called ogives, on its surface. These are due to variations in the amount of windblown dust and other superficial debris that collects in the ice.

**MER DE GLACE, HAUTE-SAVOIE, FRANCE**

EUROPE *central*

## Tschierva Glacier

**LOCATION** In the Rhaetian Alps, close to the border with Italy, southeastern Switzerland

**TYPE** Valley glacier

**TERMINUS** Terrestrial snout

**AREA** 2¼ square miles (7 square km)

**LENGTH** 3 miles (4.5 km)

**STATUS** Retreating

The Tschierva is a medium-sized alpine glacier that descends steeply from a 13,284-ft (4,049-m) peak, Piz Bernina, on the Swiss–Italian border. The glacier is notable for its large and very clearly formed lateral moraines, which mark the glacier's farthest advance during the Little Ice Age in the 18th and 19th centuries (see p.451). Because of the Tschierva Glacier's retreat over the last 120 years or so, these moraines, which are about 130 ft (40 m) high, dwarf the lower parts of the glacier itself. As the glacier has retreated, it has revealed peat beds and ancient wood. Carbon dating of these samples shows that the tree line in this region was much higher, and the glaciers far smaller, as recently as 5,000 years ago.

EUROPE *central*

## Aletsch Glacier

**LOCATION** In the Bernese Alps, 16 miles (25 km) south of the Brienzer See, southwest Switzerland

**TYPE** Valley glacier

**TERMINUS** Terrestrial snout

**AREA** 34 square miles (87 square km)

**LENGTH** 16 miles (25 km)

**STATUS** Retreating

The Aletsch is the longest and largest valley glacier in Europe, descending from a saddle between the Jungfrau and Mönch peaks at 11,300 ft (3,454 m) and winding south toward the Rhône Valley. Along the way, it passes through the Konkordiaplatz ice plateau at the base of the Jungfrau, where it merges with other glaciers. It is up to 2,950 ft (900 m) deep, and the ice moves at speeds varying from 660 ft (200 m) a year in the Konkordiaplatz to 33 ft (10 m) a year near the snout. The pattern of moraines below the snout indicates that the glacier has retreated about 2 miles (3 km) since 1860.

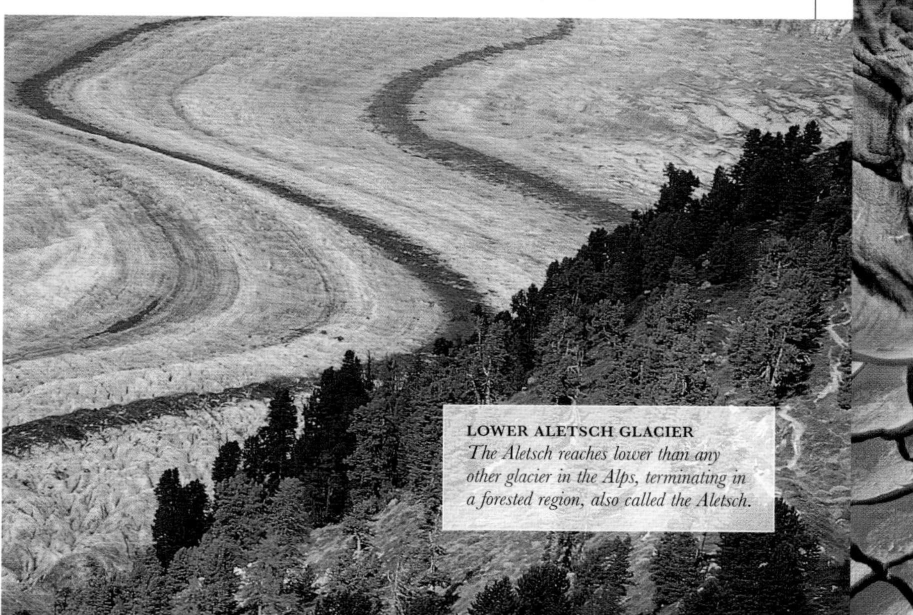

**LOWER ALETSCH GLACIER**
*The Aletsch reaches lower than any other glacier in the Alps, terminating in a forested region, also called the Aletsch.*

EUROPE *central*

## Rhône Glacier

**LOCATION** At the head of the Rhône Valley, in the eastern part of the Bernese Alps, Switzerland

**TYPE** Valley glacier

**TERMINUS** Terrestrial snout

**AREA** 6½ square miles (17 square km)

**LENGTH** 5 miles (8 km)

**STATUS** Retreating

Recognized as the source of the River Rhône, the Rhône Glacier was once the largest in the Alps. Evidence from moraines shows that, 18,000 years ago, it reached as far as Lyons, France, more than 175 miles (280 km) from its modern-day terminus. Today, it has retreated far back into the mountains and is much smaller. About 1¼ miles (2 km) wide on average, with a surface that is smooth in parts and deeply crevassed in others, the glacier was the subject of one of the earliest studies into glacial ice movement. In a classic investigation, a straight line of stakes driven into the glacier was observed between 1874 and 1882. After a few years, the stakes had formed into a bow, proving that the ice in a glacier moves and that the movement is fastest in the middle.

**GLACIER TERMINUS**
*The irregularly shaped snout of the Rhône Glacier can be seen today only by climbing up to an altitude of 7,200 ft (2,200 m).*

**SURFACE TEXTURE**
*Deep crevasses and meltwater streams have carved a mosaic of rounded blocks on the lower part of the Rhône Glacier.*

EUROPE *central*

## Allalin Glacier

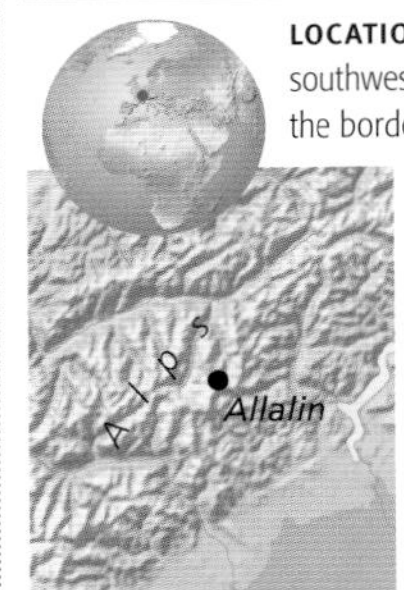


**LOCATION** In the Pennine Alps, southwestern Switzerland, close to the border with Italy

**TYPE** Cirque glacier

**TERMINUS** Terrestrial snout

**AREA** 4 square miles (10 square km)

**LENGTH** 4 miles (6 km)

**STATUS** Retreating

Occupying the eastern flank of the Allalinhorn, the Allalin is a relatively small glacier, consisting of a bowl-shaped accumulation area at its top and a steeply descending ice-fall. In recent centuries, the glacier has been a hazard to people living nearby. Between the 16th and 19th centuries, during the Little Ice Age (see p.451), it repeatedly advanced into the nearby Saas Valley, blocking the flow of the River Saaser Vispa and creating a lake called the Gletscherrandsee. Whenever the glacier retreated again, the lake would suddenly drain in a flood that devastated the valley. Since 1900, the glacier has shrunk more significantly. In 1965 a large piece of it broke off and cascaded into a dam construction site below, killing 88 people.

AFRICA *east*

## Kilimanjaro Ice Cap

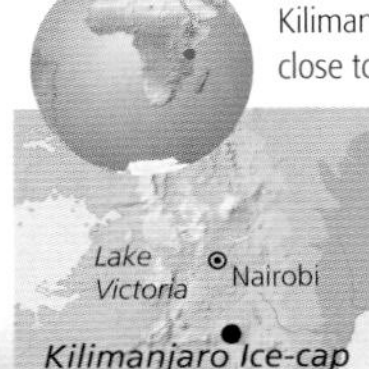

**LOCATION** At the summit of Mount Kilimanjaro, northeastern Tanzania, close to the border with Kenya

**TYPE** Ice cap

**AREA** 1 square mile (3 square km)

**STATUS** Retreating

One of the few remaining glaciated areas in Africa is at the top of Mount Kilimanjaro (see p.187). Despite the mountain's proximity to the equator, a huge continuous ice cap once covered most of its upper parts. As recently as 1912, the ice covered an area of more than 4 square miles (10 square km). Since then, it has almost all melted, leaving just a few fragments near the summit. These include a small glacier inside Kilimanjaro's volcanic crater and some ice fields around the northern flanks of the summit. Even these could soon be gone—perhaps by 2020.

**STEPPED ICE CLIFFS**
*Spectacular ice cliffs, about 80 ft (25 m) high, can still be seen at the edges of some of the ice fragments left on Kilimanjaro.*

ASIA *east*

## Bogdanovich Glacier

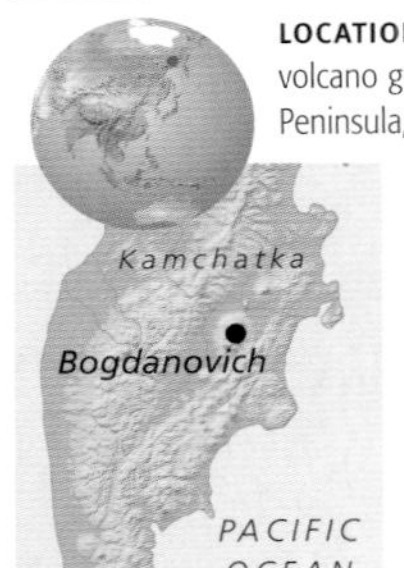


**LOCATION** Within the Kliuchevskoi volcano group, Kamchatka Peninsula, eastern Russia

**TYPE** Valley glacier

**TERMINUS** Terrestrial snout

**AREA** 6 square miles (15 square km)

**LENGTH** 6 miles (10 km)

**STATUS** Uncertain

The Bogdanovich Glacier forms part of a large glaciated area in Kamchatka, nestling in a valley surrounded by four huge volcanoes. It slopes gently down from a height of 9,200 ft (2,800 m) and suffers occasional ash falls from the Kliuchevskoi volcano (see p.189).

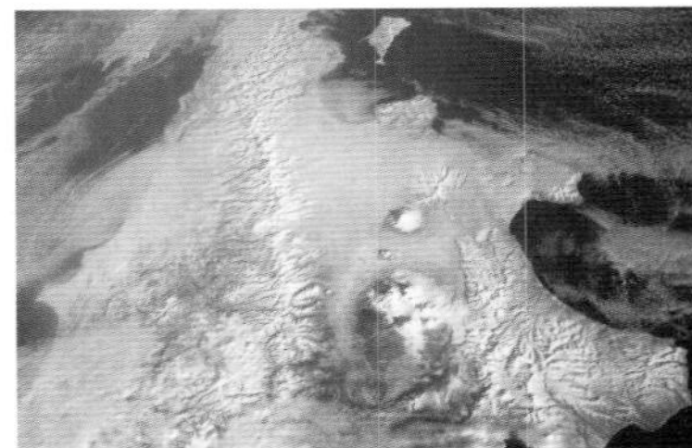

**KAMCHATKA PENINSULA**
*The Bogdanovich Glacier is located in the glaciated area visible just below and to the right of center of this satellite image of the Kamchatka Peninsula.*

ASIA *central*

## Inylchek Glacier

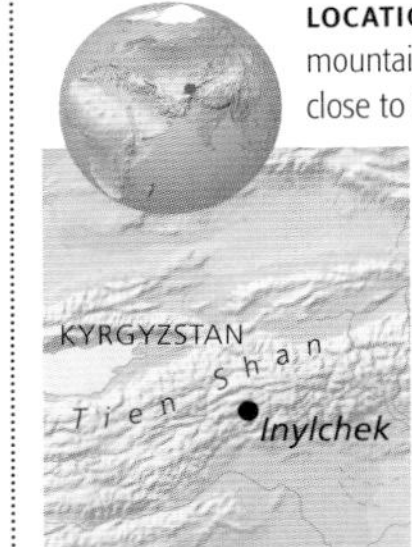


**LOCATION** In the Tien Shan mountains of eastern Kyrgyzstan, close to the border with China

**TYPE** Valley glacier

**TERMINUS** Lake-calving front

**AREA** 225 square miles (580 square km)

**LENGTH** 39 miles (62 km)

**STATUS** Uncertain

One of almost 8,000 glaciers in Kyrgyzstan, the Inylchek Glacier has two branches—North and South—that wind their way through separate valleys in the Tien Shan mountains and converge in an icy lake. The South Inylchek is one of the three longest valley glaciers in the world.

**SOUTH INYLCHEK GLACIER**

**EXPEDITION HIGHWAY**
*An expedition porter scrambles over the glacier's debris-strewn surface. Two tributary glaciers to the Baltoro can be seen in the background.*

ASIA *south*

## Baltoro Glacier

**LOCATION** In the central area of the Karakoram mountain range, northeastern Pakistan

**TYPE** Valley glacier

**TERMINUS** Terrestrial snout

**AREA** 290 square miles (750 square km)

**LENGTH** 37 miles (60 km)

**STATUS** Advancing slightly

Northern Pakistan, in and around the Karakoram mountain range, contains the greatest concentration of ice in Asia, with glaciers covering more than 5,000 square miles (13,000 square km). The Baltoro Glacier is one of the largest of these, occupying a long east–west valley running through the region. Up to 4 miles (6 km) wide in places, the Baltoro is fed by over 30 tributary glaciers, which form ice-falls where they meet the main trunk glacier. The valley walls above the glacier vary from very steep to precipitous. Much of the lower area of the glacier is covered in thick rock debris. Emerging from its snout is the fast-flowing and dangerous Braldu River, a tributary of the Indus. Unlike most Asian glaciers, the Baltoro does not seem to be in fast retreat. Between 1990 and 1997, it actually advanced slightly, and some of its tributary glaciers have displayed surge behavior. A trek along the Baltoro provides the only possible land access to the base of K2 (see pp.168–69), so the glacier has been well-trodden by mountaineers.

ASIA *south*

## Durung Drung Glacier

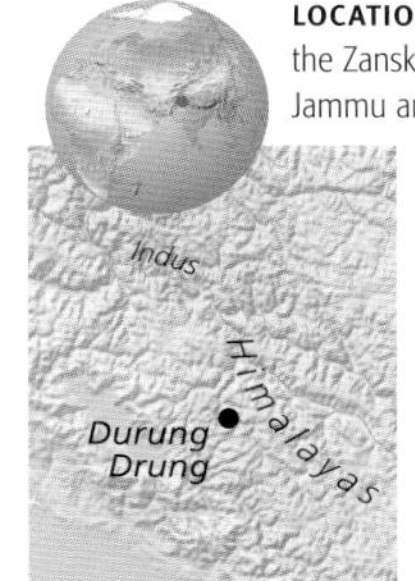


**LOCATION** In the western part of the Zanskar Range of the Himalayas, Jammu and Kashmir, northern India

**TYPE** Valley glacier

**TERMINUS** Terrestrial snout

**AREA** 12 square miles (30 square km)

**LENGTH** 9 miles (14 km)

**STATUS** Uncertain

One of the longest glaciers in the Himalayas, the Durung Drung is also one of the most accessible, from a road that crosses a pass, the Pensi-La, at 14,435 ft (4,400 m). Its surface is part clean ice, part rock debris, and features ice pinnacles, or seracs, some of which have large boulders balanced on their points. From the clifflike snout of the glacier, the fast Doda tributary of the Zanskar River arises.

ASIA *south*

## Khumbu Glacier

**LOCATION** On the southern approach to the Mount Everest region of the Himalayas, Nepal

**TYPE** Valley glacier

**TERMINUS** Terrestrial snout

**AREA** 6 square miles (15 square km)

**LENGTH** 7½ miles (12 km)

**STATUS** Retreating

The Khumbu Glacier has two main segments. The upper segment consists of a steeply descending region of clean, deeply crevassed ice, the Khumbu Ice-Fall. This is situated on the southwest flank of Mount Everest, immediately above the mountaineers' main base camp. At the bottom of the ice-fall, the glacier turns past the base camp to the south, and then descends down a straight, steep-sided valley, 5–6 miles (9–10 km) long, toward the Himalayan foothills. The glacier here has a relatively gentle downward slope and is covered in a thick layer of rock debris. Its snout has retreated a few miles since Everest was first climbed in 1953.

**ICE FINS**
*These giant finlike structures were squeezed up out of the side of the Khumbu Glacier as it was forced around a bend in its valley.*

**SERACS**
*As the East Rongbuk Glacier passed over irregular bedrock, it fractured to form isolated blocks of ice up to 80 ft (25 m) high.*

ASIA *south*

## East Rongbuk Glacier

**LOCATION** On the northern approach to the Mount Everest region of the Himalayas, Tibet

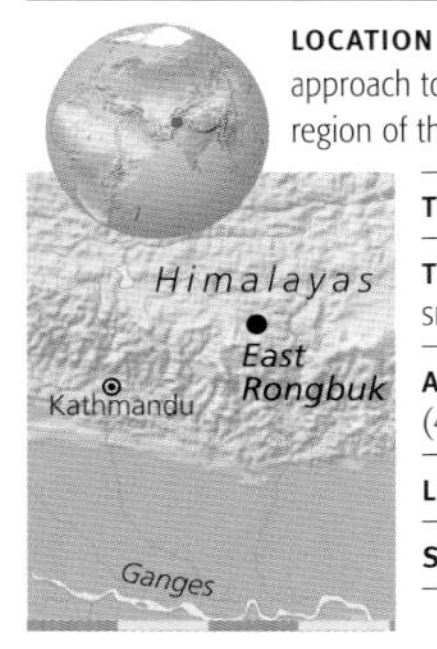


**TYPE** Valley glacier

**TERMINUS** Terrestrial snout

**AREA** 15 square miles (40 square km)

**LENGTH** 9 miles (14 km)

**STATUS** Uncertain

The East Rongbuk is part of a large, complex branching system of glaciers on the northern side of Mount Everest, and lies along one of the main routes for reaching the summit from Tibet. Its upper part occupies a large arena, at an altitude of about 22,500 ft (6,850 m), below Everest's northeast ridge. From there, it descends in a northwesterly direction, eventually merging with the Central and West Rongbuk glaciers and terminating at an altitude of about 18,000 ft (5,500 m). Its lower parts are covered in a layer of rock debris. In common with several other glaciers in the region, the East Rongbuk has a large number of seracs, or ice pinnacles, on its surface. These can be up to 100 ft (30 m) high and are caused by successive fracturing of the ice as it passes over separate steep or irregular sections of bedrock. Each section produces a series of parallel cracks in the ice, and where these intersect, the pinnacles are formed.

ANTARCTICA

# Antarctic Ice Sheet

**LOCATION** Covering most of the continent of Antarctica and extending over the sea as ice shelves

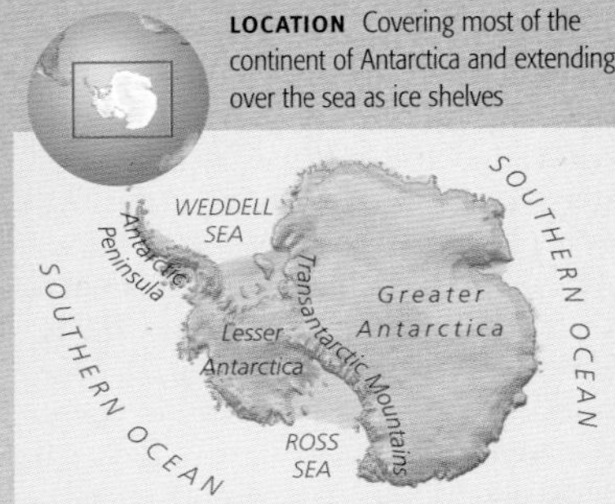


**TYPE** Ice sheet

**AREA** 5.3 million square miles (13.7 million square km)

**STATUS** Uncertain

The Antarctic Ice Sheet is the largest mass of ice on Earth. Its volume is over 7.2 million cubic miles (30 million cubic km), and it holds over 70 percent of the Earth's fresh water. It weighs so much that it depresses the Earth's crust by about 3,000 ft (900 m). The ice sheet consists of two parts, separated by the Transantarctic Mountains. The larger part, called the East Antarctic Ice Sheet (EAIS), covers most of the landmass called Greater Antarctica. It is over 2¾ miles (4.5 km) thick in places, and its base is mainly above sea level. The West Antarctic Ice Sheet (WAIS), over Lesser Antarctica, has a maximum thickness of 2¼ miles (3.5 km), and its base lies mostly below sea level. The rate of ice formation is slow (Antarctica is the driest continent, with snowfall averaging only a few inches per year), but once ice forms, very little of it melts or evaporates while it is still part of the ice sheet. Instead, under the force of its own weight, the ice slowly deforms and moves toward the coasts, where outlet glaciers and ice streams carry it down to the sea. The rate of ice movement varies from less than 3 ft (1 m) a year in parts of the ice sheet domes, to several hundred yards a year near the coasts. In many places the outlet glaciers extend over the sea as ice tongues or merge to form vast platforms of floating ice called ice shelves. Huge icebergs continually break off the ice shelves, balancing the ice input from the outlet glaciers and ice streams. Many of the outlet glaciers draining Antarctica are immense. The largest, the Lambert, is 25 miles (40 km) wide and over 250 miles (400 km) long. Other notable features include the Byrd and Beardmore glaciers, which drain

**VEHICLE CONVOY**
*These vehicles are making their way across the ice sheet close to the Adelie Coast of Greater Antarctica. The ridges of ice, called sastrugi, are caused by, and lie at right angles to, the prevailing wind.*

ANTARCTIC ICE SHEET

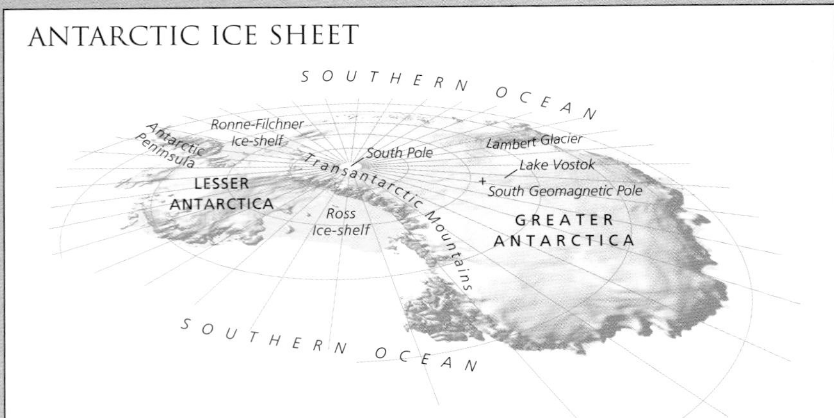


**ANTARCTIC COAST AND ICEBERG**
*Many icebergs form from glaciers flowing off the Antarctic Ice Sheet. The iceberg floating in the foreground is highly eroded, indicating that it has been drifting for several months.*

## ERNEST HENRY SHACKLETON

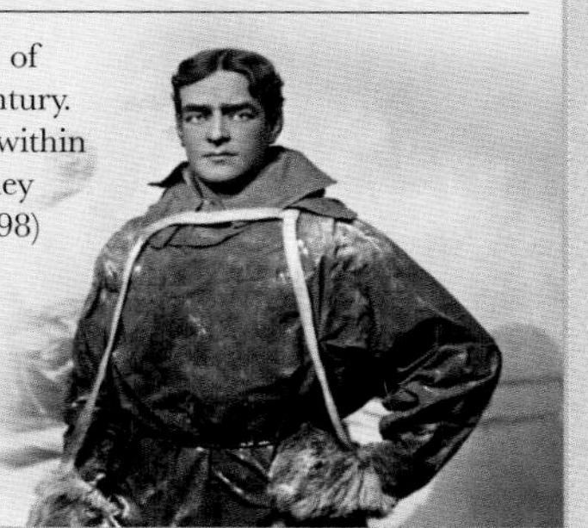

Irishman Ernest Shackleton (1874–1922) was one of the great Antarctic explorers of the early 20th century. In 1908–09, he and three companions walked to within just 97 miles (156 km) of the South Pole before they had to turn back. After Roald Amundsen (see p.398) and then Robert Falcon Scott reached the South Pole in 1911–12, he led another expedition to Antarctica in 1915, aiming to cross the continent. Unfortunately, his ship, *Endurance*, became locked in sea ice and eventually sank. However, thanks to Shackleton's leadership, no lives were lost.

the EAIS into the Ross Ice Shelf, and the Rutford Ice Stream, which drains part of the WAIS into the Ronne Ice Shelf. The only ice-free areas in Antarctica are the peaks of the highest mountains, which poke through the ice sheet as nunataks, and a few coastal areas, such as the Dry Valleys near the Ross Sea. There are concerns that the ice sheet is shrinking (see p.271), and that the WAIS is particularly vulnerable to recession through accelerated iceberg-calving because its base is below sea level. The WAIS has shown a pattern of retreat for over 10,000 years, but in the short term, some parts appear to be thinning, others getting thicker.

**EMPEROR PENGUINS**
*The tallest and heaviest of all penguins (about 82lb (37kg) in weight, and 3½ft (1.1m) tall), emperors set up their breeding colonies on the newly formed sea ice at the end of the Antarctic summer. The females lay a single egg, then return to sea to feed, while the fasting males incubate the egg through the freezing winter. The down-covered chicks have to molt into their adult plumage before they can float out to sea on the ice as it melts and breaks up in spring.*

*outline of lake visible by satellite as flat area on ice surface*

*direction of movement of ice sheet*

*upper layers of lake water freeze to the ice sheet and are carried beyond the lake edge*

*ice sheet*

*lake buried 2½ miles (4 km) beneath ice surface*

*bedrock*

**LAKE VOSTOK**
*The East Antarctic Ice Sheet overlies several lakes, the largest of which is Lake Vostok. Originally detected by ice-penetrating radar, the lake, and any organisms living in it, have been isolated for at least 1.5 million years.*

## ICE-CORE ANALYSIS

Ice accumulates in annual layers in ice sheets, somewhat like the rings of a tree trunk. By drilling down into the Antarctic Ice Sheet, extracting a long cylinder, or core, of ice, and examining how the ice changes with depth, scientists are able to find out a great deal about the Earth's past, including fluctuations in climate. The longest ice core drilled so far provides a record that goes back 420,000 years. Tiny trapped air bubbles reveal changes in the atmosphere, including rising levels of carbon dioxide. Changes in atmospheric pollution are also apparent in the ice cores.

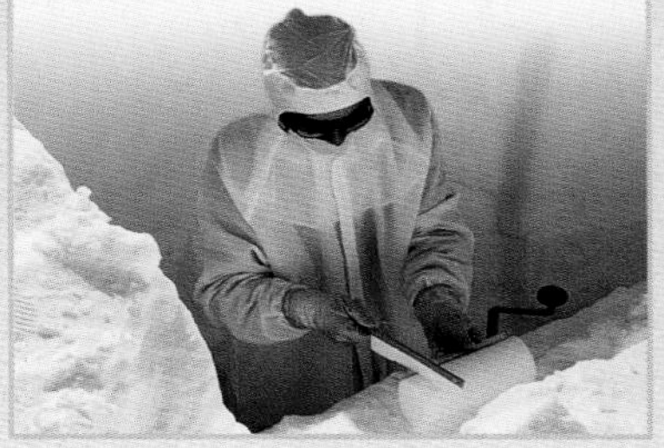

AUSTRALASIA *New Zealand*

## Franz Josef Glacier

**LOCATION** To the northeast of Mount Cook, in the Southern Alps, South Island, New Zealand

**TYPE** Valley glacier

**TERMINUS** Terrestrial snout

**AREA** 12 square miles (32 square km)

**LENGTH** 6 miles (10 km)

**STATUS** Advancing

Named in 1865, by a German geologist, after the Emperor of Austro-Hungary, the Franz Josef Glacier descends steeply from an altitude of 8,860 ft (2,700 m) in New Zealand's Southern Alps. The lower parts of the glacier almost reach sea level, flowing through a forest of ancient podocarp trees (see p.310) and other evergreen species. From the glacier's snout flows a fast and turbulent river, named the Waiho (Maori for "smoking water"; its name originates from the vapor rising from its ice-cold surface). The Franz Josef Glacier owes its existence primarily to a high rate of precipitation in the region—up to 100 ft (30 m) of snow falls on the glacier's ice-accumulation area every year. The glacier is particularly climate-responsive; that is, it retreats and advances relatively promptly (within about five years) in response to significant changes in regional precipitation and temperature.

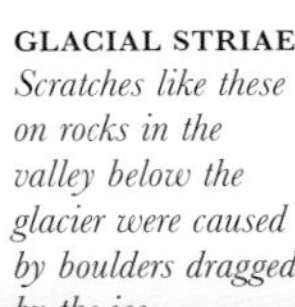

**GLACIAL STRIAE**
*Scratches like these on rocks in the valley below the glacier were caused by boulders dragged by the ice.*

**RAINFOREST AND GLACIERS**
*The combination of a temperate rainforest with large glaciers, such as the Franz Josef, flowing through it is unique to New Zealand.*

AUSTRALASIA *New Zealand*

## Tasman Glacier

**LOCATION** Immediately east of Mount Cook, in the Southern Alps, South Island, New Zealand

**TYPE** Valley glacier

**TERMINUS** Partially terrestrial snout, partially lake-calving front

**AREA** 37 square miles (95 square km)

**LENGTH** 17 miles (27 km)

**STATUS** Retreating

The Tasman is the largest glacier in New Zealand, flowing down a long valley to the east of Mount Cook (also known as Aoraki). It is up to 2,000 ft (600 m) thick and up to 2 miles (3 km) wide. It originates in a large ice-accumulation area at an altitude of 9,200 ft (2,800 m), beneath Mount Elie de Beaumont, and halfway down it is replenished by the Hochstetter Ice-Fall. This drains an enormous amphitheater of ice beneath Mount Cook and Mount Tasman. The upper Tasman Glacier receives regular heavy snowfalls and is smooth and relatively crevasse-free. Farther down are sections where it breaks up into a maze of crevasses, ice tunnels, and ice walls. The lower third of the glacier, which extends down to an altitude of about 2,300 ft (700 m), is covered in a 3-ft- (1-m-) thick layer of rock debris. This is the result of ice melting, which concentrates debris at the glacier's surface. At its bottom end, rapid melting since the 1960s has turned what were a few isolated craters and moulins in the center of the glacier, and a small elongated lake along its eastern lateral moraine, into a large melt-lake that extends for about 1½ miles (2.5 km) along the glacier's surface. Other parts of the lower glacier are heavily pitted. The main front of the glacier has retreated about 1 mile (1.5 km) since the 1960s and is continuing to retreat about 100 ft (30 m) a year. The glacier has also thinned dramatically since the beginning of the 20th century. Today, lateral-moraine walls almost 650 ft (200 m) high lie on both sides of the glacier; around 1900, the ice was above those walls. The lake at the foot of the glacier is up to 330 ft (100 m) deep and is becoming deeper, wider, and longer. Part of it still overlies a layer of ice, and its temperature stays steady at 32.5–32.9°F (0.3–0.5°C). The disintegration of the glacier's terminus has been accelerated by the onset of iceberg calving into the lake—a process that started in about 1991. Some of the icebergs, which are full of rock debris, weigh millions of tons. Occasionally, calving takes place underwater, causing icebergs to pop up in the middle of the lake. For reasons not fully understood, silt suspended in the water does not settle out, and gives the lake a uniform gray color. A demonstration of why some glaciers carry so much rock debris inside them and on their surfaces was given in 1991, when a large part of the summit of Mount Cook broke off. Millions of tons of rock cascaded down the Hochstetter Ice-Fall and then over the Tasman Glacier. The rubble is now being carried slowly down the glacier and will eventually be dumped as till at its terminus. Because insulating rock cover on a glacier reduces the rate at which it melts, this may cause some advance of the glacier in the medium term, or will perhaps slow its retreat.

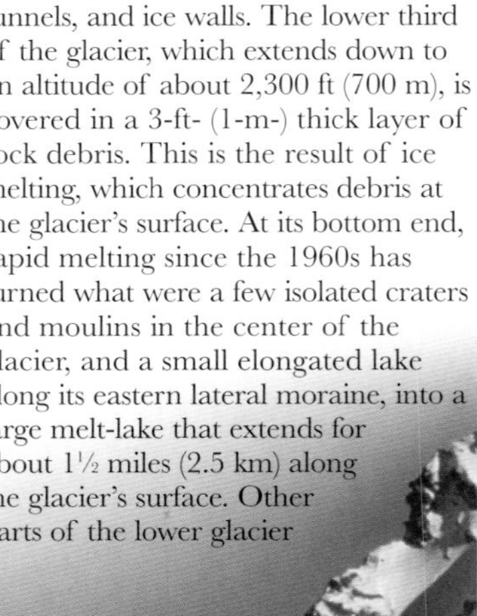

**BRAIDED STREAMS**
*Meltwater streams flow down a 9-mile- (15-km-) long outwash plain below the glacier's terminus, converging as the Tasman River.*

**EXTREME SKIING**
*The upper parts of the Tasman Glacier provide one of the longest continuous ski runs in the world—about 7 miles (12 km). These skiers were dropped off by helicopter.*

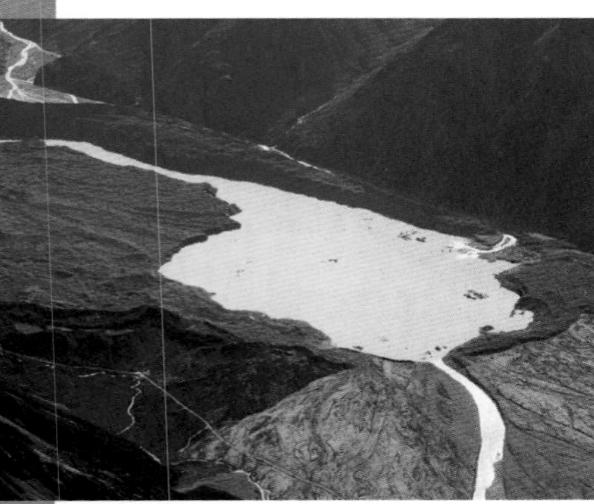

**ICY LAKE**
*This lake sits on top of part of the disintegrating lower end of the Tasman Glacier and is rapidly extending up the glacier.*

LAND

**ICEBERG IN GLACIAL LAKE**
*This huge wall, full of rock debris, is part of an iceberg that was calved into the lake at the terminus of the Tasman Glacier. Mount Cook can be seen in the background.*

**SAHARA CROSSING**
*Air travel and roads have reduced the need for camel caravans in some parts of the Sahara, but here in Mauritania near the coastal settlement of Nouakchott, they remain a vital means of communication.*

# DESERTS

THE TERM "DESERT" CONJURES UP vistas of rolling, featureless sand dunes and intense, dry heat. While these conditions are certainly found in many places, the world's deserts are far more varied in character, with some having rocky landscapes, some occupying high plateaus, and others dominated by open, salty lake beds. Nor are deserts hot all of the time. In some "cold deserts," the winters may be snowy or frosty, in strong contrast to the hot, dry summers. The temperature range between day and night may also be marked, with temperatures falling rapidly from dusk until dawn, and sometimes even dropping below freezing. Conditions for life in a desert are challenging, with water scarce most of the time, except for occasional downpours, which can lead to destructive flash floods. Windblown sand and grit is another hazard in many deserts, and it also acts as a potent agent of erosion, shaping rocks and forming dunes.

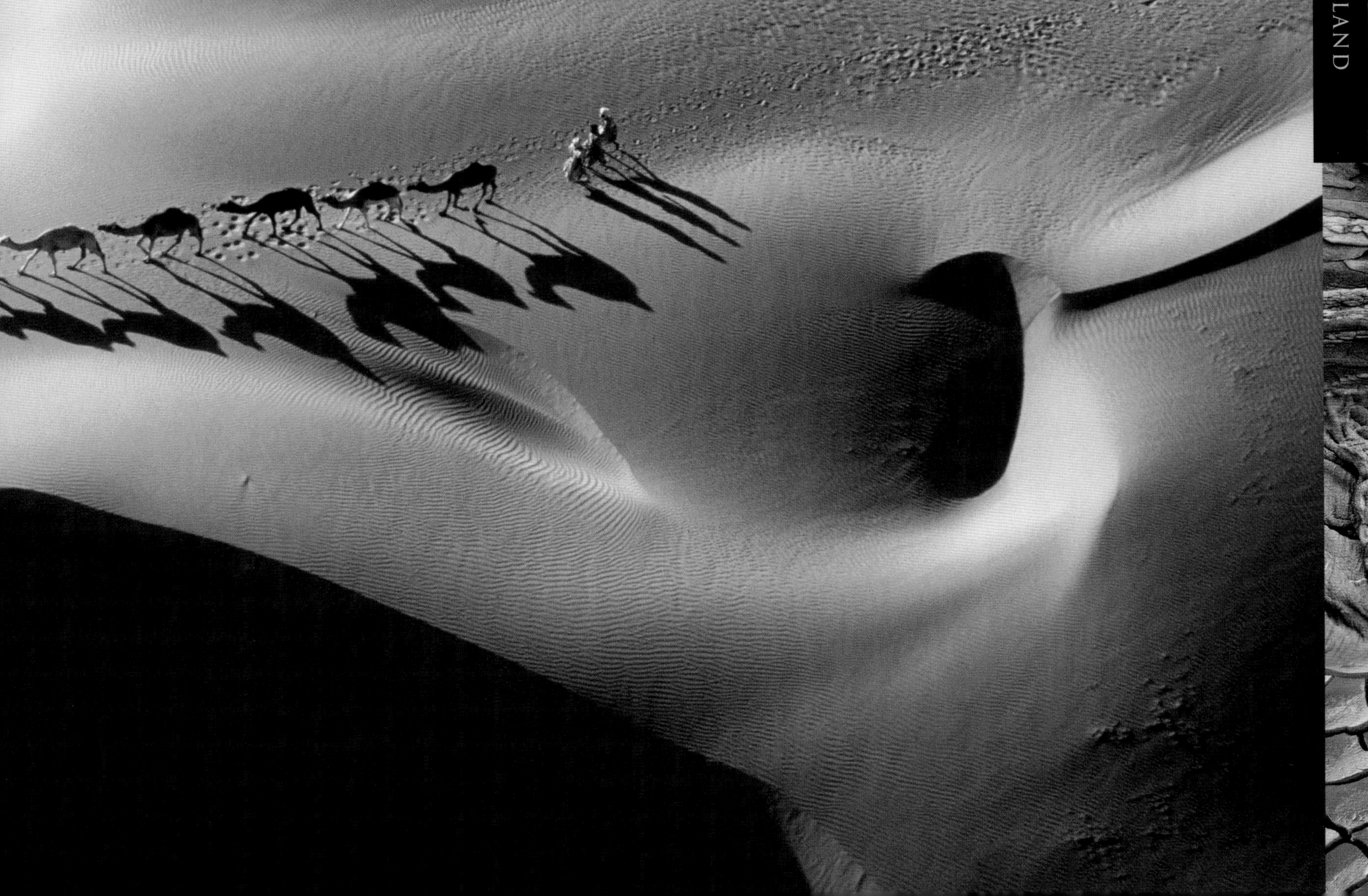

# DESERTS



DESERTS ARE AMONG THE MOST HOSTILE places on Earth. They arise in regions that receive large amounts of sunshine, or where geographical features limit rainfall. High temperatures and dry winds quickly remove any moisture by evaporation. The dry ground becomes vulnerable to processes of erosion and weathering that shift sand and sculpt rock to produce beautiful but forbidding landscapes.

## TYPES OF DESERTS

Deserts have been classified in a number of different ways, but are usually defined according to their climate or their physical characteristics. Climatically, there are large differences between deserts. Subtropical deserts, such as the Sahara, have high temperatures all year and are described as "hot," while high-altitude or continental deserts, such as the Great Basin of North America and the deserts of Central Asia, experience cold winters and are described as "cold," although the summer temperatures may be high. Deserts may be just rocky or sandy, but several contain a mixture of surface features and textures. Where the occasional surface streams have no outlet, they evaporate, sometimes forming extensive deposits of salts. Many deserts are mainly or partly sandy, and the fine sand grains are easily moved and molded by the wind, often into characteristic patterns or shaped dunes.

**SEMIARID DESERT**
*In semiarid deserts (right) the conditions are dry, but there is sufficient available moisture to sustain patches of vegetation.*

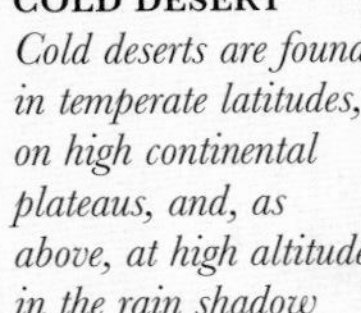

**COLD DESERT**
*Cold deserts are found in temperate latitudes, on high continental plateaus, and, as above, at high altitude in the rain shadow of the Andes.*

**ROCKY DESERT**
*Many deserts are set in rocky terrain, such as the Colorado Desert, shown at left with the rising sun illuminating bare rock pinnacles and outcrops.*

## WEATHERING AND EROSION

Desert landscapes are created and maintained by erosion and weathering acting over long periods. Both processes break down soft rocks into smaller particles, and sculpt hard rock formations. Erosion is the process of reshaping by moving agents, such as windblown sand, which scours the surface of exposed rocks and chips off small particles (see opposite). In some deserts, such erosion sculpts rounded structures called yardangs from the bedrock. They rise above the desert sands like the upturned hulls of ships. Erosion also forms pediments—gently sloping surfaces, at the foot of rocky highland areas, that are still being broken down by erosion. Sudden storms, though rare, are another major cause of erosion, since deserts tend to have little vegetation to protect their surfaces. Dry streambeds, or wadis, suddenly fill with water, and the loose gravel and sand they contain is swept away, to be deposited downstream. Weathering happens within rocks themselves—temperature variations can cause desert rocks to crack under the stress of thermal expansion and contraction. Moisture such as dew on a rock surface can also penetrate cracks, then expand or freeze, breaking the rock open. If salt water comes into contact with rock, then the growth of salt crystals within it may also crack the rock apart.

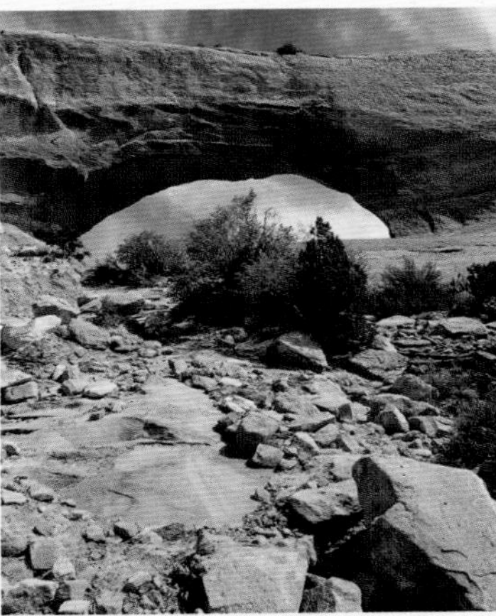

**DESERT SCREE**
*As weathering eats into exposed rocky outcrops, a scree of boulders and stones litters the desert floor below.*

**SEA OF SAND**
*The Namib Desert in Africa is famous for the vast inland sea of majestic sand dunes at Sossusvlei, which can reach heights of 650 ft (200 m).*

# LANDFORMS

The processes of erosion that happen in desert climates produce distinctive small- and large-scale landforms. Large landforms include inselbergs—isolated hills that stand out above the surrounding flat desert surface, and are all that remain of long-eroded mountain ranges. Australia's Uluru (also known as Ayers Rock) is a famous example. Mesas and buttes are towering platforms topped by flat, erosion-resistant surfaces. They mark the former ground level—their surroundings eroded after their hard cap was penetrated, perhaps by an ancient stream. Gravels of various grades are commonly found in deserts, often as alluvial fans at the feet of raised plateaus or spreading over the surface of rock pavements. Gravelly or rocky surfaces with a finer material beneath are called desert pavements. The term erg, originating in the Sahara, refers to an almost sealike expanse of sand. Flat desert basins may contain lakes, which are typically dry for part of the year, or even for many years. These are known as playas, and the soils here are usually high in a mixture of salts, mostly of calcium and sodium. Some desert rocks are covered with a black or brown glossy coating called desert varnish, which usually contains large amounts of iron and manganese oxides. How this forms is not fully understood.

**DESERT VARNISH**
*The shiny coating found on some desert rocks is called desert varnish. In some regions, the varnish has been used as a canvas for primitive rock art.*

**DESERT PAVEMENT**
*These antelope footprints in the Kalahari Desert have revealed the fine sand beneath the gravelly surface.*

**MESAS AND BUTTES**
*A mesa is the eroded remnant of a plateau, like the ones seen here rising above the flat plain of Monument Valley in the southwestern US. A butte is a small mesa.*

# WIND AND DUNES

Dunes are large accumulations of sand, forming hills, ridges, or other features in a desert landscape. Half of all dunes are linear or seif dunes—roughly parallel ridges of sand that can extend for 12 miles (20 km) or more. They are a particular feature of the Kalahari and Simpson deserts. Barchan dunes are curved, with two arms trailing downwind, and are often found around the edges of sand seas. Star dunes have three or more arms, usually irregularly arranged. They can reach heights of more than 1,000 ft (300 m), and are found mainly in the eastern Sahara, the Namib, and the deserts of Central Asia. Parabolic dunes are also large, with long trailing arms, and are a notable feature of the Thar Desert and many coastal dune regions. Crescentic dunes form ridges, with slightly wavy edges. These tend to form in areas with a narrow range of wind directions, and where there is little vegetation.

**WIND-SCULPTED ROCKS**
*Wind is a powerful erosive force, especially if it contains sand particles. Here in Arizona, veins of calcite between bands of sandstone colored with metal oxides emphasize the wavelike shapes of eroded sedimentary rock.*



**HOW DUNES FORM**
*Dune formation is a complex matter, still not fully understood, but in general, dunes form where wind loses energy and drops the sand suspended within it. Wind-shifted sand soon becomes rippled, but over a longer period of time and with varied winds and textures of sand, distinctive dune shapes are created.*

# WATER

Precipitation in deserts is scarce and unreliable. Average desert rainfall usually totals less than 10 in (250 mm) a year, but in parts of the Sahara and coastal Chile, it is often exceedingly low—less than 1/16 in (1 mm) a year. Another key factor is that the dry, hot air soon removes water by evaporation, so that even when the rains do fall, the water quickly turns to vapor, returning to the air before it can be used by animals or plants. In some deserts that experience cold winters, such as those of Central Asia, the main source of water is winter snow. Here again, availability of water is low, until the snows melt in the spring. Some coastal deserts, such as the Namib in southwest Africa and parts of the Atacama in Chile, receive moisture from fog, and this may be equivalent to as much as 5 in (130 mm) of rain annually. Dew is another important source of water, and if there is moisture in the air, it may form during the relatively cold, clear desert nights. In many deserts, such as the Thar on the India–Pakistan border, there are considerable stores of water underground, and these can be tapped by deep-rooted plants, and by humans through drilling.

**WADI**
*A wadi is a steep-sided watercourse, such as a stream, that was formed in an earlier, wetter climatic phase and now flows only irregularly.*

**WADIS IN THE YEMEN**
*This low-orbit satellite image of the Arabian Peninsula reveals a veinlike network of dry wadis, which are prone to flash floods after seasonal rains.*

## OASES

Most human activity in deserts has traditionally centered on reliable supplies of water, which guarantee plant growth for grazing livestock and enable the planting of staple crops, as well as offering drinking water. Many deserts contain large reserves of groundwater, particularly compared to surface stores, such as rivers and lakes. Sometimes, the water from these water-saturated rocks, or aquifers, comes to the surface in the form of springs. Most desert oases develop around natural permanent springs, and these produce a lush growth of plants, with palm trees a prominent feature in many regions. Even undeveloped oases provide vital supplies of water to local people, who often travel considerable distances to collect it.

**SILGAU OASIS**
*At this oasis in the Thar Desert of Rajasthan, India, local women fill metal vessels with water and carry this precious resource back to their homes.*

# ANIMAL LIFE

Although a desert may appear lifeless, night often reveals much activity from insects, reptiles, and small mammals in particular. These nocturnal animals avoid excessive water loss by sheltering from the heat of the day underground. Some species gain water from dew at night, or, in the case of the darkling beetle, from fog condensed on its body. Reptiles are perhaps the real desert specialists: they can withstand high temperatures and require little moisture, and when temperatures drop, they become torpid until it warms up again. Snakes "swim" through sand by sinuous wriggling and burrow with ease, while many desert lizards use a tiptoe stance to raise themselves above the scorching surface. Large desert mammals, such as camels, are adapted for survival. Their woolly fur helps to slow down heating, and they can tolerate higher temperatures than other large mammals. They retain water by producing concentrated urine and dry dung, but can also lose large amounts of water without ill effect.

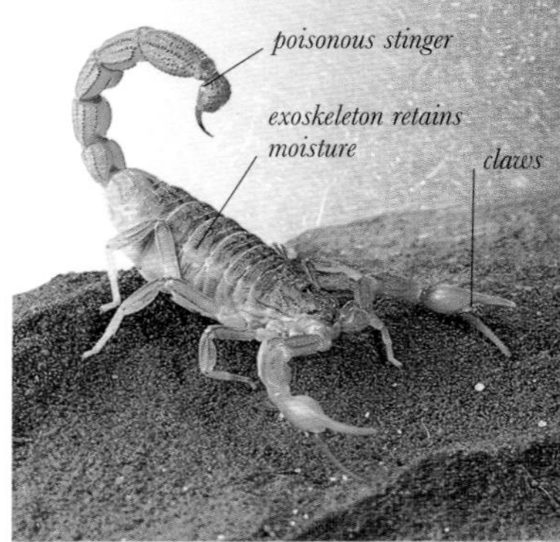


**DESERT SCORPION**
*With a tough exoskeleton to help it retain moisture, this scorpion is well adapted for the dry desert climate.*

**SIDE-WINDING ADDER**
*Almost invisible, this desert-dwelling snake lies buried in the sand, waiting for a suitable victim to approach.*

**DROMEDARY CAMEL**
*Over 90 percent of the world's camels are dromedaries. When dehydrated, this desert specialist can drink 13 gallons (50 liters) of water in a few minutes.*

**NAMAQUA SANDGROUSE**
*Male Namaqua sandgrouse absorb water in their spongelike breast feathers and carry it back to their chicks in the nest.*

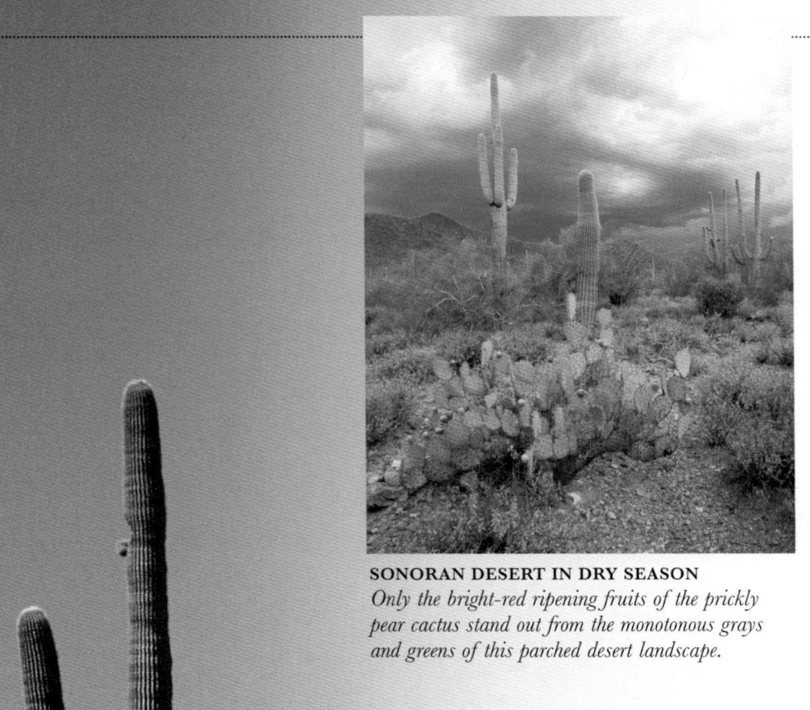

**SONORAN DESERT IN DRY SEASON**
*Only the bright-red ripening fruits of the prickly pear cactus stand out from the monotonous grays and greens of this parched desert landscape.*

## PLANT LIFE

Desert plants are generally of two main types: ephemeral annuals, which survive arid conditions as seeds, and perennials, which continue to survive and cope with periods of water scarcity. The annuals grow, flower, and set seed rapidly when conditions are more suitable, such as after rare rains, and some can complete their life cycle in as little as two weeks. The flowers are often brightly colored and large, perhaps because potential pollinators are scarce. The perennials have various strategies for enduring drought, such as sending down deep roots to tap underground moisture, storing water in swollen tissues, as many succulents and cacti do, or developing underground organs such as bulbs and sending out stems and leaves only when conditions allow. Many have tough, reduced leaves to reduce water loss through transpiration, and spines to resist grazing. Cacti are typical of the American deserts, and, as well as being spiny, many species are waxy or woolly to further reduce water loss. In the African deserts, various species of euphorbia occupy the same ecological niche as cacti—they are also succulent, and often spiny. Cacti and other succulents typically have large, attractive flowers, mainly pollinated by insects, although some, including the saguaro, are pollinated by bats. In Australian deserts, the spiky spinifex grasses can play an important role in stabilizing dunes.

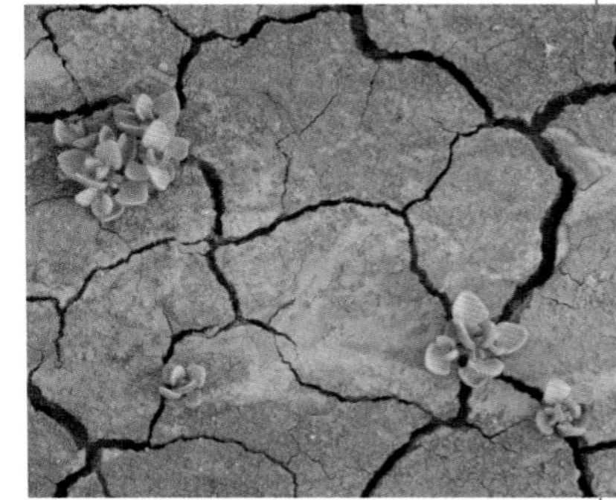

**FOURWING SALTBUSH**
*The fourwing saltbush thrives in alkaline soils, and its leaves even taste salty. The first part of its name comes from the seed, which has four paperlike wings.*

*spine (modified leaf)*

*waxy cuticle retains water*

*water-storing tissue*

*vascular cylinder takes in water*

*roots*

**GOLDEN BARREL CACTUS**
*Native to the southwestern US and Mexico, this cactus sends its roots deep into the desert soil and stores water in its fleshy tissue, as revealed in cross-section. The outer cuticle helps to prevent water loss.*



## DESERT DISTRIBUTION

The largest desert in the northern hemisphere is the Sahara. Farther east, there are extensive deserts on the Arabian Peninsula, the Thar Desert in Pakistan and India, and the deserts of Central Asia, notably the Takla Makan and Gobi. In the US, deserts are found mainly in California, Arizona, Nevada, and New Mexico, and in Mexico they occur in the northwest, especially in the states of Chihuahua and Sonora. The deserts of the southern hemisphere are generally less extensive, occupying a narrow coastal belt in Chile and Peru, a swathe of southern Africa (mainly in Namibia, Botswana, and South Africa), and large patches of mainly central and western Australia.

**SONORAN DESERT IN BLOOM**
*With the rain, the desert is transformed by the sudden flowering of ephemeral annuals, which have to bloom and reproduce in a short period.*

### DESERT PROFILES

The pages that follow contain profiles of the world's main deserts. Each profile begins with the following summary information:

**TYPE** Sandy, gravelly, stony, or rocky

**AREA** Surface area **RAINFALL** Average annual rainfall

**TEMPERATURE** Summer maximum and winter minimum temperatures or average annual temperature

NORTH AMERICA *west*

# Great Basin Desert

**LOCATION** In the states of Oregon, Idaho, Nevada, Utah, Wyoming, Colorado, and California

**TYPE** Sandy, gravelly

**AREA** 158,000 square miles (409,000 square km)

**RAINFALL** 10 in (250 mm)

**TEMPERATURE** Max: 100°F (38°C) Min: –19°F (–7°C)

The Great Basin is the largest desert in the US, lying mainly in Nevada and Utah, with minor extensions into neighboring states. Sandwiched between the Sierra Nevada range to the west and the rest of the Rocky Mountains to the east, it lies to the north of the Mojave Desert. It is classed as a cold desert, due mainly to its northerly position, but also because of its high altitude, ranging between 3,000 ft and 6,500 ft (900 m–1,980 m) above sea level, although most of it is at 4,000 ft (1,200 m). The surface consists of broad valleys on a plateau, with remnants of earlier lakes. The winters are cold, but the summers are hot and mostly dry. Because it lies in the rain shadow of the Sierra Nevada, the Great Basin receives little rain, and about 60 percent of the precipitation falls as winter snow. The gradual melting of the snow in spring seeps into the mostly fine-textured desert soils and only then is available to the plants. The Rocky Mountains also prevent most of the continental weather fronts from reaching the Great Basin, although there are occasional summer storms. Playas (temporary lakes) are a feature of the landscape, and their salty soils support vegetation dominated mainly by saltbush. Great Basin vegetation tends to be low, with shrubs growing to about 3 ft (1 m) tall. Typical species are sagebrush, blackbrush, and shadscale, but there are few cacti. Among these plants live animals including jackrabbits, the Great Basin rattlesnake, and the desert horned lizard. In some areas the concentration of salts of sodium and calcium is so high that no plants can survive.

**HARDY PLANTS**
*On the low hills of eastern Nevada, the dominant vegetation is the state flower, sagebrush. This woody shrub has narrow, gray-green leaves and small yellow blooms.*

*prominent eyes with vertical pupils*

*hind feet used to burrow 3 ft (1 m) or more into sandy soil*

**COUCH'S SPADEFOOT TOAD**
*This toad can survive a drought by lying dormant underground, encased within a watertight cocoon made of shed skin.*

NORTH AMERICA *southwest*

# Mojave and Sonoran deserts

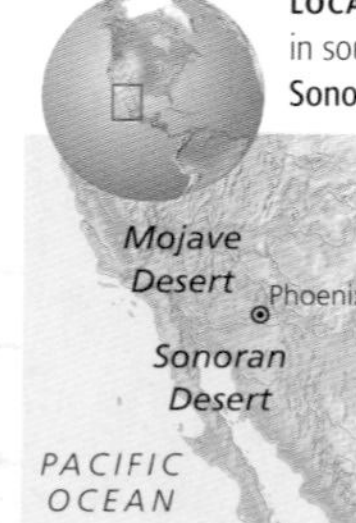


**LOCATION** Mojave: predominantly in southern California Sonoran: mostly in Arizona

**TYPE** Sandy, rocky

**AREA** Mojave 54,000 square miles (140,000 square km), Sonoran 108,000 square miles (275,000 square km)

**RAINFALL** Mojave 2–5 in (50–125 mm), Sonoran 10 in (250 mm)

**TEMPERATURE** Mojave Max: 119°F (48°C) Min: 8°F (–13°C), Sonoran Max: 119°F (48°C) Min: 119°F (–13°C)

The Mojave lies in southern California, southeast of the Sierra Nevada range. This is a high desert, lying mostly at more than 2,000 ft (600 m) above sea level. Death Valley in the north is an exception—at 282 ft (86 m) below sea level, it is the lowest place in the US, and it is also the hottest, reaching 134°F (57°C) in 1913. The plant life of the northern Mojave resembles that of the Great Basin, with a range of drought-tolerant shrubs (the extremely arid Death Valley has very little vegetation). The Sonoran Desert lies to the south of the Mojave, at lower altitude. This is a hot desert, but it has winter rains that stimulate the growth and flowering of colorful annual plants. Cacti are also a prominent feature, and include species that grow to 40 ft (12 m) high. Trees with nitrogen-fixing bacteria in their roots make an important contribution to the fertility of the soil. Fauna of the region include the desert tortoise, desert scorpion, and a large lizard called the chuckwalla.

**JOSHUA TREE**
*Something of a symbol of the Mojave Desert, this large member of the yucca family thrives in cooler sites on the higher ground.*

**DEATH VALLEY SALT FLATS**
*These strange polygons were created when high summer temperatures cracked the salt-flat crust, allowing the water trapped below to evaporate and precipitate salt crystals.*

**VOLCANIC CRATER**
*Volcanic activity toward the end of the Cretaceous Period, about 70 million years ago, created this huge caldera (volcanic crater) in the Sonoran Desert.*

## NUCLEAR STORE

Yucca Mountain, which lies within an active volcanic field in Nevada, has been earmarked as a suitable site for the storage of spent nuclear fuel and waste. Having conducted a series of scientific studies over two decades at a cost over $4 billion, experts concluded that it will be safe to store the materials, which will remain highly radioactive for thousands of years, sealed in canisters in tunnels about 650–1,000 ft (200–300 m) below the surface. However, other interest groups remain concerned about the area's tectonic stability.

**AT THE END OF THE TUNNEL**
*A specially adapted machine emerges after boring an 5-mile- (8-km-) long investigative tunnel underneath Yucca Mountain.*

NORTH AMERICA *south*

# Chihuahua Desert

**LOCATION** Between the Sierra Madre mountain ranges, Mexico, extending north into the US

**TYPE** Sandy, stony

**AREA** 200,000 square miles (518,000 square km)

**RAINFALL** 10 in (250 mm)

**TEMPERATURE** Max: 104°F (40°C) Min: –22°F (–30°C)

Named after the Mexican province near its center, the Chihuahua is a high-altitude desert, lying mostly at 3,240–4,860 ft (1,000–1,500 m). This is the largest desert in North America, sandwiched between the Sierra Madre ranges of Mexico, and extending north into New Mexico, Texas, and Arizona. Winters here are cool, with frosts occurring at night, but summers are very hot, and rain falls mainly during the summer, when it is least effective. In the lower sites, the soils are quite porous, with a calcareous base overlain by sand and gravel, and here creosote bush and tarbush cover considerable areas.

**WESTERN DIAMONDBACK RATTLESNAKE**
*This highly venomous viper locates warm-blooded prey with the two heat-sensitive pits above its nostrils.*

On higher ground, such as the smooth-sided slopes known as bajadas, the thick-leaved, frost-hardy yuccas dominate. The bajadas have a rough surface of eroded rocks, boulders, and stones, and they also trap pockets of soil, in which plants are able to thrive. Higher still, where more water is available, the vegetation undergoes a transition through desert grassland to a high-altitude scrub. These varied habitats allow the Chihuahua to support a wide range of animal life, from tarantulas to bats.

**TULAROSA BASIN**
*The shining white gypsum sand dunes of this beautiful valley at the northern end of the Chihuahua Desert cover about 275 square miles (712 square km).*

SOUTH AMERICA *south*

# Patagonia Desert

**LOCATION** East of the Andes in the southern provinces of Chubut and Santa Cruz, Argentina

**TYPE** Gravelly

**AREA** 260,000 square miles (670,000 square km)

**RAINFALL** 4–10¼ in (100–260 mm)

**TEMPERATURE** Average: 41–55°F (5–13°C)

This cold semidesert lies in the rain shadow of the Andes, and consists mainly of a series of dissected plateaus. Precipitation falls mostly in the long, bleak winters, with frosts likely any time between June and September. The region is subject to strong, steady, westerly winds, which severely restrict the vegetation and erode the soil.

**BLEACHED BONES IN THE DESERT**

SOUTH AMERICA *west*

# Atacama Desert

**LOCATION** Along the coast of northern Chile, west of the Andes, between Arica and Vallenar

**TYPE** Rocky, salty

**AREA** 40,600 square miles (105,200 square km)

**RAINFALL** Less than ½ in (15 mm)

**TEMPERATURE** Max: 95°F (35°C) Min: 25°F (–4°C)

The Atacama is the driest of all deserts, with some places having recorded no rain at all in living memory. It occupies a narrow coastal strip in northern Chile, averaging less than 100 miles (160 km) in width, but extending about 600 miles (960 km) south from the border with Peru, forming a huge strip of arid, hostile beach. Coastal fogs form in a few areas, providing moisture that allows the development of some plant communities, including cacti. On the eastern side, however, where the foothills of the Andes rise up, there is a small amount of rainfall in winter. Large deposits of salt are a feature of several parts of the Atacama, often in large salt basins, known as salars. When the salt flats flood after a storm, they can support specialist invertebrates like brine shrimp, whose eggs can survive several years of drought, and they are also home to large numbers of birds, including three species of flamingo. The Atacama has significant deposits of rare and valuable minerals, notably saltpeter (nitrates of potassium and sodium), which were formed in chemical reactions between the salty waters and volcanic debris. These are mined and extracted at a number of sites. The desert's dry climate, in which very few decomposing bacteria can exist, has preserved many features of archaeological interest, including Incan artifacts (500 years ago, this was the southern limit of their empire) and mummified remains from the Paleo-Indian civilization, thousands of years older.

**SATELLITE VIEW**
*Snow-capped volcanoes tower above the Atacama Desert in northern Chile. This is the driest place on Earth, as the dry lake beds at top right illustrate. On the left, the Atacama Fault runs from north to south.*

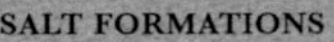

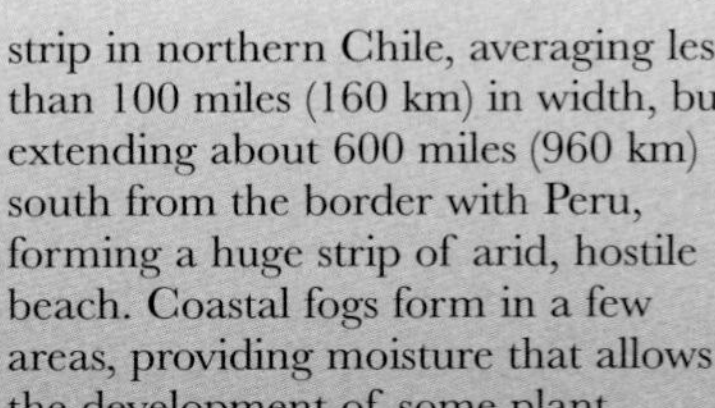

**SALT FORMATIONS**
*Many areas of the Atacama Desert are dominated by deposits of various salts, which sometimes accumulate into strange crystalline shapes.*

## COPPER MINING

Many deserts contain valuable deposits of minerals, including precious metals. Chuquicamata, the world's largest copper mine in terms of size and production, lies in the Atacama, near Calama in northern Chile. It was first mined more than 500 years ago. Blasting of terraces for most of the last 100 years has produced a huge pit, currently 1¼ by 1¾ miles (2 by 3 km) across and 2,624 ft (810 m) deep.

**CHUQUICAMATA MINE**
*As much as 670,000 tons (600,000 metric tons) of ore and waste are mined each day and removed by a fleet of over 300 trucks.*

**VALLEY OF THE MOON**
*The Valle de Luna in northern Chile is a remarkable spectacle. Here, salt, sand, and varied rock formations are set against the dramatic backdrop of the Andean peaks.*

**THE EMPTY QUARTER**
*A vast area of sand dunes lies in southern Saudi Arabia. This is the famous Ar Rub 'al Khali, or Empty Quarter, whose shifting sands buried the ancient city of Ubar.*



ASIA *west*

# Arabian Peninsula

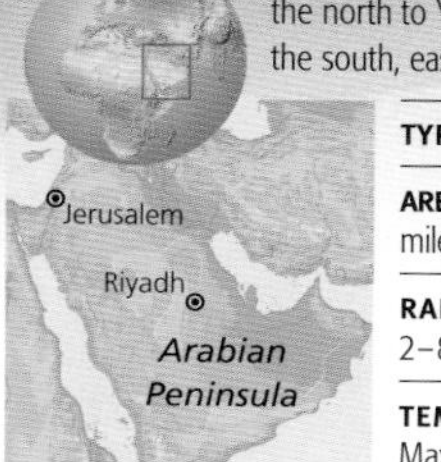


**LOCATION** Stretching from Syria in the north to Yemen and Oman in the south, east of the Red Sea

**TYPE** Sandy, gravelly

**AREA** 900,000 square miles (2.3 million square km)

**RAINFALL**
2–8 in (50–200 mm)

**TEMPERATURE**
Max: 120°F (49°C)
Min: 32°F (0°C)

In many ways, the desert of the Arabian Peninsula is an extension of the Sahara (see pp.292–93), separated from it only by the Red Sea. The desert is very arid, with less than 4 in (100 mm) of rainfall in most years, and has very hot summers and mild winters. It shares many features with the Sahara, such as large expanses of monotonous sand seas, which are difficult to traverse. Indeed, there are no permanent roads in this desert, making it impossible to drive over. The most famous expanse of sand is the Ar Rub 'al Khali, or Empty Quarter, in the south, which covers an area the size of France. Mountain ranges run along much of the Red Sea coast and also near the Gulf coast in Oman. Some gravel plains are found in the foothills, such as in the Hajar Mountains in the east. These ranges comprise some of the finest exposed rocks of the oceanic crust, and they contain fossils up to 300 million years old. The gravel plains often have a sparse cover of scrub, with species such as ghaf, a leguminous tree that is an important source of fodder and firewood. In many places, the sands lie above oil-rich sedimentary rocks on which the famous Gulf oil fields have been developed. Parts of the landscape bear the evidence of former river systems from when the area was dominated by subtropical savanna, and fossil hippos, elephants, and turtles have been recovered, notably in the Baynunah region in the United Arab Emirates. Other evidence of earlier, wetter periods is provided by fossils of creatures such as crocodiles and freshwater mollusks, which have been extracted from playa lake sediments. Some parts of the desert have extensive salt flats, known locally as sabkha, and here the surface has a hard, crystalline crust. This is often impervious and floods easily when the rare rains occur, creating temporary salty lakes. The Arabian Peninsula is used as a flight path by birds migrating between Asia and Africa, and the coastal lagoons, desert pools, and wadis provide vital refreshment.

**JERBOA**
*Exclusively nocturnal, the jerboa emerges from its burrow at night to feed on seeds, roots, and leaves. Its long hind legs enable it to hop rapidly over the desert sand—despite growing to only about 4 in (10 cm) long, it can leap as far as 10 ft (3 m).*

## WILFRED THESIGER

Born in Addis Ababa, Ethiopia, in 1910, Wilfred Thesiger studied in England before returning to Africa to work in Sudan. An intrepid explorer, he was one of the first Europeans to visit the Ar Rub 'al Khali, where between 1945 and 1950 he lived among the local nomadic people, the Bedu. He described his experiences in the desert in his first book, *Arabian Sands*.

**ROCK AND SAND**
*Remnants of a former landscape, eroded volcanic rocks rise above the shifting sands near the Ahaggar Massif in southeast Algeria, at the heart of the Sahara Desert.*

AFRICA *north*

## Sahara Desert

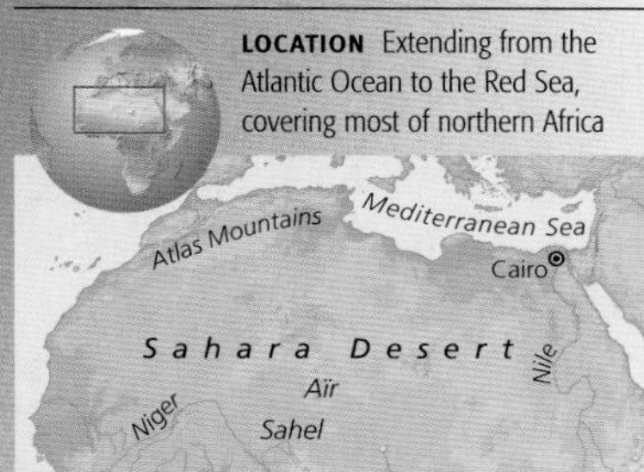


**LOCATION** Extending from the Atlantic Ocean to the Red Sea, covering most of northern Africa

**TYPE** Sandy, gravelly, stony

**AREA** 3.5 million square miles (9 million square km)

**RAINFALL** ¾–16 in (20–400 mm)

**TEMPERATURE** Average: 61–97°F (16–37°C)

The Sahara is the largest of all the deserts, and covers an area about the size of the US. Although it contains many distinct geographical features, such as gravel surfaces (regs) and rocky floors (hammadas), the most famous features of the Sahara are the large ergs, or sand seas. Saharan sand tends to have a reddish color, and the average thickness of the sand in the ergs ranges from 65 ft to over 325 ft (20–100 m). The ergs consist of dunes of a wide range of different types. Some are true giants, and these megadunes may reach 1,000 ft (300 m) in height, with a crescentic, pyramidal, or linear shape. The Sahara Desert is a landscape molded in large part by the wind. It creates the dunes and moves and shapes the sand, but it is also responsible for sculpting yardangs in the hammada areas, notably in Egypt west of the Nile, and also in Libya and Chad. The overall net movement of the sands is more or less clockwise through the Sahara, from west to east along the Mediterranean side and the reverse in the Sahel region (the boundary between desert and savanna) to the south, where it is driven mainly by the harmattan (northeasterly) winds. In the eastern, central, and southern Sahara, there is an overall loss of sand from the ergs toward the Sahel, where the sand is generally accumulating and the desert extending, partly due to overgrazing. There are regular patterns to the various prevailing winds, which can be strong and persist for days, carrying sand, soil, and dust and obscuring the landscape for miles around. Although the shifting sands have very little vegetation, oases, wadis, and other drainage channels support tough shrubs and palm trees. Several of the Sahara's oases are found in moats, which are depressions, about 130–195 ft (40–60 m) deep, around inselbergs, and are found in the Libyan Desert and elsewhere. Although there are low regions such as the Qattara Depression in northern Egypt, which is below sea level, most of the Sahara occupies a tableland with an average altitude of 1,300–1,600 ft (400–500 m), and there are clusters of high ground in the center, such as the massifs of Aïr, Ahaggar, and Tibesti. Rainfall is scant and intermittent, and some regions may see no rain at all for years at a time. Storms do develop from time to time, but these are localized, and when they do occur, erosion can be rapid. Animal species endemic to the Sahara show various adaptations to the intense heat. They include the desert horned viper, which has strong, keeled scales that enable it to slip below the sand, where it lies protected from the heat of the day. At night, it waits in ambush for lizards, birds, and rodents. Another nocturnal desert specialist is the fennec fox, whose furry soles enable it to walk on soft, hot sand.

**PLANT LIFE**
*Spiny shrubs dominate the vegetation in the rocky desert on the fringe of the Sahara in Morocco. Their thorns protect them from grazing animals.*

**CAMEL CARAVAN**
*Dromedary camels have long been domesticated in the Sahara. They eat a wide variety of plants and, given the chance, will also feed on bones.*

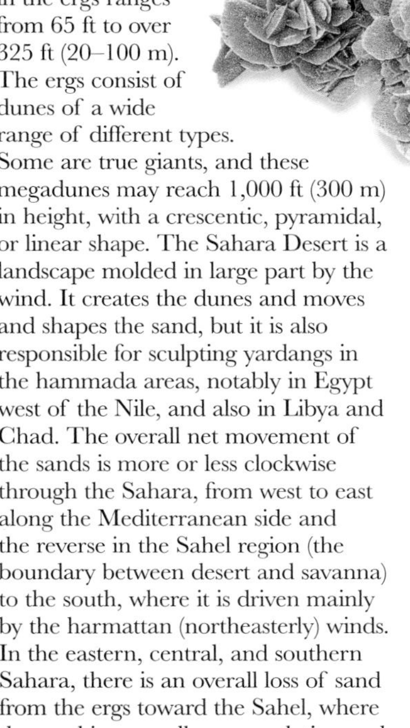

**DESERT ROSE**
*The mineral gypsum is often found in sand dunes, deposited by the evaporation of mineral-rich groundwater. The growth of the gypsum crystals is affected by the grains of sand, resulting in petal-like shapes clustered in "flowers."*

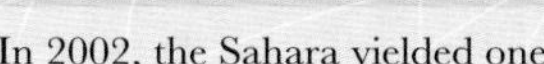

### APE OR HOMINID?

In 2002, the Sahara yielded one of its most intriguing secrets: a fossil skull found in the Djurab Desert in Chad, which some scientists are claiming to be the oldest known humanlike remains. The skull, thought to be between 6 and 7 million-years old, combines both ape and early human (hominid) features, such as a chimplike brain case and a prominent brow ridge. The new species has been named *Sahelanthropus tchadensis*, after the Sahel region of the southern Sahara where the fossil was discovered.

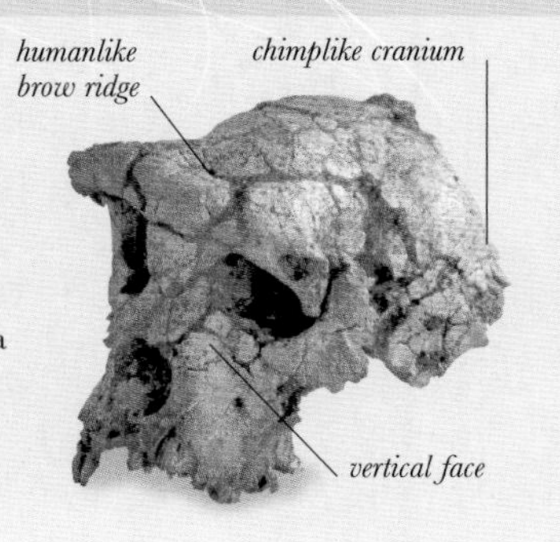

### NOMADS OF THE DESERT

The Sahara is inhabited by diverse groups, known as Tuareg, who share a common language and culture. The men wear a veil, or tagilmust, that covers almost their entire face. This is for spiritual reasons, but it also protects them from the desert sand and winds. The women wear a smaller veil, and the Tuareg refer to themselves as Kel Tagilmust, or "people of the veil." Traditionally nomadic or semi-nomadic herders of goats, sheep, cattle, and camels, for centuries the Tuareg also provided guides for the camel caravans of the trans-Saharan traders. The advent of roads and mechanized transportation means that this source of external income has largely been lost, and today many Tuareg have been forced to settle in towns.

**TUAREG OF THE SAHARA**
*The flowing blue dress of the Tuareg men absorbs more heat than white, but this creates a cooling breeze through the robes.*

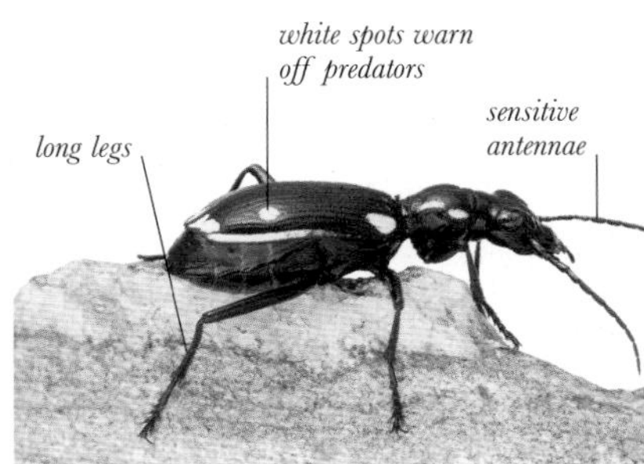


**DOMINO BEETLE**
*This ground beetle has unusually long legs, which raise its body clear of the hot desert sands when it is out hunting insects for food.*

# DESERTIFICATION

Geological evidence—such as fossilized dunes and lakebeds—shows that deserts change in the course of time. They shrink when the climate becomes more moist, and expand when it turns dry. Desertification, on the other hand, is the spread of desertlike conditions through the mismanagement of land. More than a quarter of the Earth's land surface is currently affected by this phenomenon, or at risk from it.

## LIVING WITH DROUGHT

In many parts of the world, rainfall follows a predictable pattern, with only minor variations from year to year. In arid regions, the climate is much more unreliable: occasionally, heavy rain falls in sporadic storms, but years can then go by with little or no rain at all. Wild plants and animals are well adapted to these conditions, but for humans, they are a major problem.

Where rainfall is scant and haphazard, shifting agriculture is a widespread way of life. By staying on the move, nomadic herders (see p.348) can react to changing conditions, leading their animals to places where food can be found. In regions such as the Sahel—the belt of dry land south of the Sahara—this form of shifting exploitation has continued for several thousand years, but in recent decades several factors have put extra demands on arid land worldwide. One of these is a rise in the human population. With more people to feed, animal stocks tend to rise until they strip away vegetation faster than it grows. This overexploitation leads to soil erosion, and at that point the desertification process sets in. Left unchecked, it can cause permanent changes in the landscape, and the original vegetation is unable to recover. Desertified landscapes sometimes resemble natural desert, but the remnants of soil—and the lack of balanced ecology—often reveal its recent origins.

The Sahel is particularly vulnerable to desertification, and so is much of southern Africa. But this is not solely an African problem, nor is it confined to the developing world. It currently threatens the southern US, from California to western Texas, the southern fringes of Europe, and a wide expanse of south and Central Asia.

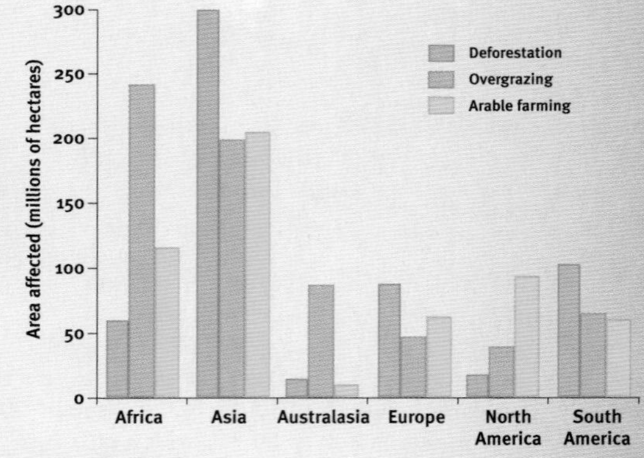


**GROWING LOSSES**
*Desertification affects all the world's inhabited continents, but typically areas near existing deserts. In Africa, the main cause is overgrazing, while in Asia, deforestation occupies first place.*

## CAUSES AND EFFECTS

Like global warming (see p.452), desertification has been the subject of intense debate among scientists. Most accept that overexploitation is a contributory factor, but opinions differ about whether it is the sole one. Some researchers think that the main trigger is local climate change, while others believe that human factors are largely to blame. The evidence is not easy to assess, because—just as with global warming—long-term trends can easily be masked by short-term change. Despite these differences, there is no disagreement about the effects: desertification can put land permanently out of action for farming.

According to the United Nations, more than 250 million people are directly affected by desertification, and 1 billion more are potentially at risk. As well as causing food shortages, desertification gives rise to other health problems, and it can displace entire populations, turning people into environmental refugees.

## COMBATING DESERTIFICATION

Despite headlines about "rolling back deserts," naturally expanding deserts cannot be stopped. Desertification, however, can be slowed and even halted by adopting measures designed to protect plant cover. Reducing livestock and planting drought-resistant trees and shrubs helps to hold the soil in place, and, in regions where wood is used for fuel, provides an alternative to trees that grow in the wild. Paradoxically, improving the water supply does not necessarily relieve the problem. Extra wells can allow livestock numbers to increase beyond the land's carrying capacity, while excessive irrigation may increase salinity of the soil. In the Fertile Crescent, from Lebanon through to Iran, where agriculture began in the Middle East, irrigation in early recorded history damaged arid land so much that it remains unproductive today.

**STORM WITHOUT RAIN**
*Caught by a sudden dust storm, a group of women huddles together in Rajasthan, northeast India. Desertification is a major threat in this part of Asia.*

**RAINFALL IN THE SAHEL**
*Records from 1930 onward show that rainfall in the Sahel is extremely variable. Since 1967, the annual departure from the long-term average has switched from being largely positive to almost entirely negative. This trend toward drier times has aggravated desertification.*

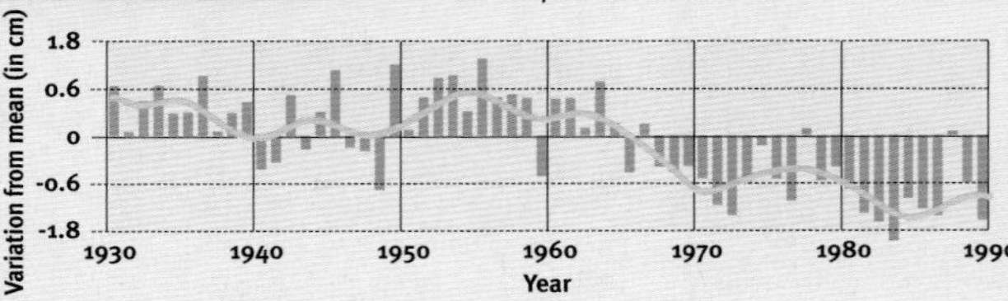

**DESERT FRONTIER**
*A giant sand dune threatens to engulf a palm plantation in Libya, on the edge of the Sahara Desert. Since the end of the last ice age, the Sahara has advanced and receded several times.*

AFRICA *south*

## Kalahari Desert

**LOCATION** Centered on southern Botswana, extending west into Namibia and south into South Africa

**TYPE** Sandy

**AREA** 275,000 square miles (712,250 square km)

**RAINFALL** 5–20 in (125–500 mm)

**TEMPERATURE** Max: 117°F (47°C) Min: 10°F (–13°C)

Ranging in altitude from about 3,000 ft (900 m) to 4,000 ft (1,200 m), the Kalahari Desert lies between the Orange River in the south and the Zambezi River in the north. Dominated by sandy ridges, the desert is also studded with lakebeds and salt pans, which are usually dry. The soil is mostly fine-grained sand—bright red in some areas and gray in others—and the dune systems are largely fixed and inactive. Parts of the Kalahari, notably the southwest, are known for their extensive linear dunes, many of which are partly vegetated. The vegetation is characterized by thorny scrub, with acacia trees and spiny shrubs, and the southern Kalahari, in particular, has a parklike landscape. Large herds of game, such as springbok, gemsbok, and wildebeest, roam the Kalahari, moving long distances in search of occasional flushes of fresh vegetation. One of the most common small mammals is the meerkat, a highly social mongoose. Expert diggers, meerkats are well adapted to the desert, creating elaborate communal burrows in the sand. The salt pans occasionally flood after heavy rains, attracting large numbers of birds, including both the greater and the lesser flamingo. The Kalahari is also home to the nomadic San people, about 40,000 in number.

**AFRICAN WHITE-BACKED VULTURE**
*This impressive scavenger is a common sight in the Kalahari, either soaring overhead on its large, broad wings or feeding on a carcass.*

**SANDSTORM IN THE KALAHARI**
*Thorn trees provide the gemsbok (center) and springbok with some shelter from sandstorms, which occur frequently in dry, sandy areas.*

AFRICA *southwest*

## Namib Desert

**LOCATION** On the Atlantic coast of Namibia, extending into southern Angola in the north

**TYPE** Gravelly, sandy

**AREA** 81,000 square miles (31,000 square km)

**RAINFALL** ½–4 in (15–100 mm)

**TEMPERATURE** Max: 77°F (25°C) Min: 50°F (10°C)

A long, narrow strip mostly less than 100 miles (160 km) wide, the Namib stretches some 800 miles (1,300 km) along the coast of southwest Africa. This is a dramatic coastline, with high dunes bordering the sea in some places. It receives very little direct precipitation—a mere ½ in (15 mm) a year near the coast—but coastal fogs provide much more moisture in the form of condensation. The dune systems are mostly active, with star dunes accounting for 10 percent of all their number. Another unusual geological feature is the presence of rounded rock yardangs. Further inland, the Namib rises across a plain to an escarpment toward the east. The flora includes a large number of lichens (some brightly colored), which are able to absorb moisture from the humid air when coastal fogs roll inland. A surprising number of animals survive in this hostile environment, ranging from small rodents to African elephants.

**GRANT'S GOLDEN MOLE**
*This unusual burrowing insectivore is found only in the Namib Desert and nearby areas. Covered in long, silky fur, it feeds on insects, such as termites and beetles, and larger prey like geckos, as above.*

*leaf shredded by wind*

*small, salmon-pink cones indicate that this is a male plant*

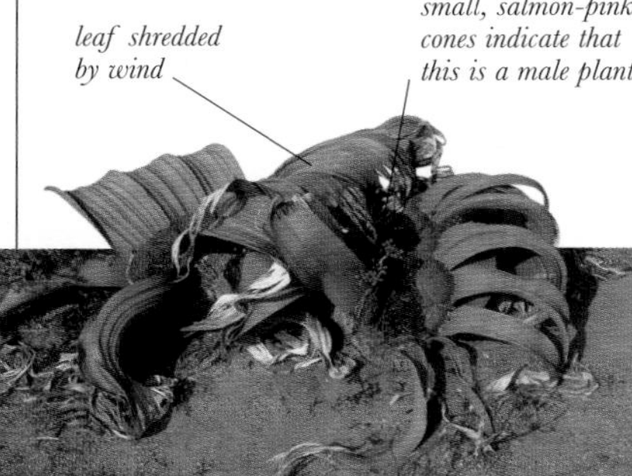

**WELWITSCHIA MIRABILIS**
*Unique to the northern Namib Desert, this plant produces just two leaves during its entire life span of several hundred years.*

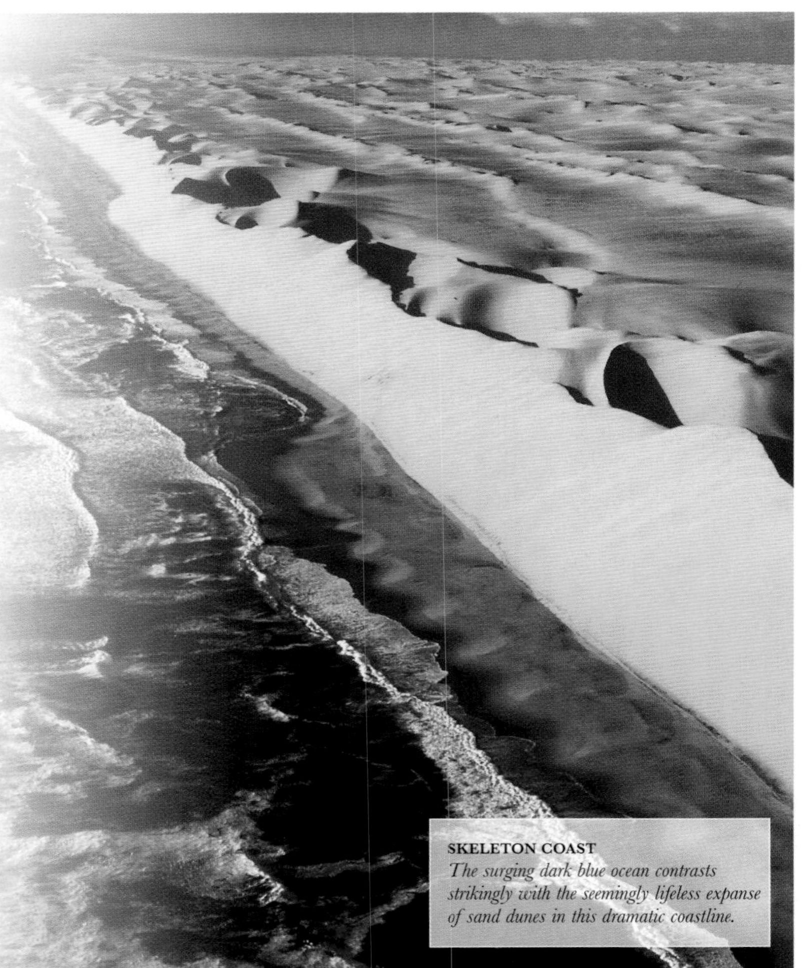

**SKELETON COAST**
*The surging dark blue ocean contrasts strikingly with the seemingly lifeless expanse of sand dunes in this dramatic coastline.*

ASIA *southwest*

## Thar Desert

**LOCATION** In the province of Rajasthan in northwest India, extending into southeast Pakistan

**TYPE** Rocky, sandy

**AREA** 172,200 square miles (446,000 square km)

**RAINFALL** 4–20 in (100–500 mm)

**TEMPERATURE** Max: 108°F (42°C) Min: 37°F (3°C)

The Thar (or Great Indian) Desert is centered on the province of Rajasthan, extending eastward to the Aravali Hills. The landscape is mixed, with areas of sandy or gravel plains, broken rocks, and dune systems. Although most of the dunes are fixed and inactive, there are some regions of shifting sands. Parabolic dunes are a special feature of this desert. U-shaped, with fronts 33–230 ft (10–70 m) high and trailing, partly vegetated, parallel arms up to 1¼ miles (2 km) long, such dunes are unusual inland and are caused by unusually consistent prevailing winds. There is a marked absence of oases, but some areas, especially in the west, have been irrigated by the Sutlej River. Vegetation is generally sparse, and tends to be dominated by tough grasses and shrubs or small trees, such as tamarisks and acacias. The Thar is one of the last haunts of a highly endangered bird, the Indian bustard.

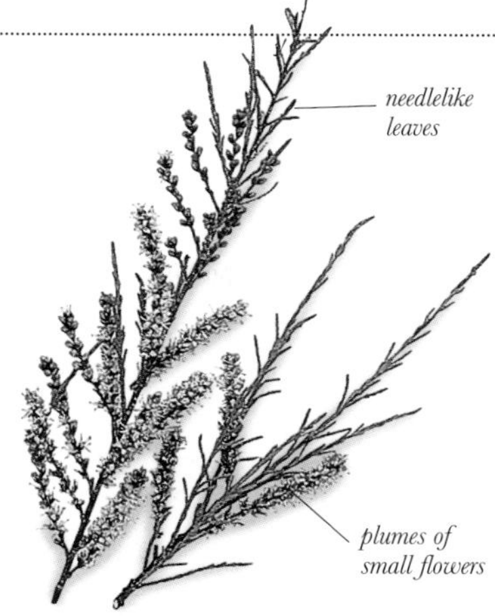

**TAMARISK**
*This heathlike small shrub is widely distributed in the Thar and many other desert regions in Asia. Its small, scalelike leaves help it conserve water.*

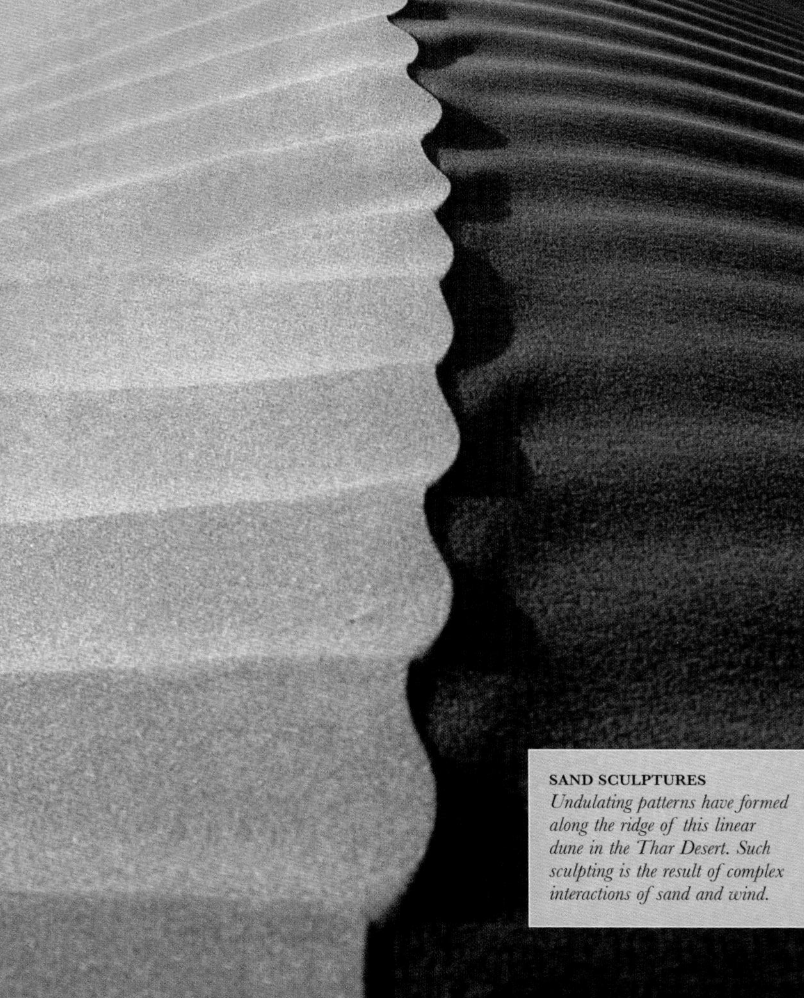

**SAND SCULPTURES**
*Undulating patterns have formed along the ridge of this linear dune in the Thar Desert. Such sculpting is the result of complex interactions of sand and wind.*

ASIA *west*

## Kara Kum Desert

**LOCATION** Occupying most of Turkmenistan, east of the Caspian Sea, south of the Aral Sea

**TYPE** Sandy

**AREA** 115,000 square miles (297,900 square km)

**RAINFALL** 4–8 in (100–200 mm)

**TEMPERATURE** Average: 7–90°F (–14–32°C)

Kara Kum means "black sands" in the language of Turkmenistan. It lies to the east of the southern end of the Caspian Sea, and west of the Amu Darya River. The landscape here is of cracked clay surfaces and crescentic dune systems. There is normally little or no rain between May and October, and dust storms are a frequent occurrence. Crops, such as cotton, are grown in oases and in areas irrigated by the Kara Kum Canal, which takes water about 500 miles (800 km) from the Amu Darya River west across the desert to Ashgabat. Black saxaul trees are part of the dominant vegetation here, their very narrow leaves helping to reduce water loss by transpiration. The sand boa snake is also especially adapted to life in the desert: it has a flat head with eyes that face upward, enabling it to keep watch for prey even when submerged in the sand.

**DESERT GRAZING**
*The edge of the Kara Kum Desert has enough grass in spring to support grazing animals, although the grass quickly becomes parched.*

ASIA *west*

## Takla Makan Desert

**LOCATION** North of the Kunlun Mountains in Xinjiang Province in western China

**TYPE** Sandy, gravelly

**SIZE** 126,250 square miles (327,000 square km)

**RAINFALL** 1½–4 in (40–100 mm)

**TEMPERATURE** Max: 62°F (38°C) Min: 16°F (–9°C)

Surrounded on three sides by snow-capped mountains, the Takla Makan Desert lies at a relatively low altitude, between 1,600 ft and 6,600 ft (500 m and 2,000 m). It is drained by the Tarim river system, which flows from west to east before finally drying up in the Lop lowland, where there are extensive deposits of gypsum and salty sand. The Tarim Basin, in particular, features yardang fields. Numerous rivers enter the desert, fed by snowmelt from the mountains. These rivers are fringed with reeds, tamarisks, and even groves of poplars. However, away from the rivers, the plant life is sparse, with drought-resistant shrubs, such as saxaul, helping to stabilize the dunes.

**DESERT MOUNTAINS**
*An inhospitable mixture of stony plains and areas of shifting sand dunes, this desert is aptly named: Takla Makan means "place of no return."*

ASIA *north*

# Gobi Desert

**LOCATION** Stretching across southern Mongolia and northern China as far as the Great Wall

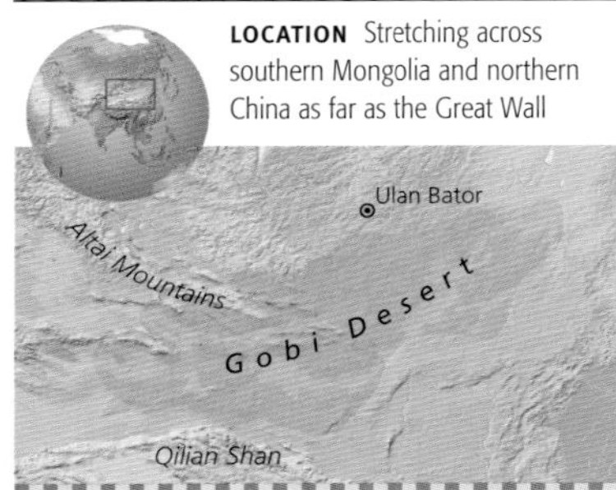


**TYPE** Stony, gravelly, sandy

**AREA** 500,000 square miles (1.3 million square km)

**RAINFALL** $^3/_8$–10 in (10–250 mm)

**TEMPERATURE** Max: 113°F (45°C) Min: –40°F (–40°C)

The Gobi is the largest desert in Asia, and gets its name from the Mongolian word for desert. To the north it grades into steppe, and stretches to the Altai Mountains in the west, while its southern limit is set by ranges such as the Qilian Shan, forming part of the northern rim of the Plateau of Tibet. The landscape ranges from rocky massifs to wide valleys and plains, but very little of the Gobi is covered by sand, unlike the very sandy Takla Makan Desert, which adjoins it to the west. The central part of the Gobi Desert is covered by stony pavement, with only very sparse vegetation, and in the west it is extremely arid, with no surface groundwater. Most rain falls during the summer, and the annual rainfall gradually diminishes across the region from about 10 in (250 mm) in the east to a mere $^3/_8$ in (10 mm) in the west. The flora of the Gobi lacks the spring ephemeral annuals of many other deserts, probably because winter and spring here are very dry. One of the most ecologically important plants of the Gobi is saxaul, a hardy, almost leafless woody shrub which is resistant to drought and grows as scattered clumps in the desert. Saxaul grows to $6^1/_2$–13 ft (2–4 m) in height, and it can survive even where the soil is sandy and unstable, helping to prevent further erosion. The shrub forms groves or woods on some of the adjacent mountain slopes, although in many places it is threatened by being overused for firewood. The sandy areas of the Gobi are home to the long-eared desert hedgehog and small rodents such as the dwarf or desert hamster. Several large species of mammals are also found in the Gobi. The last remaining wild Bactrian camels roam the desert, mainly protected in a reserve in the Southern Altai area (above right), and wild asses still live in the Gobi (below left), as do small herds of blacktailed and Mongolian gazelles. The world's only desert-dwelling bear, the mazaalai (or Gobi bear) is also found in the Altai region, but it is critically endangered, numbering only 20 to 30 individuals. Reptiles endemic to the desert include the Gobi gecko and the Tatar sand boa.

**BACTRIAN CAMELS**
*Bactrian (or two-humped) camels are native to the deserts of Central Asia and have long been domesticated, although about 300 still roam wild in a protected area of the Gobi Desert.*

**GRAVEL DESERT**
*Gravel covers large areas of the Gobi Desert. Most of the central regions, such as this area in China, are so arid that vegetation is extremely sparse and grazing animals must travel long distances to find adequate supplies of food.*

**ASIATIC WILD ASS**
*The khulan is a subspecies of Asiatic wild ass, which is the most horselike of asses. It lives in the desert and desert steppes of Mongolia, particularly in the Gobi, in herds of up to 500 animals.*

## LIVING WITH THE GOBI

Mongols are the indigenous semi-nomadic people of the Gobi Desert region. Mongol tribes once herded their grazing animals (sheep, goats, cattle, and camels) over a wide area of the desert and steppe country, moving camp several times a year to avoid using up the desert's limited resources. They depended almost entirely on these animals, which yielded milk, sour cream, and yogurt, as well as meat, skins, and wool. The Mongols also domesticated wild horses, using them to round up their flocks. The herdsmen and their families live in a type of lightweight, easily transportable tent, called a *ger*, which has a wooden frame and traditionally is covered in animal skins and woolen felt.

**BEASTS OF BURDEN**
*Camels are the favored means of transport for the Mongols, and they are also sheared for their thick wool.*

**GOBI GURVANSAIKHAN**
*The Gobi Gurvansaikhan ("Three Beauties") National Park protects a large area of desert in southern Mongolia, seen here with the eastern Altai Mountains in the background.*

AUSTRALASIA *Australia*

## Great Sandy Desert

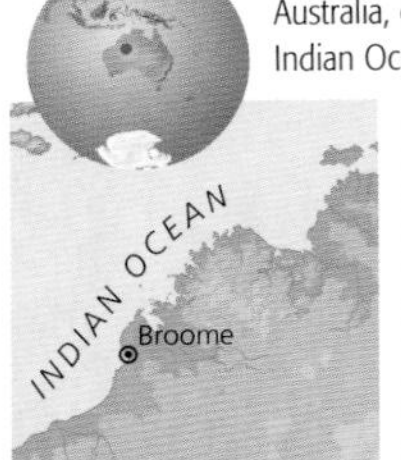


**LOCATION** In the north of Western Australia, extending as far as the Indian Ocean in the northwest

**TYPE** Gravelly, sandy

**SIZE** 130,000 square miles (340,000 square km)

**RAINFALL** 10–12 in (250–300 mm)

**TEMPERATURE** Max: 108°F (42°C) Min: 77°F (25°C)

More than 40 percent of Australia's land is classed as desert, and the Great Sandy Desert is one of the largest, occupying a vast, almost uninhabited area toward the north of Western Australia, southwest of the Kimberley Plateau. This is the only Australian desert to extend to the coast, forming the long sweep of Eighty Mile Beach where it meets the Indian Ocean. The sands here, as in most of Australia's deserts, are bright red, due to a coating of iron oxides on the sand grains. Rainfall is relatively high, with most rains arriving in the form of occasional thunderstorms. Daytime temperatures are in the range 100–108°F (38–42°C), and the short winter is warm, with temperatures averaging 77–86°F (25–30°C). The heat means that much of the moisture is lost by evaporation and is therefore not available to plants, with spinifex grasses among the few able to survive. Reserves of oil have been found beneath the sands and gravels of this desert, but these have yet to be exploited on any large commercial scale.

**GREATER BILBY**
*Also known as the rabbit-eared bandicoot, this highly endangered marsupial digs a burrow up to 10 ft (3 m) long and shelters there during the day, emerging at night to feed on seeds, fruit, and insects.*

**SHIFTING DUNES**
*Although many desert dune systems are stable, some, like these in the Great Sandy, constantly change shape due to the wind.*

### INTRODUCED SPECIES

Dingoes were introduced to Australia from Southeast Asia some 3,500 years ago. They are regarded as a pest because they attack livestock, and they have been excluded from parts of Australia by the erection of fences. Along with more recently introduced species such as foxes and domestic cats, dingoes pose a severe predatory threat to small native marsupials such as the greater bilby (see above). Another alien pest, the European rabbit, has displaced native fauna and destroyed vegetation, permanently changing the landscape in arid areas.

**DINGOES**
*Now readily associated with Australian fauna, the dingo is probably descended from the Indian wolf.*

AUSTRALASIA *Australia*

## Tanami Desert

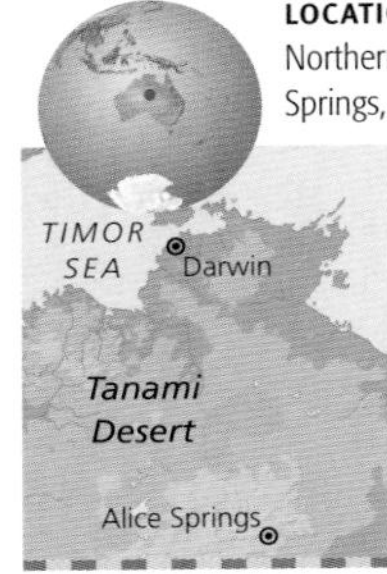


**LOCATION** In the center of the Northern Territory, north of Alice Springs, Australia

**TYPE** Rocky, sandy

**SIZE** 14,500 square miles (37,500 square km)

**RAINFALL** 8–16 in (200–400 mm)

**TEMPERATURE** Max: 108°F (42°C) Min: 77°F (25°C)

Lying to the east of the Great Sandy Desert, the Tanami Desert contains a large area of red sandy plains, with intermittent ranges of hills. The plant life here is a combination of tough spinifex grasses and small shrubs, such as saltbush, as well as occasional spiny acacias. A surprising feature of the Tanami, and the rest of Australia's arid interior, is the presence of dromedary camels, which were brought over by the early colonizers of the outback as pack animals. They were abandoned on the arrival of motorized transportation, and today's feral herds are their descendants. The Tanami also has the largest population of the endangered rufous hare wallaby. In the early 20th century, the discovery of gold led to the establishment of small mining communities, but the harsh, dry climate made large-scale extraction difficult.

AUSTRALASIA *Australia*

## Simpson Desert

**LOCATION** In the southern Northern Territory, western Queensland, and northern South Australia

**TYPE** Sandy

**SIZE** 56,000 square miles (145,000 square km)

**RAINFALL** 7–8 in (175–200 mm)

**TEMPERATURE** Max: 122°F (50°C) Min: 32°F (0°C)

The Simpson Desert, in the center of Australia, is famous for its parallel dunes of red sand, which can be 125 ft (38 m) high, although the average depth of sand is 3 ft (1 m). Some dunes are more than 75 miles (120 km) long. Summer daytime temperatures average 95–104°F (35–40°C), and winters are mild (64–74°F/18–23°C) with occasional night frosts. Rainfall is low and irregular, usually from thunderstorms, and the dominant vegetation is spinifex grasses and shrubs.

**ULURU**
*Rising 1,142 ft (348 m) above the desert floor, Uluru (also known as Ayers Rock) is a vast mass of 300-million-year-old sandstone.*

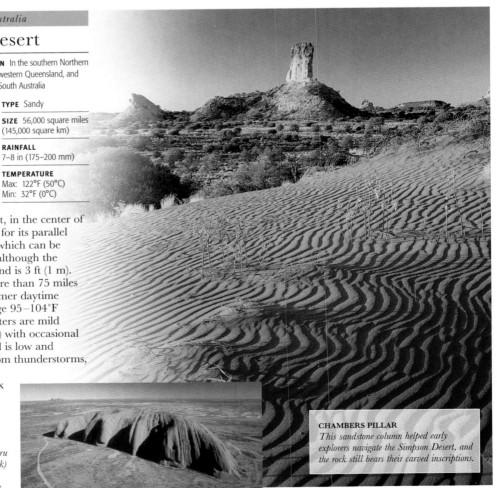

**CHAMBERS PILLAR**
*This sandstone column helped early explorers navigate the Simpson Desert, and the rock still bears their carved inscriptions.*

EASTERN GIBSON DESERT
*Rock-strewn hills, spinifex grass, and mulgas are characteristic of the eastern Gibson Desert.*

AUSTRALASIA *Australia*

## Gibson Desert

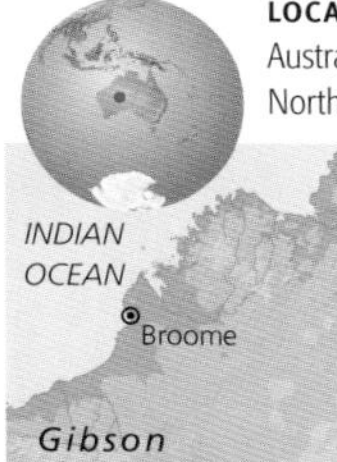


**LOCATION** In the center of Western Australia, extending eastward into Northern Territory

**TYPE** Gravelly, sandy

**SIZE** 60,250 square miles (156,000 square km)

**RAINFALL** 8–10 in (200–250 mm)

**TEMPERATURE** Max: 108°F (42°C) Min: 64°F (18°C)

The Gibson Desert lies between the Great Sandy and Great Victoria deserts. It takes its name from the Australian explorer Alfred Gibson, who attempted to cross it in 1874, but perished while searching for water. The landscape is mainly one of undulating beds of gravel, with areas of red sandhills, known as buckshot plains. Some parts are rocky, and in others, inselbergs protrude from flat plains. In the west lies Lake Disappointment, a large salt lake that attracts great numbers of waterfowl. Daytime temperatures in the Gibson average 91–108°F (33–42°C) in summer, and 64–73°F (18–23°C) in winter. The rains are very unreliable in this region, coming mostly from the occasional thunderstorm, of which there are some 20 to 30 each year, and the vegetation is patchy. Alongside the usual spinifex grasses are mulgas (species of acacia, or wattle), which lend a green tint to the landscape, and colorful flowers such as the yellow coneflower appear after the rains. Red kangaroos are a common sight.

ALEXANDRA'S PARROT
*This nomadic, slender bird is unusual among parrots in being adapted to survive in the desert. Its diet consists mainly of small seeds gathered from the ground and tree blossoms.*

AUSTRALASIA *Australia*

## Great Victoria Desert

**LOCATION** In the southeast of Western Australia, extending east to the center of South Australia

**TYPE** Sandy, gravelly

**SIZE** 150,000 square miles (338,500 square km)

**RAINFALL** 6–10 in (150–250 mm)

**TEMPERATURE** Max: 104°F (40°C) Min: 64°F (18°C)

This desert straddles the border between Western Australia and South Australia, just north of the Nullarbor Plain. It is mainly a sandy desert, with extensive dune fields, but it also includes gibber plains, on which the soil is covered by closely packed pebbles that are often glazed by iron oxides. They usually lack vegetation, although ephemeral annual species may appear after rains. In addition to the more familiar types of sand dunes, another kind, known as a lunette dune, also occurs here. Lunette dunes are a special feature of the Australian deserts, especially in the south. They are crescent-shaped, consist mostly of clay, and are formed at the downwind edges of temporary lakes from dry, windblown lake-floor sediments. Although the summers are hot, winter frosts are not infrequent in parts of the Great Victoria Desert, especially on the higher ground. But the winters are short, with temperatures rising quickly again by late September. Many parts of this desert have sufficient moisture to support open woodland, with eucalypts, common mulgas, and an understory of grass hummocks. The Great Victoria Desert is famous for its reptiles, which number over 100 species, with geckos and skinks being particularly diverse. Many of the reptiles are active during the day, such as the bizarre thorny devil, whose camouflaged pattern makes it hard to spot against the stones and sand of the desert. Equally striking are the Parentie and Gould's goanna, both of which can exceed 6 ft (1.8 m) in length. Little farming takes place here, although there are several Aboriginal reserves and some sheep and cattle ranches around the more fertile edges.

STURT'S DESERT PEA
*The floral emblem of South Australia, this striking plant is able to survive the extremes of the inland desert climate.*

THORNY DEVIL
*Despite its fearsome appearance, the desert-dwelling thorny devil is harmless to all except ants, and relies on its sharp spines for protection.*

SAND DUNES AFTER RAIN
*Although arid, the Great Victoria Desert can support scattered bushes along temporary watercourses, as here.*

LAND

**CLOUDFOREST**
*Epiphytes such as Spanish moss clothe the damp branches and trunks of trees in this Costa Rican cloudforest, providing niches for a vast number of species.*

# FORESTS

FORESTS COVER 30 PERCENT of the Earth's land surface and provide some of the richest of all habitats, offering an abundance of sites for plant germination as well as food and shelter for a wide range of woodland animals. Different types of forests are found in different regions of the world, from the dark boreal forest (taiga), which extends to the edge of the Arctic tundra, to the lush mixed broadleaf forests of temperate North America, Europe, and Asia and the dense rainforests of the humid tropics. Forests also play a vital role in the global water and carbon cycles, in improving air quality, and in preventing soil loss through erosion. They were central to early human evolution, and even today the rainforests, in particular, are still home to many native peoples. Now, many of the world's natural forests are threatened with destruction as land is cleared for housing, roads, timber, or agriculture, and some forests survive only in inaccessible areas or in enclaves on protected sites.

# FORESTS

33 Plant life
100 Soil formation
129 The global water cycle
132 Kingdoms of life
135 The carbon cycle
Deforestation 312–13

FORESTS ARE CATEGORIZED in various ways, usually according to the dominant type of tree. An important distinction is that between broadleaved and coniferous (usually needle-leaved) trees. Most coniferous trees are evergreen, which means that they retain a full complement of leaves throughout the year, but many broadleaved trees are also evergreen, and a few coniferous trees (such as larch) are deciduous. Deciduous trees shed all their leaves during the part of the year that is unfavorable for growth. This is the cold winter in temperate latitudes, but in the seasonal tropics, it may be the regular dry season.

## FOREST VEGETATION

Forest vegetation is dominated by the major tree species that form a high canopy. Conifers and their allies bear naked seeds. They have needlelike, scalelike, or blunt linear leaves, usually with a resinous coating that helps them survive cold winters. Broadleaves bear ovules in an ovary that develop into seeds only after fertilization. They have a variety of larger leaf shapes to better collect sunlight to drive photosynthesis (see p.132). Coniferous forests have less ground vegetation than broadleaved forests, since they are darker, and the resinous shed leaves produce an acid soil. In deciduous forests, the shed leaves form a nutrient-rich humus that fosters many more groundcover plants. These are often adapted to grow and flower quickly in the spring, taking advantage of the brighter conditions before the trees above unfold their foliage.

BROADLEAF

CLUSTER OF CONIFER LEAVES

**LEAF TYPES**
*Broadleaves offer a large, flat surface area for light-gathering and gaseous exchange. Narrow conifer leaves reduce water loss by transpiration.*

**CONIFEROUS CONES**
*The "fruit" of most conifers, such as this Norway spruce, is a cone, consisting of woody scales that turn brown when ripe.*

*cluster of small flowers*

*compound leaf with five distinct leaflets*

**FLOWERS AND LEAVES**

*spiky husk splits when it hits ground, releasing nut*

*shiny brown seed (or nut)*

**SEEDS**

**BROADLEAVED FLOWER AND SEED**
*Many broadleaved trees, like horse chestnut, have showy flowers to attract pollinators such as insects, and some produce large, edible seeds.*

EMERGENTS

CANOPY

UNDERSTORY

GROUND LAYER

**RAINFOREST STRUCTURE**
*Distinct layers can be seen in a mature rainforest. The midlevel canopy is overtopped by lone emergent giants, and there is a loose understory beneath.*

## TROPICAL FORESTS

Growth is rapid and lush in many tropical forests, and their complex structure offers niches for a multitude of animals and plants other than trees. Mature tropical rainforest is a multilayered habitat. The canopy is composed of trees whose crowns almost meet. Beneath it lies a more broken understory of young trees, and an often dense ground cover of saplings and undergrowth. Tall trees whose crowns stick out above the canopy are known as emergents. In many tropical rainforests, the moisture levels remain high year-round, while in others the rainfall is seasonal, as in the forests of Southeast Asia, where the monsoon brings sudden high levels of rain. Some tropical forests are dry, such as those of eastern South America.

**RAINFOREST CANOPY**
*The dense canopy of almost interlocking crowns is clearly visible in this Madagascan tropical forest.*

LAND

# TEMPERATE FORESTS

Temperate regions contain a wide variety of forest types. Many temperate forests are broadleaved deciduous, especially where the winters are hard, as in continental Europe, eastern North America, and many parts of eastern Asia. In such forests, there is a huge contrast between the forest in summer, at the height of the growing season, and in winter, when all the leaves have been shed and have accumulated on the forest floor. In China, Japan, and Korea, evergreen broadleaved trees are more common in the temperate forests, and sometimes form mixed stands with conifers. However, as these parts of the world have long been settled, many of the original forests have been destroyed, at least in part, and their rich, fertile soils used for agriculture. The remaining temperate broadleaved forests are nearly all managed in one way or another, although some are protected and are being allowed to revert to a more natural state.

**SUMMER AND AUTUMN**
*Deciduous forests change dramatically with the onset of autumn, when the trees lose all their leaves in a short time.*

**CHANGING LEAVES**
*As they die, some leaves change color from green to yellow, orange, or red, following the breakdown of the chloroplasts that contain their chlorophyll.*

# CONIFEROUS FORESTS

There are several different kinds of coniferous forest, but the most extensive, and the best known, are the boreal forests of northern North America and northern Eurasia (taiga). These are dominated mainly by evergreen coniferous trees, such as species of spruce, pine, and fir, although larches, which are deciduous, are a prominent feature of some areas. Coniferous forests are also commonly found in many mountain regions, where they tend to develop above the highest levels of broadleaved woodland, with spruces and pines being the main trees present. An unusual type of coniferous forest is found in the high-rainfall coastal zone of the Pacific Northwest of North America. This temperate rainforest boasts several very tall species, such as Douglas fir and coast redwood, while California's coniferous forest is home to the famous giant sequoia.

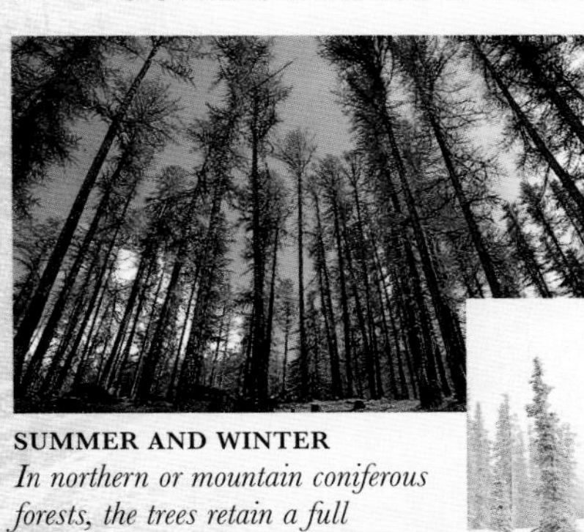

**SUMMER AND WINTER**
*In northern or mountain coniferous forests, the trees retain a full covering of needles through the harshest winters. In fact, individual leaves are shed and replaced, but not all at the same time.*

**GREEN AND GOLD**
*Larch is an unusual conifer in that it is deciduous. Here, in France, its beautiful autumn colors stand out against the dark green of the evergreen conifers.*

# FOREST ZONES

Forests are the final development of mature vegetation on average soils with an adequate water supply. Under natural conditions, forests may persist for very long periods, and individual trees often grow to be several hundred years old. This, then, is the situation in natural forests, but few of today's forests are virgin, in the sense that they are completely uninfluenced by people or their activities. If forest is cleared and the vegetation allowed to recover, ecological succession may be observed, with a slow reversion to what is known as climax forest. A snapshot of succession may also be seen, for example, in the transition from the marshy vegetation around a lake, through wet scrub to mature damp woodland. Changes in forest type are also clearly visible with a change in altitude, ascending in a series of zones from lowland to alpine levels on a mountainside (see right). In temperate forest regions, such as in the European Alps, mixed deciduous forest tends to be found in lowland sites, and the mixture of trees changes through submontane and hill forest, with an increase in coniferous trees at montane levels. At about 6,600 ft (2,000 m), the climatic limit of tree growth (the tree line) is reached, and alpine grassland and scrub begin to dominate. In the tropics, the pattern of trees is different, with a zone of bamboo often found between the montane forest and scrub bands.

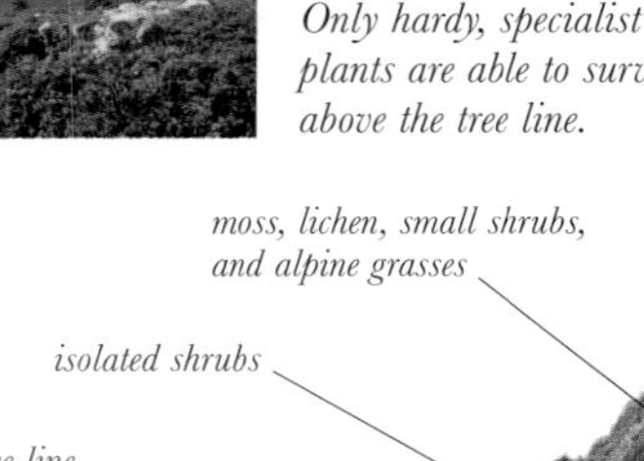

**ISOLATED SHRUBS**
*Only hardy, specialist plants are able to survive above the tree line.*

**NEEDLE-LEAF FOREST**
*Coniferous trees best tolerate the rigors of high altitude, and often form the tree line.*

**MONTANE FOREST**
*Above 3,300 ft (1,000 m), the mix changes to species such as beech and scattered conifers.*

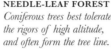

**SUBMONTANE FOREST**
*Lowland and hilly sites with good soils support a rich, mixed forest of mainly deciduous broadleaved trees.*

**ALPINE ZONATION**
*In this European alpine landscape, the lowest level of the landscape features mixed forest, with trees such as hornbeam and oaks dominant. This grades into montane forest, then a final forest zone of mainly coniferous species such as silver fir, spruce, and larch. Higher still, high winds, frost, snow, and ice discourage even coniferous trees.*

**BLUEBELL WOOD IN SPRING**
*Many deciduous woods, like this one in England, have a ground flora of species such as the bluebell that flower early, before the trees are in full leaf.*

LAND

# ANIMAL AND PLANT LIFE

Although cold and challenging for much of the year, the boreal forests can still support animal life, especially during the summer, when they are home to seed-eating birds such as crossbills, and spruce grouse and capercaillie, which feed partly on conifer needles. Many birds migrate to less harsh climates in the winter, but the mammals, which include bears and voles, either remain active beneath the snow or hibernate. In broadleaved temperate forests, conditions are more equable, and deciduous trees avoid the stresses of winter by shedding their leaves. This leaf litter is a bonus for the ground flora and fauna, and the soil of these forests is alive with invertebrates such as worms, spiders, ants, and beetles. Many woodland birds, such as warblers and tits, feed on caterpillars and other insects up in the canopy. Foxes, badgers, and deer are among the larger mammals at home in these forests. Richest of all in plant and animal life are the tropical rainforests, where researchers are still discovering new species. Many kinds of epiphytes (plants that grow on the trees without being parasitic) thrive here, including bromeliads and orchids.

*forward-facing eyes permit stereoscopic vision to judge distance*

**RUFFED LEMUR**
*This inhabitant of the rainforests of Madagascar feeds mainly on fruit, but will sometimes eat leaves, nectar, and seeds. It lives in the upper forest, and builds a nest of leaves in a tree hole or fork.*

*long tail aids balance*

**MILTONIA ORCHID**
*Many plants in moist tropical forests, such as this orchid from Brazil, grow on the trunks and branches of trees.*

**GILL POLYPORE**
*This bracket fungus is widespread in northern temperate forests, found mostly on the trunks of hardwood trees.*

**GREAT SPOTTED WOODPECKER**
*Woodpeckers use their powerful bills to drill and probe the bark of trees for grubs and bugs. The stiff tail feathers act as a sturdy prop.*

**WEAVER ANTS**
*Ants are very common insects in tropical forests. Here, worker weaver ants in the Malaysian rainforest cooperate to cut leaves, from which they fashion their nest. The tree benefits as the ants deter other plant-eaters from browsing on the foliage.*

## FOREST SIZES

The technology of satellite imaging has had an enormous impact on ecological monitoring. The comparison of satellite photographs with existing ground surveys allows detailed interpretation and analysis of such images, and enables the accurate mapping of the size of existing forests. The satellite image below, for example, clearly shows the current extent of forest in the Black Hills of South Dakota. It is vital to build up such snapshots of vegetation cover if the effects of factors such as deforestation are to be accurately monitored. Satellite imaging is also used to track forest fires, allowing their paths to be predicted and control measures mobilized.

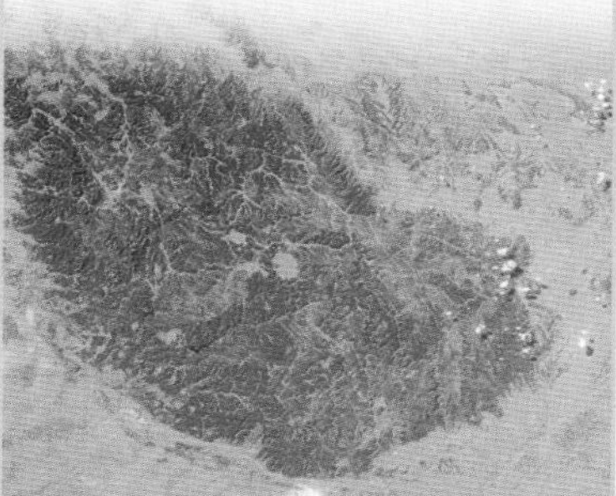

LAND

# FOREST DISTRIBUTION

In general, coniferous forests are found at higher latitudes and altitudes than temperate forests. The vast boreal forest belt, for example, lies north of about 50°N. In temperate regions, the forest climate is one with an annual rainfall of 20–60 in (500–1,500 mm). Temperate forests are found in western and eastern North America, southern South America, western and central Europe, eastern Asia, and Australia and New Zealand. Tropical rainforest requires warm conditions and an annual rainfall of 80–120 in (2,000–3,000 mm). The major belts of this forest type are in Central and South America, western and central Africa, Southeast Asia, and northeast Australia.

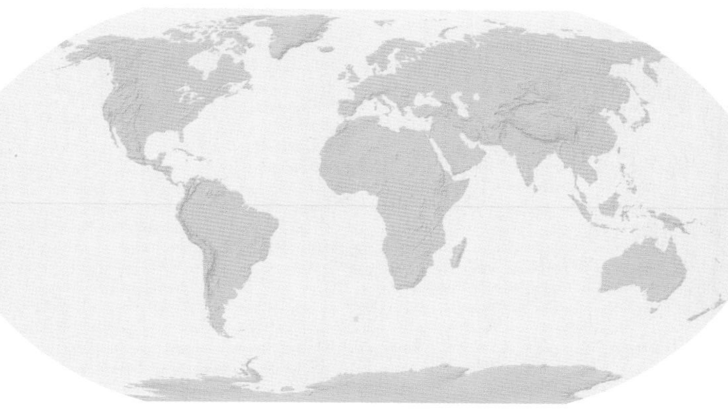

**GAP IN A FOREST**
*Gap-regeneration is a natural phenomenon in woods and forests. When an old tree dies or is felled by strong wind, as here in New Zealand's Fiordland, it decays on the forest floor, gradually returning nutrients to the soil.*

# FOREST REGENERATION

Although forests often take many decades, or even centuries, to establish themselves, death and regeneration are part of the natural dynamic of forest ecosystems. Some trees are long-lived, growing for hundreds of years, but they do die and fall eventually, creating gaps in the forest into which more sunlight can penetrate. This opening up of the canopy encourages the often rapid germination of plants on the fertile forest floor, and among these will be seedlings or suckers of trees belonging to the same species as the fallen tree. In this way, natural forests tend to display a mosaic of patches of trees of differing ages as the forest regenerates in an uneven fashion. As well as individual trees succumbing to old age or disease, large stands of trees can be destroyed by fire or strong winds. The composition of the regenerating forest may well differ from that of the surrounding mature forest, since some species of tree, such as Douglas fir, are better able to thrive in the more open conditions. Other, more shade-tolerant species such as western hemlock come to the fore as the forest returns gradually to a taller and denser—and therefore darker—state.

**FIRE AND REGENERATION**
*Fire (1) is a major regenerative factor in some habitats, such as this coniferous forest. A rich seed-bank in the ash-enriched soil (2) provides new growth, as germination follows the devastation (3).*

## FORESTS PROFILES

The pages that follow contain profiles of the world's main forested regions. Each profile begins with the following summary information:

**TYPE** Tropical rainforest, tropical dry forest, temperate rainforest, deciduous temperate forest, evergreen temperate forest, or boreal forest

**AREA** Surface area

NORTH AMERICA *north*

# North American boreal forest

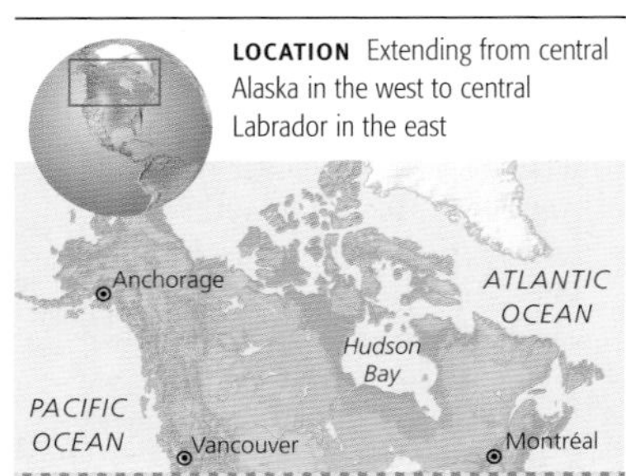


**LOCATION** Extending from central Alaska in the west to central Labrador in the east

**TYPE** Boreal forest

**AREA** 2.4 million square miles (6.25 million square km)

The boreal forest of North America dominates much of the northern part of the continent, covering around a third of Canada's landmass. This vast forested zone is the counterpart of the Eurasian boreal forest or taiga (see p.316), from which it is separated by the northern oceans. The climatic conditions under which the boreal forest develops are generally cool and humid, with very cold winters, lasting from seven to nine months. Snow covers the ground for much of the year, often into the summer months. Coniferous trees, such as black and white spruce, are well adapted to survive such rigorous conditions, and they dominate the North American boreal forest. The drooping branches shrug off the weight of heavy snow, and the waxy needles allow good control of water loss. Although conditions may be wet and boggy in summer, for much of the year water is frozen and therefore largely unavailable to the trees. To the north of its range, the boreal forest changes gradually into Arctic tundra as conditions become too harsh to support tree growth. The tall, closed coniferous forest thins out, with the trees becoming shorter (the three-month growing season limits their height to 10–15 ft/3–4.5 m) and scattered. To the south, the transition is to prairie grassland in the central part of the belt, and to mixed forest in the west and east. Many of the forests in the north are known as lichen woodland, since lichens dominate the ground flora, replacing the mosses that are more common farther south where water is more available. The Alaskan forest is the most varied, largely due to the influence of the mountain ranges and oceans on the climate. In the central region around the Canadian Shield, the forest is more uniform, with spruces tending to dominate, while in the east balsam fir is a key species. Around the Great Lakes, the boreal forest is mixed, with deciduous species including sugar maple and yellow birch.

Quaking aspens thrive on poor soil and are often found at montane levels. Mammals of the forest include the Canada lynx and its prey, the snowshoe hare (trapping records have revealed their population cycles to be closely linked), and the northern flying squirrel. Great gray owls are found here and also in the Eurasian boreal forest, and feed on voles, mice, and small birds.

**BLACK SPRUCE**
*Here, in the northern boreal forest of Alaska, scattered black spruces tower above the willow scrub. This is the dominant tree in the open woodlands at the forest–tundra edge, from Alaska across Canada to Newfoundland.*

**ALASKAN FOREST**
*The deep northern boreal forests are broken up in some places by rivers, streams, and boggy lakes, as here in Alaska.*

**QUAKING ASPENS**
*The slender white trunks of quaking aspens add to the beauty of the boreal scenery here in British Columbia.*

**CANADIAN BOREAL FOREST**
*Mount Robson Provincial Park in British Columbia protects over 840 square miles (2,170 square km) of unspoiled boreal forest close to Canada's highest peak.*

NORTH AMERICA *west*

## Pacific Northwest rainforest

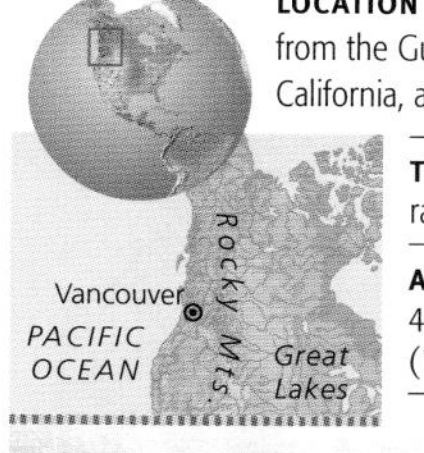


**LOCATION** Along the Pacific Coast from the Gulf of Alaska to northern California, and Canada

**TYPE** Temperate rainforest

**AREA** 480,000 square miles (1.2 million square km)

These remarkable forests, the only temperate rainforests in the northern hemisphere, occupy a relatively narrow strip along North America's northern Pacific Coast. The rainfall is high (31–118 in/800–3,000 mm, mostly between October and March) and is supplemented in some areas by coastal fogs. These conditions are ideal to support the growth of some of the giants of the plant kingdom, notably conifers such as the redwood, which can grow to over 355 ft (110 m) in height, Douglas fir, Sitka spruce, and western hemlock. Douglas firs dominate in the Coast Ranges, Olympic Mountains, and northern Cascade Range, up to about 3,300 ft (1,000 m). Redwood forests are a feature of northern California and southern Oregon.

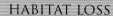

**WESTERN HEMLOCK**
*These shade-tolerant trees thrive in low light conditions, enabling them to slowly replace pioneer tree species, such as the Douglas fir.*

### HABITAT LOSS

The northern spotted owl is found only in the natural coniferous forests of the Pacific Northwest, where its habitat is under threat from logging activities, to which it is particularly sensitive. Each pair of owls requires about 4 square miles (11 square km) of unspoiled old-growth forest or heavily wooded canyons for successful breeding. Its population is now estimated at about 5,000 pairs in the US and a mere 25 pairs in British Columbia (it is Canada's rarest bird).

**HIGH RAINFOREST**
*Washington State's famous temperate rainforest receives a staggering 132 in (3,350 mm) of rain a year, resulting in lush vegetation, including many epiphytes.*

**SURVIVING THE WINTER**
*The thick, spongy bark of the giant sequoia, seen here in Sequoia National Park, protects the tree from the winter cold as well as from the effects of fire.*

NORTH AMERICA *west*

## California coniferous forest

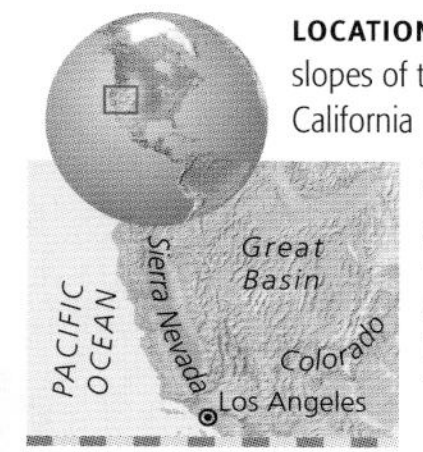


**LOCATION** Along the western slopes of the Sierra Nevada, central California

**TYPE** Evergreen temperate forest

**AREA** 16,800 square miles (43,600 square km)

These forests are famous for being the home of the world's largest tree species (in terms of bulk), the giant sequoia. This is a high-altitude community, found mainly between 4,540 ft and 7,450 ft (1,400 m and 2,300 m). There are about 75 separate groves where giant sequoias grow, often associated with white fir, sugar pine, and incense cedar. Although not quite as tall as the redwood or Douglas fir (see above), giant sequoias may grow for over 2,000 years and reach almost 325 ft (100 m) in height. Fire is an important ecological factor in these coniferous forests. The cones of giant sequoias require the heat of forest fires before they can open to shed their seeds. Fire also burns the soil surface, exposing the mineral-rich layers beneath, which are essential for seedling growth. Full-grown trees are protected by their spongy bark, up to 17½ in (45 cm) thick, which gives them an asbestos-like insulation from fire damage. Other trees in the region include Douglas fir and ponderosa and lodgepole pines, with Engelmann spruce and bristlecone pine at subalpine levels. The latter species is very slow-growing and includes the world's oldest known living plant specimen—referred to as the Old Man, it is nearly 4,770 years old. Mammals of these forests include black bears and mountain lions (cougars).

**GENERAL SHERMAN**
*This, the largest living giant sequoia, is about 2,200 years old. It is 275 ft (84 m) tall and 30 ft (9 m) across at its base.*

**YOSEMITE**
*Equally famous for its meadows, mountains, and waterfalls, Yosemite National Park also boasts three groves of ancient giant sequoias.*

NORTH AMERICA *east*

## East North American deciduous forest

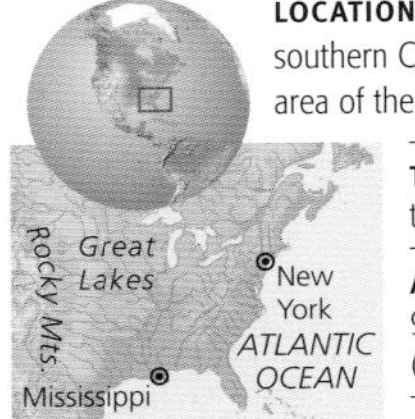


**LOCATION** From the Great Lakes in southern Canada to the east coast area of the US and south to Florida

**TYPE** Deciduous temperate forest

**AREA** 970,000 square miles (2.5 million square km)

**SPRING IN THE APPALACHIANS**
*The ridges and valleys of the Great Smoky Mountains in the Appalachians, North Carolina, are clad with several types of oak and hickory, American chestnut and beech, and basswood.*

Only scattered remnants of these once extensive forests now survive, but some of the finest are centered on the Appalachians (see p.158), whose foothills still retain considerable forest cover. On the western edge of the region, the deciduous forests are dominated by oaks and hickories, gradually giving way to prairie grassland in the drier interior. These forests are famous for their autumn colors, particularly in the north, where the leaves of the sugar maple may turn bright red.

**AUTUMN COLORS**
*The diversity of deciduous tree species on the foothills of the Appalachians leads to a colorful display in the fall.*

SOUTH AMERICA *northwest*

## Northwest South American rainforest

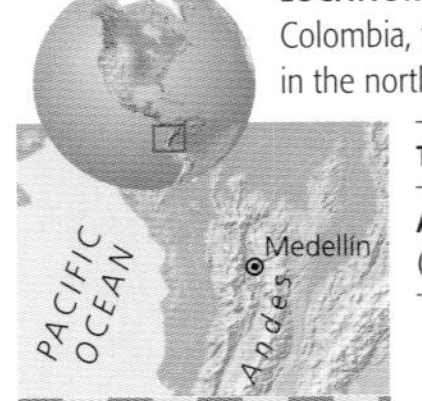


**LOCATION** On the west coast of Colombia, from the Panama border in the north, south toward Ecuador

**TYPE** Tropical rainforest

**AREA** 2,160 square miles (5,600 square km)

Of the many kinds of forest in South America, that found on the northwest coast is one of the richest, with many unique species, including 100 endemic birds. This Colombian forest, known locally as Chocó, certainly deserves the designation of rainforest—the region has one of the highest rainfalls in the world, at 465 in (11,770 mm) a year. Tall *Cecropia* trees are a feature of these forests, which are also rich in tree ferns, orchids, aroids, and bromeliads. Another prominent woody plant is the purple-flowered glory bush, but some of the most majestic, at higher altitudes, are the podocarps, conifers that can live for over 1,000 years. The rainforest is under serious threat from logging and mining operations, with a quarter having been lost already. Eight national parks now protect some areas.

**TROPICAL CLOUDFOREST**
*Lush undergrowth, rich in tree ferns, is typical of the damp cloudforest of La Planada, one of Colombia's best-known reserves.*

CENTRAL AMERICA

## Central American rainforest

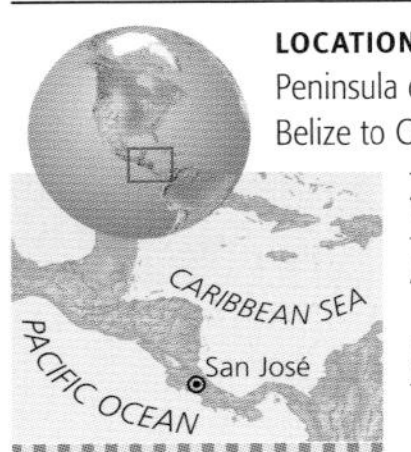


**LOCATION** From the Yucatan Peninsula of Mexico, south through Belize to Costa Rica and Panama

**TYPE** Tropical rainforest

**AREA** 200,000 square miles (520,000 square km)

The narrow land bridge connecting Mexico with South America contains important pockets of diverse tropical forest. Central America has some of the most rapid rates of deforestation in the world, and pressure on the remaining forest is high. Many sites are protected within reserves, such as the Kuna Yala reserve and the Smithsonian Tropical Research Institute on Barro Colorado Island, both in Panama, and La Tigra National Park in Honduras. Costa Rica, which lies in the middle of the land bridge, contains some of the richest rainforests, with species from the north and south intermingling with endemics. The Corcovado National Park in Costa Rica is a vital nature reserve, conserving the most important stretch of lowland tropical rainforest on the Pacific Coast of Central America. The park has over 1,500 species of higher plants, 124 mammals, 117 reptiles and amphibians, about 6,000 insect species, and 375 birds, including the rare harpy eagle. The habitats range from coastal mangroves, through swamp forests, to epiphyte-rich lowland rainforest, upland rainforest on higher ground, and finally, at the highest altitudes, to cloudforest, rich in tree ferns. Some of the trees of these forests, such as cashews, reach heights of 160 ft (50 m), with some emergents, such as vantanea, towering to 210 ft (65 m). Many of the plants are useful to humans, yielding oils, medicines, fibers, edible fruits and nuts, and even natural chewing gum from the chicle tree. Relatives of the domestic avocado provide a reserve of genetic diversity useful to commercial plant breeders.

**RED-EYED TREEFROG**
*Like all treefrogs, the red-eyed treefrog has widely spread toes with adhesive pads to help it clamber in the foliage of the Central American rainforests.*

**DANGLERS AND CLIMBERS**
*The almost constant moisture in cloudforests, such as this one in Costa Rica, allows a luxuriant growth of climbers and epiphytes to develop on the trees.*

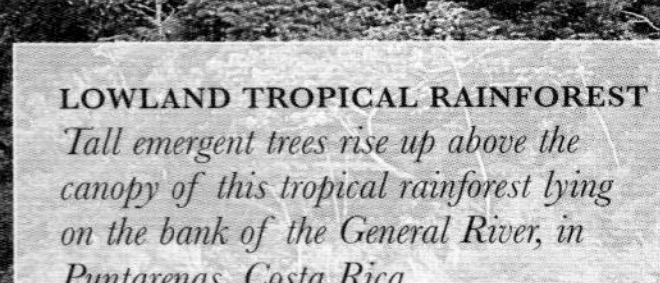

**LOWLAND TROPICAL RAINFOREST**
*Tall emergent trees rise up above the canopy of this tropical rainforest lying on the bank of the General River, in Puntarenas, Costa Rica.*

SOUTH AMERICA *north*

## Amazon Rainforest

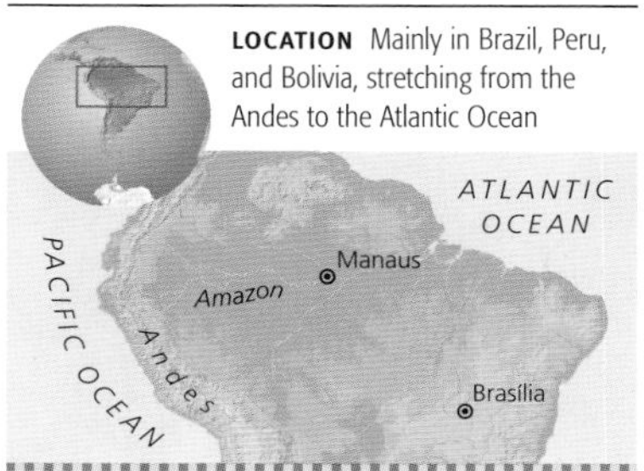


**LOCATION** Mainly in Brazil, Peru, and Bolivia, stretching from the Andes to the Atlantic Ocean

**TYPE** Tropical rainforest

**AREA** 2.3 million square miles (6 million square km)

The Amazon Rainforest is the largest area of tropical forest in the world. It occupies much of the basin of the Amazon River (see pp.222–23), which extends across most of northern Brazil, much of Bolivia, and also eastern Colombia and eastern Peru. This tropical lowland is a mosaic of different rainforest types, varying according to soil composition and the water regime.

**BROMELIADS**
*The fleshy, spirally arranged leaves of bromeliads form reservoirs that trap water, creating tiny ponds used by many animals, notably insects and frogs, as a home high up in the rainforest canopy.*

**BUTTRESSES**
*The bases of many tropical trees extend outward as distinct flanges. This gives them more stability in the often shallow, wet soil.*

Most of the basin receives an annual rainfall of about 79 in (2,000 mm), but in some places this rises to over 315 in (8,000 mm). Generally, the wetter the forest, the more diverse the flora, and the upper Amazon rainforests are as rich as those of the Colombian Chocó (see Northwest South American rainforest, left). The number of endemic plant species is estimated at 13,700—nearly 80 percent of the forest's flora. The rainforest has a complex structure, with distinct tree layers and a large number of lianas and vines, offering an array of sites for colonization by epiphytes, such as bromeliads, and various crevices and hollows that are occupied by many kinds of insects, frogs, and other animals. Some parts of the basin experience seasonal flooding, which transforms the dry forest floor into a vast lake. Such floods, which may last from 2 to 10 months, affect 38,600 square miles (100,000 square km) of forest, and the water can be as deep as 65 ft (20 m). In Brazil, the swamp forests are called igapó forests. Plant diversity studies have revealed that there are as many as 300 woody species in just a single hectare of the Amazon Rainforest, including the much-prized big-leaf mahogany. Over 2,000 plant species are used by local people as a source of food, medicine (see panel, right), and other useful products. One of the most famous is the Brazil nut, and others include natural rubber and the açaí palm, which provides palm hearts to a thriving industry. Tropical rainforest still covers a vast area of the Amazon Basin, although it is being destroyed at an alarming rate. An estimated 1,000 square miles (2,600 square km) is lost each year in Peru alone, and 1,500 square miles (4,000 square km) more in parts of Brazil.

**SWAMP FOREST**
*The Tambopata-Candamo Reserve in Peru protects swamp forest, shown here, and drier habitats on higher ground. It is home to 1,200 species of butterflies.*

**SOUTHERN TWO-TOED SLOTH**
*Sloths spend almost all their lives up in the forest canopy, moving extremely slowly, hanging from the branches of trees, and feeding on leaves and fruit.*

### MEDICINAL PLANT

One of the most important traditional medicinal plants of the Amazon region is pau d'arco (shown here). Its inner bark is used for a remarkable range of ailments, including cancer, fungal infection, diabetes, dysentery, malaria, allergies, baldness, and acne; it is also used as a painkiller. Pau d'arco is growing in importance as an herbal medicine in Europe and the US.

**ISOLATED RAINFALL**
*High temperatures and abundant rainfall in the Amazon Rainforest lead to high humidity. The shower seen here casts its shadow over just a small part of the forest's vast expanse.*

# DEFORESTATION

Humans have had a greater impact on forests than on any other land habitat. Forest clearance began with the development of agriculture, 10,000 years ago, and since then nearly a quarter of the world's tree cover has disappeared. Deforestation is now occurring most rapidly in the tropics, where it poses a major threat to plant and animal life and to the atmosphere, but the temperate world has actually lost far more of its natural forest cover; the difference is that it happened longer ago.

## TEMPERATE DEFORESTATION

Trees are the natural vegetation across most of the temperate world, apart from arid areas. Deciduous forest once covered much of Europe, eastern Asia, and eastern North America, while coniferous trees grew on mountains and in the far north. That coniferous forest is still largely intact, but almost all deciduous forest has been cut down. In Europe, clearance was gradual, but in North America it was more abrupt. When European colonists arrived in the early 1600s, forests stretched from the Atlantic coast to the Mississippi. By the mid-1800s, most of this was gone.

Forest cover is now generally stable. In North America, the deciduous forest has partially regenerated, because land fell into disuse when grain farming moved westward onto the open plains. But in most parts of the temperate world, remaining forests are managed as a resource. This is particularly true in the coniferous zone, where an area is totally cleared of trees and then replanted, a practice known as clear-cutting. As a result, natural "old growth" forest—with its varied wildlife—is becoming increasingly rare.

In the American northwest, conservationists have waged a long campaign to save the virgin coniferous forest that remains. In western Australia, a similar struggle centers on the region's unique eucalypt forests, which contain some of the continent's tallest and oldest trees.

**HISTORY IN THE HILLS**
*Scotland's Southern Uplands were once covered with natural forest. Centuries of deforestation, followed by recent replanting, have created the landscape that exists today.*

## TROPICAL DEFORESTATION

In populous parts of the tropics, such as India and southern China, most of the original tree cover was cleared several thousand years ago, and remaining trees are now heavily exploited as a source of fuel. But in the humid equatorial belt, deforestation began in earnest only in the mid-20th century, and it is now happening at an unprecedented rate, especially in rainforests.

In some regions, the principal cause is a demographic one, because expanding populations are requiring ever more land. Forest cover in Costa Rica, for example, dropped by three-quarters between 1940 and the early 1980s. But in other parts of the tropics, such as the Amazon Basin and Indonesia, deforestation is also driven by the international timber trade, and there is frequently no attempt at replanting. Rainforest soils are generally poor, so the cleared ground is often unproductive when used for agriculture. Despite many initiatives to protect it, tropical rainforest continues to disappear. At least half its original area has been cleared, and a further fourth may vanish by 2050. This extent of habitat loss will make survival difficult for many rainforest species.

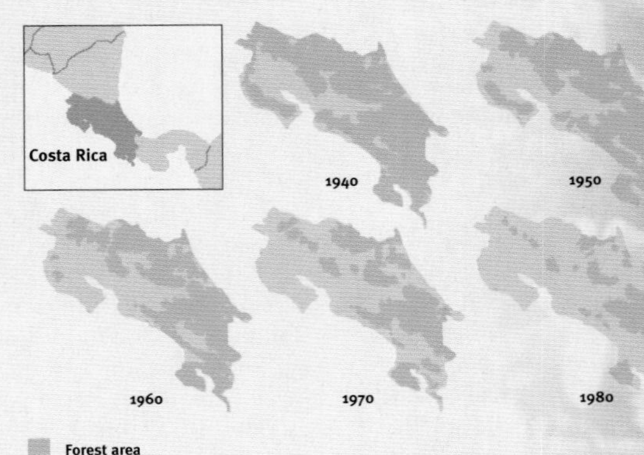


**TROPICAL TREE LOSS**
*In 1940, nearly 70 percent of Costa Rica was covered by natural forest (shaded in dark green above), yet by the early 1980s that figure was only 17 percent. Fortunately, much of the remaining forest is now protected by national parks.*

## MANGROVE FORESTS

As well as rainforests, the tropics are home to mangroves—trees that grow in the zone between high and low tides. Mangrove forests develop on muddy shores, and, depending on local topography, they can be narrow and ribbonlike, or they can reach far inland. They stabilize the coast, and they are also important breeding grounds for mollusks, fish, and wetland birds.

Mangrove wood is used for timber and charcoal, and in some parts of the tropics it is an important and sustainable resource. In recent years, the world's mangrove forests have suffered large-scale clearance, mainly to make way for aquaculture (see p.423). Mangrove removal has a direct impact on fisheries, and it also weakens the coast's natural defenses against hurricanes and typhoons. However, commercial pressure means that these unique ecosystems are disappearing even faster than forests on land.

**COASTAL RETREAT**
*Scarlet ibises use mangroves as overnight roosts and as a place to raise their young, from Venezuela southward to Brazil.*

**TREE HARVEST**
*In Canada, trees are usually removed by completely clearing an area. Although efficient, such clear-cutting harms wildlife and accelerates erosion.*

**ATLANTIC DRY FOREST**
*The Itacarambi Reserve protects a tropical dry forest near the Atlantic Coast of eastern Brazil.*

SOUTH AMERICA *central*

## South American dry forest

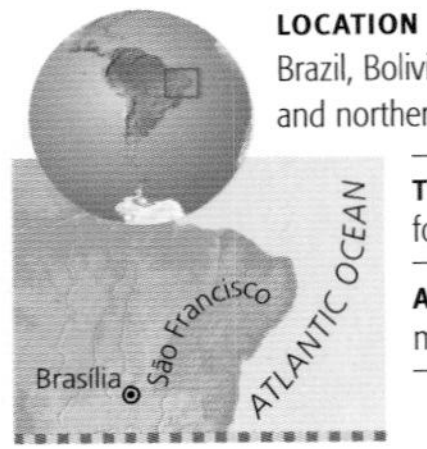


**LOCATION** Southern and eastern Brazil, Bolivia, western Paraguay, and northern Argentina

**TYPE** Tropical dry forest

**AREA** 1.5 million square miles (4 million square km)

There are two main areas of dry forest in South America: the chaco region of western Paraguay and nearby Argentina and Bolivia; and the cerrado and caatinga regions of Brazil. The chaco and caatinga are the driest, with an average annual rainfall of less than 31 in (800 mm) and a dry season of five to eight months, while the cerrado has 31–71 in (800–1,800 mm) of rain and a dry season of three to four months. Many of the trees of these forests belong to the pea family and reach heights of up to about 100 ft (30 m). Another unusual feature is that they contain large numbers of cacti, particularly in the chaco. The original area occupied by these tropical dry forests has been much reduced, and they are classed as the most threatened of all tropical vegetation. Clearing of land for cattle-ranching is one of the main causes of habitat loss.

SOUTH AMERICA *south*

## Southern Andean temperate rainforest

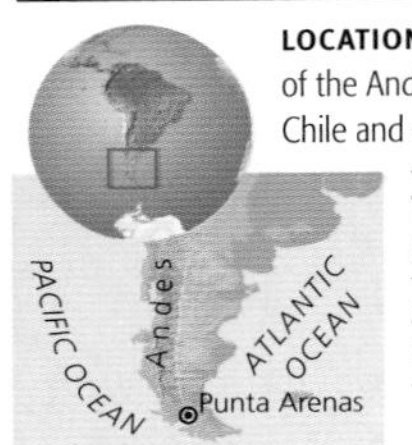


**LOCATION** On the western slopes of the Andes in central and southern Chile and some parts of Argentina

**TYPE** Temperate rainforest

**AREA** 29,340 square miles (76,000 square km)

High rainfall and sea mists produce wet conditions that are ideal for these temperate rainforests, which are found from sea level right up to the tree line of the Andes. In general, the lower altitude forests are broadleaved evergreens, with conifers dominating higher up, mixing with deciduous forest toward the tree line. Trees include lengas, podocarps, and the famous monkey puzzle. One of the more unusual trees here is the Patagonian cypress, which is valued for its timber and is also extremely long-lived, with individuals reputedly reaching 3,600 years old.

**MOSS-COVERED LENGAS**
*These mosses and lichens clothe the trunks of lengas (which belong to the beech family) in the Patagonian Andes.*

**MONKEY PUZZLES**
*These coniferous trees are found only in the northern areas of the temperate rainforest, as here in Malacahuello, Chile.*

EUROPE *southwest*

## Mediterranean evergreen forest

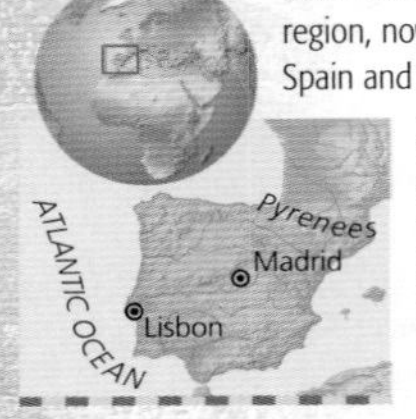


**LOCATION** In the Mediterranean region, now mostly confined to Spain and Portugal

**TYPE** Evergreen temperate forest

**AREA** 240,000 square miles (625,000 square km)

Evergreen forests once covered large areas of the Mediterranean, from Portugal and Spain to Turkey and parts of the Middle East, and also extending along the North African coast. Today, only scattered forests remain, and most of these have been significantly altered by people. In Portugal and Spain especially, managed evergreen woods, dominated by cork and holm oaks, can be found in many areas. "Dehesa" is the Spanish term for this traditionally managed habitat, called "montado" in Portugal. It is derived from the original forest and is a parklike landscape of oaks, with the land also used for crops or as pasture for sheep, goats, and pigs. The trees reduce water loss from the soil, and their acorns provide forage for pigs.

### HARVESTING CORK

Cork is probably best known for its use as a stopper for wine bottles. However, it has a wide range of other uses too, from mats to floats, and as an insulating material and sound absorber. The bark of the cork oak grows as two distinct layers, and it is the outer layer that is harvested, leaving the inner bark to regenerate more cork. The bark is stripped from the trunks of the oaks every nine years, usually during July and August.

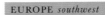

**SCATTERED OAKS**
*The Monfragüe National Park in Extremadura, Spain, protects fine expanses of dehesa. The trees here are mainly cork and holm oaks.*

**SPANISH IMPERIAL EAGLE**
*This highly endangered raptor is found only in central and southern Spain, where it hunts small mammals and reptiles in woodland and open country. There are fewer than 250 breeding pairs remaining in the wild.*

BIALOWIESKA FOREST
*Wild garlic (or ramsons) carpets the floor of this ancient forest in eastern Poland, much of which has never been felled.*



EUROPE *north*

## European mixed forest

**LOCATION** Much of lowland and submontane Europe, from the British Isles to western Russia

**TYPE** Deciduous temperate and evergreen temperate forests

**AREA** 1.6 million square miles (4 million square km)

Mixed, mainly broadleaved deciduous forest forms the natural vegetation of much of the lowland and hill country in Europe, and is particularly well developed in central Europe. The main tree species are common and sessile oaks, beech, lime, hornbeam, ash, elm, birch, and alder, with the exact mix depending on the soil type, drainage, and microclimate. Particular types of European mixed forest can be recognized according to which trees are dominant, such as oak/birch, oak/ash, oak/hornbeam, lime, or beech. In some areas—for instance, on well-drained, sandy soils—pines mingle with deciduous trees or form pure stands on their own. However, Europe has been settled for a very long time and much original forest has long been cleared, and that which remains is far from its original condition. Beech woods can be found in many parts of Europe, especially central Europe, and some of the best of these remain at submontane levels of the major mountain ranges, such as the Alps, Pyrenees, Apennines, and Carpathians. In Slovakia, for example, they form about 30 percent of the forest. At montane levels, spruce and fir often grow together with beech. Some of the forests that are closest to their natural condition, such as the Bavarian Forest straddling the German–Czech border and Bialowieska Forest in eastern Poland and adjacent Belarus, display a mosaic of woodland types within a single region, responding mainly to variations in local soil conditions. The natural processes of death and regeneration can continue unaltered by people in such places, with trees attaining heights of 130 ft (40 m) and living for 400 years before succumbing to decay or wind-felling. Dead wood is an important habitat for a range of woodland invertebrates, which in turn sustain a diverse population of birds, such as owls and woodpeckers. The mammals include wood mice, wild boar, fallow and red deer, and, in the case of the Bialowieska Forest, gray wolves and European bison. This old forest also contains over a quarter of Poland's flora (about 550 species).

BLACK FOREST
*Conifers such as spruce and silver fir mingle with stands of beech in the hills and valleys of Germany's Black Forest. Most of the forest here is secondary, and much is planted.*

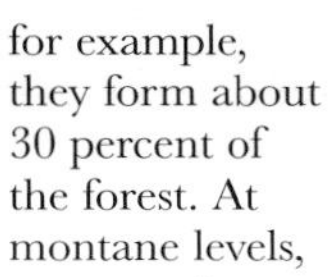

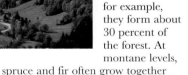

ARDENNES
*The hilly Ardennes region, which straddles the border between France and southern Belgium, is well forested. Beech, oak, and spruce are the major tree species here, mainly in managed forests.*

NEW FOREST
*This historic English hunting forest has been managed for centuries and now consists of an intricate mix of woods (as above) and heath.*

FLY AGARIC
*Common in European woods, especially under birch and oak, this toadstool is highly poisonous, but is unlikely to be mistaken for any other kind.*

### COPPICING

Traditional management of many European woods involved harvesting shoots and poles from selected trees, such as ash and hazel, a process known as coppicing. The stumps then resprout from the base, producing another crop of wood, which can be used to make fences, as below. A few trees are left to grow to their full height. Coppicing results in an open forest floor, allowing a diverse woodland flora to flourish.

EUROPE AND ASIA *north*

# Eurasian boreal forest

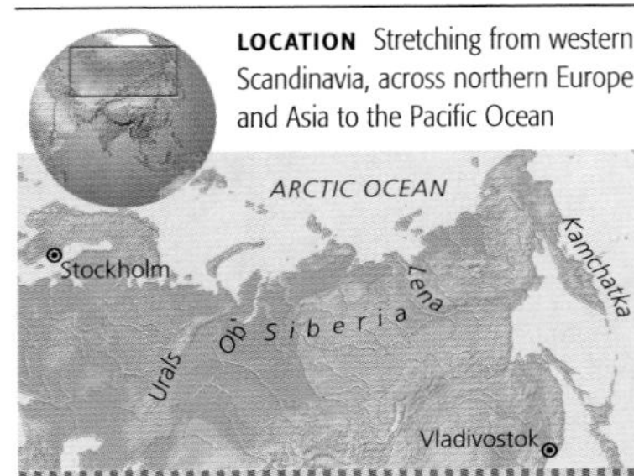


**LOCATION** Stretching from western Scandinavia, across northern Europe and Asia to the Pacific Ocean

**TYPE** Boreal forest (taiga)

**AREA** 3.4 million square miles (8.75 million square km)

The Eurasian boreal forest, often referred to by its Russian name, taiga, extends for more than 6,210 miles (10,000 km) from Scandinavia to the east coast of Russia. It is broadest in Siberia, where it stretches from 49°N to 72°N. In Europe, the main tree species is Norway spruce, although Scotch pine dominates in some parts of the boreal forest here, especially in the more oceanic west. Farther east, Siberian spruce, Siberian fir, and Siberian stone pine also occur. Although evergreen conifers cover large stretches of the boreal forest, deciduous trees flourish in some areas, often mixed with the conifers. These trees include larch, birch, alder, and rowan, with larch prevalent in a vast expanse of the Yakutiya region of north-central Russia. Toward its northern limit, the taiga starts to thin out, with trees more widely spaced or restricted to sheltered sites such as valleys, and eventually the trees give way to tundra vegetation. To the south, the taiga grades into mixed woodland or, farther east, into steppe habitats. In Russia, the northern taiga is interspersed with expanses of bogs and marshland, where impeded drainage produces conditions too wet for forest. Dwarf shrubs such as bilberry, cranberry, and crowberry flourish here. The trees of the boreal forest survive very harsh conditions, with winter temperatures in Siberia plunging to –76°F (–60°C). These mainly dense, dark forests tend to have sparse ground flora but abundant mosses and lichens. The taiga still has stable populations of mammals such as sable and Eurasian brown bears, and typical birds include capercaillie and Siberian jays.

**SIBERIAN BIRCH FOREST**
*Silver birch is especially prevalent at the interface between the taiga and tundra. Birch forest is less dense than coniferous forest, and the light reaching the forest floor encourages a thick layer of herbs to develop.*

**BOREAL FOREST, FINLAND**
*The autumn colors of birch, Finland's national tree, brighten the landscape of the boreal forest in Lapland. Birch rapidly colonizes disturbed soil, and often grows near rivers and lakes.*

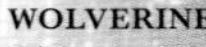

**WOLVERINE**
*This powerful member of the weasel family is found in both boreal forest belts. It takes a wide range of food, from eggs and birds, to lemmings and berries, and can even kill reindeer.*

## NENETS

The Nenet people, who number about 10,000, live in the northern boreal forests and tundra of Siberia in Russia, where they herd reindeer, on which they depend for food, clothing, tools, and transport. Each summer, they move north with the reindeer herds to fresh grazing lands in the tundra as it becomes free of ice, and then return to the protection of the taiga during the autumn and long winter. The Nenet have perfected the art of exploiting the harsh environments of the Siberian boreal forests and adjacent tundra. Their diet of reindeer meat is supplemented by sea and river fish, seals, and wildfowl. However, in the face of increased links with modern society, the Nenet are now struggling to retain their traditional lifestyle.

**TAIGA**
*These mammal tracks are preserved in snow beneath a stand of tall conifers in the Russian boreal forest. In many areas, snow covers the ground for several months.*

LAND

AFRICA *central*

## Central African rainforest

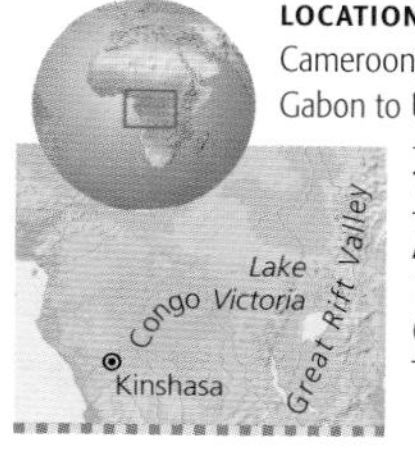


**LOCATION** From the coasts of Cameroon, Equatorial Guinea, and Gabon to Uganda and Burundi

**TYPE** Tropical rainforest

**AREA** 1.2 million square miles (1.9 million square km)

The rainforests of central Africa account for more than 80 percent of the continent's rainforest. Although they extend to the west coast, and about 90 percent of Gabon is covered by forest, they are centered on the basin of the Congo River (see p.230). The rainforests of the Democratic Republic of Congo (DRC), in particular, are the richest in all of Africa, with about 11,000 plant species, including relatives of impatiens and begonias, and over 400 species of mammals, such as the bonobo and African forest elephant. The trees include valuable timber species such as sapele, iroko, and African mahogany.

**COASTAL RAINFOREST**
*Mayombe Forest Reserve in the Congo and DRC (shown here) protects an important tract of tropical rainforest. West Africa's forests are being lost at a faster rate than any others.*

**MOUNTAIN GORILLAS**
*The mountain forests of Uganda, Rwanda, and Burundi are home to the world-famous mountain gorillas, whose dense fur keeps them warm even at 13,200 ft (4,000 m).*

AFRICA *southeast*

## Madagascan rainforest

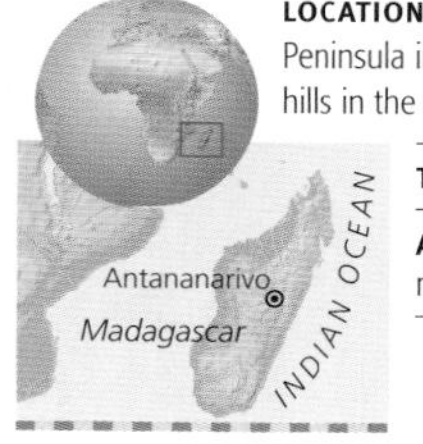


**LOCATION** On the Masoala Peninsula in the northeast, and on hills in the east, Madagascar

**TYPE** Tropical rainforest

**AREA** 14,670 square miles (38,000 square km)

Most of Madagascar's rainforests lie on the eastern side of the island and are now restricted to a narrow, broken strip following the hills just inland from the coast. Some of the best-preserved of the lowland forests are on the Masoala Peninsula in the northeast.

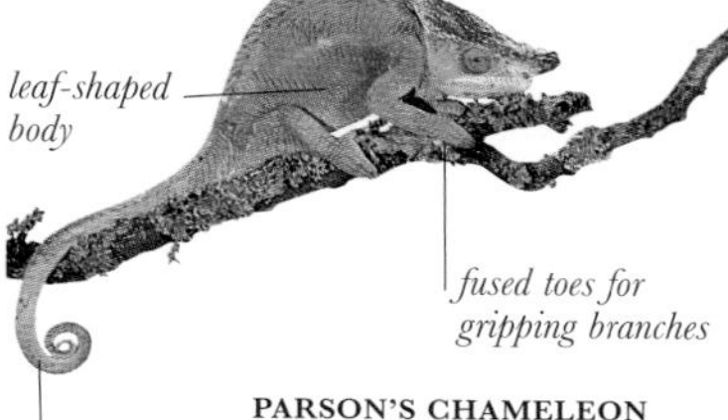


**PARSON'S CHAMELEON**
*This chameleon lives mainly in trees, is slow-moving, and so relies on camouflage to avoid predators.*

The canopy of the forest is low (about 80 ft/25 m) and, in addition to hardwood trees such as ebony, there are many palms, as well as epiphytes including orchids and ferns. Owing to the fact that Madagascar is an island, 80 percent of the plant species are endemic. Likewise, of about 300 species of reptiles identified, 95 percent are endemic, including two-thirds of the world's chameleons. Madagascar is renowned for its lemurs, which are found only there and on some nearby islands. Those species that inhabit the rainforest include the ruffed lemur (see p.307) and the nocturnal aye-aye. Over 500 species of Madagascar's plants have medicinal uses, most famously the rosy periwinkle, which is now used worldwide for treating some types of childhood leukemia.

**MASOALA NATIONAL PARK**
*This reserve, covering 840 square miles (2,175 square km), protects Madagascar's largest remaining rainforest, as well as nearby coral reefs.*

ASIA *east*

## Asian mountain mixed forest

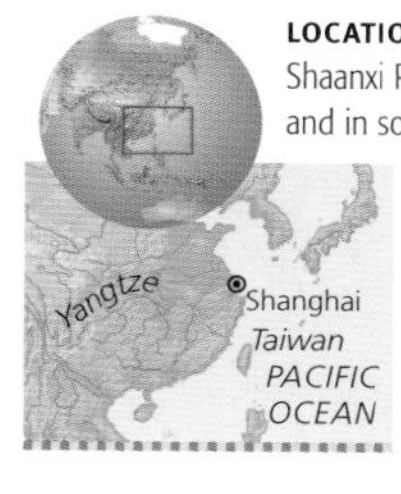


**LOCATION** In eastern China, from Shaanxi Province to the Pacific coast and in southern Japan

**TYPE** Deciduous temperate and evergreen temperate forests

**AREA** 760,000 square miles (1.9 million square km)

The mixed forests of Asia tend to be restricted to mountain sites, partly because much of the lowland has been cultivated. In China's Sichuan Province, the forests consist of a mixture of needle-leaved evergreen trees such as hemlock and spruce, with many broadleaved deciduous trees such as birch, maple, and cherry. In the Qin Ling Mountains the forests are largely of oaks, pines, and firs. A notable feature of these forests is the abundance of bamboo in the understory. In Japan, the mountain forests contain Japanese beech and cedar, while at higher altitudes conifers such as Nikko fir dominate, along with Erman's birch.

**BAMBOO FOREST**
*Bamboo forests are found in several parts of China, notably in the east. In some areas, the bamboo is cropped to provide a wide range of products, from poles to food and medicine.*

**WUHUA LAKE**
*This forest-fringed lake lies in Sichuan Province. Giant pandas are still found in these bamboo-rich mountain forests.*

**LOWLAND RAINFOREST IN BORNEO**
*Occasional tall, emergent trees rise above the general level of the canopy in this rainforest, while the gap created by the river encourages rapid growth of young trees and climbers.*

ASIA *southeast*

# Indonesian rainforest

**LOCATION** From the island of Sumatra in the west to Papua New Guinea in the east

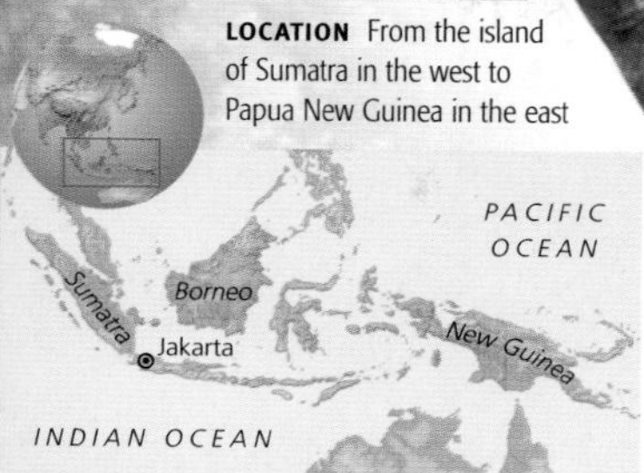


**TYPE** Tropical rainforest

**AREA** 540,540 square miles (1.43 million square km)

**ORANGUTANS**
*The orangutan, one of our closest relatives, uses its long arms and large hands to help it clamber through the tangled tropical forest.*

The rainforests of Indonesia are among the most extensive in the world, accounting for about 10 percent of the world's forest. With more than 500 species of mammals, around 1,500 bird species, 1,000 reptiles and amphibians, and 20,000 plant species, they are also some of the richest habitats in existence. In lowland areas, a mixed rainforest usually develops below about 3,100 ft (950 m). Typically this has three main layers of trees, of up to 100 different species. Many belong to the family Dipterocarpaceae, which is virtually absent from the tropical forest regions outside Asia. About 60 ft (18 m) tall, with trunks that are branch-free for most of their length, many of these trees rise majestically into the canopy, a few overtopping it as emergents. Many species flower during the same season, having been free of flowers for several years. The ground vegetation of the rainforest is made up of young trees, shrubs, palms, and ferns. The palms are more diverse than anywhere else in the world, with 975 species, but the orchids are the largest family, numbering 6,500 species. Some of the plants that grow here are remarkable, such as the rafflesia (see below). Swamp forests are widespread in Indonesia, occurring where the ground is waterlogged for much of the year, such as close to rivers that are prone to flooding or on wet, peaty ground. Many of the coasts have mangrove forests in the tidal zones, and these show a transition toward inland rainforest. Above about 3,100 ft (950 m), montane forest develops. Here the trees are mostly less than 50 ft (15 m) tall and have smaller, more leathery leaves. Such habitats are rich in epiphytes, including mosses, ferns, orchids, and lichens. Cloud and mist often form, and it is this aerial moisture that supports the epiphytes.

**MANGROVE FORESTS**
*Thriving in marine or brackish mud around many tropical coasts, some species of mangrove produce "knees" (pneumatophores), special aerial growths that bring air into the root tissues.*

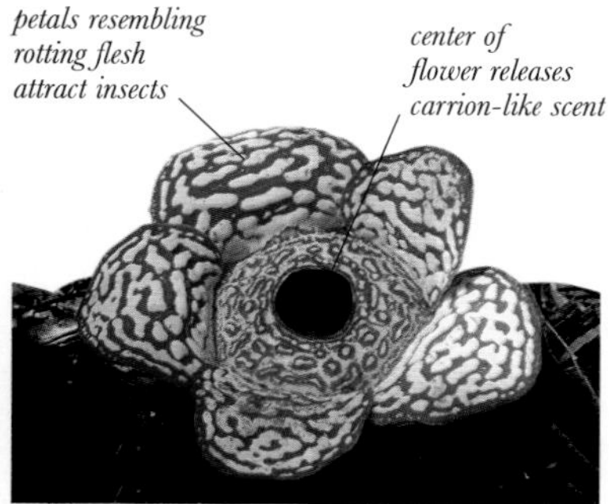


**RAFFLESIA ARNOLDII**
*This species produces a single flower that is larger than that of any other plant. The flower of one individual weighed 23 lb (11 kg) and measured 3 ft (1 m) in diameter.*

## RAINFOREST FIRES

Fires are a major threat, even to moist rainforest, and Indonesia has suffered badly, both from natural fires and from those started deliberately as an aid to logging operations or for clearing trees to make way for alternative land use. Once a forest fire has taken hold, particularly in dry conditions, such as those of the drought in 1997, the proximity of the trees allows it to spread rapidly, destroying not only the trees themselves but virtually all the associated wildlife.

**FIGHTING THE FIRE**
*This villager is attempting to douse a brush fire in Kalimantan, the Indonesian part of Borneo, in 1997.*

ASIA *northeast*

## Northeast Asian mixed forest

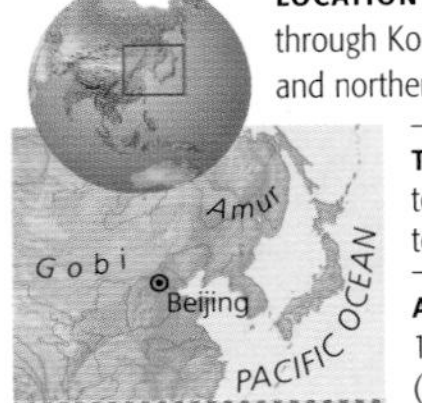


**LOCATION** From northeast China, through Korea, southeast Russia, and northern Japan

**TYPE** Deciduous temperate and evergreen temperate forests

**AREA** 1.2 million square miles (3.2 million square km)

A typical example of this type of mixed forest is that found on the foothills of the Sikhote-Alin' Range in Ussuriland, southeast Russia. It is mostly composed of cedar, pine, black fir, and local species of spruce, ash, linden, maple, and walnut. Lush undergrowth includes plants such as wild vines and ginseng. Mammals include musk deer and, in some parts, the rare Siberian tiger.

LOWLAND FOREST, USSURILAND, RUSSIA

ASIA *southeast*

## Southeast Asian rainforest

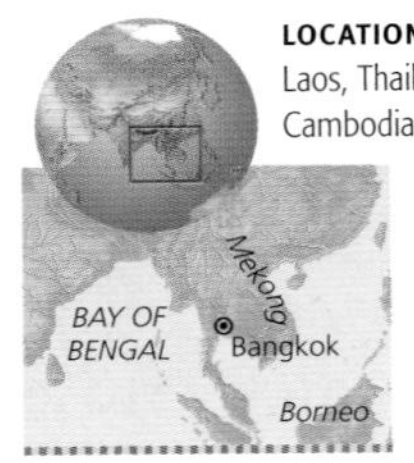


**LOCATION** In Burma (Myanmar), Laos, Thailand, peninsular Malaysia, Cambodia, and Vietnam

**TYPE** Tropical rainforest

**AREA** 655,000 square miles (1.7 million square km)

Like those of Indonesia (see p.319), the rainforests of mainland Southeast Asia are dominated by dipterocarp trees, which account for 40 percent of the understory and about 80 percent of emergents. The tallest trees reach heights of 130–162 ft (40–50 m). The valuable timber tree teak is native to Burma, Thailand, and Laos, where it is usually found mixed with other trees. But logging has greatly reduced the natural stands, and teak plantations in various Asian countries, Africa, and the Americas now help to meet the incessant demand. In peninsular Malaysia, the rainforests contain large numbers of rattans. These palms are harvested for thatching, matting, and basketry materials, and are a source of dyes and medicines. Many familiar tropical fruits originate from the Southeast Asian forests, such as bananas and mangoes, and also durian.

**RAJAH BROOKE'S BIRDWING**
*Both sexes of this large butterfly feed on flowers, and the males (shown here) often drink from wet mud on the forest floor.*

*distinctive contrast of green on black base*

*jet-black body*

*elongated forewing typical of birdwing butterflies*

**LOWLAND RAINFOREST**
*These tropical trees in southern Thailand have typically straight trunks, branching mainly toward the crown.*

**MONTANE RAINFOREST**
*The forests surrounding Thailand's highest peak, Doi Inthanon, range from tropical to a more temperate cloudforest toward the summit.*

AUSTRALASIA *New Zealand*

## New Zealand mixed forest

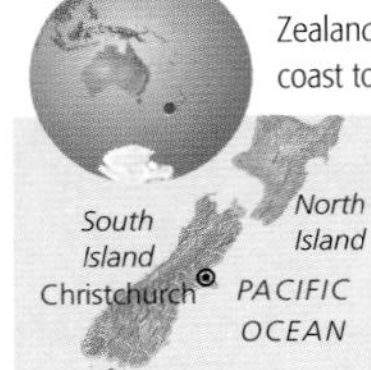


**LOCATION** Throughout New Zealand's South Island, from the coast to submontane levels

**TYPE** Deciduous temperate and evergreen temperate forests

**AREA** 44,000 square miles (110,000 square km)

Temperate forests still cover a large area of New Zealand's South Island. Common trees include southern and silver beech and members of the podocarp family, long-lived conifers that were widespread in the Mesozoic Era but which today are mostly confined to the southern hemisphere. Deciduous trees such as tawa and kamahi often dominate in the forest canopy, and these tend to be overshadowed by tall podocarps, such as kahikatea, rimu (or red pine), and totara.

**COASTAL FOREST**
*New Zealand has large areas of well-preserved coastal lowland forests, as here in Catlins, in the far south of South Island.*

AUSTRALASIA *Australia*

## Australian eucalypt forest

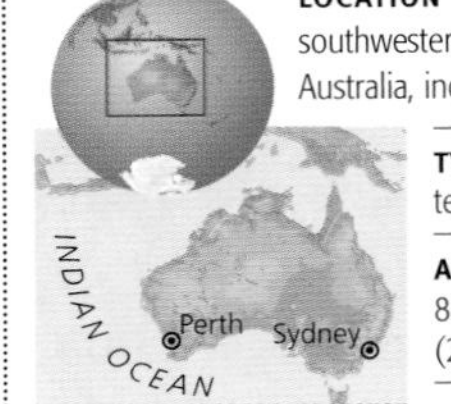


**LOCATION** In northern, eastern, southwestern, and southeastern Australia, including Tasmania

**TYPE** Evergreen temperate forest

**AREA** 888,000 square miles (2.3 million square km)

Eucalypt forest is scattered through various regions of Australia, but not in the interior, which is too arid for tree growth. Australia has about 450 species of eucalypt, nearly all of which are evergreen. Different species dominate in different areas: karri in the north and southwest; Sydney blue gum in the east; and Australian mountain ash in the southeast. Fires are frequent during dry weather, and most gum trees have fire-resistant bark. They can also regenerate rapidly from seed or by sending up suckers. Koalas are entirely dependent on the eucalypt forests of east Australia, spending nearly all their lives up in the branches. They feed at night, consuming about 1 lb (500 g) of gum leaves. One of the smallest opossums, the honey possum, is only found in the forests of the southwest. It feeds on nectar and pollen, which it gathers with its bristle-tipped tongue.

### JOSEPH BANKS

Englishman Joseph Banks (1743–1820) was one of the great botanical explorers and collectors. At age 25, he joined James Cook's expedition to the South Pacific on the *Endeavour*. In 1770, they landed in eastern Australia, where Banks amassed a vast collection of botanical specimens then unknown to Europeans. Appropriately, Captain Cook named their landing place Botany Bay.

**BLUE MOUNTAINS**
*The Blue Mountains, to the west of Sydney, get their name from the bluish haze created by sunlight filtering through a mist of eucalypt oil released by the gum trees covering their slopes.*

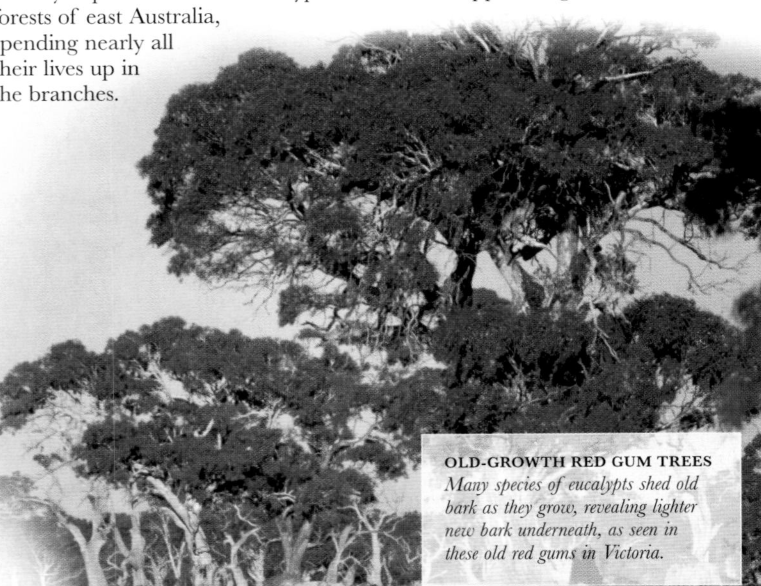

**OLD-GROWTH RED GUM TREES**
*Many species of eucalypts shed old bark as they grow, revealing lighter new bark underneath, as seen in these old red gums in Victoria.*

AUSTRALASIA *Australia*

## Northeast Australian rainforest

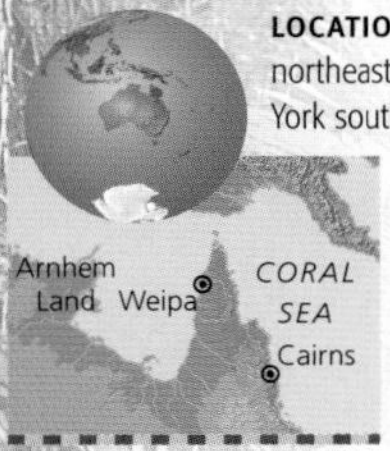

**LOCATION** In the coastal ranges of northeast Queensland, from Cape York south to the Connors Range

**TYPE** Tropical rainforest

**AREA** 14,050 square miles (0,500 square km)

The largest expanse of Australia's remaining tropical rainforest is found between Cooktown and Townsville, centered on Cairns. The forests are best developed on the eastern slopes of the coastal ranges, where the average annual rainfall is highest, at more than 59 in (1,500 mm). The canopy is uneven, varying from 65 ft to 130 ft (20 m to 40 m) in height, and the undergrowth features clambering vines and epiphytes, as well as walking stick palms, ginger, and aroids. The trees include valuable timber species such as Queensland maple, kauri pines, and red cedar (though logging is now illegal in much of the forest), and strangler figs. At the highest altitudes, over 3,300 ft (1,000 m), the trees are shorter, and these cloudforests have a dense growth of ferns and mosses. Australia's geographical isolation means that it has a high number of ancient and unique species, and Queensland's tropical rainforests account for much of the country's biodiversity. About 20 percent of Australia's cycads (see p.33) are found here, as is idiot fruit, a primitive flowering plant that originated 120 million years ago. Two species of tree kangaroo, Lumholtz's and Bennett's, live in the rainforest, along with many other endemic mammals, birds, reptiles, and frogs.

**LUMHOLTZ'S TREE KANGAROO**
*This nocturnal marsupial is a highly agile climber, using its cushioned, rough-soled feet to cling to branches and its long tail as a counterbalance.*

**SATIN BOWERBIRD**
*To attract the female, the glossy black male builds an elaborate bower on the rainforest floor, which it decorates with any blue object it can carry in its beak.*

**DAINTREE RAINFOREST**
*This rainforest, a major part of the Wet Tropics World Heritage Site created in 1988, clothes the coastal hills near Cairns and stretches right down to the sea.*

### ABOVE THE CANOPY

Skyrail is a cableway near Cairns that runs for 4⅔ miles (7.5 km) from the base of the MacAllister Range to Kuranda, a village high up in the rainforest. There are 114 gondolas, from which tourists can gaze down at the forest and its inhabitants with the minimum of disturbance. There are also two stops where the cableway descends through the canopy and people can get out and walk through the forest on boardwalks.

**FAST-FLOWING STREAM**
*This stream is in the Daintree rainforest, North Queensland. The forests in this area contain 41 percent of Australia's freshwater fish species.*

**SEASONAL FLOODING**
*Waterlilies bloom near the mouth of the Limmen River, which lies in the northeast of Australia's Northern Territory. This river has extensive areas of seasonally inundated floodplains along its banks.*

# WETLANDS

WETLANDS ARE FOUND ALL OVER the world, in places where drainage of water is impeded or in areas adjacent to the floodplains, deltas, and estuaries of large rivers. A wetland is any waterlogged or flooded terrestrial area that has a covering of water plants, either rooted or free-floating. The water is often just 3 ft (1 m) deep. Not all wetlands are flooded year-round—for example, some wetlands occur in areas that are seasonally flooded by rivers; others form where spring meltwater cannot drain away because of underlying permafrost. Wetlands are highly important ecosystems, supporting a wide variety of specially adapted plants and animals, most notably vast numbers of water birds. They may act as reservoirs of water for nearby human populations, and are also important in many regions as a source of fish and for recreational activities. Many of them are increasingly threatened by encroachment—with large areas being drained for housing or agricultural use—and also by pollution.

# WETLANDS

102 Organic soils
126–27 Salt water and fresh water
129 The global water cycle and the local water cycle
212–35 Rivers
236–51 Lakes
Tundra 342–43

WETLANDS CAN BE DIVIDED INTO freshwater wetlands, where the water is mainly derived from rainfall, and saltwater wetlands, which are generally coastal and influenced by sea water. Inland saltwater wetlands occur where evaporation has concentrated salts in the surface layers of the soil. Many freshwater wetlands occur at lake and river margins, especially along the floodplains of lowland rivers, and soil saturation can be seasonal.

## FOSSILS IN PEAT

Bogs preserve material because wet peat contains little oxygen and so slows down the processes of decay. The peat bogs of northern Europe, particularly those in Denmark, have yielded some remarkably well-preserved human bodies, about 2,000 years old, still with their hair, skin, and fingernails intact. These include the remains of the man shown here, thought to have been a victim of ritual sacrifice.

## FRESHWATER WETLANDS

The main types of freshwater wetlands are fens, bogs, swamps, and marshes. Fens and bogs, known collectively as mires, are habitats rich in wet, peaty soils. Bogs receive water from rainfall alone, and form acid peat. Fens receive groundwater as well as rainfall—they are generally wetter, and form neutral or alkaline peat. Mires are most widespread in temperate regions with high rainfall, as in the northern and western North America and Europe. They also occur at high altitudes if conditions are cool and damp. There are many gradations between fens and bogs, with mixed mires or intermediate mires occurring in some areas. The terms "swamp" and "marsh" classify wetlands by the plants that grow in them—a swamp is a wetland forest, while a marsh is a wet grassland.

**WICKEN FEN**
*This wetland reserve in Cambridgeshire protects a vital remnant of the once widespread fens of eastern England.*

**PEAT CUTTING**
*Large areas of Ireland are covered in bogs, especially in the west, as here in County Clare. The peat is traditionally cut, dried, and used for fuel, but it is a nonrenewable resource.*

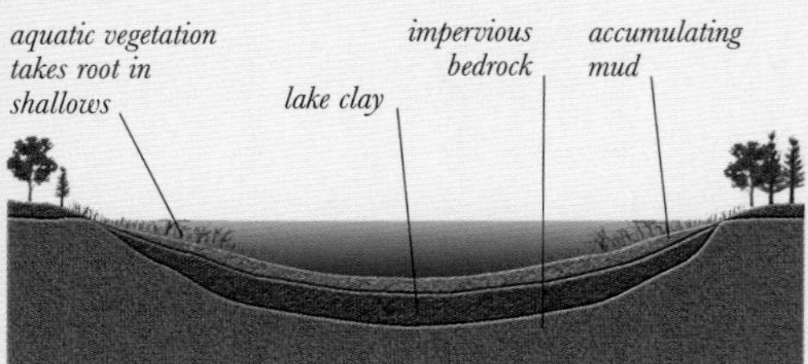


**WETLAND SUCCESSION**
*Freshwater wetlands often occur on peat bogs, which form as lakes are infilled by sediment. In the first stage of this process, mud collects on an impervious lake bed.*

**INTERMEDIATE STAGE**
*Fen peat develops above the lake mud as the partly decomposed remains of plants, such as sedges and rushes, gather under the influence of mineral-rich (alkaline) groundwater.*

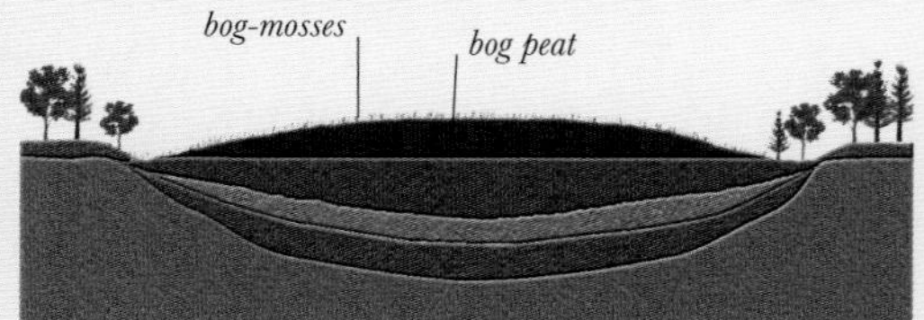


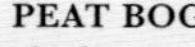

**PEAT BOG**
*As fen peat fills the lake, the surface is isolated from alkaline groundwater, and slowly becomes more acidic. This encourages the growth of bog-mosses, whose remains gather as bog peat.*

## SALTWATER WETLANDS

Most saltwater wetlands occur at the interface of land and sea, where fresh and salt water mix. A range of habitats forms where large rivers with wide floodplains braid to create extensive deltas. These include brackish (slightly salty) and saltwater lagoons, and salt marshes with salt-tolerant vegetation covering the rich estuarine mud. In warm climates, salty lagoons are often managed by people as a method of salt production. In the tropics, mangrove swamps are often found in tidal creeks and gullies. Mangrove trees germinate fast in the mud, their stiltlike roots stabilizing the soil and allowing the plants to survive tidal sea-level changes. Saltwater wetlands also occur inland, frequently in arid regions, and even in some deserts, where intermittent flooding creates temporary water bodies.

**SALT MARSH**
*Salt marshes develop in flat, coastal regions that are regularly flooded at high tide, such the Chincoteague Inlet in Virginia.*

LAND

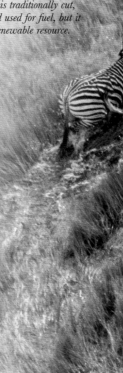

## PLANT LIFE

Wetlands present a variety challenges and opportunities to plants. Many aquatic or marsh plants have light, spongy stems and leaves, adapted to transport atmospheric oxygen to the roots, allowing the plant to keep respiring. Some, such as water hyacinth and water fern, also have waxy or hairy leaves, which resist waterlogging. By contrast, salt-marsh plants often display features shared by desert plants, such as a thick cuticle and narrow leaves. Salt water is not easily absorbed by plants (as salt solutions tend to draw water out of plant tissues by osmosis), so many salt-marsh plants suffer effective drought conditions, at least some of the time. Some plants growing at the edges of creeks, ponds, or lagoons, such as reeds, are held above the water by woody stems and can survive short-term changes in water level.

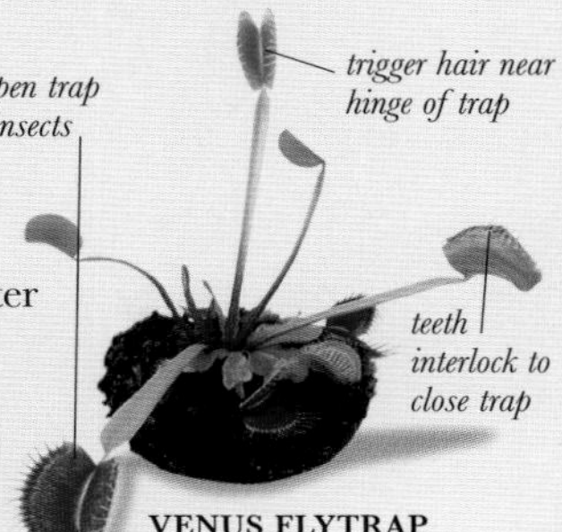

**VENUS FLYTRAP**
*Insects caught in this plant's leaves provide essential nutrients that are rare in the soils of the eastern US, where it grows.*

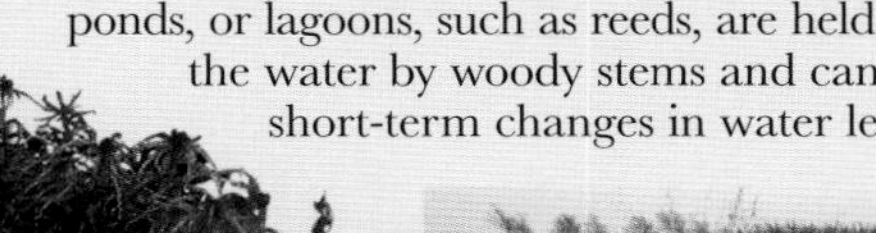

**SPHAGNUM MOSS**
*This fast-spreading inhabitant of bogs and fens grows half-submerged. Its decomposing remains form many types of peat.*

**MARSH MARIGOLD**
*The marsh marigold, which is common in temperate freshwater wetlands, produces brightly colored flowers in spring and summer.*

## ANIMAL LIFE

Fish and many aquatic invertebrates thrive in wetlands where there is sufficient water available, and in turn they provide food for a host of other animals, most notably a wide range of water birds. Herons and egrets stalk their prey in shallow water, ducks filter-feed at the surface or dive for food, while waders are adept at snatching invertebrates at or just below the surface of the wet mud. A large number of reptiles, including crocodiles, alligators, and freshwater turtles, are at home in wetlands. Grass snakes often swim in search of frogs, and tropical wetland species include the world's largest snake, the anaconda. The mammals have produced many wetland specialists, especially among the rodents, such as beavers, muskrats, and capybaras. Larger herbivores that are at home in wetlands include the Asian water buffalo and hippopotamuses.

**ATLANTIC MUDSKIPPER**
*The mudskipper is a unique fish that can climb out of water, using its fins to grasp.*

**JAGUAR**
*The jaguar's preferred habitat is swamps or seasonally flooded forests. It is a strong swimmer, and its diet includes capybaras, fish, and crocodiles.*

**COMMON BLUE DAMSELFLY**
*Damselflies are voracious aerial predators as adults, but the long juvenile stage of their life is spent underwater.*

**SEASONAL GATHERING**
*Wetlands such as Botswana's Okavango Delta burst into life in the wet season, when herds of game such as these Burchell's zebras gather.*

### WETLANDS PROFILES

The pages that follow contain profiles of some of the world's main wetlands. Each profile begins with the following summary information:

**TYPE** Marsh, bog, fen, swamp, mire, delta, or lagoon

**WATER** Fresh, acid fresh, brackish, or salt

**AREA** Surface area

NORTH AMERICA *east*

## Great Dismal Swamp

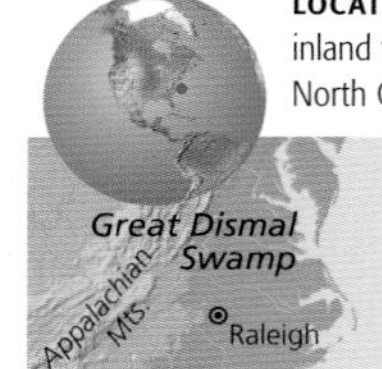


**LOCATION** About 25 miles (40 km) inland from the Atlantic Ocean, in North Carolina and Virginia

**TYPE** Swamp

**WATER** Acid fresh

**AREA** 600 square miles (1,550 square km)

**LAKE DRUMMOND**
*This lake, edged with bald cypress trees, was named after William Drummond, governor of North Carolina in the 1660s—though it lies in Virginia.*

Despite its discouraging name, the Great Dismal Swamp is a beautiful complex of wetland habitats in which tracts of wet forest are interspersed with scrub-shrub wetlands and areas of peat bog. The swamp is unusual in being located above sea level—most swamps occur in natural depressions, but the Great Dismal has been rebounding slowly since the end of the last ice age. A 150-ft- (45-m-) deep layer of clay below the swamp prevents its waters from seeping away. The variety of sites results in a rich diversity of plant and animal life. Forests of bald cypress, Atlantic white cedar, and black gum (or tupelo) dominate the marshy ground. Ground vegetation includes the rare semi-evergreen log fern. At the center of the swamp lies Lake Drummond—the lake and swamp waters support catfish, yellow perch, redfin pickerel, and the partially sighted swampfish. Almost 100 species of bird breed here, including wood ducks and barred owls, and mammals include white-tailed deer, raccoons, otters, black bears, and even bobcats. About 250 years ago, the swamp was as large as 2,200 square miles (5,700 square km), but logging, agricultural ventures (George Washington was an early investor), and roads, canals, and ditches have severely reduced its size. About 170 square miles (430 square km) of the swamp is now protected in an area designated as a National Wildlife Refuge.

**WHITE-TAILED DEER**
*Known to be good swimmers, white-tailed deer are often seen in wetland areas, feeding on green plants, including aquatic species, in the summer months.*

**DIVERSE HABITATS**
*Although it is better known for its natural stands of bald cypress and Atlantic white cedar, the Great Dismal Swamp also has areas of evergreen shrubs and marshes.*

NORTH AMERICA *southeast*

## Okefenokee Swamp

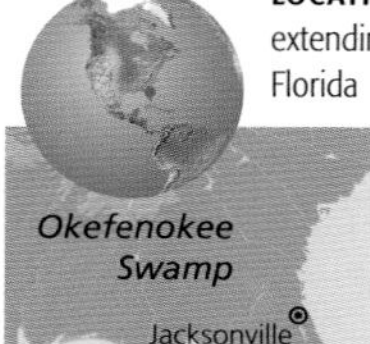


**LOCATION** In southeast Georgia, extending south into northern Florida

**TYPE** Swamp, marsh

**WATER** Acid fresh

**AREA** 685 square miles (1,770 square km)

The Okefenokee Swamp is dominated by an enormous mire lying inside a large, concave depression that is drained by the Suwannee River. Peat deposits have accumulated over the centuries to the extent that the swamp's surface is now over 100 ft (30 m) above sea level, and the peat is up to 15 ft (4.5 m) thick. In some areas the peat floats on waterlogged substrata, so that the whole surface quakes when trodden on. This is the origin of the swamp's name, from an American Indian term meaning "land of the trembling earth." Mires of this type, known as quaking bogs, are rare. The swamp's waters are sediment-free, but tinted the color of black tea by tannic acid released from decaying vegetation and peat. Although mires dominate the landscape, pines grow on drier islands, especially toward the northern margin. Another distinctive feature are large areas of marsh called water prairies. Bladderworts, sundews, and pitcher plants are all found in the swamp, compensating for the nutrient-poor soil with a carnivorous diet. Alligators thrive, as does the world's largest freshwater turtle, the alligator snapping turtle, which can weigh as much as 220 lb (100 kg). Snakes include the Florida cottonmouth, which, unlike other water snakes, swims with its head held out of the water. Birds abound, with over 230 species recorded, including the rare red-cockaded woodpecker, which nests in the pine trees. The swamp is also home to a population of Florida black bears.

**GOLDEN TRUMPET**
*This is one of three species of pitcher plant found in the Okefenokee. The "pitchers" are highly modified leaves, in which insects are trapped and digested.*

**BALD CYPRESS**
*The needles of this deciduous conifer turn a deep copper color in the autumn. The branches, draped with strands of Spanish moss, provide nesting sites for many birds.*

*scissor-sharp jaws deliver powerful bite*

*rough shell*

**ALLIGATOR SNAPPING TURTLE**
*This fearsome-looking reptile lurks in the swamp's dark, tannic waters, ready to lunge at unwary prey, which it seizes in its powerful jaws.*

WILDLIFE HAVEN
*The Everglades has a large and varied population of birds. Here, a great egret hunts in the waters beneath the buttress trunks of tall cypress trees.*



NORTH AMERICA *southeast*

# Everglades

**LOCATION** Extending south from Lake Okeechobee to Florida Bay, Florida

**TYPE** Swamp, marsh

**WATER** Fresh, salt

**AREA** 4,000 square miles (10,360 square km)

North America's only subtropical wetland, the Everglades, begins at Lake Okeechobee (see p.240). Water flowing from the lake seeps slowly through the low-lying land toward the Gulf of Mexico, creating a wetland wilderness with prairielike expanses of sawgrass. This area, known locally as the river of grass, is dependent on fires, usually ignited by lightning strikes, which burn back old grass that would otherwise impede the flow of water. Hardwoods are found on pockets of higher ground, mainly along limestone ridges, and these islands of trees are known as hammocks. The Everglades also has large areas of cypress marshes and, along the coast, mangrove swamps, salt marshes, and estuaries. Much of the original Everglades has been lost to development (see panel, right). Dikes, put in place to assist agriculture and other development, have altered the flow of water so that much of it is now diverted to the coast instead of draining into the soil. Agricultural fertilizers have also encouraged the spread of cattails, which overwhelm the sawgrass. A national park in the south protects some of the ecosystem, which is home to birds such as snail kites, purple gallinules, and anhingas. The Everglades is also famous for being the only place where crocodiles and alligators coexist.

STREAM CHANNELS
*The swamp is dissected by a multitude of channels that slowly carry the sluggish waters toward the Gulf of Mexico.*

SAWGRASS
*This plant is actually a member of the sedge family, rather than a grass, and gets its name from the serrated edges to its long leaves.*

MANGROVES
*Mangroves line many of the narrow rivers and creeks near the coast of the Everglades. Here, seedlings emerge from the rich coastal mud, putting out stilt roots for support.*

## LAND RECLAMATION

The Everglades originally occupied most of the southern tip of the Florida Peninsula. However, southern Florida's warm climate has attracted a large human population. Many areas of land that were once part of the Everglades ecosystem have been drained and reclaimed for urban use. This drainage has gradually reduced the natural flow of water through the Everglades, with devastating effects on its ecology.

HOUSING DEVELOPMENT
*These houses surround a water runoff pond, a drainage feature required by Florida law. The pond collects water from the housing development and helps to prevent possibly polluted water from reaching the Everglades.*

SOUTH AMERICA *central*

## Pantanal

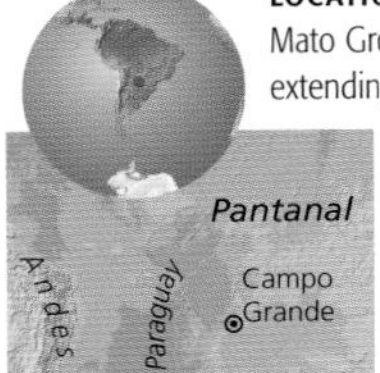


**LOCATION** In the Mato Grosso and Mato Grosso do Sul states of Brazil, extending into Bolivia and Paraguay

**TYPE** Swamp, marsh

**WATER** Fresh

**AREA** 50,000 square miles (129,500 square km)

At around 13 times the size of the Everglades, the Pantanal is the world's largest freshwater wetland, and it is also the most biologically diverse, a reflection of its tropical location. The Pantanal itself is a low-lying floodplain with rich alluvial soil, occupying about one-third of the basin of the upper Paraguay River. This is a truly dynamic ecosystem, governed by the annual flooding of the river, for which it acts as a sponge. Each year the waters rise by several yards, transforming a vast area of dry grasslands and forests into temporary swamps, islands, pools, and water channels. Floating aquatic plants, such as water hyacinth and water ferns, often carpet the water, in which piranhas, caimans, and anacondas are to be found. Caranda palms are a feature of the flooded land, while on the higher ground the mixed forest includes fig trees and acuri palms. The Pantanal is home to the endangered jaguar, and the world's largest rodent, the capybara, is also found here, as are tapirs and capuchin and howler monkeys. Birds include the rare hyacinth macaw, and there are thousands of species of butterflies.

**CAPYBARAS**
*The capybara is at home on land and in water. Its ears, eyes, and nostrils are set on top of the head, so it can remain alert while swimming.*

**VICTORIA WATERLILY**
*This magnificent aquatic lily has large floating leaves, about 3–6 ft (1–2 m) in diameter, which can support the weight of a child.*

**MARSH AND FLOODED FOREST**
*This aerial view shows the complex of interrelated habitats that make up the Pantanal, with vegetation-covered waterways cutting between forested embankments.*

LAND

SOUTH AMERICA *north*

## Llanos wetlands

**LOCATION** Centered on the Orinoco River and its tributaries in western Venezuela

**TYPE** Swamp, marsh

**WATER** Fresh

**AREA** 3,860 square miles (10,000 square km)

The Llanos is a term used to describe the huge expanse of savanna plains that stretch across Colombia and Venezuela from the Andes to the Orinoco Delta. Within this region, the Llanos wetlands comprise large areas of swamps and seasonally flooded savanna around the Orinoco River (see p.221) and its tributaries. The heaviest rains fall in May, at the end of the dry season, causing the rivers to overflow and transform the surrounding forests and grasslands into temporary wetlands. These support about 70 species of water birds, including around 90 percent of the world's population of the endangered scarlet ibis.

**SCARLET IBIS**
*This brightly colored bird breeds in treetop colonies, often over water, and feeds in flocks in lagoons, swamps, and estuaries.*

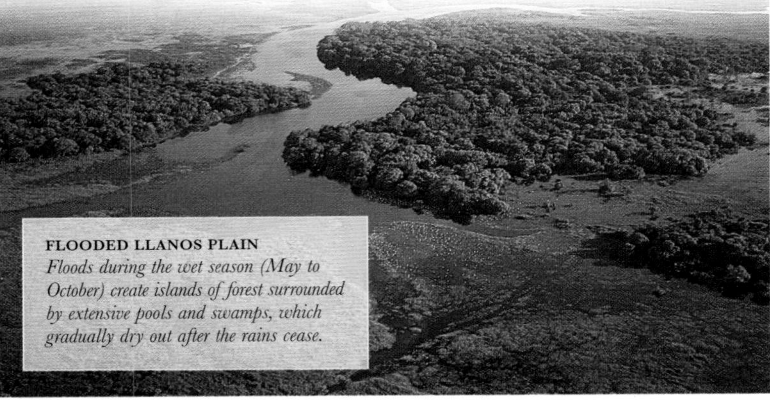

**FLOODED LLANOS PLAIN**
*Floods during the wet season (May to October) create islands of forest surrounded by extensive pools and swamps, which gradually dry out after the rains cease.*

EUROPE *northwest*

## Flow Country

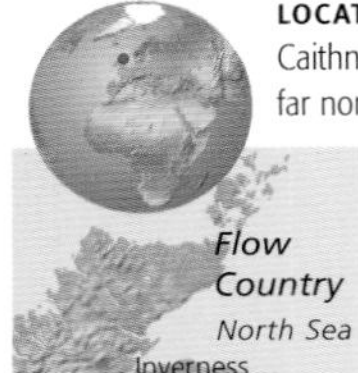


**LOCATION** In the counties of Caithness and Sutherland in the far north of Scotland

**TYPE** Bog

**WATER** Acid fresh

**AREA** 1,540 square miles (4,000 square km)

The Flow Country is a landscape of huge, rolling peatlands in the extreme north of Scotland, and it contains some of the best-developed blanket bogs in the world. Blanket bogs are wet peat formations that hug the contours of the land and grow very slowly, as little as 1/16 in (1 mm) a year, under the influence of a cool, wet climate. Sphagnum mosses, which ultimately create and sustain the bogs (see p.325), thrive, and ridges, hummocks, and small pools are rich in a variety of other plants.

**BOGBEAN**
*The white flowers of bogbean stand out from the dark, peaty waters in which it grows.*

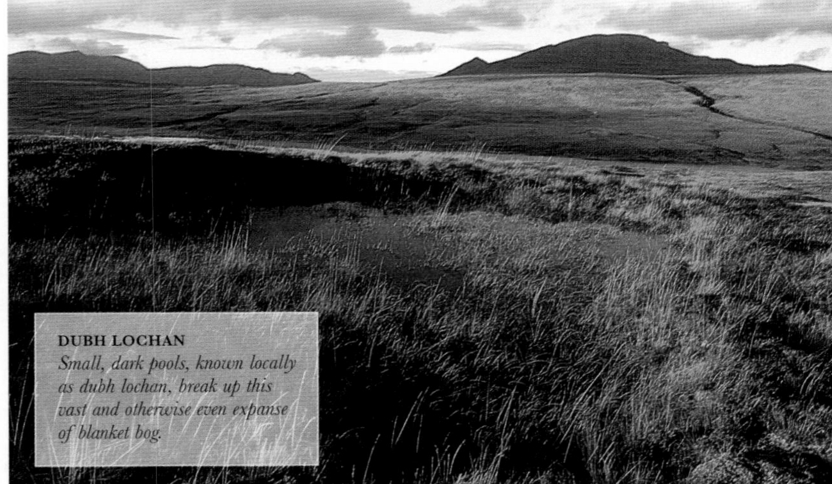

**DUBH LOCHAN**
*Small, dark pools, known locally as dubh lochan, break up this vast and otherwise even expanse of blanket bog.*

EUROPE *central*

# Biebrza Marshes

**LOCATION** Centred on the Biebrza River Valley, in Suwalki, Lomza, and Bialystok provinces, northeast Poland

**TYPE** Bog, fen, marsh

**WATER** Fresh

**AREA** 390 square miles (1,000 square km)

The Biebrza Marshes are a complex mixture of habitats, including river channels, lakes, extensive marshes, wooded areas on higher ground, and peat bogs that are some of the best preserved in the world. Many parts show a classic succession from riverside fen through raised bog, grading into wet woodland. These varied habitats support a wide range of wildlife. The birds found here include many species of waders, gulls, terns, and both white and black storks. Mammals include wetland specialists such as elk, muskrats, and Eurasian beavers.

**GREAT SNIPE**
*This wader breeds on the Biebrza Marshes, where male birds gather in groups called leks to attract females by posturing and making croaking and bubbling calls.*

**WATER MEADOWS**
*The water meadows of these marshes are regularly flooded, and the resulting rich alluvial soil supports a varied community of wetland plants.*

EUROPE *central*

# Hortobágy

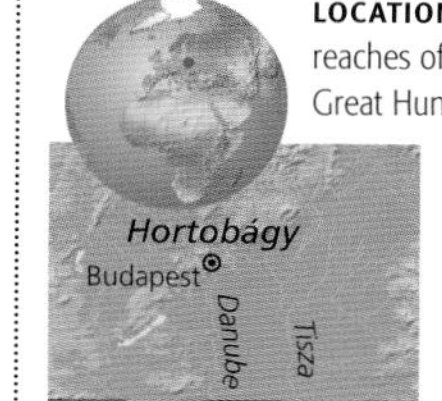

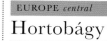

**LOCATION** Centered on the upper reaches of the Tisza River on the Great Hungarian Plain, Hungary

**TYPE** Lagoon, marsh

**WATER** Fresh, salt

**AREA** 44 square miles (115 square km )

Hortobágy lies in the Hungarian steppe region, known as the Puszta, an area consisting mainly of semiarid grassland. Regular flooding by the Tisza River has turned part of this otherwise dry landscape into a wetland encompassing streams, lakes, lagoons, salt marshes, and river-valley woodland. The region was given national park status (Hungary's first) in 1973, and today it is recognized as one of the world's best wetland reserves—one that boasts over 300 species of birds. Regulation of the flow of the Tisza has reduced flooding, and the park authorities now use canals to preserve some of the wetland habitats.

**COMMON CRANES**
*Large flocks of these migratory birds often stop over at Hortobágy to feed on insects, frogs, and plants.*

**REED BEDS**
*The lakes of Hortobágy are fringed by dense beds of reeds, which are used for shelter by ducks, herons, and other birds.*

EUROPE *west*

# Camargue

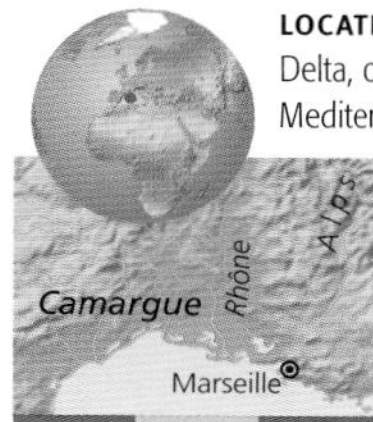


**LOCATION** Lying within the Rhône Delta, on the northwest shore of the Mediterranean Sea, in France

**TYPE** Lagoon, marsh

**WATER** Fresh, salt

**AREA** 330 square miles (850 square km)

The Camargue is based on sediments deposited by the Rhône River (see p.225) as it braids to form a huge delta, just to the west of Marseille. This diverse wetland comprises salt marshes, saltwater and brackish lagoons, freshwater ponds, rivers, grazed flooded meadows, reed beds, and dunes. The plants found here include salt-tolerant tamarisk bushes, sand crocuses, and the white-flowered sea daffodil, as well as many orchids. But, aside from its white horses and black bulls, it is for its birds that the Camargue is best known. The breeding birds include its famous colony of greater flamingos (now numbering about 10,000), as well as black-winged stilts and avocets. Its location also gives it great significance as a refueling stop for migrating birds flying between Europe and Africa. Much of the original wetland has been drained for conversion to agriculture (see panel, right), and a series of dikes and canals control the flow of the water. This, along with pollution of the Rhône, has led to a deterioration of some of the habitats. In 1970, the entire Camargue was designated a regional natural park, and some smaller areas have since been given greater protection as nature reserves.

**BRINE SHRIMP**
*These translucent crustaceans thrive in temporary salt lagoons, providing food for flamingos and other birds.*

## RICE CULTIVATION

Large areas of the Camargue, especially in the north, have been drained and are used to grow cereals, fruit, and vegetables under irrigation. Rice, which has been cultivated here since the Middle Ages, in particular thrives in the region's rich alluvial soil. The small plots of land in which the rice is grown are flooded with water from the irrigation canals in spring, and the crop is harvested mechanically during September and October, once the rice fields have been drained.

**FERAL HORSES**
*The Camargue's famous white horses are actually gray when young, only becoming white when four or five years old. For centuries they have been used by local farmers and breeders to round up the black bulls that also roam the salt marshes.*

AFRICA *south*

# Okavango Delta

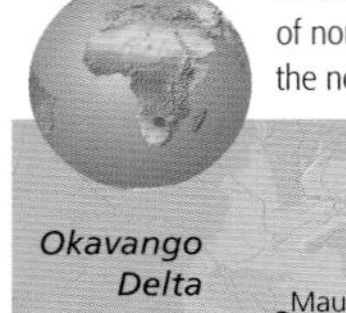


**LOCATION** In the Ngamiland District of northern Botswana, extending into the northern Kalahari Desert

**TYPE** Delta, swamp

**WATER** Fresh

**AREA** 770 square miles (2,000 square km)

The lagoons, swamps, and savannas of the Okavango Delta form what is arguably the greatest wildlife wilderness in southern Africa. The river starts its life as the Cubango in Angola, to the northwest, but instead of flowing to the coast, it spreads out through the flat land of northern Botswana, eventually dissipating in the parched savannas and desert sands of the northern Kalahari. Each year, the rains cause the Okavango River to flood, generally peaking in May, and then the waters spread, covering more than 5,000 square miles (13,000 square km). In high-flood years, the waters even reach the Makgadikgadi Pans (via the Boteti River), a huge area of salt pans about 125 miles (200 km) to the southeast, turning it into lagoons that provide a temporary haven for wildlife, most notably large flocks of flamingos. Although most of the floodwaters disappear into the dry atmosphere and arid, sandy soils, enough remains to form a network of clear streams, lagoons, and swamps, bringing a rapid greening of the landscape and a massive influx of herbivores, which in turn draw predators such as lions and leopards. The Okavango Delta's aquatic vegetation includes bladderworts (free-floating plants that trap and digest mosquito larvae), and dense beds of phragmites reeds and papyrus. These reed-beds are home to the sitatunga, a semiaquatic antelope whose widely splayed hooves are an adaptation for walking on soft, muddy ground. Kingfishers, Pel's fish-owl, and the African fish-eagle are among over 500 species of birds found in the delta, along with about 160 species of reptiles, 40 species of amphibians, and 90 species of fish.

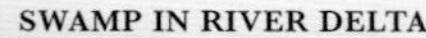

**SWAMP IN RIVER DELTA**
*The waters of the Okavango River disperse through a series of braided channels and spill over into the delta, depositing vast amounts of sediment.*

**AFRICAN ELEPHANTS**
*Although they live in a variety of habitats, African elephants will usually seek fresh water to drink—once a day, if possible, or every few days. They also bathe to help keep their skin in good condition.*

## ECOTOURISM

Ecological tourism has developed rapidly in the Okavango Delta in recent years. One of the main tourist destinations is Botswana's Moremi Wildlife Reserve, in the northeast of the delta. The reserve, established in 1963, covers about 1,160 square miles (3,000 square km), and includes large islands of dry land as well as lagoons and swamps. Ecotourism has brought welcome employment for large numbers of local people, who are skilled at identifying and tracking the wildlife.

**MOKORO SAFARI**
*These local guides are leading a party of ecotourists through the wildlife-rich waters of the delta in dugout canoes, or mokoros.*

**FLOODING IN THE KALAHARI**
*The annual flooding of the Okavango River inundates large areas of land in the northern Kalahari, creating a temporary wetland with islands of grass and palm trees.*

AFRICA *central*

## Sudd

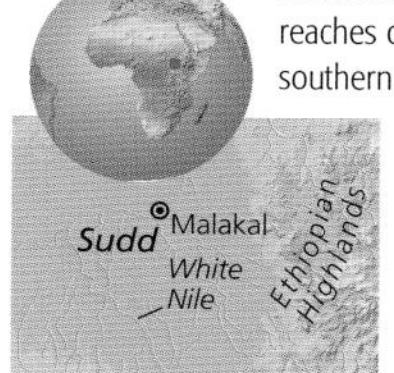


**LOCATION** Centered on the upper reaches of the White Nile river, southern Sudan

**TYPE** Marsh

**WATER** Fresh

**AREA** 13,300 square miles (34,500 square km)

One of the world's largest inland wetlands, the marshes of the Sudd cover a vast area in the upper reaches of the White Nile (see p.229). This is a landscape of reed beds and papyrus, with areas of open water that are often choked by dense mats of floating water hyacinth. The rains fall mostly between April and September, keeping the marshes wet, and flooding of nearby grasslands and woods triples the wetland's extent to over 38,600 square miles (100,000 square km). The Sudd is a haven for wildlife in an otherwise dry region, and it positively teems with birds, especially during the migration periods, when the number of species exceeds 400. The eastern edge of the Sudd is scarred by the Jonglei Canal, an ambitious and uncompleted project designed to divert water and to drain part of the marshes. Abandoned for financial and political reasons, this project has left a trough about 16 ft (5 m) deep, 245 ft (75 m) wide, and 225 miles (360 km) long, which blocks the migrations of large mammals such as giraffes, elephants, and hippopotamuses.

**SHOEBILL STORK**
*Also known as the whale-headed stork, this distinctive African bird is seldom seen far from water. It feeds mostly on lungfish, and also frogs and small mammals.*

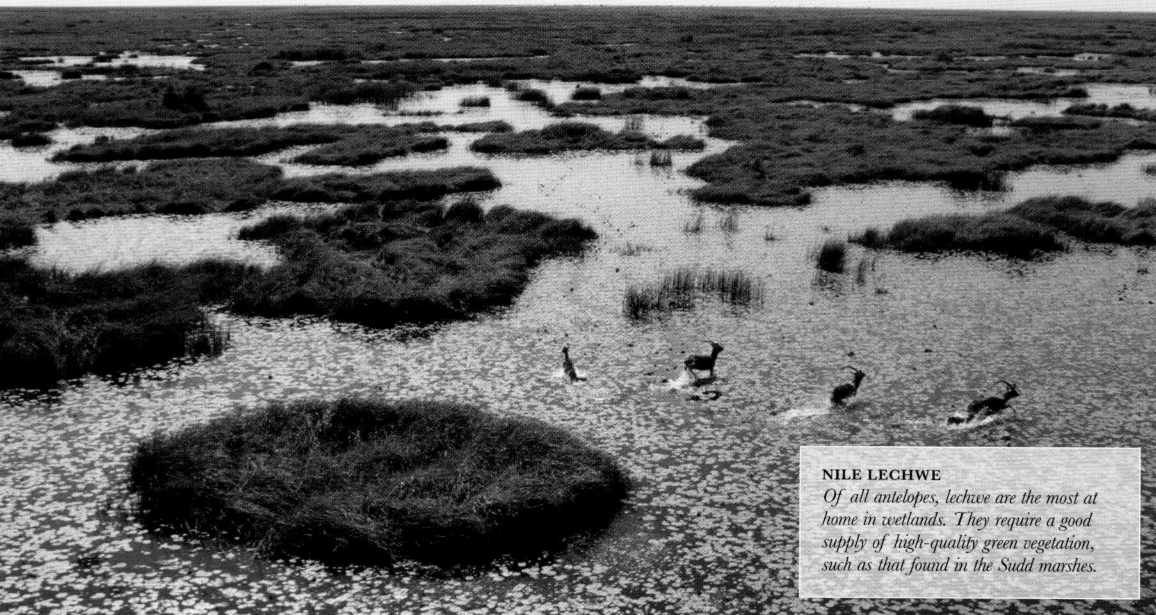

**NILE LECHWE**
*Of all antelopes, lechwe are the most at home in wetlands. They require a good supply of high-quality green vegetation, such as that found in the Sudd marshes.*

AUSTRALASIA *New Zealand*

## Waituna Lagoon

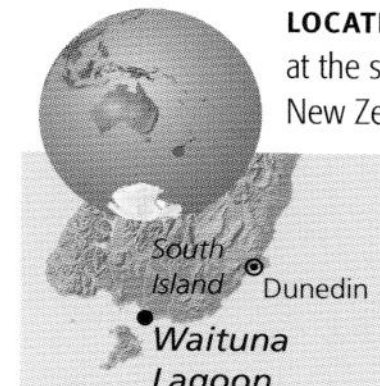


**LOCATION** In the Southland region, at the southern tip of South Island, New Zealand

**TYPE** Lagoon, bog

**WATER** Fresh, salt

**AREA** 13½ square miles (35 square km)

The southernmost of all the world's recognized wetlands, Waituna Lagoon comprises areas of salt marsh and peatland, as well as the lagoon itself. It is unusual in that it contains areas of cushion bog, which have vegetation that is normally associated with upland conditions rather than coastal regions. The cushion plant *Donatia novae-zelandiae* grows here, as do gentians, sundews, comb sedge (*Oreobolus pectinatus*), wire rush, tangle fern, and manuka. Insects are highly diverse, with over 80 species of moth having been recorded. Endemic birds include the Australasian bittern, South Island fernbird, and variable oystercatcher, while royal spoonbills and gray teal are regular visitors to the lagoon.

**MANUKA**
*Some areas of the Waituna wetlands are dominated by manuka, an evergreen shrub that yields a much-prized honey.*

LAND

AUSTRALASIA *Australia*

## Coorong

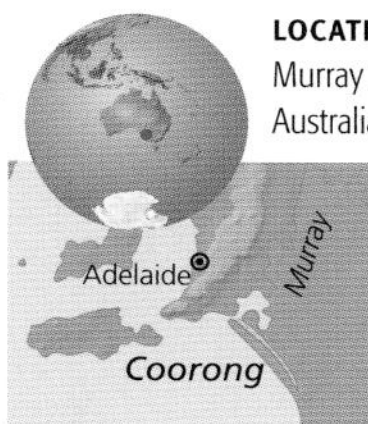


**LOCATION** At the mouth of the Murray River, southeast South Australia, Australia

**TYPE** Lagoon

**WATER** Brackish, salt

**AREA** 80 square miles (207 square km)

The Coorong lies at the mouth of the Murray River (see p.235), about 50 miles (80 km) south of Adelaide. The river opens into Lake Alexandrina and then connects with the narrow channel of the Coorong, which lies behind the Younghusband Peninsula, a narrow spit on which an impressive array of sand dunes faces the Southern Ocean. These run parallel to a set of ancient dunes (dating from the Pleistocene Epoch) on the landward side. It is between these two rows of dunes that the saltwater lagoon of the Coorong stretches southward for more than 60 miles (100 km). Other habitats found here include temporary freshwater lakes and mudflats, particularly in the south. In 1940, barrages were built between the Coorong and Lake Alexandrina, preventing seawater from reaching the lake. This altered the hydrology of the area considerably, and reduced the natural flow of the Murray into the Coorong. Water extraction for irrigation projects upstream has also reduced the flow, and if it decreases much more in the future, there is a risk that the gap connecting the Coorong to the sea will close completely. This would prevent fish and other animals from migrating between the Coorong and the sea. The Coorong is one of the best sites for viewing wildfowl in Australia. Thousands of black swans, Cape Barren geese, and countless waders flock to these waters, which are also home to the largest breeding colony of Australian pelicans. In all, 238 species of bird have been recorded in the lagoon.

**YOUNGHUSBAND PENINSULA**
*Sand dunes, partly covered by coastal scrub, stretch over 90 miles (145 km) from the mouth of the Murray River, protecting the lagoons of the Coorong from the open ocean.*

**BLACK SWAN**
*Native to Australia, the black swan is exclusively vegetarian, feeding mainly on aquatic plants and sometimes grazing on land.*

**MUD CRABS**
*Mud crabs are a prominent feature of the Coorong. These crustaceans thrive on the mudflats and in the shallow, brackish waters of the coastal lagoons.*

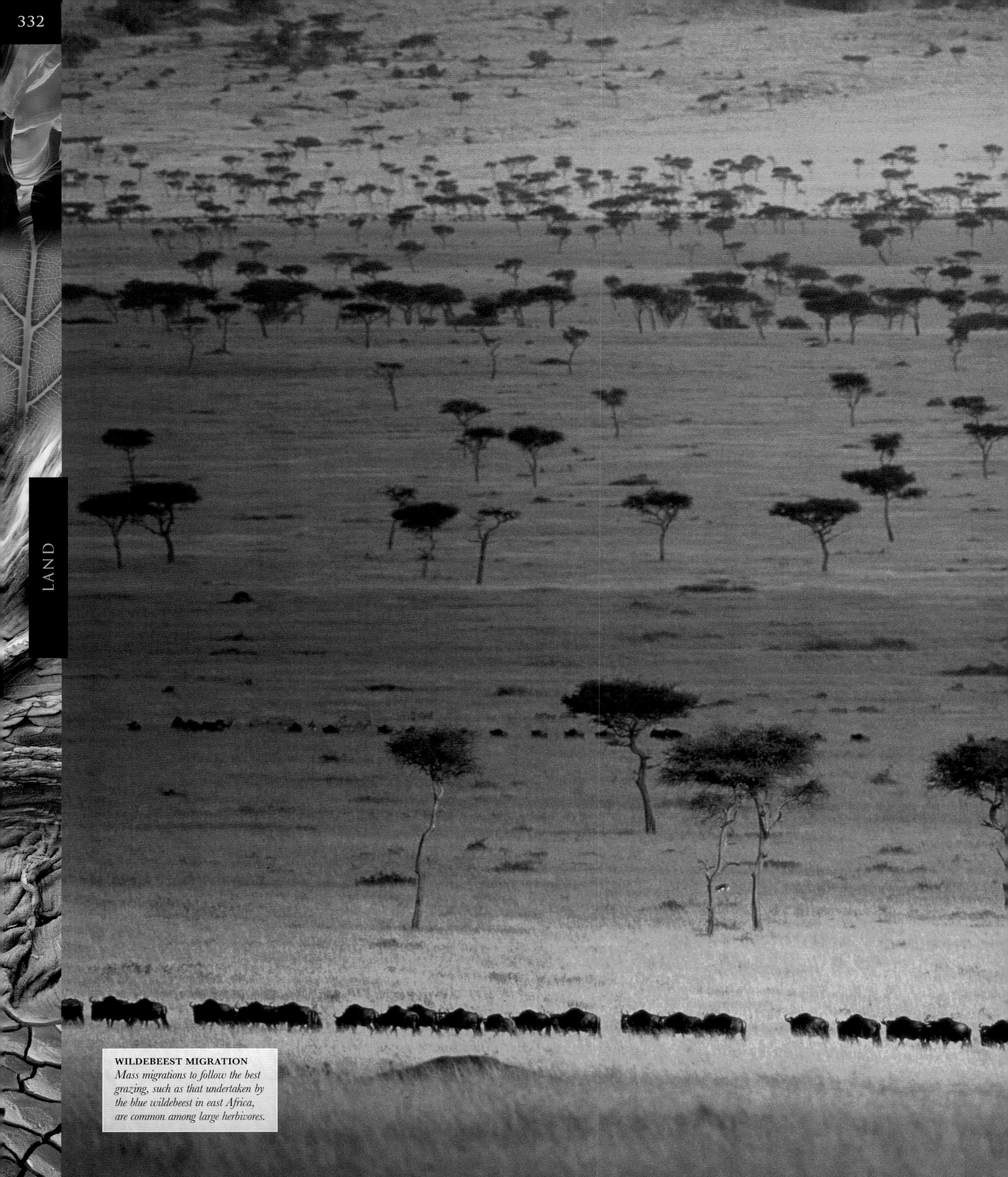



**WILDEBEEST MIGRATION**
*Mass migrations to follow the best grazing, such as that undertaken by the blue wildebeest in east Africa, are common among large herbivores.*

# GRASSLANDS AND TUNDRA

VAST AREAS OF THE EARTH'S SURFACE are covered by apparently unchanging expanses of natural grassland. In fact, this uniformity is often an illusion, for grasslands vary widely: they may be hilly or punctuated by rocky outcrops; they include both lush and semiarid areas; and some, especially in the tropics, are dotted with trees, pools, and marshes. Despite experiencing harsh weather, grasslands are one of the world's most productive biomes. Tundra is another open biome that covers a large part of the Earth's land, mainly in the Arctic. Conditions in this desolate, treeless terrain are even more extreme, with long, cold, perpetually dark winters followed in spring by an explosion of life.

# GRASSLANDS

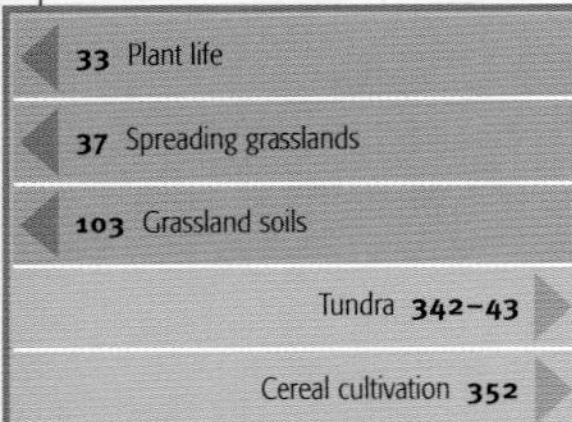


NATURAL GRASSLANDS MOSTLY DEVELOP OVER deep, fertile soils, in areas with seasonally variable precipitation and a growing season of 120–200 days. The rainfall must be too low to support forests, but too high for either scrublands or deserts to take hold. Fierce winds and large swings in temperature between day and night, and from one season to the next, are typical of many grasslands. Their boundaries are ill-defined and ever-changing due to natural climatic fluctuations and to human influence, including livestock grazing, deliberate burning, and the conversion of land to arable or pastoral crops. There are two major types of grassland, temperate and tropical, which differ both in structure and in the variety of their plant and animal species.

## TEMPERATE GRASSLANDS

Temperate grasslands are found mainly in continental interiors, far from coasts and their rain-bearing winds. Summers here are usually warm and dry, with occasional violent storms, but winters may be bitterly cold. Most rain falls in late spring and early summer. Wind has a major impact on temperate grasslands: it fans the flames of fires in late summer, and it dries the land by accelerating the process of evaporation. The vegetation is dominated by perennial grasses, which have extensive root systems that join to form a tightly woven mat known as sod. Herbs, sedges, flowers, and specialized shrubs also occur, particularly in wetter grasslands such as the tall-grass prairies of North America. Trees are generally restricted to sheltered hollows, the edges of streams and drainage channels, or marginal areas where grassland grades into parklike open woodland.

**PAMPAS GRASS**
*Many grass species grow long flower-heads in summer, which disperse clouds of pollen on the breeze. Those of the pampas grass, shown here in Uruguay, are among the most spectacular.*

**BUFFALO GRASS**
*Native to the short-grass prairies of North America's Great Plains, this species features prominently in the diet of bison (also known as buffalo).*

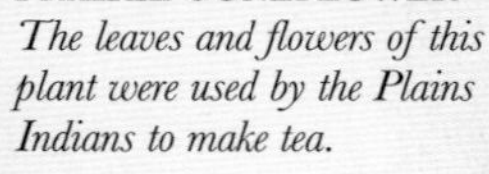

**PRAIRIE CONEFLOWER**
*The leaves and flowers of this plant were used by the Plains Indians to make tea.*

**AUSTRALIAN RANGELAND**
*This temperate grassland, found in southeast Australia, is studded with eucalyptus and paperbark trees.*

## TROPICAL GRASSLANDS

Tropical grasslands, or savannas, are typically more varied in character than temperate grasslands. There is normally a scattered cover of trees and bushes, which may be sparse in some places and dense enough to form thickets in others. Some savannas, such as those in parts of east Africa, are actually a form of open woodland, in which a ground cover of grass is always present. Many of the grasses in savannas grow very tall during periods of rain; those found in the monsoon grasslands of the Indian Himalayan foothills are the tallest on Earth. Unlike temperate grasslands, savannas are usually warm all year, but a more important distinction is the much higher rainfall, which is concentrated into distinct wet seasons. Bush fires play an even more significant role in savannas than in cooler grasslands.

**KANGAROO GRASS**
*One of the main grasses of Australian savannas, this species grows to a height of 10 ft (3 m) or more.*

**DROPSEED GRASS**
*In Africa, the edible seeds of some dropseed grasses are eaten by indigenous peoples.*

**CERRADO LANDSCAPE**
*This South American savanna is dotted with small, gnarled trees and countless red termite mounds. Its open areas are mixed with scrub and patchy dry forest.*

LAND

# ANIMAL LIFE

With the exception of the African savannas with their herds of antelope, zebra, and other large herbivores, grasslands may at first seem to be devoid of wildlife. This impression is misleading, because many grassland species are nocturnal or spend much of their life underground. A host of animals, from voles and moles to earthworms, beetles, and other invertebrates, tunnel under the sod. In doing so, they aerate the soil and create homes for non-tunneling species such as snakes and birds. Animals that remain above ground are often fast movers, since hiding places are at a premium in most grasslands; they include the ostrich and gazelles. Other strategies for avoiding predators include camouflage, the use of warning calls or signals (one example being the white rumps of certain rabbits), and herd formation. Many herbivores live in herds because predators find it harder to single out victims, and there are more animals on the lookout.

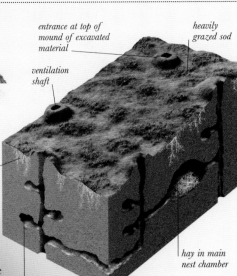

**BURROWING OWL**
*Lack of trees forces this grassland owl to nest in burrows, often those of prairie dogs.*

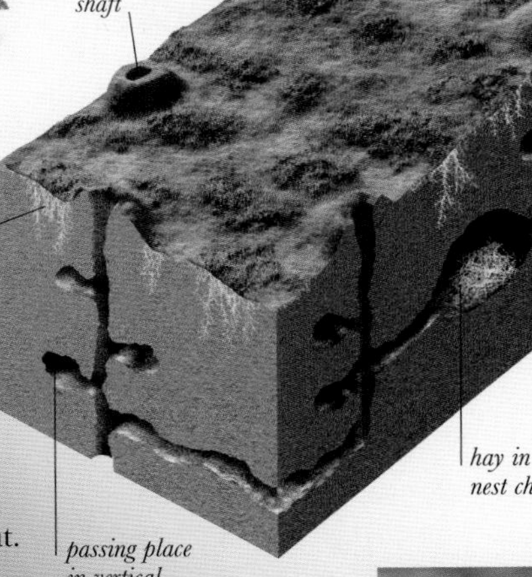


**LIFE IN A PRAIRIE-DOG TOWN**
*Prairie dogs are highly sociable ground squirrels of the North American prairies. They live in burrow systems known as towns, which provide shelter from predators and harsh winter weather, and a safe place to raise young. Prairie dogs have powerful front limbs with thick claws for excavating and scraping soil. Their sustained grazing keeps the vegetation near their colony short.*

**PRAIRIE DOGS**
*A group of prairie dogs keeps watch for danger, ready to bark the alarm and dive for cover.*

**COMMON BROWN**
*Butterflies, such as this species from Australia, are important pollinators of many grassland flowering plants.*

**HAIRY ARMADILLO**
*When threatened, this armored resident of arid South American grasslands rolls up into a tight ball.*

**BLACK-FOOTED FERRET**
*This endangered carnivore hunts prairie dogs, and lives in their abandoned burrows.*

## GRASS BURNING

Whether started naturally or by people, fire plays a crucial role in the ecology of grasslands. It removes dead plant material, and the ash created boosts the fertility of the soil, promoting new grass growth. Burning also inhibits colonization by woody shrubs and trees. Here, Masai people are undertaking a controlled burn of old grass in the Serengeti.

# GRASSLAND DISTRIBUTION

The largest temperate grasslands are the prairies, which occupy much of central North America, and the Asian steppes, which stretch in a band from easternmost Europe to northern China. Other temperate grasslands are the Pampas of southeast South America, the veld of eastern South Africa, and the rangeland of southeast Australia. The main savannas are in Brazil, in east and southern Africa and to the south of the Sahara, and in northern Australia. Smaller savannas exist in the Indian subcontinent.

**PRAIRIE LANDSCAPE**
*Temperate grasslands change with the seasons. In winter the plants lie dormant, while in the summer the grasses turn ripe yellow, as here in North Dakota.*

## GRASSLAND PROFILES

The pages that follow contain profiles of the world's main grasslands. Each profile begins with the following summary information:

**TYPE** Temperate or tropical

**AREA** Surface area

NORTH AMERICA *central*

## Great Plains

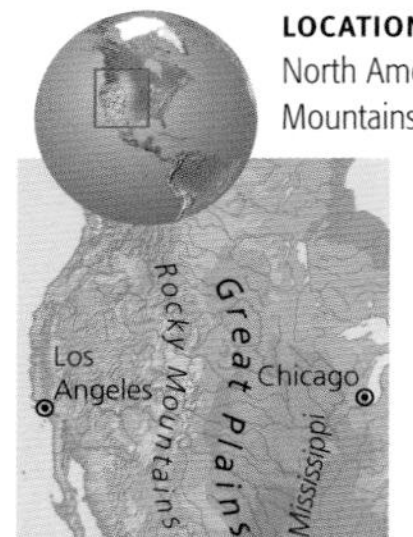


**LOCATION** Occupying much of North America between the Rocky Mountains and the Mississippi River

**TYPE** Temperate

**AREA** 1.2 million square miles (3 million square km)

By far the largest expanse of grassland in North America, the Great Plains stretch from the southern Canadian provinces of Alberta, Manitoba, and Saskatchewan, through the US Midwest, and south almost to northern Mexico. As recently as the early 19th century, this immense area was covered by grasslands known as prairies, but most of the fertile land is now under intensive agriculture. It is especially suitable for growing cereals, such as wheat and corn, and has become the main grain-producing region of North America. Only one percent of the natural grassland survives, divided into three main types: short-grass, mixed-grass, and tall-grass prairie. Short-grass prairie occurs mostly in the west in the rain shadow of the Rocky Mountains, where the annual rainfall is about 10 in (250 mm). Here, most grasses are 8–20 in (20–50 cm) high. Tall-grass prairie is found in the eastern districts of the Great Plains, with an annual rainfall of 26–39 in (650–1,000 mm). Some grasses here, such as Indian grass, big bluestem, and cordgrass, may reach 5 ft (1.5 m). In central areas, mixed-grass prairie forms a transition zone between short- and tall-grass prairie. The open prairie landscape used to be preserved partly by grazing by native herbivores, which included bison, pronghorn, and prairie dogs, but the decline of these animals has led to changes in the length and composition of the turf. Prior to European colonization, the prairies were also home to American Indian nations such as the Lakota and Pawnee.

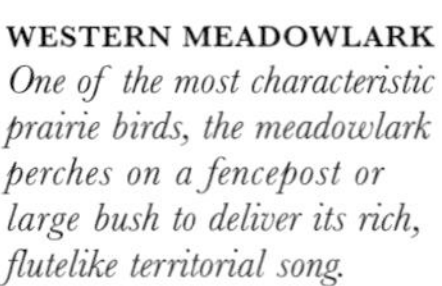

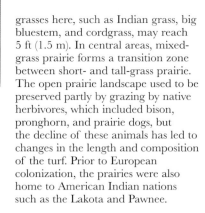

**WESTERN MEADOWLARK**
*One of the most characteristic prairie birds, the meadowlark perches on a fencepost or large bush to deliver its rich, flutelike territorial song.*

**ROLLING PLAINS**
*Undulating hills and wide, grassy plains are typical of many parts of the prairies, as here in Nebraska. The harsh winters and strong winds prevent all but a handful of trees from gaining a hold.*

### RETURN OF THE BISON

Until the early 1800s, up to 60 million bison (or buffalo) roamed the short-grass prairie in huge herds. Hunting by settlers from the East Coast reduced the bison population to just 2,000 by 1885. Today, there are more than 350,000, mainly in national parks and cattle ranches. There are plans to create a "buffalo commons," allowing herds to move freely over the prairies of Montana, the Dakotas, Nebraska, Kansas, and Oklahoma.

**COMMON SUNFLOWER**
*This is the wild ancestor of the popular garden plant. It flourishes in the rich, dark soils of the Great Plains.*

**FOXTAIL GRASS**
*In summer, foxtail grass produces long, feathery flower-heads. Later in the season, its seeds are eaten by a wide variety of rodents and birds.*

**BADLANDS NATIONAL PARK**
*Named after its strange rock formations sculpted by wind and rain, this reserve in South Dakota protects the largest area of mixed-grass prairie in the US.*

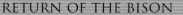

SOUTH AMERICA *north and east*

## South American tropical grasslands

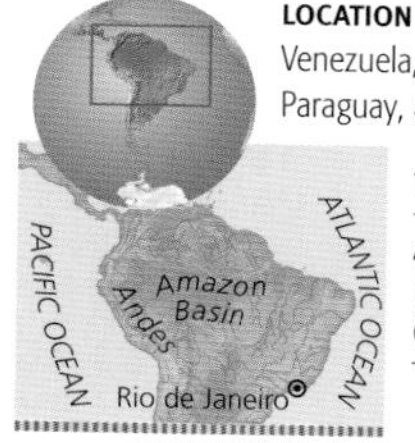


**LOCATION** In eastern Colombia and Venezuela, south-central Brazil and Paraguay, and northern Argentina

**TYPE** Tropical

**AREA** 1 million square miles (2.7 million square km)

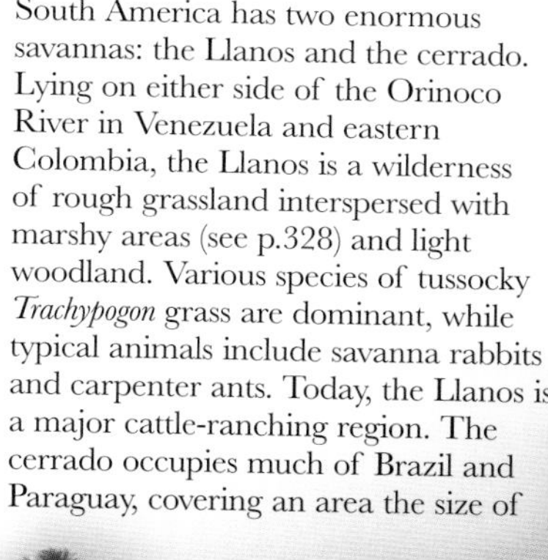

South America has two enormous savannas: the Llanos and the cerrado. Lying on either side of the Orinoco River in Venezuela and eastern Colombia, the Llanos is a wilderness of rough grassland interspersed with marshy areas (see p.328) and light woodland. Various species of tussocky *Trachypogon* grass are dominant, while typical animals include savanna rabbits and carpenter ants. Today, the Llanos is a major cattle-ranching region. The cerrado occupies much of Brazil and Paraguay, covering an area the size of western Europe, It borders the Amazon Basin (see pp.223 and 311) to the north, with the Pantanal wetland (see p.328) to the west and the Atlantic forests in the east. As many as five percent of the world's animal species live in this savanna, including the giant anteater and tall, flightless birds called rheas. The cerrado's ecological richness is due in part to its mosaic of habitats, ranging from grasslands to dry forests (see p.314). But agriculture is rapidly encroaching on the cerrado; less than two percent of it is protected. To the south, in Argentina's arid chaco region, there are sandy, desertlike grasslands.

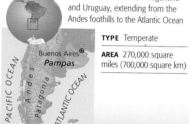

**MANED WOLF**
*Neither wolf nor fox, this member of the dog family lives in Brazil, Paraguay, and northern Argentina. Its stiltlike legs help it to move through long grass, and to leap on rodent prey.*

**BRAZILIAN CERRADO**
*Open grassland shades into scrub and palm forest in the cerrado, creating a marginal habitat rich in species.*

SOUTH AMERICA *southeast*

## Pampas

**LOCATION** In northern Argentina and Uruguay, extending from the Andes foothills to the Atlantic Ocean

**TYPE** Temperate

**AREA** 270,000 square miles (700,000 square km)

Named after a Quechua Indian word meaning "flat surface," the Pampas is a huge plain in the southeastern corner of South America. It is centered on the lowlands of the Plate River (see p.220). Apart from a few low hills, most of the region is just above sea level. Deposits of volcanic ash provide a rich soil. The eastern Pampas, which includes much of Uruguay and the Argentinian province of Buenos Aires, enjoys a mild climate, with cool summers and an annual rainfall of up to 47 in (1,200 mm). Snow and frosts are rare. The typical vegetation is coarse grasses, including pampas grass (see p.334), which is famous for its tall stems bearing silky plumes of flowers. Clusters of trees cling to the damper hollows or drainage channels. Animals of the Pampas include the endangered Pampas deer, Geoffroy's cat, and the Argentine tortoise, which spends the night in burrows for warmth. Farther west, in the Argentine provinces of La Pampa and Córdoba, the landscape is different. Here, the rainfall is half that of the eastern Pampas, and the sandier soil supports a semiarid grassland feather grasses and other species. Toward the Andes, the land rises and becomes increasingly dry, eventually turning into semidesert.

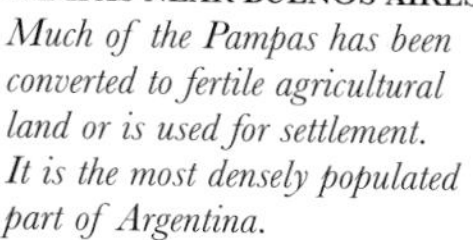

**PAMPAS NEAR BUENOS AIRES**
*Much of the Pampas has been converted to fertile agricultural land or is used for settlement. It is the most densely populated part of Argentina.*

### CATTLE RANCHING

Millions of cattle are raised on the Pampas, and gauchos, or cowboys, still follow the herds on horseback. To some extent, their livestock have replaced the ecosystem's natural grazers and fulfill a similar function. However, the mix of native grasses has largely been replaced by a smaller number of introduced fodder species.

**PATAGONIAN PAMPAS**
*Blasted by an icy, dry wind known as the pampero, this desolate plateau in southern Argentina provides a contrast to the lush Pampas farther north.*

AFRICA *east*

## Serengeti Plains

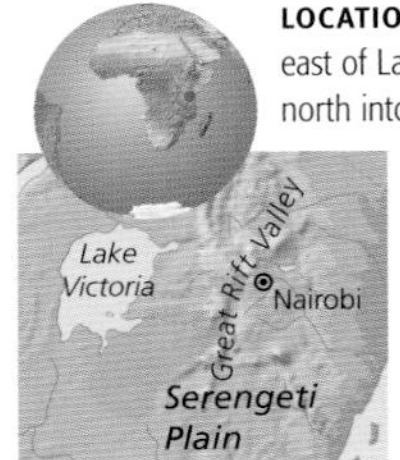


**LOCATION** In northwest Tanzania east of Lake Victoria, extending north into southwest Kenya

**TYPE** Tropical

**AREA** 8,900 square miles (23,000 square km)

Lying close to the equator in east Africa, the Serengeti Plains support an exceptionally varied collection of plants and animals, all adapted to the complex intermingling of grassland and open woodland. The main grassland areas lie in the southeast, with wooded savanna typified by flat-topped acacia trees dominating in the north and west. In some places, the undulating plains are interrupted by rocky outcrops known as kopjes, composed of hard Precambrian rocks. The Serengeti's reddish soil is derived mainly from volcanic ash. Relatively low in organic materials compared with temperate grasslands, it is replenished with vital nutrients by frequent dry-season bush fires, either natural or human-made. Billowing clouds of smoke may darken the horizon for days at a time. Another major factor shaping the landscape is grazing pressure. The Serengeti holds the largest populations of grazing and browsing mammals in the whole of Africa. During the two rainy seasons, from November to December and from March to May, these herbivores are widely dispersed over the plains,

**RESTING LIONESSES**
*A group of lionesses rests on an anthill, which offers a good vantage point over their territory. The pride's mature male will probably not be far away, but plays little role in hunting. The lionesses provide most of the food, moving as a pack to ambush their victims.*

**NGORONGORO CRATER**
*The world's largest unbroken caldera, this huge relic of past volcanic activity looms over the eastern Serengeti. Its walls are over 1,970 ft (600 m) high.*

### HUGO VAN LAWICK

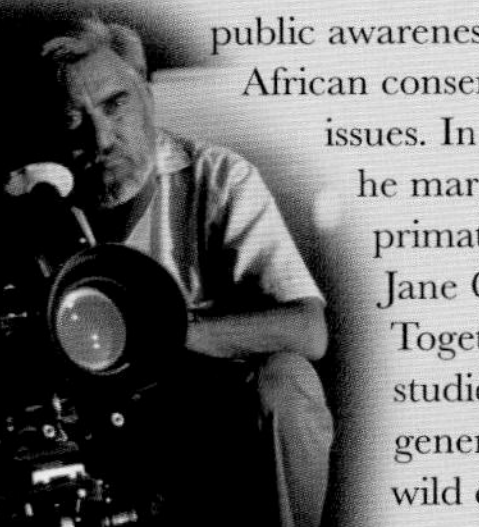

One of the earliest and most famous wildlife filmmakers, Hugo van Lawick (1937–2002) was a Dutch naturalist who first visited Africa in his twenties. His many documentaries for TV and cinema did much to increase public awareness of east African conservation issues. In 1964, he married the primatologist Jane Goodall. Together, they studied three generations of wild chimps.

**SECRETARY BIRD**
*This distinctive relative of the birds of prey stalks slowly through long grass, watching the ground for its prey—mainly insects, snakes, lizards, and small mammals.*

but when the savanna is yellow and parched, they migrate in large herds in search of fresh grass and drinking water. They include 1.3 million blue wildebeest; 200,000 zebras; and 40,000 Thomson's gazelles. Such an abundance of big game provides enough food for many predators and scavengers, including big cats, hyenas, jackals, crocodiles, and six species of vultures. Equally important are the countless dung beetles that roll away 75 percent of all the dung dropped in the Serengeti, burying it in nests for their larvae to eat. Both the Serengeti National Park and the Ngorongoro Preservation Area are World Heritage Sites, and ecotourism raises funds for conservation projects. Despite this protected status, wildlife in the region is at risk from poaching, serious overgrazing (caused by excessive numbers of cattle), and diseases spread by domestic dogs and livestock.

**BLUE WILDEBEEST AND ZEBRA**
*Although these herbivores graze side by side in large herds, they eat different foods. The zebras can consume drier, tougher grasses, while the wildebeest harvest softer grasses ignored by the zebras.*

## MASAI

The Masai live in the Serengeti and the neighboring Masai Mara. Unlike some other African nomadic peoples, they do not rely on hunting, and so historically Masai land has tended to be rich in wildlife. Instead, they herd cattle, goats, and sheep, and collect fruits and other plant products from the wild. Their main traditional food is fermented cow's milk; meat is usually eaten only during ceremonies.

**SERENGETI IN THE WET SEASON**
*With the onset of the rains, thorny shrubs and acacia trees are flushed with new leaf growth. Acacia woodland provides valuable shade and browsing for many of the Serengeti's animals.*

ASIA *central*

## Central Asian steppe grasslands

**LOCATION** Extending from Ukraine, east through southern Russia and Kazakhstan, to Mongolia and China

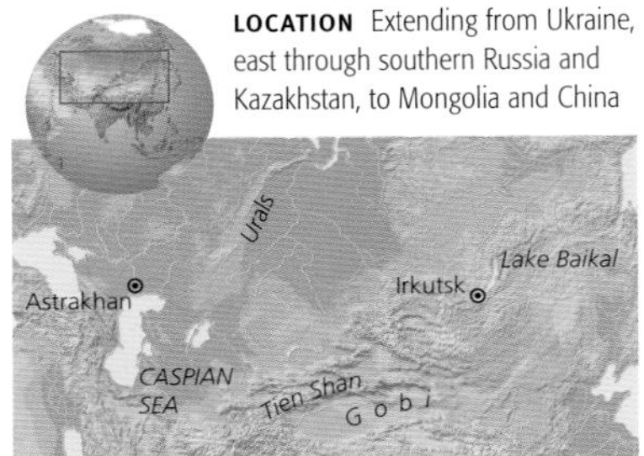


**TYPE** Temperate

**AREA** 975,000 square miles (2.5 million square km)

The steppes of Central Asia stretch almost a third of the way around the world and include some of the least populated areas on Earth. In the west, the steppes reach the northern shores of the Black Sea, while they extend east in an irregular band almost to the mountains of China's northeastern provinces on the Pacific. This immense grassland zone is bordered to the north by the Eurasian boreal forest (see p.316), and to the south by the Central Asian deserts (see pp.297–99). Dramatic swings in temperature are a defining feature of the steppes; on the grassy upland plateaus of Mongolia, it may reach 104°F (40°C) in summer and plunge to –4°F (–20°C) in winter. The annual rainfall of 8–23½ in (200–600 mm) is sporadic, with long droughts and sudden thunderstorms. Thick snow blankets the ground in winter. When the snow melts in spring, the steppes are briefly ablaze with a profusion of flowers, including irises, hyacinths, crocuses, and tulips. During the short summer, feather and fescue grasses and sedges dominate the sod. Apart from a few stunted trees in sheltered valleys, the steppes are bereft of tall vegetation, although in northern China there is a more parklike landscape, with a scattering of trees such as elms. Animals of the steppes are adapted to the extreme climate in various ways. The Saiga antelope has a bulbous, trunklike nose that warms and moistens the cold, dry air it breathes, while rodents such as the steppe lemming, susliks, and gerbils dig burrows for shelter.

### SAVING A SPECIES

Built like a heavy pony, Przewalski's horse is the only truly wild horse. At home on steppes and in semideserts, it once ranged from the Urals to northern China. It was extinct in the wild by 1969, but more than 1,000 survived in zoos. Small herds bred in captivity are now being reintroduced to suitable sites.

### MAKING HAY

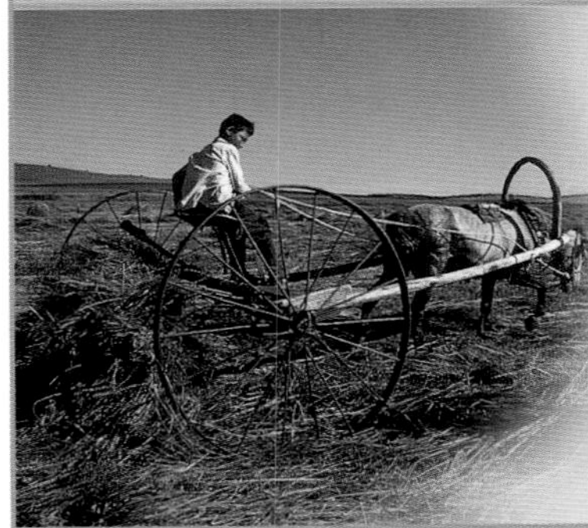

Conditions on the high steppes are too harsh for settled agriculture, but at lower levels the land is often used as pasture (few crops will grow here). Haymaking is a vital, labor-intensive activity, in which even children play a part. Hay provides fodder for cattle, sheep, goats, and horses during the long winter months, when the ground is frozen solid and grazing impossible. A poor harvest could easily lead to a crippling loss of livestock.

**SOUTHERN RUSSIAN STEPPES**
*Many of the steppes in Russia, Ukraine, and Kazakhstan have been converted to seminatural grasslands due to heavy grazing by cattle.*

**NORTHERN MONGOLIAN STEPPES**
*The mountains of southern Siberia rise above this high-altitude plateau near Ulan-Uul. A sparse covering of dusty sod is often all that distinguishes the high steppes from semidesert.*

ASIA *south*

## Indian savanna grasslands

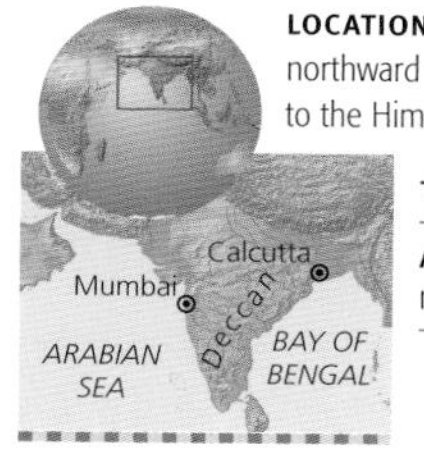


**LOCATION** Within a zone stretching northward from south-central India to the Himalayan foothills

**TYPE** Tropical

**AREA** 13,515 square miles (35,000 square km)

Forming a discontinuous area and with few large expanses remaining, Indian savanna grasslands occur mainly in inland regions, around the fringes of deserts, and in hilly country. The Terai-Duar grasslands form a narrow strip 15 miles (25 km) wide along the southern margin of the Himalayas, from Nepal east to Bhutan and the states of Uttar Pradesh and Bihar in India. The hot, humid summers and monsoon rains enable tall grasses, including giant reeds and canes, to grow luxuriantly. These "grass forests" have the highest densities of tigers in India. Farther south, the much lower rainfall of the Deccan Plateau gives rise to coarser, semiarid pastures with a scattering of trees such as sandalwood. This is the haunt of numerous reptiles, among them the highly venomous Indian cobra, which hunts rodents, small birds, lizards, and other snakes.

**GRASSLAND PLATEAU**
*The interior uplands of southern India are known as the Deccan Plateau. Pockets of wooded savanna form a patchwork with dry forest, farmland, and villages in this highly populated region.*

**DEW-COVERED GRASSES**
*Morning dew clings to grasses in the Corbett National Park in the Indian Himalayan foothills. Torrential monsoon downpours create swampy conditions in which annual grasses thrive.*

**ELEPHANT GRASS**
*In the grasslands of Nepal, this relative of sugar cane grows up to 26 ft (8 m) high. It provides the Indian rhinoceros with both shelter and the bulk of its diet.*

AUSTRALASIA *Australia*

## Australian savanna grasslands

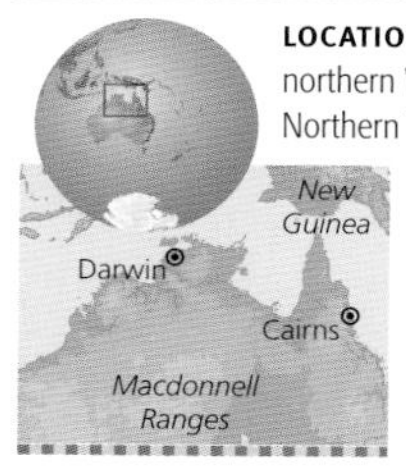


**LOCATION** Stretching from far northern Western Australia, through Northern Territory, into Queensland

**TYPE** Tropical

**AREA** 463,500 square miles (1.2 million square km)

In Australia, savanna grasslands form a wide band between the hot desert interior and the forests of the north coast, with the Great Dividing Range (see p.170) as their eastern limit. It is hot and humid during the wet season (December to March), but little or no rain falls during the cooler dry season (May to August). Wooded savanna is mainly found in the wetter north and east, but gum trees, also known as eucalypts, are present in most areas. One group of uniquely Australian grasses are the spinifex, which bristle with spearlike leaves that minimize water loss, allowing them to thrive in the desert. Aboriginal settlement of these grasslands dates back almost 40,000 years, making it one of the Earth's oldest cultures.

**AUSTRALIAN BAOBAB**
*Although mainly found in Africa, and particularly Madagascar, one species of baobab is endemic to northwest Australia.*

**RED KANGAROO**
*The largest of the kangaroos, this marsupial is common in wooded savannas. Its numbers double in years of high rainfall.*

*short forelegs*

*large, muscular hind limbs*

*very long, strong tail*

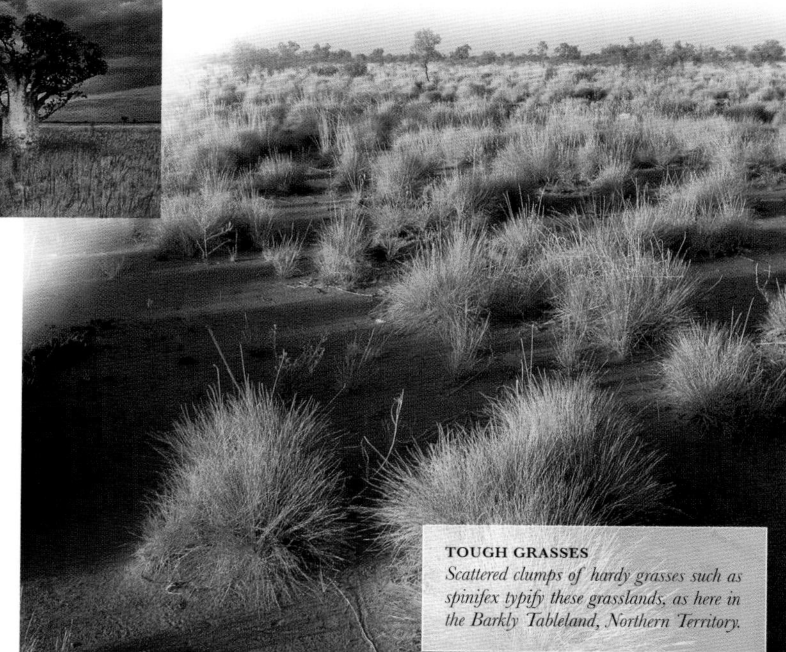

**TOUGH GRASSES**
*Scattered clumps of hardy grasses such as spinifex typify these grasslands, as here in the Barkly Tableland, Northern Territory.*

LAND

# TUNDRA

39–41 The ice ages
101 Frozen soils
110 Physical weathering
260–81 Glaciers
334–35 Grasslands

THE TERM TUNDRA IS derived from a Lappish word meaning "barren land." It is an apt description of the low-growing vegetation found in the Arctic, beyond the northern limit for tree growth. The windswept landscape here is frozen for much of the year, but its surface layers melt in spring. Summers are short and intense, with 24-hour daylight, while the Sun barely rises during the long, ferociously cold winters. Patches of tundralike habitat also exist above the tree line on some high mountains.

## PERMAFROST ZONES

Land that has been frozen for at least two years is classified as permafrost, and it has a major influence on the landscape and ecosystem of the tundra. A zone of permanent permafrost is found at the highest latitudes, spreading out from the North Pole. Immediately to the south is the zone of semi-permanent permafrost. Here, the topmost layer usually thaws out to a depth of a few inches each summer, and this so-called active layer supports plant and animal life. Still farther south lies the sporadic permafrost zone, in which the active layer is deeper and where the surface freezes less often; however, the underlying frozen soil prevents meltwater from draining away.

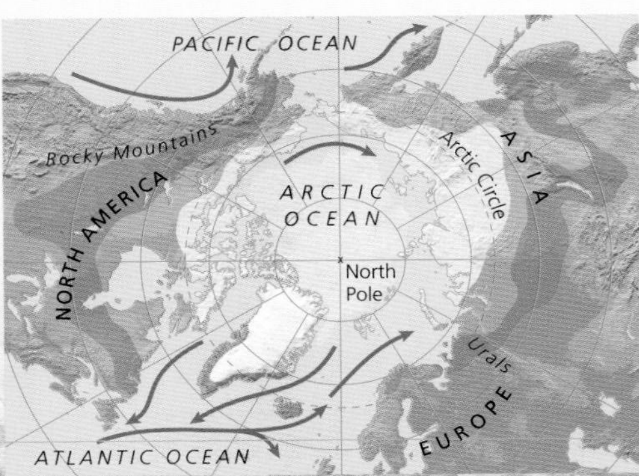


**ARCTIC PERMAFROST ZONES**
*Three zones of permafrost surround the Arctic, arranged from north to south, determined by latitude and the influence of ocean currents.*

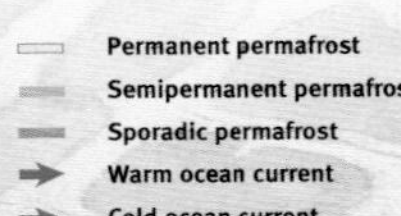


## PERIGLACIAL LANDFORMS

Periglacial landforms are those features that develop under the action of hard frosts, often in permafrost conditions. A classic periglacial feature is the polygonal patterning that appears in the surface of tundra soils, formed by the repeated freezing and thawing of water seeping along surface cracks. Rapid thawing of the active surface layers may give rise to a creeping movement of the soil known as skin flow. The accumulation of ice trapped between the surface layers of the soil and its deeper frozen layer may result in disruption of the surface, splitting and buckling it, or heaving it up into the shape of a small hill or mound. Rounded ice-heave hills of this type are called pingos, while irregularly shaped masses of ice are referred to as ice wedges. Other landforms are created by the partial thawing and refreezing of the permafrost, especially in sedimentary rocks. Where the bedrock is exposed, it may be weathered into angled fragments, to produce areas of coarse rubble called blockfields.

**ICE LENS**
*When compacted ice trapped in soil develops convex surfaces (left), it is called an ice lens.*

**FOSSIL ICE WEDGE**
*Blocks of ice sometimes leave their traces as fossils, indicating cold conditions in an earlier age.*

**SPLIT-HILL PINGO**
*A pingo is a blisterlike mound that develops as the ice at its center expands. The pressure of the expanding ice core may cause its surface to tear apart, as seen here.*

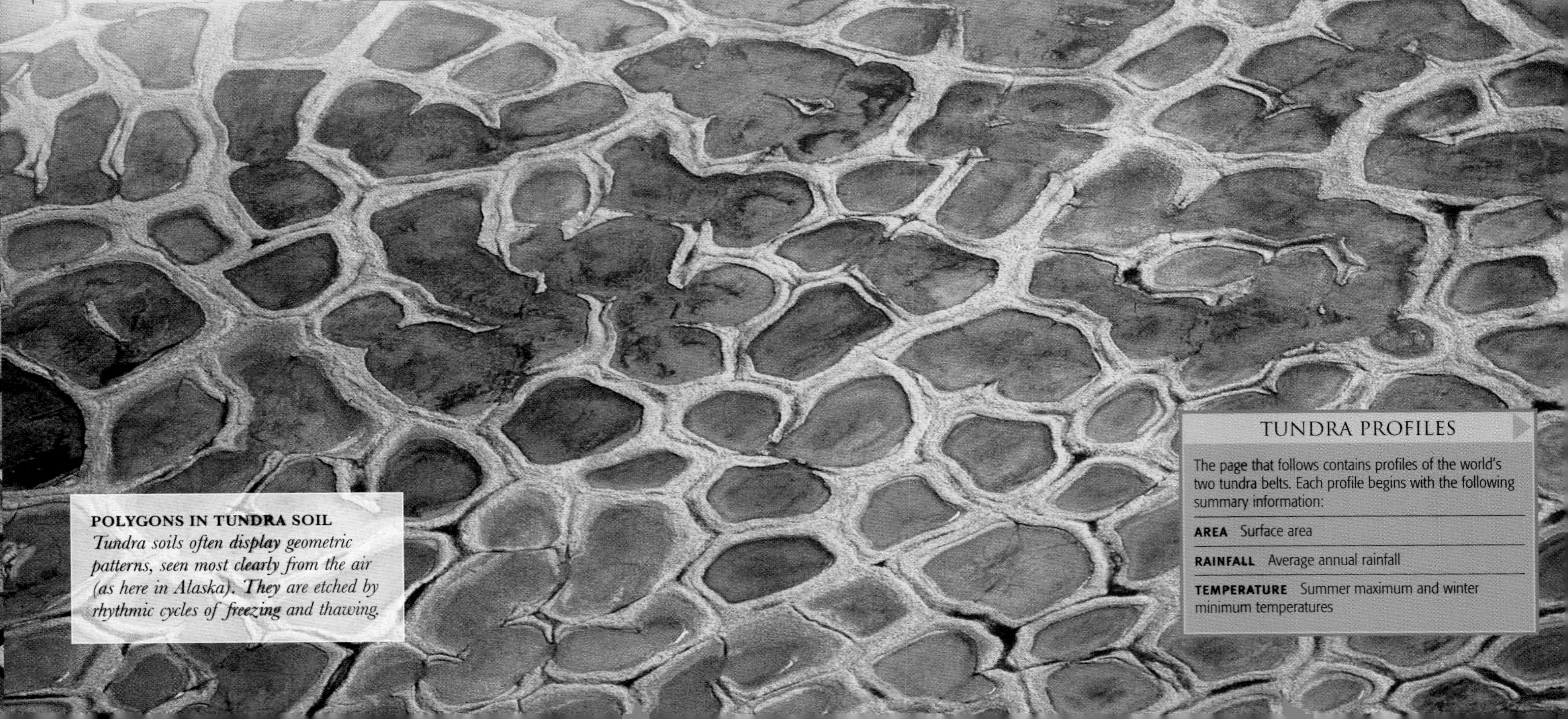

**POLYGONS IN TUNDRA SOIL**
*Tundra soils often display geometric patterns, seen most clearly from the air (as here in Alaska). They are etched by rhythmic cycles of freezing and thawing.*

### TUNDRA PROFILES

The page that follows contains profiles of the world's two tundra belts. Each profile begins with the following summary information:

**AREA** Surface area

**RAINFALL** Average annual rainfall

**TEMPERATURE** Summer maximum and winter minimum temperatures

NORTH AMERICA *north*

## North American tundra

**LOCATION** Extending from Alaska east through northern Canada; also on Greenland's coastal regions

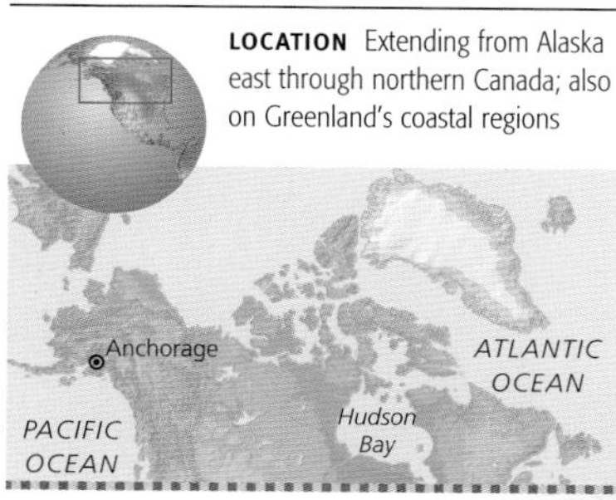


**AREA** 2 million square miles (5.3 million square km)

**RAINFALL** 2–8 in (50–200 mm)

**TEMPERATURE** –75°F–75°F (–60°C–24°C)

Tundra occupies a large swath of the North American continent, covering most land lying to the north of the North American boreal forest (see p.308). A narrow coastal strip of tundra also fringes the island of Greenland. Flat, often featureless country dominates the tundra biome, although some periglacial landforms break up the monotony (see polygons and split-hill pingo on opposite page), as do a few rugged mountain ranges. In spring, the ice and snow melt to reveal a thick mat of lichens, mosses, sedges, and specialist Arctic flowers such as saxifrages. Where the soil is deeper, there are crowberries, bilberries, and other ground-hugging shrubs. Dwarf willows and birches grow in southern areas. The tundra briefly teems with invertebrates, especially mosquitoes, beetles, and other insects, which provide food for migrant geese, ducks, and wading birds.

**DWARF WILLOW**
*This miniature tree creeps along the ground, growing less than 3 in (7.5 cm) high to avoid the drying effect of the relentless tundra wind.*

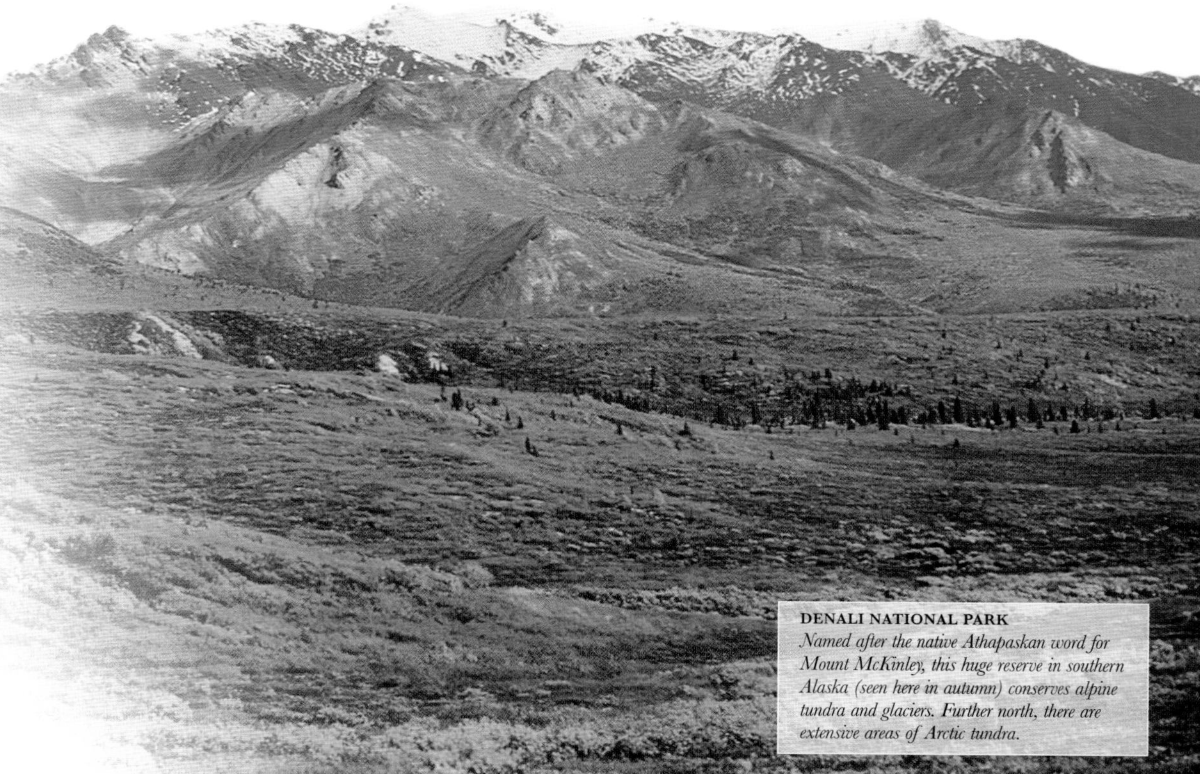

**DENALI NATIONAL PARK**
*Named after the native Athapaskan word for Mount McKinley, this huge reserve in southern Alaska (seen here in autumn) conserves alpine tundra and glaciers. Further north, there are extensive areas of Arctic tundra.*

**ARCTIC FOX**
*In winter, this tundra specialist has a white coat with long outer hairs and thick underfur for superb insulation; its summer coat is chocolate brown and much thinner.*

**ARCTIC POPPIES IN BLOOM**
*These poppies have colonized an island in Franz Josef Land, Russia. Like many tundra flowers, they concentrate warmth by turning to track the course of the Sun.*

EUROPE AND ASIA *north*

## Eurasian tundra

**LOCATION** From Iceland in the west, eastward through northern Scandinavia, Russia, and Siberia

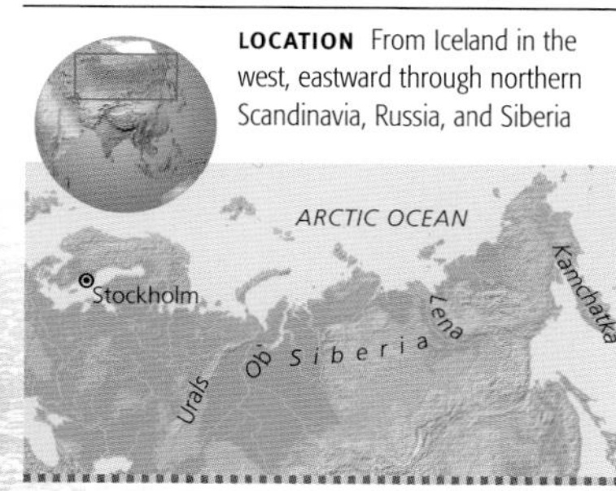


**AREA** 1.3 million square miles (3.3 million square km)

**RAINFALL** 8–12 in (200–300 mm)

**TEMPERATURE** –76°F–77°F (–60°C–25°C)

The Eurasian tundra features a range of landscapes, from the soggy plains of Siberia, crossed by numerous rivers, to the various archipelagos scattered through the southern Arctic Ocean, such as Franz Josef Land and Novaya Zemlya. In central Siberia, winters are particularly extreme, while the Scandinavian tundra is the warmest, with an average temperature of 18°F (–8°C) in winter. Most of the plants are small, long-lived perennials. Mosses, lichens, sedges, and rushes dominate in the northern tundra, with plant diversity increasing steadily from north to south. As in the North American tundra, the short but prolific 90-day growing season draws migratory animals to the region. More than 200 million birds arrive to breed, including sandpipers, ducks, and geese. Reindeer (known as caribou in North America) spend the winter in the taiga zone, then migrate north to calve and to graze lichens on the open tundra.

**SIBERIAN TUNDRA IN EARLY SPRING**
*As temperatures rise here in spring, meltwater floods the ground. Repelled by the permafrost, it forms millions of small lakes, pools, and bogs.*

**ARCTIC WILLOW**
*The fluffy seed heads of this Arctic willow have exploded to release hundreds of frost-resistant seeds, which will be spread by the wind to germinate next year.*

**FUTURE HARVEST**
*A young woman tends a rice crop in steeply terraced fields in southern China. Terracing is one solution to the global need for increased agricultural land.*

# AGRICULTURAL AREAS

WHEN FARMING FIRST BEGAN, about 10,000 years ago, it was a fringe activity carried out by small numbers of people, and it had little impact on the wider world. Since then, agriculture has developed beyond recognition. It now supports almost all of the human race, and approximately 35 percent of the world's land surface is used for farming of some kind. During its remarkably short history, agriculture has transformed the world's landscapes, creating some of the greatest physical and ecological changes the Earth has ever seen. Today, over 2 billion people are actively involved in farming. The land that they work varies enormously, from steep rice terraces to table-flat prairies, and from intensive mixed farmland to arid rangeland, where domestic animals live much like their relatives in the wild. Taken together, the world's farmland is a testament to human ingenuity—and perseverance—in the constant drive to produce food.

# AGRICULTURAL AREAS

◀ 100–03 Soils
◀ 130–31 Water resources
◀ 294–95 Desertification
Urban areas 354–71 ▶
Industrial areas 372–79 ▶
Fishing 422–23 ▶

MANY DIFFERENT FACTORS affect the way that land is farmed. Some are connected with the local soil and climate, but others relate to economics, or to local traditions. In many parts of the world, the predominant form of agriculture is arable farming, which is any kind of farming that involves growing crops. Where conditions are less suitable for crops, arable farming is often replaced by pastoralism, or raising livestock. Arable farming and pastoralism are the two principal types of agriculture, but they embrace an enormous variety of farming practices. Together, these have created the agricultural landscapes that exist today.

## SHIFTING AND SETTLED AGRICULTURE

In shifting or nomadic agriculture, farmers stay on the move and generally do not own the land that they use. This kind of agriculture was probably the first to be practiced, but it is relatively unproductive and cannot support dense populations. Today, it is restricted to regions where the climate is harsh or where the soil is not sufficiently fertile to support full-time settlement. In most other parts of the world, land is farmed on a settled or sedentary basis. Settled farmland is generally divided up by boundaries indicating ownership—a feature that has become the hallmark of agricultural land as a whole. In some regions of the world, such as northern Europe, these boundaries can date back hundreds of years.

**MOVING ON**
*Reindeer herding is one of the few forms of shifting agriculture practiced outside the tropics. The herds stay on the move in search of lichens, rather than grass.*

**FIELD SIZES**
*These fields, in Burgundy, France, are larger than their predecessors a century ago because farm machinery now needs more space.*

LAND

## SCALE AND INTENSITY

The world's earliest farmers worked on a subsistence basis, which means that they produced only enough food for themselves and their immediate dependents. By contrast, many of today's farms—particularly in the developed world—are run as commercial enterprises. They specialize in particular crops or animals, and often raise them on a substantial scale. Subsistence farms are typically small: in Southeast Asia, for example, some cover less than 2.5 acres (1 hectare). At the other extreme, some commercial farms in Australia and North America are over 100,000 times that size. From an economic standpoint, farms also differ in the intensity with which the land is worked. In intensive farming, a large amount of labor or investment is put into a relatively small amount of land, and the yields are often high. In extensive farming, inputs are more thinly spread, but, because the amount of land is greater, overall production is much higher. Most subsistence farms are run on an intensive basis, while commercial farms can be intensive or extensive.

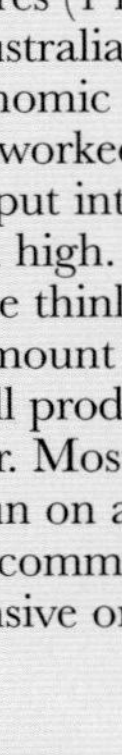

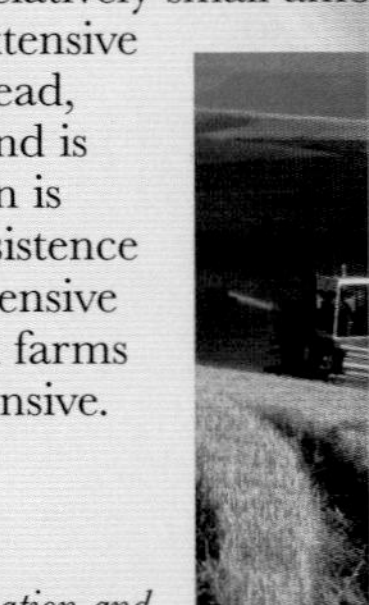

Global production (millions of metric tons)
700
600
500
400
300
200
100
0
Corn, Rice, Wheat, Potatoes, Cassava, Soybeans, Sweet potatoes, Barley, Bananas, Sorghum

**TOP TEN CROPS**
*As important food sources, corn, rice, wheat, and potatoes dwarf all other crops that are grown as primary staple foods.*

**MECHANIZED HARVESTS**
*Modern agriculture relies on mechanization and fossil fuels. For some crops, the energy used in fuel is greater than the energy produced as food.*

## FARMING AND LANDSCAPE

Shifting agriculture leaves few permanent marks on the landscape, while settled farming can bring about huge and far-reaching changes. In the Near East, for example, Mesopotamian farmers dug elaborate irrigation channels over 6,000 years ago, and in mountainous regions, from the Andes to Southeast Asia, farmers stabilized the ground by building terraces—many of which are still in use. Over time, farming can create an entirely artificial landscape that is maintained by human intervention. In northern Europe, for example, farmers cut down large expanses of forest in Neolithic times. The result is a patchwork of fields and woodland that looks semi-natural, but is almost entirely artificial.

**CHANGE OF USE**
*The slopes of this extinct volcano in Madagascar have been intensively farmed, which has led to erosion and gully formation on its flanks.*

### NORMAN BORLAUG

Born in 1914, US agriculturalist Norman Borlaug was a leading figure in the Green Revolution—an international crop-breeding program that helped to dramatically boost yields of wheat and rice from the 1960s onward. In recent years, the Green Revolution has been criticized for its reliance on agrochemicals, even though it has transformed food security in what were once famine-prone regions of the world.

## FARMING AND SOIL

In nature, soil forms very slowly, yet it erodes slowly also. When land is cultivated, this natural balance is disrupted, because erosion rates increase. One of the most catastrophic examples of this effect occurred during the 1930s, on North America's Great Plains. Here, the rapid expansion of grain farming, coupled with overgrazing, made the soil vulnerable to wind erosion. During the Dust Bowl years, millions of tons of it simply blew away (see p.103). Today's farmers are generally more alert to the danger of soil erosion, but it remains one of the major challenges facing global agriculture. The problem is worst where the climate is dry, the soil loose, and the population pressure high. In the loess plateau of northern China, these factors combine to produce the world's highest erosion rates. Soil here is lost at the rate of up to 110 tons per acre (250 metric tons per hectare) per year, much of it into the Yellow River.

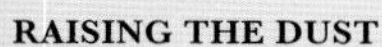

**RAISING THE DUST**
*Plants play a key role in preventing soil erosion. In dry areas, such as this farmland in Oregon, plowing is potentially risky because it exposes the soil to the wind.*

**GULLY EROSION**
*This deep gully in the African tropics is caused by loss of plant cover through overgrazing, allowing heavy rain to wash away the soil.*

## AGRICULTURE AND WILDLIFE

Farming changes natural habitats and replaces their original plants and animals with ones that live under human control. Although there are at least 2 million wild species on the Earth, the list of domesticated species includes fewer than 30 kinds of animals and about 200 kinds of food-producing plants. In farmland, these are given all the resources they need to grow, while their wild competitors are plowed up, cut down, poisoned, or fenced out. The result, after 10,000 years of farming, is a world where ecosystems are simplified. Wildlife fares worst where only one crop is grown (see Monoculture, p.349) and where high levels of agrochemicals are used. Mixed farmland and rangeland are better for wildlife, because they have more in common with the original habitats.

**DRAWN TO WATER**
*Dromedaries, seen here drinking at a well near Timbuktu, Mali, are no longer found wild in their original habitat.*

**TOTAL COMMITMENT**
*The American Great Plains are one of the world's most important grain-growing regions. Here, agriculture has replaced the original grassland habitat.*

### AGRICULTURE PROFILES

The pages that follow contain profiles of agricultural types. Each profile begins with the following summary information:

**AREA** Approximate extent (where known)

**LAND USE** Shifting or sedentary

**ECONOMIC TYPE** Extensive or intensive

**AGRICULTURAL TYPE** Arable, pastoral, or mixed

**LEADING PRODUCERS** Main producer countries, listed in order of output

SUBSISTENCE

## Hunting and gathering

**DISTRIBUTION** In the tropics and subtropics (chiefly in the southern hemisphere); also in high latitudes, including polar regions (northern hemisphere only)

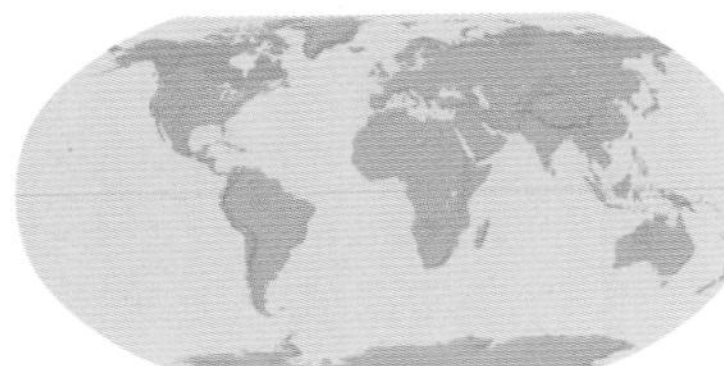

**AREA** 6,750,000 square miles (17,500,000 square km)

**LAND USE** Shifting **ECONOMIC TYPE** Extensive

**AGRICULTURAL TYPE** Not applicable

**LEADING PRODUCERS** Not known

By far the oldest form of land use, hunting and gathering stretches back to the beginning of human history. Until about 10,000 years ago, it supported the world's entire human population, but its importance waned rapidly once arable farming began.

True hunter-gatherers are now very rare. They are restricted to remote regions such as central Amazonia and parts of the Arctic, but their self-sufficient lifestyle is being eroded by contact with the outside world. Hunter-gatherers usually live in small nomadic bands and rely entirely on food that can be caught or collected in the wild. In the tropics and subtropics, this typically consists of a range of animal food, together with fruit, roots, and seeds. At high latitudes, hunting becomes increasingly important, and meat can make up almost all the diet. To survive, hunter-gatherers rely on a detailed knowledge of their environment and of the seasonal changes that affect the food supply.

**FOREST MARKSMAN**
*A hunter aims a blowpipe in the lowland rainforest of eastern Ecuador. Such traditional weapons are handmade, using locally available materials.*

**COMMUNITY ON THE MOVE**
*Among the world's last hunter-gatherers, only a minority of San people, from southwest Africa, still follows their traditional nomadic way of life.*

SUBSISTENCE AND COMMERCIAL

## Nomadic herding

**DISTRIBUTION** In mid-latitudes in Africa, the Middle East, and Central Asia, on low ground and at altitude; in high latitudes in Eurasia. Extremely rare in the Americas

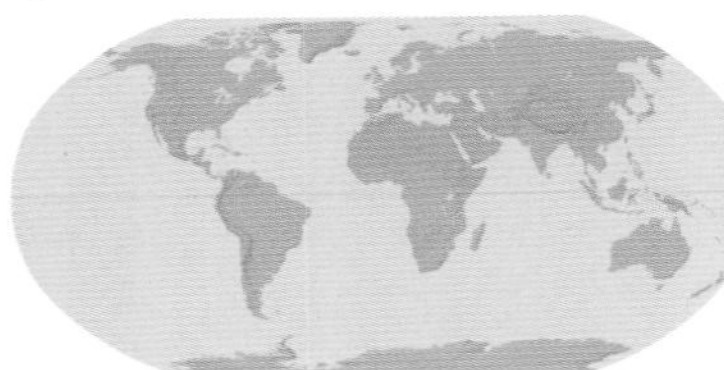

**AREA** 10,000,000 square miles (26,000,000 square km)

**LAND USE** Shifting **ECONOMIC TYPE** Extensive

**AGRICULTURAL TYPE** Pastoral

**LEADING PRODUCERS** Not known

Also known as nomadic pastoralism, this form of agriculture is an effective use of land in arid or cold climates, where the soil is unsuitable for growing crops. Animals raised include sheep, goats, cattle, yaks, and reindeer, and they can wander freely while being watched by their owners. In dry areas such as the Sahel, in central Africa, nomadic herders typically stay on the move, which helps them cope with unpredictable rainfall. In mountainous regions, such as the Himalayas, herders lead their animals between winter and summer grazing areas, a practice known as transhumance. In general, nomadic herding is a subsistence activity, with meat, milk, and animal products supplying herders and their families with most of their food. Reindeer herding, however, is carried out on a commercial scale.

**REINDEER ROUNDUP**
*The original nomadic lifestyle of reindeer herding is gradually dying out in parts of the Arctic, where it has evolved into a commercial activity.*

**MIXED HERDS**
*Sheep and goats forage for food around a Bedouin camp in the Jordanian desert.*

SUBSISTENCE

## Slash-and-burn agriculture

**DISTRIBUTION** In tropical regions worldwide including the Amazon Basin, sub-Saharan Africa, and Southern and Southeast Asia; rare or absent in Australasia

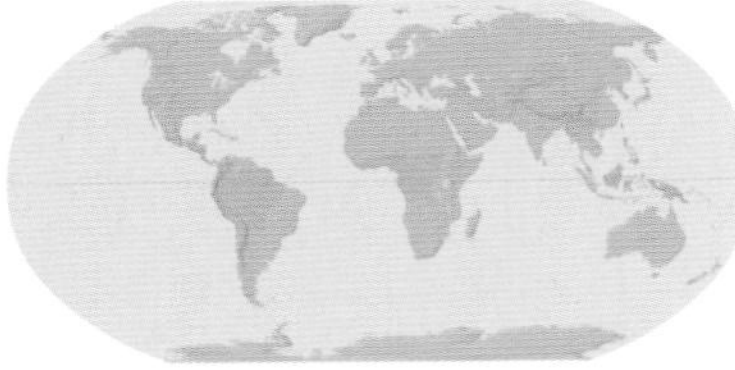

**AREA** 7,700,000 square miles (20,000,000 square km)

**LAND USE** Shifting **ECONOMIC TYPE** Extensive

**AGRICULTURAL TYPE** Arable

**LEADING PRODUCERS** Brazil, Democratic Republic of Congo, Indonesia

One in 25 of the world's population still relies on slash-and-burn for their livelihood, which has sustained them for thousands of years. In this type of cultivation, which was the original form of agriculture in the tropics, temporary clearings are hacked out of forests or woodlands, and the felled vegetation is set on fire. Crops such as cassava and sweet potatoes are planted in the newly cleared ground, drawing on the ash that fertilizes the soil. Initially, yields are good, but the soil's fertility soon declines. After a handful of years, the farmer abandons the ground and moves on to clear a new plot elsewhere. Slash-and-burn looks destructive, but it can be efficient and environmentally benign. Yields are small, but the energy returns are high, because no inputs of agricultural chemicals or fossil fuels are involved. Once land has been abandoned, it gradually reverts to its natural state, and after several decades may be used again. Slash-and-burn can support much denser populations than nomadic herding. However, it has one major weakness: if the land is worked for too long, it can become permanently infertile. This is an increasing problem in regions such as Southeast Asia, where land is in short supply.

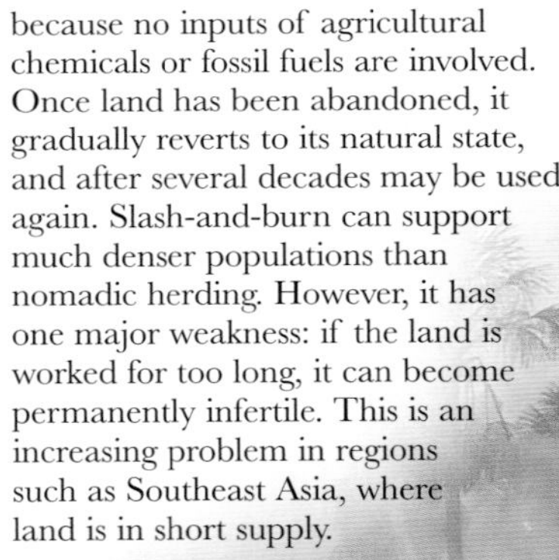

**EMERGING CROP**
*Two months after the ground has been cleared, plants begin to appear, but soil nutrients will soon be exhausted.*

**CLEARING THE LAND**
*In Roraima State, Brazil, a farmer burns off vegetation. Without the help of tractors or bulldozers, farmers often leave trees to lie where they fall, and crops are planted between them.*

LAND

COMMERCIAL

## Plantation agriculture

**DISTRIBUTION** Chiefly in the tropics, in the Americas, Africa, Asia, and Australasia, in lowlands and at altitude; common on tropical islands

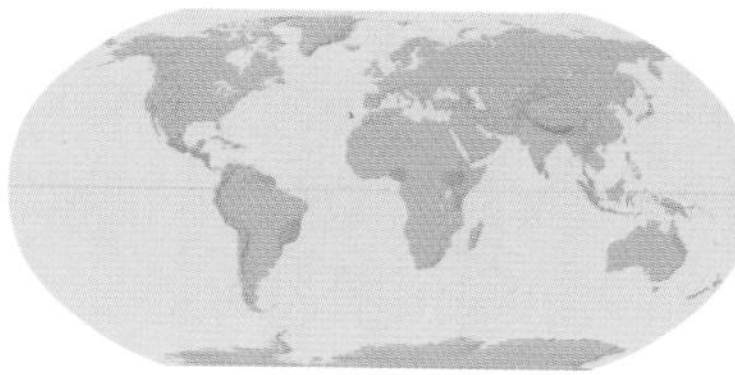

**AREA** 3,100,000 square miles (8,000,000 square km)

**LAND USE** Sedentary **ECONOMIC TYPE** Intensive

**AGRICULTURAL TYPE** Arable

**LEADING PRODUCERS** Malaysia, Brazil, Mexico, India, Cuba

Most of the large-scale commercial crops that are grown in warm climates are produced in plantations. These include tea, coffee, cocoa, bananas, palm oil, and sugar, as well as fiber crops, such as sisal and cotton. Several minor crops, such as pineapples and nuts, are also grown in this way. Three features typify this kind of agriculture. Crops are cultivated on large estates and are almost always destined for export rather than for local use. Secondly, the crops are often bushes or trees, rather than plants that have to be sown and harvested each year. Finally, much plantation agriculture had its origins in colonial times, and many plantations are still owned by corporations based overseas. For local people, plantations provide a source of employment, and often of housing and schooling. However, like all monocultures (see panel, above), their long-term nature makes them economically inflexible—a problem if prices fall.

### MONOCULTURE

In some forms of agriculture—including most plantations—the same crop is grown on a large scale year after year. This type of cultivation, called monoculture, gives important economies of scale, but it also creates a giant reservoir for pests and diseases. These have to be tackled by using high levels of agrochemicals: non-organic cotton, for example, is one of the most sprayed crops in the world.

**TEA HARVEST**
*Workers pick young shoots in a tea plantation in northern India. The tea bushes are interplanted with trees that provide timber and shade.*

**CUTTING SUGAR CANE**
*Sugar cane is one of the world's most important plantation crops. Over 1 billion tons of cane are harvested each year; much of it is cut by hand.*

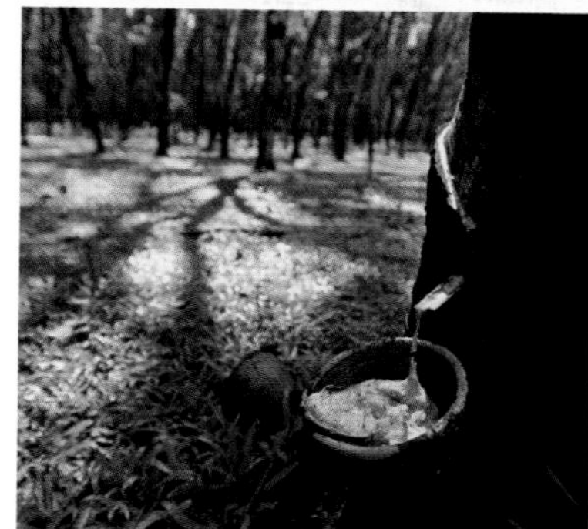

**RUBBER TAPPING**
*Latex—the source of natural rubber—trickles from a tree. Once a key commodity, natural rubber has been largely replaced by synthetic substitutes.*

SUBSISTENCE AND COMMERCIAL

## Rice cultivation

**DISTRIBUTION** Chiefly in the tropics and subtropics, in southern and Southeast Asia; also in the Americas, Africa, southern Europe, and Australasia

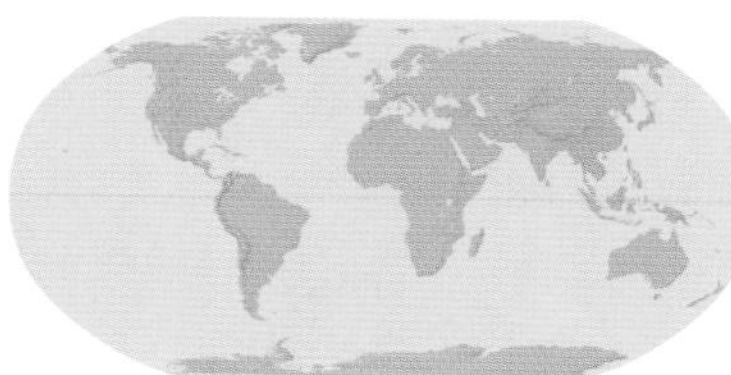

**AREA** 4,600,000 square miles (12,000,000 square km)

**LAND USE** Sedentary **ECONOMIC TYPE** Intensive

**AGRICULTURAL TYPE** Arable

**LEADING PRODUCERS** China, India, Indonesia, Bangladesh, Vietnam

Rice is the world's second most important cereal crop, with a history of cultivation dating back at least 5,000 years. It was originally domesticated in Asia, and this continent remains by far the largest producing region. Unlike corn and wheat, rice is used almost exclusively for human consumption. There are thousands of varieties of cultivated rice, and the crop can be grown in two different ways. Upland rice is planted in dry soil and raised much like other cereals. It accounts for less than 10 percent of the global harvest. Lowland rice is typically sown in nurseries. After about four weeks, it is then transplanted into paddies, or flooded fields, which have been plowed and cleared of weeds. This method is highly labor-intensive, particularly on sloping land, where terraces have to be built and maintained. It is also very productive, given sufficient water and warmth. Modern hybrid plants mature rapidly and can produce up to three crops each year. Rice cultivation spread to Europe in medieval times and from there to the Americas. In these regions, rice production is more mechanized, and pre-germinated seed is sometimes planted from the air. But this large-scale farming is still the exception. Most of the world's rice is grown on a small-scale or subsistence basis, and less than 5 percent of the total crop finds its way into global trade.

### METHANE AND RICE

Rice paddies are a major source of methane—a gas that is produced when organic matter is broken down in surroundings that are low in oxygen. Methane is an effective greenhouse gas, and is estimated to be responsible for up to 20 percent of global warming. Other forms of agriculture also contribute. Grazing livestock produce large quantities of methane when they digest their food.

**GREENHOUSE GAS**
*Flooded rice fields produce high levels of methane. It is generated when rice stalks and roots rot underwater, where oxygen cannot reach them.*

**BUFFALO POWER**
*In southern and Southeast Asia, water buffalo are important as draft animals in wet conditions. Their dung also helps to fertilize rice paddies.*

**LOWLAND RICE**
*Young rice plants grow quickly in the warmth of the tropics. In ideal conditions, rice can be ready for harvesting within 12 weeks of transplantation.*

COMMERCIAL

## Cattle ranching

**DISTRIBUTION** Principally in mid-latitudes (northern and southern hemispheres), in areas with moist or semiarid continental climate

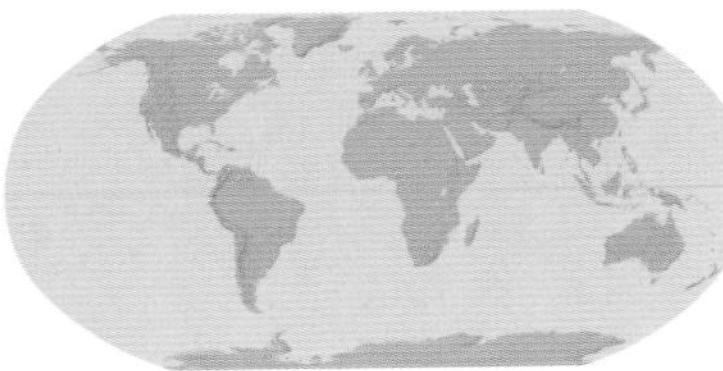

**AREA** 11,200,000 square miles (29,000,000 square km)

**LAND USE** Sedentary **ECONOMIC TYPE** Extensive

**AGRICULTURAL TYPE** Pastoral

**LEADING PRODUCERS** US, China, Brazil, Argentina

The world's most important source of meat is cattle, which is the primary product of cattle ranching. Some breeds, however, are kept chiefly for their milk (see Dairy Farming, opposite). Cattle country is normally dry, open, and relatively unproductive, and large areas are required to support cattle herds. Historically, this kind of space has always been rare in Europe, except in a few regions such as central Spain, but European colonists found it in abundance when they reached South America, North America, and Australia. Here, they developed commercial cattle ranching on a large scale, and these regions still remain key producers of beef. In Africa, the spread of cattle ranching was limited by animal diseases such as rinderpest, which spreads to cattle from wild mammals such as antelopes, buffalo, and giraffes. In the wake of deforestation, however, ranching is now expanding in tropical areas—a trend that is causing major environmental concerns, as is overgrazing (see panel, right). Cattle ranches themselves have changed with improvements in technology. In early ranches, cattle roamed freely, under the supervision of cowboys (or gauchos). But in the 1870s, the invention of barbed wire introduced cheap and effective fencing, allowing herds to be more easily controlled. In North America particularly, this was followed by improvements in transportation, which allowed animals to be carried to commercial feedlots to be fattened up for sale. The feedlot system is now highly mechanized, and animals receive a computer-controlled diet designed to maximize returns to the farmer. In energy terms, cattle farming remains an inefficient form of agriculture, because it takes more nutrients to produce a unit of beef than any other kind of animal food. However, it does enable farmers to exploit marginal land that would otherwise go unused.

### OVERGRAZING

In arid regions, vegetation is slow to recover from the effects of grazing. Wild grazers evolve a long-term balance with their food supply, but domesticated cattle can cause severe damage if they are stocked at higher densities than the land can support. The result—after several decades—is severe erosion, which can make the land worthless for farming. Rangeland erosion is a problem throughout the world, but North America is particularly badly affected.

**EXTREME EROSION**
*In Mungo, Australia, sheep and rabbits have accelerated natural erosion, creating a scarred landscape.*

**NATURAL PARTNERS**
*Originally from North America, the horse has been introduced to cattle farms worldwide, but in many regions, feral horses still live beyond human control.*

**CATTLE DRIVE**
*In North America, cattle fill the ecological role once occupied by American bison (or buffalo). Today, cattle outnumber bison by more than 5,000 times.*

SUBSISTENCE AND COMMERCIAL

## Sheep farming

**DISTRIBUTION** In the subtropics and higher latitudes (northern and southern hemispheres), in regions with widely varying climates, on low ground and mountains

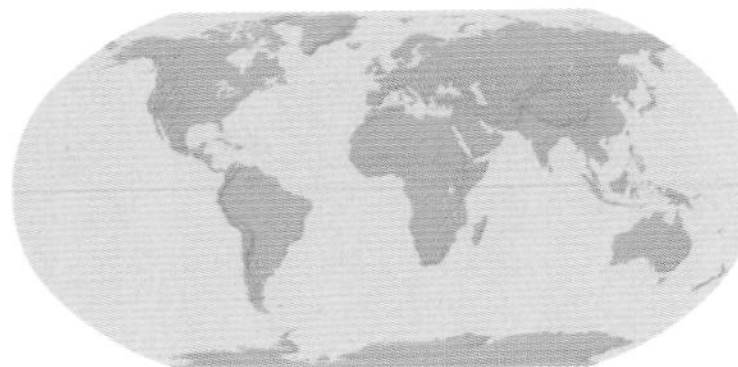

**AREA** 13,000,000 square miles (34,000,000 square km)

**LAND USE** Sedentary

**ECONOMIC TYPE** Extensive or intensive

**AGRICULTURAL TYPE** Pastoral

**LEADING PRODUCERS** China, India, Australia, Iran, New Zealand

Compared to cattle, sheep are hardy animals, with a high tolerance of heat, cold, and drought. Some are looked after by nomadic herders, but most of the world's 1 billion sheep belong to settled farmers and are raised in several different terrains. In northwest Europe—where many of today's commercial breeds were developed—sheep are nurtured on good lowland grazing and on moorland and hills. But in many other parts of the world, they are raised in open rangeland, which has a drier climate and less food. To make up for this, rangeland sheep roam over much bigger areas. In Australia, some sheep farms cover more than 385 square miles (1,000 square km), and, where grazing is poor, each sheep may need more than 50 acres (20 hectares) of land to survive. Sheep are very efficient foragers, and they can have a significant effect on the landscape. In regions with moist climates, they keep pastures closely cropped, which helps to suppress weeds and maintain the grass. In arid regions, however, they can strip away the plant cover, leading to soil erosion. Wherever they live, sheep are raised chiefly for their meat. In the past, wool was an equally important product, and breeds such as the Merino from Spain were developed for their luxuriant fleeces. Some of today's experimental breeds have relatively sparse wool, more like their relatives living in the wild.

**A CHANGE IN TRADITION**
*Sheep shearers clip a fleece so that it comes away as one piece. But with the invention of synthetic fibers, fleeces no longer command high prices.*

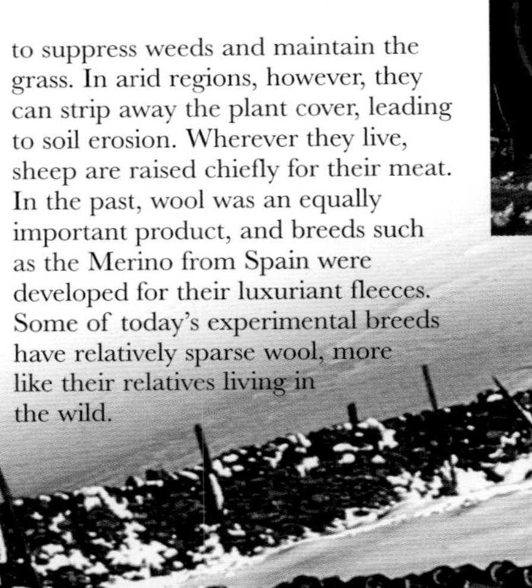

**SHEEP FARMING IN WINTER**
*Although sheep are raised in many climates, they cannot forage in deep snow and so need food supplements, such as hay, which is being provided here in northern England.*

COMMERCIAL

## Dairy farming

**DISTRIBUTION** Restricted to temperate regions, in areas with a moist climate; on low ground, and on seasonal pastures in mountains

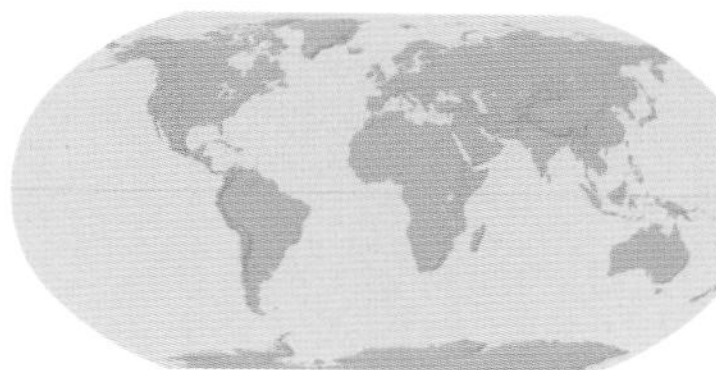

**AREA** 2,100,000 square miles (5,400,000 square km)

**LAND USE** Sedentary **ECONOMIC TYPE** Intensive

**AGRICULTURAL TYPE** Pastoral

**LEADING PRODUCERS** US, Russia, Germany, France, New Zealand

In dairy farming, animals are raised for their milk and for products that are derived from it, such as butter and cheese. The two most important milk-producing animals are cows and water buffalo, with much smaller quantities being provided by goats and sheep. However, of all dairy animals, cows are the only ones that support a complete farming system based on milk alone. Milk is highly nutritious, and cows need a high-quality diet to maintain productivity. Such food is found in the rich pastures that typify the dairy-farming landscape. Because cows have to be milked twice a day, they need easy access to a farm. As a result, fields tend to be small. In some parts of the world—such as New Zealand—the animals can be kept outside throughout the year, but in many dairy regions they are brought under cover and fed during the winter months. Milk is one of the most perishable of all agricultural products. At one time, fresh milk could be sold only for local consumption, and cities were supplied by dairy parlors housing what were known as urban cows. Today, refrigerated transportation allows milk to be carried long distances before it is processed.

**SIMPLE DIET**
*Goat's milk is an important product in many parts of the developing world. Unlike cows, goats do not need high-quality grass.*

**MILKING TIME**
*In modern milking parlors, milk is chilled to slow the growth of bacteria. Milk may also be flash-heated, or pasteurized, to extend its shelf life.*

**GRAZING COWS**
*Renowned worldwide for the richness of their milk, these Jersey cows are grazing on a farm in South Africa's Western Cape.*

COMMERCIAL

## Horticulture

**DISTRIBUTION** Widely scattered across tropical, subtropical, and temperate regions, often near centers of population; dependent on irrigation in dry areas

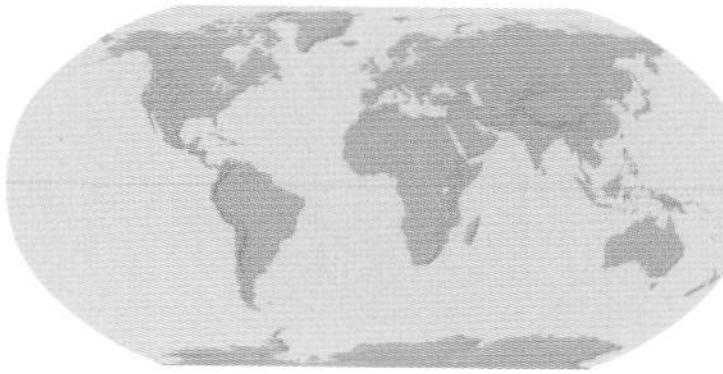

**AREA** 1,500,000 square miles (4,000,000 square km)

**LAND USE** Sedentary **ECONOMIC TYPE** Intensive

**AGRICULTURAL TYPE** Arable

**LEADING PRODUCERS** US, Netherlands, Israel, Spain, Ecuador

Fruits and vegetables are raised on an intensive basis in commercial horticulture, also known as market gardening. The actual range of crops depends on local climate and conditions, and, in many cases, several different types are grown side by side. Landholdings are often small, but in some regions—such as southern Spain, the Netherlands, and the Imperial Valley in California—they are more substantial, with major investment in glasshouses or irrigation. These centers of production include some of the most intensively worked arable land in the world (see panel, right). Traditionally, commercial horticulture is located close to towns and cities, because perishable crops must be delivered to market without delay. But with the rapid expansion of road and air freight, and improvements in refrigeration, the lines of supply grow longer each year. This has created new markets for countries such as Mali, Ecuador, and Thailand, which are able to export fresh produce during winter in the northern hemisphere. Compared with other kinds of agriculture, commercial horticulture is sensitive to changing tastes, and it frequently focuses on low-volume crops that command premium prices.

**POTATO HARVEST**
*A harvesting machine unearths potatoes in Idaho. Potatoes are a versatile crop, equally suitable for subsistence growers or large farms.*

### SALINIZATION

In intensive horticulture, and other kinds of agriculture, irrigation can cause salinization—a damaging buildup of salt in the soil. Salinization occurs when irrigation water mobilizes salts that are naturally present in the ground. Salt buildup reduces crop yields, and eventually may make land unfit for farming.

**FLOWERS UNDER COVER**
*Cut-flower cultivation is a highly intensive form of horticulture that can be very lucrative, despite high costs.*

**GOING FOR GROWTH**
*The mild winter climate here in southern Texas gives farmers a head start over those farther north.*

SUBSISTENCE AND COMMERCIAL

## Cereal cultivation

**DISTRIBUTION** Worldwide; commercial production (apart from rice growing) takes place chiefly in temperate regions

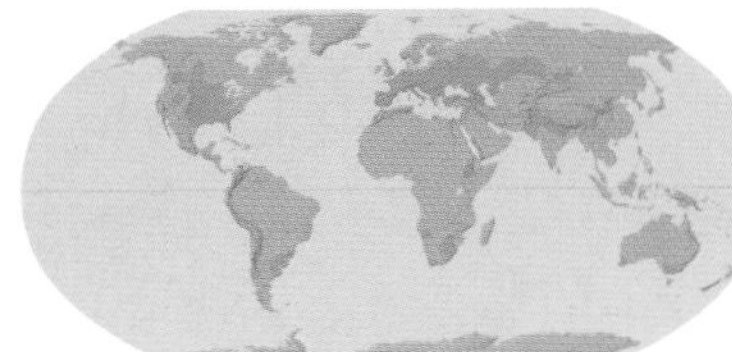

**AREA** 14,000,000 square miles (36,000,000 square km)

**LAND USE** Sedentary

**ECONOMIC TYPE** Extensive or intensive

**AGRICULTURAL TYPE** Arable

**LEADING PRODUCERS** China, US, India, Russia, France

The first plants to be domesticated were cereals, which are of supreme importance today. Cereals are grasses that are cultivated for their edible seeds, or grains). Together, just three grains—corn, rice, and wheat—account for over half the world's food, a remarkable situation given the large number of potential cereals that grow in the wild. Cereals are cultivated all over the globe, but the main centers of production are in temperate regions, except for rice (see p.349), which is grown chiefly in the tropics. As crops, cereals have two main advantages: their grain contains high levels of nutrients; and it can be stored for extended periods as long as it is kept dry. This makes grain easy to transport and to trade—a form of commerce that dates back to the world's oldest civilizations. The cereal trade is now ensuring adequate supplies of food worldwide, and over 250 million tons of grain are exported each year. In some parts of the world, cereal farming is still at subsistence level, but in developed countries, it has become a high-technology business, conducted on an increasingly large scale. Improvements in crop varieties and the use of agrochemicals have produced an extraordinary growth in yields. In the United States, for example, average corn yields stood at less than 30 bushels per acre (2 tons per hectare) in 1900. The figure is now five times higher. Productivity has also increased because more land is being plowed (see panel, left), particularly in the American Midwest, the Pampas of South America (see p.337), and the wheat belt of Australia. For the "big three" cereals, however, the growth in yields is now leveling off, and vacant land is in short supply. Many agronomists therefore believe that genetically modified crops (see panel, above) are the key to future growth, despite environmental concerns.

### JOHN DEERE

The American blacksmith and engineer John Deere (1804–88) invented the all-steel plow, which set off one of the most far-reaching changes in agriculture. Unlike wood-and-iron plows, Deere's plow could slice through densely matted soil, opening it up for arable farming. By 1867, 10,000 plows a year were sold by Deere's company.

**GLOBAL HARVEST**
*Surrounded by a sea of harvested wheat, this Australian farm lies in an area where cereal growing dates back less than 60 years.*

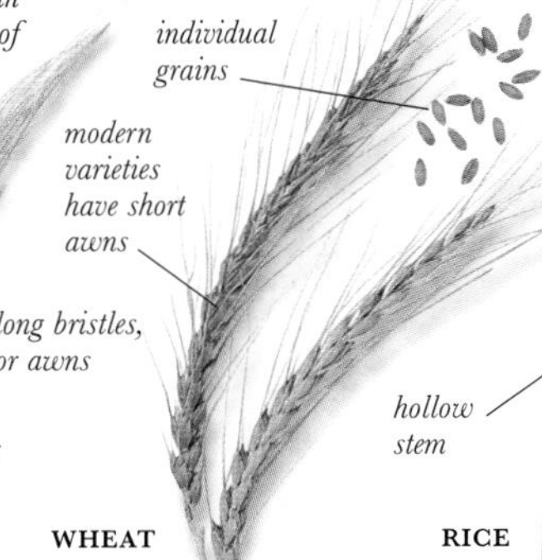



**KEY CEREALS**
*Corn, wheat, and rice are the world's major cereals. In addition, rye and barley are important in temperate regions, while sorghum and millet are popular in the tropics.*

### GM CROPS

Many agriculturists believe that genetically modified (GM) crops—particularly grains—have a vital role to play in combating world hunger. Crops can be modified to produce higher yields and to resist pests. GM crops, however, remain controversial, because of worries about their possible effect on human health, on the environment, and on the economies of poorer nations.

**WINNOWING GRAIN**
*An Egyptian separates wheat in the traditional way, by tossing it in the air. Winnowing removes the inedible husk that surrounds each grain.*

**CORN IN FLOWER**
*Multipurpose corn can be harvested while it is green as fodder for livestock, as well as when ripe for its nutritious grain.*

SUBSISTENCE AND COMMERCIAL

## Mixed farming

**DISTRIBUTION** Worldwide in the tropics and in temperate regions; absent in areas with cold or arid climates

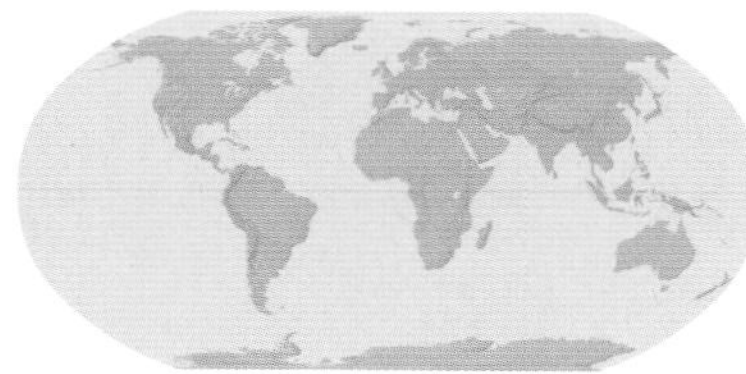

**AREA** 14,000,000 square miles (54,000,000 square km)

**LAND USE** Sedentary

**ECONOMIC TYPE** Extensive or intensive

**AGRICULTURAL TYPE** Arable and pastoral

**LEADING PRODUCERS** China, India, US, Russia, France

In agriculture, specialization can increase efficiency, but it can also create problems if crops or animals are struck by bad weather or disease. This is one of the reasons why mixed farming is so widespread throughout the world. In this form of agriculture, farmers raise a range of crops and livestock, rather than concentrating on a single product. In ecological terms, mixed farming is more balanced than specialized agriculture, because animal food is frequently grown on the farm itself, while animal manure helps to improve the fertility of the soil. Mixed farming is more variable than any other kind of agriculture. At one extreme, some mixed farms—particularly in developing countries—raise more than 20 different sources of food. In China, for example, these include grains, vegetables, fruit, and winter root crops, as well as chickens, ducks, and pigs. Many Chinese farms also have fishponds, which create high-protein animal food from kitchen scraps and farmyard waste. In other parts of the developing world, mixed farms work in a similar way, although the crops and animals raised depend on local climate and traditions. In the Bolivian Andes, for example, the traditional staple, called *chuño*, is prepared by drying potatoes, oca, or *ysaño*—three tuber-bearing crops that originated there more than 1,000 years ago. As well as poultry, Bolivian livestock includes llamas and guinea-pigs, both of which were domesticated in that part of the world. In developed countries, mixed farms are usually larger, and their produce less varied. European and North American farms typically grow grains and root crops, and most of the harvest is destined to feed livestock, which includes cattle, pigs, and poultry. Until the 1950s, it was normal for farms to include all these animals, but today's farms often concentrate on just one. This increasing specialization reaches a peak in pig and poultry farms, where hundreds or thousands of animals are raised in intensive conditions. Unlike traditional mixed farms, these industrial-style units rely solely on bought-in feed.

**PREPARING FOR WINTER**
*Haymaking allows farmers to turn surplus grass into winter feed. In many farms in eastern Poland, hay is still raked and piled by hand.*

**INTENSIVE FARMING**
*Chickens are the world's most numerous farm animals. With a population of about 15 billion birds, they outnumber people by over two to one.*

SUBSISTENCE AND COMMERCIAL

## Mediterranean-type agriculture

**DISTRIBUTION** In the Mediterranean region and areas with similar climate in North and South America, southern Africa, and Australia

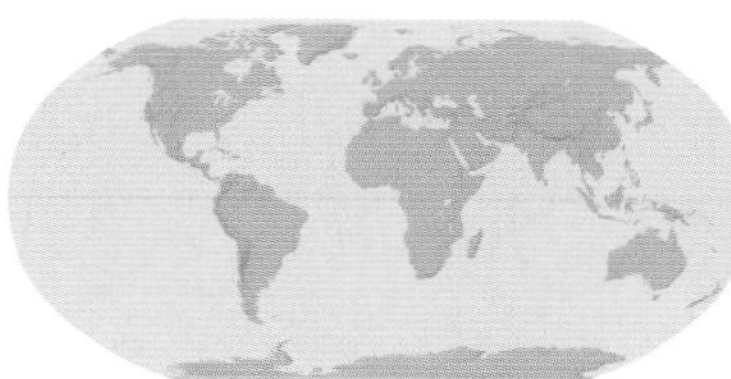

**AREA** 2,100,000 square miles (5,500,000 square km)

**LAND USE** Sedentary

**ECONOMIC TYPE** Extensive or intensive

**AGRICULTURAL TYPE** Arable and pastoral

**LEADING PRODUCERS** Spain, Italy, US, South Africa, Australia

With its distinctive climate of wet winters followed by long, dry summers, a characteristic kind of agriculture developed in antiquity in the Mediterranean region and has since spread to several other parts of the world. Although Mediterranean-type agriculture is based on a variety of plants, two of them—the olive tree and the grape vine—epitomize this kind of farming. Olives are long-lived trees, capable of fruiting for centuries. However, they cannot survive where winter temperatures drop below about 23°F (–5°C). Grape vines are much more hardy, but, like olives, they need strong sunshine to produce a useful crop. Only in a Mediterranean climate do these two plants thrive side by side. Mediterranean summers are usually too warm and dry for grains to continue growing, so cereal crops are often sown in the fall and then harvested in late spring. Where farms are still run on traditional lines, the result is a patchwork landscape of olive groves, vineyards, and grain fields. When Spanish colonists crossed the Atlantic, this kind of agriculture traveled with them, and grape vines are now grown extensively in Chile, Argentina, and California, together with many other Mediterranean-climate crops, such as citrus fruits, peaches, and figs. South Africa and Australia have also become important wine-exporting regions, as well as producers of fruits. But the spread of crops has not been only one way: tomatoes and bell peppers were unknown in Europe until the early 1500s, when they arrived from the Americas. Five centuries later, these crops are seen as a quintessential part of the Mediterranean daily diet.

**GROWING GRAPES**
*Although grape vines originated in western Asia, today's vines are grafted onto rootstocks from North American plants, which are resistant to insect pests.*

**PRESSING THE GRAPES**
*Traditionally, grapes are pressed by trampling them with bare feet. In modern wineries, presses do this work, processing thousands of tons of fruit a day.*

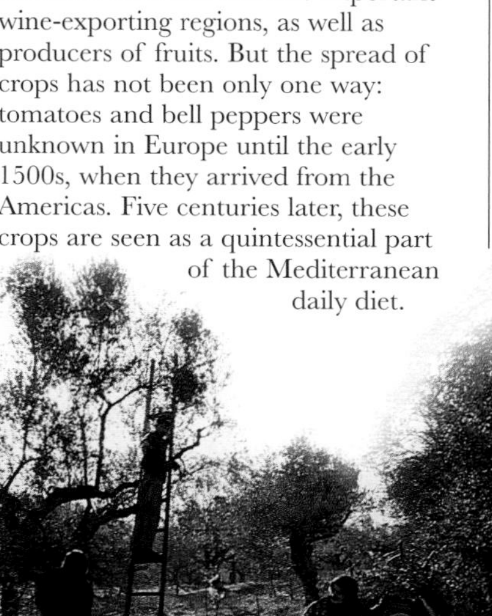

**FAMILY LABOR**
*Olives are harvested either while they are still green and semi-ripe, or when they are almost black and ripe.*

**HEIGHT ADVANTAGE**
*The skyscrapers of Chicago's central business district project above the clouds. They clearly delimit the city center, separating it from the rest of the city.*

# URBAN AREAS

TWO HUNDRED YEARS ago, most people lived in small, rural communities, and there were few large cities. Only London is known to have had over a million inhabitants, and just a handful of cities had populations in excess of 100,000 people. Today, about half of the world's population lives in cities. This rapid and continuing growth in urban populations suggests that by the year 2020, six out of every ten human beings will be city-dwellers. Accommodating large numbers of people has led to accelerated urban development, both upward, in the form of taller buildings, and outward, as cities increase in area. Cities are growing so fast that in some parts of the world, such as New York and Tokyo, they are merging into one another, covering vast areas known as megalopolises. About three percent of the Earth's land surface is already urban, and it is likely that this proportion will double over the next 20 years. Urban areas increasingly dominate economic and social life.

# URBAN AREAS

41 Human migration
344–53 Agricultural areas
Population growth 362–63
Industrial areas 372–79

THE HUMAN WORLD is now essentially urban. This is not simply because half the world's population lives in cities. Cities dominate contemporary life. They exert such enormous power that even the most remote rural regions are usually bound up in supplying their need for food, raw materials, and energy. As a result, they become focal points for transportation, trade, and technological advances. Cities are the centers of cultural life, with their museums, theaters, and other attractions. Social changes that might take centuries in the countryside can sweep through urban areas in just a few years. Cities' impact on the environment can be equally dramatic. Urban development is not only burying vast areas under concrete; it is also having a profound effect on the world's climate—at a local level, by creating heat islands, and at a global level, by filling the atmosphere with the by-products of its massive energy consumption.

## INTERNAL ZONES

The concentration of some activities in particular areas gives most cities a distinct structure. In the 1920s, it was suggested that cities were arranged in rings around the central business district, with first a transition zone of industry and poor housing, then a ring of workers' houses, a ring of suburbs, and finally an affluent commuter belt. In 1939, it was argued that cities developed not in rings, but in wedge-shaped sectors along transport routes. Another later idea was that cities develop around not just one center or nucleus, but several. Elements of all these models can be identified in most older western cities, but cities are changing, with increasing decentralization and the growth of "edge cities" on urban fringes. Analysis of cities in other parts of the world has led to a further series of models, emphasizing, for example, the central mosque and street markets for Islamic cities, and wealthy central areas and poor fringes for Latin American cities. New cities, such as Brasília in Brazil, are planned from the outset.

**A PLANNED CITY**
*Most European cities grew haphazardly. Few have a planned structure like this circular housing development near Copenhagen, Denmark.*

## LOCATION

Most cities start life as small settlements, and grow only because their location gives them some advantage over other places. In the past, cities such as London and Paris grew up at natural crossing places over major rivers. A few developed at strategic defense sites. Others, such as Johannesburg, were sited near sources of valuable minerals. In the developed world, many older urban areas, such as Pittsburgh, Pennsylvania, and the cities of the Ruhr valley, Germany, grew up where there were ready sources of coal and iron. Now, as the world becomes increasingly dominated by global trade, local resources are less important, and access to markets is the key factor. Big cities operate globally, not locally. Most of the world's major cities, such as New York and Tokyo, are close to navigable waterways. Forty percent are actually located on tidal estuaries or directly on sea coasts, and are major centers of road, rail, and air transport.

**NIGHT LIGHTS**
*The clusters of cities in the US, Europe, and Japan stand out in this composite satellite image of the Earth at night.*

**PORT OF HONG KONG**
*Inlets around the coast of Hong Kong Island make ideal harbors, such as this one at Aberdeen. They also provide shelter for local craft during typhoons.*

LAND

## POPULATION TRENDS

Urban populations are growing everywhere in the world. If current trends continue, the number of people living in cities will exceed those living in the countryside for the first time in 2007. By 2030, another 2 billion will be added to the world's urban population. In the more developed countries, already highly urbanized, over 84 percent of people will live in cities by 2030. Some of the largest cities in the developed world are already almost as big as they are ever likely to get, and Tokyo's population has probably peaked. The megacities of the future are in the developing world: by 2015, Lagos, Nigeria, may be home to at least 23 million people; and Dhaka, Bangladesh, may be nearly as big. However, both of these are likely to be eclipsed by China's Pearl River delta, centered on the city of Guangzhou (formerly Canton). Already, 40 million people live in an area smaller than Los Angeles. Soon the whole region may merge to create the biggest single urban area the world has ever seen, with a population dwarfing Tokyo's. The population of Guangzhou alone soared from from just 380,000 in 1980 to over 7 million in 2002.

**MASS MOVEMENT**
*As urban populations soar, the pressure placed on mass transit systems will grow, with underground systems moving the greatest numbers of people.*

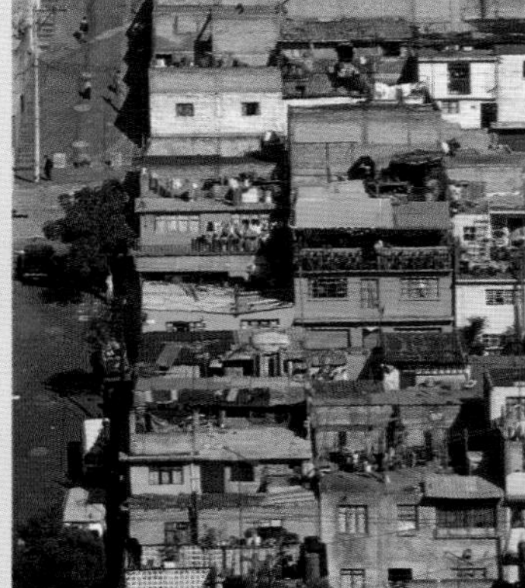

**CROWDED SHANTYTOWN**
*Shanty dwellings, such as these in Mexico City, arise when large numbers of people migrate into cities and have nowhere to live.*

## CITY CLIMATES

A large city is nearly always a little warmer than the surrounding countryside. On warm, sunny days, the difference in temperature may be 9°F (5°C) or more. The reasons for this effect, called the urban heat island, are complex. One cause is waste heat given off by city buildings, trains, and cars, which can contribute up to half as much heat as the Sun. Another cause is the way concrete and brick in the city soak up extra heat from the Sun during the day, and then release it at night. Even pollutants can add to the warming by trapping heat near the ground. Pollutants can cause other changes to the climate, too. Dense fogs called smogs were common in many industrial cities in the last century as domestic fires, industrial furnaces, and steam trains poured out soot that encouraged the condensation of moisture in the air. Clean-air legislation and changes in fuel use have cut the worst of these smogs. However, in the 1980s, a new kind of smog, called photochemical smog, began to appear in sunny cities with heavy traffic, such as Los Angeles. This smog is a result of the way emissions from car exhausts react with sunlight to produce a thick, brown haze.

**CITY SMOG**
*This policeman on traffic duty in Kuala Lumpur wears a face mask to protect himself from the fumes.*

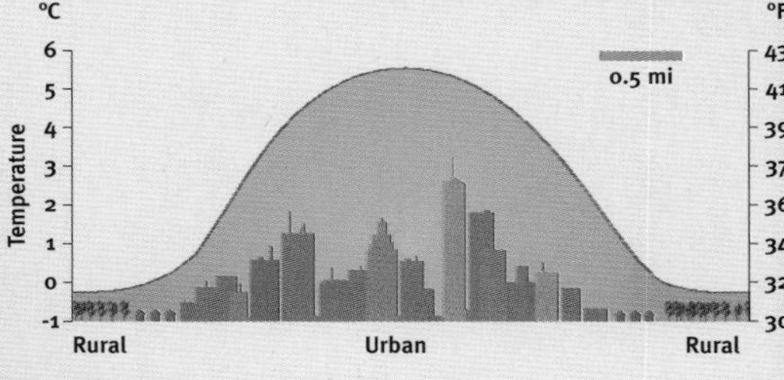


**HEAT ISLANDS**
*Large urban areas develop their own microclimates. This diagram shows the temperature gradient at night for a city in a temperate region. Downtown buildings radiate more heat than the surrounding low-density housing and parkland, keeping the center warmer.*

**BRIGHT LIGHTS**
*The centers of many cities are brightly illuminated at night. Here, Times Square in New York displays a dazzling array of neon lights.*

### URBAN AREAS CATALOG

The following pages contain profiles of selected major cities. Each profile begins with the following information:

**AREA** Area covered by the city and its suburbs

**POPULATION** People living in the city and its suburbs

**FOUNDED** Year in which city was founded

NORTH AMERICA *northeast*

## New York

**LOCATION** At the mouth of the Hudson River in New York State

**AREA** 900 square miles (2,330 square km)

**POPULATION** 18.1 million

**FOUNDED** 1609

The largest city in North America, New York is also one of the world's preeminent financial, commercial, and cultural centers. Situated on the east coast of the US, about midway between Washington, DC, and Boston, New York lies on the same latitude as Rome and Istanbul. However, it experiences a very different climate, with bitterly cold winters, often with heavy snow, and hot summers. The city's origins date back to 1609, when European colonists set up a trading post on Manhattan Island at the mouth of the Hudson River. Sheltered from the Atlantic by Long Island, and with access to the interior of the country via the Hudson River, Manhattan provided the ideal gateway to North America. Because it has the best harbor on the eastern seaboard, Manhattan attracted many European traders, and first developed as the Dutch settlement of New Amsterdam. It was a flourishing port long before it was captured by the British in 1664 and renamed New York. It is estimated that more than half of the people and goods that have ever entered the US came in through the docks of New York. Between 1892 and 1954, some 17 million immigrants landed at Ellis Island, another island in the mouth of the Hudson River, close to Manhattan and Liberty Island, where the Statue of Liberty now stands. Today, the point of arrival for many people in North America is New York's JFK airport. Until the 1950s, New York was by far the world's busiest seaport and, although it has declined in importance since the opening of the St. Lawrence Seaway in 1959, it remains the second-largest seaport in the US. As the city expanded, it

**BROOKLYN BRIDGE**
*Spanning the East River, Brooklyn Bridge was the first suspension bridge to be made of steel. This view of Manhattan is taken from its walkway.*

**ROCKS IN THE PARK**
*Despite the deposition of 10 million cartloads of stone and earth to create Central Park in 1858, outcrops of the hard rock on which Manhattan's skyscrapers are built are still visible.*

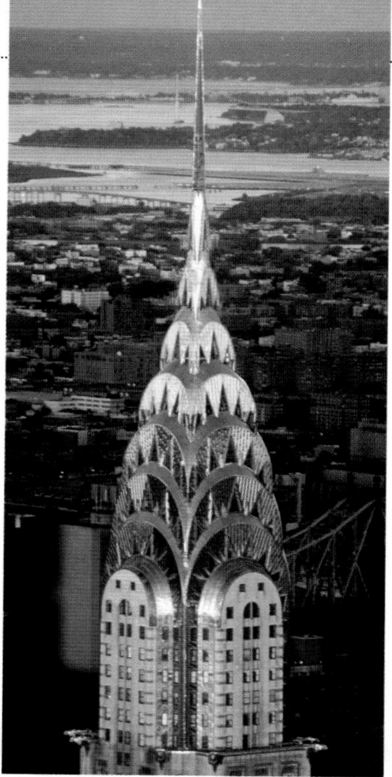

spread over adjacent islands and also onto the mainland, giving rise to Brooklyn, the Bronx, Queens, and Staten Island. Each district was connected to the other by ferries and bridges, and later by tunnels and rail links. In 1898, Manhattan and the four other boroughs united to form Greater New York, a city second only to London in size. It also adjoins other urban areas, such as New Jersey, from which it draws some of its workforce. By 1930, New York was the largest city in the world, the solid bedrock of Manhattan Island providing a firm foundation for some of the world's tallest buildings, including the Empire State Building, 1,250 ft (381 m) high, which was completed in 1931. For a long time, New York was the world's biggest industrial center, but competition from countries with cheaper land and labor had a devastating effect. Over three-quarters of a million factory jobs disappeared between 1960 and 1990. However, as both shipping and industry dwindled, the city reinvented itself as a center for stockbrokers, currency traders, and investment bankers. At the beginning of the 21st century, New York remains one of the world's most dynamic, prosperous, and influential cities, and its skyline, with its towering skyscrapers, remains one of the great icons of modern urban life.

**SYMBOL OF FREEDOM**
*The Statue of Liberty, which was unveiled by President Cleveland in 1886, stands in New York Harbor.*

**CHRYSLER BUILDING**
*Completed in 1930, the stainless-steel Art Deco tower of the 77-floor Chrysler Building is one of the best-known landmarks of central Manhattan.*

## WASTE DISPOSAL

New York generates more waste than any other city in the world. It is estimated that New York generates nearly a fourth of a ton of garbage every second, and about 24,000 tons every day. Until recently, the garbage was burned and dumped beyond the city limits in 11 landfill sites. Incineration causes pollution, and people living close to landfill sites are starting to object to the mounting tide of garbage. The 2001 closure of the giant landfill site at Fresh Kills on Staten Island brought some relief, but the problem refuses to go away.

**MANHATTAN ISLAND**
*Skyscrapers can only be built where there are solid foundations. Most of the tallest buildings in Manhattan are built on granite to the south and center of the island.*

NORTH AMERICA *northeast*

## Toronto

**LOCATION** On the northern shores of Lake Ontario, in Ontario

**AREA** 2,266 square miles (5,868 square km)

**POPULATION** 4.8 million

**FOUNDED** 1793

Toronto is the commercial and financial heart of Canada. The city owes its success to its situation on the shores of Lake Ontario, giving it direct access to Atlantic shipping via the St. Lawrence Seaway and to the major US industrial centers across the Great Lakes. The business area, close to the lake, is dominated by many tall buildings, including the CN Tower, which, at 1,814 ft (553 m), is the world's tallest free-standing structure.

**CN TOWER DOMINATES THE SKYLINE**

NORTH AMERICA *north*

## Chicago

**LOCATION** On the southwestern shores of Lake Michigan, northeast Illinois

**AREA** 228 square miles (591 square km)

**POPULATION** 6.9 million

**FOUNDED** 1816

**ELEVATED RAIL ROUTE**
*Also known as the "El", the Chicago Transit Authority train runs across the city on elevated tracks; here it is crossing the Chicago River.*

Located in the center of the North American continent, Chicago's position on the shores of the Great Lakes makes it a natural meeting point for traffic of all types. However, the flat terrain of the Midwest and the city's site beside Lake Michigan make its climate unpredictable, and severe storms are common. Chicago is a major industrial area, with massive chemical and steel-making empires. The harbor, created in 1834, allowed ships to deliver their cargoes directly onto land instead of anchoring offshore to unload. Lake Michigan has no natural harbors, but the channel that was cut through the offshore sand bar gave ships access to the Chicago River and shelter from storms. The docks now handle 90 million tons of cargo every year. Chicago's proximity to America's farming heartland also makes it a major focus of the food industry; its famous stockyards used to process 18 million head of livestock a year. At times, the city has gone into serious economic decline, yet it remains one of the world's most industrious and dynamic cities.

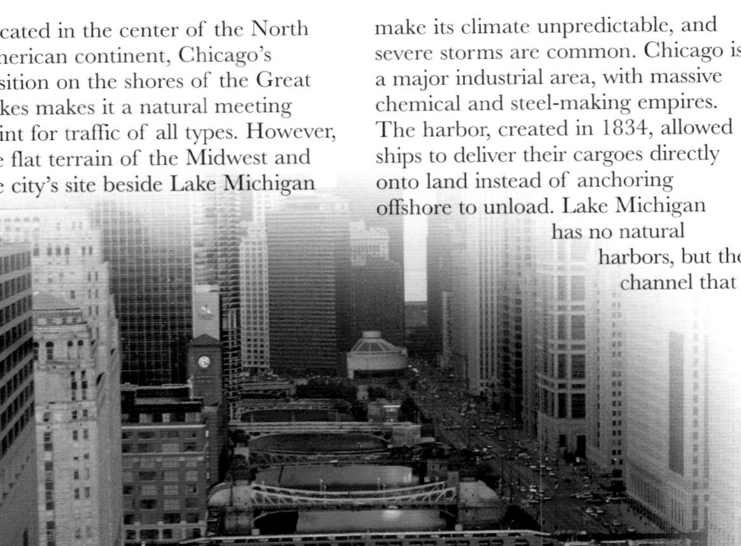

**RIVER CROSSING**
*A series of drawbridges, dwarfed by the surrounding skyscrapers, span the river in downtown Chicago.*

### FRANK LLOYD WRIGHT

Twentieth-century Chicago was the center for a new school of architecture, led by Frank Lloyd Wright (1867–1959). His Prairie Style is based on low-level houses, constructed from mass-produced materials. There are over 100 Wright-designed buildings, including the famous Robie house, in the Chicago area.

NORTH AMERICA *west*

## Los Angeles

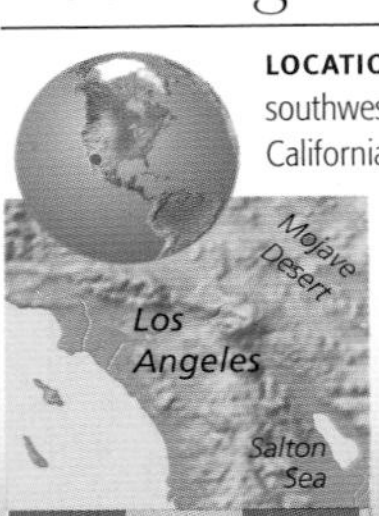


**LOCATION** On the plain of the southwestern Pacific coast of California

**AREA** 14,070 square miles (0,541 square km)

**POPULATION** 12.6 million

**FOUNDED** 1781

Sprawling across the Californian coastal plain, Los Angeles comprises 80 different towns and is the world's largest urban area. Although only the seventh-largest city by population, its cars and freeway system have made its inhabitants uniquely mobile, enabling Los Angeles to spread over a wide area. It is bounded to the east by mountain chains running north–south, and is separated from the San Fernando valley to the north by the Hollywood Hills and Santa Ana Mountains. Founded by Spanish missionaries in the 18th century, Los Angeles is now North America's leading industrial and technological center. However, the city has a serious problem with pollution. The smogs, for which Los Angeles is famous, are caused by the city's warm, dry climate. Onshore sea breezes trap the pollutants against the surrounding mountains, where temperature inversions (cool air below warm air) may prevent their dispersal.

### WATER RIGHTS

Los Angeles is warm, sunny, and very dry, with less than 15 in (380 mm) of rain a year falling in the south of the city. Providing fresh water is a major problem, and supplies were being maintained from increasingly distant sources. Today, water reclamation and conservation facilities help to avoid conflict with farmers and American Indians, who also need the water.

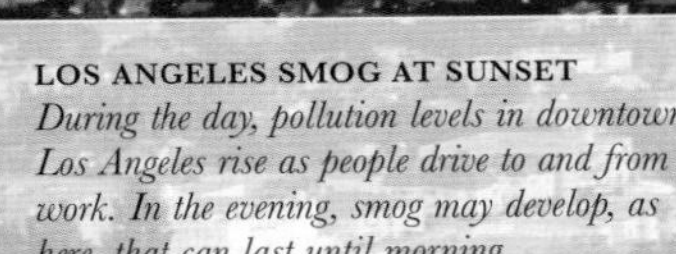

**LOS ANGELES SMOG AT SUNSET**
*During the day, pollution levels in downtown Los Angeles rise as people drive to and from work. In the evening, smog may develop, as here, that can last until morning.*

**HOLLYWOOD**
*The Hollywood sign, erected in 1923, originally advertised a housing development; nowadays it is synonymous with the film industry.*

NORTH AMERICA

## Mexico City

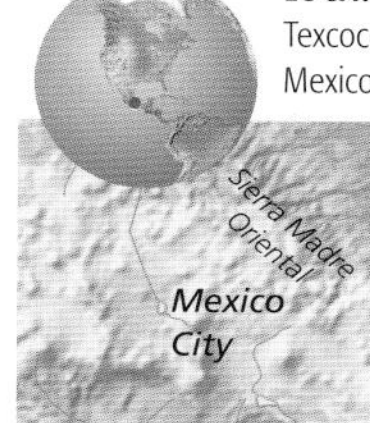


**LOCATION** On the dry lake bed of Texcoco, at the southern end of Mexico's high central plateau

**AREA** 900 square miles (2,330 square km)

**POPULATION** 18.1 million

**FOUNDED** 1325–50

Although it is 7,385 ft (2,238 m) above sea level and lies within an active earthquake zone, Mexico City is not only the capital and commercial center of Mexico, it is also one of the three largest cities in the world, along with Tokyo and New York. Due to its altitude, the climate is mild and springlike for most of the year, unlike other parts of Mexico. Surrounded by mountains, the city is built in a lake basin with no outlet channel. Drainage of Lake Texcoco, and other lakes, has lowered the water table in the area and caused buildings in the city to subside. Mexico City also has an air-pollution problem due to a combination of climatic conditions, rapid urban growth, and sulfur dioxide emissions from the nearby volcano Popocatépetl (see p.179).

**ALAMEDA PARK**
*Despite overcrowding, Mexico City has many beautiful areas, including this park, which is right in the heart of the commercial center.*

**CENTRAL LIBRARY**
*Designed by Juan O'Gorman, the Central Library is a mixture of modern and traditional styles. It is an important repository of Spanish literature.*

SOUTH AMERICA *east*

## Rio de Janeiro

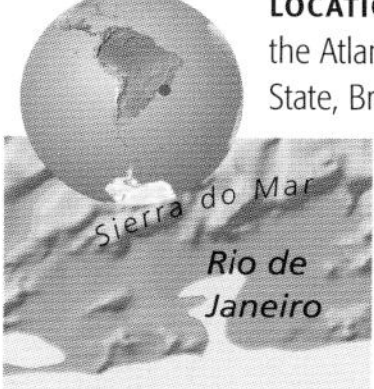


**LOCATION** On Guanabara Bay on the Atlantic coast, Rio de Janeiro State, Brazil

**AREA** 452 square miles (1,171 square km)

**POPULATION** 11.2 million

**FOUNDED** 1536

Sitting on a wide bay, with dazzling beaches and steeply domed mountains that include the Sugar Loaf (see p.199), Rio de Janeiro is one of the most dramatically sited cities in the world. It was founded by Portuguese explorers in the 16th century, who called it Rio de Janeiro ("River of January") because they thought Guanabara Bay was the mouth of a great river. Early economic activity centered around growing sugar cane and whaling, but there was more significant growth when gold and diamonds were discovered in the neighboring district of Minas Gerais. Until 1960, when the government transferred to Brasília, Rio was the capital of Brazil. It is now the second largest city after São Paulo, and one of South America's key centers of finance, trade, and transport. The mountains around Rio make it a crowded city, where the contrast between wealth and poverty is sharply apparent. Sleek beachfront suburbs are backed by shanty towns, called favelas, built on slopes that are too steep for normal housing. Despite these contrasts, Rio is a vibrant city, with a busy port that handles a huge proportion of Brazil's trade.

**FAVELA**
*Built on a steep hillside in the Catumbi district, the wooden shacks of this favela are supported on stilts.*

**IPANEMA BEACH**
*Every weekend, thousands of people head for the famous Ipanema beach to sunbathe, swim, and play volleyball.*

**OVERLOOKING THE CITY**
*The giant statue of Christ the Redeemer is one of the landmarks of Rio de Janeiro. It stands on top of the domed peak of Corcovado, which is 2,330 ft (710 m) high.*

# POPULATION GROWTH

At the beginning of the 20th century, there were about 1.5 billion people on the Earth. Since then, the world's population has quadrupled to a current total of more than 6 billion people, and, despite recent falls in the growth rate in some countries, the overall total is still rising fast. For the planet as a whole, the size of the human population is of crucial significance, because sheer numbers lie behind many of the environmental problems that affect the world today.

## RISING RATES

The current surge in human population began in the mid-1700s, with the start of the Industrial Age. At that time, the annual growth rate stood at less than 0.2 percent—a modest figure that meant the population doubled only over several centuries. From this low starting point, growth gradually accelerated, launching population levels onto an exponential path. By the early 1980s, annual growth reached a peak of nearly 2 percent, bringing doubling time down to just 35 years. This record growth fueled doomsday predictions of a world swamped by humankind.

In the last two decades, the situation has subtly changed. Humanity continues to multiply, but growth rates have eased back slightly from their 1980s peak. This fall has been due to an unexpectedly sharp drop in fertility rates—the average number of children born to each woman in her reproductive life. In developed countries, total fertility rates have now fallen below 2.1 children, which is the figure needed to keep the population stable (assuming no net immigration). In some countries, such as Italy, Spain, and Russia, the total fertility rate has now fallen below 1.5, implying a future contraction in population if present trends continue. In the world's developing nations, the picture is more complex. Some of these countries are already emerging from the same shift, known as demographic transition, as that seen in the industrialized world. China, for example, has a total fertility rate of 1.3, which is well below replacement levels, as a result of stringent legal controls on family size. But in many other developing nations, fertility levels are still high enough to maintain rapid growth. In sub-Saharan Africa, most countries have fertility rates above 4.0, while some are greater than 7.0. These differences are evident when the age profiles of national populations are compared (see above). In a typical developed country, there is a roughly even mix of people of different ages, tapering off from the age of 60. In developing countries, there is a greater bias toward younger age groups, and as today's children move up the profile into childbearing age, there will be rapid population growth for decades to come.

Number of children per woman: 0 1 2 3 4 5 6
1950 Present
Developing countries
Developed countries

**DECLINING FERTILITY**
*In the 1950s, developed and developing countries all had fertility rates well above replacement levels (2.1). Developed countries are now below replacement, while in developing countries the rate has almost halved.*

Age: 80+ 70–79 60–69 50–59 40–49 30–39 20–29 10–19 0–9
ITALY NIGERIA
Population (millions): 10 5 0 5 10 15 20 25 30 35 40

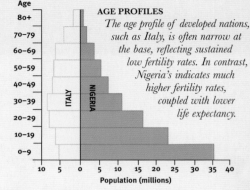

**AGE PROFILES**
*The age profile of developed nations, such as Italy, is often narrow at the base, reflecting sustained low fertility rates. In contrast, Nigeria's indicates much higher fertility rates, coupled with lower life expectancy.*

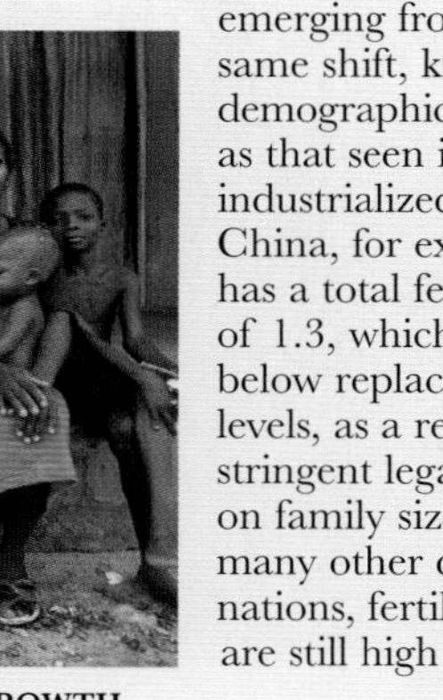

**CONTINUING GROWTH**
*Africa is the world's fastest-growing region. With a fertility rate of over 6, Nigeria is likely to more than double its current population by 2050.*

## UNCERTAIN FUTURE

According to recent UN estimates, the human population may reach a plateau of 11 billion by 2100, but a more pessimistic set of assumptions foresees continued rapid growth. In either event, humanity's impact on the Earth seems certain to increase, raising questions about whether such huge numbers can be sustained. Some economists argue that improvements in biotechnology will enable the human race to feed itself, and that similar "technological fixes" will be found for problems such as global warming (see pp.452–53). But most ecologists are not convinced and foresee widespread deprivation and environmental destruction, fueled by humanity's growing need for raw materials, water, and food.

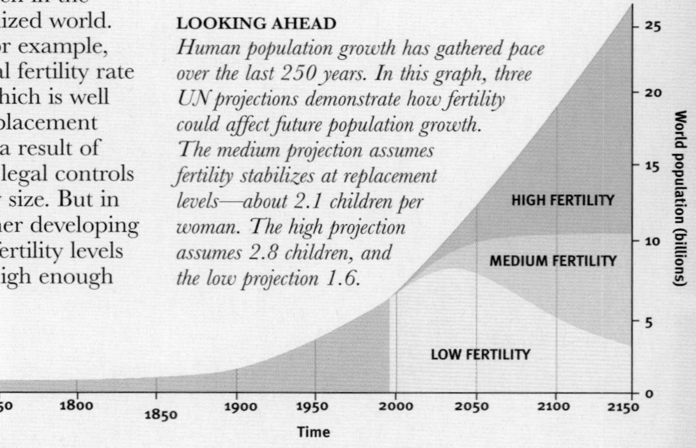


**LOOKING AHEAD**
*Human population growth has gathered pace over the last 250 years. In this graph, three UN projections demonstrate how fertility could affect future population growth. The medium projection assumes fertility stabilizes at replacement levels—about 2.1 children per woman. The high projection assumes 2.8 children, and the low projection 1.6.*

**MASS MOVEMENT**
*Commuters cycle to work in Baotou, China, the world's most populous country—and also one of the most polluted, after rapid industrialization.*

SOUTH AMERICA *east*

## São Paulo

**LOCATION** On a plateau 30 miles (50 km) from the Atlantic coast, São Paulo State, Brazil

**AREA** 3,070 square miles (7,951 square km)

**POPULATION** 17.7 million

**FOUNDED** 1554

Situated on the Tietê River, in the southeast corner of the Brazilian plateau, São Paulo is the biggest city in the southern hemisphere and one of the four largest in the world. It is also one of the fastest-growing cities, with an influx of about 350,000 people each year. The region's climate and soil are ideal for growing coffee, but there are also deposits of iron ore, which led to its industrial development. São Paulo is a major transport hub, with links to the port of Santos, 60 miles (95 km) to the east.

CITY OBELISK

SOUTH AMERICA *southeast*

## Buenos Aires

**LOCATION** On the shores of the Plate River, Buenos Aires State, Argentina

**AREA** 1,421 square miles (3,680 square km)

**POPULATION** 12.5 million

**FOUNDED** 1580

Buenos Aires, the capital of Argentina, is the second-largest city in South America and its leading port. Located on the Plate River (see p.220) 150 miles (240 km) from the Atlantic, the city sits on a gently shelving shore, so a channel was cut to admit larger ships. Exports, such as grain and meat, have made Buenos Aires a focal point for rail and road transportation. It is also the main industrial center of Argentina.

**AVENIDA NUEVE DE JULIO**
*Buenos Aires has many tree-lined boulevards, including Avenida Nueve de Julio, the widest street in the world, which is 425 ft (130 m) across.*

EUROPE *central*

## Berlin

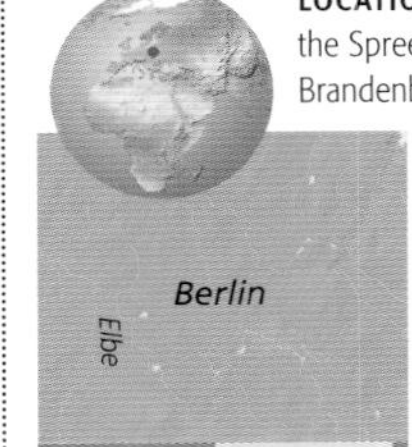


**LOCATION** At the confluence of the Spree and Havel rivers, central Brandenburg Region, Germany

**AREA** 344 square miles (891 square km)

**POPULATION** 3.4 million

**FOUNDED** before 1244

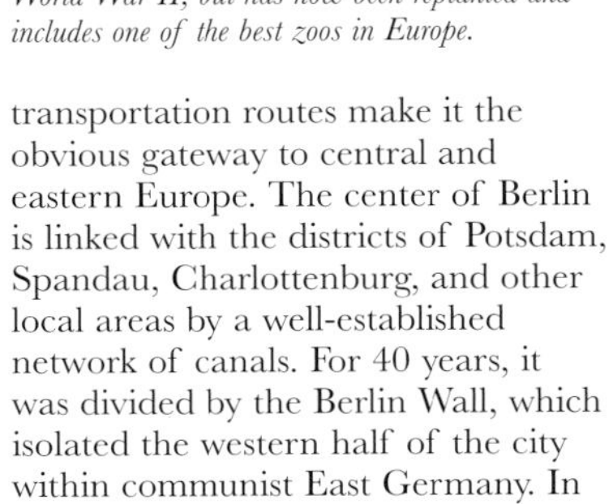

**TIERGARTEN IN WINTER**
*Berlin's famous Tiergarten park was destroyed in World War II, but has now been replanted and includes one of the best zoos in Europe.*

An amalgamation of towns around a historic center, Berlin is Germany's capital and also its largest city. It is situated in northeastern Germany, on flat land surrounding two rivers, the Spree and the Havel, which meet in the district of Spandau. Its position in the center of Europe (on a line between Vienna and Copenhagen), its recent history, and its excellent transportation routes make it the obvious gateway to central and eastern Europe. The center of Berlin is linked with the districts of Potsdam, Spandau, Charlottenburg, and other local areas by a well-established network of canals. For 40 years, it was divided by the Berlin Wall, which isolated the western half of the city within communist East Germany. In 1989, the wall was demolished, and the following year Germany reunified. Berlin soon regained its status as the capital of a reunited Germany, and is once again one of the leading cities in Europe.

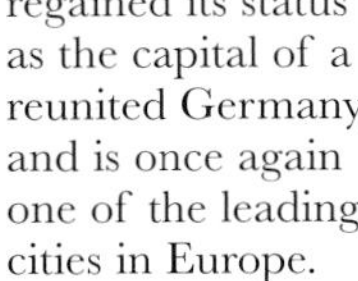

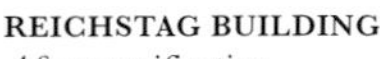

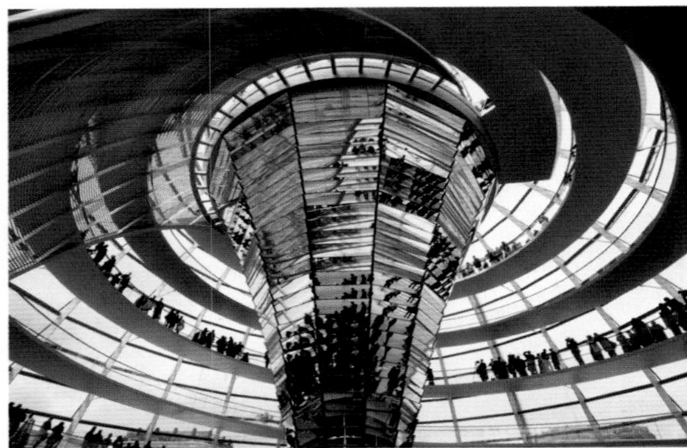

**REICHSTAG BUILDING**
*After reunification, Berlin's parliament building was restored, with the addition of this striking glass dome designed by British architect Norman Foster.*

EUROPE *east*

## Moscow

**LOCATION** On the banks of the Moskva River in the European plain of western Russia

**AREA** 1,000 square miles (386 square km)

**POPULATION** 9.1 million

**FOUNDED** Before 1147

Situated far inland, Moscow has a strategically important location on the Moskva (Moscow) River. The river links the city to seas in the north and south via the Moscow Canal and the Volga River waterways; it also connects the Baltic region to the north with the Caspian Sea to the southeast. Although Moscow shares the same latitude as Copenhagen and Glasgow, it lacks the ameliorating marine influence of the sea. Its climate is continental, with very cold winters and hot, humid summers. Despite being frozen for five months of the year, the Moskva River has been an important waterway since medieval times, and the Kremlin is built close to its banks. The Moscow Canal was built by forced labor in the Soviet era, and the city's industrial and economic might stems from that time. Moscow is the largest city in Russia, and is matched only by Paris in Europe. The city is a transport hub for a vast hinterland that covers all of Russia and the countries around it. Moscow is not only Russia's rail and air-traffic center; it is also one of its largest ports. In 1712, the country's capital was moved to St. Petersburg, and Moscow went into decline until the Communists came to power in 1917. They reinstated Moscow as capital and enforced a program of industrialization. Since the collapse of the Soviet Union in 1991, many factories have closed as their products failed to compete on the world market, but Moscow now has booming financial and service sectors in the new free market. Nowhere in Russia are contrasts of the postcommunist era more evident than in Moscow.

**ST. BASIL'S CATHEDRAL**
*With its ten colorful "onion" domes, St. Basil's Cathedral in Moscow's Red Square is like no other cathedral in the world.*

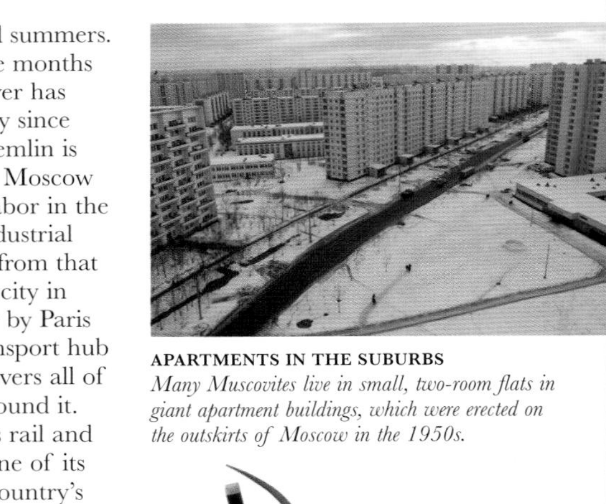

**APARTMENTS IN THE SUBURBS**
*Many Muscovites live in small, two-room flats in giant apartment buildings, which were erected on the outskirts of Moscow in the 1950s.*

**SOVIET ACHIEVEMENTS**
*This monument, celebrating industry and farming, was a centerpiece of the Soviet Exhibition of Economic Achievement. Now the building is used as a showroom for goods from around the world.*

EUROPE *west*

## London

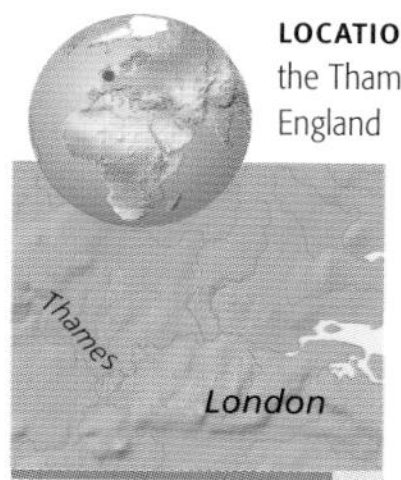


**LOCATION** On the lower reaches of the Thames River in southeastern England

**AREA** 659 square miles (1,076 square km)

**POPULATION** 7.2 million

**FOUNDED** Before AD 43

London, one of the world's oldest and most historic cities, is the capital of the United Kingdom. Although on the same latitude as Warsaw, it is influenced by the Gulf Stream, a warm oceanic current, and so has a mild, temperate climate. The city was established by the Romans over 2,000 years ago on a site now occupied by the city's financial district. During the 19th century, as the hub of the British Empire, it became the world's largest city and engulfed many small towns, including Highgate and Wimbledon. Today, despite an economic recession in the 1990s, it remains one of the financial capitals of the world. Being located on the Thames, within the tidal range of the North Sea, London is at risk of flooding from high surge tides. The threat is increasing because southeastern England is slowly sinking, while tides are rising at a rate of 2 ft (60 cm) per century. To protect the city, the Thames Barrier was built at Woolwich; it opened in 1984.

**LLOYD'S BUILDING**
*Established in 1688, the world-famous insurance company Lloyd's of London is now based in a striking modern tower designed by Richard Rogers.*

### GREEN BELTS

In the 1950s, the expansion of the London's developed area was deliberately halted by city planners, who set aside a broad band of countryside around the capital called the Green Belt. Strict controls limited building within this area. So effective have these controls been that even today, there is a green ring of open land around the capital.

**PARLIAMENT BUILDINGS**
*The Houses of Parliament and Big Ben were built in 1860 after a fire destroyed an earlier building.*

EUROPE *west*

## Paris

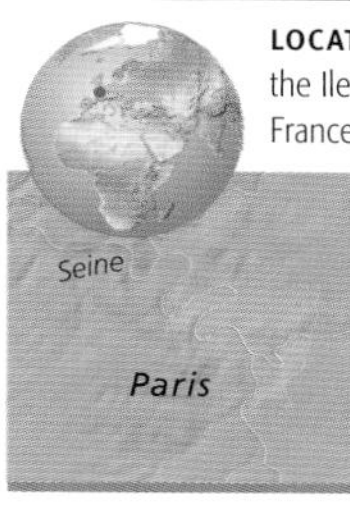


**LOCATION** On the River Seine in the Ile de France region, northern France

**AREA** 1,051 square miles (2,724 square km)

**POPULATION** 9.1 million

**FOUNDED** Before 50 BC

Paris has been one of the world's key cities for hundreds of years. Situated in the Paris Basin, an area of flat valleys and plateaus about 250 miles (400 km) across, it is divided by the Seine River into the Right and Left Banks. As the Seine passes through Paris, it flows around two islands, the boat-shaped Ile de la Cité, where Notre Dame is located, and Ile Saint-Louis, which was once swampy pasture land but is now an elegant residential area. Few parts of the old city exceed 30 ft (100 m) above sea level; Montmartre, the artists' quarter and site of the Sacré Coeur, is one of the higher parts, being located on a hill called the Butte. Paris is not only the political and cultural capital of France, but also the country's financial and commercial heart. Although the Paris region lacks tangible resources, the city's position as capital made it France's dominant industrial center up until World War II. Since then, many heavy industries have moved out. New high-tech industries have developed in the suburbs, but Paris is now more a center of commerce and one of the world's leading tourist destinations.

**EIFFEL TOWER**
*The instantly identifiable Eiffel Tower was built for the Great Exhibition in 1889. At 1,051 ft (319 m) tall, it dwarfs the buildings around it.*

**PARIS SHOP**
*Renowned for its food, Paris has not just good restaurants, but also many small shops, such as this produce stand, that offer a wide range of foodstuffs.*

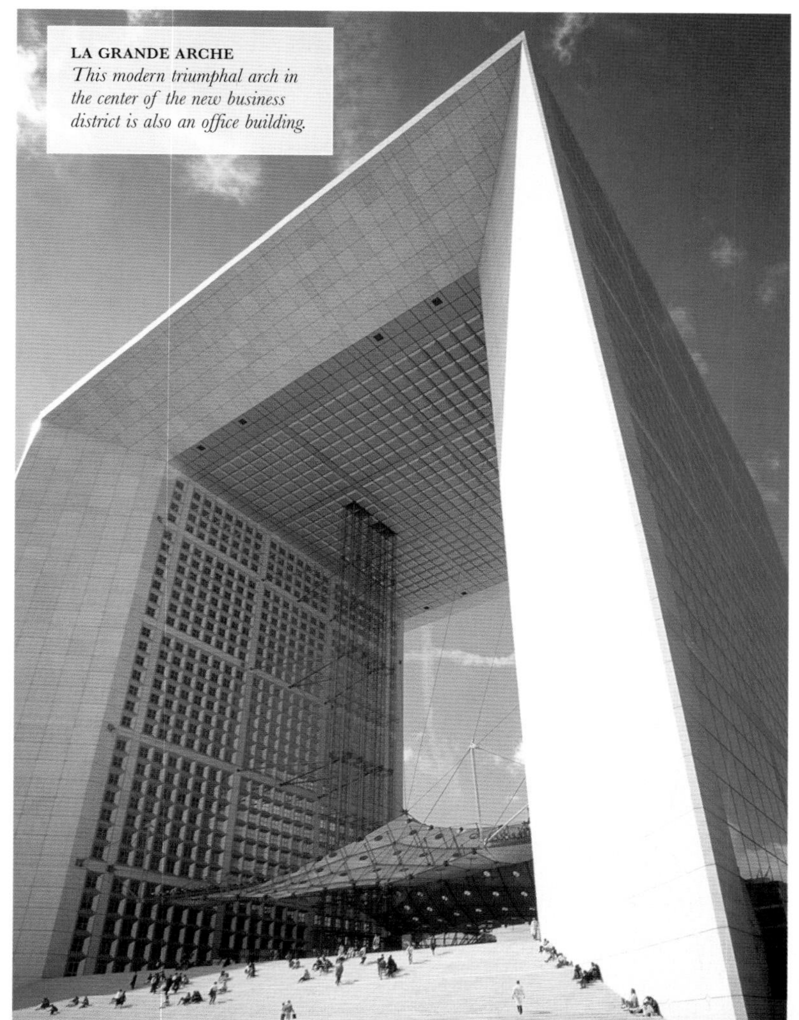

**LA GRANDE ARCHE**
*This modern triumphal arch in the center of the new business district is also an office building.*

LAND

EUROPE *south*

## Rome

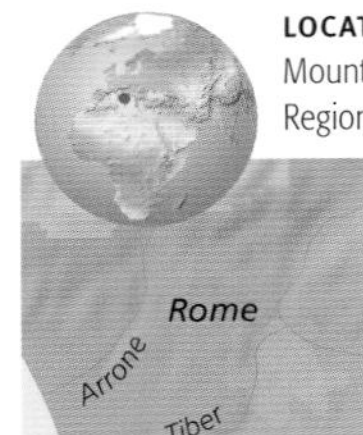


**LOCATION** West of the Abruzzi Mountains, on the Tiber River, Lazio Region, southern Italy

**AREA** 364 square miles (940 square km)

**POPULATION** 2.6 million

**FOUNDED** Before 753 BC

The center of the ancient Roman Empire and home of the Catholic church for almost 2,000 years, Rome is Italy's capital and largest city. Situated 17 miles (28 km) inland from the west coast of southern Italy, Rome originated as a shallow crossing point on the Tiber and later became a marketplace. Its low-lying site meant that until 1870, when embankments were built to contain it, the Tiber flooded regularly, and inundated parts of the city. Rome is also famous for its seven hills. The Palatine Hill, above the Forum, has been settled since Roman times, as have the Capitoline and Quirinal hills. Later expansion of the city incorporated most of the remainder. Millions of tourists come to Rome each year to marvel at its treasures; they provide the bulk of the city's income. The city is also a major administrative and financial center.

**THE SPANISH STEPS**
*A favorite place for both tourists and residents to congregate, the steps were built to link the Trinite dei Monti church with the Piazza di Spagna.*

**ST. PETER'S BASILICA**
*The magnificent Basilica of St. Peter's, with its great dome designed by Michelangelo and Bramante, is one of the greatest architectural treasures of Rome.*

### PRESERVING THE PAST

Rome's long history poses many problems for city planners. There are ancient remains buried everywhere under the center of the city, and most new buildings present a hard choice between building or preserving the past. Construction work to extend Rome's inadequate subway has been repeatedly halted as archaeologists battle to record ancient relics before they are destroyed.

EUROPE *south*

## Istanbul

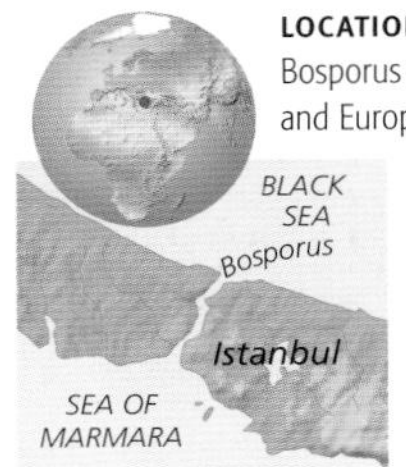


**LOCATION** On either side of the Bosporus straits, straddling Asian and European Turkey

**AREA** 2,204 square miles (5,712 square km)

**POPULATION** 8.3 million

**FOUNDED** About 650 BC

Known previously as Byzantium and Constantinople, Istanbul is the only city in the world to stand on two continents, Asia and Europe. This position has made it a unique meeting point of East and West for over 2,000 years. With the finest natural harbor in the region, it is a major trade hub and controls shipping passing into the Black Sea from the Sea of Marmara.

**VIEW OF CITY FROM BEYOGLU DISTRICT**
*Old European Istanbul, called Stambul, is visible from the modern part across a narrow channel of the Bosporus, the Golden Horn.*

AFRICA *north*

## Cairo

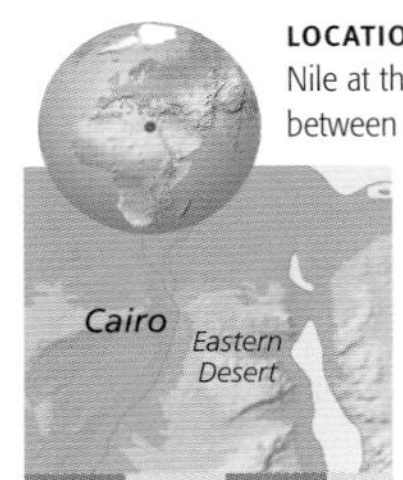


**LOCATION** On the banks of the Nile at the head of the delta, between Upper and Lower Egypt

**AREA** 175 square miles (453 square km)

**POPULATION** 14.8 million

**FOUNDED** 3011 BC

Situated about 125 miles (200 km) south of the Mediterranean Sea, Cairo is Egypt's capital and Africa's largest city. Its origins date back 6,000 years, to when the ancient Egyptians founded their capital Memphis here. The reasons for their choice of site still hold good today. Cairo is located in one of the few areas of Egypt to have fertile soil and plentiful water. Vast expanses of desert stretch away from the city fringes on all sides except the north, where the lush delta region is watered by the Nile (see p.229). The Muqattam Hills border the eastern side. Cairo receives less than 1 in (25 mm) of rain a year, and gets all its water from the river. However, the annual flood cycle of the Nile ended in 1971 with completion of the second Aswan Dam. Estimates of Cairo's size today vary, but its population seems likely to be almost 15 million, and it is growing by the year. Rapid growth has created problems such as widespread poverty, traffic congestion, and pollution. The new growth has also changed Cairo from a city that was cosmopolitan and mixed to one that is essentially Arabic and Muslim. To the west, modern skyscrapers rise along the well-irrigated Nile shoreline, while farther east there are ancient medieval streets filled with noisy souks (bazaars). Beyond these, to the southwest, rise the Great Pyramids that play a key role in the city's tourist industry.

**THE PYRAMIDS AT GIZA**
*The Giza plateau was the royal burial ground for Memphis, but today it is being extensively developed to accommodate the needs of tourists.*

**MOSQUE AND CITADEL**
*Dominating the skyline of Islamic Cairo is the heavily fortified complex of Saladin's Citadel and the Mosque of Mohammed Ali.*

AFRICA *west*

## Lagos

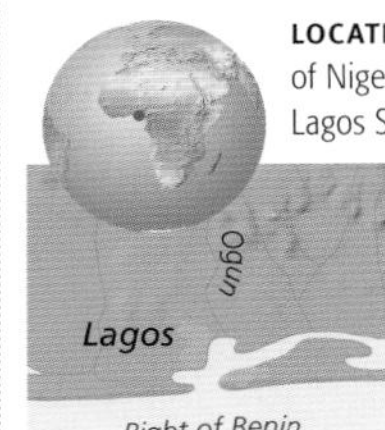


**LOCATION** On the southwest coast of Nigeria and four nearby islands, Lagos State, Nigeria

**AREA** 186 square miles (300 square km)

**POPULATION** 13.8 million

**FOUNDED** 1472

The city of Lagos is situated on the Bight of Benin, an arm of the Atlantic Ocean. It covers several islands, linked to each other by bridges and landfills, as well as parts of the mainland, such as Apapa, the main port area. Lagos is poised to become one of the world's five biggest cities by 2005, overtaking Cairo. It is Nigeria's commercial and industrial hub.

**MARKET IN LAGOS**

AFRICA *south*

## Johannesburg

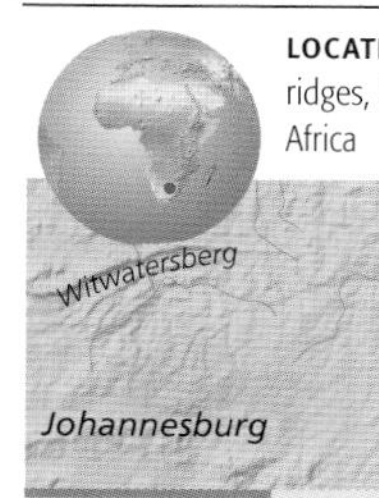


**LOCATION** On the Witswatersrand ridges, Guateng Province, South Africa

**AREA** 210 square miles (543 square km)

**POPULATION** 2.2 million

**FOUNDED** 1886

Johannesburg is South Africa's largest city and the largest in Africa south of the equator. It is built on the highest part of the interior plateau known as the Highveld, near the spot where gold was discovered in 1886. The city grew from a mining camp into a modern city, which remains the commercial and financial hub of the Witswatersrand gold-mining region. Pretoria, 35 miles (56 km) to the north, is the administrative center. Although no gold mining is carried out in Johannesburg today, white spoil heaps are littered around the city.

**SHANTY TOWN**
*Most of Johannesburg's population lives far from the center, and far from their places of work, in shanty towns on the city's outskirts.*

Soweto is situated to the southwest, sprawling across some 39 square miles (100 square km) of bare terrain.

**OVERVIEW OF CITY**
*The high-rise buildings of the central business district quickly give way to low-level urban sprawl.*

ASIA *west*

## Tehran

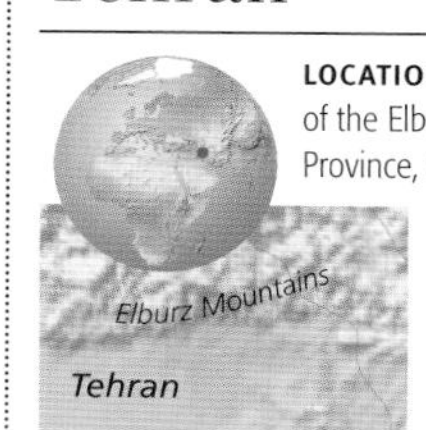


**LOCATION** On the southern slopes of the Elburz Mountains, Tehran Province, northern Iran

**AREA** 463 square miles (1,200 square km)

**POPULATION** 6.9 million

**FOUNDED** 1220

The capital of Iran, Tehran is one of the Middle East's largest cities. It lies 6,930 ft (2,100 m) above sea level at the foot of the highest peak in the Elburz Mountains, Mount Damavand. Although there are iron-ore deposits nearby, the city's rapid growth over the last 80 years is due to oil. Tehran today shows little sign of its ancient history despite its proximity to Rayy, a city built about 2,500 years ago.

**TEHRAN AND THE ELBURZ MOUNTAINS**

ASIA *southwest*

## Karachi

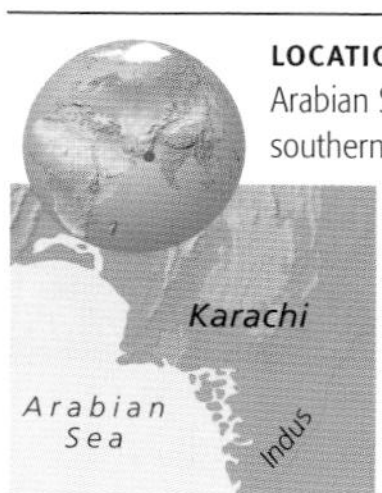


**LOCATION** On the shores of the Arabian Sea, Sindh Province, southern Pakistan

**AREA** 769 square miles (1,994 square km)

**POPULATION** 15 million

**FOUNDED** 1728

Located on the northwestern edge of the Indus river delta, on the shores of the Arabian Sea, Karachi is a major transport hub. It is also the largest city and main seaport of Pakistan. Its modern docks, situated partly on what was once the island of Kaimari, are among the busiest in Asia, handling Pakistan's own shipping and much of that from landlocked Afghanistan. The city also has a large airport and is a focus for major railroads and highways. Being on the edge of the Sind desert, Karachi has a hot, dry climate, with barely 8 in (200 mm) of rainfall a year. This creates problems in providing fresh water for the city's inhabitants. In 1947, Karachi was a small city of less than half a million people, but internal migration, and an influx of refugees after independence, greatly increased its size. Today it is overcrowded, with areas of extreme poverty and pollution problems.

**BUSTLING STREETS**
*The central market, or Sadir, is one of several bazaars in Karachi's crowded center.*

ASIA *south*

## Calcutta

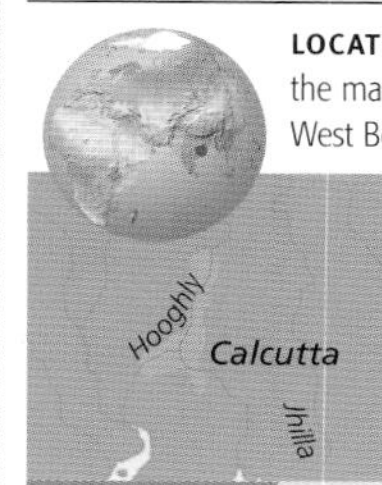


**LOCATION** On the Hooghly River, the main distributary of the Ganges, West Bengal State, India

**AREA** 533 square miles (1,380 square km)

**POPULATION** 12.1 million

**FOUNDED** 1220

**CROSSING THE HOWRAH BRIDGE**
*This is one of the most heavily used bridges in the world, carrying eight lanes of traffic, two tramways, and two busy footpaths.*

Crammed into narrow strips of land between the Hooghly River to the west and wetlands to the east, Calcutta is one of the most overcrowded cities in the world, with millions of people living in slum areas, called bustees. Its location near to the mouth of the Ganges made it India's largest port and biggest city in the days of the Raj. After independence in 1947, the city lost part of its hinterland to Bangladesh, and the Hooghly River silted up, but it remains a busy port and is a leading producer of jute.

**THE VICTORIA MEMORIAL**
*The Victoria Memorial, with its marble facade, is one of Calcutta's best known landmarks. It is now a museum, housing collections of Raj memorabilia.*

LAND

ASIA *southwest*

## Delhi

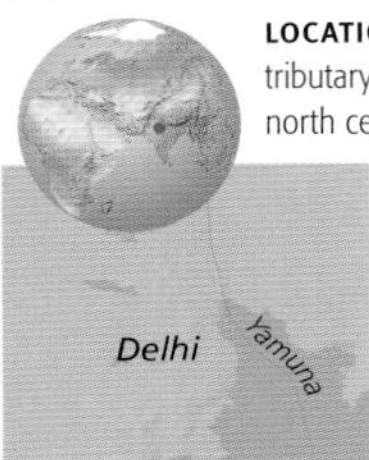


**LOCATION** On the Yamuna River, a tributary of the Ganges, Delhi State, north central India

**AREA** 572 square miles (1,283 square km)

**POPULATION** 13.8 million

**FOUNDED** 1st century BC

India's capital and second-largest city, Delhi has expanded enormously since Indian independence in 1947. There has been a city on the site since the 1st century BC, and it was the capital of many mighty empires. Because of its strategic position at the focal point of an important north–south and east–west trade route, early rulers fortified their city with an impressive array of defenses, such as the Red Fort of Shah Jahan, built in 1639. Less than a century ago, Delhi had just a quarter of a million inhabitants; now it is 50 times as big. The upsurge started with the influx of refugees in the wake of independence, but it has continued unabated ever since. The city stands pivotally between the Indus and Ganges valleys, making it a transportation hub for northern India. Five national highways converge here, and it is a key railroad junction. Old Delhi, and many of the ancient capitals, are situated on one side of the Yamuna River, while New Delhi, with its grandiose, colonial government buildings, is on the other bank. As India's capital, Delhi has a powerful administrative role, but it is also a key financial, industrial, and information-technology center.

**TRAFFIC CONGESTION IN OLD DELHI**
*The sheer volume of traffic in Old Delhi's streets, including pedicabs and bullock carts, creates some of the world's worst traffic congestion.*

**THE RED FORT**
*With its 1¼ miles (2 km) of massive red sandstone walls, the magnificent Red Fort is the largest of Old Delhi's monuments.*

### EDWIN LUTYENS

New Delhi was the grand capital of India, started by the British in 1911. Its chief architect was Edwin Lutyens (1869–1944), who was inspired by Christopher Wren's plans for London and Pierre L'Enfant's for Washington, DC. The Viceroy's House (1913–30), which is a mix of classical structure with Indian decoration, is but one of his designs.

**CULTURAL MIX**
*Delhi is a cosmopolitan city, with Hindi, Punjabi, Urdu, and English all widely spoken. There are also many different cuisines.*

ASIA *southwest*

## Mumbai

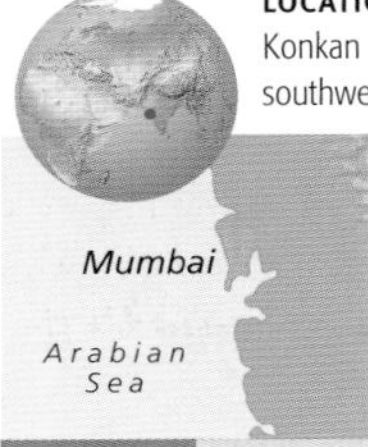


**LOCATION** On seven islands off the Konkan coast, Maharashtra State, southwestern India

**AREA** 170 square miles (438 square km)

**POPULATION** 15.7 million

**FOUNDED** 1530s

Bounded to the east by the Western Ghats and to the west by the Arabian Sea, Mumbai (formerly Bombay) is situated on a promontory at the northern end of the Konkan coast. Because of its position, it receives the full force of the southwest monsoon, and in July rainfall averages 28 in (710 mm). Established by the Portuguese in the 16th century on the site of an ancient settlement, Mumbai is now India's largest, most prosperous city. It has an excellent natural harbor that was improved in the 19th century by land reclamation, joining the seven islands and linking them with the mainland. In the late 19th century, Mumbai's proximity to the Suez Canal, across the Arabian Sea, gave access to the British market for its cotton goods. Although still important, the cotton textile industry is now shrinking; however, the city's industrial base is becoming more diversified and includes "Bollywood," the world's largest film industry. Mumbai remains India's busiest port, but it is also the center of India's economy and its commercial and financial hub.

**CONTRASTING ARCHITECTURE**
*Many of Mumbai's older buildings, dating from the 19th century, are built in a uniquely Indian and florid style of Gothic architecture.*

ASIA *east*

## Seoul

**LOCATION** On the Han River, 40 miles (60 km) from the Yellow Sea, South Korea

**AREA** 234 square miles (605 square km)

**POPULATION** 11.8 million

**FOUNDED** 1394

**BUSY STREET AT NIGHT**
*Seoul has the same traffic-congestion problems as other major cities, despite a network of highways.*

Located in a natural basin and surrounded by mountains, such as the Bukhan Range to the north, Seoul is considered one of the most attractive capital cities in the world. The Han River, which runs through the city, divides it into two distinct areas. Most people live in high-rise apartments on the south side of the river but work in the city's business district to the north. This causes massive congestion on bridges over the river as people commute to work. Founded in 1394 as the capital of the kingdom of Choson, Seoul is a city with a striking mix of ancient and modern. It grew rapidly in the 1960s and is South Korea's main financial, commercial, and industrial center, as well as being a major aviation hub for northeast Asia.

**HIGH-RISE RESIDENTIAL BLOCKS**
*Many of Seoul's inhabitants live in the huge area of high-rise apartments along the south bank of the Han River.*

ASIA *east*

## Tokyo

**LOCATION** On the shores of Tokyo Bay, on southeastern Honshu Island, Japan

**AREA** 840 square miles (620 square km)

**POPULATION** 32.2 million

**FOUNDED** 1457

Formerly known as Edo, Tokyo is Japan's capital city and the largest city in the world, with over 30 million inhabitants. Tokyo proper is smaller than this, but it merges with the cities of Yokohama, Kawasaki, and Chiba to form the largest concentration of people the world has ever seen, with over 30 million inhabitants. The ancient city of Edo was situated on land reclaimed from the Sumida estuary, but Tokyo spreads far over land reclaimed both from the river and the sea. Lying at the same latitude as Washington, DC, it has a similar climate, with hot, humid summers. Twice in the last century Tokyo was almost annihilated. Because the city lies in one of the world's most geologically active areas, where four tectonic plates converge (see p.109), earth tremors are common.

In 1923, the Kanto earthquake killed 100,000 people and flattened the center of the city. Then, in World War II, it was bombed by the US Air Force. However, within a few years, the city's economy was booming, and has been ever since. Tokyo's economy is gigantic, with over 800,000 businesses and the biggest labor force in the world. The service sector is focused mainly on the districts of Maronouchi and Nihonbashi near the old Imperial Palace, while manufacturing is centered on the shores of Tokyo Bay, with easy access to the port of Yokohama.

**OVERPASSES IN TOKYO**
*In such a crowded city, an efficient transportation system is essential. Many people travel large distances from out of town to offices in the city.*

**TOKYO CITY HALL**
*Among the many striking new buildings in Tokyo is the Tokyo Metropolitan Government Building, designed by architect Kenzo Tange.*

**TOKYO AT NIGHT**
*With more bright lights and neon than any other city, the skyline of Tokyo at night is a spectacular sight.*

### OVERCROWDING

Tokyo is one of the world's most overcrowded cities. With space at a premium, land prices and rents are among the highest in the world. A striking response to the lack of space has been Tokyo's capsule hotels with their rows of sleeping units, called pods, which are no bigger than a bed.

ASIA *east*

## Beijing

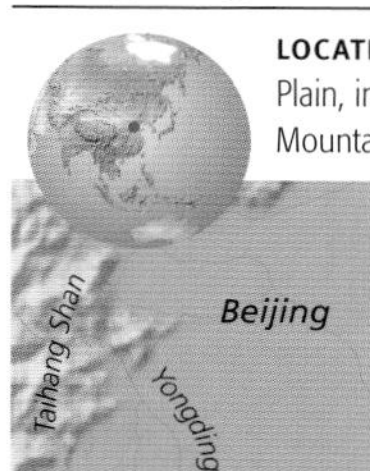


**LOCATION** On the North China Plain, in an arc of the Yen Mountains, China

**AREA** 6,490 square miles (16,810 square km)

**POPULATION** 11.4 million

**FOUNDED** 3000 BC

Set at the northern corner of the North China Plain, on the alluvial deposits between the Yongding and Chaobai rivers, Beijing is built in a massive curve in the Yen Mountains. Lying at about the same latitude as Greece, it has warm summers, but bitter winds from the north bring icy cold winters. The winds are dry, so they rarely bring snow, but they do carry fine yellow loess (see p.93). Beijing has been a political and cultural focus for longer than perhaps any other city in the world. In 1929, fossil hominids, collectively called Peking Man, were discovered to the southwest of the city. The site has been settled by modern humans for 5,000 years and has been a key trading center for at least 2,000 years. Beijing has been China's capital ever since the Mongol emperor Kublai Khan chose it for his headquarters in 1272. The Great Wall passes to the north of the city and once protected it from invasion. The Grand Canal to the south was historically important as it brought prosperity, being the only north–south route for transport and communication. Under Communist rule, Beijing has not always been sympathetic to its past. In the drive to modernize, many older buildings were demolished. Yet several old buildings survive, including the Forbidden City, once home of the Chinese emperors but now a major tourist attraction. Beijing is China's second-largest city after Shanghai, and a leading industrial center, after Shanghai and Guangzhou (formerly Canton).

**OLD BEIJING**
*Parts of old Beijing are characterized by high-density, single-story square buildings. Many are now being replaced.*

**THIRD RING ROAD AT TWILIGHT**
*Beijing is a sprawling city with an extensive transport network. Roads are busy, but many people still commute on their bicycles.*

**WALLS OF THE FORBIDDEN CITY**
*Once separating the emperor from his people, the walls of the Forbidden City today enclose a series of museums.*

ASIA *east*

## Hong Kong

**LOCATION** East of the Pearl River on the south coast of China, by the South China Sea

**AREA** 422 square miles (1,092 square km)

**POPULATION** 8.2 million

**FOUNDED** About 2,000 years ago

Vying with Singapore (see opposite page) as the world's busiest port, Hong Kong is perhaps Asia's most vibrant and successful city. Its buildings are crowded onto the northern shores of Hong Kong Island and spread some distance up the slopes of Victoria Peak and other mountains that surround the bay. Hong Kong faces Kowloon, on the mainland of China, across Victoria Harbour, one of the most spectacular stretches of natural deep water in the world. The climate is humid and monsoonal, winters are cool and dry, and summers frequently bring typhoons. Hong Kong was just a fishing village until 1842, when the British took it over; they turned it into the key trading hub of eastern Asia. It was returned to Chinese rule in 1997.

**ACROSS THE HARBOR**
*Across Victoria Harbour from Hong Kong Island lies Kowloon, to which it is linked by ferries, three road tunnels, and a subway.*

**HAPPY VALLEY RACECOURSE**
*A relic of British colonial days, Happy Valley racecourse was first established on reclaimed marshland in 1846.*

ASIA *east*

## Shanghai

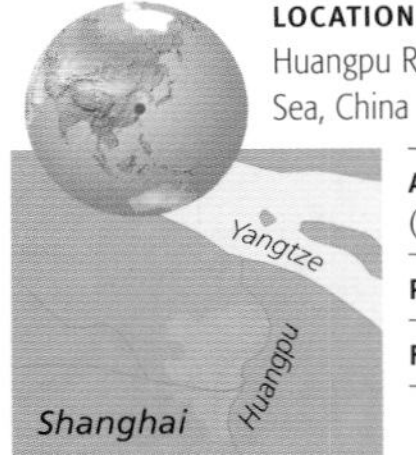


**LOCATION** On both sides of the Huangpu River near the East China Sea, China

**AREA** 794 square miles (2,057 square km)

**POPULATION** 12 million

**FOUNDED** 1074

**COLONIAL BUILDINGS ON SHORELINE**
*Known before 1949 as the Bund, Zhong Shan road is a famous avenue along the Shanghai waterfront with grand, European-style mansions.*

China's biggest city, Shanghai is one of the world's leading seaports. Sited on a bend of the Huangpu River, it spreads across the delta of the Chang Jiang River and is criss-crossed by many canals and waterways. Its mild, maritime climate is due to its position in the center of China's coastline, near the East China Sea. The city originally developed to make clothing from local cotton. When the British forced China to open to western trade in the 19th century, Shanghai prospered. By the 1930s, it was a leading seaport. Now it is a major industrial city, with shipbuilders, steelworks, and textile factories.

**TAI CHI AT ZHONG SHAN**
*Where once Europeans arrived, now Chinese citizens practice their t'ai chi.*

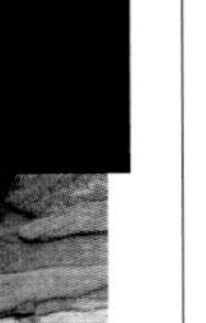

LAND

ASIA *southeast*

## Manila

**LOCATION** On the eastern shore of Manila Bay, Luzon Island, Philippines

**AREA** 245 square miles (635 square km)

**POPULATION** 13.6 million

**FOUNDED** 1574

**SMOKEY MOUNTAIN RUBBISH DUMP**
*Manila's garbage dumps are the only means of survival for many of the city's poorest people, who scour them for food and scraps to sell. This dump was leveled in the 1990s.*

Lying on the swampy delta of the Pasig River, Manila is the capital of the Philippines and its largest city. It was first settled by Spanish colonists in the 1600s because of its superb natural harbor. Sandwiched between the Pasig River and the Bataan mountains, the city is sheltered from extreme weather and has a warm, tropical climate. Metro Manila, as the Manila urban area is called, has grown rapidly since independence in 1947. Manila's setting once earned it the name "Pearl of the Orient," but today it has poor housing and traffic congestion, as well as pollution problems and inadequate sewage disposal.

**MANILA CATHEDRAL**
*The Catholic cathedral is one of the more elegant remnants of Manila's colonial past. It has been rebuilt seven times due to damage by earthquakes.*

ASIA *south*

## Kuala Lumpur

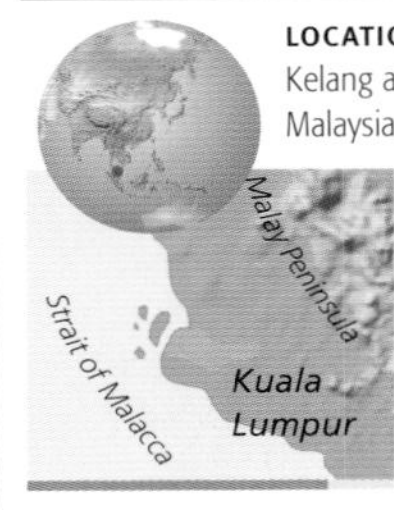


**LOCATION** On the junction of the Kelang and Gombak rivers, West Malaysia State, Malaysia

**AREA** 94 square miles (243 square km)

**POPULATION** 3.9 million

**FOUNDED** 1857

Kuala Lumpur is one of the world's youngest capitals. A century and a half ago, it was just a camp set up by Chinese tin-miners in the middle of the Malaysian jungle. The name Kuala Lumpur comes from the Malay for "muddy river mouth," a reference to the muddy riverbed from which tin was scooped by giant dredgers. Due to its strategic position at the confluence of the Kelang and Gombak rivers, Kuala Lumpur grew rapidly as a tin center, despite its humid, often malarial climate. When Malaysia became independent in 1963, Kuala Lumpur became capital and was perfectly placed to benefit from Malaysia's booming economy.

**PETRONAS TWIN TOWERS**
*One of the world's tallest buildings at 1,483 ft (452 m), the Petronas Twin Towers were designed by Cesar Peli and finished in 1996.*

ASIA *south*

## Singapore

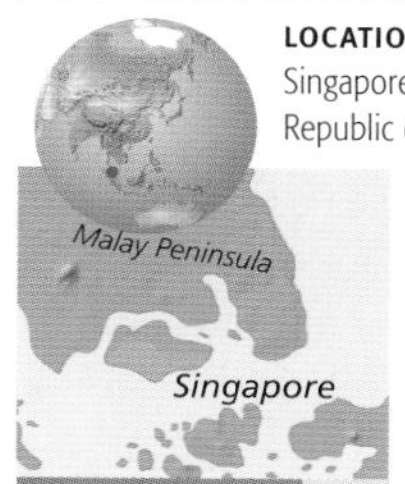


**LOCATION** On the island of Singapore, off the Malay Peninsula, Republic of Singapore

**AREA** 250 square miles (648 square km)

**POPULATION** 4.5 million

**FOUNDED** AD 100

The city state of Singapore is one of the world's smallest, yet most successful, countries. Lying off the coast of Malaysia, about 82 miles (137 km) north of the equator, it comprises 60 islands, of which the largest by far is Singapore Island. Low-lying with a humid, tropical climate and abundant rainfall, Singapore Island was once blanketed by lush jungle and mangrove swamps, but it is now all but covered by urban sprawl. Known originally as Temasek, the city got its current name, which means "Lion City" in Sanskrit, from the many tigers that once lived in the region. Singapore's success story began in 1819, when the British colonial administrator Stamford Raffles gained possession of Singapore for the British East India company. The city's strategic location between the Indian Ocean and South China Sea, and its huge deep-water harbor, rapidly made it one of the the two busiest ports in the world along with Hong Kong (see opposite page), a lead it still shares today. On any day, there are at least 600 ships in port, and a ship sails in or out of the harbor every ten minutes. In 2002, Singapore docks handled roughly a billion tons of shipping, over half of which was container ships. Singapore's most remarkable period of growth has been since its independence in 1965. The economy has boomed and includes thriving electronics and oil industries. The people of Singapore are now among the richest in the world.

**ORCHARD ROAD DISTRICT**
*This is Singapore's main shopping street and the site of some of its most expensive hotels and shops.*

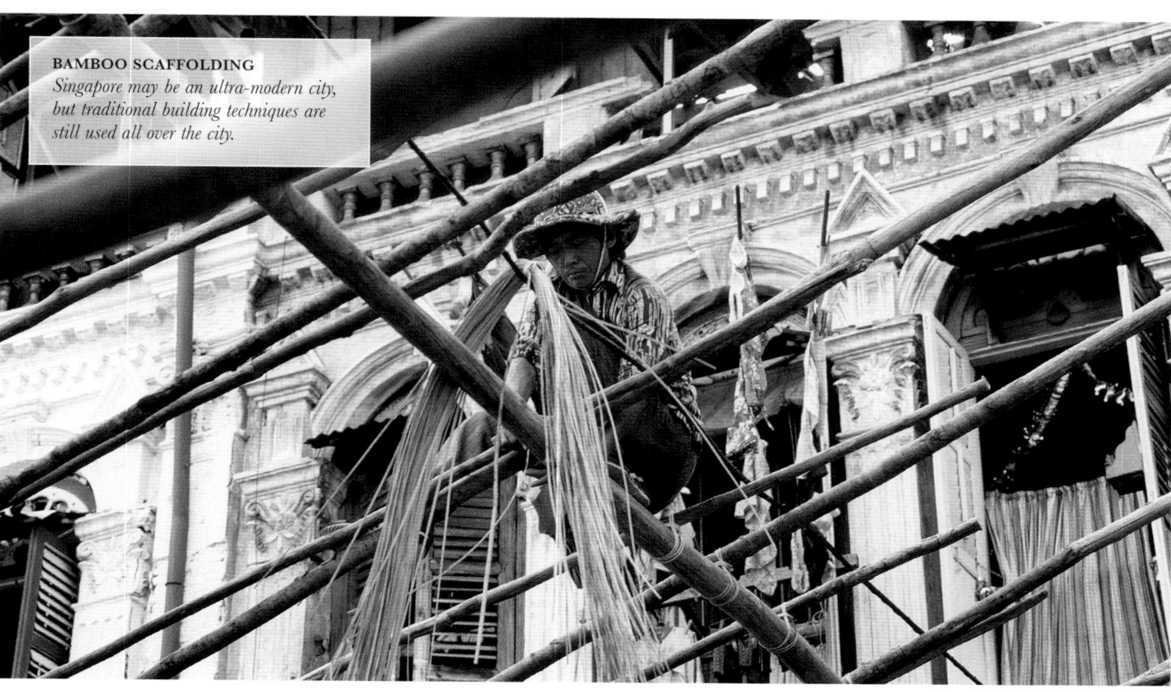

**BAMBOO SCAFFOLDING**
*Singapore may be an ultra-modern city, but traditional building techniques are still used all over the city.*

ASIA *southeast*

## Jakarta

**LOCATION** At the mouth of the Ciliwung River, Java Island, Indonesia

**AREA** 664 square miles (256 square km)

**POPULATION** 16 million

**FOUNDED** About AD 400

Situated on swampy plains at the mouth of the Ciliwung River, Jakarta is a steamy, tropical city where humidity typically rises above 85 percent. It suffers from both frequent flooding and a shortage of drinking water. The area was first settled around AD 400 by Hindu people, but got its current name from Muslims who took over in 1527. Dutch traders captured Jakarta in 1619, and for 300 years it was their trading center in Southeast Asia. The harbor still handles much of Indonesia's foreign trade, and its factories produce a wide range of goods. After Indonesia's independence in 1949, the capital grew rapidly and continues to do so. This has created a severe housing problem in the city.

**ISTIQLAL MOSQUE**
*Jakarta is a predominantly Muslim city, and the Istiqlal mosque is the largest mosque in Southeast Asia and one of the largest in the world.*

AUSTRALASIA *Australia*

## Sydney

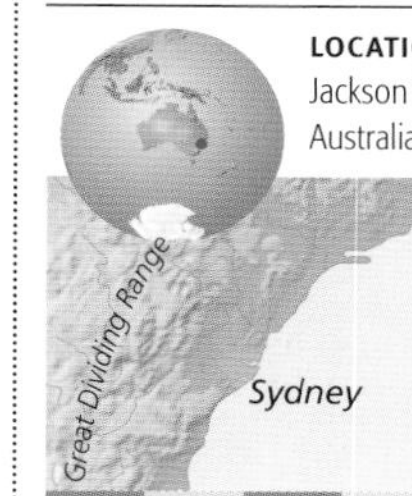


**LOCATION** On both sides of Port Jackson inlet, New South Wales, Australia

**AREA** 12,407 square miles (4,790 square km)

**POPULATION** 4 million

**FOUNDED** 1788

Built on low hills around a magnificent natural harbor, Sydney is Australia's oldest and largest city. Its waterside location and temperate, humid climate, with pleasantly warm summers, first attracted Aboriginals thousands of years ago, but in 1788 it was taken over by the British. By 1800 it was a successful commercial port, first trading in whale products, then in wool and wheat grown in Australia's interior. Sydney gradually spread north across the harbor, as well as east to include the beaches for which it is famous. The city has an excellent transport system, including the Sydney Harbour Bridge, which was completed in 1932. Sydney is the financial and industrial hub of the South Pacific and remains the main gateway to Australia.

**SYDNEY HARBOUR BRIDGE**
*Built in 1932, Sydney Harbour Bridge links downtown Sydney with the suburbs on the northern shore with a giant steel span.*

**MINING FOR GOLD**
*Open-pit mining dramatically reshapes the Earth's surface. Here, 50,000 people work the world's largest alluvial gold deposit, at Sierra Pelada, Brazil.*

# INDUSTRIAL AREAS

TWO-AND-A-HALF centuries ago, one of the most profound changes ever to affect human lifestyles began to spread across the world from northwest Europe: the Industrial Revolution. Up until that time, most people lived in the country and worked the land. The Industrial Revolution brought the first large factories, spawned the first major urban centers, and transformed landscapes. Vast areas were turned over not only to the factories and mills, but also to their support networks: homes for workers; transportation networks to supply the factories and distribute their products; and mines and quarries to feed them with raw materials. From its European roots, the tidal wave of industrialization gathered pace and spread across the world from Europe to the US, Canada, Japan, and Russia. It is now flooding into the developing world, with countries such as China industrializing at an unprecedented pace. Few parts of the world are now entirely untouched by industry.

# INDUSTRIAL AREAS

63 Ore deposits
96–97 Fossil fuels
344–53 Agricultural areas
354–71 Urban areas

THE FIRST GREAT industrial areas were located close to major sources of raw materials and power. They spread out across the Ruhr coal field in Germany, the iron fields of Pennsylvania, and many other places where coal and iron were nearby. Heavy industries still tend to develop near their source materials, especially in the developing world. But as power becomes easier to access, transportation of goods and workers more efficient, and components lighter and easier to move, so modern industry has shifted away from traditional industrial regions. Although most high-tech industries are based in metropolitan areas, these businesses do have more flexibility in choosing their locations.

## RECENT DEVELOPMENT

Europe and North America have been industrialized for so long that new developments have little impact. The most dramatic growth is now occurring in the world's most recently industrialized countries, such as Malaysia, Korea, and Taiwan. Over the last 30 years, these countries have changed rapidly from exporting raw materials to processing them. They invested first in heavy industries and then in high-tech products. They have also expanded by encouraging global companies to locate within their boundaries. Industrial development has now slowed in many Asian countries, except for China. The area along the southern coast from Shenzhen to Zhuhai, and on the Yangtze around Pudong, is currently undergoing the most rapid industrial development the world has ever seen.

**OIL REFINERY**
*The world's largest single industrial plants are oil refineries. They cover vast areas with pipelines and storage tanks.*

## LANDSCAPE IMPACT

No human activity has a more dramatic effect on the landscape than industry. It covers vast areas with concrete, brickwork, and steel, obliterating natural features. The most striking impacts are the huge holes in the ground made by open-pit mines and quarries, such as the Bingham Quarry in Utah, where over $2\frac{3}{4}$ square miles (7 square km) of the Earth's surface has been blasted away. The hills created by spoil dumps can be equally dramatic—for example, the mountains of white china clay in Cornwall, England, and the gigantic New Cornelia Tailings in Arizona.

**LANDFILL SITES**
*Industry generates a huge amount of waste, which must be disposed of. One option is to create landfill sites, such as this one in Hong Kong. Since it is a densely populated area, the site will create much-needed space for building.*

## TYPES OF INDUSTRY

Industry is divided into three types: primary, secondary, and tertiary. Primary industry uses raw materials from the land and sea: mines and quarries extract coal, oil, copper, and many other materials; while fishing, farming, and forestry exploit living resources. In the world's most advanced economies, primary industry is not as important as it once was, but it still accounts for a large share of the wealth of developing nations. Secondary industry takes raw materials and manufactures them into new products, such as cars and furniture. This type of industry was the economic backbone of the world's oldest industrial nations. Now, much of their wealth comes from tertiary industries. These do not produce goods, but instead offer a range of services, from transportation, retail sales, and customer-service operations run from call centers, to health care, education, and leisure.

**INDUSTRIAL LANDSCAPE**
*The Ruhr Valley in Germany has been industrial for over 150 years. Huge power plants like this one are needed to satisfy the demand for energy.*

**LIGHTER AND CLEANER**
*Modern, high-technology manufacturing industries (such as circuit-board assembly, seen here in China) often have less impact on the immediate environment than traditional industries.*

PRIMARY

## Power generation

**DISTRIBUTION** The world leaders in power generation are the US, China, Japan, Russia, Canada, Germany, France, and India

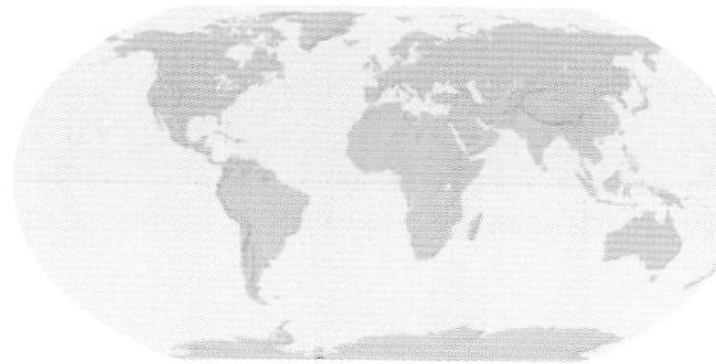

Electrical power is generated where there is demand for it, rather than close to the natural resources, so centers of power generation generally coincide with the major centers of population and wealth—that is, North America, Europe, China, and Japan. More than 70 percent of the world's electricity is generated for less than a fifth of its population. Nearly every American and European has access to electricity, while in many developing countries, only a lucky few have an electrical supply. Electricity is generated in power stations and fed out to its users across a network of cables called a grid. In some places, the cables are buried underground, but in others, electricity is carried by overhead cables hung between pylons that stretch across the landscape for mile after mile. Over two-thirds of the world's electricity is thermal power. This is generated by burning fuels to heat water, which creates steam to drive turbines that turn electrical generators. In the developed world, it is nearly always fossil fuels that are used—coal, oil, and natural gas. In developing countries, a small but significant proportion comes from burning biomass—that is, plant and animal matter converted into fuel—but in the developed world, the use of biomass is negligible. The two other major sources of power are nuclear and hydroelectric. Over the last 50 years, both these sources have been extensively developed, and they each now generate about 17 percent of the world's electricity. Alternative energy (from sources other than fossil fuels and nuclear power) and renewable sources, such as windmills, solar panels, geothermal power stations, and tidal and wave power stations, provide less than one-tenth of one percent of all the electrical energy.

**POWER STATION**
*The majority of the world's electricity is generated by burning fossil fuels in power stations, such as this one in Wakefield, England.*

**SOLAR POWER**
*Solar energy is seen here being utilized in Australia, but it can also be used in colder parts of the world.*

**WIND POWER**
*These wind turbines in California generate power, but their aesthetic value is still hotly debated.*

### DISPOSING OF NUCLEAR WASTE

Nuclear power generation creates radioactive waste, and there remains fierce debate over how to dispose of it safely. Gaseous waste is vented into the air, and in some places low-level radioactive liquid is pumped into the sea. The most dangerous waste is buried in sealed containers. In Hanford, Washington (shown here), low-level waste was dumped in canisters in open trenches over a vast area. The site is now undergoing a massive cleanup, costing $2 billion.

PRIMARY

## Fossil fuels

**DISTRIBUTION** Saudi Arabia, the US, Russia, Iran, and China are the main producers of crude oil; China, the US, and India are the main producers of coal

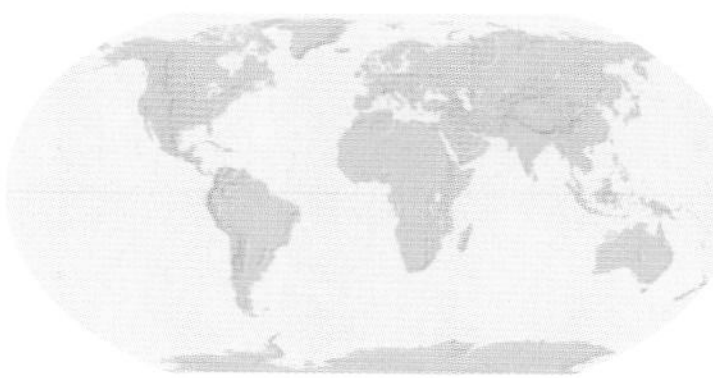

Fossil fuels are coal, oil, and natural gas. They form in the ground from the remains of organisms that lived millions of years ago. Coal is derived from plants that grew in huge, tropical swamps during the Carboniferous Period, over 300 million years ago. It is found in seams or layers, where it is compressed beneath layers of other sediments that have now turned to rock. Because continents move over the Earth's surface over time (see pp.106–107), many coal seams are no longer in the tropics, or even near coasts. Some of the world's biggest coal reserves are in the heart of Siberia, Russia. Major reserves are also found in northern Europe, the Damodar Valley in India, and the Appalachians and the Midwest in the US. Oil is made mainly of the remains of tiny organisms that lived in warm seas. Because it is liquid, oil does not settle in seams, but seeps up through rock until it meets an impervious layer. It accumulates under this layer, forming an underground reservoir. When looking for oil, prospectors search for suitable structures where oil might collect, and use seismic surveys to confirm its presence. The biggest reserves of oil are in the Middle East. Saudi Arabia is the leading producer, supplying almost 13 percent of the world's entire output. Major reserves are also found around the Caspian Sea, in Venezuela, and in Texas. Oil is often found in areas that are either distant from world markets, or in hostile environments such as the Arctic (Alaska), tropical rainforests (Nigeria and Indonesia), or under stormy seas (the North Sea). So oil exploration, extraction, and transport by pipeline or tanker is an expensive business, and spillages can cause environmental catastrophes. Natural gas is formed in much the same way oil, and is also found trapped in underground reservoirs. Because it burns cleanly and can be piped direct to homes, it has become the fastest-growing energy source of all. It now provides almost a quarter of all the world's energy needs. Gas is often found close to oil fields, so there are major reserves in the Middle East, but the biggest reserves are around the southern end of the Caspian Sea.

**COAL MINING**
*In some developing countries, such as Vietnam, extracting coal is highly labor-intensive and poorly paid.*

**UNDERSEA EXPLORATION**
*Demand for gas (and oil) has led to offshore exploration for resources under the seabed, as here in the Gulf of Mexico.*

**OPEN-PIT MINING**
*This open-pit mine in Germany is extracting lignite, a type of coal, using huge machines and few people.*

PRIMARY

# Metals

**DISTRIBUTION** Leading iron ore producers are China, Australia, Brazil, and Russia; the main producers of aluminum include Australia, Guinea, Jamaica, and India

**SHIPBUILDING**
*One of the largest demands for steel comes from the shipbuilding industry, which utilizes millions of tons of steel each year.*

Only a few metals, such as gold (see p.64), are found in a pure state. Most are impure ores, mixed with other minerals, which must be refined to produce the pure metal. Some metals are then mixed to make alloys. Steel, for example, is a mixture of iron and carbon. Most metals are found in areas of metamorphic and igneous rocks. Humans first used metals about 8,000 years ago, and they began to extract metal from ore about 6,500 years ago. Until recently, metals were usually mined and processed in the same area. Iron production, for example, was sited in places near outcrops of iron ore that also had an abundant supply of wood for fuel. With the Industrial Revolution, the need for coal for large-scale smelting shifted the iron industry to sites where iron ore was associated with coal fields, such as the Ruhr in Germany. By the late 20th century, the rising demand for metals in the developed world had outstripped supply. This meant looking farther afield for metallic ores, which often separated the processes of extraction and refinement. Ships now carry huge quantities of ore from countries such as Brazil to processing plants on distant coasts near the major markets. Iron ore is one of the most abundant metal ores in the Earth's crust, second only to aluminum. The US once led the world in steel production, but it has been hit by rising costs, and China is now the world's leading producer, followed by Japan, the US, and Russia. Aluminum is extracted mainly from bauxite, an ore that occurs in feldspars (see p.76) and other silicate minerals, which break down in tropical conditions to form a surface crust. Leading producers include Australia and Guinea, though large quantities are also mined in southeastern Europe and Russia. Base metals, such as copper and lead, are mined all around the world, but rare and precious metals are produced in just a few places. South Africa accounts for two-thirds of the world's gold and platinum supplies, while other rare and precious metals come from North America, Australia, and Russia.

## WATER POLLUTION

Mining metals can pollute nearby rivers and streams and may even cause the water to change color—as here, near Telluride, Colorado. Today this area is a famous ski resort, but in the past, zinc and copper were mined in the surrounding mountains. This has damaged some of the alpine ecosystems and contaminated streams. In places, the water does not meet minimum standards for agriculture or for drinking water.

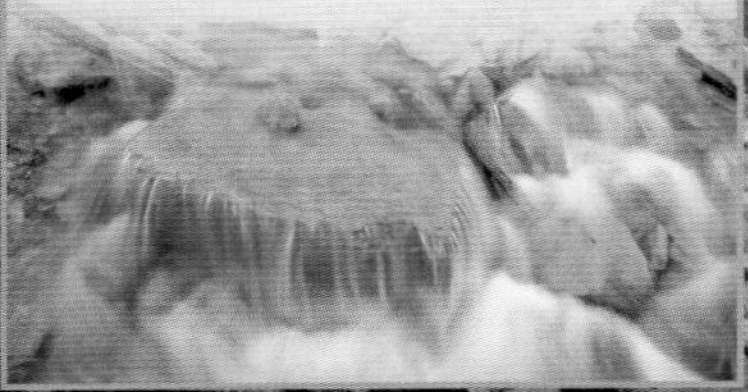

**PRODUCING NICKEL**
*Ores, such as chalcopyrite and pentlandite, are refined to produce nickel, as here at a nickel-smelting plant in Noril'sk, Russia.*

**PROSPECTING FOR GOLD**
*Gold miners in French Guiana search for alluvial gold in sedimentary rocks. It can be prospected in this way because it exists naturally in a pure form.*

**OPEN-PIT COPPER MINING**
*The huge demand for copper has led to excavation of open-pit mines, such as the Bingham copper mine in Utah, which is the largest human-made hole in the world.*

PRIMARY

## Nonmetallic minerals

**LOCATION** Nonmetallic minerals are distributed, extracted, and processed throughout the world, although some mineral ores are rare and highly localized

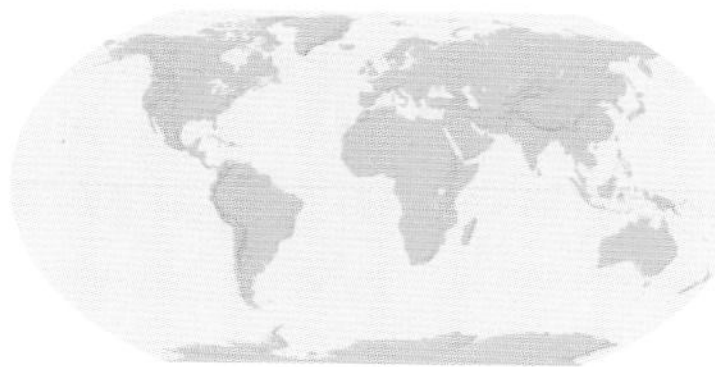

The world's most common minerals are nonmetallic. They are used to make bricks, ceramics, insulation, fertilizers, and many other chemicals, including table salt. Sediments are a rich source of building materials, such as clay, sand, gravel, and stone. Soft, dull clays, deposited on seabeds long ago, were among the first minerals ever used. Clay bricks 10,000 years old have been found under the ancient city of Jericho, Israel. Today, clays are among the most widely used, and widely distributed, of all mineral sources: kaolinite (see p.76) is used for paper-making, and pottery; montmorillonite for lubricating oil drills and cleaning wool; illite for bricks; and vermiculite (see p.76) for electrical insulation. Rare minerals such as gemstones tend to form under conditions linked to volcanic activity. So most gems are found in or near mineral veins—cracks in the Earth's surface where hot, mineral-rich fluids have risen from deep in the Earth's crust. Diamonds (see p.65), for example, formed in the crust over 3 billion years ago, and are brought to the surface by volcanic eruptions. They are found in rocks called kimberlites (see p.84), named after the area in South Africa that produces more diamonds than all other sources combined.

**SULFUR HILLS**
*Huge piles of quarried sulfur await shipment from the docks of Vancouver Harbour in British Columbia.*

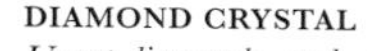

**DIAMOND CRYSTAL**
*Uncut diamonds, such as this one, can vary in color and in translucence, ranging from transparent to opaque.*

**CEMENT PLANT**
*Large, dusty cement works, such as this one in Staffordshire, England, produce cement by heating crushed limestone and clay.*

**EXTRACTION OF PHOSPHATES**
*Morocco possesses three-quarters of the world's phosphate reserves. This open-pit phosphate mine is near Casablanca.*

SECONDARY

## Chemicals

**DISTRIBUTION** The chemical industry is centered on economically advanced countries such as the US and Japan; Germany is the main producer in Europe

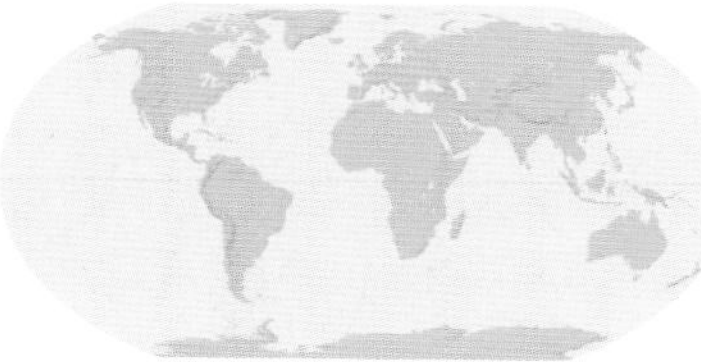

The chemical industry is one of the world's most diverse, with products ranging from pesticides to perfumes. The sources of raw materials are very diverse, but the products can be divided into two main kinds: organic (carbon-based chemicals, such as plastics) and inorganic (all non-carbon-based chemicals, most of which are used for fertilizers). Initially, the organic chemical industry was centered on coal fields, since its raw materials were obtained from coal tar. Today, most organic chemicals are oil-based, and chemical plants are sited near oil refineries or close to the markets for the products, such as cities. Inorganic chemicals include phosphate (see above) and potash, which are dug out in vast quantities and used to produce fertilizers. Since chemical plants may have a negative environmental effect, they are often sited in old industrial areas where their impact is less marked, or in developing countries, where labor is cheaper.

**PETROCHEMICAL PLANT**
*Petrochemicals, which are used mainly for producing plastics, are made in huge plants, covering thousands of acres.*

TERTIARY

## Transportation

**DISTRIBUTION** The US is the world leader in road, rail, and air transportation, followed by India, Brazil, China, and Russia; Chicago and Atlanta have the busiest airports

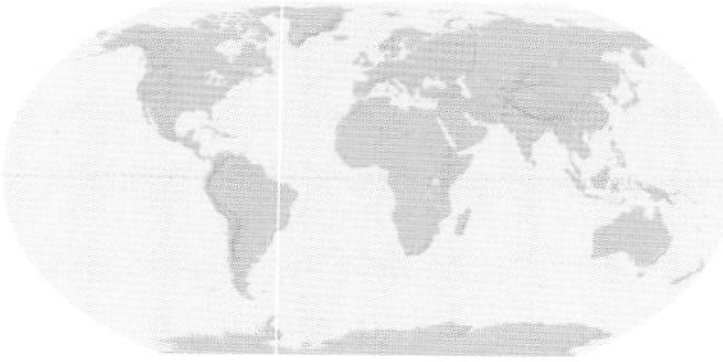

All countries have transportation networks, but their development, extent, and efficiency are variable. At a local and regional level, road and rail links are important, while rail, air, and ships are more commonly used for countrywide and international movement of people and freight. The most extensive transportation network in the world extends across the US. Second is India, due in part to the British, who built many roads and railroads during colonial times.

**CONTAINER TRAFFIC**
*Much of the world's freight is transported in containers. Victoria Harbour in Hong Kong, shown here, is the world's second-busiest container port.*

**PASSENGER TRAFFIC**
*Over 2,000 flights a day take off from the world's busiest airports, such as O'Hare in Chicago, which handles more than two planes every minute.*

SECONDARY

## Textiles

**DISTRIBUTION** China produces most cotton, followed by the US and India; Australia and China lead the way in wool production; China produces the most clothes

The raw materials used to make textiles are either natural or synthetic. Natural fibers include linen, wool, cotton, and silk, of which wool and cotton are by far the most important. Australia has 140 million sheep and is the leading wool producer. Cotton grows in subtropical areas with a long growing season and enough dry weather for the crop to be harvested. It is still the main raw material of the textile industry, but its dominance is being gradually eroded by synthetic fibers, such as acrylic, nylon, and polyester, which are not woven but mechanically, thermally, or chemically bonded together. (Four-fifths of the fibers processed in American mills 60 years ago were cotton; now it accounts for just over a third.) Meanwhile, the weaving industry, and many American and European clothing companies, have moved to developing countries where labor is cheap. Guangdong Province in southern China is now the center of the world's textile and clothing industry.

**HARVESTING COTTON**
*Over 95 percent of cotton in the US is harvested mechanically, as shown here in Mississippi. The machines are called spindle pickers or strippers.*

**WOOL USED FOR CARPET WEAVING**
*The principal source of wool used in woolen carpets is Australia. The wool is dyed into a multitude of different shades.*

**WORKERS AT SILK FACTORY IN CHINA**
*The Chinese discovered silk 3,700 years ago. Today, China produces more silk than anywhere else, manufacturing it in factories such as this one in Hotan, China.*

SECONDARY

## Food processing

**DISTRIBUTION** The world's leading food processors are the US, Germany, and the UK; China produces and consumes the most food

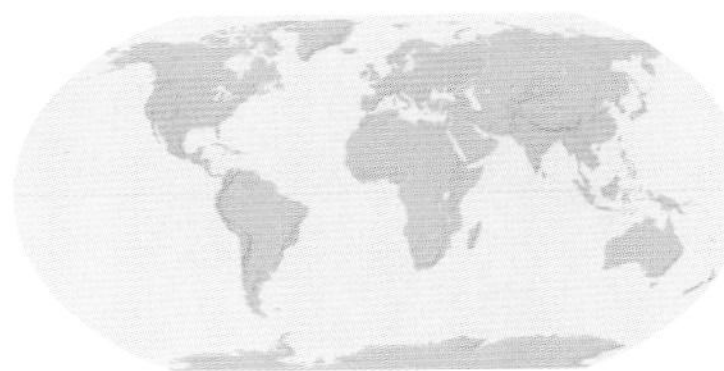

Food processing involves all stages of production, from harvesting to marketing. Whether it simply needs to be sorted and washed, or requires elaborate preparation, almost every type of food is processed, or prepared in some way, to keep it fresh, lengthen its shelf-life, or make it look more appealing. Many foods deteriorate en route to markets, so many processing plants are located close to the growing regions. The biggest food-processing centers tend to be in large cities, such as Chicago, that have good transportation links with farming areas. The cities provide labor for the plants and distribution centers for the processed food.

**SALMON PROCESSING IN RUSSIA**

SECONDARY

## High technology

**DISTRIBUTION** World leaders in advanced technology include Silicon Valley and Boston in the US, and various eastern countries, such as Japan and Korea

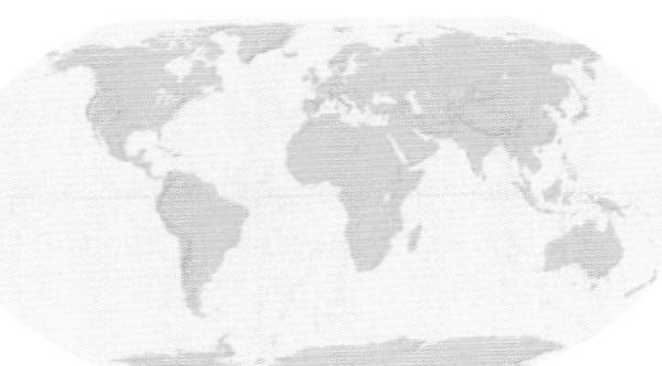

A key development of the last 50 years is the explosive growth of entirely new technologies, such as semiconductors, computer software, and robotics. The combination of low energy and raw material needs, and a lightweight product, means these high-technology industries can locate almost anywhere. They are often sited in attractive places, near universities, in order to attract the necessary skilled workers. The most famous of these sites are Silicon Valley in California and Route 128 in Boston, Massachusetts, but similar centers have appeared all around the world, in places such as Singapore, Seoul-Inch'on in South Korea, and T'aipei-Hsinchu in Taiwan.

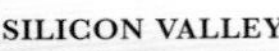

**SILICON VALLEY**
*One of the world's leading centers for high-technology industry is Silicon Valley in California.*

**CONSTRUCTING CIRCUIT BOARDS**
*Basic components of high-tech equipment are made in developing countries, where labor is cheap, as here in Guadalajara, Mexico.*

**SERVICED BY SATELLITE**
*Modern telecommunication is a high-tech industry. Today, it can provide communication services from almost anywhere on Earth. This satellite dish is in rural Bavaria, Germany.*

### RECYCLING

High-tech industries are generally considered clean, but equipment, such as computer monitors, which rapidly becomes obsolete is hard to dispose of. As a result, high-tech waste has joined the rising mountain of industrial and household waste created each year by modern society. The damage to the environment caused by disposing of this waste, and also of making the materials in the first place, has spurred efforts to recycle materials. However, as yet, only a tiny fraction of our waste is recycled.

TERTIARY

## Services

**DISTRIBUTION** Service industries are associated with urban and industrial areas; world leaders include the US, Europe, and India

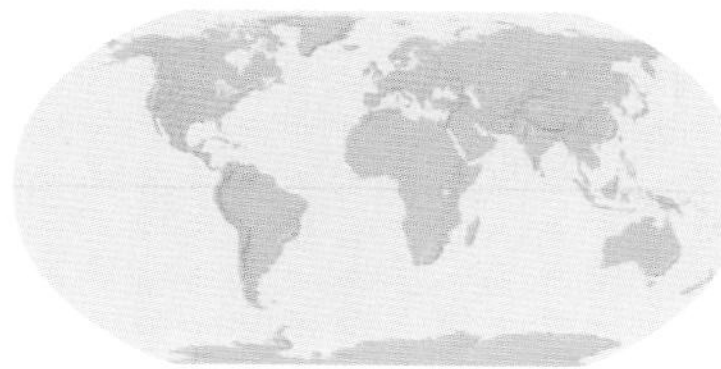

Without exception, the industrial focus in developed countries has shifted away from the traditional manufacturing industries and toward service industries, such as banking and retailing. In countries such as the US and the UK, the service sector is now by far the biggest employer, and the wealth of cities, including New York and London, is in part based upon the financial services they provide. Many developing countries are following the same path. Indeed, the proportion of the workforce involved in the service sector is often taken as a measure of their development.

Because these industries provide a service, they are located where the service is needed, typically in major urban and industrial centers. The growth of service industries is one of the most significant factors in the expansion of major megalopolises (see pp.355–57) across the world, as more and more people and businesses gather to provide services for each other. Services are normally divided into three kinds. Consumer services are provided direct to the consumer by, for example, stores, banks, and realtors. Producer services are directed toward other businesses and include financial services, warehousing, consultancy, and advertising. Finally, there are public services, such as transport, education, and health care.

**SHOPPING CENTERS**
*The Bluewater shopping complex in the UK is built on London's outskirts and therefore needs good road access and large parking lots.*

**CALL CENTERS**
*Many European call centers, like this one, are based in the UK. Their growth is due in part to fixed, low call charges.*

TERTIARY

## Tourism

**DISTRIBUTION** Leading tourist destinations are France, Spain, the US, Italy, China, and the UK; leading earners from tourism are the US, Italy, France, and Spain

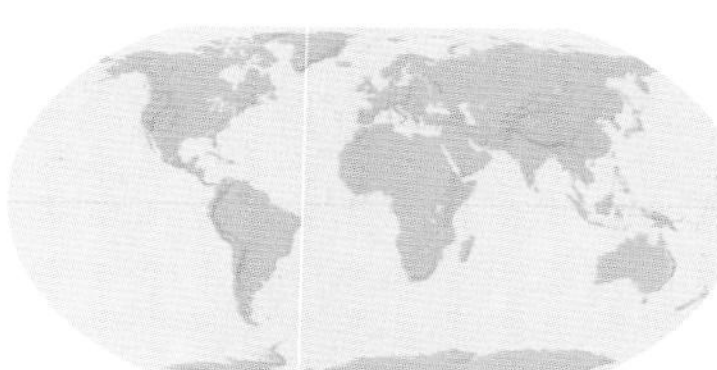

Leisure and tourism is the world's fastest-growing industry, employing 1 in 15 people. Spending on tourism is one of the main ways money moves between countries—over 8 percent of all the world's export earnings are from tourism. The biggest earners from this worldwide travel boom are North America and Europe. The biggest spenders are Americans, Germans, Japanese, and British. In recent years, tourists from these countries have ventured farther afield, and tourism is now the mainstay of the economy of countries such as Thailand, to which millions of visitors travel each year to enjoy tropical beaches. One of the fastest-growing tourist destinations is China. Before 1980, it was hard for the Chinese to travel around their own country, or for foreign tourists to visit. As travel restrictions were lifted, tourism developed rapidly, and China is poised to become the world's top tourist destination. In each tourist area, visitors tend to concentrate at a few top attractions, such as the Great Wall in China or Disneyland in Florida, and this places huge pressures on local facilities. Often the landscape is completely transformed by the construction of transportation and hotels. In the 1950s, most villages on southern Spain's Costa del Sol, for instance, were small fishing villages, but from the 1960s onward, the availability of cheap flights began to draw huge numbers of visitors from northern Europe to its warm beaches. By the 1980s, 7.5 million visitors a year were traveling to the Costa del Sol, and the villages were transformed by giant hotels and multilane highways. Ironically, all this tourist development has begun to deter tourists, and visitors are beginning to choose other destinations.

### DAMAGE TO NATURE

Geysers and hot springs, such as those seen here in Rotorua, New Zealand, are very popular, but the influx of visitors can be damaging to these fragile natural features. The damage is sometimes deliberate, as souvenir hunters and vandals attack them. More often, the problem is due to sheer numbers of people and cars. At many sites, the damage is caused by facilities built to cope with tourists, such as parking lots.

**SEEING POLAR BEARS IN THE WILD**
*Churchill, Manitoba, is one of the only places where tourists can see polar bears in their natural habitat, making it a popular "safari" destination.*

**CORAL REEF**
*Exploring coral reefs and relaxing on white sand beaches are major attractions of tropical beach resorts, such as this one in French Polynesia.*

# OCEAN

**STORMY SEAS**
*The Norwegian Sea, which has excellent fishing grounds, is a marginal sea of the Atlantic Ocean, the world's second-largest ocean.*

# OCEANS AND SEAS

WHAT CLEARLY MAKES THE EARTH unique in the Solar System is the vast expanse of ocean water that dominates its outer surface. Oceans and seas cover over two-thirds of the Earth's surface area. In fact, the volume of water contained in the oceans and seas is so great that if the Earth's surface were converted to a smooth sphere without topography, it would be entirely covered by a layer of seawater about 8,200 ft (2,500 m) deep. The floor beneath this great body of water includes such features as the most extensive mountain range on the planet, the deepest trench, and the largest structure built by living organisms. Life first evolved in the oceans, and today they support a vast diversity of species, ranging from microscopic organisms to the biggest animal in the world, the blue whale. The oceans are also the driving force that powers and modifies the world's climate, transporting huge quantities of solar-derived energy across the globe in the process.

# OCEAN WATER

27 Oceans and continents
126–29 Water
135 The carbon cycle
Coastal pollution 434–35

THE OCEANS OF THE WORLD contain about a third of a billion cubic miles (1.35 billion cubic kilometers) of seawater. This vast quantity of water is not uniform but varies in several physical attributes, including temperature, salinity, pressure, and level of illumination. The variation in these properties is mainly vertical, dividing the oceans into layers, each of which has specific attributes. However, there is also horizontal variation—between tropical and temperate latitudes, for example—and seasonal changes as well.

## LIGHT

Light cannot penetrate far through ocean water. However, different colors, or wavelengths, of light penetrate to varying extents. White light, such as sunlight, contains a mixture of wavelengths, ranging from red and orange (long wavelength) at one end of the spectrum to blue and violet (short) at the other. Ocean water strongly absorbs the longer wavelengths, and the amount of very short wavelength (violet) light in sunlight is small. As a result, only a little light in the blue area of the spectrum reaches much beyond a depth of 150 ft (45 m), and divers who descend into this zone see everything as a blue-gray color. Below a depth of about 650 ft (200 m), even this blue light has been absorbed, and so the ocean is almost completely dark. Because they rely on light to photosynthesize, phytoplankton are restricted to the upper layers of the oceans, and this in turn affects the distribution of other marine organisms. A few organisms that live at depth are able to generate their own light.

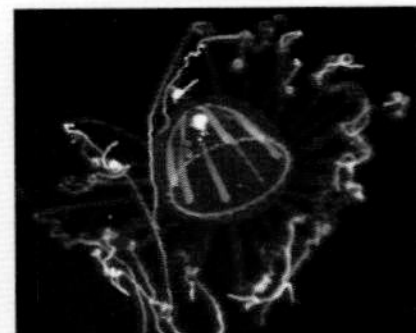

**COLOBONEMA**
*This small jellyfish lives at depths of 330–2,650 ft (100–800 m). It has bioluminescent (light-producing) tentacles.*

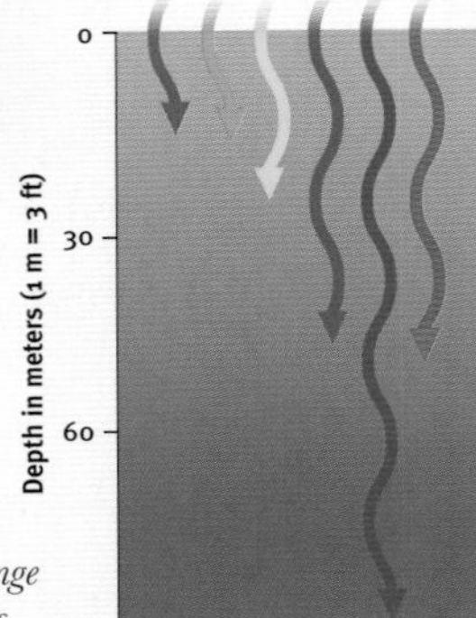


**LIGHT PENETRATION**
*Visible light in the red/orange area of the light spectrum is absorbed in the top 33–50 ft (10–15 m) of even the clearest seawater. Most other colors of the spectrum are absorbed in the next 100 ft (30 m). Only a little blue light reaches deeper than 150 ft (45 m).*

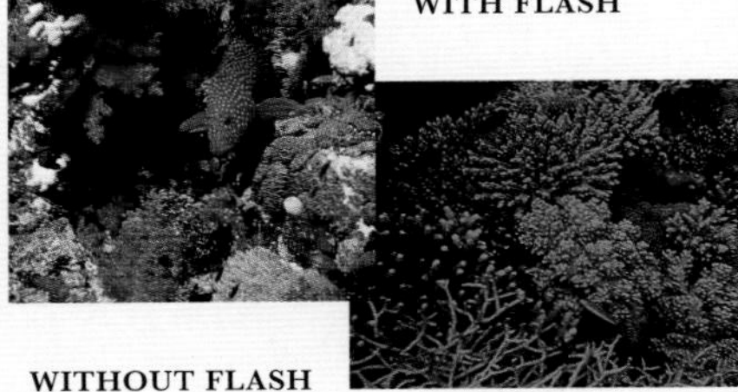

**WITH FLASH**

**WITHOUT FLASH**

**TRUE COLORS**
*Even at 65 ft (20 m) deep, features such as coral reefs appear blue-gray under ambient light conditions. Only by illuminating the reef with a lamp or photographic flash are the real colors of reef organisms revealed.*

## PRESSURE

Oceanographers measure pressure in bars (1 bar = 14.5 pounds per square inch/100,000 newtons per square meter). At sea level, all objects are subject to atmospheric pressure, which averages about one bar. Underwater, pressure increases at the rate of about one bar for each 33-ft (10-m) increase in depth, due to the weight of the overlying water. This means that at 300 ft (90 m), the total pressure is 10 bars or 10 times the surface pressure, while at 2½ miles (4 km), the pressure is over 400 times that at the surface. The huge pressures pose a considerable challenge to human exploration of the deep ocean, and animals living there exhibit special adaptations.

**COMMON OCTOPUS**

**COPING WITH PRESSURE**
*The dumbo octopus lives as deep as 5,000 ft (1,500 m), compared with a maximum of 650 ft (200 m) for the common octopus. Its adaptations include a semigelatinous body, webbed arms, and fins above the eyes.*

**DUMBO OCTOPUS**

US
Gulf of Mexico
Mexico
South America

## SALINITY

The salinity, or saltiness, of the oceans averages about 35 grams of salt per liter (1,000 grams) of water, often expressed as 35 ppt (parts per thousand). This concentration varies, being highest at the surface of warm seas, due to water evaporation from the surface, and lowest in polar oceans near the mouths of rivers, where there is a high rate of freshwater inflow. The salts in the oceans came originally from minerals that dissolved from rocks into rainwater and were then carried in rivers to the sea. The dissolved salts exist in the form of ions (charged particles), the most important being sodium, chloride, magnesium, and sulfate ions, although there are many others in smaller quantities.

**HALITE (ROCK SALT) CRYSTAL**
*Analysis of seawater trapped in ancient salt crystals has shown that its ion content has varied a little over millions of years.*

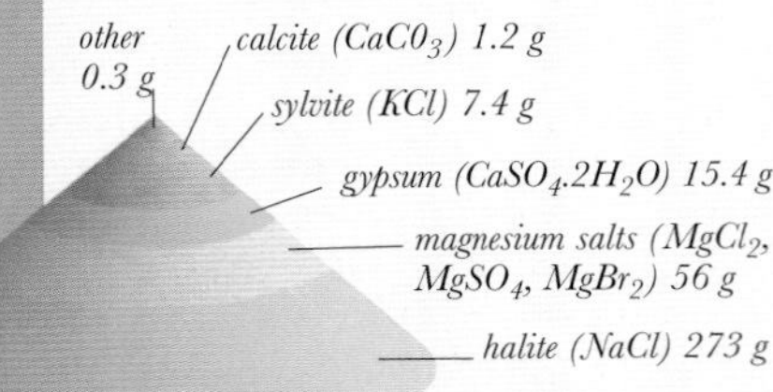


**OCEAN WATER SALTS**
*If 10 liters (about 10 quarts) of seawater were evaporated, roughly 353 g (12½ oz) of salts would be obtained, of the types shown left. The largest component is halite (sodium chloride, or table salt).*

### KEYS TO NORTH ATLANTIC MAP

This 3-D map shows the variation in early summer temperatures in part of the North Atlantic (the vertical scale has been exaggerated). The map shows that temperatures vary much more at the surface than at depth. The variation in salinity and density with depth at tropical and high latitudes is shown in the bars on the opposite page. (In the key, salinity values are expressed in practical salinity units, or psu, a measure obtained by comparing a seawater sample's electrical properties to that of a standard salt solution. The psu equates roughly to ppt, or parts per thousand.)

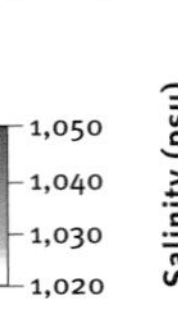


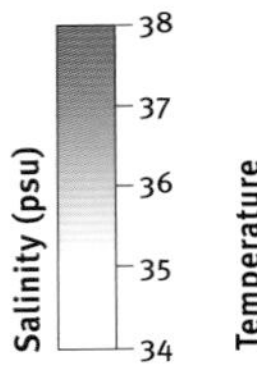


Temperature
90°F 32°C
30°C
70°F 20°C
50°F 10°C
30°F 0°C

# SOUND

Sound travels farther and faster in seawater than in air—its speed in seawater is about 5,000 ft (1,500 m) per second, more than four times faster than in air. The velocity is decreased by a decrease in temperature and increased by an increase in pressure (depth) of the water. The effects of vertical variations in temperature and pressure on sound velocity mean that in most ocean areas there is a layer of minimum sound velocity at about 3,300 ft (1,000 m) depth. This layer is called the SOFAR (Sound Fixing and Ranging) channel. It is exploited for long-distance communication underwater by people using hydroacoustic and sonar listening devices, and, it is believed, by whales and dolphins.

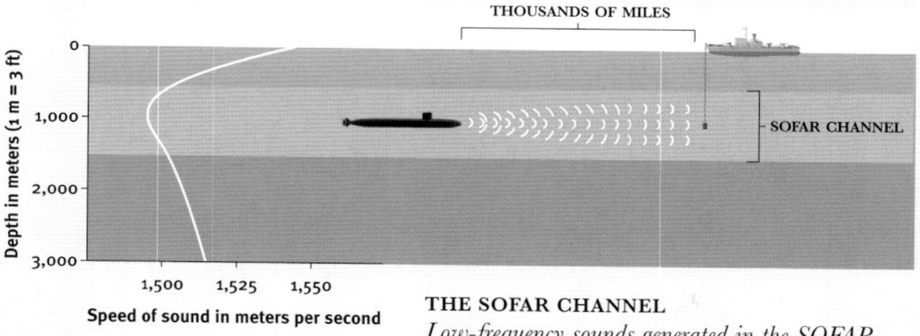

**THE SOFAR CHANNEL**
*Low-frequency sounds generated in the SOFAR channel are "trapped" in it, by refraction or bending of the sound waves toward the center of the channel. As a result, sounds can travel for long distances in this ocean layer, a phenomenon that has been exploited by submarines.*

## CETACEANS AT RISK

There is growing evidence that human use of sound underwater, such as military use of sonar for submarine detection, can damage the sensitive hearing apparatus of whales and dolphins, which forms part of their own sonar-based navigation system. This can cause the animals to become disoriented, often with fatal results.

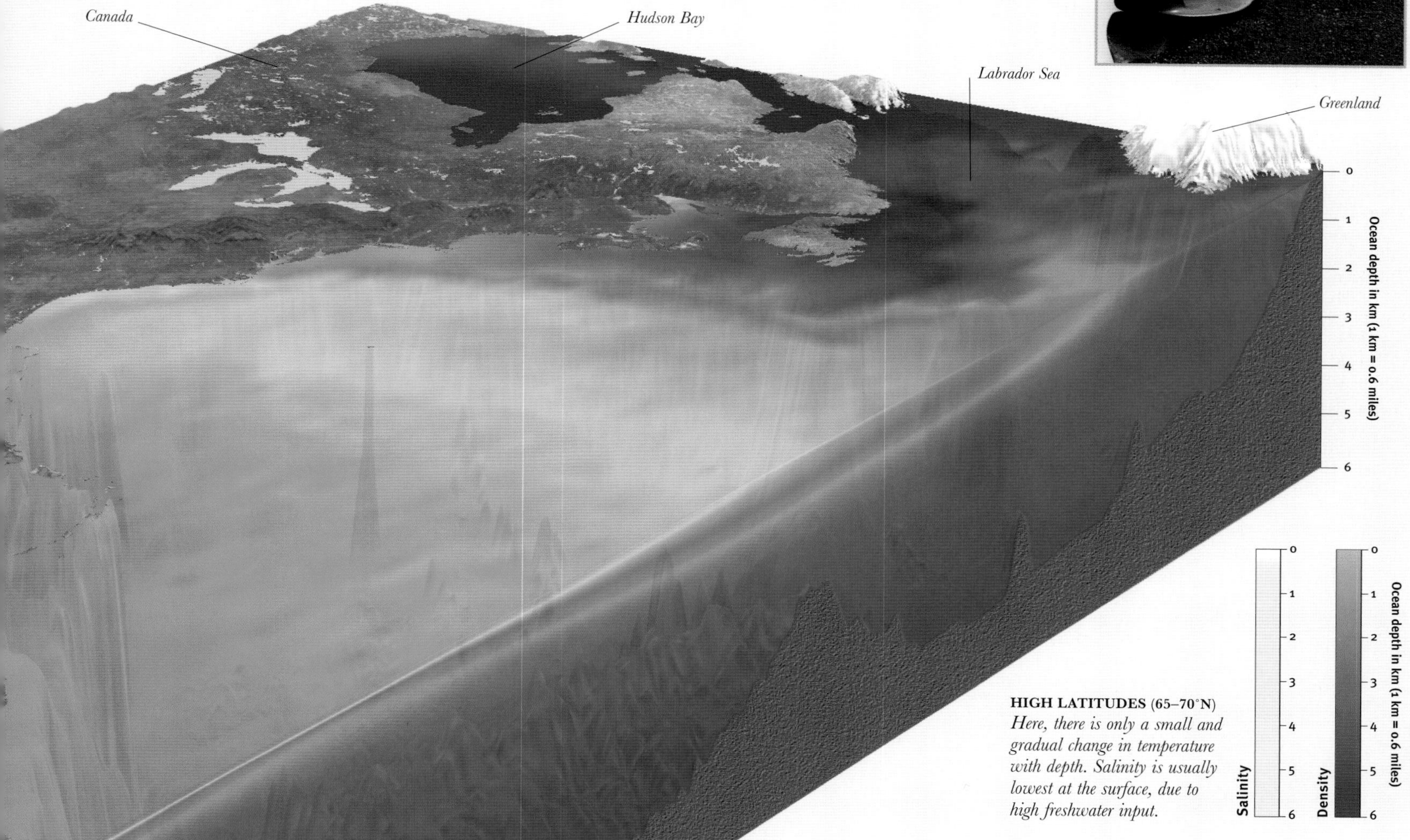

**HIGH LATITUDES** (65–70°N)
*Here, there is only a small and gradual change in temperature with depth. Salinity is usually lowest at the surface, due to high freshwater input.*

**TROPICAL LATITUDES** (0–22°N)
*Here, the ocean surface is warm and very saline due to high water evaporation. There is a sharp decline in temperature and salinity, and rise in density, down through the first 3,500 ft (1,000 m).*

# TEMPERATURE

Temperature varies significantly over the surface of the oceans, but there is little variation deeper down. In tropical and subtropical regions, solar heating warms the upper ocean to produce a broad band of water with temperatures above 77°F (25°C). Below the surface, from a depth of about 1,000 ft (300 m) down to about 3,300 ft (1,000 m), the temperature declines steeply to about 46–50°F (8–10°C). This region of steep decline is called a thermocline. Below 3,300 ft (1,000 m), the temperature decreases more gradually to a uniform 35.6°F (2°C) throughout the deep oceans. In mid-latitudes (40–50°N and S), the temperature structure of the ocean is similar to that in the tropics, except that the seasonal variation in surface temperature is much more marked, from about 63°F (17°C) in summer to 50°F (10°C) in winter, and there is a more gradual thermocline. In high latitudes and polar oceans, surface temperatures are in the range 32–41°F (0–5°C), although the temperatures in some parts and at some times may drop below the normal freezing point of water (32°F/0°C), depending on the salinity of the seawater. Below the surface, there is a gradual change to the uniform deep-water temperature of 2°C (35.6°F).

# OCEAN TECTONICS

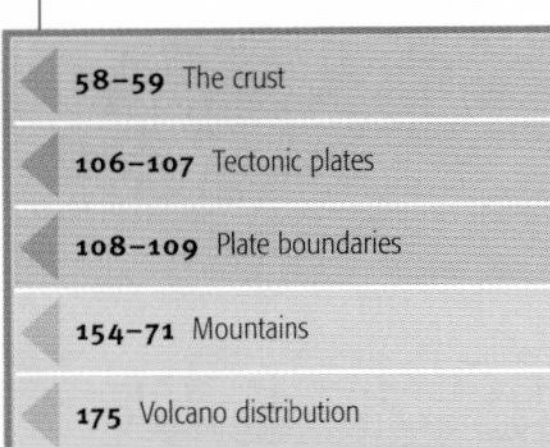
58–59 The crust
106–107 Tectonic plates
108–109 Plate boundaries
154–71 Mountains
175 Volcano distribution

THE SEA FLOOR LIES AT DEPTHS ranging from zero to more than 36,000 ft (11,000 m) below the surface of the oceans. In many places, this surface is a flat, monotonous expanse—the abyssal plains. Elsewhere, it has been heavily shaped by a combination of tectonic and volcanic activity to form features such as massive mountain ranges, seamounts, deep trenches, and basins. It is also worth noting that the ocean floor is comparatively young—no part is greater than 200 million years old—because it is continuously reworked by the plate-tectonic processes of sea-floor spreading and subduction.

## MID-OCEAN RIDGES

Mid-ocean ridges are the largest features of the sea floor. In fact, they are the largest single geological feature on the Earth's surface. They consist of an interconnected chain of mountains that twists and branches for about 40,000 miles (65,000 km) over the floor of the world's oceans. These mountains occur at divergent plate boundaries (see p.108), where new oceanic crust is created from magma welling up from the Earth's mantle, which then cools and hardens into lava. Overall, the ridges rise several thousand yards above the general level of the sea floor. In some places, volcanoes rising from the ridges break the surface of the sea; Iceland, in the North Atlantic, is an example of this. Two different types of ridges are recognized: fast- and slow-spreading. Slow-spreading ridges, such as the Mid-Atlantic Ridge, form new ocean crust at a rate of only 1–2 in (2–5 cm) a year. They have deep depressions (rift valleys), which are 6–12 miles (10–20 km) wide, running down their centers. Fast-spreading ridges, which include the East Pacific Rise, spread at 4–8 in (10–20 cm) a year and lack rift valleys.

**SURTSEY**
*A volcanic island situated directly over the Mid-Atlantic Ridge south of Iceland, Surtsey erupted out of the Atlantic Ocean in 1963. It gradually rose to a height of 555 ft (169 m) above sea level, but there has been no volcanic activity on the island since 1967.*

## HYDROTHERMAL VENTS

**LIFE AT VENTS**
*These blind rift shrimp and the vent crab are found in large numbers around mid-Atlantic hydrothermal vents.*

Located on or near the mid-ocean ridges, at an average depth of 7,000 ft (2,100 m), are some spectacular features called hydrothermal vents. These vents are like hot springs (see p.201), continuously spewing out copious amounts of mineral-rich seawater at temperatures as high as 750°F (400°C). Some of the vents have tall chimneys, formed from dissolved minerals that precipitate out when the hot vent water meets cold deep-ocean water. Black smokers are vents in which a dark cloud of sulfides precipitates out of the vent water as it exits the chimney. The minerals dissolved in the water, and bacteria that thrive on these substances, support a diverse community of organisms, including giant tubeworms and clams.

### SUBMERSIBLES

Crewed submersible vehicles have played an important part in unlocking the secrets of the sea floor. In 1977, for example, scientists aboard the submersible *Alvin* discovered the previously unknown phenomenon of hydrothermal vents in the Pacific Ocean. *Alvin* has a maximum depth of 14,764 ft (4,500 m). *Johnson Sea-Link II*, shown here, operates at shallower depths and is used to study continental slopes.

# ISLAND CHAINS AND SEAMOUNTS

At a few locations under the oceanic crust, some near mid-ocean ridges but many closer to the middle of tectonic plates, plumes of volcanic magma rise from the deep mantle. These locations, called hot spots (see p.59), are responsible for a variety of oceanic features. If oceanic crust passes over a hot spot as it is pushed away from a spreading ridge, then magma will erupt through the crust and accumulate as lava, forming a volcano that gradually enlarges and may eventually reach the ocean surface. If the crust moves in a straight line, and the hot spot stays in the same place, periodic eruptions may form a line of volcanic features called hot-spot chains. Depending on the persistence of a hot spot and the speed of plate movement, different types of hot-spot chains may develop. They include volcanic island chains, such as the Hawaiian Islands and the Azores, and chains of seamounts, the conical remains of extinct volcanoes that lie below the ocean surface (guyots, slightly different phenomena with flat tops, are volcanoes that broke above the surface before later subsiding). The Hawaiian Islands, together with a line of seamounts that extends to the northwest (the Emperor Seamounts), constitute a single hot-spot chain that stretches for 3,700 miles (6,000 km) across the Pacific.

**HAWAIIAN ISLANDS**
*This satellite photograph shows several of the Hawaiian Islands, including Hawaii at bottom right and Oahu at top left. Hawaii is composed of five merged volcanoes, only two of them still active.*

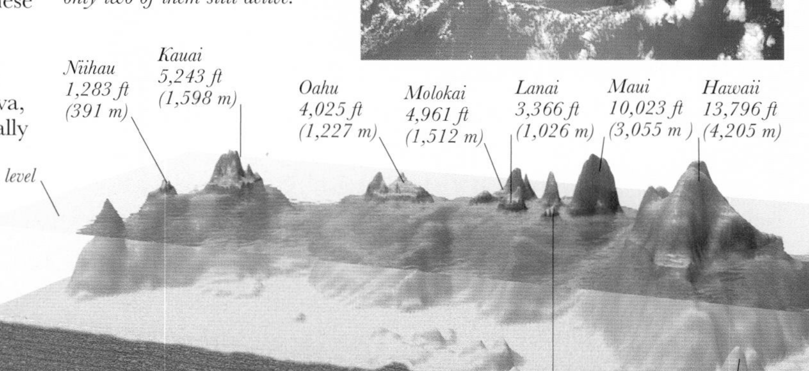


**ISLAND CHAIN**
*Each of the Hawaiian Islands is volcanic in origin and towers over the ocean floor, although the vertical scale in this depiction has been exaggerated. The chain was created by a hot spot, currently underneath Loihi, a volcanic seamount.*

**MANAGAHA ISLAND**
*This small island, close to the larger island of Saipan, is part of the Mariana Islands, a classic volcanic island arc lying to the west of the Mariana Trench on the western edge of the Pacific Ocean.*

# OCEAN TRENCHES AND ISLAND ARCS

The deepest parts of the ocean are found in arc-shaped depressions in the ocean floor called deep-sea trenches. These trenches form at convergent plate boundaries (see p.109) where one plate is subducted beneath another. Where oceanic crust is subducted beneath oceanic crust, the result is the development of an arc of volcanic islands above the overriding plate and parallel to the trench (see p.155). One example of a trench–island arc combination is the Java–Sunda Trench and Indonesian Island Arc on the edge of the Indian Ocean. Most of the world's deep-sea trenches descend to depths of over 20,000 ft (6,000 m). The very deepest, the Mariana Trench, plunges to 36,201 ft (11,034 m) at its lowest point, the Challenger Deep. Despite the total darkness, high pressure, and near-freezing temperatures, a surprising variety of life-forms are able to live in the trenches, including some limpets and clams, sea cucumbers, and barophilic (pressure-loving) bacteria.

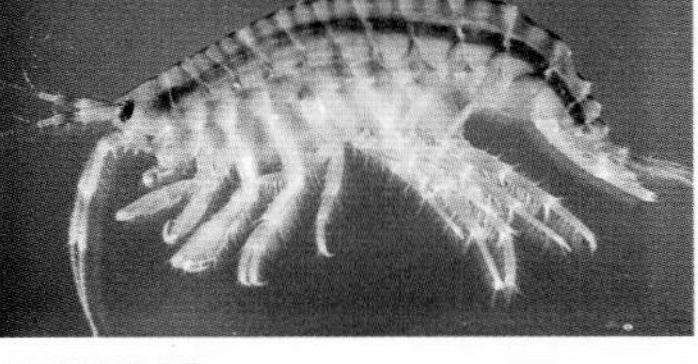

**AMPHIPOD**
*Shrimplike animals such as this small crustacean have been found in even the deepest ocean trenches, including the Mariana Trench in the western Pacific.*



**BLACK SMOKERS**
*Hydrothermal vents, such as these on the Mid-Atlantic Ridge, are among the most extreme environments in which living organisms can survive.*

# WORLD DISTRIBUTION

This map shows the distribution of mid-ocean ridges, hydrothermal vents, deep-sea trenches (marking subduction zones on convergent boundaries), and major zones in which seamounts are found. Deep-sea trenches and their associated island arcs are found mainly around the Pacific.

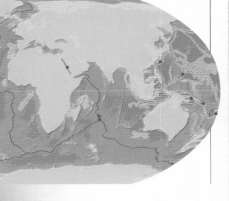

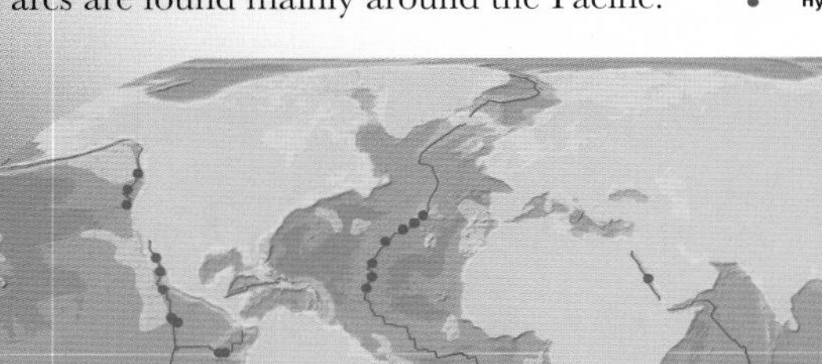

# SHELVES AND PLAINS

58–59 The crust
106–107 Tectonic plates
108–109 Plate boundaries
242–43 Freshwater quality
Fishing 422–23

AWAY FROM THE TECTONICALLY ACTIVE regions of the sea floor are other large and relatively tranquil areas of ocean, where disturbances are few apart from a constant gentle rain of sediment from above. These regions fall into two main types: the continental shelves, which themselves form part of larger zones around the edges of the world's landmasses called continental margins; and the abyssal plains, which occupy deep-sea regions between the continental margins and the mid-oceanic ridges. Both the continental margins and the abyssal plains are inhabited by a diverse range of organisms.

## LIGHT TRAPS

The sea mouse is an unusual type of floor-dwelling worm that lives in the sublittoral and bathyal zones at depths down to 6,600 ft (2,000 m). The animal has iridescent spines on its surface that are so efficient in handling light that they have attracted the attention of fiber-optics experts. The sea mouse itself is thought to use the spines for defense, their color acting as a warning to predators.

## CONTINENTAL MARGINS

The continental margins consist of three parts: the continental shelf, slope, and rise. The continental shelves are areas that were dry land during the last ice age and have since been flooded. They slope gently away from modern-day shorelines to a depth of about 660 ft (200 m). Past the edge of the continental shelf, called the shelf break, is the more steeply descending continental slope, and beyond that the continental rise, which stretches down to the abyssal plain. Many continental shelf areas are rich in deposits of oil and natural gas and also methane hydrate. The latter, a solid but volatile substance, is potentially a highly valuable energy source, but the question of how to extract it safely from the sea floor has yet to be answered.

**LOBSTERS**
*Lobsters inhabit the sublittoral zone. Some species migrate from place to place, often in groups, as here.*

**BRITTLE STAR**
*Brittle stars, relatives of starfish, inhabit the bathyal and other depth zones, often in huge numbers.*

**ANGLERFISH**
*Anglerfish swim in the bathyal and abyssal zones. Females have a luminescent organ that attracts prey.*

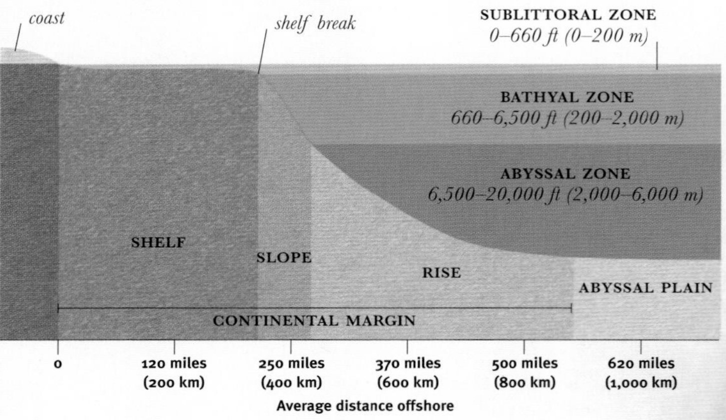


**CONTINENTAL MARGIN**
*This is a typical continental margin profile. Each part (shelf, slope, and rise) is associated with a depth zone, and each zone (sublittoral, bathyal, and abyssal) is associated with different types of organisms.*

## SUBMARINE CANYONS

At many locations on the continental margins, the continental slope has been eroded into deep, V-shaped valleys, called submarine canyons. The upper part of these canyons is often close to the point on the continental shelf where a large river runs into the sea. The canyons are excavated by a combination of sediment slumping down the slope and turbidity currents, which are large-scale movements of water and silt, like underwater avalanches, triggered by earthquakes or floods. Once they have formed, the canyons continue to act as passageways for sediment, which flows through them and finally comes to rest on the abyssal plain in a wide fan shape. The largest of these fans are associated with major sediment-carrying rivers, such as the Ganges and Amazon.

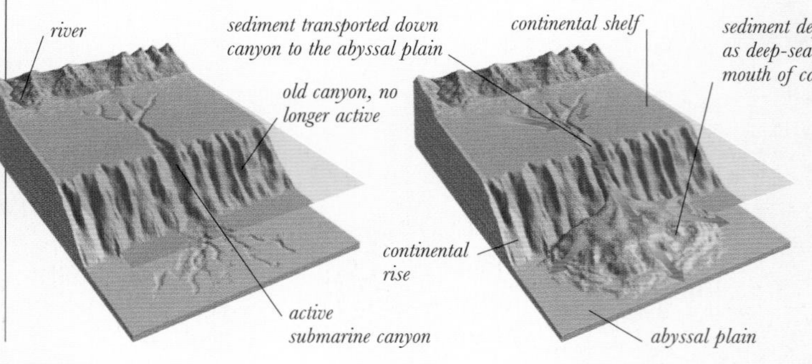

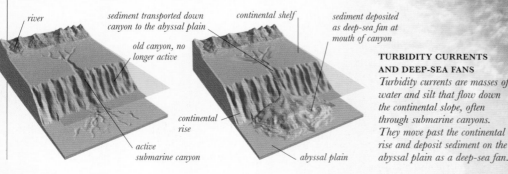

**TURBIDITY CURRENTS AND DEEP-SEA FANS**
*Turbidity currents are masses of water and silt that flow down the continental slope, often through submarine canyons. They move past the continental rise and deposit sediment on the abyssal plain as a deep-sea fan.*

**WRECK OF RMS RHONE**
*The world's continental shelves are littered with shipwrecks, such as this one in the Caribbean Sea. The ship sank in a hurricane in 1867.*

# OCEAN SEDIMENTS

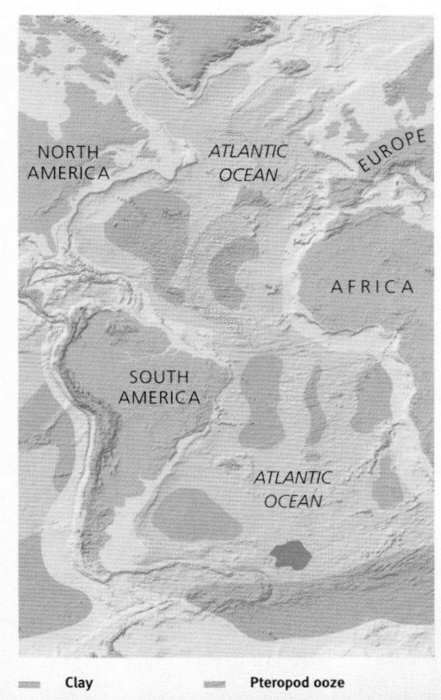


Clay
Globigerina ooze
Radiolarian ooze
Pteropod ooze
Diatom ooze

**SEABED SEDIMENTS**
*This map shows the distribution of the different sediments found on the floor of the Atlantic Ocean. Oozes are composed of the shells of countless millions of tiny organisms, such as those shown at far right.*

Most of the ocean floor is covered in layers of sediment. In some places, these are over 3 miles (5 km) thick and have taken more than 100 million years to accumulate. On the continental shelf, most sediments are gravels, sands, and muds that are derived from the erosion of rocks on land and transported by rivers to the sea. Deep-sea sediments come from many different sources. Some are fine silts that have migrated down from the continental shelf through mechanisms such as turbidity currents. Others derive from sand and dust that has been blown by winds off the continents, or from icebergs that carry rock particles away from glaciers in polar regions and then drop the particles as they melt. Authigenic sediments are precipitates of chemicals, such as iron oxide, from seawater, in forms such as manganese nodules (see below). Another important type of sediment is biogenic ooze, the skeletal remains of microscopic organisms that once lived in the ocean. There are two major varieties: calcareous oozes consisting of calcium carbonate; and siliceous oozes consisting of silica. Oozes composed of the shells of different foraminiferan species (see p.450) have been used to obtain information about ocean surface and floor temperatures and climatic conditions in the past, going back over 100 million years.

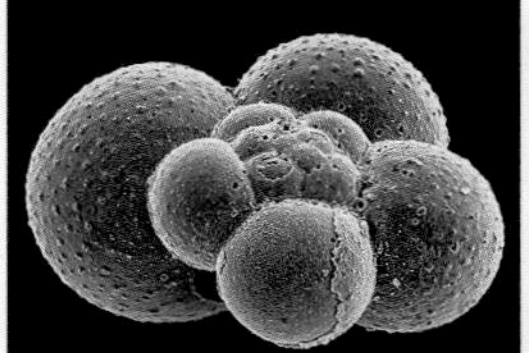

**GLOBIGERINA**
*This is a foraminiferan, a simple organism with a calcium-carbonate shell.*

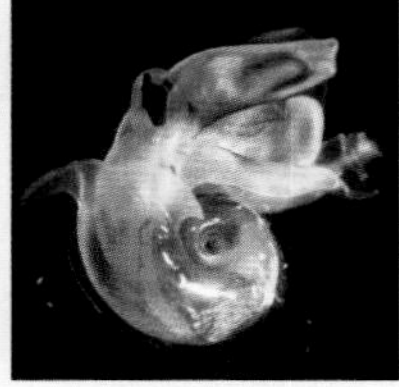

**PTEROPOD**
*Pteropods are tiny mollusks with shells of calcium carbonate.*

**RADIOLARIAN**
*Radiolarians are single-celled, floating, animal-like organisms with silica shells.*

**DIATOMS**
*Diatoms are delicate, photosynthesizing life-forms with silica shells.*

# ABYSSAL PLAINS

The abyssal plains begin where the continental margins end. They occupy extensive areas of the sea floor at depths of 13,000–20,000 ft (4,000–6,000 m) and are the deepest parts of the oceans apart from the deep-sea trenches. They are the flattest, and some of the most featureless, places on Earth, with a gradient of just 1:1,000. Despite the pitch-blackness of the water, the high pressure, and the freezing cold, many different animals live on the abyssal plains, including several species of worms, shrimp, brittle stars, sea cucumbers, and some extraordinary fish.

**MANGANESE NODULES**
*These potato-sized lumps litter wide areas of the abyssal plains. They contain many valuable metals, but so far no satisfactory way of mining them has been devised.*

# WORLD DISTRIBUTION

Areas of continental shelf (light purple on the map below) exist around the edges of all continents but vary greatly in their width, from just a few miles off the west coast of much of the Americas to up to 560 miles (900 km) off the coast of Siberia, Russia. The narrowest shelves are found at convergent plate margins. Abyssal plains (dark purple on the map) are most common in the Atlantic and Indian oceans, but are relatively rare in the Pacific.

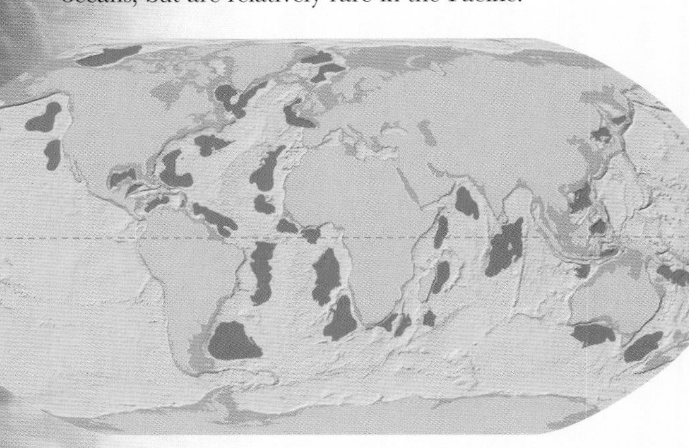

# CIRCULATION AND CURRENTS

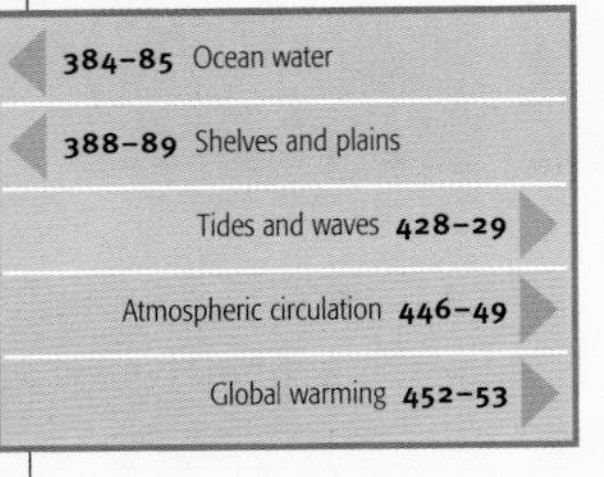
384–85 Ocean water
388–89 Shelves and plains
Tides and waves 428–29
Atmospheric circulation 446–49
Global warming 452–53

SEAWATER IS IN CONSTANT motion, and not simply due to waves and tides. Throughout the oceans, there is a constant circulation of water, both across the surface and more slowly in the ocean depths. Several interlinked processes play a part in causing and maintaining these currents. They include solar heating of the atmosphere and the surface of the oceans, the winds that this heating generates, the Earth's rotation, and various processes that affect the temperature, salinity, and density of surface waters.

## SURFACE CIRCULATION

The large-scale pattern of surface movement in the oceans is called the surface circulation. It is driven by winds, but modified by the Coriolis effect (see p.446), which results from the Earth's spin. In the northern hemisphere, the Coriolis effect deflects wind-driven surface movements of water (known as wind drag) slightly to the right (as shown below), and in the southern hemisphere to the left. However, through an accentuation of the Coriolis effect called Ekman transport, the average water motion in the top few hundred yards is almost at a right angle to the wind direction. The overall effect of the predominant winds and Ekman transport on the surface of the oceans is a pattern of large-scale circular movements of water, called gyres, which rotate clockwise in the northern hemisphere, counterclockwise in the southern. Specific components of these gyres are called boundary currents. The boundary currents on the eastern side of oceans are generally weak, cold, and move toward the equator. In contrast, those on the western sides of oceans tend to be strong, warm, and move away from the equator.

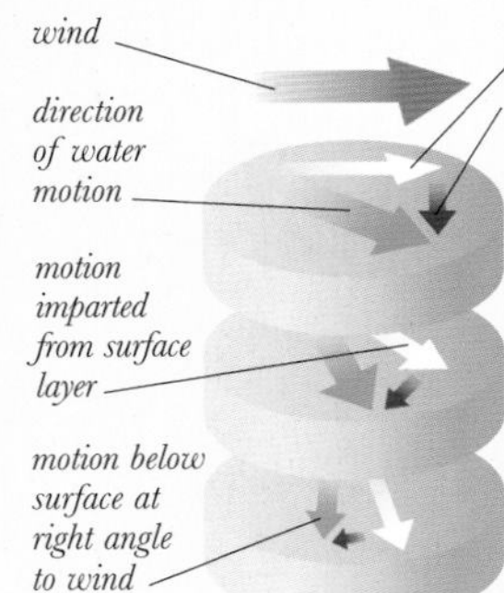


**EKMAN TRANSPORT**
*At the ocean surface, the Coriolis effect deflects water motion slightly from the direction of wind drag. This motion produces a drag in the next layer down, which is also deflected, and so on. On average, the water is pushed at 90° to the wind direction.*

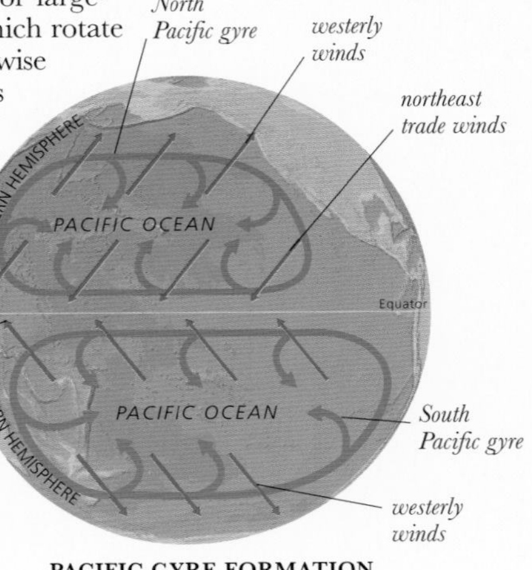


**PACIFIC GYRE FORMATION**
*Ekman transport curves water to the right of the predominant winds in the northern hemisphere, and to the left in the southern hemisphere, producing two gyres, or circular movements, of surface water.*

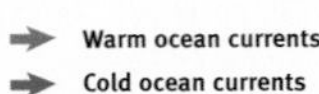


**SURFACE CURRENTS**
*This map shows the overall pattern of surface currents. In the Atlantic, Pacific, and Indian oceans, much of the circulation is in the form of circular gyres. These consist predominantly of currents carrying warm water away from the equator and colder water toward it.*

## LOCAL CURRENTS

Local currents are movements of water that are the result of localized interaction between tidal forces and the shapes of coastlines. Tides cause regular variations in sea level, which are most noticeable around coasts and are more pronounced in some locations than others. These vertical changes in water level can occur only through horizontal movement of water—into and out of bays, for example. It is these movements that bring about local or tidal currents, and they can be especially significant where large volumes of water are funneled through narrow channels, especially around promontories, between islands, and up estuaries. Because they are tide-related, these currents continuously vary in their strength and direction hour by hour over the daily tidal cycles, and also over monthly cycles (see p.428). At particular locations, they can cause phenomena such as tidal bores and whirlpools.

**BASS STRAIT**
*Bass Strait, between Tasmania and the mainland Australian state of Victoria, is famous for its strong currents, which are partly wind-driven and partly tidal.*

## VERTICAL TRANSPORT

Vertical transport refers to the movement of water from the surface to depth (downwelling) or vice versa (upwelling). One cause is an increase in the density of surface water, through cooling or an increase in its salinity. A prime example is sinking of water under sea ice in polar regions. As the water cools, it is made more saline when salt is expelled from seawater as it freezes (see p.394). Other causes include wind patterns that cause water masses to converge in an area (where they are forced down) or to diverge (where water must upwell to replace the diverging water). Where a wind blows parallel to a coast, this can also cause upwelling or downwelling. Deep waters are rich in nutrients (derived from the breakdown of the remains of organisms that have sunk from the surface), so upwelling has important biological effects because it brings these nutrients to the surface, which helps the growth of plankton.

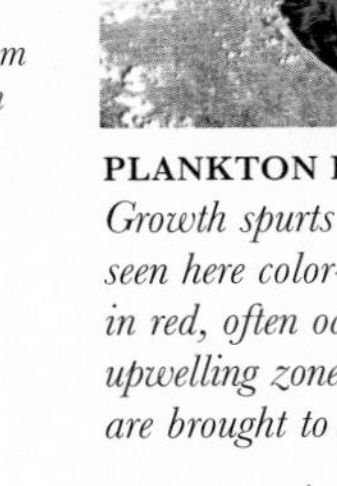

**PLANKTON BLOOM**
*Growth spurts of plankton, seen here color-enhanced in red, often occur in upwelling zones as nutrients are brought to the surface.*

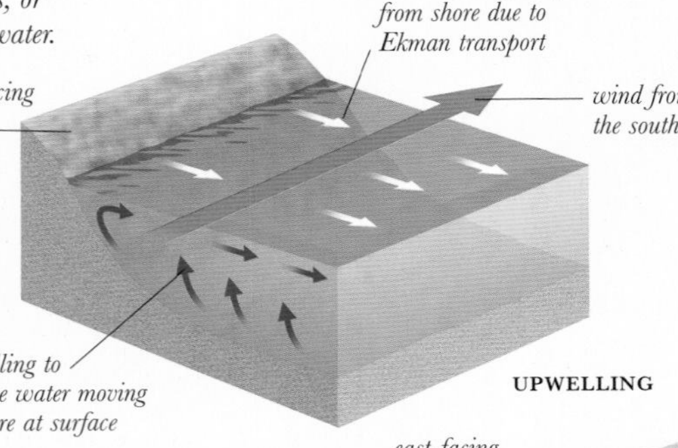


**UPWELLING**

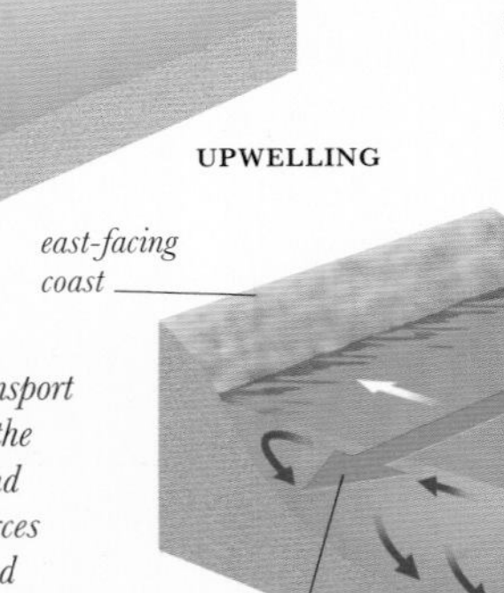

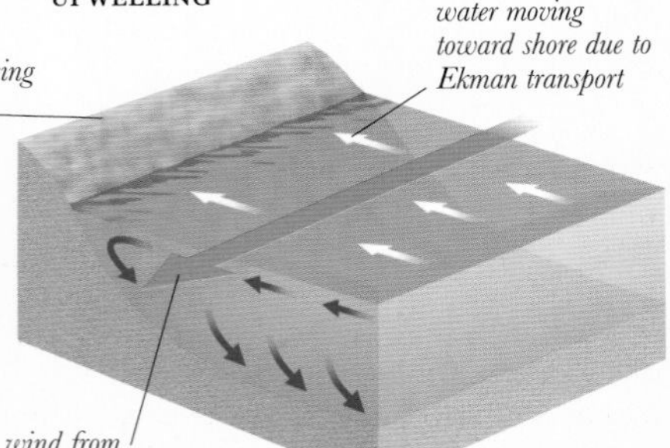


**DOWNWELLING**

**COASTAL UPWELLING AND DOWNWELLING**
*In the northern hemisphere, Ekman transport produces water motion to the right of the wind direction. When a southerly wind blows up an east-facing coast, this forces water away from the coast (above) and causes upwelling near the coast. When the wind blows from the north (right), water is pushed toward the coast and then sinks.*

**NARUTO WHIRLPOOL**
*Spectacular whirlpools caused by strong tidal currents develop several times a day in the narrow channel that links the Sea of Japan to the Pacific Ocean.*



# DEEP-WATER CURRENTS

The world's deep-water currents are set in motion by differences and changes in the density of water masses, principally by the downwelling of dense, cold, salty water in polar and subpolar regions. Once this water reaches a depth level of equal density, it spreads out, often over long distances. In other parts of the world, slightly less dense water rises to the surface as it mixes with and absorbs heat from water masses above it. This type of circulation is called thermohaline, "thermo" referring to temperature and "haline" to saltiness. Deep-water currents move slowly, no more than a few yards a day. Once a body of water sinks, it can spend hundreds of years away from the surface until it rises again to become part of the surface circulation.

**THERMOHALINE CONVEYOR BELT**
*This circulation is driven by the cooling and sinking of water masses to great depths in the North Atlantic. The cold water circulates down through the whole of the Atlantic and penetrates into the Indian and Pacific oceans, before returning as warm upper ocean currents to the South Atlantic.*

- Warm surface current
- Cold, salty, deep-ocean current

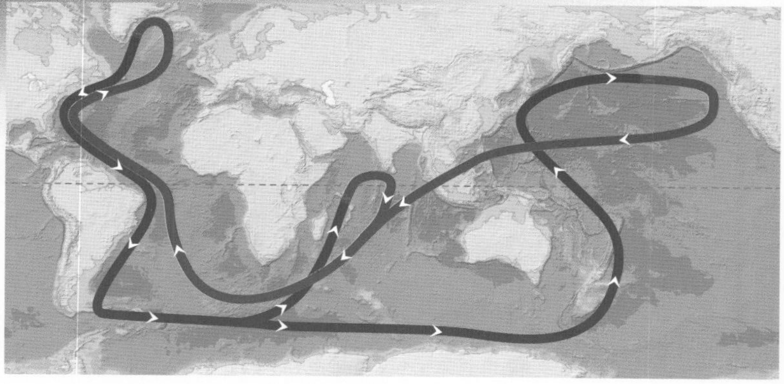

# REEFS

39–41 The ice ages
175 Volcano distribution
Coral reefs 406–407
Coastal pollution 434–35

CORAL REEFS ARE WAVE-RESISTANT structures built from the remains of small marine organisms. They are found in many shallow warm-water seas. The main bulk of a reef is a mass of limestone, consisting of the buildup of cemented skeletons and shell fragments of animals that once inhabited the reef. On the surface of this limestone is a thin skin of living organisms. When these organisms die, their hard parts remain as part of the reef structure, helping the reef to grow in size over time. Coral reefs support a highly diverse fauna and flora, with almost every group represented. They also protect shores from erosion and build new land in tropical areas, forming islands and altering continental shorelines.

### JAMES DWIGHT DANA

From his observations on various coastlines in the South Pacific, US geologist James Dana (1813–95) found evidence to support Charles Darwin's ideas on how atolls form (see right). Dana also developed a theory explaining how the shapes of atolls are related to wind and wave patterns, which affect the supply of nutrients to reef-building organisms.

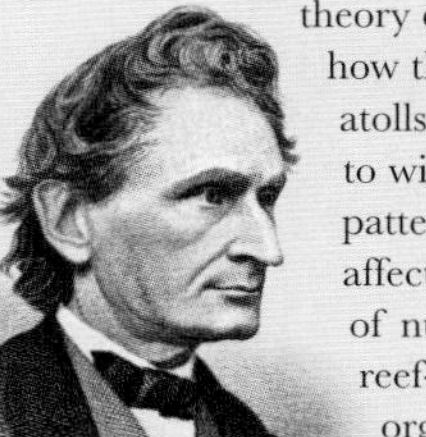

## HOW CORAL REEFS FORM

The main reef-forming organisms, known as hard or stony corals, belong to a group of animals called cnidarians. Individual animals are called coral polyps. As it grows, a coral polyp secretes limestone, building on the rock underneath. Some reef-building corals consist of single large polyps, but the majority live as colonies that, as they grow, create community skeletons in a variety of shapes. An important contributor to the life of coral polyps is the presence within their tissues of unicellular algae called zooxanthellae. Although polyps use their tentacles to capture plankton for food, most of their nutrition comes from these algae. The algae use photosynthesis to convert the carbon dioxide produced by the polyps' respiration into the nutrients that polyps need for growth and so for production of limestone. They also provide oxygen for the polyps. Other organisms that add their skeletal remains to reefs include mollusks and echinoderms. Further contributions are made by boring and grazing organisms. These fragment some of the coral skeletons into sand, which fills the gaps in the skeletons. Algae and encrusting bryozoans cement the coral fragments and other debris into a solid reef.

**POLYP OF A HARD CORAL**
*This reef-building species, called a plate or mushroom coral, consists of a single large polyp with many green, pink-tipped tentacles. Often mistaken for anemones, mushroom corals inhabit shallow reefs.*

*tentacle*
*mouth*
*gut cavity*
*connecting tissue between adjacent polyps*
*limestone exoskeleton*

**HARD CORAL POLYP ANATOMY**
*The polyp's gut cavity sits within a fluted, cuplike limestone exoskeleton. Food is taken in, and some waste products are expelled, through the mouth. The tissue around the gut cavity lays down limestone, which helps to build the reef.*

## FRINGING REEFS

Fringing reefs form a fringe of coral around a tropical island or along part of the shore of a large landmass, with little or no lagoon between the reef and shore. They are the most common type of reef worldwide. Fringing reefs consist of several zones that are characterized by their depth, structure, and coral communities. The reef crest is the part the waves break over, marked by a line of breaking surf. In front of the reef crest is an area called the buttress zone, where there are spurs of coral growing out into the sea separated by grooves. This is the region of most active coral growth. Shoreward of the reef crest is the reef flat, a shallow, flat expanse of limestone, sand, and coral fragments that may be partly exposed at low tide.

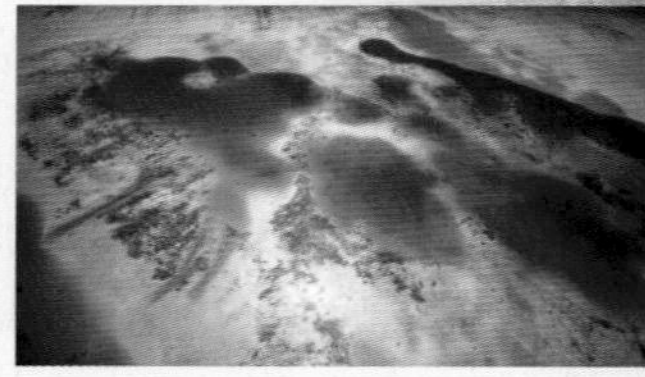

**REEFS IN PALAU**
*This aerial view shows a large area of fringing reefs (the light-colored areas) around islands and exposed coral flats in the Republic of Palau in the western Pacific.*

**CORAL SPUR**
*In this area on the surface of a spur at the front of a fringing reef, several colonies of stony coral are growing in about 3 ft (1 m) of water.*

**BORA BORA**
*Bora Bora lies in the south-central Pacific Ocean. It consists of a volcanic island surrounded by a lagoon and a necklace of submerged reefs and elongated islands that constitute a barrier reef.*

## BARRIER REEFS

A barrier reef is a coral reef that parallels the shore but is separated from it by a sizable lagoon. Large and continuous barrier reefs occur in association with continental landmasses. The largest in the world are the Great Barrier Reef off the coast of Queensland, Australia, and the Belize Barrier Reef in the Caribbean. Smaller barrier reefs can also be seen around sinking volcanic islands.
Barrier reefs contain the same zones as fringing reefs, but with some additions. In front of the buttress zone, there is frequently a steeply descending wall called a reef face, or dropoff, which is usually rich in coral formations. The main bulk of the reef behind the reef crest is often several miles wide and may include patch reefs separated by submerged areas.

**REEF FACE**
*A diver swims around the top of the reef face on part of Australia's Great Barrier Reef.*

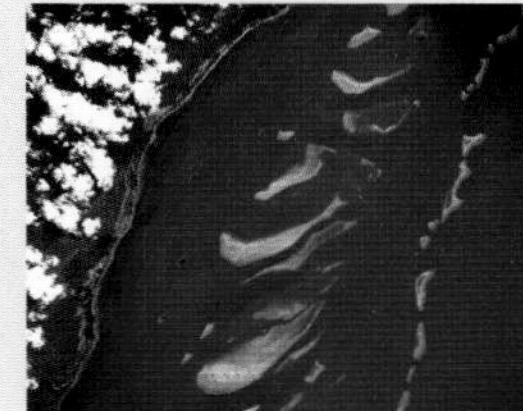

**GREAT BARRIER REEF**
*This satellite image of part of the Great Barrier Reef shows that it is an assemblage of reefs rather than a single structure.*

## ATOLLS

An atoll is a ring of coral reefs, or low-lying islands made of coral, that encloses a shallow central lagoon. Atolls are often elliptical in shape, but many are irregularly shaped. The process by which they form was first explained in the 1840s by the English naturalist Charles Darwin. Darwin proposed that fringing reefs can transform into barrier reefs and barrier reefs into atolls. He theorized that these transitions result from the upward growth of coral on the edge of a gradually sinking volcano, and that the ringlike appearance of an atoll with a central lagoon results from the total submergence of the summit of the volcano. Other scientists have contributed to Darwin's theory, especially on the importance of temperature restraints in reef formation and the contribution of wave patterns to atoll shapes. Today, it is realized that sea-level rise may contribute to atoll formation as much as volcanic subsidence.

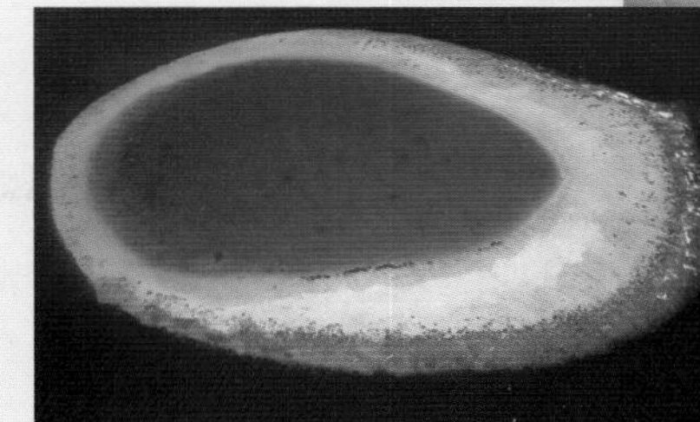

**MALDIVES ATOLL**
*This example of a small atoll is in the North Male' island group in the Indian Ocean.*

**FORMATION OF AN ATOLL**
*Darwin's explanation of atoll formation is shown here. First, the shores of an extinct oceanic volcano are colonized by corals, forming a fringing reef. As its magma chamber is depleted, the volcano sinks, but coral growth continues and forms a barrier reef, separated from the island by a lagoon. Finally, the volcano disappears entirely, leaving an atoll.*

**CROWN-OF-THORNS STARFISH**
*The large crown-of-thorns starfish, which eats coral polyps, has become a blight on coral reefs throughout the Indo-Pacific region.*

## DANGERS AND STRESSES

Many different types of stress can damage and destroy coral reefs. Some damage is due to human activity (see p.407), and a significant threat is posed by rising sea levels due to global warming (see p.453). Natural disturbances include various coral diseases, tropical storms, and an increase in predation by animals such as parrotfish, snails, and the crown-of-thorns starfish. The latter has been a particular concern since the 1960s. One theory for its emergence is that there has been a loss of some of its natural predators. Another is that the addition of nutrients from human use of the coastal zone has increased the availability of planktonic food for young starfish.

## CORAL REEF DISTRIBUTION

Reefs cover about 230,000 square miles (600,000 square km) of the world's marine areas. Reef-building corals can grow only in shallow, clear water, where there is plenty of sunlight, water temperatures of at least 64°F (18°C) but preferably 77–84°F (25–29°C), average salinity of 36 ppt, and not too much wave action, turbidity, or silting. The right conditions for reef growth are found mainly within tropical areas of the Pacific, Indian, and Atlantic oceans, and predominantly in the western parts, where the waters are warmer than in the eastern parts. Coral reefs are absent from any coastal areas where there is a large amount of sedimentation from river runoff. There are no reefs near the mouths of the large rivers of east Asia, for example.

# POLAR OCEANS

39–41 The ice ages
129 The global water cycle
264 Types of glacier
Climate change 450–51
Global warming 452–53
Cold climates 457

THE POLAR OCEANS COMPRISE THE Arctic Ocean in the northern hemisphere and the Southern Ocean in the south. They differ from other oceans in several respects, not least the vast amounts of ice that float on them in various forms, although parts of the Atlantic and Pacific oceans are also covered with ice for some of the year. This ice coverage has important stabilizing effects on world climate, insulating large areas of the oceans from solar radiation in summer and preventing heat loss in winter. The polar oceans also contain fewer temperature and density layers than the other oceans, and have different water circulation patterns.

**PANCAKE ICE**
*These scientists in Antarctica are investigating the properties of pancake ice, circular pieces of sea ice up to several yards in diameter and 4 in (10 cm) thick.*

## ICE SHELVES

Ice shelves are extensions of ice sheets and other glaciers over the sea. Although there are a few in the Canadian Arctic, most of the world's ice shelves, including two enormous ones, the Ross and Ronne-Filchner ice shelves, exist around the coasts of Antarctica (see p.279). Because they originate from snow, ice shelves contain fresh water. At their seaward edges, they are marked by cliffs that are up to 200 ft (60 m) high and may extend to 3,000 ft (900 m) underwater. Underneath the ice shelves, there is some circulation of seawater. Water drawn in from beyond the edge of the shelf freezes underneath the ice, expelling salt as it does so. This creates a layer of cold, salty, dense water that sinks and flows away from the continental shelf area beneath the ice to deeper water. Since the early 1990s, some of the smaller Antarctic ice shelves have completely broken up (see p.451). This has raised concerns for the future of the ice sheets covering Antarctica, with implications for possible accelerated sea-level rise, since ice shelves appear to provide a barrier to ice flow off the ice sheets.

**BRUNT ICE SHELF, ANTARCTICA**
*The edge of an ice shelf is usually marked by ice cliffs. Here the cliffs are fronted by sea ice on which some emperor penguins are gathered.*

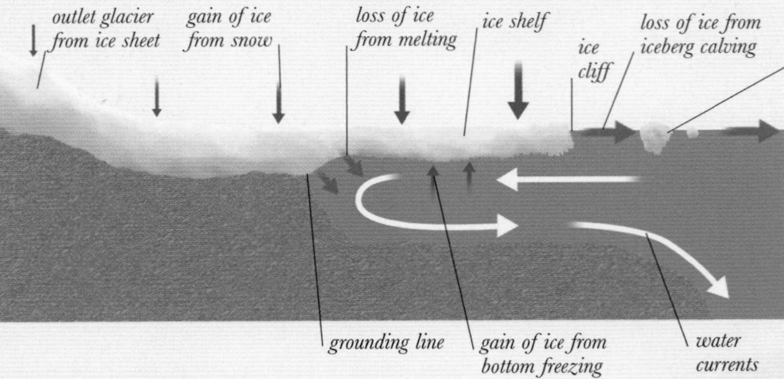


**ICE-SHELF GAINS AND LOSSES**
*Ice shelves gain ice from glaciers flowing into them, from seawater freezing to their undersurface, and from new snowfalls. They lose ice mainly by icebergs calving (breaking off) from their front edges.*

## ICEBERGS

Icebergs are pieces of ice that have broken off the edges of ice shelves or from the fronts of glaciers where they reach the sea. They range from tabular icebergs, hundreds of square miles in area, to yacht-sized chunks of ice called bergy bits and car-sized ones called growlers, which are just as dangerous to shipping. In the Southern Ocean, most icebergs have broken off ice shelves around Antarctica. They initially drift westward in a coastal current around the continent and many become trapped in bays. A few drift several hundred miles from Antarctica, some reaching as far north as 40°S. In the northern hemisphere, most icebergs have originated from glaciers flowing off Greenland, Ellesmere Island in the Canadian Arctic, or Alaska. A few reach as far south as 45°N. Because of their glacial origin, icebergs contain fresh water. The average density of icebergs is about 87 percent of the average density of seawater. As a result, an iceberg typically floats with about 87 percent of its volume underwater.

STRIATED

BLOCKY

**ICEBERG TYPES AND SHAPES**
*Icebergs come in various regular shapes, including pinnacle, pyramid, wedge, and blocky, but there are also many irregularly shaped icebergs. Striated icebergs contain layers of rock debris from the glacier from which they were calved.*

DRY DOCK

IRREGULAR

PINNACLE

# CIRCULATION IN POLAR OCEANS

The circulation of water in polar oceans is influenced by factors such as changes in the surface temperature and salinity of the water caused by the formation and melting of sea ice. Both polar oceans are sources of cold, dense bodies of water that sink toward the ocean floor and then continue toward the equator. These flows are key drivers of the Earth's deep-water circulation. In the Arctic, there is a constant flow of cold water and sea ice across the North Pole from Siberia toward Greenland in a current called the Transpolar Drift. To the east and southwest of Greenland, this cold water dips under warmer Atlantic water. Farther north, the Atlantic water plunges beneath less saline Arctic water, then moves through the Arctic Basin. An equivalent circulation operates in the Southern Ocean (see p.424).

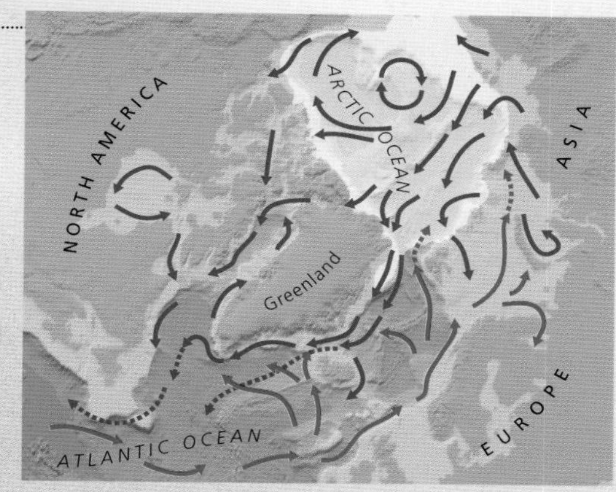


**ARCTIC CURRENTS**
*This map shows the basic pattern of surface currents and downwellings in the Arctic and North Atlantic oceans. Observations of ice drift have indicated the surface flows. One of the most pronounced currents is the southerly movement of cold water to the east of Greenland.*

# SEA ICE

Unlike icebergs and ice shelves, sea ice is seawater that has frozen. It contains little salt, since most is expelled as it forms, and it floats on the sea because it is less dense than seawater. About 7 percent of the world's oceans are covered in sea ice, nearly all of it in polar regions. New sea ice forms in distinct stages. Early on, a viscous, slushy layer of ice crystals, called grease ice, develops at the sea surface. The ice thickens and hardens, and is broken up by wave action into platelets. As these collide, the result is a mosaic of platter shapes with curling edges, called pancake ice. Later stages include first-year ice, which is more than 12 in (30 cm) thick, and multi-year ice, which has survived at least two summers' melt and can be 10–35 ft (3–10 m) thick. Fast ice is sea ice that is securely frozen to the shore or an ice shelf, while pack ice is any other area of sea ice.

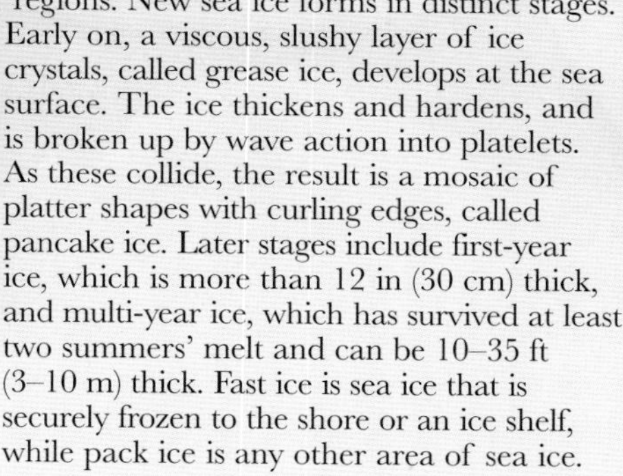

**GREASE ICE**
*A crabeater seal surfaces through grease ice—an early stage in sea-ice formation—in the Weddell Sea.*

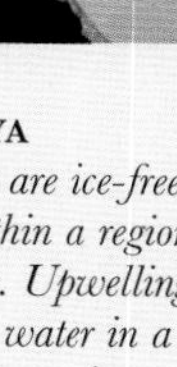

**ICE LEAD**
*A lead is a narrow fracture or passageway through sea ice, which makes the ice navigable by surface vessels and some marine mammals.*

**POLYNYA**
*Polynyas are ice-free areas within a region of sea ice. Upwelling of warm water in a localized area is one way they are formed.*

# OCEANS OF THE WORLD

IT IS GENERALLY ACCEPTED TODAY that the Earth has five oceans, although the Southern Ocean was officially recognized only in 2000. Three long-recognized oceans, the Atlantic, Indian, and Pacific, extend northward from the Southern Ocean, separating the world's major landmasses. The fifth and smallest ocean, the Arctic Ocean, caps the north polar region. Around the edges of these oceans are smaller regions called seas, bays, and gulfs. Some of the seas, such as the Sargasso Sea, are only vaguely defined, but others, such as the Mediterranean Sea and the Caribbean Sea, are almost completely surrounded by land or by arcs of islands.

**MEDITERRANEAN SEA**
*An offshoot of the Atlantic Ocean, the Mediterranean Sea was the center of the classical world. This coast is on the volcanic island of Santorini, in the Cyclades Islands, Greece.*

**ATLANTIC OCEAN**
*A major area for marine commerce and fishing, the Atlantic is known for its storms. Lighthouses like this one at Minots Ledge, Massachusetts, help to protect shipping.*

**CARIBBEAN SEA**
*The warm waters of the Caribbean Sea support large numbers of marine species, such as these bottlenose dolphins.*

**INDIAN OCEAN**
*The Indian Ocean is littered with clusters of coral islands, the best known being the Seychelles and the Maldives. The Maldive Islands lie to the southwest of Sri Lanka and consist of more than a thousand islands, of which two are shown here. Most of them are uninhabited.*

**YELLOW SEAHORSE**
*Seahorses are fish that swim upright and have prehensile tails. This species is common in the Indian Ocean.*

ARCTIC OCEAN
Greenland
Baffin Bay
Greenland Sea
Barents Sea
Kara Sea
Norwegian Sea
White Sea
Hudson Bay
Gulf of St. Lawrence
North Sea
Baltic Sea
Bay of Biscay
EUROPE
ASIA
Black Sea
NORTH AMERICA
Bay of Fundy
Mediterranean Sea
Persian Gulf
Gulf of Mexico
Sargasso Sea
ATLANTIC OCEAN
Red Sea
Caribbean Sea
AFRICA
Arabian Sea
Bay of Bengal
SOUTH AMERICA
INDIAN OCEAN
Mozambique Channel
Scotia Sea
SOUTHERN OCEAN
Weddell Sea
ANTARCTICA

**OCEANS AND SEAS**
*This map shows the range of depths in the world's oceans and seas. Surrounding all the landmasses are shallow areas of continental shelf (the lightest shade of blue). Beyond the shelves, the ocean basins fall into two main depth zones: a deeper, mostly flat region at about 16,400 ft (5,000 m) and a more elevated, rugged zone at 3,300–16,400 ft (1,000–5,000 m), lying mainly around the mid-ocean ridge areas.*

0–1,600 ft
1,600–3,300 ft
3,300–8,200 ft
8,200–16,400 ft
over 16,400 ft

**ARCTIC OCEAN**
*The Arctic Ocean consists of a deep basin surrounded by a ring of seas, which partly cover continental shelves. The volcanic island of Jan Mayen, seen here, lies on the edge of one of these marginal seas, the Greenland Sea.*

**WALRUS**
*Walruses are insulated from the Arctic cold by a thick, subcutaneous layer of blubber. When sunbathing, their heavily creased skin flushes deep pink.*

## FERDINAND MAGELLAN

The Portuguese explorer Ferdinand Magellan (1480–1521) led the first expedition to circumnavigate the globe, though he died during the voyage. Magellan embarked on the enterprise in 1519 hoping to discover a new route to the Spice Islands (now the Maluku Islands or Moluccas) by finding a passage from the Atlantic into the Pacific. In November 1520, he discovered such a passage near the tip of South America, today called the Straits of Magellan.

ARCTIC OCEAN
Laptev Sea
East Siberian Sea
Chukchi Sea
Beaufort Sea
ASIA
Hudson Bay
Sea of Okhotsk
Bering Sea
Gulf of Alaska
Yellow Sea
Sea of Japan
NORTH AMERICA
East China Sea
PACIFIC OCEAN
Gulf of Mexico
Gulf of California
South China Sea
Sulu Sea
Banda Sea
Coral Sea
Australia
Tasman Sea
PACIFIC OCEAN
SOUTHERN OCEAN
Ross Sea
ANTARCTICA

**PACIFIC OCEAN**
*By far the largest of the oceans, the Pacific is dotted with thousands of islands, many formed from (or on) the tops of undersea volcanoes, and fringed with coral reefs. This little island in the Solomon Islands group in the western Pacific is a typical example.*

**CRIMSON ANEMONE**
*Anemones are found worldwide, in all seas and at various depths. This species lives in the northeastern Pacific.*

**SOUTHERN OCEAN**
*Covered in ice for much of the year, the Southern Ocean is whipped by high winds in winter, producing mountainous seas. Despite the severe conditions, an amazing diversity of wildlife thrives there. It is home to some of the world's great whales and large numbers of birds, such as these chinstrap penguins drifting on sea ice in the Weddell Sea.*

### OCEANS AND SEAS PROFILES

The pages that follow contain profiles of the world's oceans and major seas, bays, and gulfs. Each profile begins with the following summary information:

**AREA** Surface area

**MAXIMUM DEPTH** Depth of the deepest part of the ocean, sea, or gulf

**INFLOWS** Major inflows including oceans, basins, seas, rivers, fjords and glaciers

ARCTIC OCEAN

## Arctic Ocean

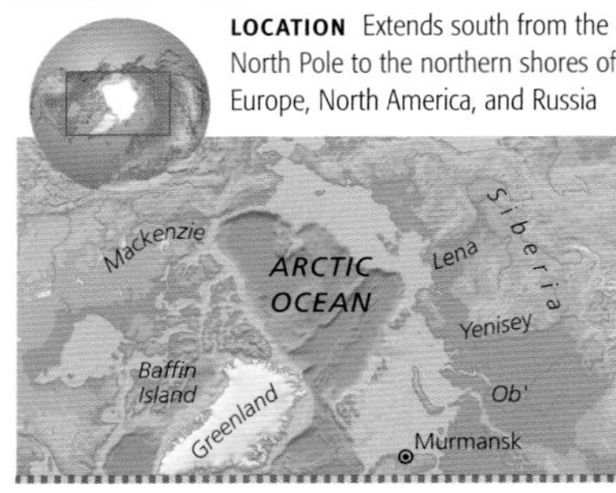


**LOCATION** Extends south from the North Pole to the northern shores of Europe, North America, and Russia

**AREA** 5.4 million square miles (14 million square km)

**MAXIMUM DEPTH** 15,305 ft (4,665 m)

**INFLOWS** Atlantic Ocean, Pacific Ocean; rivers Mackenzie, Ob', Yenisey, Lena, Kolyma

The Arctic Ocean is the smallest of the world's oceans. It is also the least studied, partly because about a third of its surface is permanently covered in ice. The ocean comprises a central region around the North Pole and several marginal seas. Most of the Arctic is in continuous darkness from November to February and continuous daylight from May to August. The floor of the Arctic Ocean has two distinct parts: an extensive depression (the Arctic Basin) in the center, with an average depth of about 13,500 ft (4,000 m), and an area of continental shelf around the edges. The central basin is divided by a long submarine ridge, the Lomonosov Ridge, into two large sub-basins. These are called the Amerasia Basin (on the North American side) and the Eurasia Basin (on the European side). The continental shelf area is much wider on the Eurasian than the Amerasian side. Many islands and archipelagoes sit on the shelf area. Sea ice covers the Arctic Ocean more or less permanently above a latitude of about 75°N. In most of the rest of the ocean, sea ice occurs in winter but retreats in summer. The permanent ice cap is composed of pack ice—pieces of sea ice that are piled up into ridges that may be more than 30 ft (10 m) thick. Wind and currents (see p.390) keep the pack ice in continuous motion, sometimes causing the formation of cracks and open pools. There is some evidence that the Arctic ice cap has thinned since the 1960s, possibly by up to 40 percent, and is reducing in area. Melting of the ice cap in itself would not affect sea levels (because it is already afloat) but would strip the ocean of its reflective cover, allowing the water to absorb and retain more solar heat. Warmer waters would, in turn, speed melting. In addition to sea ice, the Arctic Ocean contains thousands of icebergs and larger bodies called ice islands. These have broken off glaciers and ice shelves, mainly around Greenland and the Canadian Arctic. The main input of water into the Arctic Ocean comes from the Atlantic Ocean (see p.395). A second input of water enters from the Pacific Ocean through the shallow Bering Strait.

**TEMPORARY NORTH POLE**
*Although the North Pole is in a geographically fixed position, signs stuck in the ice to mark its position can be accurate only temporarily because the ice drifts a few miles per day.*

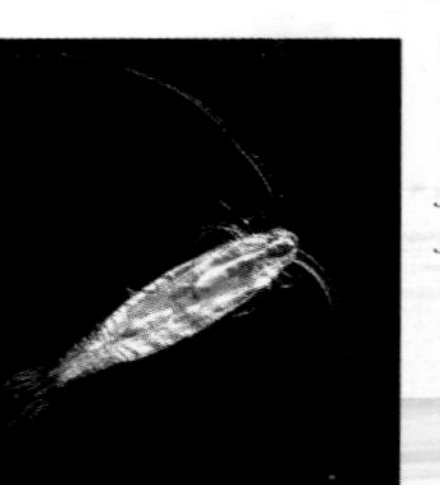

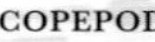

**COPEPOD**
*Small crustaceans called copepods are a key component of the Arctic food chain. They are herbivores, feeding on phytoplankton, and are themselves eaten by fish, squid, sea birds, seals, and some whales.*

**BOWHEAD WHALE BREACHING**
*The bowhead whale is found only in Arctic and sub-Arctic waters. Its broad back has no dorsal fin, which may make it easier for it to swim beneath the polar ice, to which it stays close year-round.*

### ICEBREAKERS

Icebreakers are ships that combine powerful engines, heavy displacement (weight) for their size, and reinforced hulls. A modern icebreaker can break through sea ice 6 ft (1.8 m) thick at a steady 3¾ mph (6 km/h), or through 21 ft (6.5 m) of ice by ramming. The bow of an icebreaker is shaped so that when the ship is driven forward, momentum causes it to ride up on top of the ice, which is crushed by the weight of the ship. Today, icebreakers are involved in the supply of remote settlements, creation of navigation channels for commercial ships, scientific exploration, and tourism.

### ROALD AMUNDSEN

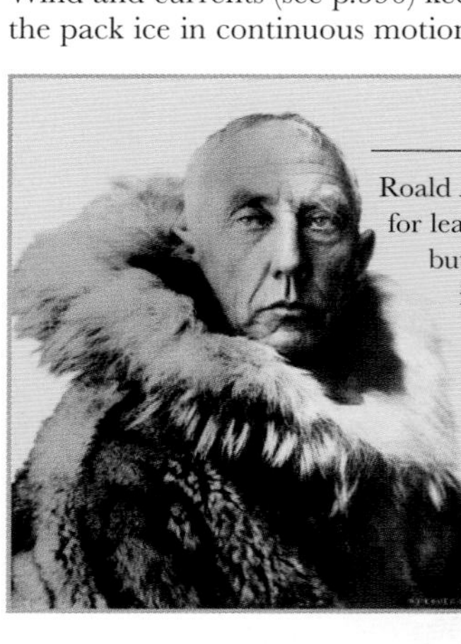

Roald Amundsen (Norway; 1872–1928) is best known for leading the first expedition to reach the South Pole, but his accomplishments in the Arctic were equally illustrious. Between 1903 and 1906, on the sloop *Gjoa*, he led the first expedition to negotiate the Northwest Passage, a long-sought-after sea route through the Canadian Arctic Archipelago, linking the Atlantic Ocean to the Pacific. As part of the expedition, Amundsen also established the position (at that time) of the North Magnetic Pole.

**ICE FACTORY**
*Disko Bay in northern Greenland, 185 miles (300 km) north of the Arctic Circle, contains one of the most productive ice floes in the northern hemisphere, averaging 20 million tons of ice per day. Icebergs are towed from here into harbors to be chipped into ice cubes.*

ARCTIC OCEAN *south*

## Barents Sea

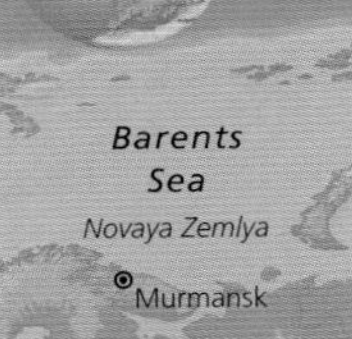


**LOCATION** North of Norway and European Russia, and south of Svalbard and Franz Josef Land

**AREA** 542,000 square miles (1.4 million square km)

**MAXIMUM DEPTH** 2,000 ft (600 m)

**INFLOWS** Norwegian Sea (Atlantic Ocean), Arctic Basin

Of all the Arctic seas, the Barents Sea is unique in that a substantial part of it is ice-free year-round. A predominantly shallow sea, it covers an area of continental shelf cut by several trenches. Ice encroaches into northern parts of the sea during the winter, and a cold current flows from the north year-round. The combination of mixing warm and cold currents and changes in surface density over the year causes frequent upwellings of nutrient-rich water, and this leads to strong plankton blooms in spring and summer. The plankton supports large populations of fish, which in turn sustain other animals and a productive fishing industry.

**UNDERWATER LIFE**
*The floor of the Barents Sea is rich in invertebrates, including feather stars, anemones, sea cucumbers, and starfish, as shown here.*

**SOUTHERN COAST**
*The Barents Sea coast of northern Norway and Russia contains many shallow bays and inlets, though in the west it is deeply indented with fjords.*

ARCTIC OCEAN *south*

## White Sea

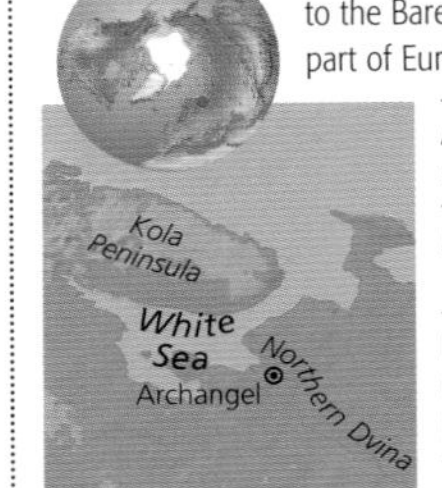


**LOCATION** Forms an inlet adjacent to the Barents Sea in the northern part of European Russia

**AREA** 35,000 square miles (90,000 square km)

**MAXIMUM DEPTH** 1,115 ft (340 m)

**INFLOWS** Barents Sea; Onega and Northern Dvina rivers

The White Sea is an almost completely landlocked body of water with an average depth of 200 ft (60 m). Its floor, which consists of continental shelf, is broken up by various troughs and ridges. The White Sea connects to the Barents Sea through a narrow strait known as the Gorlo ("throat") and to the Baltic Sea via canals and Lake Onega to its southwest. The sea can be navigated year-round, albeit with the help of icebreakers in winter. It is on a key route linking the economically active parts of northwestern Russia with the rest of the world.

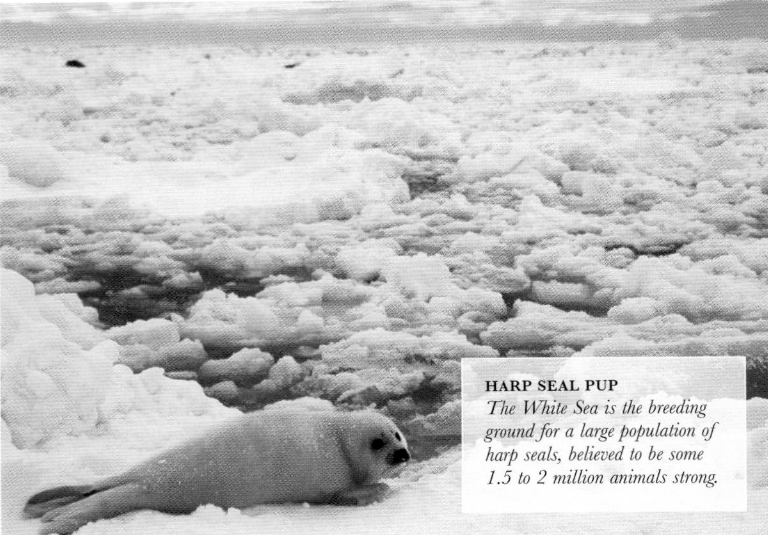

**HARP SEAL PUP**
*The White Sea is the breeding ground for a large population of harp seals, believed to be some 1.5 to 2 million animals strong.*

ARCTIC OCEAN *south*

## Kara Sea

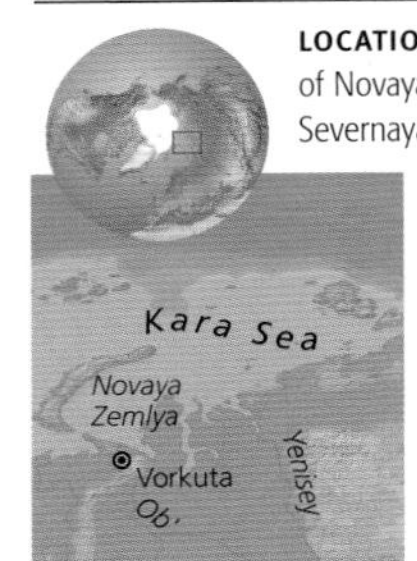


**LOCATION** Bounded by the islands of Novaya Zemlya in the west, and Severnaya Zemlya in the east

**AREA** 340,000 square miles (880,000 square km)

**MAXIMUM DEPTH** 2,035 ft (620 m)

**INFLOWS** Barents Sea, Laptev Sea; Ob' and Yenisey rivers

The Kara Sea formed as a result of deglaciation after the last ice age. Most of the sea sits over continental shelf with depths of less than 655 ft (200 m). The sea floor contains many broad terraces that step down from the southeast (where it is very shallow) to the north and west. Two deep troughs cut through the shelf. A huge volume of fresh water pours into the Kara Sea during the summer from rivers along the Siberian coast. Because of this, surface salinity varies considerably, from about 34 ppt throughout most of the sea in winter down to 10–12 ppt near the mouths of the Ob' (see p.233) and Yenisey rivers in early summer. From December to June, most of the sea is ice-covered. In contrast, between July and October, over 90 percent is ice-free, although surface temperatures never rise above 39–41°F (4–5°C). During the summer, the Kara Sea is an important route for transportation of goods into and out of western Siberia, including oil and natural gas since large deposits were discovered in the Ob'-Yenisey region. The sea is also an important fishing ground, but there are concerns about the threat to marine life posed by the dumping of nuclear waste in the recent past.

**DEEP INLETS**
*Several deep inlets of the Kara Sea cut into the mainland. Those visible at the foot of this satellite image are estuaries at the mouths of the Ob' and Yenisey rivers.*

ARCTIC OCEAN *south*

## Laptev Sea

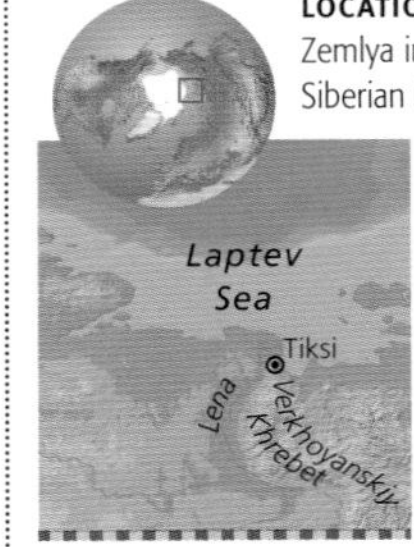


**LOCATION** Bounded by Severnaya Zemlya in the west and the New Siberian Islands in the east

**AREA** 276,000 square miles (714,000 square km)

**MAXIMUM DEPTH** 9,775 ft (2,980 m)

**INFLOWS** Kara Sea, Arctic Basin; Lena and Khatanga rivers

Of the various Arctic seas, the Laptev Sea is one of the coldest and most inhospitable. It has been called the ice factory of the Arctic because it produces most of the ice transported by the Transpolar Drift, a current that carries water and sea ice from the coasts of Siberia and Alaska across the North Pole to the Fram Strait, east of Greenland. Almost all of the Laptev Sea is ice-covered from October to May. In winter there are gales and blizzards; in the summer, frequent fogs. Much of the sea covers a shallow continental shelf, which breaks off abruptly to depths of 6,600 ft (2,000 m) or more in the far north. The sea is influenced by large amounts of runoff from large Siberian rivers, several of which form extensive deltas.

**TIKSI HARBOR**
*Tiksi is the main port on the Laptev Sea coast but is active only from June to October. For the rest of the year, the harbor is frozen over. Timber is the main cargo that goes through the port.*

ARCTIC OCEAN *south*

## East Siberian Sea

**LOCATION** Bounded by the New Siberian Islands in the west and Wrangel Island in the east

**AREA** 361,000 square miles (936,000 square km)

**MAXIMUM DEPTH** 510 ft (155 m)

**INFLOWS** Laptev Sea; Kolyma, Indigirka, and Alezeya rivers

The East Siberian Sea is the shallowest of the Arctic seas. Its average depth is just 35–65 ft (10–20 m) in its western and central parts and 100–130 ft (30–40 m) in the east. Freshwater input from the Kolyma River runs as high as 2.6 million gallons (10 million liters) per second in the spring. The salinity of the sea ranges from almost zero near the river mouth to 33 ppt in the north. It is ice-covered for most of the year, and navigation through it is possible only in August and September.

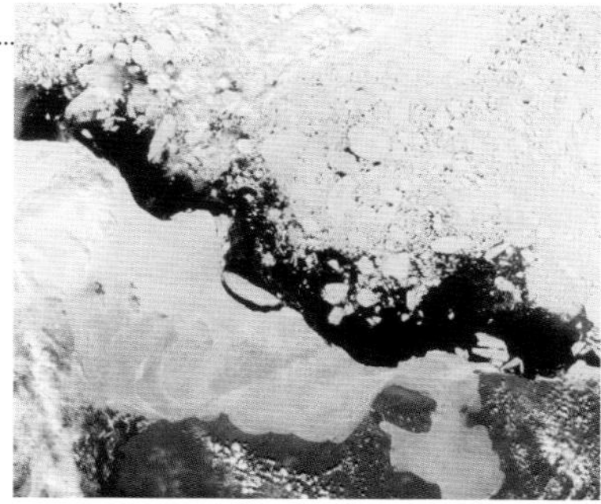

BROKEN, DRIFTING SEA-ICE (TOP RIGHT)

ARCTIC OCEAN *south*

## Chukchi Sea

**LOCATION** Lying to the northeast of eastern Siberia and to the northwest of northern Alaska

**AREA** 225,000 square miles (582,000 square km)

**MAXIMUM DEPTH** 360 ft (110 m)

**INFLOWS** Bering Sea (Pacific Ocean), East Siberian Sea, Arctic Basin

The Chukchi Sea is a mostly shallow sea situated between Alaska, Siberia, and the central Arctic Basin. The sea is fed by low-salinity waters from the Pacific Ocean via the narrow, shallow Bering Strait. This water mixes with colder and more saline water in the Chukchi Sea; the mixed water then moves into the Arctic Basin. In the northeast, a submarine channel feeds some of the mixed water into the Beaufort Sea. The Chukchi Sea is totally iced over between December and May. It supports large populations of walrus and several species of seal.

A SUMMER FOGBOW

ARCTIC OCEAN *south*

## Baffin Bay

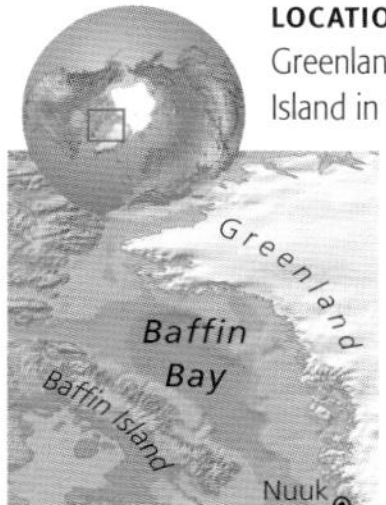


**LOCATION** Southwest of northern Greenland and northeast of Baffin Island in the Canadian Arctic

**AREA** 266,000 square miles (689,000 square km)

**MAXIMUM DEPTH** 6,900 ft (2,100 m)

**INFLOWS** Arctic Basin, Labrador Sea (Atlantic Ocean); glaciers of West Greenland

Separated from the Arctic Basin by Ellesmere Island and parts of the Canadian Archipelago, Baffin Bay is a mostly deep-water Arctic sea. From mid-August to October, the bay is largely free of sea ice, but in winter an ice cover forms and moves southeast under the prevailing winds. Combined with Arctic pack ice entering through the northern sounds, and icebergs that have broken off glaciers in northwestern Greenland, this makes the bay unnavigable for much of the year. A significant feature is the North Water polynya, an area of open water where thick ice cover would be expected. Its presence may be related to the warming effect of the West Greenland Current, which enters the bay from the south.

ICEBERGS IN DAVIS STRAIT
*Many icebergs calved from glaciers in western Greenland drift across Baffin Bay. From there, the Baffin Current moves them into the Labrador Sea.*

ARCTIC OCEAN *south*

## Beaufort Sea

**LOCATION** North of northern Alaska and northwestern Canada, bounded in the east by the Amundsen Gulf

**AREA** 184,000 square miles (476,000 square km)

**MAXIMUM DEPTH** 15,350 ft (4,680 m)

**INFLOWS** Chukchi Sea, Arctic Basin; Mackenzie and Colville rivers

The Beaufort Sea has a shallow coastal area and a deep offshore area, the Beaufort Deep. Ice covers the sea for most of the year, but it is navigable near the coast from August to October. The Mackenzie River deposits billions of tons of fresh water and 15 million tons of sediment into the sea each year.

MELTING SEA ICE

ARCTIC OCEAN *south*

## Greenland Sea

**LOCATION** Bounded by the Arctic Basin to the north, Svalbard to the northeast, and Iceland to the south

**AREA** 465,000 square miles (1.2 million square km)

**MAXIMUM DEPTH** 16,000 ft (4,800 m)

**INFLOWS** Arctic Basin, Norwegian Sea (Atlantic Ocean)

Sandwiched between the Arctic Basin to the north and the Atlantic Ocean to the south, the Greenland Sea is an important exchange region for different bodies of water moving in the global system of ocean circulation. A constant current of warm and salty Atlantic water flows north through the eastern part of the sea into the Arctic Basin. Simultaneously, enormous amounts of shallow fresh water and ice move south via the Fram Strait between Greenland and Svalbard and down the western part of the sea as the East Greenland Current. As they pass through, these waters are modified by cooling (Atlantic water) and melting (Arctic ice). In each case, the water subsequently sinks to the sea floor, when it meets the less dense water masses. From November to July, ice covers most of the sea and includes drifting Arctic pack ice, locally formed sea ice, and icebergs. Surface water temperatures range from 30°F (–1°C) in the north in February to 43°F (6°C) in the south in August. Moderately dense populations of a variety of plankton thrive in the Greenland Sea during spring and summer.

BLACK-LEGGED KITTIWAKES
*Seen here perched on an iceberg in the western part of the Greenland Sea, black-legged kittiwakes are one of the main gull species that breed on the Greenland coast.*

**NORTH ATLANTIC STORM**
*Cyclonic storms and rough seas are common in the North Atlantic above the latitude of 30°N. These wave-battered rocks are at Porto do Moniz on the island of Madeira.*

ATLANTIC OCEAN

# Atlantic Ocean

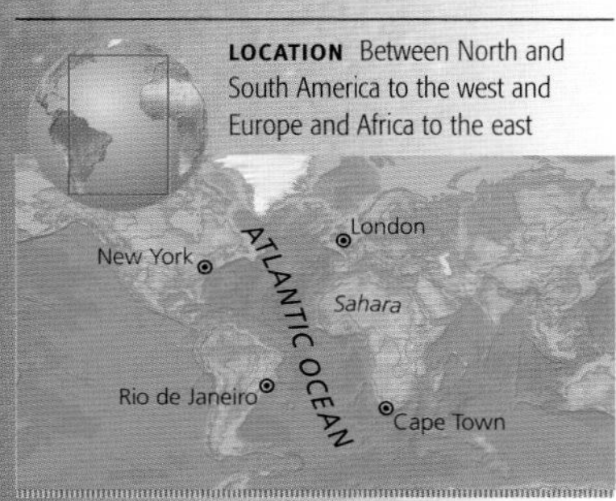


**LOCATION** Between North and South America to the west and Europe and Africa to the east

**AREA** 29.7 million square miles (77 million square km)

**MAXIMUM DEPTH** 28,230 ft (8,605 m)

**INFLOWS** Arctic Ocean, Southern Ocean; St. Lawrence, Mississippi, Orinoco, Amazon, Paraná Congo, Niger, Loire, and Rhine rivers; Mediterranean Sea

The Atlantic Ocean is the second-largest of the world's oceans, and it has several tributary seas. An immense mountain range extends from north to south along the Atlantic sea floor, covering almost a third of its area. Known as the Mid-Atlantic Ridge, this began to form about 180 million years ago, and during the Mesozoic Era, sea-floor spreading on either side of the ridge opened up the Atlantic by separating the Americas from Europe and Africa. Rising from the ridge system and its flanks are several islands, including Iceland, the Azores, and Ascension Island. On either side of the Mid-Atlantic Ridge lie basins that are smooth and flat in some areas, but disrupted in others. Some large volcanoes are found singly or in rows in the Atlantic basins. These rise to form sea mounts and, in some places, islands (such as Bermuda). The surface waters of the Atlantic Ocean move in two large gyres, a clockwise one in the North Atlantic and a counterclockwise one in the South Atlantic. In addition, there is a distinct deep-water circulation. Dense cold water formed at the poles sinks to the bottom and moves slowly at depth toward the Equator. These water masses converge and mix in the South Atlantic, where a portion of the water wells up, bringing nutrients to the surface. Prevailing weather in the Atlantic Ocean varies with latitude. The higher latitudes are characterized by variable westerly winds, changeable weather, and frequent storms. Closer to the Equator, there are belts of higher-pressure systems, and constant easterly trade winds. This region has few intense storms, though in late summer and early fall, the low latitudes of the North Atlantic are affected by hurricanes, which arise in the mid- and eastern Atlantic and move west and northwest.

**ROCKY SHORELINE**
*The Atlantic Ocean has many rocky coasts, such as this shoreline of eroded pink granite on Mount Desert Island, Maine. The glacier-sculpted island is the home of Acadia National Park.*

**ISOLATED ISLAND**
*Tristan da Cunha in the South Atlantic is the most remote inhabited island in the world. A volcano rising from 12,140 ft (3,700 m) below sea level, it is almost entirely surrounded by high cliffs.*

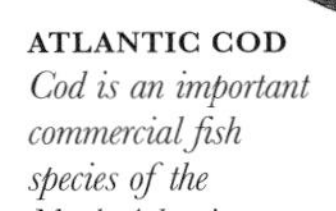

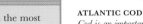

**ATLANTIC COD**
*Cod is an important commercial fish species of the North Atlantic, but overfishing has severely depleted stocks in several areas.*

**BASKING SHARK**
*Basking sharks range throughout Atlantic waters. They swim with their mouths wide open, trapping small crustaceans in their comblike gill rakers.*

## LIGHTHOUSES

The Atlantic Ocean has been the most heavily traveled of the world's oceans for hundreds of years, and many of its coasts are prone to storms and fog. To safeguard the large numbers of passing ships, thousands of lighthouses have been constructed on Atlantic coasts since the 1700s. Today, over 200 lighthouses are in active use on the US eastern seaboard alone. Lighthouses are constructed at important points on a coastline: at entrances to estuaries or harbors, and on isolated or sunken rocks or shoals. Today, most lighthouses are automated—the days of permanent lighthouse-keepers are almost gone.

ATLANTIC OCEAN *northwest*

## Hudson Bay

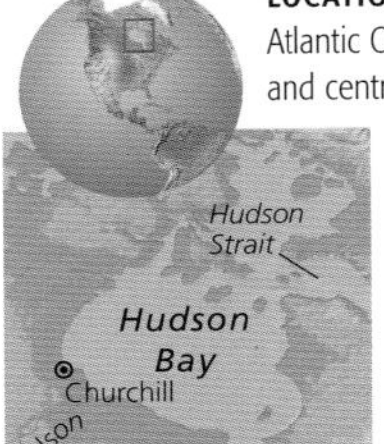


**LOCATION** A large basin from the Atlantic Ocean, extending into east and central Canada

**AREA** 316,000 square miles (819,000 square km)

**MAXIMUM DEPTH** 900 ft (270 m)

**INFLOWS** Albany, Churchill, Moose, Nelson, Severn, and Grande Rivière de la Baleine rivers

Hudson Bay is a large, shallow body of water, averaging 420 ft (128 m) in depth, which occupies a depression in the vast ancient rock mass known as the Canadian Shield. A huge area of Canada, along with small parts of Minnesota and North Dakota, drains into the bay. Because of this river input, the bay's surface waters are of low salinity. The main input of sea water comes from the northwest (through a northern part of the bay called Foxe Basin), while the main outflow is in the northeast through Hudson Strait, which connects Hudson Bay to the Atlantic. The eastern coast of the bay is rocky, with some high cliffs, but its other shores are marshy and low-lying. Hudson Bay contains a great quantity of dissolved nutrients, which allows considerable plankton growth in spring and summer. Small, shrimplike crustaceans occupy the open waters, and many invertebrates, such as mollusks and worms, live on the sea floor. Fish include Atlantic cod and Greenland halibut, as well as salmon migrating to the rivers to spawn. The shores of the bay are sparsely settled, principally by First Nations and Inuit people, who support themselves mainly by fishing and hunting.

### HENRY HUDSON

Hudson Bay is named after Henry Hudson (c.1570–1611), an English explorer who sailed into the bay in August 1610 while trying to find the Northwest Passage. After being trapped there by ice for 10 months, mutineers cast Hudson adrift in a small boat.

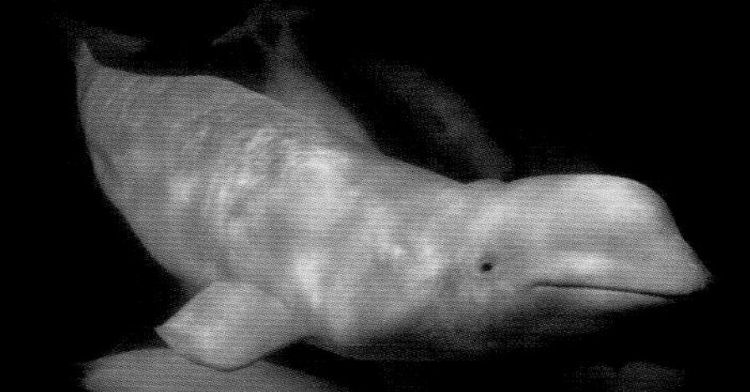

**BELUGA**
*A substantial number of belugas, or white whales, enter the bay during the summer months, where they feed on bottom-dwelling fish and invertebrates.*

**COLD WATERS**
*Ice clogs much of Hudson Bay from November until June, and surface water temperatures never rise above 49°F (9°C) even in the summer months.*

ATLANTIC OCEAN *northwest*

## Gulf of St. Lawrence

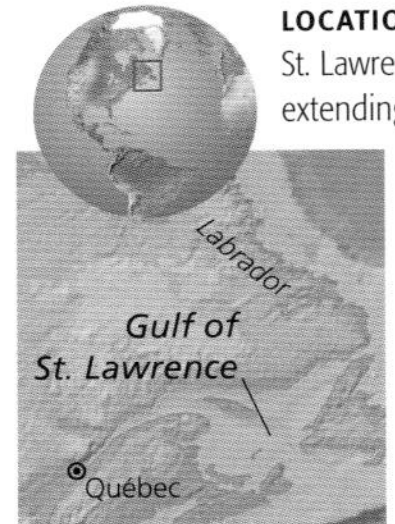


**LOCATION** At the mouth of the St. Lawrence River in eastern Canada, extending to western Newfoundland

**AREA** 60,000 square miles (155,000 square km)

**MAXIMUM DEPTH** 7,550 ft (2,300 m)

**INFLOWS** Atlantic Ocean; St. Lawrence River

The Gulf of St. Lawrence is a cold-water sea that acts as a gateway to the interior of North America. About 3.5 million gallons (14 million liters) of fresh water are discharged every second into the gulf from the St. Lawrence River. The main inflows and outflows of water from the gulf, as well as most maritime traffic, pass through Cabot Strait, which separates Newfoundland from Nova Scotia to the south. Ice floes are a prominent feature from February to April and are a hazard to ships. The deep waters of the gulf contain good fishing grounds.

**STRAIT OF BELLE ISLE**
*This northern outlet from the gulf separates the island of Newfoundland (shown here) in the south from Labrador on the Canadian mainland in the north.*

ATLANTIC OCEAN *northwest*

## Bay of Fundy

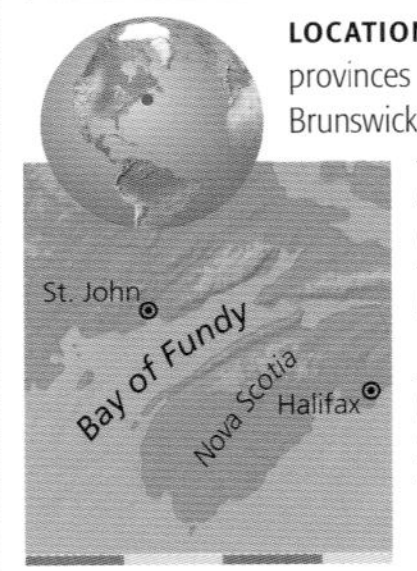


**LOCATION** Between the Canadian provinces of Nova Scotia and New Brunswick

**AREA** 3,600 square miles (9,300 square km)

**MAXIMUM DEPTH** 1,200 ft (365 m)

**INFLOWS** Atlantic Ocean; St. John River

The Bay of Fundy is best known for its fast-running tides, which can produce changes in water level as great as 70 ft (21 m), the highest in the world. High cliffs channel the bay's waters until they separate into two narrow arms at its northeastern end. In these, the tidal fluctuations are magnified by the narrowness and shape of the bay. The tidal surge in the northern arm, Chignecto Bay, produces a bore up to 6 ft (1.8 m) in height up the Petitcodiac River.

**FLOWERPOT ROCKS**
*Near Hopewell Cape, the tides have carved away the bases of cliffs, producing sandstone pillars with trees on top, known locally as the Flowerpot Rocks.*

**COASTAL SALT MARSH**
*Salt marshes, some cut through by rivers, are numerous along the 800-mile- (1,300-km-) long coastline of the Bay of Fundy.*

Over several decades, Passamaquoddy Bay on the northwest side has been the focus of studies into the feasibility of harnessing its hydroelectric potential through damming or other methods. Environmental considerations, engineering difficulties, and the huge costs involved have so far impeded any development.

OCEAN

ATLANTIC OCEAN *west*

## Sargasso Sea

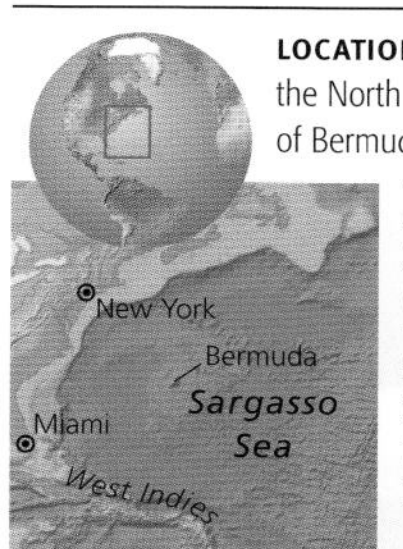

**LOCATION** In the western part of the North Atlantic Ocean, southeast of Bermuda

**AREA** 2 million square miles (5.2 million square km )

**MAXIMUM DEPTH** 23,000 ft (7,000 m)

**INFLOWS** None

The Sargasso Sea is a large, clockwise-circulating region of the North Atlantic, created by three currents around its edge: the south-moving Canaries Current to the east; the westward-moving North Equatorial Current to its south; and the Gulf Stream to its west and north. Due to the Coriolis effect (see pp.390 and 446), the movement of these currents around the perimeter of the sea forces ocean water toward its center. As a result, the middle of the Sargasso Sea is about 3 ft (1 m) above the level of water along the eastern seaboard of the US. In summer, an excess of evaporation over rainfall creates a thick lens of warm salty water in the middle of the sea. This inhibits the upwelling of nutrient-rich, colder water from the ocean bottom, and this lack of nutrients results in a sparsity of plankton. However, the mats of sargassum seaweed support a complex web of animal life.

**LIFE-SUPPORT SYSTEM**
*Eels, baby turtles, crabs, and fish such as these juvenile filefish thrive in sargassum.*

**EELS**
*All river-dwelling eels in Europe and the eastern US are born in the Sargasso Sea. Adult eels migrate there to mate when they are 10–15 years old.*

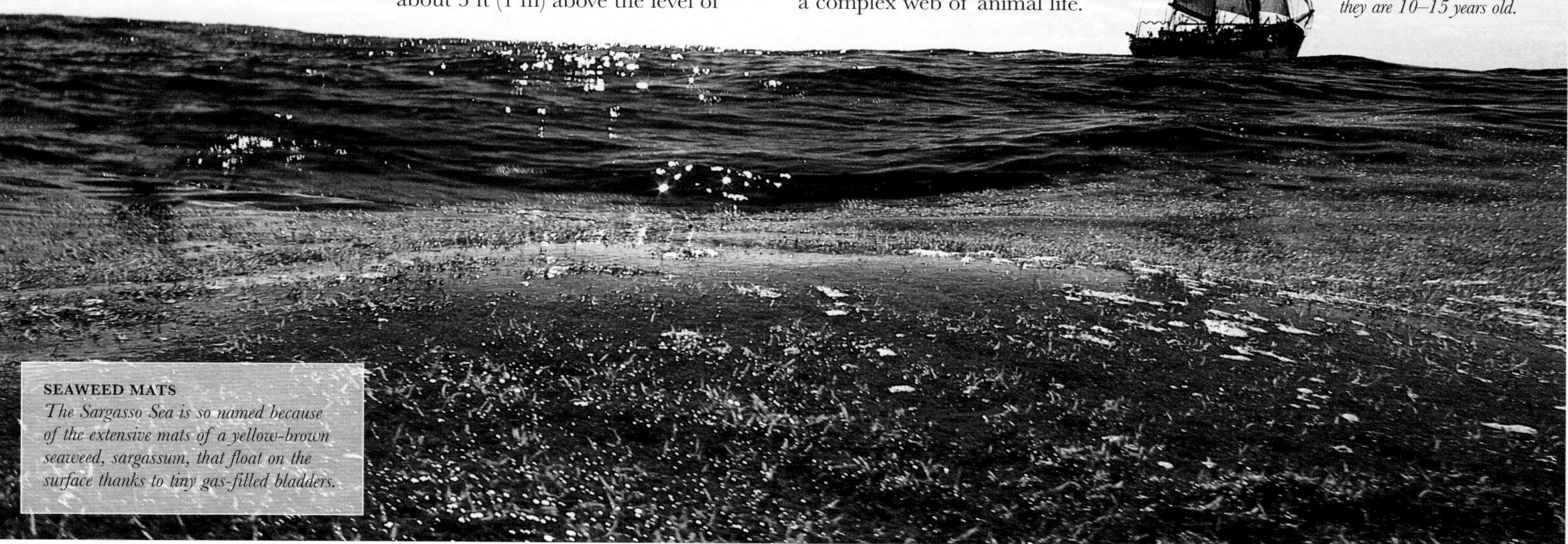

**SEAWEED MATS**
*The Sargasso Sea is so named because of the extensive mats of a yellow-brown seaweed, sargassum, that float on the surface thanks to tiny gas-filled bladders.*

ATLANTIC OCEAN *west*

## Gulf of Mexico

**LOCATION** Bounded by the southern US to the north, Florida to the east, and Mexico to the west

**AREA** 600,000 square miles (1.6 million square km)

**MAXIMUM DEPTH** 17,070 ft (5,203 m)

**INFLOWS** Caribbean Sea; rivers Mississippi, Brazos, Colorado, Alabama, Apalachicola, Rio Grande

The Gulf of Mexico is an almost landlocked body of water that is edged by a hydrocarbon-rich area of continental shelf 25–200 miles (40–320 km) wide (see panel, right). Fringing the continental shelf is a coastal zone of mangrove swamps, tidal marshes, beaches, lagoons, and estuaries. A current of warm water enters the gulf from the neighboring Caribbean Sea via the Yucatán Channel in the southeast. This exits again via the Straits of Florida into the Atlantic Ocean to the east, forming the Gulf Stream (see p.448). The Gulf of Mexico supports a large fishing industry, focused especially on shrimp but also on fish such as red snapper. A recent threat to the industry has come from over-growth of plankton in northern parts of the gulf, depleting oxygen levels in the water. The cause is thought to be heavy runoff of nitrate fertilizers from the land.

**WATERSPOUT**
*Tornadoes over water, called waterspouts, are common in the gulf. High humidity and temperatures in the range 85–95°F (30–35°C) are predisposing factors.*

### DRILLING FOR OIL

The continental shelf regions of the Gulf of Mexico contain large reservoirs of oil and gas. These resources have been explored and developed extensively since the 1940s, with numerous offshore wells drilled. They supply a substantial proportion of Mexican and US domestic energy needs.

**ROSEATE SPOONBILLS**
*These waders inhabit the gulf's coastal zone. They feed by sweeping their bills through the water and snapping them shut when they feel a shrimp or fish.*

**FLORIDA KEYS**
*The Keys are a string of low-lying islands that project into the Gulf of Mexico from the tip of Florida. They are chiefly composed of ancient coral reefs underlain by thick limestone.*

**LIVING REEF**
*Like a forest on land, this coral reef in Milne Bay, Papua New Guinea, has a complex structure that provides homes and food for a wealth of animals.*

# THREATS TO CORAL REEFS

Coral reefs are extremely rich marine habitats and by far the largest structures created by living things. However, because they grow slowly and have stringent environmental requirements, they are easily harmed by human activity. At least 20 percent of the world's coral reefs have been so badly damaged that they are effectively dead. Without urgent action, 40 percent of the reefs may disappear in the next 30 years—a situation that conservationists are fighting hard to prevent.

## REEFS IN DANGER

Corals are soft-bodied animals that protect themselves by building chalky skeletons. The skeletons last far longer than the animals themselves, and they build up to form reefs. The surface of a reef is like a living skin. It contains millions of coral animals, or polyps, and even larger numbers of microscopic algae called zooxanthellae. These live inside the polyps, and they help to nourish their hosts by harnessing the energy in sunlight (see pp.392–93).

In a healthy reef, some branching corals grow more than 8 in (20 cm) a year. Their upward spread is limited only by the sea itself, because few corals can stand more than an hour of exposure at low tide. This growth offsets the damage caused by storms and by coral-eating animals, so the reef expands or maintains its size. But if anything harms either the polyps or their zooxanthellae, this delicate balance is upset. Growth slows and the surface of the reef begins to erode. If the coral does not recover, the entire ecosystem begins to decline, because corals provide reef animals with food and habitats.

**DEAD ZONE**
*Staghorn corals are the fastest-growing coral and among the most easily damaged. They are seen here littering the sea bed in the Caribbean.*

## CAUSES OF DECLINE

Coral reefs have always been used as a local resource, and they can be seriously harmed by overfishing or by removing the coral itself for personal or commercial gain. They can also be affected by pollution from an area far away.

Coral requires clear, nutrient-poor water, because this favors their zooxanthellae and keeps competing free-living algae in check. Conversely, anything that makes seawater more turbid or more fertile can suppress their growth. In many parts of the tropics, farming and logging have increased soil erosion, producing sediment-laden runoff, which finds it way into the sea. This shades the coral, robbing it of the light that its partner zooxanthellae need to survive. Reefs are also affected by coastal and tourist development, because sewage raises water nutrient levels, causing overgrowths of algae. Sewage treatment can prevent this, but adequate funds are often lacking. Even in developed countries, many coastal communities still discharge raw sewage into the sea.

**STRANGERS ON THE SHORE**
*In coastal areas with coral reefs, tourism has to be carefully managed to avoid creating environmental damage.*

## RAISED SEA TEMPERATURES

Despite these threats, local reef damage can be reversed. In protected areas, some reefs have been quick to recover when pollution and fishing are controlled. But coral reefs are also threatened by global warming (see pp.452–53)—a worldwide phenomenon that is far harder to control.

When sea temperatures are raised, coral reefs are harmed. Coral polyps are put under stress, and so they respond by expelling their algae, in a process called coral bleaching. Bleaching reverses itself if the water then cools, but if raised temperatures continue for a prolonged period, the coral polyps die.

Measures are underway to reduce global warming, but results are still far off. For the world's coral reefs, the 21st century will be a critical time.

**SCATTERED REMAINS**
*Fragments of old coral litter a beach at Puako, Hawaii, and contrast with the black volcanic sand.*

**BLEACHING CASE HISTORY**
*At Molasses Reef in the Florida Keys, the recent record of coral bleaching has been closely studied. Here, bleaching has occurred when the maximum water temperature in summer reaches 88°F (31°C). Some bleaching years have been accompanied by outbreaks of coral disease.*

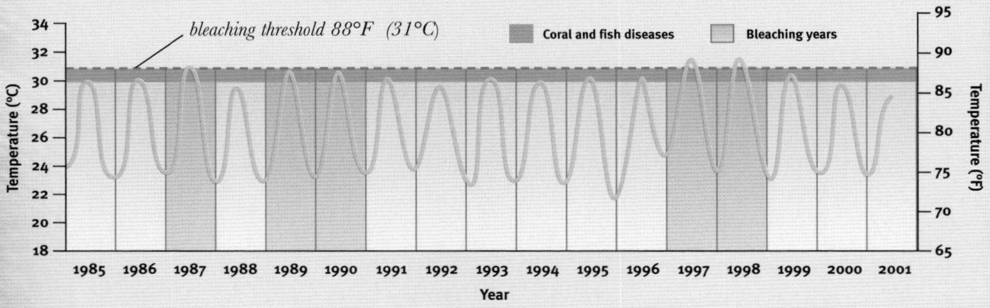

ATLANTIC OCEAN *west*

## Caribbean Sea

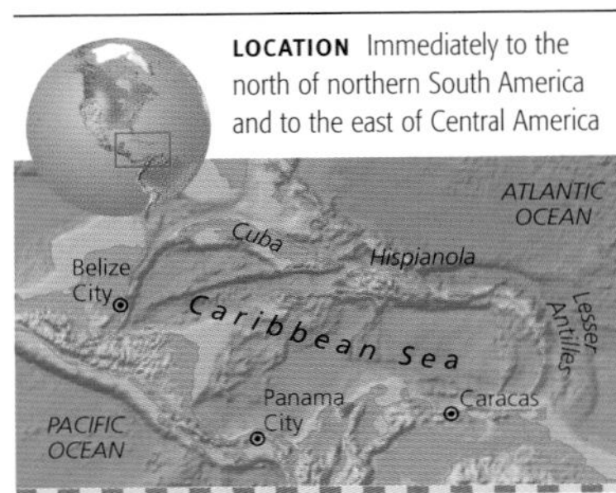


**LOCATION** Immediately to the north of northern South America and to the east of Central America

**AREA** 1.1 million square miles (2.75 million square km)

**MAXIMUM DEPTH** 25,215 ft (7,685 m)

**INFLOWS** Atlantic Ocean; Magdalena, Coco, Patuca, and Motagua rivers

A popular resort area, the Caribbean Sea is best known for its warm tropical climate, turquoise waters, and beautiful white beaches. Beneath the surface, the sea occupies five submarine basins. From southeast to northwest, these are the Grenada, Venezuelan, Colombian, Cayman, and Yucatán basins. The islands of the West Indies, which fall into three main groups, lie in a curve that extends mainly around the northern and eastern perimeter of the sea. The Lesser Antilles are a string of about 20 small volcanic islands that form an arc across its eastern boundary. The four large islands of Cuba, Hispaniola, Jamaica, and Puerto Rico in the northern part of the sea are known as the Greater Antilles. The Windward Passage, a major shipping route between the US and the Panama Canal, passes between Cuba and Hispaniola. In the southern part of the Caribbean Sea, near the coast of Venezuela, there is a separate group of three islands: Aruba, Curaçao, and Bonaire. The main surface current in the Caribbean Sea is an extension of the North Equatorial Current. Surface water enters the sea in the southeast, through channels between the Lesser Antilles, flows in a northwesterly direction, and exits through the Yucatán Channel in the northwest (between Cuba and Mexico's Yucatán Peninsula) into the Gulf of Mexico. Trade winds from the northeast dominate the region, and tropical storms occur in late summer and fall in the northern Caribbean, sometimes reaching hurricane velocity. Most of the islands, as well as some of the mainland coasts, are fringed with coral reefs that support large numbers of fish, and invertebrates including spiny lobsters and conches. These reefs have, however, suffered considerable damage over the past 30 years through a combination of storms, coral disease, increased tourism, seawater temperature rise, and coastal development.

**BEACH IN CUBA**
*One of the Caribbean's main assets is its beaches, which are important for many reasons, not just tourism. They protect coasts from wave action, especially during hurricanes, and provide boat landing sites and material for construction. In recent decades, beach erosion has become a major concern.*

### PANAMA CANAL

Much of the shipping in the Caribbean is en route to or from the Panama Canal, which links the Atlantic to the Pacific. A huge feat of engineering, the 40-mile- (64-km-) long canal opened in 1914 after decades of construction. Over 30,000 workers died during the project from malaria and yellow fever.

### MANATEE CONSERVATION

The Caribbean is home to a subspecies of manatee that was once common but is now endangered. The West Indian manatee is threatened by loss of habitat, hunting, entanglement in nets, and collisions with boats. In Belize, one of its remaining strongholds, a project is underway to track the mammal, educate local people about its conservation, and reduce destruction of its habitat.

**SOFT AND STONY CORALS**
*The Caribbean Sea contains both soft corals, such as the sea fan in the background of this picture, and stony corals, such as the purple coral (called blue crust coral) growing in front of it.*

**GREAT BLUE HOLE**
*Situated on Lighthouse Reef, east of Belize, this sinkhole (see p.253) through submerged limestone is 410 ft (125 m) deep. It formed in the last ice age, when sea levels were lower.*

ATLANTIC OCEAN *northeast*

## Norwegian Sea

**LOCATION** To the west of Norway, north of the North Sea; bordered by the Greenland Sea

**AREA** 534,000 square miles (1.4 million square km)

**MAXIMUM DEPTH** 13,020 ft (3,970 m)

**INFLOWS** Central North Atlantic Ocean; numerous Norwegian fjords

The Norwegian Sea is the most northerly of the marginal seas of the Atlantic Ocean. Although cut through by the Arctic Circle, it is kept free of ice by the warm Norway Current, a branch of the North Atlantic Current. Strong tidal currents on two areas of the Norwegian coast cause notable whirlpools: the Lofoten Maelstrom near the tip of the Lofoten Islands; and the Saltstraumen near Bodø.

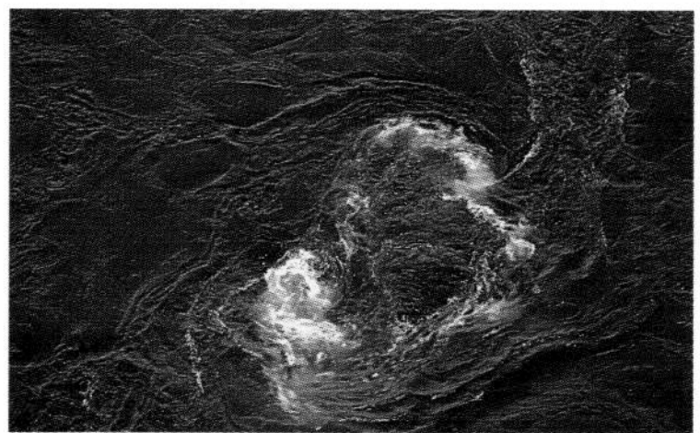

SALTSTRAUMEN WHIRLPOOL

ATLANTIC OCEAN *northeast*

## North Sea

**LOCATION** Extending from the east coast of Great Britain toward the mainland of northwestern Europe

**AREA** 220,000 square miles (570,000 square km)

**MAXIMUM DEPTH** 2,300 ft (700 m)

**INFLOWS** Central North Atlantic Ocean; rivers Elbe, Weser, Ems, Rhine, Scheldt, Thames, Humber

About 1.6 million years ago, much of the North Sea's southern half was part of the European mainland, and in the intervening period, ice sheets advanced and retreated across the region several times. Even today, on the sea floor off northern England, there are large banks of glacial moraine (see p.262), such as the Norfolk and Dogger banks. The main inflow of water is the warm North Atlantic Current from the west. Colder, less saline waters enter from the Baltic Sea via the Skagerrak channel between Norway and Denmark. The constant mixing of cold and warm waters leads to a rich supply of nutrients. As a result, the North Sea has productive fisheries, although stocks have declined markedly in the past 30 years. Since the 1950s, the North Sea has become a major site of offshore oil and gas production.

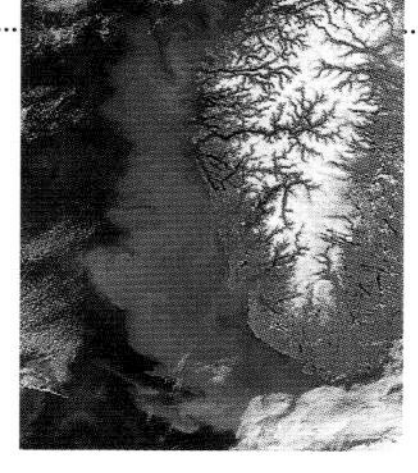

**PLANKTON BLOOM**
*A large bloom of plankton fogs the waters in a 100-mile- (160-km-) wide band off the coast of southern Norway.*

**RECLAIMED LAND**
*Areas of reclaimed land in the Netherlands, called polders, are kept safe from the sea by dikes.*

ATLANTIC OCEAN *northeast*

## Bay of Biscay

**LOCATION** In an indentation of the western European coastline, north of Spain and west of France

**AREA** 86,000 square miles (223,000 square km)

**MAXIMUM DEPTH** 15,525 ft (4,735 m)

**INFLOWS** Loire, Dordogne, Garonne, and Adour rivers

The Bay of Biscay lies partly over an area of continental shelf and partly over an abyssal plain. A number of islands dot the French area of the shelf. Navigation in the bay is often difficult because of strong northwesterly winds that cause rough seas. The coasts vary from rocky cliffs to areas of lagoon and marsh. Fishing and oyster culture are important industries here.

COAST OF BELLE ILE, FRANCE

ATLANTIC OCEAN *northeast*

## Black Sea

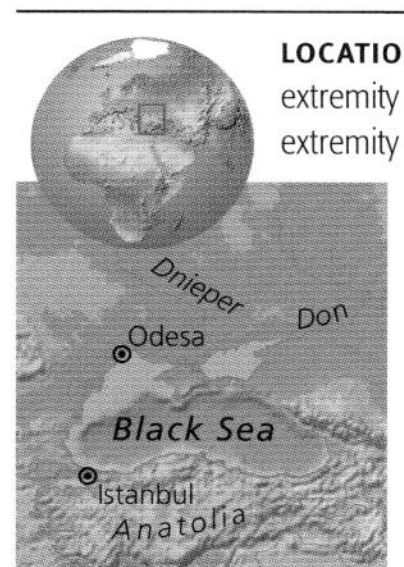


**LOCATION** At the southeastern extremity of Europe and the western extremity of Asia

**AREA** 163,000 square miles (422,000 square km)

**MAXIMUM DEPTH** 7,200 ft (2,200 m)

**INFLOWS** Mediterranean Sea, Sea of Azov; rivers Danube, Dniester, Dnieper, Kizil Irmak

An almost completely landlocked body of water, the Black Sea mostly occupies a deep, broad submarine basin separating Europe from Asia. It connects to the Mediterranean Sea in the southwest by means of two narrow channels—the Bosporus and the Dardanelles—and the intervening Sea of Marmara. A separate enclosed sea, the Sea of Azov, lies to the northeast.

BLACK SEA COAST OF TURKEY

ATLANTIC OCEAN *northeast*

## Baltic Sea

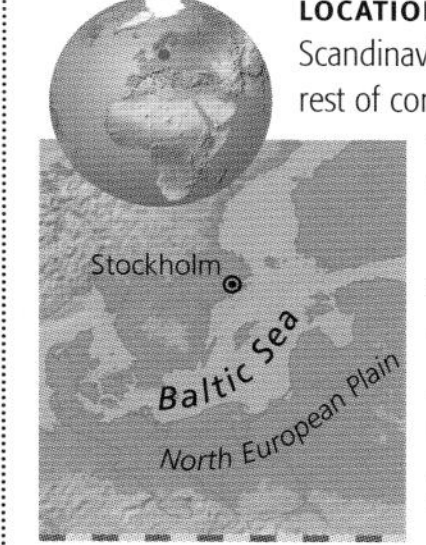


**LOCATION** Separating the Scandinavian Peninsula from the rest of continental Europe

**AREA** 149,000 square miles (386,000 square km)

**MAXIMUM DEPTH** 1,473 ft (449 m)

**INFLOWS** Vistula, Oder, and Western Dvina rivers

One of the world's largest areas of brackish (moderately salty) water, the Baltic Sea is a remnant of a huge water-covered region that formed about 8,000 years ago as the ice sheet that covered Scandinavia retreated toward the Arctic. Today, the Baltic is a semi-enclosed sea, which receives freshwater drainage from a large area of northern Europe. The Baltic Sea links to the North Sea on its western side, and there is a constant outflow of low-salinity water and, at depth, an inflow of saltier water. Winters are long and cold, summers short and cool, and storms are frequent. Pack ice accumulates in the Gulf of Bothnia in the north in winter and early spring, when navigation is suspended. With coasts on nine different countries, the Baltic Sea is of great commercial importance to northern Europe.

**CRETACEOUS LIMESTONE**
*Cliffs of limestone are a common feature in the Baltic, such as on the east coast of the Danish island of Møn, shown here.*

### WIND FARMING

The Baltic Sea has several offshore wind farms, including the world's first, built at Vindeby, Denmark, in 1991, and the Bockstigen wind farm off the Swedish island of Gotland (below). These projects have shown that offshore wind farming is an economically viable method for producing renewable energy, and many more farms are planned. The main objection to such farms is that they are a potential hazard to birds.

**ALAND, FINLAND**
*The low-lying Åland Islands rise from a narrow strait between Finland and Sweden. They mark the entrance to a northern arm of the Baltic called the Gulf of Bothnia.*

ATLANTIC OCEAN *northeast*

## Mediterranean Sea

**LOCATION** To the south of Europe and western Turkey and to the north of North Africa

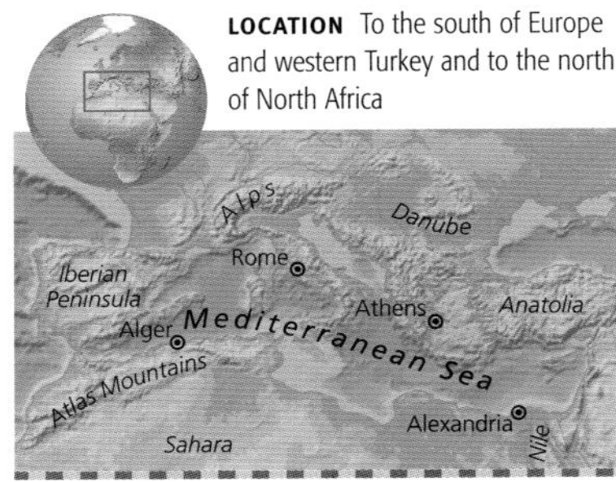


**AREA** 970,000 square miles ( 2.5 million square km)

**MAXIMUM DEPTH** 16,000 ft (4,900 m)

**INFLOWS** Atlantic Ocean; Black Sea; Nile, Rhône, Po, and Ebro rivers

**OCTOPUS ON SEA FLOOR**
*Although Mediterranean marine life has been depleted over the past 40 years through human activity, some species, such as octopuses, remain common.*

The world's largest inland sea, the Mediterranean occupies an elongated depression between Europe and Africa that originally formed tens of millions of years ago. About 6 million years ago, a fall in sea level cut the Mediterranean off from the Atlantic Ocean. During the next 2 million years, the depression repeatedly flooded and dried out, leaving some thick salt deposits. Finally, a rise in sea level allowed Atlantic waters to flood back in. Today, the Mediterranean Sea remains linked to the Atlantic Ocean via the Straits of Gibraltar in the west, a channel just 8 miles (13 km) wide and 1,050 ft (320 m) deep. It also connects to the Red Sea via the Suez Canal in the southeast (see p.414) and to the Black Sea in the northeast. Underwater, the Mediterranean sea floor is composed of a series of moderately deep basins separated by sills (see p.195). One of the main sills runs between the island of Sicily and Cape Bon in Tunisia, dividing the sea into a western half and a larger eastern half. The Mediterranean Sea also contains many smaller regions lying over shallower water, such as the Adriatic Sea to the east of Italy and the Gulf of Gabes off the eastern coast of Tunisia. Large islands in the Mediterranean include Crete and Cyprus in the eastern half, Sardinia, Corsica, and Majorca in the western half, and Sicily at the toe of Italy. The Aegean Sea, between Greece and Turkey, is dotted with the many

### JACQUES COUSTEAU

As the inventor of scuba diving, and through his films and books about the underwater world, Frenchman Jacques Cousteau (1910–97) made a huge contribution to subaquatic exploration in the 20th century and to public awareness of the marine environment. The Mediterranean Sea was the site of both Cousteau's original experiments with scuba gear and some important discoveries, such as the unearthing in 1957 of an ancient Greek wine freighter, buried beneath the sea floor off Marseilles.

**LAVA FORMATIONS**
*Along with some other Aegean Islands, Lesbos has seen considerable volcanic activity in the past, as evidenced by these lava formations on its coast.*

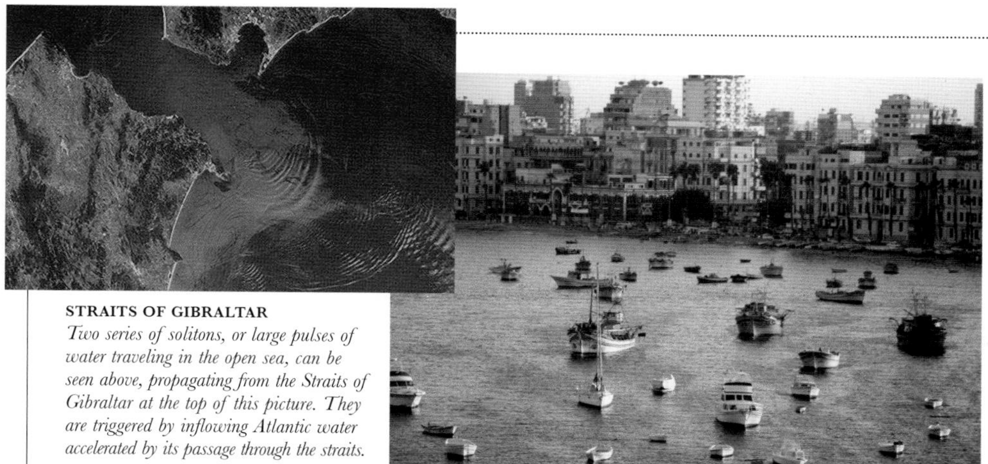

**STRAITS OF GIBRALTAR**
*Two series of solitons, or large pulses of water traveling in the open sea, can be seen above, propagating from the Straits of Gibraltar at the top of this picture. They are triggered by inflowing Atlantic water accelerated by its passage through the straits.*

**ALEXANDRIA**
*One of the busiest Mediterranean ports is Alexandria, located on the edge of the Nile Delta. A former capital of Egypt, Alexandria now handles over 80 percent of the country's exports and imports.*

islands of the Greek Archipelago. Although a large amount of fresh water flows into the Mediterranean Sea from rivers, the total amount is only about a third of the water lost by evaporation. As a result, the sea is highly saline. There is also a continuous inflow of surface waters of normal salinity into the Mediterranean from the Atlantic via the Straits of Gibraltar, and to a lesser extent, from the Black Sea. These inflows are partly offset by outflows of saltier water at depth. The dominant current in the Mediterranean Sea is eastward, flowing along the north of the African coast, a continuation of the current that brings seawater in from the Atlantic. This current is most powerful in summer, when evaporation is at its strongest. Tides are generally weak throughout the whole region, two exceptions being the Gulf of Gabes and the Adriatic Sea. The Mediterranean Sea has long been important as a route for trade, (see panel, right) and remains so today, particularly since the opening of the Suez Canal. Marine mammals, fish, sponges, corals, and other invertebrates were once plentiful. However, a rise in population and the expansion of industry and tourism over the past 40 years have led to severe pollution of many of the coastal areas. This, together with overfishing, has severely depleted some fish stocks and other marine life.

## SEA FLOOR TREASURE

Trading has been conducted by sea in the Mediterranean for over 3,000 years. As a result of navigational errors, storms, and wars, the present-day sea floor is littered with ancient wrecks, providing many sites of archaeological interest. Here, a diver brings a basket from the surface to collect the remains of bowls, bottles, and cups from a wreck that sank off the coast of Turkey in the 11th century AD. The oldest known shipwreck in the Mediterranean (at Ulu Burun, also off the coast of Turkey) dates from the 14th century BC.

INDIAN OCEAN

## Indian Ocean

**LOCATION** Between Africa and Australia, extending from southern Asia to the Southern Ocean



**AREA** 26.5 million square miles (69 million square km)

**MAXIMUM DEPTH** 23,812 ft (7,258 m)

**INFLOWS** Ganges, Indus, Tigris, Euphrates, Zambezi, Limpopo, and Murray rivers

The Indian Ocean is one of the youngest of the world's ocean basins and the most complex in terms of the structure of its floor. Both the Indian and Southern Oceans have formed over the past 120 million years, following the breakup of the ancient continent of Gondwana. The Indian Ocean narrows toward the north, where the Indian Peninsula divides it into two large bodies of water, the Arabian Sea on the west and the Bay of Bengal on the east. The most striking feature of the ocean floor is a massive mid-ocean ridge, shaped like an upside-down Y, which occupies about a fifth of the ocean's total floor area (see p.397). One arm runs to the south of Africa and joins the Mid-Atlantic Ridge. The other runs to the south of Australia, where it eventually joins the Pacific–Antarctic Ridge. The two ridge arms join roughly in the center of the ocean and continue north as the Mid-Indian Ridge. Creation of new oceanic crust at the various ridges is continuing to separate Africa, Antarctica, and Australia from each other. Along an extensive arc at its northeast margin, the Indian Plate is subducting beneath the southeast edge of the Eurasian Plate to form the Indonesian volcanic arcs of Sumatra and Java. A continental shelf extends to an average distance of about 75 miles (120 km) from the landmasses around the ocean. In places, the shelf and continental slope are penetrated by deep canyons carved by outflows from the Ganges, Indus, and Zambezi rivers, which have also deposited vast amounts of sediment onto the abyssal plains. The northern part of the Indian Ocean has a monsoon climate, characterized by rain-bearing winds that change direction twice yearly (see p.463). These changes are accompanied by twice-yearly reversals in the pattern of ocean currents, something that distinguishes the northern part of the Indian Ocean from all other oceans. The monsoon zone is also prone to destructive tropical storms that form over the open ocean and head in a westward direction. South of the

**BANDED SEA SNAKE**
*These snakes are common in shallow lagoons in the eastern part of the Indian Ocean. Their tails act like rudders, making them highly efficient swimmers.*

**MACARONI PENGUINS**
*The Indian Ocean extends as far south as 60°S. On a remote southern island, Kerguelen, there are many colonies of mammals and birds, including this colorful species of penguin.*

**SCULPTED ROCKS**
*These curious granitic rock formations, on the coast of the island of La Digue in the Seychelles, have been carved by a combination of wind, rain, and seawater erosion.*

monsoon zone lies the trade-winds zone, characterized by steady southeasterlies that prevail throughout the year. A third zone lies in the southern part of the Indian Ocean, between the latitudes of 30°S and 60°S. Here, there are light and variable winds in the northern part, and strong westerlies in the south. The warm waters that persist throughout much of the Indian Ocean provide an ideal environment for a rich variety of marine life. Some 4,000 species of fish live near the shores, many unique to this ocean. Coral reefs are abundant around all islands in the tropical areas, as well as along the western coasts of Thailand and Burma (Myanmar) and the east coast of Africa.

## ZHENG HE

One of the first explorers to travel widely across the Indian Ocean was a Chinese admiral, Zheng He (1371–1435). Accompanied by a huge fleet of ships, he visited most parts of the region, including India, Arabia, and east Africa, during 7 separate voyages in 1405–1433. His objectives included trading in goods, gathering knowledge about the region, and extending the sphere of Chinese political influence.

**STILT FISHING**
*This method of fishing is still practiced today in parts of Sri Lanka and Thailand. The fishermen and -women perch on poles in shallow water to avoid scaring away the fish.*

INDIAN OCEAN *northwest*

# Red Sea

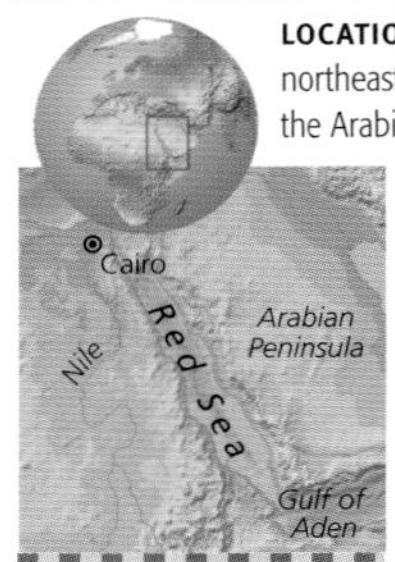


**LOCATION** Between the northeastern corner of Africa and the Arabian Peninsula

**AREA** 175,000 square miles (450,000 square km)

**MAXIMUM DEPTH** 9,975 ft (3,040 m)

**INFLOWS** Arabian Sea (via the Gulf of Aden)

**GULF OF SUEZ**
*The turquoise waters of the Red Sea are completely surrounded by desert and, behind a coastal plain, mountains that rise as high as 6,500 ft (2,000 m).*

**ANEMONE GARDENS**
*The thousands of animal species found on Red Sea reefs include many giant anemones. The clownfish swimming close by secrete a mucus that protects them from the stinging tentacles.*

One of the warmest of the world's seas, the Red Sea occupies a depression formed by the East African Rift (see p.148), the primary branch of the East African Ridge System, which has developed over the past 50 million years. Hydrothermal vents on the sea floor are evidence of continuing tectonic activity. The Sinai Peninsula splits the northern end of the sea into two arms: the shallow Gulf of Suez on the west, and the deeper Gulf of Aqaba on the east. Hardly any rain falls on the Red Sea and no rivers flow into it, but water evaporates at a high rate due to the strong solar radiation. As a result, the Red Sea is highly saline. To make up for evaporation losses, surface water flows in from the Gulf of Aden in the south and continues as a northward-moving surface current. The Red Sea contains a rich marine life based on extensive coral reefs. Since the 1970s, this marine life has attracted increasing numbers of recreational scuba divers, whose activities have caused great damage to the reefs.

**SEA ROSE**
*This curious flowerlike object, commonly seen attached to rocks at shallow depths, is actually a ribbon of eggs that was laid by a nudibranch, or sea slug.*

## SUEZ CANAL

First opened to shipping in 1869, the Suez Canal connects the Gulf of Suez at the northern end of the Red Sea to the Mediterranean Sea. One of the busiest waterways in the world, the 101-mile- (163-km-) long canal provides a short cut for maritime traffic traveling between ports in North America and Europe and those in the Middle East, southern Asia, East Africa, and Oceania. The canal has no locks because the Gulf of Suez has roughly the same sea level as the Mediterranean Sea.

INDIAN OCEAN *northwest*

## Persian Gulf

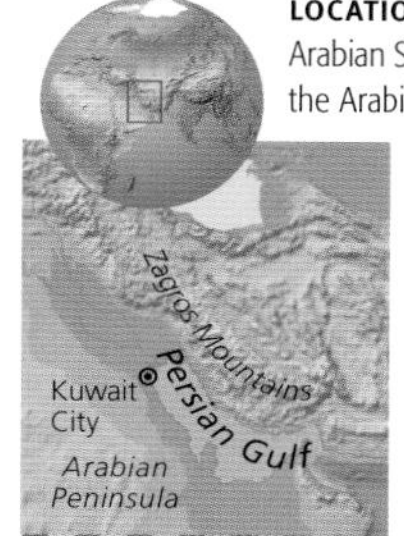


**LOCATION** Forms an inlet of the Arabian Sea, with Iran to the east and the Arabian Peninsula to the west

**AREA** 93,000 square miles (241,000 square km)

**MAXIMUM DEPTH** 360 ft (110 m)

**INFLOWS** Tigris, Euphrates, Karun rivers

Sometimes called the Arabian Gulf or just the Gulf, the Persian Gulf is best known for the large deposits of oil and natural gas that lie around its shores and beneath its thin floor. The region contains approximately two-thirds of the world's proven oil reserves and a third of its proven natural gas reserves. A warm and salty sea, the gulf receives a large inflow of both water and sediments from the Shatt al 'Arab, a river formed from the confluence of the Tigris and Euphrates rivers in the northwest (in Iraq). Its eastern shore is mountainous but its western shore is low-lying, with many islands, lagoons, and tidal flats. After oil, fishing is the most important industry in the Persian Gulf.

### DESALINATION

The Gulf region receives little rainfall so water is scarce, though there are limited groundwater reserves. To avoid water shortages, over 20 desalination plants have been built on the western side of the Persian Gulf. Together they produce over 1.3 billion gallons (5 billion liters) of desalinated water per day. They return warm, concentrated seawater back into the Persian Gulf.

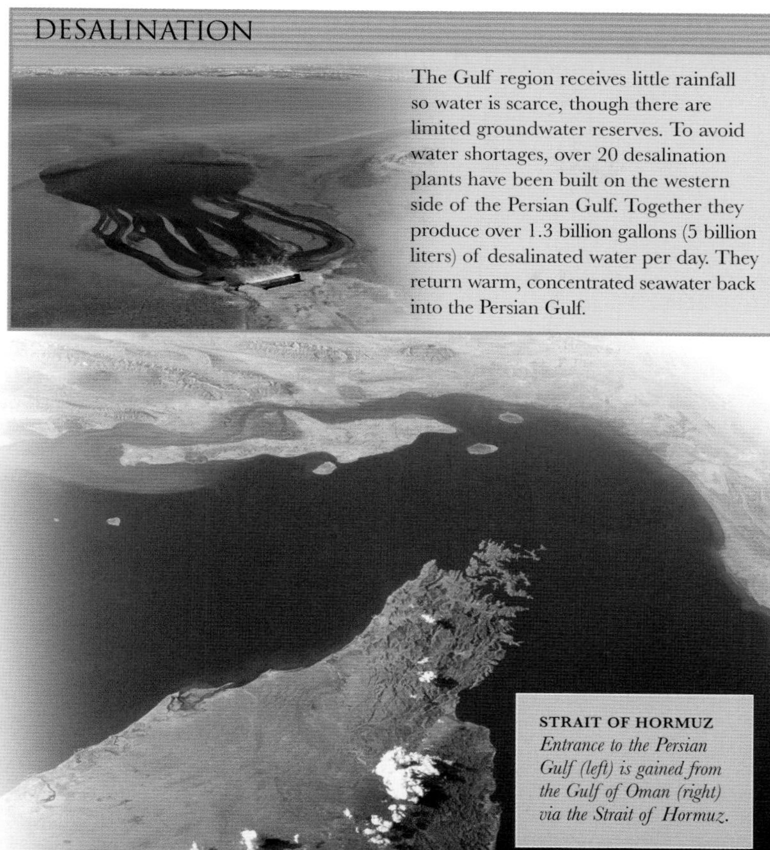

**STRAIT OF HORMUZ** *Entrance to the Persian Gulf (left) is gained from the Gulf of Oman (right) via the Strait of Hormuz.*

INDIAN OCEAN *north*

## Arabian Sea

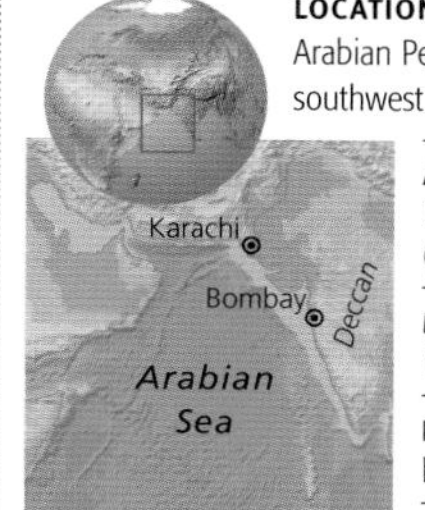


**LOCATION** To the east of the Arabian Peninsula and to the southwest of the Indian Peninsula

**AREA** 1.5 million square miles (3.9 million square km)

**MAXIMUM DEPTH** 19,038 ft (5,803 m)

**INFLOWS** Indus and Narmada rivers

The Arabian Sea occupies the western part of the northern Indian Ocean. Two northern extensions, the Gulf of Aden and Gulf of Oman, connect it to the Red Sea and Persian Gulf, respectively. Various island groups lie around its margins, including the Seychelles to the southwest and the Maldives to the southeast. The sea supports an extensive fishing industry.

**FISHERMEN ON DHOWS**

INDIAN OCEAN *west*

## Mozambique Channel

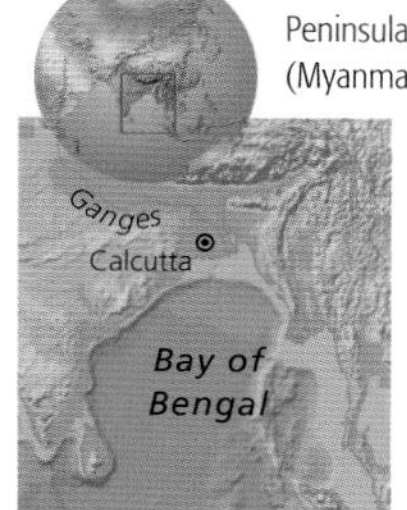


**LOCATION** Between Mozambique on the African mainland to the west and Madagascar to the east

**AREA** 386,000 square miles (1 million square km)

**MAXIMUM DEPTH** 9,800 ft (3,000 m)

**INFLOWS** Zambezi and Rio Lúrio rivers

An important shipping route for the countries of east and southern Africa, the 620-mile- (1,000-km-) long Mozambique Channel is fed by the waters of the Zambezi and all the major rivers of Madagascar. For many decades, it was believed that a continuous warm current, the Mozambique Current, passed down through the channel. Recently it has been shown that no such current exists. Rather, water movements in the area are dominated by a series of counterclockwise eddies that reach to the channel bottom.

### THE COELACANTH

The Mozambique Channel is one of just two places in the world with an identified population of coelacanths, primitive fish that were believed to have been extinct for 65 million years until, in 1938, one was caught off the east coast of South Africa. Coelacanths live at depths greater than 500 ft (150 m), where they feed on small bottom-dwelling fish and cuttlefish.

**PIROGUE BEACHED ON NOSY BE** *The island of Nosy Be lies just off mainland Madagascar, on the eastern side of the Mozambique Channel. Pirogues are boats of a simple construction, typically used by Madagascans for fishing.*

INDIAN OCEAN *north*

## Bay of Bengal

**LOCATION** To the east of the Indian Peninsula and to the west of Burma (Myanmar) and the Malay Peninsula

**AREA** 1.1 million square miles (2.9 million square km)

**MAXIMUM DEPTH** 15,400 ft (4,695 m)

**INFLOWS** Rivers Ganges, Mahanadi, Godavari, Krishna, Kaveri, Irrawaddy

The Bay of Bengal occupies the northeastern corner of the Indian Ocean. Its smaller eastern part, the Andaman Sea, is separated from the rest of the bay by two island groups, the Andaman and Nicobar islands. The Bay of Bengal is bordered by a wide area of continental shelf cut through by canyons, and its floor is a vast abyssal plain. A distinctive feature at its northern extremity is the wide and thick fan of sediments from the Ganges River (see p.234). Intense tropical storms with high winds and torrential rain occur in spring and fall. Combined with inflow from rivers during the summer monsoon, these markedly reduce the surface salinity of the bay's waters.

**LIMESTONE ROCK OUTCROPS** *On the Andaman Sea coast of Thailand, at Phang-Nga, there are many spectacular undercut limestone cliffs.*

**KO PANYEE, THAILAND** *This picturesque fishing village rests precariously on wooden piles driven into the floor of the bay.*

PACIFIC OCEAN

# Pacific Ocean

**LOCATION** Between Asia and Australia to the west and North and South America to the east

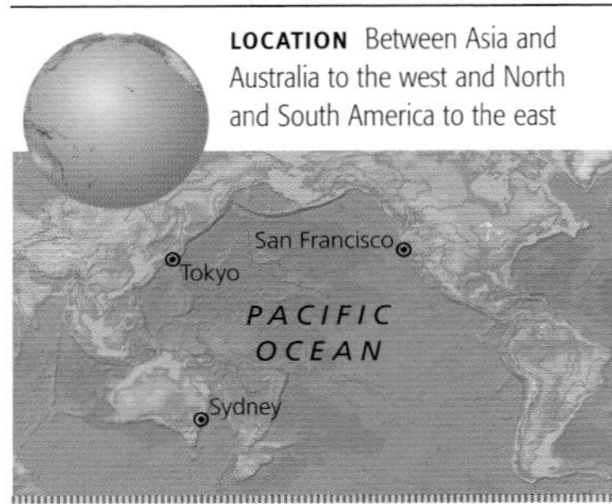


**AREA** 60 million square miles (156 million square km)

**MAXIMUM DEPTH** 35,840 ft (10,924 m)

**INFLOWS** Southern Ocean; Yukon, Columbia, Amur, Yellow, Yangtze, and Mekong rivers

By far the largest ocean, the Pacific covers more than a third of the Earth's surface and holds more than half of its liquid water. It also contains the deepest point in the world's oceans, the Challenger Deep in the Mariana Trench. About three-quarters of the Pacific Ocean lies over a single tectonic plate, the Pacific Plate. Marking the boundary between this and the adjacent oceanic Nazca Plate in the southeastern Pacific is a large mid-ocean ridge, the East Pacific Rise. Associated with plate movements away from this ridge, and occurring all around the Pacific Ocean's Ring of Fire (see p.175), are various deep-sea trenches marking subduction zones with abundant volcanic and earthquake activity. The eastern part of the Pacific Ocean has high mountain ranges on the continental side for most of its length. It has a narrow continental shelf, few large rivers discharging into it, and a regular coastline, except for fjords in the extreme north and south. In contrast, the western Pacific has an irregular continental boundary, many large marginal seas, huge rivers flowing in off the Asian mainland, and some wide areas of continental shelf. The surface waters over most of the Pacific Ocean move in two large gyres (see p.390). The main deep-water circulation is a northward spread of cold water at depth. This reaches as far north as Japan and then moves to the east. The main winds of the Pacific are twin belts of westerlies (blowing from west to east) between latitudes of 30° and 60°, both north and south of the Equator. Between the westerlies are the more predictable trade winds, which blow toward the Equator. Typhoons affect two large areas of the western Pacific in late summer and early fall. There are numerous islands in the Pacific Ocean, many of them part of volcanic island arcs. Under the Nazca Plate off Ecuador is a volcanic hot spot that created the Galápagos Islands. Over millions of years, a remarkable collection of animals and plants reached this isolated spot from South America and then diversified on the different islands (see p.134). Charles Darwin's study of this unique fauna and flora in the 1830s influenced the formation of his ideas about evolution.

**MARINE IGUANA**
*One of many unusual animals found in the Galápagos Islands, the marine iguana is the only lizard that forages underwater for food.*

## ARTIFICIAL ISLANDS

Faced with a severe shortage of suitable land for building, Japan has turned to its Pacific coastal areas as potential construction sites and created several artificial islands. There are now several in Tokyo Harbor and two near the port city of Kobe. Artificial islands are usually built by enclosing an area of shallow water with a seawall and then filling it in. Similar projects are now planned in other parts of the world, including the Baltic and Mediterranean seas.

**KANSAI AIRPORT**
*This airport was constructed on an island built some 3 miles (5 km) off the coast of Osaka, Japan, between 1987 and 1993.*

**PACIFIC BREAKERS**
*The Pacific has the world's largest fetch, the area over which wind can interact with surface water to create waves. This results in some immense breakers, like these in Hawaii.*

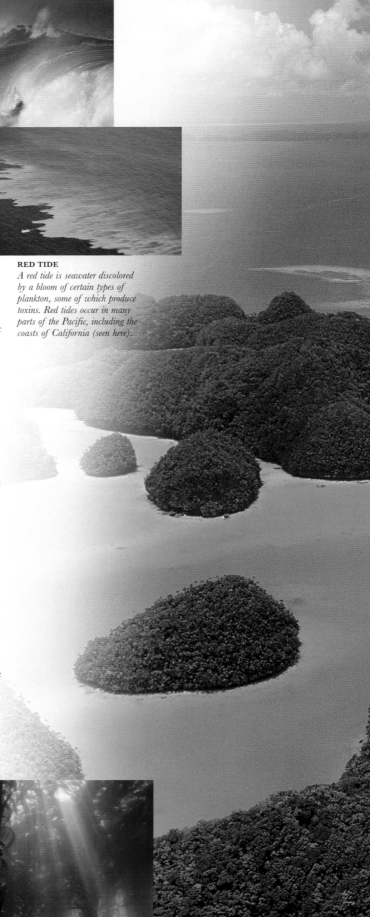

**RED TIDE**
*A red tide is seawater discolored by a bloom of certain types of plankton, some of which produce toxins. Red tides occur in many parts of the Pacific, including the coasts of California (seen here).*

**GIANT KELP FOREST**
*The northeastern coasts of the Pacific are covered in forests of giant kelp. Among the world's fastest-growing plants, they thrive along rocky reefs.*

**ISLANDS IN PALAU, MICRONESIA**
*Consisting of coral structures built on top of a submarine ridge, these are typical islands of the west Pacific, except for the way they have been uplifted and eroded into humped shapes.*

PACIFIC OCEAN *north*

## Bering Sea

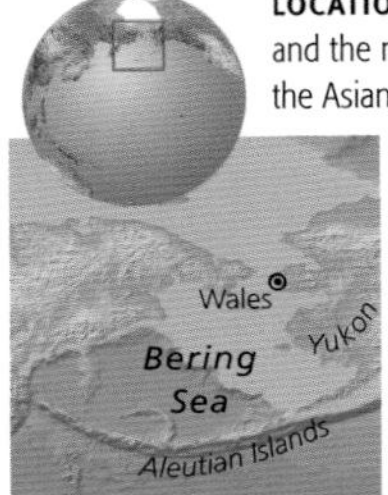


**LOCATION** Between western Alaska and the northeastern extremity of the Asian landmass

**AREA** 890,000 square miles (2.3 million square km)

**MAXIMUM DEPTH** 13,442 ft (4,097 m)

**INFLOWS** Central North Pacific Ocean; Yukon and Anadyr' rivers

The Bering Sea is the most northerly marginal sea of the Pacific Ocean, connecting to the Arctic Ocean in the north via the Bering Strait. In the south, the sea is fringed by an arc of volcanic islands, the Aleutians. In the southwest, the sea floor is separated by ridges into three basins. The Bering Sea has a severe climate, and in winter its northern part is covered in ice. It has a high biological productivity, however, with explosions of plankton growth occurring twice each year. The plankton supports large numbers of fish, mammals such as sea otters, and on the continental shelf area in the northeast, invertebrates such as mollusks, starfish, and sponges.

SEA ICE BREAKING UP IN SPRING

PACIFIC OCEAN *northeast*

## Gulf of Alaska

**LOCATION** Lying to the south of Alaska, between Kodiak Island and Cape Spencer

**AREA** 592,000 square miles (1.5 million square km)

**MAXIMUM DEPTH** 16,500 ft (5,000 m)

**INFLOWS** Susitna, Copper rivers; numerous large iceberg-producing glaciers

The Gulf of Alaska is a roughly triangular body of water, the apex of which is in Prince William Sound near the port of Valdez, North America's northernmost ice-free harbor. Two sides of the triangle extend from there to Cape Spencer (near the entrance to Glacier Bay) in the southeast and to Kodiak Island and the Alaskan Peninsula to the southwest. A warm current, the Alaska Current, moves through the gulf up the southeastern coast of Alaska, and a continuation, the Alaska Stream, runs out past the Alaskan Peninsula toward the central North Pacific. The southern part of the gulf lies over a deep abyssal plain, and in the southwest this plunges into the eastern part of the Aleutian Trench, to the southeast of Kodiak Island. There are several deep fjords and inlets along the Alaskan Gulf Coast. Many large glaciers, such as the Hubbard Glacier (see p.266), discharge icebergs into these inlets, from where they slowly float out to sea.

### THE EXXON VALDEZ

The *Exxon Valdez* disaster of 1989, caused when an oil tanker ran aground in Prince William Sound, was the world's worst-ever oil spill in terms of damage to the environment. Fourteen years later, toxic chemicals were still seeping into the Gulf of Alaska from contaminated beaches.

**BREACHING ORCA**
*Orcas, or killer whales, are a common sight in the Gulf of Alaska, preying on seals, sea lions, fish, squid, and occasionally sea birds and sea otters.*

**BAHIA DE LOS ANGELES**
*This area of the Gulf of California, surrounded by desert, has been compared to the Galápagos Islands because of its diverse marine life.*

PACIFIC OCEAN *northeast*

## Gulf of California

**LOCATION** Between the west coast of mainland Mexico and the peninsula of Baja California

**AREA** 62,000 square miles (160,000 square km)

**MAXIMUM DEPTH** 10,000 ft (3,050 m)

**INFLOWS** Fuerte, Sonora, Yaqui, Colorado rivers

Also known as the Sea of Cortés, the Gulf of California is a 750-mile- (1,200-km-) long basin from the Pacific Ocean into the west coast of Mexico. It lies over the boundary between the Pacific and North American plates and is thought to have formed by sea-floor spreading over the past 6 million years. The Gulf of California consists of two distinct regions. The northern part is mostly shallow—due to the accumulation of silt transported by the Colorado River (see p.218) over thousands of years—and has a large tidal range. The southern part is much deeper and less affected by tides, and it includes the Guaymas Basin, which has many hydrothermal vents. The large amount of sunlight in the region, combined with the nutrients provided by upwelling, allows high plankton growth, which supports some rich fishing grounds. Unfortunately, the gulf's environment is now threatened by overfishing, unregulated tourism development, and insufficient fresh-water flow (most water from the Colorado River is now diverted for residential and agricultural use).

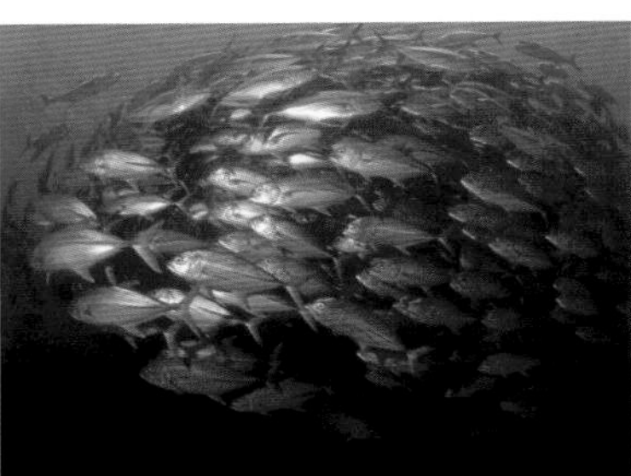

**BALL OF JACK FISH**
*Various types of jack are among the 800 species of fish found in the gulf. The balling behavior seen here is thought to be a defense against predators.*

PACIFIC OCEAN *northwest*

## Sea of Okhotsk

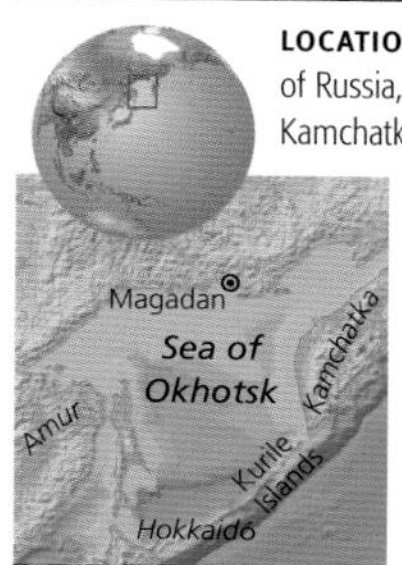


**LOCATION** Off the far eastern coast of Russia, to the west of the Kamchatka Peninsula

**AREA** 611,000 square miles (1.6 million square km)

**MAXIMUM DEPTH** 11,063 ft (3,372 m)

**INFLOWS** Sea of Japan; Amur, Uda, Okhota, and Penzhina rivers

The Sea of Okhotsk is a large, cold sea that freezes over during the winter months and is frequently covered with fog. Except where it laps the northernmost Japanese island of Hokkaido in the south, it is surrounded by Russian territory: the Kamchatka Peninsula in the east; the Russian mainland and the island of Sakhalin in the west; and the Kurile Island arc in the southeast. The floor of the sea slopes gently from north to south, then plunges to depths greater than 10,000 ft (3,000 m) to the northwest of the Kuriles. Tides and currents in the sea are strong, with a tidal range reaching 43 ft (13 m) in some locations. The Sea of Okhotsk is exceptionally important to Russia as a fishery resource. Oil and gas deposits have also been found on the continental shelf area in the northern part of the sea.

**THE WORLD'S LARGEST EAGLE**
*The coasts of the Sea of Okhotsk are a breeding ground for Steller's sea eagle.*

**ISLAND IN THE KURILES**
*The volcanic origin of Ushishur Island is clear in this view of its lake-filled central crater.*

PACIFIC OCEAN *northwest*

## Sea of Japan

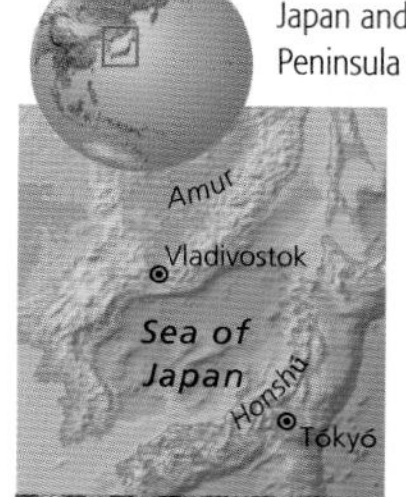


**LOCATION** To the northwest of Japan and to the east of the Korean Peninsula and southeastern Siberia

**AREA** 377,600 square miles (978,000 square km)

**MAXIMUM DEPTH** 12,276 ft (3,742 m)

**INFLOWS** East China Sea; Tumen, Ishikari, Shinano, Agano, Mogami, and Teshio rivers

The Sea of Japan is almost completely enclosed by mainland Asia and the islands of Japan. A current of warm water (the Tsushima Current) flows into the sea from the East China Sea to the south. Outflow is mainly into the Sea of Okhotsk in the north, via the La Pérouse Strait, and into the Pacific Ocean to the east through the Tsugaru Strait. The northern part of the sea freezes in winter. Occasional upwellings of nutrient-rich water encourage strong plankton blooms.

### FISHING FOR SQUID

During summer and fall, large fishing fleets set off into the Sea of Japan to fish for squid. Bright halogen lamps, specially waterproofed, are lowered underwater or floated on the surface to attract the squid, which feed at night. The lights are so brilliant that the fleets can be seen from space.

**RUGGED COASTLINE**
*The coasts around the Sea of Japan are mostly rocky, as here on the island of Honshu.*

PACIFIC OCEAN *northwest*

## Yellow Sea

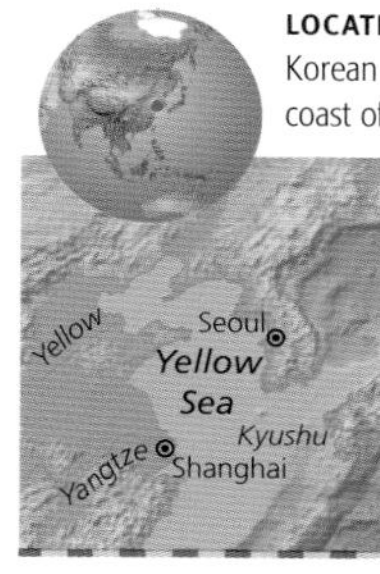


**LOCATION** To the southwest of the Korean Peninsula, extending to the coast of east-central China

**AREA** 205,000 square miles (530,000 square km)

**MAXIMUM DEPTH** 338 ft (103 m)

**INFLOWS** Yellow, Yangtze, Liao He, Luan He, Yalu, and Han rivers

The Yellow Sea is a shallow, flat-bottomed, and partly enclosed inlet of the Pacific Ocean. Its main bulk lies to the southwest of the southern part of the Korean Peninsula (see p.152). Farther north, there are two extensions of the sea: Korea Bay in the northeast; and the Gulf of Bo Hai in the northwest, into which flow the waters of the Yellow River. The Yellow Sea derives its name from the color of the silt-laden water that is discharged into it from the Yellow River, the Yangtze, and other major rivers. Most of the Yellow Sea floor is continental shelf that only 10,000 years ago was dry land or the estuary of the Yellow River. There are many sand shoals near the coast of China, and several small islands lie off the coast of Korea. A warm current flows from the southeast into the middle of the Yellow Sea, and there are southward-flowing currents along the coasts. Tides in the sea are considerable, especially along the west coast of the Korean Peninsula, with a maximum tidal range of 27 ft (8 m). The surface temperature of the sea varies remarkably, from close to freezing in the north in winter to as high as 82°F (28°C) in some of the shallower parts during the summer. The Yellow Sea is particularly rich in bottom-dwelling fish, and this is exploited by fleets of trawlers from China, Japan, and North and South Korea.

**GREEN TURTLE**
*The Yellow Sea is the northern limit of the range of the green turtle, which lives near coasts and grazes on algae and sea grass.*

**SILT-LADEN WATERS**
*In this satellite image of the Yellow Sea, the Yellow River can be seen discharging its load of silt at top left, as can the Yangtze at bottom left.*

PACIFIC OCEAN *west*

## East China Sea

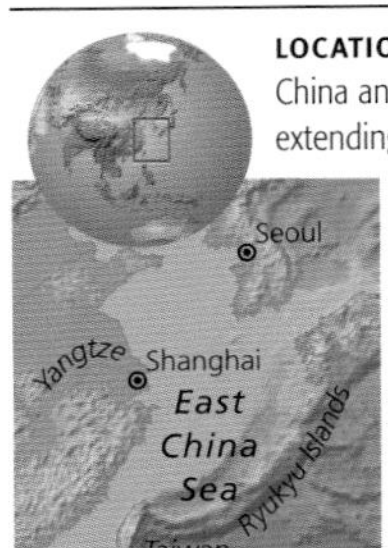


**LOCATION** Between east-central China and the Ryukyu Islands, extending south to Taiwan

**AREA** 290,000 square miles (751,000 square km)

**MAXIMUM DEPTH** 8,912 ft (2,717 m)

**INFLOWS** South China Sea, Philippine Sea; Yangtze River

The East China Sea is a shallow sea with an average depth of just 1,145 ft (349 m). It is separated from the Yellow Sea (see p.419) to the north by an imaginary line that runs from Cheju Island, off the tip of the Korean Peninsula, to the mouth of the Yangtze River in mainland China. To the south, it connects to the South China Sea (see below) through a shallow strait, the Taiwan Strait, between mainland China and the island of Taiwan. Much of the western part of the sea covers an area of continental shelf. Its deepest section, the Okinawa Trough, stretches for several hundred miles along the western side of the Ryukyu Islands in the east. A warm current (a branch of the Kuroshio Current) enters the East China Sea to the east of Taiwan and moves north, with branches penetrating into the Yellow Sea and into the Sea of Japan (see p.419) to the northeast. During summer, warm, moist winds from the western Pacific bring heavy rain to the East China Sea, sometimes accompanied by typhoons (hurricanes). In winter, cold, dry winds blow across the sea from the Asian mainland. Fishing is a major source of income in the region, the main catches being anchovy, tuna, shrimp, mackerel, and the eel-like cutlass fish. The East China Sea is the main shipping route from the South China Sea to ports in Japan and elsewhere in the North Pacific.

**TYPHOON**
*Hurricanes in the western Pacific are known as typhoons. Many form to the east of the Philippines and then move northwest over the East China Sea.*

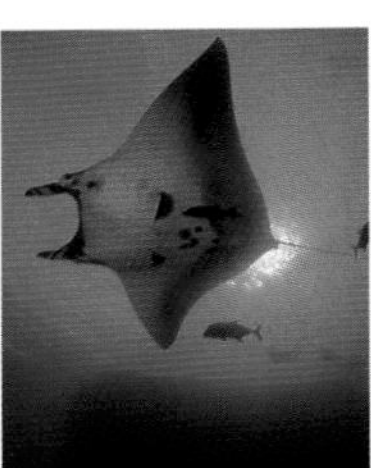

**MANTA RAY**
*Manta rays are most often encountered near the surface of lagoons. These gigantic fish, measuring up to 23 ft (7 m) across, swim by flapping their large pectoral fins.*

**HYOPCHAE BEACH ON CHEJU**
*This idyllic beach is on the west coast of Cheju Island in the northern part of the East China Sea. Offshore are some of the world's most northerly coral reefs.*

PACIFIC OCEAN *west*

## South China Sea

**LOCATION** Bounded by southern China, eastern Southeast Asia, the Philippines, and Indonesia

**AREA** 1.4 million square miles (3.7 million square km)

**MAXIMUM DEPTH** 16,457 ft (5,016 m)

**INFLOWS** Xi Jiang, Mekong, Red, Tha Chin, and Chao Phraya rivers

The largest marginal sea of the western Pacific, the South China Sea stretches over 1,700 miles (2,700 km) around the southeast edge of the Asian mainland. Two offshoots of the sea, the Gulf of Tongking and the Gulf of Thailand, indent Southeast Asia. The northeastern part of the sea, close to the Philippines, lies over a deep, diamond-shaped submarine basin, of which the most southerly part is called the South China Basin. North of this lies a broad area of continental shelf that extends for up to 150 miles (240 km) from the coast of China and Vietnam and includes the Taiwan Strait and Gulf of Tongking. To the south, off southern Vietnam, the shelf narrows and then connects with the Sunda Shelf, one of the largest areas of continental shelf in the world, which underlies the whole of the southern part of the South China Sea. This shelf is dotted with many small islands and island groups. The weather in the South China Sea is marked by monsoons and summer typhoons. Annual rainfall is as high as 160 in (4,000 mm) on parts of the Sunda Shelf. Both fishing and shipping are of high economic importance. Fish caught in the South China Sea supply as much as 50 percent of the animal protein consumed by the people of Southeast Asia. The major ports and commercial centers around the sea include Hong Kong, Manila, Hô Chi Minh City, Singapore, and Bangkok.

**WHALE SHARK**
*The world's largest fish at over 12 tons, these plankton-feeders are commonly sighted in some areas of the South China Sea, such as in western parts of the Philippines.*

### TEK SING PORCELAIN

A famous shipwreck in the South China Sea is that of the *Tek Sing*, an unusually large Chinese junk that ran aground on a coral reef 300 miles (500 km) southeast of Singapore in 1822, while en route to Java. The ship was carrying about 350,000 pieces of Chinese porcelain, much of which has been salvaged by divers and subsequently auctioned, following discovery of the wreck site in 1999.

**ANG THONG NATIONAL PARK**
*Lying in the Gulf of Thailand, this marine park consists of 42 forest-topped islands that were created by sea-level rise.*

PACIFIC OCEAN *west*

## Sulu Sea

**LOCATION** Between northeastern Borneo (Malaysia) and the central islands of the Philippines

**AREA** 100,000 square miles (260,000 square km)

**MAXIMUM DEPTH** 18,400 ft (5,600 m)

**INFLOWS** Celebes Sea

The Sulu Sea is one of several warm seas around Borneo. It connects in the northwest to the South China Sea via passages on either side of the island of Palawan, part of the Philippines. In the southeast, the Sulu Sea is fringed by the Sulu Archipelago, comprising over 4,000 small coral islands. Pearl fishing is often carried out from stilt houses built on top of the reefs.

LIMESTONE CLIFF, PALAWAN

PACIFIC OCEAN *west*

## Banda Sea

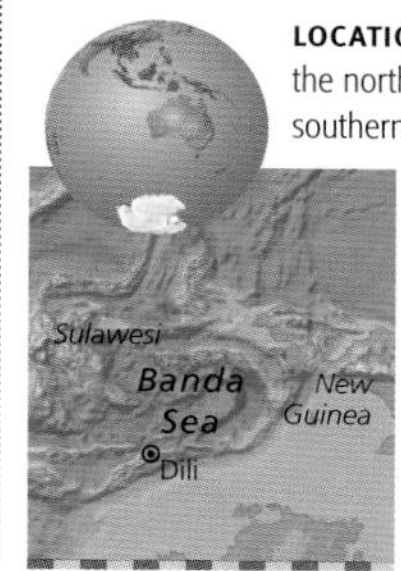


**LOCATION** Bounded by Sulawesi to the northwest and the central and southern Maluku Islands to the east

**AREA** 180,000 square miles (470,000 square km)

**MAXIMUM DEPTH** 24,410 ft ( 7,440 m)

**INFLOWS** Molucca Sea, Arafura Sea

The Banda Sea is a warm sea that laps the shores of the Maluku Islands or Moluccas, formerly the Spice Islands, of Indonesia. It connects to numerous other marginal seas, including the Molucca Sea in the north, and the Arafura Sea in the southeast, via straits between various island groups. The people of these islands are still involved in the cultivation of spices such as cloves and nutmeg, as well as fruit, coconuts, and tapioca.

NEIJALAKKA ISLAND, MALUKU, INDONESIA

PACIFIC OCEAN *southwest*

## Tasman Sea

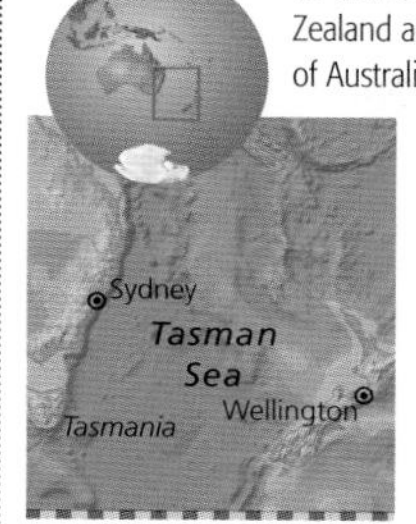


**LOCATION** Lying between New Zealand and the southeastern coast of Australia

**AREA** 900,000 square miles (2.3 million square km)

**MAXIMUM DEPTH** 19,500 ft (5,945 m)

**INFLOWS** Southern Ocean; Coral Sea

The most southerly of the marginal seas of the Pacific, the Tasman Sea lies in the belt of westerly winds known as the Roaring Forties and is noted for its storminess. The sea merges with the Coral Sea (see below) in the north, and in the southwest connects to the Indian Ocean via the Bass Strait (see p.390) between Tasmania and Australia. Its southern part overlies a broad, deep basin, the Tasman Basin, while the northeastern part covers an elongated trough, the New Caledonia Depression. These two deep areas are separated by a ridge called the Lord Howe Rise. Lord Howe Island lies near the boundary of this rise and the Tasman Basin. A dominant current in the western part of the sea is the warm, southward-moving East Australia Current. During winter and spring, colder water moves up from the south. The Tasman Sea's resources include oil deposits in the Gippsland Basin at the eastern end of Bass Strait.

VOLCANIC COAST
*Lord Howe Island is the eroded remnant of a 7-million-year-old volcano located in the northern part of the Tasman Sea. The world's most southerly coral reef surrounds the island.*

PACIFIC OCEAN *southwest*

## Coral Sea

**LOCATION** To the southeast of Papua New Guinea, east of northeast Australia and west of Vanuatu

**AREA** 1.8 million square miles (4.8 million square km)

**MAXIMUM DEPTH** 30,070 ft (9,165 m)

**INFLOWS** West Central Pacific Ocean; Fly, Purari, and Kikori rivers

The Coral Sea is a large marginal sea that extends from northeastern Australia some 800 miles (1,300 km) east to the island chain of Vanuatu (formerly the New Hebrides). A major feature is the Great Barrier Reef, the world's largest coral reef, which stretches for 1,250 miles (2,010 km) along the east coast of Queensland. However, the reef occupies only a small area of the whole Coral Sea. To its east, and north of latitude 20°S, there are many individual reefs and small islands, known collectively as the Coral Sea Islands Territory, which rise from an area called the Coral Sea Plateau. To the northeast of the Coral Sea Plateau lies a large, deep basin, the Coral Sea Basin, and there are other deep basins much farther east, toward Vanuatu. Tectonic plate boundaries lie at the northern and eastern margins of the sea, where the Indian and Australian plates are being forced beneath adjoining plates. These boundaries are marked by deep trenches, notably the Vanuatu Trench. The main current in the Coral Sea is the South Equatorial Current, which flows from the east. As it approaches Queensland, it splits into two branches, one of which moves down the east coast of Australia as the East Australia Current while the other deflects north toward Papua New Guinea and then back east across the northern part of the Coral Sea. The sea has a subtropical climate and is subject to typhoons, especially from January to April. The Coral Sea's marine life is highly diverse and includes hundreds of species of stony (reef-building) corals, over 4,000 types of mollusks, and thousands of species of fish.

CLOSED ANEMONE
*This giant anemone, with attendant clownfish, is one of several thousand species of cnidarians (corals, sea anemones, jellyfish, and hydroids) that are found in the Coral Sea.*

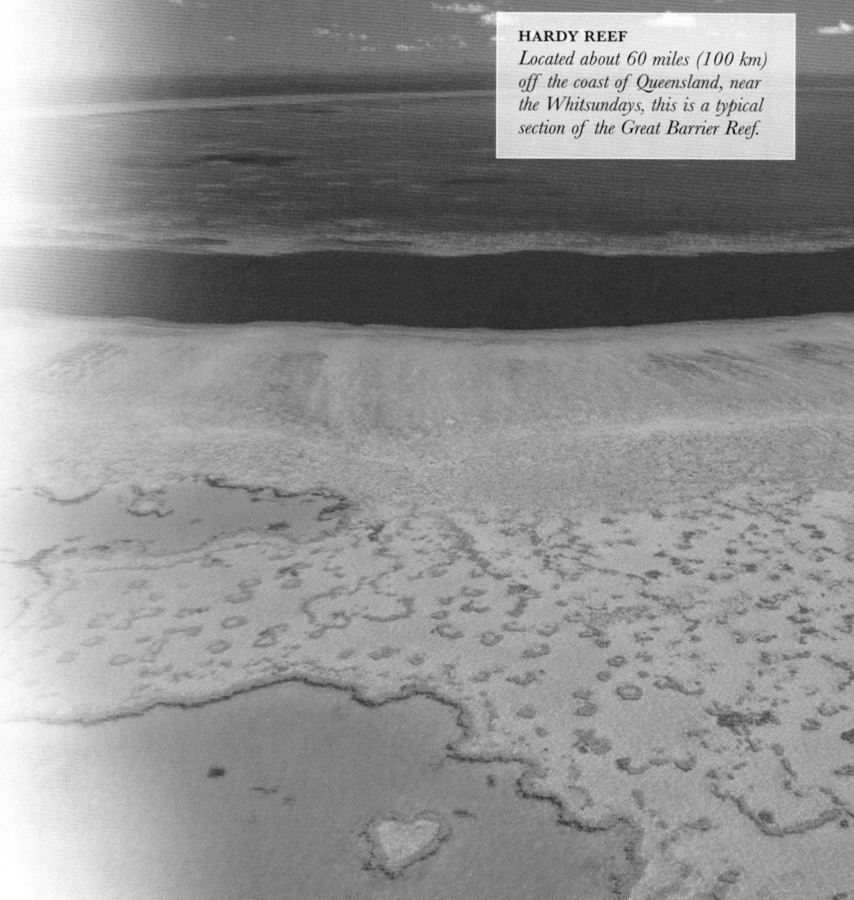

HARDY REEF
*Located about 60 miles (100 km) off the coast of Queensland, near the Whitsundays, this is a typical section of the Great Barrier Reef.*

**FINAL JOURNEY**
*A Norwegian fishing boat takes aboard a catch of herring. In the North Sea, industrial fishing of this species has brought about a collapse in stocks.*

# FISHING

Fish is the only major food that is still largely gathered from the wild. It accounts for about 10 percent of all the protein eaten by humans, which makes it an essential resource. The world's fish stocks, however, are falling rapidly, and, despite improvements in technology, the global catch remains static. As fishing nations seek to manage dwindling supplies, they are frequently torn by the conflict between short-term exploitation and long-term gain.

## FOOD FROM THE SEA

The worldwide harvest of sea fish has expanded from about 5 million tons per year in 1900 to about 90 million tons 100 years later. This huge rise has been brought about by a revolution in fishing vessels and equipment, including factory trawlers, which can catch and process several hundred tons of fish a day. Since the early 1990s, however, the total catch has leveled out, indicating that world's fisheries cannot expand any more.

Concern about overharvesting has led many coastal nations to introduce fishing quotas, as well as strict controls over the mesh sizes of nets. Setting quotas is a difficult exercise, because fish species differ in their breeding biology and also in the rate at which they grow. A catch that is sustainable for one species may drive another toward commercial extinction—the point where it becomes too rare to be worth catching by professional fishermen.

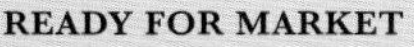

**READY FOR MARKET**
*A Senegalese woman checks fish that are drying in the sunshine. In recent years, West Africa's fishermen have faced intense competition from vast industrial trawlers.*

## BOOM AND BUST

Fish populations often undergo natural oscillations, which overfishing can disrupt. With the European herring, for example, historical records show a 50-year cycle going back several centuries. For most of this time, fishing had little effect, but during the 1960s, the catch expanded greatly, and herring stocks went into a sudden decline. They have still not fully recovered. The fate of the Peruvian anchovy was even more dramatic. Found off the west coast of South America, this finger-sized species once provided 30 percent of the world's entire fish catch. By the mid-1980s, its fishery had all but collapsed.

Although overfishing causes great ecological damage, it is not this problem but economic factors that are more likely to bring it under control. Faced with unpredictable catches and growing competition, many developing countries are turning to fish-farming as a way of meeting their food needs.

**EASTERN STYLE**
*In Kerala, India, a fisherman adjusts his net. This kind of fishing yields a sustainable harvest, because enough adult fish survive to reproduce.*

## AQUACULTURE

Practiced for centuries in Southeast Asia, fish-farming or aquaculture can be an efficient method of producing high-protein food. Unlike mammals, fish are cold-blooded, so most of their feed contributes to their growth rather than to keeping their bodies warm. China currently leads the world in this form of farming, producing more than 25 million tons of fish and crustaceans (such as shrimp) each year.

On a small scale, fish-farming dovetails well with other kinds of agriculture, because fish and other aquatic animals can be fed on a wide range of organic waste. However, large-scale fish-farming does have an environmental price. Organic waste escapes into the water, and so do preventive treatments such as antibiotics, which often damage habitats that are used by fish in the wild. But compared to sea-fishing, aquaculture is still poorly developed as a source of food, and in the future, it will almost certainly expand.

**SHORT-LIVED RETURNS**
*The Peruvian anchovy fishery began in the 1950s, as a source of fish meal and fish oil. After the boom years of the late 1960s, the fish population crashed when overfishing reinforced the natural effects of El Niño (see p.449).*

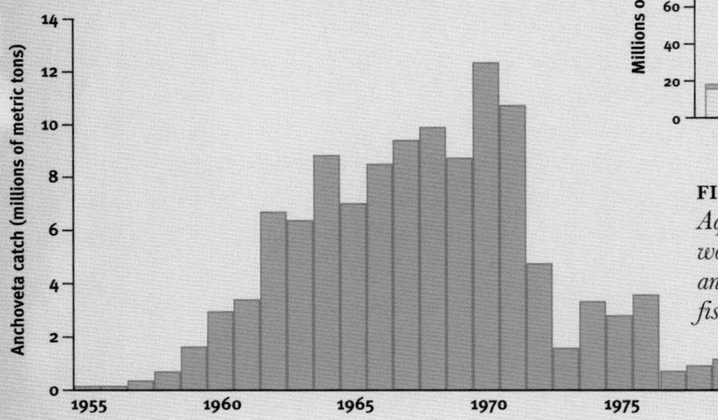


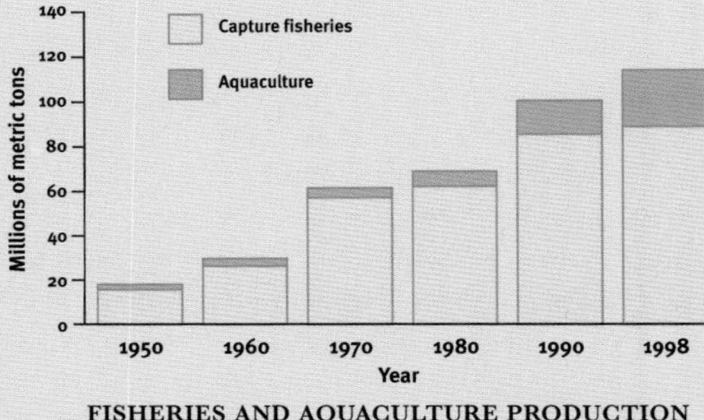


**FISHERIES AND AQUACULTURE PRODUCTION**
*Aquaculture is making an ever-larger contribution to world fish production and is currently expanding at an annual rate of 10 percent. Meanwhile, figures for fish captured in freshwater and the sea remain steady.*

SOUTHERN OCEAN

# Southern Ocean

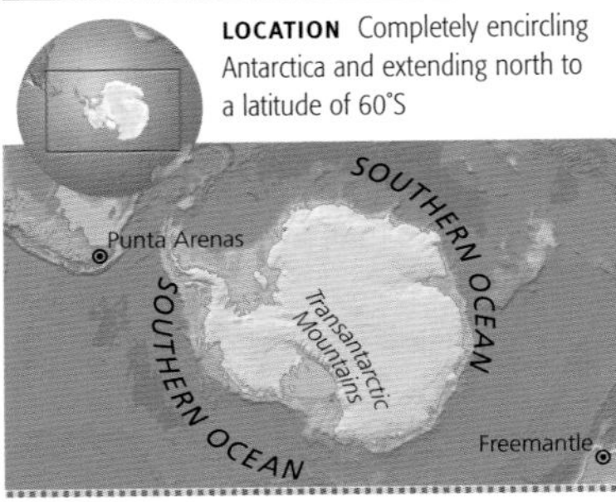


**LOCATION** Completely encircling Antarctica and extending north to a latitude of 60°S

**AREA** 7.8 million square miles (20 million square km)

**MAXIMUM DEPTH** 23,735 ft (7,235 m)

**INFLOWS** Summer melting of sea ice and icebergs calved from Antarctic ice shelves

**TABULAR ICEBERGS**
*Numerous tabular icebergs (see p.394) are present in the Southern Ocean, especially in areas close to the Antarctic landmass. After breaking off ice shelves, they initially drift westward in Antarctica's coastal current. The largest are hundreds of square miles in area.*

The Southern Ocean is the fourth-largest of the world's five oceans. Over most of its extent, it has a depth of 13,000–16,500 ft (4,000–5,000 m), its floor consisting of a series of basins separated by submarine ridges. The Antarctic continental shelf is unusually deep, as the whole of Antarctica is depressed by the weight of ice pushing down on it. The flow of currents in the Southern Ocean is complex. Close to Antarctica, highly saline cold water forms as salt is expelled from seawater freezing beneath the ice shelves. This cold salty water sinks, then moves north at depth. Farther from Antarctica, there is a northerly movement of cold but less saline surface water. This eventually sinks beneath warmer water in an area called the Atlantic Convergence, which extends all around Antarctica at latitudes between 50°S and 60°S. To replace the different bodies of cold water forming, sinking, and moving north, there is a continuous upwelling of water, called Circumpolar Deep Water, into intermediate areas of the Southern Ocean, bringing nutrient-rich waters to the surface. The world's largest ocean current, the Antarctic Circumpolar Current, lies in the same area as the Atlantic Convergence. It is driven by westerly winds, which move water perpetually eastward. In winter, much of the Southern Ocean freezes over, the overall area of sea ice around Antarctica growing seven- or eightfold. This ocean has the strongest average winds found anywhere on Earth. The combination of high winds, large waves, and ice makes ship navigation hazardous. The abundant planktonic life in the upwelling zone provides the basis for a complex food web. The most important organism in the food chain is krill, a crustacean that eats phytoplankton and sea-ice algae, and in turn provides food for fish, whales, seals, and birds. Fisheries mainly concentrate on harvesting krill and bottom-dwelling fish called toothfish.

**KRILL**
*These shrimplike animals are a key part of the food chain in the Southern Ocean. The biomass of Antarctic krill is thought to be larger than that of Earth's human population.*

## HITCHING A RIDE

One recently identified threat to Southern Ocean ecosystems comes from the increasing amount of garbage washing up on its shores. Marine biologists have established that by acting as rafts, plastic bottles and other floating objects are responsible for accelerating the spread of small marine organisms to new locations around the world, where they may displace native species. The Southern Ocean is particularly vulnerable because of its isolation.

**LIGHT-MANTLED SOOTY ALBATROSS**
*Several types of albatross inhabit the Southern Ocean. Breeding pairs of the species shown here nest on islands such as South Georgia and the Macquarie Islands.*

SOUTHERN OCEAN *south*

# Weddell Sea

**LOCATION** To the east of the Antarctic Peninsula and west of Coats Land, Antarctica

**AREA** 1.1 million square miles (2.8 million square km)

**MAXIMUM DEPTH** 10,000 ft (3,000 m)

**INFLOWS** Icebergs calved from the Ronne-Filchner Ice Shelf

The Weddell Sea is a large bay in the coast of Antarctica, forming part of the region of the Southern Ocean that lies south of the Atlantic. Ice and water moves around the sea in a clockwise direction. Overlying the southwestern portion of the Weddell Sea is the Ronne-Filchner Ice Shelf. A high proportion of the cold, dense water that lies at the bottom of the world's oceans was originally formed beneath this ice shelf. On its west side, along the coast of the Antarctic Peninsula, the sea overlies a 150-mile- (240-km-) wide area of continental shelf, which also extends for about 300 miles (480 km) beyond the edge of the Ronne-Filchner Ice Shelf. The part of the Weddell Sea beyond the ice shelf is more or less permanently covered in thick pack ice, which has greatly restricted exploration.

**MIDNIGHT SUN**
*The southern part of the Weddell Sea has continuous daylight from November to February.*

**YOUNG WEDDELL SEAL**
*Weddell seals live around the pack ice of the entire Southern Ocean. They swim beneath the ice and can break through to the surface to create breathing holes.*

**SASTRUGI**
*A research scientist surveys the thick pack ice at Cape Norvegia at the eastern end of the Weddell Sea, which has had its surface formed into ridges (sastrugi) by wind action.*

SOUTHERN OCEAN *south*

# Ross Sea

**LOCATION** To the east of Victoria Land and to the west of Marie Byrd Land, Antarctica

**AREA** 370,000 square miles (960,000 square km)

**MAXIMUM DEPTH** 8,300 ft (2,500 m)

**INFLOWS** Icebergs calved from the Ross Ice Shelf

The Ross Sea is an indentation into the coast of Antarctica, forming a part of the section of the Southern Ocean lying south of the western Pacific. Its southerly arm is overlain by the Ross Ice Shelf. The sea is generally shallow, its floor extending northward as a broad continental shelf before plunging into the depths of the Pacific–Antarctic Basin. Of all the seas around Antarctica, the Ross Sea is the least prone to sea ice, being free of pack ice in summer, and so is the most accessible to shipping.

**ICEFISH**
*The ghostly-looking icefish, in common with other Antarctic fish species, contains a substance in its tissues that prevents it from freezing.*

**SEA FLOOR UNDER ICE**
*Under the sea ice in shallow water near McMurdo Sound, the marine life is dominated by mobile organisms, such as starfish.*

SOUTHERN OCEAN *north*

# Scotia Sea

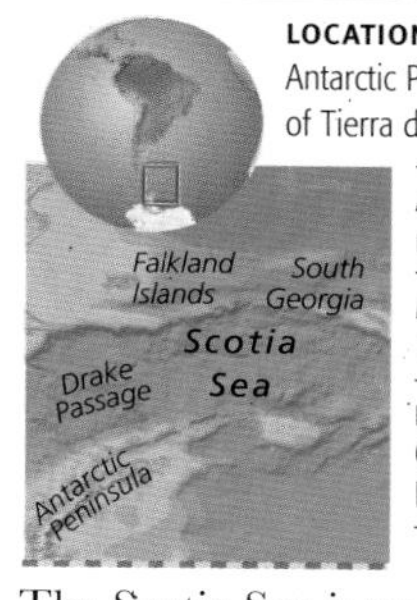


**LOCATION** Northeast of the Antarctic Peninsula and southeast of Tierra del Fuego, Argentina

**AREA** 350,000 square miles (900,000 square km)

**MAXIMUM DEPTH** 13,000 ft (4,000 m)

**INFLOWS** Southern Ocean to the west of Drake Passage

The Scotia Sea is an elongated body of water that stretches east from the 600-mile- (965-km-) wide channel, Drake Passage, that runs between the southern tip of South America (Cape Horn) and the northern tip of the Antarctic Peninsula. The Scotia Sea occupies its own small tectonic plate, sandwiched between the South American Plate to the north and the Antarctica Plate to the south. Underwater, the sea consists of two adjoined deep basins that are almost completely surrounded by a ridge, the Scotia Ridge, from which rise various islands. These include South Georgia to the northeast, the South Sandwich Islands (a volcanic island arc) to the east, and the South Shetland Islands to the south.

**CAPE HORN**
*The area south of Cape Horn is notorious for its rough seas, caused partly by strong currents as water funnels through Drake Passage.*

**LOOKOUT POINT, ELEPHANT ISLAND**
*A huge iceberg drifts past the coast of Elephant Island, the easternmost of the South Shetland Islands on the southern edge of the Scotia Sea.*

**EROSIONAL COASTLINE**
*These cliffs pounded by Atlantic waves are near the village of Port, County Donegal, Ireland. They show features typical of long-term marine erosion.*

# COASTS

COASTS ARE THE PLACES WHERE the Earth's oceans and seas meet the land. Within these regions can be found a great variety of different landforms, which are shaped by the interaction of breaking waves, the rise and fall of tides, the effect of currents, and slow changes in sea level. Coasts can also be affected by land-based phenomena—such as glacier advance and retreat, lava flows, and the discharge of sediment from rivers—and weathering by wind, rain, or frost. Variations in these processes and factors over time and along the length of shorelines cause constant change, sometimes eroding coasts and sometimes building them up. Last but not least, the activities of people also influence coastlines—about half of the world's human population lives on or within 60 miles (100 km) of a coast. These activities include the building of coast defenses such as sea walls and groynes, the reclaiming of land from the sea, and the creation of artificial islands.

# TIDES AND WAVES

48–49 The Earth and the Sun
50–51 The Earth and the Moon
Coasts and sea level 430–31
Erosional and depositional coastlines 432–33

TIDES AND WAVES ARE TWO PHENOMENA that are both highly noticeable in coastal areas and can have a profound effect on coastal landforms. Tides are regular rises and falls in sea level that occur all over the oceans but are most obvious near coasts, where they can cause strong currents and are important to coastal life. Waves are wind-generated and are instrumental in erosional and depositional processes, including the formation and maintenance of beaches.



## TIDES

Tides are produced by gravitational interactions between the Moon, the Sun, and the Earth. The daily tides are mainly caused by the Moon, which produces two bulges in Earth's oceans (see p.51). As the Earth rotates, these bulges sweep over the oceans, causing peaks and troughs in sea level, called high and low tide. Different coastal locations may have semi-diurnal tides (two high and two low tides per day), diurnal tides (one high and one low tide per day), or a mixture of these patterns. On top of the basic daily tidal pattern, there is a 28-day cycle in the difference in height between high and low tide (known as the tidal range). This cycle is caused by the interaction between tidal forces caused by the Moon and the somewhat weaker forces caused by the Sun. The strength of tides at a particular spot on the Earth's surface also depends on other complex factors, such as the shape of nearby land. In some narrow estuaries, for example, tides can cause strong surges of water, known as tidal bores.

**INTERTIDAL ZONE**
*Many different types of organisms have adapted to living in the zone between high and low tides.*

**TIDAL BORE**
*The tidal bore up China's Qiantang river estuary reached an exceptional height in September 2002.*

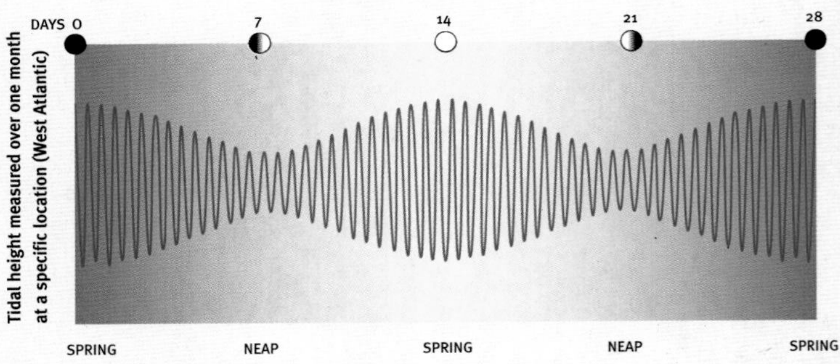


**SPRING AND NEAP TIDES**
*Twice every 28 days (around days 0 and 14), the Sun and the Moon's gravitational pull combine to produce extra-large bulges in the Earth's oceans, leading to a high tidal range called spring tides. At other times (around days 7 and 21), the Sun partly cancels out the effects of the Moon, causing a smaller tidal range called neap tides.*

**SPRING TIDES**
*Spring tides occur when the Sun, the Earth, and the Moon are aligned, at the times of new moon and full moon.*

**NEAP TIDES**
*Neap tides occur at lunar first and last quarters, when the Sun's and Moon's pull on Earth are at right angles.*

## WAVES

Ocean waves are caused by the action of wind blowing across the surface of the sea. Their size depends on the speed of the wind, the length of time the wind blows (wind duration), and the distance over which the wind blows (the fetch). A series of waves can be characterized by their wavelength (distance between wave crests), frequency (number of crests that pass each minute), and height (amplitude). In the area where the wind is blowing, the sea often has a chaotic form because the wind generates waves with many different frequencies that interfere with each other. As the waves travel away from the generation area, they become sorted by frequency into a steady pattern called a swell. Overall, waves transfer energy, not water, across the sea surface. When they reach a coast, this energy is released in the form of breakers.

movement of individual particles of seawater
INTERMEDIATE WAVES
WAVELENGTH
SHALLOW WATER WAVES
BREAKER ZONE
SURF ZONE
SWASH ZONE
water movement occurs to depth of half the wavelength
waves shorten in length and decrease in speed, but increase in height (amplitude)
waves reach critical ratio of wave height to wavelength and begin to break

**MOTION IN WAVES**
*The particles of water in a choppy sea do not move forward with the waves. Instead, they gyrate in little circles or loops. Underwater, the particles move in smaller and smaller loops, until at a depth equal to half the distance between the wave crests, there is hardly any movement at all.*

**CHAOTIC AND REGULAR WAVES**
*Where they are generated, waves give the sea a chaotic form (far right). Far from the generation area, waves come onshore in a regular pattern (right).*

# BEACHES

A beach is a deposit of sedimentary particles, varying in size from mud to boulders, but mostly sand that occupies a coastal area between high and low tide. The particles are derived from erosion of the land, or fragmentation of marine structures such as corals or shells, which are brought to the beach by rivers and waves. Beach deposits never stay in the same place for long but are continually moved around—onshore, offshore, or along the shore—through the action of waves, tides, and currents. The average intensity of wave action is important in determining the nature of a beach. Low-energy waves tend to build beaches, while high-energy waves tend to erode them. There are several zones on a beach. The nearshore zone extends from where waves break to the upper end of the swash zone (the area covered and uncovered by each wave surge). Further up most beaches is a berm, an accumulation of beach material with a flat top surface and a relatively steep seaward slope.

**SANDY BEACH**
*Beaches of fine sand or coral fragments are associated with low-energy waves and a calm sea.*

**GRAVEL BEACH**
*Pebble or gravel beaches, such as this one in Dorset, England (left), tend to be associated with higher-energy waves.*

**HUGE DUNES**
*Coastal dunes, such as these in Arcachon, France, are caused by wind blowing sand off the dry part of a beach.*

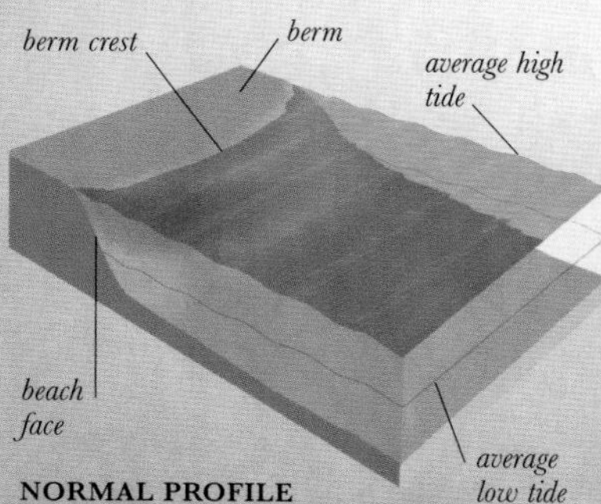


**NORMAL PROFILE**
*Fair weather and a low-energy swell (regular waves) for several days results in accumulation of beach material, giving a steep beach face. This is sometimes called a swell profile.*

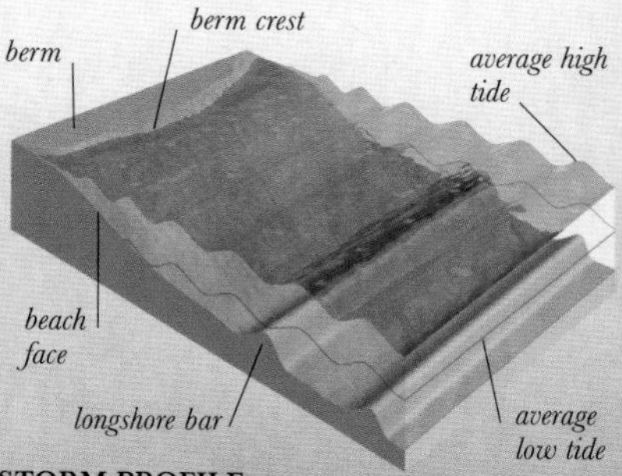

berm
berm crest
average high tide
beach face
longshore bar
average low tide

**STORM PROFILE**
*During stormy weather, higher-energy waves erode the beach, flattening the beach face and carrying material offshore, where it is deposited as longshore bars. The result is a beach with a storm profile.*

# TSUNAMIS

A tsunami is a powerful pulse of wave energy that can propagate for a long distance, and at high speed, through the surface of an ocean. A tsunami has a long wavelength but low amplitude, so that in the open ocean its passage is hardly noticed. However, when a tsunami reaches shallow water, its amplitude increases dramatically. Sometimes it forms a huge wave, as high as 100 ft (30 m), which can cause tremendous amounts of damage as it breaks on a shore. Tsunamis are usually set off by submarine earthquakes or the sudden slumping of large masses of sediment underwater, and they occur regularly in some parts of the world. In recent years, scientists have recognized the rarer phenomena of massive megatsunamis, which produce waves many times higher than those generated by normal tsunamis.

**AFTERMATH**
*When a tsunami hits a shore, it can smash buildings and carry boats far inland. The damage shown here occurred on Okushiri Island, Japan, in 1993.*

## BEACH PROTECTION

Groins are structures built out from a beach into the sea with the object of slowing beach erosion. They are supposed to work by trapping sand particles that are being moved along the beach by wave action (longshore drift). Ultimately, groins often prove ineffective, as they starve other beach areas of sand. The beach shown below, on the Hel Peninsula on the Baltic Sea coast of Poland, reveals several past attempts to slow erosion.

**WAVE ENERGY**
*The intensity of wave energy release can have a profound effect on the form of a coast. These waves are breaking in a rough sea on the coast of Oregon.*

# COASTS AND SEA LEVEL

39–41 The ice ages
59 Isostasy
386–87 Ocean tectonics
388–89 Shelves and plains
Erosional and depositional coastlines 432–33
Coastal pollution 434–35
Global warming 452–53

A COAST IS A STRIP OF LAND in immediate contact with the sea, extending from the shoreline inland to the first major change in terrain. Coasts are dynamic rather than static. They quickly alter in response to natural forces such as wave action and tides, land-based processes, and sea-level change. These forces and processes constantly push and pull at coastlines, and, as a result, their shape and location changes. There are also many different types of coasts. The types and the processes that shape them are in turn determined by the various causes of sea-level change.

## EMERGENT COASTS

Emergent coastlines result from land being uplifted faster than sea level has risen since the last ice age. The causes are either tectonic activity (for example, on the US Pacific coast) or slow glacial rebound (such as in parts of Scandinavia and New England). On emergent coasts, areas that were formerly sea floor may become exposed above the shoreline, and former beaches may become the tops of cliffs. Emergent coasts are typically rocky, but usually have a smooth shoreline.

**RAISED BEACH**
*The tops of these cliffs in the Bay of Fundy, Canada, are former beaches that were raised by glacial rebound.*

## SEA-LEVEL CHANGE

Changes in sea level are of two different types, regional and global. The most important cause of a global change in sea level is an increase or decrease in the extent of the world's ice sheets and glaciers (see below). When these grow larger, sea level drops because water becomes locked up in the ice; and when they melt, sea level rises. Two other parameters that affect global sea level are ocean temperature and the size of ocean basins (see right). Regional sea-level change occurs when a specific area of land rises or falls relative to the general sea level. Two of the main causes are tectonic uplifting of land, which is common in regions where oceanic crust is being forced beneath continental crust, and glacial rebound, which is a slow rise of an area of land after an ice sheet that once weighed it down has melted. Sea levels have constantly fluctuated throughout the Earth's history, not least in the recent past. For example, about 15–18,000 years ago, during the last ice age, sea levels were about (400 ft (120 m) lower than they are today.

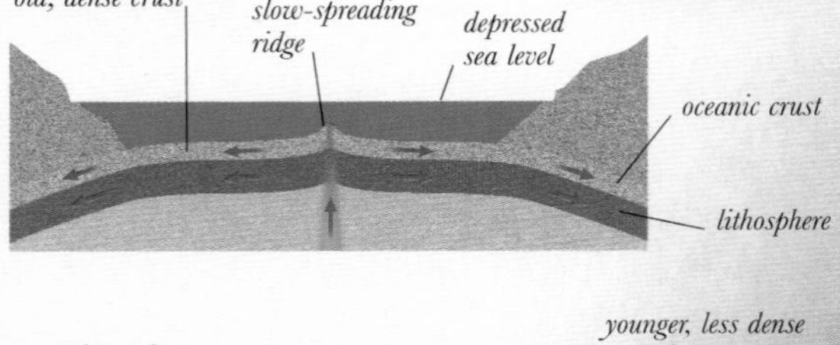


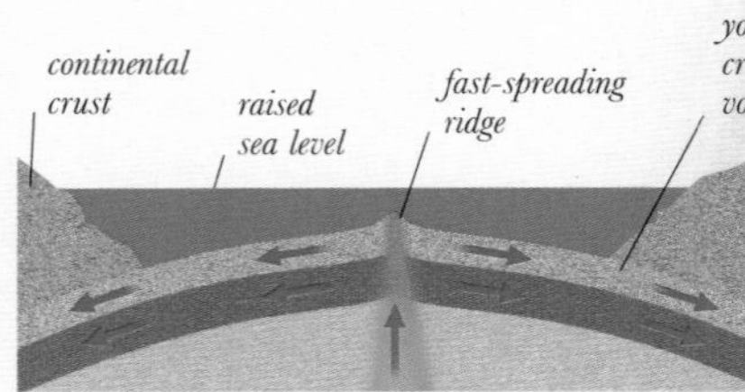


**OCEAN BASIN CHANGE**
*A rise in sea level can occur when new crust is produced at a fast-spreading ridge. The new crust swells because it is relatively hot and buoyant, pushing the ocean water upward.*

*overall drop in sea level relative to land*
*continental crust depressed through loading with ice*
*water locked in ice sheet*
*worldwide warming and deglaciation*
*overall increase in sea level relative to land*
*oceanic crust rises due to unloading of seawater*
*worldwide cooling and glaciation*
*oceanic crust depressed through loading with seawater*
*continental crust rises due to unloading of ice*

**CHANGES IN SEA LEVEL OVER GLACIAL CYCLES**
*When temperatures rise, glacial ice melts and flows into the sea and seawater expands, raising sea levels. The rise is partially offset by depression of oceanic crust and, in some regions, by land rising following glacial rebound.*

## TYPES OF COASTS

Coasts are either primary or secondary. Primary coasts are relatively young ones that have been created by terrestrial processes or by sea-level change. Examples of primary coasts are volcanic coasts, where lava has flowed to the sea and solidified, and deltas, where a river deposits sediment as it enters the sea. Coasts in which sea-level change has played a significant part include drowned and emergent coasts. Other types of primary coasts are those dominated by glaciers or glacial moraines, and karst coasts, where a region of limestone has been eroded and later inundated by the sea. Secondary coasts are older ones that have been shaped by marine erosional and depositional processes (see pp.432–33) or the activities of marine organisms (see Reefs, pp.392–93).

**VOLCANIC COAST**
*This coast on the Snaefellsnes Peninsula, Iceland, was created by lava from a nearby volcano. Since the coast formed, it has started to undergo some wave erosion.*

**DROWNED KARST**
*This coast of eroded limestone in Ha Long Bay, Vietnam, has been inundated by a rise in sea level.*

**DELTA COAST**
*A delta, such as this one, the Wadi al-Mujib on the Dead Sea, Jordan, is the result of sediment deposited by a river.*

# DROWNED COASTS

Drowned coasts (sometimes called submergent coasts) are formed as a result of a global or regional rise in sea level. Several types are recognized. A ria coast is a river-eroded hilly landscape, with valleys running perpendicular to the coast, that has been partly drowned. The coast usually has deep bays or estuaries, intervening headlands, and many offshore islands. There are examples in southwest Ireland and Great Britain, northwest Spain, and the eastern US (such as Chesapeake Bay, see p.122). A fjord coast, seen in Norway, Chile, and New Zealand's South Island, for example, develops when a glaciated mountain landscape becomes partly drowned, forming long, deep bays with steep sides. A Dalmatian coast, named after the coast of Dalmatia, Croatia, is a series of valleys formed by folds running parallel to the coast that has been almost completely submerged.

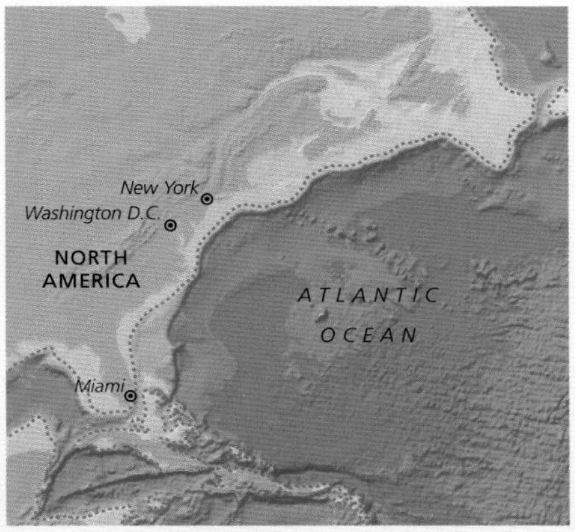

**COASTLINE PAST AND PRESENT**
*The east coast of North America has changed in the past 15,000 years. Sea-level rise has drowned many ancient valleys.*

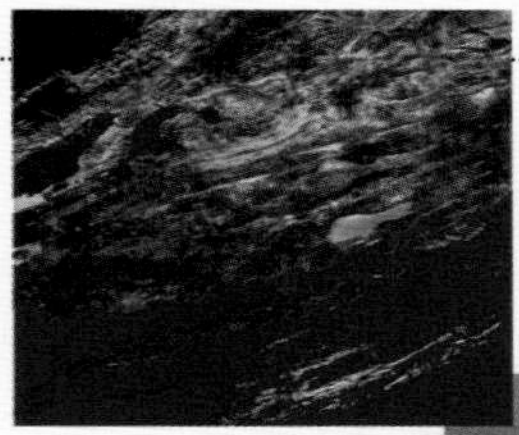

Eastern coastline of North America 15,000 years ago

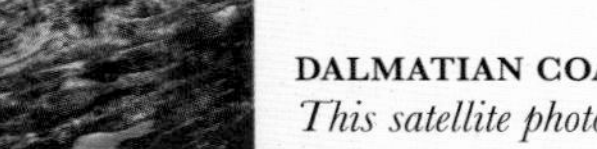

**DALMATIAN COAST**
*This satellite photograph shows the classic drowned Dalmatian coast in the Adriatic Sea.*

**RIA COAST**
*Southwest Ireland, in the counties of Cork and Kerry, contains a series of elongated bays typical of a ria coast.*

**FJORD COAST**
*This coast, a type of drowned coast, is located in the Lofoten area of Norway. Fjords are narrow, steep-sided, and, unlike rias, penetrate far inland.*

# EROSIONAL AND DEPOSITIONAL COASTLINES

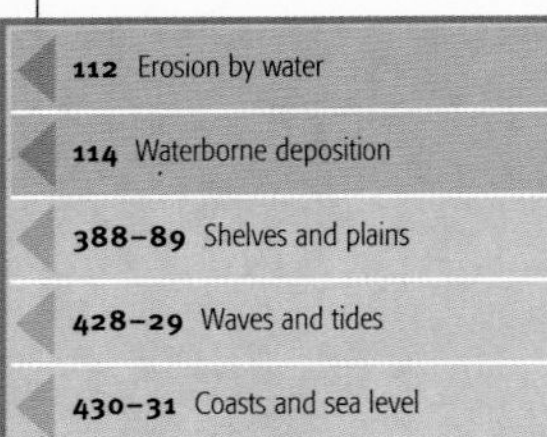
112 Erosion by water
114 Waterborne deposition
388–89 Shelves and plains
428–29 Waves and tides
430–31 Coasts and sea level

EROSIONAL AND DEPOSITIONAL coastlines are two forms of secondary coasts, which have been shaped primarily by marine processes. Erosional coasts occur where land is being eroded by the sea faster than sediment can be deposited. They are typically associated with emergent coasts (see p.430), where the land is being lifted by tectonic processes. Depositional coasts occur where sediment is being deposited on a coast faster than it is eroded. They are often associated with tectonically inactive continental margins, where the coast is slowly subsiding.

## PROCESSES OF EROSION

Energy for the processes of coastal erosion is supplied mainly by waves and currents. As they beat against cliffs, waves compress air within cracks in rocks, and on re-expansion the air can shatter the rock. Mechanical erosion refers to waves hurling beach material against cliffs and abrading the rock. Where waves encounter headlands, a significant factor is refraction (bending) of the wave fronts, which causes their erosive energy to be directed toward the headlands and away from the bays. Waves arriving on a coast at an angle also generate a longshore drift (movement of sediment along a shore), which shifts material eroded from the headland toward adjacent bays. Groundwater that has emerged at the surface and surface-water runoff also gradually wear channels through the headland.

**REFRACTING WAVE**
*The bending of the wave front toward the headland can be seen clearly here in this bay in southern Ireland.*

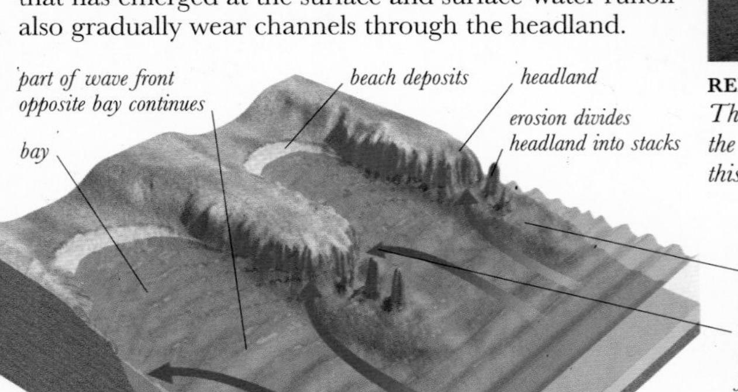


**CONCENTRATION OF WAVE ENERGY**
*When a wave front reaches a shore, the parts opposite headlands slow down before the parts opposite bays. As a result, the wave front refracts (bends), and this concentrates its energy on the headlands.*

## EROSIONAL LANDFORMS

At headlands subjected to wave erosion, various distinctive features develop in a classic sequence. First, sea caves form at the bases of the cliffs on the sides of the headlands, usually along faults and joints (planes of weakness in a rock; see p.142). Wave action gradually deepens and widens the caves until they penetrate through the headland to form an arch. Next, the roof of the arch collapses to leave an isolated rock pillar called a stack. These are ultimately worn down to stumps. Along cliffs, wave action cuts into the base of the cliff. This produces a horizontal platform at sea level called a wave-cut platform, which provides some protection from further erosion by dissipating wave energy. Tectonic uplift sometimes raises the land in short bursts, and it is then worn into a series of wave-cut platforms called marine terraces.

**CAVES**
*The coast of Cornwall, England, is dotted with numerous erosional features, including these caves at Nanjizel, near Land's End.*

**ARCH**
*This spectacular example of a wave-eroded arch is at Loch Ard Gorge (named after a ship that was wrecked there) on the coast of Victoria, Australia.*

# PROCESSES OF DEPOSITION

Like coastal erosion, coastal deposition is brought about primarily by wave action. However, while erosion occurs in areas of high wave energy, deposition happens in environments where most of the energy has been dissipated before it reaches the shore, allowing sand or other material in the water to settle out. The deposited sediment comes mostly from rivers, but some comes from eroded headlands or from offshore. An important mechanism of transport and deposition is that of longshore drift. When wind and waves strike a shore obliquely, the movement of surf (swash) propels water and sand up the shore at an angle, but backwash drags them back down at a right angle to the shore. The overall effect is to move water and sand gradually along the shore. Eventually, where the sediment-carrying water arrives at a lower-energy environment, the sediment settles out and builds up to form features such as spits, sand bars, and barrier islands. Some depositional processes modify one type of landform into another. For example, the development of a beach may lead to a spit, which may eventually extend across the mouth of a bay to create a baymouth bar.

**SPIT**
*This spit at Hurst Castle is one of the most prominent depositional features on the southern coast of England.*

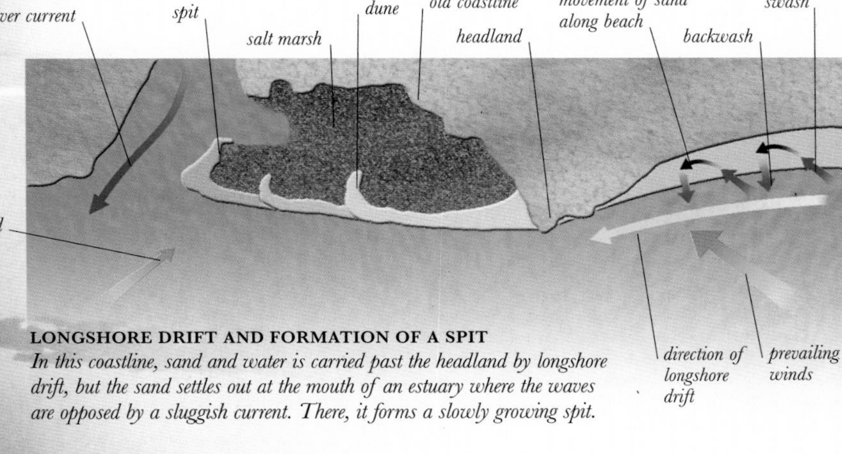


**LONGSHORE DRIFT AND FORMATION OF A SPIT**
*In this coastline, sand and water is carried past the headland by longshore drift, but the sand settles out at the mouth of an estuary where the waves are opposed by a sluggish current. There, it forms a slowly growing spit.*

# DEPOSITIONAL LANDFORMS

Depositional processes form a number of characteristic features around coasts. A spit is a sandy peninsula with one end attached to land. An offshore bar is an unattached sand bar (a general name for any sand deposit) that lies parallel to the coast. About 13 percent of the world's coasts have offshore bars. Where a spit extends all the way across the mouth of a bay, or two spits have merged, the result is called a baymouth bar. A tombolo is a sand bar that connects an island to land or to another island. Barrier islands are sand bars that coalesce to form long, linear islands parallel to the coast. They are typical of the eastern and Gulf coasts of the US, and often slowly migrate shoreward. Other types of depositional landform include salt marshes, which often develop on the landward sides of spits and bars, and mud flats, which form from the deposition of fine silt in sheltered tidal water, particularly in estuaries.

**TOMBOLOS**
*This example of a tombolo (a sand bar between two islands) is in Puget Sound, near Seattle, Washington.*

**ALASKAN MUD FLAT**
*Large mud flats on the Alaskan coast are the result of glacial erosion of nearby mountains.*

**STACKS**
*Stacks—such as the Twelve Apostles off Victoria, Australia—represent an advanced stage in the erosion of a headland.*

**BEACHED**
*Cargoes that have leaked from wrecked ships are major sources of coastal pollution. Synthetic organic chemicals can have particularly damaging effects.*

# COASTAL POLLUTION

Biologically as well as physically, coasts are frontiers between land and open sea. They teem with life, but are also highly vulnerable to pollution from contaminated rivers as well as waste that has been discharged at sea and washed up by inshore currents. For humans, coastal pollution may be a minor nuisance only, or a major economic threat, but for coastal wildlife, its effects are far-reaching and often unseen.

## MARINE LITTER

More than 8 million pieces of litter are estimated to enter the seas and oceans every day. Fifty years ago, wood was the main ingredient, but it has now been superseded by plastic. The lightness and durability of plastic enable it to reach some of the remotest coastlines on the Earth. During the 1990s, Chilean scientists identified 1,500 pieces of plastic waste on the shores of Livingston Island, in Antarctica, while in Ducie Island, in the South Pacific, a researcher found nearly 1,000 pieces of litter on a stretch of beach under 2 miles (3 km) long. They included 189 buoys, 71 plastic bottles, 44 pieces of rope, and 29 lengths of plastic pipe. Ducie Island is uninhabited, and the nearest continental coast is more than 2,800 miles (4,500 km) away.

Plastic is chemically inert, but it causes physical problems by not rotting. Marine animals, such as dolphins, seals, and turtles, become entangled in plastic debris, while seabirds often mistake plastic fragments for food. To counter this threat, an international convention forbids the disposal of plastics at sea—but, as yet, there is no equivalent agreement about litter that comes from land.

**LIFE SENTENCE**
*This young herring gull has plastic tangled in its beak. Gulls are naturally curious, and their feeding habits put them at risk from waste.*

## OIL SPILLAGE

Few sights evoke the idea of pollution more vividly than coastal wildlife covered with oil. Birds are foremost among the victims, but casualties include all kinds of inshore animals, from seals to fish and mollusks. Oil clogs up fur, feathers, and gills, and it is toxic if it is swallowed. When the densest part of crude oil coalesces and sinks, it can remain in the seabed sediment for years. In a study conducted off the coast of Massachusetts in the year 2000, oil contamination was found from a tanker accident that had occurred in 1969.

Large-scale oil spills generate intense coverage in the media and calls for emergency action. However, scientists are divided on the best way to respond. Although detergents break up visible slicks, the dispersed oil can then enter marine food chains, where it may cause hidden environmental damage. As with all forms of pollution, prevention is better than cure, and here the record is more promising. Double-hulled tankers, together with better navigation systems, have helped cut major oil spills by four-fifths in the last 30 years.

Surprisingly, accidental spills actually account for less than 20 percent of oil that finds its way into the seas. Coastal areas are also polluted by land-based discharges, and by the oil from small craft to large ships. Badly maintained engines in recreational boats probably create more oil pollution than all the world's oil spills combined. But because this pollution comes from many small sources, it is largely invisible and thus very hard to control.

**CLEANING UP**
*Off the coast of South Africa, rescue workers treat penguins that are covered in oil. Rapid action is needed to treat oil-damaged birds.*

**SOURCES OF OIL POLLUTION AT SEA**
*Shipping accounts for nearly half of the oil waste found at sea, while roughly a third is from urban and industrial waste. However, at least 8 percent of sea pollution comes from natural sources, such as oil seepages on the sea bed.*

Shipping
Waste and runoff
Atmospheric pollution
Natural sources
Offshore oil production

## SEWAGE AND INDUSTRIAL WASTE

Coastal waters are common dumping grounds for urban and industrial waste. Sewage is sometimes screened (in a process called primary treatment), but in most coastal communities it is discharged without any treatment at all. In time, sewage decomposes, but industrial pollutants are often more persistent. Heavy metals—such as cadmium and lead—may remain in seabed sediment for a century or more.

Until relatively recently, coastal pollution was a local problem. Estuaries were often contaminated, but nature's recycling systems meant that there was little human impact farther out to sea. Today, the sheer volume of waste produced by modern society means that this is no longer the case. In New York and New Jersey, for example, nearly 10 million tons of sewage are dumped each year, creating a submarine desert that stretches miles from the shore. However, because it affects the seabed, and not the shore itself, this form of pollution often goes unnoticed.

**OUT OF SIGHT**
*When waste is discharged on the shore, it often returns with each tide. Long underwater outfall pipes aid successful off-shore dispersal of such waste.*

# ATMOSPHERE

**DARK SKIES OVER THE DESERT**
*Some climates are strongly influenced by physical features. The hot, dry climate of Death Valley is due to its position in the rain shadow of the Sierra Nevada.*

# CLIMATE

THE WEATHER CHANGES from one day to the next and is evident as variations in temperature, precipitation, wind, and clouds. Examine the weather over many years, however, and a pattern emerges: winters are cold, with snow and ice; summers are warm; and rainfall in June is about the same as in November. This pattern of weather, repeated over many years, constitutes the climate of a particular region. Obviously, different parts of the world have different climates. In equatorial regions, the weather is always warm and usually wet. Deserts are dry, and polar regions are cold. The different climate regions result partly from the Sun, which shines more strongly at the tropics than anywhere else, and partly from the way the atmosphere and oceans transfer the Sun's warmth away from the equator. If the Earth had no atmosphere or oceans, it could have no climates. The way in which the air and water produce climates is a complicated but fascinating story.

# STRUCTURE OF THE ATMOSPHERE

8–17 Dynamic Earth
28 Oxygenating the atmosphere
Energy in the atmosphere 442–43
Atmospheric circulation 446–49
Global warming 452–53
Weather 458–77

THE ATMOSPHERE FORMS distinct layers around the Earth. These layers are remarkably uniform in chemical composition, but their density decreases with altitude. Within each layer there is a constant temperature change with height that alters abruptly at its boundaries. The lowest layer, called the troposphere, is where life exists and weather occurs because of the heating effect of the Sun. The Sun's rays pass through the atmosphere and warm the Earth's surface, causing the air to move and water to evaporate or condense. This causes weather and creates the different climate regions. Harmful ultraviolet rays are filtered out by the ozone layer, a gas that forms a thin layer in the stratosphere.

## LAYERS

The lowest layer in the atmosphere is the troposphere, where air moves vertically and horizontally, and is thoroughly mixed. Air temperature decreases with height. As warm air rises, it loses some of its energy and becomes cooler. Eventually, it reaches a level at which it cools no further. The air above it is no denser, and so it rises no further. This level marks a boundary, called the tropopause. It is the first of several such boundaries that divide our atmosphere into its distinct layers. It separates the troposphere from the stratosphere, which itself ends at the stratopause. Above lies the mesosphere, and the thermosphere. At the top of the thermosphere, the temperature rises to about 1,800°F (1,000°C) due to the absorption of ultraviolet radiation. The upper limit of the thermosphere, the thermopause, extends up to 625 miles (1,000 km). This outermost layer gradually merges with the vacuum of the Sun's atmosphere (which we call space), so Earth's atmosphere has no distinct upper boundary.

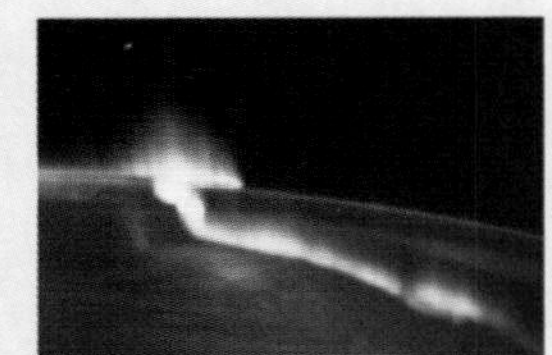

**AURORA**
*The Northern and Southern Lights, also known as the Aurora Borealis and Australis, appear in the thermosphere.*

**RADIATION FOG**
*The troposphere is warmed by the Sun during the day, but at night it cools again by radiating heat. The cool early morning air may contain fog.*

### LIVING AT HIGH ALTITUDE

Air pressure decreases with altitude, and above about 13,200 ft (4,000 m) there is insufficient oxygen for people to sustain strenuous activity—unless they are acclimatized. These tea pickers in Darjiling, India, have greater lung capacity and breathe more deeply than lowlanders. Their blood has a greater affinity for oxygen and can absorb it faster. These unique adaptations are genetic and are therefore passed on from one generation to the next. It allows the tea pickers to work on the upper slopes and carry heavy loads all day without becoming unduly breathless.

**LAYERS OF THE ATMOSPHERE**
*The atmosphere consists of layers, principally the troposphere, stratosphere, mesosphere, and thermosphere. Each is defined by the way temperature changes within its limits. The layers are separated by clear boundaries.*

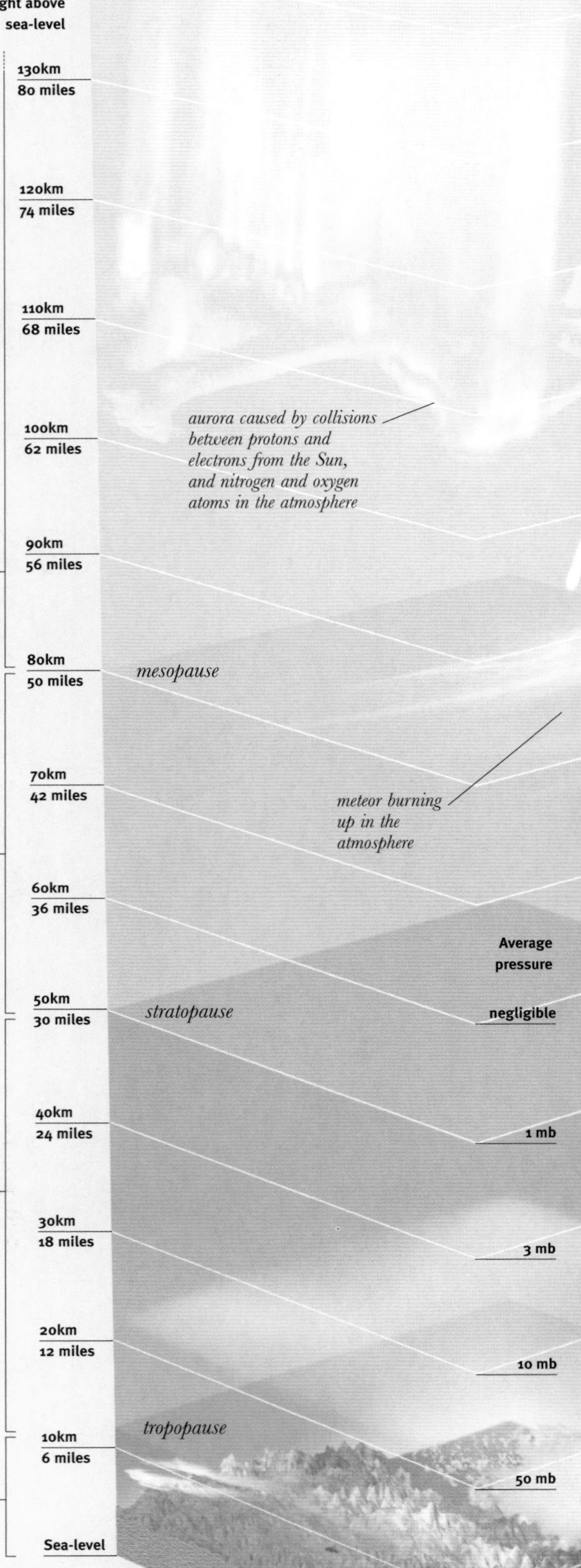


**THERMOSPHERE**
*This layer extends up to the thermopause, as high as 625 miles (1,000 km). The temperature in the lower part of this layer remains constant with height, but increases rapidly above 55 miles (88 km).*

**MESOSPHERE**
*The temperature of the mesosphere remains constant with height through the lower mesosphere, but above about 35 miles (55 km) it decreases with height to about –112°F (–80°C) at the mesopause.*

**STRATOSPHERE**
*Temperature in the stratosphere remains stable up to about 12 miles (20 km), then increases, due to absorption of ultraviolet radiation. This layer's upper boundary, called the stratopause, is at about 28 miles (48 km).*

**TROPOSPHERE**
*The upper limit, called the tropopause, is at about 10 miles (16 km) at the equator, and 5 miles (8 km) over the poles. Temperature at high altitude drops to about –22°F (–30°C) at the poles and –85°F (–65°C) at the equator.*

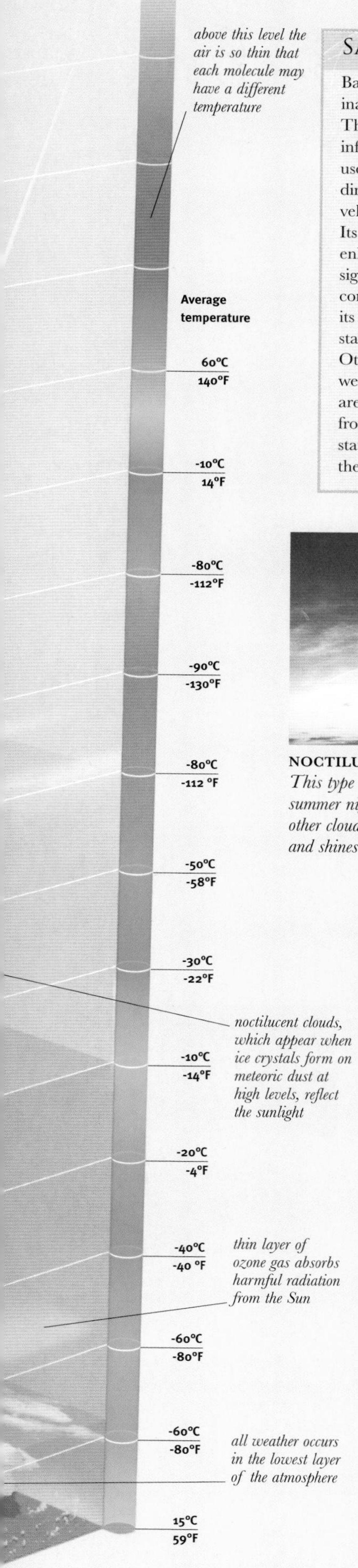

*above this level the air is so thin that each molecule may have a different temperature*

Average temperature

60°C 140°F

-10°C 14°F

-80°C -112°F

-90°C -130°F

-80°C -112 °F

-50°C -58°F

-30°C -22°F

*noctilucent clouds, which appear when ice crystals form on meteoric dust at high levels, reflect the sunlight*

-10°C -14°F

-20°C -4°F

*thin layer of ozone gas absorbs harmful radiation from the Sun*

-40°C -40 °F

-60°C -80°F

-60°C -80°F

*all weather occurs in the lowest layer of the atmosphere*

15°C 59°F

## SAMPLING THE AIR

Balloons are used to sample the more inaccessible parts of the atmosphere. The Jimsphere seen here is an inflatable weather balloon, which is used to determine wind speed and direction prior to the launch of space vehicles, and missiles. Its shiny surface enhances radar signals, and the conical "spikes" on its surface help to stabilize its flight. Other types of weather balloons are released daily from weather stations all over the world.

# COMPOSITION

Air consists mainly of three gases: nitrogen, oxygen, and argon, in constant proportions. It also contains water vapor, but in variable amounts. In addition to these principal constituents, there are minute quantities of 10 other gases. These are carbon dioxide, neon, helium, methane, krypton, hydrogen, nitrous oxide, carbon monoxide, xenon, and ozone. These form about 0.04 percent of the atmosphere by volume. Air also contains even smaller amounts of gases that come from the surface, such as ammonia, nitrogen dioxide, hydrogen sulfide, and sulfur dioxide. Finally, there are dust and smoke particles.

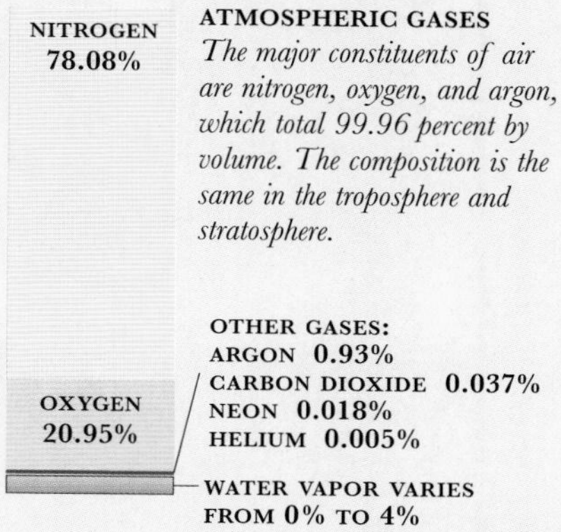

NITROGEN 78.08%

OXYGEN 20.95%

OTHER GASES:
ARGON 0.93%
CARBON DIOXIDE 0.037%
NEON 0.018%
HELIUM 0.005%

WATER VAPOR VARIES FROM 0% TO 4%

**ATMOSPHERIC GASES**
*The major constituents of air are nitrogen, oxygen, and argon, which total 99.96 percent by volume. The composition is the same in the troposphere and stratosphere.*

# AIRBORNE WATER

Water evaporates from the surfaces of oceans, lakes, and rivers, and from wet ground. It also evaporates from cloud droplets and raindrops, many of which fail to reach the ground. Water vapor is a gas, and it is present in all air. The amount of water vapor varies widely. Desert air often contains almost no water vapor, but even the moist air of the humid tropics seldom contains more than about four percent water vapor by volume. The humidity of the air is a measure of the amount of water vapor it contains. However, the most often cited measure is relative humidity, which is the amount of water vapor present expressed as a percentage of the amount of water vapor needed to saturate the air. This varies according to the amount of water vapor and the temperature of the air. Warm air can hold more water vapor than cold air.

**NOCTILUCENT CLOUDS**
*This type of cloud is sometimes seen on summer nights in high latitudes. Unlike other clouds, it forms in the mesopause, and shines by reflecting light from the Sun.*

**THE SNOWLINE**
*At altitude, low temperatures stop snow from melting, even in summer. The lower limit of the permanent snow, seen here on Cotopaxi in Ecuador, is called the snowline.*

**RAINBOW**
*This rainbow in Idaho is produced when light is refracted on entering raindrops, then reflected from their rear, and refracted again as it leaves.*

# ATMOSPHERIC PARTICLES

Air contains solid particles, called aerosols, that are so small they remain temporarily suspended. They fall so slowly that upcurrents repeatedly carry them aloft. Nevertheless, individual particles do not remain airborne for very long. Large particles return to the ground within minutes, and small particles, such as smoke, within a few hours. Some fall to the ground and some collide with surfaces as they are carried horizontally by the wind, but most are washed to the ground by rain and snow. Many particles are lifted into the air as a result of natural events. Volcanic eruptions eject vast quantities of ash and dust, while desert winds carry sand and dust over great distances—for example, Saharan dust occasionally falls in North America. Fires, ignited by lightning or humans, release smoke and ash that can be carried upward by rising hot air. Even pollen and spores released from plants, fungi, and bacteria become airborne, along with the aerosols produced by various agricultural and industrial practices. Water vapor in the atmosphere condenses onto certain minute particles, including dust, smoke, and sulfate. Some aerosols reflect sunlight back toward space, partially offsetting the effects of global warming. These cloud condensation nuclei, as they are known, allow clouds to form.

**POLLUTION**
*Clouds of aerosol particles are seen here in the skies along the southern edge of the Himalayas in northern India, streaming south over Bangladesh. This type of industrial and urban pollution now affects much of southern Asia.*

**VOLCANIC ASH**
*This is a magnified image of a particle of volcanic ash. Many tons of ash are produced during a volcanic eruption.*

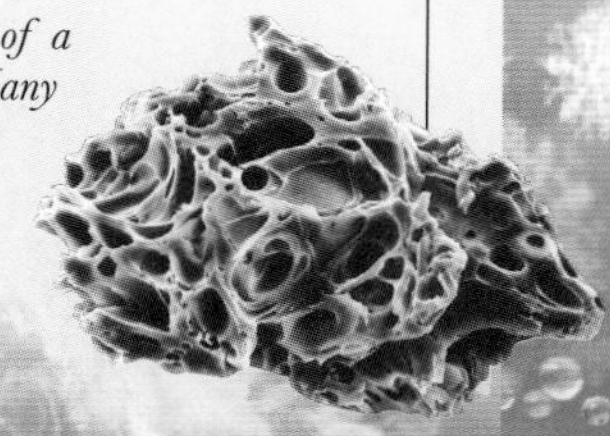

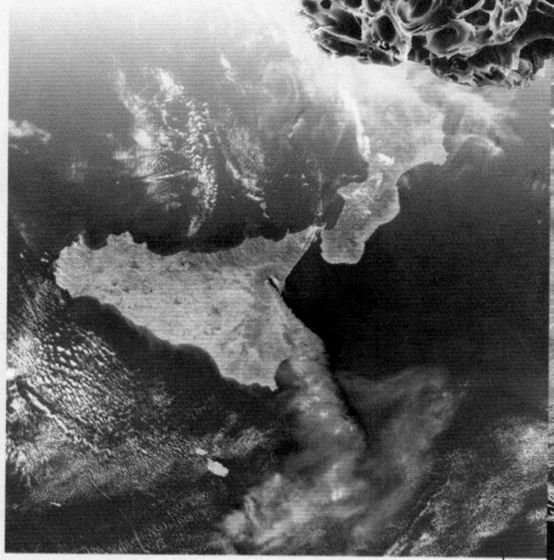

**ETNA ERUPTION**
*Mount Etna, Sicily, erupted in 2002. This satellite image shows a cloud of ash traveling southward across the Mediterranean Sea.*

# ENERGY IN THE ATMOSPHERE

46–47 The Earth and the Sun
Atmospheric circulation 446–49
Climate change 450–51
Global warming 452–53
Climate regions 454–57
Weather 458–77

THE SUN SUPPLIES THE ENERGY to produce our weather. The Sun's radiant heat is absorbed by the land and sea surface, and air is warmed from below by contact with it. Air movements transport this heat throughout the troposphere. Warmth from the Sun also provides energy for water to evaporate. Water vapor enters the air, and when it condenses to form clouds, the water molecules release the latent heat that was absorbed when the water first evaporated. A huge amount of energy is involved in this process: an average summer thunderstorm releases as much energy as burning 7,000 tons of coal in less than one hour; and a tornado releases enough energy to light the streets of New York City for one night.

## SOLAR POWER

The Sun's energy can be used to heat homes and generate power. Solar collectors on the roof can heat water, and solar cells can convert sunlight into electrical power. A solar furnace, seen here, uses hundreds of mirrors to focus light and heat onto a target in a central tower. It heats to more than 2,000°C (3,600°F), raising steam for power generation or for research.

## SOLAR BUDGET

The balance between the amount of solar energy received by the Earth's surface, its atmosphere, or the clouds, and the amount of energy reflected or radiated back into space, is known as the solar budget. The temperature at the visible surface of the Sun is about 10,472°F (5,800°C). It radiates its heat into space, but in all directions. The Earth, which is about 93 million miles (150 million km) away, receives only a tiny proportion of it. The amount of solar energy arriving as shortwave radiation at the topmost level of the atmosphere is known as the solar constant. Of the solar energy that reaches the top of the atmosphere, about 30 percent is lost due to backscatter by air molecules and reflection by clouds, land, and sea. Reflection will vary with both the amount of cloud cover and the nature of the Earth's surface (see opposite). About 70 percent of all incoming radiation is absorbed, mostly by the land and sea, but also by the air and clouds. The surface radiates the energy it has absorbed as longwave radiation, maintaining the balance between heat gains and losses. About 45 percent of the radiation reaching the Earth is visible as light.

**THE SOLAR BUDGET**
*Just over half the sunlight reaching the Earth penetrates to the surface, and almost one-fifth is absorbed by the atmosphere. The remainder is reflected by clouds and the surrounding surface or scattered back into space by air molecules.*

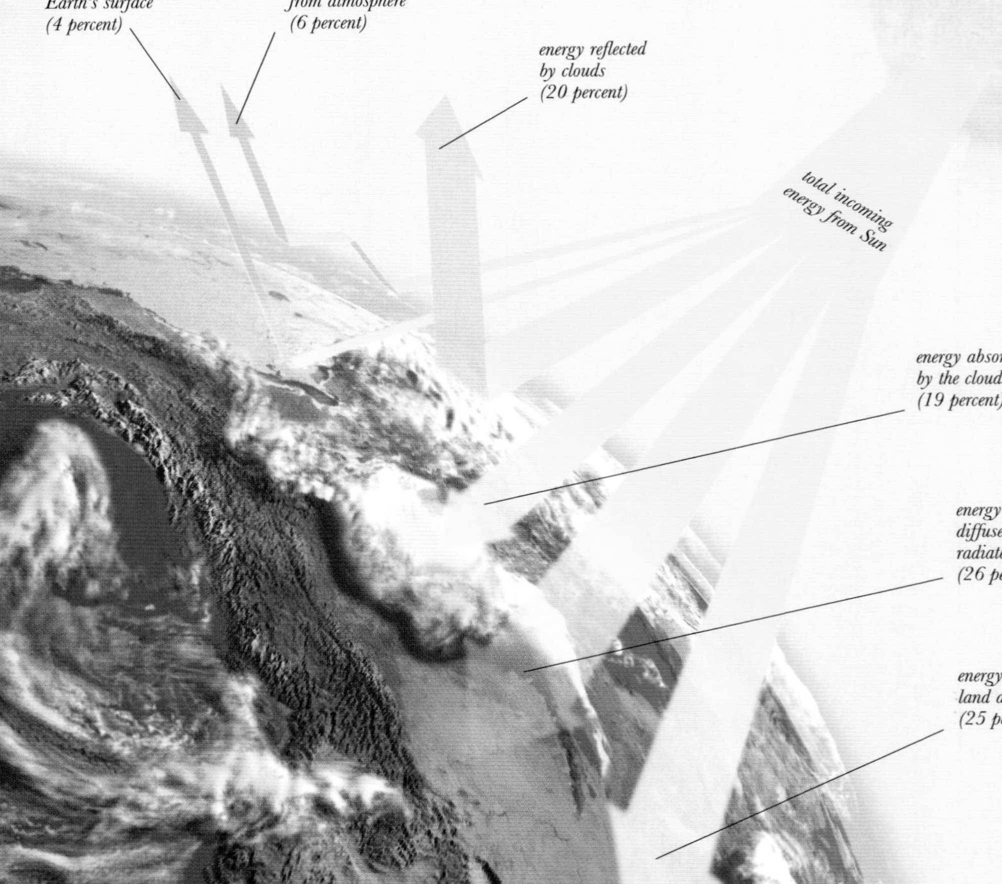

**SOLAR ACTIVITY**
*The solar flare, shown right, is produced by a sudden release of the Sun's energy. It creates a huge amount of radiated heat, which has a direct effect on our climate.*

**WAVELENGTHS**
*Incoming solar radiation is emitted at shorter wavelengths than that which is radiated back into space. Ultraviolet wavelengths are absorbed by oxygen and ozone, and the longer infrared wavelengths are absorbed by carbon dioxide and water vapor.*

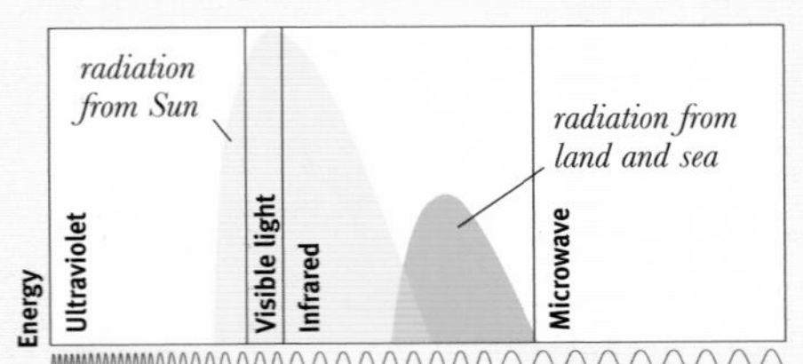

**SUNFLOWERS**
*The Sun's energy is important to most living things. Green plants, such as these sunflowers, need sunlight to photosynthesize food. The flowers turn to follow the path of the Sun.*

# GREENHOUSE EFFECT

Radiation from the Sun is most intense at short wavelengths. The atmosphere is almost completely transparent to this radiation, which passes through the air and is absorbed by the Earth's surface. When the land and sea are warm, they radiate the energy they have absorbed, but at longer wavelengths. Some of the longwave radiation is absorbed and re-radiated back to the Earth by atmospheric gases, including water vapor, carbon dioxide, methane, and ozone. This warms the gases, and although the Earth's radiation eventually escapes into space, it is retained long enough to warm the atmosphere. This is the greenhouse effect, and the gases causing it are called greenhouse gases. Without the natural greenhouse effect, temperatures would be 54–72°F (30–40°C) cooler. However, since 1900, the concentration of carbon dioxide in the atmosphere has increased by about 30 percent due to an increase in the burning of fossil fuels, atmospheric pollution, and deforestation (trees absorb carbon dioxide). Many scientists agree that the addition of carbon dioxide and other greenhouse gases to the air is changing the temperature balance, producing an enhanced greenhouse effect and increasing average temperatures over the Earth as a whole.

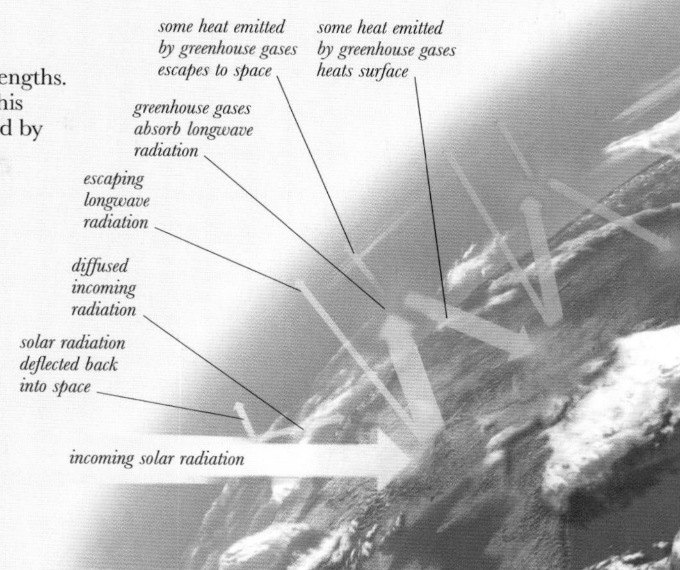


**GLOBAL WARMING**
*The lower atmosphere gradually becomes warmer when longwave solar radiation emitted by the Earth's surface is trapped in the atmosphere by carbon dioxide, water vapor, methane, and other greenhouse gases.*

**DARK CANOPY**
*This forest canopy in the Amazon basin is dark and will absorb sunlight. This affects the air temperature in the forest and the rate at which water evaporates.*

# ABSORPTION VARIABLES

Different surfaces reflect different proportions of the radiation falling on them. This is because pale colors reflect light and heat, and dark colors absorb them. The percentage of the radiation that a surface reflects is known as the albedo of that surface. Freshly fallen snow is highly reflective. It has an albedo of 75–95 percent, usually written as 0.75–0.95. Bright cloud is also very reflective. Its albedo is 0.70–0.90. So surfaces with a high albedo absorb very little heat, while dark surfaces absorb solar radiation much better. A broadleaf forest, with an albedo of 0.10–0.20, absorbs 80–90 percent of the radiation falling on it. A coniferous forest absorbs even more. If the surface of a large area is changed, the albedo may be altered. Removing forest and replacing it with farm crops, for example, increases the albedo from about 0.10 to about 0.20, and so more sunlight is reflected. When grassland is replaced with concrete, the albedo increases from about 0.15 to 0.22. Changes of this kind can affect the local climate, because increased reflection reduces convection in the air, and this in turn affects the amount of cloud cover.

**ICE AND SNOW**
*Most of Antarctica is covered by ice and snow. This reflects almost all of the sunlight, so the layers beneath the surface remain cold.*



# CHANGES IN STATE

Water exists in three different forms (or phases). It occurs as liquid water, as solid ice, and as water vapor, which is an invisible gas. Most substances can exist in these three phases, but water is remarkable in that it can exist in all three states at temperatures commonly experienced at the Earth's surface. For example, in winter, the surface of a pond may partially freeze over. Here water is present as ice (solid) and water (liquid), and in the air above the pond, there is water vapor (gas). All three are present in the same place, at the same time. When ice melts or changes directly into water vapor, and when liquid water evaporates, energy is needed to break the bonds holding the water molecules together. The molecules absorb this energy, but it does not change the temperature of the ice or water. The energy is latent (or hidden) heat. The same amount of heat is released when water vapor condenses, or changes directly to ice, and when liquid water freezes into ice. The release of latent heat when water vapor condenses provides the energy that fuels the growth of huge storm clouds. These clouds usually have great vertical depth due to unstable atmospheric conditions. This instability occurs because latent heat is released from water vapor as it condenses from the air, making the rising air warmer than the surrounding air, so it continues to rise.

**HEAVY RAINSTORM**
*Warm air rises and cools over the hills around Lake Shastina, California, causing water vapor to condense and resulting in a rainstorm.*

**WATERSPOUT**
*This waterspout, off the Bahamas, is visible because low pressure inside the wind funnel causes water vapor to condense into droplets.*

**TROUBLED SKIES**
*High-altitude polar clouds—seen here over Lapland—indicate unusually low temperatures in the stratosphere. Ozone thins rapidly in these conditions.*

# THE OZONE LAYER

Ozone is a natural component of the atmosphere, and a gas that plays a crucial part in the survival of life on Earth (see p.441). In the stratosphere, the ozone layer screens out short-wave ultraviolet (UV) radiation—a form of solar energy that can damage or kill living cells. During the 20th century, the ozone layer was seriously thinned by atmospheric pollutants, but following concerted international action in the late 1980s, the damage is gradually being reversed.

## OZONE AND LIFE

Toxic and highly reactive, ozone is a relatively rare form of oxygen that has three atoms in each of its molecules ($O_3$), rather than the normal two ($O_2$), which many organisms breathe. Most of the world's ozone is found in the ozone layer, which is 12–16 miles (20–25 km) above the Earth's surface. Here, ozone is continually formed from $O_2$, and it reverts back into $O_2$ when it breaks down again.

In the ozone layer, ozone molecules absorb UV-B and UV-C radiation—short-wavelength forms of ultraviolet that carry high levels of energy. They re-emit this energy as heat. As a result, the ozone layer acts as a screen, preventing most of this radiation from reaching the ground. Any depletion of the ozone layer is potentially dangerous for life, because UV-B and UV-C radiation can disrupt organic molecules, producing cancerous changes in living cells.

## OZONE DEPLETION

The ozone layer has been stable around the Earth for millions of years, but that situation ended in the late 1920s, when chemists synthesized the first chlorofluorocarbons (CFCs). These non-flammable gases turned out to have many uses: as solvents, aerosol propellants, and industrial cleaners, and also as coolants in refrigerators and air-conditioners. Unfortunately, CFCs can destroy ozone if they escape into the atmosphere. They can continue to do damage for decades or even centuries after they have been released, because they are stable and do not dissolve in rain.

During the early 1970s, two American chemists, Mario Molina and Sherwood Rowland, identified the theoretical threat of ozone depletion. Initially, reaction was skeptical, but when in 1983 scientists at the British Antarctic Survey discovered an ozone "hole" over the South Pole, the abstract threat suddenly became real.

**SOUTHERN EXPOSURE**
*The Antarctic ozone hole forms during the southern winter and reaches a maximum each spring. In this composite satellite map, the zone of greatest deficiency is shown in red.*

## INTERNATIONAL ACTION

The reality of ozone depletion prompted a rapid response. In 1987, two dozen industrialized countries signed the Montreal Protocol, which committed them to phasing out CFC production by 1996. In the period leading up to this deadline, "ozone-friendly" products were developed to take their place. In 1990, a further agreement was drawn up to include most developing nations, with a ten-year extension to allow them to adopt ozone-friendly technology. As a result of these measures, atmospheric CFC levels have reached a plateau and may already be starting to fall.

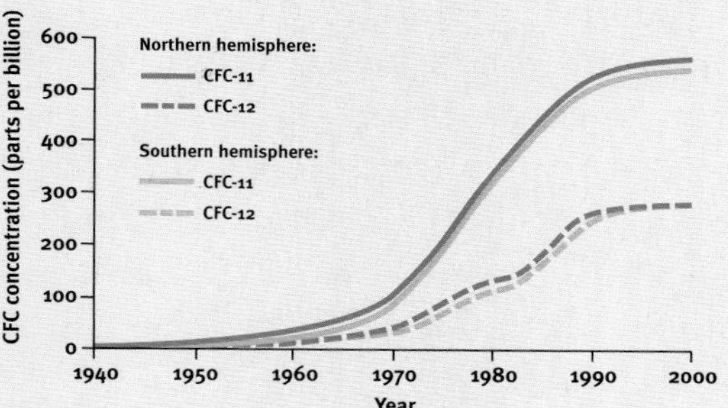


**ATMOSPHERIC CFC CONCENTRATIONS**
*Since the 1950s, levels of CFC-11 and CFC-12 have grown steeply before flattening out in response to a production ban. The concentrations are highest in the northern hemisphere, where most of the world's production was based.*

How long the ozone layer takes to repair itself will depend on the residence time of different CFCs—the period that an average molecule stays in the atmosphere—and also on the amounts originally released. For example, CFC-11, which was widely used in refrigeration, aerosols, and plastic foams, has a residence time of 45 years, while that of CFC-115 is 500 years. Current data suggests that the ozone layer might be near normal by 2050, though scientists remain wary of newly synthesized substances that may also have ozone-depleting effects.

**TASK FORCE**
*To prevent CFCs from escaping into the atmosphere, old refrigerators and freezers in southern England have their coolant removed before being scrapped.*

## GROUND-LEVEL OZONE

Ozone can be dangerous when it is near the ground. Such ozone is created by lightning, by any kind of machinery that creates electric sparks, and also by internal combustion engines. The exhaust gases of internal combustion engines include nitrogen oxides and organic compounds, and these can form ozone when they react—particularly in the presence of sunlight. In traffic-dense urban areas, such photochemical reactions can raise the air's ozone content to ten times its normal background level.

Ground-level ozone is toxic to plants, and it irritates the lining of the lungs, causing asthma and bronchitis. This form of pollution can be reduced by improvements in engine design, and by the use of catalytic converters in car exhausts.

# ATMOSPHERIC CIRCULATION

46–47 The Earth and the Sun
390–91 Circulation and currents
440–41 Structure of the atmosphere
442–43 Energy in the atmosphere

AIR IS CONSTANTLY MOVING. Worldwide, this movement constitutes the general circulation of the atmosphere, transporting warmth from equatorial areas to high latitudes, and returning cooler air to the tropics. It comprises three sets of "cells." These cells produce wind systems called prevailing winds, which drive the surface waters of the ocean, creating currents. The winds are deflected in opposite directions north and south of the equator by the Coriolis effect. In the upper troposphere, fast-moving jet streams form due to differences in temperature and pressure at the boundaries of air masses. They can increase the intensity and movement of low-pressure systems, resulting in climate cycles, or oscillations. El Niño is one such oscillation that affects weather patterns on a global scale. The behavior of the atmosphere is unpredictable, so it is difficult to know when these climatic events will take place.

*polar easterlies blowing away from high-pressure system over the North Pole*

*inner core of jet stream flows much faster than outer shell*

*polar-front jet stream*

*subtropical jet stream flows at about 30°N year-round*

*cloud indicating position of intertropical convergence zone*

*direction of Earth's rotation*

*cool air subsides at the South Pole, flowing toward mid-latitudes as the polar easterlies*

## CORIOLIS EFFECT

Air flowing toward or away from the equator invariably follows a curved path that swings it to the right in the northern hemisphere and to the left in the southern hemisphere. The reason for this was discovered in 1835 by Gustave-Gaspard de Coriolis (see panel, below). It used to be called the Coriolis force and is still abbreviated as CorF, but no force is involved, and nowadays it is known as the Coriolis effect. It occurs because the Earth is rotating counterclockwise on its axis, so as air moves across the surface, the surface itself is also moving beneath it but at a different speed. The magnitude of the Coriolis effect depends on the latitude, and the speed of the moving air. The Coriolis effect is greatest at the poles.

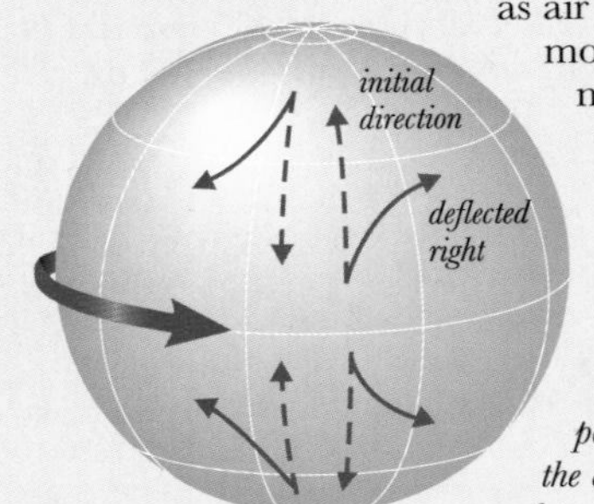


**HOW CORIOLIS EFFECT WORKS**
*All points on the Earth's surface complete one revolution every 24 hours, so a point at the equator travels farther and faster than a point at higher latitude. This rotation causes the air to shift to the right north of the equator and to the left south of the equator.*

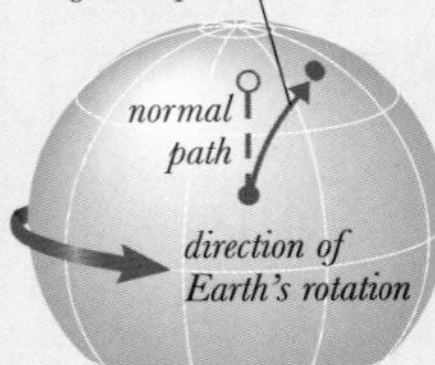


**DEFLECTING AIR**
*Air moving over the Earth's surface travels at a different speed from the Earth's rotation, so when plotted on a map, it appears to be deflected.*

**STORM BUILDUP**
*These storm clouds, forming off the southeast coast of the US, are in the northern hemisphere, and so rotate counterclockwise.*

### GUSTAVE-GASPARD DE CORIOLIS

Mathematician and mechanical engineer Gustave-Gaspard de Coriolis (1792–1843) was born in Paris, France. In 1816, he became a tutor at the École Polytechnique, where he carried out research on friction and hydraulics. He went on to discover the Coriolis effect, publishing his results in 1835. He died in Paris at age 51.

## FLUID CIRCULATION

The air in the Earth's atmosphere moves in response to heating by the Sun. When a fluid is heated, its molecules absorb energy. This allows them to move faster, so they move farther apart, and the fluid expands. Expansion makes it less dense, because a given volume contains fewer molecules than before. Denser fluid, which is heavier, then sinks beneath the less dense fluid, pushing it upward. Air acts like a fluid when it is heated by contact with the Earth's surface that has been warmed by the Sun. As warm air rises, air pressure decreases, since there is a smaller amount of air above it pressing down. This allows the air to expand, but that uses energy, so it also cools. When the air reaches a level where its density equals that of the air immediately above it, it stops rising. If the air then subsides, the pressure on it increases. It absorbs energy and its temperature rises. These changes give rise to the different cells of the Earth's circulation system.

**HOW AIR MOVES IN CELLS**
*Atmospheric circulation in each hemisphere consists of three cells. At the equator, warm air rises, moves north and south, and subsides at the tropics. The air that flows back to the equator creates the Hadley cells. Cold air subsides over the poles, flows to mid-latitudes, where it meets air from Hadley cells and rises, and returns to the poles, forming polar cells. Rising mid-latitude air divides, flowing to the poles and to the equator, forming Ferrel cells.*

*polar-cell air circulation is caused by subsiding air at the pole flowing south toward the equator. On meeting Hadley-cell air, it rises*

*air in Ferrel cells rises to the tropopause and then divides, some air flowing to the pole and some toward the equator*

*winds in mid-latitudes produced by Ferrel cells flow from the west*

*northeasterly trade winds deflected to right by Coriolis effect*

*air within Hadley cells rises moist at the equator and subsides dry at the subtropics*

*southeasterly trade winds*

*Roaring Forties*

# JETS AND WAVES

Jet streams are narrow, winding ribbons of strong wind in the upper troposphere. They occur at the top of fronts, marking the boundary between air masses at different temperatures (see pp.460–61). Since they are produced by a temperature difference, they are also known as thermal winds. Sometimes waves form in jet streams, and travel along with them. Wind speed at the center of a jet stream is about 65 mph (105 km/h), but speeds of 310 mph (500 km/h) have been recorded. Polar-front jet streams flow in temperate latitudes, between 30° and 50°N and S, depending on the season. The subtropical jet stream is at about 30°N and S year-round. These jet streams blow with cold air on their left in the northern hemisphere, and on their right in the southern hemisphere. They therefore blow from west to east in both hemispheres. In summer, there is a jet stream that blows from east to west at about 20°N, crossing Asia, southern Arabia, and northeastern Africa.

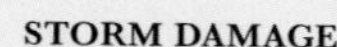

**JET STREAM**
*These bands of cirrus cloud mark the position of the jet stream over the Red Sea and Egypt. The cloud forms in air that is lifted as it is drawn into the core of the jet stream.*

**STORM DAMAGE**
*During the night of October 16, 1987, there was a severe storm in the UK, caused by an unusual northward shift of the jet stream.*

# PREVAILING WINDS

Although the wind may blow from any direction, when directions at a particular place are compared over a long period, it is usually found that the wind blows from one direction more than it blows from any other. This is the prevailing wind. The general circulation of the atmosphere produces belts of prevailing winds around the world. In the tropics, wind blowing toward the equator, as part of the Hadley cells, is deflected to the right, and forms the northeasterly and southeasterly trade winds. Air flowing away from the high pressure over the poles is deflected to the right and forms the polar easterlies. Between these, in middle latitudes, the Ferrel cells produce westerly winds. Over the world as a whole, the force of the easterly winds is precisely balanced by the force of the westerlies. If this were not so, the rotation of the Earth would either speed up or slow down. In sandy deserts, and in polar regions, prevailing winds produce dunes, and drifts, with characteristic shapes. Sand and snow grains are blown up a slope, and tumble down the far side. This creates high dunes and snow drifts, with sharp, sinuous crests, extending for great distances. Knowing about prevailing winds is important for locating structures such as airport runways. Sailors used to depend on prevailing winds to cross oceans and would avoid places such as the doldrums, where there is little wind.

**HIGH DUNES**
*The sinuous crest of these Namibian Desert dunes shows that the prevailing wind flows along the dune, sometimes a little to the left or to the right.*

**WIND SYSTEMS**
*Global wind belts result from the way air circulates. Seasonal differences in the northern hemisphere are due to large landmasses where pressure systems change from high in January to low in July.*

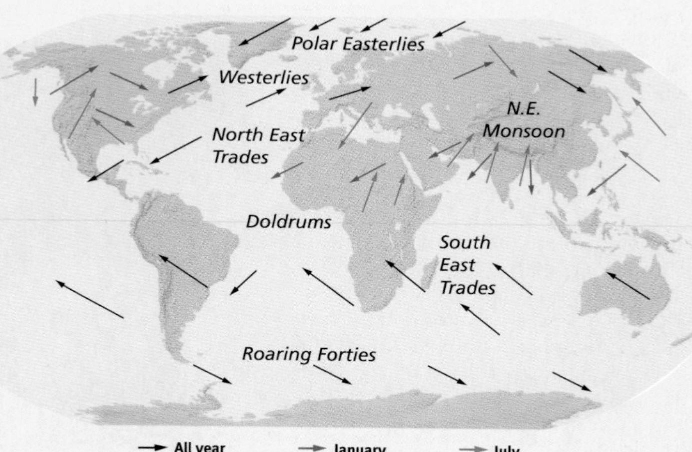

**INTERTROPICAL CONVERGENCE ZONE**
*In equatorial regions, the meeting of trade winds at the equator is known as the intertropical convergence zone. Here, a band of cloud indicates its position.*

# MOISTURE AND THE OCEANS

Oceans cover about 70 percent of the surface of the Earth, and they play a major part in the formation and regulation of climates. Because water warms and cools much more slowly than dry land, the oceans moderate temperatures, making summers cooler and winters warmer than they would be on a completely dry planet. Ocean currents (see p.390) also transport heat from the equator into high latitudes. The North Pole has a much warmer climate than the South Pole because it lies at the center of the Arctic Ocean, rather than the center of the continent of Antarctica, and there is always liquid water beneath the sea ice. Precipitation—rain, snow, hail, fog, dew, and frost—consists of water that came originally from the ocean, and that returns there through rivers and groundwater. This is an obvious contribution to climate, but it is not the only one. The condensation of water vapor in rising air releases latent heat (see p.443) that makes the air continue to rise. This is how storm clouds form. The energy of tornadoes and hurricanes is provided mainly by condensation. Clouds also shade the surface by reflecting incoming sunlight, which has a cooling effect. The climatic importance of the oceans is most evident in years with low rainfall and crops wither in a drought.

**WATERSPOUT**
*When a tornado moves over water, it is called a waterspout. This one occurred off the coast of Spain and killed six people. The spray ring at the base consists of water whipped up from the surface.*

**GULF STREAM**
*This satellite image of the east coast of the US shows temperature differences in the ocean due to warm (orange and yellow) and cool (blue and green) currents. The warm current is the Gulf Stream, which transfers heat poleward.*

**ICEBERGS**
*These icebergs near Cuverville Island, Antarctica, provide information about ocean currents as they move north into warmer waters and then melt.*

# WINDS AND CURRENTS

Prevailing winds, blowing over the ocean, drive the surface water along particular paths to form ocean currents. The trade winds drive equatorial currents westward in both hemispheres. As they approach land, they turn away from the equator into higher latitudes. Here the Coriolis effect swings the currents further, until they enter the middle latitudes, where prevailing winds come from the west. They then form currents flowing eastward, but the Coriolis effect continues to deflect them. Finally, they head back toward the equator. This circular movement forms gyres, one in each major ocean, apart from the Southern Ocean. On the western sides of oceans, currents are narrow, deep, fast-flowing, and carry warm water; on the eastern side, they are wide, shallow, slow-moving, and carry cool water. In the Southern Ocean, there is no land to deflect the Antarctic Circumpolar Current. It is the only current to flow all the way around the world. Currents do not flow precisely in the direction of the wind. This is because friction between the surface layer of water and the layer below, combined with the Coriolis effect, deflects the current at about 45° to the right of the wind direction in the northern hemisphere, and to its left in the southern hemisphere.

**TYPHOON**
*Typhoon Odessa (left) formed in the tropical trade-wind belt of the western Pacific. It moved northward until it met the mid-latitude westerlies.*

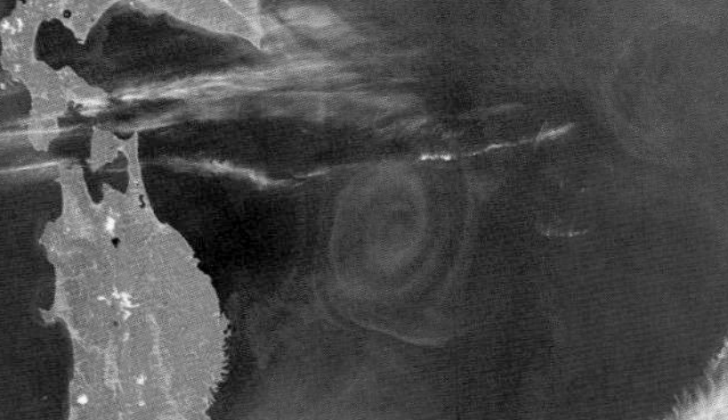

**EDDY OFF JAPAN**
*The winds around pressure systems drive eddies (small currents), such as this one, visible as a green spiral below the white streaks of clouds.*

## VARIATION AND OSCILLATION

Climates change due to cyclical variations in the distribution of pressure over varying periods. An example of a climate cycle is the North Atlantic Oscillation (NAO), which results from pressure differences between the Azores high and Iceland low. When pressure is higher than average over the Azores, and lower over Iceland, the NAO index is said to be positive (high). At this time the jet stream flows strongly over the Atlantic, bringing mild, wet winters to Europe and warm, dry winters to the Mediterranean. When the pressure difference is small, the NAO index is said to be negative (low). This blocks the jet stream, so in winter cold air enters Europe from Asia, and rainfall increases in the Mediterranean region. The NAO is unpredictable, but tends to occur every couple of years.

**HARSH WINTER IN CANADA**
*Tuktoyaktuk, northwest Canada, shown here, experiences harsher-than-normal winters due to climatic oscillations.*

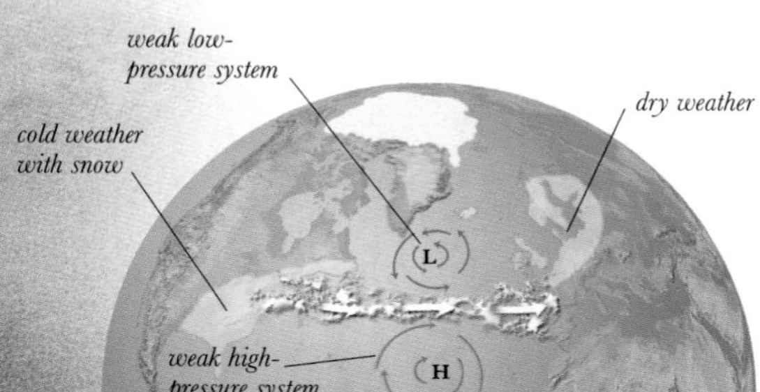


**NORTH ATLANTIC OSCILLATION (NEGATIVE)**
*A negative (low) North Atlantic Oscillation exists when pressure systems are weak, bringing mild winters to the northwestern Atlantic, cold weather with snow to northeastern North America, cold, dry winters to northern Europe, and wet weather to the Mediterranean.*

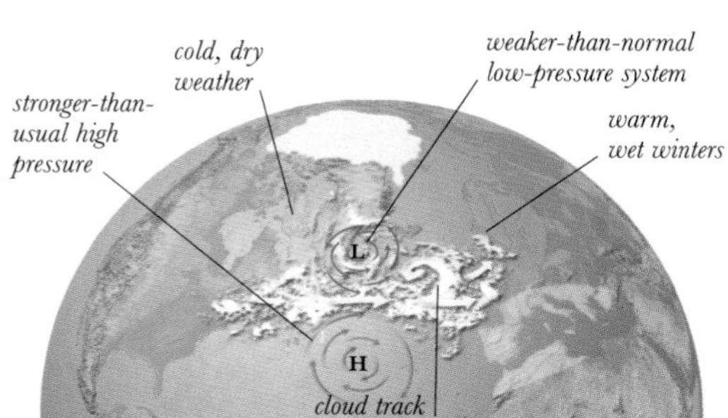


**NORTH ATLANTIC OSCILLATION (POSITIVE)**
*A positive (high) North Atlantic Oscillation index brings cold winters to the northwestern Atlantic and northeastern North America, mild winters and higher rainfall to northern Europe, and dry weather to the Mediterranean region.*

## EL NIÑO

El Niño is a reversal in the normal flow of the South Equatorial Current in the Pacific that brings dramatic changes in the weather at intervals of between two and seven years. The effects of El Niño become evident in late December, and are associated with a change in the distribution of air pressure over the South Pacific. This change is known as the Southern Oscillation. The full cycle is called an El Niño–Southern Oscillation (ENSO) event and includes La Niña, the opposite of El Niño. Ordinarily, pressure is high over the eastern South Pacific and low in the west. This produces the trade winds that drive the South Equatorial Current, carrying warm water away from South America and toward Indonesia. This pattern produces heavy rain over Indonesia and extremely arid conditions along the coast of Peru and northern Chile. During an ENSO event, the pressure difference weakens or reverses, as do the trade winds. Heavy rain falls over Peru and Chile, causing deserts to bloom, while Indonesia experiences drought, sometimes leading to serious forest fires.

**FLOODING**
*Heavy rains caused by El Niño cause extensive damage. Here, a mudslide in California has flowed down a slope and covered a section of road, blocking it to traffic. Other risks include floods and landslides.*

rising warm, moist air associated with heavy rainfall and low pressure
southeast trade winds
descending warm air associated with dry conditions and high pressure
accumulation of warm water
South Equatorial Current
upwelling of cold, nutrient-rich waters, since warm surface waters are shallow

**NORMAL CLIMATIC CONDITIONS**
*Ordinarily, a low-pressure system over Australia draws the southeast trade winds across the eastern Pacific from a high-pressure system over South America. These winds drive the warm South Equatorial Current toward the coast of Australia. Off the coast of South America, the warm water layer is only 300 ft (100 m) deep, so upwellings of cold water bring nutrients to the surface.*

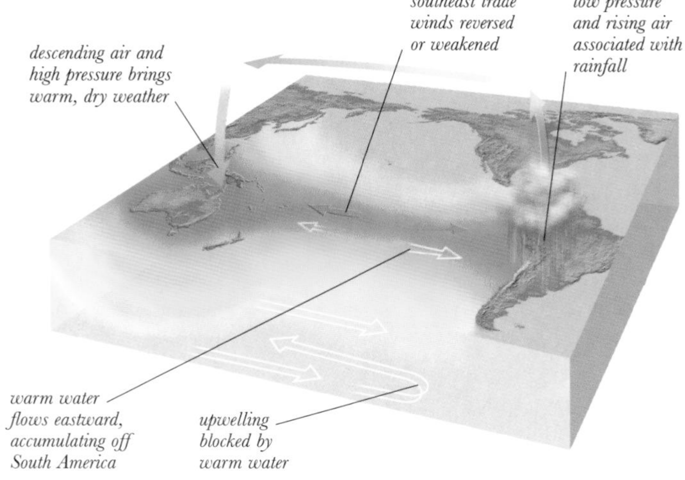
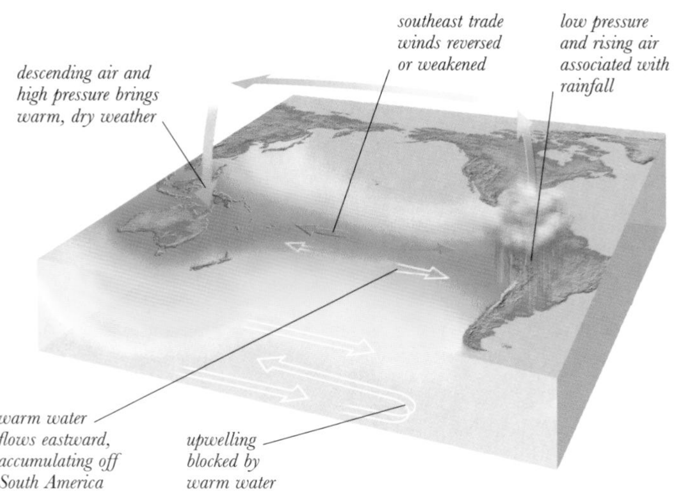


**EL NIÑO EFFECT**
*During El Niño, the pressure systems that normally develop over Australia and South America are much weaker or reversed, which is reflected by the flow of the trade winds and ocean currents. Warm water to a depth of about 500 ft (152 m) flows eastward, blocking the normal upwelling of nutrients along the west coast of the Americas, and devastating fish stocks.*

### EL NIÑO HUMAN IMPACT

El Niño brings drought to Indonesia and northeastern South America. Soils dry out and crack. This farmer must hope there is enough irrigation water to keep his crop alive, and that the rains will soon return. Indonesian farmers commonly clear the ground by burning off old, dead vegetation in preparation for planting. In El Niño years, the fires may blaze out of control, as they did during the unusually severe El Niño of 1997–98. When subsistence crops, such as rice and corn, fail, the government has to purchase the shortfall of food from elsewhere. In 1997, when over 105 million people in Indonesia were still dependent on agriculture to make a living, over five billion tons of rice was imported. The cost puts considerable additional strain on the country's economy.

# CLIMATE CHANGE

26–41 The Earth's past
48–49 The Earth and Sun
121 Meteorite impacts
262–63 Glaciers
442–43 Energy in the atmosphere
Global warming 452–53

THE WORLD'S CLIMATES are changing constantly. In the past they have been both warmer and cooler than they are now, due to catastrophic events and cycles that have natural causes. At present, the average global temperature is increasing, but the rise is not spread evenly around the Earth and some regions are becoming cooler. Several factors are contributing to the present warming. Most climatologists agree that part of it is due to an enhanced greenhouse effect, which is caused by the release of certain gases when fossil fuels—coal, oil, and natural gas—are burned.

## PAST EVIDENCE

Reliable written climate records are available for about the last 200 years, but until recently there were few weather stations in much of Africa, Asia, the Middle East, and South America, so scientists must reconstruct past climates from other types of evidence. Tree rings, for example, vary in width, depending on growing conditions, and details of climate over more than 8,000 years have been obtained from bristlecone pines in California. Many species of beetles live only within a narrow temperature range, and their wing cases, which are sometimes preserved in the soil, can provide evidence about the climate in which the insects lived. Plant pollen also preserves well, and can reveal the plant communities that inhabited a particular area and thus the climate at the time. Scientists also study cores drilled from seabed and lakebed sediments, and from polar ice sheets, in search of clues to water temperatures, rainfall, and the composition of the gases in the atmosphere.

**UNICELLULAR FORAMINIFERAN**
*This is a magnified image of the shell of a temperature-sensitive animal found in sedimentary rocks.*

**POLLEN GRAIN**
*Scientists use pollen to reconstruct past climates. It is extracted from cores drilled in sediment or peat.*

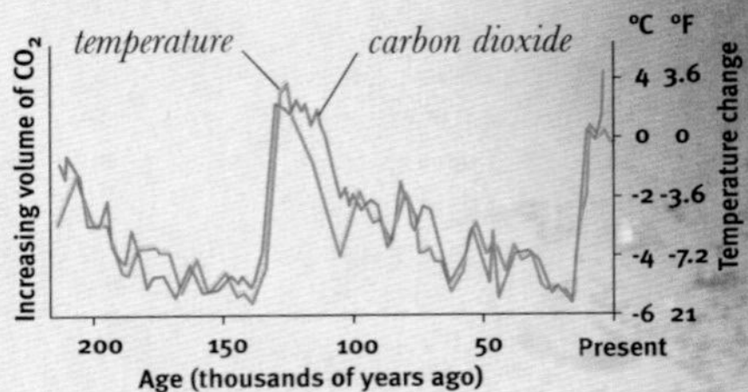


**CARBON DIOXIDE VARIATION**
*From this graph, using data from Vostok Station in Antarctica, it can be seen that the amount of carbon dioxide in the atmosphere and temperature have varied together over the past two ice age cycles.*

**DESERTIFICATION**
*Deserts increase in area when climates become hotter and drier, as here in the Sahel, where the desert has encroached on a village.*

## ORBITAL CYCLES

Fluctuations in the Earth's orbit around the Sun and rotation on its axis over time are reflected by cyclical changes in the climate. When these fluctuations coincide, temperatures fall sufficiently to trigger an ice age. The Earth's orbital path extends from almost circular to slightly elliptical over a cycle of about 100,000 years, altering the distance between the Earth and the Sun. The tilt of the Earth's axis varies over 42,000 years, altering the area directly exposed to the Sun. The rotational axis of the Earth wobbles over 25,800 years, causing the dates of the solstices and equinoxes to move.

**ORBITAL VARIATIONS**
*The orbit and rotation of the Earth is not constant, but changes cyclically over time. Its orbit fluctuates from elliptical to circular (when the orbit is described as more or less eccentric), and it is also subject to tilt on its axis, and wobble through movement of the axis. Over time, these changes in the Earth's orbit affect temperature and can be correlated with specific events, such as ice ages.*

**ORBITAL ECCENTRICITY**
*100,000-year cycle*

*circular Earth orbit*
*Sun*
*Earth*
*elliptical Earth orbit*
Present day
Eccentricity
200 100 0 -100
Thousands of years

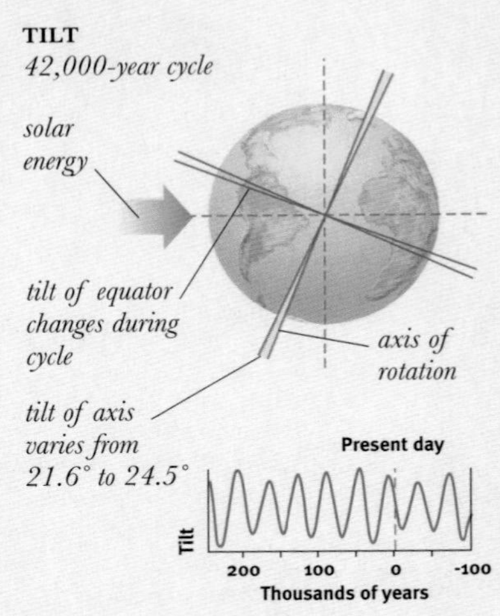


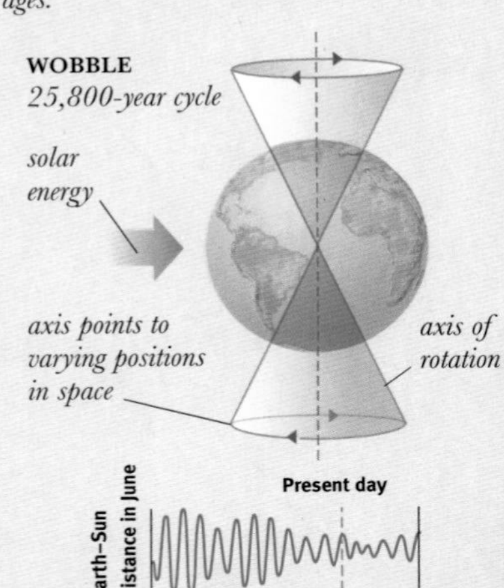


### MILUTIN MILANKOVICH

Serbian mathematician and climatologist Milutin Milankovich (1879–1958) was determined to test a suggested link between climate and the levels of solar radiation reaching the Earth. He identified three cyclical changes (see left) that might be relevant, and calculated the dates when these combined to minimize and maximize solar radiation over hundreds of thousands of years. The dates coincided with the ice ages.

# OTHER CAUSES

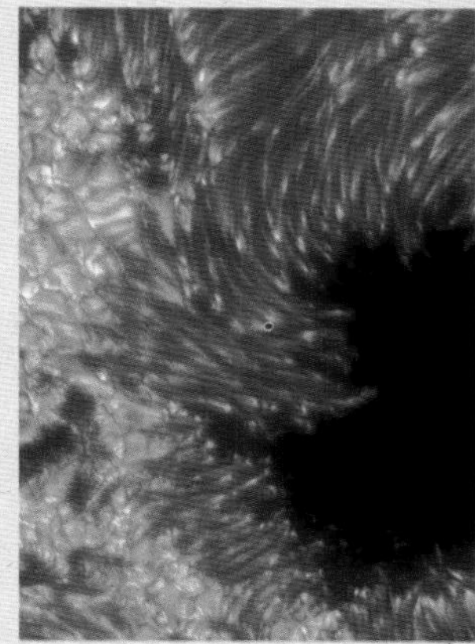

**SOLAR ACTIVITY**
*Sunspots reflect variations in solar activity. They have two distinct regions: a dark center, called the umbra, and a brighter, hotter surrounding area, the penumbra.*

The present climatic warming may be due, at least in part, to changes other than higher concentrations of greenhouse gases. It may be a natural trend as the Earth emerges from the Little Ice Age, a cold period that lasted from the 15th to the mid-19th century. Warming may also be caused by natural oscillations. The North Atlantic Oscillation (see p.449) remained strongly positive for much of the 1990s. It brought mild winters to Europe, and increased the flow of warmer water into the Arctic Basin, melting the sea ice in places. The warmest year in recent times (1998) was due entirely to a strong El Niño event. The retreat of glaciers in tropical Africa, often blamed on global warming, is due to reduced snowfall; no rise in temperature has been recorded. Changes in solar output may also affect climate. Throughout history, periods when few sunspots were recorded have coincided with cold conditions. A period of low sunspot activity called the Maunder Minimum was the coldest part of the Little Ice Age. In contrast, when solar output increases, as it did in the 1990s, the solar wind, a stream of particles from the Sun, intensifies. This deflects the cosmic radiation, comprising charged particles that react with atmospheric molecules to trigger cloud formation. Increased solar output therefore results in reduced cloudiness. Less sunlight is reflected from cloud tops, so more heat is absorbed by the surface, raising air temperature. Some scientists believe this explains the warming experienced during the 1990s, but others remain unconvinced.

**FOREST FIRES**
*Scientific evidence suggests that there is a link between climate and fire. Burning vegetation releases carbon dioxide into the atmosphere, and the smoke screens out solar radiation. Here, flames and smoke engulf Northfork, Wyoming.*

**VOLCANIC ERUPTIONS**
*Clouds of fine dust and ash particles rise from Cerro Negro, a volcano in Nicaragua. Most of the particles will fall to the ground in a matter of hours, but any that enter the stratosphere will remain airborne much longer, shading the Earth's surface and slightly reducing the temperature.*

# THE CURRENT TREND

At present, the average global temperature is rising. Calculating the size of the rise is difficult, because the change is very small. The temperature is recorded in three ways. Surface weather stations record temperatures at least twice each day—there are about 10,000 weather stations on land and 7,000 on ships. Weather balloons monitor the temperature of the upper air and the air pressure at different altitudes. If the pressure at the surface and at a specified altitude are known, the temperature at that altitude can be worked out precisely. Finally, orbiting satellites monitor the temperature of the atmosphere above about 6,500 ft (2 km). The most reliable data indicates that the average global temperature rose by about 1.08°F (0.6°C) between the late 19th century and 2000. The best present estimate is that the average temperature is increasing by about 3.1°F (1.7°C) every 100 years.

**BREEDING EARLIER**
*The North American red squirrel is now breeding earlier in the year as a result of global warming. Due to mild winters, plants germinate sooner, increasing food availability.*

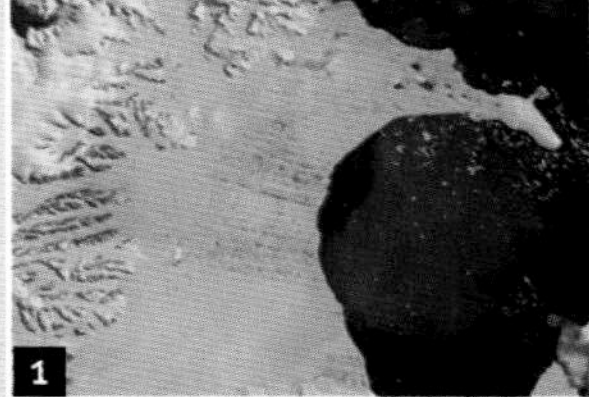


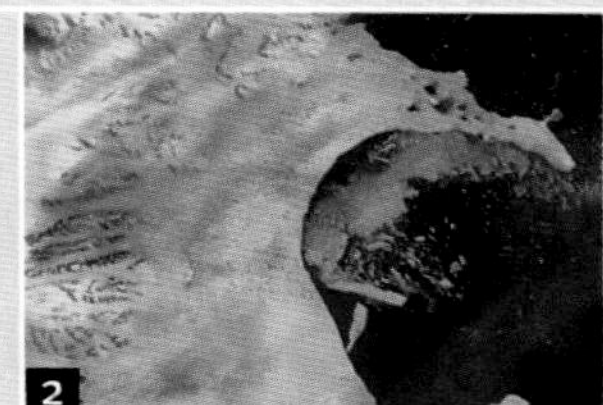


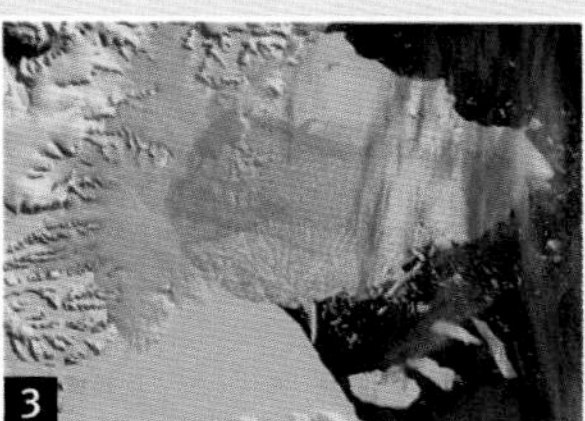


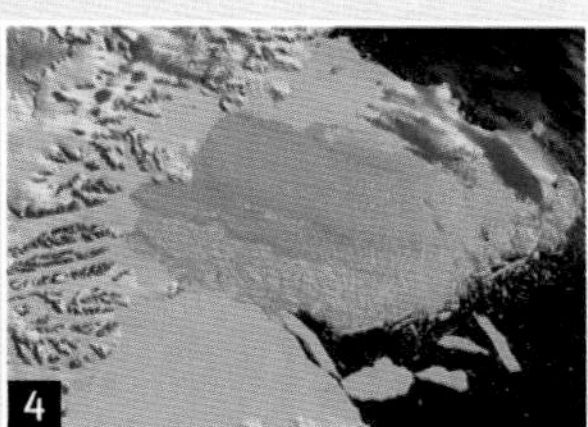


**LARSEN B ICE SHELF BREAKUP**
*Over the last 50 years, the Antarctic Peninsula has warmed by 2.5°C (4.5°F), probably due to a change in water flow in the Southern Ocean. This has caused part of the Larsen B ice shelf to disintegrate.*

**A SHRINKING WORLD**
*One-fifth of the Arctic's summer sea ice could disappear by 2050. Because sea ice is used by polar bears when they hunt, its decline threatens their survival.*

# GLOBAL WARMING

Since the early 1800s, the atmosphere's carbon dioxide levels have climbed by nearly half, chiefly due to the increased use of fossil fuels. This has accentuated the greenhouse effect (see p.443), making the Earth warm up, and during the coming century global warming is likely to accelerate still more. On a local scale this is significant enough, but for an entire planet, it represents a huge and momentous change.

## CHANGING CLIMATES

Scientists generally agree that global warming is a human-made phenomenon, not the result of natural climate change. Increased carbon dioxide levels are the main cause, but other greenhouse gases such as methane and chlorofluorocarbons (CFCs; see pp.444–45) also play an important part. Forecasting the effect of global warming is extremely difficult, because so many variables are involved. However, Earth's dynamic systems—such as the atmospheric circulation and ocean currents—mean that warming will not be felt equally all over the world.

According to current computer forecasts, areas at high latitudes will experience the most rapid warming over the next 50 years. This phenomenon can already be seen in the Antarctic Peninsula, where summer temperatures have risen by 3.5°F (2°C) since the 1960s. Paradoxically, warmer parts of the world may also experience abnormally cold periods as weather patterns shift.

## RISING SEA LEVELS

Since the early 20th century, most of the world's glaciers have retreated, and their meltwater has helped to raise sea levels by a fraction of an inch each year. If melting breaks up the West Antarctic Ice Sheet (see p.278), the rise would be far faster. On a warming planet, sea levels will also increase through thermal expansion, because, above 39°F (4°C), warm water is less dense than cold water.

Projections of sea-level rise vary widely, because some climatic conditions may accelerate the change, while others could slow it. Early estimates of a rise of 13 ft (4 m) or more by 2100 now seem alarmist, but even 20 in (50 cm) could have dramatic effects on coastal cities and island nations throughout the world.

**ISLANDS AT RISK**
*Atolls are often less than 3 ft (1 m) above sea level, which makes them, and their surrounding coral reefs, prone to global-warming damage.*

## RAIN AND DROUGHT

As the Earth's surface temperature rises, the additional heat increases the amount of seawater that evaporates, which in turn creates more rain clouds and triggers storms. Already there is evidence that storms are becoming more frequent, although short-term changes do not necessarily become long-term trends. Recent storms also seem to leave more damage in their wake, although again, this has to be viewed with caution, because some of this effect may be caused by the increasing density of the human population throughout the world.

As precipitation patterns change and pressure systems shift, some parts of the Earth will become wetter, while others may become more prone to drought. According to recent projections, one of the drier regions will be the American Midwest—an area where rainfall is already a factor limiting the growth of food.

**FAILED HARVEST**
*Hit by drought, corn withers in Texas. In a warmer world, this could become a familiar sight in grain-growing areas.*

## WARMING AND WILDLIFE

By changing weather patterns, global warming will alter natural habitats also. Where it becomes warmer and wetter, woodland will encroach into grassland, and coniferous forest in the far north is likely to spread into the tundra zone. Mountain habitats will extend to higher altitudes as average temperatures increase.

For some plants and animals, these changes will open up new opportunities, but for many species, such as polar bears (see left), survival will become more difficult.

## THE FUTURE

Global warming may turn out to be the largest environmental change ever triggered by human activities. The Kyoto Protocol of 1997 sought to bring carbon dioxide emissions under control, but even if all countries agree to ratify it, global warming will still be hard to reverse. A major switch from fossil fuels will be needed, and renewable sources—such as solar power—will have to play a much more significant role. According to the Intergovernmental Panel on Climate Change, emissions could start to drop by 2050, but, even if this happens, decades will pass before the warming trend comes to an end.

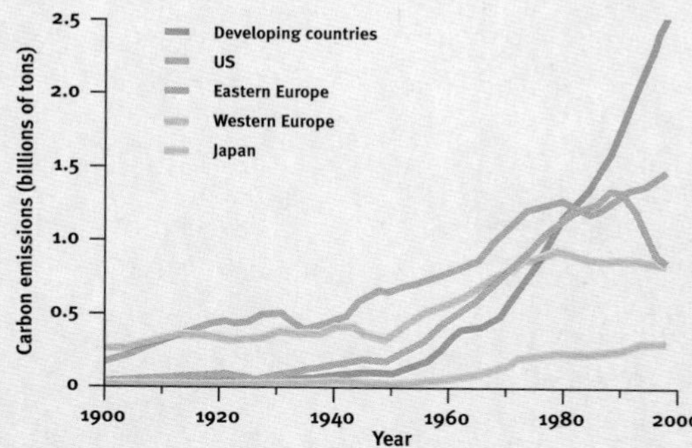


**WORLDWIDE CARBON EMISSIONS**
*Carbon emissions have climbed steeply in the last 100 years, but emissions are now stabilizing in developing nations. Unfortunately, they are still growing in the developing world.*

# CLIMATE REGIONS

282–301 Deserts
302–21 Forests
322–31 Wetlands
334–41 Grasslands
342–43 Tundra
Weather 458–77

SCIENTISTS DIVIDE THE WORLD into regions according to their climates. The first system, devised by Aristotle, based the regions on the height of the Sun above the horizon. This method produced three belts, the torrid, temperate, and frigid zones, separated by the tropics of Cancer and Capricorn and the Arctic and Antarctic circles. Modern climate maps show many more regions, the distribution of which is determined only partly by latitude. The elevation above sea level and distance from the ocean are just as important. Continental interiors have climates very different from those of ocean islands in the same latitude, and mountains have colder and wetter climates than those of the adjacent lowland plains.

## CLASSIFICATION

Climate classifications arrange climates according to their main characteristics, and give each climate a short, simple name to identify it. Modern classifications are of two types: generic and genetic. Generic classifications are based mainly on aridity and temperature, and relate these to plant growth. The most widely used generic classifications were devised by Wladimir Köppen (Germany) and Charles Thornthwaite (US). There are fewer genetic classifications. These are based on the physical causes of particular types of climates and are related to the general circulation of the atmosphere and its implications. Hermann Flohn (Germany) and Arthur Strahler (US) have proposed generic classifications.

**NEW YORK**
*In winter, the climate of New York is influenced by cold, continental air masses, so it has low temperatures and snow.*

**VENICE**
*Winters in Venice are mild, even though the city is slightly farther north than New York (see above). The Alps protect it from the cold, continental air masses of Europe.*

## WORLD CLIMATES

One of the most significant differences between climate regions is the amount of solar radiation they receive. Polar regions receive the least energy, so arctic and antarctic climates are characterized by extreme cold. With increasing distance from the poles, these conditions give way to temperate climates, with warm summers, cool or cold winters, and moderate precipitation. Exceptions are continental interiors, where distance from the ocean produces cool deserts. Subtropical deserts, such as the Sahara, are arid. Equatorial climates are hot and, since oceans cover most of the equatorial belt, they are wet. The map below (and the profiles of climate types on the pages that follow) are based on the classification devised by Wladimir Köppen.

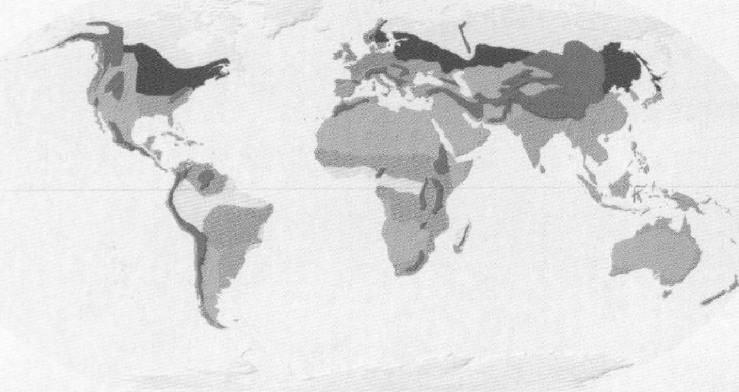

Hot climates with year-round rain
Hot climates with monsoon rain
Hot climates with seasonal rain
Hot, dry climates
Cool, temperate maritime climates
Warm, temperate climates
Cool, temperate continental climates
Cold climates
Mountain climates

**CLIMATIC BOUNDARY**
*The Sneffels Mountains in Colorado are visibly colder than the San Juan Forest in the foreground.*

### CLIMATE PROFILES

The pages that follow contain profiles of the world's main climate regions. Each profile begins with the following summary information:

**ANNUAL REGIMEN FOR** Name of example locality

**TOTAL PRECIPITATION** Total annual rainfall

**TEMPERATURE OF WARMEST MONTH** Average figure for the warmest month of the year

**TEMPERATURE OF COLDEST MONTH** Average figure for the coldest month of the year

TROPICS

## Hot climates with year-round rain

**DISTRIBUTION** In equatorial regions, comprising the northern part of South America, southern Central America, and central Africa

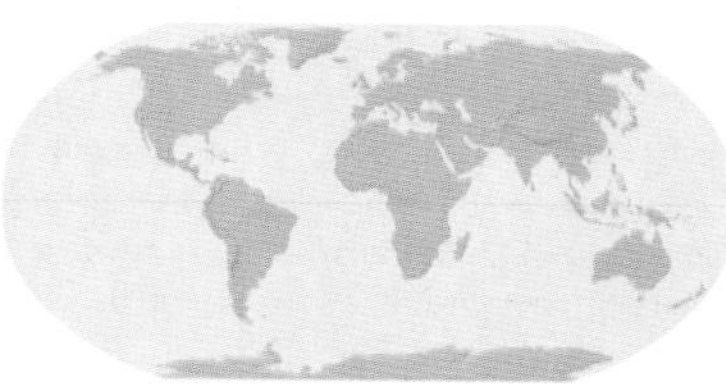

**ANNUAL REGIME FOR** Belém, Brazil

**TOTAL PRECIPITATION** 96 in ( 2,439 mm)

**TEMPERATURE OF WARMEST MONTH** 90°F (32°C)

**TEMPERATURE OF COLDEST MONTH** 86°F (30°C)

**FOREST ANTELOPE**
*The bongo is the largest, and most distinct, of the forest antelopes. It lives in the lowland, montane, and highland forests of tropical Africa.*

This type of climate is hot—the temperature at sea level rarely falls below 64°F (18°C)—and wet, with abundant rainfall distributed evenly through the year. The noon Sun is always overhead somewhere within the tropics, as it moves between the tropics of Cancer (June) and Capricorn (December) during its annual cycle. As a result, seasonal differences are small and there is little variation in day length. The climate supports luxuriant, evergreen rainforest in the lowlands, on mountains, giving way, with increasing height, to other types of forests, culminating in cloud forest. It is estimated that half of the world's animal species live in these regions.

**KENYAN GIANT FIG**
*Fig trees, such as this one in Kakamega Forest, western Kenya, are typical plants of tropical rainforests. Some species of figs are parasites that use other trees for support and then strangle them.*

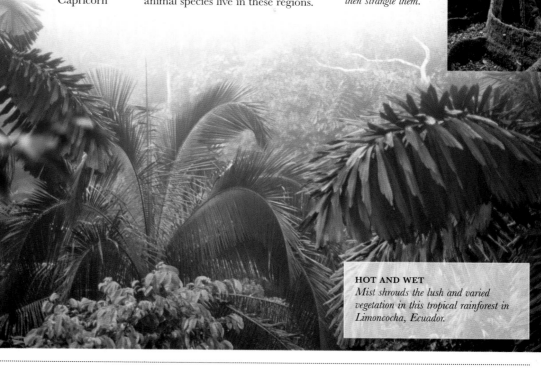

**HOT AND WET**
*Mist shrouds the lush and varied vegetation in this tropical rainforest in Limoncocha, Ecuador.*



TROPICS

## Hot climates with monsoon rain

**DISTRIBUTION** In Southeast Asia, Korea, southern Japan, and the northern tip of Australia; also in localized areas of west Africa, and northeastern South America

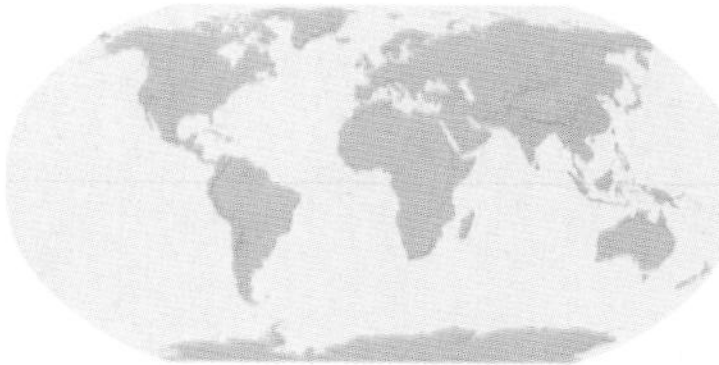

**ANNUAL REGIME FOR** Darjiling, India

**TOTAL PRECIPITATION** 120 in (3,037 mm)

**TEMPERATURE OF WARMEST MONTH** 66°F (19°C)

**TEMPERATURE OF COLDEST MONTH** 46°F (8°C)

Monsoon climates have two seasons. In the Indian subcontinent, where the monsoon is most extreme (see p.463), the winter season is very dry. The prevailing winds blow outward, from land to sea, carrying dry air from a large area of high pressure over Central Asia. In spring, the humidity begins to rise, but there is no rain. The wind direction reverses as a low-pressure system develops over Tibet and the intertropical convergence zone shifts northward. In mid-June the weather changes rapidly. There are torrential rains which, by the end of July, have reached all of southern Asia, extending as far as Japan.

**CAUGHT IN THE MONSOON RAINS**
*In India, as here in Simla, people eagerly await the onset of the monsoon rains to relieve the hot, humid conditions.*

TROPICS

## Hot climates with seasonal rain

**DISTRIBUTION** In the Sahel region of Africa, much of east Africa, coastal parts of southern Central America, and parts of South America

**ANNUAL REGIME FOR** Dodoma, Tanzania

**TOTAL PRECIPITATION** 22 in (570 mm)

**TEMPERATURE OF WARMEST MONTH** 88°F (31°C)

**TEMPERATURE OF COLDEST MONTH** 79°F (26°C)

Many parts of the tropics experience constant high temperatures, but have highly seasonal rainfall, with over 95 percent of their annual total falling during the wet season. Regions with this type of climate, such as east Africa, lie between the intertropical convergence zone and the subtropical deserts. The intertropical convergence zone moves with the seasons. As it approaches in summer, it brings heavy rain and there are many violent thunderstorms. As the intertropical convergence zone retreats in winter, the dry season commences. It affects plants and animals in much the same way as the cold winters of higher latitudes. When there is no water, plants become dormant and large animals migrate. The wet season causes rapid regrowth of the plants, and the wildlife returns.

**SERENGETI PLAINS**
*Animals graze near some acacias during the wet season in the Serengeti National Park, Tanzania.*

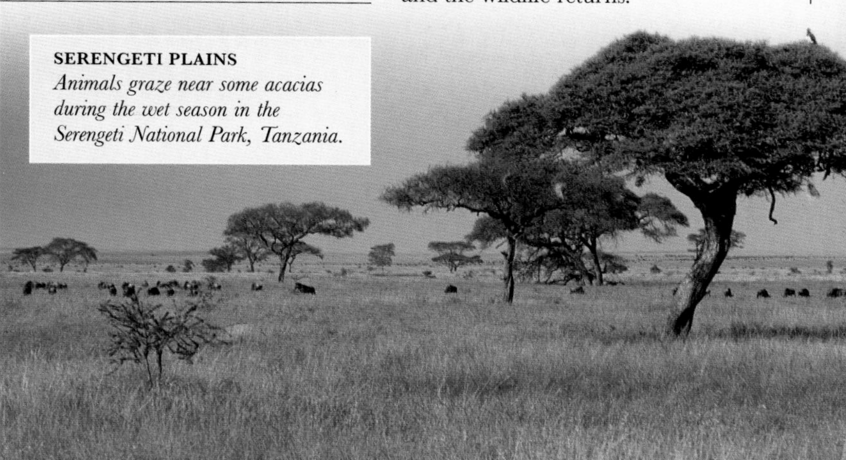

MID-LATITUDES

## Hot, dry climates

**DISTRIBUTION** In mid-latitude deserts and adjacent semiarid regions in the subtropics of the northern and southern hemispheres

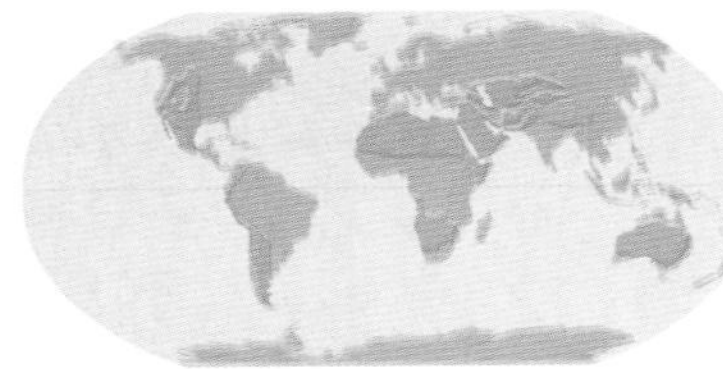

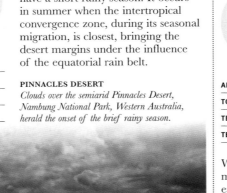

**ANNUAL REGIMEN FOR** I-n-Salah, Algeria

**TOTAL PRECIPITATION** ¼ in (17 mm)

**TEMPERATURE OF WARMEST MONTH** 113°F (45°C)

**TEMPERATURE OF COLDEST MONTH** 70°F (21°C)

The Sahara is an example of a mid-latitude desert, with a hot, dry climate. It has areas that receive no rain for several years but then experience a violent, short-lived storm. This is because hot air rises over the equator and sinks over the subtropics, creating stable areas of high pressure and very dry conditions. In contrast, semiarid regions, such as the Sahel, usually have a short rainy season. It occurs in summer when the intertropical convergence zone, during its seasonal migration, is closest, bringing the desert margins under the influence of the equatorial rain belt.

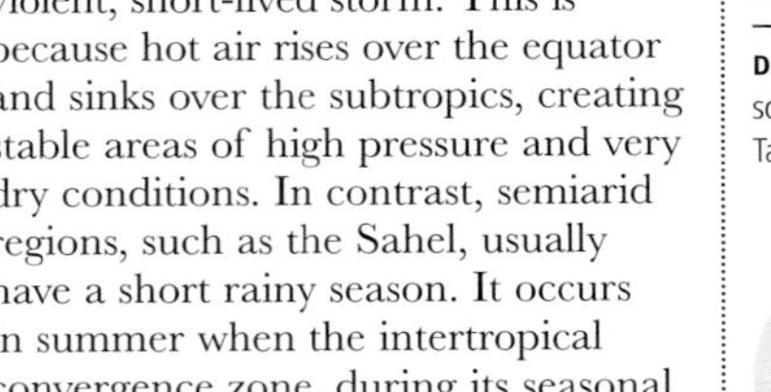

**PINNACLES DESERT**
*Clouds over the semiarid Pinnacles Desert, Nambung National Park, Western Australia, herald the onset of the brief rainy season.*

MID-LATITUDES

## Cool, temperate maritime climates

**DISTRIBUTION** In the mid-latitudes of North America, southern South America, the western margin of Europe, Tasmania, and New Zealand

**ANNUAL REGIMEN FOR** Cork, Ireland

**TOTAL PRECIPITATION** 41 in (1,049 mm)

**TEMPERATURE OF WARMEST MONTH** 68°F (20°C)

**TEMPERATURE OF COLDEST MONTH** 48°F (9°C)

Weather systems in the middle latitudes most often travel from west to east, driven by the prevailing winds and the overlying jet stream. Air reaching the western coasts of the continents has crossed the ocean, where it acquired large quantities of moisture. When the air rises to cross the coast, much of this moisture falls as precipitation. Frontal depressions develop between tropical and polar air, and they also cross the ocean, delivering more precipitation. This regimen produces mild climates, with winter temperatures falling only slightly below freezing and summer temperatures that rise above 50°F (10°C), usually with only brief spells of hotter weather. These climates are moist, with rain distributed through the year, though often with slightly more rain in winter than in summer. Cool maritime climates provide good conditions for farming. The wetter western areas favor pasture and are used for dairy and beef production, with sheep on the higher ground. Fruit and vegetables are also grown. Farther east, it is drier, and the climate is suitable for growing grains.

**COUNTY KERRY, IRELAND**
*Prevailing winds from the Atlantic Ocean rise to cross the rocky cliffs on the northern coast of the Dingle Peninsula, resulting in precipitation.*

MID-LATITUDES

## Warm, temperate climates

**DISTRIBUTION** In southeastern US and California, parts of South America and Australia, countries around the Mediterranean, and Cape Province, South Africa

**ANNUAL REGIMEN FOR** Rome, Italy

**TOTAL PRECIPITATION** 29 in (730 mm)

**TEMPERATURE OF WARMEST MONTH** 86°F (30°C)

**TEMPERATURE OF COLDEST MONTH** 52°F (11°C)

Regions with mild, wet winters and warm or hot, dry summers have a warm, temperate, or Mediterranean climate. Los Angeles, for example, has daytime temperatures ranging from 64°F (18°C) in January to 82°F (28°C) in August. Rainfall averages 15 in (381 mm) a year, falling mainly between December and March, with no rain at all in August. The natural vegetation is adapted to the dry conditions. It comprises certain species of coniferous trees, such as the Aleppo pine of the Mediterranean, and evergreen shrubs with tough, leathery leaves, such as heathers and broom. Taller shrubs and small trees, such as the olive, form a type of vegetation known as maquis, while smaller shrubs and herbs are known as garrigue or chaparral. Many plants, such as rosemary, produce aromatic oils and are cultivated as herbs. The climate is excellent for growing a wide variety of fruits, especially grapes.

**BEE-EATER**
*This medium-sized Mediterranean bird feeds on insects caught in flight, including bees, hence its common name.*

**MEDITERRANEAN VEGETATION**
*This view of a valley between Cogolin and Collobrières, southern France, shows the typical vegetation of the region.*

**HUNTER VALLEY**
*Warm, temperate climates provide ideal conditions for vineyards, such as these in the Hunter Valley, New South Wales, Australia.*

MID-LATITUDES

## Cool, temperate continental climates

**DISTRIBUTION** In the continental interiors of North America and Europe, parts of Sweden, eastern Asia, and northern Japan

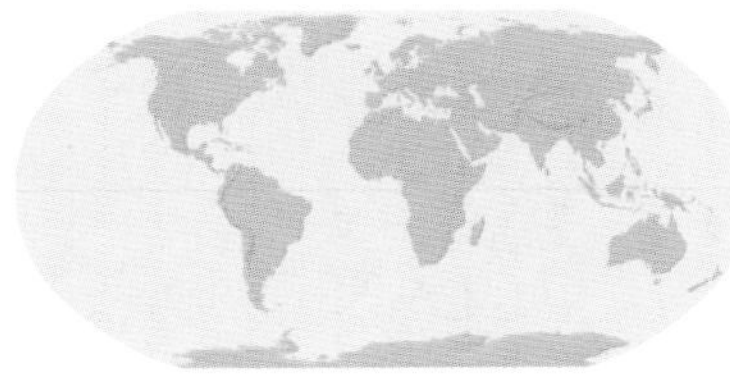

| ANNUAL REGIMEN FOR | Minneapolis, Minnesota |
|---|---|
| TOTAL PRECIPITATION | 28½ in (725 mm) |
| TEMPERATURE OF WARMEST MONTH | 82°F (28°C) |
| TEMPERATURE OF COLDEST MONTH | 21°F (–6°C) |

Regions with continental climates experience much more extreme temperatures than those areas with maritime climates. This is because the land heats and cools more rapidly than the sea. Summers are warm or hot, and winters are cold and often severe. They are also influenced by high- and low-pressure weather systems. Precipitation is distributed evenly through the year. The central European climate, though more severe than that farther to the west, is moderated by the fact that air reaching the region from the Atlantic is still relatively mild and moist. Vladivostok, on the Pacific coast, receives air that has crossed the entire landmass of Eurasia and so experiences more severe conditions. Regions with temperate continental climates are extensively cultivated and support large-scale grain production.

**SNOW-COVERED LANDSCAPE**
*Snow-covered corn stubble and grain silos are characteristic of grain-growing areas such as here in Ellsworth, Wisconsin.*

POLAR REGIONS

## Cold climates

**DISTRIBUTION** In the Arctic and Antarctic circles, above 66.5° north and 66.5° south, including Greenland, and the extreme north of North America and Eurasia

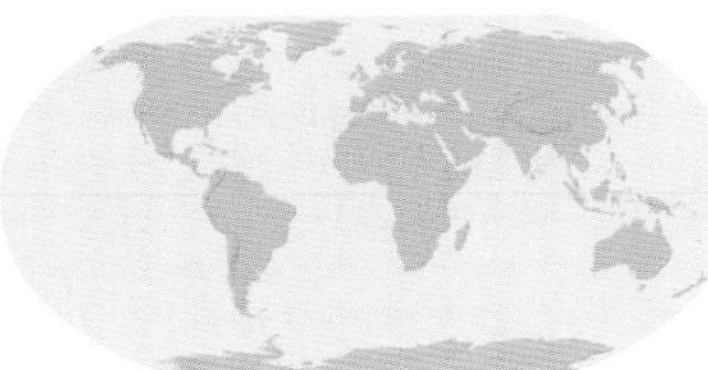

| ANNUAL REGIMEN FOR | Vostok Station, Antarctica |
|---|---|
| TOTAL PRECIPITATION | ³⁄₁₀ in (4.5 mm) |
| TEMPERATURE OF WARMEST MONTH | –26°F (–32°C) |
| TEMPERATURE OF COLDEST MONTH | –89°F (–67°C) |

Latitudes above 66.5° are extremely cold and very dry. At Qaanaaq, on the coast of northern Greenland, average daytime temperatures in July reach 46°F (8°C). The temperature in February, the coldest month, averages –4°F (–20°C). Inland, temperatures are much lower. Most of the Arctic is covered by the Arctic Ocean, which is fed with water from farther south. Although ice-covered for most of the year, there are breaks in the ice where sea and air meet, preventing the temperature from falling as low as it would otherwise. Antarctica is colder because it is a continent and high above sea level. Surface temperatures measured above the ice sheet, about 14,750 ft (4,500 m) above sea level, average –56°F (–49°C) at the South Pole. Stable high pressure over the Antarctic and Greenland ice sheets produces winds that blow outward, preventing milder, moister air from entering. The low temperatures mean that the air can hold very little moisture. Consequently, the climate is extremely arid, and all precipitation falls as snow that remains on the ground and accumulates gradually over time.

**GREENLAND ICE SHEET**
*Air in contact with this huge ice sheet is cooled, creating an area of high pressure. Winds blowing from it are both cold and extremely dry.*

WORLDWIDE

## Island and coastal climates

**DISTRIBUTION** On the coasts of continents, especially the west coasts of continents in temperate latitudes, and oceanic islands worldwide

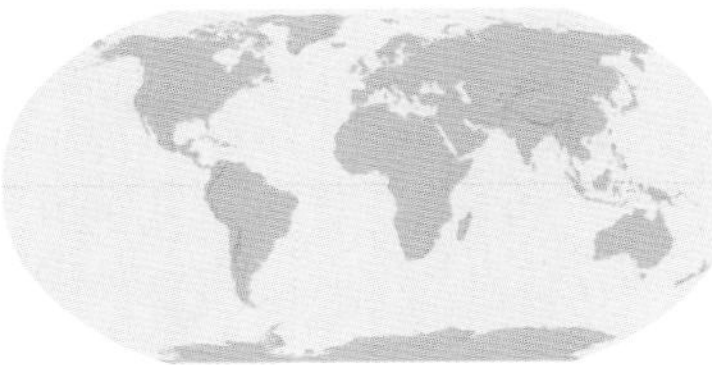

| ANNUAL REGIMEN FOR | Honolulu, Hawaii |
|---|---|
| TOTAL PRECIPITATION | 25 in (643 mm) |
| TEMPERATURE OF WARMEST MONTH | 82°F (28°C) |
| TEMPERATURE OF COLDEST MONTH | 75°F (24°C) |

Climates dominated by the sea are cooler in summer and warmer in winter, compared with continental interiors in the same latitude. The maritime air also brings abundant precipitation, but there are exceptions. The very dry coastal Atacama and Namib deserts occur in latitudes 15–30° south, near the edges of the subtropical high-pressure regions, where dry air flows away from the continental interior. Some oceanic islands are relatively dry for the same reason. For example, Las Palmas in the Canary Islands has only 7 in (176 mm) of rain a year, with a temperature range of about 63°F (17°C) to 75°F (24°C). The Solomon Islands, at 6.2° south, have a tropical climate, with temperatures ranging from 84°F (29°C) in winter to 90°F (32°C) in summer, with 120 in (3,000 mm) of rain a year.

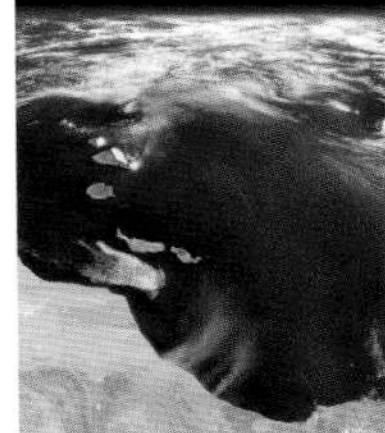

**CANARY ISLANDS**
*This satellite image shows dust being blown from the African mainland by seasonal sirocco winds. They flow toward the Canary Islands, but are dry and carry little rain.*

**STEEP COASTAL CLIFFS**
*The Na Pali coast of Kauai, Hawaii, is strongly influenced by the sea, which brings plentiful rain.*

WORLDWIDE

## Mountain climates

**DISTRIBUTION** Mountains, plateaus, and mountain ranges throughout the world, starting at about 2,000 ft (600 m) above sea level

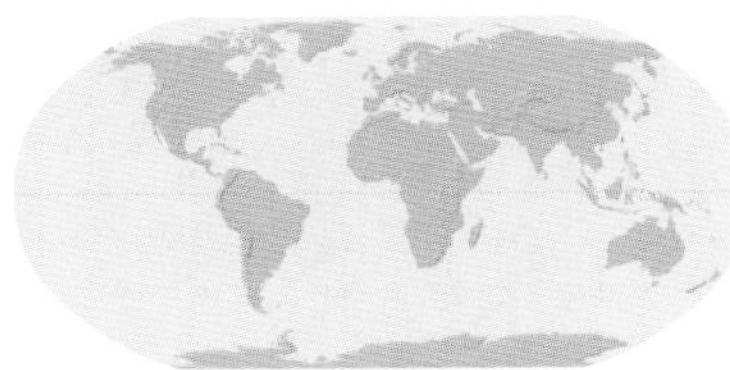

| ANNUAL REGIMEN FOR | Les Escaldes, Andorra |
|---|---|
| TOTAL PRECIPITATION | 32 in (808 mm) |
| TEMPERATURE OF WARMEST MONTH | 79°F (26°C) |
| TEMPERATURE OF COLDEST MONTH | 43°F (6°C) |

The temperature of the air decreases with height by about 3.6°F per 1,000 ft (6.5°C per km). This has two consequences. The first is that the temperature in mountain climates decreases with increasing elevation. The second is that mountain climates are wetter than the adjacent lowland climates. This is because air is forced to rise over mountains, the water vapor condenses to form clouds, and these deliver precipitation on the windward slopes. The lee side is often dry, because it is in the rain shadow. Although temperature decreases with height everywhere, actual mountain climates vary according to latitude, because the temperature beginning at sea level varies in different locations. Mountain vegetation changes with height in much the same way that it changes with latitude (see p.306).

**SNOW-CAPPED MOUNTAINS**
*The snow-covered peaks of the La Sal Mountains, Utah, indicate that temperatures are lower above the snow line than below it, preserving the snow.*

**CHANGING WEATHER**
*Unstable atmospheric conditions can produce rapid changes in weather. This storm cloud in New Mexico will probably cause a violent thunderstorm and heavy rainfall.*

# WEATHER

"WEATHER" IS A GENERAL TERM for the constantly changing atmospheric conditions that prevail at a particular place and time. It is a result of the physical interactions between sunshine, air, and water. Sunlight heats the air more in some places than others, leading to differences in air pressure, which in turn cause the air to move in the form of wind to eliminate the differences. The Sun's heat also causes water to evaporate, only to condense again to form clouds as the air carrying the water vapor rises and cools. Changes in the weather can be rapid, as when a warm, sunny morning gives way to a cold, wet afternoon, or slow, as in the seasonal change from summer to winter. Over longer periods, weather is less changeable: there are warm summers and cool ones, and hard or mild winters, but over the years these seasonal variations cancel each other out. The average weather over many years makes up the climate for a particular place.

# AIR MASSES AND WEATHER SYSTEMS

390–91 Circulation and currents
446–49 Atmospheric circulation
Living with hurricanes 464–65
Precipitation and clouds 466–69

AN AIR MASS is a body of air with uniform characteristics. Different air masses meet and interact along boundaries called fronts. There is a close link between air masses and patterns of circulating air known as pressure systems: warm air masses are associated with low-pressure systems and cool air masses with high-pressure systems. Over the course of a year, the distribution of air masses and pressure systems changes. These shifts can have significant effects on the weather. For example, continental air is hot in summer, causing low pressure over the land, and bitterly cold in winter, resulting in high pressure.

## AIR MASSES

An air mass is a volume of air that covers most of a continent or ocean and acquires characteristics from the surface below it. Air over a continent will be drier than air over the ocean, and it will be warmer in summer and cooler in winter, because of the different rates at which the land and sea warm and cool with the seasons. Once it has acquired these characteristics, the almost homogeneous air constitutes an air mass. Air masses extend from the surface all the way up to the tropopause (see p.440). Air masses have names that describe their origins: those that form over land are continental, and those forming over the sea are maritime. Depending on the latitude in which they form, air masses may be arctic, polar, tropical, or equatorial. These names can then be combined to produce six types of air mass: continental arctic; continental polar; continental tropical; maritime arctic; maritime polar; maritime tropical; and maritime equatorial. There is no continental equatorial type, because there is no continental land mass in equatorial regions that is large enough to produce continental equatorial air. Arctic and polar air are essentially identical at surface level, but they differ in the upper atmosphere. When two air masses meet, they do not simply merge, because they are at different temperatures. This means that the density of the air is different in the two air masses. Instead of mixing, the denser air moves beneath the less dense air, lifting it from the surface. The boundary between two air masses is known as a front, named for the air behind it: as a warm front passes, warm air replaces cool air; a cold front brings cooler air.

polar continental air mass
polar maritime air mass
polar maritime air mass
tropical continental air mass
tropical maritime air mass
tropical maritime air mass

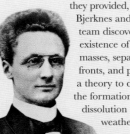

**NORTH AMERICAN AIR MASSES**
*The weather at a particular place is usually influenced by several air masses. For example, North American weather is influenced by the six air masses shown above: cool polar air flows from the north, and warm tropical air from the south. In winter, tropical maritime air comes in from the Pacific Ocean.*

### VILHELM BJERKNES

Vilhelm Bjerknes (Norway; 1862–1951) was one of the founders of scientific meteorology. He established the Bergen Geophysical Institute in 1917, and set up a series of weather stations throughout Norway. Using the data they provided, Bjerknes and his team discovered the existence of air masses, separated by fronts, and proposed a theory to describe the formation and dissolution of weather fronts.

## PRESSURE SYSTEMS

Warm, rising air produces an area of low surface pressure (or cyclone). Dense, subsiding, cool air forms an area of high surface pressure called an anticyclone. The difference in pressure between the center of a pressure system and the air outside it is called a pressure gradient. Air flows along this gradient from areas of high to low pressure—the steeper the gradient, the stronger the wind. The moving air is deflected by the Coriolis effect (see p.446), so in the northern hemisphere, air flows counterclockwise around cyclones and clockwise around anticyclones.

**HIGH AND LOW PRESSURE**
*Air spirals into a low-pressure area (cyclone), then rises, perpetuating the low pressure below, but creating high pressure above, where air diverges. In a high-pressure area (anticyclone), the situation is reversed.*

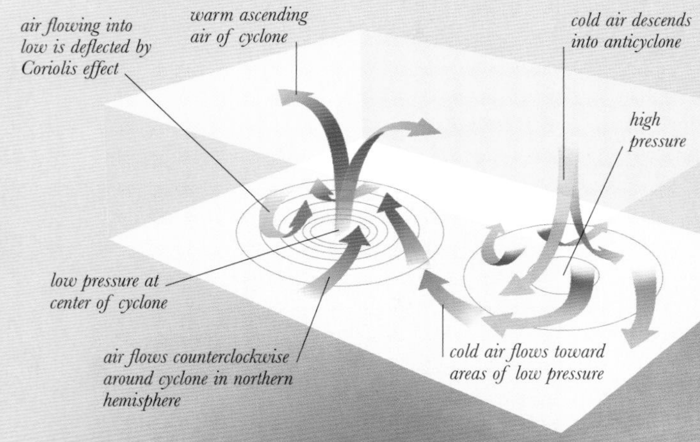


**DEEP CLOUD**
*When two air masses meet along a cold front, air is forced upward and becomes unstable, resulting in the formation of deep cumulus clouds.*

# FRONTS

On a weather map, the position of a front is indicated where it meets the ground, but fronts also extend upward, often all the way to the tropopause, the boundary between the troposphere and the stratosphere (see p.440). Fronts are 60–120 miles (100–200 km) wide, and they form along the edges of air masses. They also slope, so a frontal system, comprising a warm and cold front with warm air between them, is shaped rather like a bowl, but with uneven sides. A warm front slopes at about 0.5°–1°, and a cold front at about 2°. Cold fronts travel at an average speed of 22 mph (35 km/h), and warm fronts at 15 mph (24 km/h). At a cold front, cool air undercuts warm air, forcing it steeply upward along the line of the front. Strong convection currents develop, leading to the formation of storm clouds and heavy rainfall. Cold fronts are often associated with low-pressure systems and unsettled conditions blowing away from a high-pressure region. At a warm front, warm air rises over cold air more gradually. Moisture condenses in the rising air, producing clouds and precipitation, the type depending on how rapidly the air rises. Because it moves more quickly, a cold front will eventually overtake a warm one, lifting it clear of the ground. This is known as an occlusion. A warm occlusion occurs when the cold front rises over the warm one, and a cold occlusion occurs when the cold front undercuts the warm front.

**GRADUAL SLOPE OF A WARM FRONT**
*This warm front over Illinois has produced layered, stratus-type cloud. It meets the ground about 186 miles (300 km) away.*

**STORM FRONT**
*A fast-moving cold front may "shovel up" warm air, producing huge storm clouds. Each cloud dissipates quickly, but another forms, creating a squall line, as seen here.*

**COLD FRONT**
*A cold front advances into the warm air ahead of it, forcing air to rise up the relatively steep frontal slope. This triggers the formation of cumulus-type clouds.*



**WARM FRONT**
*Air rises more slowly up the shallow slope of a warm front and produces stratus-type clouds. The clouds warn people of its approach as the front overhangs the region ahead of it.*



# JET STREAMS

Jet streams are narrow ribbons of fast-moving air that circle the Earth, close to the tropopause, at 60–310 mph (100–500 km/h). They are produced by the difference in temperature between air masses. Because warm air is less dense than cold air, pressure decreases with height more rapidly in a column of warm air than in cold air. This difference in temperature is reflected in a difference in pressure that increases with height. It is this difference that generates the wind, which blows with the cold air on its left in the northern hemisphere and on its right in the southern hemisphere. This means the jet streams blow from west to east in both hemispheres. The exception is a jet stream that develops in summer over India and Africa and flows in the opposite direction.

**JET-STREAM VARIATION**
*Jet streams do not blow in a constant direction. Shallow waves, known as Rossby waves, develop, becoming bigger over about three weeks and then disappearing as a steady flow resumes.*

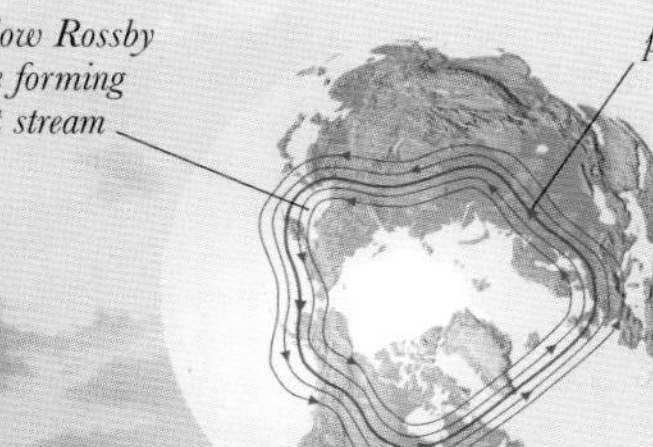



**EARLY WAVES**

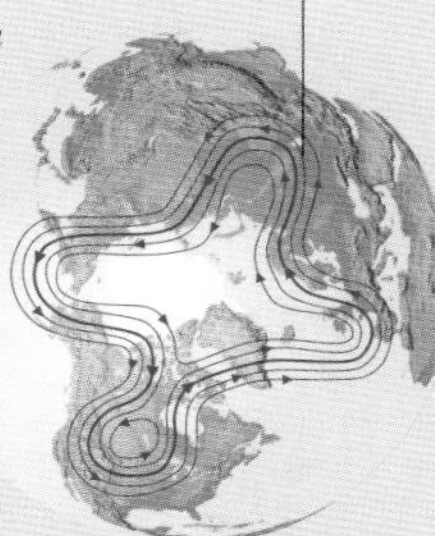



**ENLARGED WAVES**

# HIGH- AND LOW-PRESSURE WEATHER

As cold air sinks, it becomes warmer, creating an area of high pressure at the surface and bringing settled weather with cloudless skies. It is usually dry and sunny in summer and cold and frosty in winter. However, if the descending air traps warmer air beneath it, fog may form. Warmth at the surface causes air to rise, producing low surface pressure. Moisture condenses in the rising air, producing both cloud and precipitation, the type depending on how rapidly the air rises. Winds blowing into low-pressure regions are generally stronger than the winds blowing away from high-pressure regions because the pressure gradient is usually steeper around low-pressure systems. The very high winds and heavy rain of a tropical cyclone are extremely destructive. In temperate areas, low-pressure cyclones bring less severe weather and are known as depressions.

**HIGH-PRESSURE SYSTEMS**
*Clear skies, dry air, and fine weather indicate anticyclonic (high-pressure) systems, such as seen here in the Namib Desert. Air flowing outward prevents moister air from entering.*

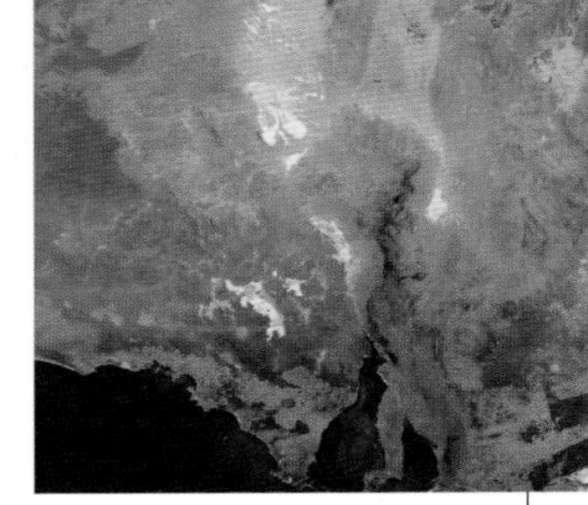

**CLEAR SKIES**
*This satellite image of South Australia shows anticyclonic conditions in summer, which bring high temperatures and almost no rain.*

**STORMY WEATHER FRONT**
*A cold front is lifting warm, moist, unstable air, causing these storm clouds to form over Norfolk, England. The clouds will produce heavy rain.*

**LOW-PRESSURE SYSTEMS**
*Cyclonic (low-pressure) weather, as seen here in Brazil, is usually dull, with gray cloud, often accompanied by persistent, light rain.*

# MID-LATITUDE DEPRESSIONS

Between polar and tropical regions, low-pressure areas, called mid-latitude depressions, travel from west to east, usually with one depression following another. They are caused by horizontal waves in the jet stream, called Rossby waves, at the top of the front between tropical and polar air. Rossby waves produce similar waves at surface level. A crest in a Rossby wave, projecting into cold air on the polar side of the jet stream, produces a ridge of high pressure. A trough in the jet stream, projecting into warm air on the equatorial side, produces low pressure. These ridges and troughs are where the winds are fastest. At the surface, anticyclones usually occur downwind of the ridges and cyclones downwind of the troughs. A mid-latitude depression tends to develop a characteristic arrangement of warm and cold fronts. The system begins as a simple boundary between warm and cold air, with the air on either side flowing in opposite directions. As it travels eastward, separate fronts develop. Because cold fronts move more quickly than warm ones, they eventually merge, form an occlusion, and then dissipate.

**DEPRESSION DIARY**
*These illustrations show the development, movement, and decay of a mid-latitude depression. A front develops, and then deepens. Over time, the cyclonic circulation results in occlusion of the front, which then dissipates.*

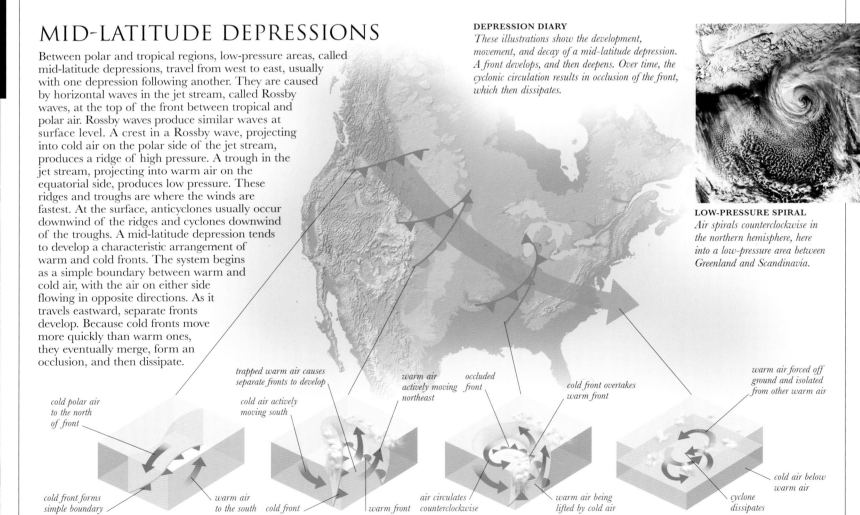

**LOW-PRESSURE SPIRAL**
*Air spirals counterclockwise in the northern hemisphere, here into a low-pressure area between Greenland and Scandinavia.*

*cold polar air to the north of front*
*cold front forms simple boundary*
*warm air to the south*

**FRONT FORMS**

*trapped warm air causes separate fronts to develop*
*cold air actively moving south*
*warm air actively moving northeast*
*cold front*
*warm front*

**CYCLONE ESTABLISHED**

*occluded front*
*cold front overtakes warm front*
*air circulates counterclockwise*
*warm air being lifted by cold air*

**FRONT OCCLUDES**

*warm air forced off ground and isolated from other warm air*
*cold air below warm air*
*cyclone dissipates*

**CYCLONE DISSIPATES**

# TROPICAL CYCLONES

A tropical cyclone, known also as a hurricane or typhoon, is a low-pressure system that develops only where a large area of the sea's surface has a temperature of at least 80°F (27°C) and the wind direction or speed changes with height, to remove rising air. The Coriolis effect makes the winds spiral. These conditions occur mainly in late summer, when the sea is warm, between latitudes 5° and 20°. A disturbance in the wind pattern starts the air turning, producing low surface pressure with air accelerating into it. The high temperature causes high evaporation, so the rising air is very moist. As it cools, condensation releases latent heat, increasing instability and generating towering storm clouds.

**HURRICANE PAULINE**
*This hurricane struck Mexico in October 1997. Wind speeds in the cloud wall surrounding the eye exceeded 125 mph (200 km/h).*

**TYPHOON ODESSA**
*The eye of typhoon Odessa, over the western Pacific, is evident from the "hole" where warm, dry air is sinking, causing cloudless skies.*

**ANATOMY OF A HURRICANE**
*Warm, sinking air forms the calm area known as the eye, around which there is a cloud wall where the strongest winds and the heaviest rain occur. Rising air produces high pressure aloft, while subsiding air produces spiral bands of cloud with clear skies in between. Strong convection and condensation inside the clouds power the hurricane.*



**HURRICANE TRACK**
*Hurricanes travel westward, and swing to the right as they approach land. How soon this happens will determine whether or not they cross land. This is the track of Mitch, which struck Central America in October 1998. Its winds blew continuously at 180 mph (290 km/h) for 15 hours.*

# MONSOON SYSTEMS

Monsoon weather occurs in some parts of the tropics due to a seasonal change in the direction of the prevailing wind. During the winter, very dry air flows outward from high pressure in the continental interior. In summer, as the land warms, the high pressure gives way to low pressure. Over the ocean, however, pressure remains relatively high because water warms much more slowly. The winds reverse direction, bringing moist air over the land. As it crosses onto the land, the air rises, producing huge clouds and the torrential rains of the monsoon.

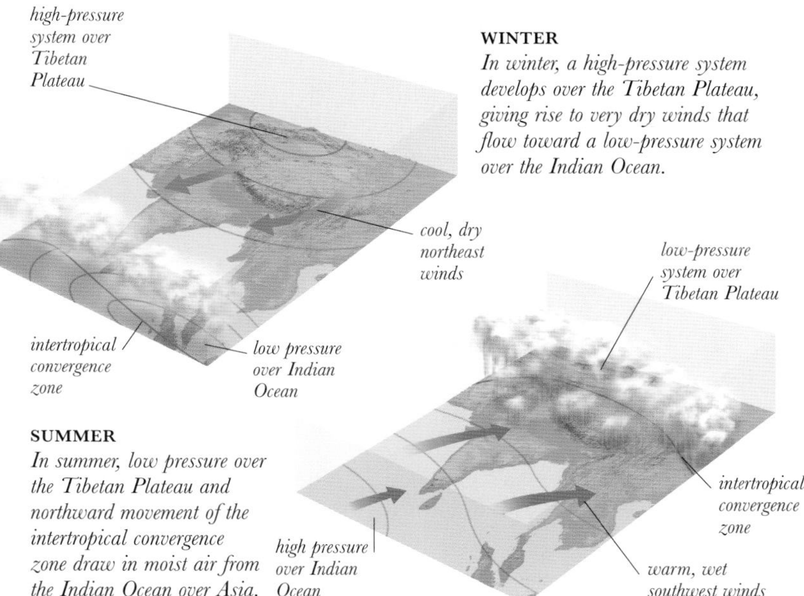


**WINTER**
*In winter, a high-pressure system develops over the Tibetan Plateau, giving rise to very dry winds that flow toward a low-pressure system over the Indian Ocean.*

**SUMMER**
*In summer, low pressure over the Tibetan Plateau and northward movement of the intertropical convergence zone draw in moist air from the Indian Ocean over Asia.*

## THE ONSET OF THE MONSOON

People in India and elsewhere in Asia are dependent on the monsoon for growing their crops. If the monsoon is early, the fields may not be ready for planting, but if it is late, planting is delayed and the crop yield is poor. This can be devastating for subsistence farmers. In drought years, people go hungry and their stock starves. Today, the increasing accuracy with which the onset of the monsoon is predicted allows some farmers to plan which crops to plant. If they know when the rains will arrive, they also know the length of the growing season. The monsoon starts with a spectacular thunderstorm, and torrential rain that lasts for a week or more. Heavy rain can cause flooding, but people still celebrate the first rains of the monsoon.

# LIVING WITH HURRICANES

Although tropical cyclones—also known as hurricanes and typhoons—are generated at sea, they can inflict widespread devastation and loss of life if they reach coasts and travel inland. These storms also trigger storm surges, flooding, and catastrophic landslides. Attempts to defuse hurricanes by cloud-seeding—a method of inducing rain—have met with limited success. In recent years, however, improvements in hurricane and typhoon warning systems have saved thousands of lives.

## FOURFOLD IMPACT

Hurricane strength is measured using the Saffir-Simpson scale, which has a maximum rating of five. In a category-five hurricane, wind speeds can reach over 155 mph (240 km/h), and, because the wind's impact increases exponentially with its speed, it can tear off roofs and smash wooden buildings. This destructive power is made worse by the fact that the wind blows in the opposite directions on either side of the hurricane's center, or eye. As the eye passes overhead, the wind stops and then reverses, ripping out anything that has been loosened in the initial blast.

Violent though it is, wind takes second place to water as a cause of cyclone-related deaths. Where land is low-lying, seawater is forced inland in storm surges, which can be more than 25 ft (8 m) high. Such surges are a hazard in the Gulf of Mexico, where they played a deadly role in the hurricane that hit the Texas port city of Galveston in 1900, but they can also strike the Atlantic coast from Florida to the Carolinas. They are also a particular risk in the Ganges delta, in the northern reaches of the Bay of Bengal. This region is densely populated, and the land is only a few yards above sea level during a normal high tide. To make matters worse, the gentle slope of the seabed increases the chance of that large storm surges will form when winds blow toward the coast. Since 1980, storm surges in the Bay of Bengal have claimed over half a million lives.

When Hurricane Mitch struck Central America in 1998, most of the fatalities were due to heavy rain. The storm faltered after it reached land over Honduras and Nicaragua, and then shed more than 50 in (125 cm) of rain in the mountainous interior. As the runoff poured toward the coast, towns and villages were smothered by water carrying huge amounts of loose volcanic soil.

**STORM SURGE**
*This town in the Dominican Republic was flooded in 1998 by Hurricane George, which caused millions of dollars' worth of damage.*

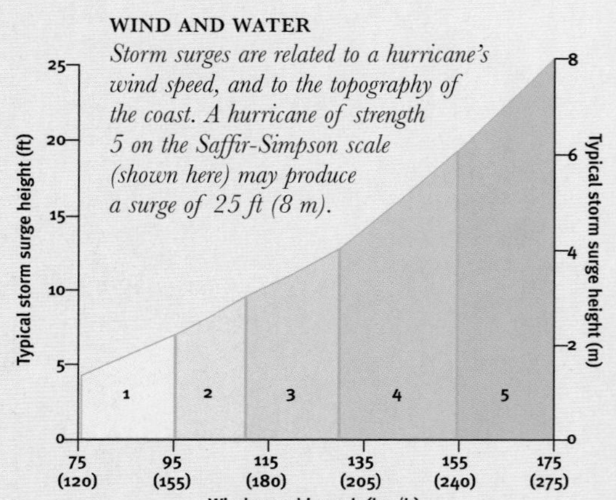


**WIND AND WATER**
*Storm surges are related to a hurricane's wind speed, and to the topography of the coast. A hurricane of strength 5 on the Saffir-Simpson scale (shown here) may produce a surge of 25 ft (8 m).*

## ADVANCE WARNING

Geostationary and polar-orbiting satellites are used to track developing storms in cyclone-generating regions. Once a potential threat has been identified, regional forecasting systems estimate a storm's strength, path, and eventual arrival on land. Hurricanes and typhoons make slow progress across the sea, so storm alerts can give several days' advance warning—vital preparation time for people likely to be affected.

Unfortunately, as Hurricane Mitch demonstrated, it is relatively easy to track storms, but much more difficult to forecast their future path. Hurricanes and typhoons follow curving tracks, moving outward from the tropics, but local conditions such as surface-temperature variations often make them deviate from their original course. Hurricane Mitch was particularly erratic: after moving northeast through the Caribbean, it abruptly turned south before doubling back across Mexico and heading out to sea.

**CLOSING IN**
*Seen by satellite, Hurricane Fran spirals near the US Atlantic coast in 1996. The hurricane inflicted extensive damage in North Carolina.*

## WARMING OF THE SEAS

Hurricanes develop best in hot conditions, and form in regions where the sea's surface temperature exceeds about 79°F (26°C). In the North Atlantic, heat stored during the summer produces a peak hurricane season from August to October, while off northern Australia the cyclone season reaches its height between January and March. Once the sea starts to cool, violent cyclonic storms are rare, but with an increase in global warming (see pp.452–53), zones of warm surface water will become larger and more persistent. As a result, the threat from hurricanes and typhoons is almost certain to grow.

With more heat available to generate rising air, cyclonic storms are likely to become more frequent, and the strength of each storm may increase. The amount of moisture that it carries will grow. Hurricanes may also wander farther from the tropics than they do now, threatening areas as far north as New England relatively often. At present, this danger is still hypothetical – the next decades will show whether or not it is real.

ATMOSPHERE

**THE AFTERMATH**
*A broken window reveals a scene of devastation in Puerto Vallarta, Mexico. This Pacific coast resort was hit by Hurricane Kenna in October 2002.*

# PRECIPITATION AND CLOUDS

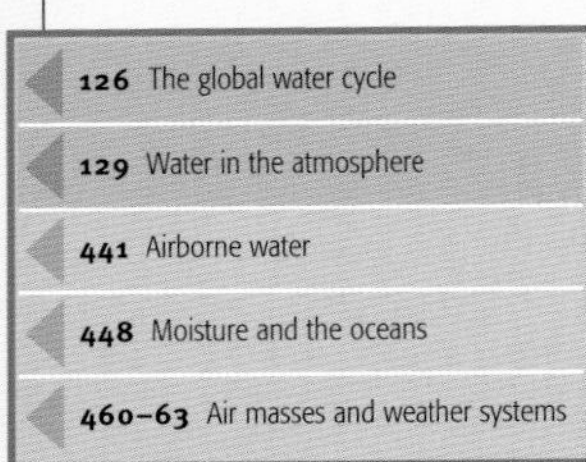


CLOUDS FORM WHEN water vapor in rising air condenses into droplets or freezes directly into ice crystals. The height at which this occurs depends on the stability of the air and the amount of moisture present. A typical cloud droplet, or ice crystal, is about 0.0004 in (0.01 mm) across. Cold clouds formed at high altitude contain only ice crystals, lower-altitude warm clouds contain only water droplets, and mixed clouds contain both. Snowflakes form if ice crystals and water droplets cool to below freezing. Water evaporates from the droplets and is deposited on the ice crystals, which collide and form snowflakes. Fog and dew form when condensation occurs at ground level, but if temperatures fall below freezing, dew is replaced by frost and rime.

**STORM CLOUD**
*Clouds like this form in unstable air. Water vapor condenses and releases latent heat, so the air continues to rise. Violent vertical currents in the cloud may produce a thunderstorm.*

## STABLE AND UNSTABLE AIR

Air is warmed when it is in contact with the ground. This causes it to become less dense than the cooler air above, and so it starts to rise. As air moves upward, its temperature falls at a set rate, the lapse rate, making it denser and heavier as its height increases. Eventually the rising air reaches a point at which it is more dense than the surrounding air, and it starts to sink again. This air is said to be stable. As air rises and cools, the amount of water vapor it can hold decreases until it becomes saturated and water droplets form. The temperature at which this happens is referred to as the dew point. The dew point is not fixed and varies according to air temperature. When water vapor condenses into water droplets, it releases latent heat (see below). If the saturated air is cooling at a slower rate than the surrounding air, it will also be warmer and so it will continue to rise. This air that rises without being forced is referred to as unstable. Clouds form at the dew point and may attain great depth under these conditions.

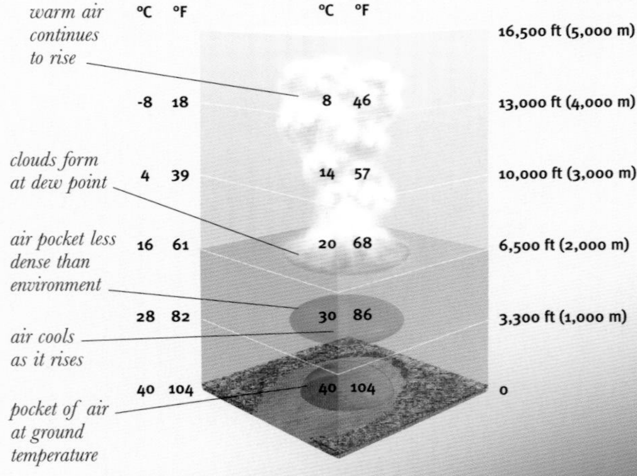


**UNSTABLE CONDITIONS**
*In these conditions, a rising pocket of air is always warmer than the air surrounding it. At the dew point, the air becomes saturated with water vapor, and clouds start to form.*

## CONDENSATION

Condensation is the change that takes place when a gas, such as water vapor, changes into a liquid, such as water. A water-vapor molecule consists of atoms— two hydrogen and one oxygen—that move freely among the molecules of the other atmospheric gases. When two water molecules collide, they bounce off each other. However, as the temperature falls, the molecules have less energy and move more slowly. When they reach the dew-point temperature, at which water condenses, water-vapor molecules merge when they meet. A hydrogen bond forms between one of the hydrogen atoms of one molecule and the oxygen atom of another, forming them into short strings. The water vapor has then condensed into a liquid. As they merge, the molecules fall into a state that requires less energy to sustain it, and so they give up some of their energy as heat. This is called latent heat. Its release warms the surrounding air and may be absorbed by liquid molecules, allowing them to vaporize. Thus groups of water molecules are constantly breaking and re-forming. If the water droplets grow large enough, they eventually fall as rain.

**RAINBOW IN HAWAII**
*Light is refracted (bent) as it enters a raindrop and again as it leaves it. This separates the light into its constituent colors and forms the rainbow.*

**STORM CLOUD**
*Cumulonimbus clouds are among the largest of all clouds and have great vertical height, but they are short-lived, rarely lasting for more than an hour.*

ATMOSPHERE

# LIFTING AND RAINFALL

Clouds form when moist air cools to its dew-point temperature. Air becomes cooler if it is forced to rise, and so cloud often forms in rising air. The height at which water vapor condenses in rising air marks the cloud base. Air can be forced to rise in three ways. If it is heated from below, due to contact with a warm surface, air rises by convection, forming small fair-weather cumulus or cumulonimbus clouds. If moving air has to pass over a mountain, it must rise. This is called orographic lifting and results in orographic cloud formation. Warm air also rises when cooler air pushes beneath it at a weather front. This is frontal lifting, producing frontal clouds. If clouds are thick enough, water droplets and ice crystals link together until they are heavy enough to fall as precipitation. Convection produces showers and brief storms. Orographic and frontal clouds produce more persistent rain and drizzle.

**LAKE EFFECT CLOUDS**
*In winter, large lakes, such as Lake Superior, are slow to freeze. Air crossing them gathers moisture that condenses into clouds as it passes over cold land.*

**OROGRAPHIC CLOUD**
*Clouds form in the rising air passing over this mountain in Greenland. Orographic clouds often form a cap shrouding the mountain summit and extending a short distance downwind.*

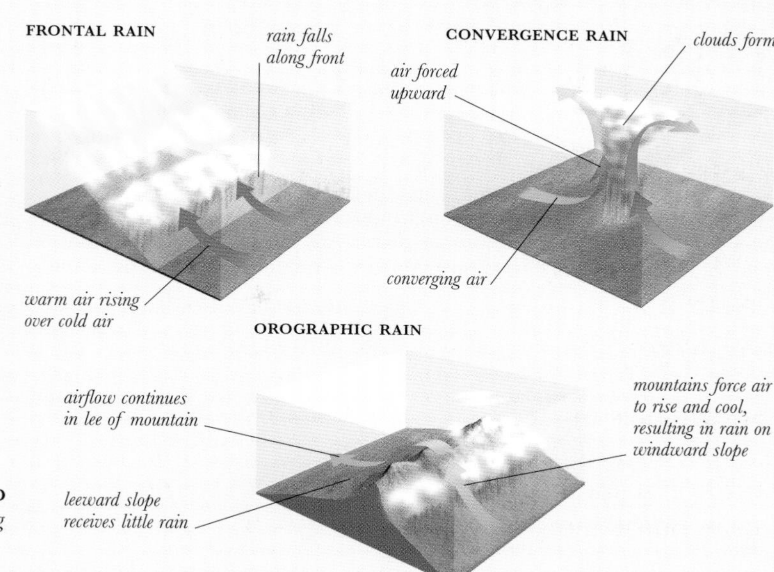


**RAIN-CLOUD FORMATION**
*Moist air that is forced to rise above the dew point forms clouds. They form where warm air rises over cool air at a front (frontal rain); where air rises by convection over warm ground, producing a local area of low pressure into which air converges and rises (convergence rain); and where air rises over mountains (orographic rain).*

# NO-FREEZE RAINFALL

Rain feels warm if it falls from a cloud that consists entirely of water droplets. Such clouds, which form only in air that is above freezing point, are known as warm clouds, and the rain they produce is called no-freeze rainfall. The lower parts of clouds are often warm, but outside the tropics, the freezing level is rarely high enough for deep warm clouds to form and produce rain. Raindrops form in warm clouds by the merging of cloud droplets, which vary greatly in size. Bigger ones, being heavier, fall faster than small ones and overtake them. When droplets of similar size collide, they merge and break in two again, but when a big droplet meets a small one, they coalesce. The droplet grows bigger, collects more small droplets, and finally falls as rain.

**HEAVY RAIN**
*Warm clouds produce big, heavy raindrops. Regular, heavy rainfall is a feature of regions with tropical climates.*

**TROPICAL RAINSTORM**
*Water evaporates quickly in the tropics. It rises rapidly, forming warm clouds that produce frequent rainstorms.*

## SATELLITE DATA

This is a false-color image produced as part of a project called the Tropical Rainfall Measuring Mission. The aim of the project is to measure the energy released by tropical rainfall, and to calculate the effect of this energy on the global climate. Data is gathered using instruments carried on satellites, including radar to measure raindrop formation in clouds, and microwave receivers to measure air temperature. When the images are received, computers color parts of them differently to enhance any variation. In this image, pale blue indicates light rainfall. Areas of heavy rainfall are colored orange and red.

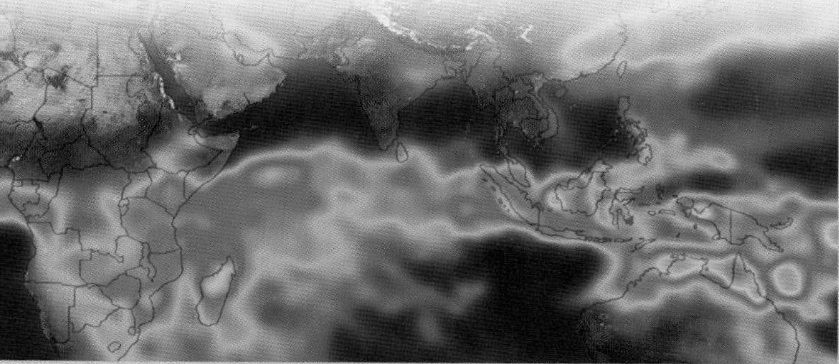

# RAIN

Water that reaches the ground as droplets is known as rain or drizzle. Drizzle consists of drops that are less than 1/50 in (0.5 mm) in diameter and very close together. These droplets form near the base of stratus cloud, in which there are no upcurrents that would allow them to coalesce with other droplets and grow larger. They sink slowly, but because the cloud base is so low, they have no time to evaporate before reaching the ground. Drops larger than those of drizzle fall as rain. Raindrops vary in size, but within limits. The smaller the drop, the more slowly it falls, and therefore the longer it has to evaporate in the air between the cloud base and the ground. This means that a raindrop must be at least 1/50 in (0.5 mm) in diameter if it is to reach the ground. Most raindrops are 1/25–1/5 in (2–5 mm) in diameter. This is 1 million to 2.5 million times bigger than an average cloud droplet, so many cloud droplets must merge together before they are big enough to fall as rain. Raindrops fall at 14–20 mph (23–33 km/h). Friction with the air breaks up drops over 1/5 in (5 mm).

**RAINY SEASON**
*Seasonal rainfall, such as this in the Serengeti in Tanzania, triggers the regrowth of plants, which in turn supports a wide variety of wildlife.*

**DISTANT RAINFALL**
*Rain seen falling in the distance looks like a gray curtain under cloud. Sometimes the curtain does not reach the ground because raindrops evaporate in the dry air beneath the cloud.*

## SAHEL DROUGHT

The semiarid Sahel region lies along the southern edge of the Sahara Desert. Usually there is a short rainy season in summer, when the intertropical convergence zone moves north, which revives the pasture for livestock and allows some crop cultivation. However, the rains are unreliable, and the region has endured many droughts. The worst drought of modern times began with light rains in the 1960s and complete failure of the rains in 1972 and 1973. Rainfall did not return to normal until the 1980s. Up to 200,000 people and 4 million cattle died.

# HAIL AND SNOW

Most hailstones are hard, roughly spherical pellets of ice, 1/5–2 in (5–50 mm) across. Soft hail forms when rime (see opposite page) collects on snow crystals, producing pellets up to 1/5 in (5 mm) in diameter that flatten when they hit the ground. A hailstone begins as a raindrop near the bottom of a large cumulonimbus cloud. Carried aloft by air currents, it freezes when it is above the freezing level, then falls in downcurrents near the top of the cloud. The hailstone grows by collecting supercooled water droplets that freeze onto it. Its final size depends on the number of circuits it has made between the top and bottom of the cloud. Snowflakes form when ice crystals adhere to one another, and they are 1/25–4/5 in (1–20 mm) across. The biggest snowflakes develop between 32° and 23°F (0° and –5°C). At this temperature, water coats the crystals and freezes as soon as another crystal touches it. All snowflakes are six-sided, but the random way they form ensures that no two are identical. Ice pellets, or sleet, are melted snowflakes or raindrops, 2/25–1/5 in (2–5 mm) across, that freeze between the cloud base and the ground in air that is below freezing. Ice particles smaller than 1/25 in (1 mm) are called snow grains.

**SHAPED BY THE WIND**
*An ice storm in New Hampshire has decorated this sign. Freezing raindrops, driven by a strong wind, struck the sign, spread out, and then froze. Later drops spread over earlier ones, producing this shape.*

**SNOWFLAKE**
*Snowflakes, such as the one shown here, form when ice crystals collide with one another. Each has a unique shape, but all are six-sided.*

**INSIDE A HAILSTONE**
*In this section through a hailstone, viewed in polarized light, the alternating layers of clear ice and opaque rime that form the hailstone are clearly evident.*

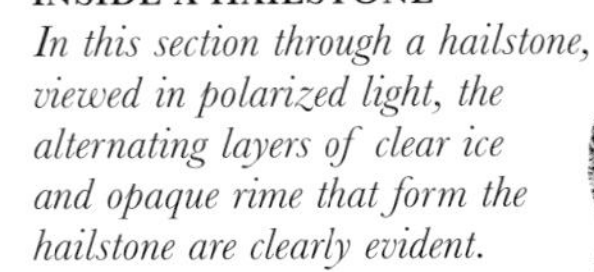

**HAILSTONES**
*Hailstones form only in large cumulonimbus clouds, where they are carried up and down repeatedly, growing all the time. The bigger the cloud, the larger the hailstones become.*

**BLIZZARD CONDITIONS**
*Heavy snow, low temperatures, and strong winds are characteristic of blizzards. Here, emperor penguins sit out a blizzard in Antarctica.*

## FOG AND DEW

Fog and mist occur when a layer of moist air, near ground level, cools to below its dew-point temperature. On high ground, fog and mist are often stratus cloud with a base lower than the hilltops. Dew forms late at night, when objects on the ground lose the warmth they absorbed by day. If the air is moist, contact with cold surfaces can cause water vapor to condense, coating them with liquid droplets. Cooling the surface also produces radiation fog because the ground cools by radiating away its warmth. Valley fog is a type of radiation fog. Advection fog forms if a wind carries warm, moist air over a cold surface.

**RADIATION FOG**
*Early-morning fog on the Vermilion Lakes, Alberta, formed by condensation when the Earth's surface cooled and reached the dew point.*

**ICE DEPOSITS**
*Freezing raindrops fell onto these pine needles, which were below the freezing point, and froze almost instantly into clear ice.*

## FROST AND RIME

Frost is a layer of ice crystals that covers objects such as plants, parked cars, and windows. It also forms on the surface of fallen snow and in the air spaces inside it. If the dew-point temperature is below freezing, water vapor will be deposited on surfaces as ice crystals. This is how hoarfrost, the most common type of frost, forms. Dew may also freeze if the temperature falls. It may form clear ice, with rounded surfaces, or fern frost, which covers windowpanes with patterns resembling fern fronds. Fern frost develops when the dew on the window cools below freezing, but remains liquid. Ice crystals form between the droplets, then the droplets freeze onto the crystals, rapidly covering the window with frost. Rime frost is white and rough to the touch, forms thicker layers than hoarfrost, and may grow into feathery shapes. It forms in cold, foggy conditions when supercooled water droplets in freezing fog or drizzle freeze on contact with a very cold surface, or when water vapor is deposited directly as ice crystals. Once the surface is coated with ice crystals, more water freezes onto the existing crystals. If a wind is blowing, rime frost forms only on the side of structures, such as radio towers and ships' masts, that are facing into the wind.

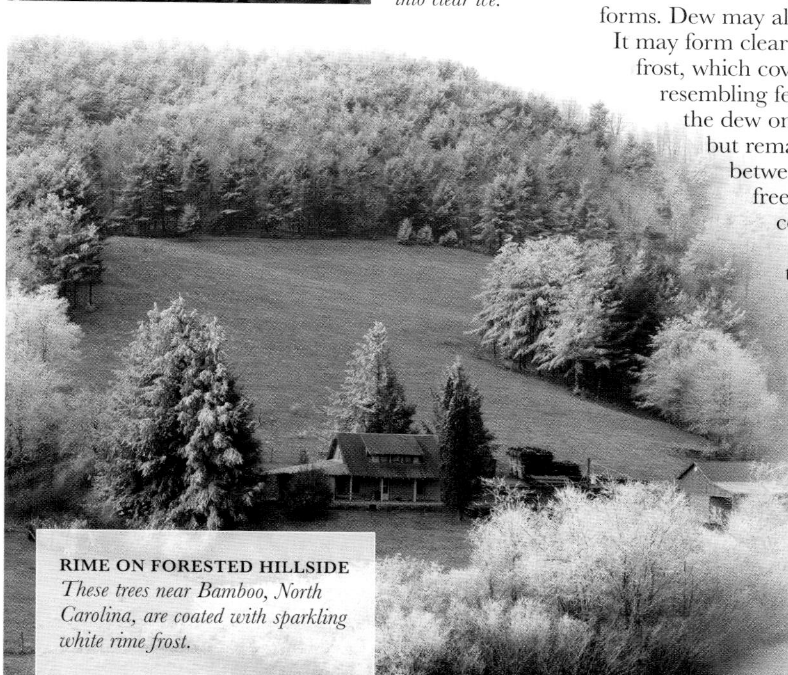

**RIME ON FORESTED HILLSIDE**
*These trees near Bamboo, North Carolina, are coated with sparkling white rime frost.*



### LUKE HOWARD

Luke Howard (1772–1864) was an English pharmacist with an avid interest in meteorology. He began keeping weather records when he was 11 years old. Howard wrote the first study of urban climate and the first book on meteorology. His cloud classification was strongly influenced by the system of two-part names used for biological classification.

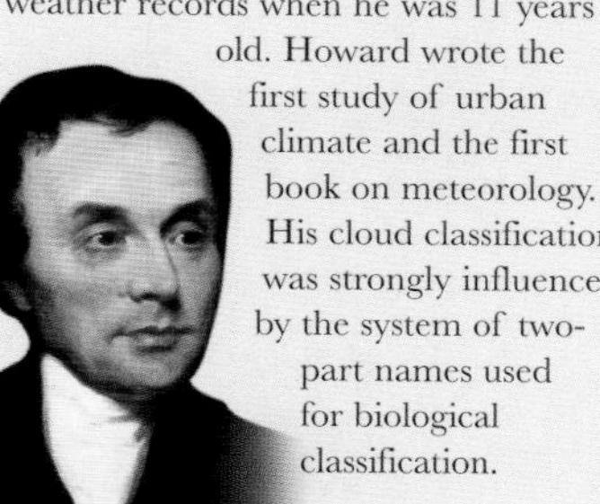

## CLOUD TYPES

Clouds differ in appearance and in the height at which they develop. These differences provide the basis for an international system for classifying them. The classification is derived from the one proposed in 1803 by Luke Howard (see panel, left), and it uses the names he coined: combinations of cirrus (curl), stratus (layer), nimbus (rain), and cumulus (heap). It groups clouds as high-, middle-, or low-level, according to the height at which the cloud base most often occurs. In middle- and high-level clouds, this varies with latitude, the height and range being lower at the poles than in temperate latitudes and highest at the equator. Clouds are grouped into ten basic types: cirrus, cirrostratus, and cirrocumulus (high clouds); altocumulus and altostratus (medium clouds); and nimbostratus, cumulonimbus, cumulus, stratus, and stratocumulus (low clouds). Some cloud types are further subdivided using 14 specific terms, such as humilis and congestus (with cumulus), which refer to the character of the cloud, or lenticularis and undulatus (with altocumulus), which further define the shape.

**GLOBAL CLOUD COVER**
*This satellite image shows that clouds occur worldwide, but are not uniform. Different clouds form under different climatic conditions.*

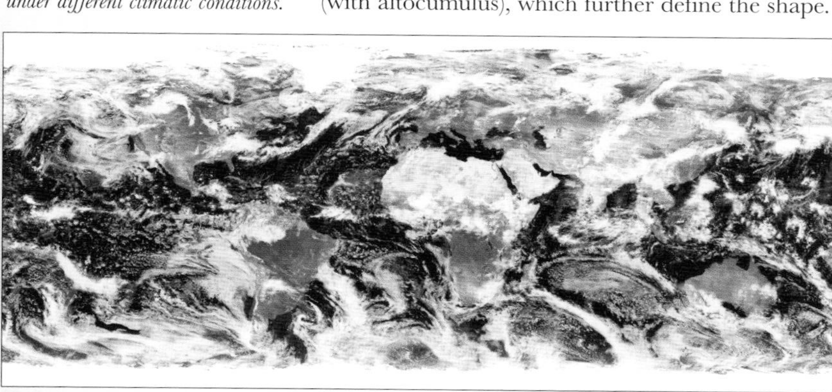

cirrus
cirrocumulus
cirrostratus
altocumulus
altostratus
stratocumulus
stratus
cumulus
nimbostratus
cumulonimbus

Tropopause
high level, above 20,000 ft (6,000 m)
middle level, 6,500–20,000 ft (2,000–6,000 m)
low level, 0–6,500 ft (0–2,000 m)
Height of cloud base

**CLOUD TYPES**
*The ten basic types of clouds, which are classified by their appearance, are shown here. There is a clear division between high-, middle-, and low-level clouds. Only the cumulonimbus cloud has great vertical height.*

### CLOUD PROFILES

The pages that follow contain profiles of the ten basic cloud types. Each profile begins with the following summary information:

**LEVEL** Cloud base: low level, 6,500 ft (0–2,000 m); middle level, 6,500–20,000 ft (2,000–6,000 m); high level, above 20,000 ft (6,000 m)

**DISTRIBUTION** Global distribution

**OCCURRENCE** Season and conditions in which it forms

LOW LEVEL

## Stratus

**DISTRIBUTION** Worldwide, more common near mountains and coasts

**OCCURRENCE** All year, more common in winter

Low cloud that forms a featureless, gray layer covering most or all of the sky is called stratus. It consists entirely of liquid droplets, typically containing 0.00005–0.0005 ounces of water per cubic foot (0.05–0.5 grams per cubic meter). The cloud base is usually lower than 6,500 ft (2,000 m). Stratus at ground level occurs as fog. Hill fog has a base lower than the hilltop it covers. Fog clears by the evaporation of its lowest layer, and any remaining fog becomes stratus cloud. Because there is no vertical air movement within stratus, it cannot produce rain, snow, or hail, but drizzle or snow grains sometimes fall. If the cloud is thin enough for the Sun or Moon to be visible, they appear as sharply defined disks. Stratus often forms overnight in fine weather, especially over water. It dissipates quickly during the morning as the temperature rises and the cloud droplets evaporate. Stratus also forms in valleys. Frontal stratus forms along warm fronts, where warm, stable air is being slowly raised up by cooler air. Cirrostratus and altostratus appear first, followed by stratus, which is the most likely to produce precipitation.

### ACID RAIN

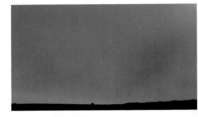

Rain is slightly acid because carbon dioxide and other atmospheric gases dissolve in cloud droplets. Industrial emissions, especially of sulfur dioxide, increase its acidity. The acid reaches the surface as acid rain, snow, or mist, or as dry particles. It can damage plants directly, as shown here, or indirectly by altering chemical reactions in the soil. Mist and dry deposition cause more harm to plants than rain, since rain tends to run off leaves.

**STRATUS OPACUS**
*This type of stratus forms an opaque layer of cloud that is thick enough to obscure the Sun and Moon completely.*

**FOG OVER THE GOLDEN GATE BRIDGE**
*Moist air approaching San Francisco from the ocean crosses the cold California current. This chills the air, and its moisture condenses to produce advection fog, which is identical in composition to stratus cloud.*

**STRATUS TRANSLUCIDUS**
*This sheet of stratus cloud forms a thin, translucent layer through which the Sun or Moon is visible as a sharply defined disk.*

LOW LEVEL

## Stratocumulus

**DISTRIBUTION** Worldwide

**OCCURRENCE** All year

Stratocumulus is low cloud, usually with a base below 6,500 ft (2,000 m), composed entirely of liquid droplets. It forms patches, sheets, or extensive layers of white or gray cloud. Rolls, or rounded masses of darker cloud, give stratocumulus a textured appearance. Stratocumulus forms when rising air meets a layer of warmer air and is flattened against its underside. In winter this produces an overcast sky that lasts for several days, but in summer the sky usually clears quickly. Gaps in the clouds allow the Sun to shine through at intervals. At dusk and dawn, the sunlight shining through these gaps may illuminate dust particles, producing rays that converge onto the cloud from below.

**BANDS OF STRATOCUMULUS CLOUDS**

LOW LEVEL

## Cumulus

**DISTRIBUTION** Worldwide, common in humid regions

**OCCURRENCE** All year, but more common in summer

This cloud forms by convection, with a base that is usually below 6,500 ft (2,000 m). Cumulus varies greatly in size and can be up to 6 miles (10 km) across and 12 miles (20 km) high. The cloud is fleecy and very bright when seen from above, because of the amount of sunlight it reflects. The clouds are usually separate from each other, with blue sky in between. This allows the Sun to shine directly on the clouds, so they appear very white with clearly defined edges. Cumulus can be embedded in clouds of other types, making them more difficult to identify. Cumulus clouds form in columns of rising air, or thermals. Their base marks the level at which water vapor condenses, which is why all cumulus clouds in one area are at the same height. As it rushes upward, a thermal draws in surrounding air at its sides and at its base. This mixes the cloud with drier and cooler air. Cloud droplets evaporate in the drier air, and fragments of cloud sink in the cooler air. This prevents the cloud from growing wider, and also prevents other clouds from forming close to it.

**CUMULUS CONGESTUS**
*This large cumulus cloud is growing rapidly, mainly by expanding upward, giving the top a cauliflower-like appearance.*

**SCATTERED CUMULUS CLOUDS**
*These detached, dense cumulus clouds in the Great Karoo, South Africa, have a typical horizontal cloud base and extend upward due to convection.*

LOW LEVEL

## Cumulonimbus

**DISTRIBUTION** Worldwide except Antarctica; some types are common in tropical regions

**OCCURRENCE** All year, associated with tropical cyclones and thunderstorms

Low-level cloud that extends vertically to a great height, sometimes as far as the tropopause or beyond, is called cumulonimbus. It is composed of liquid cloud droplets in the lower regions and ice crystals near the top. These clouds produce thunderstorms, hailstorms, and tornadoes, as well as torrential rain or snow. Because of its great depth, cumulonimbus is very dark when viewed from below. Light encounters many ice crystals and cloud droplets in passing through the cloud. These reflect and scatter it, reducing the amount of light that reaches the ground. Cumulonimbus cloud is created by convection currents in very unstable air. Warm, moist air rises rapidly. As it cools, its moisture condenses, releasing latent heat (see p.466) that warms the air, so it continues to rise. Air is drawn into the cloud to feed the upcurrents, which can reach a vertical speed of 100 mph (160 km/h). Ice crystals and hailstones falling through the cloud produce downcurrents that exit the base of the cloud as strong winds, cooling the air in the adjacent upcurrents. Eventually this suppresses the upcurrents, convection ceases, and the cloud dissipates. Cumulonimbus cloud rarely survives more than an hour. As it dissipates, it may release all of its moisture at once—up to 275,000 tons in a very large storm cloud. In some cumulonimbus, the upcurrents and downcurrents separate. This produces a "supercell" in which the downcurrents no longer interfere with upcurrents, allowing the cloud to exist for several hours. If the wind speed or direction changes with height, the center of the cloud may start rotating and extend downward to become a tornado.

**LIT UP BY LIGHTNING**
*This cumulonimbus cloud in Wyoming is illuminated by lightning sparking between two regions within it.*

**THREAT FROM THE AIR**
*Cumulonimbus cloud is typically deep and large, with a dark, menacing appearance. Here, there are scattered cumulus clouds in the background.*

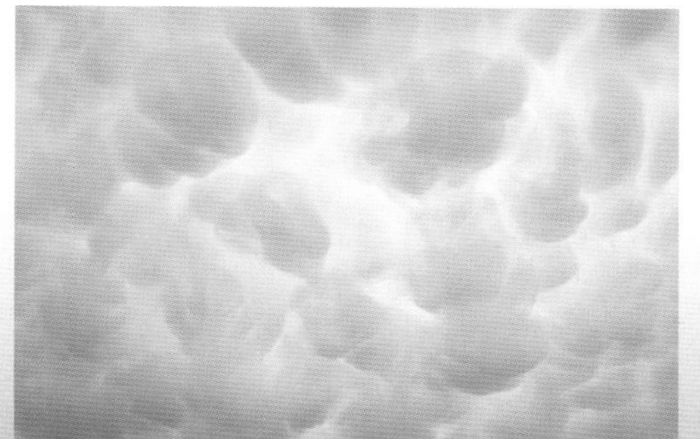

**MAMMATUS CLOUDS**
*Distinctively shaped mammatus clouds, like these in Utah, form on the underside of the "anvil" (see below) in cumulonimbus clouds.*

### CLOUD SEEDING

Rain can be induced from clouds by "seeding" from an aircraft, such as the one shown below. Silver iodide is burned above a cloud. As the smoke cools, the silver iodide crystals re-form and fall into the cloud. The water vapor freezes onto the crystals, which may induce the clouds to produce rain when otherwise they would not.

**TOWERING STORM CLOUD**
*The height of this cumulonimbus cloud over the European Alps is readily apparent. The top has spread out into an anvil shape, and it is being lit from within by lightning.*

LOW LEVEL

## Nimbostratus

**DISTRIBUTION** Worldwide

**OCCURRENCE** All year except in Antarctica

Nimbostratus is dark and shapeless, but may appear to be illuminated from inside because gaps in the cloud reveal the paler stratus cloud above. It has a low base, but usually extends upward to more than 6,500 ft (2,000 m), and so is thick enough to obscure the Sun and Moon completely, resulting in dull days and dark nights. It may produce steady, persistent rain, or snow. If the cloud temperature is above 14°F (–10°C) throughout, the cloud will consist mainly of water; from 14°F to –4°F (–10° to –20°C), it will be a mixture of water at the bottom of the cloud and ice at the top; and below –4°F (–20°C) it will consist mainly of ice. The mixture of water and ice favors the formation of larger ice crystals that fall as snow, or melt and fall as rain.

**NIMBOSTRATUS PRAECIPITATIO**
*This nimbostratus cloud has precipitation falling from it. The precipitation is clearly visible, resembling a pale curtain beneath the dark cloud.*

**NIMBOSTRATUS CLOUDS**
*This picture, taken over Utah, reveals the vertical extent of the cloud and the darkness of its base.*

MIDDLE LEVEL

## Altostratus

**DISTRIBUTION** Worldwide, most common in mid-latitudes

**OCCURRENCE** All year, associated with warm fronts, especially in temperate climates

A mid-level cloud, called altostratus, contains ice crystals near the top, but water droplets lower down. A layer of this cloud can be 6,500–10,000 ft (2,000–3,000 m) thick and cover a very large area, appearing as a uniform, gray or slightly blue sheet of cloud, or as a thin veil of fibrous cloud. The Sun and Moon may be faintly visible through it, but more often the cloud is thick enough to obscure them completely. Altostratus forms at warm fronts, where warm, moist air is being lifted above cooler air. Cirrostratus (see opposite page) usually appears ahead of it and merges with the altostratus as the cloud base becomes lower. Its appearance usually heralds rain or snow, and the altostratus itself may produce light snow or drizzle.

**ALTOSTRATUS CLOUDS**

MIDDLE LEVEL

## Altocumulus

**DISTRIBUTION** Worldwide

**OCCURRENCE** All year, associated with cold fronts especially in temperate climates

Altocumulus is a mid-level cloud, with a base ranging from 6,500 to 13,000 ft (2,000–4,000 m) over polar regions to 20,000 ft (6,000 m) over the tropics. It is composed mainly of water droplets. Its appearance is very variable, but it often tends to form rolls, arranged in lines or waves or distinctive rounded masses. The clouds are sometimes shaded around the edges, which defines them clearly, but they may also form so close together that they merge into a continuous sheet. Altocumulus cloud is white, gray, or a mixture of both. Its structure results from small, vertical air movements. It absorbs warmth at its base and also at its top. This causes air to rise and water vapor to condense, but when the cloud droplets reach the top of the cloud, where they are exposed to direct sunshine, they evaporate again. Altocumulus often forms at night, when the radiation balance is different from its daytime state. The cloud still absorbs heat radiated from the surface below, but it also radiates away its warmth at the cloud top. Air rises, water vapor condenses into droplets, and at the top these are chilled and sink back into the cloud. This type of altocumulus dissipates in the morning, when the Sun warms the top of the cloud and the droplets evaporate.

**LAMMERGEIER**
*Many of the larger carnivorous birds use gliding flight, which relies on the same thermal currents that produce altocumulus clouds to give them lift.*

**MACKEREL SKY**
*Waves in the air sometimes produce cloud near their crests, with clear sky in the troughs. This produces a pattern in altostratus and cirrocumulus clouds reminiscent of the markings on a mackerel.*

**ALTOCUMULUS CLOUD COVER**
*These clouds often indicate fair weather in temperate latitudes. If enough of these clouds form, they may merge so that little blue sky is visible.*

**TOPOGRAPHIC BARRIER**
*Altocumulus lenticularis cloud forms when air rises rapidly to pass over a topographic barrier, such as Mount Rainier in Washington State.*

HIGH LEVEL

## Cirrus

**DISTRIBUTION** Worldwide

**OCCURRENCE** All year

THIN AND WISPY
*Cirrus clouds are the most common type of high-level cloud. Winds can draw out these clouds into wisps, known as mares' tails.*

High-level cloud that consists entirely of ice crystals is called cirrus. It is thin and wispy, and has a fibrous appearance. Once it starts forming, its ice crystals grow until they reach a maximum size of about 1⁄500 in (0.05 mm) across. Once the crystals are heavy enough to start falling, they sometimes drift downward into a wind flowing from a different direction. This stretches the cloud into long tails that often curve—cirrus means "curl." They are then known as mares' tails or cirrus uncinus. Their appearance usually means that the ground-level wind will soon strengthen. When the top of a towering cumulonimbus cloud reaches the tropopause and spreads out beneath the boundary of air that is less dense than itself, it forms an incus, or anvil shape, composed of ice crystals. This is a type of cirrus, and when the storm cloud dissipates, its incus often remains. Cirrus is most often produced by the lifting of stable air. This happens along a weather front. As the air rises, different cloud types form in sequence as more and more moisture condenses. Cirrus is the last to form and represents almost the last of the water vapor. The air above the cirrus is extremely dry. Although the last to form, cirrus is the first cloud to be seen as a front approaches. Lifting also occurs close to the core of a jet stream. The location of a jet stream is sometimes revealed by the cirrus along its track. This type of cloud also forms as rapidly moving air masses cross high mountains.

CIRRUS FIBRATUS
*The cirrus fibratus clouds shown in this photograph are among the highest of all clouds. They rarely cover the sky, allowing sunlight to shine through.*

### CONTRAILS

Condensation trails, or contrails, form behind aircraft (particularly jet aircraft) flying above a certain height. Water vapor is produced whenever fossil fuel burns, and it is one of the exhaust gases discharged by engines. The engine exhaust cools in the air behind an aircraft. If the air is close to saturation, the water vapor condenses as a trail that starts some distance behind the aircraft. The trail dissipates, but on major routes there are often many contrails, and these may spread out to form a layer of cirrus.

HIGH LEVEL

## Cirrocumulus

**DISTRIBUTION** Worldwide

**OCCURRENCE** All year, associated with cold fronts, especially in temperate climates

Cirrocumulus is a high cloud, consisting entirely of ice crystals, that appears as white patches, or approximately spherical masses, called elements. These are arranged in regular patterns, often resembling ripples on a sandy beach or, less frequently, straight lines. Small vertical air movements can also produce a mackerel sky (see opposite page) made of cirrocumulus. Like mares' tails, a cirrocumulus mackerel sky is an indication of approaching strong surface-level winds. Like all cirrus-type cloud, cirrocumulus has a fibrous appearance. This betrays its origin, because cirrocumulus is usually cirrus or cirrostratus that has been reshaped by air movements.

CIRROCUMULUS UNDULATUS
*The ripples seen here are due to cloud forming at the crests of air waves. The clouds below cirrocumulus are small patches of altocumulus.*

CIRROCUMULUS CLOUDS
*Occurring at high altitude, cirrocumulus appears as thin sheets or layers with varying degrees of regularity.*

HIGH LEVEL

## Cirrostratus

**DISTRIBUTION** Worldwide

**OCCURRENCE** All year, associated with approaching warm fronts

When the Sun shines on high-level cirrostratus at dawn and dusk, it sometimes makes the cloud appear pink, producing spectacular sunrises and sunsets. At other times of day, thin cirrostratus appears as a white veil that makes the sky look milky. If the cloud is thicker, it looks like many tangled filaments. Cirrostratus consists entirely of ice crystals. It is not thick enough to obscure the Sun or Moon, but it can produce a halo around them, which is caused by the refraction (bending) of light as it passes through the ice crystals. Haloes are quite large: most have a diameter of 22° (measured as an angle at the eye of the observer, formed by two lines drawn from opposite sides of the halo), but less commonly 46°, depending on the route taken by the light through the ice crystals. A halo is mainly white, but sometimes it is faintly colored, changing from yellow at the center to green, white, and blue at the outer edge. A halo often means rain is approaching, because the cirrostratus occurs at the top of a warm front. This is especially likely if cirrus appears first, and cirrostratus gradually replaces it. If a halo is seen around the Moon, it often means rain will arrive by morning. The cloud forms when an entire layer of air is lifted, as colder air moves beneath it at a front. The air rises slowly, at no more than 8 in (20 cm) per second. When cirrostratus is associated with either nimbostratus or altostratus, prolonged rainfall is likely.

CIRROSTRATUS CLOUDS
*This uniform and featureless sheet of cloud is characteristic of high-level cirrostratus. Its appearance is usually a sign of approaching rain.*

# WIND

390–91 Circulation and currents
446–49 Atmospheric circulation
460–63 Air masses and weather systems
464–65 Living with hurricanes
466–69 Precipitation and clouds

WIND IS THE MOVEMENT of air away from regions of high pressure toward regions of low pressure. The air continues to move until the pressure difference is eliminated and an equilibrium is reached. Pressure differences are associated with large-scale weather systems, but they can also occur and produce winds on a smaller scale. A towering storm cloud, for example, draws in air to feed its upcurrents and expels air from its downcurrents, producing winds with gusts that can reach gale or even hurricane force.

## THEODORE FUJITA

Tetsuya Theodore Fujita (1920–98) was a Japanese meteorologist who, with the help of his wife and a colleague, devised the six-point tornado-intensity scale that bears his name. He worked first on tornadoes triggered by firestorms resulting from the atomic explosions of World War II. He moved to the US in the early 1950s and continued his research on tropical cyclones at the University of Chicago.

## PRESSURE AND WIND SPEEDS

Wind speed is proportional to the pressure gradient, which is the rate of pressure change over a horizontal distance. However, it flows parallel to the gradient due to deflection by the Coriolis effect (see p.446). Air spirals into a low-pressure center. The closer it gets, the smaller the radius of the spiral. The conservation of its angular momentum accelerates the wind in the same way a pirouetting dancer speeds up by drawing her arms in to her sides. This means that deep depressions, drawing air from a large area toward a small center, generate stronger winds than mild ones. It is also why cyclones and tornadoes, with small cores, generate such strong winds.

**HIGH WINDS IN THE TROPICS**
*In the tropics, strong surface heating produces intense low pressure and violent weather, as seen here in Kerala state, India.*



### THE BEAUFORT WIND FORCE SCALE

This wind force scale was devised by Francis Beaufort, an English naval officer, in 1805. Initially devised for sailors, who used visual features of the sea to judge wind strength, it was later modified for use on land by substituting features such as trees and cars.

| Beaufort number | Wind speed MPH | KM/H | Description |
|---|---|---|---|
| 0 | 0–1 | 0–2 | Calm—smoke rises vertically, air feels still |
| 1 | 1–3 | 2–6 | Light air—smoke drifts |
| 2 | 4–7 | 7–11 | Slight breeze—wind detectable on face, some leaf movement |
| 3 | 8–12 | 12–19 | Gentle breeze—leaves and twigs move gently |
| 4 | 13–18 | 20–29 | Moderate breeze—loose paper blows around |
| 5 | 19–24 | 30–39 | Fresh breeze—small trees sway |
| 6 | 25–31 | 40–50 | Strong breeze—difficult to use an umbrella |
| 7 | 32–38 | 51–61 | High wind—whole trees bend |
| 8 | 39–46 | 62–74 | Gale—twigs break off trees, walking into wind is difficult |
| 9 | 47–54 | 75–87 | Strong gale—roof tiles blow away |
| 10 | 55–63 | 88–101 | Whole gale—trees break and are uprooted |
| 11 | 64–74 | 102–119 | Storm—damage is extensive, cars overturn |
| 12 | 75+ | 120+ | Hurricane—widespread devastation |

**HIGH-ALTITUDE WINDS**
*At altitude, wind speed and strength can be especially high. This storm cloud over Mount Chephren in the Canadian Rockies is being sculpted by strong winds.*

# OCEANIC WINDS

The wind is always stronger over the sea than it is over land. This is because there is less friction over water. Hills, trees, buildings, and other obstructions slow the wind on land. The distance the wind blows without interruption is called the fetch, and the long fetch over an ocean alters the characteristics of air masses crossing it from a continent. Continental air is dry, and water will evaporate into it as it crosses the ocean. If the air were still, evaporation would be slow, since the lowest layer would soon become saturated. But because the wind constantly replaces the surface layer with dry air drawn down from above, it greatly increases the rate at which the air gathers moisture, and changes it from a continental to a maritime air mass. The longest fetch on Earth is in the Southern Ocean, where the wind blows, uninterrupted by land, all around the world, and the wind and waves are the strongest on the planet. Sailors refer to the latitudes south of South America as the Roaring Forties, Furious Fifties, and Screaming Sixties.

**TREE SCULPTURE**
*Salt-laden winds have dried out and destroyed the buds on the exposed side of this tree on the British coast.*

**POLAR WINDS**
*The Greenland and Antarctic ice sheets are higher at the center than the edges. Sinking cold air flows down to the coast, producing almost constant wind, as here at Marsh Base, Antarctica.*

# SEASONAL WINDS

Some winds are associated with particular seasons. A few such winds are so constant that they are given individual names. Most occur because the high- and low-pressure systems that create wind change position during the year. The sirocco, for example, most often occurs in spring, and is drawn northward by the low-pressure systems that develop in the Mediterranean. It blows ahead of depressions, gathering dust in the Sahara and moisture from the sea, and arrives over southern Europe as a warm, moist wind that sometimes delivers dust-laden rain. In contrast, the mistral may occur at any time, but is most common in winter when there is a high-pressure system in the North Atlantic and low pressure over Europe. It is a cold, northerly wind that blows over southern Europe, increasing in speed as it is funneled down the lower Rhône Valley. Other cold, winter winds include the buran, a fierce blizzard that blows across the Russian plains.

**SEASONAL WINDS IN EUROPE**
*Depending on the season, different winds occur in the Mediterranean. The cold mistral, bora (blue arrows), and warm föhn (orange) flow in winter; the hot leveche and sirocco (red) flow in summer.*

**CLOUD ARCH**
*This arch in the altostratus cloud over the Rocky Mountains in Montana heralds the arrival of the warm chinook wind.*

# DAILY WINDS

Some local winds occur at particular times of day. These daily winds, as they are known, include land–sea breezes (see right), and katabatic (see below) and anabatic winds, which flow up and down the sides of valleys due to changes in temperature. The haboob is a dust storm that occurs in northern Sudan, most often late on a summer day, when a squall, producing strong downdrafts, moves over ground covered with loose sand and dust. It can raise pillars of dust to a great height, and when these merge, they form a wall of dust up to 15 miles (24 km) long, advancing at about 35 mph (56 km/h). The harmattan blows from the north or northeast, over west Africa to the south of the Sahara, at any time of year but only by day. At night, the wind dies down. The northeasterly trade wind becomes very hot, dry, and dusty as it blows across the desert. Ordinarily, hot air rises by convection, drawing in cooler air, but if the wind blows under a temperature inversion, where warmer air lies below cool air, the hot air is trapped, which is when it becomes a harmattan.

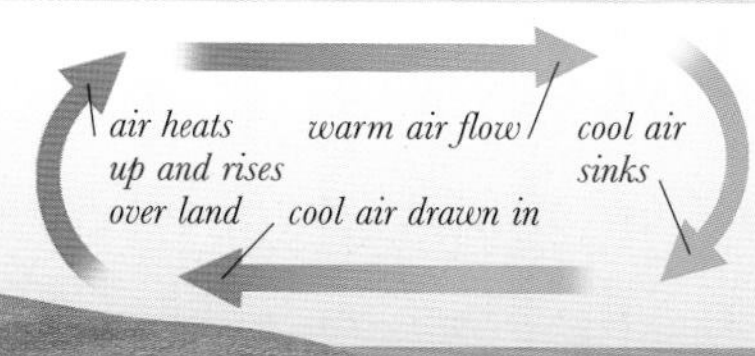


**DAYTIME SEA BREEZE**

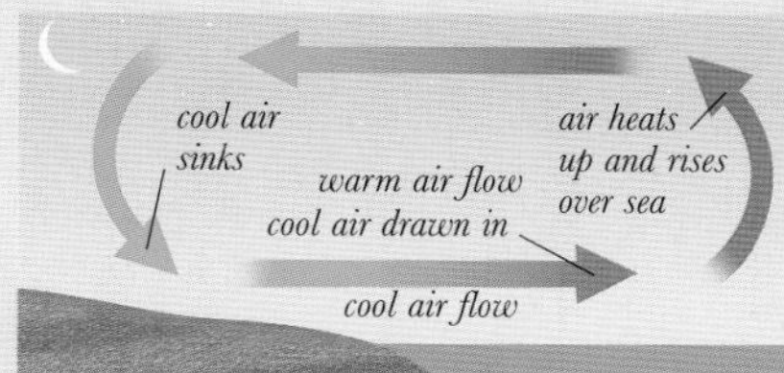


**NIGHTTIME LAND BREEZE**

**LAND–SEA BREEZES**
*During the day, land heats faster than water. Warm air rises over land, and cool air flows in to replace it, creating a sea breeze. At night, land cools more quickly than water, and the airflow is reversed. Similar winds develop over large lakes, such as the Great Lakes.*

**KATABATIC WIND FOG**
*At night, the ground cools, the temperature drops, and moisture condenses. The cold fog creates a katabatic wind by sinking into the valley, displacing warmer air, as here in the Grand Canyon.*

# THUNDERSTORMS

Thunderstorms are generated inside large cumulonimbus clouds (see p.471). Hailstones and ice crystals, colliding as they move up and down through the cloud, cause particles with a positive electric charge to accumulate near the top of the cloud. Those with a negative charge accumulate near the base and then induce a positive charge on the ground below. When the charge builds beyond a certain point, it discharges as a spark of lightning. Lightning flashing between different regions inside a cloud or between two clouds is called sheet lightning. Forked lightning flashes between a cloud and the ground. Lightning heats the adjacent air by up to 54,000°F (30,000°C) in less than a second, causing it to explode violently. Thunder is the sound of that explosion.

**STORM CLOUDS**
*Frontal rain is a feature of temperate climates in spring. This front, passing over Colorado, is accompanied by a line of thunderstorms.*

**THUNDERSTORM OVER BRAZIL**
*The high humidity and warm temperatures found near the equator provide ideal conditions for the formation of cumulonimbus cloud that results in thunderstorms.*

# SQUALL LINES

A squall line develops when active cumulonimbus clouds (see p.471) merge to form a continuous belt of storms up to 600 miles (1,000 km) long. The storms advance at a right angle to the squall line. They begin along a cold front that is moving rapidly beneath warm, moist air. The warm air is lifted and becomes unstable, producing towering clouds. Wind speed increases with height, pushing the tops of the clouds forward so they overhang their bases as anvils. Air is drawn in ahead of the line, and cold air leaves as downdrafts in the rain behind the storms. Some of the cold air flows beneath the cloud, undercutting the updraft and producing what is known as a gust front. This front is so vigorous that it makes the squall line move faster than the cold front that gave rise to it. The two then become separated, with the squall line traveling ahead of the cold gust front.

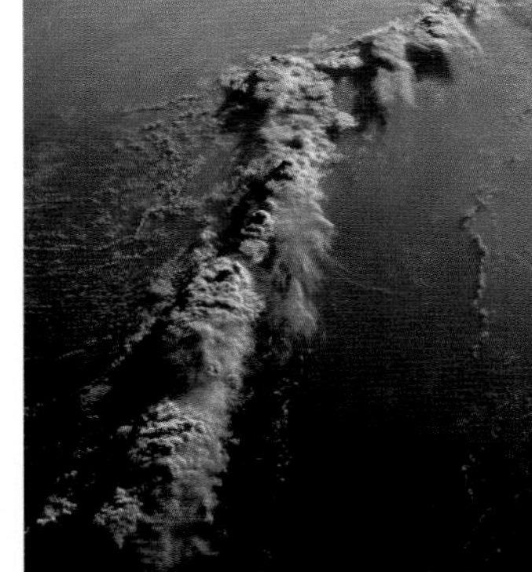

**STORMY WEATHER**
*A squall line in the Atlantic Ocean, bringing heavy precipitation, gale-force winds, and thunderstorms. It may travel a huge distance before finally expending all of its energy.*

**FULGURITE**
*This curiously shaped mineral is a type of glass made from sand grains fused together by the heat of a lightning strike.*

# DUST DEVILS AND SAND STORMS

A dust devil is a column of rapidly spinning air that rises without warning and dies down just as suddenly. Most are about 100 ft (30 m) tall, but they can rise to 300 ft (100 m) and occasionally even to 6,000 ft (1,800 m). They last for only a few minutes, snaking and moving erratically over the ground, but as one dies, another often rises nearby. Dust devils are caused by the uneven heating of the ground. Rocks heat at different rates, so by early afternoon in a desert, some patches of ground can be up to 30°F (17°C) hotter than others. Air rises over the hot patches, and cooler air spirals in to replace it. The wind lifts dust and sand, making the dust devil visible. As the cool air reduces the surface temperature, the temperature of the hot patch soon drops to match that of its surroundings. Deprived of energy, the dust devil disappears. A sand storm occurs when a wind of 35 mph (55 km/h) or more coincides with very unstable air, usually at a front. The wind raises dust and sand, but not high enough for it to travel very far. However, if the air is unstable, vertical air currents lift the dust and sand to a great height, over a vast area. In March 1998, a sandstorm that passed over Egypt and parts of Lebanon and Jordan reduced visibility to about 600 ft (180 m). It meant that both Cairo Airport and the Suez Canal had to be closed.

**DUST STORM OVER THE SAHARA**
*This dust storm, seen from space, covers a vast area of Libya and Algeria. Visibility on the ground will be reduced to almost zero.*

**DUST DEVIL**
*Pockets of circulating air become visible as they collect fine particles of dust. This dust devil has developed on a hot patch of ground in the savanna region of Kenya.*

**WALKING INTO THE WIND**
*This male lion in South Africa walks into the wind and dust, possibly to maximize his chance of scenting prey and stalking it successfully.*

# TORNADOES

A tornado is the fiercest wind on Earth, reaching speeds of up to 300 mph (480 km/h). Tornadoes can happen anywhere, but most occur on the Great Plains of the US. When the wind changes speed or direction near the top of a cumulonimbus cloud (see p.471), it tilts the upcurrents. Cold downcurrents then fall to the side of the rising air without cooling it. This allows the cloud to grow bigger and last longer than an ordinary storm cloud. The rising air starts rotating slowly, starting near the top of the cloud and extending downward, getting narrower as it does so. As the radius of the spiral decreases, the rate of rotation increases. The rotating air continues downward, forming a funnel below the cloud. Air drawn into the funnel enters an area of much lower pressure, so it expands, cools, and gives up its moisture. A funnel becomes a tornado on touching the ground.

**TORNADO**
*The funnel of this developing tornado is snaking down from a cumulonimbus cloud over Colorado. If the funnel touches the ground, it will become a tornado, changing color as dust and debris are drawn into the vortex.*

**TORNADO FORMATION**
*A tornado begins to form as a funnel, emerging beneath a cloud (1). It extends downward (2) and narrows at the base (3). On touching the ground (4), it raises a cloud of dust and debris.*

**TORNADO APPROACHING**
*This tornado is in the region of the Great Plains called Tornado Alley, which has more tornadoes than anywhere else in the world.*

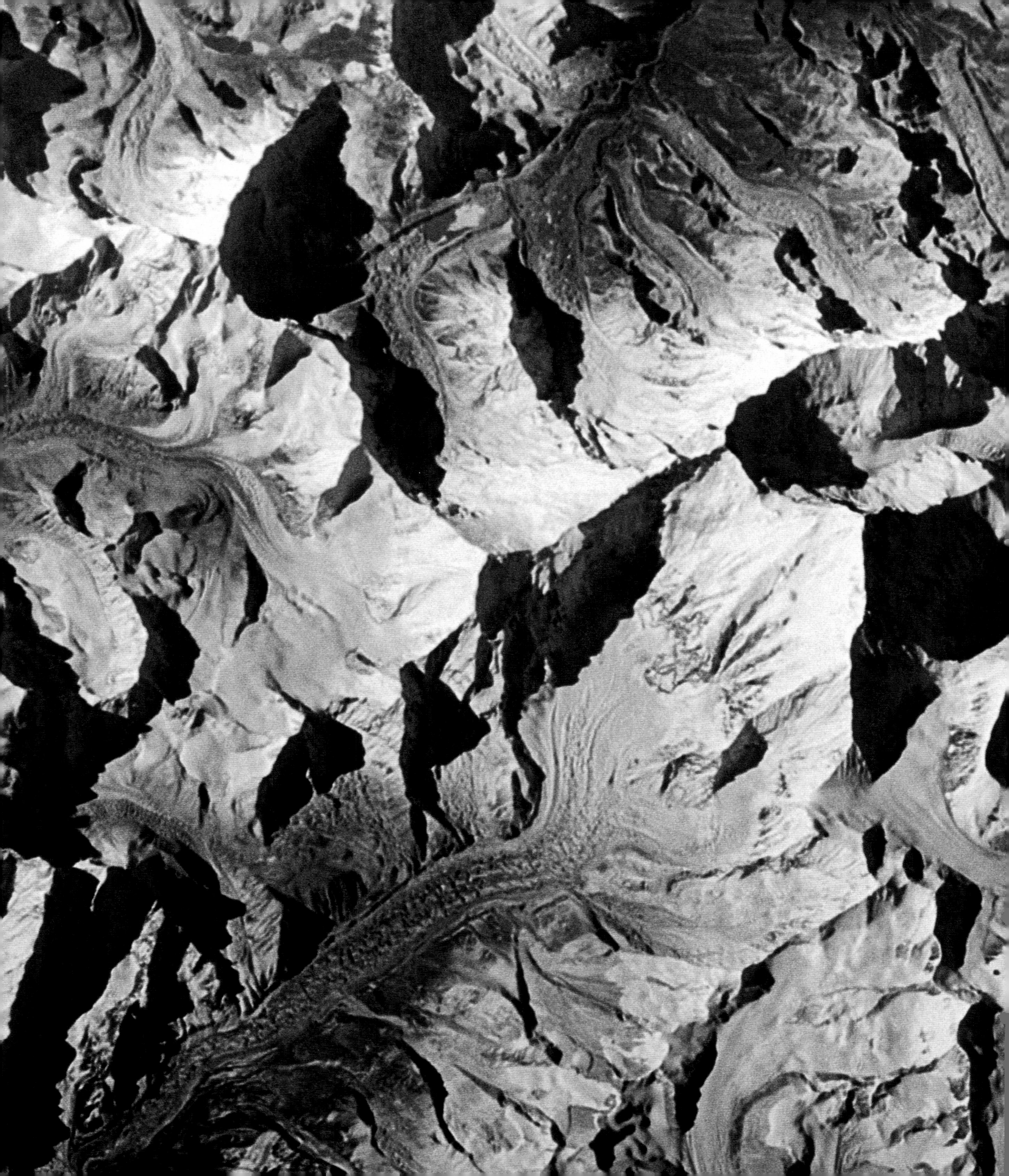

# TECTONIC EARTH

# THE EARTH'S PLATES

58–59 The crust
106–107 Tectonic plates
108–109 Plate boundaries
142–53 Fault systems
154–71 Mountains
172–93 Volcanoes
386–87 Ocean tectonics
396–97 Oceans of the world

THE SHAPE AND GEOLOGY of the Earth's surface is mainly a result of billions of years of slow shifting, making, and breaking of the Earth's tectonic plates. There are seven large plates (the Pacific, African, North American, South American, Eurasian, Australian, and Antarctic) and a dozen or so smaller ones. Six of the major plates carry a continent, meeting adjacent plates under the ocean. Oceanic margins have spreading ridges, where plates grow, and subduction zones, where one plate is consumed. Of the large plates, only the Pacific lies entirely underwater, as do most of the smaller plates.



## PLATE MOVEMENT

Tectonic plates have been moving over the Earth's surface for at least hundreds of millions of years. Today, they are moving at about ¾–7¾ in (2–20 cm) a year, the rate at which fingernails grow, but over geological time, the familiar pattern of the continents will change. The most striking movements are the widening of the Atlantic, which is pushing the Americas west; the splitting of Arabia from Africa down the Red Sea; and the pushing of India up against Asia, which has thrown up the Himalayas. Most dramatic of all is the rapid subduction of the western edge of the Pacific plate, which makes eastern Asia uniquely prone to violent volcanic eruptions and earthquakes.

**VOLCANIC ACTIVITY**
*Volcanoes, such as these in the Galápagos Islands, are fed by a hot spot, the lava rising from within the deep mantle.*

**FOLDING**
*Rock strata buckle and fold when plates collide, as here in Utah. Such collisions throw up long mountain chains at continental margins.*

**FAULT LINE**
*This ancient fault line in Arizona is not close to the edge of a tectonic plate; rather, it is in a broad zone of distributed tectonic activity.*

**EARTH GEOLOGY AND TECTONICS**
*Although tectonic activity constantly alters the Earth's surface, each continent has a stable core of ancient, crystalline rock, called a craton, forming a base onto which the younger rocks are welded. Where exposed at the surface, cratons often form vast, flat plains in continental interiors. They are usually covered by a thin layer of younger sediments. Around the margins of continents, volcanoes add new igneous rocks, and mountain ranges are uplifted as rock strata buckle under pressure caused by the collision of tectonic plates.*

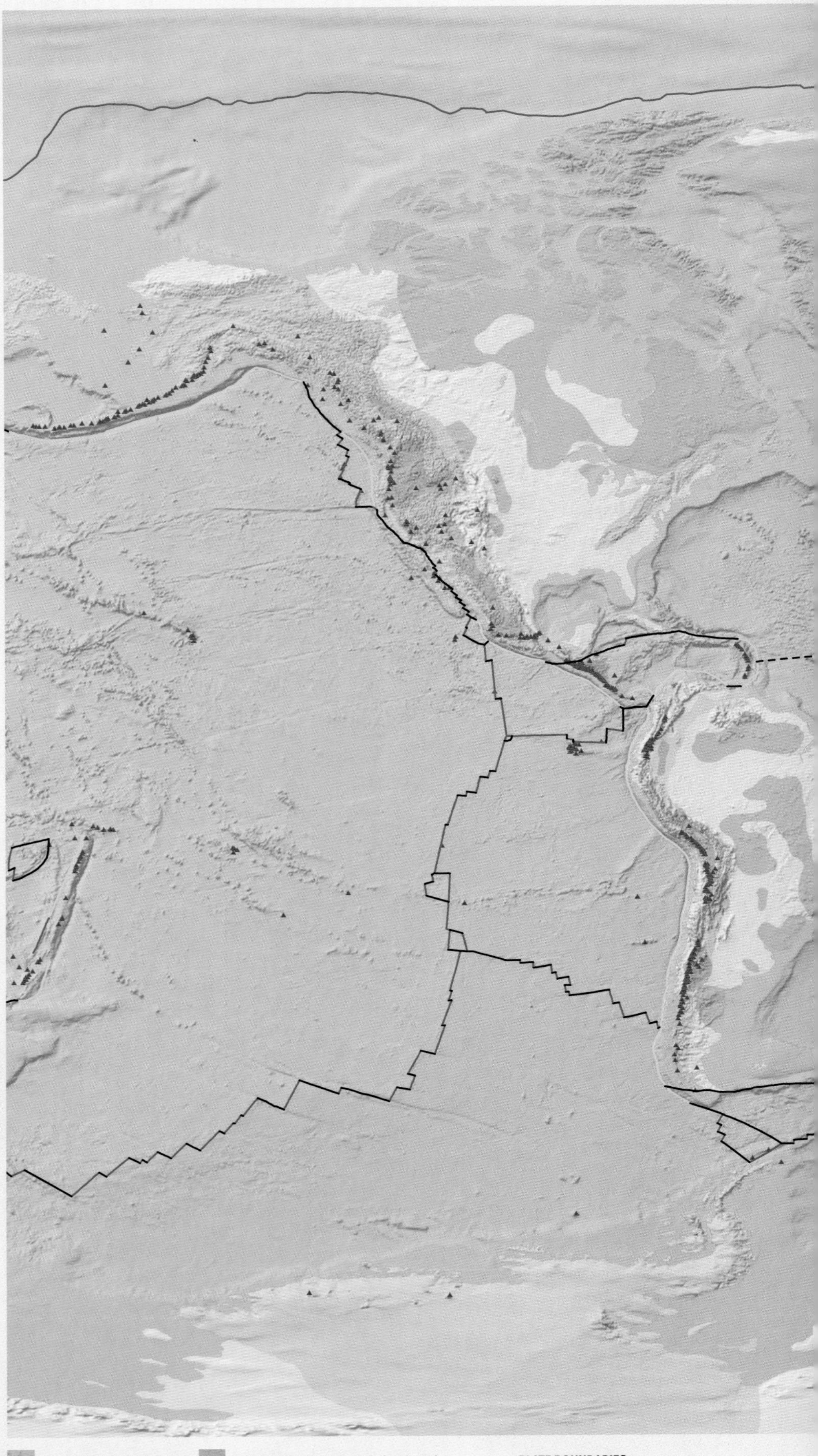

Precambrian shields

AREAS OF FOLDING
Paleozoic
Mesozoic
Cenozoic

Mesozoic and Cenozoic volcanic rocks
Sedimentary cover
▲ Active volcanoes

PLATE BOUNDARIES
Convergent
Divergent
Transform
Uncertain

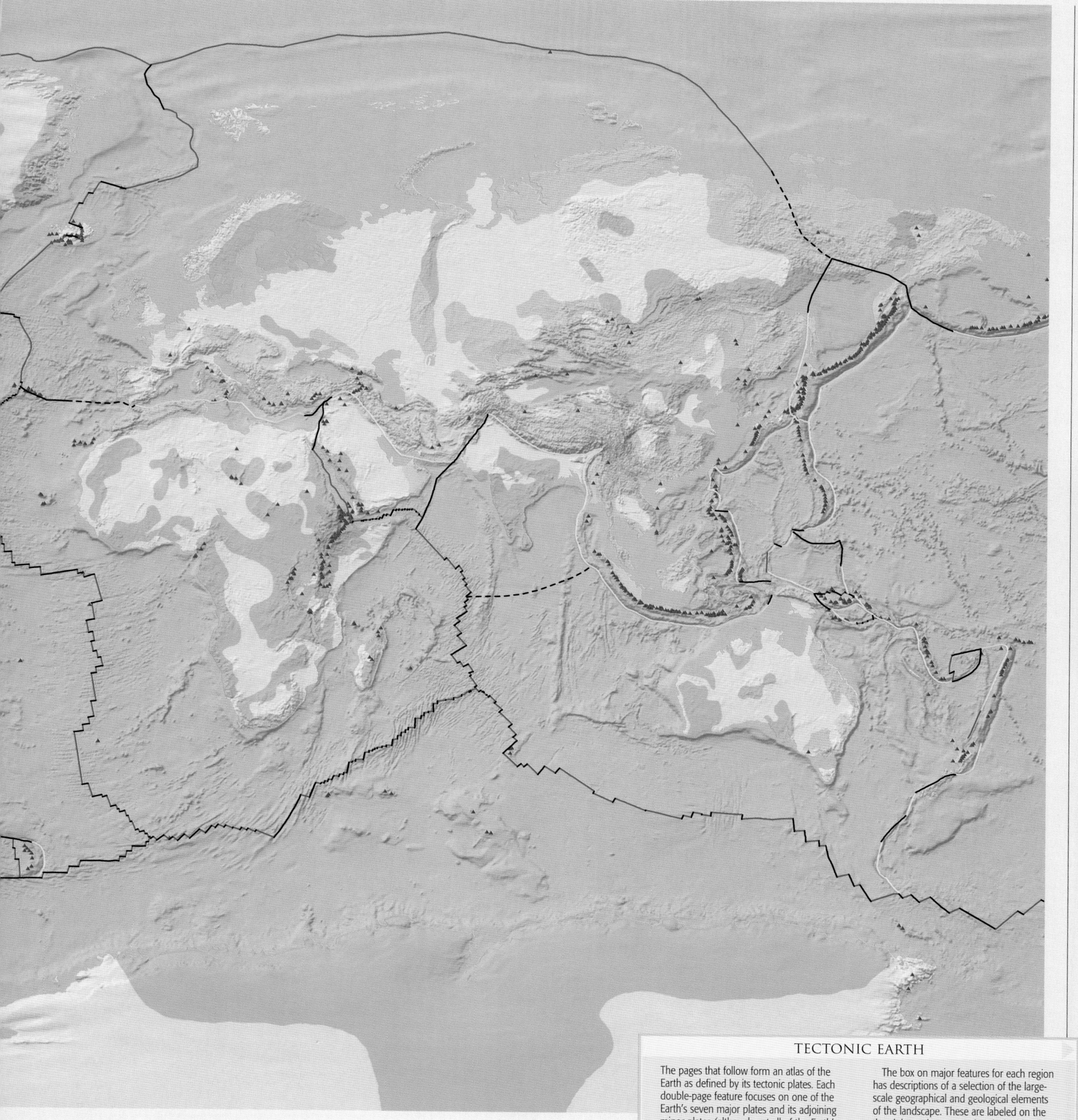



## TECTONIC EARTH

The pages that follow form an atlas of the Earth as defined by its tectonic plates. Each double-page feature focuses on one of the Earth's seven major plates and its adjoining minor plates (although not all of the Earth's many minor plates have been mapped here).

The data presented for each region refers to the combined area of the plates shown. For example, the highest point on the five plates described in the feature on North America is Mount McKinley, Alaska, and the largest city is New York.

The box on major features for each region has descriptions of a selection of the large-scale geographical and geological elements of the landscape. These are labeled on the the globe at the center of each double-page feature. Around the globe are numbered descriptions of other features of interest, which can also be located on the globe.

For clarity, boundaries between plates have been colored and exaggerated. Important boundary features, such as ocean ridges and trenches, are also labeled on the map.

# NORTH AMERICAN

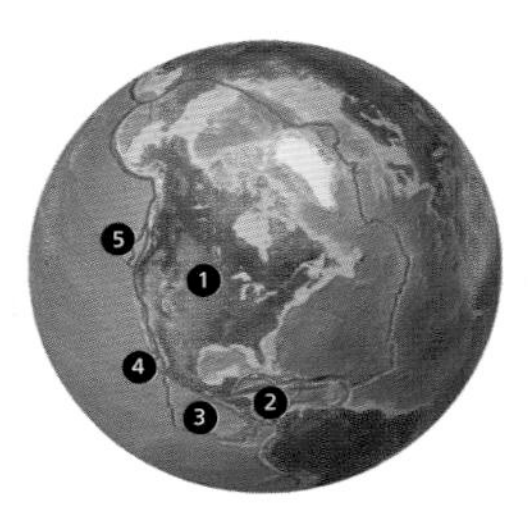

**PLATES** ❶ North American Plate ❷ Caribbean Plate ❸ Cocos Plate ❹ Rivera Plate ❺ Juan de Fuca Plate

**COMBINED AREA** 24 million square miles (62 million square km)

**HIGHEST POINT** Mount McKinley, Alaska, 20,320 ft (6,194 m)

**LOWEST POINT** Puerto Rico Trench 28,374 ft (8,648 m) below sea level

**LARGEST CITY** New York (18.1 million)

THE SIXTH-LARGEST PLATE, the North American Plate was joined to the Eurasian, African, and South American Plates until the former supercontinent of Pangea started to break up and the Atlantic Ocean began to open. The relatively young mountains of western North America reflect ongoing subduction and collision with the Pacific Plate. The more ancient Appalachian mountains of eastern North America reflect older Paleozoic collisions in a pre-Pangean Atlantic Ocean. The Cocos and Rivera plates are currently subducting beneath the southern edge of North America to produce the active Mexican Volcanic Belt.

① **MOUNT MCKINLEY**
*The highest peak in North America is Mount McKinley in the Alaska Range. Its granite outline has been shaped by glaciers that have blanketed the region for the last 2 million years.*

② **BADLANDS**
*This part of South Dakota is made up of ridges, separated by deep ravines and gullies, that formed as violent cloudbursts eroded away the soft clay rock. The area covers 1,000 square miles (2,600 square km).*

③ **BRYCE CANYON**
*At Bryce Canyon in Utah, rivers cut deep into the Paunsaguant Plateau, and carved out tall, thin ridges, known as fins. Over time they were eroded into spectacular sandstone pinnacles and spires called hoodoos.*

④ **YOSEMITE**
*The granite mountains of the North American Sierra Nevada were shaped by glaciers in the last ice age, creating spectacular rock formations and cliffs up to 3,300 ft (1,000 m) high.*

⑤ **DEATH VALLEY**
*This is the lowest, hottest, and driest place in North America, descending to 282 ft (86 m) below sea level, with shade temperatures up to 134°F (57°C) and years with no rain at all.*

ALASKA PENINSULA
CANADIAN SHIELD
ROCKY MOUNTAINS
GREAT PLAINS
MISSISSIPPI
GREAT LA
APPALA
GULF OF MEXICO
MIDDLE AMERICAN TRENCH
ISTH OF PA
San Francisco
Los Angeles
Chicago
Detroi
Dallas
Houston
Atlanta
Guadalajara
Mexico City

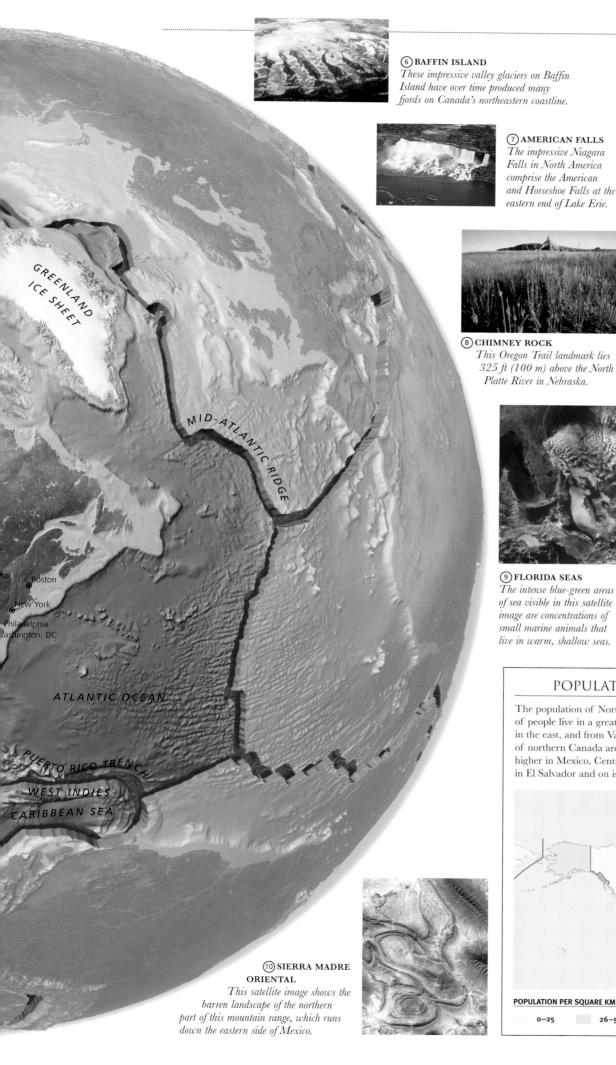

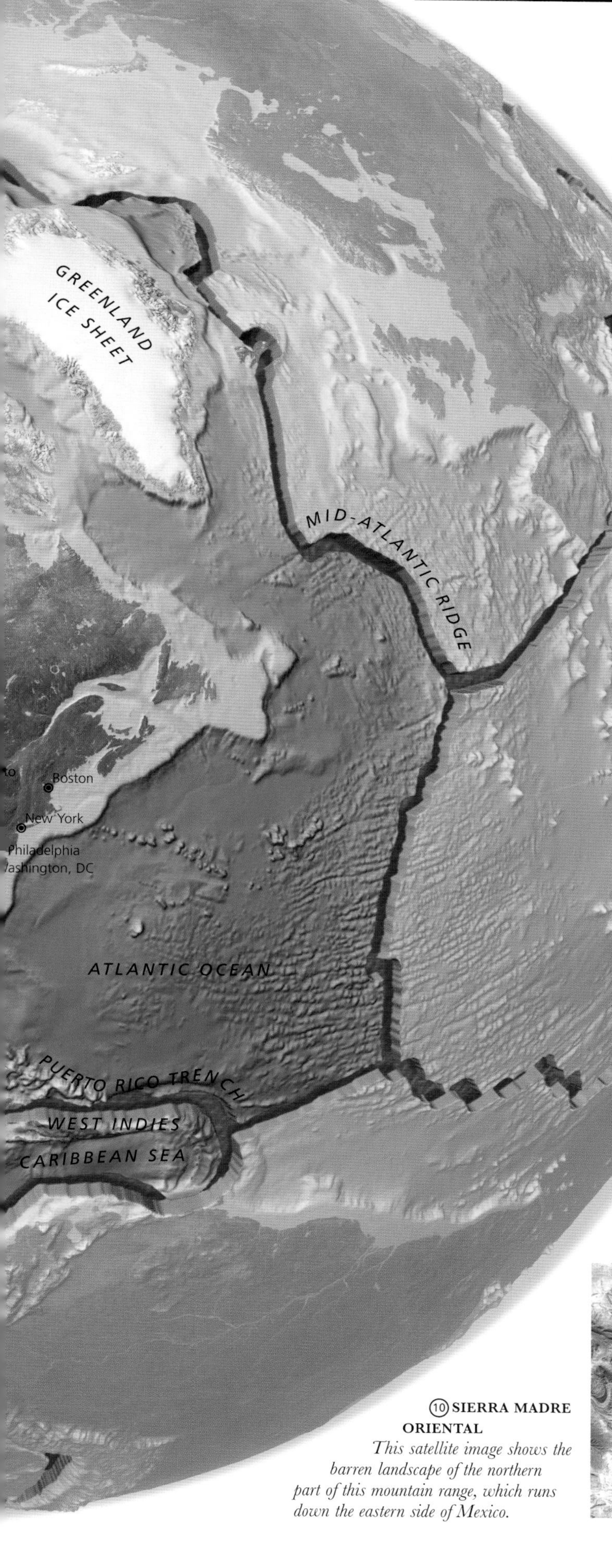

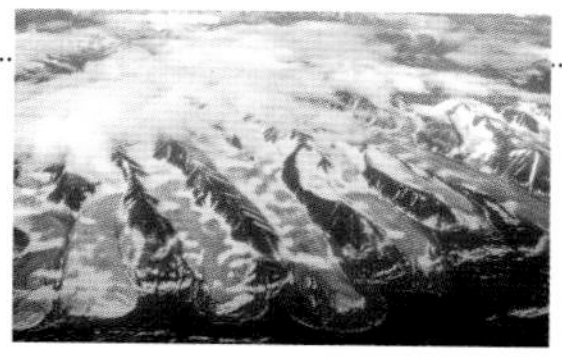

⑥ **BAFFIN ISLAND**
*These impressive valley glaciers on Baffin Island have over time produced many fjords on Canada's northeastern coastline.*

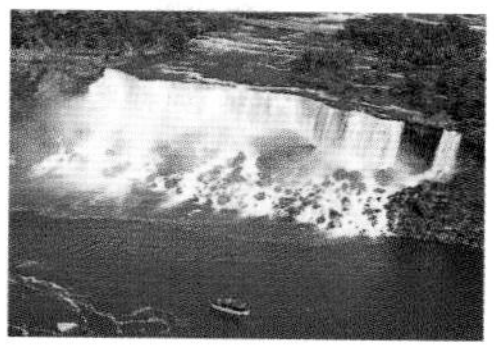

⑦ **AMERICAN FALLS**
*The impressive Niagara Falls in North America comprise the American and Horseshoe Falls at the eastern end of Lake Erie.*

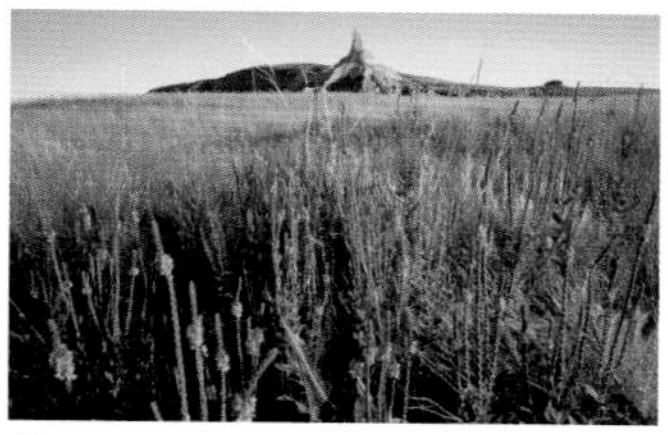

⑧ **CHIMNEY ROCK**
*This Oregon Trail landmark lies 325 ft (100 m) above the North Platte River in Nebraska.*

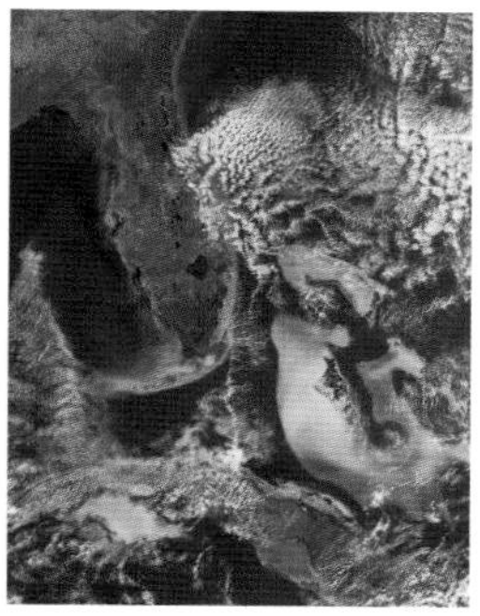

⑨ **FLORIDA SEAS**
*The intense blue-green areas of sea visible in this satellite image are concentrations of small marine animals that live in warm, shallow seas.*

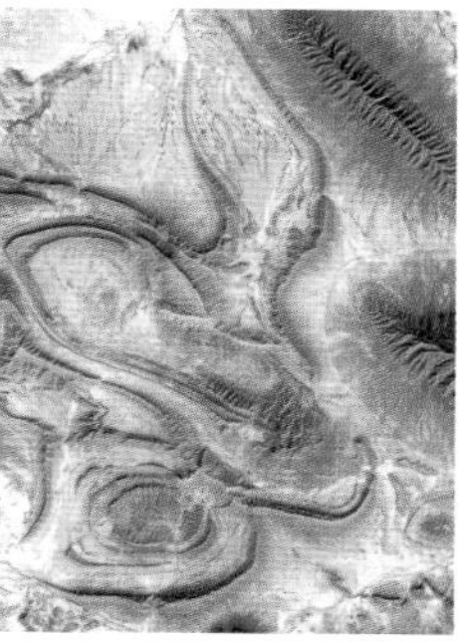

⑩ **SIERRA MADRE ORIENTAL**
*This satellite image shows the barren landscape of the northern part of this mountain range, which runs down the eastern side of Mexico.*

## MAJOR FEATURES

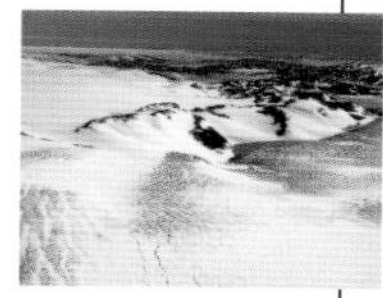

**GREENLAND ICE SHEET**
Containing a tenth of the world's ice, the Greenland Ice Sheet is 14,000 ft (4,300 m) thick in places.

**CANADIAN SHIELD**
Formed some three billion years ago, the ancient crystallized rocks of the Canadian Shield cover a vast area of North America.

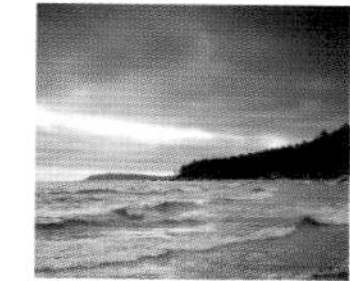

**GREAT LAKES**
Water from glaciers created the Great Lakes at the end of the last ice age, 15,000 years ago.

**GREAT PLAINS**
Up to 625 miles (1,000 km) wide, the Great Plains are a vast area that stretches down the middle of the continent.

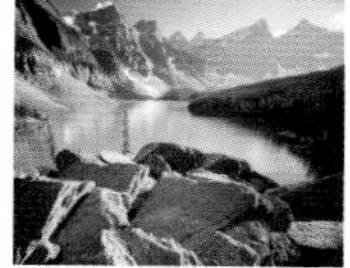

**ROCKY MOUNTAINS**
These mountains run from California to Alaska, a distance of about 3,700 miles (6,000 km).

**APPALACHIANS**
These are among the oldest of North America's mountain ranges, formed in Paleozoic times, over 250 million years ago.

**MISSISSIPPI**
The second longest river in the US after its tributary the Missouri, the Mississippi flows 2,350 miles (3,780 km) from Minnesota to the Gulf of Mexico.

**ISTHMUS OF PANAMA**
A strip of land just 36 miles (58 km) wide that appeared about 3 million years ago, linking North and South America.

## POPULATION AND BOUNDARIES

The population of North America is unevenly spread. Huge numbers of people live in a great arc of cities from Chicago to Washington, DC in the east, and from Vancouver to Los Angeles in the west. Vast areas of northern Canada are virtually uninhabited. Average densities are higher in Mexico, Central America, and the Caribbean, being highest in El Salvador and on islands such as Barbados and Haiti.

**POPULATION PER SQUARE KM** (1 SQUARE KM = ⅓ SQUARE MILE)

0–25 | 26–50 | 51–100 | 101–200 | 200+

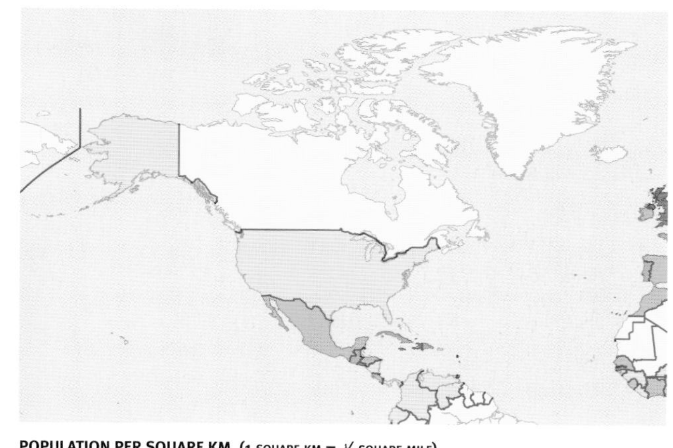

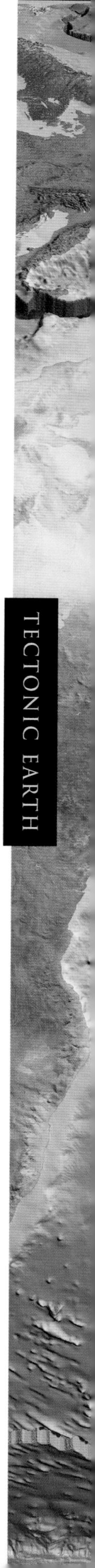

# SOUTH AMERICAN

THE SMALLEST OF THE SEVEN large tectonic plates, half of the South American Plate is under the Atlantic. Its northern boundary is washed by the waters of the Caribbean, but its southern tip at Cape Horn is less than 600 miles (1,000 km) from the Antarctic Circle, where it touches the Scotia Plate. Down the west coast are the Andes, thrust up as the South American Plate moves west over the subducting Nazca Plate that lies beneath the southeastern Pacific. The Nazca Plate is the fastest-moving of all plates, and as it dives steeply under South America, it generates earthquakes and volcanoes along the length of the Andes.

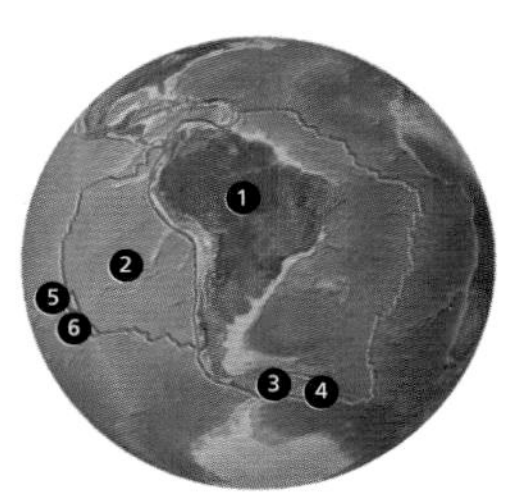

**PLATES** ❶ South American Plate ❷ Nazca Plate ❸ Scotia Plate ❹ Sandwich Plate ❺ Easter Plate ❻ Juan Fernandez Plate

**COMBINED AREA** 23 million square miles (60 million square km)

**HIGHEST POINT** Mount Aconcagua, Argentina, 22,834 ft (6,960 m)

**LOWEST POINT** South Sandwich Trench, 30,615 ft (9,334 m) below sea level

**LARGEST CITY** São Paulo (17.7 million)

## MAJOR FEATURES

**AMAZON BASIN**
The rainforest in the Amazon basin covers an area larger than western Europe.

**ANDES**
Winding all the way down the west side of South America, the Andes are the world's longest mountain range.

**ATACAMA DESERT**
This windswept plain, 3,300 ft (1,000 m) above sea level, is the world's driest desert. Some areas have no record of any rainfall.

**PAMPAS**
The extensive plains in Argentina and Uruguay are used for growing corn and for grazing.

**PATAGONIA**
Bitterly cold and windswept, Patagonia is virtually a desert, where almost nothing grows except in the river valleys.

**TIERRA DEL FUEGO**
The southern tip of this island, called Cape Horn, is South America's most southerly point.

**SCOTIA SEA**
Along the northern edge of the Scotia Sea, the Scotia Plate is moving beneath South America, creating a trench in the ocean floor and an arc of volcanic islands.

**PERU–CHILE TRENCH**
Where the Nazca Plate meets the South American Plate, the Peru–Chile trench has developed. It is 3,700 miles (6,000 km) long and up to 26,460 ft (8,065 m) deep.

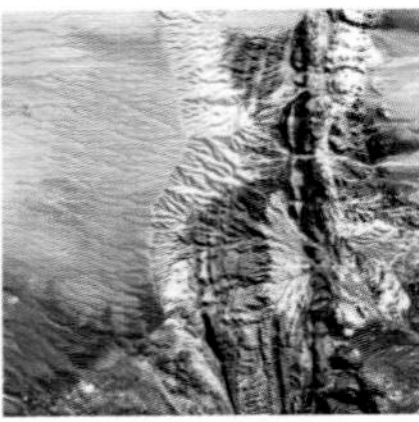

① **ALTIPLANO**
*This high plateau (seen from directly overhead in this satellite image) extends from Peru through Bolivia. Its rolling plains lie between the two mountain ranges of the Andes at about 12,000 ft (3,660 m) and are covered with pyroclastic deposits.*

② **ACONCAGUA**
*The world's highest peak outside the Himalayas, Aconcagua in Argentina is a volcano that formed in a rift, resulting from the folding and rising of the Andes.*

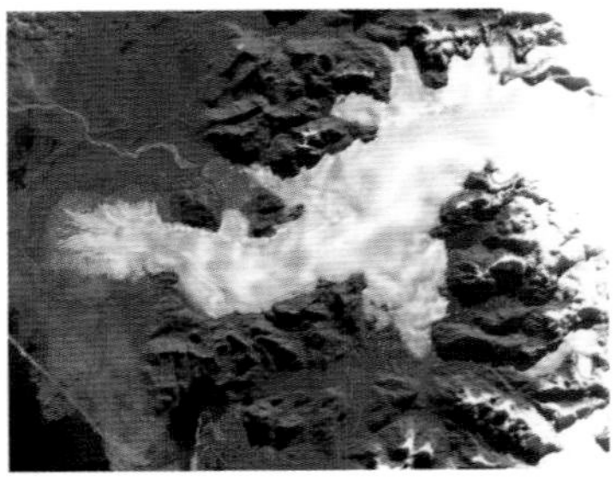

③ **PATAGONIA GLACIER**
*This thermal satellite image shows part of southern Patagonia. A vast glacier, colored white, has carved deep valleys in the landscape.*

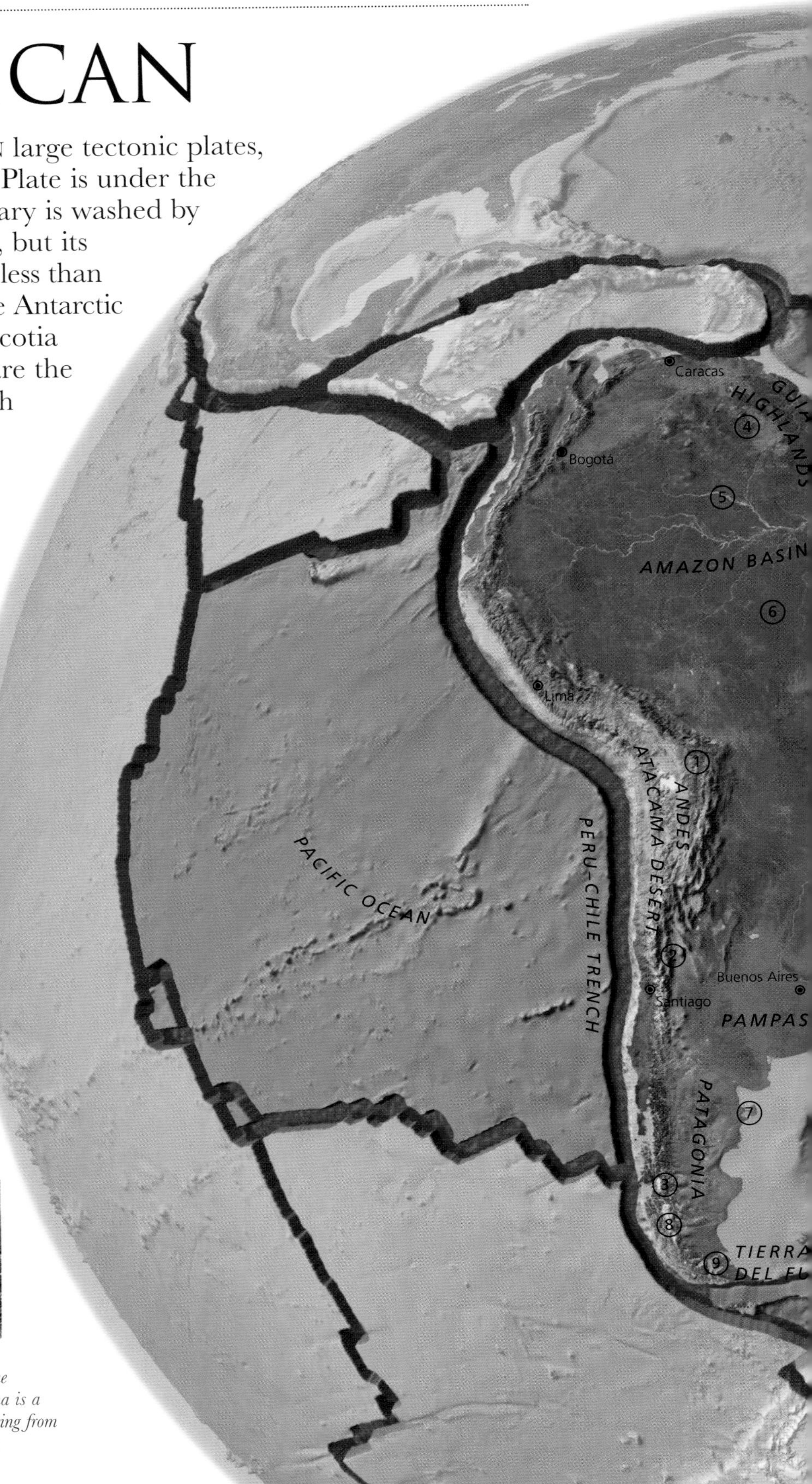

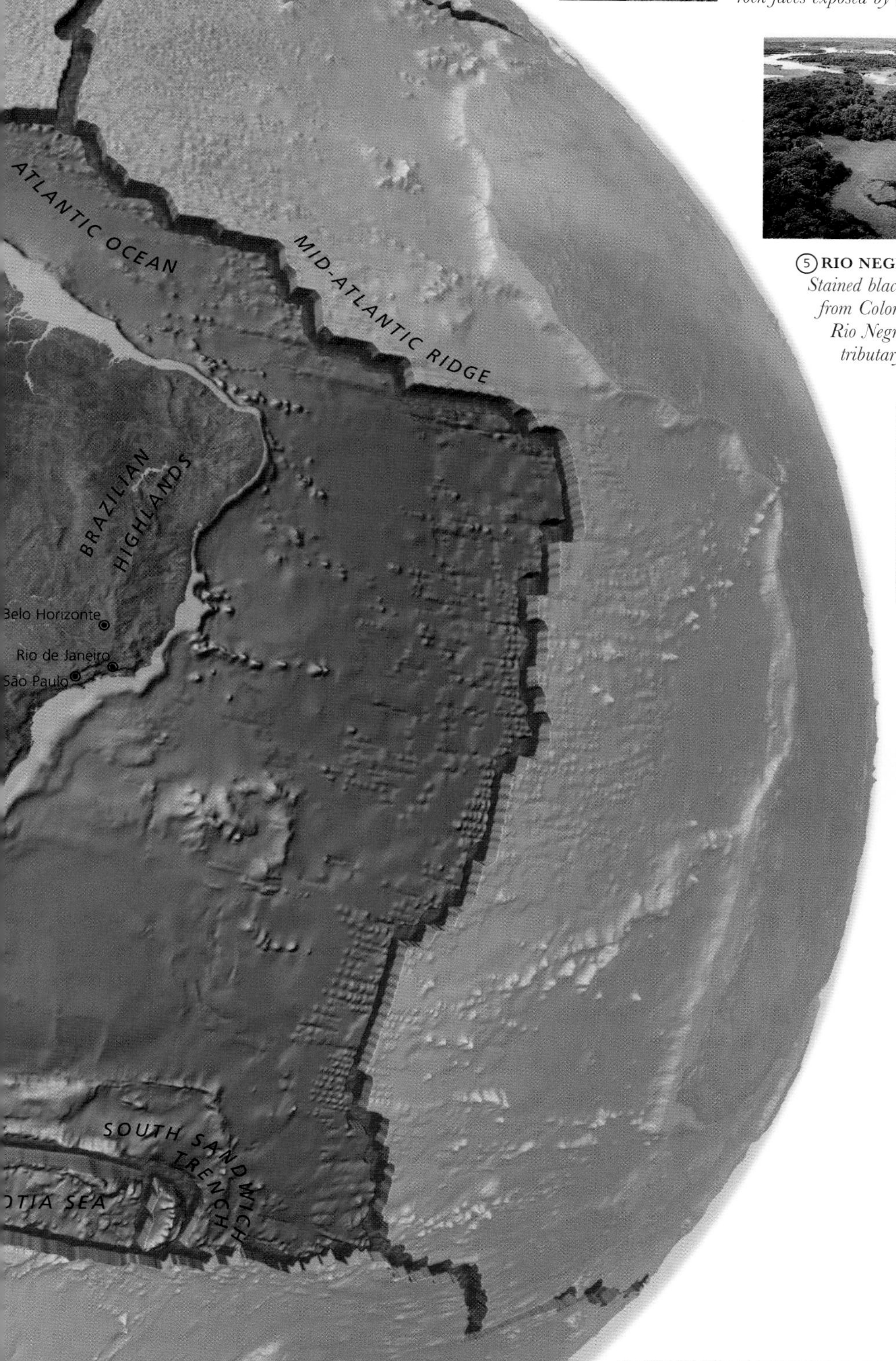

④ **TEPUIS**
*In Venezuela, ancient rocks form gigantic tablelands, known as tepuis. Spectacular waterfalls plunge over steep cliffs, which typically have pink sandstone rock faces exposed by erosion.*

⑤ **RIO NEGRO**
*Stained black by rotting vegetation from Colombian swamps, the Rio Negro is the main tributary of the Amazon.*

⑥ **DEFORESTATION**
*Although the Amazon basin is the world's greatest region of tropical rainforest, it is being destroyed for its timber, by mining activities, and to make way for farms. In this picture, one side of the river is dominated by farmland (upper half), while the other side is still covered in rainforest (lower half).*

ATLANTIC OCEAN

MID-ATLANTIC RIDGE

BRAZILIAN HIGHLANDS

Belo Horizonte

Rio de Janeiro

São Paulo

SOUTH SANDWICH TRENCH

OTIA SEA

⑦ **VALDES PENINSULA**
*Projecting out into the Atlantic from Patagonia in Argentina, the Valdes Peninsula is bounded by steep cliffs. Inland there is a salt flat 131 ft (40 m) below sea level, which is the lowest point in South America.*

⑧ **TORRES DEL PAINE**
*One of the world's last remaining wildernesses, the Torres del Paine in southern Chile has a unique range of wildlife, including guanacos, nandus, and condors. The spectacular Torres del Paine are twin peaks of towering granite.*

⑨ **STRAIT OF MAGELLAN**
*This channel links the Atlantic and Pacific oceans, and separates Tierra del Fuego from South America.*

## POPULATION AND BOUNDARIES

South America has the third-smallest population of any of the continents (behind Australia and Antarctica), but the cities of São Paulo, Rio de Janeiro, and Buenos Aires are among the world's most populous. It also has the fewest countries—just 12 (compared with 53 in Africa), including Brazil, one of the world's largest countries. Overall, Ecuador has the densest population, followed by Colombia and Venezuela.

**POPULATION PER SQUARE KM** (1 SQUARE KM = ⅓ SQUARE MILE)

0–25 | 26–50 | 51–100 | 101–200 | 200+

TECTONIC EARTH

# EURASIAN

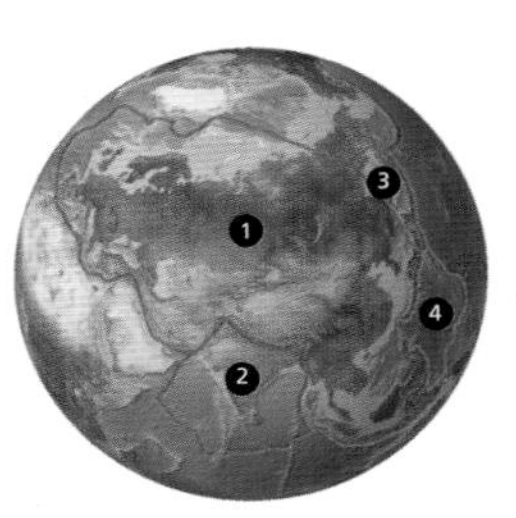

**PLATES** ❶ Eurasian Plate ❷ Indian Plate ❸ Okhotsk Plate ❹ Philippine Plate

**COMBINED AREA** 35 million square miles (90 million square km)

**HIGHEST POINT** Mount Everest, Nepal, 29,035 ft (8,850 m)

**LOWEST POINT** Galathea Deep, 34,580 ft (10,540 m) below sea level

**LARGEST CITY** Tokyo (32.2 million)

THE EURASIAN PLATE is the largest tectonic plate. It contains some of the world's oldest rocks, which are exposed in east Siberia, under the world's widest plains. To the west, the North American Plate is pulling away to open up the Atlantic Ocean. To the east, the Pacific and Philippine plates are subducting beneath the Eurasian Plate, creating an arc of volcanic islands that includes southern Japan and the Philippines. To the south, the Indian and Australian plates are moving north. This is the only place in the world where two continents are actively colliding—and the impact is pushing up the world's highest mountain range, the Himalayas.

## MAJOR FEATURES

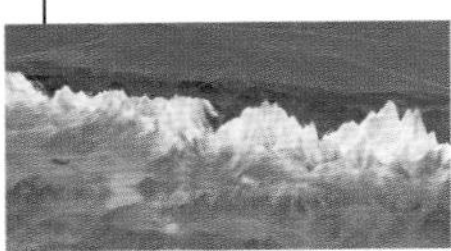

**HIMALAYAS**
These mountains started to form about 60 million years ago.

**ALPS**
The Alps are young in geological terms. They formed 10–25 million years ago as the African Plate collided with Europe.

**MEDITERRANEAN SEA**
About 5 million years ago, the Straits of Gibraltar appeared; water flowed in from the Atlantic to form the Mediterranean.

**YANGTZE**
Asia's longest river flows 3,900 miles (6,300 km) from the Tibetan Highlands to the China Sea, near Shanghai.

**DECCAN PLATEAU**
The basaltic rocks covering the Deccan Traps erupted about 65 million years ago.

**GOBI DESERT**
This is named after the small stones, called gobi, that remain on the surface after the wind has removed any finer material.

**ISLAND ARC OF SUMATRA, JAVA, AND THE PHILIPPINES**
Along the southeastern edge of the Eurasian Plate is an island arc with some of the most violent subduction-related volcanoes in the world, such as Krakatau.

**BOREAL FOREST**
Also known as taiga, this is the climax vegetation of northern Asia.

① **MOSELLE**
*The valley of the Moselle River follows a narrow, winding course through western Germany to join the Rhine at Koblenz.*

② **METEORA**
*The rock pinnacles of Metéora in Greece are all that remains of a high sandstone plateau that was traversed by innumerable faults and eroded over millions of years.*

③ **MOUNT EVEREST**
*Satellite measurements of the world's highest peak show that it is still being uplifted due to tectonic plate movement.*

④ **TIBETAN PLATEAU**
*The crust beneath this vast plateau is over 50 miles (80 km) thick, making it the thickest part of the Earth's crust.*

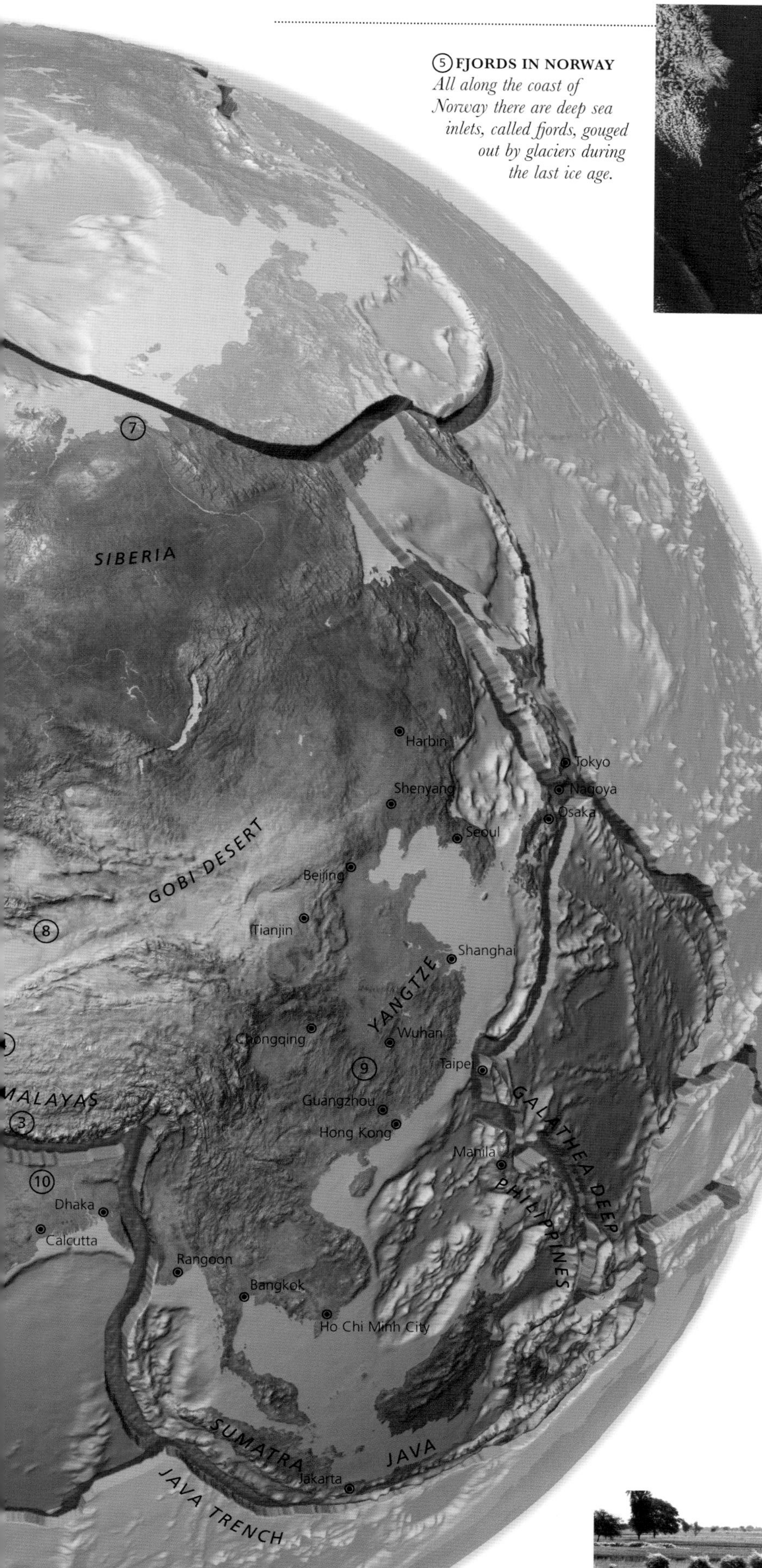

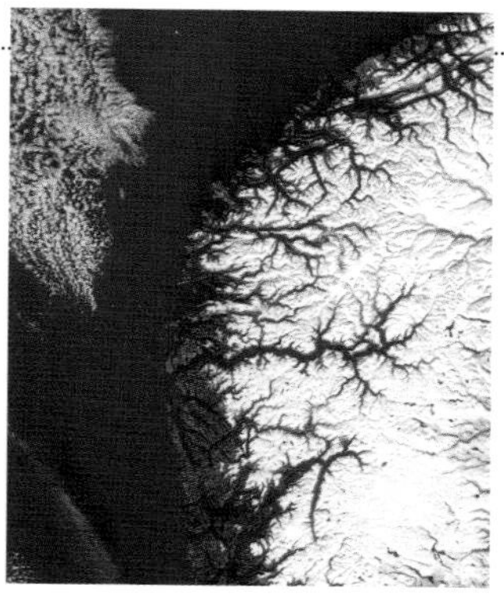

**⑤ FJORDS IN NORWAY**
*All along the coast of Norway there are deep sea inlets, called fjords, gouged out by glaciers during the last ice age.*

**⑥ BIALOWIESKA FOREST**
*The Bialowieska Forest on Poland's border with Belarus is one of the last remnants of the vast primeval forests that once covered much of Europe. It provides a refuge for wolves, bears, deer, wild boar, and European bison.*

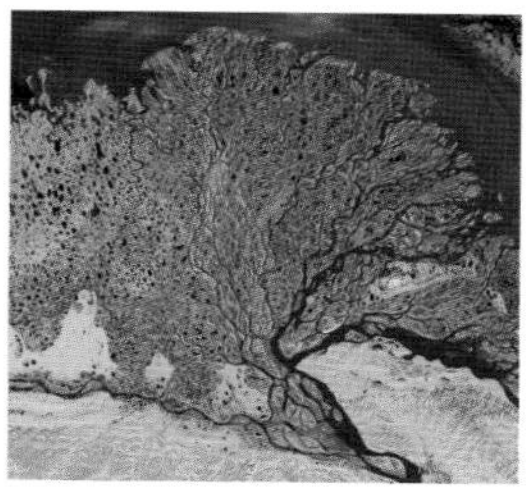

**⑦ LENA DELTA**
*Frozen solid during the bitter Siberian winter, the Lena River is transformed in June into a surging flood as the ice melts, and water flows through its green delta into the Laptev Sea.*

**⑧ TURPAN DEPRESSION**
*Created as a large block of the Earth's crust gradually slipped down between parallel faults, this rift valley is northern Asia's lowest point, descending to 505 ft (154 m) below sea level.*

**⑨ GUILIN KARST**
*The bewitching pinnacles of the Guilin Hills in China are the world's best example of karst landscape. They were formed as the limestone rock was gradually dissolved by naturally acidic rainwater.*

## POPULATION AND BOUNDARIES

About 75 percent of the world's population lives on the Eurasian Plate, but its distribution is very uneven. The vast majority live in Europe, the Indian subcontinent, southeast Asia, or China, and these are the most densely populated parts of the world. Up until the 1990s, much of northern Asia fell within the boundaries of the Soviet Union, but with its collapse, an array of independent states has emerged.

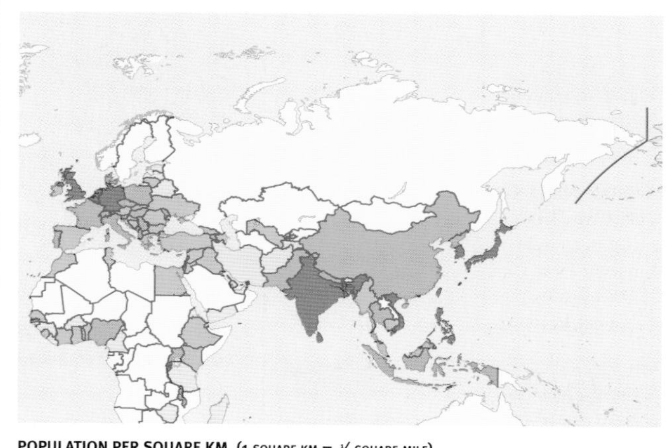

**POPULATION PER SQUARE KM** (1 SQUARE KM = ⅓ SQUARE MILE)

0–25 | 26–50 | 51–100 | 101–200 | 200+

**⑩ GANGES FLOODPLAIN**
*Flooding of the Ganges has created this fertile floodplain, which is intensively cultivated.*

# AFRICAN

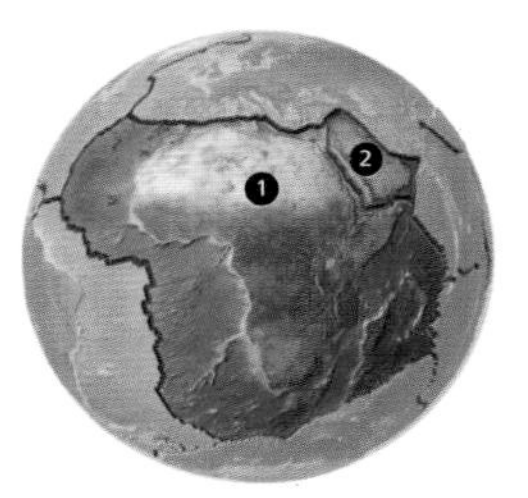

**PLATES** ❶ African Plate
❷ Arabian Plate

**COMBINED AREA** 32 million square miles (83 million square km)

**HIGHEST POINT** Mt. Kilimanjaro, Tanzania, 19,340 ft (5,895 m)

**LOWEST POINT** Somali Basin, 19,335 ft (5,826 m) below sea level

**LARGEST CITY** Cairo (14.8 million)

AS WELL AS THE AFRICAN CONTINENT, the African Plate includes parts of the Atlantic, Indian, and Southern oceans. Unlike other major plates, it has no great range of fold mountains, only the small Atlas range in the northwest. Much of the continent has been warped into saucer-shaped basins, and highlands. The south and east is mostly a vast high plateau, broken by a few mountain regions and rift valleys, notably the East African Rift. The north and west is much lower and flatter. To the northeast lies the Arabian Plate, recently split from the African Plate by the opening of the Red Sea, which is growing wider all the time.

## MAJOR FEATURES

**ATLAS MOUNTAINS**
These mountains run parallel to the northwestern edge of Africa, passing through Morocco, Algeria, and Tunisia.

**NILE**
The second-longest river in the world, the Nile flows north from several main sources, including the Luvironza River in Burundi, to the Mediterranean Sea.

**SAHARA DESERT**
The largest desert in the world, the Sahara covers almost a third of the African continent.

**ARABIAN PENINSULA**
Almost the entire Arabian Plate is a vast, sandy desert. The Red Sea borders its western edge.

**EAST AFRICAN RIFT**
This valley in east Africa has opened up over the last 30 million years as the continent has slowly started to split apart.

**CONGO BASIN**
These forest elephants live in the tropical rainforests of the Congo basin in central Africa.

**MADAGASCAR**
Separated from Africa 95 million years ago, Madagascar has a unique fauna, including its lemurs.

**MID-ATLANTIC RIDGE**
This long seafloor-spreading ridge marks the western boundary of the African Plate.

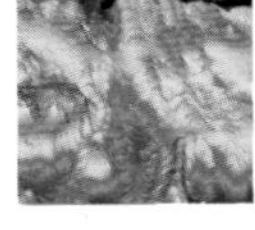

① **NEFTA OASIS**
*A green island in the desert, the Nefta oasis in Tunisia has 152 natural springs that are artesian wells.*

② **AHAGGAR MOUNTAINS**
*This vast lunar landscape in Algeria is formed from rocks about 2 billion years old—some of the oldest in Africa.*

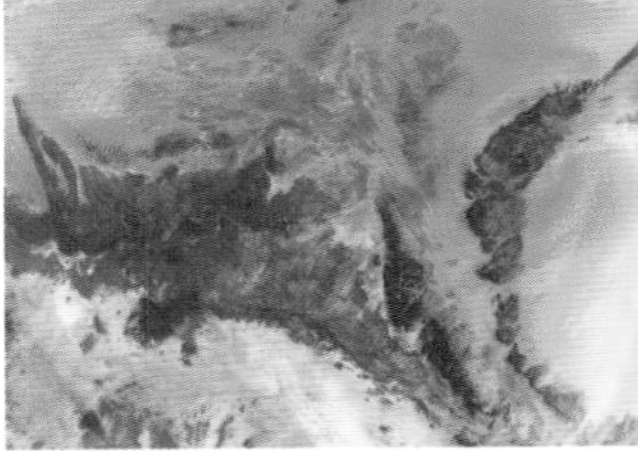

③ **ACACUS–AMSAK REGION**
*This satellite image shows three large rock massifs exposed between orange-colored sand dunes: Tassili (left), Tadrart–Acacus (center), and the crescent-shaped Amsak (right).*

④ **ORANGE RIVER**
*This is the longest river in southern Africa, flowing from the Drakensberg Plateau in Lesotho west into the Atlantic.*

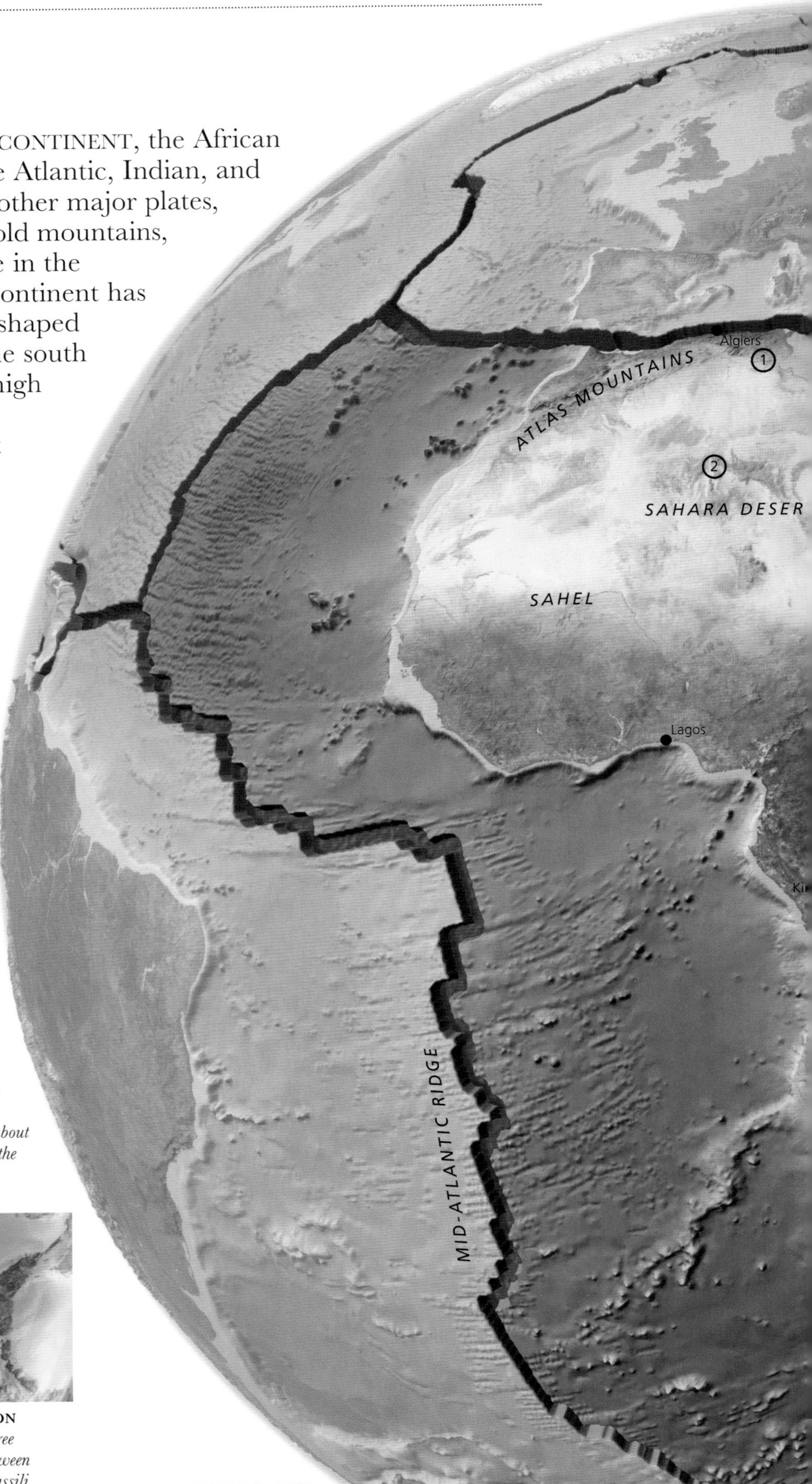

Algiers ① ATLAS MOUNTAINS ② SAHARA DESER SAHEL Lagos MID-ATLANTIC RIDGE

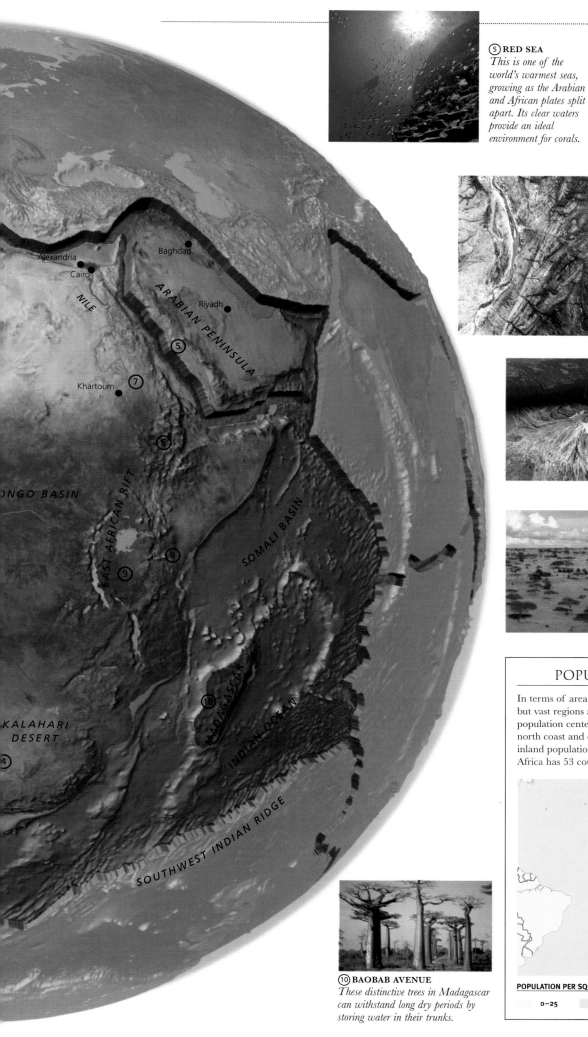

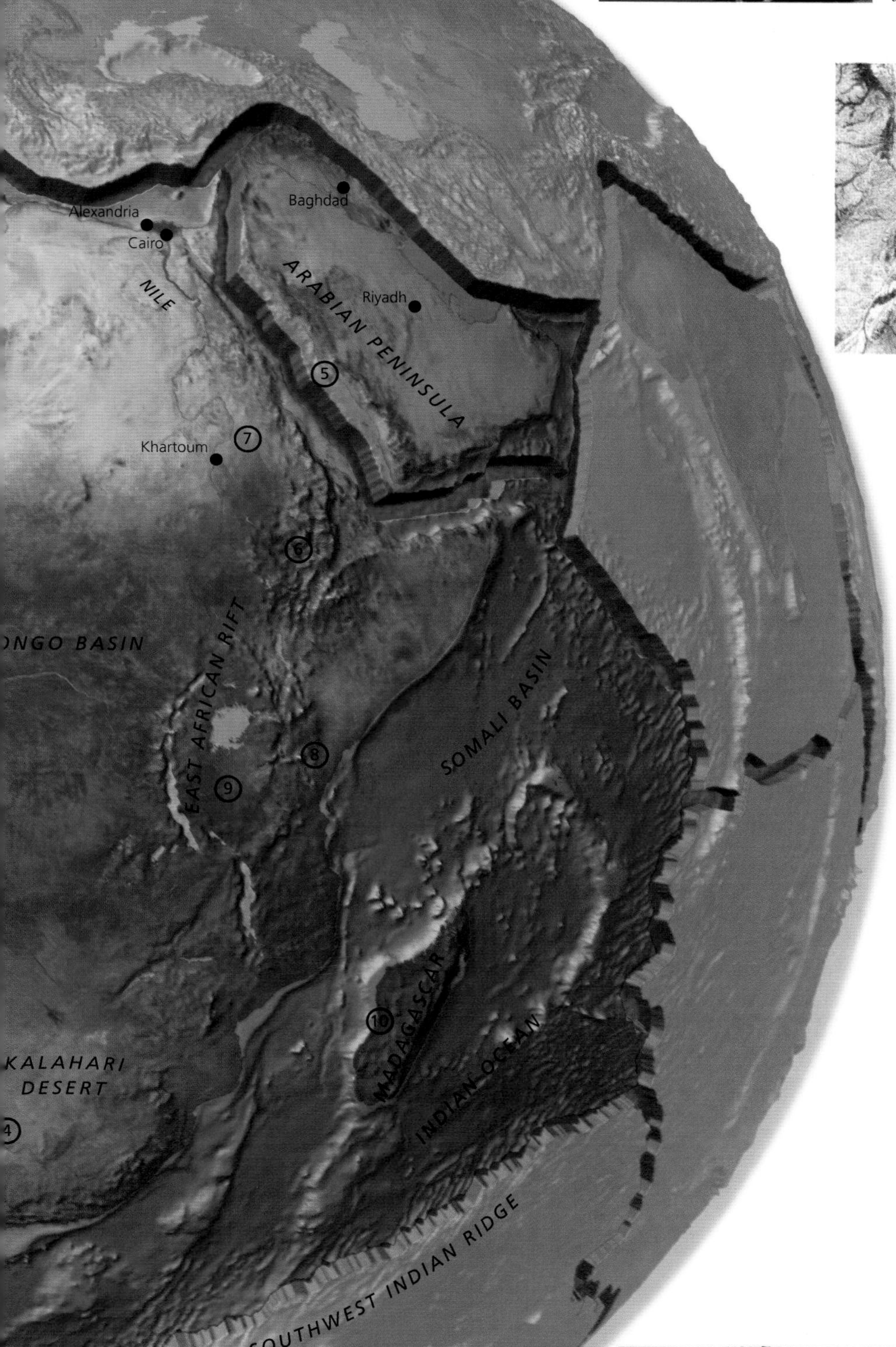

⑤ **RED SEA**
*This is one of the world's warmest seas, growing as the Arabian and African plates split apart. Its clear waters provide an ideal environment for corals.*

⑥ **ETHIOPIAN HIGHLANDS**
*Although split by the East African Rift, the Ethiopian highlands form the largest mountainous region in Africa. Many areas are higher than 13,000 ft (4,000 m).*

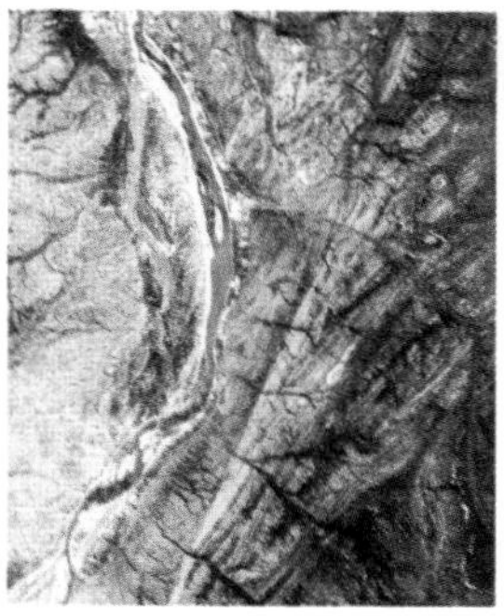

⑦ **KERAF SUTURE**
*This radar image reveals an approximately north–south line, known as the Keraf Suture, in northern Sudan. It shows where two ancient continents collided about 650 million years ago.*

⑧ **MOUNT KILIMANJARO**
*The highest peak in Africa, Mount Kilimanjaro formed about 1.8 million years ago, growing to 19,340 ft (5,895 m). Today there are three volcanoes, Shira, Mawenzi and Kibo, nested in its summit.*

⑨ **EAST AFRICAN SAVANNA**
*The tropical grasslands that cover much of east Africa are famed for their wildlife, which includes lions, elephants, and vast herds of antelope and zebra.*

## POPULATION AND BOUNDARIES

In terms of area and population, Africa is the second-largest continent, but vast regions are almost completely empty. Most of the main population centers are in countries near the sea—for example, along the north coast and down the Nile Valley, including Cairo. The one major inland population center is around lakes Victoria and Tanganyika. Africa has 53 countries, more than any other continent.

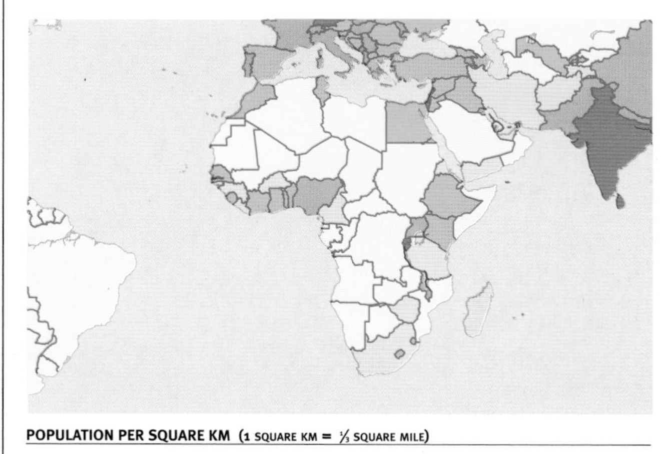

**POPULATION PER SQUARE KM** (1 SQUARE KM = ⅓ SQUARE MILE)

0–25 | 26–50 | 51–100 | 101–200 | 200+

⑩ **BAOBAB AVENUE**
*These distinctive trees in Madagascar can withstand long dry periods by storing water in their trunks.*

# AUSTRALIAN

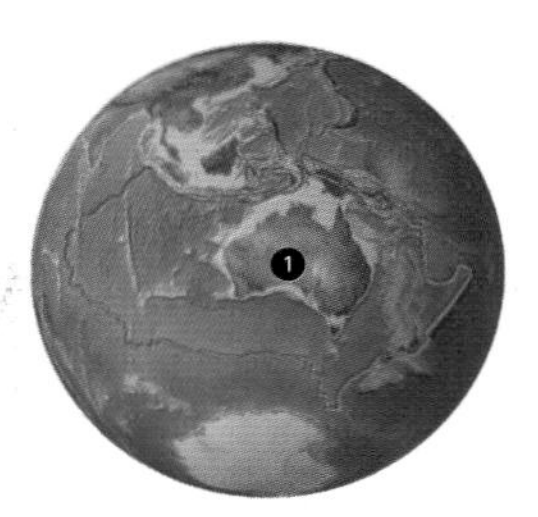

**PLATES** ❶ Australian Plate

**AREA** 18 million square miles (46 million square km)

**HIGHEST POINT** Mount Wilhelm, Papua New Guinea, 14,794 ft (4,509 m)

**LOWEST POINT** Java Trench, 24,440 ft (7,450 m) below sea level

**LARGEST CITY** Sydney (4 million)

AUSTRALIA, ANTARCTICA, and New Guinea split away from the other continents about 200 million years ago. Then Antarctica broke away from Australia and New Guinea about 50 million years ago, leaving them isolated. The gap between them is still widening, along the Indian Ocean Ridge. Due to its isolation, Australia is one of the most stable of all the continents, made mostly of ancient rocks. The western plateau is a Precambrian shield, 570 million–3 billion years old. Even the Great Dividing Range of mountains, down the east side, formed hundreds of millions of years ago and is now extremely eroded.

① **LAKE CARNEGIE**
*Australia's interior is so flat that rivers drain into inland lakes, rather than to the sea. With barely enough rainfall to keep rivers flowing, many of these lakes are ephemeral. This satellite image of Lake Carnegie in Western Australia shows one of the rare occasions when the lake contains water.*

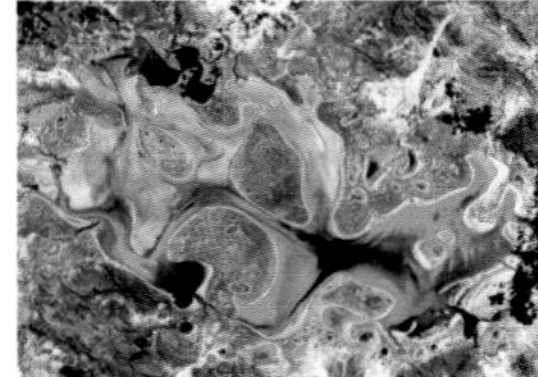

② **PINNACLES**
*The dry conditions of Western Australia have produced extraordinary rock formations, none more so than the Pinnacles, 150,000 limestone stacks up 13 ft (4 m) tall near Cervantes.*

## POPULATION AND BOUNDARIES

Australia is the most sparsely populated of all the continents except Antarctica, with a total of 31 million people and an average of less than 4 people per square kilometer, or about ⅓ square mile. The interior of Australia is almost empty; most people live near the southeast coast or in a small area around Perth in the southwest. The western part of the island of New Guinea is more densely populated.

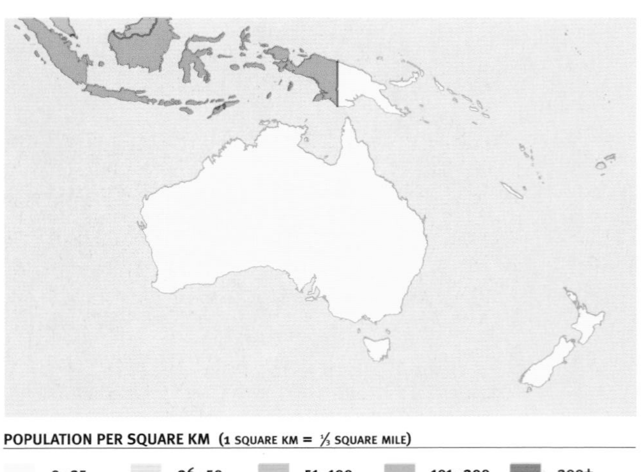

**POPULATION PER SQUARE KM** (1 SQUARE KM = ⅓ SQUARE MILE)

0–25 | 26–50 | 51–100 | 101–200 | 200+

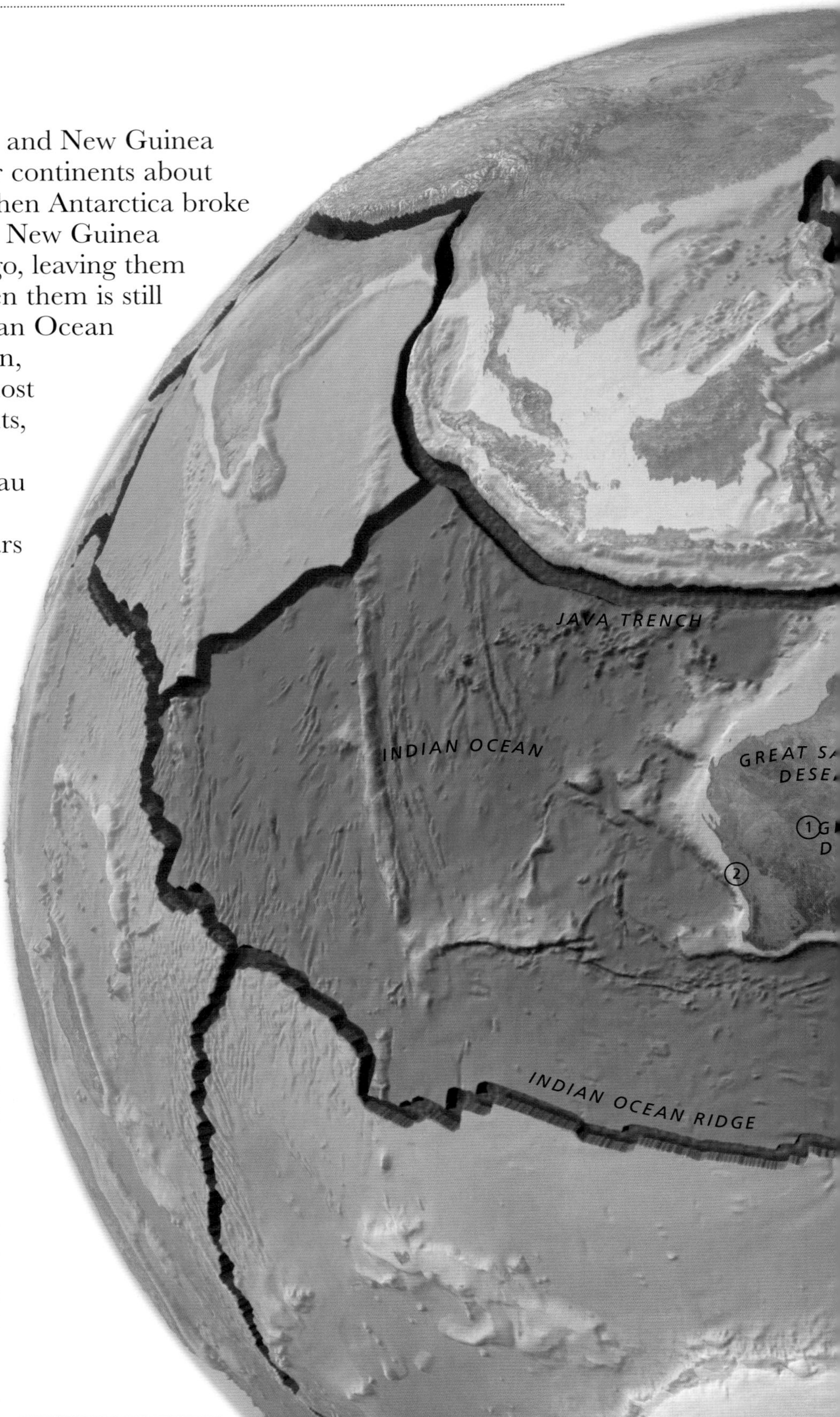

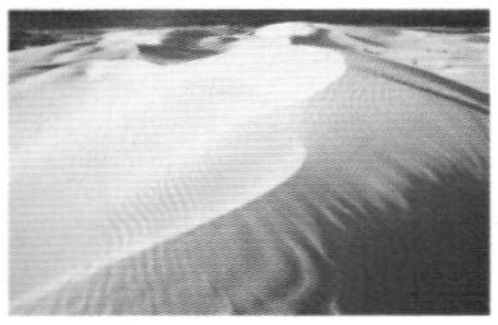

③ **SAND DUNES**
*As well as being famous for its caves and karst scenery, the Nullarbor Plain has some spectacular sand dunes along its southern edge, such as these on the coast of South Australia at Eucla.*

④ **BUNGLE BUNGLES**
*In Western Australia, there is an extraordinary landscape of beehive-shaped, orange-and-black sandstone pinnacles, up to 660 ft (200 m) tall, called the Bungle Bungles.*

⑤ **COPPER AND GOLD**
*Papua New Guinea has some of the world's richest copper and gold deposits. This giant quarry is in the Star Mountains.*

⑥ **CAPE YORK**
*The northernmost point of the Australian mainland, Cape York is at the head of a peninsula covered in dense tropical rainforest on the east and grassland to the west.*

⑦ **ULURU (AYERS ROCK)**
*The world's largest freestanding rock, Uluru formed when beds of sandstone about 450 million years old were tilted 90 degrees by crustal movement, then eroded by wind and water.*

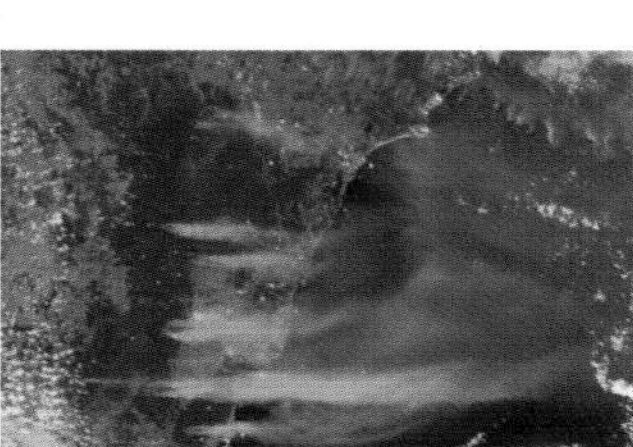

⑧ **BUSH FIRES**
*Bush fires, here fanned by prevailing winds, spread rapidly through the dry eucalypt vegetation of New South Wales. People living on forest fringes are particularly at risk.*

⑨ **WHITE ISLAND**
*White Island in the Bay of Plenty, off the coast of New Zealand's North Island, is the tip of a huge volcano that rises from the sea floor.*

⑩ **MOUNT COOK**
*New Zealand's highest peak, Mount Cook is in the Southern Alps on South Island. Standing 12,284 ft (4,744 m) high, it was forced up by movements of the Australian and Pacific plates.*

## MAJOR FEATURES

**BARKLY TABLELAND**
This flat, grassland region in the Northern Territory has become Australia's foremost area for rearing beef cattle.

**GREAT DIVIDING RANGE**
Running down the east side of Australia, the Dividing Range is so called because it separates rivers that flow east to the coast from those that run west to the interior.

**GREAT SANDY DESERT**
This is one of the many deserts of Australia's interior. Rainfall is patchy, falling mainly in an isolated downpour, and there are many drought years.

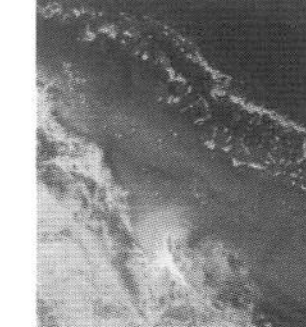

**GREAT BARRIER REEF**
The world's largest living structure, this reef extends 1,600 miles (2,570 km) along the Queensland coast.

**MURRAY–DARLING**
Rising in the Snowy Mountains, the Murray is Australia's longest river. The Darling flows only sporadically.

**NULLARBOR PLAIN**
This vast, flat, limestone plain in South Australia is an ancient sea floor that was uplifted 25 million years ago.

**TASMAN SEA**
This is one of the world's stormiest seas, due to its proximity to a wind belt called the Roaring Forties.

# PACIFIC

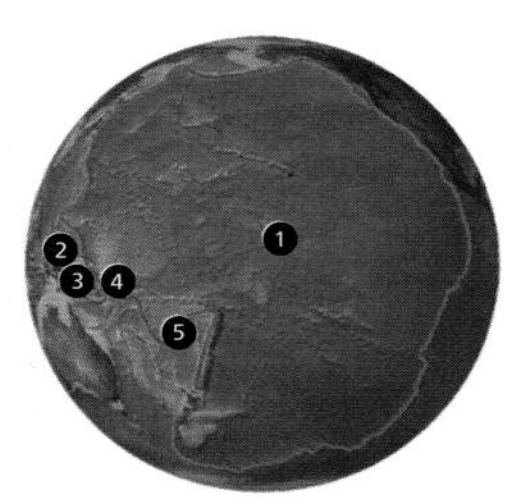

**PLATES** ❶ Pacific Plate
❷ Solomon Plate ❸ Bismarck Plate
❹ Caroline Plate ❺ Fiji Plate

**AREA** 42 million square miles (108 million square km)

**HIGHEST POINT** Mauna Kea, Hawaii, 13,796 ft (4,205 m)

**LOWEST POINT** Challenger Deep, 36,198 ft (11,033 m) below sea level

**LARGEST CITY** Honolulu (372,000)

ABOUT 85 MILLION YEARS AGO, the Pacific Plate was one of several in the Pacific Ocean, but as it widened and spread northwestward, the other plates were mostly subducted under the Americas. The Pacific Plate moves northwest at about 4 in (100 mm) a year. Its leading edge turns down sharply under the Eurasian, Philippine, Australian, Caroline, and Bismarck plates, creating major subduction zones and the world's deepest ocean trenches. Much of this plate is featureless and flat, but volcanic islands, such as Hawaii, and undersea volcanoes, such as the Emperor Seamounts, occur where plumes of magma have burst through the plate.

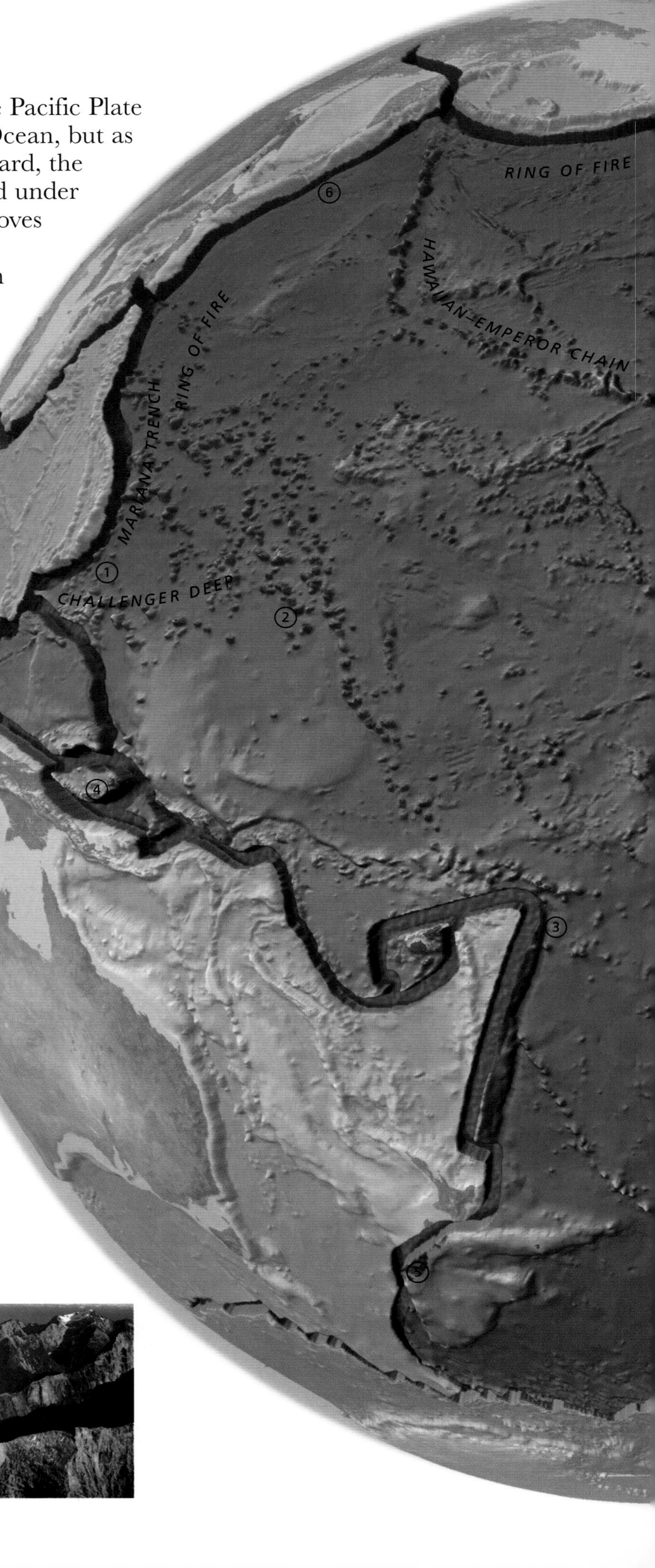

① **MOUNT JUMULLONG**
*This mountain is inland from Cetti Bay in Guam, one of the Mariana Islands. These volcanic islands have formed along the edge of the Philippine Plate as the Pacific Plate is thrust down beneath it.*

② **MARSHALL ISLANDS**
*Many of the 34 Marshall Islands are ring-shaped reefs, called atolls, that form as corals grow in the waters around a volcanic island, now submerged beneath the ocean surface. The islands include some of the world's largest atolls.*

③ **TONGA**
*This is the Vivau Island Group to the east of Tonga, composed of coral islands. To the west, volcanic islands formed along the edge of the Australian Plate as the Pacific Plate subducted beneath it.*

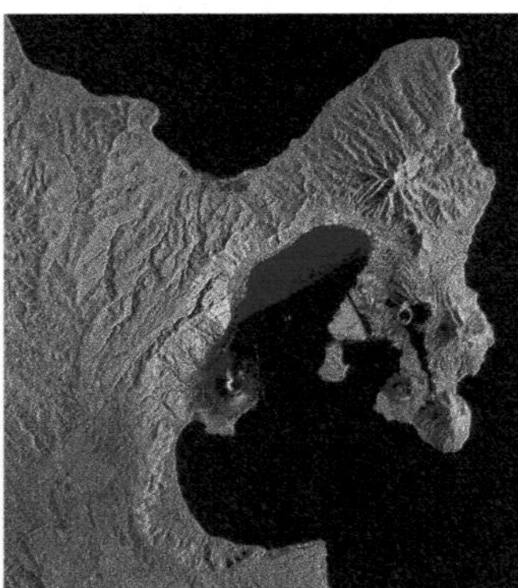

④ **RABAUL VOLCANO**
*This satellite image shows Blanche Bay, New Britain, a semicircular body of water formed from the submerged caldera of Rabaul volcano. New cones have developed and are visible around the margin of the old caldera.*

⑤ **NEW ZEALAND FJORD**
*Milford Sound, on the southwest coast of the South Island, is the most spectacular of New Zealand's fjords. It is a drowned valley that was carved out by a glacier about 15,000 years ago.*

RING OF FIRE
HAWAIIAN-EMPEROR CHAIN
RING OF FIRE
MARIANA TRENCH
CHALLENGER DEEP

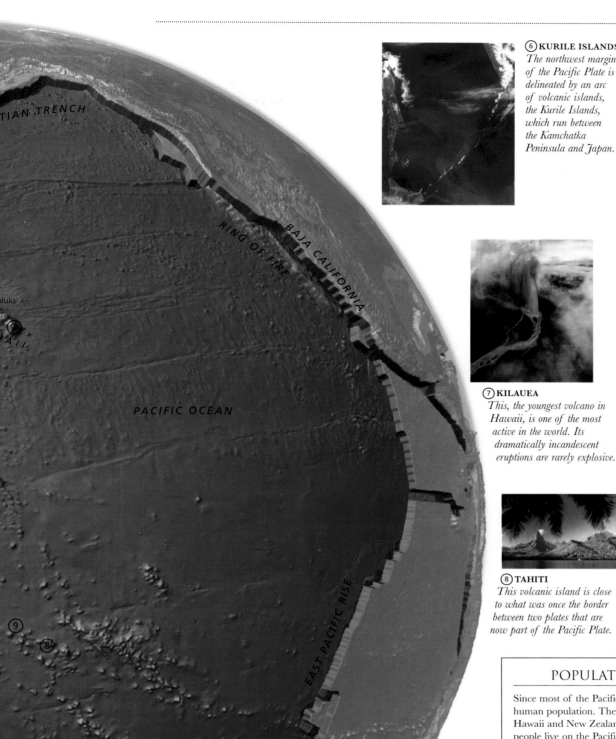

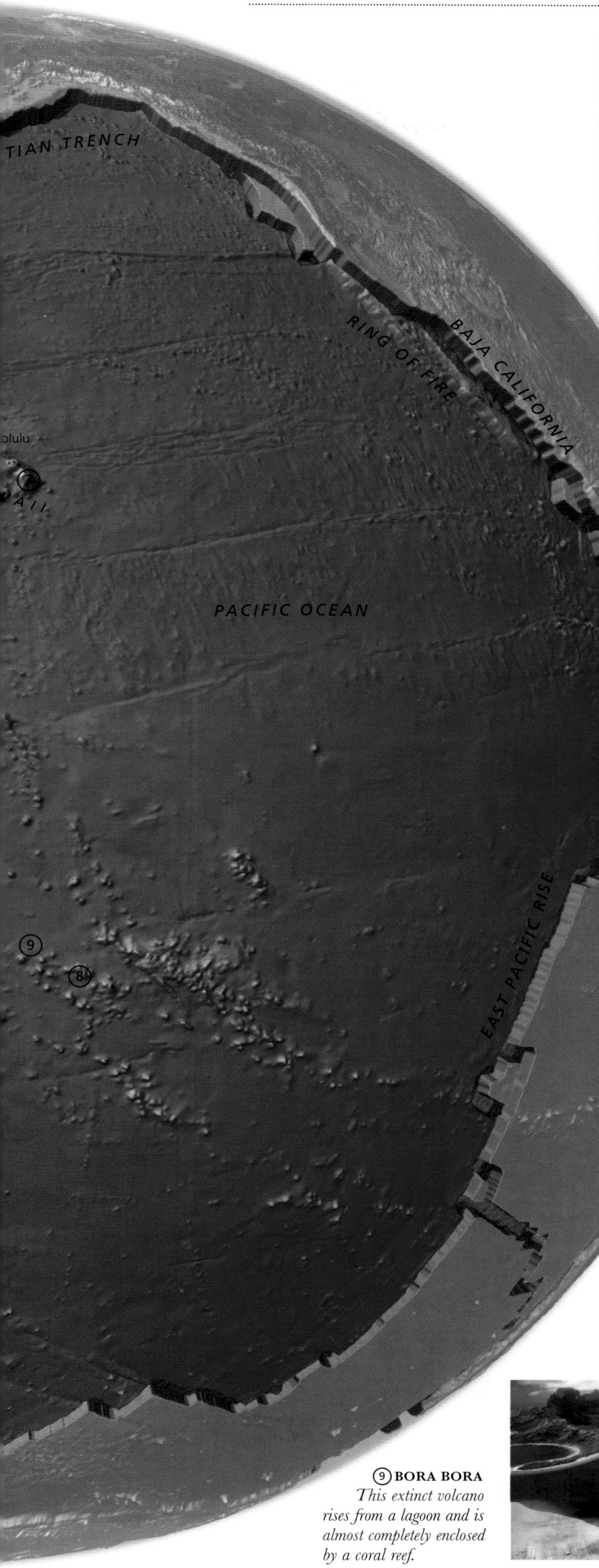

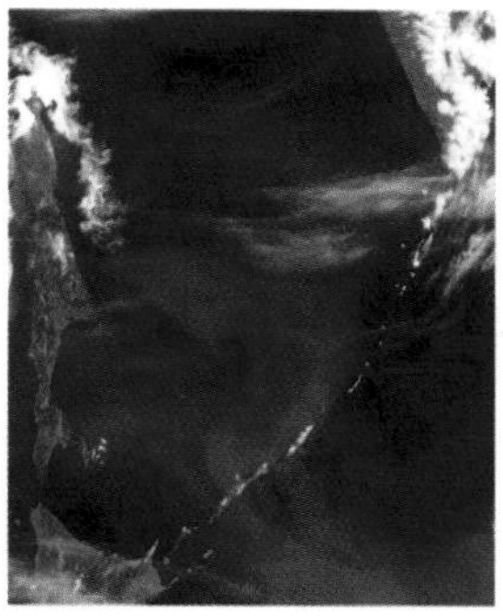

**⑥ KURILE ISLANDS**
*The northwest margin of the Pacific Plate is delineated by an arc of volcanic islands, the Kurile Islands, which run between the Kamchatka Peninsula and Japan.*

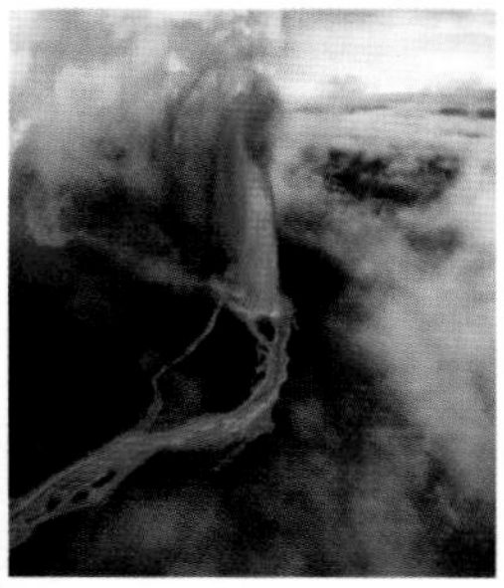

**⑦ KILAUEA**
*This, the youngest volcano in Hawaii, is one of the most active in the world. Its dramatically incandescent eruptions are rarely explosive.*

**⑧ TAHITI**
*This volcanic island is close to what was once the border between two plates that are now part of the Pacific Plate.*

**⑨ BORA BORA**
*This extinct volcano rises from a lagoon and is almost completely enclosed by a coral reef.*

## MAJOR FEATURES

**RING OF FIRE**
The Pacific plate is nearly encircled by a string of active, often violently explosive volcanoes called the Ring of Fire.

**BAJA CALIFORNIA**
Extending 800 miles (1,300 km) south into the Pacific, this rugged peninsula is mainly desert.

**HAWAIIAN–EMPEROR CHAIN**
About 100 volcanic islands and seamounts developed over a fixed mantle hot spot, as the Pacific Plate moved northwest over it.

**CORAL REEFS**
The warm, shallow waters of the South Pacific are ideal for coral growth. There are both long fringing reefs and atolls.

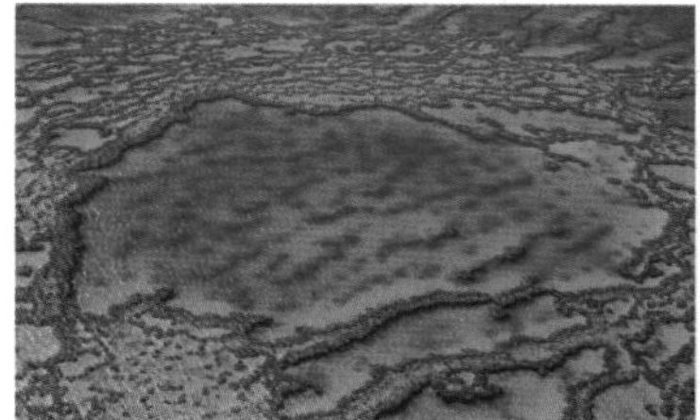

**SOUTH ISLAND, NEW ZEALAND**
The Pacific Plate is riding over the Australian Plate to create this island.

**EAST PACIFIC RISE**
Situated between the Pacific and Nazca plates, this rise contains one of the world's fastest-spreading ridge systems.

**MARIANA TRENCH**
The lowest point on Earth is in the Mariana Trench, where the Pacific Plate starts to pass under the Philippine Plate.

## POPULATION AND BOUNDARIES

Since most of the Pacific Plate is covered by ocean, it has a very small human population. The only islands with any sizeable population are Hawaii and New Zealand's South Island. However, a huge number of people live on the Pacific Plate's margins along the California coast. The borders between island groups, because they include areas of ocean, are not well defined other than by a reference map.

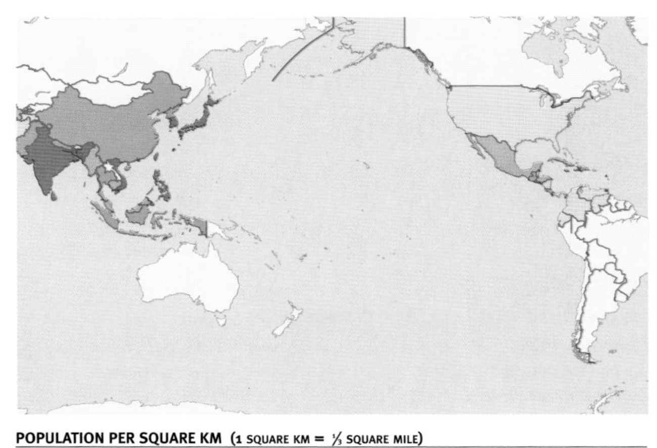

**POPULATION PER SQUARE KM** (1 SQUARE KM = ⅓ SQUARE MILE)

0–25 | 26–50 | 51–100 | 101–200 | 200+

# ANTARCTIC

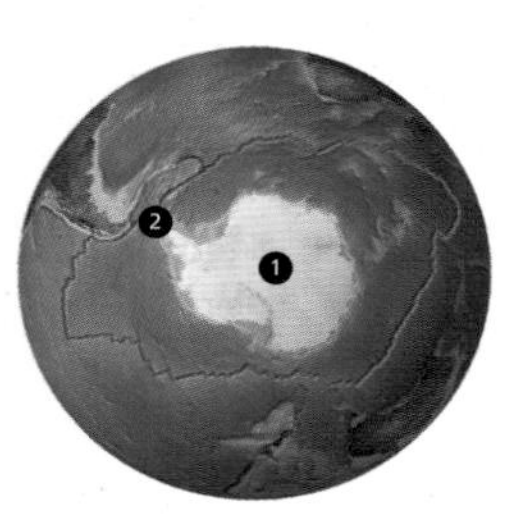

**PLATES** ❶ Antarctic Plate
❷ Shetland Plate

**COMBINED AREA** 22 million square miles (58 million square km)

**HIGHEST POINT** Vinson Massif, 16,067 ft (4,897 m)

**LOWEST POINT** Bentley Subglacial Trench, 8,327 ft (2,538 m) below sea level

**LARGEST CITY** None

LARGER THAN AUSTRALIA, the continent of Antarctica lies in the center of the Antarctic Plate. High mountain peaks are visible, but the continent is almost entirely covered by ice, formed over 15 million years and now extending beyond the land to form floating ice shelves. Beyond the the ice, the Antarctic plate extends into the Southern Ocean for 1,000 miles (1,600 km) or more in every direction. In fact, Antarctica is entirely surrounded by oceanic spreading ridges. The Antarctic continent was once two plates, West Antarctica and East Antarctica, which have now bonded along a line joining the Ross and Weddell Seas.



## MAJOR FEATURES

**ANTARCTIC PENINSULA**
This peninsula stands out as a thin tail of land, projecting from an otherwise almost circular mass of land and ice.

**TRANSANTARCTIC MOUNTAINS**
These mountains include two (Erebus and Buckle Island) of Antarctica's three volcanoes that have erupted since 1900.

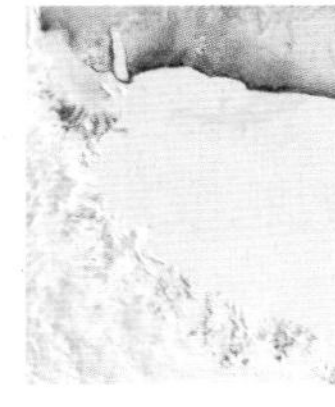

**ROSS ICE SHELF**
This is the largest of the ice shelves floating around the coast of Antarctica. It is about the same size as France.

**GEOMAGNETIC SOUTH POLE**
Situated near Russia's Vostok Station, the geomagnetic south pole is where magnetic currents in the upper atmosphere are vertical. The South Pole is at 90° south.

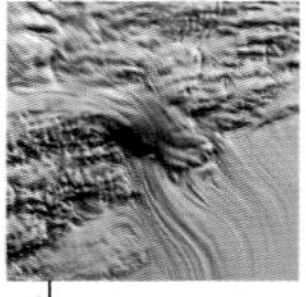

**LAMBERT GLACIER**
Not only is the Lambert Glacier the world's largest glacier, it is also the fastest-moving, flowing 4,000 ft (1,200 m) or more a year.

**ELLSWORTH MOUNTAINS**
The highest peak in Antarctica, Vinson Massif, is found in these mountains at the head of the Ronne Ice Shelf.

**OZONE HOLE**
The ozone hole forms each spring and is increasing in size. Gases released by human activity may be the cause.

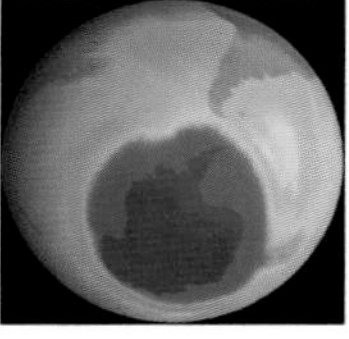

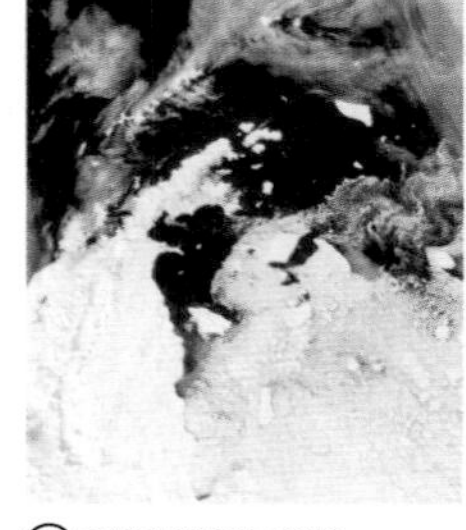

**① PALMER LAND**
*The narrow Antarctic Peninsula is formed of low mountains, such as those of Palmer Land. They were created by the same tectonic events that threw up the Andes mountains. These low mountains are permanently covered by a thin sheet of ice.*

**② MOUNT EREBUS**
*The world's most southerly active volcano, Erebus is the largest of three volcanoes on Ross Island. Daily, it ejects lava bombs and jets of steam from its icebound summit.*

**③ MCMURDO BASE**
*The largest settlement in Antarctica, McMurdo station is like a small town, with stores and banks. Even in winter, several hundred people stay here, and in summer, visitors push the population up to 3,000.*

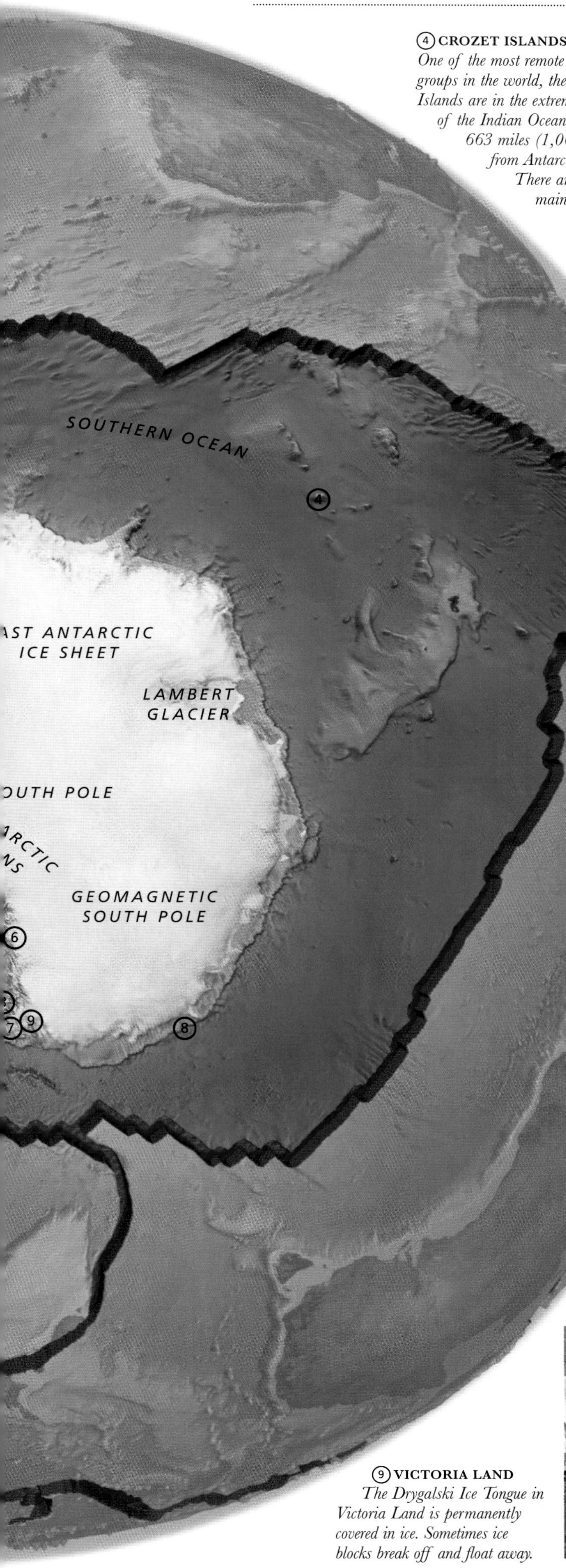

④ **CROZET ISLANDS**
*One of the most remote island groups in the world, the Crozet Islands are in the extreme south of the Indian Ocean, only 663 miles (1,067 km) from Antarctica. There are five main islands.*

⑤ **WEDDELL SEALS**
*The Weddell Sea lies between the Antarctic Peninsula and the mainland. The Weddell seal lives farther south than any other mammal, diving through holes in the ice to catch fish.*

⑥ **WRIGHT VALLEY**
*One of very few areas of land in Antarctica that is sometimes completely free of ice, the Wright Valley may appear to be devoid of life, but algae, bacteria, and fungi are all found there.*

⑦ **TERRA NOVA**
*Located at the foot of the Transantarctic Mountains in Victoria Land, Terra Nova overlooks a bay with the same name in the Ross Sea. The area is renowned for its spectacular ice caves.*

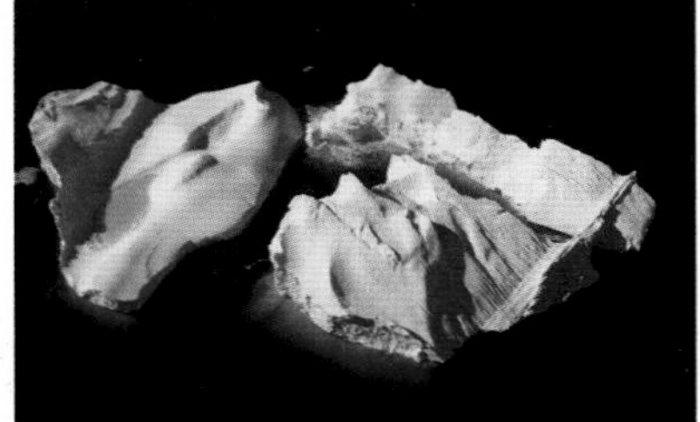

⑧ **ICEBERGS**
*These icebergs floating off the Adelie coast are huge. Usually bigger than Arctic icebergs, they last much longer, surviving up to ten years. The largest to date was 435 miles (700 km) long and broke away from the Ross Ice Shelf in 2002.*

⑨ **VICTORIA LAND**
*The Drygalski Ice Tongue in Victoria Land is permanently covered in ice. Sometimes ice blocks break off and float away.*

## POPULATION AND BOUNDARIES

Antarctica does not belong to any one country. Seven nations claim territory, but no claim has met with international agreement. Since 1961, the continent has been administered under the Antarctic Treaty, an international agreement between 43 countries to preserve the continent for peaceful scientific study. In 1991, the members of the treaty agreed to a 50-year ban on the exploitation of minerals.

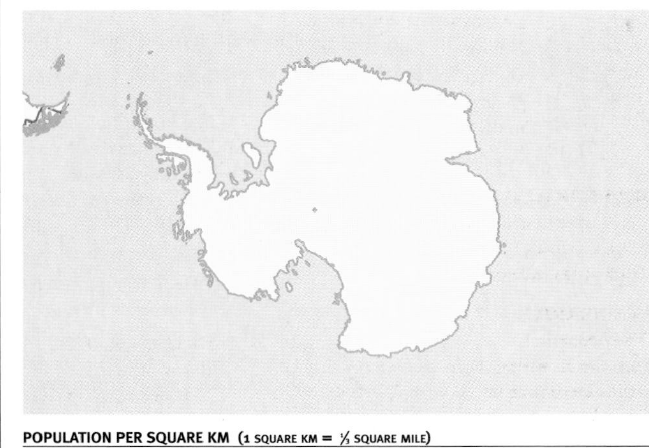

**POPULATION PER SQUARE KM** (1 SQUARE KM = ⅓ SQUARE MILE)

0–25 | 26–50 | 51–100 | 101–200 | 200+

# GLOSSARY

In this glossary, terms defined within a larger entry are highlighted in **bold** type, while references to terms defined in other entries are identified with *italic* type.

## A

**ABLATION** The loss of ice from a glacier due to melting, evaporation, or calving. The region of a glacier where there is a net loss of ice is known as the **ablation area**. See also *accumulation area*.

**ABYSSAL** Relating to the deep ocean floor and its environment, beyond the continental shelves and slopes. Abyssal depths are deeper than bathyal depths but not as deep as deep-sea trenches. The **abyssal plain** is the flat, sediment-covered plain at about 13,000–20,000 ft (4,000–6,000 m) depth that forms the bed of most of the world's oceans. See also *bathyal*, *deep-sea trench*.

**ACCESSORY MINERAL** A mineral present in only small quantities in a particular rock.

**ACCRETION** The gravitational coming-together of small particles—for example, in the early Solar System—resulting in the formation of the planets.

**ACCUMULATION AREA** The part of a glacier where snowfall exceeds losses by melting, evaporation, and calving. See also *ablation*.

**ACID RAIN** Any precipitation (snow as well as rain) that contains dissolved acids. Most rain is slightly acidic due to dissolved carbon dioxide. More strongly acidic rain occurs due to atmospheric pollution, or sometimes from gases released by volcanic eruptions.

**AEROSOL** Tiny particles (about a millionth of a millimeter across) of dust or liquid suspended in the air.

**AGNATHAN** A primitive jawless fish. Present-day agnathans include lampreys and hagfish.

**AIR MASS** A mass of air of uniform characteristics that may extend over thousands of miles.

**ALBEDO** The extent to which a part of the Earth's surface reflects incoming radiation from the Sun, usually expressed as a decimal fraction.

**ALLUVIUM** Sedimentary material deposited by rivers. Any particular accumulation of alluvium is called an **alluvial deposit**. An **alluvial fan** is a deposit laid down where a stream leaves a highland area and spreads out over a plain. **Alluvial soil** is soil derived from alluvium.

**ANGIOSPERM** A flowering plant (as distinct from a conifer, fern, moss, or other plant). Angiosperms also include many tree species. See also *coniferous*.

**ANNUAL** A plant that lives and dies within one year.

**ANOXIA** Lack of molecular oxygen ($O_2$), which many (but not all) organisms require for respiration.

**ANTICLINE** An archlike, upward fold of originally flat strata due to horizontal compression. See also *syncline*, *fold*.

**ANTICYCLONE** A weather system in which winds circle around an area of high pressure. See also *cyclone*.

**AQUIFER** A layer of porous rock that can hold and transmit water.

**ARCHIPELAGO** A group of islands forming a chain or cluster.

**ARÊTE** A narrow mountain ridge separating two adjacent cirques. See also *cirque*.

**ASTEROID** One of thousands of bodies of rocky material orbiting the Sun that are smaller in size than the planets.

**ASTHENOSPHERE** The layer of the mantle immediately below the rigid lithosphere. It is sufficiently nonrigid to flow slowly in a solid state and plays a key part in the movement of tectonic plates. See also *lithosphere*, *tectonic plate*, *mantle*.

**ATOLL** see *coral reef*.

**AUREOLE** The area around an igneous intrusion where thermal alteration (metamorphism) of country rock has taken place. See also *igneous rock*, *metamorphism*, *country rock*.

**AURORA BOREALIS** Flickering lights in the sky that are sometimes visible in northern polar regions. They are caused by the interaction between high-energy particles from the Sun and the Earth's magnetic field. Their southern-hemisphere equivalent is called the **aurora australis**.

## B

**BACKWASH** The movement of water down a beach after waves have broken. See also *swash*.

**BAR** A long, sandy deposit built up by sea or river action. An **offshore bar** is a coastal bar not attached to the land. A **baymouth bar** is a spit that has grown so it extends across the mouth of a bay. A **longshore bar** lies between the high- and low-water marks, parallel to the shoreline. A **barrier bar** is a long bar parallel to the coastline. A **point bar** is a bar deposited by a river on the inside of a meander. See also *spit*, *meander*.

**BARCHAN** see *dune*.

**BARRIER BAR** see *bar*.

**BARRIER ISLAND** A long island of sediment parallel to a coastline. It is normally higher than a barrier bar. See also *bar*.

**BASALT** The Earth's most common volcanic rock, which usually originates as solidified lava. Basaltic lava can erupt on continents, but basalt also forms the oceanic crust. Basalt is glassy to fine-grained (composed of very small crystals). See also *igneous rock*, *flood basalt*, *crust*.

**BATHOLITH** A very large igneous intrusion, 60 miles (100 km) or more across, which originates deep underground. The Earth's great batholiths are known to science only if exposed on the surface by weathering. See also *igneous rock*.

**BATHYAL** A term referring to ocean depths of about 660–6,500 ft (200–2,000 m). See also *abyssal*.

**BAYMOUTH BAR** see *bar*.

**BEACH** An accumulation of sand, shell debris, or larger particles along a shoreline. A **beach face** is the steeply sloping part of a beach below a berm. See also *berm*.

**BEDDING** The manner in which a sedimentary rock was originally laid down in layers. A **bedding plane** is any plane in the rock that is parallel to these layers. See also *sedimentary rock*.

**BERGSCHRUND** A deep crack formed at the back of a cirque glacier, caused by the glacier moving away from the cirque wall. See also *cirque*.

**BERM** A ridge of beach gravel, usually marking the point of the highest high tides. See also *beach*.

**BIODIVERSITY** The total variety of living things, either on the Earth as a whole or in the region described.

**BIOLUMINESCENCE** The production of light by living organisms.

**BIOME** A biological community existing on a large scale and defined mainly by vegetation type (for example, rainforest). The extent of any particular biome is determined by climatic conditions.

**BLACK SMOKER** see *hydrothermal system*.

**BLOCKFIELD** An area of broken rocks, usually in a mountainous region.

**BOG** A peat-accumulating wetland that receives water mainly from rainfall. Bogs are acidic, extremely poor in nutrients, and most support an abundance of *Sphagnum* mosses. The soils are composed almost entirely of dead plant matter, and are waterlogged rather than completely covered by water. See also *fen*, *marsh*.

**BOREAL** Relating to or coming from the colder parts of the northern hemisphere, between the Arctic and temperate zones.

**BOSS** An igneous intrusion that is roughly circular in horizontal cross-section. See also *igneous rock*.

**BRAIDED STREAM** A stream or river where water flows in many shallow channels that constantly separate and rejoin.

**BRECCIA** A sedimentary rock made up of angular fragments of minerals and other rocks.

**BROADLEAF** Of trees: having broad leaves (such as oak and chestnut), as distinct from needlelike leaves. Broadleaved trees are mainly angiosperms rather than conifers, and often deciduous rather than evergreen. See also *angiosperm*, *coniferous*, *deciduous*, *evergreen*.

## C

**CAATINGA** A thorny, dry, tropical woodland found in northeast Brazil.

**CALCITE** A common mineral form of calcium carbonate.

**CALDERA** A bowl-shaped volcanic depression larger than a crater, typically greater than 0.6 miles (1 km) in diameter, caused by the collapse of a volcano into its emptied magma chamber after it has been evacuated by a volcanic explosion.

**CALVE** Of a glacier: to create icebergs by shedding ice blocks into a sea or lake.

**CANYON** A deep valley, usually not as narrow as a gorge; also a deep, irregular passage in a cave system. See also *gorge*, *tube*.

**CARAT** A unit of weight equal to about 0.2 grams, used for diamonds and other gemstones.

**CERRADO** A type of South American savanna dotted with small, gnarled trees. See also *savanna*.

**CHEMOSYNTHESIS** The biological process of extracting food energy from inorganic (nonliving) chemicals. Many microorganisms rely on chemosynthesis. See also *photosynthesis*.

**CHLOROPHYLL** The green pigment that plants require for photosynthesis. See also *photosynthesis*.

**CHLOROPLAST** A structure, found within the cells of green plants, in which chlorophyll is found and photosynthesis takes place. See also *photosynthesis*.

**CINDER CONE** see *volcano*.

**CIRQUE** A steep-sided, rounded hollow carved out by a glacier. Many glaciers have their origin in a mountain cirque, from which they flow to lower ground.

**CLAST** A fragment of rock, especially when incorporated into another newer rock. See also *grain*.

**CLAY** Mineral particles smaller than about 0.00008 in (0.002 mm) in diameter that are found in soils and other earthy deposits. A clay soil contains an abundance of clay particles and transmits water slowly.

**CLEAVAGE** The characteristic plane (or planes) along which a particular rock or mineral splits. See also *foliation*.

**CLIMAX VEGETATION** The vegetation that in time would naturally dominate any given site or region.

**CLOUD CONDENSATION NUCLEUS** A tiny particle in the air around which a raindrop or snow crystal may begin to form.

**CLOUDFOREST** A damp forest almost constantly under clouds, especially in highland regions.

**CLOUD-SEEDING** A method of creating rain by scattering tiny particles onto clouds.

**COALESCENCE** In meteorology, the fusion of tiny droplets within clouds until they are big enough to fall as rain.

**COAST** The boundary between land and sea. A **drowned coast** occurs where, over the centuries, the sea level has risen relative to the land; its opposite is an **emergent coast**.

**COLD DESERT** A desert in high-altitude, high-latitude, or temperate regions that may become very cold during the winter months.

**COMET** A body composed of rock and ice that orbits the Sun. Vaporization of the ice produces the spectacular tail seen during passage close to the Sun.

**CONDENSATION** The conversion of a substance from the gaseous to the liquid state—for example, water vapor condensing to liquid water. See also *cloud condensation nucleus*.

**CONFLUENCE** A place where two rivers, streams, or glaciers meet.

**CONIFEROUS** Cone-bearing. Typical coniferous trees, such as pines and firs, have needlelike leaves and are usually evergreen. See also *evergreen*.

**CONSERVATIVE BOUNDARY** see *plate boundary*.

**CONSTRUCTIVE BOUNDARY** see *plate boundary*.

**CONTINENTAL MARGIN** The continental shelf and continental slope taken together.

**CONTINENTAL RISE** The slightly sloping area around the edge of the deep sea floor where it meets the continental slope.

**CONTINENTAL SHELF** The gently sloping, submerged portion of continental crust seaward of most continental coasts. See also *crust*.

**CONTINENTAL SLOPE** The sloping ocean bottom between the edge of the continental shelf and the continental rise.

**CONVECTION** The movement and circulation of fluids (gases, liquids, or hot and ductile rocks) in response to differences in temperature, which in turn cause density variations between different parts of the fluid.

**CONVERGENCE** In meteorology, the situation where two air masses are moving toward one another. See also *plate boundary*.)

**CONVERGENT BOUNDARY** see *plate boundary*.

**COOLING JOINT** see *joint*.

**CORAL** Any one of various simple animals related to sea anemones. They are often colonial and can secrete skeletons to support themselves. **Hard corals** (sometimes called true corals) eventually create reefs out of the calcium-carbonate skeletons they lay down.

**CORAL REEF** A structure in shallow tropical seas built up by the activity of coral animals and other organisms over many years. A **fringing reef** is attached to the shore with little or no water between the reef and shore. A **barrier reef** is parallel to the shore but separated from it by a lagoon. A reef that has grown on top of a sunken extinct volcano forms a circular island called an **atoll**.

**CORE** The innermost part of the Earth. It consists of a liquid outer core and a solid inner core, both made of nickel-iron. See also *mantle, crust*.

**CORIOLIS EFFECT** The tendency for winds and currents moving in a northerly or southerly direction to be deflected and move at an angle because of the effect of the Earth's rotation. The deflection is to the right in the northern hemisphere and to the left in the southern hemisphere.

**CORRASION** Erosion of rocks by scraping—for example, by rock-laden glacial ice.

**COUNTRY ROCK** Existing rock into which a new body, such as an igneous intrusion, is emplaced. See also *igneous rock*.

**CRATER** A bowl-shaped depression through which an erupting volcano discharges gases, pyroclasts, and lava. The crater walls form by accumulation of ejected material. This term also refers to a circular depression in the landscape caused by a large meteorite impact.

**CRATON** A stable area of the Earth's continental crust, made of old rocks largely unaffected by mountain-building activity since the Precambrian eon. Also called a shield.

**CRUST** The rocky outermost layer of the solid Earth. The continents and their margins are made of thicker but less dense **continental crust**, while thinner, denser **oceanic crust** underlies the deep ocean floors. See also *mantle, tectonic plate*.

**CRYOTURBATION** The churning up of the land surface by the action of ice below ground level.

**CRYPTOCRYSTALLINE** Made of crystals too small to be seen with the naked eye.

**CRYSTAL** Any solid in which the individual atoms or molecules are arranged in a regular geometrical pattern. Most pure substances can form crystals. There are seven basic patterns of crystal growth, called **crystal systems**.

**CURRENT** In oceanography, a flow of ocean water. Both surface and deep-sea currents exist. Some are driven by the wind; others (**thermohaline currents**) by differences in temperature and salinity that create differences in density between water masses.

**CYCLONE** A pressure system in which air circulates around an area of low pressure. A **tropical cyclone** is another name for a hurricane or typhoon. See also *pressure system, hurricane, anticyclone*.

# D

**DECIDUOUS** Of trees and shrubs: having leaves that fall at a particular time of year, such as winter or a dry season. See also *evergreen*.

**DEEP-SEA FAN** A fan-shaped deposit on the ocean floor built up by turbidity currents. See also *turbidity current, alluvial fan*.

**DEEP-SEA TRENCH** A canyonlike, linear depression in the ocean floor, the site of subduction of one tectonic plate below another. Trenches are the deepest regions of the oceans. See also *subduction, plate boundary*.

**DEFORMATION** The ductile flow or brittle fracture of existing rocks due to geological movements.

**DELTA** The area of gently sloping sediment built up by many rivers when they enter the sea or a lake. Its overall shape depends on factors such as the amount of sediment brought down by the river, and the currents, wave action, and tides that it meets. See also *estuary*.

**DENDRITE** A mineral occurring naturally as a branching, treelike form within rocks. A dendritic crystal is a crystal that has grown in this way.

**DEPOSITION** The laying down of material such as sand and gravel in new locations, usually by wind or moving water or ice.

**DESERT PAVEMENT** A rocky or stony surface layer found in many deserts. A **hammada** is a solid-rock surface with weathered rock fragments on top.

**DESERT VARNISH** A black or brown glossy coating sometimes found on rock surfaces in a desert that have been exposed to the atmosphere for long periods of time.

**DESERTIFICATION** The transformation of a formerly more fertile region into desert.

**DESTRUCTIVE BOUNDARY** see *plate boundary*.

**DEW POINT** The temperature at which, under given conditions, liquid water will start to condense out of the air.

**DIKE** A sheet of intrusive igneous rock at a steep angle to the surface. A large number of dikes in a given area is termed a **dike swarm**. See also *country rock, sill*.

**DINOSAUR** A member of a dominant group of reptiles that died out between 145—65 million years ago. Their closest living relatives are birds and crocodiles.

**DIPTEROCARP** A member of a family of tall, mainly evergreen trees native to Africa and south Asia.

**DISTRIBUTARY** A branch of a river that flows away from the main stream and does not reconnect with it later.

**DIVERGENT BOUNDARY** see *plate boundary*.

**DOLINE** A depression in the surface of a karst landscape. It often leads down to an underground drainage system. Also called a sinkhole. See also *karst*.

**DOME VOLCANO** see *volcano*.

**DRAINAGE BASIN** The area within which all precipitation is drained by a single river system.

**DROWNED COAST** see *coast*.

**DRUMLIN** A hill-sized, usually streamlined mound of debris left behind by a retreating glacier. See also *moraine*.

**DRY FOREST** A forest growing in an area with a predominantly dry climate.

**DUNE** A mound or hill of sand, either in a desert, on a river bed, or by the shore of a sea or lake. There are several varieties of desert sand dunes. A **barchan** is a crescent-shaped dune. A **linear** or **seif dune** can run for long distances roughly parallel to the prevailing wind. **Crescentic dunes** form ridges with slightly wavy edges. **Parabolic dunes** have long, trailing arms, while **star dunes** have arms stretching in several directions.

# E

**ECOSYSTEM** All of the living and nonliving features of a region and the interactions among them. Organisms include plants, animals, and microorganisms; nonliving features include light, water, nutrients, and the environment. An ecosystem can be exceedingly small or as large as the Earth itself.

**EKMAN SPIRAL** The tendency of deep-ocean currents to change direction as they rise to the surface. The term also refers to the tendency of winds lower in the atmosphere to blow at an angle compared with winds higher up. The causes of the Ekman spiral include frictional effects and the Coriolis effect.

**EL NIÑO** The phenomenon in which the normally strong currents of the equatorial Pacific weaken, and warm water that has piled up in the western Pacific floats eastward, making water in the eastern Pacific warmer than usual. It is part of a larger phenomenon that can trigger changes to weather patterns worldwide. See also *La Niña*.

**EMERGENT** Of a forest tree: growing taller than most other species of tree in the forest. Of a wetland plant: growing taller than the water surface.

**EMERGENT COAST** see *coast*.

**ENDEMIC** Of animals and plants: native to a particular region, and found nowhere else.

**EPEIROGENESIS** The raising or lowering of parts of the Earth's crust due to vertical movements only, rather than the sideways movements involved in orogenesis. See also *orogenesis*.

**EPHEMERAL** Appearing for only a short time. In deserts, both plant life and rivers are frequently ephemeral.

**EPIPHYTE** A plant, especially a nonparasitic one, that grows on another plant.

**ERG** A large expanse of sand dunes within a desert.

**EROSION** The processes by which rocks or soil are loosened and worn or scraped away from a land surface. The main agents of erosion are wind, water, and moving ice, and the sand grains and other rock particles that they carry. See also *transport, weathering*.

**ERRATIC** A rock that has been transported from its original location, usually carried by ice.

**ERUPTION** The discharge of lava, pyroclasts, gases, and other material from a volcano. **Hawaiian eruptions** involve large flows of very liquid lava but little explosive activity. **Strombolian eruptions** involve frequent ejections of incandescent clasts and gases, but without major explosions. **Surtseyan eruptions** take place in shallow water and are partly powered by the conversion of water to steam. **Vulcanian eruptions** involve a series of explosive events that feed significant eruption columns. **Plinian eruptions** involve massive explosions to form towering eruption columns, in which much of the volcanic cone may be destroyed. **Fissure eruptions** take place through long cracks in the ground rather than via a volcanic crater. See also *volcano*.

**ESCARPMENT** The steep slope at the edge of a plateau, or at the edge of an area of exposed, gently sloping strata. An escarpment is also called a scarp.

**ESKER** A long line of raised debris left behind by a glacier. Eskers are thought to mark the position of former meltwater channels. See also *meltwater*.

**ESTUARY** The broad, funnel-shaped stretch of water found where many large rivers meet the sea. Most present-day estuaries owe their form to the drowning of river valleys when sea level rose at the end of the last ice age. The term is also used more broadly to include all bays and inlets where seawater and fresh water mix and muddy sediments are deposited.

**EUKARYOTE** Any living organism built out of a complex cell or cells containing nuclei (which are enclosed in a membrane). All complex animals and plants are eukaryotes. See also *prokaryote*.

**EUTROPHICATION** The enrichment of natural waters with extra nutrients, especially as a result of human action. It often has drastic deleterious effects on aquatic ecosystems.

**EVAPORITE** A chemical sedimentary deposit created by the progressive evaporation of salty water to form various minerals such as halite and gypsum.

**EVERGREEN** Having leaves year-round. See also *coniferous*, *deciduous*.

**EXFOLIATION** A weathering process that involves the splitting off of outer layers of rocks, like the layers of an onion.

# F

**FAULT** A fracture where the rocks on either side have moved relative to one another. If the fault is at an angle to the vertical, and the overhanging rock has slid downward, it is called a **normal fault**. If the overhanging rock has slid upward (in relative terms) it is called a **reverse fault**. A **strike-slip fault** is one where movement is horizontal. A **transform fault** is a large-scale strike-slip fault associated with oceanic spreading.

**FELSIC ROCK** An igneous rock rich in silica. The name reflects the abundance of the mineral feldspar and the high silica content. See also *silica*, *igneous rock*.

**FEN** A peat-accumulating wetland that receives water mainly from groundwater seepage. Fens are typically alkaline and nutrient-poor (although not extremely so). They support a variety of vegetation types including *Sphagnum* mosses, sedges, and conifer trees. The soils are composed almost entirely of dead plant matter, and they are sometimes completely covered by water.

**FERREL CELL** A large-scale circulation within the atmosphere, in which air rises at around 60° North or South and flows toward the equator before descending at around 30°. See also *Hadley cell*.

**FETCH** The extent of open water across which a wind or water wave has traveled.

**FJORD** A former glacial valley on the coast that has become an inlet of the sea.

**FISSURE ERUPTION** see *eruption*.

**FLASH POINT** The point at which superheated water changes to steam. See also *geyser*.

**FLASH FLOOD** A sudden flood occurring after heavy rain.

**FLOOD BASALT** An extensive area of basalt resulting from massive volumes of lava erupting and flowing widely over the landscape in relatively short intervals of geological time. It is resistant to erosion and commonly forms extensive plateaus after its formation.

**FLOODPLAIN** A flat plain next to a river that is likely to be covered with water when the river floods. See also *alluvium*.

**FLOWSTONE** A smooth coating of younger limestone precipitated over surfaces in caves. See also *calcite*.

**FOG** Condensed droplets of water forming in the air at ground level. **Radiation fog** is fog caused by the ground surface radiating heat at night, becoming cooler and also cooling the air above it.

**FOLD** A geological structure in which originally flat-lying rocks appear to have been flexed and bent. In reality, the rocks behaved in a ductile manner and flowed into these shapes, like a warm caramel bar. They may bend upward in the middle to form a ridge (anticline) or downward to form a trough (syncline). A **symmetrical fold** has slopes that are the same steepness on either side of the vertical; an **asymmetrical fold** does not. A **recumbent fold** is a fold that is lying on its side. In an **overturned fold**, a syncline is partly tucked underneath a neighboring anticline. See also *anticline*, *syncline*.

**FOLIATION** The arrangement of platy minerals in parallel bands in some deformed metamorphic rocks. See also *cleavage*, *lamination*.

**FOSSIL** The remains or traces of an organism that have been buried and often petrified (with its original tissues replaced by minerals). Some fossils, such as preserved footprints and coprolites (fossilized droppings), also record activities.

**FOSSIL FUEL** Any fuel such as coal or oil that is derived from the remains of once-living organisms that were deeply buried beneath the ground.

**FRONT** In meteorology, the forward-moving edge of an air mass. See also *air mass*.

**FUMAROLE** In volcanic regions, a small opening in the ground through which hot gases can escape.

**FUNNEL** A tube of whirling air descending from a cloud. It becomes a tornado when its lower end touches the ground. See also *tornado*.

# G

**GALAXY** An aggregation of millions of stars, often formed into an immense spiral shape. The Solar System lies within the Milky Way Galaxy.

**GEODE** A hollow cavity found within a rock and lined with crystals.

**GEOTHERMAL ENERGY** Energy obtained by tapping the heat generated in the Earth's interior.

**GEYSER** A jet of boiling water and steam that rises at intervals from the ground. It is powered by hot rocks heating groundwater. See also *hot spring*.

**GLACIER** A mass of semi-permanent ice capable of flowing downhill. Glaciers come in many varieties. The largest are **ice sheets**, such as the Antarctic ice sheet. Similar, but slightly smaller, are **ice caps**. A **valley glacier** is smaller and flows down a valley, eroding rock as it does so. An **outlet glacier** is one that flows down from an ice cap or ice sheet. A **piedmont glacier** spreads out from higher ground onto a broad plain. A **surge-type glacier** is one that periodically increases its speed much above normal. A **cirque glacier** does not spread behind the mountain hollow (cirque) that it has carved out. In a **polythermal glacier**, a cold surface layer of ice overlies much warmer ice.

**GORGE** A narrow, deep valley, usually with vertical cliffs on either side. See also *canyon*.

**GRABEN** see *rift valley*.

**GRAIN** A term referring to the texture of a rock: rocks can be either fine- or coarse-grained. The term also refers to a single small particle (for example, of sand).

**GREASE ICE** see *sea ice*.

**GREENHOUSE EFFECT** The tendency of the atmosphere to contribute to warming of the Earth, by letting through the Sun's radiation but absorbing some of it that is re-radiated by the Earth. A **greenhouse gas** (such as water vapor, carbon dioxide, or methane) is any gas that promotes this process, whether naturally or via human action.

**GROUNDMASS** The fine-grained material surrounding larger crystals or clasts. The term is most commonly applied to volcanic rocks and some heterogeneous sedimentary rocks. See also *matrix*.

**GROUNDWATER** Water occurring below the surface of the land and held between grains or in the interstices of rocks. The upper limit of such a groundwater zone is called the water table. Water held in the soil is not usually counted as groundwater. See also *aquifer*.

**GUST FRONT** An area of strong winds that travels ahead of a line of storm clouds. See also *front*.

**GUYOT** A flat-topped seamount. See also *seamount*.

**GYRE** A large-scale circular movement of ocean currents.

# H

**HABITAT** Any area that can support a particular group of living things.

**HADLEY CELL** A major element in the Earth's atmospheric circulation, in which warm air rises at the equator and flows north and south at high level before sinking and flowing at the surface, mainly back to the equator.

**HALO** A hazy ring sometimes seen around the Sun or Moon, caused by refraction of light by high-level clouds.

**HAMMOCK** A small area of vegetation growing on raised ground. Hammocks are common in the Florida Everglades.

**HANGING VALLEY** A valley, usually carved by a glacier, that enters high up the side of a larger valley.

**HEADWATER** The upper portion of any river or stream, close to its source.

**HORN, GLACIAL** A steep-sided mountain left behind after erosion by glaciers occupying cirques. See also *cirque*.

**HOT SPOT** A long-lived zone of volcanic activity thought to originate deep in the Earth's mantle. Tectonic plates passing over hot spots are marked by linear chains of volcanoes that become progressively older with increased distance from the hot spot. See also *mantle plume*.

**HOT SPRING** A bubbling up of hot water and steam from the ground, caused by hot rocks heating the groundwater beneath. See also *geyser*.

**HUMUS** A dark-colored substance found in soils, derived from dead plants, microorganisms, and animals.

**HURRICANE** A large-scale low-pressure system of tropical regions, involving powerful circulating winds and torrential rain. It is also termed a **tropical cyclone** and (especially in east Asia) a **typhoon**. Its energy comes from the latent heat of water that has evaporated from warm oceans and later condenses. See also *latent heat*.

**HYDROCARBON** A chemical compound of carbon and hydrogen. Petroleum is a complex mixture of hydrocarbons.

**HYDROTHERMAL SYSTEM** Any natural system involving the heating and circulation of underground water and steam, powered by nearby hot or molten rocks. A **hydrothermal vein** is a mineral vein laid down by past hydrothermal action. A **hydrothermal vent** is an outlet for heated water from a hydrothermal system, especially one at the bottom of an ocean. When the heated water is colored with dark material, sulfides, and other minerals precipitated when the hot fluid comes into contact with cold seawater, the vent is known as a **black smoker**. See also *vein*, *geyser*, *hot spring*.

# I

**ICE LEAD** A open channel formed in sea ice.

**ICEBERG** A floating mass of ice derived from a glacier.

**ICE CAP** see *glacier*.

**ICE FIELD** A large expanse of ice at high altitude, partly hemmed in by mountains.

**ICE HEAVE** The growth of ice within the soil in periglacial regions, leading to disruption of the land surface and the creation of ice wedges and pingos. See also *periglaciation*, *ice wedge*, *pingo*.

**ICE SHEET** see *glacier*.

**ICE SHELF** A large area of floating ice attached to land and originally derived from a glacier, especially from an ice sheet.

**ICE STREAM** A region of an ice field or ice sheet within which faster flow is taking place.

**ICE WEDGE** An irregular mass of ice growing within a soil layer in a periglacial region. See also *ice heave*.

**IGNEOUS ROCK** Rock that originates from the solidification of molten magma. **Extrusive** (volcanic) igneous rocks have solidified on the Earth's surface after volcanic activity and commonly have small, hardly visible crystals. **Intrusive** (plutonic) igneous rocks have solidified below the surface, cooling slowly enough to allow larger crystals to form. A body of intrusive igneous rock is called an **igneous intrusion**.

**INSELBERG** An isolated, steep-sided hill rising up out of a flat plain.

**INTERTROPICAL CONVERGENCE ZONE** The region where the main air masses north and south of the equator come into contact. See also *Hadley cell*.

**INTRUSION** see *igneous rock*.

**INVERTEBRATE** An animal without a backbone, such as an insect, snail, or worm.

**IRRIGATION** The artificial supply of water to agricultural crops.

**ISLAND ARC, VOLCANIC** A line of volcanic

islands associated with an ocean–ocean subduction zone. See also *subduction.*

**ISOBAR** A line on a weather map that joins points with equal atmospheric pressure.

**ISOSTASY** The concept that all columns of rock on the Earth have the same weight at some depth in the mantle. Continents rise high above the sea floor because continental crust is less dense than oceanic crust.

# JK

**JET STREAM** Any of several winds that blow for long distances high in the troposphere. See also *troposphere.*

**JOINT** A crack running through a rock. Unlike a fault, the rocks on either side of a joint remain in the same position with respect to each other. A **cooling joint** results from an igneous rock shrinking as it cools. See also *fault.*

**KARAT** A term used to describe the proportion of gold in a gold alloy (for example, 24-karat gold is pure gold).

**KARST** A characteristic landscape formed in regions built of water-soluble rocks, especially limestone. A **karst coast** occurs where limestone has been eroded and later inundated by the sea.

**KATABATIC WIND** A wind that blows downward from a glacier or cold valley at night.

**KETTLE LAKE** A lake occupying a **kettle hole**, a depression left after an ice block from a former glacier melted.

**KT BOUNDARY** The boundary between the Cretaceous and Tertiary Periods. It dates to about 65 million years ago and coincides with the extinction of some dinosaurs and many other life-forms.

# L

**LA NIÑA** A situation in which the waters of the eastern Pacific become unusually cold; the opposite of the El Niño phenomenon. See also *El Niño.*

**LAGOON** An area of sheltered seawater, almost cut off from the open ocean. The term includes both shallow coastal lagoons and the lagoons of atolls.

**LAHAR** A muddy flow of water mixed with volcanic ash and other volcanic debris. See also *mass movement.*

**LAMINATION** The occurrence of thin, distinct, and generally parallel layers within rocks.

**LAPSE RATE** The rate at which air temperature decreases with height. See also *latent heat.*

**LATENT HEAT** The heat that is input when a liquid such as water turns into vapor. It is released again when the vapor condenses.

**LAVA** Molten rock that has reached the Earth's surface. Basaltic **a'a lava** (45–52 percent by weight of silica) forms a rough-textured rock when it cools. **Pahoehoe lava**, of similar basaltic composition, flows easily and creates a smooth or ropy-textured surface. Basaltic **pillow lava** is lava ejected underwater, forming pillow-shaped mounds of rock. More silica-rich and andesitic-dacitic lavas (more than 57 percent by weight of silica) typically form **block lava**, jostling masses of angular blocks, some of them many yards across.

**LEACHING** The removal of minerals from soil or rock by downward-percolating water.

**LEVEE** A raised bank of a river that is higher than the surrounding active floodplain. Levees commonly mark former floodplains prior to major downcutting events (when the river rapidly erodes and lowers its bed).

**LIGHTNING** The visible flash that accompanies electrical discharges from storm clouds. Discharges within or between clouds are **sheet lightning**; those between clouds and the ground are **forked lightning**.

**LITHOSPHERE** The Earth's crust together with the rigid uppermost layer of the underlying mantle. Each of the Earth's tectonic plates is made of a section of lithosphere. See also *crust, mantle, asthenosphere, tectonic plate.*

**LITTORAL** Relating to the shoreline, especially between the high- and low-water marks. See also *tide.*

**LONGSHORE BAR** see *bar.*

**LONGSHORE DRIFT** A current that flows parallel to a shoreline, usually picking up and transporting sediment as it does so.

# M

**MAAR** A wide volcano crater formed by explosive eruptions that have excavated into underlying bedrock, which is exposed in the crater walls.

**MAGMA** Molten rock rising from the interior of the Earth.

**MAGMA CHAMBER** A region just below the Earth's surface where magma (molten rock) has collected.

**MAGMATIC DIFFERENTIATION** The various processes by which magma changes composition, typically toward higher silica contents. The processes include gravitational separation of early formed low-silica minerals and assimilation of high-silica crustal rocks.

**MANTLE** The rocks lying between the Earth's crust and its core. It makes up 84 percent of the Earth's volume. See also *crust, core.*

**MANTLE PLUME** A column of hot rocks rising through the mantle and crust, giving rise to a hot spot at the Earth's surface. See also *hot spot.*

**MARSH** A freshwater or saltwater wetland that is dominated by soft-stemmed plants such as grasses or reeds, but not by woody plants such as trees or shrubs. Marshes are characterized by frequent or continual flooding. See also *swamp, salt marsh.*

**MASS MOVEMENT** The movement of rocks, soil, or mud down a slope in response to gravity. See also *transport.*

**MASSIF** A well-defined mountainous region whose rocks and landforms tend to be similar across the region.

**MASSIVE** Of a rock: having a structure that does not show layering or other divisions into smaller segments.

**MATRIX** The mass of relatively fine-grained material that binds together larger particles in some heterogeneous sedimentary rocks and larger crystals in volcanic rocks. See also *groundmass.*

**MEANDER** A loop or bend in a river. Meandering rivers gradually change their course, as erosion occurs on the outside of the bend and deposition on the inside.

**MELTWATER** Water flowing within or from a glacier. A **meltwater channel** is a channel carved by meltwater running under or near a glacier.

**MESA** A tall, usually flat-topped outcrop of rock, especially in desert regions, left behind after the rest of the former land surface has been eroded. It is usually capped by a layer of resistant rock.

**MESOSPHERE** The layer of the Earth's atmosphere between the stratosphere and thermosphere, at an altitude of about 30–50 miles (50–80 km).

**METAL** Any of numerous chemical elements that are typically hard, shiny, malleable in the solid state, and good conductors of heat and electricity; also any alloy (mixture) of such elements.

**METAMORPHIC ROCK** A rock that has been transformed underground by heat or pressure to a new texture or new set of minerals. For example, marble is metamorphosed limestone.

**METAMORPHISM** The processes by which rocks are transformed by heat, pressure, or chemical reactions underground. In **thermal metamorphism**, the agent is heat. **Dynamic metamorphism** involves differential stresses that can impart new textures and structures to the rock. **Regional metamorphism** affects broad areas the size of mountain belts. **Contact metamorphism** results from the effects of hot magma on surrounding rocks.

**METASOMATIC ROCK** A rock that has been transformed by hot fluids penetrating it and changing its bulk composition and mineral assemblage.

**METEOR** A small mass of rock from space that vaporizes completely as it falls through the Earth's atmosphere, glowing as it does so.

**METEORITE** A rock from space that has fallen to the Earth's surface without completely burning up.

**METEOROID** A potential meteor or meteorite before it has entered the Earth's atmosphere.

**MICROCLIMATE** The distinctive climate of a particular place, such as a narrow valley or hillside.

**MID-OCEAN RIDGE** Any of the submerged mountain ranges running across the floors of the major oceans (although not all occur exactly in mid-ocean). They are sites where two tectonic plates are spreading apart and material is rising up from the mantle to form new oceanic crust. Mid-ocean ridges vary widely in their spreading rates, from fast- to slow-spreading ridges. See also *plate boundary, tectonic plate.*

**MINERAL** Any solid, naturally occurring inorganic material in the Earth that has a characteristic crystal structure and well-defined chemical composition. Most rocks are mixtures of more than one mineral. See also *rock.*

**MIRE** A peat-accumulating wetland such as a bog or fen.

**MONSOON** A pattern of winds, especially in southern Asia, that blow from one direction for about half the year, and from the opposite direction for the other half. The term is also used to refer to the heavy rains carried by these winds at certain times of the year.

**MONTANE** Of a forest: a type of woodland found in mountainous areas that includes conifers and other species such as beech.

**MORAINE** An accumulation of rock debris resulting from glacial action. Active valley glaciers create **lateral moraines** at their edges, **medial moraines** where two glaciers merge, and **terminal moraines** at their tips. These features often remain after the glacier disappears. See also *till.*

# N

**NATIVE ELEMENT** A chemical element that is found in its pure (uncombined) state in nature. Gold is always found native; other elements, such as sulfur and copper, sometimes occur in native form.

**NEEDLE LEAF** A needle-shaped leaf, typical of pine trees and other conifers. See also *coniferous.*

**NOCTILUCENT CLOUD** A very high-level cloud of ice crystals or dust particles that has a shiny appearance at night.

**NUNATAK** A mountaintop that rises above an ice sheet that is otherwise covering the land.

# O

**OASIS** A fertile area within a desert, normally fed by a spring arising from an aquifer. See also *aquifer.*

**OCCLUSION** In meteorology, the situation where a moving mass of colder air catches up with a warmer air mass, pushing the latter away from the surface of the Earth. See also *front.*

**OFFSHORE BAR** see *bar.*

**OOZE** The fine sediment that covers much of the deep sea floor. The names of some kinds of ooze refer to the particular tiny organisms whose skeletons or remains dominate: for **globigerina ooze**, single-celled organisms (foraminifera) of the genus *Globigerina*; for **diatom ooze**, certain algae (diatoms); for **radiolarian ooze**, other single-celled organisms called radiolarians; and for **pteropod ooze**, the shells of certain free-swimming snail relatives.

**ORE** A rock from which a metal can be profitably mined.

**OROGENESIS** The process of mountain building, resulting from pressures generated by the horizontal movement of tectonic plates. An **orogeny** is a particular episode of mountain building. See also *epeirogenesis.*

**OROGRAPHIC LIFTING** The rising of an air mass as it moves over mountains, often resulting in clouds being formed.

**OUTLET GLACIER** see *glacier.*

**OXIDE** A compound of oxygen with one or more chemical elements. Examples of oxides include ferrous oxide (FeO) and silicon dioxide ($SiO_2$).

# PQ

**PACK ICE** see *sea ice*.

**PANCAKE ICE** see *sea ice*.

**PEAT** Dead plant material that accumulates in bogs, fens, and other wetland habitats, remaining largely undecomposed. Peat soils are composed almost entirely of plant matter, particularly *Sphagnum* mosses, with little mineral content. See also *bog*.

**PEGMATITE** A plutonic igneous rock made up of unusually large crystals. **Pegmatitic veins** are narrow sheets or lenses of such rock. They are sometimes the source of valuable ores. See also *vein*.

**PENINSULA** An area of land jutting out into the sea or a lake.

**PERENNIAL** A plant that lives for three or more years. See also *annual*.

**PERIGLACIATION** Any of various features peculiar to landscapes that are subject to very cold conditions, while not having a permanent ice cover. See also *permafrost*, *pingo*, *ice wedge*, *tundra*.

**PERMAFROST** Technically, land that remains frozen for at least two years. It is characteristic of glacial and many periglacial regions. In tundra, the top layer of the soil generally thaws in spring and summer, while permafrost may remain deeper down. See also *periglaciation*, *tundra*.

**PHENOCRYST** A large crystal surrounded by smaller ones in an igneous rock. See also *porphyry*.

**PHOTOSYNTHESIS** The process by which green plants and some microorganisms use the energy of sunlight in the presence of chlorophyll to convert water and carbon dioxide into food.

**PHREATIC ERUPTION** A type of volcanic eruption or geyser activity where groundwater is turned to steam by contact with hot rocks near the surface.

**PHYTOPLANKTON** The tiny, mostly single-celled plants (algae) that float near the surface of oceans and lakes and are the foundation of most aquatic food chains.

**PIEDMONT GLACIER** see *glacier*.

**PILLOW LAVA** see *lava*.

**PINGO** In periglacial areas, a hillock with a core of ice that has pushed up above the surrounding landscape.

**PLACER DEPOSIT** An accumulation of sediments deposited by running water that contains minerals that can be profitably mined.

**PLANET** A large astronomical body that orbits a star and, unlike a star, is not heated by thermonuclear reactions taking place within it.

**PLANETESIMAL** One of millions of rocky objects of variable size believed to have been present in the early history of the Solar System and which later came together to create the planets.

**PLATE** see *tectonic plate*.

**PLATE BOUNDARY** A boundary between two tectonic plates. The plates concerned can be diverging (**constructive boundary**), converging (**destructive boundary**), or sliding past one another (**conservative boundary**).

**PLATEAU** A large area of relatively flat land that stands topographically above its surroundings.

**PLATY** A term used to describe thin, flat crystals.

**PLAYA** A level plain in a desert basin, sometimes occupied by a temporary lake that fills and empties seasonally.

**PLINIAN ERUPTION** see *eruption*.

**PLUTON** A large body of igneous rock that solidifies slowly underground to form a coarse-grained (plutonic) rock such as granite. A mass of many hundreds of adjacent plutons is called a batholith. See also *batholith*.

**PLUTONIC** Relating to igneous processes or materials formed deep below the Earth's surface. The term is derived from Pluto, the Roman god of the underworld.

**PODOCARP** A member of a family of coniferous trees found mainly in the southern hemisphere.

**POINT BAR** see *bar*.

**POLDER** An area of low-lying land reclaimed from the sea.

**POLYMORPH** A substance that can exist in different mineral or crystalline forms, all sharing the same chemical composition. For example, pure carbon can exist as both diamond or graphite.

**POLYNYA** A stretch of open water in an ice-covered sea.

**POLYTHERMAL GLACIER** see *glacier*.

**PORPHYROBLAST** A large crystal surrounded by smaller crystals found in some metamorphic rocks.

**PORPHYRY** Igneous rock that contains large phenocrysts, embedded in finer-grained material. See also *phenocryst*.

**PRECIPITATE** Any material that has come out of solution to form a solid deposit.

**PRECIPITATION** Water that reaches the Earth's surface from the atmosphere, including rain, snow, hail, and dew.

**PRESSURE SYSTEM** Any pattern of weather in which air circulates around an area of high or low pressure. The **pressure gradient** is the pressure difference between two given points. See also *cyclone*, *anticyclone*, *isobar*.

**PREVAILING WIND** The source direction of wind that tends to blow most commonly at a particular location.

**PRIMARY MINERAL** A mineral that derives from the cooling of molten igneous rock, without being altered subsequently. See also *secondary mineral*.

**PROKARYOTE** Any single-celled organism, such as a bacterium, whose cell has a simple structure that lacks a well-defined nucleus. See also *eukaryote*.

**PUMICE** A light-colored, low-density volcanic rock containing innumerable bubbles formed by expanding gases, ejected during some volcanic eruptions.

**P-WAVE** see *seismic wave*.

**PYROCLASTIC** Consisting of, or containing, volcanic rock fragments. **Pyroclastic flows** are fast-moving, sometimes deadly clouds of hot gases and debris. See also *pumice*, *scoria*, *tuff*.

# R

**RADIATION** A flow of high-energy particles or waves. **Electromagnetic radiation** consists of electromagnetic waves: listed from long-wavelength to short-wavelength forms, these include radio waves, microwaves, infrared rays, visible light, ultraviolet light, X-rays, and gamma rays. Short-wavelength electromagnetic radiation has the highest energy.

**RADIATION FOG** see *fog*.

**RAINBOW** A large-scale, prismlike light effect caused by the Sun's rays being refracted (bent) as they pass through raindrops. See also *refraction*.

**RAIN SHADOW** The occurrence of low rainfall downwind of a mountain range, caused by the air shedding its moisture as it passes over the mountains.

**RAINFOREST** A forest, usually tropical, with high rainfall, high humidity, and high temperatures year-round.

**REEF** see *coral reef*.

**REFRACTION** The tendency for waves of any kind to be bent when they pass from one medium to another of different properties. For example, light waves can be refracted by passage from air to water. The amount of refraction depends on wavelength.

**REGOLITH** A general term for all the materials that are not solid rock covering the surface of the Earth, Moon, and other planetary bodies. On Earth, it includes soil and debris from glaciers. The lunar regolith refers to the thick blanket of rock dust produced by meteorite impacts.

**RELATIVE HUMIDITY** The amount of water vapor in the atmosphere, relative to the total amount that the atmosphere can hold at that particular temperature. Above 100 percent relative humidity, water vapor will tend to condense as liquid water. Warm air can hold more water vapor than cold air. See also *dew point*.

**REPLACEMENT DEPOSIT** A mineral deposit in which the original material has been replaced by other minerals.

**RESURGENCE** The emergence of water from an underground aquifer at the surface as a spring.

**RIA COAST** A type of drowned coast indented by former river valleys, known as **rias**, that are now inlets of the sea. See also *coast*.

**RIDGE OF HIGH PRESSURE** A long, narrow area of high pressure extending out from an anticyclone. See also *anticyclone*.

**RIFT VALLEY** A large block of land that has dropped vertically downward compared with the surrounding regions, as a result of horizontal extension and normal faulting. A rift valley is also called a graben. See also *fault*.

**RIMSTONE** A deposit of calcite around an underground pool in a cave. See also *calcite*, *flowstone*.

**ROCK** Any solid material made up of one or more minerals and occurring naturally on the Earth or other planetary bodies. See also *mineral*.

**ROSSBY WAVE** A wave that forms in a jet stream at temperate latitudes, giving rise to areas of high and low pressure. See also *jet stream*, *pressure system*.

# S

**SALINITY** The concentration of dissolved salts in, for example, water or soil.

**SALT MARSH** A wetland that supports salt-tolerant plants that grow on flat coastal areas regularly inundated by the tides.

**SARGASSUM** A floating seaweed common in the Sargasso Sea in the North Atlantic.

**SASTRUGI** The sculpted appearance of some ice surfaces, caused by erosion from windblown ice crystals.

**SAVANNA** A general term for all tropical natural grasslands. Most savannas also have scattered trees.

**SCARP** see *escarpment*.

**SCORIA** A dark-colored volcanic rock containing innumerable bubbles formed by expanding gases ejected during some basaltic volcanic eruptions. Scoria is also known as cinder. See also *pyroclastic*.

**SEA ARCH** A natural arch formed in a sea cliff by erosion.

**SEA CAVE** A cave eroded into a sea cliff by the action of the sea.

**SEA ICE** Ice that arises from seawater freezing. (It does not include icebergs, which originally come from land.) The sea freezes in several stages. At first, separate crystals form. These begin to join together, forming a thick, oily texture called **grease ice**. Thin slabs of ice called **pancake ice** may form. When the sea ice forms a continuous sheet it is called **fast ice**. Winds and storms may later break this up into fragments called **pack ice**.

**SEAMOUNT** An undersea mountain, usually volcanic in origin. See also *guyot*.

**SEA STACK** A tall column of rock rising out of the sea near a coastline. It is a resistant remnant of a formerly higher land surface that has otherwise disappeared due to erosion.

**SECONDARY MINERAL** A mineral that has replaced a primary igneous mineral, one originally solidified from a molten state. See also *primary mineral*.

**SEDIMENT** Solid particles that have been transported by water, wind, volcanic processes, or mass movement, and later deposited.

**SEDIMENTARY ROCK** Rock formed when small particles are deposited—by wind, water, volcanic processes, or mass movement—and later harden.

**SEISMIC WAVE** A shock wave generated by an earthquake. **P-waves** can travel through both solid and liquid portions of the Earth's interior, **S-waves** only through the solid parts.

**SEMIDESERT** A dry region that has enough precipitation to support some plant life.

**SERAC** A pinnacle of ice in a glacier.

**SHIELD** see *craton*.

**SHIELD VOLCANO** see *volcano.*

**SILICA** Silicon dioxide ($SiO_2$), the most common mineral form of which is quartz, a hard mineral that is the dominant component of sand.

**SILICATE** Any rock or mineral composed of groups of silicon and oxygen atoms in chemical combination with atoms of various metals. Silicate rocks make up most of the Earth's crust and mantle.

**SILL** A roughly horizontal, sheetlike igneous intrusion that usually forms when igneous rock forces its way between layers of existing sedimentary rocks. See also *dike.*

**SINKHOLE** see *doline.*

**SKIN FLOW** Creeping movements of soil in periglacial regions following rapid thawing of the surface layers. See also *periglaciation.*

**SOFAR CHANNEL** Short for Sound Fixing and Ranging channel. Sound achieves a much longer range at particular depths in the ocean, typically at 3,300 ft (1,000 m). This layer allows scientists with underwater microphones to detect sounds thousands of miles away.

**SOIL HORIZON** A particular layer of a soil with distinctive appearance and physical or chemical properties.

**SOIL PROFILE** A vertical section through the soil showing all of its horizons.

**SOLAR BUDGET** The overall flow of radiant energy from the Sun to the Earth, and its ultimate re-radiation into space. See also *radiation, greenhouse effect.*

**SOLAR CONSTANT** The amount of energy from the Sun that reaches a unit area at the top of the Earth's atmosphere. It is not truly constant, but nearly so.

**SPECIFIC GRAVITY** A measure of the density of a substance, in terms of its weight for a given volume compared to the same volume of a reference material, usually pure water.

**SPIT** A peninsula of sand or shingle projecting from a shore, usually where the coastline changes direction. It is created by longshore drift. See also *bar, tombolo, longshore drift.*

**SPUR** A mountain ridge cut off as a result of past glacial action or tectonic faulting.

**STALACTITE** A deposit of calcite hanging down from the roof of a cave or underground passage. See also *calcite.*

**STALAGMITE** A deposit of calcite rising up from the floor of a cave or underground passage. See also *calcite.*

**STAR** A self-luminous astronomical body, such as the Sun, that generates energy by nuclear reactions, mainly involving the conversion of hydrogen to helium.

**STEPPE** Temperate grassland habitats, especially in regions with hot, dry summers and cold winters.

**STRATOSPHERE** A layer of the Earth's atmosphere extending from the top of the troposphere, at 5–10 miles (8–16 km), up to about 30 miles (50 km). See also *troposphere, mesosphere.*

**STRATOVOLCANO** see *volcano.*

**STRATUM** (pl. STRATA) A layer of sedimentary rock.

**SUBDUCTION** The descent of an oceanic tectonic plate under another plate when two plates converge. Subduction zones can be classified as either ocean–ocean or ocean–continent depending upon the nature of the two converging plates. See also *deep-sea trench, plate boundary, tectonic plate, crust.*

**SUBLITTORAL** Below the low-tide mark.

**SUBMONTANE** Of a forest: a type of woodland found in lowland and hilly sites that consists mainly of deciduous, broadleaved trees.

**SUCCESSION** A gradual change with time from one community of plants to another in a particular location—for example, from grassland to forest.

**SUCCULENT** A plant having thick, juicy leaves or stems for storing water.

**SURGE-TYPE GLACIER** see *glacier.*

**SWAMP** A freshwater or saltwater wetland that is dominated by trees. See also *marsh.*

**SWASH** The movement of turbulent water up a beach when a wave breaks. See also *backwash.*

**S-WAVE** see *seismic wave.*

**SYNCLINE** A downward fold of originally flat strata as a result of horizontal compression. See also *anticline, fold.*

# T

**TAIGA** The coniferous forest (also called boreal forest) that covers much of Europe, Asia, and northern North America.

**TECTONIC PLATE** Any of the large rigid sections into which the Earth's lithosphere is divided. The relative motions of different plates lead to earthquakes, volcanic activity, continental drift, and mountain-building. See also *plate boundary, subduction, deep-sea trench, mid-ocean ridge, lithosphere.*

**TEKTITE** A glassy particle sometimes formed when a large meteorite strikes the Earth. Tektites originate when melted target rock is thrown great distances from the site of impact.

**TEMPERATE** Relating to the regions of the Earth between the tropics and the polar regions.

**TERRACE** A flat region in a river valley that is higher than the present floodplain. It represents a former floodplain created when the river ran at a higher level. See also *floodplain.*

**THERMOHALINE** see *current.*

**THERMOPAUSE** The upper edge of the thermosphere, about 400 miles (640 km) above the surface.

**THERMOSPHERE** The layer of the Earth's atmosphere above the mesosphere. It extends over altitudes of about 50–400 miles (80–640 km).

**TIDAL BORE** A single large wave sometimes created when a rising tide enters a narrowing channel such as an estuary.

**TIDE** The rise and fall (usually twice each day) of water on the shore caused by the gravitational pull of the Moon and Sun. The difference in level between high and low tide is called the **tidal range**, and tends to vary on a monthly basis: **spring tides** (when the effects of the Moon and Sun reinforce each other) produce the highest high tide and the lowest low tide; **neap tides** (when the effects of the Moon and Sun oppose each other) have the smallest tidal range.

**TILL** Solid material laid down or left behind by a glacier. It typically consists of rock fragments of many sizes. Till left behind by glaciers from the last ice age covers many regions today. See also *moraine.*

**TOMBOLO** A type of spit connecting a small island to the mainland. See also *spit.*

**TORNADO** A narrow, rapidly whirling tube-shaped column of air, which is often highly destructive.

**TRANSPIRATION** The process by which water is drawn through plants. The evaporation of water vapor from leaves causes more water to be drawn in through the roots to compensate.

**TRANSPORT** The conveying of weathered and eroded material to another location, for example by wind, water, or ice action. See also *weathering, erosion.*

**TRAVERTINE** A rock, mainly consisting of calcium carbonate, that is deposited around the edges of hot springs.

**TREE LINE** The latitude or altitude above which conditions are too harsh for trees to grow in a particular location.

**TRIBUTARY** Any river or stream that flows into a larger river.

**TROPICAL** Relating to the warm regions of the Earth that lie between the equator and the tropics of Cancer and Capricorn, at latitudes of 23.5° North and South, respectively.

**TROPOPAUSE** The boundary between the troposphere and the stratosphere. Air temperature starts to increase with height above the tropopause, which varies from about 10 miles (16 km) at the equator to 5 miles (8 km) over the poles.

**TROPOSPHERE** The lowest, densest layer of the atmosphere, where most weather phenomena occur. The height of the upper limit of the troposphere (known as the tropopause) varies from the equator to the poles. See also *tropopause.*

**TROUGH** In meteorology, a long, relatively narrow area of low pressure. See also *front.*

**TSUNAMI** A fast-moving, often destructive sea wave generated by earthquake activity; popularly but incorrectly known as a tidal wave. It rises in height rapidly as it reaches shallow water.

**TUBE** A tunnel-shaped passage within a cave system through which water flows or has flowed. See also *canyon.*

**TUFF** A rock made of fine-grained pyroclastic material produced by an explosive volcanic eruption. See also *pyroclastic.*

**TUNDRA** A treeless habitat of low-growing, cold-tolerant plants widespread in far northern North America and Siberia.

**TURBIDITY CURRENT** A flow of sediment-containing water from the continental margin down onto the ocean floor. Sediment deposited by such a current is known as **turbidite**.

**TWINNING** A form of crystal growth in which the same crystal grows in more than one orientation, giving the appearance of two interpenetrating crystals separated by a mirror plane of symmetry.

**TYPHOON** see *hurricane.*

# UV

**ULTRAVIOLET** see *radiation.*

**UNDERSTORY** The layer of smaller trees and shrubs growing beneath the main trees of a forest.

**VALLEY FOG** A type of radiation fog occurring in valleys. See also *fog.*

**VALLEY GLACIER** see *glacier.*

**VEIN** A thin, sheetlike body of rock penetrating through other rocks. See also *hydrothermal system.*

**VENTIFACT** A rock or pebble polished by windblown sand in a desert.

**VERTICAL TRANSPORT** The upwelling or downwelling of nutrient-rich water in the oceans.

**VISCOSITY** Resistance to flow in fluids. The higher the viscosity of a fluid, the more sluggishly it flows.

**VOLATILE** In volcanology: a term used to describe water, carbon dioxide, and other potentially gaseous compounds that are dissolved in molten rocks.

**VOLCANO** An opening in the Earth's crust where magma reaches the surface; also, a mountain that contains such an opening. A **shield volcano** has shallow slopes and is built from lava that flowed easily. A **stratovolcano** is steeper and built from alternate layers of ash and lava. A **dome volcano** is rounded, steep, and built from viscous lava. A **cinder cone** is built from scoria that has fallen from the explosion clouds of an eruption.

**VULCANIAN ERUPTION** see *eruption.*

# WXYZ

**WADI** A usually waterless river valley in a desert.

**WATER TABLE** The upper surface of the groundwater zone in places where it is not confined by impermeable rocks. Water soaking into the ground will tend to sink down until it reaches the water table. See also *groundwater.*

**WATERSHED DIVIDE** The divide between two drainage basins.

**WAVE** A regular motion or disturbance that transfers energy. For a wave crossing the open ocean, the water itself does not move significantly except up and down as the wave passes. The high point of a wave is its **crest** and the low point its **trough**. For waves breaking on shores (**breakers**), water motion becomes more complex and chaotic (**turbulent**).

**WAVE REFRACTION** see *refraction.*

**WAVELENGTH** The distance between successive crests (or troughs) of a wave.

**WEATHERING** The alteration of rocks caused by their being exposed at or close to the Earth's surface. Usually the original rock eventually crumbles or becomes weakened. See also *erosion.*

**XENOLITH** A fragment of older rock incorporated within a younger igneous rock. See also *igneous rock.*

**YARDANG** A wind-sculpted ridge of rock in a desert.

# INDEX

Page numbers in **bold** indicate feature profiles or extended treatments of a topic. Page numbers in *italic* indicate pages on which the topic is illustrated.

## A

# Q



# R

# T

# U



# V

# ACKNOWLEDGMENTS

**Dorling Kindersley** would like to thank the following people for their help in the preparation of this book: Vanessa Hamilton for original design work and Rebecca Milner for design assistance; Cathy Meeus for editorial help; Gemma Casajuana and Mark Bracey for additional DTP support; Erin Richards for administrative support; Julia Lunn and Rob Stokes for additional cartography; Robyn Bissette and Ellen Nanney of the Smithsonian Institution; Dr. A.G. Smith and Lawrence Rush of Cambridge University for plotting bases for the maps of the Earth's past plate distributions; Philip Eales, Andrew Wayne, and Kevin Tildsley of Planetary Visions Ltd; and Sean Mulshaw and Geoffrey Gilbert for advice on the selection of features.

**The Smithsonian Institution** would like to thank Ellen Nanney and Robyn Bissette of the Office of Product Development & Licensing; the National Museum of Natural History Department of Mineralogy, Department of Botany, Department of Paleobiology, and Department of Anthropology; the Smithsonian Environmental Research Center; and the Center for Earth and Planetary Sciences.

**PICTURE CREDITS**
Dorling Kindersley would like to thank the following for their help with images: Romaine Werblow in the DK Picture Library; Giovanni Cafagna at Corbis; Tony Waltham at Geophotos; Caroline Thomas and Mariana Sonnenberg.

**Key: L = Left, R = Right, C = Centre, T = Top, B = Bottom, a = Above, b = Below.**

**Abbreviations**
**B&C:** Bryan and Cherry Alexander Photography; **DK:** DK Picture Library www.dkimages.com; **FLPA:** FLPA – Images of Nature; **Geo:** Tony Waltham Geophotos; **Getty:** Getty Images; **RHPL:** Robert Harding Picture Library; **NG:** National Geographic Image Collection; **NHM:** The Natural History Museum, London; **OSF:** Oxford Scientific Films; **SI:** Smithsonian Institution; **SPL:** Science Photo Library; **Still:** Still Pictures; **WWI:** Woodfall Wild Images.

**Sidebar Images** (from top to bottom)
**Earth - NASA**: Hubble Space Telescope Center (PR96-27B); **Corbis:** NASA, L Clarke, Ken Straiton, Michael T Sedam; **DK.**
**Land – Corbis**: Darrell Gulin, Walter Hodges, Wolfgang Kaehler, Gary Braasch, José F Poblete, David Muench.
**Oceans – Corbis**: FLPA / Winifred Wisniewski, David Pu'u, Darrell Gulin, Stuart Westmorland, Jeffrey L Rotman, Robert Yin.
**Atmosphere – Corbis**: Stuart Westmorland, Paul A Souders, William James Warren, Matthias Kulka, Craig Tuttle, William A Bake.

**1 Still**: UNEP / D Stanfill. **2/3 Getty**: Taxi / Bav. **4 Corbis**: Darrell Gulin TC. **4/5 Still**: Pascal Kobeh b. **5 alamy.com**: ImageState / Adam Jones TL; **Corbis**: William James Warren C; **Galaxy Contact**: Johnson Space Center (STS066-208-205) TCRb; **Getty**: Image Bank / Peter Cade TC. **8/9 Corbis**: Firefly Productions. **10/11 Getty**: Stone / G Brad Lewis. **12/13 Magnum**: Stuart Franklin. **14/15 Corbis:** Ralph A Clevenger. **16/17 NG:** Panoramic Images / David Lawrence. **18/19 Corbis:** Darrell Gulin. **20/21 SPL:** John Reader. **22 Geo:** CLb; **University of Edinburgh, Grant Institute:** TRb. **22/23 Corbis:** Tom Bean B. **23 Corbis:** George HH Huey Ca; James L Amos TCL; **NG:** Louis Mazzatenta TCbR; **SPL:** James King-Holmes CRb. **24 SPL:** NASA BR; Patrice Loiex, Cern BL. **24/25 SPL:** Roger Harris. **25 NASA:** Chandra X-Ray Observatory (MSFC-9905485) C; Hubble Space Telescope Center (PR96-38B) CRa. **26 NHM:** C. **26/27 Corbis:** Ric Ergenbright. **27 Courtesy of the Archives, California Institute of Technology:** BR. NASA: Saturn Apollo Program (MSFC-690096) CL. **SPL:** Dr Kari Lounatmaa Cb. **28 Roger Buick:** Department of Earth and Space Sciences & Astrobiology Program, University of Washington Cb, BC; **Corbis:** Roger Garwood BL. **28/29 Corbis:** Ric Ergenbright. **29 Dr Peter Crimes:** BR; **Dr Bruce Runnegar:** Centre for Astrobiology, Institute of Geophysics, University of California. **30 alamy.com:** Stephen Frink Collection / Masa Ushioda BL; **Corbis:** Bettmann CR; Ric Ergenbright L; **DK:** Colin Keates BC. **30/31 FLPA:** L West. **31 DK:** Royal Museum of Scotland / Harry Taylor C; NHM / Harry Taylor CR. **32 British Antarctic Survey:** Pete Bucktrout C; **Corbis:** Jonthan Blair CbR; **FLPA:** L West (main); **Geo:** CLb. **33 Corbis:** Douglas Peebles BC; Kevin Schafer BL; Layne Kennedy R (main); **DK:** Hunterian Museum C; Jon Hughes BR. **34 Corbis:** Layne Kennedy (main); Tom Bean BL; **DK:** Senekenberg Nature Museum / Andy Crawford C; **NHM:** BC. **35 Corbis:** Richard Cummins CR; Roger Ressmeyer CLb; Royalty-Free R (main). **36 alamy.com:** Ukraft C; **Nature Picture Library:** Jurgen Freund CR; **Corbis:** Jonathan Blair BCL; **DK:** Harry Taylor BCR. **36/37 Corbis:** Royalty-Free. **37 Corbis:** Gallo Images CR; Torleif Svensson BC; **SPL:** Geospace C. **38 Corbis:** Hulton-Deutsch Collection C; **DK:** BLR; Museum of London / Dave King CRb; Pitt Rivers Museum / Dave King Cra; **NHM:** BL; **SPL:** John Reader CLb, BC. **38/39 Corbis:** Royalty Free. **39 DK:** Gables CL **Dr Sandy Tudhope:** Institute of Geology and Geophysics, Edinburgh University. **40 NHM:** BL; **Geo:** BR. **40/41 Corbis:** Royalty-Free. **41 Corbis:** William Findlay BL; **DK:** Royal British Columbia Museum / Andrew Nelmerm CL. **42/43 Galaxy Contact:** C. **44 NASA:** The Hubble Heritage Team (STScI/AURA) BCL; ESA, R de Grijs (Institute of Astronomy, Cambridge UK) BCR; Hubble Space Telescope (PR 99-25) BR; M Clampin (STScI), H Ford (JHU), G Illingworth (UCO/Lick Observatory), J Krist (STScI), D Ardila (JHU), D Golimowski (JHU), The ACS Science Team, J Bahcall (IAS) and ESA BL; NSSDC / Hubble Space Telescope BLa; Robert Williams and the Hubble Deep Field Team (STScI) C. **44/45 Galaxy Contact. 45 Corbis:** Bettmann CR; Dennis di Cicco TR; **NASA:** Hubble Space Telescope Center (PR00-10) TCb; Hubble Space Telescope Center (PR93-01) CRb; Hubble Space Telescope Center (PR99-30B) CbL; Michael Rich, Kenneth Mighell and James D Neill (Columbia University), and Wendy Freedman (Carnegie Observatories) BL. **46 NASA:** NSSDC / David Crisp and the WFPC2 Science Team (Jet Propulsion Laboratory / California Institute of Technology) *mars*; NSSDC / Jet Propulsion Laboratory / California Insitute of Technology *jupiter*; NSSDC / Johnson Space Center / Apollo 17 *earth*; NSSDC / Mariner 10 *mercury*; NSSDC / Pioneer Venus Orbiter *venus*. Courtesy of **SOHO/EIT Consortium.** SOHO is a project of international cooperation between ESA and NASA: L. **47 Corbis:** Aaron Horowitz BR; Bettmann BL; Lorenzo Lovata: BC. **NASA:** Dr R Albrecht, ESA / ESO Space Telescope European Coordinating Facility *pluto*; Erich Karkoschka (University of Arizona) *uranus*; John Hopkins University Applied Physics Laboratory CRb; NSSDC *neptune*; NSSDC / Jet Propulsion Laboratory / California Institute of Technology *saturn*. **48 Corbis:** Gary Braasch Ca; Roger Ressmeyer TRb; William James Warren TCR. **48/49** Courtesy of **SOHO/EIT Consortium.** SOHO is a project of international cooperation between ESA and NASA: C. **49 Corbis:** Jonathan Blair CRb; NASA: Marshall Space Flight Center (MSFC-9513744) BC; Courtesy of **SOHO/EIT Consortium.** SOHO is a project of international cooperation between ESA and NASA: BRa. **50 NASA:** Jet Propulsion Laboratory / California Institute for Technology (PIA00342) TR. **50/51 Michael Light / www.projectfullmoon.com:** C; **U.S. Navy Observatory:** Antonio Cidadao B. **51 Corbis:** Roger Ressmeyer CRa; Sandro Vannini TCb; Sygma / Colin McPherson CRb. **NASA:** Johnson Space Center (AS16-114-18423) TL. **52/53 OSF:** Olivier Grunewald. **54 Hutchison Library:** Patricia Goycolea CLb; **NASA:** NSSDC / Apollo 8 BLa. **55 Corbis:** Stuart Westmorland CbR; **DK:** Clive Streeter BCR; **SPL:** Alfred Pasieka BR. **56 alamy.com:** Stock Connection Inc / Jagdish Agarwal TCRb; **Corbis:** Raymond Gehman CLb; **DK:** CaR; **Dr Sally Gibson:** Department of Earth Sciences, University of Cambridge BL. **57 Corbis:** Roger Ressmeyer BR; **SPL:** J Lees & P Malin, UC Santa Barbara CRa; **Department of Geophysics, Faculty of Science, University of Zagreb:** TR. **58 alamy.com:** David Noton Photography / David Noton Ca; Stephen Bond BCRa; Worldwide Picture Library / John Fowler CLb; **Still:** Staffan Andersson TRb; **Geo:** BCL, CbL, CbR. **59 Corbis:** Guy Motil CLa; Philip James Corwin Ca; Tiziana and Gianni Baldizzone Cb; **SPL:** George Bernard BR; **Geo:** TR, TCbR. **60 Philip E Batson:** CLa; **Corbis:** Kevin Schafer BR. **60/61 Corbis:** Frank Lane Picture Agency / Maurice Nimmo C; **61 DK:** *diamond*; Colin Keates *hematite, rock crystal*; Harry Taylor *azurite, calcite, galena, halite, kaolinite, magnetite, milky quartz, muscovite, smokey quartz, talc*; NHM / Colin Keates *opal*; NHM / Harry Taylor *gold*; **SI:** BC. **62 DK:** *stibnite*; Colin Keates *gold, olivine*; Harry Taylor *anhydrite, apatite, borax, calcite, chromite, green fluorite, wulfenite*; **Geo:** Harry Taylor B. **63 Corbis:** Robert Garvey TCb. **DK:** Colin Keates *burmese ruby crystal, cusioned mix-cut*; Harry Taylor *amethyst quartz, fluorite, sphalerite*; **NHM:** CL. **OSF:** Nick Gordon CRa. **SPL:** TCL; Dr Jeremy Burgess CbL. **64 Corbis:** Bettmann TR. **DK:** CLb; Harry Taylor CLA, TCL; NHM / Harry Taylor BCL; **SI:** BR; **Geo:** Cb. **65 DK:** Colin Keates TC; Harry Taylor CRa, C, BL; NHM / Harry Taylor BCL; **OSF:** Mary Plage CLa; Nick Gordon CRb; **P A Photos:** Peter Jordan Bla; **SI:** BR. **66 DK:** Harry Taylor CRa, CLb, CRb, BR, CaL; **NHM:** BCL. **67 AISA - Archivo Iconográfico S. A., Barcelona:** TC; **Corbis:** Fulvio Roiter CRb; **DK:** Harry Taylor TR, CLa, Cb, BL, BCR; **Geo:** Ca. **68 Corbis:** Elio Ciol Ca; José Manuel Sanchis Calvete BCL; **DK:** Colin Keates CR; Harry Taylor TL, TC, CLb; NHM / Harry Taylor CRa (inset); **NHM:** Cb; **SPL:** Astrid & Hanns-Freider Michler BR. **69 Corbis:** Tim Thompson BL; **Paul Deakin:** TRb; **DK:** Harry Taylor CLa, C, BR, CaR, CbR, TCL. **70 Tony Dickson:** Department of Earth Sciences, University of Cambridge CRa; **DK:** Harry Taylor TR, CLa, CLb, BR, BCL, BCLa, CaL, CbR. **71 Corbis:** Scott T Smith CL; **DK:** Colin Keates BCR; Harry Taylor CRa, BL, BRa, CbR, TCL, TCRb; NHM / Harry Taylor BCL. **72 Corbis:** Jim Sugar Photography BR; **DK:** Colin Keates Ca; Harry Taylor CRa, Cb, CRb, BL, BC (*andradite*), BC (*pyrope*), BC (*spessartine*), TCR; **NHM:** TCb; **SPL:** Andrew Syred CaL; **Geo:** TLb. **73 DK:** Colin Keates CRa, CaR; Harry Taylor TL, CL, CRb, BL, BCR, CaL, TCLb; NHM / Colin Keates BCLa; **SI:** TR. **74 Corbis:** Mike Zend TCR; **DK:** Harry Taylor Ca, CL, CR, CRb, BL, BR, BCR, CaR; **SI:** BCL, CbR. **75 Corbis:** Ecoscene / Ian Harwood TR; Lowell Georgia BR; **DK:** Harry Taylor CLa, BL, CaR, CbR, TCL; NHM / Colin Keates C; **SI:** CaL. **76 DK:** Harry Taylor TR, CLa, CR, BL, BC, CbL (*labradorite*), CbL (*microline*), CbL (*yellow*), TCLb; NHM / Harry Taylor CbL (*moonstone*); **SPL:** Philippe Gontier BRa. **77 Corbis:** David Muench B; George HH Huey CRb; Gianni Dagli Orti TR. **DK:** Harry Taylor CLa, CbR, TCL; NHM / Harry Taylor Ca. **78 DK:** Harry Taylor CL. **83 SI:** BL; **Dr AG Tindle:** Open University Cb. **78/79 OSF:** Paul Kay C. **79 Corbis:** Roger Ressmeyer Ca; **OSF:** Stuart & Jan Bebb BR; **US Geological Survey:** BC; **Geo:** TR. **80 Corbis:** Hubert Stadler Ca; **Landform Slides:** Ken Gardner BCL; **OSF:** Michael Fogden CbR; **SPL:** Tony Craddock TL; **Geo:** CRb. **80/81 Corbis:** Lester Lefkowitz T. **81 B&C:** Ann Hawthorne BCL; **SPL:** BRa; Martin Dohrn CRa; Mike McNamee TCR; **SI:** BCa. **82 DK:** Clive Streeter CR; Harry Taylor CRa, CaR; **SI:** CL; **Geo:** TR, B, CaL. **83 British Antarctic Survey:** Brian Storey BL; **DK:** Harry Taylor TR, CL, CRb, BCR, TCL; **SPL:** CbL; **Geo:** CaR. **84 Corbis:** Raymond Gehman BL; **DK:** Harry Taylor Cb, BR; **SI:** PD & L CaR, TCLb; **Geo:** BC, TRb. **85 DK:** Harry Taylor TL, BL, CaL; **OSF:** Richard Packwood BR; **SI:** Cra; **Geo:** TCRb. **86 Corbis:** Douglas Peebles BR; **DK:** Colin Keates CLa, BC, TCL; **SPL:** CRb; **Geo:** CRa, BL. **87 Corbis:** Jason Hawkes Ca; **DK:** Dinesh Khama BCL; Harry Taylor TR, CLa, BC; **OSF:** George Gardener BR. **88 Corbis:** Tiziana and Gianni Baldizzone BCL; **DK:** Harry Taylor TC, TR, CLa, Ca, Cb, CRb; **FLPA:** Maurice Nimmo BRa. **89 DK:** Harry Taylor TR, C; **FLPA:** E & D Hosking Cra; **OSF:** Niall Benvie BL; **SI:** CR; **Geo:** BR. **90 DA Carswell:** University of Sheffield BC; **DK:** Andreas Einsiedel BL; Clive Streeter CbR; Harry Taylor Cla; **NHM:** CaL; **SI:** BR; **Geo:** TR, CaR. **91 British Antarctic Survey:** Brian Storey TL; **DK:** Harry Taylor CLb, BR; **SI:** CRa, BC, CbR. **92 alamy.com:** Worldwide Picture Library / Ivan J Belcher BC; **DK:** Harry Taylor CL, CR, BL, BR, TCb. **93 DK:** Andreas Einsiedel BL; Harry Taylor CRa, CRb, BCR, CbL; Peter Wilson Cla; **Geo:** TR. **94 Corbis:** Richard Cummins BL; **DK:** Harry Taylor CLa, BC, CbL; NHM TC; **SPL:** Bill Bachman BR; **Geo:** Ca. **95 Corbis:** TCL; Michael St Maur Sheil Cra; **DK:** Andreas Einseidel CRb; Colin Keates CL; **SPL:** TR; Sinclair Stammers B. **96 Still:** UNEP / S Compoint. **97 Still:** Chris James BC. **98 Corbis:** Wolfgang Kaehler Cra; **DK:** Harry Taylor TC, BR, BCL; Steve Gorton Cb; **SI:** CLb; **Geo:** CLa. **99 DK:** Colin Keates CRb; Harry Taylor BL; NHM: CL; **SI:** BCR; CaR. **100 Corbis:** Phil Schermeister CR; Stephanie Maze C; **Geo:** B. **101 alamy.com:** Bernd Mellmann BR; **Corbis:** Yann Arthus-Bertrand Cra; **DK:** Geoff Dann BL; **Geo:** BC; **SSSA Marbut Memorial Slide Set:** TR, BC, CaL; **University of Idaho:** CLa. **102 Corbis:** Macduff Everton Cla; **SPL:** James King-Holmes Ca; **Geo:** CRa, BR; Reprinted from *Soil Teaching Aid* by Andrew R Aandahl by permission of the University of Nebraska Press. **Copyright © 1979 by the University of Nebraska Press:** BC, TCR; **SSSA Marbut Memorial Slide Set:** BL; **University of Idaho:** CRb. **103 Hulton Archive/Getty:** Three Lions Cra; **OSF:** Berndt Fischer CLa; Geo: BC; **SSSA Marbut Memorial Slide Set:** BL; **University of Idaho:** BR, CRb, TCR. **104/105 Corbis:** Yann Arthus-Bertrand. **106 Corbis:** L Clarke CbL; Pat O'Hara BL; SPL: TRb. **106/107 Visible Earth:** Image Courtesy Jacques Descloitres, MODIS Land Rapid Response Team, NASA / GSFC B. **107 Corbis:** Roger Ressmeyer BR; **DK:** Peter Anderson CRb; **SPL:** Mehau Kulyk Cb. **108 Corbis:** Dewitt Jones BC; Galen Rowell BCaL; **SPL:** Dr Ken MacDonald BL. **108/109 Corbis:** Yann Arthur-Bertrand B. **109 Corbis:** Hubert Stadler TCb; **Ontario Science Centre:** BR. **110 Corbis:** James Randklev TR; **Geo:** B. **111 alamy.com:** Jon Arnold Images / James Montgomery CaL; **Corbis:** David Muench BCR; Paul A Souders TCR; Royalty-Free BR; Wolfganf Kaehler CbL; **SPL:** Adam Hart-Davis CRa. **112 Association of American Geographers:** CL; **Corbis:** David Muench Cb; Michael S Yamashita BR; Paul A Souders BL. **112/113 Corbis:** Royalty-Free C. **113 Corbis:** Galen Rowell TR; Jon Sparks BRa; Wolfgang Kaehler CRa; **Geo:** BCR. **114 Corbis:** Charles Krebs Photography CL; Lowell Georgia BC; Travel Ink / Derek M Allan TRb; **SPL:** William Ervin Cb. **114/115 Corbis:** Yann Arthus-Bertrand B. **115 NASA:** Johnson Space Center (STS026-43-98) TR; Johnson Space Center / Image courtesy of Earth Science and Image Analysis Laboratory (STS043-22-23) BCR; Johnson Space Center / Image courtesy of Earth Science and Image Analysis Laboratory (STS41D-32-14) BCL; **SPL:** Dr Juerg Alean CbR; **Geo:** TCb. **116 Corbis:** Hans Strand Ca; Lloyd Cluff BC; **Geo:** BL. **116/117 Corbis:** SP Gillette C. **117 Corbis:** Jon Sparks BC; ML Sinibaldi BRa; **DisasterMan Ltd:** Bill McGuire CR; **Geo:** TR, CaL. **118 Corbis:** David Lees C; Natalie Fobes BC. **119 Rex Features:** Sipa Press. **120 Corbis:** Roger Ressmeyer CbL; **DK:** Harry Taylor BR; **Getty:** Photodisc CLa; **Images Of Africa Photobank:** Vanessa Burger CL; **SPL:** Dr David King BC. **120/121 Corbis:** Yann Arthur-Bertrand C. **121 Corbis:** Roger Ressmeyer TC; Sygma / Duthei / Didier BCRa; **Department of Geology, University of Witwatersrand:** BCL. **122 Aurora & Quanta Productions Inc.:** Peter Essick CL; **Corbis:** Paul A Souders BL; Royalty-Free BR; **NASA:** Haughton-Mars Project 2002 TCL; Marshall Space Flight Center (MSFC-0201920) CR. **123 AirPhoto:** Jim Wark Ca; **The Barringer Crater Company:** TRb; **Corbis:** NASA TC; Image courtesy of **Virgil L Sharpton, University of Alaska – Fairbanks / Lunar and Planetary Institute and NASA:** Cb; **SPL:** Prof Walter Alvarez BR. **124 Corbis:** Michael S Lewis Cb; **Landsat:** CR; **NASA:** Johnson Space Center (STS51I-33-56A) BCL; **Stadt Noerdlingen:** TL. **125 Geological Survey of Western Australia:** BL; **Global Impact Studies Project:** TR; **NASA:** Johnson Space Center (STS41D-14-41-028) CRb; **Novosti (London):** CLA; **Skyscan:** photo by skycam.com.au BR; Geo: Ca. **126 Corbis:** Lowell Georgia BCL; Natalie Fobes Cb; Ricki Rosen BLA; Science Pictures Limited CLa; **126/127 Getty:** Image Bank / Harald Sund C. **127 Corbis:** David Muench TC; Galen Rowell CRa; Papilio / Bob Marsh BCR; **OSF:** Ben Osbourne CRb; **SPL:** David Parker C. **128 Corbis:** Tom Bean TCR; **DK:** Brian Cosgrove Collection CaL, TLb; **Getty:** Taxi / Andrew Mounter B. **129 Corbis:** Doug Wilson TCR; Sygma / Aim Patrice BRa; **Galaxy Contact:** CLa; **Getty:** Stone / David Muench CLb. **130 alamy.com:** Peter Bowater C; **Corbis:** Tony Roberts CLb. **131 Still:** UNEP / A Ishokon. **132 Corbis:** Michael & Patricia Fogden CL; **DK:** BL (*vertebrate*); Christine M Douglas BL (*non flowering*) pl; Dave King TR; Kim Taylor BL (*invertebrate*); Neil Fletcher BCL (*fungi*); Steve Wooster BCL (*flowering plant*); **OSF:** CbL; Gary Gaugler / OKAPIA BC (*monerans*); London Scientific Films BC (*protist*); SPL: CDC CaL. **133 Gerald D Carr:** Department of Botany, University of Hawaii TC; **DK:** Colin Keates CLa, TLb. **OSF:** Richard Packwood B. **134 alamy.com:** Jan Baks CLb; **Corbis:** Agliolo / Sanford CLa; Eric and David Hosking BRa; **DK:** Geoff Dann BL; Ken Findlay BC; Peter Anderson BCL; **FLPA:** David Hosking CL. **135 alamy.com:** David Boag CRb. **136 Corbis:** James L Amos BCa; Tom Bean BL; Wolfgang Kaehler TCb. **137 FLPA:** Minden Pictures/Frans Lanting. **138/139 alamy.com:** ImageState / Adam Jones. **140/141 Corbis:** Roger Ressmeyer. **142 Corbis:** Bettmann CRa; Roger Ressmeyer BCL; **DK:** Joe Cornish CaL; **James Jackson:** Department of Earth Sciences, University of Cambridge CLb. **142/143 Corbis:** Yann Arthur-Bertrand B. **143 Martin Miller:** Department of Geological Sciences, University of Oregon TL; **SPL:** Fred McConnaughey TR; Massonnet et al / CNES BRa. **144 Corbis:** TCL; Bettmann CLb; Lloyd Cluff B; **School of Earth Sciences, Stanford University:** TRb. **145 Corbis:** BRa; Progressive Image / Bob Rowan TR **Martin Miller:** Department of Geological Sciences, University of Oregon C, BCR, CLa. **146** Reproduced by permission of the **British Geological Survey** © NERC All Rights Reserved IPR/38-31c CL, BL; **SPL:** Robert Brook CR; **Art Directors & TRIP:** C Sanders BR; **WWI:** Patricia McDonald TR. **147 Agence France Presse:** Sergei Chirikov Ca; **FLPA:** Walter Rohdich Cb; **NASA:** Johnson Space Center (STS41G-34-6) TCR; **SOA Photo Agency:** Blue Box BR. **148 alamy.com:** ImageState / Georgette Douwma C; **Corbis:** Bettmann BC; Stephen Frink Ca; **DK:** NHM / Harry Taylor BL; **Images Of Africa Photobank:** David Keith Jones CL; **NG:** Chris Johns TR. **149 OSF:** NASA / Johnson Space Center (STS40-152-180). **150 Corbis:** Michael S Yamashita. **151 NG:** Reza BCR; **SPL:** James Stevenson TRb. **152 Corbis:** Lloyd Cluff CLb; **Art Directors & TRIP:** TCb; M Fairman BCL; **Visibile Earth:** Image courtesy Jacques Descloitres, MODIS Rapid Response Team, NASA/GSFC TR. **153 Institute of Geological and Nuclear Sciences, New Zealand:** TC, CR, B. **154 Corbis:** Galen Rowell TR, Yann Arthus-Bertrand C. **154/155 Corbis:** Galen Rowell B. **155 Corbis:** CRb; NASA BCR; Yann Arthus-Bertrand TR. **156 Corbis:** Hulton-Deutsch Collection CL; **Getty:** Stone / Joe Cornish BL; **Magnum:** Stuart Franklin TCLb; **SI:** CR. **157 Bruce Coleman Ltd:** Johnny Johnson TCL; **Corbis:** Galen Rowell Ca; Lester Lefkowitz B (main); **RHPL:** Patrick Endres BCL; Scott Darsney TCR; **Still:** Peter Arnold Inc / Alan Maichrowicz TR. **158 Corbis:** Ecoscene / Graham Neden CR; Galen Rowell C; Hubert Stadler CRa, CRb; Jonathan Blair BCL; Lynda Richardson TCL; NASA CLb; William Manning CLa. **159 AKG London:** Alte Nationalgalerie, Berlin TRb; **Corbis:** Brian

Vikander BCR; **DK:** TL; **South American Pictures:** Kathy Jarvis B (main); **Still:** Alan Watson TCR. **160 Bruce Coleman Ltd:** Rob Jordan Cb; **Corbis:** James L Amos CRa; John Noble BR; **OSF:** Konrad Wothe / SAL CLa. **161 Corbis:** Jon Sparks Ca; Michael Busselle CL; Roger Ressmeyer BL; **FLPA:** BB Casals B (main); J&C Sohns TR; R Wilmshurst TRb. **162 Corbis:** Galen Rowell TCLb; Patrick Johns TCLa; Randy Wells BRa; Sandro Vannini BL, BC; **P A Photos:** European Press Agency CbL; **SPL:** TR. **163 Getty:** Taxi / Pascal Tournaire. **164 Corbis:** A Alamary & E Vicens CLa; Bojan Brecelj TR; Natalie Fobes CbL; **NG:** John Eastcott and Yva Momatiuk BL; Randy Olsen CRa; **Konstantin Mikhailov:** BR. **165 Corbis:** Christine Osborne CaL; Richard Bickel CRa; Yann Arthus-Bertrand CLa, CbL; **FLPA:** Panda Photo TC; **NG:** Steve McCurry B. **166 Corbis:** CLa; Earl & Nazima Kowall CaL; Gallo Images / Roger de la Harpe TCR; **DK:** Frank Greenaway CbR; Ken Findlay TCRb; **FLPA:** Fritz Polking BR; **NPA Group:** CbL; **OSF:** JC Stevenson / AA BL. **167 alamy.com:** Michael Grant BL; **Corbis:** NASA BCL; **DK:** Gables CRa; **Getty:** Taxi / Walter Bibikow T; **RHPL:** Gavin Hellier Cb. **168 Corbis:** Bettmann BL; David Keaton TRb; Galen Rowell TCL; Howard Davies TCb. **168/169 Mountain Images:** Ian Evans B; **169 Corbis:** Ecoscene / Robert Weight TR; **NASA:** Johnson Space Center (STS41G-120-22) CL; **OSF:** Dinodia Picture Agency TCL. **170 Corbis:** Eric & David Hosking CRb; **DK:** Rob Reichenfeld CLa; **Getty:** Image Bank / John William Banagan TR; **NG:** Paul Chesley B; **WWI:** Joe Cornish CR. **171 British Antarctic Survey:** David Cantrill CRa; **Corbis:** Galen Rowell B, TCLb; **FLPA:** Eric & David Hosking TCR; **OSF:** Kim Westerskov Ca. **172 Corbis:** NASA CL. **172/173 Bruce Coleman Ltd:** Gerald S Cubitt C. **173 Corbis:** Gary Braasch Cb; Roger Ressmeyer CR; **DK:** Rob Reichenfeld C; **SI:** CRb; Ca; **Geo:** TCR, TRb. **174 Agence France Presse:** TC; **Corbis:** TRb; Yann Arthur-Bertrand CaL; **US Geological Survey:** JD Griggs B; **Geo:** Ca. **175 Agence France Presse:** Romeo Gacard CLa; **Corbis:** Sygma / Patrick Robert CR; **Geo:** TR, CL. **176 Corbis:** David Muench BL; Terry W Eggers CRa; **NG:** Bates Littlehales BR; **Geo:** Cla, CRb. **177 Corbis:** Gary Braasch B (main); **RHPL:** Calvin W Hall CR; **NG:** James C Richardson TRb; **Gary Rosenquist:** B (inset); **SI:** TCb; **US Geological Survey:** TC. **178 Corbis:** BLa; Roger Ressmeyer TL, Ca, CRb, BR; **NG:** James Amos TRb. **179 Corbis:** Nik Wheeler BLa; Sygma / Chloe Harford BR; **NG:** Sarah Leen Ca, TCb; **Rex Features:** Sipa Press CRb; **SI:** Olger Aragon BC; William I Rose, Michigan Tech University, Department of Geological and Mining Engineering and Science CRa. **180 Corbis:** Galen Rowell BL; Jeremy Horner CL; Roger Ressmeyer CRa; **Rex Features:** Sipa Press TCb; **SI:** Oscar Gonzalez-Ferran, Departmento de Geolgia y Geofisica, Universidad de Chile CRb; **US Geological Survey:** CLa. **181 Corbis:** Ric Ergenbright TR; Roger Ressmeyer CRa; **Magnum:** Patrick Zachmann BL; **Geo:** BR, TCL; **Mats Wibe-Lund:** CL. **182 Agence France Press:** TCR; **Corbis:** Roger Ressmeyer CL; **Getty:** Image Bank / Guido Alberto Rossi CbL; **Tom Pfeiffer:** B (inset). **183 Still:** Otto Hahn. **184 Corbis:** Yann Arthur-Bertrand. **185 Corbis:** Michael S Yamashita CRa; Ric Ergenbright CbL; Roger Ressmeyer CRb. **186 Corbis:** Bettmann TR; Roger Ressmeyer CLa, BR, TCL; **NG:** Sisse Brimberg CRb; **N.H.P.A.:** Kevin Schafer CLb. **187 Corbis:** Ecoscene / Chinch Gryniewicz Cb; Torleif Svensson B; Yann Arthus-Bertrand CLa; **Magnum:** Bruno Barbey TR, CRa; **NG:** George F Mobley CRb. **188 alamy.com:** RHPL BL; **Corbis:** Peter Turnley TC; **Katz/FSP:** Gamma CLa; **NG:** Chris Johns CRb; Medford Taylor BR; **SI:** Krafft Collection CRa. **189 Corbis:** Yann Arthur-Bertrand CLa; **NG:** Steve Raymer TR; **OSF:** NASA BR; **SI:** Alexander Belousov BC. **190 Bruce Coleman Ltd:** Orion Press Ca; **Corbis:** Michael S Yamashita BC; **Magnum:** Chris Steele-Perkins CLa; **SI:** Liu Xiang, Changchun University of Science and Technology, Department of Geology BL; **US Geological Survey:** Jack Lockwood BR. **191 Corbis:** Roger Ressmeyer CaR; Saba / Alberto Garcia C; **Katz/FSP:** Van Cappellen / Rea TCR; **SI:** Chris Newhall, Department of Earth and Space Sciences, University of Washington BR; Rizal Dasoeki / Volcanological Survey of Indonesia Cb. **192 Corbis:** Dean Conger TC; Roger Ressmeyer CLa; **NG:** Paul Chesley CR. **Popperfoto:** CRa; **SI:** Krafft Collection BL; **Still:** Alain Compost BR. **193 Corbis:** Galen Rowell C; **Hulton Archive/Getty:** CRb; **OSF:** Doug Allan B; Tarnmy Peluso TCL; Tui de Roy TR. **194 Geo:** BCL, Bla; **PJ Fleisher:** R; **Martin Miller:** Department of Geological Sciences, University of Oregon CbL. **195 Martin Miller:** Department of Geological Sciences, University of Oregon Cb; **PJ Fleisher:** CRa; **Dr Parvinder S Sethi:** BCL; **SPL:** Alfred Pasieka TCL; **Geo:** CRb. **196 Jens C Andersen:** Camborne School of Mines CLa; **Corbis:** Charles E Rotkin Cb; Kelly Mooney Photography BL; **Kurt Hollocher:** Union College TC; **Geological Survey of Canada:** Dr Robert H Rainbird CRa. **197 Corbis:** Bill Ross C; Galen Rowell TR; Phil Shermeister CRa; Robert Holmes BL. **198 AirPhoto:** Jim Wark TL; **Corbis:** Charles O'Rear BRa; David Muench CRa; Jonathan Blair C; **Still:** Peter Arnold Inc / Robert MacKinlay BL. **199 alamy.com:** Neil Cameron CRa; RHPL CLa; **Art Directors & TRIP:** T Bognar BC; **Geo:** Tim Fogg BRa; Reproduced by permission of the **British Geological Survey** © NERC All Rights Reserved IPR/39-5c BL. **200 Corbis:** Yann Arthus Bertrand BL, BC; **DK:** Joe Cornish CLa; **Art Directors & TRIP:** B Gadsby TCL; **METI / ERSDAC:** TR. **201 Getty:** Photodisc Green CR; **ImageState/Pictor:** Premium Stock BL; **Geo:** BRa; **National Park Service:** BC. **202 Corbis:** Jeff Vanuga CL; Kevin R Morris BL; **Still:** Peter Arnold Inc / John Keiffer TC; **WWI:** Mark Hamblin Ca. **202/203 Corbis:** Raymond Gehman C. **Still:** Peter Arnold Inc / John Keiffer TC; **WWI:** Mark Hamblin Ca. **203 Getty:** Image Bank / Michael Melford CRa. **204 Corbis:** Charles O'Rear BC; Dave G Houser BL; Nik Wheeler BR; **Getty:** Stone / Jack Dykinga Ca; **NG:** E.C. Kolb CLa. **205 Corbis:** Macduff Everton TR; **SPL:** Klaus Guldbrandsen BC; **Geo:** BR; **WWI:** Andreas Leemann Ca. **206 Corbis:** Eye Ubiquitous / Mike Powles CaL; Michael Freeman CRa; Michael S Yamashita BL; Robert Holmes Cb, BR. **207 Getty:** Stone / John Lamb C; Taxi / Travel Pix BL; **Geo:** CRb, BR; **Waimangu Volcanic Valley Ltd, New Zealand:** TR, TLb. **208 Corbis:** Roger Ressmeyer. **209 Corbis:** Bob Krist Cb. **210/211 Corbis:** Yann Arthus-Bertrand. **212 Getty:** Stone / Hans Strand BL. **212/213 Corbis:** Lester Leftowitz C. **213 Getty:** Image Bank / China Tourism Press Bca; **RHPL:** Simon Harris CR; **Still:** Peter Arnold Inc / Aldo Brando CRb; **Louise Thomas:** C; **Geo:** BR. **214 © 2002 Environment Agency, National Centre for Environmental Data & Surveillance:** TR; **NG:** Paul Nicklen Cla; Philip Schermeister TCLb. **214/215 Getty:** Image Bank / China Tourism Press B. **215 Corbis:** Nathan Benn CL; **Getty:** Image Bank / NASA Cb; **Still:** Peter Arnold Inc / Jim Wark TRb; **Visible Earth:** NASA / GSFC / JPL, MISR Team BRa. **216 Corbis:** Charles E Rotkin BL; Lowell Georgia TR; Paul A Souders CLa; **RHPL:** Maurice Joseph CRb; **FLPA:** Steve McCutcheon TCL. **217 AirPhoto:** Jim Wark TCL; **Corbis:** Richard Hamilton Smith CL; **Still:** Peter Arnold Inc / Alex S MacLean B; **Telegraph Herald:** CaR. **218 AirPhoto:** Jim Wark TC. **Corbis:** Annie Griffiths Belt C; Buddy Mays BC; **Russ Finley Stock Photography:** BL. **219 Aurora & Quanta Productions Inc.:** Peter Essick. **220 Getty:** Image Bank / Jack Dykinga TR; Image Bank / Michael Melford TC; Stone / Donald Nausbaum B; **Landsat:** CR. **221 Corbis:** Bettmann BC; James Marshall R; **DK:** Cyril Laubscher CLb; **WWI:** David Woodfall BL. **222 Corbis:** Yann Arthus-Bertrand BCLa, TCL; **Getty:** Stone / Will & Deni McIntyre TCR; **South American Pictures:** Tony Morrison CL, CbL. **223 Still:** Luiz C Marigo. **224 Corbis:** Adam Woolfitt BR; Eye Ubiquitous / Tim Hawkins CaL; **Getty:** Image Bank / Daniel Barbier CR; Taxi / David Noton BL. **225 Corbis:** Papilio / Pat Jerrold CbR; **Getty:** Image Bank / Werner Dieterich BC; **RHPL:** Michael Busselle BL; **Magnum:** Thomas Hoepker C; **P A Photos:** European Press Agency TRb. **226 Corbis:** Joe McDonald BC; Lester Lefkowitz CL. **227 Still:** Jim Wark. **228 Agence France Presse:** Pavel Neubauer CL; **Getty:** Stone / Simeone Huber TR; **Landsat:** CRb; **Russia and Eastern Images:** Mark Wadlow BR. **229 DK:** Peter Hiscock TC; **Getty:** Image Bank / NASA B; **Images Of Africa Photobank:** David Keith Jones TR; **NG:** David S Boyer CR; **Still:** Chris Caldicott CL. **230 Aurora & Quanta Productions Inc.:** Robert Caputo / IPN B, TCL; **Corbis:** Dave G Houser TR. **231 Corbis:** CL; Ed Kashi BRa; Galen Rowell CR; Nik Wheeler B; **NG:** Chris Johns TR. **232 Corbis:** Dean Conger B; How-Man Wong TC; Tom Nebbia TRb; **Still:** Dera CR; Mark Carwardine Ca. **233 Corbis:** Liu Liqun BL; Michael Freeman CR; Michael S Yamashita BR; Photowood Inc TCL; **RHPL:** Gina Corrigan CL. **234 Getty:** Taxi / Gavin Hellier B; **RHPL:** CL; **NASA:** Johnson Space Center (STS087-707-092) TR; **Still:** Andre Maslennikov TCL; K McCullough C. **235 Australian Portraits:** David Hancock BR; **Corbis:** Rik Ergenbright CL; **DK:** Cyril Laubscher BCL; The British Museum, London / Peter Hayman TRb; **RHPL:** TC. **236 Getty:** Image Bank / Art Wolfe B; **NG:** Emory Kristof Ca; **Geo:** CR; **SC Porter:** University of Washington Cb. **237 DK:** Dave King TR; **Mike Grandmaison:** BR; **RHPL:** Paolo Koch Ca; **FLPA:** GT Andlewartna BL. **238 Corbis:** Visions of America / Joseph Sohm CLb; **Getty:** Stone / Vito Palmisano BL; **NG:** Medford Taylor TC; Raymond K Gehman CaR. **239 NG:** Medford Taylor. **240 alamy.com:** Masrdis CaL; **Corbis:** Scott T Smith BR; **RHPL:** S Grandadam CaR; **Leland Howard:** BL. **241 Corbis:** Craig Lovell CR; Kevin Schafer TR; Ronan Soumar B; **South American Pictures:** Tony Morrison CL. **242 Still:** Glen Christian. **243 Corbis:** Ian Harwood / Ecoscene CbL; **Greenpeace:** Shailendra Yashwant CR; **Still:** Gilles Corniere TCbR. **244 DK:** Paul Harris BC; Steve Shott CLA; **FLPA:** F Ardito / Panda TR; **Art Directors & TRIP:** O Semenenko BL; **WWI:** David Woodfall CRb. **245 Corbis:** Blaine Harrington III TRb; Bo Zaunders B; **Getty:** Image Bank / Macduff Everton CR; Image Bank / Werner Dieterich TCL. **246 Corbis:** Bernard and Catherine Desjeux TCb; Jonathan Blair BL; **Images Of Africa Photobank:** David Keith Jones CR, BR; **OSF:** Richard Packwood TR. **247 Corbis:** Caroline Penn Ca; **Getty:** Image Bank / Guido Alberto Rossi BL; **Images Of Africa Photobank:** David Keith Jones TR; Friedrich von Hörsten BR. **248 Corbis:** Galen Rowell BC; Hulton-Deutsch Collection TRb; **DK:** Jerry Young TC; **NG:** Barry Tessman BRa; Sarah Leen BL; **OSF:** Mark Deeble & Victoria Stone Ca. **249 Corbis:** David Turnley TL, CR; Yann Arthur-Bertrand B. **250 Aurora & Quanta Productions Inc.:** Evan Roberts BR; **Corbis:** Diego Lezama Orezzoli TR; Mark Garanger TCLb; NASA BL; **OSF:** Tony Bomford CaL. **251 Corbis:** Galen Rowell BL; Ted Spiegel BR; **RHPL:** CCD Tokeley BC.; **ImageState/Pictor:** David South TR; **NG:** Reza Ca. **252 Geo:** CLb, TRb. **252/253 Tom Till Photography:** B; **253 Paul Deakin:** TCb; **Geo:** CRa, CR, CbR; **Jenolan Caves Reserve Trust:** TCL. **254 Corbis:** Craig Lovell TR; David Muench CR; **Russ Finley Stock Photography:** TCL; **OSF:** Rodger Jackman CL; **Terra Galleria Photography:** T Luong B. **255 Aurora & Quanta Productions Inc.:** Stephen Alvarez C; **Corbis:** Bojan Brecelj CR; **NG:** Wes Skiles BL; **OSF:** David M Dennis TCR; **Arthur Palmer:** CL; **Geo:** BR. **256 Caving Club of Touraine:** Jean-Luc Roch BC; **RHPL:** TRb. **NG:** Sisse Brimberg Ca; **Geo:** CLa, BL, BR. **257 Cango Caves:** Steve Mouton BL, CbL; **Geo:** TC, BR, CaL. **258 Corbis:** Robert Holmes CR; **Getty:** Image Bank / Nevada Wier TCL; **OSF:** Clive Bromhall CLb; **Geo:** TR, B. **259 Geo:** CLa, CRa; **Jenolan Caves Reserve Trust:** CRb, BR; **Koonalda Caves:** Peter Bell BL. **260/261 SPL:** Bernhard Edmaier. **262 Corbis:** David Muench BCL; Marc Muench Ca; **RHPL:** Andrew Sanders BL. **262/263 Corbis:** Tom Bean C. **263 Corbis:** Phil Schermeister BCR; **Getty:** Image Bank / Darrell Gulin CLa; **Greenpeace:** BRaL,CRb; **RHPL:** Tony Waltham TR. **264 AirPhoto:** Jim Wark B; **Corbis:** CR; **Geo:** CL, CRR; **Tom Lowell:** University of Cincinnati CLL. **265 AirPhoto:** Jim Wark CRa; **Geo:** TR, Cb, BLA, TCL. **266 Corbis:** Tom Bean CRa, BR; **RE Johnson Photography:** CRb; **Dan Stone:** CLa; **US Geological Survey:** Rod March BL. **267 Anthony Arendt:** University of Alaska CR; **Jamie Buscher:** Department of Geological Sciences, Virginia Tech R; **Corbis:** Neil Rabinowitz CLa; **Still:** Peter Arnold Inc / Jim Wark B. **268 Corbis:** Danny Lehman TR; David Muench CLa; David Samuel Robbins CRa; Jim Zuckerman BL; The Purcell Team BR. **269 Corbis:** Tom Bean Ca; **NASA:** Goddard Space Flight Center / USGS EDC B; **Still:** Peter Arnold Inc / Jim Wark TR. **270 B&C. 271 Jason E Box:** Polar Research Center, Ohio State University CRa; **Corbis:** Staffan Widstrand Cb; **Geo:** BR. **272 Josh Beck:** BL; **Peter Burden:** CLa; Simon Heyes, **Last Frontier Ltd:** BCL; **NASA:** Goddard Space Flight Center BR; **Still:** Peter Arnold Inc / SJ Krasemann CRa; **Nozomu Takeuchi:** Research Institute for Humanity and Nature (RIHN), Kyoto TCbR. **273 Corbis:** Eye Ubiquitous / James Davies C; Yann Arthus-Bertrand TR; **NG:** Steve Winter BL; **Freysteinn Sigmundsson:** Nordic Volcanological Institute, Reykjavik BR; **Still:** Alan Watson TL. **274 Kim Holmén:** Department of Meteology, Stockholm University Ca; **Dr Miriam Jackson:** Norweigan Water Resources & Energy Directorate Hydrology Department BR; **Dr John Wood:** Department of Geography and Earth Sciences, Brunel University TR; **Jøran Zahl Marken:** Bca. **275 Corbis:** Marc Garanger CaR; **Still:** Peter Arnold Inc / Bill O'Connor BL; Peter Arnold Inc / Gordon Wiltsie TCLb; Peter Arnold Inc / Helmut Gritscher BR. **276 Corbis:** NASA BC; **Jamie McPherson:** BL; **Klaus E Schwartz:** BR; **Still:** Theresa de Salis TC, CLa. **277 Corbis:** Galen Rowell BL. **Thomas E Dietz:** International Society for Mountain Medicine CR; **NG:** Bobby Model TL. **278 Corbis:** Yann Arthus-Bertrand TCL. **278/279 Getty:** Image Bank / Joseph van Os B. **279 Corbis:** Galen Rowell TR; Hulton-Deutsch Collection TCL; **SPL:** JG Paren CR. **280 alamy.com:** Stock Connection Inc / James Kay BC; **Corbis:** Paul A Souders TCR. Tom Lowell: University of Cincinnati BR; **Still:** Peter Arnold Inc / Walter H Hodge CLb; **Ulrich Walthert Coast Photography, New Zealand:** BL. **281 David Summers. 282/283 Corbis:** Yann Arthus-Bertrand. **284 Corbis:** Kevin Schafer CL; Scott T Smith CbL; **DK:** Linda Whitwarm C; **Getty:** PhotoDisc BL. **284/285 alamy.com:** Geoffrey Morgan C. **285 Corbis:** Ric Ergenbright CRa; **Getty:** Taxi / Josef Beck Cb; **Geo:** TCL, TCR. **286 Corbis:** CaL; David Muench TLb; Gallo Images / Hein von Horsten BC; Jeremy Horner CaR; **DK:** Alistair Duncan BL; Frank Greenaway Cb; **NG:** Des and Jen Bartlett BCLa. **286/287 WWI:** David Woodfall C. **287 Corbis:** Ecoscene / Andrew Brown TCL; George HH Huey CRa; **DK:** Geoff Dann C. **288 DK:** Dave King CRb; **Getty:** Stone / RGK Photography BC; Taxi / Alan Kearney BL; Jim Harding: CLa. **288/289 Scott T Smith:** T; **289 Agence France Presse:** John Gurzinski / STR CRa; **Corbis:** B.S.P.I. BL; Marko Modic BR; **DK:** Frank Greenaway Cb. **290 D Donne Bryant Stock Picture Agency:** LLC / David Ryan B; **Corbis:** Charles O'Rear CLa, CRa. **SPL:** CNES, Distribution Spot Image tcl. **291 FLPA:** A Christiansen Cb; **Royal Geographical Society Picture Library:** BR; **Still:** Michael Gunther T. **292 Still:** M&C Denis-Huot. **293 DK:** Andy Crawford CaR; Frank Greenaway BR; **Getty:** Image Bank / Jean du Boisberranger BCR; **NG:** Michael S Lewis CbL; **Still:** Cyril Ruoso CLa; **© M.P.F.T.:** TR. **294 Magnum:** Steve McCurry BLA. **295 Still:** UNEP / Voltchev. **296 Corbis:** Gallo Images / Peter Lillie Ca; Michael & Patricia Fogden CbL; Papilio / Robert Gill BL; **Gerald Cubitt:** BR; **DK:** Frank Greenaway TR. **297 DK:** Colin Walton TCL; **RHPL:** Gina Corrigan BR; **Russia and Eastern Images:** Mark Wadlow CRa; **Wax Visual Stock Photography:** BL. **298 Nature Picture Library:** Gertrud Neumann-Denzau TC; **Corbis:** Dean Conger BC; Steve Kaufman BL; **Geo:** Cb. **299 Getty:** Image bank / Art Wolfe. **300 Corbis:** Australian Picture Library BC; Eric & David Hosking CLa; Gallo Images / 318 TR; **WWI:** Ted Meads CRa, BR. **301 Corbis:** Gallo Images / 318 Cb; Michael & Patricia Fogden BL; **DK:** Cyril Laubscher CRa; **WWI:** Ted Mead TL, BR. **302/303 Magnum:** Stuart Franklin. **304 DK:** Matthew Wark CaL; Neil Fletcher & Matthew Ward CLb, CbL; Steve Gorton CaR; **FLPA:** David Hosking CL. **Still:** Roland Seitre BCR.**304/305 Natural Image:** Bob Gibbons C. **305 DK:** Peter Chadwick CaR; **FLPA:** B Borrel Casals TCb; Chris Mattison CaL; Fritz Polking BRa; **Still:** Frank Vidal Cb. **306 Corbis:** Ecoscene / Andrew Brown TCR; **Getty:** Stone / Jane Gifford CL; **Natural Image:** Bob Gibbons B; **Rex Features:** Sipa Press TCLb; **World Wildlife Fund:** MedPo / Pedro Regato CaL. **307 Corbis:** Jonathan Blair BCL; Raymond Gehman BL, BC; **DK:** Eric Crichton TLb; Neil Fletcher CaL; Peter Anderson TCR; **FLPA:** M Moffett / Minden Pictures CL; Martin B Withers CaR; **NASA:** Johnson Space Center (STS41G-45-40) CRa; **Art Directors & TRIP:** W Jacobs CLb. **308 Corbis:** Gunter Marx Photography CR, TRb; **FLPA:** Mark Newman B; **OSF:** Lon E Lauber TCL. **309 Corbis:** Bob Rowan / Progressive Image CRb; Galen Rowell TRb; Marc Muench BL; **Still:** Art Wolfe Ca; David Drain TC; **Art Directors & TRIP:** M Barlow BCR. **310 Corbis:** David Muench TCL; Mark Muench CLa; **DK:** Jerry Young BL; **OSF:** CRA, CRb; **WWI:** Alan Watson BR. **311 Corbis:** Brian Vikander CL; **DK:** Jerry Young TR; Pau D'Arco CR; **South American Pictures:** Tony Morrison TCR; **Still:** Juan Carlos Munoz TCL; **WWI:** David Woodfall B. **312 Corbis:** Macduff Everton CLb; Wayne Bennett BC. **313 Corbis:** Robert Semeniuk. **314 FLPA:** Roger Tidman BC; T de Roy / Minden Pictures CR; **Natural Image:** Bob Gibbons BL, BR; **South American Pictures:** Tony Morrison TL; **Still:** Alan Watson Ca. **315 Corbis:** Adrian Arbib T; Michael Busselle BL; **DK:** Neil Fletcher BCL; **Getty:** Stone / Stephen Studd C; **Natural Image:** Bob Gibbons BR; **Rex Features:** Isopress CR. **316 NG:** Maria Stenzel BCL; **OSF:** Daniel Cox CbL; **Russia and Eastern Images:** Mark Wadlow TC; **Still:** Patrick Bertrand Ca; Roland Seitre R. **317 Corbis:** Keren Su CR; **Gerald Cubitt:** BL; **DK:** Kim Taylor CL; **Nigel Hicks:** BR; **OSF:** Andrew Plumptre TR; **Still:** Michael Gunther TCL. **318 FLPA:** Frans Lanting. **319 Associated Press AP:** Yenni Kwok, Stringer bc; **Getty:** Stone / Manoj Shah TR; **RHPL:** Louise Murray BR; **SPL:** Geoff Tompkinson C. **320 Corbis:** Hulton-Deutsch Collection BC; Paul A Souders CbR; Peter Johnson BR; **DK:** NHM / Frank Greenaway TCR; **FLPA:** Terry Whittaker CRa; Tim Rushforth BL. **Still:** Mark Edwards TR; **World Wildlife Fund:** Steve Nelson & Zovtaigi c/o World Wise Ecotourism Network CLa. **321 Corbis:** FLPA / Pam Gardner CRa; Paul A Souders L; Yann Arthus-Bertrand Cb; **N.H.P.A.:** Dave Watts TR; **Skyrail, Cairns, Australia:** BR. **322/323 WWI:** Ted Mead. **324 Corbis:** David Muench BL; Ecoscene / Ian Harwood CaL; Nik Wheeler TCb; **NG:** courtesy Moesgård Museum, Arhus, Denmark / Ira Block TRb. **325 alamy.com:** Steve Bloom Images / Steve Bloom B; **Corbis:** FLPA / Fritz Polking CaL; **DK:** Frank Greenaway TLb; Geoff Dann TC; Kim Taylor & Jane Burton TR; Philip Dowell CaR; **FLPA:** Tony Hamblin CR. **326 Corbis:** Patrick Johns Cb; Raymond Gehman CL, BL; **DK:** Karl Stone BR; **Dave Liebman:** TC; **Virginia Museum of Natural History:** Susan B Felkner TR. **327 AirPhoto:** Jim Wark CbR; **Corbis:** David Muench BL; Kevin Fleming BR; Tony Arruza BC; **NG:** Farrell Greham T. **328 Corbis:** Tom Brakefield CLa; **FLPA:** Jurgen & Christine Johns CRb; **OSF:** Chris Catton CRa; Niall Benvie BR; **Still:** Gunter Ziesler TCL; Roland Seitre TRb; **WWI:** M Powles BL. **329 Getty:** Image Bank / Art Wolfe BL; **FLPA:** Terry Andrewartha TCL; **Natural Image:** Bob Gibbons CLA; **OSF:** John Cooke Cb; Roland Mayr TR; **Still:** Nigel Dickinson BRa; Peter Arnold Inc / Gunter Ziesler CRa. **330 Corbis:** Wolfgang Kaehler TR; Yann Arthus-Bertrand CLA; **FLPA:** Minden Pictures / Frans Lanting B; **Still:** Paul Springett CRA. **331 Aurora & Quanta Productions Inc.:** Robert Caputo / IPN CL; **Australian Portraits:** David Hancock BR; **Corbis:** Gallo Images / 318 TC; **DK:** Andrew Butler CR; Exmoor Zoo / Peter Cross CbR; **NewsPhotos:** Chris Crerar BCL. **332/333 Still:** Fritz Polking. **334 Corbis:** Michael S Yamashita CL; Peter Johnson BCLa; Tom Bean CbL; **OSF:** Mills Tandy Ca; Rafi Ben-Shahar BCR; **South American Pictures:** Tony Morrison BL, TCb. **334/335 Corbis:** Tom Bean B. **335 Corbis:** D Robert & Lorri Franz CaR; Staffan Widstrand CR; W Perry Conway TRb; **DK:** Frank Greenaway TC; Jerry Young CaL; NHM / Frank Greenaway CLa.

**336 Corbis:** David Muench B; Joe McDonald TCL; Tom Bean CR; **FLPA:** Daphne Kinzler CL; **OSF:** Scott Camazine CLL; **Still:** Roland Seitre TR. **337 Corbis:** CRb; Macduff Everton C; **DK:** Jerry Young TLb; **South American Pictures:** Mike Harding B; **Still:** Roland Seitre TRb. **338 Corbis:** Sygma TRb; Wolfgang Kaehler TC; **Art Directors & TRIP:** J Farmar CLa. **338/339 Corbis:** Brian A Vikander B; **339 DK:** Mike Dunning TL; **Images Of Africa Photobank:** David Keith Jones TRb; **FLPA:** Terry Andrewartha TC. **340 DK:** Jerry Young CL; **Russia and Eastern Images:** Mark Wadlow CRa; **Still:** Adrian Arbib B; Stephen Pern TCR. **341 Corbis:** O Alamany & E Vicens BR; **DK:** Geoff Dann BCL; **FLPA:** E&D Hosking C; **Natural Image:** Bob Gibbons TCb; **NG:** Thad Samuels Abell II Cb; **Still:** Mark Edwards TR. **342 AirPhoto:** Jim Wark B; **Geo:** CRa, C, CaR. **343 B&C:** TR, BL, BCR; **Corbis:** Tom Brakefield CL; **Hutchison Library:** Andrey Zvoznikov CRb; **NG:** Rich Reid CRa. **344/345 Corbis:** Keren Su. **346 Corbis:** Sylvain Saustier CbL; W Wayne Lockwood M.D. BCR; **NG:** Maria Stenzell CLa. **346/347 Corbis:** Charles O'Rear C. **347 Corbis:** Kennan Ward CR; Pallava Bagla TRb; Yann Arthus-Bertrand BC, TCb; **Getty:** Stone / Bruce Forster C. **348 Corbis:** Joe McDonald CaL; Peter Johnson CL; **NG:** Jodi Cobb BCL; Maria Stenzel TR; **Still:** Nigel Dickinson BR; **Geo:** CR. **349 Corbis:** Fulvio Roiter CLa; Macduff Everton CRa; Michael Freeman TR; Richard Hamilton Smith TCLb; **Images Of Africa Photobank:** Ivor Migdoll BR; **NG:** Michael Nicols BL; **Still:** UNEP / C.K. Au CRb. **350 Corbis:** Macduff Everton CLa; **NG:** William Albert Allard CRa; **SPL:** Francois Sauze CRb; Gordon Garradd TR; **Geo:** BR. **351 Corbis:** David Muench BL; Dean Conger Cb; Jonathan Blair TCL; Morton Beebe, SF TR; Pablo Corral BR; **Images Of Africa Photobank:** Friedrich von Hörsten Ca; **SPL:** Kaj R Svensson CRb. **352 Corbis:** Lanz von Horsten / Gallo Images TC; Photo courtesy of **Deere & Company, Moline, Illinois, USA:** CLb; **DK:** Andrew McRobb *wheat*; Frank Greenaway *rye*; Neil Fletcher & Watthew Ward *rice*; Steve Gorton *barley*, *maize*; **Getty:** Taxi / Walter Bibikow B; **NG:** Kenneth Garrett CRb; **Rex Features:** Ray Tang TRb. **353 alamy.com:** Cephas Picture Library / Mick Rock BL; **Corbis:** Dennis Degnan Cb; Raymond Gehman CLa; Tony Arruza TR; **NG:** James L Stanfield BR. **354/355 NG:** Steve Raymer **356 Corbis:** Michel Setboun Ca; Tom Nebbia BL; **NASA:** Goddard Flight Center / C Mayhew & R Simmon TR. **356/357 Corbis:** Lester Lefkowitz B; **357 Corbis:** Sygma / Kontos Yannis TCb; **Still:** Mark Edwards TR; **Art Directors & TRIP:** ASK Images CRb. **358 Corbis:** Bob Krist TR; Michael S Yamashita CRa. **358/359 Magnum:** Hiroji Kubota B. **359 Corbis:** Owen Franken CRa; **RHPL:** TC; **Magnum:** Dennis Stock TL. **360 Corbis:** Alan Schein Photography Inc CLa; Bettmann TRb; Kelley Mooney Photography TCR; Lloyd Cluff CRb; Sandy Felsenthal Ca; **DK:** Neil Setchfield BR; **Getty:** Taxi / Walter Bibikow BL. **361 Corbis:** Danny Lehman TR; Paul Almasy CR; Stephanie Maze CbR; Yann Arthus-Bertrand B; **Art Directors & TRIP:** ASK Images CRa. **362 Rex Features:** John Sutcliffe BL. **363 Magnum:** Hiroji Kubota. **364 Corbis:** Bojan Brecelj CaR; David Turnley CRb; John & Dallas Heaton BL; Yann Arthus-Bertrand CLa; **Getty:** Image Bank / Jorg Greuel TR; Stone / Chad Ehlers CaL; Stone / John Lamb BR. **365 Corbis:** Craig Aurness BR; Owen Franken BL; WildCountry CLa, Ca; **Getty:** Image Bank / Pete Seaward Photography CbL.; **Art Directors & TRIP:** H Rogers TR. **366 Corbis:** Archivo Iconográfico S. A TC; Nik Wheeler CRa; Yann Arthus-Bertrand BL; **Magnum:** Steve McCurry CLa; **Panos Pictures:** Mark Henley Cb; **Art Directors & TRIP:** A Tovy Ca; J Highet BR. **367 alamy.com:** Peter Bowater BL; **Corbis:** Jeremy Horner BR; **RHPL:** Robert Harding CRa; **Images Of Africa Photobank:** David Keith Jones TCR; **Magnum:** Harry Gruyaert CRb; **Art Directors & TRIP:** M Barlow CLa. **368 Corbis:** Bettmann TRb; Lindsay Hebberd CLa; Ted Spiegel TCR; **Getty:** Stone / Markus Amon CRa; **NG:** Steve McCurry BL; **Rex Features:** Sipa Press CRb, BR. **369 Corbis:** Jose Fuste Raga CLa; Liu Liqun BL; Roger Ressmeyer CRa; Tibor Bognár TC; **RHPL:** P Scholey BR; R McLeod TR; **Still:** Andy Crump CRb. **370 Corbis:** Bob Krist CRa; Ed Wheeler BR; Travel Ink / Derek M Allan CLa; **Panos Pictures:** Jeremy Horner CbL; **Art Directors & TRIP:** P Treanor BL; **WWI:** Nigel Hicks TR, TCLb. **371 Corbis:** Dean Conger Ca; Penny Tweedie BR; Sergio Dorantes BL; **RHPL:** Gavin Hellier TR. **372/373 Magnum:** Miguel Rio Branco. **374 Corbis:** Bohemian Nomad Picturemakers / Kevin R Morris BR; Manfred Vollmer CRb; Michael S Yamashita CRa; **Getty:** Taxi / Benelux Press BL. **375 Corbis:** David H Wells CbL; Derek Croucher CLa; Kevin Schafer TRb; Roger Ressmeyer CRA; Stock Photos TCR; **Magnum:** Gueorgui Pinkhassov BL; **Still:** Thomas Rampack BR. **376 Corbis:** Galen Rowell CLb; Jim Richardson B; **Still:** Harmut Schwarzbach TCL; Thierry Montford CRa; **Art Directors & TRIP:** A Kuznetsov TR. **377 Corbis:** Gunter Marx CLa; Paul Hardy BL; Yann Arthus-Bertrand Ca; **DK:** Harry Taylor TC; **Getty:** Image Bank / Andy Caulfield BR; Stone / Paul Chesley CRb; **Still:** Nigel Dickinson TR. **378 Corbis:** Christopher Morris BL; Michael S Yamashita CL; Natalie Fobes CR; Owen Franken BC; Richard Hamilton Smith TCL; Saba / Keith Dannemiller Cb; Sygma / Bisson Bernard BR; **Magnum:** Patrick Zachmann TCR. **379 Corbis:** London Aerial Photo Library CL; Staffan Widstrand CR; Tim McKenna BR; **Magnum:** David Hurn TRb; Peter Marlow BL. **380/381 Getty:** Image Bank / Peter Cade. **382/383 Getty:** Stone / Arnulf Husmo. **384 Robert Dinwiddie:** CLb, CLLb; **DK:** Andreas Einseidel BCLa; Frank Greenaway TCbR; **© 1999 MBARI:** CaR, CaL. **385 Corbis:** Brian A Vikander TRb. **386 Corbis:** Chris McLaughlin BCL; **FLPA:** S Jonasson Ca; **Richard Lutz:** CLb; **386/387 NG:** Image from Volcanoes of the Deep, a giant screen motion picture, produced for IMAX theaters by the Stephen Low Company in association with Rutgers University. Major funding for the project is provided by the National Science Foundation C. **387 NG:** Paul Chesley CL; **SPL:** ER Degginger CR; **Visible Earth:** Provided by the SeaWiFS Project, NASA/Goddard Space Flight Center, and ORBIMAGE TR. **388 OSF:** David Fox TRb; Howard Hall Ca; Joyce & Frank Burek / AA C; Neil Bromhall Cb. **388/389 Corbis:** Jonathan Blair B. **389 SPL:** Eric Grave CaR; Institute of Oceanographic Sciences / NERC CbR; Jan Hinsch CRa; Manfred Cage TCR; Sinclair Stammers TR. **390 Corbis:** Paul A Souders Ca; **SPL:** Geospace CRb. **391 Japanese National Tourist Organization:** T. **392 Corbis:** Bettmann TRb; Jeffrey L Rotman Cb; Robert Yin Ca; **Southampton Oceanography Centre:** Dr Alex Mustard BCL. **392/393 Corbis:** Jose Fuste Raga C; **393 Corbis:** Yann Arthus-Bertrand TR; **FLPA:** Minden Pictures / Fred Bavendam Cb; **NASA:** Image courtesy of Earth Sciences and Image Analysis Laboratory, Johnson Space Center (STS046-77-31) TCb; **Southampton Oceanography Centre:** Dr Alex Mustard TCL. **394 Corbis:** George Lepp CRb; Paul A Souders BC; Wolfgang Kaehler BCa; **FLPA:** Fritz Polking Cb; **NG:** Maria Stenzel T; **OSF:** Ben Osborne BR; Doug Allan Ca. **395 Corbis:** Rick Price BL; Wolfgang Kaehler BR, BCa. **396 Corbis:** Jon Hicks TR; Onne van der Wal CLa; Stuart Westmorland CL; **DK:** BLa; **FLPA:** Silvestris CLb. **397 B&C:** Wendy Else TL; **Corbis:** Gianni Dagli Orti TRb; Peter Johnson BRA; Stephen Frink CRa; **DK:** Frank Greenaway TCb; **OSF:** David Fleetham CR. **398 B&C:** TCL; **Corbis:** Bettmann BL; **NG:** Emory Kristof TCb; Paul Nicklen CbL; **NOAA:** CL. **399 Getty:** Stone / Hans Strand. **400 alamy.com:** F.R. CLa; **B&C:** BR; **Corbis:** Staffan Widstrand CRa; **NOAA:** V Juterzenka, Piepenburg, Schmid TCL; **Visible Earth:** Provided by the SeaWiFS Project, NASA/Goddard Space Flight Center, and ORBIMAGE BCL. **401 B&C:** TRb; **Corbis:** Bruce Burkhardt CR; Raymond Gehman BL; Wolfgang Kaehler BR; **Visible Earth:** Image courtesy Jacques Descloitres, MODIS Rapid Response Team, NASA/GSFC TCL. **402 Corbis:** Jonathan Blair. **403 Corbis:** Neil Rabinowitz TR; Richard Cummins BCL; **DK:** Frank Greenaway BRa; **NG:** Nick Caloyianis BR; **David Robinson:** Cb. **404 Corbis:** Kevin Schafer CLa; Raymond Gehman BL; The Purcell Team CRb; **Hulton Archive/Getty:** Stock Montage / Archive Photos TRb; **NG:** James P Blair BR; Norbert Rosing Ca. **405 Corbis:** B; Darren Gulin CLb; Kevin R Morris CRb; Lynda Richardson TR; Stephen Frink TC; Tony Arruzal Cb; **Henry Genthe:** Ca. **406 Still:** Fred Bavendam. **407 Corbis:** Phil Schermeister CRb; Stephen Frink Cb; **Getty:** Taxi / Ron Whitby CRa. **408 Corbis:** TRb; Gavriel Jecan TCL; Macduff Everton B; Stephen Frink CR; **NG:** Wes Skiles CL. **409 Gunnar Britse:** windpowerphotos.com CRb; **Corbis:** Chris Hellier BL; Nik Wheeler BR; Patrick Ward CRa; **Geological Survey of Denmark and Greenland:** Peter K Warna-Moors Cb; **Geo:** CLa, Ca; **Visible Earth:** Provided by the SeaWiFS Project, NASA/Goddard Space Flight Center, and ORBIMAGE TCR. **410 Agence France Presse:** TRb; **Corbis:** Jeffrey L Rotman TCL. **410/411 Getty:** Image Bank / Macduff Everton B. **411 Corbis:** Jonathan Blair TRb; **Magnum:** Stuart Franklin TC; **NASA:** Johnson Space Center (STS41G-17-34-081) TL. **412 B&C:** CbR; **Corbis:** Brandon D Cole CbL; Charles & Josette Lenars BR. **412/413 NG:** Gilbert M Grosvenor T. **413 42 Degrees South:** Rob Walls BC. **414 Corbis:** Dean Conger BL; Lawson Wood R; **Robert Dinwiddie:** CLb; **Geo:** TC. **415 Corbis:** Ca; Bojan Brecelj CRa; Kevin R Morris CRb; Yann Arthus-Bertrand BR, TCL; **Karen Hissman:** CbL; **Images Of Africa Photobank:** Brian Charlesworth BL. **416 Corbis:** BC; Sygma / Haruyoshi Yamaguchi BL; Tim McKenna TC; **Robert Dinwiddie:** CL; **Professor Peter Franks:** University of California, San Diego TCb; . **417 Corbis:** Amos Nachoum. **418 alamy.com:** RHPL CRb; **Corbis:** Annie Griffiths Belt BL; Natalie Fobes Ca, TRb; Ralph A Clevenger CL. **419 Corbis:** BL; Joel W Rogers CRa; Michael S Yamashita CLa; Rick Price TCL; Hosik Kim: TR; **Visible Earth:** Provided by the SeaWiFS Project, NASA/Goddard Space Flight Center, and ORBIMAGE BR. **420 Corbis:** Catherine Karnow Ca; Jeffrey L Rotman Cb; Nik Wheeler BL; **P A Photos:** European Press Agency BR; **Visible Earth:** Provided by the SeaWiFS Project, NASA/Goddard Space Flight Center, and ORBIMAGE TR; **© 2003 Norbert Wu:** www.norbertwu.com CLa. **421 Corbis:** FLPA / Pam Gardner CRa; Stuart Westmorland BCL; Yann Arthus-Bertrand BR; **Jezohare.com:** Ca; **NG:** Paul Chesley CLa. **422 Magnum:** Jean Gaumy. **423 Corbis:** Hans Georg Roth CRa; Margaret Courtney-Clarke Cb. **424 Corbis:** Peter Johnson CL; Yann Arthus-Bertrand TCL; **Getty:** Stone / Kim Westerskov B; **SPL:** Simon Fraser TRb. **425 British Antarctic Survey:** D Allan CbL. **Corbis:** Galen Rowell TR, CaL; Rick Price C; Wolfgang Kaehler CRb, BR; **© 2003 Norbert Wu:** www.norbertwu.com BL. **426/427 Geo. 428 Associated Press AP:** Zhan Xiaoding CaR; **Corbis:** Brandon D Cole CaL; **Getty:** Stone / Natalie Fobes BC; Stone / Philip Long BCL. **428/429 Getty:** Stone / H Richard Johnston C. **429 Corbis:** B.S.P.I. TC; Barry Davies TCbL; Sygma / Frederick Astier CaR; **OSF:** Richard Packwood CRb; **Rex Features:** Timepix / G Davies CRA. **430 Corbis:** Catherine Karnow BCaR; Paul A Souders CaR; Yann Arthus-Bertrand BL, BR. **431 Corbis:** NASA TCR; **Getty:** Image Bank / Peter Hendrie B; **Geo:** TRb. **432 Corbis:** John Heseltine BCA; Micahel S Yamashita BCR; **DJ Sauchyn:** Prairie Research Collaborative C. **432/433 Corbis:** Dallas and John Heaton C. **433 Corbis:** Carol Havens BCa; Ecoscene / John Farmer TRb; Joel W Rogers Cb. **434 Corbis:** Otto Rogge. **435 Corbis:** Gallo Images / Martin Harvey BC; Joe McDonald Cb; **Getty:** Stone / Jeremy Walker BR. **436/437 Corbis:** William James Warren. **438/439 alamy.com:** IMAGINA / Atsushi Tsunoda. **440 Corbis:** Eye Ubiquitous / Bennett Dean BCL; **Getty:** Image Bank / Kevin Kelley C; **NG:** Thad Samuels Abell II CLb. **441 Corbis:** Ron Watts CR; **Tom Ekland:** CL; **Getty:** Image Bank / Guido Alberto Rossi C; **P A Photos:** European Press Agency BR. **SPL:** David Parker TCLb; **US Geological Survey:** AM Sarna-Wojcicki CRb; **Visible Earth:** Image courtesy Jacques Descloitres, MODIS Rapid Response Team, NASA/GSFC BC. **442 Corbis:** Grafton Marshall Smith CRa; **Transition Region and Coronal Explorer (TRACE)** – a mission of the Stanford-Lockheed Institute for Space Research, part of the NASA Small Explorer Program CRb. **443 Corbis:** Sygma / Herve Collart CL; **Getty:** Image Bank / Stephen Frink BC; Stone / Kim Blaxland TL; Stone / Kim Westerskov CR; Stone / Mark Muench BR. **444 SPL:** David Hay Jones. **445 DK:** NASA BCa; **Rex Features:** Brian Harris CRb. **446 Bridgeman Art Library, London / New York:** Giraudon. *Portrait of Gustave Gaspard Coriolis* (1792-1843) engraved by Zephirin Felix Jean Marius Belliard (b.1798) (litho), Academie des Sciences, Paris, France BL; **Corbis:** Stocktrek Cb. **447 alamy.com:** Geoffry Morgan CRb; **Corbis:** NASA TR; **National Trust Photographic Library:** Mike Howarth CRa; **Visible Earth:** Image courtesy GOES Project Science Office BCL. **448 Corbis:** Paul A Souders CL; **FLPA:** H Hoflinger TR; **Visible Earth:** Provided by the SeaWiFS Project, NASA/Goddard Space Flight Center, and ORBIMAGE BR; Image courtesy Liam Gurnley, MODIS Atmosphere Team, University of Wisconsin – Madison CLa; Image courtesy Jacques Descloitres, MODIS Rapid Response Team, NASA/GSFC BCL. **449 Corbis:** Lowell Georgia TR; Sygma / Telegram Tribune / Jason Mellom CL; **P A Photos:** European Press Agency BCL. **450 Corbis:** Frank Lane Picture Agency / Ron Boardman CL; **SPL:** BR; Eye of Science Ca. **450/451 Magnum:** Steve McCurry **451 Corbis:** CRb; Jonathan Blair C; Richard Hamilton Smith Cb; **NASA:** Goddard Space Flight Center Scientific Visualization Studio B; **Royal Swedish Academy of Sciences:** TR. **452 Corbis:** Hans Strand. **453 Corbis:** Douglas Faulkner Cb; **NG:** Michael S Lewis CRa **454 Corbis:** ChromoSohm Inc / Joseph Sohm B; Saba / James Leynse Ca; **Magnum:** Bruno Barbey CbL. **455 Corbis:** Adrian Arbib TR; Brian Vikander BR; Eye Ubiquitous / Ben Spencer BL; The Purcell Team Ca; **DK:** Marwell Zoological Park / Frank Greenaway CLa. **456 Corbis:** Maurizio Lanini CbL; Michael Freeman CLa; **DK:** Joe Cornish CRa; Max Alexander BL; **Getty:** Stone / Oliver Strewe BR. **457 Corbis:** Bob Krist BCL; James L Amos CaR; Richard Hamilton Smith CLa; Yann Arthus-Bertrand BR; **Visible Earth:** Provided by the SeaWiFS Project, NASA/Goddard Space Flight Center, and ORBIMAGE Cb. **458/459 Corbis:** Tom Bean. **460 SPL:** BL. **460/461 Corbis:** A&J Verkaik B. **461 Corbis:** Richard Hamilton Smith CLa; **OSF:** Bill Pike / SAL TCR. **462 Corbis:** Paul A Souders TCR; Roger Tidman CLa; **Getty:** Image Bank / Pascal Perret CRa; **Ferdinand Valk:** CRb; **Visible Earth:** Image courtesy Jacques Descloitres, MODIS Rapid Response Team, NASA/GSFC TR. **463 Agence France Presse:** STR BCR; **Corbis:** Stocktrek CLa; **NASA:** Johnson Space Center (STS51I-37-83) TR. **464 Corbis:** Sygma / Diaro Listen CLb; **NASA:** Goddard Space Flight Center (GPN-2000-001331) Cb. **465 Popperfoto:** Reuters / Andrew Winning. **466 Getty:** Stone / David Olsen BL; **OSF:** Warren Faidley TR. **466/467 Getty:** Taxi / Ron Chapple B. **467 Getty:** Photodisc Green CbR; **OSF:** Michael Leach C; **SPL:** Simon Fraser CaL; **Visible Earth:** Provided by the SeaWiFS Project, NASA/Goddard Space Flight Center, and ORBIMAGE TLb; Provided by the SeaWiFS Project, NASA/Goddard Space Flight Center, and ORBIMAGE BR. **468 Corbis:** Galen Rowell B; Wolfgang Kaehler TC; **Magnum:** Steve McCurry CRa; **NG:** George Grall TCbL; Medford Taylor C; Patricia Rasmussen: **CbL; SPL:** Astrid & Hanns-Frieder Michler CRb; National Center for Atmospheric Research CR. **469 Corbis:** Raymond Gehman CLA; The Purcell Team TCL; William A Bake Ca; **Royal Meteorological Society:** CLb; **Visible Earth:** NASA Goddard Space Flight Center Image by Reto Stöckli. Enhancements by Robert Simmon BC. **470 Corbis:** Charles O'Rear BR; Ted Spiegel CLa; Wolfgang Kaehler TR; **DK:** Brian Cosgrove Cloud Collection Ca, CRa, BL; **Getty:** Stone / Donovan Reese CRb. **471 Corbis:** Scott T Smith CRa; **Getty:** Image Bank / Alan R Moller CLa; **North American Weather Consultants Inc, Utah:** BL; **Still:** Francois Suchel BR; Keith Kent TR. **472 Corbis:** Galen Rowell CLa; Gallo Images Cb; John McAnulty BCL; Richard Hamilton Smith CRb; **DK:** Brian Cosgrove Cloud Collection CRa, BR; **Getty:** Stone / Paul A Souders TCR. **473 Corbis:** Maurice Nimmo CRa; Royalty-Free CaL; Tom Bean CbL; **DK:** Brian Cosgrove Cloud Collection TR, BL, BR. **474 Corbis:** Bettmann TCRb; Darrell Gulin B; **NG:** Priitt Esilind CL. **475 Corbis:** Ecoscene / Nick Hawkes TC; Kit Kittle TR; **OSF:** Stan Osolinski CRa; **SPL:** Picture Researchers Inc BC. **476 AirPhoto:** Jim Wark CL; **Corbis:** Inge Yspeert BL; NASA CRb; **NASA:** Johnson Space Center (STS41B-41-2347) TC; Johnson Space Center (STS51G-46-5) CRa; **NG:** Chris Johns BCR; **SPL:** Peter Menzel CR. **477 Corbis:** Jim Zuckerman B; W Perry Conway TCR; **Gene E. Moore:** CLa, CRa, CaL, CaR. **478/479 Galaxy Contact:** Johnson Space Center (STS066-208-025). **480 Corbis:** James L Amos Cb; Tom Bean BL; Yann Arthus-Bertrand CLb. **482 alamy.com:** Peter Haigh C; **Corbis:** David Muench Cb, BC; **Getty:** Stone / James Balog Ca; **Hutchison Library:** Nigel Smith BL. **483 alamy.com:** Winston Fraser TCL; **Corbis:** David Muench Ca, CR; Galen Rowell CaR; James L Amos TR; **RHPL:** Roy Rainford TCb; **Landsat:** BC; **Visible Earth:** Image courtesy Jacques Descloitres, MODIS Rapid Response Team, NASA/GSFC C. **484 Corbis:** Wolfgang Kaehler CL; **Hutchison Library:** T Moling Cb; **Magnum:** Burt Glinn CLb; **Visible Earth:** NASA / GSFC / MITI / ERSDAC / JAROS, and US / Japan ASTER Science Team BC, C. **485 Corbis:** FLPA Ca; Larry Lee BC; Richard List TCL; **Getty:** Stone / Glen Allison Cb; **Still:** Michel Roggo TCb; UNEP / S Rocha TR. **486 alamy.com:** B&C BL; **Corbis:** Michael S Yamashita BCR; Yann Arthus-Bertrand Cb; **Getty:** Image Bank / Weinberg / Clark C; **Hutchison Library:** Trevor Page BCLa; **Geo:** CbL; **Visible Earth:** MISR Instrument Team, MISR Project CL. **487 Landsat:** Ca; **Magnum:** Hiroji Kubota C, Cb; **Rex Features:** Richard Gardner TR; **Geo:** BC; **Visible Earth:** Jeff Schmaltz, MODIS Rapid Response Team, NASA / GSFC TC. **488 Nature Picture Library:** Bruce Davidson BRa; **Corbis:** Charles O'Rear BCR; Owen Franken Cb; Wolfgang Kaehler C; **Getty:** Taxi / Hans Christian Heap CbL; **SPL:** Dr Ken MacDonald BCL; **Visible Earth:** Image Courtesy Luca Pietranera Telespazio, Rome, Italy BCa. **489 Corbis:** FLPA / Winifred Wisniewski Cb; Jon Hicks TR; Stephen Frink TC; **Getty:** Image Bank / AEF BC; **NASA:** Goddard Space Flight Center Scientific Visualization Studio and USGS Ca; **Visible Earth:** NASA / JPL TCb. **490 Getty:** Image Bank / Theo Allofs Cb; Image Bank / Thomas Schmitt BC; **Landsat:** C. **491 Corbis:** Ecoscene / Wayne Lawler TCb; Hubert Stadler CR; O Alamany & E Vicens TR; Yann Arthus-Bertrand TC; **Getty:** Image Bank / Guido Alberto Rossi C; Stone / Art Wolfe Cb; Stone / Kim Westerkov BC; **WWI:** Joe Cornish BR; **Visible Earth:** Image courtesy Jacques Descloitres, MODIS Rapid Response Team, NASA/GSFC Bca; Image courtesy NASA / GSFC / LaRC / JPL / MISR Team CaR. **492 Corbis:** James L Amos BC; Michael S Yamashita C; Neil Rabinowitz BCLa; **NG:** Emory Kristof Cb; **Visible Earth:** NASA JPL BLa. **493 alamy.com:** David Noton Photography / David Noton C; **Corbis:** Douglas Peebles Ca; Eye Ubiquitous / David Batterbury CR; Jim Zuckerman BC; Kevin Schafer TRb; **Visible Earth:** Image courtesy Jacques Descloitres, MODIS Rapid Response Team, NASA/GSFC TC. **494 Corbis:** Anne Hawthorne BC; Galen Rowell Cb; **Landsat:** Bla; **Visible Earth:** Image courtesy Jacques Descloitres, MODIS Rapid Response Team, NASA/GSFC CL, C; Image courtesy Greg Shirah, GSFC Scientific Visualization Studio, based on data from the TOMS science team BL. **495 Corbis:** Galen Rowell Ca; Kevin Schafer BC; Sygma TC; Wolfgang Kaehler TR; Yann Arthus-Bertrand C, Cb.

**Endpapers AirPhoto:** Jim Wark.

**Jacket pictures**
Front jacket: **OSF:** Olivier Grunewald
Back jacket: **Still:** UNEP / R Bernardo
Front Flap: **Joe Cornish**

**Additional illustrations**
Centrepiece artworks on the following pages by **Planetary Visions Ltd:** 54–55, 56–57, 58–59, 384–385, 440–441, 442–443, 446–447, 482–483, 484–485, 486–487, 488–489, 490–491, 492–493, 494–495.